Adaptation in Wireless Communications

Edited by

Mohamed Ibnkahla

ADAPTIVE SIGNAL PROCESSING in WIRELESS COMMUNICATIONS

ADAPTATION and CROSS LAYER DESIGN in WIRELESS NETWORKS

THE ELECTRICAL ENGINEERING AND APPLIED SIGNAL PROCESSING SERIES

Edited by Alexander Poularikas

The Advanced Signal Processing Handbook: Theory and Implementation for Radar, Sonar, and Medical Imaging Real-Time Systems
Stergios Stergiopoulos

The Transform and Data Compression Handbook
K.R. Rao and P.C. Yip

Handbook of Multisensor Data Fusion
David Hall and James Llinas

Handbook of Neural Network Signal Processing
Yu Hen Hu and Jenq-Neng Hwang

Handbook of Antennas in Wireless Communications
Lal Chand Godara

Noise Reduction in Speech Applications
Gillian M. Davis

Signal Processing Noise
Vyacheslav P. Tuzlukov

Digital Signal Processing with Examples in MATLAB®
Samuel Stearns

Applications in Time-Frequency Signal Processing
Antonia Papandreou-Suppappola

The Digital Color Imaging Handbook
Gaurav Sharma

Pattern Recognition in Speech and Language Processing
Wu Chou and Biing-Hwang Juang

Propagation Handbook for Wireless Communication System Design
Robert K. Crane

Nonlinear Signal and Image Processing: Theory, Methods, and Applications
Kenneth E. Barner and Gonzalo R. Arce

Smart Antennas
Lal Chand Godara

Mobile Internet: Enabling Technologies and Services
Apostolis K. Salkintzis and Alexander Poularikas

Soft Computing with MATLAB®
Ali Zilouchian

Wireless Internet: Technologies and Applications
Apostolis K. Salkintzis and Alexander Poularikas

Signal and Image Processing in Navigational Systems
Vyacheslav P. Tuzlukov

Medical Image Analysis Methods
Lena Costaridou

MIMO System Technology for Wireless Communications
George Tsoulos

Signals and Systems Primer with MATLAB®
Alexander Poularikas

Adaptation in Wireless Communications - 2 volume set
Mohamed Ibnkahla

ADAPTATION AND CROSS LAYER DESIGN IN WIRELESS NETWORKS

Edited by
Mohamed Ibnkahla

CRC Press
Taylor & Francis Group
Boca Raton London New York

CRC Press is an imprint of the
Taylor & Francis Group, an **informa** business

CRC Press
Taylor & Francis Group
6000 Broken Sound Parkway NW, Suite 300
Boca Raton, FL 33487-2742

Printed in the United States of America on acid-free paper
10 9 8 7 6 5 4 3 2 1

International Standard Book Number-13: 978-1-4200-4603-8 (Hardcover)

Library of Congress Cataloging-in-Publication Data

Adaptation and cross layer design in wireless networks / editor, Mohamed Ibnkahla.
p. cm. -- (Electrical engineering and applied signal processing series ; 21)
Includes bibliographical references and index.
ISBN 978-1-4200-4603-8 (alk. paper)
1. Wireless communication systems. I. Ibnkahla, Mohamed. II. Title. III. Series.

TK5103.2.A355 2008
621.384--dc22 2008025417

Visit the Taylor & Francis Web site at
http://www.taylorandfrancis.com

and the CRC Press Web site at
http://www.crcpress.com

Contents

Preface

Adaptive techniques play a key role in modern wireless communication systems. The concept of adaptation is emphasized in the *Adaptation in Wireless Communications Series* across all layers of the wireless protocol stack, ranging from the physical layer to the application layer.

This book covers the concept of adaptation at the data link layer, network layer, and application layer. It presents state-of-the-art adaptation techniques and methodologies including cross layer adaptation, joint signal processing, coding and networking, selfishness in mobile ad hoc networks, cooperative and opportunistic protocols, adaptation techniques for multimedia support, self-organizing routing, and tunable security services. The book offers several new theoretical paradigms and analytical findings, which are supported by various simulation and experimental results, and contains more than 170 figures, 25 tables, and 650 references.

I would like to thank all the contributing authors for their patience and excellent work. The process of editing started in June 2005. Each chapter has been blindly reviewed by at least two reviewers (more than 50% of the chapters received three reviews or more), and I would like to thank the reviewers for their time and valuable contribution to the quality of this book.

Finally, a special thank you goes to my parents, my wife, my son, my daughter, and all my family. They all have been of a great support for this project.

Mohamed Ibnkahla
Queen's University
Kingston, Ontario, Canada

Editor

Dr. Mohamed Ibnkahla earned an engineering degree in electronics in 1992, an M.Sc. degree in signal and image processing in 1992, a Ph.D. degree in signal processing in 1996, and the Habilitation à Diriger des Recherches degree in 1998, all from the National Polytechnic Institute of Toulouse (INPT), Toulouse, France.

Dr. Ibnkahla is currently an associate professor in the Department of Electrical and Computer Engineering, Queen's University, Kingston, Canada. He previously held an assistant professor position at INPT (1996–1999) and Queen's University (2000–2004).

Since 1996, Dr. Ibnkahla has been involved in several research programs, including the European Advanced Communications Technologies and Services (ACTS), and the Canadian Institute for Telecommunications Research (CITR). His current research is supported by industry and government agencies such as the Ontario Centers of Excellence (OCE), the Natural Sciences and Engineering Research Council of Canada (NSERC), the Ontario Ministry of Natural Resources, and the Ontario Ministry of Research and Innovation.

He is currently leading multidisciplinary projects designing, implementing, and deploying wireless sensor networks for various applications in Canada. Among these applications are natural resources management, ecosystem and forest monitoring, species at risk tracking and protection, and precision agriculture.

Dr. Ibnkahla has published a significant number of journal papers, book chapters, technical reports, and conference papers in the areas of signal processing and wireless communications. He has supervised more than 40 graduate students and postdoctoral fellows. He has given tutorials in the area of signal processing and wireless communications in several conferences, including IEEE Global Communications Conference (GLOBECOM, 2007) and IEEE International Conference in Acoustics, Speech, and Signal Processing (ICASSP, 2008).

Dr. Ibnkahla received the INPT Leopold Escande Medal for the year 1997, France, for his research contributions in signal processing; the Prime Minister's Research Excellence Award (PREA), Ontario, Canada in 2000, for his contributions in wireless mobile communications; and the Favorite Professor Award, Queen's University in 2004 for his excellence in teaching.

Contributors

Dharma P. Agrawal
Center for Distributed and Mobile Computing
Department of Computer Science
University of Cincinnati
Cincinnati, Ohio

Ali Alemdar
Electrical and Computer Engineering Department
Queen's University
Kingston, Ontario, Canada

Giuseppe Anastasi
Department of Information Engineering
University of Pisa
Pisa, Italy

Farid Benbadis
Université Pierre et Marie Curie
Paris, France

John Bigham
Department of Electronic Engineering
Queen Mary, University of London
London, UK

Anna Brunstrom
Department of Computer Science
Karlstad University
Karlstad, Sweden

Marco Conti
IIT-CNR
National Research Council
Pisa, Italy

Sajal K. Das
Department of Computer Science and Engineering
University of Texas at Arlington
Arlington, Texas

Marcelo Dias de Amorim
CNRS
Paris, France

Zoltán Faigl
Mobile Innovation Center
Budapest University of Technology and Economics
Budapest, Hungary

Fethi Filali
Eurecom Institute
Sophia-Antipolis, France

Hossam Hassanein
School of Computing
Queen's University
Kingston, Ontario, Canada

Ekram Hossain
Department of Electrical and Computer Engineering
University of Manitoba
Winnipeg, Manitoba, Canada

Mohamed Ibnkahla
Electrical and Computer Engineering Department
Queen's University
Kingston, Ontario, Canada

Massimiliano Lenardi
Hitachi Europe
Sophia-Antipolis, France

Yuxia Lin
Department of Electrical and Computer Engineering
University of British Columbia
Vancouver, British Columbia, Canada

Stefan Lindskog
Department of Computer Science
Karlstad University
Karlstad, Sweden
and
Centre for Quantifiable Quality of Service in Communication Systems
Norwegian University of Science and Technology
Trondheim, Norway

Yonghe Liu
Department of Computer Science and Engineering
University of Texas at Arlington
Arlington, Texas

Ning Lu
MPI-QMUL
Information Systems Research Centre
Macao Polytechnic Institute
Macao

Hong Luo
School of Computer Science and Technology
Beijing University of Posts and Telecommunications
Beijing, China

Hamid Menouar
Hitachi Europe
Sophia-Antipolis, France

Nidal Nasser
Department of Computing and Information Science
University of Guelph
Toronta, Ontario, Canada

Dusit Niyato
Department of Electrical and Computer Engineering
University of Manitoba
Winnipeg, Manitoba, Canada

Andrea Passerella
IIT-CNR
National Research Council
Pisa, Italy

Luciana Pelusi
IIT-CNR
National Research Council
Pisa, Italy

Mihail S. Sichitiu
North Carolina State University
Raleigh, North Carolina

Sameh Sorour
Department of Electrical and Computer Engineering
University of Toronto
Toronto, Ontario, Canada

Ananthram Swami
Communications and Information Sciences Directorate
U.S. Army Research Laboratory
Adephi, Maryland

Alberto Lopez Toledo
Telefonica Research
Barcelona, Spain

Shahrokh Valaee
Department of Electrical and Computer Engineering
University of Toronto
Toronto, Ontario, Canada

Tom Vercauteren
INRIA
Sophia-Antipolis, France

Aline Carneiro Viana
INRIA
Paris, France

Yannis Viniotis
North Carolina State University
Raleigh, North Carolina

Xiaodong Wang
Department of Electrical Engineering
Columbia University
New York, New York

Vincent W. S. Wong
Department of Electrical and Computer Engineering
University of British Columbia
Vancouver, British Columbia, Canada

Yunnan Wu
Microsoft Research
Redmond, Washington

Chi-Hsiang Yeh
Department of Electrical and Computer Engineering
Queen's University
Kingston, Ontario, Canada

Younghwan Yoo
Ubiquitous Computing Laboratory
School of Computer Science and Engineering
Pusan National University
Busan, Korea

Tiantong You
School of Computing
Queen's University
Kingston, Ontario, Canada

Guohua Zhang
Department of Computer Science and Engineering
University of Texas at Arlington
Arlington, Texas

Qing Zhao
Department of Electrical and Computer Engineering
University of California
Davis, California

1

Adaptive Optimization of CSMA/CA MAC Protocols Based on Bayesian State Estimation

Alberto Lopez Toledo
Telefonica Research Barcelona

Tom Vercauteren
INRIA

Xiaodong Wang
Columbia University

1.1 Introduction

The IEEE 802.11 protocol [1] has become the predominant technology for wireless local area networks (WLANs). One of the most important elements of the 802.11 in terms of performance is the medium access control (MAC). The MAC protocol is used to provide arbitrated access to a shared medium, in which several terminals access and compete for the radio spectrum. The design of the MAC protocols is often application dependent, and it is closely linked to the characteristics of the medium in which it operates. It also determines the performance and quality of service (QoS) metrics of the network, such as throughput, stability, and delay. The IEEE 802.11 wireless networks employ the distributed coordination function (DCF) as a primary medium access mechanism. It is based on the carrier-sensing multiple-access with collision avoidance (CSMA/CA) protocol and binary exponential backoff [2, 3].

Several studies have shown that the DCF protocol is very sensitive to the number of competing terminals that access the wireless channel [2–7], and that a way to optimize the network performance is to make the parameters of the backoff window depend on the number of terminals competing for the medium. However, estimating the number of competing terminals is not an easy task. While a terminal could cache the identity of the past senders in the network, the number of *competing* terminals is the number of terminals that *have data to send* at any given time, so a simple list of neighbors is not sufficient. The estimation of the number of competing terminals faces two problems. First, the number of competing terminals is a non-Gaussian nonlinear dynamic system that is difficult to track accurately with conventional filters. Advanced estimators such as the extended Kalman filter (EKF)-based one from [3] provide better results, but they are subject to critics due to their complexity [8]. Second, the performance of the IEEE 802.11 DCF is extremely sensitive to the number of competing terminals [2], particularly in the typical operating point of one to fifteen terminals. This makes approximate methods such as [5, 9, 10] to yield suboptimal operation of the protocol compared with the theoretical optimum. Hence, there is a need for an accurate estimation algorithm that is able to efficiently track the number of competing terminals in an IEEE 802.11 network and, at the same time, is easy to implement. As we will see, sequential Monte Carlo methods are appropriate for this purpose.

Bayesian Monte Carlo signal processing techniques [11, 12] offer a paradigm for tackling challenging signal processing problems for which traditional methods are difficult to apply. Among them, the sequential Monte Carlo (SMC) methodology [12] has been shown to be extremely powerful in dealing with filtering problems in non-Gaussian and nonlinear complex dynamic systems, where conventional approaches fail to work. In order to tackle the above estimation of the competing terminals, we develop several SMC-based adaptive estimators that outperform the existing classical estimator, such as those based on the extended Kalman filter (EKF) [3]. In particular, we develop a deterministic variant of the SMC estimator that is simpler to implement and offers superior performance, by including a set of sufficient statistics about the unknown parameters in the deterministic sample filter setting proposed in [13]. The use of sufficient statistics is pushed one step further than in [14] because this information about the parameters is now integrated out so that no Monte Carlo approximation needs to be performed. We further extend it into a maximum a posteriori (MAP) estimator whose computational load and memory requirements are equivalent to those of the well-known Viterbi algorithm. We will see that our SMC-based algorithms overcome both of the problems mentioned above: accuracy and ease of implementation. We show that the accuracy of the Bayesian algorithms is particularly good at small timescales, which makes our proposal attractive to optimize the protocol when the terminals are in a nonsaturation regime, a problem usually not addressed in the literature. Finally, we propose an optimization mechanism that uses the predictive distribution of the number of competing terminals to adapt the parameters of the IEEE 802.11 DCF protocol and maximize its throughput.

1.2 The IEEE 802.11 Distributed Coordination Function

1.2.1 The DCF Protocol

The DCF defines two distinct techniques to access the medium: *basic access* and *RTS/CTS access.*

1.2.1.1 Basic Access

In basic access, the terminals implement a two-way handshake mechanism (Figure 1.1). A terminal senses the channel to be idle before starting a transmission. If the channel is idle for at least a period of distributed interframe space (DIFS), then thc terminal is allowed to transmit. If during this sensing time the channel appears to be busy at any time, the terminal defers the transmission and enters into the collision avoidance (CA) mode. In CA mode the terminal generates a random backoff interval during which it waits before attempting another transmission. This random backoff is used to minimize the probability of collision between terminals accessing the medium. The idle time after waiting a DIFS interval is slotted, and the terminals are only allowed to transmit at the beginning of the slot time. The slot time size σ accounts for the time the signal is propagating, and is set equal to the time needed for any terminal to detect the transmission of a packet from any other terminal [2]. If this time were not accounted for, a terminal could assess the channel as idle when the data sent by another terminal have not yet arrived.

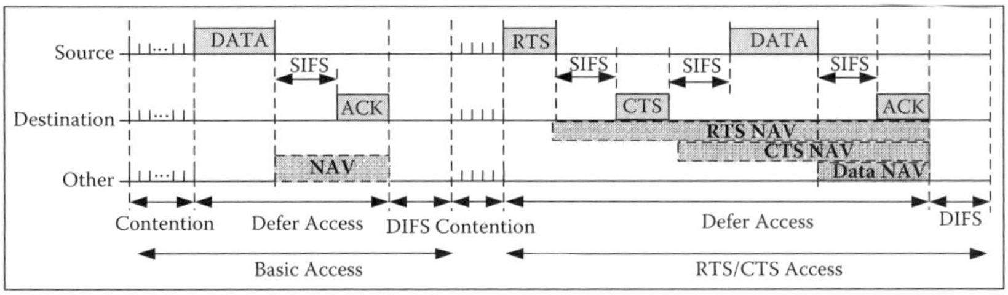

FIGURE 1.1 IEEE 802.11 MAC access mechanisms.

The random backoff timer is uniformly chosen between $[0,v)$, where v is called the *contention window*, and it is such that it satisfies $v \in [CW_{min}, CW_{max}]$, where CW_{min} and CW_{max} are called *minimum* and *maximum* contention window, respectively. At the first transmission attempt the value of the contention window v is set to CW_{min}. The backoff timer is decremented while the channel is idle (i.e., it only counts the *idle time*). If at any time the channel is sensed busy, the backoff timer is paused until the channel is sensed idle again *after* the corresponding DIFS time. When the backoff timer reaches 0, the terminal is allowed to transmit. Following the successful reception of the data, the receiving terminal waits a short interframe space (SIFS) interval and transmit an acknowledgment (ACK) to the transmit terminal. Because the SIFS interval is shorter than the DIFS interval, the destination terminal has priority in sending the ACK. Such two-way handshake-based ACK is necessary because the CSMA/CA protocol does not assume the terminals have the capability to detect collisions. Upon reception of the ACK, the backoff stage is reset to 0 and $v = CW_{min}$. This is referred as a "heavy decrease" in [4]. If the source terminal does not receive the ACK after a timeout period (ACK_timeout) or it detects the transmission of any other frame in the channel (collision), the frame is assumed to be lost. After each unsuccessful transmission the value of v is doubled up to a maximum of $CW_{max} = 2^m\, CW_{min}$, where m is usually referred to as the *maximum backoff stage* [4]. The values of CW_{min}, CW_{max}, and slot size σ are determined by the characteristics of the physical layer.

As shown in Figure 1.1, when another terminal is transmitting, the rest of the terminals set up the network-allocation vector (NAV) timer, which acts as a virtual carrier sense. When hearing a data frame, the rest of the terminals set up the duration of the NAV to that specified in the header of the transmitted data frame. The NAV vector includes the SIFS and the duration of the ACK transmission. All the terminals defer their access to the medium until the NAV timer expires.

1.2.1.2 RTS/CTS Access

RTS/CTS access is similar to *basic access*, but makes use of a four-way handshake protocol in which prior to data transmission a terminal transmits a special short request-to-send (RTS) frame to try to "reserve" the transmission and reduce the cost of collisions. The receive terminal responds with a short special clear-to-send (CTS) frame, as shown in Figure 1.1.

1.2.2 Analytical Throughput of IEEE 802.11 DCF

Let us consider an IEEE 802.11 network with DCF operating in the basic access mode as described in section 1.2.1. We assume that the number of terminals using the network at a given time is finite. We also assume that the terminals transmit in a saturation regime, i.e., they always have something to send. In this saturation regime, it is shown in [2] that the normalized throughput of the system can be analytically derived.

From the point of view of a terminal, the time can be slotted into variable-length slots. Specifically, in the DCF operation, a time slot will correspond to an idle slot σ, or a busy slot that has the duration of a successful transmission $T_s(L)$ or the duration of a collision

$T_c(L^*)$, where L is the time length of the packet and L^* is the time length of the largest of the packets involved in the collision. The normalized throughput is then given by

$$S=\frac{\mathbb{E}[L]}{T_s-T_c+\dfrac{T_c+\sigma(1-P_{tr})}{P_s}}, \tag{1.1}$$

where P_{tr} is the probability of a terminal transmitting in the slot and $\mathbb{E}[L]$, $T_s = T_s(\mathbb{E}[L])$, $T_c = T_c(\mathbb{E}[L^*])$, and σ are constants denoting the average packet payload length, the average time of a busy slot with successful transmission, the average time of a busy slot with collision, and the duration of an empty slot, respectively. $\mathbb{E}[L^*]$ is the average length of the longest packet involved in a collision.

Let x_t be the number of competing terminals in the network at time t (discrete). Here the term *competing* refers to a terminal that is either transmitting or backlogged (i.e., it has data to send). Let q be the probability that a terminal transmits in a given slot. Then the probability P_{tr} that at least one terminal transmits in a given time slot t is given by

$$P_{tr}=1-(1-q)^{x_t}. \tag{1.2}$$

In [2] it is shown that

$$q=\frac{2(1-2p_c)}{(1-2p_c)(CW_{min}+1)+p_cCW_{min}\left(1-(2p_c)^m\right)}, \tag{1.3}$$

where CW_{min} and m are the minimum contention window and maximum backoff stage, respectively. Then, the collision probability p_c and the probability of a successful transmission P_s for a terminal are given by

$$p_c=1-(1-q)^{x_t-1}, \qquad P_s=\frac{x_t\cdot q(1-q)^{x_t-1}}{1-(1-q)^{x_t}}. \tag{1.4}$$

This analysis shows that the throughput is a function of the number of competing terminals x_t and the probability of a terminal transmitting q. Given x_t, CW_{min}, m, equations (1.3) and (1.4) can be solved and a unique solution can be found [2]. Therefore, the normalized saturation throughput only depends on the number of competing terminals and the backoff parameters, i.e.,

$$S=S(x_t, CW_{min}, m). \tag{1.5}$$

Once the number of competing terminals x_t is estimated, the optimization problem involves selecting the other parameters CW_{min} and m to maximize the system throughput S.

1.3 Bayesian Estimation of the Number of Competing Terminals in IEEE 802.11 DCF

1.3.1 Problem Formulation

It is shown in [2] that when the terminals are in saturation regime, i.e., they always have a packet to send, and when the system reaches a steady state, the number of competing terminals x_t can be expressed as a function of the collision probability p_c as

$$x_t = f(p_c) \underline{\Delta} 1 + \frac{\log(1-p_c)}{\log\left(1 - \dfrac{2(1-2p_c)}{(1-2p_c)(CW_{min}+1) + p_c CW_{min}\left(1-(2p_c)^m\right)}\right)}. \tag{1.6}$$

The above function is monotonic increasing in p_c, and hence an inverse function exists, i.e., $p_c = h(x_t)$, where $h(\cdot) = f^{-1}(\cdot)$. The problem of estimating the number of competing terminals then involves estimating x_t based on a noisy observation of p_c, which each terminal can acquire by monitoring the channel activity. The observation variable of the collision probability y_t can be defined at each time step t as [3]

$$y_t = \sum_{i=(t-1)B}^{tB-1} C_i, \tag{1.7}$$

where $C_i = 0$ if the ith time slot is empty or corresponds to a successful transmission (i.e., no collision), and $C_i = 1$ if the ith basic time slot is busy or corresponds to an unsuccessful transmission (i.e., would result in collision); B is the number of slots that compose the *observation slot* for the measurement. It is easy to see that y_t follows a binomial distribution $\mathcal{B}(B, p_c)$ with B trials and probability of success p_c, i.e.,

$$p(y_t = b) = \binom{B}{b} p_c^b (1-p_c)^{B-b}, \quad b = 0, 1, 2, \ldots, B. \tag{1.8}$$

The number of competing terminals x_t at time $t = 1, 2, \ldots,$ takes value from the set $\mathcal{X}$. In wireless LAN systems, admission control is always performed to maintain certain quality of service (QoS), and thus $\mathcal{X}$ is a finite set. Typically we have $\mathcal{X} = [1, \cdots, N]$, with N being the maximum number of users, so that we do not need to differentiate between the index of the states and the states themselves. We assume that x_t evolves according to a first-order Markov chain with a transition probability matrix $\boldsymbol{A} = [a_{i,j}]$, i.e., $p(x_{t+1} = j \mid x_t = i) = a_{i,j}$, where $a_{i,j} \geq 0$ and

$$\sum_{j=1}^{N} a_{i,j} = 1.$$

We denote the initial probability vector as $\boldsymbol{\pi} = [\pi_1, \cdots, \pi_N]$, i.e., $p(x_0 = i) = \pi_i$.

1.3.1.1 The Inference Problem

From the above discussion, we can cast our problem into a hidden Markov model (HMM) with unknown parameters:

$$x_t \sim \mathcal{MC}(\boldsymbol{\pi}, \boldsymbol{A}), \tag{1.9}$$

$$y_t \sim \mathcal{B}(B, h(x_t)), \tag{1.10}$$

where $\mathcal{MC}(\boldsymbol{\pi},\boldsymbol{A})$ denotes a discrete-time Markov chain with the initial probability distribution $\boldsymbol{\pi}$ and the transition probability matrix $\boldsymbol{A}$; x_t is the state realization of the Markov chain at time instant t; $\mathcal{B}(B, p)$ denotes a binomial distribution with B trials and probability of success p; y_t is the observation; and $h(\cdot) \triangleq f^{-1}(\cdot)$, where $f(\cdot)$ is given in (1.6).

Denote the observation sequence up to time t as $\boldsymbol{y}_t \triangleq [y_1, y_2, \cdots, y_t]$ and the network state sequence up to time t as $\boldsymbol{x}_t \triangleq [x_1, x_2, \cdots, x_t]$. Let the model parameters be $\boldsymbol{\theta} = \{\pi, \boldsymbol{A}\}$. Given the observations $\boldsymbol{y}_t$ at time t, we are interested in estimating the current state x_t.

1.3.2 Sequential Monte Carlo Estimation

We consider a generic dynamic model described by

$$\begin{aligned} \text{initial eqn:}\ & p_{\boldsymbol{\theta}}(x_0); \\ \text{state eqn:}\ & p_{\boldsymbol{\theta}}(x_t \mid x_{t-1}), \forall t \geq 1; \\ \text{measurement eqn:}\ & p_{\boldsymbol{\theta}}(y_t \mid x_t), \forall t \geq 1, \end{aligned} \tag{1.11}$$

where x_t and y_t are respectively the state and observation at time t, and $p_{\boldsymbol{\theta}}(\cdot)$ are some probability density functions (p.d.f.) depending on some static parameters $\boldsymbol{\theta}$ assumed known for the moment. Suppose we want to make online inference of the unobserved states $\boldsymbol{x}_t = (x_0, x_1, \ldots, x_t)$. That is, at current time t we wish to make an estimate of a function of the state variable, say $\psi(\boldsymbol{x}_t)$, based on the currently available observations, $\boldsymbol{y}_t = (y_0, y_1, \ldots, y_t)$. The optimal solution (in terms of any common criterion) only depends on the conditional p.d.f. $p_{\boldsymbol{\theta}}(\boldsymbol{x}_t \mid \boldsymbol{y}_t)$ (e.g., the minimum mean squared error estimator is the conditional mean).

Monte Carlo methods provide an approximation of this p.d.f. based on K random samples $\{\boldsymbol{x}_t^{(k)}\}_{K=1}^{K}$ from the distribution $p_{\boldsymbol{\theta}}(\boldsymbol{x}_t \mid \boldsymbol{y}_t)$. Since sampling directly from $p_{\boldsymbol{\theta}}(\boldsymbol{x}_t \mid \boldsymbol{y}_t)$ is often not feasible or computationally too expensive, the idea of importance sampling can be used to approximate $p_{\boldsymbol{\theta}}(\boldsymbol{x}_t \mid \boldsymbol{y}_t)$ by employing some trial sampling density $q_{\boldsymbol{\theta}}(\boldsymbol{x}_t \mid \boldsymbol{y}_t)$ from which we can easily draw samples. Suppose a set of random samples $\{\boldsymbol{x}_t^{(k)}\}_{K=1}^{K}$ has been drawn according to $q_{\boldsymbol{\theta}}(\boldsymbol{x}_t \mid \boldsymbol{y}_t)$. By associating the *importance weight*

$$w_t^{(k)} = \frac{p_{\theta}\left(\boldsymbol{x}_t^{(k)} \mid \boldsymbol{y}_t\right)}{q_{\theta}\left(\boldsymbol{x}_t^{(k)} \mid \boldsymbol{y}_t\right)}$$

to the sample $\boldsymbol{x}_t^{(k)}$, the posterior distribution of interest is approximated as

$$\hat{p}_\theta(\boldsymbol{x}_t \mid \boldsymbol{y}_t) = \frac{1}{W_t} \sum_{k=1}^{K} w_t^{(k)} \mathbb{I}\left(x_t - x_t^{(k)}\right). \tag{1.12}$$

where

$$W_t = \sum_{k=1}^{K} w_t^{(k)},$$

and the set

$$\left\{\boldsymbol{x}_t^{(k)}, \boldsymbol{w}_t^{(k)}\right\}_{k=1}^{K}$$

is called a set of *properly weighted samples* with respect to the target distribution $p_\theta(\boldsymbol{x}_t \mid \boldsymbol{y}_t)$ [12].

Suppose a set of properly weighted samples

$$\left\{\boldsymbol{x}_{t-1}^{(k)}, \boldsymbol{w}_{t-1}^{(k)}\right\}_{k=1}^{K}$$

with respect to $p_\theta(\boldsymbol{x}_{t-1} \mid \boldsymbol{y}_{t-1})$ is available at time $(t - 1)$. The SMC procedure generates a new set of samples and weights

$$\left\{\boldsymbol{x}_t^{(k)}, \boldsymbol{w}_t^{(k)}\right\}_{k=1}^{K},$$

properly weighted with respect to $p_\theta(\boldsymbol{x}_t \mid \boldsymbol{y}_t)$, from the previous set. In particular, if we choose the optimal trial distribution

$$q_{\boldsymbol{\theta}}\left(x_t \mid \boldsymbol{x}_{t-1}^{(k)}, \boldsymbol{y}_t\right) = p_{\boldsymbol{\theta}}\left(x_t \mid \boldsymbol{x}_{t-1}^{(k)}, \boldsymbol{y}_t\right)$$

and if x_t can only take values from a finite set, say $\mathcal{X} = \{b_1, b_2, \cdots, b_{|\mathcal{X}|}\}$, assuming $\boldsymbol{\theta}$ is known, then the SMC procedure is as follows [12]:

Algorithm 1.1: Sequential Monte Carlo (SMC)

1. For $i = 1, \cdots, \mathcal{X}$, compute

$$q_{\boldsymbol{\theta}}\left(x_t = b_i \mid \boldsymbol{x}_{t-1}^{(k)}, \boldsymbol{y}_t\right) = p_{\boldsymbol{\theta}}\left(x_t = b_i \mid \boldsymbol{x}_{t-1}^{(k)}, \boldsymbol{y}_t\right)$$

up to a normalizing constant

$$q_{\boldsymbol{\theta}}\left(x_t = b_i \mid \boldsymbol{x}_{t-1}^{(k)}, \boldsymbol{y}_t\right) \propto p_{\boldsymbol{\theta}}\left(y_t \mid x_t = b_i\right) p_{\boldsymbol{\theta}}\left(x_t = b_i \mid x_{t-1}^{(k)}\right). \tag{1.13}$$

2. Normalize these values such that

$$\sum_{b\in\mathcal{X}} q_{\boldsymbol{\theta}}\left(b \mid \boldsymbol{x}_{t-1}^{(k)}, \boldsymbol{y}_t\right)=1.$$

3. Draw a sample $x_t^{(k)}$ from the trial distribution

$$q_{\boldsymbol{\theta}}\left(\cdot \mid \boldsymbol{x}_{t-1}^{(k)}, \boldsymbol{y}_t\right)$$

   and let

$$\boldsymbol{x}_t^{(k)}=\left(\boldsymbol{x}_{t-1}^{(k)}, x_t^{(k)}\right).$$

4. Update the importance weight

$$w_t^{(k)} \propto w_{t-1}^{(k)} p_{\boldsymbol{\theta}}\left(y_t \mid x_{t-1}^{(k)}\right) \propto w_{t-1}^{(k)} \sum_{b\in\mathcal{X}} p_{\boldsymbol{\theta}}\left(y_t \mid x_t=b\right) p_{\boldsymbol{\theta}}\left(x_t=b \mid x_{t-1}^{(k)}\right). \tag{1.14}$$

5. Normalize the importance weights for them to sum up to 1.

A common problem with the SMC algorithm is known as the degeneracy phenomenon. In [12] it is shown that the variance of the importance weights can only increase over time, which makes the degeneracy problem ineluctable. After a few iterations, some samples will have very small weights. Such samples are said to be ineffective. If there are too many ineffective samples, the Monte Carlo procedure becomes inefficient. The *resampling* scheme is a useful method for reducing ineffective samples and enhancing effective ones. One simple resampling scheme can be described as follows (cf. [15] for other schemes):

Algorithm 1.2: Simple Resampling Scheme

1. Draw K sample streams $\{\overline{\boldsymbol{x}}_t^{(j)}\}_{j=1}^K$ from $\{\boldsymbol{x}_t^{(k)}\}_{k=1}^K$ with probabilities proportional to the weights $\{w_t^{(k)}\}_{k=1}^K$.
2. Assign equal weights to each stream, $\overline{w}_t^{(k)} = K^{-1}$.

The degeneracy of the samples can be measured by the effective sample size, which is defined and approximated respectively by [16]

$$K_{\mathit{eff}} \triangleq K\left[1+\mathrm{Var}\left(\frac{p_{\boldsymbol{\theta}}\left(\boldsymbol{x}_t^{(k)} \mid \boldsymbol{y}_t\right)}{q_{\boldsymbol{\theta}}\left(\boldsymbol{x}_t^{(k)} \mid \boldsymbol{x}_{t-1}^{(k)}, \boldsymbol{y}_t\right)}\right)\right]^{-1}, \widehat{K_{\mathit{eff}}}=\left(\sum_{k=1}^{K}\left(w_t^{(k)}\right)^2\right)^{-1}. \tag{1.15}$$

Heuristically, $\widehat{K_{eff}}$ reflects the equivalent size of a set of independent and identically distributed (i.i.d.) samples for the set of K weighted ones. It is suggested in [15, 16] that resampling should be performed whenever the effective sample size becomes small, e.g., $\widehat{K_{eff}} \le \frac{K}{10}$.

1.3.2.1 Prior and Posterior Distributions

Modeling realistic priors for a particular application is a difficult task. It is therefore common to choose priors that convey little prior knowledge or ease the calculations. A well-known strategy for Monte Carlo Markov Chain (MCMC) computation is to choose the prior distributions with a suitable form so that the posteriors belong to the same functional family as the priors. The priors and posteriors are then said to be conjugate [17]. The choice of the functional family depends on the likelihood. We will use the following *conjugate* strategy:

1.3.2.1.1 Prior Distributions

Denote $\boldsymbol{a}_i \triangleq [a_{i,1}, \cdots a_{i,N}]$ the ith row of the state transition probability matrix $\boldsymbol{A}$, $i = 1, \cdots, N$. It can be seen here that the discrete states x_t are drawn from multinomial distributions. For this kind of likelihood function it is well known that the Dirichlet distribution provides conjugate priors. We will therefore assume multivariate Dirichlet priors for both the initial probability vector $\boldsymbol{\pi}$ and $\boldsymbol{a}_i$.

The multivariate Dirichlet distribution $\mathcal{D}(\gamma_1, \cdots, \gamma_N)$ with strictly positive shape parameters $\gamma_1, \gamma_2, \cdots, \gamma_N$ has the following density function,

$$p(\boldsymbol{u}) = \left(\Gamma\left(\sum_{i=1}^{N} \gamma_i \right) \Big/ \prod_{i=1}^{N} \Gamma(\gamma_i) \right) \cdot \prod_{i=1}^{N} u_i^{\gamma_i - 1}, \tag{1.16}$$

where $\Gamma(\cdot)$ is the Gamma function and $\boldsymbol{u} = [u_1, \cdots, u_N]$ is such that $u_i \ge 0$ and

$$\sum_{i=1}^{N} u_i = 1.$$

It is easy to draw samples from such a distribution by using Gamma distributed samples. The prior distributions for the unknown parameters and network states are thus as follows:

1. The prior distribution for the initial probability vector π is given by

$$\pi \sim \mathcal{D}(\rho_1, \rho_2, \cdots, \rho_N). \tag{1.17}$$

2. The prior distribution for the ith row of the transition probability matrix A is given by

$$\boldsymbol{a}_i \sim \mathcal{D}(\alpha_{i,1}, \alpha_{i,2}, \cdots, \alpha_{i,N}) \quad i = 1, \cdots, N. \tag{1.18}$$

Note that for large N, it is common to assume that the matrix $\boldsymbol{A}$ is banded, i.e., $a_{i,j} = 0$ for $|i - j| > \delta$. Correspondingly, with some notational abuse, in the prior distribution of $\boldsymbol{a}_i$, we set $\alpha_j = -\infty$, if $|i-j| > \delta$. The resulting Dirichlet distribution then has a reduced dimension, i.e., $(a_{i,i-\delta},\cdots,a_{i,i},\cdots,a_{i,i+\delta}) \sim \mathcal{D}(\alpha_{i,i-\delta},\cdots,\alpha_{i,i},\cdots,\alpha_{i,i+\delta})$.

3. Finally, the prior distribution for the network state, x_t, is imposed by our choice of (1.17) and (1.18). It can be sampled from its prior distribution by using the samples $\pi^{(0)}$ and $\boldsymbol{A}^{(0)}$ of $\boldsymbol{\pi}$ and $\boldsymbol{A}$,

$$x_t \sim \mathcal{MC}(\pi^{(0)}, \boldsymbol{A}^{(0)}). \tag{1.19}$$

1.3.2.1.2 *Conditional Posterior Distributions*

Based on our model and our choice of prior distributions, we get the following conditional posterior distributions (using Bayes rules and Markovian assumptions):

1. The conditional posterior distribution of the initial probability vector $\boldsymbol{\pi}$:

$$\begin{aligned} p\left(\pi \mid y_T, \mathbf{A}, x_T\right) &= p\left(\pi \mid x_0\right) \propto p\left(x_0 \mid \pi\right) p(\boldsymbol{\pi}) \propto \pi_{x_0} \prod_{j=1}^{N} \pi_j^{\rho_j - 1} = \prod_{j=1}^{N} \pi_j^{\rho_j + \mathbb{I}(x_0 - j) - 1} \\ &= \mathcal{D}\left(\pi; \rho_1 + \mathbb{I}(x_0 - 1), \rho_2 + \mathbb{I}(x_0 - 2), \cdots, \rho_N + \mathbb{I}(x_0 - N)\right), \end{aligned} \tag{1.20}$$

where $\mathcal{D}(\cdot;\boldsymbol{\gamma})$ is the p.d.f. of the Dirichlet distribution with parameters $\boldsymbol{\gamma}$.

2. The conditional posterior distribution of the ith row $\boldsymbol{a}_i$ of the transition probability matrix:

$$\begin{aligned} p\left(a_i \mid \boldsymbol{y}_T, \pi, \eta, x_T, a_{-i}\right) &= p\left(a_i \mid x_T, \pi\right) \propto p\left(x_T \mid a_i, \pi\right) p(a_i) \\ &\propto p\left(x_0 \mid \pi\right) \left[\prod_{t=1}^{T} p\left(x_t \mid x_{t-1}, a_i\right)\right] \prod_{j=1}^{N} a_{i,j}^{\alpha_{i,j}-1} \\ &\propto \pi_{x_0} \prod_{j=1}^{N} a_{i,j}^{n_{i,j}} \prod_{j=1}^{N} a_{i,j}^{\alpha_{i,j}-1} \propto \prod_{j=1}^{N} a_{i,j}^{\alpha_{i,j}+n_{i,j}-1} \\ &= \mathcal{D}\left(\boldsymbol{a}_i; \alpha_1 + n_{i,1}, \alpha_2 + n_{i,2}, \cdots, \alpha_N + n_{i,N}\right), \end{aligned} \tag{1.21}$$

where $n_{i,j}$ is the number of state transitions from state i to state j in x_T and $\boldsymbol{a}_i = (\boldsymbol{a}_1,\cdots,\boldsymbol{a}_{i-1},\boldsymbol{a}_{i+1},\cdots,\boldsymbol{a}_N)$. Note that if $\boldsymbol{A}$ is banded, then, $n_{i,j} = 0$ for $|i-j| > \delta$.

3. The conditional posterior distribution of network state x_t:

$$\begin{aligned} p\left(x_t = i \mid \boldsymbol{y}_T, \pi, \boldsymbol{A}, \boldsymbol{x}_{-t}\right) &\propto p(y_t \mid x_t = i) \cdot p\left(x_t = i, \boldsymbol{x}_{-t}, \pi, \boldsymbol{A}\right) \\ &\propto \begin{cases} \mathcal{B}\left(y_t; B, h(i)\right) \cdot a_{x_{t-1},i} \cdot a_{i,x_{t+1}}, & \text{if } t \in [2, T-1] \\ \mathcal{B}\left(y_t; B, h(i)\right) \cdot \pi_i \cdot a_{i,x_{t+1}}, & \text{if } t = 1 \\ \mathcal{B}\left(y_t; B, h(i)\right) \cdot a_{x_{t-1},i}, & \text{if } t = T \end{cases}, \end{aligned} \tag{1.22}$$

where $\boldsymbol{x}_{-t} \triangleq \left[x_1, \cdots, x_{t-1}, x_{t+1}, \cdots, x_T\right]$, and $\mathcal{B}(\cdot; B, p)$ denote a binomial p.d.f. with B trials and probability p.

1.3.2.2 Sequential Monte Carlo Estimation with Unknown Static Parameters

If the parameters $\boldsymbol{\theta}$ are unknown, the usual approach is to include the parameters into the state vector. Because of the static evolution of the parameters, the space of parameters is only explored during initialization, which is obviously inefficient. Several works have proposed better algorithms for dealing with static parameters within an SMC framework [18, 19]. Gilks and Berzuini [20] proposed to use MCMC moves within the SMC framework. Such a procedure avoids the usual sample depletion problem but requires the storage of the complete path of the particles and increases the computational load. In [21], it is shown that sufficient statistics can be used to perform these MCMC steps without the growing storage requirement. Here we use the approach developed in [14], which also uses sufficient statistics. In the model under consideration, the posterior distribution of $\boldsymbol{\theta} = \{\pi, \boldsymbol{A}\}$ given $\boldsymbol{x}_t$ and $\boldsymbol{y}_t$ has been shown in (1.20) and (1.21) to be defined by Dirichlet distributions. It therefore depends on some sufficient statistics $\boldsymbol{T}_t = \boldsymbol{T}_t(\boldsymbol{x}_t, \boldsymbol{y}_t)$ that can easily be updated.

We consider a generic case where the parameters of the probability density function depend on some sufficient statistics, $p(\boldsymbol{\theta}|\boldsymbol{x}_t, \boldsymbol{y}_t) = p(\boldsymbol{\theta}|\boldsymbol{T}_t(\boldsymbol{x}_t, \boldsymbol{y}_t))$. Since $p(\boldsymbol{\theta}|\boldsymbol{x}_t, \boldsymbol{y}_t)$ is easily updated, we are interested in having a Monte Carlo approximation of $p(\boldsymbol{x}_t|\boldsymbol{y}_t)$. Suppose a set of properly weighted samples $\{\boldsymbol{x}_{t-1}^{(k)}, w_{t-1}^{(k)}\}_{k=1}^K$ with respect to $p(\boldsymbol{x}_{t-1}|\boldsymbol{y}_{t-1})$ is available at time $(t-1)$,

$$\hat{p}\left(\boldsymbol{x}_{t-1} \mid \boldsymbol{y}_{t-1}\right) = \frac{1}{W_{t-1}} \sum_{k=1}^{K} w_{t-1}^{(k)} \mathbb{I}\left(\boldsymbol{x}_{t-1} - \boldsymbol{x}_{t-1}^{(k)}\right). \tag{1.23}$$

The main idea is to get a Monte Carlo approximation of $p(\boldsymbol{x}_t, \boldsymbol{\theta}|\boldsymbol{y}_t)$ from (1.23) and the set of sufficient statistics

$$\left\{\boldsymbol{T}_t^{(k)}\right\}_{k=1}^{K} = \left\{\boldsymbol{T}_t(\boldsymbol{x}_t^{(k)}, \boldsymbol{y}_t)\right\}_{k=1}^{K}.$$

The approximation of the marginal distribution $p(\boldsymbol{x}_t|\boldsymbol{y}_t)$ is then simply obtained by discarding the samples $\boldsymbol{\theta}^{(k)}$. Therefore, only the samples $\boldsymbol{x}_t^{(k)}$ and the corresponding sufficient statistics $\boldsymbol{T}_t^{(k)}$ are stored, but samples for $\boldsymbol{\theta}$ are drawn jointly to $x_t^{(k)}$ to simplify the computations. Specifically, this approach is based on the following identity:

$$\begin{aligned} p(\boldsymbol{x}_t, \boldsymbol{\theta} \mid y_t) &\propto p(\boldsymbol{x}_t, \boldsymbol{\theta}, y_t \mid \boldsymbol{y}_{t-1}) \\ &\propto p(\boldsymbol{x}_{t-1} \mid \boldsymbol{y}_{t-1}) p(\boldsymbol{\theta} \mid \boldsymbol{x}_{t-1}, \boldsymbol{y}_{t-1}) p(x_t \mid \boldsymbol{x}_{t-1}, \boldsymbol{y}_{t-1}, \boldsymbol{\theta}) p(y_t \mid \boldsymbol{x}_t, \boldsymbol{y}_{t-1}, \boldsymbol{\theta}) \\ &\propto p(\boldsymbol{x}_{t-1} \mid \boldsymbol{y}_{t-1}) p(\boldsymbol{\theta} \mid \boldsymbol{T}_{t-1}) p(x_t \mid x_{t-1}, \boldsymbol{\theta}) p(y_t \mid x_t, \boldsymbol{\theta}). \end{aligned} \tag{1.24}$$

Based on the importance sampling paradigm, a Monte Carlo approximation of (1.24) can be obtained by keeping the past simulated streams $\{\boldsymbol{x}_{t-1}^{(k)}, w_{t-1}^{(k)}\}_{k=1}^K$ unmodified and drawing $(\boldsymbol{\theta}^{(k)}, x_t^{(k)})$ from a proposal distribution $q(\boldsymbol{\theta}, x_t | \boldsymbol{x}_{t-1}^{(k)}, \boldsymbol{y}_t) = q_1(\boldsymbol{\theta} | \boldsymbol{x}_{t-1}^{(k)}, \boldsymbol{y}_t) \cdot q_2(x_t | \boldsymbol{x}_{t-1}^{(k)}, \boldsymbol{y}_t, \boldsymbol{\theta})$. The weights are updated according to the usual rule:

$$w_t^{(k)} \propto w_{t-1}^{(k)} \frac{p\left(\boldsymbol{\theta}^{(k)} | \boldsymbol{T}_{t-1}^{(k)}\right) p\left(x_t^{(k)} | x_{t-1}^{(k)}, \boldsymbol{\theta}^{(k)}\right) p\left(y_t | x_t^{(k)}, \boldsymbol{\theta}^{(k)}\right)}{q_1\left(\boldsymbol{\theta}^{(k)} | \boldsymbol{x}_{t-1}^{(k)}, \boldsymbol{y}_t\right) q_2\left(x_t^{(k)} | \boldsymbol{x}_{t-1}^{(k)}, \boldsymbol{y}_t, \boldsymbol{\theta}^{(k)}\right)}. \tag{1.25}$$

The sufficient statistics are then updated and the samples for $\boldsymbol{\theta}$ discarded. Estimation of $\boldsymbol{\theta}$ is done through Rao-Blackwellization as follows [22]:

$$\mathbb{E}\{\boldsymbol{\theta} | \boldsymbol{y}_t\} = \mathbb{E}_{\boldsymbol{x}_t | \boldsymbol{y}_t}\left\{\mathbb{E}\{\boldsymbol{\theta} | \boldsymbol{y}_t, \boldsymbol{x}_t\}\right\} \approx \frac{1}{W_t} \sum_{k=1}^{K} w_t^{(k)} \mathbb{E}\left\{\boldsymbol{\theta} | \boldsymbol{T}_t^{(k)}\right\}. \tag{1.26}$$

Resampling can be performed as usual.

1.3.2.3 An Online SMC Estimator

We next outline the SMC algorithm for online estimation of the number of competing terminals when the system parameters $\boldsymbol{\theta} = [\pi, \boldsymbol{A}]$ are unknown. The prior distributions (1.17) and (1.18) for the parameters will be used hereafter. At time $(t - 1)$, the posterior distribution of $\boldsymbol{\theta} = \{\pi, \boldsymbol{A}\}$ given $\boldsymbol{x}_t$ and $\boldsymbol{y}_t$ has been shown in (1.20) and (1.21) to be given by some Dirichlet distributions. Let us denote $\rho_{i,t-1}$ and $\alpha_{i,j,t-1}$ the parameters of these distributions:

$$\begin{aligned} p\left(\pi | \boldsymbol{x}_{t-1}, \boldsymbol{y}_{t-1}\right) &= \mathcal{D}\left(\pi; \rho_{1,t-1}, \rho_{2,t-1}, \cdots, \rho_{N,t-1}\right), \\ p\left(\boldsymbol{a}_i | \boldsymbol{x}_{t-1}, \boldsymbol{y}_{t-1}\right) &= \mathcal{D}\left(\boldsymbol{a}_i; \alpha_{i,1,t-1}, \cdots, \alpha_{i,N,t-1}\right). \end{aligned} \tag{1.27}$$

The posterior distribution of the parameters therefore only depends on the sufficient statistics $\boldsymbol{T}_{t-1} = \{\alpha_{i,j,t-1}, \rho_{m,t-1}\}_{(i,j,m)\in[1,N]^3}$. Furthermore, we have

$$\begin{aligned} p\left(\pi | \boldsymbol{x}_t, \boldsymbol{y}_t\right) &= p\left(\pi | \boldsymbol{x}_{t-1}, \boldsymbol{y}_{t-1}\right), \\ p\left(\boldsymbol{a}_i | \boldsymbol{x}_t, \boldsymbol{y}_t\right) &= p\left(\boldsymbol{a}_i | \boldsymbol{x}_t\right) \propto p\left(x_t | x_{t-1}, \boldsymbol{a}_i\right) p\left(\boldsymbol{a}_i | \boldsymbol{x}_{t-1}\right) \end{aligned} \tag{1.28}$$

$$= \mathcal{D}(\boldsymbol{a}_i; \underbrace{\alpha_{i,1,t-1} + \mathbb{I}(x_{t-1} - i)\mathbb{I}(x_t - 1)}_{\alpha_{i,1,t}}, \cdots, \underbrace{\alpha_{i,N,t-1} + \mathbb{I}(x_{t-1} - i)\mathbb{I}(x_t - N)}_{\alpha_{i,N,t}}), \tag{1.29}$$

so that $\boldsymbol{T}_t$ is easily updated. It can also be seen that the trial distribution $q_2(x_t | \boldsymbol{x}_{t-1}^{(k)}, \boldsymbol{y}_t, \boldsymbol{\theta})$ has no reason to depend on π since x_{t-1} is given. Therefore, we only need to consider the transition matrix $\boldsymbol{A}$ in the parameters. The initialization step is derived in a straightforward manner from the present discussion.

We consider the optimal proposal distribution for the number of terminals:

$$q_2\left(x_t = i \mid \boldsymbol{x}_{t-1}^{(k)}, \boldsymbol{y}_t, \boldsymbol{\theta}\right) = p\left(x_t = i \mid \boldsymbol{x}_{t-1}^{(k)}, \boldsymbol{y}_t, \boldsymbol{\theta}\right) \propto \mathcal{B}\left(y_t; B, h(i)\right) \cdot a_{x_{t-1}^{(k)}, i}. \tag{1.30}$$

Sampling $\boldsymbol{A}$ is a little bit more involved if we want to include the latest observation in the proposal distribution. It is shown in the appendix that the posterior distribution of $\boldsymbol{a}_i$ given $\boldsymbol{x}_{t-1}$ and $\boldsymbol{y}_t$ is a mixture of Dirichlet distributions, which we use as a proposal distribution:

$$\begin{aligned} q_1\left(\boldsymbol{A} \mid \boldsymbol{x}_{t-1}^{(k)}, \boldsymbol{y}_t\right) &= \prod_{i=1}^{N} p\left(\boldsymbol{a}_i \mid \boldsymbol{x}_{t-1}^{(k)}, \boldsymbol{y}_t\right) \propto \prod_{i=1}^{N}\left[\sum_{j=1}^{N} \beta_{i,j,t}\, p\left(\boldsymbol{a}_i \mid \boldsymbol{T}_t(x_t = j, \boldsymbol{x}_{t-1}^{(k)}, \boldsymbol{y}_t)\right)\right] \\ &\propto \sum_{j=1}^{N} \beta_{x_{t-1}^{(k)}, j, t}\, p\left(\boldsymbol{a}_{x_{t-1}^{(k)}} \mid \boldsymbol{T}_t(x_t = j, \boldsymbol{x}_{t-1}^{(k)}, \boldsymbol{y}_t)\right) \prod_{i \neq j} p\left(\boldsymbol{a}_j \mid \boldsymbol{T}_t(\boldsymbol{x}_{t-1}^{(k)}, \boldsymbol{y}_{t-1})\right), \end{aligned} \tag{1.31}$$

where $\beta_{i,j,t} = \mathcal{B}\mathbb{I}(y_t; B, h(j))\alpha_{i,j,t-1}^{\mathbb{I}(x_{t-1}-i)}$. The weight update formula (1.25) can now be computed and its derivation can be found in the appendix:

$$w_t \propto w_{t-1} \frac{\sum_{i=1}^{N} \mathcal{B}\left(y_t; B, h(i)\right) \alpha_{x_{t-1}, i, t-1}}{\sum_{i=1}^{N} \alpha_{x_{t-1}, i, t-1}}. \tag{1.32}$$

Interestingly, the weight update formula does not depend on the actual values sampled at time t, so it is possible to compute the weights before sampling. This is very attractive since it allows to perform resampling before sampling, which lowers the loss of diversity occurring during the usual resampling scheme. The same idea appears in the auxiliary sample filter [23], where a part of the weight can be computed before sampling and the rest is roughly estimated. The approximate weights are then used to perform the resampling as a prior step. The major difference here is that the complete weights can be precomputed, and we advocate to use resampling only if necessary, e.g., whenever $\widehat{K_{eff}} \le \frac{K}{10}$. The complete SMC estimator with unknown parameters is summarized in algorithm 1.3.

Algorithm 1.3: Online SMC Estimator

1. Draw the initial samples of prior distributions $\pi^{(k)}$ according to $p(\pi \mid y_1)$, and $x_1^{(k)}$ according to $p(x_1 \mid y_1, \pi^{(k)})$. The corresponding weights are all equal.
2. **for** $t = 1, 2, \cdots$ **do**
3. Compute the new weights according to (1.32).
4. Compute $\widehat{K_{eff}}$ according to (1.15). If $\widehat{K_{eff}} \le \frac{K}{10}$, perform resampling.
5. **for** $k = 1, \ldots, K$ **do**

6. Sample $\boldsymbol{A}^{(k)}$ from (1.31).
7. Sample $x_t^{(k)}$ from (1.30).
8. Update the sufficient statistics $\boldsymbol{T}_1^{(k)} = \boldsymbol{T}_t(\boldsymbol{x}_t^{(k)}, \boldsymbol{y}_t)$.
9. **end for**
10. If necessary, estimate the posterior probability distribution of x_t and compute an estimate of $\boldsymbol{A}$ according to (1.26).
11. **end for**

1.3.3 Deterministic Sequential Monte Carlo Estimators

Algorithm 1.3 provides the basis for online Bayesian estimation. However, the discrete characteristic of the number of terminals and the fact that a set of sufficient statistics for the posterior probability of the unknown transition matrix can easily be updated allows us to significantly improve the basic SMC procedure. We will derive two new algorithms: a deterministic sequential sampling algorithm that outperforms the SMC estimator in terms of computational load, robustness, and accuracy; and a novel approximate maximum a posteriori algorithm that trades the accuracy of the deterministic sampling algorithm for computational load.

1.3.3.1 Deterministic Sequential Sampling

The online SMC estimator discussed in section 1.3.2.3 randomly generates samples according to $p(\boldsymbol{x}_t | \boldsymbol{x}_{t-1}^{(k)} \boldsymbol{y}_t)$ for $x_t \in \mathcal{X}$. Consequently, some information is distorted if the number of Monte Carlo samples is not sufficiently large. Furthermore, the use of the optimal sampling distribution implicitly led us to consider all possible extensions of a sample. The mixture densities indeed arose from this fact. This was possible because x_t can only take values from the finite set $\mathcal{X}$. An alternative deterministic approach, developed in [13, 24], and extended here to the case of unknown parameters and Markov state processes, consists of explicitly considering all K_{ext} possible extensions of the K samples and then performing a selection step so as to avoid the exponential increase of the number of samples and keep a constant number K of them. Another idea in this context is that there is no point in keeping different samples representing the same path. Therefore, the selection step should not rely on the usual resampling scheme. Here we rely on the simplest (but effective) idea, which is to select the K most likely samples at each time step. Strictly speaking, we drop the properly weighted characteristic by cutting out the tails of the p.d.f. during the selection step. To avoid this, a more sophisticated scheme is developed in [25, 26]; some of the most likely samples are kept, and resampling without replacement is performed on the remaining ones.

We consider again the state-space model (1.11) where the parameters $\boldsymbol{\theta}$ are assumed known. Suppose a set of weighted samples $\{\boldsymbol{x}_{t-1}^{(k)}, w_{t-1}^{(k)}\}_{k=1}^{K}$ representing $p_{\theta}(\boldsymbol{x}_{t-1} | \boldsymbol{y}_{t-1})$ is available at time $(t-1)$. We assume that this set does not contain any duplicate samples. The posterior distribution of $\boldsymbol{x}_{t-1}$ is approximated as

$$\hat{p}_{\boldsymbol{\theta}}\left(\boldsymbol{x}_{t-1} | \boldsymbol{y}_{t-1}\right) = \frac{1}{W_{t-1}} \sum_{k=1}^{K} w_{t-1}^{(k)} \mathbb{I}\left(\boldsymbol{x}_{t-1} - \boldsymbol{x}_{t-1}^{(k)}\right), \tag{1.33}$$

where

$$W_{t-1} = \sum_{k=1}^{K} w_{t-1}^{(k)}.$$

From Bayes theorem we have

$$\begin{aligned} p_{\boldsymbol{\theta}}(\boldsymbol{x}_t \mid \boldsymbol{y}_t) &\propto p_{\boldsymbol{\theta}}(y_t \mid \boldsymbol{x}_t, \boldsymbol{y}_{t-1}) p_{\boldsymbol{\theta}}(\boldsymbol{x}_t \mid \boldsymbol{y}_{t-1}) \\ &\propto p_{\boldsymbol{\theta}}(y_t \mid \boldsymbol{x}_t, \boldsymbol{y}_{t-1}) p_{\boldsymbol{\theta}}(x_t \mid \boldsymbol{x}_{t-1}, \boldsymbol{y}_{t-1}) p_{\boldsymbol{\theta}}(\boldsymbol{x}_{t-1} \mid \boldsymbol{y}_{t-1}), \end{aligned} \tag{1.34}$$

and the state transition distribution can be written as

$$p_{\boldsymbol{\theta}}(x_t \mid \boldsymbol{x}_{t-1}, \boldsymbol{y}_{t-1}) = p_{\boldsymbol{\theta}}(x_t \mid x_{t-1}) = \sum_{i=1}^{N} a_{x_{t-1},i} \mathbb{I}(x_t - i). \tag{1.35}$$

The posterior distribution of $\boldsymbol{x}_t$ can be approximated by

$$\hat{p}_{\boldsymbol{\theta}}^{ext}(\boldsymbol{x}_t \mid \boldsymbol{y}_t) = \frac{1}{W_t^{ext}} \sum_{k=1}^{K} \sum_{i=1}^{N} w_t^{(k,i)} \mathbb{I}\left(\boldsymbol{x}_t - \left[\boldsymbol{x}_{t-1}^{(k)}, i\right]\right), \tag{1.36}$$

where

$$W_t^{ext} = \sum_{k,i} w_t^{(k,i)}$$

and

$$w_t^{(k,i)} \propto w_{t-1}^{(k)} p_{\boldsymbol{\theta}}(y_t \mid x_t = i) p_{\boldsymbol{\theta}}(x_t = i \mid x_{t-1}^{(k)}). \tag{1.37}$$

The initialization steps of the algorithm proceed exactly as stated above except that no selection needs to be done until the total number of samples exceeds the maximum number allowed, N.

We now extend this approach to the case where the system parameters $\boldsymbol{\theta}$ are unknown but their posterior distribution, given $\boldsymbol{x}_t$ and $\boldsymbol{y}_t$, only depends on a set of sufficient statistics that can easily be updated, such as considered in section 1.3.2.2. Similarly to (1.34) we have

$$\begin{aligned} p(\boldsymbol{x}_t \mid \boldsymbol{y}_t) &\propto p(y_t \mid \boldsymbol{x}_t, \boldsymbol{y}_{t-1}) p(x_t \mid \boldsymbol{x}_{t-1}, \boldsymbol{y}_{t-1}) p(\boldsymbol{x}_{t-1} \mid \boldsymbol{y}_{t-1}) \\ &\propto p(x_t \mid \boldsymbol{x}_{t-1}, \boldsymbol{y}_{t-1}) \int p_{\boldsymbol{\theta}}(y_t \mid x_t) p(\boldsymbol{\theta} \mid x_t, \boldsymbol{x}_{t-1}, \boldsymbol{y}_{t-1}) \mathrm{d}\boldsymbol{\theta} \int p_{\boldsymbol{\theta}}(x_t \mid x_{t-1}) p(\boldsymbol{\theta} \mid \boldsymbol{T}_{t-1}) \mathrm{d}\boldsymbol{\theta}. \end{aligned} \tag{1.38}$$

Depending on the specific state-space (1.11) under consideration, evaluating (1.38) can be an easy or very difficult task. If no analytical form is available, it is possible to approximate these integrals. Monte Carlo sampling or the unscented transform [27] are some of the options, but one could also simply evaluate $p_{\bar{\boldsymbol{\theta}}}(\cdot)$, where $\bar{\boldsymbol{\theta}}$ could be the mean, mode, or any other likely value of $\boldsymbol{\theta}$ under $p(\boldsymbol{\theta}|x_t = i, \boldsymbol{x}_{t-1}, \boldsymbol{y}_{t-1})$ and $p(\boldsymbol{\theta}|\boldsymbol{T}_{t-1}^{(k)})$, respectively. Such an approximation is used, for example, in the auxiliary particle filter [23] during the auxiliary weights computation. This rough approximation should often be sufficient when $p_{\boldsymbol{\theta}}(\cdot)$ is smooth with respect to $\boldsymbol{\theta}$.

In our case, the emission probabilities $p_{\boldsymbol{\theta}}(y_t|x_t = i) = \mathcal{B}(y_t; B, h(i))$ do not depend on the parameters $\boldsymbol{\theta}$. Thanks to the Dirichlet prior, the integral with respect to $p(\boldsymbol{\theta}|\boldsymbol{T}_{t-1})$ can be computed analytically,

$$
\begin{aligned}
p\left(x_t \mid \boldsymbol{x}_{t-1}, \boldsymbol{y}_{t-1}\right) &= \int p_{\boldsymbol{\theta}}\left(x_t \mid x_{t-1}\right) p\left(\boldsymbol{\theta} \mid \boldsymbol{T}_{t-1}\right) \mathrm{d}\theta \\
&= \mathbb{E}_{p(\boldsymbol{\theta}|\boldsymbol{T}_{t-1})}\left\{a_{x_{t-1},x_t}\right\} = \frac{\alpha_{x_{t-1},x_t,t-1}}{\sum_{j=1}^{N} \alpha_{x_{t-1},j,t-1}}.
\end{aligned}
\tag{1.39}
$$

The recursion (1.38) can thus be computed analytically as well. If, at time $(t - 1)$, $\{\boldsymbol{x}_{t-1}^{(k)}, w_{t-1}^{(k)}\}_{k=1}^{K}$ represents $p(\boldsymbol{x}_{t-1}|\boldsymbol{y}_{t-1})$, then, as in (1.36), $p(\boldsymbol{x}_t|\boldsymbol{y}_t)$ can be approximated by

$$
\hat{p}^{ext}\left(\boldsymbol{x}_t \mid \boldsymbol{y}_t\right) = \frac{1}{W_t^{ext}} \sum_{k=1}^{K} \sum_{i=1}^{N} w_t^{(k,i)} \mathbb{I}\left(\boldsymbol{x}_t - \left[\boldsymbol{x}_{t-1}^{(k)}, i\right]\right), \tag{1.40}
$$

where the weight update is given by

$$
w_t^{(k,i)} \propto w_{t-1}^{(k)} \mathcal{B}\left(y_t; B, h(i)\right) \frac{\alpha^{(k)}_{x_{t-1}^{(k)},i,t-1}}{\sum_{j=1}^{N} \alpha^{(k)}_{x_{t-1}^{(k)},j,t-1}}. \tag{1.41}
$$

As one can see, such a procedure has many benefits since the parameters are analytically integrated out, no random sampling has to be performed, and no computation needs to be done twice. The complete procedure is summarized in algorithm 1.4.

Algorithm 1.4: Online Deterministic Estimator

1. Initialization: Enumerate the N possible samples and compute their weights.
2. **for** $t = 2, 3, \ldots$ **do**
3. **for** $k = 1, 2, \ldots$ **do**
4. Enumerate all possible sample extensions: $\boldsymbol{x}_t^{(k,i)} = [\boldsymbol{x}_{t-1}^{(k)}, i]$
5. $\forall i$, compute the weights $w_t^{(k,i)}$ according to (1.41)
6. **end for**

7. If necessary, estimate the posterior probability distribution of x_t.
8. Select and preserve N distinct sample streams $\{\boldsymbol{x}_t^{(k)}\}_{k=1}^N$ with the highest importance weights $\{w_t^{(k)}\}_{k=1}^N$ from the set $\{\boldsymbol{x}_t^{(k,i)}, w_t^{(k,i)}\}_{k,i}$.
9. $\forall k$, update the sufficient statistics $\boldsymbol{T}_t^{(k)} = \boldsymbol{T}_t(\boldsymbol{x}_t^{(k)}, \boldsymbol{y}_t)$.
10. If necessary, compute an estimate of $\boldsymbol{A}$ according to (1.26).
11. **end for**

A more accurate estimate of $\boldsymbol{A}$ can easily be obtained by updating the sufficient statistics and estimating $\boldsymbol{A}$ before the selection step. However, this would induce a heavier computational load.

1.3.3.2 Approximate MAP Estimator

For HMM with known parameters, the Viterbi algorithm provides a recursive solution to get the best state sequence estimation in terms of the maximum a posteriori (MAP) [28]. When the parameters are unknown, the most common procedure is to use an EM algorithm, which only converges to some local maximum of the a posteriori density, but above all, it is a batch procedure and thus cannot be used in our setting. Online estimation of the HMM parameters has been studied in [29–33]. In this section, another approach based on the use of sufficient statistics developed above is taken. An approximate MAP algorithm is presented whose computational load and memory need are equivalent to a usual Viterbi algorithm.

We are interested in recursively maximizing $p(\boldsymbol{x}_t \mid \boldsymbol{y}_t)$ with respect to $\boldsymbol{x}_t$. In order to do that, the Viterbi algorithm makes use of the quantity

$$\delta_t(i) = \max_{\boldsymbol{x}_{t-1}|x_t=i} p(\boldsymbol{x}_t \mid \boldsymbol{y}_t). \tag{1.42}$$

From (1.38) we have

$$\delta_t(i) = p(y_t \mid x_t = i) \max_{x_{t-1}|x_t=i} \max_{\boldsymbol{x}_{t-2}|x_{t-1},x_t=i} [p(\boldsymbol{x}_{t-1} \mid \boldsymbol{y}_{t-1}) p(x_t \mid \boldsymbol{x}_{t-1}, \boldsymbol{y}_{t-1})], \tag{1.43}$$

which can recursively be computed by taking $p(x_t \mid \boldsymbol{x}_{t-1}, \boldsymbol{y}_{t-1}) = a_{x_{t-1},x_t}$ out of the inner max so that $\delta_t(i) = p(y_t \mid x_t = i)\max_j[\delta_{t-1}(j) \cdot a_{j,i}]$. The estimate $\boldsymbol{x}_t$ of $\boldsymbol{x}_t$ at time t is then given by $\max_i[\delta_t(i)]$. When the transition matrix is unknown, even if the probability of any path can be analytically computed, such a recursion cannot directly be used because $p(x_t \mid \boldsymbol{x}_{t-1}, \boldsymbol{y}_{t-1})$ depends on $\boldsymbol{x}_{t-2}$. However, if we make the approximation that $p(\boldsymbol{x}_{t-1} \mid \boldsymbol{y}_{t-1})$ $p(x_t \mid \boldsymbol{x}_{t-1}, \boldsymbol{y}_{t-1})$ is maximized when only $p(\boldsymbol{x}_{t-1} \mid \boldsymbol{y}_{t-1})$ is maximized, we then get,

$$\begin{aligned} &\max_{\boldsymbol{x}_{t-2}|x_t=i,x_{t-1}=j} \left[p\left(\boldsymbol{x}_{t-1} \mid \boldsymbol{y}_{t-1}\right) p\left(x_t \mid \boldsymbol{x}_{t-1}, \boldsymbol{y}_{t-1}\right)\right] = \\ &p\left(\boldsymbol{x}_{t-2}^*, x_{t-1} = j \mid \boldsymbol{y}_{t-1}\right) p\left(x_t = i \mid \boldsymbol{x}_{t-2}^*, x_{t-1}, \boldsymbol{y}_{t-1}\right) \end{aligned}, \tag{1.44}$$

where $\boldsymbol{x}_{t-2}^* = \arg\max_{\boldsymbol{x}_{t-2}|x_t=i,x_{t-1}=j} p(\boldsymbol{x}_{t-1} \mid \boldsymbol{y}_{t-1})$. This allows us to derive an approximate MAP algorithm. The rationale behind the assumption above is that, as time goes on, our estimation of the transition matrix will stabilize. The impact of the transition probability should therefore be lower than that of the observation probability. Our simulations showed that (1.44) provides rather good results given the low complexity of the resulting algorithm. Our approximation $\hat{\delta}_t(i)$ of $\delta_t(i)$ can thus be recursively computed by keeping, for every possible value j, only the best path ending at this particular value j together with the corresponding set of sufficient statistics

$$\boldsymbol{T}_t^{(j)} = \left\{\alpha_{i,k,t}^{(j)}, \rho_{m,t}^{(j)}\right\}_{(i,k,m)\in[1,N]^3}.$$

The recursion is then given by

$$\hat{\delta}_t(i) = p\left(y_t \mid x_t = i\right) \max_j \left[\hat{\delta}_{t-1}(j) \cdot \frac{\alpha_{j,i,t-1}^{(j)}}{\sum_{k=1}^{N} \alpha_{j,k,t-1}^{(j)}}\right]. \tag{1.45}$$

Let $\boldsymbol{x}_t$ be our estimate of $\boldsymbol{x}_t$ at time t. The approximate MAP algorithm is presented in algorithm 1.5.

Algorithm 1.5: Approximate MAP Algorithm

1. Initialization: For $i = 1, \cdots, N$, set the weight of each point to

$$\hat{\delta}_1(i) = p\left(x_1 = i \mid y_1\right) \propto \mathcal{B}\left(y_1; B, h(i)\right) \frac{\rho_i}{\sum_j \rho_j}.$$

2. **for** $t = 2, 3, \cdots$ **do**
3.   **for** $i = 1, \cdots, N$ **do**
4.   Set $\hat{\delta}_t(i) = \mathcal{B}(y_t; B, h(i)) \max_j \left[\hat{\delta}_{t-1}(j) \cdot \frac{\alpha_{j,i,t-1}^{(j)}}{\sum_{k=1}^{N} \alpha_{j,k,t-1}^{(j)}}\right].$
5.   Set $\hat{\psi}_t(i) = \text{argmax}_j \left[\hat{\delta}_{t-1}(j) \cdot \frac{\alpha_{j,i,t-1}^{(j)}}{\sum_{k=1}^{N} \alpha_{j,k,t-1}^{(j)}}\right].$
6.   Set $\boldsymbol{x}_t^{(i)} = \left[\boldsymbol{x}_{t-1}^{(\hat{\psi}_t(i))}, i\right].$
7.   Update the sufficient statistics: $\boldsymbol{T}_t^{(i)} = \boldsymbol{T}_t(\boldsymbol{x}_t^{(i)}, \boldsymbol{y}_t)$.
8.   **end for**
9.   If necessary, get the approximate maximum likelihood (ML) estimate $\hat{\boldsymbol{x}}_t$ of $\boldsymbol{x}_t$ by using the sequence that maximizes $\delta_t(\cdot)$.
10.   If necessary, get an estimate of $\boldsymbol{A}$ from $\hat{\boldsymbol{A}} = E[\boldsymbol{A} \mid \boldsymbol{T}_t^{(\hat{x}_t)}]$.
11. **end for**

1.3.4 Performance of the SMC Estimators

We assume an 802.11 network as modeled in [2], where the relation between the number of competing terminals and the probability of collision is given by (1.6). Our scenario is composed of a variable number of competing stations x transmitting in saturation conditions. As in [2], only DCF basic access is considered, with no capture or hidden terminals. The arrival and departure of competing terminals from the network follow a random Markov chain as in (1.10). The exponential backoff parameters are CW_{min} = 16,32 and 64, with m = 4,5 and 6, respectively, i.e., $CW_{max} = 2^m\, CW_{min} = 1024$.

For the model-based simulations, we generate noisy observations from (1.10), and each station monitors the medium and estimates the probability of collision by counting the number of busy slots as indicated in (1.7) with $B = 100$. For each estimator that provides an approximation of the filtering density, we first make a hard estimate by taking the mode of the output distribution. The different estimators are then compared by using this hard estimate. For the real data simulations, we use the ns-2 network simulator version 2.27 [34] with the parameters described in section 1.5.1. For testing the estimators in a realistic scenario, we used the curves obtained in Figure 1.2(a) instead of the analytical model in (1.6). Our simulation scenario is composed of a variable number of competing stations x transmitting in saturation conditions. The terminals use $B = 100$ for estimating the collision probability. The arrival and departure of competing terminals to the network (to attach to the corresponding access point) follow a Markov process in one case, and an on-off exponential process in continuous time in the other. In our irregular time grid, this is not Markovian any more.

The effectiveness of the proposed estimators for all the parameters is summarized in Table 1.1. The SMC algorithms substantially outperform the CUSUM-EKF algorithm [3]. Both the deterministic algorithm and the approximate MAP algorithm perform in a manner similar to that of the SMC estimator. As we can see, the approximate MAP algorithm appears as an excellent option for real-time implementation given its lower complexity and excellent performance.

1.4 Adaptive Optimization of IEEE 802.11 DCF Based on the SMC Estimators

Once the number of competing terminals in an IEEE 802.11 DCF is estimated using the SMC methods proposed in the previous section, we can optimize the algorithm based on the estimations. In order to simplify the problem, we impose the number of competing terminals to be less than 40* and set m to be fixed such as $CW_{max} = 2^m\, CW_{min} = 1024$, i.e.,

$$m = \log_2\left(\frac{1024}{CW_{min}}\right),$$

* See section 5.1 for the rationale behind this assumption.

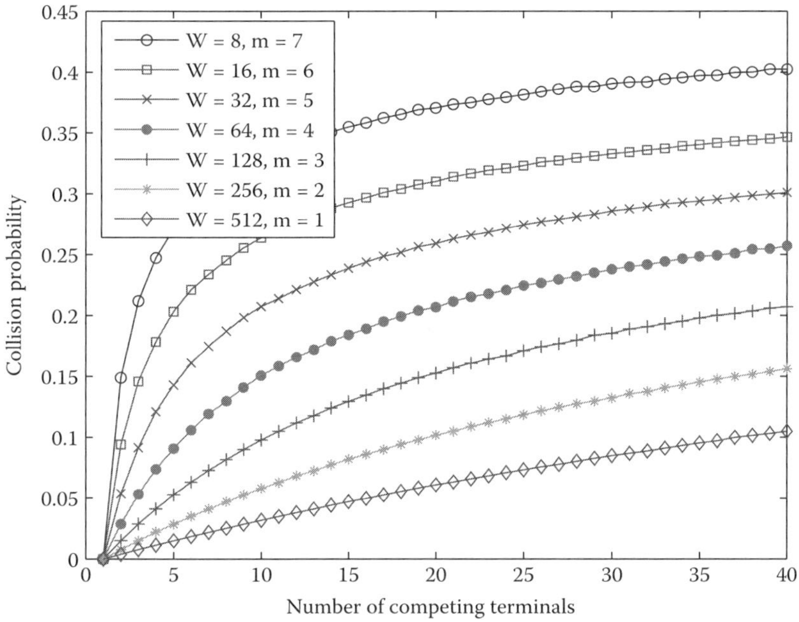

FIGURE 1.2 Collision probabilities and throughput in ns-2. (a) Collision probabilities from ns-2 used in the estimator. (b) Saturation throughput of the implementation of IEEE 802.11 in n2-s.

and CW_{min}* takes values from a set $\mathcal{W}$. This set can be fixed or it can be constructed, for example, using the method in section 1.4.2. Then, assuming m is no longer a variable, a simple formulation of the backoff window choice is given by

$$W_{t+1}^{*} = \arg\max_{W \in \mathcal{W}} \mathbb{E}_{p(x_{t+1}|\mathbf{y}_t)} \left\{ \Psi_u \left(S(x_{t+1}, W, m) - S(x_{t+1}, W_t, m) \right) \right\}, \tag{1.46}$$

where Ψ_u is a utility function of the difference in throughput, and $S(\cdot)$ is given in (1.5). Ψ_u will typically be a nondecreasing function and should be convex on the positive part and concave on the negative part.

Considering the choice of the cost function, we are interested in studying the case in which a change in CW_{min} negatively affects the normal operation of the protocol. Let v be the actual window size for an IEEE 802.11 terminal. Then by the operation of the protocol we know that $v(t) \in [W_{min}(t), 1024]$ and, in the next observation slot,

$$v(t+1) = \begin{cases} CW_{min}(t), & \text{if success in } t, \\ \min(2 \times v(t), 1024), & \text{if collision in } t. \end{cases} \tag{1.47}$$

---

* For ease of notation, we would use the term CW_{min} and W interchangeably.

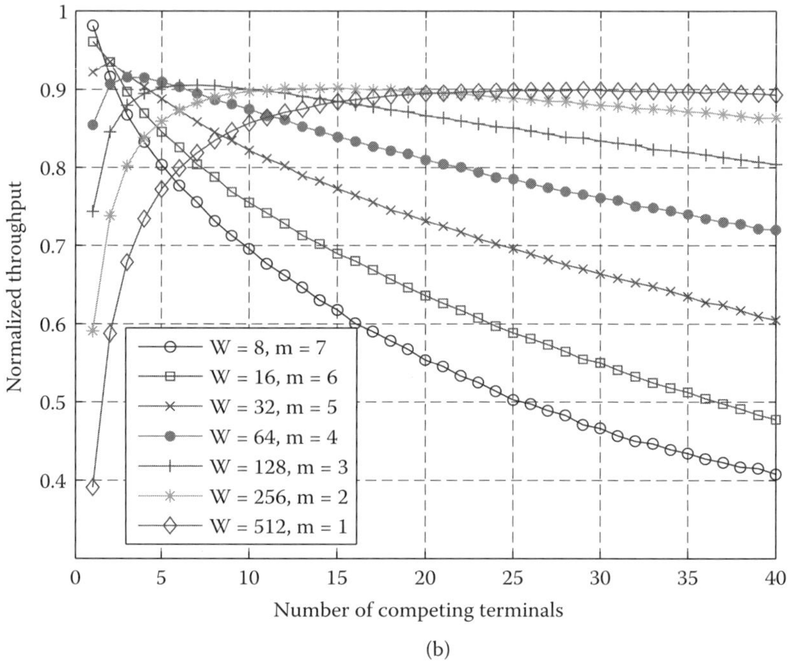

FIGURE 1.2 (continued).

TABLE 1.1 Average MSE of the hard estimation obtained from 100 data sets

Estimator	Model Based			ns-2 (Markov)		ns-2 (on-off exponential)		
CW_{min}, m	16, 6	32, 5	64, 4	32, 5	64, 4	16, 6	32, 5	64, 4
SMC	0.4635	0.4351	0.5420	1.2232	1.0213	1.6042	1.1111	0.9132
Deterministic	0.6533	0.5961	0.8056	1.1663	1.0733	1.4797	1.0614	1.0181
EKF + CUMSUM	1.1528	1.1820	1.2133	1.5236	2.8760	1.8903	2.3663	2.5957
Approximate MAP	0.6079	0.5180	0.6557	1.0680	0.9108	1.5338	1.0842	0.9100

Note that because the protocol dictates that $v(t) \in [CW_{min}(t), CW_{max}(t)]$, a change in CW_{min} may produce a jump in the value of v if $v(t) < CW_{min}(t+1)$. Figure 1.4(a) shows the average evolution of $v(t)$ with the number of competing terminals for different backoff parameters. As $x \rightarrow 20$, the optimal $CW_{min}(t) = 256$, and the average value for $v(t) \approx 330$. If $x(t+1) > 20$, the optimal value of $CW_{min}(t+1)$ becomes 512, forcing $v(t+1)$ to be at least 512, as indicated by the arrow. If the estimate $x(t+1) > 20$ is spurious and $x(t+2) < 20$ (so $CW_{min}(t+2)$ is back to 256 again), it would take a terminal an extra successful transmission to return v to the average correct level (i.e., 256), incurring an average delay of 128 slots. For this reason, the utility function Ψ_u must penalize oscillations of CW_{min}, so the change of window is not made for small differences.

Note that our utility function in (1.46) is not based on a hard decision on the number of competing terminals but makes use of its distribution if available. In [8, 35], a similar optimization scheme was introduced, but a hard estimate of the number of terminals was used to make a range estimation. To prevent frequent switching, the authors proposed use of overlapping ranges. We believe that this Bayesian criterion is more natural to make a soft decision.

1.4.1 Predictive Distribution Based on SMC Samples

As shown in our criterion (1.46), we need to have access to the predictive distribution $p(x_{t+1}|\mathbf{y}_t)$ in order to perform an optimal control of the protocol. Given a set of samples and weights $\{\mathbf{x}_t^{(k)}, w_t^{(k)}\}_{k=1}^K$ representing $p(\mathbf{x}_t|\mathbf{y}_t)$ at time t, (1.46) can be approximated as

$$
\begin{aligned}
\hat{W}_{t+1}^{SMC} &= \arg\max_{W\in\mathcal{W}} \sum_{k=1}^{K}\sum_{i=1}^{N} \Psi_u\left(\Delta S(x_{t+1}=i,W)\right) p\left(x_{t+1}=i,\mathbf{x}_t^{(k)} \mid \mathbf{y}_t\right) \\
&= \arg\max_{W\in\mathcal{W}} \sum_{k=1}^{K}\sum_{i=1}^{N} \Psi_u\left(\Delta S(i,W)\right) \frac{\alpha_{x_t^{(k)},i,t}}{\sum_{j=1}^{N}\alpha_{x_t^{(k)},j,t}} w_t^{(k)} \\
&= \arg\max_{W\in\mathcal{W}} \sum_{k=1}^{K} \frac{w_t^{(k)}}{\sum_{j=1}^{N}\alpha_{x_t^{(k)},j,t}} \sum_{i=1}^{N} \Psi_u\left(\Delta S(i,W)\right)\alpha_{x_t^{(k)},i,t},
\end{aligned}
\tag{1.48}
$$

where $\Delta S(x_{t+1},W) = S(x_{t+1},W,m) - S(x_{t+1},W_t,m)$.

For the case in which we only have access to a hard estimate of the number of competing terminals, the backoff window choice (1.46) is simply approximated by

$$
W_{t+1}^{MAP} = \arg\max_{W\in\mathcal{W}} \Psi_u\left(S\left(\hat{x}_{t+1|t},W,m\right) - S\left(\hat{x}_{t+1|t},W_t,m\right)\right), \tag{1.49}
$$

where $\hat{x}_{t+1|t} = \operatorname{argmax}_{x_{t+1}} p(x_{t+1} \mid \hat{x}_t, \mathbf{y}_t) \approx \operatorname{argmax}_{x_{t+1}} p(x_{t+1} \mid \mathbf{x}_t, \mathbf{y}_t)$ is an approximate MAP estimate of x_{t+1}, with $\hat{\mathbf{x}}_t$ being the current MAP estimate of $\mathbf{x}_t$.

For comparison with the EKF algorithm in [3], $p(x_t \mid y_t, y_{t-1},\ldots,y_1)$ is approximated by a Gaussian $p(x_t \mid y_t, y_{t-1},\ldots,y_1) \approx N(x_t; h(x_t), P_t)$. This would involve complex numerical integrations, so we use the hard estimate of the number of competing terminals, as in (1.49).

1.4.2 Choice of Backoff Window Size Set $\mathcal{W}$

Having discussed how to perform an optimal choice of the backoff window within a given set $\mathcal{W}$, we can now give some insight on the choice of this set. It will be chosen such that the optimal throughput can always be approached, and such that its cardinality remains low. Indeed, a small number of configurations will allow a more stable system and an easier implementation. Our design criterion can be written as

$$\forall i \in [1,\cdots,N], \quad \left| \max_{W \in \mathbb{N}^*} S(i,W,m) - \max_{W \in \mathcal{W}} S(i,W,m) \right| < \Delta S_{max}, \tag{1.50}$$

where ΔS_{max} is the maximum throughput loss to optimality we allow. ΔS_{max} will typically be chosen small, for instance, 2.5%. Within this constraint, we would like to have as few points in $\mathcal{W}$ as possible. Because of the regularity of $S(\cdot)$, such a set can be constructed by performing the following operations:

1. Let $i_{mid} = 1$.
2. Choose the greatest integer j_{ref} such that $S_{opt}(i_{mid}) - S(i_{mid},W_{opt}(j_{ref}),m) < \Delta S_{max}$, where $W_{opt}(k) = \text{argmax}_{W \in \mathbb{N}*} S(k,W,m)$ and $S_{opt}(k) = S(k,W_{opt}(k),m)$. Let $W_{opt}(j_{ref})$ be in $\mathcal{W}$.
3. Find the smallest integer $i_{mid} = j_{ref}$ such that $S_{opt}(i_{mid}) - S(i_{mid},W_{opt}(j_{ref}),m) > \Delta S_{max}$.
4. If $i_{mid} < N$ and $j_{ref} < N$, go back to step 2.
5. If $i_{mid} \geq N$ and $j_{ref} \geq N$, remove $W_{opt}(j_{ref})$ from $\mathcal{W}$ and let N be in $\mathcal{W}$.

1.5 Simulation Results

1.5.1 Simulation Setup

For the simulations we use the ns-2 network simulator version 2.27 [34]. We modified the 802.11 implementation so the terminals measure the observation slots as in (1.7) for the estimates of the collision probability. The parameters used in the simulation are classical for a 1 Mbps WLAN and are taken after [3] for a fair comparison. No packet fragmentation occurs, and the terminals are located close to each other to avoid capture or hidden terminal problems. The propagation delay is 1 μs. The packet size is fixed with a payload of 1024 bytes. The MAC and PHY headers use 272 and 128 bits, respectively. The PHY preamble takes 144 bits. The ACK length is 112 bits with an ACK timeout of 300 μs. The Rx/Tx turnaround time is 20 μs and the busy detect time 29 μs. The short retry limit and long retry limit are set to 7 and 4, respectively. Finally, the slot time is 50 μs, the SIFS 28 μs, and the DIFS 130 μs. The RTS/CTS threshold was increased so that only basic access was used.

For the cases in which an analytical model is not available, empirical models can also be used. Figure 1.2(a) shows the collision probability versus the number of competing terminals obtained empirically in the ns-2 simulator. Each point was obtained by simulating a fixed number of terminals transmitting under saturation conditions and measuring the total probability of collision. The simulation time for this empirical measurement lasted 3,000 s to provide better accuracy. To avoid including address resolution protocol (ARP) packets in the measurement, an initial 20 s transmission was used to ensure all the terminals had updated automatic repeat request (ARQ) tables. Finally, an additional 100 s transmission was added before measurements to allow the system to reach the steady state.

We assume that 40 is a reasonable upper limit for the number of competing terminals. To select the appropriate set $\mathcal{W}$ of backoff parameters, we measured the utilization of IEEE 802.11 for different values of x_t. Our simulations showed that there is almost no impact in performance for $CW_{max} > 1024$, so we fixed m such that $m = \log_2(1024/CW_{min})$. Because the number of parameters needs to be finite, we selected for CW_{min} the powers of 2 lower than 1,024, i.e., $\mathcal{W} = \{(8,7),(16,6),(32,5),(64,4),(128,3),(256,2),(512,1)\}$.

For the EKF estimator we used the parameters suggested in [3]: the state variance Q_t is set to 0 except when a change is detected, where Q_t is set to $Q_{max} = 10$. The initial error variance $P_0 = 100$. R_t is known and given by the observation model. For the change detection filter we used $v = 0.5$ and $h = 10$. These parameters were used by the authors in [3] for both the saturation and nonsaturation schemes, and thus we used the same values in our simulations.

Our simulation scenario is composed of a variable number of competing terminals x_t transmitting in saturation conditions. Each ns-2 simulation run lasts between 300 and 1,000 s. The arrival and departure of competing terminals to the network (to attach to the corresponding access point) follows an on-off exponential process in continuous time.

1.5.2 Effect of the Adaptive Choice of Parameters on the DCF Optimization

As discussed in section 1.1, the optimization of the IEEE 802.11 DCF based on the estimation of the number of competing terminals often trades off accuracy for complexity. However, we believe that given the sensitivity of the throughput to the number of competing terminals, as shown in Figure 1.2(b), especially in the 1–15 range, both the speed and accuracy of the estimator are crucial in order to rapidly select the optimal parameters to increase the network utilization. Figure 1.2(b) shows the saturation throughput for our backoff parameters set $\mathcal{W}$.

We implemented in ns-2 the optimization algorithm described in section 1.4 using the estimation algorithms described in section 1.3. For comparison purposes we also implemented an optimized version of PDCF [36] where the reset of the window to CW_{min} after successful transmission is 0.5 (on average the best option for the 1–40 terminals range). For ease of discussion, and because it is the most appropriate algorithm for a real implementation of the SMC estimators presented here, we only show the results of the MAP approximation described in algorithm 1.5.

Figure 1.3(a) shows the normalized throughput of the optimization DCF with respect to the standard DCF of $W = 32$, $m = 5$ and the PDCF implementation. As the figure shows, the increase in efficiency is dramatic. The benefit of the optimized algorithm with respect to the regular IEEE 802.11 is as high as 40% for large values of x. While a nonadaptive protocol can never outperform an adaptive one, it is interesting to see how simple modification of the existing vanilla DCF protocol results in a considerable throughput benefit. The actual average evolution of the actual window size v is shown in Figure 1.4(a), and the normalized throughput of the network wasted in collisions is shown in Figure 1.4(b).

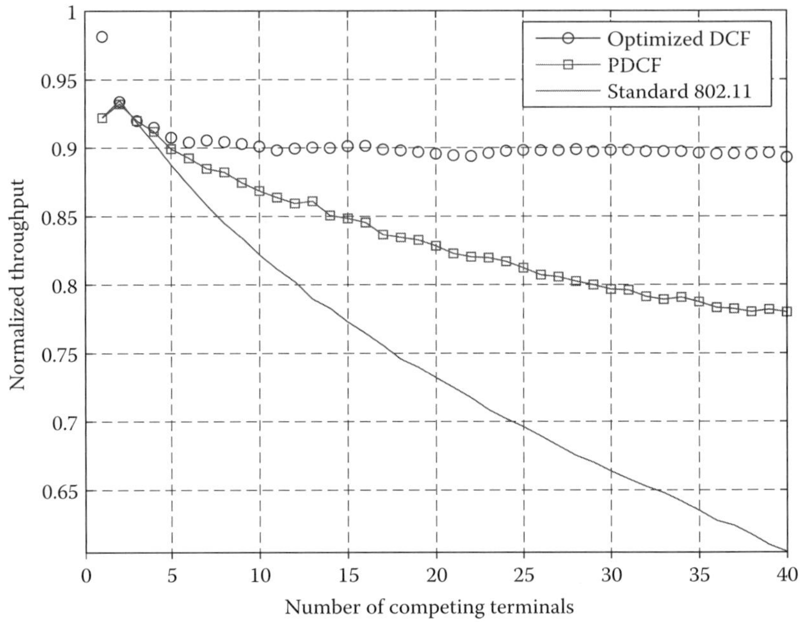

(a)

FIGURE 1.3 Performance of the optimized algorithm with respect to the standard DCF. (a) Total throughput of the network. (b) Observed probability of collision.

1.5.3 Instantaneous Network Utilization

Figure 1.5(a) shows the instantaneous network utilization of the optimization protocols when the terminals follow the step arrival shown in Figure 1.5(b). Figure 1.5(b) also shows the actual estimates for both the approximate MAP and the EKF algorithms. The estimation window size B is 100. Note that the accuracy of the estimate of the probability of collision is directly related to the value of B; a large B means more accuracy, but also greater delay in the estimation. A smaller B provides a noisy measurement of y_t but a faster reactive estimation. The speed of the estimator may be crucial when the number of competing terminals oscillates in the 1–15 range, as the decision regions for the optimal contention window size are narrower, and the estimator may miss the optimal points. In the step case, the algorithms have time to detect and estimate the number of competing terminals. The expected result is a flat line of maximum throughput, like the one shown in the perfect estimator, where the algorithm is fed with the actual number of competing terminals and not an estimate. The nonadaptive algorithms fall in throughput after the increment in the number of terminals. The approximate MAP algorithm outperforms the EKF algorithm in the estimates, and hence the positive effect in the network performance.

On the other hand, Figure 1.6(a) shows the instantaneous utilization of the protocols when the terminals follow an exponential on-off activation with a parameter of 10 s. A

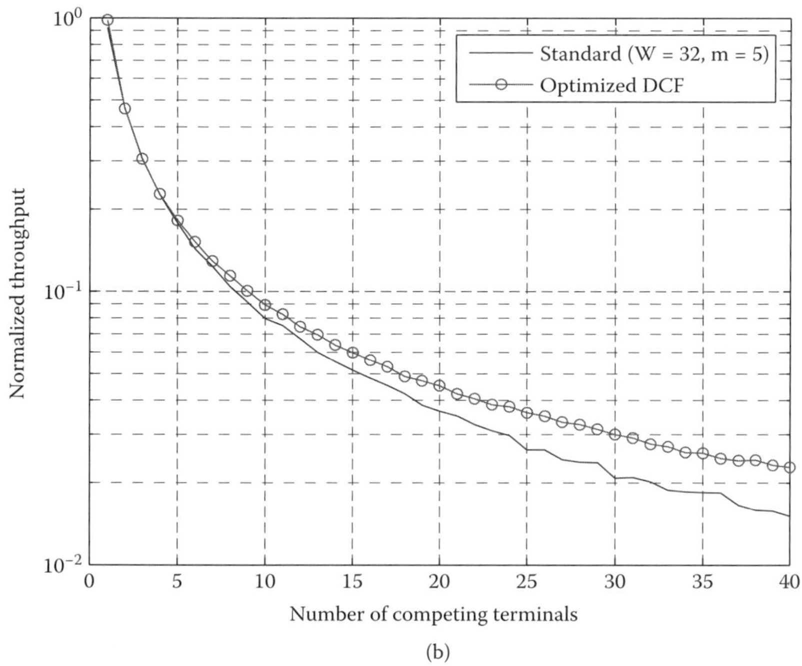

FIGURE 1.3 (continued).

terminal is active for an exponential time with $\lambda = 10$ s and then deactivates for an exponential time of $\lambda = 10$ s. The evolution of the number of competing terminals and the estimates of both the approximate MAP and EKF algorithms are shown in Figure 1.6(b). The fast and accurate tracking capabilities of the approximate MAP are evident, and its MSE is 14.1483, while the MSE of the EKF algorithm is 28.3253. We want to compare the effect of the estimation in time for $B = 50$, to keep the estimation within the granularity of the change in the number of terminals. We used an optimized version of the PDCF protocol for the range of 1–10 terminals (reset probability is 0.9). Note that the estimation algorithms never outperform the perfect estimation at any point. This is an indication of the benefit of the accurate estimation when optimizing the DCF. A protocol that simplifies the estimations will necessarily fall apart from the perfect curve. Moreover, the estimation protocols take some time to converge to their optimal operation. At time $t = 100s$ the optimization algorithms start their operation and take between 10 and 20 s to converge. Note that the approximate MAP algorithm outperforms both the EKF and the modified PDCF algorithm at all times, which is an indication of the benefits of its accuracy in the IEEE 802.11 operation.

1.5.4 Results under Nonsaturated Network Conditions

A common problem of the estimation mechanisms described in section 1.3.2.1 is that they base their estimations on the fact that the network is in saturation mode, i.e., at any

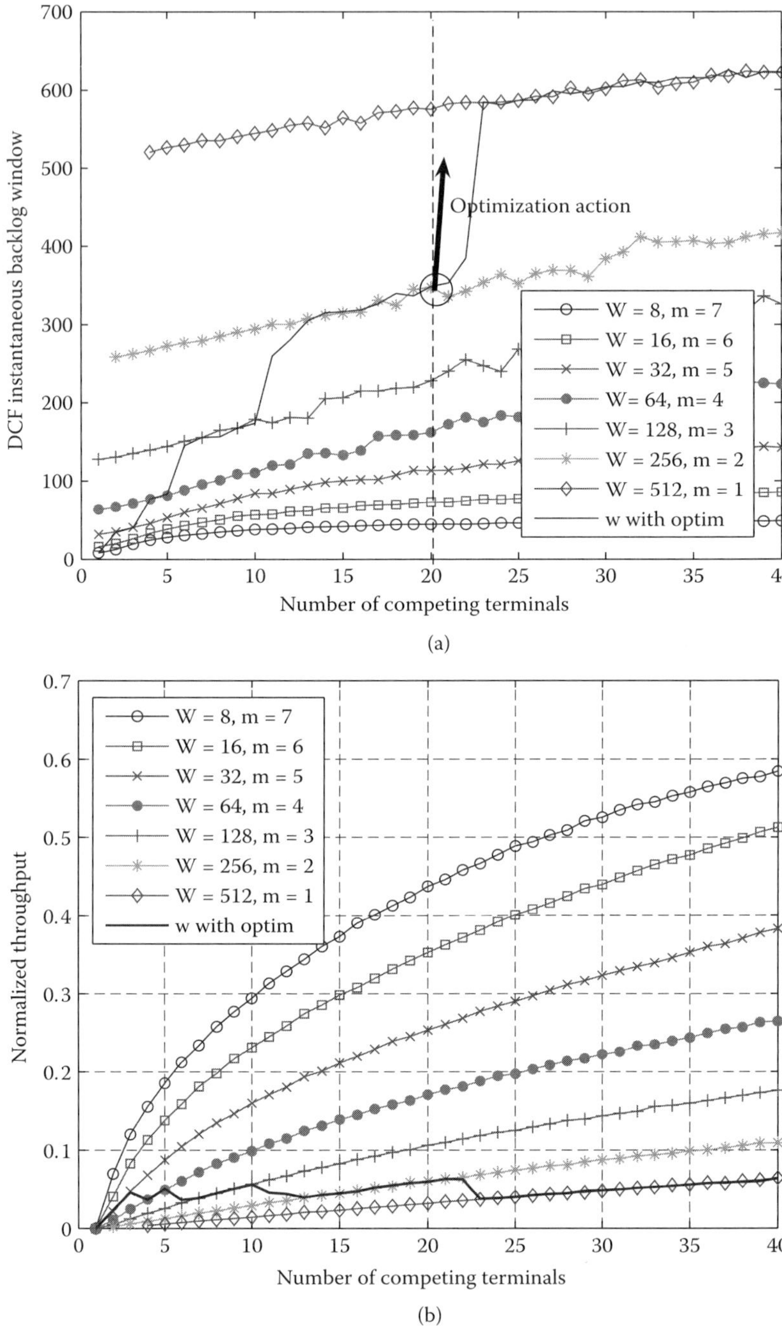

FIGURE 1.4 Performance of the optimized algorithm. (a) Evolution of the instantaneous backoff window versus the number of computing terminals for fixed backoff parameters. (b) Normalized throughput wasted in collisions.

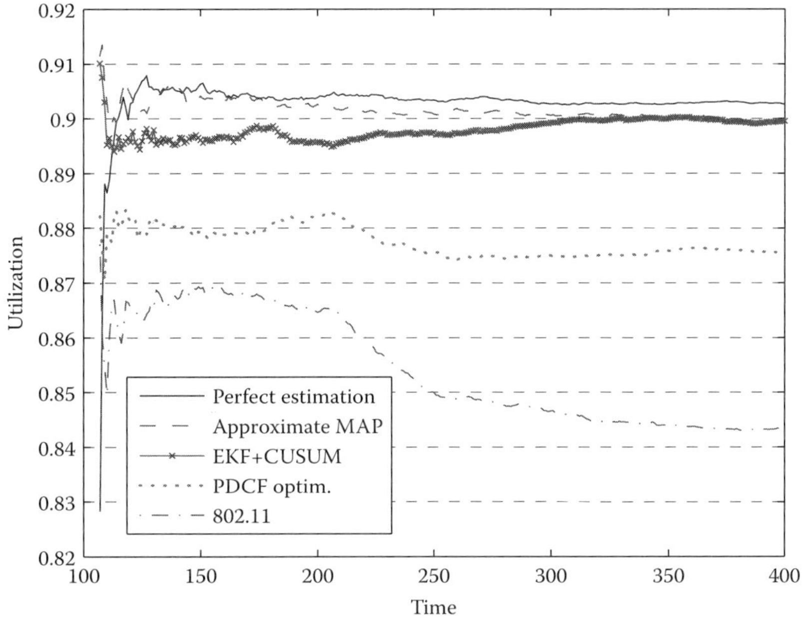

(a)

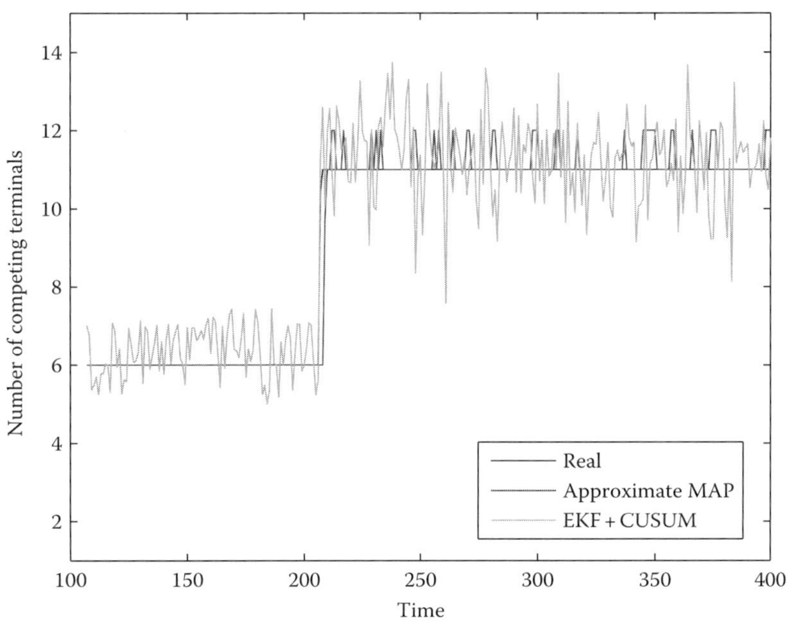

(b)

FIGURE 1.5 Instantaneous utilization when terminal's arrival has a step form. (a) Instantaneous utilization. (b) Evolution of the number of competing terminals.

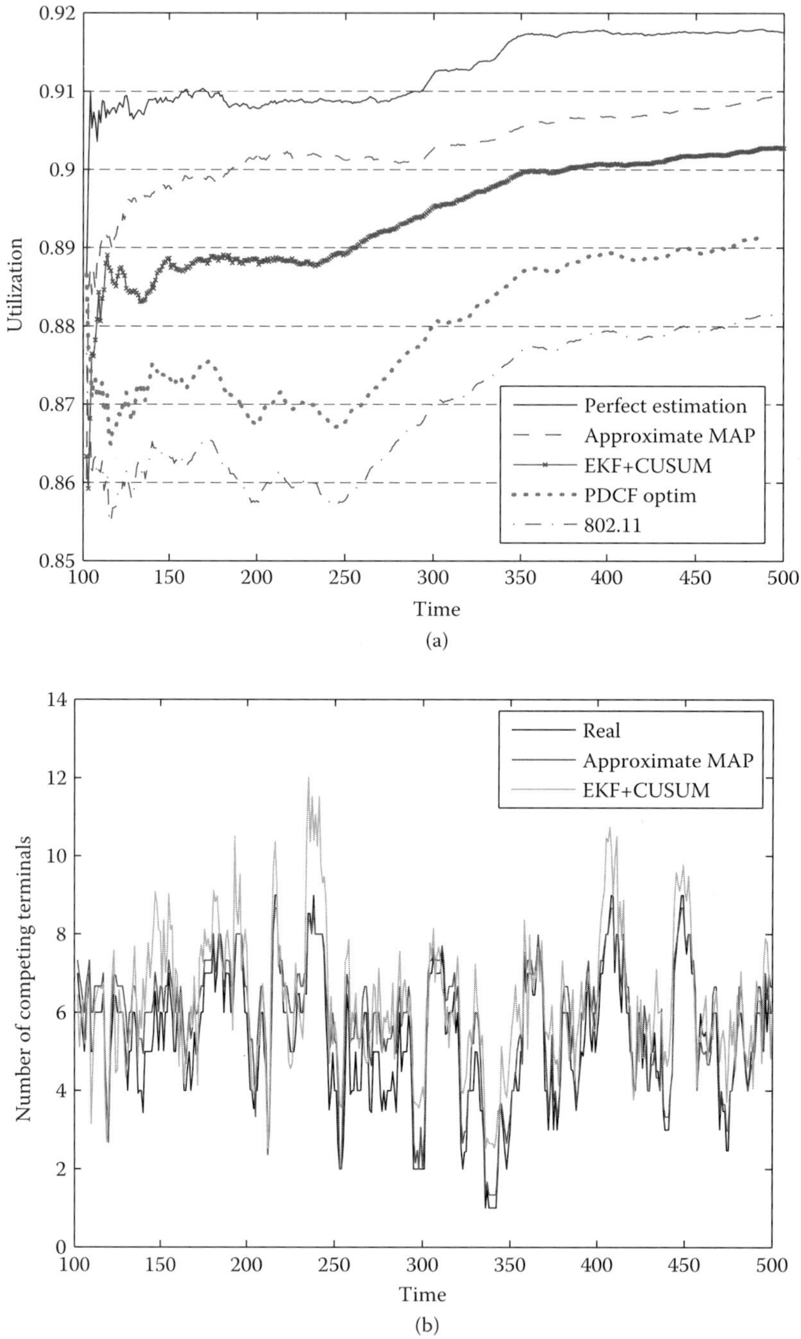

FIGURE 1.6 Instantaneous utilization when terminals arrive exponentially. (a) Instantaneous utilization. (b) Evolution of the number of competing terminals.

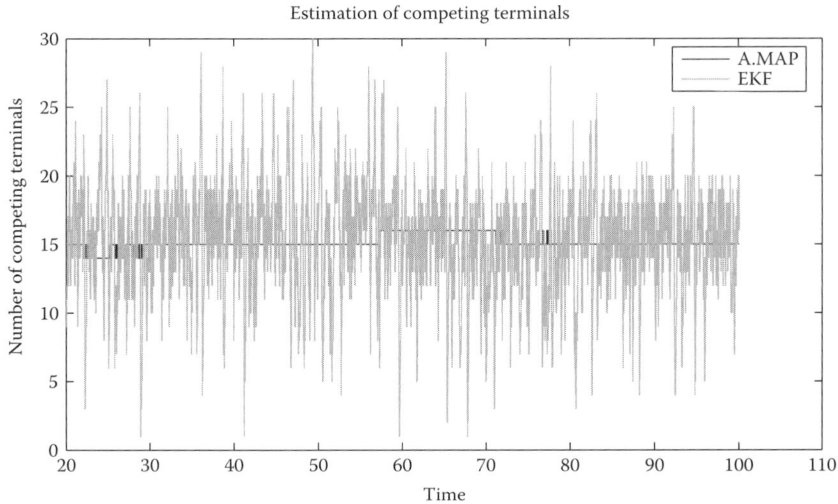

FIGURE 1.7 Accuracy of the estimation algorithms for the extreme case of very noisy measurements ($B = 10$). The number of competing terminals is 15.

given time the terminals always have something to transmit. As [3] shows, the number of competing terminals fluctuates heavily under nonsaturation conditions. As a rude approximation, and intuitively, we can think of n terminals in the nonsaturation regime as a process of $x(t)$ saturating terminals (those that have something to transmit in the allowed slots) that fluctuates very fast. In this scenario, the effect of a highly accurate and fast estimate of the number of competing terminals may be crucial to the optimal operation of the protocol. We tested the accuracy of both the EKF and the approximate MAP estimator in a very simple scenario: the number of competing terminals is fixed to 15, all of them saturating, and we reduced the observation slot $B = 10$. Note that $B = 10$ means the average time for which the terminals measure the channel before estimation averages less than 300 ms. Figure 1.7 shows that the approximate MAP estimator is more accurate than the EKF estimator at very low timescales, and potentially able to better track fast fluctuations. Figure 1.8(a) and (b) shows the instantaneous utilization and evolution of the number of competing terminals when twenty terminals are not in the saturation regime. Each terminal randomly picked a throughput between 70 and 100% of one-twentieth of the network saturation throughput. As we see in Figure 1.8(b), both estimators have problems in tracking the small fluctuations in the number of competing terminals. However, the approximate MAP estimator clearly does a better job, with a MSE of 154.4874 against a MSE value of 322.4615 for the EKF estimator. This difference makes our algorithm clearly superior in the nonsaturation regime.

1.6 Conclusions

We have presented several algorithms for the problem of estimating the number of competing terminals in an IEEE 802.11 wireless network under the framework of Bayesian

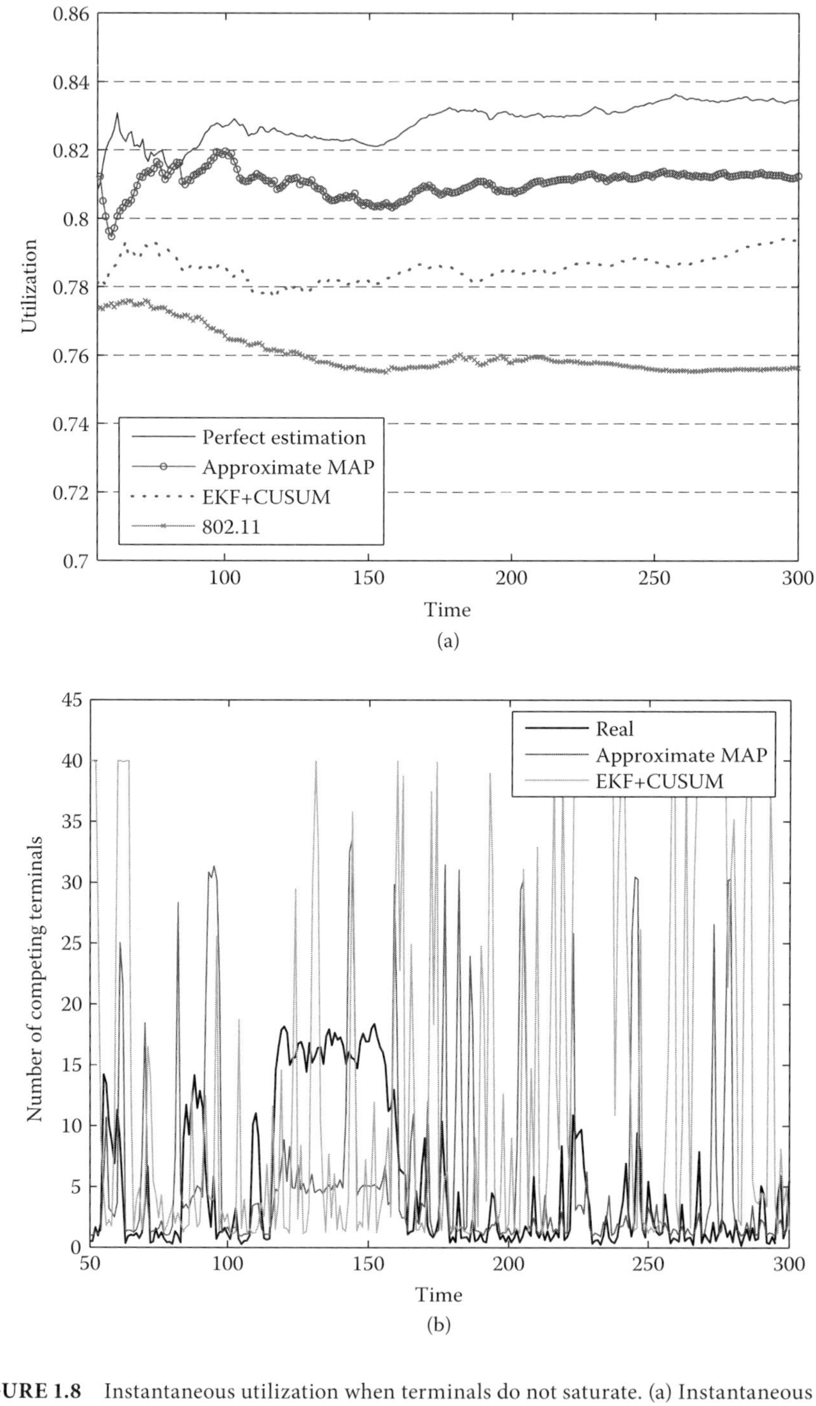

FIGURE 1.8 Instantaneous utilization when terminals do not saturate. (a) Instantaneous utilization. (b) Evolution of the number of competing terminals.

Monte Carlo signal processing. We have employed a powerful yet computationally simple online algorithm to estimate the number of competing terminals, based on the sequential Monte Carlo method. The online estimators can be applied to any hidden Markov chain with unknown transition probabilities and unknown prior distributions, which makes them appropriate for an 802.11 protocol where, from the terminal point of view, there is very little knowledge of the state of the system. Moreover, its low computational requirements make it a good candidate for introduction in an actual IEEE 802.11 network. We then use the estimates from those protocols to optimize the operation of the IEEE 802.11 DCF by adjusting the contention window parameters based on the number of competing terminals in the network. We have provided extensive ns-2 simulation results and have shown that the proposed technique outperforms existing state-of-the-art approaches in all cases.

We have shown that the accuracy of the estimation of the number of competing terminals in an 802.11 network has a significant impact on the network performance—in terms of overall network utilization and in terms of observed delay due to collisions. This accuracy is shown to be extremely important when the number of competing terminals fluctuates heavily in small timescales, as in the case when the network is in a nonsaturation regime. Consequently, a fast and accurate estimation of the number of competing terminals, such as the one provided by the SMC techniques, offers a great benefit toward optimizing the operation of an IEEE 802.11 DCF by adjusting the contention window parameters to the existing network conditions.

References

[1] ANSI/IEEE. 2003. *Wireless LAN medium access control (MAC) and physical layer (PHY) specifications.* ANSI/IEEE Std. 802.11, 1999 ed. (R2003), pp. i–503.

[2] G. Bianchi. 2000. Performance analysis of the IEEE 802.11 distributed coordination function. *IEEE J. Select. Areas Commun.* 18:535–47.

[3] G. Bianchi and I. Tinnirello. 2003. Kalman filter estimation of the number of competing terminals in an IEEE 802.11 network. In *Proceedings of Infocom 2003*, vol. 2, San Francisco, pp. 844–52.

[4] C. Wang, B. Li, and L. Li. 2004. A new collision resolution mechanism to enhance the performance of IEEE 802.11 DCF. *IEEE Trans. Veh. Technol.* 53:1235–46.

[5] F. Cali, M. Conti, and E. Gregori. 2000. Dynamic tunning of the IEEE 802.11 protocol to achieve a theoretical throughput limit. *IEEE/ACM Trans. Netw.* 8:785–99.

[6] M. Natkaniec and A. R. Pach. 2000. An analysis of the backoff mechanism used in IEEE 802.11 networks. In *Proceedings of IEEE ISCC 2000*, vol. 1, pp. 444–49.

[7] Z. J. Haas and J. Deng. 2003. On optimizing the backoff interval for random access schemes. *IEEE Trans. Mobile Comput.* 51:2081–90.

[8] H. Ma, X. Li, H. Li, P. Zhang, S. Luo, and C. Yuan. 2004. Dynamic optimization of IEEE 802.11 CSMA/CA based on the number of competing stations. In *Proceedings of the IEEE International Conference on Communications* (ICC'04) pp. 121–132.

[9] F. Cali, M. Conti, and E. Gregori. 2000. IEEE 802.11 protocol: Design and performance evaluation of an adaptive backoff mechanism. *IEEE J. Select. Areas Commun.* 18:1774–86.

[10] Y. Kwon, Y. Fang, and H. Latchman. 2004. Design of MAC protocols with fast collision resolution for wireless local area networks. *IEEE Trans. Wireless Commun.* 3:793–807.

[11] W. R. Gilks, S. Richardson, and D. J. Spiegelhalter. 1996. *Markov chain Monte Carlo in practice.* London: Chapman & Hall.

[12] A. Doucet, N. de Freitas, and N. Gordon, Eds. 2001. *Sequential Monte Carlo methods in practice.* New York: Springer-Verlag.

[13] E. Punskaya. 2003. Sequential Monte Carlo methods for digital communications. PhD dissertation, University of Cambridge.

[14] G. Storvik. 2002. Particle filters for state-space models with the presence of unknown static parameters. *IEEE Trans. Signal Process.* 50:281–89.

[15] J. S. Liu and R. Chen. 1998. Sequential Monte Carlo methods for dynamic systems. *J. Am. Stat. Assoc.* 93:1032–44.

[16] S. Arulampalam, S. Maskell, N. Gordon, and T. Clapp. 2002. A tutorial on particle filters for online non-linear/non-Gaussian Bayesian tracking. *IEEE Trans. Signal Process.* 50:174–88.

[17] J. M. Bernardo and A. F. M. Smith. 1994. *Bayesian theory.* New York: Wiley.

[18] C. Andrieu and A. Doucet. 2003. Online expectation-maximization type algorithms for parameter estimation in general state space models. In *Acoustics, Speech, and Signal Processing, ICASSP'03*, vol. 6, pp. 69–72.

[19] J. Liu and M. West. 2001. Combined parameter and state estimation in simulation-based filtering. In *Sequential Monte Carlo methods in practice*, ed. A. Doucet, N. de Freitas, and N. Gordon. New York: Springer-Verlag.

[20] W. R. Gilks and C. Berzuini. 2001. Following a moving target—Monte Carlo inference for dynamic Bayesian models. *J. R. Stat. Soc. B* 63:127–46.

[21] P. Fearnhead. 2002. MCMC, sufficient statistics and particle filters. *J. Comput. Graphical Stat.* 11:848–62.

[22] A. Doucet, S. J. Godsill, and C. Andrieu. 2000. On sequential Monte Carlo sampling methods for Bayesian filtering. *Stat. Comput.* 10:197–208.

[23] M. K. Pitt and N. Shephard. 1999. Filtering via simulation: Auxiliary particle filters. *J. Am. Stat. Assoc.* 94:590–99.

[24] J. Tugnait and A. Haddad. 1979. A detection-estimation scheme for state estimation in switching environment. *Automatica* 15:477–81.

[25] P. Fearnhead. 1998. Sequential Monte Carlo methods in filter theory. PhD dissertation, University of Oxford.

[26] P. Fearnhead and P. Clifford. 2003. On-line inference for hidden Markov models via particle filters. *J. R. Stat. Soc. B* 65:887–99.

[27] S. J. Julier and J. K. Uhlmann. 2002. The scaled unscented transformation. In *Proceedings of the IEEE American Control Conference '02*, pp. 4555–4559.

[28] L. R. Rabiner. 1989. A tutorial on hidden Markov models and selected application in speech recognition. *Proc. IEEE* 77:257–85.

[29] V. Krishnamurthy and J. B. Moore. 1993. Online estimation of hidden Markov model parameters based on the Kullback-Leibler information measure. *IEEE Trans. Signal Process.* 41:2557–73.

[30] V. Krishnamurthy and G. G. Yin. 2002. Recursive algorithms for estimation of hidden Markov models and autoregressive models with Markov regime. *IEEE Trans. Inf. Theory* 48:458–76.

[31] J. J. Ford and J. B. Moore. 1998. Adaptive estimation of HMM transition probabilities. *IEEE Trans. Signal Process.* 46:1374–85.

[32] F. LeGland and L. Mevel. 1997. Recursive estimation in hidden Markov models. In *Proceedings of the Conference on Decision and Control*, vol. 4, pp. 3468–3473.

[33] T. Rydén. 1997. On recursive estimation for hidden Markov models. *Stoch. Proc. Appl.* 66:79–96.

[34] Network simulator 2. Accessed 2007 from http://www.isi.edu/nsnam/ns.

[35] H. Ma, X. Li, H. Li, P. Zhang, S. Luo, and C. Yuan. 2004. Range estimation and performance optimization for IEEE 802.11 based on filter. In *Proceedings of the IEEE Wireless Communications & Networking Conference (WCNC'04)*, vol. 3, Atlanta, GA, pp. 1469–1475.

[36] C. Wang and W. Tang. 2004. A probability-based algorithm to adjust contention window in IEEE 802.11 DCF. *ICCCAS Commun. Circuits Syst.* 1:418–22.

Appendix

Derivation of (1.31)

$$p(\boldsymbol{a}_i \mid \boldsymbol{x}_{t-1}, \boldsymbol{y}_t) \propto p(y_t \mid \boldsymbol{a}_i, \boldsymbol{x}_{t-1}, \boldsymbol{y}_{t-1})\, p(\boldsymbol{a}_i \mid \boldsymbol{x}_{t-1}, \boldsymbol{y}_{t-1})$$

$$\propto \sum_{j=1}^{N} p(y_t \mid x_t = j)\, p(x_t \mid x_{t-1}, \boldsymbol{a}_i)\, p(\boldsymbol{a}_i \mid \boldsymbol{x}_{t-1}, \boldsymbol{y}_{t-1})$$

$$\propto \sum_{j=1}^{N} \mathcal{B}(y_t; B, h(j))\, a_{i,j}^{\mathbb{I}(x_{t-1}-i)} \prod_{m=1}^{N} a_{i,m}^{\alpha_{i,m,t-1}-1}$$

$$\propto \sum_{j=1}^{N} \mathcal{B}(y_t; B, h(j)) \frac{\prod_{m=1}^{N} \Gamma\left(\alpha_{i,m,t-1} + \mathbb{I}(x_{t-1}-i)\mathbb{I}(j-m)\right)}{\Gamma\left(\mathbb{I}(x_{t-1}-i) + \sum_{m=1}^{N} \alpha_{i,m,t-1}\right)} \times$$

$$\mathcal{D}\left(\boldsymbol{a}_i; \alpha_{i,1,t-1} + \mathbb{I}(x_{t-1}-i)\mathbb{I}(j-1), \cdots, \alpha_{i,N,t-1} + \mathbb{I}(x_{t-1}-i)\mathbb{I}(j-N)\right)$$

$$\propto \sum_{j=1}^{N} \mathcal{B}(y_t; B, h(j)) \frac{\Gamma\left(\alpha_{i,j,t-1} + \mathbb{I}(x_{t-1}-i)\right)}{\Gamma(\alpha_{i,j,t-1})} \times$$

$$\mathcal{D}\left(\boldsymbol{a}_i; \alpha_{i,1,t-1} + \mathbb{I}(x_{t-1}-i)\mathbb{I}(j-1), \cdots, \alpha_{i,N,t-1} + \mathbb{I}(x_{t-1}-i)\mathbb{I}(j-N)\right)$$

$$\propto \sum_{j=1}^{N} \mathcal{B}(y_t; B, h(j))\, \alpha_{i,j,t-1}^{\mathbb{I}(x_{t-1}-i)}\, \mathcal{D}\left(\boldsymbol{a}_i; \alpha_{i,1,t-1} + \mathbb{I}(x_{t-1}-i)\mathbb{I}(j-1), \cdots, \alpha_{i,N,t-1} + \mathbb{I}(x_{t-1}-i)\mathbb{I}(j-N)\right).$$

The posterior distribution of $\boldsymbol{a}_i$ given $\boldsymbol{x}_{t-1}$ and $\boldsymbol{y}_t$ is thus a mixture of Dirichlet distributions that can be rewritten as

$$p(\boldsymbol{a}_i \mid \boldsymbol{x}_{t-1}, \boldsymbol{y}_t) = \frac{1}{\sum_{m=1}^{N} \beta_{i,m,t}} \sum_{j=1}^{N} \beta_{i,j,t}\, p\left(\boldsymbol{a}_i \mid \boldsymbol{T}_t(x_t = j, \boldsymbol{x}_{t-1}, \boldsymbol{y}_t)\right),$$

where $\beta_{i,j,t} = \mathcal{B}(y_t; B, h(j))\, \alpha_{i,j,t-1}^{\mathbb{I}(x_{t-1}-i)}$.

Derivation of (1.32)

From (1.6), we can see that

$$\frac{p(\boldsymbol{a}_i \mid \boldsymbol{x}_{t-1}, \boldsymbol{y}_t)}{p(\boldsymbol{a}_i \mid \boldsymbol{x}_{t-1}, \boldsymbol{y}_{t-1})} = \frac{1}{\sum_{m=1}^{N} \beta_{i,m,t}} \sum_{j=1}^{N} \beta_{i,j,t} \frac{\left(\sum_{m=1}^{N} \alpha_{i,m,t-1}\right)^{\mathbb{I}(x_{t-1}-i)}}{\alpha_{i,j,t-1}^{\mathbb{I}(x_{t-1}-i)}} a_{i,j}^{\mathbb{I}(x_{t-1}-i)}$$

$$= \begin{cases} 1, & \text{if } x_{t-1} \neq i \\ \dfrac{\sum_{m=1}^{N} \alpha_{i,m,t-1}}{\sum_{m=1}^{N} \beta_{i,m,t-1}} \sum_{j=1}^{N} \beta_{i,j,t-1} \dfrac{a_{i,j}}{\alpha_{i,j,t-1}}, & \text{if } x_{t-1} = i \end{cases}$$

$$= \begin{cases} 1, & \text{if } x_{t-1} \neq i \\ \dfrac{\sum_{j=1}^{N} \alpha_{i,j,t-1}}{\sum_{j=1}^{N} \beta_{i,j,t-1}} \sum_{j=1}^{N} \mathcal{B}\big(y_t; B, h(j)\big) a_{i,j}, & \text{if } x_{t-1} = i \end{cases}$$

Therefore, we get

$$\frac{p(\boldsymbol{A} \mid \boldsymbol{x}_{t-1}, \boldsymbol{y}_t)}{p(\boldsymbol{A} \mid \boldsymbol{x}_{t-1}, \boldsymbol{y}_t)} = \frac{\sum_{i=1}^{N} \alpha_{x_{t-1},i,t-1}}{\sum_{i=1}^{N} \beta_{x_{t-1},i,t-1}} \sum_{i=1}^{N} \mathcal{B}\big(y_t; B, h(i)\big) \cdot a_{x_{t-1},i}.$$

The rest of the weight update follows from the usual formula:

$$\frac{p(x_t \mid x_{t-1}, \boldsymbol{\theta})\, p(y_t \mid x_t, \boldsymbol{\theta})}{p(x_t \mid \boldsymbol{x}_{t-1}, y_t, \boldsymbol{\theta})} = p(y_t \mid x_{t-1}, \boldsymbol{\theta}) = \sum_{i=1}^{N} \mathcal{B}\big(y_t; B, h(i)\big) \cdot a_{x_{t-1},i}.$$

The weight update is thus performed by

$$w_t \propto w_{t-1} \frac{\sum_{i=1}^{N} \beta_{x_{t-1},i,t-1}}{\sum_{i=1}^{N} \alpha_{x_{t-1},i,t-1}}.$$

2

A Survey of Medium Access Control Protocols for Wireless Local and Ad Hoc Networks

Tiantong You
Queen's University

Hossam Hassanein
Queen's University

Chi-Hsiang Yeh
Queen's University

2.1 Introduction

In recent years, wireless networks have undergone a spectacular boost in popularity, evidenced by tremendous research and commercialization efforts. Being a competitive approach of wireless data provisioning, wireless local area networks (WLANs) or wireless fidelity (Wi-Fi) has gained increasing interest due to the exponential growth in the number of various wireless devices and the dramatic drop in their prices.

2.1.1 Background

From the perspective of network structure, wireless networks can be categorized into two types: infrastructure-based wireless networks and ad hoc networks [1]. An infrastructure-based network conceptually comprises two levels: a stationary level and a mobile level. The stationary level consists of fixed access points (APs) that are interconnected through wired or wireless media. The mobile level refers to mobile terminals (MTs), which communicate with APs or each other through wireless links. The APs are permanently in service at fixed locations to coordinate the operation of the MTs. When an MT is turned on, it first associates with an AP. Thereafter, the AP tracks the device and provides service to MTs.

In the ad hoc type of wireless network, portable devices are brought together to form a network on the fly. Ad hoc wireless networks have no infrastructure; that is, they do not have fixed points and usually every node is able to communicate with every other node when all nodes are spread in a relatively small geographic range. As an example, consider a meeting in a conference room where employees bring laptop computers together to communicate and share a design or financial information. However, nodes may spread over a geographic range larger than the communication signal can reach. In such cases, nodes may have to communicate over multiple hops. Ad hoc networks do not have a central control, but do have the advantage of being relatively inexpensive to deploy, as they do not require a communication infrastructure. Applications of such networks also exist in military and emergency scenarios [1].

In general, there are certain unique issues that need to be addressed in wireless transmission. First, the transmission medium is shared. Signals sent from one MT to another experience propagation delay and attenuate with distance. Furthermore, when more than two packets from different sources in the same radio frequency reach an MT, the MT is not able to decode the message correctly. This is commonly referred to as a collision. Normally, the hardware of the transceiver is incapable of sending and receiving simultaneously because the overwhelming sideband interference from the concurrent transmission will make the attenuated signal at the destination undecodable.

Wireless Medium Access Control (MAC) protocols are used to coordinate MTs' access to the shared medium in WLANs. One goal of the MAC protocols is to decrease the collision rate of transmission while increasing the medium utilization. There are two typical MAC paradigms in the literature. The first comprises authority-driven MAC protocols, which require the existence of at least one authority center. The second comprises distributed MAC protocols. In wireless ad hoc networks, with or without APs, MTs have to contend for medium access in a distributed manner. Currently, the implementations of authority-driven MAC protocols have not gone beyond academic study due to complexity and centricity requirements, even though they can potentially provide quality of service (QoS) support, especially for ad hoc wireless networks.

Because the wireless medium is shared by all MTs in the same geographic range, MAC protocols aim to prevent concurrent data transmission from different MTs that require coordination. Carrier Sense Multiple Access (CSMA) [2] is currently the main mechanism to implement distributed medium access. In CSMA, an idle medium must be sensed before any data transmission. With MTs spread in a large area, the communication environment may also change from single hop to multihop. A MAC protocol based on CSMA that works well in a single-hop environment may suffer performance degradation in a multihop environment due to the hidden terminal problem [3]. Given an active transmission between an arbitrary sender and a receiver, a hidden terminal is an MT that resides in the interference range of the receiver but outside the sensing range of the sender. The issue is that the hidden terminal cannot sense the current data transmission from the sender and may send signals that interfere with the data reception at the receiver. Another problem also related to CSMA in a multihop wireless system is the exposed terminal problem. Exposed terminals [3] are MTs that are located in the sensing range of a sender and are hence inhibited from transmission during the sender's transmission. Recalling that a data collision, by definition, only occurs at the receiver's side, the existence of exposed terminals results in a reduced medium utilization.

2.1.2 Design Objectives of MAC Protocols

In a multihop environment, solving the hidden terminal problem can decrease the probability of collision in data transmission, indirectly resulting in an increased system throughput. On the other hand, solving the exposed terminal problem can directly increase the system throughput. The distributed coordination function (DCF) defined in the Institute of Electrical and Electronics Engineers (IEEE) 802.11 standard [4, 5] is currently the most popular commercial MAC protocol for WLANs. However, the performance of DCF in multihop environments is far from optimal. This is because DCF was originally designed for centralized, single-hop wireless networks. With the continuing development of wireless networks, to save energy and increase the spatial reuse of medium resources, the communication range of an individual MT is reduced. To exploit this feature, new MAC protocol paradigms are desirable. Such protocols must be efficient and operate in a distributed manner in multihop environments.

The following are specific design objectives of the MAC protocols under study. *High achievable network throughput*, which is the key objective of most new MAC protocols [4–30]. The higher network throughput normally results in short expected packet delays on arrival. *Low collision probability (or collision-free) communication* is one way to achieve higher network performance. In the single-hop environment, a MAC that reduces concurrent data transmissions is highly desirable. However, solving the hidden terminal problem has been the main objective of recent research [12–30]. To transmit data with fewer collisions will reduce the data retransmission and, accordingly, achieve a higher network throughput. Another key method to raise network throughput in multihop environments is by increasing parallel transmission through enlarging the spatial reuse rate. Achieving parallel transmissions becomes more possible if the exposed terminal challenge is overcome.

Quality of service (QoS) support recently became an important objective in designing new wireless MAC protocols. With the increase of wireless communication bandwidth, a variety of new applications have different service requirements. QoS requires that high-priority packets be delivered within a hard time bound, in most cases. Likewise, service differentiation enables and requires MAC protocols to provide differentiated throughput distribution and average packet delay for packets of different priorities. Supporting service differentiation can result in more overhead, limiting the number of possible priority classes.

Within the same priority level, *starvation prevention and fairness* are crucial objectives when designing a wireless MAC protocol. Without authority center, MAC protocols based on medium competition could sacrifice the resource-sharing principle for a substantial duration of time. In such cases, the system throughput can be optimized at the cost of starving certain nodes, i.e., depriving them of transmission. Thus, a good MAC protocol needs to consider not only the system throughput, but also the achievable fairness among all nodes. Fairness can be satisfied by giving higher medium access priority to nodes suffering starvation. The fairness issue is often related to QoS support and service differentiation. In a system that supports service differentiation, low-priority packets normally suffer from starvation when the traffic load for high-priority packets is high. An issue examined in this thesis is how to prevent low-priority packets from starving while maintaining the proper service differentiation among different priority classes.

Energy preservation is another objective under consideration in this thesis. The greatest advantage of wireless communication is the convenience of mobility. Users like to hold and move with their devices without wired connection for communication and energy supply. The powering of the mobile device hence becomes temporarily supported by a battery. Accordingly, to design an energy-efficient MAC protocol is elementary to sustaining the convenience of mobility for a longer duration of time.

Efficient broadcasting in multihop wireless networks is another objective in designing a MAC protocol. In an ad hoc network, broadcast transmission is an unavoidable type of data transmission that provides synchronization, routing, and other messages within neighboring devices. To guarantee that all neighboring MTs receive a broadcast message is a hard task for a wireless network that operates in a distributed manner, especially when the network is managing a heavy traffic load. Little has been done on this topic in the literature. Previous work [4, 5] only focuses on the infrastructure-based network where APs are in charge of broadcasting and the neighboring geometry ranges apply different radio frequencies to prevent collisions due to hidden terminals. To propose an innovative MAC protocol that possesses efficient broadcasting is highly desirable from both the academic and industrial perspectives.

Simple hardware requirements is another objective of proposed MAC protocols. The more complicated the required hardware, the more expensive it will be to implement the MAC protocol. To solve the exposed terminal or hidden terminal problem, some MAC protocols split the communication band into several channels and require the node to simultaneously receive (or sense) and send data. Theoretically, however, these MAC protocols can achieve higher network throughput. However, the complex hardware requirements hinder their commercial implementation and deployment.

Based on the challenges of the wireless MAC protocol design, we organize this chapter as follows. In section 2.2, we will discuss the basic MAC protocol's design ideas that pursue high system throughput and low average delay in the single-hop environment, targeted at the concurrent data transmission problem. We extend the single-hop environment to multihop in section 2.3 to discuss the protocols that deal with hidden terminal and exposed terminal problems. In section 2.4, we give a review on the MAC protocols that focus on energy efficiency. In section 2.5, MAC protocols dealing with QoS and service differentiation are discussed further. In section 2.6, we study the fairness issue. In section 2.7, we give a survey on the wireless MAC protocols that claim a collision-free medium access, as well as discuss their shortcomings. Finally, the summary will be drawn in section 2.8.

2.2 Medium Access Control (MAC) in Wireless LANs

In a single-hop environment, the shared medium can only allow for one collision-free transmission at any given time. Therefore, we can normalize the theoretical maximum system throughput to be 1. Two factors degrade the system throughput from the ideal case. The first factor is due to collisions, which occur in a single-hop environment when there is an overlap time period between any two data transmissions. The second factor is the overhead brought by a MAC protocol, which in many cases is introduced in order to reduce the probability of data collision. Any MAC protocol for the single-hop environment aims at reducing the probability of concurrent data transmissions while using reasonable overhead. In infrastructure-based wireless networks, an MT usually communicates with the access point with the larger communication range. The MAC protocols for the single-hop environment work well for their purpose. Normally, MAC protocols with complicated mechanisms can reduce the concurrent transmission rate but bring about larger overhead. The achievable network throughput is the most important criterion to evaluate the overall performance of any MAC protocol. Carrier Sense Multiple Access (CSMA) is the basic technique for wireless MAC protocols in a distributed manner. It requires an MT to have medium sensing ability. Before transmitting a packet, an MT is required to sense the medium and if the medium is idle, transmit the data; otherwise, it either keeps sensing the medium until it becomes idle again, which is what happens in persistent CSMA, or defers transmission for a period of time with random duration, then senses again, which is the operation of the nonpersistent CSMA. In the persistent CSMA, the time right after a busy period is vulnerable to collisions. In the following, three MAC protocols, namely, CSMA/CA, HIPERLAN, and CSMA/IC, all based on the persistent CSMA technique, will be discussed in more detail.

2.2.1 Carrier Sense Multiple Access (CSMA)/ Collision Avoidance (CA)

In wireless networks, nodes cannot detect collisions during a data transmission. One way to reduce the collision rate is to try to avoid collisions before they happen. Note that in CSMA, collision occurs with higher probability at the time point when the medium changes from busy to idle while several competitors attempt to access the medium. CSMA/CA [6] can ease this high collision probability problem, in which terminals wait for random time periods before data transmission even when the medium becomes idle. The distributed coordination function (DCF) in the IEEE 802.11 standard [4, 5], a discrete-time version of the CSMA/CA, is currently the most popular MAC protocol for WLANs. The standard defines three interframe spaces (IFSs) of different lengths: DCF interframe space (DIFS), point coordination function interframe space (PIFS), and short interframe space (SIFS), with DIFS > PIFS > SIFS. DCF requires the nodes to sense the medium idle for DIFS to send a data packet, and for SIFS to send control packets such as Acknowledgment (ACK) messages. When a data packet and an ACK packet simultaneously compete for the medium, the medium sensing for the ACK is shorter than for other common packet competitors. ACK packets are thus transmitted earlier, blocking data packets, as the medium will be sensed busy. Note that since an ACK is sent in response to a successful data transmission, it is impossible for two or more ACKs to concurrently compete for the medium. To avoid the collision among data packets, DCF applies a binary exponential back-off (BEB) mechanism that extends several fixed-length time slots. The number of time slots is randomly created after the DIFS and takes values between zero and the contention window (CW). CW varies between minimum contention window (CW_{min}) and maximum contention window (CW_{max}). More specifically, CW is initialized to CW_{min}; whenever a packet suffers collision, the value of the CW is doubled with CW_{max} as an upper bound. Whenever a packet is sent out successfully, i.e., when the respective ACK is received, the CW is reset to CW_{min}. With the BEB scheme, as the traffic load becomes heavy, the collision rate increases. More collisions result in larger values of CW. It should be noted that, on average, increasing traffic load will result in a larger CW. In this manner, congestion control is achieved.

To reduce the cost of collision for large data packets, a small-size packet can be used as a preamble to competition to prevent the large data packets suffering collision. The virtual sense mechanism in the DCF implements this notion and further improves the system throughput. DCF defines that if the size of a data packet is larger than a certain threshold, instead of sending the data packet directly, the node sends out a small-size packet called Request to Send (RTS) after sensing the medium idle for DIFS plus the required time slots. The destination responds back using another small-size packet called Clear to Send (CTS) after sensing the medium idle for SIFS. When the CTS is successfully received, the sender transmits the large data packet. With the four-way handshake mechanism, i.e., RTS-CTS-Data-ACK, CSMA/CA achieves even better system throughput. In the DCF of IEEE 802.11, the virtual sense mechanism is a powerful and efficient mechanism to alleviate the hidden terminal problem in a multihop environment, which will be further discussed in the next subsection.

2.2.2 High-Performance Radio Local Area Networks

The Europe Telecommunication Standardization Institute (ETSI) formed a working group called HIPERLAN/1 [7, 8] to define a standard for local area networks. HIPERLAN is proposed for distributed networks in which a backlogged node first senses the medium for a fixed period. If the medium is idle during this whole period, the node transmits its packet immediately. If the channel is busy during the required sensing period, it triggers the competition mechanism, which lasts for three phases. The three phases of competition are prioritization, contention (including an elimination subphase and a yield subphase), and transmission, as shown in Figure 2.1.

The priority of a packet, for example, can be derived from the duration of the packet time in the queue divided by a time unit parameter. The longer a packet is delayed, the higher its priority. In the prioritization phase, the node senses the medium and quits the competition if the medium is sensed busy. After sensing the medium idle for the priority phase, the node transmits a noise signal for a random number of slots, dictating the elimination phase, at the end of which the node senses the medium again. If the medium is busy, the node quits competition. During the yield phase, the node further senses the medium and defers for a random number of slots. If no transmission is detected, the node completes its data transmission. Ideally, only one node can complete the whole process and send a packet.

2.2.3 Carrier Sense Multiple Access/ID Countdown (CSMA/IC)

As shown in Figure 2.2, instead of continually sensing the media silently until sending the packet in CSMA/CA, CSMA/IC [31–33] sends the so-called buzz signal during the time slot unit in order to lead other sensing nodes out of competition. Whether to send the buzz signal or sense the media silently in one specific time slot is decided by the

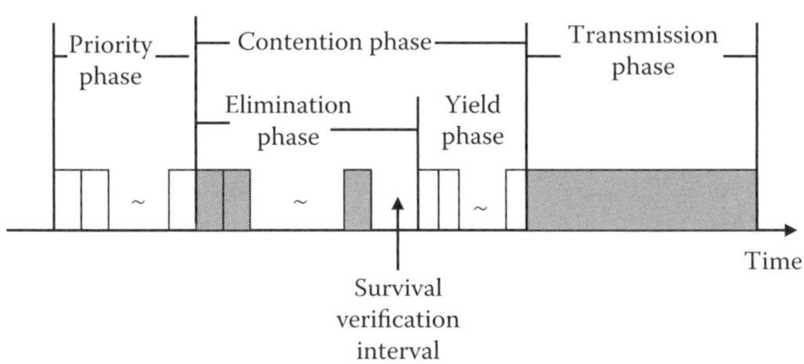

FIGURE 2.1 HIPERLAN/1 MAC.

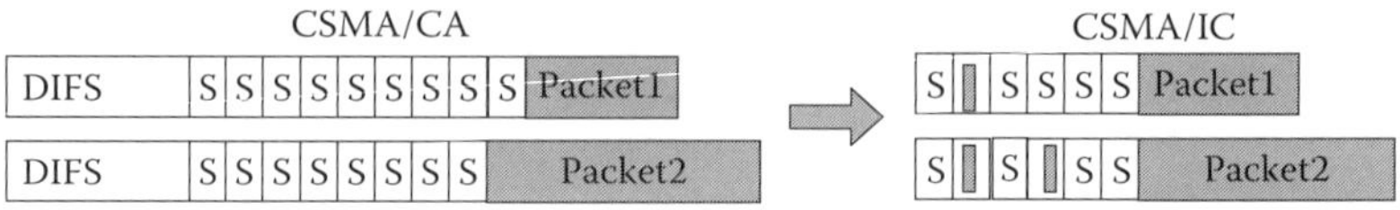

FIGURE 2.2 The slots preceding the data transmission in CSMA/CA and CSMA/IC.

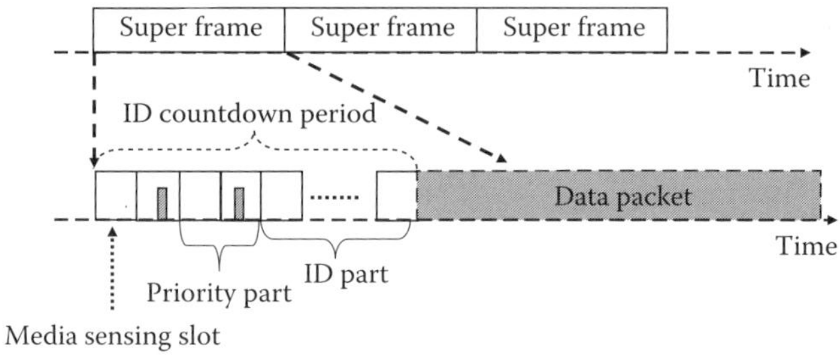

FIGURE 2.3 Time format in the CSMA/IC with variable data size.

pattern of competition ID codes the node owns for medium access. For example, the three-digit ID 010 means sensing silently in the first slot, buzzing in the second slot, and sensing silently again in the third slot. If the node is in the time slot of medium sensing and senses a buzzing signal, it quits the competition process. In CSMA/IC, if two nodes use different IDs to compete for medium access, then starting from the first time slot to the last time slot there must be a certain time slot in which one node sends the buzz (1 bit) signal while the other node senses (0 bit) the medium. This will let the latter quit. Hence, the ID uniqueness in CSMA/IC will guarantee the uniqueness of the winner if all the nodes begin processing the competition at the same time (i.e., synchronization is required) and the number of time slots between start of competition and actual data transmission is identical.

Figure 2.3 shows the time format of CSMA/IC in a variable-data-size system. The time period is formatted through mutual agreement into equal-length superframes. The MTs can periodically begin competition processing at the start point of the superframe. The size of common data is varied and normally larger than a superframe. To avoid collisions, each superframe begins with a time slot called the media sensing slot. Only when the medium is sensed idle during the media sensing slot can the rest of competition processing be continued. Otherwise, the node will quit the current competition and resume at the next superframe. The higher value of the competition ID means higher priority. Thus, the packet's priority (packet type and waiting time) could be encrypted to the slots (priority part) preceding the slots (ID part) for processing the competition ID code.

2.3 MAC Protocols for Multihop Environments

When the communication range of each MT is reduced, not all MTs can directly communicate with an AP. In multihop environments, MTs need other MTs to relay the packets. To clarify the problems existing in the multihop communication, three ranges related to a transmitter can be defined: the communication range, the interference range, and the sensing range:

Communication range refers to the range within which a node can receive the signal from a transmitting node if there is no interference from other nodes.

Interference range refers to the range within which an ongoing data reception can be interfered with by other transmitters.

Sensing range refers to the range within which a node can recognize the existence of an ongoing data transmission, and thus restrain itself from any transmission. The maximum sensing range reflects the signal detection ability of the node and is normally much larger than the interference and communication ranges.

MAC protocols designed for the multihop environment intend to solve or alleviate the hidden terminal and exposed terminal problems. Consider Figure 2.4, where a circle illustrates the communication range of the node at the circle's center. During data transmission from node A to node B, nodes like g, h, f, e, d, and c are in the hidden position of node A; i.e., they cannot sense the data transmission of node A, but are all in the interference range of node B. If any of these hidden nodes sends a signal during the data transmission from A to B, the data received at node B will suffer collision. When the traffic becomes heavy, the probability of data collision from the hidden terminals becomes higher. Solving the hidden terminal problem can extend to multihop environments the relatively low data collision rate that CSMA-based MAC protocols achieve in single-hop environments.

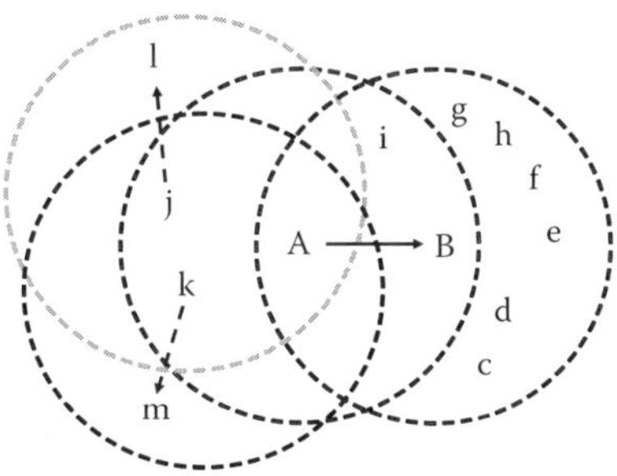

FIGURE 2.4 The hidden terminal and exposed terminal problems.

In a multihop environment, it is possible to concurrently transmit data without suffering collisions, making the theoretical maximum network throughput higher than 1. The exposed terminal problem holds the network throughput from reaching the optimal point. As illustrated in Figure 2.4, during the data transmission from node A to B, nodes j and k sense the medium is busy and accordingly block themselves from transmitting. From the view of some receiver such as node l or m, the medium is idle. It is therefore viable for node j to send data to l, and node k to send data to node m, without collisions. Alleviating the exposed terminal problem increases the number of successful concurrent transmissions and increases the network throughput.

Recently, many MAC protocols have been proposed to alleviate the hidden terminal problem based on the assumption that the sizes of the communication range, interference range, and sensing range are the same, and that the size is fixed for all nodes. In practice, the interference range is likely to be larger than the communication range, and the sensing range could be even larger than the interference range. Accordingly, the hidden terminal problem is extended to the interference range problem [9–11], which extends the hidden terminal to nodes that can potentially interfere with the data reception at the destination but cannot be contacted directly by either the sender or the receiver. The traditional way to solve the hidden terminal problem is by reserving the medium around the destination by transmitting a small-size packet preceding the actual data transmission. In the interference range problem, the preceding small-size message cannot reserve the entire interference range of the destination. This degrades the performance of dialogue-based MAC protocols.

We can categorize MAC protocols proposed for multihop environments into two groups according to their channel requirements: protocols working in the single-channel system and protocols using dual or multiple channels in Frequency Division Multiple Access (FDMA)-based systems. The hardware requirements for single-channel systems are much simpler than those of dual- or multichannel systems. The dual- or multichannel system assumes individual nodes either have two transmitters or have the ability to switch the single transmitter among the different channels.

2.3.1 MAC Solutions for Hidden Terminal Problem

For a single-channel system, many researchers have proposed a dialogue-based solution [12–17] to ease the hidden terminal problem in multihop environments. Based on this, the IEEE 802.11 embraced the virtual sense mechanism to solve the hidden terminal problem. The virtual sense mechanism discussed in the last section could increase the throughput in a single-hop environment. Another purpose of the RTS and CTS, which is actually the main objective of the virtual sense proposal, is to reserve the medium at the hidden terminal site and alleviate the hidden terminal problem. The preamble small RTS packet sent by the sender and the CTS sent by the destination contain information on the length of the packet to follow. All the nodes in the transmission range of the source node and destination node hearing the RTS or CTS will set a value in the network allocation vector (NAV) to defer the competition for the medium until the completion of the ACK. A backlogged node is required to check the NAV first to see whether it is zero; if it is nonzero, which indicates that another node already reserved the medium for a time period equal to the value in the NAV, it will wait. When the NAV is equal to zero, the sender begins to sense the medium.

Figure 2.5 compares the data transmission with and without the virtual sense mechanism. In Figure 2.5, the section above the time line shows the data transmission from node A to node B without the virtual sense mechanism, while the section below the time line shows the data transmission with virtual sensing. In the former case, if node C sends in the time period t_0 to t_5, a collision will happen. With virtual sensing, instead of sending the data packet directly, a node sends an RTS packet that includes a time that covers the following data transmission period length. The destination responds with a CTS packet, which reserves the medium in its neighborhood. Collisions will happen when node C sends data in the period from t_0 to t_3. Due to the small size of RTS and CTS packets, the time interval t_0 to t_3 is usually much smaller than time interval t_0 to t_5. Thus, the collision probability with virtual sensing is largely decreased over direct data packet transmission. Another advantage of data transmission with virtual sensing is that retransmitting small-size RTSs is more efficient than retransmitting large-size data packets.

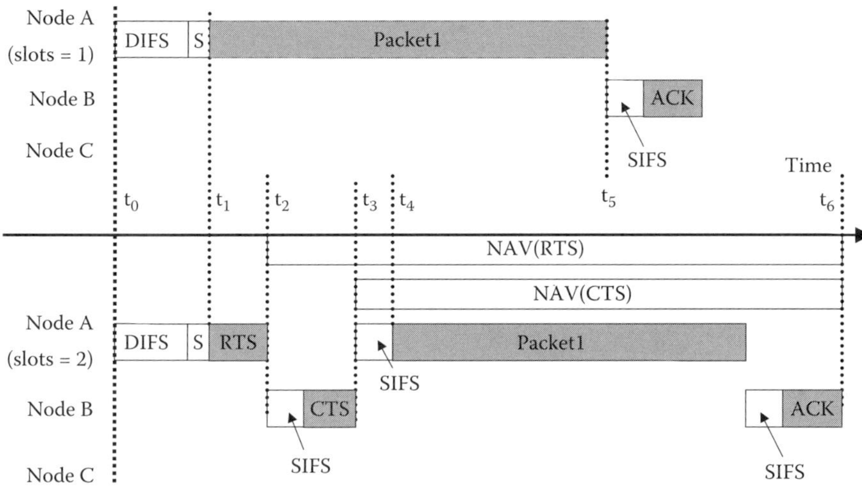

FIGURE 2.5 Packet transmissions with virtual sense.

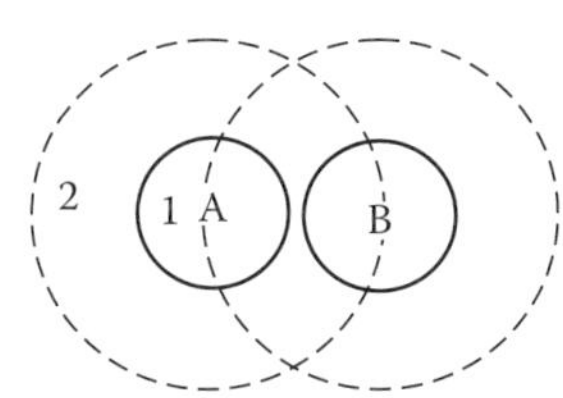

FIGURE 2.6 The transmission range and preventing range.

CSMA/IC [31–33] solves the hidden terminal problem without the RTS-CTS dialogue by blocking an enlarged area. This is called the preventing range, and has a radius of at least twice the radius of the communication range. As Figure 2.6 shows, the solid circles indicate the maximum transmission range of nodes A and B; the nodes outside this range or on the edge of this range cannot receive the data clearly. The collision radius is supposed to be the same as the transmission radius. The dashed circles indicate the prevented areas of correspondent nodes. An enlarged preventing range could be achieved by simply setting the noise signal threshold lower to make the sensing range match the preventing range. The nodes inside the sensing radius but outside the transmission radius can sense (not hear) the buzz signal or data transmission signal of the sender and accordingly suppress transmission and avoid collision. This approach is similar to the one in [9]. An optional method would be enlarging the transmission range of the buzz signal to match the preventing range, as shown in Figure 2.7. To avoid collisions, the stronger buzz signal must be sent in a channel different from the one used for data transmission (using TDMA or FDMA).

2.3.2 MAC Solutions for Exposed Terminal Problem

A single-channel, traditional dialogue-based protocol can alleviate the hidden terminal problem to a certain degree but does not solve the exposed terminal problem. With a single channel, an exposed terminal cannot receive the control packet, such as CTS, from

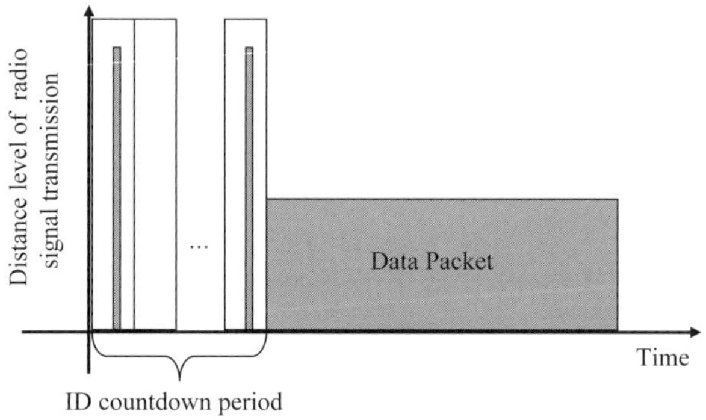

FIGURE 2.7 Strong buzz signal versus data packet signal.

the destination. However, if a control message like CTS for the exposed terminal can go through a channel other than the data channel, it becomes possible for the exposed terminal to receive CTS without collision. Thus, the exposed terminal problem is solved or alleviated. In [18–23], multichannel systems are proposed to improve the network performance in multihop environments. The systems proposed assume that the entire radio bandwidth is split into two types of channels: control channels and data channels. The preceding dialogue messages are transmitted in the control channels while the data packets are transmitted in the data channels. The Medium Access Control to access the control channels is based on one of the MAC protocols we discussed in the last section (e.g., Aloha, CSMA based, etc.), while the medium access for the data channel is based on the result of preceding negotiation that took place in the control channel. In [19], Choi et al. proposed using the single network interface card (NIC) in the multichannel system. Since there is only one NIC with every node, the nodes that are currently sending or receiving the data in the data channel are not able to record the transmission schedule negotiated in the control channel. This is recognized as the hidden multichannel problem. There is no efficient way to solve this problem. Normally, multichannel systems require two or more NICs.

The other direction in multihop environments is applying the busy tone [3, 24–28] or busy-tone-like signal [29, 30] to improve the network performance. Busy Tone Multiple Access (BTMA) [24] is the first MAC protocol that targeted the hidden terminal problem. In BTMA, the entire bandwidth is divided into two channels, i.e., a data channel and a busy-tone channel. Whenever a terminal has a packet to send, it senses the busy-tone channel first. If the busy-tone channel is idle, the MT sends the packet; otherwise, the MT backs off for a random duration of time. If a node senses the data channel is busy, it will send the busy-tone signal in the busy-tone channel and stop sending the busy-tone signal when the data channel becomes idle. Therefore, when a sender begins to transmit the data, all nodes in its sensing range will send a busy-tone signal in the busy-tone channel, and thus block the entire enlarged area. To address the exposed terminal problem and save energy from the busy-tone signal, the receiver-initiated BTMA [24] was proposed, where only the intended destination sends the busy-tone signal, and thus energy is saved and the reservation area is reduced. Before sending the data packet,

BTMA requires the sender to send a preamble mini-packet, RTS, which contains the destination ID information. The destination begins to send the busy-tone signal when it receives the RTS. The sender begins to send the packet when it senses the busy-tone signal in the busy-tone channel. The Dual Busy Tone Multiple Access (DBTMA) [25–28] also applies a busy tone to decouple the transmission. It claims a collision-free medium access, and will hence be discussed in section 2.6.

The multiple accesses with reservation lag time (RLT) [34–38] or detached dialogue approaches [39, 40] utilize a duration gap between the completion of the preceding control messages dialogue and the following data transmission. The duration gap is used by other MTs to complete the preceding dialogue. RLT-based MAC protocols [36, 37] can solve the heterogeneous hidden and exposed terminal problems with a single channel for both control messages and data packets in a distributed manner. Specifically, power-controlled Medium Access Control (POWMAC) presented in [41] and power-controlled binary countdown Request to Send, Clear to Send, and Ensure to Send (PBRCE) presented in [42] are applications of the RLT technique. To solve the exposed terminal problem, the authors suggest deferring the data transmission from the preceding dialogue, thus allowing a chance for the other parallel communications to complete their preceding dialogues. They both also work in single-channel systems.

Pair-wise ID Countdown (PIDC) [43] for multihop WLANs is motivated by the group activation approach [44, 45], where transmissions that can be processed concurrently without collision are categorized in the same group and scheduled to access the medium at the same time. PIDC stands as a novel direction that is different from the two above, and solves the exposed terminal problems without relying on either a busy tone or a preceding dialogue. Different from CSMA/IC, PIDC assigns the ID to the communications. The communications that could cause collision are encouraged to assigned different codes, while those that could transmit concurrently (i.e., the communications blocked by the exposed terminal problem) are encouraged to assign the same code.

2.3.3 Multicasting and Broadcasting in Multihop Environments

Due to the error-prone characteristics of wireless communications, an ACK packet is adopted to confirm a successful point-to-point data transmission. A multicast or broadcast model should first address the problem of how the successful receptions can be acknowledged. If all the destinations respond back with ACKs at the same time, they will all suffer collision because the IFSs prior to any ACK transmission are exactly the same. If a multicast/broadcast model ignores the ACK mechanism, then the multicast/broadcast becomes error-prone. Another problem with multicast/broadcast models is how to implement the virtual sense mechanism in a multihop environment. Very little research relative to this matter has been reported in the literature. In wireless networks, most multicast/broadcast models are based on a unicast model. In [46], the Multiple Access Collision Avoidance Protocol for Multicast Service (MACAM) is proposed to deal with multicast service in ad hoc networks. In MACAM, the RTS can contain several destinations' addresses. If the receiver's address is in the k position of the receiver list, then it will respond back with a CTS at the time k*SIFS + (k – 1)*CTS. The node that does not answer back with CTS can be further asked to receive packets in the next round of RTS.

The original CSMA/IC [31–33] is targeted at the collision-free at both unicast and broadcast data communication. For every data communication, the CSMA/IC blocks the entire preventing range; this is necessary for the broadcast data communication. Compared to other reported competition-based protocols, CSMA/IC is an efficient MAC protocol for broadcasting. The CSMA/IC is driven by competition code, which is processed in the competition period before actual data transmission. The competition period is regarded as the overhead that CSMA/IC carries. Notice that the broadcast is necessary for the ad hoc network to maintain connectivity among the MTs. The character of these kinds of packets is periodical created and may be small in size. With the small-size broadcast packet, the overhead of competition period in CSMA/IC is relatively huge compared to the duration of data transmission. Sequential MAC (SeMAC) [47] is proposed to broadcast the periodical management packets with small overhead. SeMAC is a reservation-based MAC protocol that ideally fits broadcasts with periodical message arrivals. In SeMAC, some periods of time are dedicated for broadcast. In these periodical broadcast periods, time is further sliced to slots. Each MT tries to reserve a slot for its periodical broadcast.

2.4 Energy-Efficient MAC

There are mainly two categories of energy-saving schemes for MAC protocol design in WLANs: power saving and transmission power control (TPC). Power saving involves letting the wireless device switch to power-saving mode when it is not engaged in transmission. TPC involves reducing the signal transmission power to a minimum level that enables the destination to clearly receive the message. Theoretically, TPC not only reduces the energy consumed for each data packet transmission, but also improves the network throughput by means of increasing the channel spatial reuse.

2.4.1 Power-Saving MAC Protocols

The IEEE 802.11 specification [4, 5] details a power-saving mechanism. In the specification, time is divided into beacon intervals. The front part of the beacon interval is called ad hoc traffic indication message (ATIM) window. Nodes exchange their packets' communication plans in the ATIM window with ATIM frame and ATIM-ACK. Apart from the nodes that will be involved in data transmission, all the nodes turn into "doze mode" for the rest of the beacon interval. The following should be noted: First, nodes that are scheduled to transmit or receive in the last small portion of the beacon interval will stay awake during the entire beacon interval. Second, the size of beacon interval and ATIM window are fixed. Third, the power-saving mechanism (PSM) in the IEEE 802.11 is regulated in the fixed maximum power communication even for destinations close to the source.

The dynamic power-saving mechanism (DPSM) proposed in [48] is aimed at achieving a dynamic variation of the ATIM window. It attempts to enlarge the dozing time by switching the nodes that complete the duty to the doze mode before the end of the beacon interval. A Power-Efficient MAC (PEM) protocol is proposed in [49]. The PEM attempts to reduce the transmission power besides the energy preservation mechanism proposed by IEEE 802.11. The proposed protocol works in the single-hop environment. Assuming that the broadcast of frames in the ATIM window could be heard by all the nodes, by collecting the information in the ATIM window the nodes will know the ongoing transmissions in the following beacon interval. Due to reduced power for the transmission, it is possible for parallel communication to take place in the single-hop system. By the proposed maximum independent set algorithm, all nodes can get the same scheduling order, on the justifiable assumption that all nodes know the same information.

The power-saving mode is normally designed for the multihop wireless ad hoc network in the situation that the traffic load is light. It is more suitable for wireless sensor networks where energy efficiency is much important than the network throughput.

2.4.2 Power Control MAC Protocols

The network performance of the interference-aware TPC protocols has been analytically studied in [50], where it is shown that TPC can increase the network throughput and reduce the energy consumption. Many TPC schemes for mobile ad hoc networks (MANETs) have been proposed in the literature [11, 12, 34–41 51–72].

In [52–55], TPC is used to control the network topology in order to maintain the connectivity of a certain node with its neighboring nodes. Power control is aimed at distinguishing the proper transmission power for all the nodes in their set of one-hop neighbors, which are previously selected. The TPC mechanisms proposed in [34–41, 56–72] are aimed at selecting the proper transmission power for individual peer-to-peer communication. When applying the TPC mechanism to the CSMA technique for MAC protocols, the performance of ad hoc networks will be considerably degraded due to the heterogeneous terminal problem [36, 39, 69, 73]. This problem may be further divided into heterogeneous hidden terminal problem and heterogeneous exposed terminal problem, both of which can be viewed as a new form of hidden terminal and exposed terminal problems uniquely due to heterogeneous communication ranges observed in ad hoc networks. Figure 2.8 shows the heterogeneous terminal problems in the heterogeneous ad

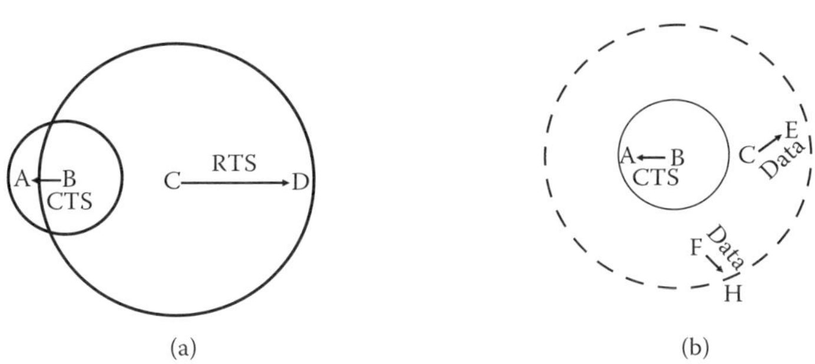

FIGURE 2.8 The heterogeneous terminal problems. (a) Heterogeneous hidden terminal problem. (b) Heterogeneous exposed terminal problem.

hoc networks with CSMA and RTS/CTS-based MAC protocol. As shown in Figure 2.8a, with TPC applying for the CTS message, the CTS message sent from B cannot reach the MT C, which means that B cannot block C from transmitting the signal. During the data transmission of A to B, if MT C intends to send data to MT D, which is farther away, it needs stronger power, and the RTS message from C interferes with the ongoing data reception at B. This can be viewed as a new form of the hidden terminal problem that is unique in heterogeneous ad hoc networks. On the other hand, if the ad hoc network applies the fixed maximum transmission range for the CTS message, the floor area it blocks could be far bigger than the necessary ones, blocking potential concurrent data transmissions. As shown in Figure 2.8b, with the maximum transmission power, the CTS sent from B will block the area depicted with a dash circle instead of a solid circle. This blocks the following data transmission of C to E and F to H. This could be viewed as an exposed terminal problem for the heterogeneous ad hoc networks.

TPC MAC protocols can be further divided into two subgroups: energy-oriented TPC and throughput-oriented TPC. Energy-oriented TPCs, presented in [56–59, 61–65], primarily aim at reducing energy consumption, with network throughput being a secondary factor. In the BASIC scheme in [58], nodes exchange the preceding dialogue packets that contain the power and interference information of the sender and the destination with maximum transmission power. After the preceding dialogue packets, the sender transmits the large-size data packet with negotiated transmission power. When facing the aforementioned heterogeneous terminal problem, these schemes could alleviate the problem in homogeneous ad hoc networks. They have no effect on the heterogeneous exposed terminal problem. These MAC protocols only attempt to reduce their power consumption rather than utilizing smaller transmission power to increase the spatial reuse, and hence the network throughput.

On the other hand, the throughput-oriented TPC schemes reported in [34–40, 60, 66–73] aim mainly at increasing the spatial reuse. The schemes proposed in [60, 66, 67] apply the preceding dialogue with maximum transmission power to alleviate the heterogeneous hidden terminal problem. To alleviate the heterogeneous exposed terminal problem and achieve parallel communication, these schemes use a multichannel system to enable the completion of multiple preceding dialogues. These schemes also impractically assume that nodes are able to simultaneously transmit data on one channel while receiving on the other. Specifically, Monks, Bharghavan, and Hwu propose the Power Controlled Multiple Access (PCMA) [66] protocol based on a busy tone. In PCMA, the data receiver senses the channel and measures the current noise and interference level at its location, and then calculates the additional interference it can tolerate. Simultaneously, the data receiver will transmit a discrete busy-tone signal in the busy-tone channel. The power level of busy tone reflects the additional interference it can tolerate. An MT that intends to transmit data will first collect all the busy-tone signals sent by all of its neighboring MTs. Based on this information, the MT will determine the maximum power it can use to transmit, and such that it does not interfere with an ongoing data reception. Although PCMA can solve the aforementioned heterogeneous terminal problems, a PCMA device requires two transceivers and hardware for signal strength measurement. As a result, the hardware cost and power consumption of PCMA will be increased.

One way to increase the spatial reuse in the single-channel system is to apply the TPC for the preceding dialogue, but this will lead to the aforementioned heterogeneous hidden terminal problem. Poon and Baochun [68] propose to dynamically vary the transition power between the maximum and the minimum to reach the destination. As indicated in [68], the heterogeneous hidden terminal problem often leads to unfairness because the higher-power communication has a higher success rate of transmission than the lower-power communication. This is because a higher-power communication has more chance to interfere with the data reception of a lower-power communication. The authors propose to refer the success rate of transmission as feedback for the transmission power.

POWMAC [41] introduces a power-controlled MAC protocol modified from the DCF of IEEE 802.11 that not only aims at saving energy, but also improves the network throughput. To save power, the authors suggest controlling the transmission power of data following the preceding control packet. The goal of POWMAC is similar to the PBRCE proposed in [42], which aims not only at energy efficiency, but also at a higher network throughput. They both also work in single-channel systems. Energy efficiency in PBRCE is achieved by reducing the collision rate—thus saving energy through avoiding retransmission of the same packets, controlling the signal transmission power, and forcing terminals to go into the sleep mode to avoid unnecessary passive listening.

2.5 QoS MAC Protocols

As mentioned earlier, IEEE 802.11 is currently the most widely accepted and commercialized WLAN standard. It defines DCF and point coordination function (PCF). The PCF is designed to be driven by the access points in the infrastructure-based wireless networks, while the DCF can work in both ad hoc and infrastructure-based wireless networks. Due to its simplicity and adaptability, the DCF is now a completely implemented and commercialized MAC protocol. On the other hand, the PCF, which can potentially provide QoS support, is still limited to academic study due to its complexity and centricity requirements. With the transmission speed of the WLANs becoming faster and faster and the applications in the wireless networks becoming more varied and popular, QoS becomes a general concern of the MAC protocol design. Several adverse factors make such a design a challenge. Since the IEEE 802.11 DCF is fully commercialized, many researches focus on the QoS enhancements of the IEEE 802.11 DCF, while a few other efforts are dedicated to make the PCF work for the wireless ad hoc environment, e.g., Mobile Point Coordinator MAC (MPC-MAC) [74] protocol. In this subsection, we give an overview of the recently emerging QoS MAC protocols designed for wireless ad hoc networks. In these QoS MAC protocols, when two MTs with different priority compete for the medium, the probability (p) of the higher-priority MT winning the medium Is higher than 50% but less than or equal to 100%. Based on the value of p, we can categorize QoS MAC protocols into two groups: (1) MAC protocols supporting absolute priority ($p = 100\%$) and (2) MAC protocols supporting relative priority ($50\% < p < 100\%$).

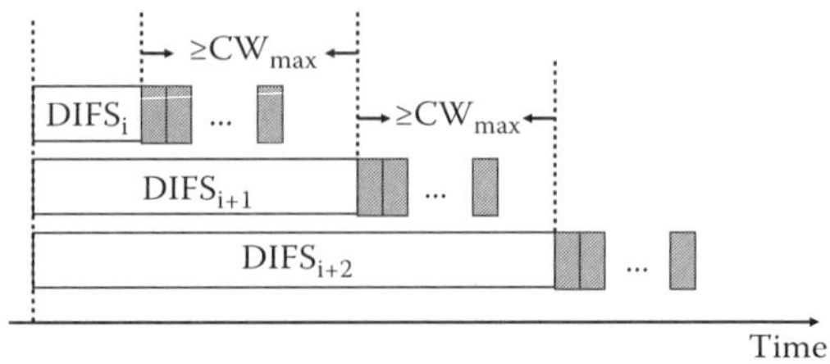

FIGURE 2.9 Differentiated DIFSs mechanism for absolute priorities.

2.5.1 MAC Protocols Supporting Absolute Priority

To support packets with different priorities for different services, it is intuitive to define and differentiate the values of relative system parameters in DCF. Modified DCF MAC protocols can give higher-priority packets an absolute priority or a relative priority. The work in [75] proposes to enlarge the duration difference between the IFSs; that is, making the DIFS for the lower-priority packet longer than the sum of the DIFS and contention window size of the higher-priority packet, as shown in Figure 2.9. In effect, the higher-priority packets have absolute medium access privilege compared to lower-priority packets. The disadvantages of this design include: (1) The CW for high-priority packets is fixed, which cuts the congestion control ability of the original DCF for all priority packets except the lowest-priority packet. (2) The difficulty to optimize the length difference among the DIFSs. Small differences between DIFSs lead to the limitation of the back-off space for the higher-priority packets. In this case, heavy traffic load of high-priority packets leads to a high collision probability. On the other hand, a large difference will cause a long DIFS, which might result in capacity wastage if traffic load for the high-priority packet is light. (3) The number of priority classes is limited.

To overcome the disadvantages mentioned above, Banchs et al. [76] proposed to use a jamming signal after the carrier sensing period to gain absolute priority for high-priority packets. As Figure 2.10 shows, different priority packets will use different lengths of DIFS for sensing the medium, and the difference between the neighboring priorities is only one time slot. After sensing the medium for the duration of a DIFS, the nodes send the first buzz signal for the duration of a random number of slots, then sense the medium for one time slot while sending the second buzz signal for the duration of another random number of time slots. Before sending the data packet, the node needs

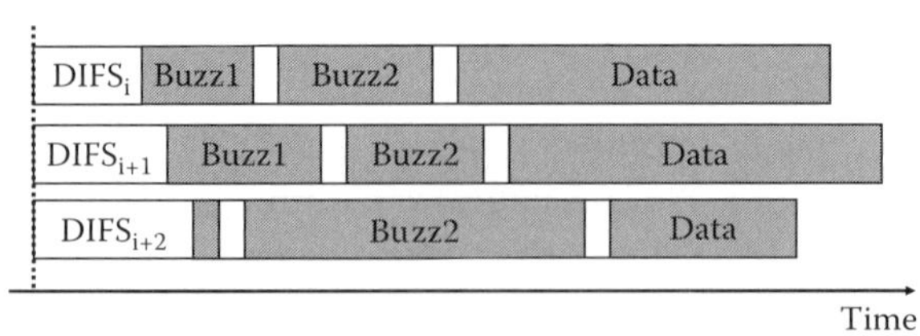

FIGURE 2.10 Another MAC protocol that supports absolute priorities.

to sense the medium for another time slot. If the node sensing the medium is busy in any sensing period, the node quits processing the next step and backs off. The first buzz signal is at least one slot. If the same priority packets happen to create the same length of first and second buzz signals, these packets will be transmitted concurrently and may suffer collision.

2.5.2 MAC Protocols Supporting Relative Priority

The enhanced DCF (EDCF) defined in IEEE 802.11e [77] gives an example of supporting relative priority differentiation by modifying the parameters of DCF. In EDCF, the length difference in arbitrary IFSs (AIFSs) between the neighboring priorities is only one time slot. A higher-priority packet would use a shorter AIFS to compete for the medium. Combined with the following back-off time slots, the higher-priority packets have a relatively higher chance to transmit the data packet first, and thus win the competition. The lower-priority packets also have a chance to beat the higher-priority packet, but with a lower probability. Therefore, the priority mechanism defined in the EDCF can only provide statistical QoS differentiation rather than strict QoS guarantees.

To further increase these winning probability differences, the back-off scheme following the DIFC could be modified to support service differentiation. This can be achieved through giving the higher-priority packet a higher probability to create a smaller number of back-off time slots and making it reach the data transmission state first. Accordingly, higher-priority packets could be given smaller values for CW_{min}, CW_{max}, and the CW enlargement factor. With the smallest CW_{min} and CW_{max}, higher-priority packets are more likely to choose a smaller back-off counter and finish the back-off process earlier. Upon a collision, the CW of the higher-priority packet will be increased at a smaller rate than the lower-priority packet for the next round of competition, and would thus be more likely to create smaller values of back-off counter for the higher-priority packets.

An alternative way to differentiate the back-off counter is a fairness mechanism addressed in [78] that combines the p-persistent CSMA with the DCF. In this protocol, the CW remains the same all the time for all the different priority packets. However, a station would send the RTS packet with probability p after the back-off period, and different priority packets have different values of p, where different values of p indicate different retransmission factors.

These differentiated back-off schemes can also be used in the length definition of the jamming signal reported in [76, 79, 80]. Contrary to the differentiation priority defined in the back-off interval, a longer jamming signal corresponds to higher priority. The differentiated jamming signal can coexist with the DCF protocol to support the QoS for the real-time packets. The Black Burst (BB) protocol proposed in [79] adds a jamming signal mechanism to the commercialized DCF protocol in order to support absolute priority for voice traffic. The BB protocol uses PIFS for the medium access of voice data, while nonvoice data are proposed to use exactly the same mechanism as DCF. The highest-priority response-type data, like ACK or CTS, use SIFS as medium access IFS. They thus have higher priority than voice data. However, medium access for the voice data utilizes the additional BB protocol that uses PIFS combined with a jamming signal, providing the voice data with absolute priority over nonvoice data. The disadvantages of the jamming signal scheme include: (1) transmission of the jamming signal consumes additional energy; (2) the jamming signal may collide with ongoing packet transmission/receptions, especially in a multihop environment; and (3) in the jamming-based scheme, a larger counter indicates a higher priority, and a previously at-loss node tends to get the larger counter to compete for the medium. Upon heavy traffic load, the voice data will tend to have a larger jamming signal counter, which enlarges the average packet delay.

2.5.3 MAC Protocols Supporting Controllable Priority

Normally, in the wireless ad hoc networks, with the increase of traffic load of higher-priority packets, the throughput of low-priority packets will decrease. Controllable Fair ID Countdown (CFIC) proposed in [81, 82], which is developed from CSMA/IC, achieves controllable relative throughput differentiation among different priority levels and can guarantee starvation-free access for all nodes. In CFIC, the priority is controlled by a single priority slot. Figure 2.11 illustrates the timing format in the ID countdown period of CFIC, where the priority slot is in front of the ID segment. Through setting the priority bit, CFIC attempts to achieve differentiated service by assigning different back-off waiting times to packets with different priorities. Under CFIC, low-priority packets may suffer longer waiting times, but they will not experience starvation.

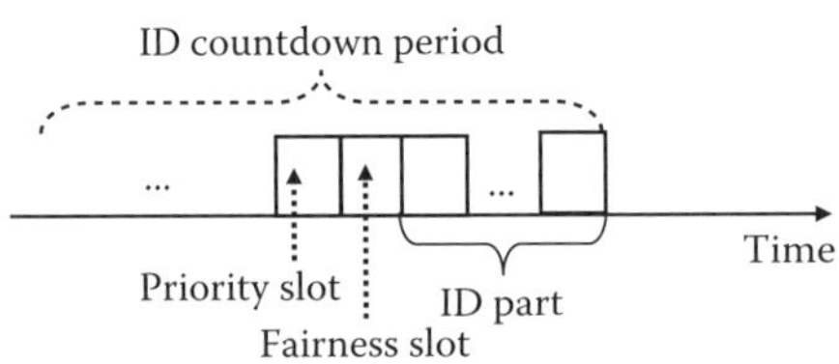

FIGURE 2.11 Illustration of the priority and fairness slots in CFIC.

Although there is only a single priority bit in CFIC, several priority levels may be defined and supported. Theoretically, however, there is no upper bound for priority levels in CFIC. If there are totally *k* different priority levels, where smaller values indicate higher priorities, each node will maintain a series of *k* different parameters called waiting threshold (WT), $\{WT_1, WT_2, \ldots, WT_k\}$, for each priority level of packets. Whether to set the priority bit strictly depends on whether the waiting time T_i (counted from the time this priority *i* packet becomes the first packet in the buffer) exceeds the corresponding WT_i. Once $T_i > WT_i$, the priority bit will be set. There are a number of ways to define the length of WT_i for the priority *i* packet.

2.6 Fairness and Starvation Prevention Issues

The fairness problem is relative to the problem of prioritized medium access. In a distributed system, it is easier and more efficient for a node to occupy the medium for a long time. Therefore, the fairness between the same priority packets is an important point for consideration when designing a MAC protocol for wireless networks. The method for achieving fairness is to increase the medium access priority for each competition loss.

2.6.1 Fairness in the Back-off Mechanism

The back-off mechanism in the DCF of IEEE 802.11 achieves fairness among the competing nodes by processing the back-off counter, the value of which is inherited from the previous round instead of creating a new value. Therefore, for the new competition, the previous competition losers are more likely to have smaller counter values than the newly joining senders. Figure 2.12 shows an example of the DCF back-off that achieves the relative fairness. In the scenario, node A has a packet to send to node B, and node C has a packet to send to D. Both nodes A and C begin to sense the medium idle for DIFS and go into the back-off stage at the same time. Suppose node A creates a back-off value 3 while node C creates a back-off value 5. Then, at time t_2 the counter of node A reaches 0, while node C still has 2 left. Node A wins the medium and sends the packet while node C stops processing. At the next round of competition beginning at t_4, node C will process the 2 back-off value while node A will re-create a back-off value for its new packet, which is more likely to be larger than that of node C. Accordingly, there will be a higher probability that node C will win the medium. However, node A still has some probability to create a back-off value that is smaller than 2 and win the medium again. The DCF back-off mechanism hence only achieves relative fairness.

The back-off mechanism can alleviate the unfairness of the nodes that lost the last medium competition in the DCF. Considering another important aspect, we know that collisions are the main reason behind long transmission delays. However, in a distributed system, congestion control is normally achieved by the feedback of individual collision status. The congestion is the result of aggregate behavior of all nodes. Nevertheless, distributed systems attempt to overcome collision in the individual behavior. This unavoidably leads to unfairness of the individual node. The collided packet already suffers longer delay, but the back-off mechanism still requires it to back off for a longer duration to retransmit. The BEB in Medium Access with Collision Avoidance (MACA) [15], the multiplicative increase and linear decrease (MILD) in Medium Access with Collision Avoidance for Wireless (MACAW) [17], and variations of BEB for the CW in the DCF all apply congestion control with the individual collision status, unavoidably leading to unfairness. MILD seems better than BEB because change in the value of

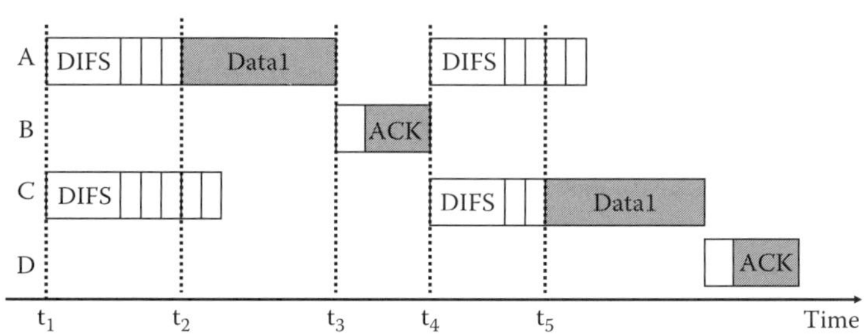

FIGURE 2.12 The back-off mechanism of DCF.

back-off (BO) in MILD is softer than those in BEB. This somewhat alleviates the situation where nodes undergoing collision become worse and the successful nodes become better off. We strongly believe that the splitting algorithms discussed in section 2.2 may potentially be used here. This is because splitting algorithms transmit the collided packets before other nodes that did not undergo a collision can participate into contention.

Some unfairness is caused in a multihop setting where it is possible for the destination to block itself from sending the feedback message. This mechanism can also increase the value of BO in the back-off mechanism of the MACA protocol. For example, in Figure 2.13, there are two streams creating a saturated traffic load: from node C to D and from node B to A. Suppose node B tries to negotiate the next data transmission by sending the RTS package during the data transmission of the packet from node C to D. However, the response message, i.e., the CTS from A, cannot be heard by B because of the interference caused by C. In this case, node B increases its BO, and this leads to unfairness between these two streams in MACA. MACAW solves this problem by introducing another control message called Data Sending (DS). It requires that when the sender receives the CTS, it needs to broadcast the DS message to ask all the exposed terminals to block themselves from sending the RTS, and thus protect the values of BO of the exposed terminals from unnecessary increase. In the CSMA/CA, the medium reservation by the RTS and the carrier sensing before sending the RTS has the same function.

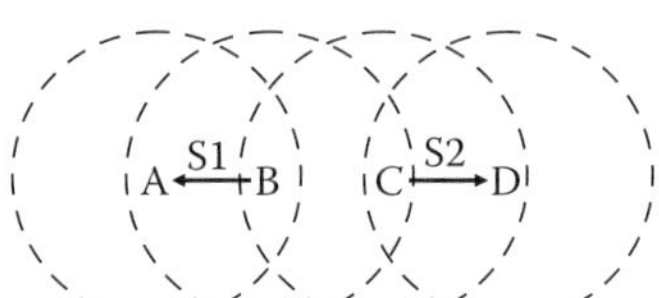

FIGURE 2.13 An example shows the necessity of the DS in MACAW.

Figure 2.14 shows another unfairness scenario of MACA that results in a multihop network topology. During the data transmission of one stream, say S1, node C remains silent when it hears the RTS from D for the duration indicated by the previous CTS from B. It is hard for node D to predict the complement of the data transmission, and it will hence often miss the chance to win the medium when the current data transmission is completed. This will lead to unfairness for the second stream, S2. Furthermore, the BO for node D will unnecessarily increase, worsening the situation of S2. MACAW introduces another control message called receiver-initiated RTS (RRTS), which is initiated by the receiver to require the sender to send the RTS to let S2 win the medium access competition in the future.

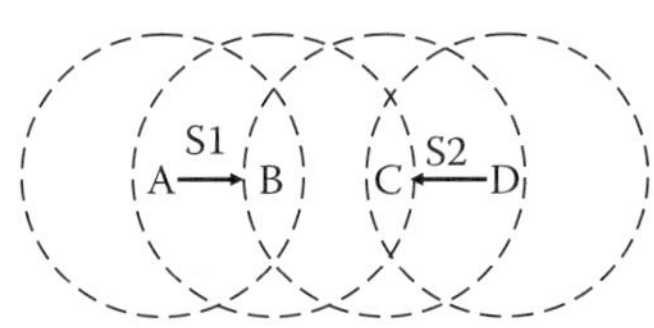

FIGURE 2.14 An example shows the necessity of the RRTS in MACAW.

2.6.2 Fairness Solution Based on Queuing Delay

A more effective fairness scheme assigns a longer queuing delay packet a higher priority to access the medium. The Black Burst mechanism for the voice packets is an example that applies queuing delay to the priority of medium access. Nodes with real-time packets contend for access to the channel after the medium IFS rather than after the long IFS used by data nodes. Thus, real-time nodes as a group have priority over data nodes.

Real-time nodes first sort their access rights by jamming the channel with pulses of energy, denominated Black Burst signals (BBs), before sending real-time packets. The length of BB transmitted by a real-time node is an increasing function of the contention delay experienced by the node, measured for the instant ($T_{duration}$), when an attempt to access the channel has been scheduled until the channel becomes idle for the medium IFS. Similar to the notion of slot time in DCF, an integer number of BB slot units form the BB. The minimum BB slot number is 1, as it needs at least one BB slot unit to gain the channel first from the non-real-time data nodes. The BB duration is given by $BB_{duration} = 1 + \lfloor T_{duration}/T_{unit} \rfloor \times BB_{slot}$. Here, $BB_{duration}$ is the period of time that the competing MTs are sending the BBs, $T_{duration}$ is the period of time that the MTs are waiting for the medium to clear, T_{unit} is a constant parameter defining the length of time used to count for the MTs waiting for the medium to clear, and BB_{slot} is another constant parameter that defines the length of unit BB slot time.

After the BB period, a node with real-time data will sense the medium for the sensing time period to check if the medium is clear. If the channel is clear, it will send its real-time packet; otherwise, it will quit the competition. Therefore, instead of a shorter slot time extension winning the medium in the non-real-time data group, longer BB duration will win access to the medium.

Figure 2.15 shows an example of the BB mechanism. The wireless network consists of four mobile terminals, A, B, C, and D. They can all communicate with each other. In this case, node C has a series of data packets to send to node D. As node C sends one packet, node A plans to send a real-time packet at time t_0, and node B plans to send a real-time packet at time t_1. They both find the medium is busy and wait until time t_2. At time t_2, node D will respond with ACK, which has highest priority. Therefore, node D will win the competition over nodes A and B, then send the ACK. When node D finishes sending the ACK at time t_4, node C schedules to send its next non-real-time data packet, still

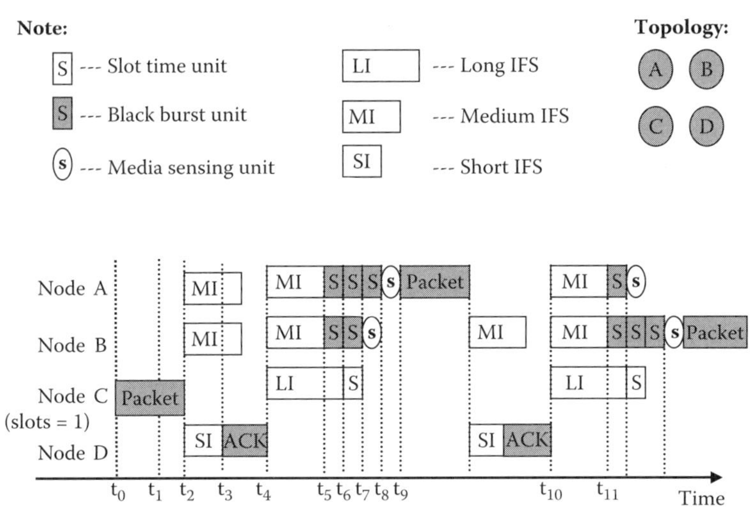

FIGURE 2.15 Black Burst contention mechanisms.

using the long IFS to compete, while both nodes A and B are using the medium IFS. At time t_5, real-time nodes A and B sense that the medium is clear and begin to send the BB signal, causing node C to quit the competition. Simultaneously, using the BB equation, nodes A and B calculate their respective BB durations. Node B calculates a value of 2, but node A calculates a value of 3 because it has been waiting longer, i.e., time duration $t_5 - t_0$ is longer than time duration $t_5 - t_1$. After two BB time slots, i.e., at time t_7, node B senses the medium and finds the medium busy, so it quits the competition. After BB duration, node A senses the medium is clear and sends its real-time packet. The next competition occurs at time t_{10}, when node B has now been waiting for a longer time: its BB duration will be calculated from time interval time t_1 to time t_{11}, but node A will only calculate from time t_{10} to t_{11}. Accordingly, node B will win access to the medium. From this example, we can see that the BB mechanism does not cause starvation for real-time data.

2.6.3 Fairness Based on Traffic Weight

Another objective of fairness scheduling is based on the weight of the flow and can be described by an idealized fluid flow model with equation (2.1). In equation (2.1), r_i denotes the bandwidth used by flow *i* and φ_i denotes the weight of the flow, i.e., the service share of each flow should be proportional to the flow weight during any period of time. It has been proven that this fair scheduling provides end-to-end delay bounds for any well-behaved traffic flows [83]. In the real world, the wireless MAC protocol is often cooperating with certain routing protocols and a Call Admission Control (CAC) mechanism. With a certain routing protocol and CAC mechanism, the traffic load is uneven for each individual node. This fairness concept is the objective of some research, such as the Distributed Fair Scheduling (DFS) proposed by Vaidya et al. in [84] and Distributed Weighted Fair Queuing (DWFQ) proposed by Banchs and coworkers in [76, 80, 85]:

$$\frac{r_i}{\varphi_i} = \frac{r_j}{\varphi_j} \forall i, \forall j. \tag{2.1}$$

In the DFS, the back-off interval for the *k*th packet of flow $i(B_i^k)$ is obtained by using equation (2.2), where the value of *Scaling_Factor* is properly chosen to make B_i^k at a certain suitable scale, L_i^k denotes the size of the *k*th packet of flow *i*, and φ_i denotes the weight of the flow *i*:

$$B_i^k = \left\lfloor Scaling\_Factor \times \frac{L_i^k}{\varphi_i} \right\rfloor. \tag{2.2}$$

To reduce the probability of collision, the B_i^k above is randomized as B_i^k = $rand(0.9,\ 1.1) * B_i^k$. In the DWFQ, each node estimates the value of r_i/φ_i and piggybacks it to its transmission. All other nodes retrieve the corresponding value from the packet and compare the value received to their own estimated values to decide whether to decrease or increase their contention windows.

2.6.4 Fairness in CSMA/IC

In the binary countdown MAC protocol, a node with a higher binary code always beats nodes with smaller values of binary competition code. When the network traffic load is saturated, it can be expected that the stations with higher addresses always occupy the channel, while stations with lower addresses suffer starvation. The traditional way to achieve fairness for binary countdown MAC protocol in wired networks is using ID circulation. The binary code used for the medium access is continuously changed. That is, the winner will change its ID to a value equal to 0, while all other stations with lower ID increase their ID by 1. This algorithm works well for static wired LANs. However, in multihop wireless ad hoc networks, nodes are mobile and may shut down or be turned on, all in an unpredictable manner. Such specific features make the ID circulation scheme hard to implement. The fatal point is that some nodes with lower ID may be located at hidden positions and may not increase their IDs by 1. This leads to future collisions by duplicate ID holders. To illustrate this point, during the first round of competition in Figure 2.16, node A wins and its ID is changed to 000. This makes node B increase its ID from 101 to 110. However, node C is in a hidden position from node A and keeps its original ID. Hence, nodes B and C will hold the same ID after the first round of competition and will cause collision in the following round.

A practical way to solve the fairness problem for binary countdown in ad hoc networks is to use a randomly created ID [86–88] to replace or insert in front of the static ID for each medium competition. Inserting a randomly created ID before the original static ID [87, 88] can maintain the uniqueness of the entire competition code and at the same time alleviate the fairness problem for the binary countdown MAC protocol. This, however, increases the number of competition slots, making the total number of time slots before the data transmission even longer. Replacing the static competition ID with a randomly created ID [38, 87, 88] solves the fairness problem but cannot guarantee collision-free operation. Fair ID Countdown (FIDC) [89] or Differentiated Fair ID Countdown (DFIC) [81, 82] is extended from CSMA/IC; it focuses on fair medium access for multihop wireless ad hoc networks. The basic idea of FDIC or DFIC is inserting one slot, called the fairness slot, between the priority segment and the ID segment.

(a)

Node	A	B	C	D
1	111	101	110	100
2	000	110	110	100

(b)

FIGURE 2.16 A scenario that shows ID circulation in multihop wireless networks. (a) Network topology. (b) First two rounds of competition.

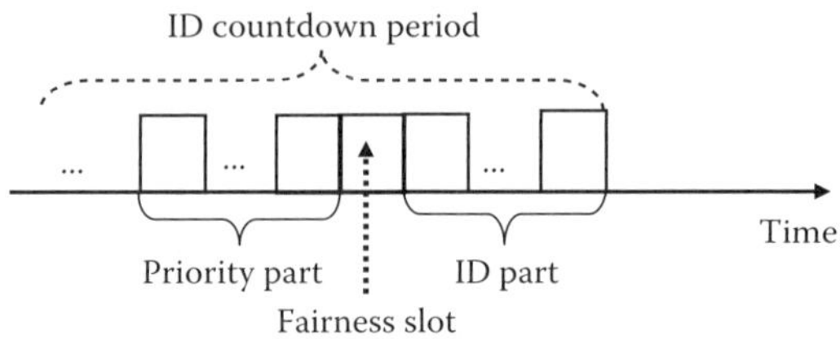

FIGURE 2.17 Illustration of fairness slot position in DFIC.

Figure 2.17 illustrates the position of the fairness slot in DFIC. Because the priority segment comes before the fairness slot and ID segment, all the medium contenders with lower-priority packets will be pushed out in this field. Consequently, after processing the priority segment, the medium competition will be actually restricted among the nodes with the current highest-priority packets.

The fairness bit is processed before the bits in the ID segment. Accordingly, once the fairness bit is set to 1, the node can beat the node with the highest ID value and is hence used to achieve fairness among nodes within the same priority level. There are two situations that can cause a node to set its fairness bit in DFIC: by either detecting an unfair situation or being beaten by the same priority packets enough times to exceed a certain threshold value. An unfair situation is recognized by the node if it was beaten by a same priority node with higher ID whose fairness bit is also clear. It is simple for a node to realize this unfair situation. Once a node begins processing bit patterns in the ID segment, the only reason for a node to lose the competition is having an ID value lower than that of one or more other competitors.

2.7 Collision-Free MAC

Several MAC protocols [25–28, 90–114] have been shown to be collision-free under some network and traffic conditions. We can generally classify these MAC protocols into four categories. The following subsections provide a survey for these MAC protocols.

2.7.1 CDMA-Based Collision-Free MAC Protocols

In a wireless network system that applies Code Division Multiple Access (CDMA) [115], each MT uses an assigned binary code to spread each bit of original digital data (e.g., using code –1010... to represent 1 of the digital data and the inverse of the code, –0101..., to represent 0). At the receiver side, the destination MT applies the same code to decode the received signal and recovers the original digital information. Many researchers proposed CDMA-based MAC protocols and claimed that they are collision-free, as in [90–98].

To apply CDMA for the ad hoc network, several mechanisms must be established. First is an effective spreading code assignment mechanism. Second is a coordination method of the pairs of senders and receivers such that they apply the same spreading code to spread and decode the signal. Third is the way to avoid the same spreading code being applied for concurrent data transmission, and hence avoid collisions. The above three mechanisms are not easy to achieve for the dynamically changing ad hoc network. The CDMA-based MAC protocols [90–98] propose ways to achieve the above-mentioned mechanisms. Collision-free medium access is thus claimed when different spreading codes are applied for different concurrent data transmissions. While this may be true for some cases, in the following we argue the generality of such claims.

In CDMA, signals spread by different codes have nonzero cross-correlations [116] at the receiver side because of different propagation delay from different source MTs. Based on the mathematical analysis in [115, 117, 118], to achieve a 10^{-6} bit error probability, the total interference power must satisfy

$$\frac{\sum p_i}{w \times p_o} \leq 0.4743,$$

where w is the ratio of the bit rate of spread digital signal to the bit rate of original digital data information, and p_o and p_i represent the intended signal power and interference power at the receiver side. If the interference power is over the threshold, the receiver also cannot decode the received data correctly. This is called CDMA secondary collision [116]. The so-called near-far problem [116] increases the probability of CDMA secondary collision. As shown in Figure 2.18, the distance of A to B and C to D is much longer than the distance of A to D or B to C. Suppose A transmits data to B concurrently with a transmission of data from C to D. No matter how the transmission power is controlled, on the receiver side (B or D) the p_i/p_o cannot be guaranteed to be under the threshold, possibly leading to a secondary collision.

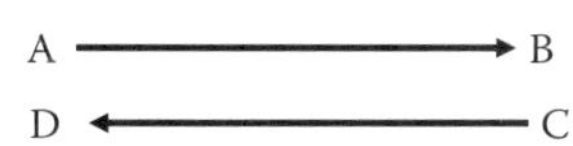

FIGURE 2.18 Near-far problem.

2.7.2 Collision-Free MAC Protocols Based on FHSS

In [99–105], MAC protocols based on the frequency hopping spreading spectrum (FHSS) technique are proposed to achieve collision-free data communication. Specifically, [105] focuses on spread spectrum-based techniques for solving the interference range hidden/exposed terminal problem.

With the FHSS technique, all the MTs in the system process a predefined sequence of radio frequency hopping. Time is sliced to fixed size slots as shown in Figure 2.19. If an MT is not active, i.e., neither transmitting nor receiving, in the current frequency, it will hop to another frequency according to the predefined sequence in the next time slot. The transmission-intended MTs need a certain way to reserve the current frequency for the following data transmission.

The Hop Reservation Multiple Access (HRMA) in [99, 100] applies the RTS-CTS dialogue between the ongoing sender and receiver to reserve and stay in the current frequency. If the destination is currently not ready to receive data or RTS is collided with at

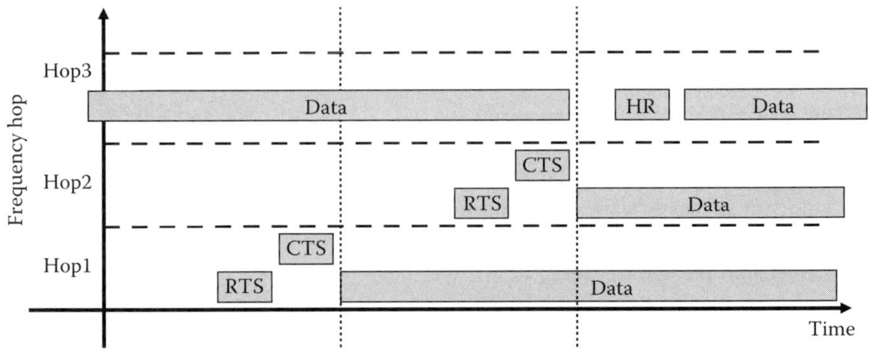

FIGURE 2.19 Hop Reservation Medium Access.

the destination for some reason, the sender will not receive the expecting CTS message, and this will cause the sender to back off for a random duration. The unsuccessful sender will return back to the normal channel hopping process. To avoid the possible collision of the overlength data packet with the next round of RTS message at the same hop, the HRMA reserves a time space before each RTS message for the hop reservation (HR) message. As shown in Figure 2.19, when current frequency jumps to hop 3, the transmitting MT transmits an HR message before any other RTS messages, and this continues the uncompleted transmission.

The receiver-initiated channel hopping (RICH) [101–104] applies the polling mechanism from the receiver to the possible sender. According to this protocol, if an MT has a packet to transmit for another MT, it transmits a Ready to Receive (RTR) message instead of an RTS message. The destination of RTR will respond with transmitting a data packet if it happens to have a packet for the polling source. Otherwise, the destination of RTS will respond with a CTS message to let the polling source transmit data. If the polling MT does not receive the expected data message or CTS message, it will return back to the normal frequency hopping and back off itself for another try.

In HRMA and RICH, the medium access for the control message is based on the slotted ALOHA mechanism and cannot avoid collisions. Although the collision of preceding control messages for both HRMA and RICH does not lead to further collisions of the larger-size data packet, they will result in resource wastage of channel for one round, from the time the common frequency hops away to the time it comes back.

2.7.3 TDMA-Based Collision-Free MAC Protocols

There have been many studies applying Time Division Multiple Access (TDMA) to ad hoc networks for collision-free medium access [106–112]. In TDMA-based MAC protocols, time is synchronized and divided into equal-duration time slots, where a number of continuing time slots form a time frame. The collision-free TDMA-based MAC protocols aim at assigning any two communications that may lead to collision with two different time slots. The schemes proposed in [119–122] apply a fixed-length time frame and assign each MT in the system one TDMA time slot in each time frame. Poor channel utilization is shown in these schemes because the length of time frame must be set to a value large enough to accommodate the maximum possible MTs in the system, even in scenarios with low nodal density. This leads to many time slots being unassigned. To reduce the number of unassigned slots in a time frame, Kanzaki et al. in [108] propose an algorithm that applies dynamic frame sizes for TDMA. That is, the frame length is dynamically set as a power of two slots according to the needs of existing MTs.

In a multihop environment, parallel communications become possible. Many MAC protocols based on TDMA aim at maximizing spatial reuse and minimizing the TDMA frame length. The Spatial TDMA (STDMA) [107] works under the assumption that the network topology is static and is known to all the MTs in the network. The communications that can be processed simultaneously without collision form a clique [107]. Each transmission joins one or more cliques. Each clique reserves a time slot in the frame, where the frame length is set to be equal to the number of different cliques. An MT that intends to transmit a packet must wait until a slot that is reserved for the clique that contains the intended communication. The authors in [109] proposed to apply power control for the STDMA to increase the size of an individual clique, and reduce the size of a frame that improves the network throughput. In a typical ad hoc network, however, all MTs are in the state of mobility, and thus the network topology is in a constant state of change. This results in a large overhead for the Spatial TDMA that is required to maintain the slot assignment to the cliques. To make STDMA match the characteristics of ad hoc networks, the authors in [110, 112, 123–127] present different algorithms to optimize the STDMA. In general, obtaining an optimal solution to the problem of time slot assignment of STDMA is NP-hard [128, 129]. The necessary condition for a collision-free time slot scheduling for TDMA is the accurate knowledge of the surrounding network topology and the slot assignment status, and this is typically impractical. In mobile ad hoc networks, collision-free time slot scheduling in the last frame may still lead to collisions in data transmission due to the change of network topology. So, TDMA-based MAC protocols are not completely collision-free.

2.7.4 Competition-Based Collision-Free MAC Protocols

2.7.4.1 Dual Busy Tone Multiple Access (DBTMA)

Dual Busy Tone Multiple Access (DBTMA) [25–28] uses two out-of-band tones to decouple communications in two directions, as shown in Figure 2.20. The entire channel is divided into a control channel and a data channel. Data packets are transmitted over the data channel, while control packets (e.g., RTS and CTS) are transmitted over the control channel. Two extra-narrow-band channels are reserved for the transmit busy

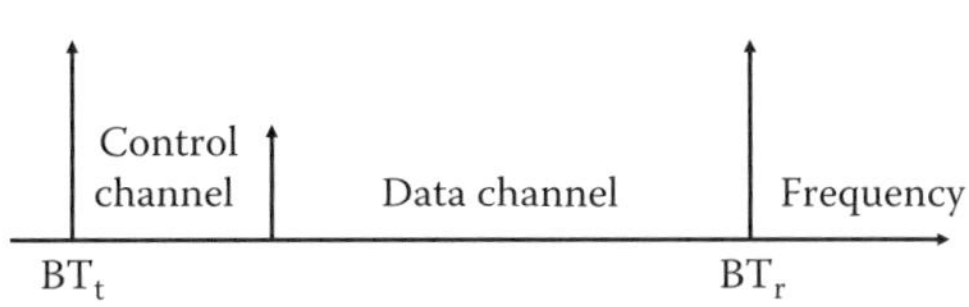

FIGURE 2.20 DBTMA frequency chart. Adapted from [28].

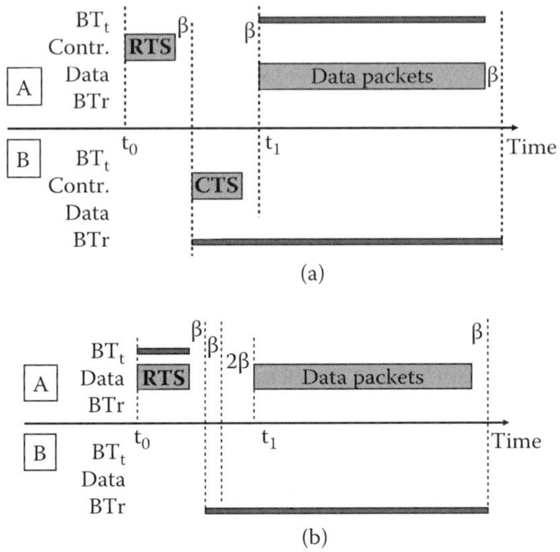

FIGURE 2.21 Channel diagram of DBTMA. (a) Original DBTMA. (b) Modified DBTMA.

tone (BT_t) and the receive busy tone (BT_r). When the destination node receives the RTS packet and finds that it is able to receive the packet, e.g., the medium is clear, it transmits its BT_r signal and replies with a CTS packet. Upon the source receiving the CTS packet, it starts to transmit its BT_t signal and sends the data. All the other nodes in the transmission range of the receiving node can sense the BT_r signal and defer from transmitting. All the other nodes in the transmission range of the sender will determine that they cannot receive the other node's data. However, if a node cannot sense the BT_r signal, it is capable of transmitting data. The original DBTMA protocol requires two separate channels plus two additional busy-tone channels. The modified version of DBTMA [28] adopts a shared channel for the control packet and the data packet, and uses the BT_r as the response signal from destination, as shown in Figure 2.21. When sending RTS, the sender also sends BT_t. At the same time, the sender is required to sense both busy-tone channels. If any one of the busy-tone channels becomes busy, the RTS sender is required to stop sending immediately.

Both the hidden terminal problem and the exposed terminal problem for the data packets are solved by the DBTMA. The collision-free medium access is achieved for data packets, but collisions between the control packets are unavoidable. The disadvantage of DBTMA is that it wastes the battery capacity, as the wireless node continuously senses the medium for the BT_r and BT_r signals. A further frequency is reserved for BT_r and BT_t, resulting in more bandwidth wastage. The DBTMA requires the transceiver having the ability to transmit and receive signals simultaneously, which is impractical in the real world.

2.7.4.2 Medium Access Collision Avoidance by Invitation (MACA-BI)

To deal with the hidden terminal problem, Talucci and coworkers propose MACA by Invitation (MACA-BI) [113, 114], in which the data transmission is initiated by the receiver. It assumes that the receiver can predict its future reception time in a network with periodic data traffic. The reception is initiated by sending out a Ready to Receive (RTR) message to request data. All other nodes that hear the RTR will defer their transmissions. However, since most transmission instances cannot be predicted due to the burst characteristic of data traffic, this mechanism does not take effect in most cases. However, once it predicts that its neighbors have packets to send (i.e., piggyback "have more bits" by the previous packet), it will trigger the MACA-BI protocol to improve efficiency. Incorrect predictions lead to unnecessary resource reservation by the surrounding terminals. Furthermore, the RTR itself can suffer collision at the data sender side, but some nodes still hear the RTR and block themselves unnecessarily.

2.7.4.3 A Distributed Dynamic Channel Assignment MAC

In [106], Cidon and Sidi proposed an algorithm to dynamically assign time slots for nodes that could transmit data simultaneously without collisions. Figure 2.22 shows the time format for the algorithm where time is divided into equal-length slots. Each slot is further separated to a control segment, which is used to determine the medium access right of active MTs and a transmission segment that is dedicated for the transmission. The control segment is further divided into a front request segment and a confirmation segment, and both segments are further separated into N tiny slots (N ≥ number of nodes). Each node is assigned a tiny slot and can transmit control signals only in its tiny slot. The tiny slot in the confirmation segment is further separated into two parts; the first part is used to decline, while the second part is used for confirming transmission.

Each node that plans to transmit data will transmit a control message in the corresponding tiny slot of the request segment to broadcast the intended communication. After request segment, each node will know the status of surrounding medium competitions. In the confirmation segment, for each slot, the active MT will transmit a confirmation signal in the second half if it does not sense the signal in the first half. The

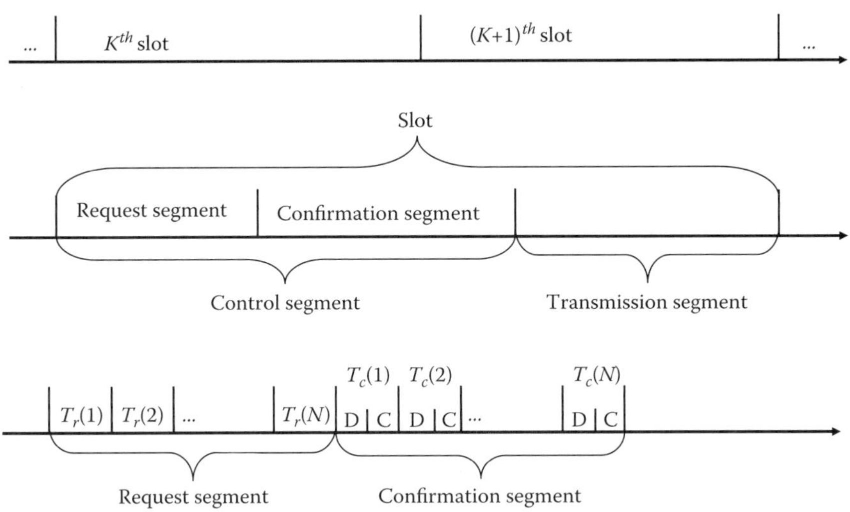

FIGURE 2.22 Time format of the collision-free MAC protocol proposed in [106].

communication that can cause collision with previously confirmed communication will be declined by a transmitting signal in the front half of the corresponding mini-slot. The MTs that transmit the decline signal could be either the confirmed receivers, which are in charge of declining all the later transmissions in their communication range, or the undergoing receiver, which finds that the underconfirming data reception is in the interference range of previously confirmed data transmission. After the confirmation segment, the confirmed communications will be processed simultaneously in the following transmission segment.

This algorithm can lead to collision-free, parallel transmissions on the assumption that the interference range is the same as the communication range. However, in practical operation, the interference range is normally larger than the communication range. In this case, the collision-free transmission cannot be guaranteed because it is impossible for the previously confirmed receiver to contact the unconfirmed transmitter that is located in its interference range but beyond its communication range.

2.7.4.4 Binary Countdown Application

As mentioned earlier, CSMA/IC is an innovative MAC protocol for multihop ad hoc wireless networks. The medium access mechanism for CSMA/IC is based on the binary countdown technique, a collision-free medium access technique originally designed for wired networks. To make the binary countdown technique work properly in wireless ad hoc networks, a frame synchronization mechanism is proposed in [31–33]. The distributed ID claiming (DIDC) algorithm is proposed in [130] for an MT to claim a unique but small competition ID automatically in the distributed manner. It is shown that CSMA/IC can achieve complete collision freedom by solving the hidden terminal problem and concurrent transmission problems. Note that the hidden terminal problem is solved by applying stronger transmission power for the buzz signal to enlarge the blocking range of a data sender. If there is an obstruction blocking the signal propagated from the sender, it could not be blocked by the strengthened signal, while the signal from it could interfere with data reception at the destination. This is the shortcoming of the hidden terminal solution that depended purely on the activity of the sender. Both PBRCE [42] and BROADEN [33, 131] are collision-free MAC protocols that apply CSMA/IC on the preceding dialogue for scheduling. With the proposed proper scheduling algorithm, PBRCE can achieve collision freedom for data transmission.

2.8 Summary

In this chapter, we provide a review of MAC protocols categorized by their goal and functionality, namely, collision avoidance, energy efficiency, quality of service (QoS), fairness, and collision freedom, respectively (as illustrated in Figure 2.23). Collision avoidance techniques are further categorized into techniques that prevent concurrent data transmissions in the single-hop environment, techniques that solve the hidden terminal and exposed terminal problems, and techniques that broadcast in the multihop environment. Energy-efficient MAC protocols can be categorized into two types: power

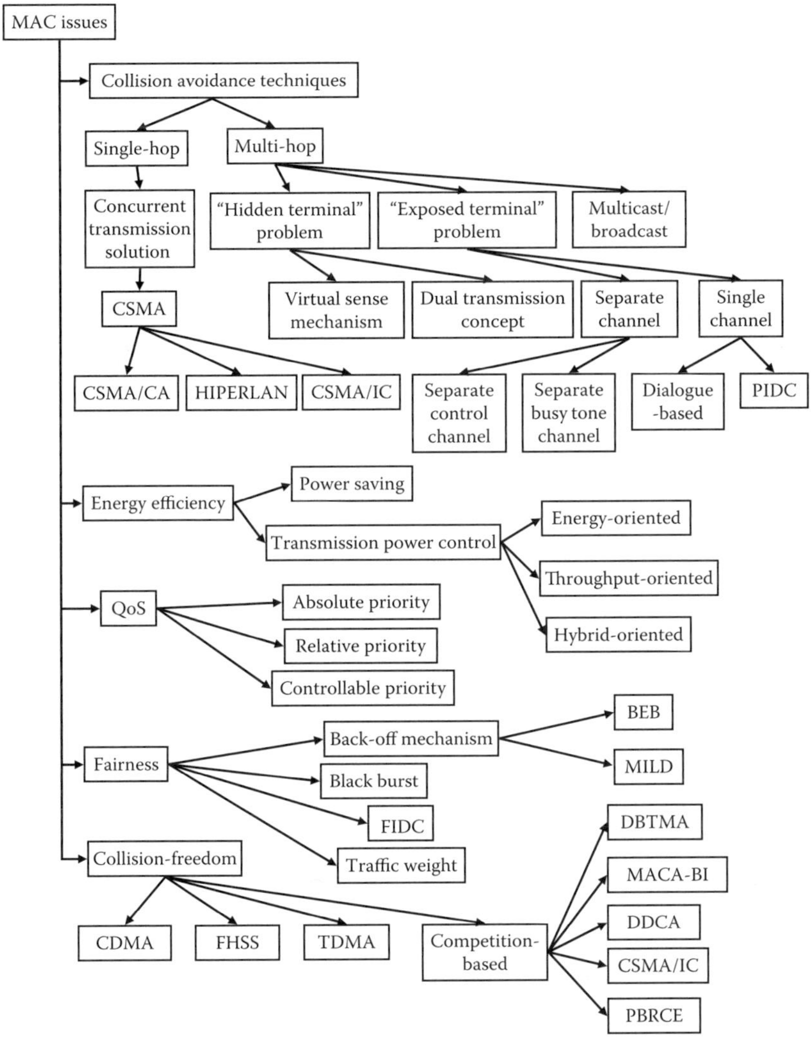

FIGURE 2.23 A classification of WLAN MAC protocols.

saving and transmission power control. Specifically, the Transmission Power Control MAC protocols can be further divided into those focusing on energy saving, throughput, or both. For QoS, some MAC techniques support absolute priority packets, while others provide relative medium access power for the higher-priority packets. Protocols that address fair medium access include the back-off mechanism, Black Burst mechanism, and schemes based on traffic load to decide the medium access. Lastly, we give a brief overview of MAC protocols that were claimed to be collision-free for wireless networks. They could be further categorized into four groups: CDMA, FHSS, TDMA, and competition-driven MAC.

References

[1] C. Toh. 2002. *Ad hoc mobile wireless networks protocols and systems.* Upper Saddle River, NJ: Prentice Hall.

[2] L. Kleinrock and F. A. Tobagi. 1975. Packet switching in radio channels. Part I. Carrier sense multiple access modes and their throughput-delay characteristics. *IEEE Trans. Commun.* 23:1400–16.

[3] F. A. Tobagi and L. Kleinrock. 1975. Packet switching in radio channels. Part II. The hidden terminal problem in CSMA and busy-tone solution. *IEEE Trans. Commun.* 23:1417–33.

[4] IEEE 802.11 WG. 1997. *IEEE standard for wireless LAN medium access control (MAC) and physical layer (PHY) specifications.* IEEE Standard 802.11.

[5] Institute of Electrical and Electronics Engineers. 1999. *Wireless LAN medium access control (MAC) and physical layer (PHY) specifications.* IEEE Standard 802.11, 1999 ed.

[6] A. Colvin. 1983. CSMA with collision avoidance. *Comput. Commun.* 6:227–35.

[7] T. Wilkinson, T. G. C. Phipps, and S. K. Barton. 1995. A report on HIPERLAN standardization. *Int. J. Wireless Inform. Networks* 2(2).

[8] G. Anastasi, L. Lenzini, and E. Mingozzi. 1998. Stability and performance analysis of HIPERLAN. In *Proceedings of the IEEE Joint Conference of the IEEE Computer and Communications Societies (INFOCOM)*, vol. 1, pp. 134–41.

[9] C. Yeh, H. Zhou, and H. Mouftah. 2002. A time space division multiple access (TSDMA) protocol for multihop wireless networks with access points, for routing and resource management in ad hoc wireless networks. In *Proceedings of the IEEE Vehicular Technology Conference (VTC)*, pp. 1185–1189.

[10] C. Yeh. 2003. IPMA: An interference/power-aware MAC scheme for heterogeneous wireless networks. In *Proceedings of the IEEE International Symposium on Computer Communications (IEEE ISCC'03)*, pp. 849–854.

[11] C. Yeh. 2003. The advance access mechanism for differentiated service, power control, and radio efficiency in ad hoc MAC protocols. In *Proceedings of the IEEE Vehicular Technology Conference (VTC)*, vol. 3, pp. 1652–1657.

[12] C. L. Fullmer and J. J. Garcia-Luna-Aceves. 1995. Floor acquisition multiple access (FAMA) for packet-radio networks. In *Proceedings of the ACM Conference on Applications, Technologies, Architectures, and Protocols for Computer Communication (SIGCOMM)*, vol. 25, pp. 262–273.

[13] C. L. Fullmer and J. J. Garcia-Luna-Aceves. 1995. FAMA-PJ: A channel access protocol for wireless LANs. In *Proceedings of the 1st Annual International Conference on Mobile Computing and Networking*, pp. 76–85.

[14] C. L. Fullmer and J. J. Garcia-Luna-Aceves. 1997. Solutions to hidden terminal problems in wireless networks. In *Proceedings of the Conference on Applications, Technologies, Architectures, and Protocols for Computer Communication, ACM SIGCOMM Comput. Commun. Rev.*, 27:39–49.

[15] P. Karn. 1990. MACA—A new channel access method for packet radio. In *Proceedings of the ARRL/CRRL Amateur Radio 9th Computer Networking Conference*, pp. 134–40.
[16] K. Biba. 1992. *A hybrid wireless MAC protocol supporting asynchronous and synchronous MSDU delivery services*. IEEE 802.11 Working Group Paper 802.11/91-92.
[17] V. Bharghavan, A. Demers, S. Shenker, and L. Zhang. 1994. MACAW: A media access protocol for wireless LAN's. In *ACM SIGCOMM Comput. Commun. Rev., Proceedings of the Conference on Communications Architectures, Protocols and Applications*, 24:212–225.
[18] F. A. Tobagi and L. Kleinrock. 1976. Packet switching in radio channels. Part III. Polling and (dynamic) split-channel reservation multiple access. *IEEE Trans. Commun.* 24:832–45.
[19] N. Choi, Y. Seok, and Y. Choi. 2003. Multi-channel MAC protocol for mobile ad hoc networks. In *Proceedings of the IEEE Vehicular Technology Conference (VTC)*, vol. 2, pp. 1379–82.
[20] A. Muqattash and M. Krunz. 2003. Power controlled dual channel (PCDC) medium access protocol for wireless ad hoc networks. In *Proceedings of the 22nd International Annual Joint Conference of the IEEE Computer and Communications Societies (INFOCOM)*, vol. 1, pp. 470–80.
[21] J. Zhang, Y. Wang, and J. Wang. 2003. DCC-MAC: A new MAC protocol for ad-hoc networks based on dual control channel. In *Proceedings of the IEEE Personal, Indoor and Mobile Radio Communications (PIMRC)*, pp. 1341–45.
[22] F. H. P. Fitzek, D. Angelini, G. Mazzini, and M. Zorzi. 2003. Design and performance of an enhanced IEEE 802.11 MAC protocol for multihop coverage extension. *Wireless Commun.* 10:30–39.
[23] T. You, C.-H. Yeh, and H. S. Hassanein. 2005. DRCE: A high throughput QoS MAC protocol for wireless ad hoc networks. In *Proceedings of the IEEE International Symposium on Computer Communications (ISCC)*, pp. 671–76.
[24] C. Wu and V. Li. 1987. Receiver-initiated busy-tone multiple access in packet radio networks. In *ACM SIGCOMM Comput. Commun. Rev., Proceedings of the ACM Workshop on Frontiers in Computer Communications Technology*, 17:336–342.
[25] J. Deng and Z. J. Haas. 1998. Dual busy tone multiple access (DBTMA): A new medium access control for packet radio networks. In *Proceedings of the IEEE International Conference on Universal Personal Communications (ICUPC)*, vol. 2, pp. 973–77.
[26] Z. J Haas and J. Deng. 1999. Dual busy tone multiple access (DBTMA)-performance evaluation. In *Proceedings of the IEEE Vehicular Technology Conference (VTC)*, vol. 1, pp. 314–19.
[27] Z. J. Haas, J. Deng, and S. Tabrizi. 1999. Collision-free medium access control scheme for ad-hoc networks. In *Proceedings of the IEEE Military Communications Conference (MILCOM)*, vol. 1, pp. 276–80.
[28] Z. J. Haas and J. Deng. 2002. Dual busy tone multiple access (DBTMA)—A multiple access control scheme for ad hoc networks. *IEEE Trans. Commun.* 50:975–85.

[29] S. Singh and C. S. Raghavendra. 1998. PAMAS—Power aware multi-access protocol with signaling for ad-hoc networks. *ACM Comp. Commun. Rev.* 28:5–6.

[30] S. Ye, Y. Wang, and Y.-C. Tseng. 2003. A jamming-based MAC protocol for wireless multihop ad hoc networks. In *Proceedings of the IEEE Vehicular Technology Conference (VTC).*

[31] T. You, H. S. Hassanein, and C.-H. Yeh. 2005. CSMA/IC: A new paradigm for medium access control in wireless LANs and ad hoc wireless networks. *Int. J. Wireless Mobile Comput.* No. 6.

[32] T. You., C.-H. Yeh, and H. S. Hassanein. 2003. CSMA/IC: A new class of collision-free MAC protocols for ad hoc wireless networks. In *Proceedings of the IEEE International Symposium on Computer Communications (ISCC)*, vol. 2, pp. 843–48.

[33] T. You., C.-H. Yeh, and H. S. Hassanein. 2003. A new class of collision-prevention MAC protocols for ad hoc wireless networks. In *Proceedings of the IEEE International Conference on Communications (ICC)*, vol. 2, pp. 1135–40.

[34] C.-H. Yeh. 2001. Medium access control with differentiated adaptation for QoS management in wireless networks. In *Proceedings of the IEEE International Conference on Mobile and Wireless Communication Networks*, pp. 208–19.

[35] C.-H. Yeh. 2001. ROC: A wireless MAC protocol for solving the moving terminal problem. In *Proceedings of the IEEE International Conference on Wireless LANs and Home Networks*, pp. 182–89.

[36] C.-H. Yeh. 2002. ROAD: A class of variable-radius MAC protocols for ad hoc wireless networks. In *Proceedings of the IEEE Vehicular Technology Conference*, pp. 399–403.

[37] C.-H. Yeh and T. You. 2003. A power-controlled multiple access scheme for differentiated service and energy efficiency in mobile ad hoc networks and wireless LANs. In *Proceedings of the 14th IEEE Proceedings on Personal, Indoor and Mobile Radio Communications (PIMRC)*, vol. 1, pp. 765–71.

[38] C.-H. Yeh. 2004. High-throughput interference-aware MAC protocols for heterogeneous ad hoc networks and multihop wireless LANs. In *Proceedings of the International Conference on Parallel Processing (ICPP) Workshops*, pp. 447–55.

[39] C.-H. Yeh. 2004. The heterogeneous hidden/exposed terminal problem for power-controlled ad hoc MAC protocols and its solutions. In *Proceedings of the IEEE Vehicular Technology Conference (VTC)*, vol. 5, pp. 2548–54.

[40] C.-H. Yeh. 2004. Interference-controlled multiple access for mobile ad hoc networks and multihop wireless LANs. In *IEEE Vehicular Technology Conference (VTC)*, vol. 4, pp. 2863–67.

[41] A. Muqattash and M. Krunz. 2005. POWMAC: A single-channel power-control protocol for throughput enhancement in wireless ad hoc networks. *IEEE J. Select. Areas Commun.* 23:1067–84.

[42] T. You, H. S. Hassanein, and C.-H. Yeh. 2006. PBRCE: Energy efficient MAC protocol for wireless ad hoc networks. In *2nd IEEE International Conference on Wireless and Mobile Computing, Networking and Communications (WiMob).*

[43] T. You, H. S. Hassanein, and C.-H. Yeh. 2005. PIDC: Towards an Ideal MAC protocol for multihop wireless LANs. In *Proceedings of the International Conference on Wireless Networks, Communications, and Mobile Computing (WIRELESSCOM)*, vol. 1, pp. 655–60.

[44] C.-H. Yeh. 2004. A new scheme for effective MAC-layer DiffServ supports in mobile ad hoc networks and multihop wireless LANs. In *Proceedings of the IEEE Vehicular Technology Conference (VTC)*, vol. 4, pp. 2149–55.

[45] C.-H. Yeh. 2004. Collision-controlled multiple access for mobile ad hoc networks and multihop wireless LANs. In *IEEE 60th Vehicular Technology Conference (VTC)*, vol. 5, pp. 3080–84.

[46] K. Lee and D. Cho. 2003. A multiple access collision avoidance protocol for multicast service in mobile ad hoc networks. In *Proceedings of the 57th IEEE Semiannual Vehicular Technology Conference (VTC)*, vol. 3, pp. 1793–97.

[47] T. You, H. S. Hassanein, and C.-H. Yeh. 2006. SeMAC: Robust broadcast MAC protocol for multihop. In *Proceedings of the 25th IEEE International Performance Computing and Communications Conference (IPCCC)*, pp. 71–78.

[48] E. Jung and N. H. Vaidya. 2002. An energy efficient MAC protocol for wireless LANs. In *Proceedings of the Twenty-First Annual Joint Conference of the IEEE Computer and Communications Societies (INFOCOM)*, vol. 3, pp. 1756–64.

[49] K. Shih, C. Chang, C. Chou, and S. Chen. 2005. A power saving MAC protocol by increasing spatial reuse for IEEE 802.11 ad hoc WLANs. In *Proceedings of the 19th International Conference on Advanced Information Networking and Applications (AINA)*, vol. 1, pp. 420–25.

[50] P. Gupta and P. R. Kumar. 2000. The capacity of wireless networks. *IEEE Trans. Inform. Theory* 46:388–404.

[51] P. Karn. 1990. MACA—A new channel access method for packet radio. In *Proceedings of the ARRL/CRRL Amateur Radio 9th Computer Networking Conference*, pp. 134–40.

[52] T. A. ElBatt, S. V. Krishnamurthy, D. Connors, and S. Dao. 2000. Power management for throughput enhancement in wireless ad hoc networks. In *Proceedings of the IEEE International Conference on Communications (ICC)*, pp. 1506–13.

[53] R. Ramanathan and R. Rosales-Hain. 2000. Topology control of multihop wireless networks using transmit power adjustment. In *Proceedings of the 19th International Annual Joint Conference of the IEEE Computer and Communications Societies (INFOCOM)*, pp. 404–13.

[54] V. Rodoplu and T. Meng. 1999. Minimum energy mobile wireless networks. *IEEE J. Select. Areas Commun.* 17:1333–44.

[55] R. Wattenhofer, L. Li, P. Bahl, and Y.-M. Wang. 2001. Distributed topology control for power efficient operation in multihop wireless ad hoc networks. In *Proceedings of the 20th International Annual Joint Conference of the IEEE Computer and Communications Societies (INFOCOM)*, pp. 1388–97.

[56] J. Gomez, A. T. Campbell, M. Naghshineh, and C. Bisdikian. 2003. PARO: Supporting dynamic power controlled routing in wireless ad hoc networks. *ACM/Kluwer J. Wireless Networks* 9:443–60.

[57] T.-W. Chen et al. 1998. A comparison of MAC protocols for wireless local area networks based on battery consumption. In *Proceedings of International Annual Joint Conference of the IEEE Computer and Communications Societies (INFOCOM)*, pp. 150–157.

[58] E. S. Jung and N. H. Vaidya. 2002. A power control MAC protocol for ad hoc networks. In *Proceedings of ACM/IEEE International Conference on Mobile Computing and Networking (MOBICOM)*, pp. 36–47.

[59] M. B. Pursley, H. B. Russell, and J. S. Wysocarski. 2000. Energy-efficient transmission and routing protocols for wireless multiple-hop networks and spread spectrum radios. In *Proceedings of the EUROCOMM Conference*, pp. 1–5.

[60] S.-L. Wu, Y.-C. Tseng, and J.-P. Sheu. 2000. Intelligent medium access for mobile ad hoc networks with busy tones and power control. *IEEE J. Select. Areas Commun.* 18:1647–57.

[61] S. Agarwal, S. Krishnamurthy, R. H. Katz, and S. K. Dao. 2001. Distributed power control in ad-hoc wireless networks. In *Proceedings of the 12th IEEE International Symposium on Personal, Indoor and Mobile Radio Communications (PIMRC)*, vol. 2, pp. F-59–F-66.

[62] J.-P. Ebert and A. Wolisz. 1999. Combined tuning of RF power and medium access control for WLANs. In *IEEE International Workshop on Mobile Multimedia Communications*, pp. 74–82.

[63] J.-P. Ebert, B. Stremmel, E. Wiederhold, and A. Wolisz. 2000. An energy efficient power control approach for WLANs. *J. Commun. Networks* 2:197–206.

[64] J. Gomez, A. T. Campbell, M. Naghshineh, and C. Bisdikian. 2001. Conserving transmission power in wireless ad hoc networks. In *International Conference on Network Protocols.*

[65] S.-L. Wu, Y. C. Tseng, C.-Y. Lin, and J.-P. Sheu. 2002. A multi-channel MAC protocol with power control for multihop mobile ad hoc networks. *Comput. J.* 45:101–10.

[66] J. Monks, V. Bharghavan, and W.-M. Hwu. 2001. A power controlled multiple access protocol for wireless packet networks. In *Proceedings of the 20th International Annual Joint Conference of the IEEE Computer and Communications Societies (INFOCOM)*, pp. 219–28.

[67] A. Muqattash and M. Krunz. 2003. Power controlled dual channel (PCDC) medium access protocol for wireless ad hoc networks. In *Proceedings of the 22nd International Annual Joint Conference of the IEEE Computer and Communications Societies (INFOCOM)*, vol. 1, pp. 470–80.

[68] E. Poon and L. Baochun. 2003. SmartNode: Achieving 802.11 MAC interoperability in power-efficient ad hoc networks with dynamic range adjustments. In *Proceedings of the 23rd International Conference on Distributed Computing Systems*, pp. 650–57.

[69] C.-H. Yeh and H. Zhou. 2002. A new class of collision-free MAC protocols for ad hoc wireless networks. In *Proceedings of the International Conference on Advances in Infrastructure for e-Business, e-Education, e-Science, and e-Medicine on the Internet.*

[70] H. Zhou, C.-H. Yeh, and H. Mouftah. 2004. A power efficient medium access control protocol for heterogeneous wireless networks. In *IEEE 60th Vehicular Technology Conference (VTC)*, vol. 7, pp. 5021–25.

[71] C.-H. Yeh, H. Zhou, P.-H. Ho, and H.T. Mouftah. 2003. A variable-radius multichannel MAC protocol for high-throughput low-power heterogeneous ad hoc networking. In *IEEE Global Telecommunications Conference (GLOBECOM)*, vol. 3, pp. 1284–89.

[72] C.-H. Yeh. 2001. Variable-radius routing protocols for high throughput, low power and small latency in ad hoc wireless networks. In *Proceedings of the IEEE International Conference on Wireless LANs and Home Networks*, pp. 215–27.

[73] H. Zhou, C.-H. Yeh, and H. T. Mouftah. 2002. A solution for multiple access in variable-radius mobile ad hoc networks. In *Proceedings of the IEEE International Conference on Communications, Circuits and Systems (ICCCAS) and West Sino Expositions*, pp. 150–54.

[74] T. You, H. Hassanein, and H. T. Mouftah. 2002. Infrastructure-based MAC in wireless mobile ad-hoc networks. In *Proceedings of the IEEE Local Computer Networks (LCN)*, pp. 821–30.

[75] I. Aad and C. Castelluccia. 2001. Differentiation mechanisms for IEEE 802.11. In *Proceedings of the IEEE Joint Conference of the IEEE Computer and Communications Societies (INFOCOM)*, vol. 1, pp. 209–18.

[76] A. Banchs, M. Radimirsch, and X. Perez. 2002. Assured and expedited forwarding extensions for IEEE 802.11 wireless LAN. In *Proceedings of the IEEE/ACM International Workshop on Quality of Service (IWQoS)*, pp. 237–46.

[77] IEEE 802.11 WG. 2004. *Draft supplement to standard for telecommunications and information exchange between systems LAN/MAN specific requirements. Part II. Wireless access control (MAC) and physical layer (PHY) specifications: Medium access control (MAC) enhancements for quality of service (QoS)*. IEEE 802.11e/D8.0.

[78] T. Ozugur, M. Naghshineh, P. Kermani, and J. A. Copeland. 1999. Fair media access for wireless LANs. In *Proceedings of the IEEE Global Telecommunications Conference (GLOBECOM)*, vol. 1b, pp. 570–79.

[79] J. L. Sobrinho and A. S. Krishnakumar. 1999. Quality-of-service in ad hoc carrier sense multiple access networks. *IEEE J. Select. Areas Commun.* 17:1353–68.

[80] A. Banchs and X. Perez. 2002. Providing throughput guarantees in IEEE 802.11 wireless LAN. In *Proceedings of the IEEE Wireless Communications and Networking Conference (WCNC)*, vol. 1, pp. 130–38.

[81] T. You, H. S. Hassanein, and C.-H. Yeh. 2005. A paradigm for controllable QoS-based medium access control in local and ad hoc wireless networks. *Int. J. Parallel Emergent Distributed Syst.* 20:281–302.

[82] T. You, H. S. Hassanein, and C.-H. Yeh. 2004. Controllable fair QoS-based MAC protocols for ad hoc wireless networks. In *Proceedings of the IEEE International Workshop on Mobile and Wireless Networking (MWN)*, pp. 21–28.

[83] A. K. Parekh and R. G. Gallager. 1993. A generalized processor sharing approach to flow control in integrated services networks: The single-node case. *IEEE/ACM Trans. Networking* 1:344–57.

[84] N. H. Vaidya, P. Bahl, and S. Gupta. 2000. Distributed fair scheduling in a wireless LAN. In *Proceedings of the ACM/IEEE International Conference on Mobile Computing and Networking (MOBICOM)*, pp. 167–178.

[85] A. Banchs and X. Perez. 2002. Distributed weighted fair queuing in 802.11 wireless LAN. In *Proceedings of the IEEE International Conference on Communications (ICC)*, vol. 5, pp. 3121–27.

[86] H. Wu, A. Utgikar, and N. Tzeng. 2004. SYN-MAC: A distributed medium access control protocol for synchronized wireless networks. *ACM Mobile Networks Appl. J.*

[87] C. Yeh and T. You. 2003. A QoS MAC protocol for differentiated service in mobile ad hoc network. In *Proceedings of the International Conference on Parallel Processing (ICPP)*, pp. 349–56.

[88] C.-H. Yeh. 2004. A collision-controlled MAC protocol for mobile ad hoc networks and multihop wireless LANs. In *IEEE Global Telecommunications Conference (GLOBECOM)*, vol. 2, pp. 867–71.

[89] T. You, H. S. Hassanein, and C.-H. Yeh. 2004. FIDC: A fair MAC protocol for ad hoc wireless networks. In *Proceedings of the IEEE Global Telecommunications Conference (GLOBECOM)*, vol. 5, pp. 2775–80.

[90] Q. Liu, Z. Zhang, and R. A. Scholtz. 1996. Analysis of the throughput of random-access multichannel CDMA systems with multiuser detection. In *5th IEEE International Conference Record on Universal Personal Communications*, vol. 2, pp. 881–85.

[91] D. I. Kim. 1995. Analysis of a direct-sequence CDMA mobile radio system with reduced set of code sequences. *IEEE Trans. Vehicular Technol.* 44:525–34.

[92] G.-S. Kuo and P.-C. Ko. 1999. A collision free medium access control protocol for flow-oriented ad hoc wireless LAN. In *IEEE 49th Vehicular Technology Conference (VTC)*, vol. 1, pp. 325–31.

[93] X. H. Chen and J. Oksman. 1992. Busy code broadcasting and sensing protocol for collision-free CDMA packet radio networks and its performance analysis. In *IEE Proceedings I of Communications, Speech and Vision*, vol. 139, pp. 613–19.

[94] J. Nishimura, T. Haruki, K. Mutsuura, and H. Okada. 2001. CDMA with code sense for collision free access in packet-typed wireless networks. In *IEEE 54th Vehicular Technology Conference (VTC)*, vol. 2, pp. 1143–47.

[95] J. Nishimura, T. Haruki, K. Mutsuura, and H. Okada. 2001. A novel access scheme for packet-typed wireless networks CDMA/CS-CF. In *IEEE Global Telecommunications Conference (GLOBECOM)*, vol. 1, pp. 115–20.

[96] A. Butala and L. Tong. 2002. Dynamic channel allocation and optimal detection for MAC in CDMA ad hoc networks. In *Conference Record of the Thirty-Sixth Asilomar Conference on Signals, Systems and Computers*, vol. 2, pp. 1160–64.

[97] D. D. Lin and T. J. Lim. 2004. Subspace-based active user identification for a collision-free slotted ad hoc network. *IEEE Trans. Commun.* 52:612–21.

[98] Y. S. Su, S. L. Su, and J. S. Li. 2004. Topology-transparent link activation scheduling schemes for multihop CDMA ad hoc networks. In *IEEE Global Telecommunications Conference (GLOBECOM)*, vol. 6, pp. 3563–67.

[99] Z. Tang and J. J. Garcia-Luna-Aceves. 1998. Hop reservation multiple access (HRMA) for multichannel packet radio networks. In *Proceedings of the 7th International Conference on Computer Communications and Networks*, pp. 388–95.

[100] Z. Tang and J. J. Garcia-Luna-Aceves. 1999. Hop-reservation multiple access (HRMA) for ad-hoc networks. In *Proceedings of the Eighteenth Annual Joint Conference of the IEEE Computer and Communications Societies (INFOCOM)*, vol. 1, pp. 194–201.

[101] A. Tzamaloukas and J. J. Garcia-Luna-Aceves. 2001. A receiver-initiated collision-avoidance protocol for multi-channel networks. In *Proceedings of the Twentieth Annual Joint Conference of the IEEE Computer and Communications Societies (INFOCOM)*, vol. 1, pp. 189–98.

[102] A. Tzamaloukas and J. J. Garcia-Luna-Aceves. 2000. Receiver-initiated channel-hopping for ad-hoc networks. In *IEEE Wireless Communications and Networking Conference (WCNC)*, vol. 3, pp. 1262–67.

[103] A. Tzamaloukas and J. J. Garcia-Luna-Aceves. 2000. Channel-hopping multiple access. In *IEEE International Conference on Communications (ICC)*, vol. 1, pp. 415–19.

[104] A. Tzamaloukas and J. J. Garcia-Luna-Aceves. 2000. A channel-hopping protocol for ad-hoc networks. In *Proceedings of the Ninth International Conference on Computer Communications and Networks*, pp. 142–47.

[105] C.-H. Yeh. 2004. Spread spectrum techniques for solving MAC-layer interference issues in mobile ad hoc networks. In *IEEE Vehicular Technology Conference (VTC)*, vol. 3, pp. 1339–44.

[106] I. Cidon and M. Sidi. 1989. Distributed assignment algorithms for multihop packet radio networks. *IEEE Trans. Comput.* 38:1353–61.

[107] R. Nelson and L. Kleinrock. 1985. Spatial TDMA: A collision-free multihop channel access protocol. *IEEE Trans. Commun.* 33:934–44.

[108] A. Kanzaki, T. Uemukai, T. Hara, and S. Nishio. 2003. Dynamic TDMA slot assignment in ad hoc networks. In *17th International Conference on Advanced Information Networking and Applications (AINA)*, pp. 330–35.

[109] O. Somarriba and T. Giles. 2002. Transmission control for spatial TDMA in wireless radio networks. In *4th International Workshop on Mobile and Wireless Communications Network*, pp. 394–98.

[110] Z. Fang, B. Bensaou, and J. Yuan. 2004. Collision-free MAC scheduling algorithms for wireless ad hoc networks. In *IEEE Global Telecommunications Conference (GLOBECOM)*, vol. 5, pp. 2770–74.

[111] K. Oikonomou and I. Stavrakakis. 2004. Analysis of a probabilistic topology-unaware TDMA MAC policy for ad hoc networks. *IEEE J. Select. Areas Commun.* 22:1286–300.

[112] J. Gronkvist, J. Nilsson, and D. Yuan. 2004. Throughput of optimal spatial reuse TDMA for wireless ad-hoc networks. In *Proceedings of the IEEE 59th Vehicular Technology Conference (VTC)*, vol. 4, pp. 2156–60.

[113] F. Talucci and M. Gerla. 1997. MACA-BI (MACA by invitation). A wireless MAC protocol for high speed ad hoc networking. In *IEEE 6th International Conference Record on Universal Personal Communications*, vol. 2, pp. 913–17.

[114] F. Talucci, M. Gerla, and L. Fratta. 1997. MACA-BI (MACA by invitation)—A receiver oriented access protocol for wireless multihop networks. In *8th IEEE International Symposium on Personal, Indoor and Mobile Radio Communications (PIMRC), 'Waves of the Year 2000*, vol. 2, pp. 435–39.

[115] J. G. Proakis. 2001. *Digital communications*. New York: McGraw-Hill.

[116] R. L. Pickholtz, D. L. Schilling, and L. B. Milstein. 1982. Theory of spread spectrum communications—A tutorial. *IEEE Trans. Commun.* 30:855–84.

[117] T. Rappaport. 2002. *Wireless communications: Principles and practice*. Upper Saddle River, NJ: Prentice Hall.

[118] E. Sousa and J. A. Silvester. 1990. Optimum transmission ranges in a direct-sequence spread-spectrum multihop packet radio network. *IEEE J. Select. Areas Commun.* 8:762–71.

[119] H. Lee, J. Yeo, S. Kim, and S. Lee. 2001. Time slot assignment to minimize delay in ad-hoc networks. In *IST Mobile Communications Summit*.

[120] L. C. Pond and V. O. K. Li. 1989. A distributed time-slot assignment protocol for mobile multihop broadcast packet radio networks. In *Proceedings of the IEEE MILCOM*, vol. 1, pp. 70–74.

[121] C. D. Young. 1996. USAP: A unifying dynamic distributed multichannel TDMA slot assignment protocol. In *Proceedings of the IEEE MILCOM*, vol. 1, pp. 235–239.

[122] C. D. Young. 1999. USAP multiple access: Dynamic resource allocation for mobile multihop multichannel wireless networking. In *Proceedings of the IEEE MILCOM*, vol. 1, 271–275.

[123] J. Gronkvist. 2000. Assignment methods for spatial reuse TDMA. In *First Annual Workshop on Mobile and Ad Hoc Networking and Computing (MobiHOC)*, pp. 119–24.

[124] J. Gronkvist. 1998. Traffic controlled spatial reuse TDMA in multihop radio networks. In *Ninth IEEE International Symposium on Personal, Indoor and Mobile Radio Communications*, vol. 3, pp. 1203–7.

[125] J. Ju and V. O. K. Li. 1998. An optimal topology-transparent scheduling method in multihop packet radio networks. *IEEE/ACM Trans. Networking* 6:298–306.

[126] J. Ju and V. O. K. Li. 1999. TDMA scheduling design of multihop packet radio networks based on latin squares. *IEEE J. Select. Areas Commun.* 17:1345–52.

[127] J. Ju and V. O. K. Li. 1999. TDMA scheduling design of multihop packet radio networks based on latin squares. In *Proceedings of the Eighteenth Annual Joint Conference of the IEEE Computer and Communications Societies (INFOCOM)*, vol. 1, pp. 187–93.

[128] A. Ephremides and T. V. Truong. 1990. Scheduling broadcasts in multihop radio networks. *IEEE Trans. Commun.* 38:456–60.

[129] G. Wang and N. Ansari. 1997. Optimal broadcast scheduling in packet radio networks using mean field annealing. *IEEE J. Select. Areas Commun.* 15:250–60.

[130] T. You, C.-H. Yeh, and H. S. Hassanein. 2004. An adaptive distributed ID assignment scheme for CSMA/IC. In *Proceedings of the IEEE Local Computer Networks (LCN)*, pp. 634–41.

[131] T. You, C.-H. Yeh, and H. S. Hassanein. 2003. BROADEN: An efficient collision-free MAC protocol for ad hoc wireless networks. In *Proceedings of the IEEE Local Computer Networks (LCN)*, pp. 698–707.

Appendix: List of Acronyms

ACK	Acknowledgment
AIFS	Arbitrary IFS
AP	Access point
ATIM	Ad hoc traffic indication message
BB	Black Burst protocol
BCBS	Busy code broadcasting and sensing
BEB	Binary exponential back-off
BO	Back-off
BTMA	Busy Tone Multiple Access
BT_r	Receive busy tone
BT_t	Transmit busy tone
CAC	Call Admission Control
CDMA	Code Division Multiple Access
CDMA/CS-CF	CDMA/Code Sense for Collision-Free Access
CFIC	Controllable Fair ID Countdown
CSMA	Carrier Sense Multiple Access
CSMA/CA	Carrier Sense Multiple Access/Collision Avoidance
CSMA/IC	Carrier Sense Multiple Access/ID Countdown
CTS	Clear to Send
CW	Contention window
CW_{max}	Maximum contention window
CW_{min}	Minimum contention window
DBTMA	Dual Busy Tone Multiple Access
DCA	Dynamic channel allocation
DCF	Distributed coordination function
DFIC	Differentiated Fair ID Countdown
DFS	Distributed Fair Scheduling
DIDC	Distributed ID claim
DIFS	Distributed coordination function interframe space
DPSM	Dynamic power-saving mechanism
DS	Data Sending
DSC	Distributed slot claimed
DTS	Decide to Send
DWFQ	Distributed Weighted Fair Queuing
EDCF	Enhanced DCF
ETS	Ensure to Send
ETSI	Europe Telecommunication Standardization Institute
FCC	Federal Communications Commission
FDM	Frequency Division Multiplex

FDMA	Frequency Division Multiple Access
FHSS	Frequency hopping spreading spectrum
GPS	Global Positioning System
HIPERLAN	High-Performance Radio Local Area Network
HR	Hop reservation
HRMA	Hop Reservation Multiple Access
IEEE	Institute of Electrical and Electronics Engineers
IFS	Interframe space
ISM	Industrial, scientific, and medical
L-hello	Long hello message
LAN	Local area network
MAC	Wireless Medium Access Control
MACA	Medium Access with Collision Avoidance
MACA-BI	Medium Access Collision Avoidance by Invitation
MACAM	Multiple Access Collision Avoidance Protocol for Multicast Service
MACAW	Medium Access with Collision Avoidance for Wireless
MANET	Mobile ad hoc network
MILD	Multiplication increase and linear decrease
MPC	Mobile Point Coordinator
MPC-MAC	Mobile Point Coordinator–based Medium Access Control
MT	Mobile terminal
NAV	Network allocation vector
NIC	Network interface card
PBRCE	Power-control binary countdown RTS/CTS/ETS
PCF	Point coordination function
PCMA	Power Controlled Multiple Access
PEM	Power-Efficient MAC
PIDC	Pair-wise ID Countdown
PIFS	Point coordination function interframe space
POWMAC	Power-controlled Medium Access Control
PRCD	Power-controlled RTS/CTS/DTS
PSM	Power-saving mechanism
QoS	Quality of service
RICH	Receiver-initiated channel hopping
RLT	Multiple accesses with reservation lag time
RRTS	Receiver-initiated RTS
RTR	Ready to Receive
RTS	Request to Send
S-hello	Short hello message
SB	Sync-beacon
SeMAC	Sequential Medium Access Control

SIFS	Short interframe space
SN	Sequential number
SPCCS	Spreading Code Control Server
STDMA	Spatial TDMA
TDM	Time Division Multiplex
TDMA	Time Division Multiple Access
TPC	Transmission Power Control
Wi-Fi	Wireless fidelity
WLAN	Wireless local area network
WT	Waiting threshold

3

Adaptive Scheduling for Beyond 3G Cellular Networks

Sameh Sorour
University of Toronto

Shahrokh Valaee
University of Toronto

3.1 Introduction

The main target of future cellular network designers is to improve the broadband wireless access of these networks by achieving high-rate, high-reliability packet-based transmission over the wireless channel, the bottleneck against this achievement. Three of the most important wireless channel impairments are:

- Frequency selective fading due to the multipath impulse responses of wireless channels
- Intersymbol interference (ISI)
- Time-varying properties of mobile wireless channels due to user mobility

To tackle the first two problems, orthogonal frequency division multiple access (OFDMA) has been proposed as the multiple-access scheme for the wireless channel in each cell of the cellular network. This choice is explained by the high immunity of OFDMA against these two impairments. Although frequency selective fading is removed or substantially reduced by proper spacing of subcarriers in an OFDMA signal, flat fading still remains a hindering impairment. Consequently, the key point to solve this problem is the use of channel-adaptive resource allocation techniques that can exploit the time-varying nature of the wireless channel by adaptively allocating the OFDMA subcarriers or subchannels (which represent a group of successive subcarriers) to the users that can exploit them best. This concept is called multiuser diversity and is considered a very crucial aspect for all new high-speed wireless network designs, including beyond third-generation (3G) cellular networks [1].

In wired networks, scheduling is generally defined as the process that determines the order of allocation of network resources to different users in order to provide them with their required services. This allocation order is generally based on these users' traffic needs and the queuing status of their packets. Consequently, in the wireless cellular context described above, adaptive scheduling can be defined as a scheduling process that has the ability of resource allocation according to the instantaneous wireless channel conditions of different users as well as their traffic requirements. It thus aims to exploit multiuser diversity by dynamically changing the allocation of the different cell resources to its users in order to achieve global targets for cell performance, such as high overall throughput, low transmission power, low average delay, etc., while providing users with their individual quality-of-service (QoS) requirements.

Studies to implement efficient adaptive scheduling techniques for OFDMA follow two main chains, each of which has its importance and complementarities to the other chain. These two chains can be defined as follows:

- *Adaptive physical layer scheduling*: In this study chain, the designed scheduling techniques consider only the conditions of the user physical channels in the adaptation process. In this case, the traffic of all users is regarded as a continuous stream of bits. The scheduling algorithms are thus designed to adaptively allocate the OFDMA subcarriers or subchannels to different users, then adaptively load the subcarriers granted by each user with coded bits from its traffic stream.
- *Adaptive cross-layer scheduling*: In this study chain, the designed scheduling algorithms consider both channel and traffic conditions of users in the adaptation process. However, these schedulers are mainly designed to adaptively allocate the subcarriers or subchannels to users without paying much attention to the bit loading issue.

In this chapter, we aim to introduce some of the scheduling algorithms designed in both chains and describe their classifications and targets in some detail.

3.2 OFDMA: Prospective Multiple-Access Scheme in Beyond 3G Cellular Networks

Before launching our description of adaptive scheduling techniques for future cellular networks, we will first describe OFDMA, in some detail, since it is the main candidate for multiple access in all future wireless networks, as previously mentioned. This description is essential in the understanding of our illustration of the scheduling techniques.

OFDMA has greatly attracted attention as an efficient multiple-access and modulation technique for wireless communications since it is based on orthogonal frequency division multiplexing (OFDM) and inherits its very high immunity to ISI and frequency selective fading, the conventional problems of wireless transmission. Moreover, OFDMA is very flexible as a multiple-access scheme since it has the ability to schedule resources in two dimensions: frequency and time. This property makes it more suitable for the exploitation of multiuser diversity through scheduling in the frequency dimension, as well as adaptable to the time-varying nature of wireless mobile channels through scheduling in the time dimension.

It has been shown in various researches that providing an efficient and flexible resource allocation in the two dimensions of OFDMA can result in the achievement of both high transmission rates and high transmission reliability. Consequently, OFDMA has been adopted in the issued versions of the IEEE 802.16 standard and is still the major candidate for beyond 3G cellular standards, as declared by the 3rd Generation Partnership Project (3GPP), the organization responsible for standardizing future cellular communication systems [1].

In elementary OFDMA schemes, the M subcarriers of the OFDM symbol are distributed among the users in the cell based on the scheduler decision. The value of M, in future wireless standards, is expected to grow up to two thousand subcarriers [1]. Consequently, this scheme becomes very complicated from the complexity viewpoint, if adaptive scheduling is adopted, since the scheduler is required to allocate resources on the subcarrier level in the frequency domain and on the OFDM symbol in the time domain. This complexity will result in prohibitive scheduling delays and overheads that are not acceptable for high-speed wireless transmission—the top target for future wireless networks. Another drawback of this small scheduling granularity is the need of packet fragmentation into many small fragments to fit the transmission on these small resource units. This adds another delay due to the fragmentation and reconstruction processes.

Consequently, new OFDMA schemes were developed to solve these problems. Most of these schemes adopt the following approach:

- The time axis of the wireless channel is partitioned into large scheduling frames, generally termed transmission time intervals (TTIs). The duration of these TTIs is determined carefully to allow a good track of user channel variations in the scheduling process.
- Each TTI is internally divided into N subchannels in the frequency domain and S time slots (we will call them slots for simplicity) in the time domain. Each subchannel contains N_f consecutive subcarriers and each slot contains N_s

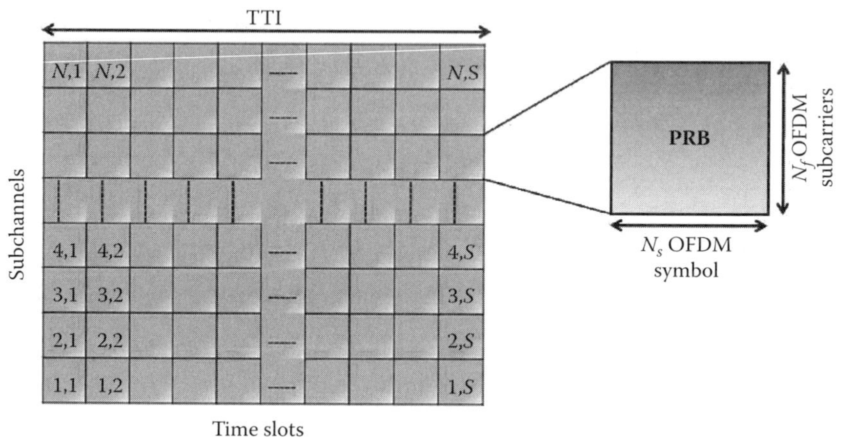

FIGURE 3.1 OFDMA TTI structure.

consecutive OFDM symbols. The intersection of a subchannel with a slot, in the frequency-time resource plane, corresponds to the smallest resource unit that can be allocated to a user. This unit is called the physical resource block (PRB).

- Each PRB is unique in its (n, s) coordinates in the frequency-time resource plane, with $n \in (1, \ldots, N)$ and $s \in (1, \ldots, S)$ representing the frequency domain and time domain components, respectively. The overall number of subcarriers in a PRB in both dimensions is $N_f \times N_s$.

The resulting TTI structure is depicted in Figure 3.1.

In order to solve the complexity and delay problems of the elementary scheme, the following measures are taken:

- Scheduling is performed only once for each TTI and at least on the PRB level.
- A constant modulation and coding scheme is employed for all the subcarriers of each PRB.
- The granularity of the PRBs ($N_f \times N_s$) is chosen to match the minimum packet size and at the same time large enough to prevent numerous packet fragmentations.

It is important to note that, unlike the flexibility in the choice of the number of symbols per slot (N_s), the choice of the number of subcarriers per subchannel (N_f) is restricted by the fact that the channel quality for the N_f grouped subcarriers must be highly correlated.

3.3 Adaptive Physical Layer Scheduling

We defined the adaptive physical layer scheduling techniques as the ones in which physical resources of each cell are distributed among different users only based on their physical channel conditions. These scheduling techniques should perform adaptive subchannel allocation and adaptive bit loading while aiming to both achieve a global cell target and provide different users with a certain level of physical layer performance in terms of bit rate and bit error rate.

The adaptive physical layer scheduling techniques are subdivided according to the global target of the adaptation procedure into two main categories:

- *Margin adaptation* (MA): In which the adaptive scheduler aims to minimize the overall power needed for the transmission of a per user constant throughput, in each TTI, while achieving a minimum transmission reliability, i.e., maximum allowable bit error rate (BER).
- *Rate adaptation* (RA): In which the adaptive scheduler aims to maximize the overall throughput, achieved in each TTI, under a maximum transmission power constraint and while achieving both fairness and a minimum transmission reliability.

Our description of the adaptive physical layer scheduling techniques will follow this categorization.

It is important to note that since these techniques allocate resources according to channel quality only, two assumptions are always considered in the design of their schedulers:

- All active users always have traffic to be transmitted. This traffic is assumed, from the physical layer viewpoint, to be a simple stream of bits obtained from MAC layer packets.
- Since all users always have traffic to transmit, and since scheduling is performed only according to channel qualities, which are variant only between different subchannels, there is no meaning of scheduling PRBs belonging to the same subchannel to different users. Thus, the smallest resources assignable to users are subchannels.

The physical channel quality of different users on different cell subchannels is expressed in terms of the channel quality indicator (CQI). Different designed scheduling algorithms utilize different physical channel parameters as CQI. The main two parameters are the received signal-to-interference-and-noise ratio (SINR) and the average subchannel power gain (ASPG).

It is evident that the scheduler of the cell is located in the base station and is used to allocate resources for both downlink and uplink transmissions. For uplink transmission, the values of the CQIs for different uplink channels and different users are measured at the base station. Thus, the scheduler has full knowledge of the uplink channel characterization. However, CQIs for the downlink channels are measured at the mobile stations of the different users. These CQIs should thus be reported to the base station in order to be utilized by the scheduler in the downlink scheduling process. However, if the number of users and subchannels is too large, the feedback of the CQIs of all users over all subchannels will result in a huge overhead that may limit the possibility of achieving high transmission throughputs. Thus, limited feedback reporting techniques were developed to overcome this problem. However, the scheduling algorithms based on limited knowledge of the channel will strongly depend on the methodology used to reduce this feedback, and thus may strongly change from the scheduling algorithms that operate with full channel knowledge.

In this section, we will focus on formulating and describing the two main categories (MA and RA) of adaptive physical layer scheduling techniques that operate with full channel knowledge. As we mentioned, these techniques are of importance for uplink resource scheduling. In section 3.4, we will illustrate the modifications that could be made in these scheduling techniques to be operable with limited channel knowledge in downlink resource scheduling.

3.3.1 Margin Adaptation Scheduling

The adaptive multiuser subcarrier and bit allocation problem for MA is formulated in [2] and [3]. However, we will make slight modifications to this problem formulation to fit to the new OFDMA schemes described in section 3.2.

3.3.1.1 Margin Adaptation Problem Formulation

Suppose that the cell serves K active users, each of which requires to transmit or receive a throughput denoted R_k in each TTI, where $k \in (1, \ldots, K)$. Also assume that each of the available modulation and coding schemes (MCSs) results in a certain throughput per subcarrier and that all these per-subcarrier throughputs are grouped in a set denoted D. The per-subcarrier throughput that the kth user can achieve on all the subcarriers of the nth subchannel is denoted $c_{k,n}$. Clearly, the value of $c_{k,n}$ should be one of the available values in D. We then define a subchannel assignment indicator, denoted $\rho_{k,n}$, such that:

$$\rho_{k,n} = \begin{cases} 1, & \text{if the } n^{\text{th}} \text{ sub-channel is assigned to the } k^{\text{th}} \text{ user} \\ 0, & \text{otherwise} \end{cases}. \tag{3.1}$$

The CQI used in this formulation is the ASPG. The ASPG of the nth subchannel for the kth user is denoted $\alpha_{k,n}^2$ and is defined as

$$\alpha_{k,n}^2 = \left| H_{k,n} \right|^2, \tag{3.2}$$

where $H_{k,n}$ is the channel frequency response coefficient of the nth subchannel for the kth user. We finally define $f_k(c_{k,n})$ as the required received power by the kth user on a subcarrier in the nth subchannel, for reception of $c_{k,n}$ per-subcarrier throughput using the corresponding MCS, in the presence of additive white Gaussian noise (AWGN), with a certain target bit error rate. In general, the higher the modulation level and the lower the coding rate, the higher the value of $f_k(c_{k,n})$.

For uncoded square constellation quadrature amplitude modulation (QAM), $f_k(c_{k,n})$ can be calculated using the following close form [2]:

$$f_k(c_{k,n}) = \frac{N_0}{3}\left[Q^{-1}\left(\frac{BER_k}{4}\right)\right]^2 (2^{c_{k,n}} - 1), \tag{3.3}$$

where BER_k is the target bit error rate desired for traffic of the kth user and N_0 is the AWGN power spectral density (PSD). Consequently, if the kth user is granted the nth subchannel in the TTI of concern, the transmission power that should be allocated to that subchannel for proper detection over all its subcarriers is equal to

$$P_{k,n} = \frac{m\, f_k(c_{k,n})}{\alpha_{k,n}^2}, \tag{3.4}$$

where $m = N_f \times N_s \times S$. Thus, the total required transmission power for the whole TTI becomes

$$P_T = \sum_{k=1}^{K}\sum_{n=1}^{N} \frac{m\, f_k(c_{k,n})}{\alpha_{k,n}^2}\rho_{k,n}. \tag{3.5}$$

We can finally formulate the MA optimization problem over $c_{k,n}$ and $\rho_{k,n}$ as follows:

$$\min_{\substack{c_{k,n}\in D\\ \rho_{k,n}\in\{0,1\}}} P_T = \min_{\substack{c_{k,n}\in D\\ \rho_{k,n}\in\{0,1\}}} \sum_{k=1}^{K}\sum_{n=1}^{N} \frac{m\, f_k(c_{k,n})}{\alpha_{k,n}^2}\rho_{k,n} \tag{3.6a}$$

$$\text{subject to} \quad \sum_{n=1}^{N} m\, c_{k,n}\,\rho_{k,n} = R_k \qquad \forall k \tag{3.6b}$$

$$\sum_{k=1}^{K} \rho_{k,n} = 1 \qquad \forall n. \tag{3.6c}$$

It is obvious that the constraints in (3.6b) guarantee the achievement of user throughput requirement, whereas the constraints in (3.6c) guarantee the assignment of each subchannel to a single user.

The optimization problem, defined in (3.6), is a nonlinear optimization problem since $f_k(c_{k,n})$ is generally a nonlinear function in $c_{k,n}$. The solution of this problem is very complex and requires a prohibitive delay for real-time applications. In [4], simplifications of the optimization problem were developed by exploiting the fact that $c_{k,n}$ takes only restricted values. Hence, the values of $f_k(c)$ for all $c \in D$ can be precalculated for all k and n. A new joint subchannel and per-subcarrier throughput assignment indicator ($\gamma_{k,n,c}$) is defined as follows:

$$\gamma_{k,n,c} = \begin{cases} 1, & \text{if the } n^{\text{th}} \text{ sub-channel is both assigned to the } k^{\text{th}} \\ & \text{and loaded with } c \text{ per-subcarrier throughput} \\ 0, & \text{otherwise} \end{cases} . \tag{3.7}$$

Hence, $f_k(c_{k,n})$ can be expressed in terms of $\lambda_{k,n,c}$ as follows:

$$f_k(c_{k,n}) = \sum_{c \in D} f_k(c)\gamma_{k,n,c}. \tag{3.8}$$

Since the values of $f_k(C)$, for all $c \in D$, are constants depending on the corresponding MCS, (3.8) states that $f_k(c_{k,n})$ is a linear combination in the indicator variables ($\gamma_{k,n,c}$). From the definitions of the indicators $\rho_{k,n}$ and $\gamma_{k,n,c}$, we can infer that $\forall c$, $\gamma_{k,n,c} = 1 \Rightarrow \rho_{k,n,c} = 1$. Substituting (3.8) in (3.5), the MA cost function becomes

$$P_T = \sum_{k=1}^{K} \sum_{n=1}^{N} \sum_{c \in D} \frac{m\, f_k(c)}{\alpha_{k,n}^2} \gamma_{k,n,c}\, \rho_{k,n}. \tag{3.9}$$

Thus, the MA cost function can be finally written as

$$P_T = \sum_{k=1}^{K} \sum_{n=1}^{N} \sum_{c \in D} \frac{m\, f_k(c)}{\alpha_{k,n}^2} \gamma_{k,n,c}. \tag{3.10}$$

Hence, the cost function becomes linear in $\gamma_{k,n,c}$ and the final MA optimization problem can be defined as

$$\min_{\gamma_{k,n,c} \in \{0,1\}} P_T = \min_{\gamma_{k,n,c} \in \{0,1\}} \sum_{k=1}^{K} \sum_{n=1}^{N} \sum_{c \in D} \frac{m\, f_k(c)}{\alpha_{k,n}^2} \gamma_{k,n,c} \tag{3.11a}$$

$$\text{subject to} \quad \sum_{n=1}^{N} \sum_{c \in D} m\, c\, \gamma_{k,n,c} = R_k \qquad \forall k \tag{3.11b}$$

$$\sum_{k=1}^{K} \sum_{c \in D} \gamma_{k,n,c} = 1 \qquad \forall n. \tag{3.11c}$$

The constraints in (3.11b) and (3.11c) are modified versions of constraints (3.6b) and (3.6c), respectively, that match with the new definition of the problem.

The above optimization problem in $\gamma_{k,n,c}$ can be solved using integer programming (IP). It is quite known that the solution of an IP problem requires considerably less computations than the nonlinear optimization problem [5, 6]. However, in general, the complexities of IP algorithms, and thus their execution times, grow exponentially with the number of integer variables and constraints. In practical cases, the time needed to solve the problem will still be prohibitive for real-time applications. In [7], it was shown that the relaxation of integer constraints in the above problem, by using linear programming (LP) instead of IP, leads to the same solution. This results in a considerable decrease in the algorithm complexity and execution time. However, more decrease in the resource allocation time is still needed.

3.3.1.2 Margin Adaptation Heuristic Scheduling Algorithms

In this section, we will present some MA heuristic algorithms that have proven to achieve near-optimal scheduling with less complexity and running time.

3.3.1.2.1 LP-Based Subchannel Allocation–Greedy Bit Loading Algorithm

The LP-based subchannel allocation–greedy bit loading (LPSA-GBL) algorithm, proposed in [4], splits the scheduling process into two steps:

- *Step 1*: Subchannels are adaptively assigned to different users based on the solution of the following optimization problem:

$$\min_{\rho_{k,n}\in\{0,1\}} \sum_{k=1}^{K}\sum_{n=1}^{N} \frac{\rho_{k,n}}{\alpha_{k,n}^{2}} \tag{3.12a}$$

$$\text{subject to} \quad \sum_{n=1}^{N} \rho_{k,n} = N_k \qquad \forall k \tag{3.12b}$$

$$\sum_{k=1}^{K} \rho_{k,n} = 1 \qquad \forall n, \tag{3.12c}$$

- where N_k is the number of subchannels to be assigned to the kth user. This number generally depends on the ratio of the kth user throughput requirement (R_k) to the overall throughput required by all the users. In [4], the subchannel distribution is assumed to be equal on all users (i.e., $N_k = \lfloor N/K \rfloor$), assuming that they all have the same rate requirements.
- Although this problem should be solved via IP, LP is used, since it was proven in [4] that the LP relaxation of IP for this problem leads to the same results.

- *Step 2*: Coded bits of each user are adaptively loaded on its assigned subchannels using one of the MA single-user greedy adaptive bit loading algorithms such as the Hughes-Hartogs algorithm [8] or the Chin-Wang algorithm [9]. The bit loading algorithm is run for each user until its required throughput (R_k) is reached.

The benefits achieved in this algorithm are:

- Reduction in the number of variables and constraints in the LP step ⇒ complexity reduction
- Simple bit loading (especially if the Chin-Wang algorithm is implemented)

3.3.1.2.2 *Constrained Greedy Subchannel Allocation–Greedy Bit Loading Algorithm*

The constrained greedy subchannel allocation–greedy bit loading (CGSA-GBL) algorithm [3, 10] splits the scheduling process into three steps:

- *Step 1*: Subchannels are assigned to users by a constrained greedy approach. For each TTI, the allocation progresses one subchannel at a time such that it is assigned to the kth user if it both has the highest average power gain for this subchannel and has not yet reached its maximum limit in the number of assigned subchannels (N_k) in this TTI. Obviously, this step is much simpler than solving an LP problem [3, 10]. However, the subchannel allocation obtained is likely to be trapped in a local minimum due to the greedy strategy followed in this step.
- *Step 2*: The allocation, obtained in step 1, can be improved by iterative swapping of the subchannels between users to reduce the total transmit power. In each iteration, the subchannels allocated to two users are swapped if this swapping will result in the reduction of the following cost function:

$$C = \sum_{k=1}^{K} \sum_{n=1}^{N} \frac{\rho_{k,n}}{\alpha_{k,n}^2}. \tag{3.13}$$

  It is obvious that this cost function corresponds to the objective function in (3.18) of the LP subchannel allocation problem described for the LPSA-GBL algorithm. It is claimed in [3] and [10] that the complexity of both subchannel allocation steps (steps 1 and 2 in the CGSA-GBL algorithm) is still lower than the complexity of step 1 in the LPSA-GBL algorithm.
- *Step 3*: Coded bits of each user are adaptively loaded on its assigned subchannels using the Hughes-Hartogs or Chin-Wang algorithm.

The benefit achieved in this algorithm, compared to the previous one, is the further complexity reduction obtained in the subchannel allocation step.

It has been shown, through simulation in [7], that the CGSA-GBL algorithm, without its second step, can already achieve a near-optimal scheduling, if the average CQIs of different users over all their subchannels are in the same order of magnitude. Consequently, a new algorithm that can switch between running the algorithm with or without its second step can be implemented. The switching between these two modes

TABLE 3.1 Comparison among MA Scheduling Algorithms

Algorithm	Complexity	Performance
IP	Exponential complexity	Optimal
LP	Lower than IP	Optimal
LPSA-GBL	Lower than LP	Near optimal
CGSA-GBL	Lower than LPSA-GBL	Near optimal

should be based on the maximum of pair-wise differences in the average CQI of all users in each TTI.

3.3.1.3 Comparison between MA Scheduling Algorithms

Table 3.1 depicts a comparison between the illustrated MA scheduling algorithms. The performance metric used for comparison is the overall transmission power achieved by the algorithm.

3.3.2 Rate Adaptation Scheduling

The RA problem aims to maximize the overall throughput per TTI under a maximum transmission power (P_{max}) constraint and while achieving both a fair service between users and a minimum transmission reliability. In the physical layer scheduling context, fairness of service among users refers to either achieving proportional throughputs to certain weights assigned to them or guaranteeing a weighted minimum throughput for each of them.

3.3.2.1 Rate Adaptation Problem Formulation

3.3.2.1.1 Max-Min Formulation

The adaptive multiuser subcarrier and bit allocation problem for RA is formulated in [11]. However, we will make slight modifications on the problem formulation to fit to the new OFDMA scheme described in section 3.2. Based on the ideology of [11] and the same notation defined in section 3.1.1, the RA problem can be formulated as follows:

$$\max_{\substack{c_{k,n}\in D \\ \rho_{k,n}\in\{0,1\}}} \min_{k} \frac{R_k}{\varphi_k} = \max_{\substack{c_{k,n}\in D \\ \rho_{k,n}\in\{0,1\}}} \min_{k} \sum_{n=1}^{N} \frac{m\, c_{k,n}\, \rho_{k,n}}{\varphi_k} \tag{3.14a}$$

$$\text{subject to} \quad \sum_{k=1}^{K}\sum_{n=1}^{N} \frac{m\, f_k(c_{k,n})}{\alpha_{k,n}^2}\rho_{k,n} \le P_{max} \tag{3.14b}$$

$$\sum_{k=1}^{K} \rho_{k,n} = 1 \qquad \forall n, \tag{3.14c}$$

where $\rho_{k,n}$ is the subchannel assignment indicator defined in (3.1), P_{max} is the maximum allowed transmission power, and φ_k is the weight of the kth user traffic. The max-min optimization in (3.14a) tends to assign more subchannels, and thus bits and transmission power, to users with poorer channel conditions in proportion to their weights. As a result, the allocated throughputs to all users via this procedure usually become proportional to these weights, thus achieving the required service fairness. The optimization problem in (3.14) is nonlinear due to the nonlinear power constraint in (3.14b).

In [4], authors used the same linearization process described in section 3.1.1 to transform the nonlinear power constraint into a linear one, hence reducing the complexity of the solution and its execution time. With using (3.8) in (3.14b), the power constraint becomes:

$$\sum_{k=1}^{K}\sum_{n=1}^{N}\sum_{c\in D}\frac{m\,f_k(c)}{\alpha_{k,n}^2}\gamma_{k,n,c}\le P_{max}, \tag{3.15}$$

where $\gamma_{k,n,c}$ is the joint subchannel and per-subcarrier throughput assignment indicator defined in (3.7). Thus, the simplified optimization problem becomes:

$$\max_{\gamma_{k,n,c}\in\{0,1\}}\min_{k}\frac{R_k}{\varphi_k}=\max_{\gamma_{k,n,c}\in\{0,1\}}\min_{k}\sum_{n=1}^{N}\sum_{c\in D}\frac{m\,c\,\gamma_{k,n,c}}{\varphi_k} \tag{3.16a}$$

$$\text{subject to}\quad\sum_{k=1}^{K}\sum_{n=1}^{N}\sum_{c\in D}\frac{m\,f_k(c)}{\alpha_{k,n}^2}\gamma_{k,n,c}\le P_{max} \tag{3.16b}$$

$$\sum_{k=1}^{K}\sum_{c\in D}\gamma_{k,n,c}=1\qquad\forall n. \tag{3.16c}$$

Thus, we obtain an IP optimization problem that can be solved with less complexity than the nonlinear problem. LP relaxation of IP may also be used for further complexity reduction. However, IP and LP max-min problems, employed to achieve service fairness, are considerably complex compared to simple minimization or maximization IP and LP problems, respectively.

3.3.2.1.2 Formulation with Minimum Rate Constraints

The RA problem with minimum rate constraint (MRC) is described in [12]. The objective was to formulate the RA problem to maximize the aggregated throughput. The authors developed a new formulation based on simple maximization with additional constraints that guarantee for each user the achievement of a minimum throughput per TTI regardless of its channel conditions.

The problem is defined as follows:

$$\max_{\gamma_{k,n,c}\in\{0,1\}}R_T=\max_{\gamma_{k,n,c}\in\{0,1\}}\sum_{k=1}^{K}\sum_{n=1}^{N}\sum_{c\in D}m\,c\,\gamma_{k,n,c} \tag{3.17a}$$

$$\text{subject to} \quad \sum_{k=1}^{K}\sum_{n=1}^{N}\sum_{c\in D}\frac{m\,f_k(c)}{\alpha_{k,n}^2}\gamma_{k,n,c} \le P_{max} \tag{3.17b}$$

$$\sum_{k=1}^{K}\sum_{c\in D}\gamma_{k,n,c} = 1 \qquad \forall n \tag{3.17c}$$

$$\sum_{n=1}^{N}\sum_{c\in D} m\,c\,\gamma_{k,n,c} \ge R_k^{min} \qquad \forall k, \tag{3.17d}$$

where R_T is the total cell throughput and R_k^{min} is the minimum throughput guaranteed for the kth user per TTI. This minimum throughput is determined by the scheduler for each user based on its traffic weight and average CQI over all subchannels. Obviously, this formulation added only K constraints while using maximization instead of max-min IP or LP optimization. This results in a faster solution.

3.3.2.1.3 Formulation with Proportional Rate Constraints

Another RA formulation can be obtained by maximizing the aggregated cell throughput while achieving for each user a throughput proportional to its traffic weight. We refer to this formulation as the proportional rate constrained (PRC) RA formulation. The definition of this problem can be obtained from (3.17) by replacing the constraints in (3.17d) by the following constraints:

$$\left|\frac{R_i}{\varphi_i} - \frac{R_j}{\varphi_j}\right| \le \varepsilon \qquad \forall i,j \in \{1,\cdots,K\}, \tag{3.18}$$

where R_k is the throughput of the kth user in the current TTI and ε is the allowed weight mismatch factor. The number of added constraints to this problem formulation is

$$\binom{K}{2}.$$

3.3.2.2 Rate Adaptation Heuristic Scheduling Algorithm

Despite all the simplifications in the formulation of the RA problem, explained in the previous section, further complexity reduction is required. In this section, we will present some RA heuristic algorithms that have proven to achieve near-optimal scheduling with less complexity and running time.

3.3.2.2.1 LP-Based Subchannel Allocation–Greedy Bit Loading Algorithm

This algorithm is the RA version of the LPSB-GBL algorithm introduced in section 3.1.2 and is also proposed in [4]. The scheduling process is split into two steps:

- *Step 1*: Subchannels are adaptively assigned to different users based on the solution of the same problem defined in (3.12). In [4], the subchannel distribution was assumed to be equal on all users assuming that they all have the same weights and their average CQIs over all subchannels are in the same order of magnitude. Although this problem should be solved via IP, LP can be used instead, since it has been proven in [4] that the LP relaxation of IP for this problem leads to the same results.
- *Step 2*: Coded bits of each user are adaptively loaded on its assigned subchannels using one of the single-user RA greedy adaptive bit loading algorithms, such as the Hughes-Hartogs algorithm [8] or Campello algorithm [13]. The loading algorithm is run until the overall allowed transmission power (P_{max}) is reached.

The benefits achieved in this scheduling algorithm are:

- Reduction in the number of variables and constraints in the LP step ⇒ complexity reduction
- Simple bit loading (especially if the Campello algorithm is implemented)

3.3.2.2.2 Joint Scheduling Algorithm Based on Equal Subchannel Power Allocation

The joint scheduling algorithm based on equal subchannel power allocation (JS-ESP) is introduced in [12] with MRC as fairness constraint. We will first describe algorithm strategy and then describe its two steps.

The algorithm assumes equal power distribution on all subchannels, i.e., each subchannel is loaded with a maximum transmission power equal to P_{max}/N. Consequently, we can calculate the maximum number of per-subcarrier throughput that each user can transmit on each subchannel from the inverse of the $f_k(c_{k,n})$ function. Note that in general, the chosen value for $c_{k,n}$ will be the one having the highest $f_k(c_{k,n})$ lower than $P_{max}/(m\ N)$.

The two steps of the algorithms can be described as follows:

- *Step 1*: Initial subchannel allocation and bit loading are executed jointly by assigning each subchannel to the user that can transmit the maximum number of bits on it.
- *Step 2*: The initial scheduling step is greedy and does not provide any fairness guarantees. Therefore, the initial assignment needs to be readjusted so that more subchannels are allocated to the users whose MRCs have not been satisfied yet. This process is called subchannel reallocation. This reallocation process should satisfy the following conditions:
    1. The nth subchannel that is originally assigned to user k_n^* cannot be reallocated to another user if this reallocation will cause the violation of user k_n^*'s minimum rate constraint (i.e., if $R_{k_n^*} - c_{k_n^*,n} < R_{k_n^*}^{min}$).
    2. Each subchannel reallocation should cause the least reduction in the overall throughput.
    3. The number of reallocation operations should be kept as low as possible.

The first condition ensures that the number of users satisfying the MRC increases monotonically during the reallocation process, guaranteeing that the number of reallocation operations will converge. Conditions 2 and 3 can be realized by using the reallocation cost function of the nth subchannel to user k', instead of its original owner k_n^*, defined as follows:

$$e_{k',n} = \frac{c_{k_n^*,n} - c_{k',n}}{c_{k',n}} \quad \forall n \text{ not owned by } k' . \tag{3.19}$$

Obviously, this cost function is directly proportional to the decrease in the overall throughput and inversely proportional to the increase in the throughput of user k'. Therefore, reallocating the subchannel having the least value of $e_{k',n}$ to user k' instead of user k_n^* will achieve both the least overall throughput reduction, fulfilling condition 2, and the greatest increase in user k' throughput. The latter achievement lowers the probability of further reallocation to user k', fulfilling condition 3. The reallocation procedure ends when all users meet their minimum rate requirements.

The benefit achieved in this algorithm over the previous one is that it performs joint subcarrier allocation and bit loading without involving any LP solving. However, two problems can be found in this algorithm:

- The reallocation step complicates the algorithm, especially when some of the users have bad CQI over most of the subchannels. Thus, this algorithm may be implemented only if the users' average CQIs over all subchannels are in the same order of magnitude.
- There is an amount of power wasted in each subchannel due to the difference between the amount of power assigned to that subchannel and the actual amount of power required for proper reception of the chosen number of bits loaded on the subcarriers of that subchannel.

In [14] and [7], some modifications were proposed to solve these drawbacks. The modified algorithm is described in the next section.

The JS-ESP algorithm can be generalized to the PRC. This can be done by modifying the reallocation steps such that subchannel n^* owned by user k^* is reallocated to user k' iff:

$$k' = \arg\min_k \left\{ \frac{R_k}{\varphi_k} \right\} \tag{3.20a}$$

$$k^* = \arg\max_k \left\{ \frac{R_k}{\varphi_k} \right\} \tag{3.20b}$$

$$\frac{R_{k^*}}{\varphi_{k^*}} - \frac{R_{k'}}{\varphi_{k'}} > \varepsilon \tag{3.20c}$$

$$\frac{R_{k'} + c_{k',n^*}}{\varphi_{k'}} - \frac{R_{k^*} - c_{k^*,n^*}}{\varphi_{k^*}} \leq \varepsilon \,. \tag{3.20d}$$

Condition (3.20d) is added to prevent the reallocation steps from diverging. If it is not satisfied in a certain reallocation iteration, user k^* is temporarly removed from the list and the above computations are rechecked for the remaining users. User k^* is added back to the list when this reallocation iteration is terminated. After each reallocation iteration, the throughtputs of both users (k' and k^*) are updated. Reallocations continue until no further improvement can be achieved.

3.3.2.2.3 Modified Joint Scheduling Algorithm Based on Equal Subchannel Power Allocation

This algorithm, proposed in [14] and denoted by M-JS-ESP, is a modified version of the JS-ESP algorithm with MRC. It similarly assumes equal distribution of power among all subchannels. Conversely, it tends, in a single step, to achieve service fairness along with subchannel and bit allocations. It schedules the cell resources in one step that operates as step 1 of the JS-ESP algorithm with insertion of a user throughput-checking step before additional subchannel and bit assignment. In other words, one subchannel after the other is allocated to the user whose MRC is not yet satisfied and is capable of transmitting the maximum number of per-subcarrier throughput on that subchannel after omitting all the users that already satisfied their MRC from the comparison.

Finally, an additional step was suggested in [14] to exploit the amounts of power wasted in each subchannel. The idea is to group these amounts of power and exploit the resulting amount in increasing the overall spectral efficiency of the cell by running a greedy modulation-level increment procedure, similar to the RA Hughes-Hartogs algorithm, until the exhaustion of this residual amount power. This is called the residual power exploitation procedure.

Although it seems that this algorithm may be trapped in a local minimum, [14] showed that this is not true for the case where the average CQIs of different users on all subchannels are in the same order of magnitude. Thus, we can say that this algorithm is able to achieve a better performance than the JS-ESP algorithm with a lower complexity, as it replaces the entire reallocation step by simple comparisons in the subchannel and bit allocation step. It also improves the overall spectral efficiency of the cell through the residual power exploitation step.

3.3.2.3 Comparison between RA Scheduling Algorithms

Table 3.2 depicts a comparison between the illustrated RA scheduling algorithms. The performance metric used in the comparison is the total cell throughput achieved by the algorithm. It is also important to note in this comparison that MRC-based algorithms may fail in achieving the minimum rate required by different users if some of them have poor average channel conditions. PRC-based algorithms do not suffer from this problem.

TABLE 3.2 Comparison between RA Scheduling Algorithms

Algorithm	Complexity	Performance	Fairness	Power Efficiency
Max-Min	High complexity	Optimal	Proportional	Yes
MRC	Lower than Max-Min	Optimal	Minimum rate	Yes
PRC	Higher than MRC	Optimal	Proportional	Yes
LPSA-GBL	Lower than MRC	Near optimal	Proportional	Yes
JS-ESP-MRC	Lower than LPSA-GBL	Suboptimal	Minimum rate	No
JS-ESP-PRC	Lower than LPSA-GBL	Suboptimal	Proportional	No
M-JS-ESP	Lower than JS-ESP	Near optimal	Proportional	Yes

3.4 Adaptive Physical Layer Scheduling with Limited Feedback

In the previous section, we assumed that the scheduler has full knowledge of all the CQIs for all users over all subchannels, which is acceptable for uplink transmission. However, in practical downlink scenarios, this huge amount of information cannot be fed back to the base station due to the high overhead that this feedback will result in. As a solution, several limited CQI feedback mechanisms were developed to reduce this overhead. Consequently, the scheduling techniques in such schedulers must rely only on this available limited information, and thus are expected to be simpler than the full-channel knowledge techniques. In this section, we will describe several limited feedback techniques and the scheduling approach that can be adopted with each of them.

3.4.1 Scheduling Based on CQI Quantization

The main idea behind this feedback technique can be described as follows: Each user compares its CQI for each subchannel to 2^l quantization levels and reports, in l bits/subchannel, the number of the level closest to that CQI (whether ASPG or SINR). Thus, this scheme needs only $N \cdot l$ bits per user to report a quantized CQI version of all its subchannels. A special case is obtained when $l = 1$. In this case, if the CQI is above a certain threshold, 1 is transmitted back to the base station; otherwise, 0 is transmitted. Consequently, only N bits per user are required for the feedback. Methods for optimizing and adapting comparison thresholds in order to achieve the maximum sum-rate capacity can be found in [15] and [16].

Upon receipt of the feedback bits from all users, scheduling can be performed as follows:

- Subchannel allocation is carried out as in step 1 of the MA or RA heuristic algorithms described in sections 3.3.1.2 and 3.3.2.2 by replacing the real CQI values by their quantized ones.
- If two or more users have the same quantized CQI value for the same subchannel (these subchannels are termed subchannels in conflict), it is assigned to the one among them that has not yet reached or is furthest from reaching

its constraining parameter, like MRC. If they are all at the same position, the subchannel is assigned randomly. For PRC, the subchannels are first greedily allocated, then reallocation iterations are carried out as described in the JS-ESP algorithm with PRC.

- In case of RA with MRC, if all users reach their minimum throughputs before the allocation of all subchannels, the subsequent subchannels in conflict are distributed randomly or in round-robin.

When all subchannels are allocated, bit loading is then performed as in the algorithms described in sections 3.3.1.2 and 3.3.2.2.

Several researches show that quantization of CQI does not lead to a significant reduction of the system gain on certain mean CQI ranges [15, 16]. The number of quantization levels should increase as the number of users increases, in order to reduce throughput degradation.

3.4.2 Scheduling Based on Contention Feedback with User Splitting

The feedback mechanism, in this technique, can be described as follows: For each subchannel, users are divided into *k* equal-size static groups. Users of each group are allowed to contend on a specific mini-slot to send their feedback if they are eligible. To determine whether it is eligible, each user compares its CQI against a quantile threshold forwarded to all users by the base station and denoted *q*. The user is eligible and transmits a feedback only when its CQI for that subchannel is greater than *q*. If only one user contends during a given slot, the base station stores its CQI for this subchannel. If either no or more than one user of the same group sends its feedback, the base station stores no feedback or collision, respectively.

When the contention phase is terminated, scheduling of subchannels is performed as follows:

- Subchannels for which more than one single mini-slot reporting is detected are assigned to the user that has the highest CQI and has not yet reached its constraining parameter.
- Subchannels for which no single mini-slot reporting is detected but one or more collisions occur are assigned to the user belonging to one of these groups that is furthest from reaching its constraining parameter.
- Subchannels with no reporting or with one reporting from a user that reached its constraining parameter are assigned to the furthest user from reaching its constraining parameter.

When all subchannels are allocated, bit loading is performed as in the MA or RA heuristic algorithms according to the scheduling technique.

3.4.3 Scheduling Based on L-Best Subchannels Feedback

The main idea of this feedback-scheduling mechanism is the reporting of the CQI of only the L-best subchannels by each of the users.

Scheduling of subchannels is performed as follows:

- Subchannels with more than one single reporting are assigned to the user that has the highest CQI and has not yet reached its constraining parameter.
- Subchannels with no reporting or one reporting from a user that reached its constraining parameter are assigned to the furthest user from reaching its constraining parameter.

Finally, bit loading is performed as described in section 3.3.1.2 or 3.3.2.2.

3.5 Adaptive Cross-Layer Scheduling

We define cross-layer scheduling as the process in which cell resources are distributed among users based on both their traffic information (e.g., traffic needs, queuing information, etc.) and their channel qualities. The aim of the scheduler is to achieve both QoS requirements and fairness of service among users from the link layer viewpoint as well as maximizing the physical throughput of the cell.

Since the scheduler is in the base station, it has all the information about downlink traffic information of all users. For uplink, users can either piggyback their traffic information in their uplink transmission or send it as a separate feedback in order to be used in uplink scheduling. Users starting to be active can also send their traffic information along with their CQIs in their session setup procedures.

Before proceeding in classifying and introducing cross-layer scheduling techniques, it is important to clarify that the two assumptions declared for physical layer scheduling in section 3.3 are no longer valid. In other words:

- Since cross-layer scheduling considers traffic and queuing information, which is the MAC layer information that deals with packet arrivals and departures, it is no longer valid to assume that user traffic is simply a stream of bits and that all active users always have bits to transmit. A packet-based queuing environment is considered instead.
- Since cross-layer scheduling does not allocate resources only according to CQIs, it is no longer suitable to assume scheduling on the subchannel level, as other information is considered in the resource allocation decision. Thus, scheduling is performed on the PRB level.

We can distinguish two major subclasses of cross-layer scheduling techniques according to the delay sensitivity of the traffic to be scheduled. In this sense, the two scheduling subclasses can be defined as:

- *Delay-insensitive scheduling*: Scheduler, belonging to this class, should consider queuing delays, fairness, and traffic priorities along with channel qualities in the scheduling process. This class of schedulers is more dedicated to traffic that has a best-effort nature, such as multimedia messaging, web browsing, etc.
- *Delay-sensitive scheduling*: In this class, schedulers should add to the concerns of the delay-insensitive case, the packet expiry times beyond which their transmission becomes of no use to the receivers. This class of schedulers is mainly

dedicated to real-time traffic such as packet-based voice telephony and multimedia streaming applications.

Although other subclassifications of cross-layer schedulers might be found, we will rely on the aforementioned one to discuss some cross-layer scheduling algorithms in this section. In section 3.6, adaptive cross-layer scheduling algorithms based on the limited feedback mechanisms described in section 3.4 are introduced for downlink scheduling.

3.5.1 Delay-Insensitive Traffic Scheduling

The cross-layer delay-insensitive traffic scheduling policies aim to allocate transmission resources to different backlogged sessions based on a fairness notion and according to their queuing delays and their priorities while considering the channel qualities of their corresponding users.

It is important to clarify, at this point, that the term *delay-insensitive traffic* does not practically mean that the scheduler should completely ignore the delays experienced by traffic flows in the scheduling process. It only means that the packets of this traffic family do not have deadlines beyond which their transmission becomes useless. This means that packets of this family should be transmitted even if they experienced high queuing delay. However, the scheduler is required to reduce their queuing delay as much as possible.

Several wireless scheduling algorithms described in the literature, for delay-insensitive traffic, were designed for the single-channel model. However, several modifications were proposed to develop versions of some of these algorithms that are suitable for multichannel transmission, such as OFDMA. In our context, we will describe these modified algorithms.

3.5.1.1 Maximum Rate Scheduling Algorithm

The maximum rate (MR) scheduling algorithm simply schedules PRBs to the users that can transmit the highest throughput on them, thus obtaining the highest achievable overall cell throughput.

Obviously, such a scheduling algorithm does not consider any queuing or fairness notions in its allocation procedure. Consequently, it is highly prone to uncontrollable backlog increases and huge packet delays for the users that temporarily undergo bad channel conditions on most of their PRBs. Thus, this algorithm is not of practical use but is always considered for comparison purposes as a reference of the maximum achievable throughput in a cell.

3.5.1.2 Proportional Fair Scheduling Algorithm

The proportional fair (PF) scheduling algorithm was introduced in [17] and [18] for high-data-rate (HDR) services in CDMA systems. In [19], modifications were proposed to allocate the PRBs of OFDMA systems based on the proportional fair scheduling algorithm. In general, proportional fairness is a simple yet effective fairness notion. It heuristically tries to balance the services of the sessions, while implicitly maximizing the system throughput in a greedy manner. This is achieved by an algorithm that assigns the channel to the user that can best exploit it while having a relatively low average service within a certain, but relatively long, past time window. The computation method of this windowed average service depends on the PRB allocation scheme. Three allocation schemes were proposed in [19]:

- Parallel PRB allocation
- Serial PRB allocation
- Serial PRB allocation with per-slot update

For a better understanding of the three versions of the algorithm, some variables should be first defined as follows:

- $D_{k,n}$ is the current rate (or throughput) that could be supported by the base station for the kth user on the nth subchannel in the current TTI. This value reflects the current channel state of the kth user on the nth subchannel. This parameter can be calculated through a physical layer scheduling algorithm.
- T_c is the averaging window parameter. In our multichannel multiple-access model described in section 3.2, T_c is expressed in terms of a number of either PRBs or TTIs, depending on the updating granularity of the window. Generally, this number is very large, which results in a relatively long averaging window.
- R_k is the average rate (or throughput) previously assigned to the kth user on all PRBs averaged over a sliding window of length T_c.
- $R_{k,n}$ is the average rate (or throughput) previously assigned to the kth user only on the PRBs of the nth subchannel averaged over a sliding window of length T_c.

In this section, we will first describe the three schemes presented in [19] for the OFDMA version of the PF scheduling algorithm, then will discuss its performance and drawbacks.

3.5.1.2.1 Parallel PRB Allocation Scheme

In this scheme, PRBs of each subchannel are scheduled independently. Thus, the number of schedulers is equal to the number of subchannels. For each slot, each PRB is assigned by its specified scheduler to the user k^*, satisfying the following allocation rule:

$$k^* = \arg\max_k \left\{ \frac{D_{k,n}}{R_{k,n}} \right\}. \tag{3.21}$$

After the parallel allocation of all PRBs of one slot, the windowed average rate is updated as follows:

$$R_{k,n}^{new} = \left(1 - \frac{1}{T_c}\right) R_{k,n}^{old} + \frac{1}{T_c} \rho_{k,n,s} D_{k,n} \qquad \forall\, k \text{ and } n, \tag{3.22}$$

where $\rho_{k,n,s}$ is the PRB assignment indicator defined as

$$\rho_{k,n,s} = \begin{cases} 1, & \text{if the } (n,s)^{\text{th}} \text{ PRB is assigned to the } k^{\text{th}} \text{ user} \\ 0, & \text{otherwise} \end{cases}. \tag{3.23}$$

If different traffic flows are to be treated with different priorities (differential service), the allocation rule can be modified to be

$$k^* = \arg\max_k \left\{ \varphi_k \frac{D_{k,n}}{R_{k,n}} \right\}, \tag{3.24}$$

where φ_k is the priority index (also called weight) of the kth user traffic.

3.5.1.2.2 Serial PRB Allocation Scheme

This scheme allocates PRBs to users using only one scheduler in a serial iterative manner. In each iteration, a PRB is assigned to the user k^* that satisfies the allocation rule defined as

$$k^* = \arg\max_k \left\{ \frac{D_{k,n}}{R_k} \right\}. \tag{3.25}$$

After the allocation of each PRB, the windowed average rates of all users are updated as follows:

$$R_k^{new} = \left(1 - \frac{1}{T_c}\right) R_k^{old} + \frac{1}{T_c} \rho_{k,n,s} D_{k,n} \qquad \forall k. \tag{3.26}$$

Afterwards, the next iteration starts for the following PRB. In general, the PRBs of each slot are scheduled serially until they are all scheduled before moving to the following slot. However, several other schemes can be used. The above procedure continues until all PRBs in the TTI are allocated. Again, different priorities can be assigned to different traffic flows by modifying the allocation rule to

$$k^* = \arg\max_k \left\{ \varphi_k \frac{D_{k,n}}{R_k} \right\}. \tag{3.27}$$

3.5.1.2.3 Serial PRB Allocation Scheme with Per-Slot Update

The serial allocation scheme updates the windowed average rates for different users after the allocation of each PRB, which results in a high complexity.

To reduce this complexity, this third scheme updates the windowed average rates for all users only after the allocation of all the PRBs of each slot. In this case, the allocation rules are the same rules defined in (3.25) and (3.27) for nonprioritized and prioritized scheduling, respectively, while the update formulation is changed to

$$R_k^{new} = \left(1 - \frac{1}{T_c} \sum_{n=1}^{N} \rho_{k,n,s}\right) R_k^{old} + \frac{1}{T_c} \sum_{n=1}^{N} \rho_{k,n,s} D_{k,n} \qquad \forall k, \tag{3.28}$$

where

$$\sum_{n=1}^{N} \rho_{k,n,s}$$

is the number of PRBs allocated to the kth user in the sth slot.

3.5.1.2.4 *Performance and Drawbacks of the PF Scheduling Algorithm*

We previously mentioned that, in general, the three schemes of the PF scheduling algorithm heuristically try to balance the services of different user sessions, while implicitly maximizing the system throughput in a greedy manner. It was found in [19] that the first scheme achieves a slightly higher throughput than the third one, while the second scheme has the worst throughput performance. It was also shown that the first and third schemes better exploit the variations of the channels for a bigger window size, while the additional gain of the second scheme is almost saturated when the window size grows above a certain value. However, two main drawbacks were noticed for the PF scheduling algorithm:

- It does not guarantee fairness in a strict sense. For example, consider the situation where a session has experienced a prolonged period of poor channel states for most of its PRBs (hence, has a small value of R_k); it may not for some time get enough service after the improvement of its channel conditions, if there is a dominant session having a very good channel state (i.e., very large $D_{k,n}$ values). It was shown in [19] that this phenomenon, termed throughput jitter, is more likely to occur as the window size increases.
- It does not give any consideration for the packet waiting times and other queue properties in the allocation rules.

These two drawbacks, either exclusively or combined together, may cause an uncontrollable increase in the delay experienced by the session packets.

3.5.1.3 Queue Size and Delay-Controlled Scheduling Algorithm

The design objective of the queue size and delay-controlled (QSDC) scheduling algorithm [20] is to find a new allocation priority metric that can increase the number of users achieving minimum bit rate requirements assigned to their traffic flows depending on their nature. It was shown in [20] that the number of satisfied users can be increased if the following allocation priority rule was used:

$$k^* = \arg\max_k \left\{ \frac{D_{k,n}}{R_k} \cdot \frac{W_k}{L_k} \right\}, \tag{3.29}$$

where W_k is the waiting time experienced by the kth user head-of-line (HoL) packet and L_k is the total number of bits for all the backlogged packets in the kth user queue at the scheduling instant of the $(n,s)^{\text{th}}$ PRB.

The allocation rule in (3.29) is obtained from the allocation rule of the PF scheduling algorithm in (3.25) with two new parameters, W_k and L_k, added to it. These two parameters impose the following effects:

- If W_k is large and equal for two users, the user that has the lowest expected backlog depletion time ($L_k/D_{k,n}$) gets a higher priority in seizing the PRB. This property resembles the ideology of the well-known weighted fair queuing (WFQ) algorithm for wired networks that assigns the channel to the packet that has the lowest virtual departure time.
- If, for a queue, W_k is small and L_k is large, then the queue has recently received a large number of packets or few large-size packets. They thus need not have higher priority for transmission compared to higher waiting time, smaller-size packets.

The allocation rule for the prioritized traffic mode can be expressed as

$$k^* = \arg\max_k \left\{ \varphi_k \frac{D_{k,n}}{R_k} \cdot \frac{W_k}{L_k} \right\}. \tag{3.30}$$

In the larger scope, the scheduling algorithm serially allocates different PRBs to the users according to the allocation rule. Updates for W_k and L_k are performed after each PRB allocation, while R_k can be updated either after each PRB allocation or after the allocation of all the PRBs of each slot.

In [20], it has been shown that the number of users satisfying their minimum rate requirements with QSDC scheduling is higher than the number of users satisfying the same minimum rate requirements with PF scheduling. Furthermore, the QSDC scheduling algorithm outperforms the PF scheduling algorithm from the overall cell throughput viewpoint. It is also obvious that the former algorithm is more unlikely to suffer from high uncontrollable delays since the HoL packet waiting times are taken in consideration of the allocation procedure.

3.5.1.4 Queue Arrival and Delay-Controlled Scheduling Algorithm

The queue arrival and delay-controlled (QADC) scheduling algorithm [21] performs, in the large scope, as the QSDC scheduling algorithm but with a different allocation rule, defined as

$$k^* = \arg\max_k \left\{ \frac{A_k}{R_k} D_{k,n} W_k \right\}, \tag{3.31}$$

where A_k is the mean windowed arrival rate of the kth user traffic. In this case, the priority of assigning a PRB to the kth user increases with the waiting time of its HoL packet, the ratio of its mean traffic arrival rate to its mean service rate (A_k/R_k), and indeed its channel condition for that PRB. Priorities of different traffic classes can also be included by modifying the rule to

$$k^* = \arg\max_k \left\{ \varphi_k \cdot \frac{A_k}{R_k} \cdot D_{k,n} \cdot W_k \right\}. \tag{3.32}$$

In [21], it was shown that the QADC scheduling algorithm can achieve an overall throughput close to that of the MR scheduling algorithm but with a much higher queue stability. It also allocates resources fairly among the users.

3.5.1.5 Queue Left-Over and Delay-Controlled Scheduling Algorithm

The idea of the queue left-over and delay-controlled (QLDC) scheduling algorithm, proposed in [22], is to initially allocate subchannels to the users that can best utilize them, as in physical layer scheduling, and then refine this allocation by reallocating some PRBs in order to minimize a cost function that reflects the queuing properties of different users. This cost function is defined as follows:

$$C = \sum_{k=1}^{K} J_k, \tag{3.33a}$$

where

$$J_k = \left[\frac{L_k - \sum_{s=1}^{S} \sum_{n=1}^{N} \rho_{k,n,s} D_{k,n}}{Z_k} \exp\left(\frac{-\tau}{W_k} \right) \right]^{+}, \tag{3.33b}$$

where L_k is defined as in section 3.5.1.3, Z_k is the average backlogged packet size in bits/packet, and τ is a constant time offset decided after the scheduling termination of the previous TTI. $[x]^{+}$ is defined as $\max\{0,x\}$.

It was clearly stated in [22] that the function J_k is not optimum in any sense but is just a candidate function. Obviously, J_k is an increasing function in both the number of left-over traffic bits (i.e., the number of nonscheduled bits in the queue of kth user after an initial allocation phase) and the waiting times of the HoL packets. It is also a decreasing function in the average packet sizes of different queues. The algorithm is performed in two steps:

- *Step 1*: Assignment of subchannels to the users that can best utilize them, as in physical layer scheduling.
- *Step 2*: Reallocation of some PRBs in order to minimize the cost function C. This is done iteratively as follows:
  1. The values J_k for all k are computed and a PRB iterative reallocation phase is launched based on these values.
  2. In each iteration, the user having the maximum value for J_k is allocated, as a test, a PRB in the best subchannel not currently assigned to him. If this test leads to a reduction in C, the reallocation is executed and the values of J_k for the two concerned users (i.e., the user that granted the PRB and the one that lost it) are recomputed.
  3. These reallocation trials are carried out until either no further reduction in C can be achieved or a maximum allowed number of reallocations is

TABLE 3.3 Comparison between Delay-Insensitive Scheduling Algorithms

Algorithm	Throughput	Queue Stability	Throughput Fairness
MR	Highest	Uncontrolled	Worst
PF	Lower than MR	Better than MR	Good
QSDC	Better than PF	Better than PF	Good
QADC	Close to MR	Better than PF	Good
QLDC	Better than PF	Better than PF	Good

reached. This maximum limit is chosen to prevent the reallocation phase from diverging.

Despite the use of a nonoptimized function, it has been shown in [22] that performing reallocations based on this function in the QLDC scheduling algorithm makes its throughput and delay performance better than that of the PF scheduling algorithm.

3.5.1.6 Comparison between Delay-Insensitive Scheduling Algorithms

Table 3.3 depicts a comparison between the illustrated delay-sensitive scheduling algorithms.

3.5.2 Delay-Sensitive Traffic Scheduling

The QoS requirements of delay-sensitive traffic are usually defined in terms of a delay bound, generally termed deadline, before which the packets should be delivered to the receiver. Otherwise, the information contained in these packets will be of no use to the receiver, and thus should be dropped. Therefore, the cross-layer delay-sensitive scheduler should be able to minimize the number of packet deadline violations while maintaining its other previously explained duties (channel awareness and fairness among users).

The main scheduling ideology for most delay-sensitive traffic schedulers relies on modified versions of the PF scheduling algorithm, explained in section 3.5.1.2, as well as channel-aware versions of the well-known earliest-deadline-first (EDF) scheduling technique. Due to time-varying properties of the wireless channel, EDF scheduling cannot always meet the deadlines of all packets. Consequently, a portion of these packets expires and is dropped. A good delay-sensitive scheduler is one that has the ability to minimize this portion. In other words, if W_k, T_k, and δ_k are the respective waiting time, deadline, and expiry probability (also termed violation probability) for the packets of the kth user, then a good delay-sensitive traffic scheduler should guarantee that

$$P\{W_k > T_k\} < \delta_k. \tag{3.34}$$

Before illustrating some delay-sensitive scheduling algorithms, we will introduce some more variables:

- a_k is a weighting parameter reflecting the QoS requirements of the kth user. For large values of T_k and small values of δ_k, a_k can be defined as [23]

$$a_k = -\frac{1}{T_k}\log(\delta_k). \tag{3.35}$$

- W_k is the delay encountered by the kth user HoL packet at the PRB scheduling instant. By definition, W_k is always less than T_k since any packet whose W_k becomes greater than T_k is dropped.
- d_k is the time to expire of the kth user HoL packet at the PRB scheduling instant. Obviously, $d_k = T_k - W_k$.
- V_k is the number of violation (packet expiry) occurrences for the kth user at the scheduling instant.
- V is the average number of violations for all users, defined as

$$V = \frac{1}{K}\sum_{k=1}^{K} V_k. \tag{3.36}$$

Delay-sensitive traffic is inclusively classified if different values of T_k and δ_k are assigned to different traffic types. Thus, no further prioritization will be introduced in the algorithm description.

Finally, before starting the description of delay-sensitive traffic scheduling algorithms, it is important to note that some of the algorithms that will be described were initially designed for TDMA systems. Consequently, our illustration of these algorithms will follow the serial allocation schemes adopted by [19] to match the PF scheduling algorithm to OFDMA systems. We will term our illustrations by the OFDMA versions of the algorithms when this modification is done. To deduce the formulations for the parallel scheme, $R_{k,n}$ should replace R_k in the allocation rules and should be updated according to (3.22) after the parallel allocation of the PRBs of each slot.

3.5.2.1 Modified Largest Weighted Delay First Algorithm

The modified largest weighted delay first (M-LWDF) algorithm was described in [23] and [24] for TDMA-based systems. An OFDMA version of the algorithm can be described as follows. The algorithm schedules PRBs serially to users according to the following allocation rule:

$$k^* = \arg\max_k \left\{ a_k \frac{D_{k,n}}{R_k} W_k \right\}. \tag{3.37}$$

W_k is updated after each PRB allocation, while R_k can be updated either after each PRB allocation or after the allocation of all the PRBs of each slot. Obviously, the assignment rule aims to reduce the packet delays for all users and to exploit at the same time the multiuser diversity by providing users with their good-quality PRBs. It has been shown in [25] that M-LWDF is throughput optimal: it renders the queues at the base station stable if any other rule can do so.

Another version of the M-LWDF algorithm is introduced in [26]. This version considers the relative number of violation occurrences in the scheduling process, and thus is called violation-fair M-LWDF (VF-M-LWDF) algorithm. The new allocation rule is defined as

$$k^* = \arg\max_k \left\{ a_k \frac{D_{k,n}}{R_k} \cdot \frac{\max\{1, V_k\}}{V} W_k \right\}. \tag{3.38}$$

Consequently, this version adds the achievement of a fair distribution of the violation occurrences among all users to the benefits of the M-LWDF algorithm.

3.5.2.2 Channel-Dependent Exponential Rule Scheduling Algorithm

The OFDMA version of the channel-dependent exponential rule (CD-ER) scheduling algorithm, described in [24], assigns PRBs to users in a serial or parallel scheme according to an exponential allocation rule defined as

$$k^* = \arg\max_k \left\{ a_k \frac{D_{k,n}}{R_k} \exp\left(\frac{a_k \cdot W_k - W_{av}}{1 + \sqrt{W_{av}}} \right) \right\}, \tag{3.39}$$

where W_{av} is defined as

$$W_{av} = \frac{1}{K} \sum_{k=1}^{K} a_k W_k. \tag{3.40}$$

This algorithm attempts to equalize the weighted delays ($a_k W_k$) in user queues when their differences are large. If one of the queues has a larger weighted delay than the others by more than the order of $\sqrt{W_{av}}$, the exponent term becomes very large and overrides the channel considerations, hence giving priority to that queue as long as its channel can support a nonzero throughput. On the other hand, for small weighted delay differences, the exponential term is close to unity, and the rule becomes a PF allocation rule with a_k representing different class priorities (as φ_k in equation (3.27)). Hence, the exponential rule policy adapts from a proportionally fair one to one that balances delays [24]. Simulation results in [27] showed that, for HDR systems, the exponential rule scheduling exhibits better delay tails than any other scheduling policy in the sense that the delays of all users are about the same and are all reasonably small. This occurs, however, for large values of T_k, which is not desired practically, and very small values of δ_k.

As for the M-LWDF algorithm, a violation-fair version of the CD-ER scheduling algorithm (VF-CD-ER) can be obtained by modifying the assignment rule to be

$$k^* = \arg\max_k \left\{ a_k \frac{D_{k,n}}{R_k} \cdot \frac{\max\{1, V_k\}}{V} \exp\left(\frac{a_k \cdot W_k - W_{av}}{1+\sqrt{W_{av}}} \right) \right\}. \tag{3.41}$$

This guarantees the fair distribution of the violation occurrences among all the users.

3.5.2.3 Channel-Dependent Earliest-Deadline-Due Algorithm

The channel-dependent earliest-deadline-due (CD-EDD) algorithm, described in [26] and [28] for TDMA-based systems, operates for the OFDMA-based systems similarly to the previous serially iterative algorithms using a new assignment rule defined as

$$k^* = \arg\max_k \left\{ a_k \frac{D_{k,n}}{R_k} \cdot \frac{W_k}{d_k} \right\}. \tag{3.42}$$

The behavior of the CD-EDD algorithm can be explained as follows: When a certain queue has its HoL packet waiting in the system for a relatively long period (but has not expired yet), its time to expire will decrease significantly. In such a situation, the term W_k/d_k will grow significantly until it overcomes other terms in (3.42). This results in reducing the number of dropped packets due to deadline violations. On the other hand, if the delay characteristics of all users are about the same (i.e., their time to expire and waiting times are close), the term W_k/d_k will be in the same order for all users, and the algorithm then reduces to a prioritized PF algorithm.

An important feature of the CD-EDD algorithm is its weak dependency on the variance of the value of QoS weights a_k, and thus it can be used for a wide variety of QoS requirements.

Finally, if the proposed violation-fair technique is applied to the proposed CD-EDD, we can get a scheduling algorithm that explicitly provides delay-sensitive traffic with excellent fairness characteristics for transmission rates, queuing delays, and violation occurrences. The assignment rule of the VF-CD-EDD algorithm is expressed as

$$k^* = \arg\max_k \left\{ a_k \frac{D_{k,n}}{R_k} \cdot \frac{\max\{1, V_k\}}{V} \cdot \frac{W_k}{d_k} \right\}. \tag{3.43}$$

3.5.2.4 Channel-Dependent Direct Prioritized Assignment Algorithm

Unlike all previous algorithms, the channel-dependent direct prioritized assignment (CD-DPA) algorithm, described in [29], was mainly designed for OFDMA systems, and thus schedules the resources in a direct, noniterative scheme. The assignment of PRBs among users is performed in two steps with a proportional prioritization to the number of violation occurrences of different users.

The algorithm steps can be explained as follows:

- *Step 1: PRB allocation*: The target of this step is to determine the number of PRBs (B_k) to be allocated to each user. To calculate B_k for all k, we first define $\mu_{k,n}$ as the maximum capacity of the k_{th} user on the n_{th} subchannel and μ_k as the average subchannel capacity of the k_{th} user, expressed as

$$\mu_k = \frac{1}{N}\sum_{n=1}^{N} \mu_{k,n} \qquad \forall k. \tag{3.44}$$

- Defining r_k as the average rate of the k_{th} user traffic, the number of PRBs allocated to the k_{th} user is calculated as

$$B_k = \left\lceil \frac{r_k}{\frac{1}{K}\sum_{j=1}^{K} r_j} \cdot \frac{\mu_k}{\frac{1}{K}\sum_{j=1}^{K} \mu_j} \right\rceil \qquad \forall k. \tag{3.45}$$

- Three situations may result from the previous equation:
  1. The sum of B_k over all k is equal to the overall number of PRBs in one TTI (B). In this case, the PRB allocation step is terminated.
  2. The sum of B_k over all k is greater than B. In this case, the active users are listed in a descending order of their HoL times to expire, and the algorithm iterates through this list, subtracting a PRB in each iteration until equality is obtained.
  3. The sum of B_k over all k is less than B. In this case, the remaining number of PRBs is distributed among the users directly proportional to their numbers of packet dropping occurrences and inversely proportional to the time to expire of their HoL packets, as follows:

$$B_k \leftarrow B_k + \left\lceil \left(B - \sum_{i=1}^{K} B_i \right) \cdot \frac{\frac{\max\{1, V_k\}}{d_k}}{\sum_{j=1}^{K} \frac{\max\{1, V_j\}}{d_j}} \right\rceil \qquad \forall k. \tag{3.46}$$

- *Step 2: Prioritized PRB assignment*: This step aims to assign the PRB shares determined in step 1 to all users in a descending order of their priorities. The user priority is initialized to unity once it becomes active. This priority is then incremented by 1 whenever a deadline violation occurs to a packet of this user.
- The PRB assignment is performed as follows:
  1. In each TTI, the users are sorted in a descending order of their priorities. If more than one user shares the same priority level, ties are broken by giving a higher priority to the user with the best average CQI over all

TABLE 3.4 Comparison between Delay-Sensitive Scheduling Algorithms

Metric	M-LWDF	CD-ER	CD-EDD	VF Versions	CD-DPA
Allocation method	Serial	Serial	Serial	Serial	Direct
Total throughput	High	Low	Highest	Lowest	High
Delay fairness	Depends on QoS	Depends on QoS	Good	Good	Good
Throughput fairness	Good	Bad	Best	Good	Good
Violation fairness	Bad	Bad	Good	Best	Good

subchannels. Let the user at the head of the list be denoted user 1, the following one user 2, and so on.

2. The highest-priority user (user 1) is allowed to pick its B_1 best PRBs (i.e., from the subchannels for which it has the highest CQIs) from the set of all PRBs.
3. The PRBs assigned to user 1 are removed from the set of available PRBs.
4. The next-higher-priority user (user 2) is granted its B_2 best PRBs from the set of remaining PRBs. The PRB set is reupdated accordingly.
5. This procedure continues until all users are granted their allocated number of PRBs.

Hence, this algorithm enhances the fairness performance of the scheduler with respect to violation occurrences while exploiting multiuser diversity in assigning the available PRBs. Simulation results in [29] showed that this algorithm can achieve a high overall data rate while maintaining low levels of violation occurrences and achieving fairness among different users.

3.5.2.5 Comparison between Delay-Sensitive Scheduling Algorithms

Table 3.4 depicts a comparison between the illustrated delay-sensitive scheduling algorithms.

3.6 Adaptive Cross-Layer Scheduling with Limited Feedback

In section 3.5, we introduced several adaptive cross-layer scheduling techniques. These techniques require the knowledge of the CQIs of different users over all subchannels to compute the channel-dependent parameter ($D_{k,n}$ or $\mu_{k,n}$). As previously explained in section 3.4, the full CQI reporting, for downlink scheduling, is very exhaustive from the overhead viewpoint. Consequently, we will focus, in this last section, on exploring the changes that are needed in the scheduling algorithms, described in section 3.5, to fit with the limited CQI feedback mechanisms explained in section 3.4. Unlike the physical layer scheduling case, it is important to note that the changes will be minor for adaptive cross-layer scheduling algorithms since they depend on several factors in their PRB allocation rules other than the CQIs.

3.6.1 Scheduling Based on CQI Quantization

The operation of the feedback mechanism is the same as the one explained in section 3.4.1. Upon receipt of the feedback bits from all users, scheduling can be performed as follows:

- PRB allocation is carried out as in ordinary cross-layer scheduling algorithms described in section 3.5 by replacing the values of $D_{k,n}$ or $\mu_{k,n}$, calculated based on the accurate CQI values, with their quantized ones.
- PRBs located within a subchannel in conflict are ordinarily allocated through the same allocation rules. In such situations, the other scheduling parameters will break the ties.

3.6.2 Scheduling Based on Contention Feedback with User Splitting

When the contention phase of the feedback mechanism, explained in section 3.4.2, is terminated, scheduling of PRBs is performed as follows:

- PRBs located within a subchannel for which more than one single mini-slot reporting is detected are allocated among these successfully contending users through the ordinary allocation rule of the adopted scheduling algorithm, as described in section 3.5.
- PRBs located within a subchannel for which no single mini-slot reporting is detected but one or more collisions occurred are allocated among the users belonging to these groups, again through the allocation rule of the scheduling algorithm of concern, after removing the parameter $D_{k,n}$ or $\mu_{k,n}$ from the rule.
- PRBs located within a subchannel with no reporting are allocated among all users through the allocation rule of the adopted algorithm after removing $D_{k,n}$ or $\mu_{k,n}$ from the calculation.

3.6.3 Scheduling Based on L-Best Subchannel Feedback

As previously mentioned in section 3.4.3, the main idea of this feedback-scheduling mechanism is the CQI reporting of the L-best PRBs by each of the users. Scheduling of PRBs is performed as follows:

- PRBs located within a subchannel for which more than one single reporting is detected are allocated among these successfully contending users through the ordinary allocation rule of the adopted scheduling algorithm, as described in section 3.5.
- PRBs located within a subchannel with no reporting are allocated among all users through the allocation rule of the adopted algorithm after removing $D_{k,n}$ or $\mu_{k,n}$ from the calculations.

3.7 Summary

In this chapter, we illustrated the two major classes of adaptive scheduling techniques suitable for implementation in beyond 3G OFDMA-based cellular networks, namely, adaptive physical layer scheduling and adaptive cross-layer scheduling. The former class schedules network resources based on channel qualities of different users, while the latter class combines user channel qualities with traffic and queue information in the scheduling process.

We first described OFDMA as a prospective multiple-access technique for future cellular communications.

The MA and RA formulations of adaptive physical layer scheduling were then described, and we showed that they are too complex to implement in practical real-time scenarios. Consequently, we presented and compared several heuristic algorithms that can achieve near-optimal performance with less complexity.

Afterwards, we described adaptive cross-layer scheduling techniques with their two subclasses: delay-insensitive and delay-sensitive subclasses. For each subclass, we illustrated several algorithms and provided a comparison of their performance.

Finally, we illustrated some solutions to modify the presented scheduling algorithms for operation in limited channel knowledge, as in the case of downlink transmission in cellular networks. We introduced three limited feedback mechanisms and described the modifications required for each of them. We explained that modifications for physical layer scheduling techniques are major, since this class relies only on channel quality in the resource allocation process. Oppositely, the required modifications for cross-layer scheduling techniques were limited, as they rely on several parameters in the scheduling decisions.

Acknowledgments

This work has been done with partial support from LG Electronics, Korea.

References

[1] 3rd Generation Partnership Project. 2006. *Technical Specification Group Radio Access Network—Physical layer aspects for evolved Universal Terrestrial Radio Access (UTRA) (Release 7)*. 3GPP-LTE TR 25.814 V7.0.0.

[2] C. Y. Yong, R. S. Cheng, K. Ben Letaief, and R. D. Murch. 1999. Multiuser OFDM with adaptive subcarrier, bit and power allocation. *IEEE J. Select. Areas Commun.* vol. 17, pp. 1747–1758.

[3] C. Y. Wong, C. Y. Tsui, R. S. Cheng, and K. B. Letaief. 1999. A real-time subcarrier allocation scheme for multiple access downlink OFDM transmission. In *IEEE Proceedings VTC '99*, vol. 2, pp. 1124–1128.

[4] I. Kim, H. L. Lee, B. Kim, and Y. H. Lee. 2001. On the use of linear programming for dynamic subchannel and bit allocation in multiuser OFDM. In *IEEE GLOBECOM '01*, vol. 6, pp. 3648–52.

[5] H. P. Williams. 1993. *Model solving in mathematical programming*. Chichester: Wiley.

[6] D. A. Wismer and R. Chattergy. 1979. *Introduction to nonlinear optimization: A problem solving approach*. New York: North-Holland Publishing.

[7] S. Sorour. 2006. Adaptive resource allocation for multiuser wireless networks based on OFDMA. MSc thesis, Faculty of Engineering, Alexandria University, Department of Electrical Engineering.

[8] D. Hughes-Hartogs. Ensemble modem structure for imperfect transmission media. U.S. Patents 4,679,277 (July 1987) and 4,731,816 (March 1988).

[9] C. I. Wang and Y.-S. Chang. An optimal discrete loading algorithm for DMT modulation. In research work in Hsinchu, Taiwan (1999).

[10] S. Pietrzyk and G. J. M. Janssen. 2002. Multiuser subcarrier allocation for QoS provision in the OFDMA systems. In *Proceedings of VTC 2002I*, vol. 2, pp. 1077–81.

[11] W. Rhee and J. M. Cioffi. 2000. Increase in channel capacity of multiuser OFDM system using dynamic subchannel allocation. In *IEEE Proceedings of VTC'00*, pp. 1085–89.

[12] Y. J. Zhang and K. Ben Letaief. 2004. Multiuser adaptive subcarrier and bit allocation with adaptive cell selection for OFDM systems. *IEEE Trans. Wireless Commun.* 3(5).

[13] J. Campello. 1999. Practical bit loading for DMT. In *Transactions of IEEE International Conference on Communications*, Vancouver, Canada, pp. 801–5.

[14] O. Abdel-Alim, S. Shabaan, and S. Sorour. 2006. Achieving fairness, simplicity, and near optimal performance in rate adaptation algorithms for wireless multiuser OFDM systems. In *Proceedings of the 23rd NRSC Conference*, pp. C33:1–C33:7.

[15] M. Johansson. 2004. On scheduling and adaptive modulation in limited feedback channel. Accessed March 24, 2008 from http://www.signal.uu.se/Staff/mj/pub/MJTComm20040405.pdf.

[16] S. Sanayei, A. Nosratinia, and N. Aldhahir. 2004. Opportunistic dynamic subchannel allocation in multiuser OFDM networks with limited feedback. In *Proceedings of the IEEE Information Theory Workshop*, pp. 182–186.

[17] P. Bender, P. Black, M. Grob, R. Padovani, N. Sindhushayana, and A. Viterbi. 2000. CDMA/HDR: Bandwidth efficient high-speed wireless data service for nomadic users. *IEEE Commun. Mag.* 38:70–77.

[18] A. Jalali, R. Padovani, and R. Pankaj. 2000. Data throughput of CDMA-HDR: A high efficiency high data rate personal communication wireless system. In *Proceedings of VTC'2000*, vol. 3, pp. 1854–58.

[19] A. Wang, L. Xiao, S. Zhou, X. Xu, and Y. Yao. 2003. Dynamic resource management in the fourth generation wireless systems. In *International Conference on Communication Technology Proceedings*, vol. 2, pp. 1095–98.

[20] D. Kim, B. Ryu, and C. Kang. 2004. Packet scheduling algorithm considering a minimum bit rate for non-real-time traffic in an OFDMA-FDD-based mobile Internet access system. *ETRI J.* 26(1).

[21] P. Parag, S. Bhashyam, and R. Aravind. 2005. A subcarrier allocation algorithm for OFDMA using buffer and channel state information. In *IEEE VTC'05*, vol. 62, pp. 622–25.

[22] L. Sidiropoulos, S. Ben Slimane, and G. K. Karagiannidis. 2006. An improved resource allocation algorithm for the downlink of OFDM networks. In *Proceedings of IST Mobile and Wireless Communications Summit.*

[23] A. L. Stolyar and K. Ramanan. 2001. Largest weighted delay first scheduling: Large deviations and optimality. *Ann. Appl. Probability* 11:1–48.

[24] M. Andrews, K. Kumaran, K. Ramanan, A. L. Stolyar, R. Vijayakumar, and P. Whiting. 2000. *CDMA data QoS scheduling on the forward link with variable channel conditions.* Bell Laboratories Technical Report.

[25] M. Andrews, A. Stolyar, K. Kumaran, R. Vijayakumar, K. Ramanan, and P. Whiting. 2001. Providing quality of service over a shared wireless link. *IEEE Commun. Mag.* 39:150–54.

[26] A. Khattab and K. Elsayed. 2004. Channel-quality dependent earliest deadline due fair scheduling schemes for wireless multimedia networks. In *Proceedings of the 7th ACM International Symposium on Modeling, Analysis and Simulation of Wireless and Mobile Systems (MSWiM 2004)*, Venice, pp. 31–38.

[27] S. Shakkottai and A. L. Stolyar. 2001. Scheduling algorithms for a mixture of real-time and non-real-time data in HDR. In *Proceedings of ITC-17*, Salvador da Bahia, Brazil, pp. 793–804.

[28] K. Elsayed and A. Khattab. 2006. Channel-aware earliest deadline due fair scheduling for wireless multimedia networks. *J. Wireless Personal Commun.* 38:233–252.

[29] A. Khattab and K. Elsayed. 2006. Opportunistic scheduling of delay sensitive traffic in OFDMA based wireless networks. In *Proceedings of the International Symposium on a World of Wireless, Mobile and Multimedia Networks,* pp. 279–288.

4

Adaptive Resource Allocation in CDMA Cellular Wireless Mobile Networks under Time-Varying Traffic: *A Transient Analysis-Based Approach*

Dusit Niyato
University of Manitoba

Ekram Hossain
University of Manitoba

4.1 Introduction

Next-generation cellular wireless mobile networks based on code division multiple-access (CDMA) technology will support different types of data services along with traditional voice service. These data services can be either elastic or inelastic. For inelastic service, the transmission rate is fixed (which is suitable for constant-bit-rate [CBR] applications), while for elastic data service, the transmission rate of the ongoing calls can be adjusted, for example, to serve more users. The call-level quality of service (QoS) for the different types of services is generally measured by new call blocking and handoff call dropping probabilities. Also, the average transmission rate is an important QoS metric for calls under the elastic service class. These performance measures need to be optimized in a CDMA cellular wireless mobile network.

A system exhibits transient behavior when it is not in the steady state, i.e., during the transition period until the system reaches an equilibrium/steady state. The transient behavior is important especially for a time-varying system since the system may never reach the steady state. In the context of CDMA cellular networks, in the transient state, the network parameters (e.g., call arrival rate) can vary depending on time. In such an environment, adaptive optimized resource allocation and admission control are required to ensure that the system utilization is maximized (i.e., results in the highest revenue for the service provider) while the QoS performances are maintained at the target level. The system can be modeled as a continuous-time Markov chain (CTMC) to capture the evolution of the number of ongoing calls. Various performance measures (e.g., new call blocking, handoff call dropping probabilities, and average transmission rate) in both the steady and transient states can be obtained by using an analytical model.

For a dynamic system, the performance measures need to be obtained in both transient and steady states. Mathematically, the system performance measure can be written as [1]

$$P(t)=P_{tr}(t)+P_{ss}(t), \tag{4.1}$$

where $P(t)$ is the performance measure at time t, and $P_{tr}(t)$ and $P_{ss}(t)$ denote the transient state and steady state performance measures, respectively. The system performance approaches the steady state performance as time approaches infinity (i.e., $t \rightarrow \infty$).

Transient analysis is used to evaluate the dynamic response of a system under time-varying system parameters. It is useful for a system for which most of the time system parameters (e.g., inputs) are time varying due to both internal and external factors, and therefore, the steady state may never be reached. Based on transient analysis, it is possible to investigate system characteristics as follows:

- *Fluctuation in the transient behavior*: When the system parameters change, the system performance measures will vary, which can be different from those in the steady state. For example, in a second-order dynamic system, the performance measures may oscillate before reaching the steady state. It is desirable that the transient fluctuations of the system performance measures do not go beyond the respective upper and lower bounds.
- *Time-dependent system behavior*: In contrast to the steady state analysis, which gives the system performance measure when the system has been observed for a long period of time (theoretically as time approaches infinity), transient analysis can provide a set of performance measures associated with a particular time.
- *Time to reach the steady state*: This is one of the crucial characteristics for a dynamic system. In general, the faster the system reaches the steady state, the better is the performance. Transient analysis can be used to investigate the rate of convergence to the desirable level of the performance measure after the system parameters have changed or been adjusted.
- *Impact of initial condition*: All of the aforementioned system characteristics depend strongly on the initial condition of the system. The system may reach the steady state quickly if the initial condition is set properly. Otherwise, the system behavior may fluctuate and never reach the steady state, for instance, if the initial condition is chosen too far from the region of convergence.

An example of transient and steady state performance measures (i.e., new call blocking and handoff call dropping probabilities in a cellular wireless network) are shown in Figure 4.1 [2]. Here, the system is in transient state for a certain period of time (i.e., about 15 minutes). In the transient state, new call blocking and handoff call dropping probabilities are smaller than those in the steady state (e.g., at $t = 4$, these probabilities in transient state are only half of those in steady state). Therefore, to obtain the accurate system performance measures for a dynamic system, the transient state performances need to be considered. Based on the system characteristics, an adaptive control can be applied (Figure 4.2), for example, to maintain the output in the operational range or to increase the speed of convergence to the system steady state. In the context of the system model considered in [2], the inputs to the control system (in Figure 4.2) are the new call and handoff call arrival rates, and the transient state performance measures (i.e., outputs) are new call blocking and handoff call dropping probabilities. The objective of the controller is to minimize the difference between new call blocking probabilities in two consecutive time periods, while the handoff call dropping probability is maintained below the target level. In this case, the feedback control is used to reserve some channel capacity for handoff calls.

In this chapter, we present an analytical model to investigate the transient QoS performances in a multiservice CDMA mobile network where the call admission control mechanism prioritizes handoff calls over new calls by reserving a portion of radio

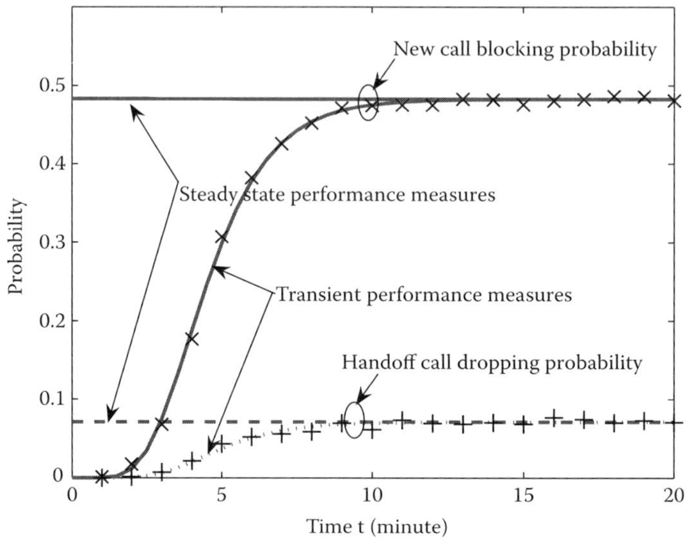

FIGURE 4.1 Example of transient and steady state performance [2].

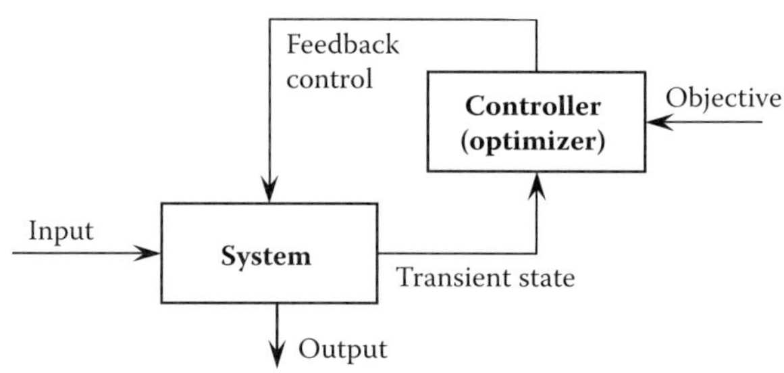

FIGURE 4.2 Adaptive control.

resources (e.g., transmission power) in a cell for handoff calls. The model assumes different channel holding times for new calls and handoff calls. Since the traffic in a cell is time dependent in nature, the system may not reach the steady state within a radio resource allocation update interval. In such an environment, the transient analysis model is able to provide time-dependent performance measures that are crucial to design, analysis, and engineering of CAC and dynamic resource allocation mechanisms. The transient performance measures analyzed in this chapter are call blocking and handoff call dropping probabilities, and average transmission rate (for elastic service). To this end, an application of the analytical model is presented to adaptively optimize the radio resource management parameters so that the system revenue can be maximized. Specifically, an optimization problem is formulated to maximize system revenue over time while maintaining the ratio of new call blocking probability and handoff call dropping probability at the desired level.

4.2 Related Works

4.2.1 Resource Allocation in CDMA System

CDMA has emerged as the preferred radio transmission technology for the next-generation cellular wireless networks. Resource allocation (i.e., power and transmission rate allocation) in CDMA networks is a crucial problem and needs to be carefully designed to maximize network capacity. Also, admission control is necessary to ensure that the system is not overloaded due to the acceptance of too many calls, and that the QoS requirements of ongoing and accepting calls can be guaranteed. Radio resource allocation and admission control problems in CDMA cellular networks were studied extensively in the literature [3–11]. These problems were also addressed in [12–16] considering multiservice scenarios and different QoS requirements for different types of services. In addition, the optimal resource allocation problem in an integrated CDMA cellular network and wireless LAN (WLAN) was addressed in [17].

4.2.2 Applications of Transient Analysis

Transient analysis based on Markov models was used to compute the different performance measures for reliability analysis [18, 19]. Transient analysis for cellular systems was used in [20] to investigate time-dependent packet-level performance measures (e.g., packet blocking probability). In [2], transient analysis based on the *uniformization method* was used to investigate the performance of hard-capacity cellular mobile networks under time-varying traffic. Also, an adaptive resource reservation scheme for handoff calls was proposed based on the transient analysis.

Transient QoS performances in nonadaptive CDMA cellular networks were analyzed in [21–23]. An admission control mechanism based on transient outage probability was also proposed in [21]. These transient performance measures were obtained by using a fluid-flow approximation model and were used to adjust the call admission control policy. A similar approach was used in [22] for CDMA systems with a smart antenna. The user traffic was assumed to follow an on-off model. Again, transient outage probability was used as a performance measure to decide whether an incoming call could be accepted into the system or not. In [23], for a CDMA system with video/voice/data services, an access control scheme was proposed. The transient state performance was used to predict the residual capacity of the system, and the access control was performed based on this capacity information.

However, the analytical models for CDMA systems used in all of the above works were developed assuming identical channel holding times for both new calls and handoff calls, which may not be true due to the the mobility of the users [24]. Also, these works did not consider any prioritization between handoff calls and new calls for radio resource allocation and call admission control. Again, the problem of optimizing the radio resource management parameters was ignored. Performance evaluation of call admission control and resource reservation schemes for CDMA cellular networks was carried out in [25, 26]. In [27], an integrated analytical model was proposed to obtain both the call-level and packet-level performance measures in a rate-adaptive CDMA system.

In such a system, the transmission rates of ongoing calls can be reduced to achieve larger system capacity so that the call blocking and call dropping probabilities are minimized. However, transient performance analysis was not carried out in this work.

4.2.3 Analytical Models for Optimized Resource Allocation in CDMA Networks

Based on an analytical formulation for QoS performance evaluation, optimization of resource allocation in a CDMA network can be performed. In [27], user utility, which is a function of perceived performances, was optimized to maximize users' satisfaction. A cross-layer strategy for optimal admission control for multimedia services in the CDMA system was proposed in [28]. This optimal admission control was designed based on stochastic optimization—specifically, as a Markov decision process (MDP). In this MDP formulation, the system state was defined as the number of ongoing calls in each service class, and the actions were defined in terms of accepting or rejecting an incoming call. The objective of this optimization formulation was to minimize the connection blocking probability (i.e., maximize network utilization) while maintaining the target level the SIR performances for both ongoing and incoming calls. Similar optimization formulations can be found in [29, 30].

4.3 Mathematical Preliminaries: Transient Analysis of a Markov Process

4.3.1 Continuous-Time Markov Chain

A continuous-time Markov chain (CTMC) is generally used to model and analyze the behavior of a system that possesses the Markov property. This Markov property states that the future state of the system depends only on the current state, not on any past state. The evolution of a CTMC is defined as

$$\mathbb{X}=\{\mathcal{X}(t), t \geq 0\}, \tag{4.2}$$

where $\mathcal{X}(t)$ is the state of the system at time t. This state $\mathcal{X}(t)$ is in a set of state space Ψ that could be finite or infinite. The transition probability from state i at time t to state j at time $t + 0$, where 0 is a small increment in time, can be defined as follows:

$$Pr\{\mathcal{X}(t+\Delta)=j \mid \mathcal{X}(t)=i\}=q_{i,j}\Delta+O(\Delta), \tag{4.3}$$

where $O(.)$ is the order notation, and $q_{i,j}$ denotes the rate of transition from state i to state j where $j \neq i$. The transition matrix of CTMC with a finite state space $\Psi = \{1, \cdots, N\}$ can be defined as follows:

$$\mathbf{Q}=\begin{bmatrix} q_{1,1} & q_{1,2} & \cdots & q_{1,N} \\ q_{2,1} & q_{2,2} & \cdots & q_{2,N} \\ \vdots & \vdots & \cdots & \vdots \\ q_{N,1} & q_{N,2} & \cdots & q_{N,N} \end{bmatrix}. \tag{4.4}$$

The probability that the system is in state j at time t is denoted by

$$\pi_j(t) = Pr\{\chi(t) = j\}. \tag{4.5}$$

The matrix corresponding to the probability that the system is in any state can be expressed as follows:

$$\pi(t) = \begin{bmatrix} \pi_1(t) & \pi_2(t) & \cdots & \pi_N(t) \end{bmatrix}. \tag{4.6}$$

At the steady state (i.e., when $t \to \infty$), the probability that the system is in state j is given by

$$\pi_j = \lim_{t\to\infty} \pi_j(t). \tag{4.7}$$

The steady state probability matrix can be expressed as

$$\pi = \begin{bmatrix} \pi_1 & \pi_2 & \cdots & \pi_N \end{bmatrix}. \tag{4.8}$$

Based on the transition matrix **Q**, the steady state probability matrix can be obtained by solving the following set of equations:

$$\pi\mathbf{Q} = 0 \quad \text{and} \quad \sum_{j=1}^{N} \pi_j = 1. \tag{4.9}$$

4.3.2 Transient Analysis of CTMC

Let us consider a system with finite state space. The transient state probability vector $\pi(t)$ can be obtained from a transition function, which is denoted by matrix $\mathbf{M}(t)$. This transition function $\mathbf{M}(t)$ satisfies both Kolmogorov's backward and Kolmogorov's forward equations [31], defined as follows:

$$\frac{\mathrm{d}\mathbf{M}(t)}{\mathrm{d}t} = \mathbf{Q}\mathbf{M}(t) \tag{4.10}$$

$$\frac{d\mathbf{M}(t)}{dt} = \mathbf{M}(t)\mathbf{Q}, \tag{4.11}$$

and the transient state probability vector $\pi(t)$ can be obtained from this transition function as follows:

$$\pi(t) = \pi(0)\mathbf{M}(t). \tag{4.12}$$

The popular methods used to obtain the transition function $\mathbf{M}(t)$ are based on ordinary differential equation, exponential of a transition matrix, and uniformization [32].

4.3.2.1 Ordinary Differential Equation Approach

An ordinary differential equation (ODE) can be defined as

$$\frac{dm(t)}{dt} = m'(t) = f(t,m). \tag{4.13}$$

An ODE can be written in matrix form as $\mathbf{M}'(t) = \mathbf{QM}(t)$, which is Kolmogorov's backward equation.

Euler's method [33] can be used to obtain a solution of this ODE. In this method, the observation interval $(0,t)$ is divided into subintervals of length h, and the value at the current subinterval (i.e., $\mathbf{M}(ih)$, where $i = 1, 2, \ldots$ denotes the index of subinterval) is approximated from the value at the previous subinterval (i.e., $\mathbf{M}((i-1)h)$) and its derivative. This approximation can be expressed as follows:

$$\mathbf{M}(t) = (\mathbf{I} + h\mathbf{Q})\mathbf{M}(t-h), \tag{4.14}$$

where $\mathbf{M}(0)$ is an initial transition function and generally chosen to be $\mathbf{M}(0) = \mathbf{I}$. Equation (4.14) denotes a single-step approximation. The accuracy of the computation can be improved by using a multistep approximation, which can be defined as follows:

$$\mathbf{M}(t) = \left(\mathbf{I} + h\mathbf{Q} + \frac{h^2}{2!}\mathbf{Q}^2 + \frac{h^3}{3!}\mathbf{Q}^3 + \cdots\right)\mathbf{M}(t-h). \tag{4.15}$$

The advantage of Euler's method is the implementation simplicity, since the solution can be obtained directly from (4.14) and (4.15). However, the major disadvantage of this method is the computational complexity, which is relatively high, since the length h has to be small so that the acceptable accuracy of the solution can be maintained.

4.3.2.2 Matrix Exponential Approach

The simplest solution of the Kolmogorov-backward equation can be defined as

$$\mathbf{M}(t) = e^{\mathbf{Q}t}, \tag{4.16}$$

where e^{QT} can be defined as follows:

$$e^{\mathbf{Q}t} = \sum_{i=0}^{\infty} \mathbf{Q}^i \frac{t^i}{i!}. \tag{4.17}$$

There are several methods to obtain the exponential of a matrix [34]. For example, the simplest method is to compute (4.17) and truncate the summation to a large number (e.g., K). In this method, the transition function can be expressed as follows:

$$\mathbf{M}(t) = \sum_{i=0}^{K} \mathbf{N}_i(t) + \mathbf{E}(K), \tag{4.18}$$

where the error due to truncation of summation is determined by $\mathbf{E}(K)$, and $\mathbf{N}_i(t)$ is recursively defined by

$$\mathbf{N}_i(t) = \mathbf{N}_{i-1}(t)\mathbf{Q}\frac{t}{i}. \tag{4.19}$$

An alternative to compute the exponential of a matrix is to use transformation of $\mathbf{Q}$. In this method, it is assumed that $\mathbf{Q}$ is diagonalizable, and can be written in Jordan canonical form [35] as follows:

$$\mathbf{Q} = \mathbf{U}\mathbf{V}\mathbf{U}^{-1}, \tag{4.20}$$

where $\mathbf{V}$ is a diagonal matrix of eigenvalues λ_i of transition matrix $\mathbf{Q}$. Then, the transition function can be written as follows:

$$\mathbf{M}(t) = \mathbf{U}\left(\sum_{i=0}^{\infty} \mathbf{V}\frac{t^i}{i!}\right)\mathbf{U}^{-1} \tag{4.21}$$

$$= \mathbf{U}e^{\mathbf{V}t}\mathbf{U}^{-1}. \tag{4.22}$$

Since $\mathbf{V}$ is diagonal, the solution can be obtained as follows:

$$e^{\mathbf{V}t} = \begin{bmatrix} e^{\lambda_1 t} & & & \\ & e^{\lambda_2 t} & & \\ & & \ddots & \\ & & & e^{\lambda_N t} \end{bmatrix}. \tag{4.23}$$

However, most of the existing techniques to compute the exponential of a matrix suffer from the truncation and numerical round-off errors, which could be problematic for practical applications that require a high degree of accuracy for the solution.

4.3.2.3 Uniformization Approach

The uniformization method [36] (also called randomization) is the most attractive technique to analyze the transient behavior of a continuous-time Markov chain due to its numerical stability and the fact that the computation error can be well controlled and determined in advance.

The transient state probability can be obtained from

$$\pi(t)=\pi(0)\sum_{i=0}^{\infty} e^{-\Lambda t}\frac{(\Lambda t)^i}{i!}\mathbf{P}^i, \tag{4.24}$$

where $\mathbf{P} = \mathbf{I} + \mathbf{Q}/\Lambda$ and $\Lambda \geq \max_i(q_{i,i})$. One of the most attractive features of the uniformization method is that the error (which is controllable) can be computed from

$$\text{error}=1-e^{\Lambda t}\sum_{i=0}^{K}\frac{(\Lambda t)^i}{i!}. \tag{4.25}$$

Therefore, K can be chosen so that this error remains smaller than the target value.

The uniformization technique was further improved in many works [37, 38]. For example, an enhanced computation method for a large continuous-time Markov chain was proposed in [39]. This method is based on a randomization technique and the Kronecker product representation of probability vectors, and it results in significant reduction of the space and time requirements for computation of the transient performance measures. Adaptive uniformization methods were proposed to analyze the transient state behavior of a Markov chain in [40, 41]. The major advantage of adaptive uniformization is that fewer iterations are required to obtain the solution compared with the standard uniformization method. Adaptive and standard uniformization techniques were combined to improve the performance and accuracy of the solution further in [42].

4.4 Adaptive Resource Allocation in CDMA Cellular Wireless Mobile Networks Based on Transient Analysis: Methodology and Modeling Assumptions

4.4.1 Methodology

Transient analysis can be used for resource allocation and admission control in a cellular CDMA network. The general approach is shown in Figure 4.3. First, the parameters (e.g., new call and handoff call arrival rates) of the system are estimated. These parameters, which are generally time dependent, will be used to construct the Markov chain. Then, the estimated system parameters are used in the transient analysis. The time-dependent transition matrix will be established and solved by using a standard technique (e.g., uniformization) to obtain the transient performance measures. These performance measures will be used in an optimization formulation for resource allocation to make decisions on reserved radio resources and admission decisions so that the target QoS can be achieved in a specific period of time.

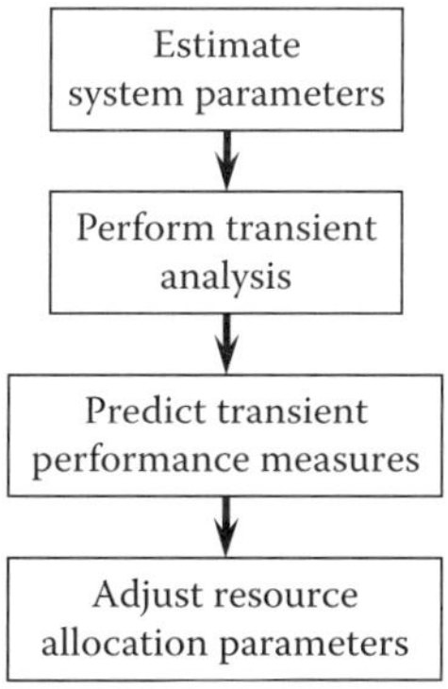

FIGURE 4.3 Steps in adaptive resource allocation based on transient analysis.

4.4.2 Service and Traffic Models

We consider an uplink transmission scenario in a cell in a cellular CDMA network that supports elastic and inelastic types of services. For inelastic service, the transmission rate is fixed during the entire duration of the call. On the other hand, for calls using elastic type of service, the transmission rate can be adaptively adjusted according to the traffic load condition in a cell. This will allow more calls to be admitted into the system.* At the base station, a call admission control mechanism is used to decide whether an incoming call (new call or handoff call) can be accepted. Note that admission control and rate adaptation algorithms are executed independently for each type of service.

We assume that the call arrival process in a cell for both new calls and handoff calls follows Poisson distribution with respective average rates $\rho_n^{(c)}$ and $\rho_h^{(c)}$ for service class *c*. The channel holding times for new calls and handoff calls are assumed to be exponentially distributed with means $1/\mu_n^{(c)}$ and $1/\mu_h^{(c)}$, respectively. However, since the call arrival rates and the channel holding times may depend on the time of the day (e.g., the same call arrival rate during a particular time of the day), we consider time-dependent call arrival rate and channel holding time in which their patterns periodically repeat every day [43]. The average call arrival rates and channel holding times at a particular time (or time interval) *t* of the day are denoted by $\rho_n^{(c)}(t)$, $\rho_h^{(c)}(t)$, $1/\mu_n^{(c)}(t)$, and $1/\mu_h^{(c)}(t)$, respectively. Since the call arrival rate and the channel holding time are time varying, using steady state analysis may not provide accurate results [2]. Therefore, transient state analysis would be required for a time interval that considers the variations in system parameters (e.g., arrival rate and resource allocation) in the previous time interval. In addition, to adapt to the variation in traffic load, the call admission control parameters should be

* This is due to the fact that as the transmission rate is decreased, for a fixed transmission power, the achieved SIR at the receiver becomes higher than that required to sustain a certain bit error rate (BER) requirement.

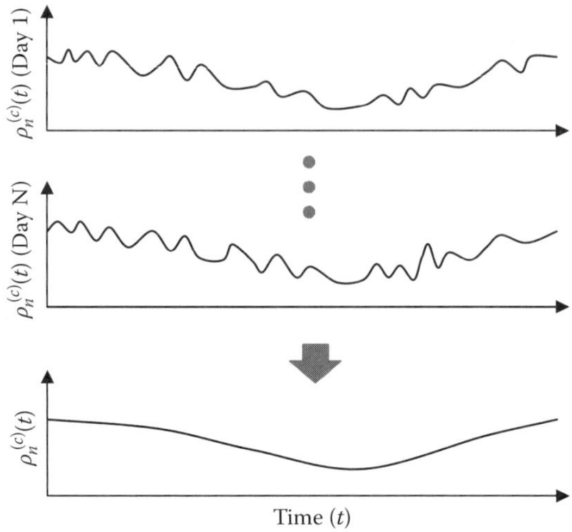

FIGURE 4.4 Periodic traffic pattern.

tuned accordingly. Note that the traffic (i.e., call) arrival parameters for a particular time interval can be obtained by averaging the call arrival rate during that particular time interval for several days (e.g., as shown in Figure 4.4).

4.4.3 Modeling Cell Capacity

In a multiservice scenario, the ratio $\tilde{\delta}_c$ of the received power from a mobile in service class c ($c \in \{1, 2, \ldots, C\}$) to the total interference power received at the base station can be expressed as follows [44]:

$$\tilde{\delta}_c = \frac{Pow(c)}{I_N + I_{own} + I_{other} - Pow(c)} \quad c \in \{1, 2, \ldots, C\}, \tag{4.26}$$

where $Pow(c)$ is the received power (at base station) corresponding to a call in service type c, I_N is the background noise, I_{own} is the total power received from other mobiles within the same cell, and I_{other} is the total power received from mobiles in other cells. I_{own} can be obtained from

$$I_{own} = \sum_{c=1}^{C} M_c Pow(c),$$

in which M_c denotes the number of ongoing calls in service class c. I_{other} can be simplified as follows [45]: $I_{other} = \zeta \times I_{own}$, where ζ denotes the outer-cell interference factor, which can be obtained from measurements. Again, to sustain a target BER, the SIR requirement at the base station receiver can be expressed as

$$\tilde{\delta}_c = \frac{E_b^{(c)}}{I_o^W} R_c,$$

where $E_b^{(c)}$ is the energy per bit, I_o is the interference-plus-noise power density, W is the spread spectrum bandwidth, and R_c is the transmission rate (in bits per second) for a call in service class c.

Let us define another term, δ_c (which was referred to as *nominal capacity* in [46]), as follows:

$$\delta_c = \frac{Pow(c)}{I_N + I_{own} + I_{other}}, \quad \text{where} \quad \delta_c = \frac{\tilde{\delta}_c}{1+\tilde{\delta}_c} \Longleftrightarrow \tilde{\delta}_c = \frac{\delta_c}{1-\delta_c}. \tag{4.27}$$

Rewriting (4.26), we have

$$Pow(c) = \frac{I_N \delta_c}{1-(1+\zeta)\sum_{j=1}^{C} M_j \delta_j}.$$

In order to prevent the transmission power for a call in service class c from going to infinity, the condition

$$\sum_{c=1}^{C} M_c \delta_c < \frac{1}{1+\zeta} \tag{4.28}$$

needs to be satisfied. Therefore, the *pole capacity*, which is defined as the maximum number of ongoing calls for the different service types, can be expressed as a vector $\mathbf{M}^*$ as follows:

$$\mathbf{M}^* = \left\{ M_c^* : 1 = (1+\zeta)\sum_{c=1}^{C} M_c \delta_c \right\}.$$

We define $\gamma_c = M_c^* \delta_c$ as the *normalized power*, which represents the amount of transmission resource allocated to service class c. Note that the total available normalized power is completely partitioned among the different service classes.

4.4.4 Call Admission Control (CAC) and Transmission Rate Adaptation

A CAC algorithm is used to prevent a cell from becoming overloaded as well as to prioritize handoff calls over new calls. For the former condition, the CAC decision is based on the condition in (4.28). For the latter condition, a guard channel scheme [47] is used to reserve a portion of the normalized power for handoff calls. Let $\gamma_c^{(h)}$ denote the amount of normalized power reserved for handoff calls in service class c.

For elastic service, decreasing (i.e., slowing down) the transmission rate can increase the number of ongoing calls in that service class (and hence the pole capacity). Let a_c ($a_c \in \{1, 2, \ldots, A_c\}$), where $A_c > 1$, indicate the level of slow-down in the transmission rate of elastic calls in service class c. The slow-down factor and the transmission rate are related by

$$R_c = \frac{R_p^{(c)}}{a_c},$$

where $R_p^{(c)}$ is the maximum transmission rate for calls in elastic service class c. Therefore, the nominal capacity can be represented as a function of the slow-down factor as follows:

$$\delta_c(a_c) = \frac{\tilde{\delta}_c}{a_c + \tilde{\delta}_c} = \frac{\delta_c(1)}{a_c\left(1-\delta_c(1)\right)+\delta_c(1)}, \tag{4.29}$$

where $\delta_c(a_c)$ is the nominal capacity when the slow-down factor is a_c and $\delta_c(1)$ is the nominal capacity for the highest transmission rate.

For an elastic service class, the slow-down factor a_c is initially set to 1. When a call arrives, the call admission control and rate adaptation procedure (in Figure 4.5) is invoked. In this algorithm, N_c denotes the number of ongoing calls in service class c. When a call terminates or is handed over to another cell, the number of ongoing calls in the cell decreases. Therefore, if service class c is elastic and the condition $N_c \times \delta_c(a_c - 1) \leq \gamma_c(t)$ holds, the transmission rate for elastic calls can be increased by adjusting a_c as follows: $a_c \leftarrow a_c - 1$.

4.5 Queueing Analytical Model for Transient Analysis and Its Application

4.5.1 Transition Matrix

Based on the aforementioned system model, we can model the evolution of the number of ongoing calls in class c as a *quasi-birth and death* (QBD) process, and the two-dimensional state space is defined as follows:

$$\Psi = \left\{(\mathcal{N}_c(t), \mathcal{H}_c(t)); 0 \leq \mathcal{N}_c(t)\delta_c(A_c) \leq \gamma_c(t) - \gamma_c^{(h)}(t), 0 \leq \mathcal{H}_c(t)\delta_c(A_c) \leq \gamma_c(t)\right\}, \tag{4.30}$$

where $\mathcal{N}_c(t)$ and $\mathcal{H}_c(t)$ denote the number of ongoing new calls and handoff calls, respectively, at a particular time t of the day. The transition diagram of this Markov chain is shown in Figure 4.6.

```
Call Admission Control and Rate Adaptation Algorithm
if ((N_c + 1) × δ_c(a_c) ≤ γ_c(t)) then
  if (incoming call is a handoof call) then
    accept the incoming call
    return
  else
    if ((N_c + 1) × δ_c(t) ≤ γ_c(t) - γ_c^(h)(t)) then
      accept the incoming new call
      return
    end if
  end if
end if
if ((service class c is elastic) AND (a_c < A_c)) then
  a_c = a_c + 1
else
  reject incoming call
  return
end if
Go to 1
```

FIGURE 4.5 Call admission control and rate adaptation algorithm.

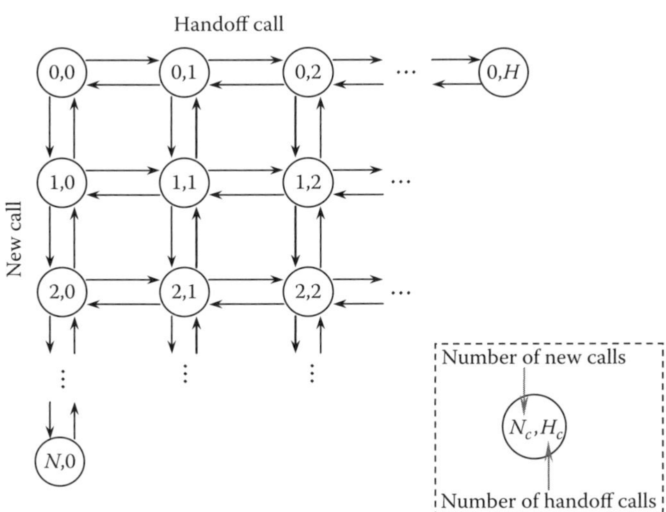

FIGURE 4.6 Transition diagram of the Markov chain.

The time-dependent transition matrix of this model can be expressed as follows:

$$\mathbf{Q}(t)=\begin{bmatrix} \mathbf{q}_{0,0}(t) & \mathbf{q}_{0,1}(t) & & & \\ \mathbf{q}_{1,0}(t) & \mathbf{q}_{1,1}(t) & \mathbf{q}_{1,2}(t) & & \\ \ddots & \ddots & \ddots & & \\ & \mathbf{q}_{N-2,N-1}(t) & \mathbf{q}_{N-1,N-1}(t) & \mathbf{q}_{N-1,N}(t) \\ & & \mathbf{q}_{N,N-1}(t) & \mathbf{q}_{N,N}(t) \end{bmatrix}, \tag{4.31}$$

where N denotes the maximum number of ongoing new calls

$$\left(\text{i.e., } N=\left\lfloor \frac{\gamma_c(t)-\gamma_c^{(h)}(t)}{\delta_c(A_c)} \right\rfloor\right)$$

in service class c. The element $\mathbf{q}_{u,u'}(t)$ corresponds to the change in the number of ongoing new calls from u to u'. The elements inside $\mathbf{q}_{u,u'}(t)$ represent the number of ongoing handoff calls. This matrix $\mathbf{q}_{u,u'}(t)$ can be classified into three cases. First, $\mathbf{q}_{u,u-1}$ represents the case when a new call departs the cell. The diagonal elements of this matrix are obtained as follows:

$$\left[\mathbf{q}_{u,u-1}(t)\right]_{e+1,e+1}=u.\mu_n^{(c)}(t), \quad 0\leq e\leq H-u, \tag{4.32}$$

where H denotes the maximum number of ongoing handoff calls

$$\left(\text{i.e., } H=\left\lfloor \frac{\gamma_c(t)}{\delta_c(A_c)} \right\rfloor\right),$$

and

$$\left[\mathbf{q}_{u,u-1}(t)\right]_{e+1,e'+1}$$

denotes the element at row $e + 1$ and column $e' + 1$ of matrix $\mathbf{q}_{u,u-1}(t)$, and it corresponds to the state where the number of ongoing handoff calls changes from e to e'. Second, $\mathbf{q}_{u,u+1}(t)$ represents the case that a new call arrives. The diagonal elements of these matrices are obtained as follows:

$$\left[\mathbf{q}_{u,u+1}(t)\right]_{e+1,e+1}=\rho_n^{(c)}(t), \quad \text{for } 0\leq e\leq H-u. \tag{4.33}$$

Third, $\mathbf{q}_{u,u}(t)$ represents the case that the number of ongoing new calls does not change. This matrix also represents the change in the number of ongoing handoff calls and can be expressed as in (4.34) to (4.36), for the cases when a handoff call departs, arrives, and the number of ongoing handoff calls does not change, respectively:

$$\left[\mathbf{q}_{u,u}(t)\right]_{e+1,e} = e.\mu_h^{(c)}(t), \quad \text{for } 1 \le e \le H-u \tag{4.34}$$

$$\left[\mathbf{q}_{u,u}(t)\right]_{e+1,e+2} = \rho_h^{(c)}(t), \quad \text{for} \quad 0 \le e \le H-1-u \tag{4.35}$$

$$\left[\mathbf{q}_{u,u}(t)\right]_{e+1,e+1} = -\sum_{j \ne e+1}\left[\mathbf{q}_{u,u}(t)\right]_{e+1,j} - \sum_{\forall j}\left[\mathbf{q}_{u,u+1}(t)\right]_{e+1,j} - \sum_{\forall j}\left[\mathbf{q}_{u,u-1}(t)\right]_{e+1,j} \tag{4.36}$$

for $0 \le e \le H - u$.

4.5.2 Transient Analysis

The transient state probabilities at time t depend on the initial state probability at time zero. For a service class, let $P_{n,h}(t) = \Pr\{\mathcal{N}_c(t) = n, \mathcal{H}_c(t) = h | \mathcal{N}_c(0) = n0, \mathcal{H}_c(0) = h0\}$ denote the probability that there are n ongoing new calls and h ongoing handoff calls at time t given the initial state $P_{n_0,h_0}(0)$ at time zero. Let $\pi(t) = [\pi_0(t) \ \dots \ \pi_i(t) \ \dots \ \pi_Q(t)]$ denote a row vector of state probabilities at time t, where Q is the size of the matrix $\mathbf{Q}(t)$. The probability of having n and h number of ongoing new calls and handoff calls in a service class at time t can be obtained from this matrix as follows:

$$P_{n,h}(t) = \pi_{col(n,h)}(t), \text{ where } col(n,h) = \left(\sum_{i=0}^{n} H - i + 1\right) + h. \tag{4.37}$$

To calculate $\pi(t)$, we solve the following Kolmogorov-forward equation:

$$\frac{d\pi_i(t)}{dt} = -\pi_i(t)\left[\mathbf{Q}\right]_{i,i}(t) + \pi_i(t)\left(\sum_{j \ne i}\left[\mathbf{Q}\right]_{i,j}\right), \tag{4.38}$$

which can be expressed in the matrix form as follows:

$$\frac{d\pi(t)}{dt} = \pi(t)\mathbf{Q}(t). \tag{4.39}$$

By solving (4.39) based on the assumption that the elements in the generator matrix are constant during time interval $[0,t)$ (i.e., $\mathbf{Q}(t) = \mathbf{Q}$), we have $\pi(t) = \pi(0)e^{\mathbf{Q}t}$.

To obtain the transient performance measures using the *uniformization* technique [48] (also called *Jensen's method* or *randomization method*) the new transition probability matrix $\mathbf{P}$ is obtained from transition matrix $\mathbf{Q}$ as follows:

$$\mathbf{P}=\frac{\mathbf{Q}}{\Lambda}+\mathbf{I}, \tag{4.40}$$

where $\Lambda \geq \min_i(|q_{i,i}|)$ and $\mathbf{I}$ is an identity matrix. Therefore, we have

$$\pi(t)=\pi(0)e^{(\mathbf{P}-\mathbf{I})\Lambda t}=\pi(0)e^{\mathbf{P}\Lambda t}e^{-\Lambda t}=\pi(0)\sum_{i=0}^{\infty}\mathbf{P}^i\frac{(\Lambda t)^i}{i!}e^{-\Lambda t}. \tag{4.41}$$

However, to reduce the computational time, the index of the summation above is limited to $i = K$ so that the error remains below ε. That is,

$$\pi(t)=\pi(0)\sum_{i=0}^{K}\mathbf{P}^i\frac{(\Lambda t)^i}{i!}e^{-\Lambda t}, \quad \text{where} \quad 1-e^{-\Lambda t}\sum_{i=0}^{K}\frac{(\Lambda t)^i}{i!}\leq\varepsilon. \tag{4.42}$$

With time-dependent call arrival rates and channel holding times, let $\mathbf{P}(t_i)$ denote the transition probability matrix of the process $\mathbf{Q}(t_i)$, whose elements are constant during $[t_i,t_{i-1})$. After uniformization, we obtain the transient probability at time t_i conditioned on the previous time t_{i-1} as follows:

$$\pi(t_i)=\pi(t_{i-1})\sum_{i=0}^{K}\mathbf{P}^i(t_i)\frac{(\Lambda t)^i}{i!}e^{-\Lambda t}. \tag{4.43}$$

Based on the transient state probability $\pi(t)$, various transient state performance measures can be obtained.

4.5.3 Performance Measures

4.5.3.1 Average Number of Ongoing Calls

The average number of ongoing new calls $\bar{n}_c(t)$ and handoff calls $\bar{h}(t)$ in a service class at a particular time t can be obtained from

```
\begin{equation}
\overline{n}(t) = \sum_{n=1}^{N_n} \sum_h n \times P_{n,h} (t), \quad
\overline{h}(t) = \sum_{h=1}^N_h \sum_n h \times P_{n,h}(t).
\end{equation}
where $N_n = \left \lfloor \frac{ \gamma_c (t) - \gamma^{(h)}_c(t)
}{\delta_c (A_c)} \right \rfloor$ and $N_h = \left \lfloor \right
\frac{\gamma_c(t)}{\delta_c(A_c)} \rfloor$.
```

(4.44)

4.5.3.2 New Call Blocking and Handoff Call Dropping Probabilities

For a service class, these performance measures at time t can be obtained as follows:

$$P_{bl}(t)=\left(\sum_{n+h\geq\left\lfloor\frac{\gamma_c(t)-\gamma_c^{(h)}(t)}{\delta_c(A_c)}\right\rfloor} P_{n,h}(t)\right),\quad P_{dr}(t)=\left(\sum_{n+h\geq\left\lfloor\frac{\gamma_c(t)}{\delta_c(A_c)}\right\rfloor} P_{n,h}(t)\right). \tag{4.45}$$

Based on $P_{bl}(t)$ and $P_{dr}(t)$ the observed grade of service for a service class at time t is defined as

$$GoS(t)=P_{bl}(t)+\omega P_{dr}(t), \tag{4.46}$$

where the parameter ($\omega > 1$) can be adjusted based on the priority given to handoff calls over new calls.

4.5.3.3 Average Transmission Rate

The average transmission rate for a call in an elastic service class with maximum rate R_p can be obtained from

$$\bar{R}(t)=R_p\sum_{a=1}^{A}(A-a+1)\left(\sum_{S(a)}P_{n,h}(t)\right),$$

where

$$S(a)=\left\{(n,h)\left\lfloor\frac{\gamma(t)(a-1)}{\delta}\right\rfloor<n+h\leq\left\lfloor\frac{\gamma(t)(a)}{\delta}\right\rfloor\right\}. \tag{4.47}$$

4.5.4 Application of the Transient Analysis Model to Optimize the CAC Parameters

We illustrate an application of the transient analysis model to obtain the optimal normalized power allocation at time t. An optimization problem is formulated as follows:

$$\text{Maximize:}\sum_{c=1}^{C}\phi_c(t)\left(\bar{n}_c(t)+\bar{h}_c(t)\right)\frac{\bar{R}_c(t)}{R_p^{(c)}} \tag{4.48}$$

$$\text{Subject to } P_{bl}^{(c)}(t)\geq\omega P_{dr}^{(c)}(t) \tag{4.49}$$

where the objective is to maximize average revenue during interval t, and $\phi_c(t)$ denotes the revenue for a call in service class c. Also, the average revenue is calculated by considering the normalized transmission rate

$$\left(\text{i.e., } \frac{\bar{R}_c(t)}{R_p^{(c)}}\right).$$

Specifically, if service class c is elastic, the average revenue is proportional to the call's transmission rate. The constraint in (4.49) ensures that the ratio between handoff call dropping probability and new call blocking probability remains below a target threshold. This optimization problem can be solved by enumeration.

4.6 Performance Evaluation

4.6.1 Parameter Setting

We consider three service classes (i.e., $C = 3$): voice, constant-bit-rate (CBR) data, and best-effort data (which are indicated by indices (v), (d), and (b), respectively). While voice and CBR data services are inelastic, best-effort data service is elastic. We assume $\zeta = 0.02$, $\delta_{(v)} = 0.01$, $\delta_{(d)} = 0.05$, $\delta_{(b)} = 0.20$, and $E_b/I_0 = 5$ dB for voice service, and $E_b/I_0 = 6.99$ dB for CBR data and best-effort data service (at the maximum transmission rate). The transmission rate for a CBR data call is 50 Kbps. The maximum transmission rate for a best-effort data call is 200 Kbps, and the maximum slow-down factor is $A_{(b)} = 10$. We assume that the average channel holding times for a new voice call, a CBR data call, and a best-effort data call are 4, 9, and 14 minutes, respectively, and those for the handoff calls are 5, 12, and 16 minutes, respectively.

The new call arrival rates (number of calls per minute) for the different service classes are defined as functions of time (for an observation period of 300 minutes) as follows: $\rho_n^{(v)}(t) = 0.4 + 0.5(1 + cos(2\pi t/350))$, $\rho_n^{(d)}(t) = 0.05 + 0.1(1 + cos(2\pi t/500 + 350))$, and $\rho_n^{(d)}(t) = 0.2 + 0.4(1 + cos(2\pi t/200 + 300))$. We assume that the handoff call arrival rate is half of the new call arrival rate. Each time interval during the observation period has a length of 15 minutes; therefore, the total number of intervals for which the transient state performances are measured is 20. Also, we assume that $\phi_{(v)} = 1$, $\phi_{(d)} = 4$, and $\phi_{(b)} = 2$.

4.6.2 Numerical Results

We investigate the transient behavior of the normalized power adaptation for voice call. The new voice call arrival rate is defined as $1.5\cos\left(\frac{2\pi t}{100}\right)+1.5$, and handoff call arrival rate is half of the new call arrival rate. Normalized power $\gamma_{(v)}^{(h)}(t) = 0.02$ is reserved for the handoff call. At time zero, there is no ongoing call (i.e., $\pi_1(0) = [1 \quad 0 \quad \ldots \quad 0]$). We adjust $\gamma_{(y)}(t)$ to observe the transient state performance measures. In particular, $\gamma_{(v)}(t)$ is initialized to be 0.2 at time zero. Since the arrival rate decreases during time period $t = [30, 70]$, the normalized power allocated for voice service can be reduced and given to other service classes. In the first scenario, we set $\gamma_{(v)}(t) = 0.19$ for $t = 40, \ldots, 60$. We adjust

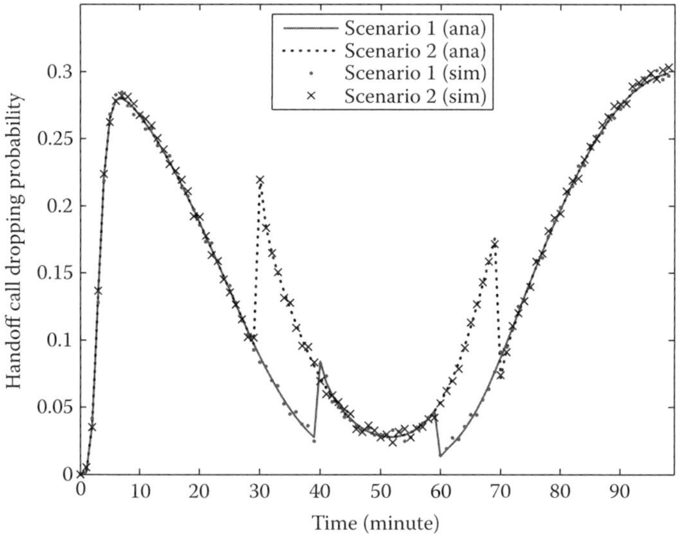

FIGURE 4.7 Transient behavior resulting from adjustment of normalized power.

$\gamma_{(v)}(t) = 0.19$ for $t = 30, \ldots, 70$. Figure 4.7 shows the variation in handoff call dropping probability with time.

We observe that reducing the normalized power allocated to voice service results in higher handoff call dropping probability, and also, changing normalized power allocation causes abrupt variations in the handoff call dropping performance (e.g., the spike in Figure 4.7, scenario 2). These variations imply that all of the handoff calls arriving after increase in the allocated normalized power cannot be admitted until some of the ongoing calls leave the cell. Therefore, the normalized power adjustment should be delayed to avoid (or at least reduce) the sudden increase in handoff call dropping probability. Also, when the normalized power allocated to voice service increases, the handoff call dropping probability decreases rapidly since most of the calls arriving after adjustment of the allocated normalized power can be admitted.

In order to validate the transient analysis, we also use an event-driven simulation to obtain the performance measures of interest. In this case, we simulate the admission control process for 100 minutes (i.e., for the same time period as that of transient analysis), and the simulation is replicated five thousand times. We observe that transient state performance measures obtained from the proposed model are very close to those obtained from simulations.

Due to the time-dependent nature of the traffic load, the normalized power allocation among the different service classes needs to be adaptively adjusted to achieve the desired performance objectives. Figure 4.8 shows variations in the normalized power allocation in each interval during the observation period. In particular, the highest amount of normalized power is allocated to the best-effort service. By reducing the transmission rate, more best-effort data calls are admitted, which increases the system revenue. The highest amount of system revenue is achieved by allocating the highest amount of normalized power to the best-effort service. The normalized power allocated to the voice

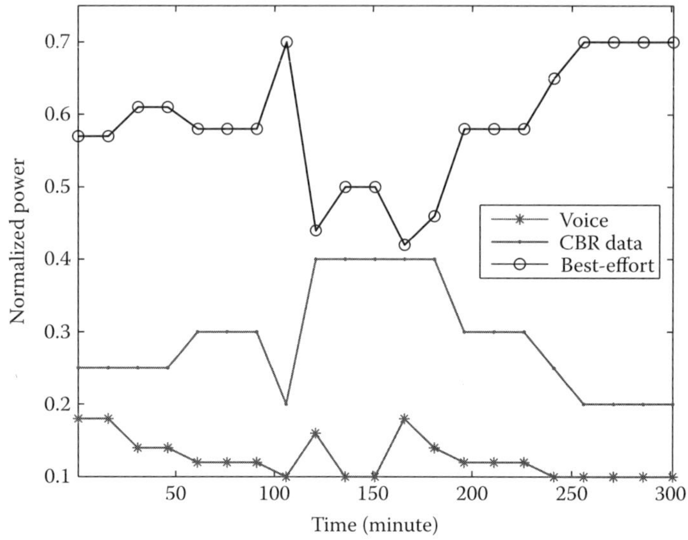

FIGURE 4.8 Normalized power allocation to each service class.

service is the smallest in this case, since each voice call has the smallest SIR requirement and the revenue per voice call is the smallest as well. For all the service classes, the normalized power allocation is gradually adjusted. In particular, $\gamma_{(c)}$ ($c \in \{v,d,b\}$) is increased/decreased by 0.01 every 30 seconds until it becomes equal to the optimized value for that period. This gradual adaptation avoids abrupt variations in call blocking/call dropping probabilities when the normalized power has to be increased/decreased.

We observe that the performance measures depend strongly on traffic arrival rate (which is time-dependent) and adaptation of the normalized power allocation. For example, voice call blocking and dropping probabilities are high during intervals 10–110 and 250–200 when traffic load is high (Figure 4.9). Due to the higher SIR requirement for CBR data calls, call blocking and call dropping probabilities are higher than those for voice and best-effort data calls (Figure 4.10). For the best-effort service, call blocking and dropping probabilities (Figure 4.11) are the smallest since the transmission rate in this case can be reduced to decrease the SIR requirement. Consequently, more calls can be admitted.

Typical variation in the normalized transmission rate for a best-effort data call is shown in Figure 4.12. In this case, the transmission rate is adjusted according to the traffic load. During high traffic load periods (e.g., during intervals 10–110 and 200–300), transmission rate is reduced. while during light traffic load period (e.g., during interval 110–200), the transmission rate is increased to achieve higher system revenue.

4.7 Conclusion

In this chapter, we have shown the importance of transient analysis in a time-varying system for which the steady state analysis may not provide very accurate performance

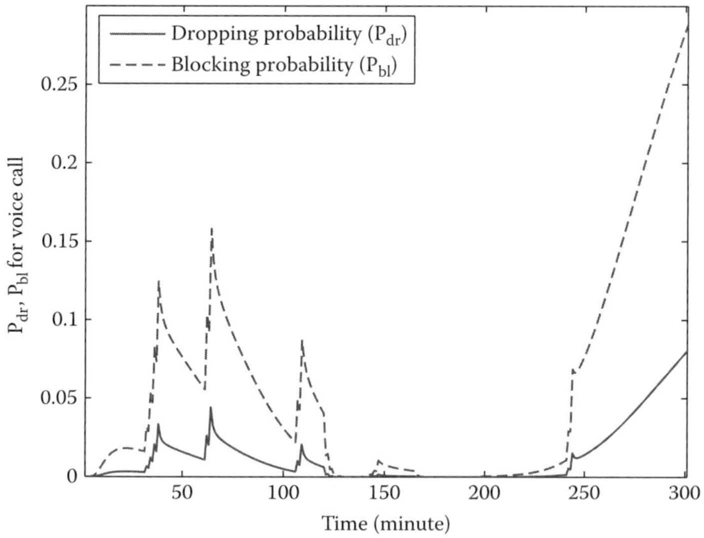

FIGURE 4.9 New call blocking and handoff call dropping probabilities for voice calls.

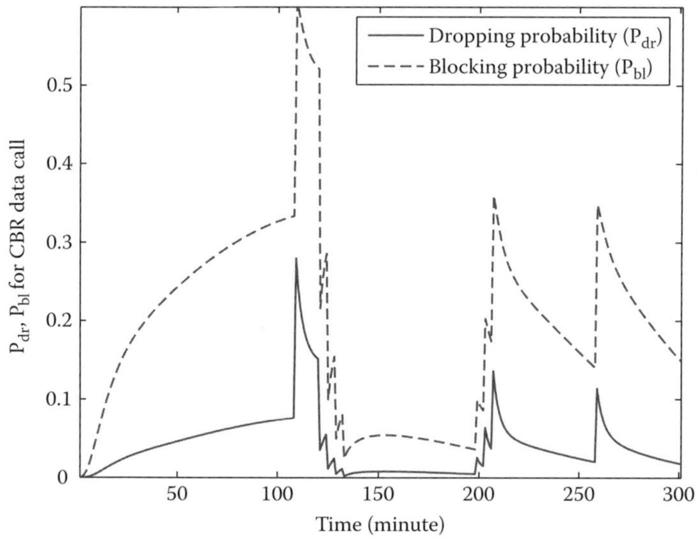

FIGURE 4.10 New call blocking and handoff call dropping probabilities for CBR data calls.

measures. For many systems with Markov property, transient analysis can be performed based on the continuous-time Markov chain. We have reviewed three different approaches (i.e., ordinal differential equation, matrix exponential, and uniformization) for transient analysis of a continuous-time Markov chain. To this end, we have presented a framework for resource allocation and admission control in a CDMA cellular network

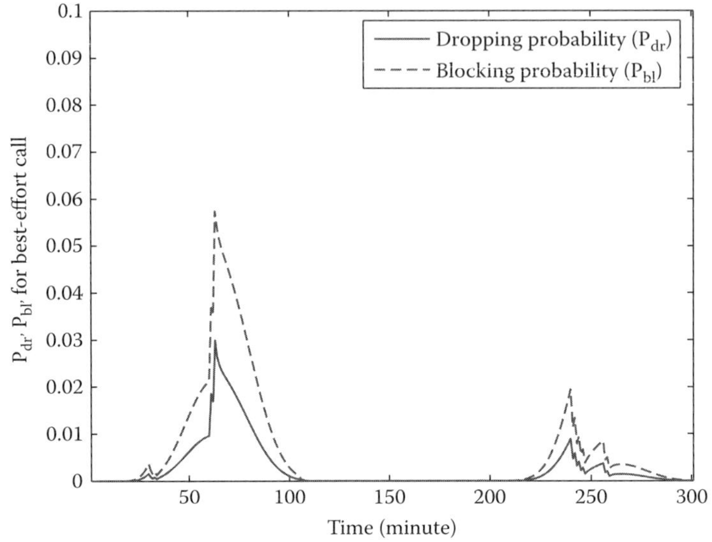

FIGURE 4.11 New call blocking and handoff call dropping probabilities of best-effort data calls.

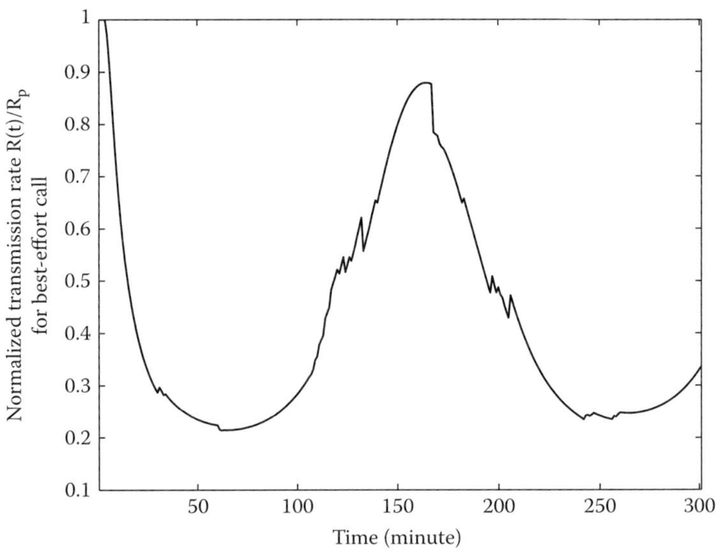

FIGURE 4.12 Normalized transmission rate for best-effort data calls.

based on the transient performance analysis. The *uniformization* method has been used to obtain the transient state performance measures under time-dependent call arrival rates and channel holding times. An application of the transient analysis model to obtain adaptive optimal normalized power allocation among the different service classes has been illustrated, with which the system revenue can be maximized.

The framework presented in this chapter can be extended to investigate both packet-level and call-level performance measures and their relationships in the transient state. The temporal optimization technique (e.g., dynamic programming) can be applied to obtain optimal resource allocation and admission control policy considering multiple time periods.

References

[1] K. Ogata. 2001. *Modern control engineering.* Upper Saddle River, NJ: Prentice Hall.

[2] D. Niyato, E. Hossain, and A. S. Alfa. 2005. Performance analysis and adaptive call admission control in cellular mobile networks with time-varying traffic. In *Proceedings of IEEE ICC'05*, vol. 5, pp. 3324–28.

[3] M. Cheung and J. W. Mark. 2000. Resource allocation in wireless networks based on joint packet/call levels QoS constraints. In *Proceedings of IEEE GLOBECOM'00*, vol. 1, pp. 271–75.

[4] C.-J. Chang, S. Shen, J.-H. Lin, and F.-C. Ren. 2000. Intelligent call admission control for differentiated QoS provisioning in wideband CDMA cellular systems. In *Proceedings of IEEE VTC'00*, vol. 3, pp. 1057–63.

[5] X. Tang and A. Goldsmith. 2001. Admission control and adaptive CDMA for integrated voice and data systems. In *Proceedings of IEEE VTC'01*, vol. 1, pp. 506–10.

[6] D. Zhao, X. Shen, and J. W. Mark. 2003. Radio resource management for cellular CDMA systems supporting heterogeneous services. *IEEE Transactions on Mobile Computing* 2:147–60.

[7] P. Liu, P. Zhang, S. Jordan, and M. L. Honig. 2004. Single-cell forward link power allocation using pricing in wireless networks. *IEEE Transactions on Wireless Communications* 3:533–43.

[8] J. Yao, J. W. Mark, T. C. Wong, Y. H. Chew, K. M. Lye, and K.-C. Chua. 2004. Virtual partitioning resource allocation for multiclass traffic in cellular systems with QoS constraints. *IEEE Transactions on Vehicular Technology* 53:847–64.

[9] X. Wang, R. Ramjee, and H. Viswanathan. 2005. Adaptive and predictive downlink resource management in next-generation CDMA networks. *IEEE Journal on Selected Areas in Communications* 23:1219–32.

[10] D. Liu, Y. Zhang, and H. Zhang. 2005. A self-learning call admission control scheme for CDMA cellular networks. *IEEE Transactions on Neural Networks* 16:1219–28.

[11] O. Yu, E. Saric, and A. Li. 2006. Fairly adjusted multimode dynamic guard bandwidth admission control over CDMA systems. *IEEE Journal on Selected Areas in Communications* 24:579–92.

[12] J. S. Evans and D. Everitt. 1999. Effective bandwidth-based admission control for multiservice CDMA cellular networks. *IEEE Transactions on Vehicular Technology* 48:36–46.

[13] J.-G. Choi and S. Bahk. 2001. Multiclass call admission control in QoS-sensitive CDMA networks. In *Proceedings of IEEE ICC'01*, vol. 2, pp. 331–35.

[14] S. Singh, V. Krishnamurthy, and H. V. Poor. 2002. Integrated voice/data call admission control for wireless DS-CDMA systems. *IEEE Transactions on Signal Processing* 50:1483–95.

[15] J.-G. Choi, Y.-J. Choi, and S. Bahk. 2007. Power-based admission control for multiclass calls in QoS-sensitive CDMA networks. *IEEE Transactions on Wireless Communications* 6:468–72.

[16] J. Yao, D. T. C. Wong, and Y. H. Chew. 2006. Capacity balancing between the reverse and forward links in multiservice CDMA cellular networks with cross-layer design. *IEEE Transactions on Vehicular Technology* 55:1397–411.

[17] F. Yu and V. Krishnamurthy. 2007. Optimal joint session admission control in integrated WLAN and CDMA cellular networks with vertical handoff. *IEEE Transactions on Mobile Computing* 6:126–39.

[18] J. A. Carrasco. 2003. Solving dependability/performability irreducible Markov models using regenerative randomization. *IEEE Transactions on Reliability* 52:319–29.

[19] J. A. Carrasco. 2004. Transient analysis of some rewarded Markov models using randomization with quasistationarity detection. *IEEE Transactions on Computers* 53:1106–20.

[20] J. Zhang and E. J. Coyle. 1990. The transient performance analysis of voice/data integrated networks. In *Proceedings of IEEE INFOCOM'90*, vol. 3, pp. 963–68.

[21] Y. W. Jang and J. Ahn. 2000. A connection admission control using transient outage probability in CDMA systems. In *Proceedings of IEEE VTC'00*, vol. 3, pp. 1412–16.

[22] H. S. Kim, Y. M. Jang, and G. J. Jeon. 2001. A transient analysis approach for connection admission control in multi-cell CDMA networks with smart antennas. In *Proceedings of IEEE GLOBECOM'01*, vol. 1, pp. 690–94.

[23] T. S. Randhawa and R. H. S. Hardy. 2003. Transient-state analysis based access control in wireless CDMA networks supporting integrated services. In *Proceedings of IEEE ICT'03*, vol. 1, pp. 799–805.

[24] Y. Fang and Y. Zhang. 2002. Call admission control schemes and performance analysis in wireless mobile networks. *IEEE Transactions on Vehicular Technology* 51:371–82.

[25] A. Sampath and J. M. Holtzman. 1997. Access control of data in integrated voice/data CDMA systems: Benefits and tradeoffs. *IEEE Journal on Selected Areas in Communications* 15:1511–26.

[26] M. Casoni, G. Immvilli, and M. L. Merani. 2002. Admission control in T/CDMA systems supporting voice and data applications. *IEEE Transactions on Wireless Communications* 1:540–48.

[27] D. Niyato and E. Hossain. 2006. Call-level and packet-level quality of service and user utility in rate-adaptive cellular CDMA networks: A queuing analysis. *IEEE Transactions on Mobile Computing* 5:1749–63.

[28] F. Yu, V. Krishnamurthy, and V. C. M. Leung. 2006. Cross-layer optimal connection admission control for variable bit rate multimedia traffic in packet wireless CDMA networks. *IEEE Trans Signal Processing* 54:542–55.

[29] F. J. Vazquez-Abad and V. Krishnamurthy. 2002. Self learning call admission control for multimedia wireless DS-CDMA systems. In *Proceedings of the IEEE International Workshop on Discrete Event Systems*, pp. 399–404.

[30] C. Comaniciu and H. V. Poor. 2003. Jointly optimal power and admission control for delay sensitive traffic in CDMA networks with LMMSE receivers. *IEEE Transactions on Signal Processing* 51:2031–42.

[31] E. Cinlar. 1975. *Introduction to stochastic processes*. Englewood Cliffs, NJ: Prentice-Hall.

[32] W. J. Stewart. 1994. *Introduction to the numerical solution of Markov chains*. Princeton, NJ: Princeton University Press.

[33] A. Iserles. 1996. *A first course in the numerical analysis of differential equations*. Cambridge: Cambridge University Press.

[34] Roger A. Horn and Charles R. Johnson. 1991. *Topics in matrix analysis*. Cambridge: Cambridge University Press.

[35] A. N. Michel and C. J. Herget. 1993. *Applied algebra and functional analysis*. New York: Dover.

[36] A. Jensen. 1953. Markoff chains as an aid in the study of Markoff processes. *Skandinavsk Aktuarietidskrift* 36:87–91.

[37] W. K. Grassman. 1977. Transient solutions in Markovian queuing systems. *Computers and Operations Research* 4:47–53.

[38] D. Gross and D. R. Miller. 1984. The randomization technique as a modeling tool and solution procedure for transient Markov processes. *Operations Research* 32:343–61.

[39] P. Buchholz and W. H. Sanders. 2004. Approximate computation of transient results for large Markov chains. In *Proceedings of IEEE QEST'04*, pp. 126–35.

[40] A. P. A. van Moorsel. 1993. Performability evaluation concepts and techniques. PhD thesis, University of Twente, The Netherlands.

[41] A. P. A. van Moorsel and W. H. Sanders. 1994. Adaptive uniformization. *Communications in Statistics—Stochastic Models* 10:619–47.

[42] A. P. A. van Moorsel and W. H. Sanders. 1997. Transient solution of Markov models by combining adaptive and standard uniformization. *IEEE Transactions on Reliability* 46:430–40.

[43] Y. Shu, M. Yu, J. Liu, and O. W. W. Yang. 2003. Wireless traffic modeling and prediction using seasonal ARIMA models. In *Proceedings of IEEE ICC'03*, vol. 3, pp. 1675–79.

[44] E. Altman. 2002. Capacity of multi-service cellular networks with transmission-rate control: A queueing analysis. In *Proceedings of ACM Mobicom'02*, pp. 205–214.

[45] J. Laiho and A. Wacker. 2000. Radio network planning process and methods for WCDMA. *Annals of Telecommunications*. 56:317–331.

[46] L. Xu, X. Shen, and J. W. Mark. 2004. Dynamic fair scheduling with QoS constraints in multimedia wideband CDMA cellular networks. *IEEE Transactions on Wireless Communications* 3:60–73.

[47] D. Hong and S. S. Rappaport. 1986. Traffic model and performance analysis for cellular mobile radio telephone systems with prioritized and nonprioritized hand-off procedures. *IEEE Transactions on Vehicular Technology* 35: 77–92.
[48] S. M. Ross. 1983 *Stochastic processes.* New York: John Wiley & Sons.

5

Utility-Based Bandwidth Adaptation for Multimedia Wireless Networks

Ning Lu
Macao Polytechnic Institute

John Bigham
Queen Mary University of London

Nidal Nasser
University of Guelph

5.1 Introduction

Over the past few years the rapid development of wireless communications technologies has greatly enriched the diversity of wireless applications. Wireless services are evolving from the traditional voice service to a wide range of multimedia services, including data, voice, and video. Different multimedia services over wireless networks have different bandwidth requirements. For example, some applications like telephone call and video conference require strict end-to-end performance guarantees. Hence, it is critical for the networks to provide reliable and timely packet transmission. On the other hand, applications such as file transfer and video streaming can adapt their bandwidth to various network loads since they can tolerate certain delays.

As a result, providing quality of service (QoS) to multimedia applications according to their bandwidth requirements is becoming an important resource management issue for wireless networks. However, due to the limited and highly varying bandwidth resource, the QoS provisioning for multimedia traffic in wireless networks is much more challenging than in their wired counterpart. Despite the relatively high data rate provided by some current wireless networks, such as 3G/3.5G and wireless local area network (WLAN), bandwidth is still the major bottleneck in wireless networks. Moreover, compared to wireline networks, the resource availability in wireless networks is highly varying due to channel fading and user mobility. Although the effect of channel fading can be mitigated by rich-function transmission/reception wireless subsystems [17], mobility may cause severe fluctuations of network resources.

The above observations motivate the study of efficient resource management to support multimedia services in wireless networks. Bandwidth adaptation is one of the most promising resource management methods to provide QoS guarantees to multimedia traffic in wireless networks. The main rationale behind bandwidth adaptation is that it can explore the adaptive nature of multimedia applications to deal with network congestion and increase the resource utilization. In the traditional non-adaptive network environment, once a call is admitted into the network, its allocated bandwidth is kept fixed throughout the lifetime of the call. When a new or handoff call requests a certain amount of bandwidth, the network rejects the call if there is not sufficient bandwidth available. In contrast with adaptive bandwidth allocation paradigms, when a new or handoff call arrives at a congested network, the bandwidth allocated to the ongoing calls can be degraded to smaller values to accept the new or handoff call. When an ongoing call is terminated due to its completion or handoff to another cell, the released bandwidth can be utilized to upgrade other ongoing calls. Some examples of these adaptive multimedia services include the International Organization for Standardization's (ISO) Motion Picture Experts Group (MPEG)-4 [21] and the International Telecommunication Union's (ITU) H.263 [33, 37], which are expected to be used extensively in future wireless communication networks.

In this chapter we explore an efficient utility-maximization bandwidth adaptation scheme to provide QoS support for multimedia wireless networks. The structure of the chapter is as follows. Section 5.2 surveys related work in the area of bandwidth adaptation for wireless networks and shows that the proposed scheme is new and original. Section 5.3 presents the utility-based adaptive multimedia traffic model. Section 5.4 introduces the implementation of multimedia adaptation in wireless networks. Section 5.5 gives a detailed description and formulation of the utility-maximization bandwidth adaptation problem. Following this, section 5.6 analyzes design issues for the utility-maximization algorithm and proposes an efficient search-tree-based maximization algorithm and its corresponding call admission control (CAC) and bandwidth reservation mechanisms. Section 5.7 presents the wireless network simulation model. Section 5.8 evaluates the performance of the proposed utility-maximization bandwidth adaptation scheme by comparing it with two existing schemes under various traffic loads. Finally, concluding remarks are given in section 5.9.

5.2 Previous Work on Wireless Network Bandwidth Adaptation

Recently some bandwidth adaptation schemes have been proposed for QoS provisioning in wireless networks. In [34], the authors propose an adaptive bandwidth reservation scheme to provide QoS guarantees for multimedia traffic in wireless networks. The scheme allocates bandwidth to a call in the cell where the call request originates and reserves bandwidth dynamically in all neighboring cells according to the network conditions. Bandwidth reservation in all neighboring cells guarantees the QoS of handoff calls, but it often results in the underutilization of network resources as the mobile user hands off to only one of the cells. Reference [19] presents an adaptive bandwidth allocation scheme for QoS support in broadband wireless networks consisting of three service classes with different handoff dropping requirements. The scheme includes the measurement-based CAC and bandwidth reservation algorithms to adaptively allocate bandwidth to the calls so that the target handoff dropping probability can be met. The main disadvantage of the scheme is that the allocated bandwidth of the call is kept fixed during the stay in the cell, and it can only be changed when handoff happens. This is also the case for [34].

Nasser and Hassanein [32] describe an adaptive bandwidth allocation framework that can adjust the bandwidth of ongoing calls during their stay in the cell whenever there are resource fluctuations in wireless networks. When a new or handoff call arrives to an overloaded network, the bandwidth adaptation algorithm can reduce the allocated bandwidth of ongoing calls to free some bandwidth for the new or handoff call. The bandwidth adaptation algorithm minimizes the number of calls receiving lower bandwidth than that requested. In [23], a bandwidth adaptation scheme is developed for wireless networks to guarantee the upper bound of the call degradation probability. The CAC measures the state of the network and reflects the observed system history on making call admission decisions. The adaptation algorithm adjusts the bandwidth of multimedia calls to minimize the call degradation probability. In the work of El-Kadi et al. [15], a rate-based borrowing scheme (RBBS) is provided for multimedia wireless networks. In case of insufficient bandwidth, in order not to deny service to requesting calls, bandwidth can be borrowed on a temporary basis from existing calls to accept the new or handoff call. When enough bandwidth becomes available due to call completion or outgoing handoff, the bandwidth is returned to the ongoing calls. To reduce handoff dropping probability, a fixed amount of bandwidth is reserved for handoff calls in each cell. Reference [8] proposes a borrowing-based adaptive bandwidth allocation scheme to improve the work in [15]. The scheme makes adaptive decisions for bandwidth allocation by employing an attribute-measurement mechanism and service-based bandwidth borrowing policy. A dynamic time interval reservation strategy is introduced to provide QoS guarantees for handoff calls by adjusting the amount of reserved bandwidth in each cell according to the online traffic information. Compared to [19, 34], the bandwidth adaptation schemes proposed in [8, 15, 23, 32] provide more flexibility in bandwidth allocation since they can change the bandwidth of ongoing calls during their stay in the cell. However, these schemes have one common drawback: they have not provided any mechanism to measure the degradation of calls.

The bandwidth adaptation scheme for wireless networks described in [11] measures the bandwidth degradation of multimedia calls. Two bandwidth degradation metrics, i.e., bandwidth degradation ratio and bandwidth degrade frequency, are taken into account in the bandwidth degradation process. Similar bandwidth degradation measurements can also be found in [44]. The bandwidth adaptation schemes introduced in [11] and [44] evaluate the application-level QoS using bandwidth instead of a quantitative measure that can be perceived by end users. Hence, the consequence of bandwidth degradation, namely, the decrease of the satisfaction degree of end users, and the adaptive characteristics of ongoing calls cannot be reflected. For example, a small portion of bandwidth degradation on a non-real-time data call may result in unnoticeable perceived QoS change on the end users, while the same bandwidth degradation on a real-time multimedia call may cause the application to be dropped. The quantitative QoS measure is also a missing factor in other bandwidth adaptation schemes mentioned above. To address such problems, some schemes apply utility (revenue) to bandwidth adaptation to provide both connection-level and application-level QoS to multimedia traffic in wireless networks.

Bharghavan et al. [3] present the TIMELY adaptive resource management architecture and algorithms for resource reservation and adaptation in wireless networks. The architecture has four layers—link, reservation, adaptation, and transport—all of which perform resource adaptation in a coordinated manner to solve the problems introduced by the scarce and dynamic network resources. A revenue model for resource usage is introduced and a rate adaptation algorithm is presented to distribute resources among the adaptable flows to maximize network revenue. However, the multilayer architecture has made the bandwidth adaptation work at the expense of high message overhead. In [13], the authors introduce a revenue-based bandwidth degradation model for both real-time multimedia traffic and non-real-time data traffic in wireless networks. An effective bandwidth degradation algorithm is presented to maximize the net revenue of the network. Reference [24] describes a near-optimal bandwidth allocation algorithm for multimedia wireless networks. When the network is overloaded, bandwidth can be degraded from ongoing calls to accept the new or handoff call. The bandwidth degradation algorithm is based on a greedy approach and seeks to achieve maximum network revenue with polynomial time complexity. The limitation of the above two schemes is that they only work in wireless networks with one single cell and do not take into account the basic characteristic of wireless networks, i.e., user mobility. The authors in [16] formulate the QoS provisioning in a wireless network with multiple cells as a constrained Markov decision problem and propose a bandwidth adaptation scheme using Q-learning to maximize network revenue and meet the QoS constraints. The problem of employing Q-learning is that the learning process can be time consuming, making the scheme not suitable for real-time QoS support in wireless networks. Moreover, as in [24], the proposed scheme only applies to adaptive real-time traffic, and therefore it is not clear how to deal with multiclass multimedia traffic with different adaptive characteristics. Another bandwidth adaptation scheme using QoS satisfaction function for multimedia wireless network QoS provisioning can be found in [1]. When some bandwidth becomes available due to the termination of an ongoing call, the released bandwidth is used to upgrade other ongoing calls that need more bandwidth allocation. Bandwidth is upgraded in a way to maximize the total satisfaction degree of end users in the network. An algorithm based on the Lagrangean relaxation is developed to support the real-time bandwidth adaptation. The drawback of the algorithm is that it does not address the problem of bandwidth degradation when the network is overloaded. In the work of Curescu and Nadjm-Tehrani [12], a time-aware bandwidth allocation scheme is introduced to maximize the accumulated utility of the network. Interestingly, the scheme identifies how bandwidth reallocation affects the utility of the applications. To integrate this information into the bandwidth allocation algorithm, it is assumed that the duration of every connection can be estimated. However, in a real-time wireless network environment where the connections' durations are updated dynamically with bandwidth reallocation, the feasibility of estimating connections' durations is doubtful. Moreover, the scheme does not reserve bandwidth for real-time handoff calls; thus, it risks an inability of meeting their QoS requirements when the network is heavily overloaded.

The utility-maximization bandwidth adaptation scheme proposed in this chapter provides QoS support in wireless networks containing both real-time multimedia traffic and non-real-time data traffic. With the proposed scheme, each call is assigned a utility function according to its adaptive characteristics. Bandwidth adaptation is divided into two processes: bandwidth degrades and bandwidth upgrades. Depending on the network load, the allocated bandwidth of ongoing calls is degraded or upgraded dynamically so that the achieved utility in each individual cell of the network is maximized. Appropriate call admission control (CAC) and bandwidth reservation mechanisms have also been integrated into the bandwidth adaptation scheme to strike the appropriate performance balance among multiple QoS requirements, i.e., maximizing network utility, reducing the call blocking and handoff dropping probabilities, and maintaining high bandwidth utilization.

5.3 Utility-Based Multimedia Traffic Model

In multimedia wireless networks, different applications have different bandwidth requirements. To provide QoS to multimedia applications according to their bandwidth needs under the wireless networks environment featuring limited and varying bandwidth resources, an explicit traffic model is needed to reflect the QoS sensitivity of the applications to bandwidth allocation. For example, references [2, 31] use a single-class traffic model and references [15, 34] apply a multiclass traffic model including real-time traffic and non-real-time traffic for the bandwidth adaptation problems. In this section a utility-based multiclass multimedia traffic model is proposed to differentiate the multimedia traffic according to their adaptive characteristics.

5.3.1 The Definition of Utility Functions

A utility function is defined as a curve mapping the amount of bandwidth received by the application to the performance as perceived by the end users. Utility function is monotonically nondecreasing; in other words, more bandwidth allocation should not lead to degraded application performance [6, 27, 36, 39]. The key advantage of the utility function is that it can inherently reflect the QoS requirements of the end users and quantify the adaptability of the application. The shape of the utility function varies according to the adaptive characteristics of the application.

5.3.2 Utility Functions Formulation for Multimedia Traffic

In the past few years utility functions have been widely applied by many bandwidth adaptation schemes for QoS support in multimedia wireless networks [1, 3, 12, 13, 16, 24, 28]. However, the formulation of utility functions for multimedia traffic remains a problem since none of these schemes provide an explicit method to capture the adaptive nature of the applications and map their allocated bandwidth to the utilities. For instance, reference [13] adopts the Sigmoid utility functions, reference [24] uses linear and convex utility functions, and reference [12] constructs utility functions using subjective values from the authors' experiments. The work presented in [5, 27, 28, 42] introduces methods to generate utility functions for multimedia applications by evaluating their qualities subjectively in a discrete bandwidth set. The drawback of these methods is that they need to create the utility function for each multimedia application individually. Usually wireless networks have a large amount of applications; thus, it is not practical for them to be applied to real-time environments. To solve the above problem, this section categorizes multimedia traffic into different classes according to their adaptive characteristics and formulates the utility function for each class of traffic to reflect its nature of adaptability.

According to the bandwidth requirements, the multimedia traffic used in this chapter can be classified into two broad classes:

- Class I: Real-time traffic
- Class II: Non-real-time traffic

Class I traffic can be further classified into two subclasses: adaptive real-time traffic and hard real-time traffic.

5.3.2.1 Adaptive Real-Time Traffic

Adaptive real-time traffic refers to the applications that have flexible bandwidth requirements. In case of congestion, they can gracefully adjust their transmission rates to adapt to various network conditions. However, such applications have an intrinsic bandwidth requirement b^{intr} because the data generation rate is independent of the network congestion. Thus, the quality starts dropping sharply as soon as the bandwidth is reduced below b^{intr}, and becomes unacceptable when the bandwidth is reduced below b^{min} [39]. Typical examples are interactive multimedia services and video on demand [34]. The utility function of adaptive real-time traffic is modeled as follows:

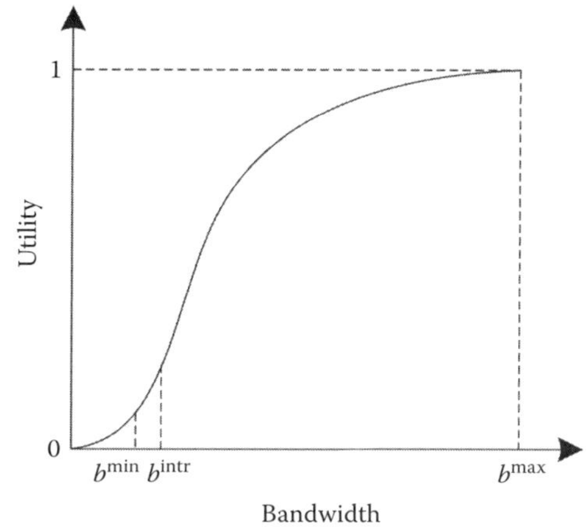

FIGURE 5.1 Adaptive real-time traffic utility function.

$$u(b) = 1 - e^{-\frac{k_1 b^2}{k_2 + b}}, \tag{5.1}$$

where k_1 and k_2 are two positive parameters that determine the shape of the utility function and ensure that when the maximum bandwidth requirement $b^{\max}$ is received, the achieved utility $u^{\max}$ approximately equals 1. A similar utility function has also been used to model adaptive real-time traffic in [6, 35]. The general shape of the utility function is depicted in Figure 5.1. At high bandwidth values the marginal utility of additional bandwidth is very slight because the signal quality is much better than humans need. At very small bandwidth values, the marginal utility is also very slight because the signal quality is unbearably low [39]. The utility function is convex in the neighborhood around $b^{\min}$ and starts becoming concave after b^{intr}.

To determine the exact shape of the adaptive real-time traffic utility function, parameters k_1 and k_2 need to be calculated. It is obvious that for a utility function, when its allocated bandwidth b reaches $b^{\max}$, the corresponding utility u equals $u^{\max}$. Thus, there is the following equation:

$$1 - e^{-\frac{k_1 (b^{\max})^2}{k_2 + b^{\max}}} = u^{\max}. \tag{5.2}$$

From equation (5.2) the relationship between k_1 and k_2 can be derived as follows:

$$k_1 = \frac{\ln(1 - u^{\max}) \cdot (k_2 + b^{\max})}{-(b^{\max})^2}. \tag{5.3}$$

Now one more equation is needed to calculate k_1 and k_2. The intrinsic bandwidth requirement b^{intr} is defined as the bandwidth before which the utility function is convex and after which the function becomes concave. This happens at the point where the second-order derivative of the utility function is equal to zero [36], i.e.,

$$\left(1-e^{-\frac{k_1(b^{\text{intr}})^2}{k_2+b^{\text{intr}}}}\right)''=0. \tag{5.4}$$

After calculating the second-order derivative there is the following equation:

$$2k_1\cdot\frac{e^{-\frac{k_1(b^{\text{intr}})^2}{k_2+b^{\text{intr}}}}}{(k_2+b^{\text{intr}})^4}\cdot\left(k_2^3+\left(b^{\text{intr}}-2(b^{\text{intr}})^2k_1\right)k_2^2-2(b^{\text{intr}})^3k_1k_2-\frac{(b^{\text{intr}})^4k_1}{2}\right)=0. \tag{5.5}$$

Since both k_1 and k_2 are positive numbers, equation (5.5) is equal to zero only when the cubic polynomial in the brackets is zero, i.e.,

$$k_2^3+\left(b^{\text{intr}}-2(b^{\text{intr}})^2k_1\right)k_2^2-2(b^{\text{intr}})^3k_1k_2-\frac{(b^{\text{intr}})^4k_1}{2}=0. \tag{5.6}$$

After substituting k_1 with

$$\frac{\ln(1-u^{\max})\cdot(k_2+b^{\max})}{-(b^{\max})^2},$$

equation (5.6) becomes:

$$\begin{aligned}&\left(1+2\ln(1-u^{\max})\cdot\left(\frac{b^{\text{intr}}}{b^{\max}}\right)^2\right)\cdot k_2^3+\\&\left(b^{\text{intr}}+2\ln(1-u^{\max})\cdot\frac{(b^{\text{intr}})^2}{b^{\max}}+2\ln(1-u^{\max})\cdot\frac{(b^{\text{intr}})^3}{(b^{\max})^2}\right)\cdot k_2^2+\\&\left(2\ln(1-u^{\max})\cdot\frac{(b^{\text{intr}})^3}{b^{\max}}+\ln(1-u^{\max})\cdot\frac{(b^{\text{intr}})^4}{2(b^{\max})^2}\right)\cdot k_2+\\&\ln(1-u^{\max})\cdot\frac{(b^{\text{intr}})^4}{2b^{\max}}=0.\end{aligned} \tag{5.7}$$

Since b^{intr}, $b^{\max}$, and $u^{\max}$ are all predefined by the network operator, end users, or both, equation (5.7) is a cubic polynomial in k_2 with all coefficients constant. The equation can be solved easily, and its positive cube root is the value for k_2. After achieving k_2, k_1 can then be calculated using equation (5.3).

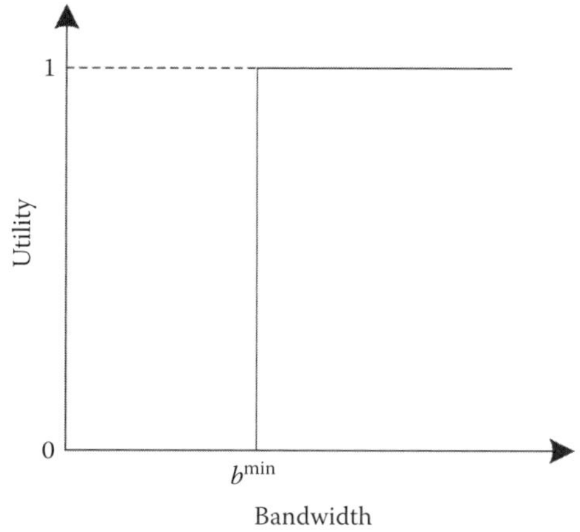

FIGURE 5.2 Hard real-time traffic utility function.

5.3.2.2 Hard Real-Time Traffic

Hard real-time traffic refers to the applications with stringent bandwidth requirements. A call belonging to hard real-time traffic requires strict end-to-end performance guarantees and does not show any adaptive properties. It is not allowed to enter the network if its minimum bandwidth requirement $b^{\min}$ cannot be met. Once accepted, the maximum $u^{\max}$ utility is achieved. The bandwidth of the call cannot be changed during its lifetime, and any bandwidth decrease will cause the utility to drop to zero. Examples include audio or video phone, video conference, and telemedicine [34, 35]. The following utility function is used to model hard real-time traffic:

$$u(b) = \begin{cases} 1, \text{ when } b \geq b^{\min} \\ 0, \text{ when } b < b^{\min} \end{cases}. \tag{5.8}$$

The shape of the utility function is depicted in Figure 5.2.

5.3.2.3 Non-Real-Time Traffic

Non-real-time traffic refers to the applications that are rather tolerant of delays. In case of congestion, it is acceptable to buffer non-real-time applications at a network node (such as a base station) and transmit them at a slower rate. For non-real-time traffic, it is assumed that there is no minimum required bandwidth since non-real-time traffic can tolerate relatively large delays. Most traditional data applications such as e-mail, file transfer, and remote login [35, 39] belong to non-real-time traffic and can work without guarantees of timely packet delivery. The following utility function is used to model non-real-time traffic:

$$u(b) = 1 - e^{-\frac{kb}{b^{\max}}}, \tag{5.9}$$

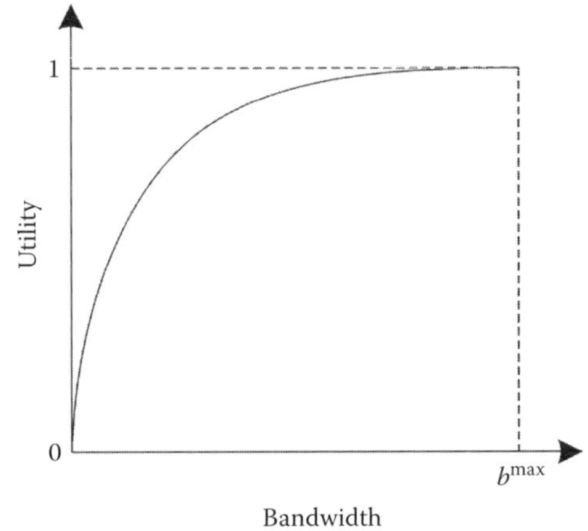

FIGURE 5.3 Non-real-time traffic utility function.

where k is a positive parameter that determines the shape of the utility function and ensures that when the maximum bandwidth requirement b^{max} is allocated, the achieved utility u^{max} approximately equals 1. The utility function of non-real-time traffic has the general shape shown in Figure 5.3. From the figure it can be seen that there is a diminishing marginal rate of performance enhancement as bandwidth increases, so the utility function is strictly concave everywhere.

To determine the exact shape of the non-real-time traffic utility function, parameter k needs to be calculated. For a utility function when its allocated bandwidth b reaches b^{max}, the corresponding utility u equals u^{max}. Thus, there is the following equation:

$$1-e^{-\frac{kb^{max}}{b^{max}}}=u^{max}. \tag{5.10}$$

Parameter k can be derived from equation (5.10) as follows:

$$k=-\ln(1-u^{max}). \tag{5.11}$$

Since u^{max} is a constant predefined by the network operator, end users, or both, k can be easily calculated using equation (5.10).

5.3.3 Utility-Based Adaptive Traffic Model

The utility functions with various shapes proposed above can accurately reflect the adaptive characteristics of multimedia applications. While in practice, utility functions should be simple enough to support the design of bandwidth adaptation schemes in wireless networks. Hence, there must be a trade-off between the accuracy and simplicity

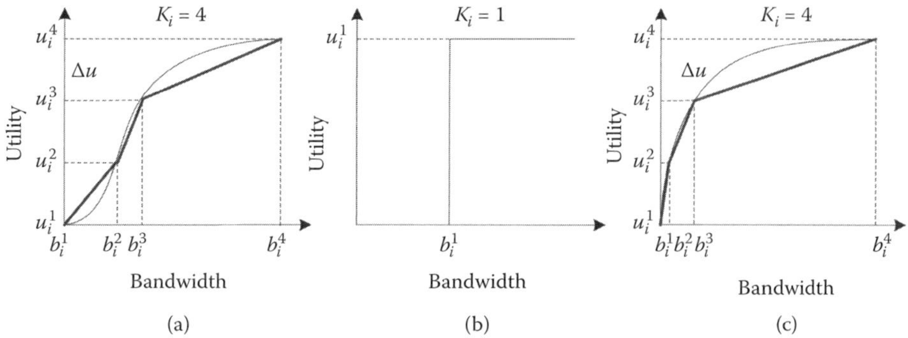

FIGURE 5.4 Utility function quantization using equal utility interval. (a) adaptive real-time traffic; (b) hard real-time traffic; (c) non-real-time traffic.

of utility functions. It has been proved that the use of linear piecewise functions can greatly simplify utility functions while still maintaining acceptable accuracy [12, 28]. A utility function can be quantized into linear piecewise functions by dividing its utility range into a number of equal intervals. After quantization, utility function $u_i(b_i)$ is approximated to a continuous linear piecewise function represented by a list of <bandwidth, utility> points in increasing order of bandwidth, i.e.,

$$u_i(b_i) = (<b_i^1, u_i^1>, <b_i^2, u_i^2>, \ldots, <b_i^{K_i}, u_i^{K_i}>),$$

where 1 is the application's lowest acceptable operating level corresponding to the minimum bandwidth requirement $b_i^{\min}$, and K_i is the application's optimal operating level corresponding to the maximum bandwidth requirement $b_i^{\max}$. Figure 5.4 demonstrates the quantization of utility function $u_i(b_i)$ using equal utility range Δu. Note that due to the strict bandwidth requirements of hard real-time traffic, the utility functions of hard real-time traffic are the same before and after quantization, containing only one <bandwidth, utility> point, i.e., $u_i(b_i) = <b_i^1, u_i^1>$).

When a call requests a connection to the network, it is assumed to provide the following information:

- Traffic class (real-time or non-real-time traffic)
- Bandwidth requirements
- Utility function (quantized)

With adaptive bandwidth allocation paradigms, if there is enough bandwidth available in the network, the call is allocated its maximum requested bandwidth $b_i^{K_i}$; otherwise, depending on how much the network is overloaded, the call is allocated a bandwidth ranging from to b_i^1 to $b_i^{K_i}$. If the call belongs to hard real-time traffic, once admitted its allocated bandwidth is fixed during the lifetime. Hard real-time traffic is regarded as a special case of adaptive real-time traffic; in the rest of the chapter no distinction will be made between them.

5.4 Multimedia Adaptation Implementation in Wireless Networks

5.4.1 Multimedia Adaptation Architecture

In the past few years, tremendous efforts have been made in designing the adaptation architecture for multimedia applications in wireless networks. Although there has not been a standard overall architecture for the end-to-end implementation, some partial solutions have been proposed. As introduced in [9, 36, 38], the adaptation of multimedia applications in wireless networks can be achieved at different Open System Interconnection (OSI) layers, and the function of each OSI layer is as follows:

- **Physical layer:** At the physical layer, adaptability can be achieved by choosing appropriate modulation and power control techniques.
- **Data link layer:** At the data link layer, error control mechanisms, e.g., retransmission, can be used to protect against the varying error rates of wireless links.
- **Network/transport layer:** At the network/transport layer, routing methods can be used to adapt the applications when there is user mobility.
- **Application layer:** At the application layer, most multimedia applications can adapt to the changing networking conditions using multimedia coding techniques such as layered coding and fine-granular scalable (FGS) coding.

5.4.2 Multimedia Adaptation Techniques

Based on the granularity of bandwidth allocation, multimedia adaptation can be divided into two types: discrete adaptation and continuous adaptation. Discrete adaptation limits the bandwidth choice of the applications to a set of discrete bit rates between their minimum and maximum bandwidth requirements, whereas continuous adaptation allows the allocated bandwidth of the applications to be adjusted to any bit rate between their minimum and maximum bandwidth requirements.

5.4.2.1 Discrete Adaptation

With the discrete adaptation approach, the multimedia content is encoded in the form of different layers to adapt to the varying network resource conditions [18, 26, 30, 41]. The base layer contains critical information for decoding the multimedia content at the lowest quality. The higher layers improve the application quality progressively. When the network is congested, only the base layer is transmitted; when more bandwidth becomes available, better application quality can be obtained by transmitting the higher layers.

Figure 5.5 illustrates an example of layer encoded multimedia. The bandwidth of multimedia application can only take discrete values in the set $\{b^1, b^2, b^3\}$. If the network is overloaded, the application is allocated its base layer bandwidth requirement b^1. When more bandwidth becomes available, additional bandwidth Δb^1 or $(\Delta b^1 + \Delta b^2)$ can be allocated, increasing the total bandwidth to b^2 or b^3, where Δb^1 and Δb^2 are the first and second enhanced layers of the multimedia application in addition to the base layer bandwidth b^1, respectively.

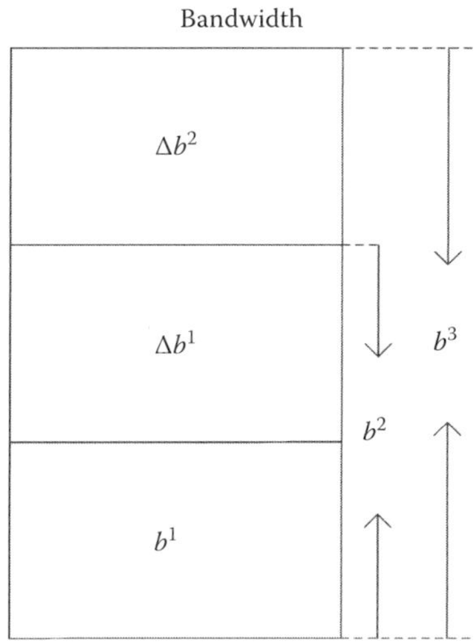

FIGURE 5.5 Example of layer encoded multimedia.

5.4.2.2 Continuous Adaptation

Discrete adaptation is easy to implement, but it has scalability limitation since the bandwidth of the multimedia applications can only be adapted on a number of discrete values.

To overcome this drawback, FGS coding techniques have been proposed to support continuous multimedia adaptation [7, 10, 14, 40, 45]. With FGS, multimedia applications can specify the range of acceptable bandwidth, and their allocated bandwidth can be adapted to any value in the bandwidth range. An example is the wavelet-based Joint Photographic Experts Group 2000 (JPEG-2000) image coding standard [20]. With the embedded coding system, the image can be encoded at any desirable bit rate within the specified bandwidth range. The new-generation multimedia communications coding standard MPEG-4 also uses FGS coding to support continuous adaptation [21].

Continuous adaptation provides more control over multimedia applications than discrete adaptation. However, it is usually more complicated to implement. The choice between continuous and discrete adaptation depends on the needs of the network operator, end users, or both. The utility functions proposed earlier assume that the bandwidth of multimedia applications is continuous and can take any value between the minimum and maximum bandwidth requirements. Therefore, without loss of generality, as in [4, 12, 28], continuous multimedia adaptation is considered throughout this chapter.

5.5 Utility-Based Bandwidth Adaptation Formulation

In wireless networks containing multiclass multimedia traffic, bandwidth adaptation should be performed selectively. For example, calls belonging to hard real-time traffic have stringent bandwidth requirements and their allocated bandwidth cannot be changed

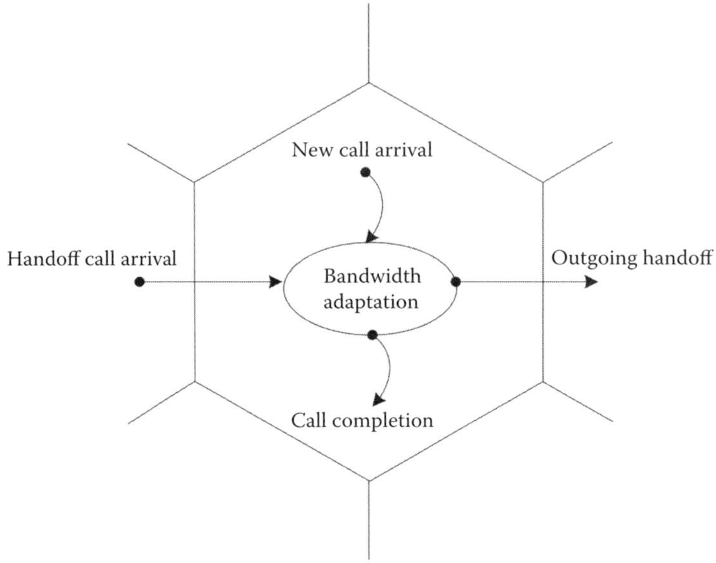

FIGURE 5.6 Bandwidth adaptation trigger events.

during their lifetime. Therefore, bandwidth adaptation can only be applied to adaptive ongoing calls, i.e., calls belonging to adaptive real-time traffic or non-real-time traffic.

To avoid the complexity of central coordination, bandwidth adaptation is performed based on each base station (cell) of wireless networks in a distributed manner. It consists of two processes—bandwidth degrades and bandwidth upgrades, which are triggered by call arrival events and call departure events, respectively (see Figure 5.6). Call arrival events include new call arrival events (a new call is generated within the cell) and handoff call arrival events (a handoff call arrives at the cell). Call departure events include call completion events (a call within the cell terminates) and outgoing handoff events (a call leaves its current cell) [2, 43].

5.5.1 Bandwidth Degrades

Consider a saturated cell containing n adaptive ongoing calls. When a new or handoff call arrives, the allocated bandwidth of ongoing calls can be degraded to smaller values to accommodate the new or handoff call. Denote the utility function of the ith ongoing call as $u_i(b_i)(1 \leq i \leq n)$ and its current allocated bandwidth as β_i; thus, the degradable utility function of the ith ongoing call can be written as $u_i^{\downarrow}(b_i^{\downarrow}) = u_i\ (\beta_i - b_i^{\downarrow})\ (0 \leq b_i^{\downarrow} \leq \beta_i - b_i^{\min}$ where $b_i^{\downarrow}$ and $b_i^{\min}$ are the bandwidth degrades and the minimum bandwidth requirement of the call, respectively. Figure 5.7 illustrates the degradable utility function of the ith ongoing call when it belongs to non-real-time traffic and adaptive real-time traffic, respectively.

Also denote the utility function of the new or handoff call as $u_{n+1}(b_{n+1})$. The objective of bandwidth degrades is to find the bandwidth degrades profile $\{b_i^{\downarrow}\}$ for all adaptive

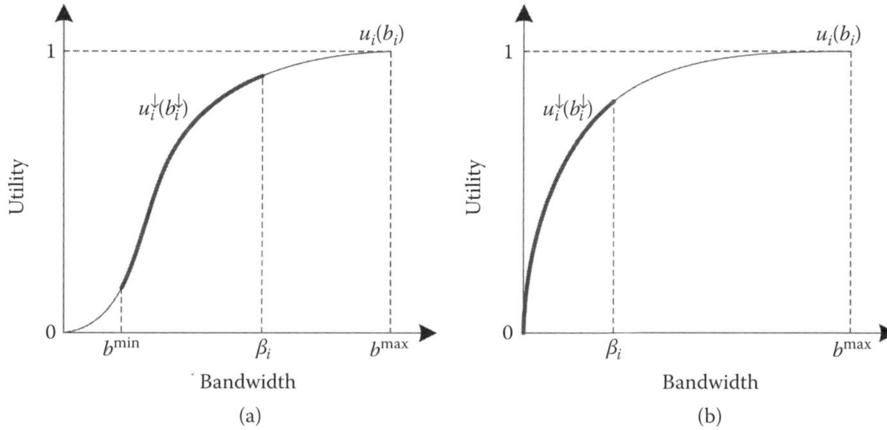

FIGURE 5.7 The degradable utility function of the ith ongoing call. (a) adaptive real-time traffic; (b) non-real-time traffic.

ongoing calls and the allocated bandwidth b_{n+1} for the new or handoff call to maximize the utility sum of all calls subject to bandwidth constraints, i.e.,

$$\text{maximize} \left(\sum_{i=1}^{n} u_i^{\downarrow}(b_i^{\downarrow}) \right) + u_{n+1}(b_{n+1}) \tag{5.12}$$

$$\text{subject to } 0 \le b_i^{\downarrow} \le \beta_i - b_i^{\min}, \tag{5.13}$$

$$\left(\sum_{i=1}^{n} (\beta_i - b_i^{\downarrow}) \right) + b_{n+1} \le B, \tag{5.14}$$

where B is the bandwidth capacity of the cell.

Figure 5.8 illustrates the procedure of bandwidth degrades, where b_i^{cur} and b_i^{adapt} are the current bandwidth allocation and the bandwidth allocation after bandwidth adaptation of the ith adaptive ongoing call, respectively.

5.5.2 Bandwidth Upgrades

Assume that in an overloaded cell when a call is terminated or handed off to another cell, there are n adaptive ongoing calls that have not received their maximum bandwidth requirements. The released bandwidth from the terminated or outgoing handoff call (denoted by β) can be used to upgrade these ongoing calls to enhance their quality and increase network bandwidth utilization. Again, denote the utility function of the ith ongoing call as $u_i(b_i)$ $(1 \le i \le n)$ and its current allocated bandwidth as β_i; thus, the

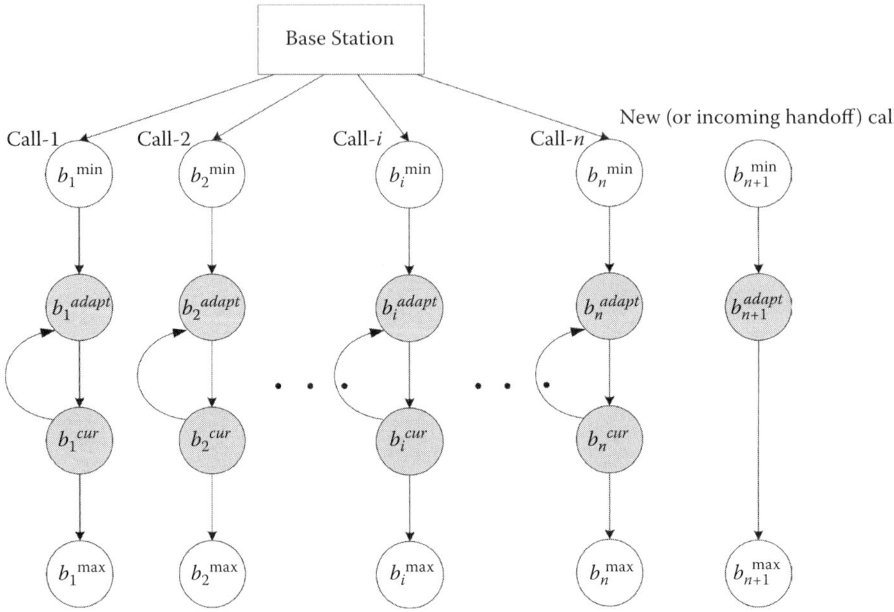

FIGURE 5.8 Bandwidth degrades procedure.

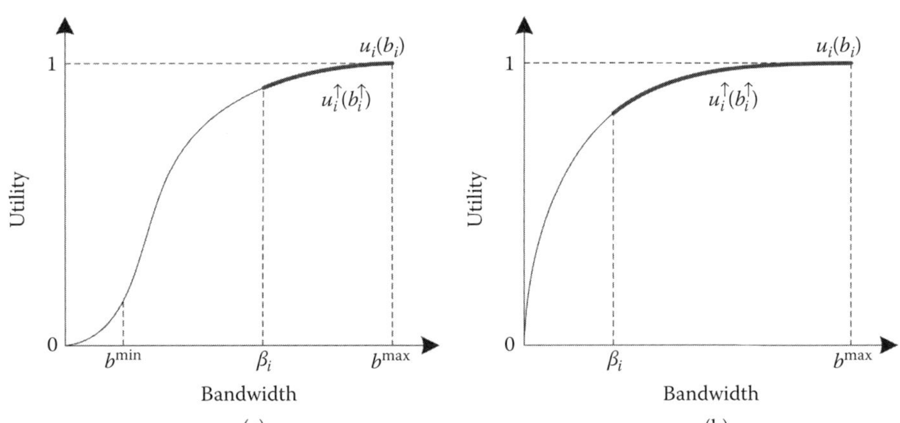

FIGURE 5.9 The upgradable utility function of the ongoing call. (a) Adaptive real-time traffic; (b) non-real-time traffic.

upgradable utility function of the ith ongoing call can be written as $u_i^{\uparrow}(b_i^{\uparrow}) = u_i\,(\beta_i + b_i^{\uparrow})$ $(0 \le b_i^{\uparrow} \le b_i^{max} - \beta_i)$, where $b_i^{\uparrow}$ and b_i^{max} are the bandwidth upgrades and the maximum bandwidth requirement of the call, respectively. Figure 5.9 illustrates the upgradable utility function of the ith ongoing call when it belongs to non-real-time traffic and adaptive real-time traffic, respectively.

The objective of bandwidth upgrades is to find the bandwidth upgrades profile $\{b_i^{\uparrow}\}$ for all ongoing calls to maximize the sum of their utilities subject to bandwidth constraints, i.e.,

$$\text{maximize} \sum_{i=1}^{n} u_i^{\uparrow}(b_i^{\uparrow}) \tag{5.15}$$

$$\text{subject to } 0 \le b_i^{\uparrow} \le b_i^{\max} - \beta_i, \tag{5.16}$$

$$\sum_{i=1}^{n} b_i^{\uparrow} \le \beta. \tag{5.17}$$

Figure 5.10 illustrates the bandwidth upgrades procedure when a call is terminated in a cell containing n adaptive ongoing calls, where b_i^{cur} and b_i^{adapt} are the current bandwidth allocation and the bandwidth allocation after bandwidth adaptation of the ith adaptive ongoing call, respectively.

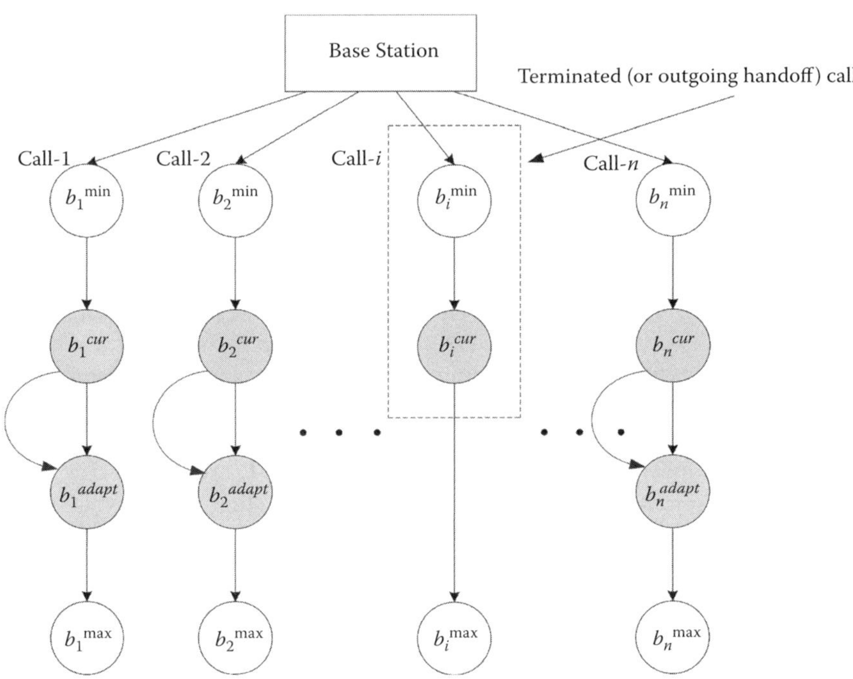

FIGURE 5.10 Bandwidth upgrades procedure.

5.6 The Proposed Utility-Maximization Algorithm

The essence of utility-maximization bandwidth adaptation is to maximize the utility sum of n utility functions subject to bandwidth constraints. After quantization, each utility function becomes a function represented by a list of <bandwidth, utility> points in the increasing order of bandwidth, i.e.,

$$u_i(b_i) = \left(< b_i^1, u_i^1 >,\ < b_i^2, u_i^2 >,\ \ldots,\ < b_i^{K_i}, u_i^{K_i} > \right),$$

where K_i is the maximum <bandwidth, utility> level. Thus, bandwidth adaptation maximizes the utility sum of n quantized utility functions, which can be arranged as follows:

$$u_1(b_1) = \left(< b_1^1, u_1^1 >,\ < b_1^2, u_1^2 >,\ \ldots, < b_1^{K_1}, u_1^{K_1} > \right)$$

$$u_2(b_2) = \left(< b_2^1, u_2^1 >,\ < b_2^2, u_2^2 >,\ \ldots, < b_2^{K_2}, u_2^{K_2} > \right)$$

$$\vdots$$

$$u_n(b_n) = \left(< b_n^1, u_n^1 >,\ < b_n^2, u_n^2 >,\ \ldots, < b_n^{K_n}, u_n^{K_n} > \right)$$

The utility-maximization problem is NP-hard, and finding optimal solutions has exponential time complexity [22, 25]. In wireless networks bandwidth adaptation needs to be performed in real time to support the frequent bandwidth fluctuations. Therefore, in this section, an efficient search-tree-based utility-maximization algorithm is designed to reduce the computational complexity.

The algorithm can be formulated by a tree as illustrated in Figure 5.11. The quantized utility function of each call is represented by a branch in the tree, and the <bandwidth, utility> points of each utility function are represented by the nodes in the branch. The nodes of each branch are laid out downwards to reflect the bandwidth allocation order, i.e., the second <bandwidth, utility> point of a utility function is connected to the first <bandwidth, utility> point, and so on. The tree contains n branches, and the number of branches is the same as the number of utility functions. Each branch (utility function) is associated with ten variables b_i^{cur}, u_i^{cur}, b_i^{next}, u_i^{next}, b_i^{temp}, u_i^{temp}, b_i^{req}, r_i^{cur}, r_i^{temp}, and $r_i^{\max}$ (all notation can be found in Table 5.1). The algorithm allocates the bandwidth in a greedy fashion based on utility generation ratio, i.e., it gives priority to the <bandwidth, utility> points with higher utility generation ratio. The utility generation ratio r is calculated by dividing the utility increase by bandwidth increase between two <bandwidth, utility> points. The pseudo-code of the algorithm is as follows:

```
(1) b^avail = B
    for each call-i
      initialize b_i^cur, u_i^cur, b_i^next, u_i^next, b_i^temp and u_i^temp to be at
    the first level
      b_i^req = b_i^cur+1 - b_i^cur
```

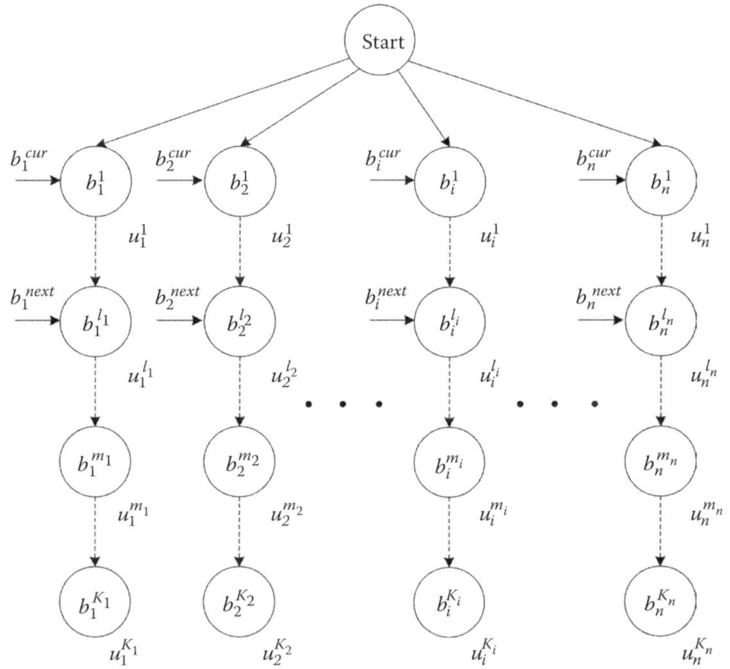

FIGURE 5.11 Bandwidth allocation tree.

TABLE 5.1 Notation for the Utility-Maximization Algorithm

B	The total available bandwidth to be allocated
b^{avail}	The current available bandwidth to be allocated
b_i^{cur}	The current bandwidth allocation of the ith call
u_i^{cur}	The current achieved utility of the ith call
b_i^{next}	The next possible bandwidth allocation of the ith call
u_i^{next}	The next possible achieved utility of the ith call
b_i^{temp}	The temporary bandwidth allocation of the ith call
u_i^{temp}	The temporary achieved utility of the ith call
b_i^{req}	The required bandwidth for upgrading the current bandwidth allocation of the ith call to its next-higher bandwidth level
$b^{req,\max}$	The maximum required bandwidth for upgrading the current bandwidth allocation of the call to its next-higher bandwidth level among all n calls, i.e., $b^{req,\max} = \max\{b_i^{req}\}$ $(1 \le i \le n)$
r_i^{cur}	The current utility generation ratio of the ith call
r_i^{temp}	The temporary utility generation ratio of the ith call
$r_i^{\max}$	The maximum utility generation ratio of the ith call

$r_i^{cur} = (u_i^{cur+1} - u_i^{cur})/b_i^{req}$
$r_i^{temp} = 0$
$r_i^{\max} = 0$

(2) for each call-i
  while $(b_i^{temp} < b_i^{K_i})$
    $b_i^{temp} = b_i^{temp+1}$
    $r_i^{temp} = (u_i^{temp} - u_i^{cur})/(b_i^{temp} - b_i^{cur})$
    if $(r_i^{temp} > r_i^{\max})$
      $b_i^{next} = b_i^{temp}$
      $r_i^{\max} = r_i^{temp}$

(3) among all calls find the largest b_i^{req} denoted by $b^{req,\max}$
  if $(b^{avail} \leq b^{req,\max})$
    among all calls with $b_i^{req} \geq b^{avail}$ find the call with the highest r_i^{cur} denoted by call-k
    $b_k^{cur} = b_k^{cur} + b^{avail}$
    return b_i^{cur} as the bandwidth allocation for each call

(4) among all calls find the call with the largest $r_i^{\max}$ denoted by call-j
  if $(b^{avail} \geq (b_j^{next} - b_j^{cur}))$
    $b^{avail} = b^{avail} - (b_j^{next} - b_j^{cur})$
    $b_j^{cur} = b_j^{next}$
    $b_j^{temp} = b_j^{next}$
    $r_j^{temp} = 0$
    $r_j^{\max} = 0$
    if $(b_j^{cur} < b_j^{K_j})$
      $b_j^{req} = b_j^{cur+1} - b_j^{cur}$
      $r_j^{cur} = (u_j^{cur+1} - u_j^{cur})/b_j^{req}$
    else
      $b_j^{req} = 0$
      $r_j^{cur} = 0$
  else
      $b_j^{cur} = b_j^{cur} + b^{avail}$
      return b_i^{cur} as the bandwidth allocation for each call

(5) for call-j found in step (4)
  while $(b_j^{temp} < b_j^{K_j})$
    $b_j^{temp} = b_j^{temp+1}$
    $r_j^{temp} = (u_j^{temp} - u_j^{cur})/(b_j^{temp} - b_j^{cur})$
    if $(r_j^{temp} > r_j^{\max})$
      $b_j^{next} = b_j^{temp}$
      $r_j^{\max} = r_j^{temp}$

(6) Go to step (3)

In step (1), for each call the algorithm initializes its associated variables and calculates the required bandwidth b_i^{req} for upgrading the call to the next-higher bandwidth level and the current utility generation ratio r_i^{cur}.

In step (2), for each call the algorithm increases its temporary bandwidth allocation b_i^{temp} by one level, and then calculates its temporary utility generation ratio r_i^{temp}. If r_i^{temp} is greater than the maximum utility generation ratio $r_i^{\max}$, the algorithm upgrades its next possible bandwidth allocation b_i^{next} to b_i^{temp} and assigns r_i^{temp} to $r_i^{\max}$. The above process is repeated until every node of the call has been investigated and the node with the maximum utility generation ratio is the next possible bandwidth allocation node in the corresponding call.

Step (3) checks if the current available bandwidth b^{avail} is less than or equal to the maximum required bandwidth $b^{req,\max}$ for upgrading the call to the next-higher bandwidth level among all calls. If the answer is yes, then b^{avail} is allocated to the call with the highest current utility generation ratio r_i^{cur} among all calls with $b_i^{req} \geq b^{avail}$, and the algorithm terminates.

The algorithm keeps track of the maximum utility generation ratio $r_i^{\max}$ for each call. In step (4), it chooses the call with the highest $r_i^{\max}$, denoted as call-j. If there is enough bandwidth available, the algorithm upgrades its current bandwidth allocation b_j^{cur} and temporary bandwidth allocation b_j^{temp} to its next possible bandwidth allocation b_j^{next}; otherwise, the current available bandwidth b^{avail} is allocated to call-j and the algorithm terminates.

Subsequently, for call-j it has the same current and next possible bandwidth allocation. Therefore, in step (5) the algorithm updates its next possible bandwidth allocation using the same approach described in step (2). After finding the new next possible bandwidth allocation for call-j, the algorithm goes back to step (3) to execute the above procedure repeatedly until it terminates due to insufficient bandwidth (it is assumed that the total available bandwidth B cannot satisfy the maximum bandwidth requirements of all calls).

To provide QoS guarantees for multimedia wireless networks, apart from the bandwidth adaptation algorithm, two supplementary bandwidth management functions, i.e., CAC and bandwidth reservation, have also been integrated into the proposed bandwidth adaptation scheme.

With CAC policy, when a new call requests admission into the network, the cell first attempts to allocate the maximum bandwidth requirement to the new call. If there is enough bandwidth available in the cell, the CAC accepts the new call by assigning it the maximum bandwidth requirement. If there is not enough bandwidth, the bandwidth adaptation algorithm is invoked to free some bandwidth from the existing ongoing calls. After bandwidth adaptation, if the sum of the available bandwidth in the cell plus the freed bandwidth according to the bandwidth adaptation algorithm is greater than or equal to the desired bandwidth requirement of the new call, the new call is admitted; otherwise, the new call is blocked.

The objective of CAC is to admit new calls and handoff calls as much as possible under the guidance of the utility-maximization bandwidth adaptation algorithm. However, new calls and handoff calls are competing for the usage of the finite network bandwidth;

the number of admitted new calls and handoff calls cannot be increased simultaneously under network congestion—it is a matter of trade-off. It is well known and obvious that, from the end users' perspective, the dropping of a handoff call during its service session is more undesirable than the blocking of a new call at its beginning. Based on this fact, some bandwidth can be reserved for handoff calls to reduce their dropping probability. But bandwidth reservation should be used carefully since it may decrease the bandwidth utilization and increase the new call blocking probability. The proposed bandwidth reservation policies differentiate between class I (real-time) and class II (non-real-time) traffic. A certain amount of bandwidth is reserved exclusively for class I traffic; the motivation is that real-time traffic would suffer an actual loss by being dropped.

When a handoff call requests admission into the network, the cell first attempts to allocate the maximum bandwidth requirement to the handoff call. If there is enough bandwidth available in the cell, the CAC accepts the handoff call by assigning it the maximum bandwidth requirement. If there in not enough bandwidth, the bandwidth adaptation algorithm is invoked to free some bandwidth from the existing ongoing calls. After bandwidth adaptation, the CAC checks the traffic class to which the handoff call belongs. If the handoff call is a class I call and the sum of the available bandwidth in the cell plus the freed bandwidth according to the bandwidth adaptation algorithm plus the available reserved bandwidth is greater than or equal to the minimum bandwidth requirement of the handoff call, the handoff call is admitted. Otherwise, the handoff call is dropped. If the handoff call is a class II call, it is accepted as long as there is some bandwidth available in the cell that the call is moving in to. It will only be dropped when there is not bandwidth available at all after bandwidth adaptation. The reserved bandwidth is not available to class II traffic because it is assumed that a class II call, although inconvenienced by being dropped, would be able to resume its transmission at a later time without any significant loss due to its elastic characteristics.

Note that to provide higher priority to handoff calls over new calls, apart from reserving bandwidth for handoff calls, the proposed scheme also admits new calls more strictly than handoff calls. New calls are only accepted if enough bandwidth is available to accommodate them at the desired level. When the available bandwidth in the cell is less than the desired amount of bandwidth, a new connection is always rejected, whereas a handoff connection may be accepted if the minimum required bandwidth can be provided.

When a call is completed or handed off from a current cell to another, if the call is a new call when it is admitted in the current cell, then its released bandwidth is used to upgrade other ongoing calls or saved for future usage, depending on whether there are any ongoing calls served with a bandwidth less than b^{max}. If the terminated call is a handoff call, and when admitted in the current cell it has used the reserved bandwidth, then the released reserved bandwidth is returned to the reserved bandwidth pool for future incoming handoffs and the released cell bandwidth is used to upgrade other ongoing calls or saved for future usage, depending on whether there are any ongoing calls served with a bandwidth less than b^{max}.

The pseudo-code for the integrated utility-maximization bandwidth adaptation scheme is as follows, and the notation can be found in Table 5.2:

TABLE 5.2 Notation for the CAC

b^{avail_cell}	The current available cell bandwidth to be allocated
$b^{avail\_reserved}$	The current available reserved bandwidth to be allocated
b^{min}	The minimum bandwidth requirement of the call
b^{max}	The maximum bandwidth requirement of the call
$b^{desired}$	The desired bandwidth requirement of the call
$b^{degrade}$	The freed bandwidth after performing bandwidth degrade
$b^{upgrade}$	The used bandwidth after performing bandwidth upgrade
$b^{released\_cell}$	The released cell bandwidth when a call is terminated due to its completion or outgoing handoff
$b^{released\_reserved}$	The released reserved bandwidth when a call is terminated due to its completion or outgoing handoff

New Call Arrival

```
if (b^avail_cell ≥ b^max)
  assign b^max to the new call;
  b^avail_cell = b^avail_cell - b^max;
else
  perform bandwidth degrade;
  if (b^avail_cell + b^degrade ≥ b^desired)
    assign min(b^avail_cell + b^degrade, b^max) to the new call;
    b^avail_cell = 0;
  else
    reject the new call request;
```

Handoff Call Arrival

```
if (b^avail_cell ≥ b^max)
  assign b^max to the handoff call;
  b^avail_cell = b^avail_cell - b^max;
else
  perform bandwidth degrade;
  if (is class I call)
    if (b^avail_cell + b^degrade ≥ b^min)
    assign min(b^avail_cell + b^degrade, b^max) to the handoff call;
    b^avail_cell = 0;
  else
    if (b^avail_cell + b^degrade + b^avail_reserved ≥ b^min)
      assign b^min to the handoff call;
      b^avail_cell = 0;
      b^avail_reserved = b^min - (b^avail_cell + b^degrade);
    else
      reject the handoff call
  else // is class II call
    if (b^avail_cell + b^degrade ≠ 0)
      assign min(b^avail_cell + b^degrade, b^max) to the handoff
call;
```

```
b^avail_cell = 0;
    else
      reject the handoff call request;
```

Call Departure

```
if (is new call)
  if (every call has received b^max)
      b^avail_cell = b^avail_cell + b^released;
  else
    perform bandwidth upgrade;
    b^avail_cell = b^avail_cell + b^released - b^upgrade;
else // is handoff call
  b^avail_reserved = b^avail_reserved + b^released_reserved;
  if (every call has received b^max)
      b^avail_cell = b^avail_cell + b^released_cell;
  else
    perform bandwidth upgrade;
    b^avail_cell = b^avail_cell + b^released_cell - b^upgrade;
```

5.7 Simulation Modeling

A wireless network simulator has been developed to evaluate the performance of the proposed bandwidth adaptation scheme.

5.7.1 Network Model

The simulated network consists of 36 (6 × 6) hexagonal cells. The diameter of each cell is 1 km, and each cell has a total bandwidth capacity of 30 Mbps. The layout of the simulation model is shown in Figure 5.12. To avoid the edge effect of the finite network size,

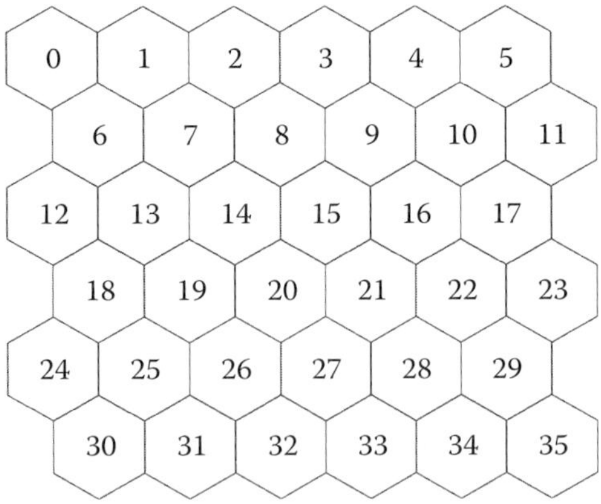

FIGURE 5.12 The layout of the wireless network model.

wrap-around is applied to the edge cells so that each cell has six neighboring cells. The similar network model has also been used by previous work [15, 29].

5.7.2 Traffic Model

Six representative groups of multimedia traffic belonging to the two traffic classes, i.e., real-time traffic (class I) and non-real-time traffic (class II), are considered. They are typical traffic seen in multimedia wireless networks, and similar traffic has been used by previous work [1, 8, 12, 15, 29, 34]. Each group of traffic is associated with an appropriate utility function, and all calls belonging to the same traffic group are assumed to have the same bandwidth requirements and utility function. The exact characteristics of the traffic are shown in Table 5.3.

From Table 5.3 it can be seen that traffic groups 0 and 1 belong to hard real-time traffic, and according to the utility functions formulation, their utility functions can be achieved from b^{min} and u^{max} directly. Traffic group 2 belongs to adaptive real-time traffic, and the two parameters k_1 and k_2 of its utility function can be calculated from b^{min}, u^{max}, and b^{intr}. Traffic groups 3, 4, and 5 belong to non-real-time traffic, and the

TABLE 5.3 Traffic Characteristics for the Simulation

App. Group	Traffic Class	Bandwidth Requirement (Mbps)	Average Connection Duration	Example	Utility Function (b is Mbps)
0	I (hard real-time)	$b^{min} = 0.03$ $b^{desired} = 0.03$	3 minutes	Voice service and audio phone	$\begin{cases} 1, & b \geq 0.03 \\ 0, & b < 0.03 \end{cases}$ $u^{max} = 1$
1	I (hard real-time)	$b^{min} = 0.25$ $b^{desired} = 0.25$	5 minutes	Video phone and video conference	$\begin{cases} 1, & b \geq 0.25 \\ 0, & b < 0.25 \end{cases}$ $u^{max} = 1$
2	I (adaptive real-time)	$b^{min} = 1$ $b^{intr} = 1.5$ $b^{desired} = 2$ $b^{max} = 6$	10 minutes	Interactive multimedia and video on demand	$1 - e^{-\frac{1.8b^2}{8.3+b}}$ $u^{max} = 0.99$
3	II (non-real-time)	$b^{min} = 0$ $b^{desired} = 0.003$ $b^{max} = 0.002$	30 seconds	E-mail, paging, and fax	$1 - e^{-\frac{4.6b}{0.02}}$ $u^{max} = 0.99$
4	II (non-real-time)	$b^{min} = 0$ $b^{desired} = 0.1$ $b^{max} = 0.5$	3 minutes	Remote login and data on demand	$1 - e^{-\frac{4.6b}{0.5}}$ $u^{max} = 0.99$
5	II (non-real-time)	$b^{min} = 0$ $b^{desired} = 1.5$ $b^{max} = 10$	2 minutes	File transfer and retrieval service	$1 - e^{-\frac{4.6b}{10}}$ $u^{max} = 0.99$

parameter of their utility functions can be calculated from b^{max} and u^{max}. Each group of traffic is also associated with a desired amount of bandwidth $b^{desired}$, which is defined as the bandwidth that enables its utility function to receive half of u^{max}. The traffic is generated in the following way. The new call arrival of a group-i ($i = 1, \ldots, 6$) traffic is assumed to follow Poisson distribution with mean rate λ_i^{new} (calls/second/cell), and all six groups of traffic are generated with equal probability, i.e., $\lambda_1^{New} = \lambda_2^{New} = \cdots = \lambda_6^{New}$. Handoff call arrivals of group-i traffic are assumed to be proportional to the new call arrivals by $\lambda_i^{handoff} = \alpha\lambda_i^{new}$, where α is set to be 0.5 in the experiments. The call holding time (CHT) of group-i traffic is assumed to follow exponential distribution with mean $1/\mu_i$. For the mobility characterization, the cell residence time (CRT), i.e., the amount of time during which a group-i call stays in a cell before handoff, is assumed to follow an exponential distribution with mean $1/\eta_i$ [32, 46]. Here, η_i is also called the handoff rate, which means how fast the calls hand off.

5.8 Numerical Results

To highlight the performance of the proposed utility-maximization bandwidth adaptation scheme, it is compared with a non-adaptive bandwidth allocation scheme and RBBS [15].

The non-adaptive scheme is simulated assuming that a call must have its maximum bandwidth allocated to be admitted, and once accepted, its bandwidth cannot be changed throughout the lifetime of the call. If such bandwidth is not available, the call is either blocked or dropped, depending on whether the call is a new or handoff call. RBBS is a well-known adaptive bandwidth allocation scheme for providing QoS in multimedia wireless networks. With RBBS, each call has a minimum bandwidth requirement b^{min} and a maximum bandwidth requirement b^{max}. The actual borrowable bandwidth (*ABB*) of the call is calculated as a fraction of the difference between the minimum and maximum bandwidth, i.e., $ABB = f \times (b^{max} - b^{min})$, where f is a local parameter of the cell and $f = 0.5$. *ABB* is divided into a number of equal levels, i.e., ABB/λ, where λ is also a local parameter of the cell and $\lambda = 10$. When a new or handoff call arrives and the network is overloaded, some bandwidth can be freed from the existing ongoing calls to accept the new or handoff call. To ensure a smooth bandwidth adaptation and guarantee the bandwidth fairness criteria defined by the authors, every time the bandwidth adaptation happens, only one share of the *ABB*, i.e., ABB/λ, can be borrowed from each of the ongoing calls. The CAC policy of RBBS treats new calls more strictly than handoff calls. New calls are only accepted if enough bandwidth is available to accommodate them at the same bandwidth level as the existing ongoing calls, while handoff calls can be accepted as long as their minimum bandwidth requirements can be met. When enough bandwidth becomes available due to call termination or outgoing handoff, if there are calls degraded below the level of the existing ongoing calls due to handoff, these calls will be upgraded first. If all ongoing calls in the cell are operating in the same bandwidth level, the available bandwidth is used to upgrade each of these ongoing calls to its next-higher bandwidth level.

In order to make an objective and fair comparison, the non-adaptive and RBBS experiments use the same network and traffic model as that used in the proposed scheme. In the experiments, apart from the traditional performance metrics including call blocking probability and handoff dropping probability, a newly introduced performance metric (i.e., average cell utility) is used. Cell utility means the sum of the utilities of all ongoing calls within a cell. Average cell utility is calculated as follows: every time a bandwidth adaptation occurs, the achieved cell utility is re-calculated and added to the total accumulated cell utility. Thus at the end of the simulation the average cell utility can be obtained by dividing the total accumulated cell utility by the bandwidth adaptation frequency.

Figure 5.13 illustrates the average cell utility of the three schemes as a function of the call arrival rate. As expected, the proposed utility-maximization scheme achieves higher utility than the other two schemes. The reason behind this is that the proposed scheme works in a fashion to maximize the utility sum of all ongoing calls every time the bandwidth adaptation happens, whereas the other two schemes do not.

Figures 5.14 and 5.15 demonstrate the handoff dropping probabilities for class I and class II traffic, respectively. For class I handoff calls the non-adaptive scheme shows an extremely high dropping ratio since it does not utilize the bandwidth flexibility of ongoing calls to free bandwidth to accept the handoff calls under network congestion. The handoff dropping probabilities of the proposed scheme and RBBS are much lower than that of the non-adaptive scheme (RBBS slightly outperforms the proposed scheme) in both moderate and high traffic load because these two schemes not only degrade ongoing calls to free bandwidth for handoff calls when the network is overloaded, but also give class I traffic handoff calls exclusive use of the reserved bandwidth to protect them from the actual loss by being dropped. In terms of class II traffic dropping probabilities, the non-adaptive scheme still features the highest dropping ratio. The dropping ratios of the proposed scheme and RBBS have been reduced to a negligible level even though the reserved bandwidth is not available to the handoff calls of class II traffic because class II calls do not have minimum bandwidth requirements and can be accepted as long as there is some free bandwidth available in the network.

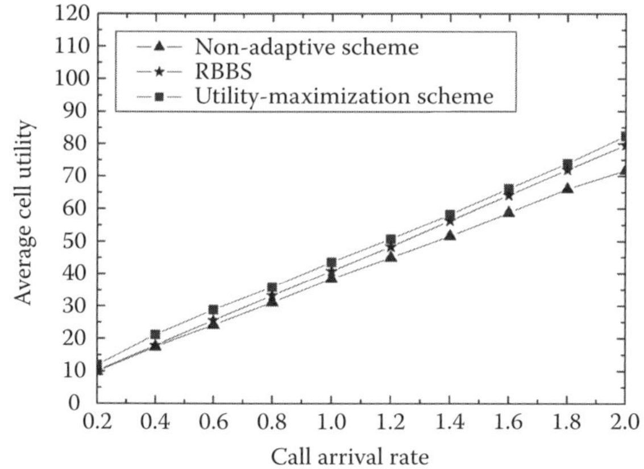

FIGURE 5.13 Average cell utility.

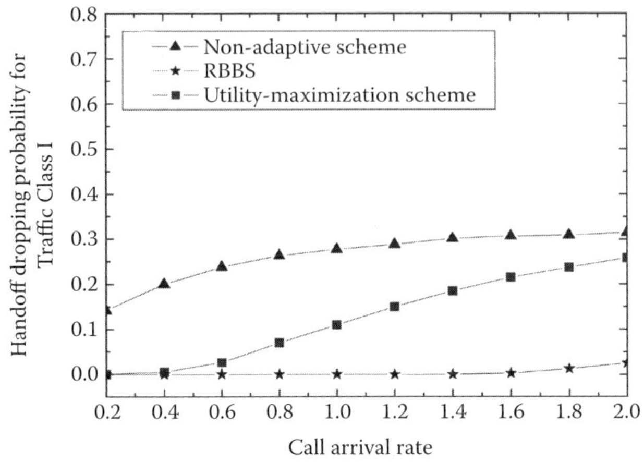

FIGURE 5.14 Handoff dropping probability for traffic class I.

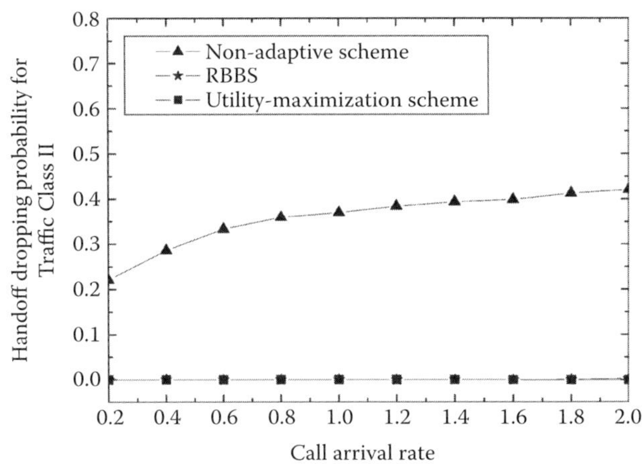

FIGURE 5.15 Handoff dropping probability for traffic class II.

Figure 5.16 compares the call blocking probabilities of the three schemes. It can be observed that the proposed scheme allows a significant improvement in the blocking ratio compared with the other two schemes while keeping the handoff dropping probability low. These results indicate that the proposed scheme can serve many more calls than the other two schemes. The proposed scheme outperforms RBBS partially because with RBBS only half of the adaptive bandwidth is degradable from each ongoing call, and every time the bandwidth adaptation happens, only one share (a share is a predefined parameter) of the degradable bandwidth can be borrowed, while with the proposed scheme the allocated bandwidth of the adaptive ongoing calls can be degraded down to the minimum level in one step to accommodate the new calls. It can be observed

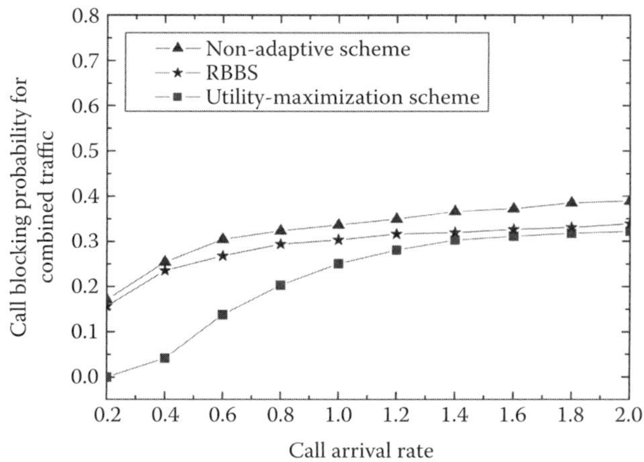

FIGURE 5.16 Call blocking probability for combined traffic.

that at the call arrival of 0.2, the blocking ratio of the proposed scheme is significantly lower than that of RBBS, and the advantage becomes smaller when the call arrival rate increases. The underlying reason is that RBBS maintains an operating bandwidth level among ongoing calls, and a new call request can only be accepted if the available bandwidth can bring it to the same bandwidth level as other ongoing calls. At low traffic load, the operating bandwidth level is low since the ongoing calls are only degraded moderately, thus making the CAC threshold high. Therefore, new calls are more likely to be rejected due to insufficient bandwidth. When the traffic load increases, the operating bandwidth level becomes higher, i.e., the CAC threshold becomes lower, and thus new calls are more easily admitted.

Figure 5.17 shows the values of bandwidth utilization for the three schemes under various traffic loads. Bandwidth utilization is the percentage of the total bandwidth

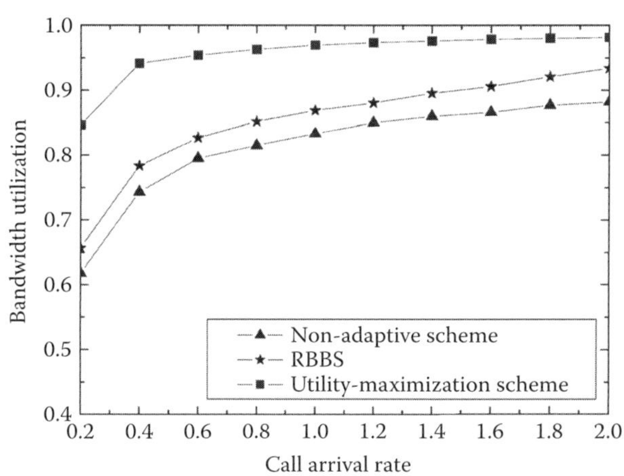

FIGURE 5.17 Bandwidth utilization.

actually used by all calls in a cell. From the diagram it can be seen that for all schemes, the bandwidth utilization becomes higher as the call arrival increases, and the proposed scheme uses the bandwidth much more efficiently than the other two schemes. The reason why the proposed scheme outperforms RBBS is that to maintain the operating bandwidth level, RBBS may waste the bandwidth resource when it rejects a new call when there is free bandwidth, but the available bandwidth cannot bring the new call to the operating bandwidth level of the cell.

5.9 Conclusions

This chapter introduces the utility-based bandwidth adaptation for multimedia wireless networks. To deal with multimedia applications with different QoS requirements, multimedia applications are classified into various classes according to their adaptive characteristics. Based on the availability of the utility-based adaptive multimedia traffic model, a utility-maximization bandwidth adaptation scheme is proposed for QoS provisioning in multimedia wireless networks. With the proposed scheme, each call in the network is assigned a utility function according to its adaptive characteristics. When there are bandwidth fluctuations, the allocated bandwidth of the ongoing calls can be adapted dynamically to maximize the total network utility. After mathematical formulation for the bandwidth adaptation processes is given, including bandwidth degrades and bandwidth upgrades, an efficient search-tree-based algorithm is proposed to solve the utility-maximization problem. To provide QoS guarantees to the new and handoff calls, CAC and bandwidth reservation policies have also been incorporated into the proposed scheme. Simulation experiments have been carried out to highlight the performance of the proposed bandwidth adaptation scheme compared to that of a non-adaptive scheme and RBBS. Numerical results have clearly demonstrated the superior performance of the proposed scheme.

References

[1] Ahn, K. M., and Kim, S. 2003. Optimal bandwidth allocation for bandwidth adaptation in wireless multimedia networks. *Computers and Operations Research* 30:1917–29.

[2] Aljadhai, A., and Znati, T. F. 2000. A bandwidth adaptation scheme to support QoS requirements of mobile users in wireless environments. In *Proceedings of the 9th International Conference on Computer Communications and Networks*, pp. 34–39.

[3] Bharghavan, V., et al. 1998. The timely adaptive resource management architecture. *IEEE Personal Communications* 5:20–31.

[4] Bianchi, G., Campbell, A. T., and Liao, R. R. F. 1998. On utility-fair adaptive services in wireless networks. In *Proceedings of the 6th International Workshop on Quality of Service (IWQOS '98)*, pp. 256–67.

[5] Bocheck, P., Nakajima, Y., and Chang, S. F. 1999. Real-time estimation of subjective utility functions for MPEG-4 video objects. In *Proceedings of the 9th International Packet Video Workshop (PacketVideo '99).*

[6] Breslau, L., and Shenker, S. 1998. Best-effort versus reservations: A simple comparative analysis. In *Proceedings of the ACM SIGCOMM '98 Conference on Applications, Technologies, Architectures, and Protocols for Computer Communication*, pp. 3–16.

[7] Chakrabarti, S., and Wang, R. 1994. Adaptive control for packet video. In *Proceedings of the International Conference on Multimedia Computing and Systems*, pp. 56–62.

[8] Chang, J. Y., and Chen, H. L. 2006. A borrowing-based call admission control policy for mobile multimedia wireless networks. *IEICE Transactions on Communications* E89-B:2722–32.

[9] Chen, H., et al. 2003. *Radio resource management for multimedia QoS support in wireless networks*. Dordrecht, The Netherlands: Kluwer Academic Publishers.

[10] Cheng, P. Y., Li, J., and Kuo, C. C. J. 1997. Rate control for an embedded wavelet video coder. *IEEE Transactions on Circuits and Systems for Video Technology* 7:696–702.

[11] Chou, C. T., and Shin, K. G. 2004. Analysis of adaptive bandwidth allocation in wireless networks with multilevel degradable quality of service. *IEEE Transactions on Mobile Computing* 3:5–17.

[12] Curescu, C., and Nadjm-Tehrani, S. 2005. Time-aware utility-based resource allocation in wireless networks. *IEEE Transactions on Parallel and Distributed Systems* 16:624–36.

[13] Das, S. K., et al. 2003. A framework for bandwidth degradation and call admission control schemes for multiclass traffic in next-generation wireless networks. *IEEE Journal on Selected Areas in Communications* 21:1790–1802.

[14] Duffield, N. G., Ramakrishnan, K. K., and Reibman, A. R. 1998. SAVE: An algorithm for smoothed adaptive video over explicit rate networks. *IEEE/ACM Transactions on Networking* 6:717–28.

[15] El-Kadi, M., Olariu, S., and Abdel-Wahab, H. 2002. A rate-based borrowing scheme for QoS provisioning in multimedia wireless networks. *IEEE Transactions on Parallel and Distributed Systems* 13:156–66.

[16] Fei, Y., Wong, V. W. S., and Leung, V. C. M. 2006. Efficient QoS provisioning for adaptive multimedia in mobile communication networks by reinforcement learning. *Mobile Networks and Applications* 11:101–10.

[17] Gomez, J., Campbell, A. T., and Morikawa, H. 1998. A systems approach to prediction, compensation and adaptation in wireless networks. In *Proceedings of the 1st ACM International Workshop on Wireless Mobile Multimedia*, pp. 92–100.

[18] Hartung, J., et al. 1998. A real-time scalable video codec for collaborative applications over packet networks. In *Proceedings of the 6th ACM International Conference on Multimedia*, pp. 419–26.

[19] Huang, L., Kumar, S., and Kuo, C. C. J. 2004. Adaptive resource allocation for multimedia QoS management in wireless networks. *IEEE Transactions on Vehicular Technology* 53:547–58.

[20] ISO/IEC/JTC1/SC29/WG1. 2000. *JPEG 2000*. Part 1. *Final committee draft version 1.0*. ISO/IEC International Standard N1646R.

[21] ISO/IEC/JTC1/SC29/WG11. 2000. *Overview of the MPEG-4 standard.* ISO/IEC N3747.

[22] Kellerer, H., Pferschy, U., and Pisinger, D. 2004. *Knapsack problems.* Heidelberg: Springer.

[23] Kwon, T., et al. 1999. Measurement-based call admission control for adaptive multimedia in wireless/mobile networks. *IEEE Wireless Communications and Networking Conference (WCNC '99)* 2:540–44.

[24] Kwon, T., Choi, Y., and Das, S. K. 2002. Bandwidth adaptation algorithms for adaptive multimedia services in mobile cellular networks. *Wireless Personal Communications* 22:337–57.

[25] Lee, C., et al. 1999. On quality of service optimization with discrete QoS options. In *Proceedings of the 5th IEEE Real-Time Technology and Applications Symposium*, pp. 276–86.

[26] Li, X., Paul, S., and Ammar, M. 1998. Layered video multicast with retransmissions (LVMR): Evaluation of hierarchical rate control. In *Proceedings of the 17th Annual Joint Conference of the IEEE Computer and Communications Societies (INFOCOM '98)*, vol. 3, pp. 1062–72.

[27] Liao, R. R. F., Boukelee, P., and Campbell, A. T. 1999. Dynamic generation of bandwidth utility curves for utility-based adaptation. In *Proceedings of the 9th International Packet Video Workshop (PacketVideo '99).*

[28] Liao, R. R. F., and Campbell, A. T. 2001. A utility-based approach for quantitative adaptation in wireless packet networks. *Wireless Networks* 7:541–57.

[29] Malla, A., et al. 2003. A fair resource allocation protocol for multimedia wireless networks. *IEEE Transactions on Parallel and Distributed Systems* 14:63–71.

[30] McCanne, S., Vetterli, M., and Jacobson, V. 1997. Low complexity video coding for receiver-driven layered multicast. *IEEE Journal on Selected Areas in Communications* 15:983–1001.

[31] Nasser, N. 2005. Real-time service adaptability in multimedia wireless networks. In *Proceedings of the 1st ACM International Workshop on Quality of Service and Security in Wireless and Mobile Networks*, pp. 144–49.

[32] Nasser, N., and Hassanein, H. 2007. Enabling seamless multimedia wireless access through QoS-based bandwidth adaptation. *Wireless Communications and Mobile Computing* 7:53–67.

[33] Ngan, K. N., Chai, D., and Millin, A. 1996. Very low bit rate video coding using H.263 coder. *IEEE Transactions on Circuits and Systems for Video Technology* 6:308–12.

[34] Oliveira, C., Kim, J. B., and Suda, T. 1998. An adaptive bandwidth reservation scheme for high-speed multimedia wireless networks. *IEEE Journal on Selected Areas in Communications* 16:858–74.

[35] Rakocevic, V., Griffiths, J., and Cope, G. 2001. Performance analysis of bandwidth allocation schemes in multiservice IP networks using utility functions. In *Proceedings of the 17th International Teletraffic Congress (ITC '17).*

[36] Rakocevic, V. 2002. Dynamic bandwidth allocation in multi-class IP networks using utility functions. PhD thesis, Queen Mary, University of London.

[37] Rijkse, K. 1996. H.263: Video coding for low-bit-rate communication. *IEEE Communications Magazine* 34:42–45.

[38] Semeia, A. L. I. 2003. Wireless network performance analysis for adaptive bandwidth resource allocations. PhD thesis, Stevens Institute of Technology.

[39] Shenker, S. 1995. Fundamental design issues for the future Internet. *IEEE Journal on Selected Areas in Communications* 13:1176–88.

[40] Taubman, D., and Zakhor, A. 1996. A common framework for rate and distortion based scaling of highly scalable compressed video. *IEEE Transactions on Circuits and Systems for Video Technology* 6:329–54.

[41] Vickers, B. J., Albuquerque, C., and Suda, T. 1998. Adaptive multicast of multi-layered video: Rate-based and credit-based approaches. In *Proceedings of the 17th Annual Joint Conference of the IEEE Computer and Communications Societies (INFOCOM '98)*, vol. 3, pp. 1073–83.

[42] Wang, Y., Kim, J. G., and Chang, S. F. 2003. Content-based utility function prediction for real-time MPEG-4 video transcoding. In *Proceedings of the International Conference on Image Processing (ICIP '03)*, vol. 1, pp. 189–92.

[43] Xiao, Y., Chen, C. L. P., and Wang, Y. 2001. Optimal admission control for multi-class of wireless adaptive multimedia services. *IEICE Transactions on Communications* E84-B:795–804.

[44] Xiao, Y., et al. 2005. Proportional degradation services in wireless/mobile adaptive multimedia networks. *Wireless Communications and Mobile Computing* 5:219–43.

[45] Yeadon, N., et al. 1996. Filters: QoS support mechanisms for multipeer communications. *IEEE Journal on Selected Areas in Communications* 14:1245–62.

[46] Yeung, K. L., and Nanda, S. 1996. Channel management in microcell/macrocell cellular radio systems. *IEEE Transactions Vehicular Technology* 45:601–12.

6

An Extensive Survey and Taxonomy of MAC Protocols for Vehicular Wireless Networks

Hamid Menouar
Hitachi Europe

Fethi Filali
Eurecom Institute

Massimiliano Lenardi
Hitachi Europe

6.1 Introduction

Vehicular ad hoc networks (VANETs) represent a rapidly emerging, particularly challenging class of mobile ad hoc networks (MANETs). VANETs are distributed, self-organizing communication networks built up from traveling vehicles, and are thus characterized by very high speed and limited degrees of freedom in nodes' movement patterns. In contrast to classical nodes, in a vehicle there is no problem of storage and power resources limitation. Thus, in VANETs, wider transmission ranges and longer communication lifetimes are possible [1, 2]. In VANETs, vehicles are supposed to be equipped with some satellite geopositioning system, like the global positioning system (GPS) [3] or the future European positioning satellite system (GALILEO) [4], which allows each vehicle in the network to get its own physical position in real time. This position information is as useful for the application layer as it is for the Medium Access Control (MAC) and routing layers. These satellite geolocalization systems provide another source of useful information: a common clock that can be used for time synchronization in the network. Another advantage in VANETs is the nonrandom mobility of the nodes; roads are mapped and digitally available, and driving rules can be electronically represented as well.

The main disadvantage in VANETs, when compared to MANETs, belongs in the fact that vehicles' speed is relatively high, which makes the network topology change very fast and frequently and further decreases the stability of the transmission wireless channel.

The above particular features often make standard MANET protocols inefficient or unusable in VANETs, and this, combined with the huge impact that the deployment of VANET technologies could have on the automotive market, explains the growing effort in the development of communication protocols that are specific to vehicular networks.

The applications for VANETs can be roughly divided into two main categories: the ones related to active safety on the roads* and those dedicated to improve the comfort of drivers and passengers (imagine, for example, a roadside unit wanting to contact a vehicle for downloading a previously requested set of data). Both of these categories are important, but the first one is very sensitive, since human lives will depend on it.

For any application, in a wireless network the communicating terminal needs access to the radio channel in order to transmit or receive data. The radio channel (say, medium) is considered a common resource that several terminals in the same neighborhood can attempt to access at the same time. Since only one terminal should transmit on the same radio channel at the same time, some Medium Access Control (MAC) protocols are needed in wireless networks to efficiently share this common medium.

The existing and successful MANET MAC protocols are not suitable (as they are) for vehicular communications [5, 6]. Our goal in this chapter is not to design a novel MAC solution for VANETs, nor is it to adapt MANET existing solutions for VANETs. We rather describe some MAC schemes and protocols that can be adapted for VANETs, and expose the VANET-related application requirements from the MAC layer. We also discuss the VANET MAC protocols proposed in the open literature.

The remainder of this chapter is organized as follows. In section 6.2, we enumerate and discuss the main issues to consider when designing a new MAC protocol for wireless networks. Section 6.3 reviews the MAC protocols for MANETs. The MAC protocols proposed for wireless vehicular networks is the subject of section 6.4, and a qualitative comparison is offered in section 6.5. Section 6.6 provides a summary of this chapter.

---

* Active in informing drivers or acting on the vehicle in order to avoid accidents, instead of alleviating their consequences, as do airbags, for example.

6.2 Issues in Designing MAC Protocols

In mobile ad hoc networks (MANETs), which are totally distributed, nodes communicate through wireless communication by transmitting the information on a radio channel. This radio channel is limited in terms of bandwidth; only limited transmissions can be done on it at the same time. Thus, only a limited number of nodes are allowed to transmit on it at the same time. In wireless networks, two nodes within the communication range of each other should never transmit on the same radio channel at the same time; otherwise, transmission collision occurs. To avoid this problem, a centralized infrastructure can efficiently share the medium between nodes in its communication coverage. In MANETs we do not have any centralized infrastructure, which makes managing the medium-sharing process more complicated. It should be solved in a distributed manner. Many solutions, named Medium Access Control (MAC), have already been proposed for solving this problem in wireless networks [7]. When designing a MAC protocol for MANETs, there are important issues that have to be taken into account, knowing that some of these issues are more or less important when considering vehicular ad hoc networks (VANETs) [2]. VANETs have specific characteristics (see section 6.4.1), such as the high movement speed of vehicles, which concludes in fast and frequent topology changes.

In the following are listed the most important issues to be taken into account when designing a MAC protocol for MANETs.

6.2.1 Bandwidth Efficiency

The radio spectrum is a limited resource that concludes to a limited bandwidth. This bandwidth should be used in an efficient way by decreasing the control overhead, knowing that the bandwidth efficiency can be defined as the ratio of the bandwidth used for data transmission to the total available bandwidth. In VANETs we may have more control overhead caused by the frequent and the fast changes in the topology network due to the high mobility of vehicles.

6.2.2 Quality-of-Service Support

Many applications, such as video and voice communication, need some quality of service (QoS) to be guaranteed by the network in order to insure proper functionality. One good way to provide a good QoS consists of the bandwidth reservation, which is difficult to manage in MANETs because nodes in such networks are almost mobile. In VANETs, providing a certain QoS may be more difficult and more complex to manage than in MANETs, because of the relatively higher mobility of nodes (vehicles). Thus, the bandwidth reservation is not so easy to manage in VANETs.

6.2.3 Synchronization

Synchronization time between nodes in a wireless network is very important. In a centralized network, providing a synchronization time is easy to manage since the network

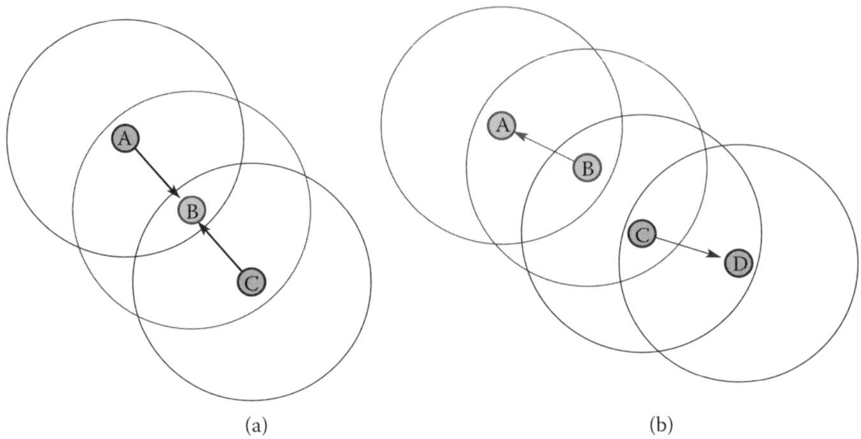

FIGURE 6.1 The hidden and exposed terminals problem.

has a common clock, the clock of the centralized infrastructure. This is not as easy in MANETs, which are totally distributed. Thus, synchronization should be provided in a distributed way in MANETs, as in VANETs. Even in VANETs this is less difficult to manage since vehicles should be equipped with positioning systems, like Global Positioning System (GPS) [3], which already provides a common clock that can be used for network synchronization.

6.2.4 Hidden and Exposed Terminal Problems

The hidden terminal problem is easy to understand but not easy to resolve. Figure 6.1(a) shows three wireless nodes: A, B, and C. A and C cannot communicate directly via their physical layer because they are not within the communication range of each other. But, both of them can communicate with B, which is in their communication range at the same time. Now, suppose A is transmitting data to B. At that time, C cannot hear this transmission; thus, it can transmit at any time, which can cause a transmission collision on B with the ongoing transmission from A. This is what we call the hidden terminal problem, where A and C are each hidden from the other.

The exposed terminal problem is similar to the hidden terminal problem in the sense that the problem is caused by the limitation of the communication coverage range and the common medium. Figure 6.1(b), which explains this problem, shows four wireless nodes: A, B, C, and D. When B is transmitting data to A, C is prevented from transmitting to D, as it believes that it will interfere with the ongoing transmission from B to A. In reality, C can transmit to D without any risk of interfering with the ongoing transmission between B and A. C is the exposed node.

6.2.5 Error-Prone Shared Broadcast Channel

Because of the transmission radio nature, when a source node is receiving a transmission from a sender node, no other node in its neighborhood should transmit; otherwise, a transmission interference can occur. That is because when a node is transmitting, all nodes in its neighborhood hear this transmission. Since it often happens that nodes in the same neighborhood attempt to access the medium at the same time, transmission collision probability is quite high in MANETs, as in VANETs. So, the role of a MAC protocol is reducing as much as possible these communication collisions.

6.2.6 Distributed Nature and No Central Coordination

One of the main characteristics of MANETs and VANETs is the lack of any centralized infrastructure or any centralized coordination. In such kinds of networks, nodes should interact in a purely distributed way. Thus, the MAC protocol should carry out a good control of the channel access, based on some control packet exchanges that may decrease the bandwidth.

6.2.7 Mobility of Nodes

If nodes were not mobile in MANETs, scheduling the channel access would be done in a static manner where the medium is allocated to communicating nodes in advance. Of course, this is not the case in MANETs, where nodes are mobile most of the time. In VANETs, this mobility is relatively higher and should be taken more seriously into account when designing MAC protocols.

6.3 MAC Protocols for MANETs

In wireless networks, two nodes located within the communication range to each other should not transmit on the same radio channel at the same time; otherwise, a transmission collision occurs. In such kinds of networks, a Medium Access Control (MAC) protocol should process in order to guarantee an efficient medium sharing between communicating nodes, knowing that the medium here refers to the communication radio channels.

In MANETs, as in VANETs, we have almost the same problems to face when controlling the medium access. In such kinds of networks, nodes operate in a total distributed way without any centralized management. Therefore, the MAC protocol that operates in these networks should be able to guarantee an efficient medium sharing, and in a distributed manner, which is not so easy to manage.

Thousands of works have been done around the medium access control in ad hoc networks, and many solutions have been proposed for this purpose. But unfortunately, still no solution is able to offer complete reliability.

6.3.1 Medium-Sharing Methods

There are three main basic medium access methods:

- **Time Division Multiple Access (TDMA):** To allow several nodes to share the same frequency channel (medium), TDMA divides the medium into different

time slots. Nodes use the medium one after another, by using a specific allocated slot in a repeated time frame. This technique allows several nodes to transmit on the same medium without transmission collisions.
- **Frequency Division Multiple Access (FDMA):** This is another way to allow different nodes to transmit on the same medium. It consists of dividing the medium into different radio frequencies and making each node use an associate frequency. This is another technique to allow many transmissions at the same time without collisions.
- **Code Division Multiple Access (CDMA):** This is a form of multiplexing allowing multiple nodes to access the medium. It divides up a radio channel not by time as in TDMA, nor by frequency as in FDMA, but instead by using different pseudo-random code sequences for each mobile.

These three different manners to split the medium can also be coupled and used at the same time, for example, by applying TDMA on each radio frequency in FDMA.

6.3.2 Classification of MAC Protocols

The existing MAC protocols can be classified in different ways based on several criteria, such as time synchronization and initiation approach. In the rest of this section we give a short overview on the main existing MAC protocols, taking care to classify most of them into these three big categories:

- **Contention-based access:** In this access mode, a node, when having a packet to send, has to contend with nodes in its neighborhood for access to the medium. In this mode no QoS can be guaranteed since no medium reservation is possible.
- **Contention-based access with reservation:** MAC protocols classified in this mode work as in contention-based access, but with resources reservation. Here a node is able to make a reservation of certain bandwidth for a certain time, allowing the support of real-time and other QoS-based applications.
- **Contention-based access with scheduling:** In this type of protocol, the MAC process is based on packet scheduling at the nodes and scheduling of the nodes. The target in this scheduling is to provide a certain ordering mechanism to guarantee some QoS differentiation or fairness, which means each packet is processed following a defined order.

6.3.2.1 Contention-Based Access

ALOHA [8] was the first MAC protocol proposed for packet radio networks. Its name comes from the world *Aloha*, used especially in Hawaii as a greeting meaning "hello" and "good-bye." This MAC protocol is very simple in its process; it is based on random access and processes like that: when a node has data to send, it transmits them immediately, and then if any collision occurs, it tries again after a random time. The maximum throughput that this MAC protocol can reach is only around 18%. To improve this throughput, a slotted version of ALOHA (S-ALOHA) [8] was proposed. In this slotted version the medium is divided into several time slots. And when a node wants to access the medium, it attempts to transmit at the beginning of a time slot. When compared to ALOHA, the vulnerable period of a transmission is halved in S-ALOHA, which doubles the system efficiency (maximum throughput).

As a refinement of ALOHA, the Carrier Sense Multiple Access (CSMA) [9] scheme has been proposed for wired networks. In CSMA a node senses if the medium is idle or not before transmitting. If the medium is sensed to be idle, the node can transmit; otherwise, it attempts to transmit again after a random time. A collision detection was integrated to CSMA in CSMA with Collision Detection (CSMA/CD) [10]. In CSMA/CD a node is able to detect transmission collisions when they occur. Therefore, if a node detects during its transmission any collision with another transmission, it stops transmitting and attempts to transmit again after a random time T. If the same transmission collides again, the sender waits twice the time T before attempting to transmit again. Even with collision detection, CSMA/CD suffers from charged networks because the probability of transmission collisions increases, which decreases network performances. Another disadvantage in CSMA/CD is that collisions are detected by the sender and not by the receiver, which make it not able to solve the hidden terminal and exposed terminal problems (see section 6.2.4).

Multiple Access with Collision Avoidance (MACA) [11, 12] was proposed as an alternative to CSMA. It tries to overcome the hidden terminal and exposed terminal problems based on establishing a handshake transmission between the transmitter and the receiver. In MACA, when a transmitter has a data packet to send, it asks the receiver if it is free to receive its transmission by sending a Request to Send (RTS) packet. When the receiver gets the RTS packet, if it is free to receive the corresponding transmission, it replies to the transmitter by sending back a Clear to Send (CTS) packet. Once the transmitter receives this CTS packet, it can start the transmission of its data packet without any risk of collision since all neighbors, in both its vicinity and the receiver's vicinity, are aware of the ongoing transmission through the RTS and CTS packets. The nodes in the vicinity of the transmitter, when hearing the RTS packet, do not transmit for a long enough period of time so that the sender can receive the CTS packet. The expected duration time of the data packet transmission is indicated in both CTS and RTS packets. All nodes within the vicinity of the receiver, when hearing the CTS packet, defer their transmission until the receiver receives the data packet. Based on this RTS/CTS exchange mechanism, MACA overcomes the hidden terminal problem.

In MACA, the node that receives only the RTS and not the CTS should be free to transmit even when the node from the RTS is transmitting. So, this node is considered an exposed terminal (see section 6.2.4). For example, as shown in Figure 6.2, we suppose node A is transmitting to node C, and node B has received the RTS corresponding to this transmission. Node B does not receive the CTS sent by node C, and thus it is free to transmit. When node B wants to transmit to node D, it sends node D an RTS. When receiving this RTS, node D, if it is ready to receive the corresponding transmission, sends back a CTS to node B. This CTS packet can collide with the ongoing transmission between node A and node C. Thus, node B cannot get it, and cannot start the transmission to node C.

FIGURE 6.2 Example of the exposed terminal problem in MACA.

MACA Wireless (MACAW) [13] has been proposed to improve the MACA performances. In addition to RTS and CTS control packets used in MACA, an Acknowledgment (ACK) packet is used in MACAW. In MACAW, a receiver node, when receiving the data packet, sends an ACK packet to the sender to acknowledge the good reception of the data packet. If the sender does not receive this ACK packet, it again sends the RTS packet, and when receiving this RTS packet, the receiver sends back an ACK, since the corresponding data packet has already been received.

Because of the exposed terminal problem caused in MACA, MACAW proposes to use another new control packet, Data Sending (DS). The DS packet, which is transmitted by the sender after receiving a CTS and before starting the transmission, makes the exposed node aware of the expected time duration of the corresponding transmission. For example, in Figure 6.2, after receiving a DS packet from node A, node B becomes aware of the expected time duration of the ongoing transmission between nodes A and C. Thus, when having data to transmit to node D, node B defers this transmission until the ongoing transmission between nodes A and C is expected to be finished. Figure 6.3 summarizes the packet exchange mechanism in MACAW, including control and data packets.

The Floor Acquisition Multiple Access (FAMA) [14] protocol is similar to MACA. The main difference is that it makes a carrier sensing before sending each control packet in order to avoid control packet collisions.

In the Busy Tone Multiple Access (BTMA) MAC protocol [15] a new way to overcome the hidden terminal problem is proposed, by splitting the medium into two channels:

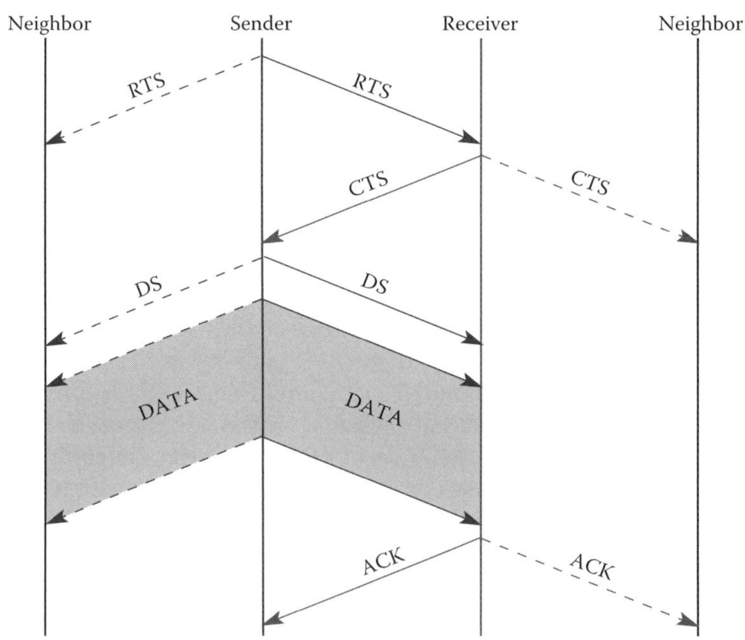

FIGURE 6.3 Packet exchange in MACAW.

control channel (say, busy-tone channel) and data channel. The first channel is used for transmitting the busy-tone signal, and the second one for transmitting data packets. In BTMA, when a sender wants to transmit, it first senses the busy-tone channel. If and only if it is free, it transmits the busy-tone signal on it and then starts the data transmission on the data channel. All neighbors that sense the busy-tone signal transmit it as well. Thus, all nodes in the two-hop neighborhood of the sender are not allowed to transmit, which avoids the collisions and the hidden terminal problem. An extension of BTMA was proposed in Dual BTMA (DBTMA) [16]. DBTMA uses two busy tones. These two busy-tone channels are used to inform nodes within the neighborhood about the ongoing transmission. The first busy tone is used by the sender node and the second one by the receiver node.

Right now, all MAC protocols we have presented are sender initiated. In MACA by Invitation (MACA-BI) [17], it is the receiver who initiates the data transmission by transmitting a Ready to Receive (RTR) control packet. And then, if the sender is ready to transmit, it responds by starting the data packet transmission. By doing that, MACA-BI removes the RTS packet, which makes the control packet overhead decrease.

6.3.2.2 Contention Based with Reservation

In MANETs, as in VANETs, some applications like voice and video communication need some QoS to be guaranteed. As said in section 6.3.1, MAC protocols can split the medium into different manners to make the medium seen as several physical or logical channels. Some MAC protocols allow the reservation of a channel for a dedicated transmission to provide some QoS in the network.

The Five Phase Reservation Protocol (FPRP) [18] is a contention-based MAC protocol with reservation. It uses a five-phase reservation process to establish TDMA slot assignments (see section 6.3.1). As shown in Figure 6.4, FPRP proposes to split the medium into two types of frames: reservation frame (RF) and information frame (IF). Each RF is followed by a sequence of IFs. In each IF there are N information slots (ISs). To each IS corresponds a reservation slot (RS) in the previous RF. When a node wants to reserve an IS in the following IFs, it contends in the corresponding RS. The reservation schedule generated in a RF is used in the next subsequent IFs until the next RF.

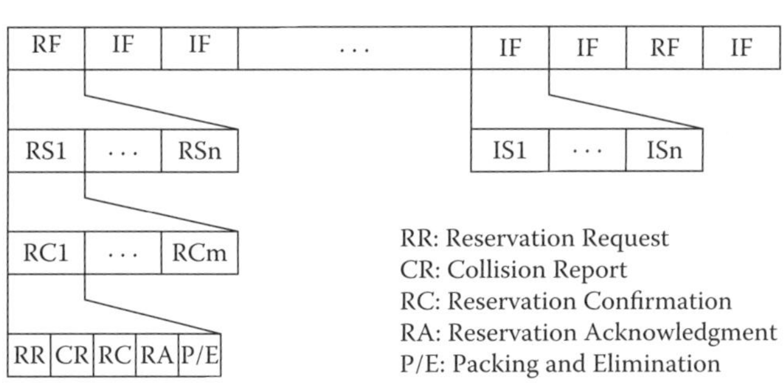

FIGURE 6.4 Frame structure of FPRP.

Each RS is composed of M reservation cycles (RCs). Within each RC, one IS reservation is done through a five-phase dialog between the contending node and its neighbors, as follows:

- Reservation Request (RR) phase: A node sends an RR packet when a reservation is needed.
- Collision Report (CR) phase: When a collision is detected a CR packet must be broadcast.
- Reservation Confirmation (RC) phase: If no CR packet has been received in the previous phase, then the source node is considered to have won the contention for the corresponding SI. Thus, it sends reservation confirmation to the destination node.
- Reservation Acknowledgment (RA) phase: The destination, after receiving the RC packet, sends back to the source a reservation acknowledgment. All nodes in the neighborhood of the destination hear this RA, which make them aware of the reservation done by the source node. By doing that, the hidden terminal problem is avoided.
- Packing and Elimination (P/E) phase: In this phase, two kinds of packets are transmitted: a packing packet, which serves to make the broadcasting pattern denser in a given slot, and an elimination packet, which is used to remove possible deadlocks (DLs) between adjacent broadcast nodes (see [18] for more details on this phase).

FPRP is totally distributed. For a given node, the reservation process only involves nodes within two-hop neighborhoods. This local reservation process makes FPRP scalable in terms of network size. FPRP is well robust in a rapidly changing topology because it does not need any prior information about the network.

Distributed Packet Reservation Multiple Access (D-PRMA) [19] splits the medium into several time frames (TFs), each TF into several time slots (TSs), and each TS into several mini-TSs (MTSs). When a node wants to transmit, it contends for an MTS in a TS, and if it wins, it uses the same MTS in the following TS sequence until it completes its transmission.

The Soft Reservation Multiple Access with Priority Assignment (SRMA/PA) [20] is another TDMA-based MAC protocol that resembles the previously presented contention-based with reservation MAC protocol, FPRP, with the main difference that it supports integrated services of real-time and non-real-time applications at the same time. SRMA/PS splits the medium into several TFs, each TF into several STs, and each ST into six small fields: SYNC, Soft Reservation (SR), Reservation Request (RR), Reservation Confirmation (RC), Data Sending (DS), and Acknowledgment (ACK). The SYNC field is used for synchronization; SR, RR, RC, and ACK are used for transmitting the corresponding control packets; and DS is used for the data transmission. The SR packet contains the information of the according priority. This priority information allows classification of nodes into high-priority and low-priority nodes. High-priority nodes are first served during the time slots reservation. This scheme makes SRMA/PA able to provide at the same time both QoS-based services, for voice applications, for example, and non-QoS-based services, for data transmission, for example.

Different contention-based MAC protocols with reservation schedules have been proposed, and most of them use the TDMA slot assignments. As in TDMA-based MAC schemes, in both CDMA-based and FDMA-based MAC schemes, the channel reservation can be provided. In CDMA-based MAC schemes, the medium is divided into several channels by using different orthogonal codes. In FDMA-based MAC schemes, by using different radio frequencies, the medium is slotted into several channels. Having several channels available in the network makes the reservation scheme possible, where each node is able to reserve a specific channel for a specific period of time. Multi-Code MAC (MC MAC) [21] is a CDMA-based MAC protocol. Among the different available codes, it uses one common code for transmitting control packets, and the other codes for transmitting data packets. In MC MAC, a sender node indicates in the RTS the orthogonal code it is going to use for the data transmission. When receiving this RTS packet, if there is no code conflict with other ongoing transmissions, the receiver replies by sending back a CTS packet; otherwise, it proposes to the sender the available and usable orthogonal codes, among which the sender will select one and then retransmit the RTS packet. The sender can start the data packet transmission when receiving a CTS from the destination. The destination acknowledges to the sender the good reception of the data packet by sending an ACK packet. MCSMA [22] is a good example of FDMA-based MAC protocols. It uses the CSMA mechanism on each frequency channel.

6.3.2.3 Contention Based with Scheduling

Distributed Priority Scheduling MAC (DPS-MAC) [23] uses distributed priority scheduling based on the basic RTS/CTS/DATA/ACK packet exchange mechanism used in the IEEE 802.11 distributed coordination (DCF) (see section 6.4.2.1). The example in Figure 6.5 shows how this distributed scheduling mechanism works. When the source node, node 1, has data to send, it sends an RTS packet including the information of the priority index corresponding to the data packet to be sent. Once receiving this RTS packet, the destination node, node 2, if ready to receive the data packet, sends a CTS packet including the same priority index indicated in the RTS packet. All nodes in the neighborhood, including the hidden nodes, when hearing the RTS or CTS packets, retrieve the priority index of the data packet to be sent and create the corresponding entree in their local scheduling table (ST). Node 3, for example, which is a neighbor of both nodes 1 and 2, initially has its ST as shown in ST(a) in Figure 6.5, and when receiving the RTS sent by node 1, it adds the related entree in its ST as shown in ST(b). As shown in ST(c), another entree is added to the ST of node 3 after receiving the data packet sent by node 1. Finally, after receiving the ACK packet from node 1, node 3 removes the corresponding entree from its ST as shown in ST(d).

Based on this mechanism, each node in the network is able to evaluate its priority in relation to other nodes' priority.

The Distributed Wireless Ordering Protocol (DWOP) [24] is another contention-based MAC protocol with scheduling. The scheduling proposed here is that the packet arriving first is to be transmitted first, based on the first in, first out (FIFO) scheme. Each node in the network keeps a local scheduling table (ST), which is ordered according to the arrival times. Having this ST table updated allows each node to evaluate its priority in relation with other nodes in its communication range. According to the FIFO scheme,

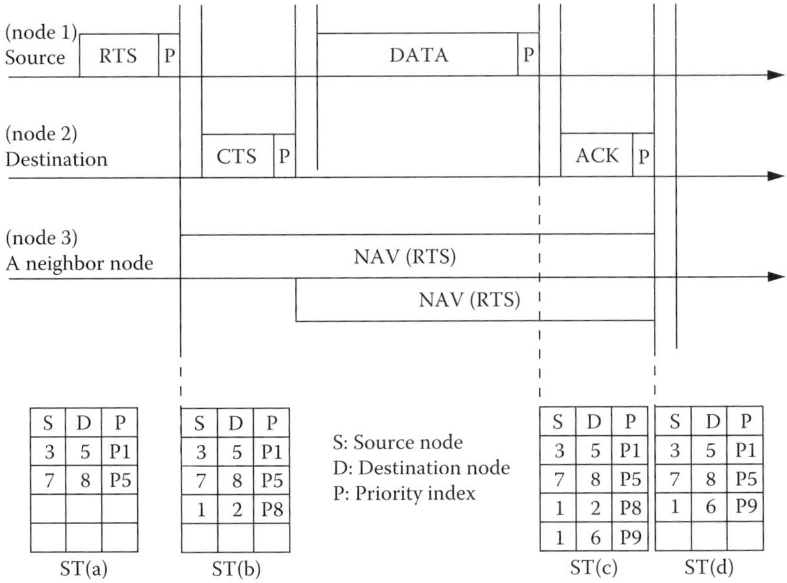

FIGURE 6.5 Scheduling table update in DPS-MAC.

the node (or the packet) that has the highest priority is the one that has the lowest arrival time. Thus, in DWOP a node contends to access the channel only when it has the lower arrival time compared to its neighbors.

In DWOP, as in DPS-MAC, once receiving an Acknowledgment (ACK), it removes from its ST the entree related to the corresponding transmission. It can happen that a node does not hear the ACK packet; in that situation, the entree corresponding to this ACK will be removed from the ST when hearing other packets with higher arrival times (lower priority) being transmitted.

So far we have talked about MAC solutions proposed for MANETs, but, as said in the beginning of this chapter, because of their specific characteristics (see section 6.4.1) compared to MANETs. These MAC solutions are not as suitable as they are for VANETs. In the rest of this chapter we will note proposed MAC solutions or adaptations for VANETs, but we will just select among the existing MAC solutions the ones that have good potential to be implemented for VANETs, and the ones already proposed for such kinds of networks. VANETs are still in progress, and no large real implementation has been done, which makes the selection and comparison of MAC protocols for these networks difficult.

6.4 MAC Protocols for VANETs

6.4.1 VANETs Characteristics and Issues for MAC Protocols

Vehicular ad hoc networks (VANETs) [2], also called Vehicle-to-Vehicle Communication (V2VC) or Intervehicle Communication (IVC) networks, are considered a specific instance of MANETs, where nodes are the vehicles. Thus, one of the main differences in VANETs compared to MANETs is defined in the characteristics of the vehicles, such as the high speed and the nonlimitation of energy and storage resources.

The vehicle's mobility speed, which is relatively high in VANETs, makes such kinds of networks suffer on frequent and fast topology changes. This is one of the disadvantages of VANETs compared to traditional MANETs. Regarding advantages in these networks, we have no limitation of energy and data resources, since vehicles are already equipped with batteries and can be equipped with powerful computers. Vehicles can be equipped with positioning systems as well, such as GPS [3] or Galileo [4], which allow getting geographic positions. In contrast to nodes in MANETs, vehicles do not have large freedom in their movement in VANETs. They move on roads following driving rules. Both roads and corresponding driving rules can be electronically represented.

All these specific characteristics make existing MAC solutions already proposed for MANETs not as suitable as they are for VANETs. Therefore, we do not have the same issues when designing a MAC protocol for VANETs as when designing a MAC protocol for MANETs. In section 6.2 we present the main issues for designing a MAC protocol for MANETs. We notice that some of these issues are less important and some more important when considering MAC protocols for VANETs.

For both MANETs and VANETs, when designing a MAC protocol we have to think about how to guarantee an efficiency bandwidth use, how to reduce or avoid the hidden and exposed terminal problems, and of course, how to make the proposed solution able to work in a distributed way. The network synchronization is also very important in VANETs as in MANETs, but it is easier to guarantee in VANETs since vehicles should be equipped with a satellite positioning system, like GPS, which is able to provide a common clock in the network.

What we have to take more care about when designing a MAC protocol for VANETs is discussed in the following sections.

6.4.1.1 Quality-of-Service Support

The quality-of-service importance in any communication network depends directly on the application use. In VANETs, the important candidate applications would be around driver and passenger safety. Human lives depend on these safety applications, which depend on the communication network efficiency, which depends on the MAC layer. Thus, any MAC protocol proposed for VANETs should take into account quality of service, at least for safety applications.

6.4.1.2 Mobility of Vehicles

A vehicle's mobility pattern can be the main difference when comparing VANETs to MANETs. Because of the high-speed movement specification of vehicles, the topology network in VANETs changes frequently and rapidly. Therefore, vehicles that want to exchange some data have relatively small time to access the channel and establish the desired communication. If two vehicles move in opposite directions with a speed of 130 km/h, the relative speed between these two vehicles becomes double, i.e., 260 km/h.

And if the radio communication range is assumed to be 300 m for each vehicle, we conclude that these vehicles have less than 8 s to access the channel, establish the communication, and transfer the data.

Thus, any MAC protocol proposed for VANETs should take this into account and reduce as much as possible the medium access/allocation time.

6.4.2 Candidate MAC Protocols for VANETs

6.4.2.1 IEEE 802.11 Standards

The wireless communication standard IEEE 802.11 [25], which was brought out in 1997, is very interesting and famous in the domain of wireless LANs (WLANs). The most famous versions of this standard are IEEE 802.11b and IEEE 802.11g, commercially known as Wi-Fi (wireless fidelity). These two standards have obtained much interest from researchers in the domain of VANETs, in both theoretical and application sides (prototypes and test beds).

The IEEE 802.11 standards work in two modes:

1. Centralized mode, where mobile terminals communicate with one or different fixed and centralized infrastructures already deployed in the network.
2. Ad hoc mode, where mobile terminals are connected to each other without any centralized infrastructure, and in which two mobile terminals are able to communicate directly through the physical layer when they are within the communication range of each other, or by multihop communication through the routing layer when they are not close enough to each other. In ad hoc mode, the network should be autoconfigurable and totally distributed.

In VANETs, the IEEE 802.11 standard can be used in both modes: in centralized mode for vehicle-to-infrastructure (V2I) communications and in ad hoc mode for vehicle-to-vehicle (V2V) communications. Ad hoc mode can be used for V2I communications as well, when infrastructure spots, a repeater, for example, play the role of an ad hoc terminal in the network.

In the Open Systems Interconnection (OSI) model, the IEEE 802.11 standard addresses the last two layers: data link and physical layers. The data link layer is divided into a Media Access Control (MAC) layer and a Logical Link Control (LLC) layer. In this chapter, we focus only on the MAC layer, which is almost the same as that used in interactions with different physical layers in all IEEE 802.11 standards.

Considering VANET environment specifications (see section 6.4.1), the IEEE 802.11 working group is working on a new standard version, IEEE 802.11p [26]. This new standard should deal with VANET characteristics. It uses a modified IEEE 802.11a with slight changes, mainly on the physical layer. It operates in the licensed 5.9 GHz band, and has almost the same MAC layer as the other standards in the family of IEEE 802.11, based on CSMA/CA technology. In the following, we give a description of the IEEE 802.11 MAC layer and a short overview on the new IEEE 802.11p physical layer.

6.4.2.1.1 IEEE 802.11 MAC Layer

The MAC layer mechanism used by IEEE 802.11 wireless LANs (WLANs) is based on Carrier Sense Multiple Access with Collision Avoidance (CSMA/CA), which is provided by the distributed coordination function (DCF). If contention-free service is required, it can be provided by the point coordination function (PCF), which is built on top of the DCF. The contention-free services are provided only in networks with centralized infrastructures. The carrier sensing is used to determine whether the shared medium is free for use or already in use by another candidate terminal in the neighborhood. It can be managed in two manners: physical carrier sensing or virtual carrier sensing.

- The physical carrier sensing function is provided by the physical layer, where the radio channel is sensed physically. This method needs specific hardware and cannot overcome the hidden terminal problem since the hidden terminal can never be heard physically.
- The virtual carrier sensing is provided by the network allocation vector (NAV), as shown in Figure 6.6. The NAV is a timer that indicates the duration time of a transmission. Each terminal should indicate the amount of time it expects to use the medium. Terminals in its neighborhood count down the corresponding NAV from this amount of time until zero. The carrier sensing function indicates that the medium is busy until the NAV reaches zero.

The IEEE 802.11 WLANs use some interval spaces, named interframes spacing (IFSs). These IFSs are set between two successive transmission frames in order to coordinate access to the common transmission medium. IEEE 802.11 uses four different IFSs, as shown in Figure 6.7. Having different IFSs makes it possibile to have different priority levels for different types of traffic, which means that after the medium becomes idle, traffic with lower priority waits more than traffic with higher priority before attempting to access it.

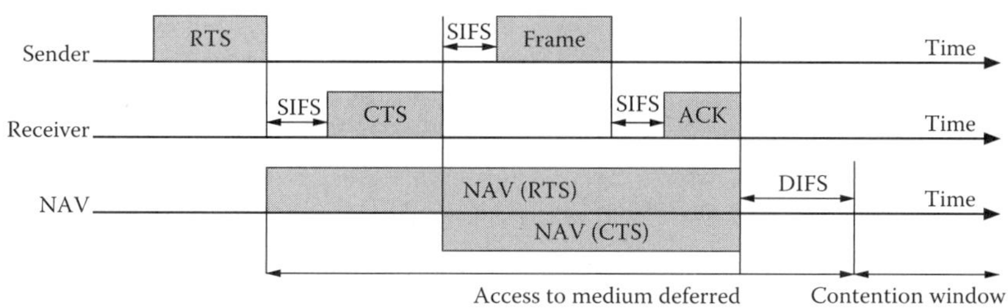

FIGURE 6.6 Virtual carrier sensing by the NAV.

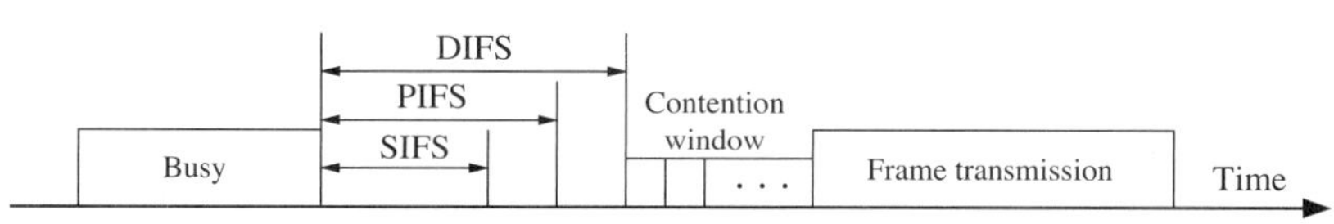

FIGURE 6.7 Interframe spacing in 802.11.

More details on these IFSs are given below:

- Short interframe space (SIFS) is the shortest. It is used for the highest-priority transmission, such as RTS/CTS/ACK messages. The medium should be sensed idle for a minimum time period of at least SIFS before any transmission can be done.
- PCF interframe space (PIFS) is used for real-time services. When the medium has been sensed free of any traffic for a period greater than the PIFS, stations operating in a contention-free mode may have immediate access to the medium.
- DCF interframe space (DIFS) is longer than PIFS, and is used in the case of DCF transmission mode. A terminal can access immediately a medium if it senses it free for a period of time longer than DIFS.
- Extended interframe space (EIFS) is the longest, and it is used when an error in frame transmission occurs.
- When having a data packet to send in an IEEE 802.11-based VANET, a vehicle first checks whether the medium is idle. If it is sensed to be idle for a time duration of DIFS, the vehicle can transmit. Otherwise, it backs off and attempts again after an amount of time chosen within a contention window (CW).

The IEEE 802.11 MAC layer is principally based on the RTS/CTS/ACK packet exchange, as shown in Figure 6.8, and interframes spacing, as shown in Figures 6.6 and 6.7. A vehicle, when having a data packet to transmit, checks whether the medium is idle. If the medium is sensed idle for a time duration of DIFS, an RTS packet is transmitted, after a random back-off time, including the sender ID and the expected transmission duration time. If the destination vehicle is ready to receive the transmission, when receiving the corresponding RTS packet, it waits for a SIFS duration time and then sends back a CTS packet to the sender. Once the sender receives this CTS packet, it waits again

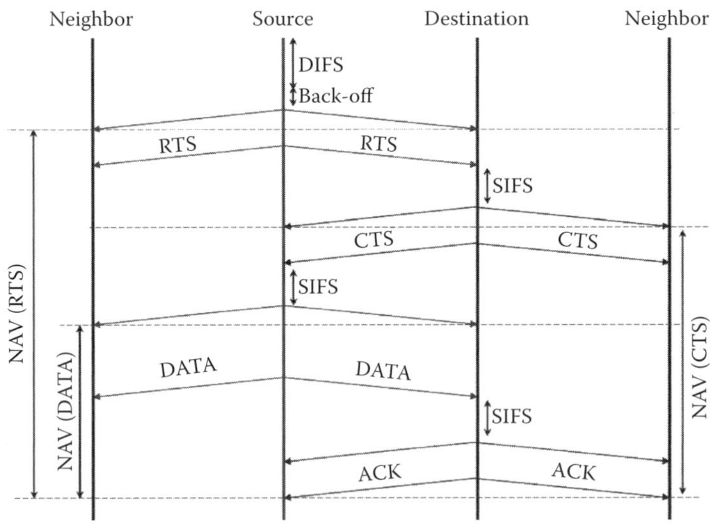

FIGURE 6.8 Control packet exchange in IEEE 802.11.

for a DIFS duration time and then starts the transmission of the data packet to the destination. After receiving the data packet, the destination waits for a DIFS time and then sends an ACK packet to the sender, and sets its NAV to zero. When receiving the ACK packet, the vehicle sets its NAV to zero. To reduce the risk of collision and the hidden terminal problem, all vehicles that hear the RTS and CTS packets set their NAVs according to the information provided in these packets.

Thanks to the RTS/CTS/ACK packet exchange and the interframes spaces, 802.11 minimizes the risk of transmission collisions and the hidden terminal problems.

6.4.2.1.2 Toward an IEEE 802.11 Physical Layer for VANETs

Several IEEE 802.11 standards have been proposed. Each version is dedicated for a specific domain of application and for a specific environment. The most famous versions are IEEE 802.11b and the IEEE 802.11g, as well as IEEE 802.11a, thanks to the large availability in the market of devices based on those technologies. Some versions have been proposed just as extensions or enhancement. IEEE 802.11i, for example, takes care of security in network communications, and IEEE 802.11e takes care of providing a certain QoS in the network.

Both b and g versions of IEEE 802.11 are popular and known under the name of Wi-Fi. Both operate on top of an almost identical MAC layer. Like IEEE 802.11g, IEEE 802.11b uses the unlicensed 2.4 GHz band, where interferences are possible with cordless phones, microwave ovens, wireless IP (Internet protocol) cameras, and other devices using the same band. Theoretically, IEEE 802.11b data rates can reach 11 Mbps, but in practice, due to CSMA/CA protocol overhead, reach only about 7.5 Mbps.

Wireless devices equipped with IEEE 802.11b and IEEE 802.11g communication technologies are compatible and can communicate to each other. IEEE 802.11a operates on the 5 GHz frequency band, which makes it incompatible with IEEE 802.11b and IEEE 802.11g. Theoretically, the maximum throughput of IEEE 802.11a is up to 54 Mbps, but in reality it goes up to 25 Mbps at most. The 5 GHz band lets 802.11a have the advantage of less interferences, but unfortunately, it does not let it penetrate walls and other obstacles.

IEEE 802.11g can reach the same higher bit rate technology as IEEE 802.11a, which lets it operate at a maximum raw data rate of 54 Mbps, which is about 25 Mbps maximum net throughput. Super G is a new proprietary feature used by some products in the market, allowing network speed to reach up to 108 Mbps by using channel bonding over IEEE 802.11g, which can bond two 20 MHz channels together.

6.4.2.1.3 WAVE (IEEE 802.11p)

As said above, an IEEE 802.11 working group is working on a new PHY/MAC amendment of the 802.11 standard, named IEEE 802.11p, and also referred to as Wireless Access in Vehicular Environments (WAVE). It will deal with Intelligent Transportation Systems (ITS) applications requirements, including data exchange between high-speed vehicles and between vehicles and roadside infrastructure in the licensed ITS band of 5.9 GHz (5.85–5.925 GHz). Requirements for this amendment are mostly coming from vehicular active safety concepts and applications (communications between vehicles or from vehicles to road infrastructures), where reliability and low latency are extremely important.

In terms of MAC operations, because WAVE is based on IEEE 802.11a, the MAC layer used is almost the same, and based on CSMA/CA and interframes spacing as described above.

At the PHY layer, WAVE should work around the 5.9 GHz spectrum, using Orthogonal Frequency Division Multiplexing (OFDM) technology and providing vehicular wireless communications over distances up to 1000 m in theory, while taking into account the environment, i.e., absolute and relative high velocities (up to 200 km/h), fast multipath fading, and different scenarios (rural, highway, and city). It will probably operate on 10 MHz channels, but optionally on 20 MHz channels.

6.4.2.2 ADHOC MAC

ADHOC MAC [27] was basically proposed for vehicular networks. It was proposed within the European research project CarTALK2000, which has the purpose to design novel solutions for intervehicle communications. ADHOC MAC has been designed to provide a reliable single- and multihop broadcast. It is totally distributed and based on a dynamic TDMA mechanism that can be easily adapted to the UMTS Terrestrial Radio Access Time Division Duplex (UTRA-TDD), which has been chosen as the physical target system in the CarTALK2000 project. ADHOC MAC is able to provide QoS based on time slots reservation.

To achieve the dynamic TDMA mechanism, ADHOC MAC uses an extension of the Reservation ALOHA (R-ALOHA) [28] protocol, named Reliable R-ALOHA protocol (RR-ALOHA) [29]. R-ALOHA is capable of achieving the dynamic TDMA, but in centralized networks, where a central repeater provides the status information (busy, free, or collided) of time slots. It proposes that a node that has a data packet to send contends to access a time slot by transmitting on it, and then if the transmission is recognized as successful, the slot is reserved for that node in the next subsequent frames. To apply R-ALOHA in a distributed way, RR-ALOHA was proposed. RR-ALOHA makes each vehicle in the network able to reserve a slot time within two-hop neighborhoods, which allows it to transmit without any risk of collision and with avoiding the hidden terminal problem.

In RR-ALOHA each vehicle periodically transmits the perceived status of slots in the preceding period (frame), called frame information (FI). In more detail, the medium is divided into repeated successive frames, and each frame is composed by N time slots. A node that wants to transmit has to reserve a time slot that we call basic channel (BCH), which will be used for its data packet transmissions during the next subsequent frames. The statute of a time slot (channel) is considered busy when a transmission in this slot is recognized as successful; otherwise, it is considered free. Each vehicle in the network listens first to the FI broadcast by all active vehicles during one frame (N slots), and then marks each slot as free or busy, depending on the transmissions heard during this one-frame period. When a time slot is heard as busy, a vehicle puts in its FI's corresponding slots the ID of the corresponding transmitter. This FI is sent periodically every one-frame period. When receiving an FI, a vehicle is able to know which time slot is reserved by which vehicle within two-hop neighborhoods. When a new vehicle reaches the network, it first listens during one time frame before attempting to transmit on one free time slot. Then, if in the next frame period the same time slot is marked by its ID in the whole received FI, the time slot is considered as reserved for it by all vehicles within

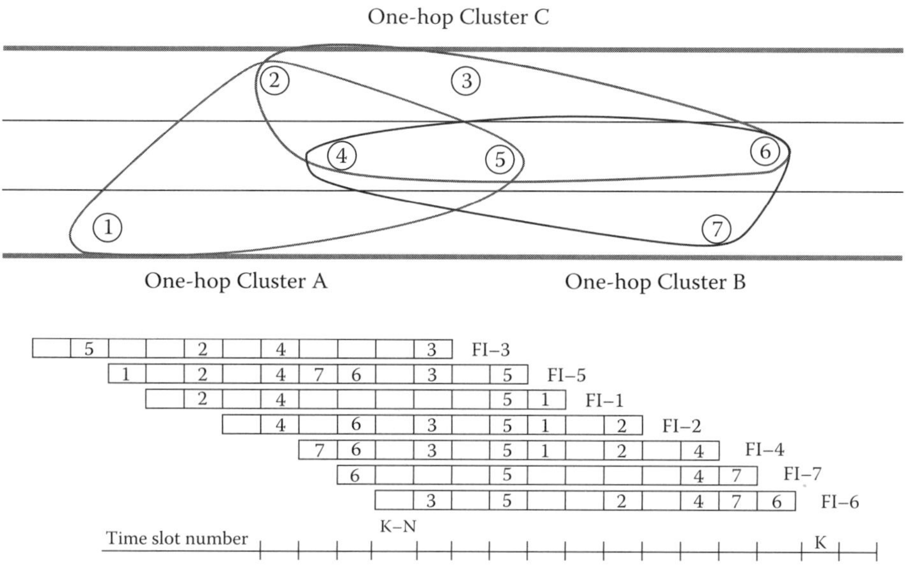

FIGURE 6.9 FIs propagation within TH cluster in RR-ALOHA.

its two-hop neighborhood. Thus, this time slot can be considered as its BCH, and can be used for its data packet transmissions.

Figure 6.9 shows an example of FIs transmitted by vehicles in a two-hop (TH) cluster, knowing that a TH cluster is a union of one-hop (OH) clusters having a common subset of vehicles. A OH cluster is a set of vehicles that are within the communication range of each other, meaning all vehicles are able to communicate with each other directly at the physical layer.

By hearing the medium, each vehicle in the network knows which time slot is reserved by which one-hop neighbor. And based on the information provided by its one-hop neighbors through their FIs, each vehicle in the network is able to know which time slot is reserved by which vehicle within two-hop neighborhoods. Thus, each vehicle is aware of all ongoing transmissions in its two-hop neighborhoods, which allows RR-ALOHA to easily overcome the hidden terminal problem, and so to reduce transmission collisions. And based on the dynamic TDMA mechanism with slot reservation, it can guarantee a relatively good QoS in VANETs.

6.4.2.3 Directional Antenna–Based MAC Protocols

Directional antennas have many important benefits in MANETs as well as in VANETs. Directional antennas can provide higher system capacities by directing narrow beams toward the users of interest, while nulling users not of interest. This allows for lower transmission interferences, lower power levels, and more channel reuse within the same terminal neighborhood. Figure 6.10 shows how a vehicle is able to have a limited communication coverage space when using directional antennas.

In this chapter, we consider each directional antenna-based communicating vehicle as a vehicle equipped with N directional antennas, and each antenna covers an angular

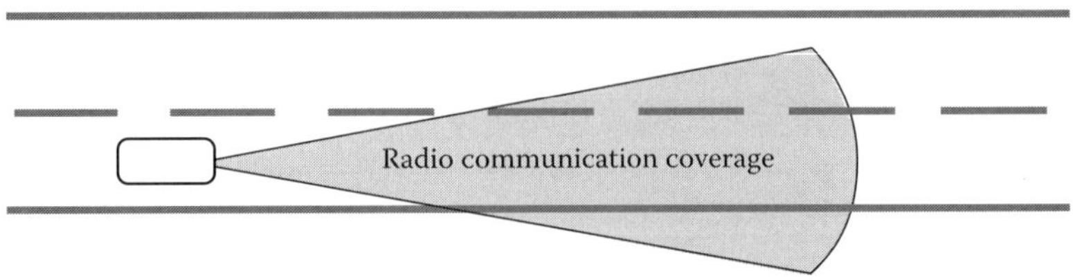

FIGURE 6.10 Communication coverage when using directional antenna.

space of (360/N) degrees around it. Based on such a system, a transmitter can optimize its transmission by limiting the coverage space, which allows channel reuse and makes transmission collisions and hidden terminal problems decrease in the network. Thus, it increases the network throughput and reliability.

In some vehicular communication scenarios, directional antenna-based systems are very welcome. Vehicles move along roads, and in general cases within the same or opposite direction. Therefore, limiting radio transmission to roads on which we are driving can be beneficial, in particular for MAC issues, when transmission collisions with vehicles moving on neighboring roads are avoided. For example, in a highway scenario, based on such a system, transmission may be limited to only the highway in terms of coverage space, which avoids collisions with other ongoing radio communications among vehicles driving on other neighboring roads and highways.

In the following we present some MAC solutions among the many solutions that have been proposed based on this directional antenna scheme.

In [30], the authors propose an ad hoc network of terminals equipped with M multiple directional antennas, as shown in Figure 6.11. Each antenna spans an angle of $2\Pi/M$ radians. The N antennas are fixed on each terminal with nonoverlapping beam directions, so as to collectively span the entire plane. It is supposed that we can switch any one or all of the antennas to active or passive modes. When transmitting, if all antennas are active, the radio signal is transmitted in all directions, like in omnidirectional antennas. A node can receive a transmission from all antennas, and selects the one receiving the maximum power.

The MAC protocol proposed here adapts the same scheme used in MACA [11, 12] for use with directional antennas. It proposes that a terminal that has a data packet to transmit has to find the direction of the destination in order to transmit the packet on the corresponding directional antenna. Same for the destination, which has to find the direction of the sender in order to know on which directional antenna the data packet will be received. To manage this task, the following process is used.

Every terminal in the network listens to ongoing transmissions in its neighborhood on all its antennas. When having a data packet to transmit, a sender first transmits an omnidirection RTS to the destination. When receiving the RTS packet, if the destination is ready to receive the transmission, it sends back an Omnidirectional CTS packet to the sender. Once the sender receives the CTS packet, it starts transmitting the data packet on the antenna that points toward the destination. The destination should receive this transmission on the antenna that is in the direction of the sender. To explain in more

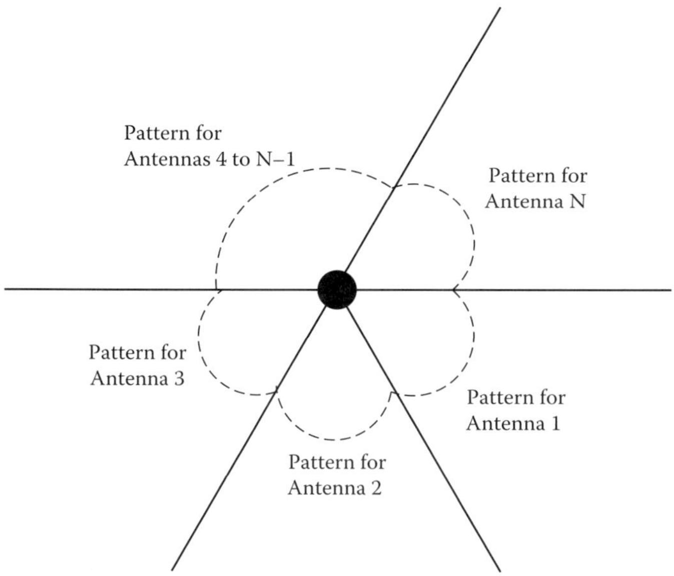

FIGURE 6.11 Illustration of deployment of directional antennas.

detail the packet exchanges in this directional antenna-based MAC protocol, an example is shown in Figure 6.12. In this example each terminal is supposed to be equipped with four directional antennas, numbered from 1 to 4. The sender S, when having a data packet to send to D, first transmits an RTS packet to D on all four antennas. D receives this RTS packet on directional antenna 2. If D is ready to receive the data packet, it transmits a CTS packet to S on all four antennas. Once receiving the CTS packet, S transmits the data packet on antenna 4, on which the CTS sent by D has been received. All of the neighbors of S and D that hear this RTS-CTS use the related information to prevent interfering with the ongoing data transmission.

Directional Busy Tone-Based MAC (Directional BTMA) [31] is another MAC protocol based on directional antennas. It adapts the DBTMA [16] protocol described in section 6.3 to directional antennas. The same busy-tone concept used in DBTMA is again used in Directional BTMA. As in the previously described directional MAC protocol [30], in Directional BTMA, each terminal is supposed to have M multiple directional antennas, as shown in Figure 6.11. Each antenna spans an angle of $2\Pi/M$ radians. It is supposed that each terminal, when having no ongoing transmission, keeps all its antennas sensing the channel.

In order to not interfere with ongoing transmissions in the neighborhood, a terminal that needs to send data senses if the busy-tone channel is free before sending the RTS packet on all antennas (Omnidirectional RTS). When receiving an RTS packet, if ready to receive the corresponding data transmission, the destination sends a CTS packet on all antennas (Omnidirectional CTS), and activates the busy tone on the antenna that points toward the sender. Upon receiving the CTS packet, the sender activates the busy tone on

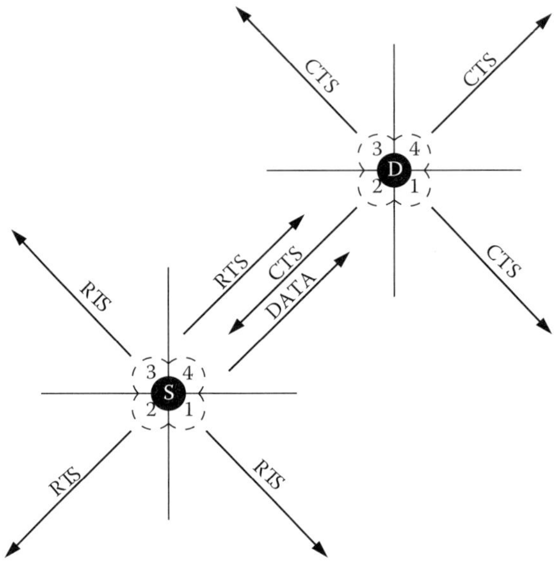

FIGURE 6.12 Illustration of the packet transmissions.

the antenna that points toward the destination and starts the data packet transmission on the same antenna. All neighbors within the communication coverage of the directional antennas having the busy tone active are not allowed to transmit. This scheme can avoid transmission collisions but not in certain scenarios. For example, in Figure 6.13, terminal A is supposed to be transmitting data to B on its directional antenna. B is receiving this transmission on directional antenna 4, which has the busy tone activated. Terminal C is not in the transmission coverage of directional antenna 4 of terminal B; thus, it is free to transmit at any time. If terminal C gets a data packet to send to E, it can transmit it on directional antenna 1 without any risk of collision with the ongoing transmission between A and B. Now, if this same terminal C gets a data packet to send to D, it is also free to transmit it on directional antenna 3, which points toward D and B. But this transmission causes a collision on B with the ongoing transmission between A and B, even though the busy tone on antenna 1 of B is inactive. Therefore, Directional BTMA, as it is, is not collision-free.

In both directional antenna-based MAC protocols described above, no localization system is needed to determine in which direction a neighbor is localized in order to know the antenna that points toward it. The Directional MAC (D-MAC) protocol [32] requires that each terminal knows its neighbors' locations as well as its own location. This assumption does not cause any problem in case of VANET applications. In VANETs it is already supposed that each vehicle knows the geographic locations of itself and its neighbors. A positioning system like GPS [3] and Galileo [4] can be used for that.

D-MAC uses the same handshake mechanism based on RTS-CTS packet exchange as in the IEEE 802.11 MAC layer [25]. It works in two different schemes: in the first one, RTS packets are directional transmitted only, and in the second, they are both directional and omnidirectional transmitted.

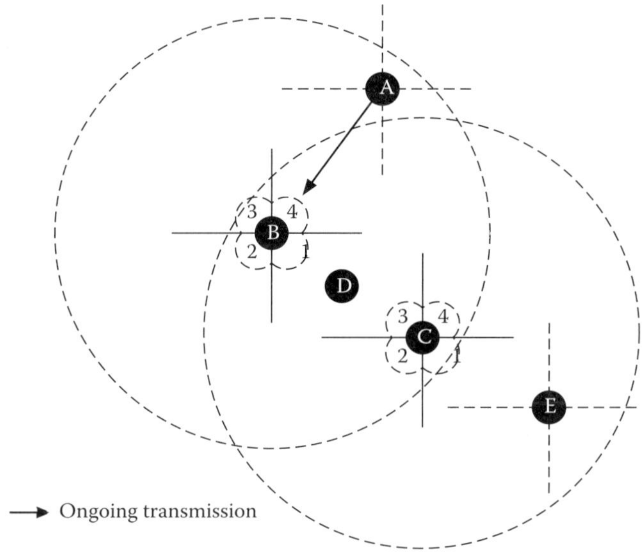

FIGURE 6.13 Directional BTMA: transmission collision example.

6.4.2.3.1 Scheme 1: Using DRTS Packets

When terminal B, as shown in Figure 6.14, has a data packet to send to C, first it sends a Directional RTS (DRTS) in the direction of C. Since A and C are not in the same direction, A will not hear this DRTS packet. After receiving the DRTS, C replies to B by sending an Omnidirectional CTS (OCTS). Once receiving this OCTS, B starts the transmission of its data packet using a directional antenna. When B receives the data packet, it immediately transmits a Directional ACK (DACK).

Now, suppose that during the ongoing transmission between B and C, D has a data packet to send to E. By using a directional antenna, D can transmit to E without any risk

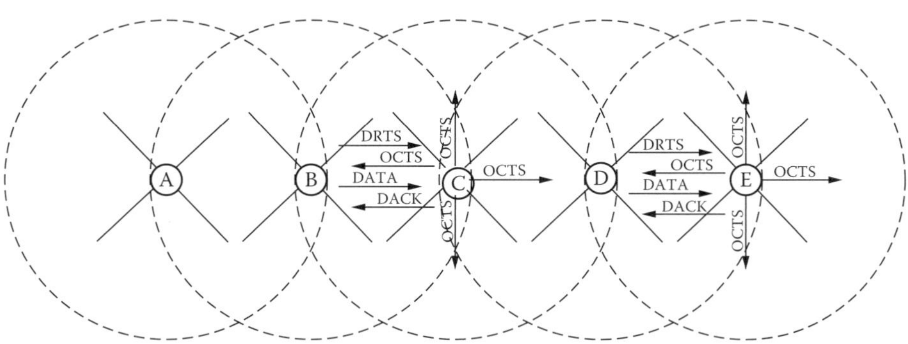

FIGURE 6.14 The process of the first scheme of D-MAC.

of collision on C, which is impossible when not using directional antennas. Note that when receiving the OCTS from C, D blocks its directional antenna that points toward C. So, D first sends a DRTS in the direction of E. Then, if ready, E sends an OCTS that authorizes D to start the transmission of its data packet using the antenna that points toward E. Thus, D can transmit to E when B is transmitting to C without any collision.

6.4.2.3.2 *Scheme 2: Using Both DRTS and ORTS Packets*

In this scheme D-MAC uses both Directional and Omnidirectional RTS depending on the state of the antennas (blocked or not). This scheme is proposed to improve the first, which may increase the probability of control packet collisions in some cases. For example, in Figure 6.14, terminal A cannot hear the DRTS sent by B to C; thus, it is free to transmit to B. Therefore, if A has a data packet to send to B, it is free to send the DRTS to B, which can collide on B with the OCTS or DACK packets sent by C. To avoid this problem, in this second scheme of D-MAC, Omnidirectional RTS (ORTS) is sent instead of DRTS when possible. In more detail, when a transmitter has a data packet to send, if all its antennas are not blocked, it transmits an ORTS; otherwise, it transmits a DRTS if the desired antenna is not blocked. If the desired antenna is blocked, the transmission is deferred until the antenna becomes unblocked.

Let us suppose that all antennas at terminal B (see Figure 6.14) are unblocked. When having a data packet to send to terminal C, terminal B transmits an ORTS. Since this ORTS is sent in all directions, A hears it this time and blocks its antenna that points toward C until the ongoing transmission between B and C finishes. Thus, when having data to send to B, A should defer its transmission until the desired antenna becomes unblocked.

D-MAC, based on directional antennas and the two schemes, increases channel reuse by decreasing packet transmission collisions, which improves the network throughput.

The directional antenna-based mechanism can decrease the probability of transmission collisions in the network, and can increase channel reuse as well. Thus, it can improve network performances in terms of throughput. In VANETs, vehicle movement is limited by roads, and in general cases, vehicles are moving in the same or opposite direction of each other. This makes directional antenna systems easier to be adapted for VANETs than for MANETs.

6.4.2.4 Other MAC Solutions and Improvements for VANETs

Much work has been done around communication technologies in vehicular networks, and in the context of MAC solutions, as far as we know, all proposed solutions are just adaptations or extensions of the existing MAC protocols already proposed for VANETs. Earlier in this section we presented some protocols and schemes that have been or may be adapted for VANETs. In the following we present slight MAC extensions that have been proposed for VANETs.

In [33] a MAC extension layer is proposed for safety message transmission. The main goal of this extension layer is to maximize the probability that a safety message is received by all vehicles within the message range and within the message lifetime. So, the strategy is to repeat the message a certain number of times within its lifetime. The design of this MAC extension layer is specified as a state machine that can be overlayed on IEEE 208.11 DCF.

Vehicle safety applications generate messages to be delivered to destinations with a specified probability that has the following meaning. Each message has an associated intended communication range and useful lifetime (say, τ). A safety message should reach all vehicles within the specified range within the specified lifetime. The packet transmission delay (say, t_{trans}), which is a function of the packet size and data rate of the radio, can be calculated. Authors in [33] propose to divide the lifetime into n = $\lfloor t/t_{trans} \rfloor$ slots. And then they pick any $k(1 < k < n)$ slots to transmit the packets. If at least one message is correctly received at the destination, the transmission is considered successful; otherwise, the transmission fails. The scheme proposed here can increase the probability that a safety message is received by all concerned vehicles within the desired delay, which is very important in safety-related applications.

A Location-Based Channel Access (LCA) protocol is proposed in [34]. LCA exploits the position information of the vehicle, easily provided by GPS or any other geolocalization system, to move the centralized scheme into an ad hoc scheme. In centralized networks, the medium access is managed by the base station. Each base station has a set of channels that it allocates to terminals within its communication coverage. Each base station covers a specific geographic plan, and two stations near each other should have different channels to allocate in order to avoid transmission collision. This is purely incompatible with ad hoc networks, which are totally distributed. A great idea was inspired from this centralized scheme to be applied in LCA. In LCA, the geographic area is divided into a cellular structure, as in centralized networks. To each cell, a unique channel is associated, and used by the vehicle located in it. So, when a vehicle wants to transmit, it first gets its own position, and then determines the cell that corresponds to this position. Once having the ID of the cell in which it is located, the vehicle obtains the corresponding channel from a locally stored cell-to-channel mapping.

This scheme allows use of the cellular concept without having any base station in the network, and any centralized management. It allows each vehicle to get a unique channel within its neighborhood. Having each vehicle with a different physical position makes the LCA scheme scalable in terms of network size because of channel reuse. A channel can be reused by more than one cell if these cells are far enough from each other.

The main problem in this scheme is that it is supposed that the cell size should be small enough so that two vehicles never happen to be in the same cell. If we take into account this assumption, which is important for the smooth working of LCA, it will be very difficult to manage the channel allocation process. In cellular networks, there is a special mechanism to manage the transfer of the terminal from one cell to another, say, handover. In VANETs, vehicle movement speed is relatively high, and having the cell size so small makes the handover problem very difficult, if not impossible, to manage.

6.5 Qualitative Comparison of VANET MAC Protocols

The MAC protocols that we have selected as candidates for VANETs in section 6.4 can be classified into three main categories: contention based, contention based with reservation, and directional antenna based.

A good MAC protocol should be selected in relation with the network characteristics, as well as in relation with the network requirements. Network requirements refer here to the network-related applications requirements. The characteristics of a VANET are not always the same since the network environment is not always the same. For example, the characteristics of a VANET composed by a set of a vehicles moving on a highway are not the same as those for a VANET composed by a set of a vehicles moving in a city. The application requirements also change from one application to another. Safety applications, for example, do not have the same requirements as video and voice communication applications. Thus, in the rest of this chapter, we will try to select which scheme of MAC access is better for use for which VANET and for which application.

The CSMA/CA scheme used in IEEE 802.11 standards uses contention-based access, which is suitable only for burst data traffic where the available bandwidth is used effectively. Such kinds of MAC schemes are not suitable for real-time communications, like video and voice communications. Safety applications need the medium access delay to be reduced as much as possible, which CSMA/CA cannot guarantee when using a contention-based system and interframe spacings. CSMA/CA suffers from the network density in terms of number of active communicating terminals, which makes efficiency of the network decrease when the density increases. Wireless local area networks (WLANs), which already employ CSMA/CA in ad hoc mode, have the same limitations as mentioned earlier.

ADHOC MAC uses a contention scheme with time slots reservation based on the TDMA concept. Time slots reservation makes this MAC protocol able to guarantee a certain level of QoS, which is very welcome for real-time communications and any QoS-based application. One of the problems from which ADHOC MAC or any other TDMA-based system can suffer is time synchronization. A vehicular network is considered an ad hoc network, which is totally distributed. In centralized networks, it is the role of the base station to provide a common clock time for synchronizing the network. Of course, this is not possible in an ad hoc network, but in VANETs there is a solution to provide such a common clock. In VANETs each vehicle is supposed to be equipped with a geo-positioning system, like GPS, which is able to provide, in additional to the physical position, a common clock that can be used to synchronize the network. Thus, TDMA-based MAC solutions are welcome in VANETs.

The medium access delay in ADHOC MAC depends on the duration and number of time slots within one time frame. To reserve a time slot in ADHOC MAC, a vehicle has to wait at least one frame time before attempting to reserve its time slot. And sometimes it needs to attempt many times before getting a free time slot. Thus, in case of a dense network with an important number of active vehicles, where many vehicles can attempt at the same time to reserve a time slot, the medium access delay can increase considerably because of collisions. Therefore, to be suitable for safety applications, ADHOC MAC, as it is, needs some improvements in the sense of reducing the medium access delay in cases of networks with high density. Just an optimal slotting way may be enough, i.e., finding the optimal number and duration of time slots in one frame time.

Directional antenna–based MAC protocols adapted existing MAC protocols by using direction antennas. Thus, what we discuss here is this directional antenna concept. This concept is of course useful in the theoretical field. It is supposed to be able to provide higher system capacities by directing narrow beams toward the vehicles of interest, while nulling vehicles not of interest, which provides lower transmission interferences, lower power levels, and more channel reuse within the same terminal neighborhood. This is how it is seen from the theoretical side, which may be the same from the practical side. But what is sure is that such a concept needs specific devices (directional antennas), which are expensive and very difficult to manage. Also, using directional antenna makes the coverage for each antenna very limited, which makes the network topology changes faster in vehicular networks.

So, we believe that medium access is still a big open issue in vehicular networks. Existing proposals are not suitable as they are to cover all requirements in VANET-related applications. Safety is an important application field in VANETs. Human lives depend on the efficiency of these applications, which directly depends on the network and the MAC layers' efficiency. Thus, a good MAC layer should take care of both safety application requirements and other application requirements. Another idea that can be employed is using two different MAC layers, the first used by safety applications and the second by other applications.

Since the main difference in VANETs when compared to MANETs is the relative high speed of vehicles, we believe that vehicle movement prediction in the network can be very useful for improving the medium access process, knowing that this concept has already been used in [35] to improve data routing in VANETs.

6.6 Summary

Vehicular networks will certainly emerge as communication networks dedicated to specific purposes. Research and demonstration of vehicular networks could show the power of different sorts of architecture and help shape the future. Vehicular networks can improve navigation safety using wireless car-to-car and car-to-curb communications to rapidly propagate unsafe road conditions, and report accidents to oncoming cars, unsafe drivers in the proximity, and imminent intersection crashes. These networks can also consume location-aware resource services, and can be used as an emergency communications network. However, vehicular networks have very different characteristics from many of the other networks: large-scale, temporary network disconnections; correlation between motion patterns and performance; and rapidly changing connectivity to the fixed network infrastructure.

This chapter focused on MAC protocols proposed in the open literature for VANETs. We first enumerated and discussed the most important metrics to take into account when designing new MAC protocols for wireless networks in general. Then we presented a classification of the MAC protocols of MANETs, and finally we emphasized the proposals for vehicular networks that take into account the inherent characteristics of these networks.

References

[1] J. Luo and J.-P. Hubaux. 2004. *A survey of inter-vehicle communication. Technical Report IC/2004/24*. School of Computer and Communication Sciences, Lausanne, Switzerland.

[2] *Proceedings of the ACM International Workshops on Vehicular Ad Hoc Networks (VANET)*. 2004–2005.

[3] The GPS system. Accessed March 27, 2008 from http://en.wikipedia.org/wiki/GPS.

[4] The Galileo system. Accessed March 27, 2008 from http://en.wikipedia.org/wiki/Galileo_positioning_system.

[5] J. J. Blum, A. Eskandarian, and L. J. Hoffman. 2004. Challenges of inter-vehicle ad hoc networks. *IEEE Transactions on Intelligent Transportation Systems* 5(4).

[6] C. Lochert, H. Hartenstein, J. Tian, H. Fussler, D. Hermann, and M. Mauve. 2003. Routing strategy for vehicular ad hoc networks in city environments. In *58th IEEE Semiannual Vehicular Technology Conference VTC 2003*, Orlando, FL, pp. 156–61.

[7] C. Siva Ram Murthy and B. S. Manoj. 2004. *Ad hoc wireless networks—Architectures and protocols*. Upper Saddle River, NJ: Pearson/Prentice Hall.

[8] A. C. V. Gummalla and J. O. Limb. 2000. Wireless medium access control protocols. *IEEE Communications Surveys and Tutorials* 3(2).

[9] L. Kleinrok and F. A. Tobagi. 1975. packet switching in radio channels. Part I. Carrier sense multiple-access models and their throughput-delay characteristics. *IEEE Transactions in Communications* 23:1400–16.

[10] IEEE/ANSI. 1985. *Carrier sense multiple access with collision detection*. IEEE/ANSI standard.

[11] P. Karn. 1990. MACA—A new channel access method for packet radio. In *ARRL/CRRL Amateur Radio 9th Computer Networking Conference*, pp. 134–40.

[12] K. Biba. 1992. *A hybrid wireless MAC protocol supporting asynchronous and synchronous MSDU delivery services*. IEEE 802.11 Working Group paper 802.11/91-21.

[13] V. Aharghavan, A. Memers, S. Shenker, and L. Zhang. 1994. A media access protocol for wireless LAN's. In *ACM SIGCOMM'94*, pp. 212–25.

[14] C. L. Fullmer and J. J. Garcia-Luna-Aceves. 1995. Floor acquisition multiple access (FAMA) for packet-radio networks. In *ACM SIGCOMM'95*, pp. 262–73.

[15] F. A. Tobagi and L. Kleinrock. 1975. Packet switching in radio channels. Part II. The hidden terminal problem in carrier sense multiple-access modes and the busy-tone solution. *IEEE Transactions in Communications* 23:1417–33.

[16] J. Deng and Z. J. Hass. 2002. Dual busy tone multiple access (DBTMA): A new medium access control for packet radio networks. *IEEE Transactions in Communications* 50:975–85.

[17] F. Talucci, M. Gerla, and L. Fratta. 1997. MACA-BI (MACA by invitation): A receiver-oriented access protocol for wireless multi-hop networks. In *Waves of the Year 2000, PIMRC'97, the 8th IEEE InternationaI Symposium on Personal, Indoor and Mobile Radio Communications*, vol. 2, pp. 435–39.

[18] C. Zhu and M. Scott Corson. 2001. A five-phase reservation protocol (FPRP) for mobile ad hoc networks. *ACM/Baltzer Journal of Wireless Networks* 7:371–84.
[19] S. Jiang, J. Rao, D. He, and C. C. Ko. 2002. A simple distributed PRMA for MANETs. *IEEE Transactions on Vehicular Technology* 51:293–305.
[20] C. W. Ahn, C. G. Kang, and Y. Z. Cho. 1999. Soft reservation multiple access with priority assignment (SWPA): A novel MAC protocol for QoS-guaranteed integrated services in mobile ad-hoc networks. In *Proceedings of IEEE INFOCOM 1999*, vol. 1, pp. 194–201.
[21] K. T. Jin and D. H. Cho. 2002. Multi-code MAC for Multi-hop wireless ad hoc networks. In *Proceedings of IEEE Vehicular Vechnology Conference (VTC 2002)*, vol. 2, pp. 1100–4.
[22] S. Agrawal et al. 2001. Distributed power control in ad hoc wireless networks. In *Personal, Indoor and Mobile Radio Communications*, vol. 2, pp. F-53–F-58.
[23] V. Kanodia, C. Li, A. Sabharwal, B. Sadeghi, and E. Knightly. 2002. Distributed priority scheduling and medium access in ad hoc networks. *ACM/Baltzer Journal of Wireless Networks* 8:455–66.
[24] V. Kanodia, C. Li, A. Sabharwal, B. Sadeghi, and E. Knightly. 2002. Ordered packet scheduling and wireless ad hoc networks: Mechanisms and performance analysis. In *Proceedings of ACM MOBIHOC 2002*, pp. 58–70.
[25] M. S. Gast. 2002. *802.11 wireless networks: The definitive guide*. Sebastopol, CA: O'Reilly.
[26] IEEE 802.11p. 2006. *Trial use standard wireless access in vehicular enviroments (WAVE) multichannel operation*. For Trial Operation.
[27] F. Borgonovo et al. 2003. ADHOC MAC: A new, flexible and reliable MAC architecture for ad-hoc networks. In *Proceedings of IEEE WCNC 2003*, vol. 4, pp. 965–70.
[28] W. Crowther et al. 1973. A system for broadcast communications: Reservation ALOHA. In *Proceedings of Hawaii International Conference on Systems Sciences*, pp. 596–603.
[29] F. Borgonovo et al. 2002. RR-ALOHA, a reliable R-ALOHA broadcast channel for ad-hoc inter-vehicle communication networks. In *Proceedings of Med-Hoc-Net.*
[30] A. Nasipuri et al. 2000. A MAC protocol for mobile ad hoc networks using directional antennas. In *Proceedings of IEEE WCNC 2000*, vol. 1, pp. 1214–19.
[31] Z. Huang et al. 2002. A busy tone-based directional MAC protocol for ad hoc networks. In *Proceedings of IEEE MILCOM 2002*, vol. 2, pp. 1233–38.
[32] Y. B. Ko et al. 2000., Medium access control protocols using directional antennas in ad hoc networks. In *Proceedings of IEEE INFOCOM 2000*, vol. 1, pp. 13–21.
[33] Q. Xu et al. 2007. Medium access control protocol design for vehicle-vehicle safety messages. *IEEE Transactions on Vehicular Technology* 56(2).
[34] S. Katragadda et al. 2003. A decentralized location-based channel access protocol for inter-vehicle communication. In *Proceedings of the 57th IEEE Vehicular Technology Conference (VTC)*, vol. 3, pp. 1831–35.
[35] H. Menouar, M. Lenardi, and F. Filali 2005. A movement prediction-based routing protocol for vehicle-to-vehicle communications. In *V2VCOM 2005, 1st International Vehicle-to-Vehicle Communications Workshop*, San Diego.

7

Network Coding for Wireless Networks

Yunnan Wu
Microsoft Research

Network coding refers to a scheme where a node is allowed to generate output data by mixing (i.e., computing certain functions of) its received data. The unique characteristics of wireless medium renders network coding particularly useful. For instance, network coding can be used to achieve the minimum energy per bit for multicasting in a wireless ad hoc network. In addition to optimizing energy efficiency, the network coding-based scheme has only polynomial time complexity, breaking through the NP-hardness barrier of the conventional routing approach. As another example, recently network coding has been developed into a link layer enhancement scheme. The network coding engine in the link layer can opportunistically mix the outgoing packets to reduce the transmissions in the air. This chapter provides an overview of some recent developments about using network coding in wireless networks, including (1) network coding for end-to-end multicasting, (2) network coding in the link layer, and (3) network coding in the physical layer.

7.1 Introduction

In today's practical communication networks such as the Internet and wireless networks, information delivery is performed by routing, i.e., having intermediate routers store and forward data. *Network coding* is a recent generalization of routing in which

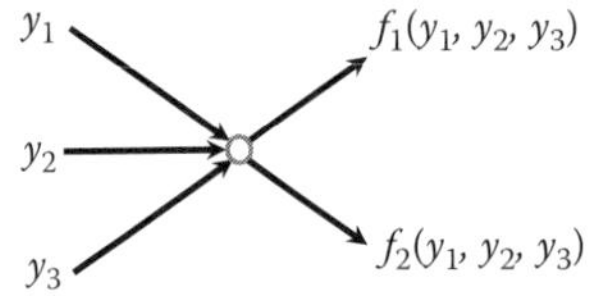

FIGURE 7.1 The concept of network coding: network nodes can compute functions of input messages.

nodes can generate output data by encoding (i.e., computing certain functions of) previously received input data. As illustrated by Figure 7.1, in network coding each node in a network can perform some computation, whereas in routing the output messages can only be copies of the received messages. Intuitively, network coding allows information to be "mixed" at a node.

The potential advantages of network coding over routing include resource (e.g., bandwidth and power) efficiency, computational efficiency, and robustness to network dynamics. As shown by the pioneering work of Ahlswede et al. [1], network coding can increase the possible network throughput, and in the multicast case can achieve the maximum data rate theoretically possible. This is illustrated by Figure 7.2. Each link in the graph can carry 1 bit/second. Using network coding, we can multicast information from the source node s to the two receivers t_1 and t_2 at rate 2.0 bits/second, which cannot be achieved using traditional routing.

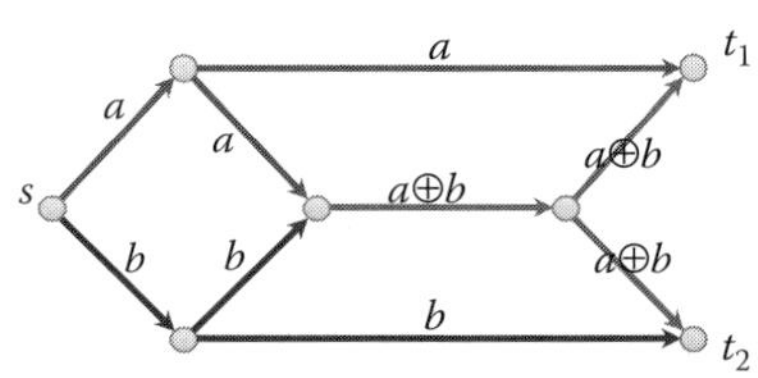

FIGURE 7.2 An example showing that network coding can achieve the multicast capacity, whereas routing cannot. Here $a \oplus b$ stands for the bit-wise XOR of the two bits. This example was introduced by Ahlswede et al. [1].

In addition to maximizing throughput, network coding can also maximize the energy efficiency. Consider the example wireless network shown in Figure 7.3. Assume each node is equipped with a transmitter operating at a fixed transmission range, which is just sufficient to reach its lateral neighbors, but not the diagonal ones. Under this setting, each physical layer transmission consumes a unit amount of energy. Using routing, the minimum number of transmissions required to deliver one message from s to $\{t_1,t_2\}$ is 5; a solution is shown in Figure 7.3(a), where the first transmission is a broadcast transmission. As shown in Figure 7.3(b), using network coding, we can multicast two messages using nine transmissions, resulting in a better energy efficiency. More generally, for multicasting in a multihop wireless network, it has been shown that under a layered model of wireless networks, the minimum energy per bit can be found by a linear program; the minimum energy per bit can be attained by performing linear network coding [2]. In contrast, minimum-energy multicast routing is NP-hard to compute and may not achieve the minimum possible energy per bit.

The advantages of network coding go beyond multicasting. Consider the example shown in Figure 7.4, where the left node wants to send packet a to the right node and the right node wants to send packet b to the left node. Using conventional routing, this requires four transmissions, as shown in Figure 7.4(a); using network coding, two packets can be exchanged using three transmissions [3, 4]. It looks as if the two packets a and b are sharing a ride in the air!

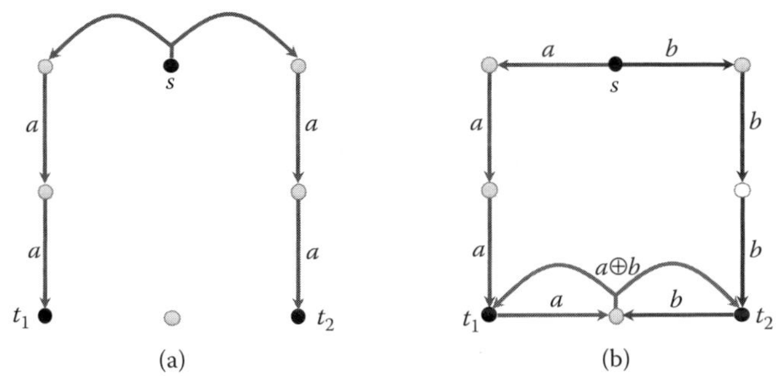

FIGURE 7.3 (a) Minimum-energy routing solution uses five transmissions to multicast one message. (b) Minimum-energy network coding solution uses nine transmissions to multicast two messages. (Adapted from Wu et al., 2005. Copyright © 2005 IEEE. With permission.)

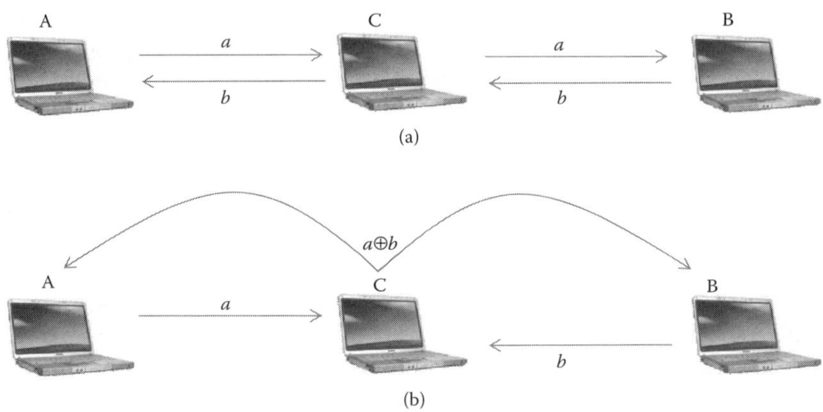

FIGURE 7.4 (a) The conventional solution requires four transmissions to exchange two packets between ***A*** and ***B*** via a relay node ***C***. (b) Using network coding, two packets can be exchanged in three transmissions.

The three-node information exchange scenario mentioned above has been investigated and extended from various perspectives; see, e.g., [3–12]. For instance, the technique depicted in Figure 7.4 has been generalized and developed into a generic technique in the link layer of the protocol stack, which mixes packets belonging to different communication sessions [6]. As another example, it has been shown that by mixing and demixing signals in the physical layer (as opposed to mixing and demixing message bits), further performance gain can be achieved; see, e.g., [7–12].

In the above we have seen some examples that demonstrate the usefulness of network coding in wireless networks. Indeed, in recent years, significant progress has been made regarding the theory and practice of network coding in wireless networks. This chapter provides an overview of some recent development. We start by discussing the use of network coding for end-to-end multicasting in section 7.2. Then we explain how network coding can be applied as a link layer technique in section 7.3. Section 7.4 reviews some techniques that perform mixing and demixing in the physical layer.

7.2 Network Coding for End-to-End Multicasting

7.2.1 Theory

In this section we present a theoretical model for analyzing the performance of multicasting information in wireless networks using network coding. We cast this problem as a mathematical optimization, where the objective function is some function of the end-to-end multicast throughput and the overall resource (e.g., energy) consumed in providing such throughput. For simplicity, we assume there is only a single multicast session in the network, from a source node s to a set of destination nodes T.

To establish such an optimization, we use a popular layered model of wireless networks (see, e.g., [13]), which is a mathematical abstraction of the well-known layered network architecture. The basic assumptions of the layered model can be explained as follows. The lower and upper layers are respectively abstracted as supply and demand of communication resources. The interface between the supply and demand is a network of lossless channels with rate limits, which can be described as $G = (V, E, \mathbf{c})$ where V and E are sets of vertices and edges, respectively, and c is a length $|E|$ vector assigning to each edge $e \in E$ a bit-rate limit $c(e)$.

The physical and link layers can supply many possible graphs $G = (V, E, \mathbf{c})$, by applying communication mechanisms such as scheduling, modulation, and channel coding over the underlying noisy and interfering channels. We call these graphs *realizable graphs* because they correspond to feasible ways of operating the wireless network, e.g., arranging different subsets of nodes to communicate for different time fractions. In section 7.2.1.1, we discuss how to obtain the realizable graphs.

Using a realizable graph G as the available communication resource, the network layer coordinates the information flow from the source to the destinations such that certain end-to-end throughput is achieved. We then discuss how to characterize the achievable end-to-end throughput via network coding.

Each realizable graph $G = (V, E, \mathbf{c})$ provides certain end-to-end throughput and has an associated resource (e.g., energy) consumption that represents the cost in supplying the bit-rate resources. To find a realizable graph that achieves a good trade-off between throughput and resource consumption, we can use an optimization that has the following abstract form:

$$\text{maximize} \quad U(r) - \lambda p, \tag{7.1}$$

$$\text{subject to: } G \text{ can provide multicast rate } r; \tag{7.2}$$

$$G \text{ is a realizable graph with power consumption } p, \tag{7.3}$$

where $\lambda \geq 0$ is a weighting coefficient reflecting the desired emphasis on throughput versus resource consumption. In section 7.2.1.1, we explain how to concretely characterize constraint (7.3). In section 7.2.1.2, we explain how to concretely characterize constraint (7.2). In section 7.2.1.5, we shall examine some concrete optimization formulations for wireless networks.

7.2.1.1 Realizable Graphs for Wireless Networks

In this section, we discuss the structure of realizable graphs for wireless ad hoc networks. The set of realizable graphs represents all possible supplies of bit-rate resources arising from power control and scheduling in the physical and medium access layers (under certain simplifying assumptions about the physical layer). For simplicity, we assume the nodes are static and the link conditions do not change over time.

A wireless ad hoc network can operate in many different *physical states*, where each physical state represents a "snapshot" of all nodes in the physical layer, such as which nodes are transmitting, what transmitting powers are used, and what the channel conditions are. A physical state may support a collection of concurrent *links*, which are assumed to be point-to-multipoint in general. Let V_0 denote the set of nodes in the network. A *link* can be described as $u \xrightarrow{c} Y_u$, where $u \in V_0$ is the transmitter, $Y_u \subseteq V_0$ is its receiver set, and c is the associated bit rate in a reliable communication.

Each collection of links supported by a certain physical state corresponds to an *elementary (realizable) graph*. Loosely speaking, an elementary graph refers to a graph that can be directly realized in the physical layer. For example, Figure 7.5 shows two physical states and the corresponding elementary graphs. Figure 7.5 corresponds to the special case of omnidirectional transmissions. In a first state shown in Figure 7.5, only node s is transmitting, and the transmission power is just enough to reach a and b. This supports the link $s \xrightarrow{1.0} \{a,b\}$. The corresponding elementary graph is shown as G_1. In a second state, only node s is transmitting, and the transmission power is just enough to reach a, b, c, and d. This supports the link $s \xrightarrow{1.0} \{a,b,c,d\}$. The corresponding elementary graph is shown as G_2.

Generally, an elementary graph consists of a number of concurrent broadcast links. To reflect the broadcasting of common information, we use the following graph model for broadcast links. For each link $u \rightarrow Y_u$, we add to the associated elementary graph a

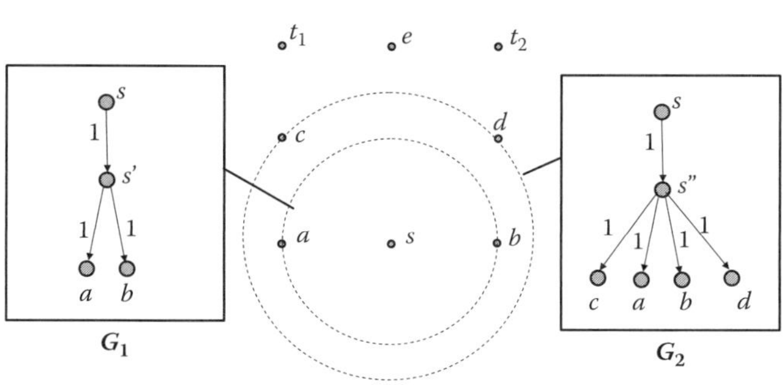

FIGURE 7.5 Two example elementary graphs G_1 and G_2. G_1 corresponds to the physical state where s is transmitting at a power just enough to reach a and b. G_2 corresponds to the physical state where s is transmitting at a power just enough to reach a, b, c, and d. (From Wu et al., 2005. Copyright © 2005 IEEE. With permission.)

*distinct** virtual vertex (e.g., u') and unit capacity edges $uu', u'v, v \in Y_u$. Two examples are given in Figure 7.5 as G_1 and G_2, in which s' and s'' are the virtual vertices introduced. The virtual vertex plays the role of an artificial bottleneck that constrains the rate of new information going out of the transmitter. To see this, consider the example elementary graph G_1 in Figure 7.5. If we instead represent the broadcast of s to $\{a, b\}$ by two unit capacity links, sa and sb, then in this graph representation s can send new information to the set $\{a, b\}$ at rate 2, which is not supported by the underlying broadcast transmission.

By timesharing among different physical states, it is possible to achieve any convex combination of the elementary graphs. That is, if λ_k is the relative share of time for the elementary graph $G_k = (V_k, E_k, \mathbf{c}_k)$, then it is possible to achieve on average the graph

$$G = \left(\cup_k V_k, \cup_k E_k, \sum_k \lambda_k \mathbf{c}'_k \right).$$

Here the edge capacities $\boldsymbol{c}_k$ are each extended to a length $|\cup_k E_k|$ vector $\boldsymbol{c}'_k$ in the obvious way. Denote such combinations by

$$G = \sum_k \lambda_k G_k .$$

Let the set of all elementary graphs be $\mathcal{B}_0$. The number of elementary graphs, $|\mathcal{B}_0|$, generally grows exponentially with the number of network nodes. For analytic tractability, we shall identify a limited number of promising elementary graphs that provide a reasonably good span for a specific application. For details on how to do so in a general setting, see, for example, [13]. For the minimum-energy multicast problem, however, we only need to examine a polynomial number (in the number of nodes) of elementary graphs, or corresponding physical states. This is because separating interfering transmissions into different time slots improves the energy efficiency; thus, we can restrict our attention to those physical states involving only a single transmitter. It is this fact that results in the polynomial solvability of the minimum-energy multicast problem for a wireless ad hoc network.

With a finite set of elementary graphs $\mathcal{B} \subseteq \mathcal{B}_0$, the set of realizable graphs is

$$\mathcal{G}(\mathcal{B}) = \left\{ G \,\middle|\, G = \sum_k \lambda_k G_k, \sum_k \lambda_k \leq 1, \lambda_k \geq 0 \; \forall k, G_k \in \mathcal{B} \right\},$$

where the dependence on $\mathcal{B}$ is explicitly shown. The power consumption of a composite graph

$$G = \sum_k \lambda_k G_k \text{ is } \sum_k \lambda_k p(G_k),$$

* To see why it can be problematic to treat the virtual vertices associated with the same transmitter as the same in different elementary graphs, please refer to [14].

where $p(G_k)$ is the power consumption of elementary graph G_k. The power consumption reflects one possible metric that measures the cost of providing the bit-rate resources at the physical and medium access layers.

7.2.1.2 Characterizing End-to-End Throughput of Network Coding in a Given Graph

The lower layers in a wireless network can support many possible realizable graphs. In this section, we assume we are given a network of lossless links represented by $G = (V, E, \mathbf{c})$; the graph can be a particular realizable graph supported by the lower layers. We characterize the achievable end-to-end multicast throughput using network coding in G via a set of linear inequalities.

7.2.1.3 Unicasting and Max-Flow-Min-Cut

Before talking about multicasting, let us begin by reviewing the results for unicasting from a source node s to a destination node t. We are interested in the *unicast capacity*, which refers to the maximum rate at which s can communicate information to t.

An upper bound of the unicast capacity can be obtained by examining the *s–t cuts*. Given two nodes $s,t \in V$, an s–t cut $(U,\bar{U})$ refers to a partition of the nodes $V = U + \bar{U}$ with $s \in U$, $t \in \bar{U}$. The *capacity* of the cut refers to the sum of the edge capacities for edges going from U to $\bar{U}$. An s–t cut with minimum capacity is called a *minimum s–t cut*. Let $\rho_{s,t}(\mathbf{c})$ denote the capacity of a minimum s–t cut for graph (V,E) with link capacities $\mathbf{c}$. The significance of an s–t cut comes from the fact that it exhibits a bottleneck for communication from s to t. It is intuitively clear that all information t can get from s must be derived from the information flowing across the cut. Consequently, the maximum rate at which s can transfer information to t cannot exceed the minimum s–t cut capacity $\rho_{s,t}(\mathbf{c})$.

A fundamental theorem in graph theory, the max-flow-min-cut theorem, shows that the cut bound $\rho_{s,t}(\mathbf{c})$ is achievable by routing along parallel paths. To explain the max-flow-min-cut theorem, we now review the notion of *flow*.

An *s–t flow* is a nonnegative vector $\boldsymbol{f}$ of length $|E|$ satisfying the *flow conservation constraint*:

$$\text{excess}_v(\mathbf{f}) = 0, \quad \forall v \in V - \{s,t\}, \tag{7.4}$$

where

$$\text{excess}_v(\mathbf{f}) \equiv \sum_{e \in In(v)} f_e - \sum_{e \in Out(v)} f_e \tag{7.5}$$

is the *flow excess of* v, viz., the amount of incoming traffic less the amount of outgoing traffic for node v. The flow excess is not required to be zero at s and t. The flow excess $\text{excess}_t(\mathbf{f})$ at the destination node t is called the *value* of the flow.

A flow from a source to a destination can be decomposed into a sum of several *path flows* and *cycle flows*. It turns out that the cycle flows can be eliminated without affecting

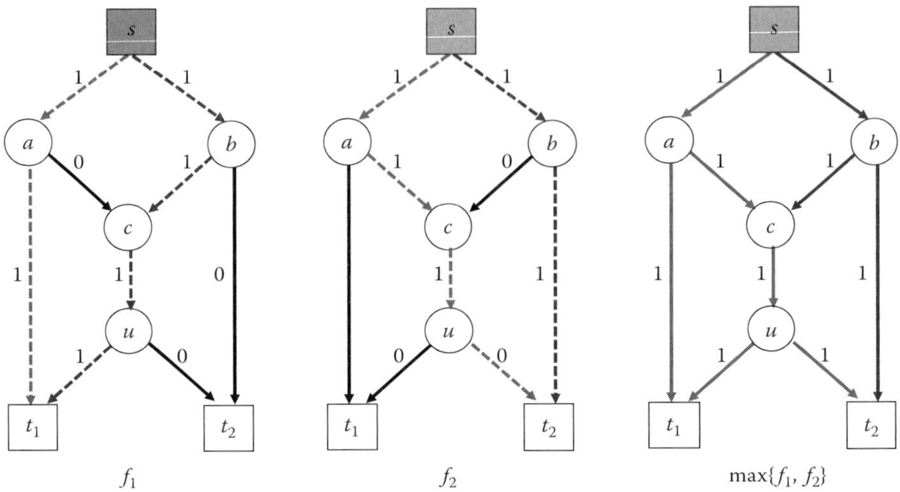

FIGURE 7.6 (a) An s–t_1 flow $\mathbf{f}_1$ on graph (V, E). The corresponding two parallel s–t_1 paths are shown in dotted lines. (b) An s–t_2 flow f_2 on graph (V, E). The corresponding two parallel s–t_2 paths are shown in dotted lines. (c) The edge-wise max of flows $\max\{f_1, f_2\}$, which can provide a multicast rate of 2 from s to $\{t_1, t_2\}$. (From Wu et al., 2005. Copyright © 2005 IEEE. With permission.)

the value of the flow. Each path flow corresponds to a path from the source with the destination with an associated rate. The value of the flow is the sum of the rates carried by the individual path flows. Thus, a flow prescribes a way for information to be routed from the source to the destination along parallel paths; the communication rate achieved by such a routing scheme is the value of the flow. For an illustration of a flow, see Figure 7.6(a).

Let $\mathcal{F}_{s,t}(r)$ denote the set of s–t flows, each with its flow value equal to r. Then $\boldsymbol{f} \in \mathcal{F}_{s,t}(r)$ if and only if

$$\begin{aligned}
&\boldsymbol{f} \geq 0, \\
&\text{excess}_s(\boldsymbol{f}) = -r, \\
&\text{excess}_t(\boldsymbol{f}) = r, \\
&\text{excess}_v(\boldsymbol{f}) = 0, \quad \forall v \in V - \{s,t\}.
\end{aligned}$$

Note that the above inequalities are linear in $\boldsymbol{f}$ and r; for this reason, $\mathcal{F}_{s,t}(r)$ is called the *s–t flow polyhedron*. A useful property of $\mathcal{F}_{s,t}(r)$ is its linearity in r, i.e.,

$$\mathcal{F}_{s,t}(r) = r\mathcal{F}_{s,t}(1) \equiv \{r\boldsymbol{f} \mid \boldsymbol{f} \in \mathcal{F}_{s,t}(1)\}. \tag{7.6}$$

In order for a flow to correspond to a feasible routing arrangement, we often need to enforce that the assigned flow on an edge fits in the available bit-rate resource, i.e., $\boldsymbol{f} \leq \boldsymbol{c}$

(in the element-wise sense). The max-flow-min-cut theorem says that the minimum cut capacity $\rho_{s,t}(\mathbf{c})$ is equal to the maximum value of an s–t flow within $G = (V, E, \mathbf{c})$. It follows then

$$r \leq \rho_{s,t}(\mathbf{c}) \Leftrightarrow \exists \mathbf{f}_t \in \mathcal{F}_{s,t}(r), \mathbf{f}_t \leq \mathbf{c}. \tag{7.7}$$

7.2.1.4 Multicasting and Edge-wise Maximum of Flows

Given $G = (V, E, \mathbf{c})$, a source node s, and a set of destination nodes T, the *multicast capacity* refers to the maximum multicast throughput. Since the capacity of any s–t cut is an upper bound on the rate at which information can be transmitted from s to t, $\min_{t\in T} \rho_{s,t}(\mathbf{c})$ is an upper bound of the multicast capacity. Ahlswede et al. [1] showed that $\min_{t\in T} \rho_{s,t}(\mathbf{c})$ can be achieved by performing network coding. Hence, it is the multicast capacity. Now we have arrived at a characterization of the achievable multicast throughput in a graph $(V, E, \mathbf{c})$: end-to-end throughput r can be achieved if and only if

$$r \leq \min_{t\in T} \rho_{s,t}(\mathbf{c}). \tag{7.8}$$

From (7.7), we obtain an equivalent formulation of (7.8): end-to-end throughput r can be achieved in $(V, E, \mathbf{c})$ if and only if

$$\exists \mathbf{f}_t \in \mathcal{F}_{s,t}(r), \mathbf{c} \geq \max_{t\in T} \mathbf{f}_t. \tag{7.9}$$

We call $\max_{t\in T} \mathbf{f}_t$ in (7.9) an *(edge-wise) max of flows*. Just as a flow is the critical structure for unicasting from a source to a destination, a max of flows plays a fundamental role for multicasting using network coding. Figure 7.6 illustrates the structure of a max of flows, using the classical example of network coding introduced in [1]. Figure 7.6(a) shows an s–t_1 flow, which prescribes two parallel paths from s to t_1 (shown in dotted lines); similarly, Figure 7.6(b) shows an s–t_2 flow. Figure 7.6(c) shows the edge-wise maximum of these two flows, which is sufficient to provide a multicast rate of 2.

The (edge-wise) max of flows formulation (7.9) is especially important because it can be written as a set of linear inequalities, and integrated with other constraints in an optimization formulation. We will see some concrete examples in the next section.

7.2.1.5 Optimization Formulations

Having characterized the supply and demand sides of wireless networks, we are now ready to formulate various optimizations for network coding–based multicasting in wireless networks. For instance, if we want to find the maximum throughput, we can use the following linear optimization:

$$\text{maximize } r, \tag{7.10}$$

$$\text{subject to: } \mathbf{c} \geq \mathbf{f}_t, \ \forall t \in T, \tag{7.11}$$

$$f_t \in \mathcal{F}_{s,t}(r), \forall t \in T, \tag{7.12}$$

$$\boldsymbol{c} = \sum_k \lambda_k \boldsymbol{c}_k, \tag{7.13}$$

$$\sum_k \lambda_k \leq 1, \tag{7.14}$$

$$\lambda_k \geq 0, \quad \forall k. \tag{7.15}$$

Here we use a finite collection of elementary graphs $\{G_k = (V_k, E_k, \boldsymbol{c}_k)\}$. Constraints (7.11) and (7.12) enforce that $G = (V, E, \boldsymbol{c})$ must be able to support a rate of r, or equivalently, that the s–T multicast capacity in $G = (V, E, \boldsymbol{c})$ must be greater than or equal to r. Constraints (7.13)–(7.15) enforce that $G = (V, E, \boldsymbol{c})$ must be physically realizable by operating the wireless network in different physical states, each for a certain fraction of time.

As another example, we can minimize the energy per bit for multicasting by using the following optimization, where r, $\boldsymbol{c}'$, λ_k are treated as variables:

$$\varepsilon^* = \min \frac{\sum_{k:G_k \in \mathcal{B}} \lambda'_k p(G_k)}{r}$$

$$\text{subject to: } \boldsymbol{c}' \geq \boldsymbol{f}_t, \quad \forall t \in T, \tag{7.16}$$

$$\boldsymbol{f}_t \in \mathcal{F}_{s,t}(r), \forall t \in T, \tag{7.17}$$

$$\boldsymbol{c}' = \sum_k \lambda'_k \boldsymbol{c}_k, \tag{7.18}$$

$$\sum_k \lambda'_k \leq 1, \lambda'_k \geq 0, \forall k, \tag{7.19}$$

$$r > 0. \tag{7.20}$$

Note that here we only need to include a polynomial number of elementary graphs, each containing a single transmitter operating at a certain power level (see the discussion in section 7.2.1.1).

At first glance, the objective function of the above optimization is nonlinear in the variables. However, we can renormalize the above optimization to arrive at a linear program. Specifically, by a variable change $\lambda_k = \lambda'_k/r$, $\boldsymbol{c} = \boldsymbol{c}'/r$, we have the following linear program:

$$\varepsilon^* = \min \sum_{k:G_k \in \mathcal{B}} \lambda_k p(G_k)$$

$$\text{subject to: } \boldsymbol{c} \geq \boldsymbol{f}_t, \quad \forall t \in T, \tag{7.21}$$

$$\boldsymbol{f}_t \in \mathcal{F}_{s,t}(1), \forall t \in T, \tag{7.22}$$

$$\boldsymbol{c} = \sum_k \lambda_k \boldsymbol{c}_k, \tag{7.23}$$

$$\lambda_k \geq 0, \forall k. \tag{7.24}$$

7.2.2 Practice

In this section we examine how to practically apply network coding for multicasting in multihop wireless networks. Recently, several practical protocols have been proposed [15–18]. These protocols can be understood using a single framework, which contains four pillar ideas: random linear coding, packet tagging, buffering, and output pacing.

The first three ideas, random linear coding, packet tagging, and buffering, are not unique to wireless networks. Random linear coding allows the encoding to proceed in a distributed manner. Tagging each packet with the corresponding coding vector allows the decoding to proceed in a distributed manner. Buffering allows for asynchronous packet arrivals and departures with arbitrarily varying rates, delay, and loss. These ideas together constitute a practical way to implement network coding in real-world packet networks. Simulation results in [19] have shown that a practical network coding system based on these ideas can achieve a multicast throughput close to the multicast capacity. In addition, theoretical studies in [20, 21] have shown that the use of random linear coding and buffering gives rise to an asymptotically capacity achieving scheme. Recall from section 7.2.1 that some inequalities in the optimization formulations impose the constraint that the desired multicast rate r in $(V, E, \boldsymbol{c})$ cannot exceed the s–T multicast capacity in $(V, E, \boldsymbol{c})$. We now know that via random linear coding, packet tagging, and buffering, we can approach the multicast capacity for given network resource $\boldsymbol{c}$.

In section 7.2.1, we showed that for wireless networks there are many ways to provision the network resource $\boldsymbol{c}$, and we can use various optimization formulations to determine an efficient resource allocation $\boldsymbol{c}$. Accordingly, in practice, we need a distributed method for deciding the resource allocation. Such a role is played by the fourth pillar idea, *output pacing*. A proper output pacing mechanism judiciously decides when to generate an output packet, to ensure efficient use of the network resource and coordinate the nodes in moving the information toward the destinations. The output pacing mechanisms to be discussed below can be interpreted as heuristic methods for optimizing the resource allocation.

We now describe these pillar ideas one by one.

7.2.2.1 Random Linear Coding

While Ahlswede et al.'s work showed that the multicast capacity can be achieved using network coding, their result is mainly an existence proof, established via information theoretical techniques. Hence, the first problem is to find a practical way of assigning network codes. The work by Li, Yeung, and Cai [22] showed that the maximum multicast capacity can be achieved by using linear encoding functions at each node. With linear encoding, decoding at a receiver amounts to solving linear equations. Linear encoding and decoding represent an important step toward making network coding practical. However, determining the proper coefficients in a distributed manner is another practical challenge. The studies by Ho et al. [23] and Sanders et al. [24] addressed this challenge, by showing that random linear network coding over a sufficiently large finite field can (asymptotically) achieve the multicast capacity. This leads to a simple and distributed encoding scheme, where each node chooses its own encoding coefficients at random for each output packet, without any coordination with other nodes. Random linear coding is the first key idea for making network coding practical for multicasting.

We now explain the idea of random linear coding in the context of a packet network. For concreteness, suppose $\mathbb{F}=GF(2^8)$ and each packet contains 1,000 bytes. Each packet can thus be viewed as a row vector of length 1,000, with elements in $GF(2^8)$. Taking Figure 7.1 as an example, with random linear coding, $f_1(y_1,y_2,y_3)=\alpha_1 y_1+\alpha_2 y_2+\alpha_3 y_3$, and $f_2(y_1,y_2,y_3)=\beta_1 y_1+\beta_2 y_2+\beta_3 y_3$, where $\alpha_1,\alpha_2,\alpha_3,\beta_1,\beta_2,\beta_3$ are randomly and independently drawn from the field $GF(2^8)$. Note that all (1,000) symbols within one packet are mixed in the same way.

7.2.2.2 Packet Tagging

Suppose there are h source packets, denoted by $\mathbf{x}_1,\ldots,\mathbf{x}_h$. Linear coding is applied throughout the network so that each packet flowing in the network is a linear combination of the source packets. For example,

$$\mathbf{y}=\sum_{i=1}^{h} q_i\mathbf{x}_i, \tag{7.25}$$

where $\mathbf{q}=[q_1,\ldots,q_h]$ is the *global coding vector* that shows how this packet $\mathbf{y}$ relates to the source packets.

A critical practical issue is to inform the destinations how the coding is done so that they can decode the original packets. To address this issue, Chou et al. [19] and Ho et al. [23] propose to explicitly record the global coding vector in the packet header. The cost of this scheme is the overhead of transmitting h extra symbols in each packet; note that this is amortized over the number of data symbols in the packet (e.g., 1,000). On the other hand, the benefits of the scheme are significant. It allows the system to be

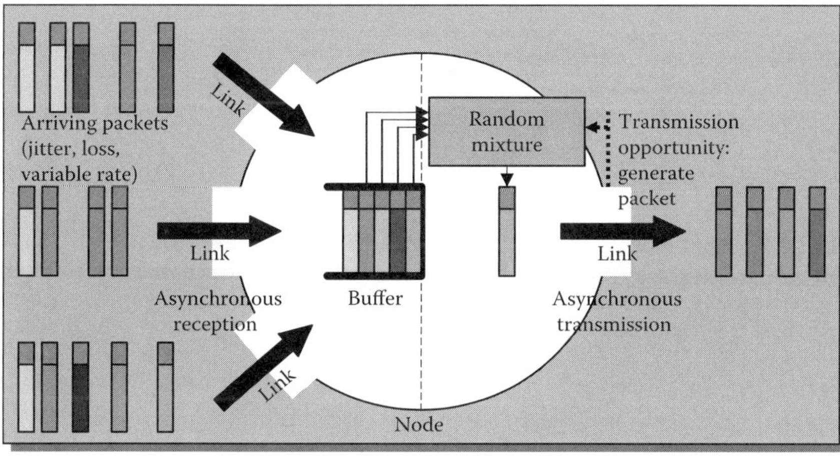

FIGURE 7.7 Illustration of the random linear network coding scheme with buffering.

completely decentralized: destinations can decode without knowing the network topology or the encoding rules; destinations can decode even if nodes or links are added or removed in an ad hoc fashion.

7.2.2.3 Buffering and Generations

In many practical networks, synchronization among nodes is difficult and expensive, if not impossible. To make network coding practical, we must handle the asynchronous arrivals of packets and other dynamics. Another important idea is the use of buffering to eliminate the need for distributed synchronization [19]. As illustrated by Figure 7.7, each node in the system maintains a buffer. Whenever a node receives a packet from one of its incoming links, it stores the packet in its buffer if the packet is "innovative." A mixture packet is said to be innovative if it is not a linear combination of the packets in the buffer. Noninnovative packets do not provide any new information to the node, and hence are immediately discarded. To efficiently test whether a packet is innovative, we can maintain the packets in the buffer in the standard row echelon form. Notice that the testing can be done by using only the global coding vectors of the packets in the buffer, not involving the payloads.

Whenever a transmission opportunity arises on one of its outgoing links, a node generates an output packet by linearly mixing the packets in the buffer using random coefficients in $\mathbb{F}$, as illustrated in Figure 7.8.

7.2.2.4 Output Pacing

With random linear coding, generating an output packet is easy: simply combine the buffered packets using a set of randomly generated coefficients. But when should we generate an output packet? This is what we call the output pacing problem. We now review two example approaches to output pacing.

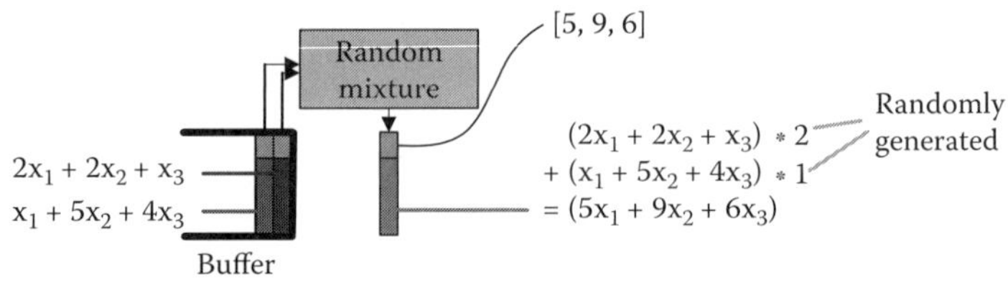

FIGURE 7.8 Illustration of the random mixing of packets in the buffer. Suppose $h = 3$ and $\mathbb{F} = GF(11)$. The two packets residing in the current buffer are linear combined with randomly generated coefficients, 2 and 1, respectively. The global coding vector is recorded within each packet to describe how the packet relates to the original packets.

Widmer, Fragouli, and Le Boudec [15] studied broadcasting in a multihop wireless network. In their output pacing protocol, each node maintains a send counter, which is initialized to 0. A new source packet is always broadcast once in the original form (i.e., without mixing with other packets). Each time a node receives an innovative packet, the send counter is incremented by a value called the forwarding factor d. A node broadcasts a mixture packet when its send counter is positive. Each time it broadcasts a packet, its send counter is decremented by 1. A subsequent paper by the authors [16] extends this scheme by allowing the nodes to have different forwarding factors. Let d_v denote the forwarding factor of v. Here the forwarding factor d_v is adjusted heuristically based on the local (two-hop) density information. A specific rule for setting d_v that works well in the simulations is to set the forwarding factor to be:

$$d_v = \frac{k}{\min_{v' \in N(v)} |N(v')|}, \tag{7.26}$$

where k is a network-wide constant parameter and $N(v)$ is the set of direct neighbors of node v. The intuition is that if v has a neighbor whose only neighbor is v, then v must forward the information, regardless of how many neighbors v has.

Chachulski et al. [17] proposed a multicasting protocol called MORE. A unique feature of MORE is its use of distance information. Each node is labeled with the distances to the destinations. Only nodes that are closer to at least one destination than the source are involved in the distribution of the session data. Intuitively, this ensures that information is not flowing in the opposite direction.

Similar to the broadcast protocol discussed above, in MORE, each node performs output pacing by making use of a credit counter. When the MAC allows transmission of a packet, the node generates a mixture when the credit counter is positive, broadcasts it, and reduces the credit counter by 1. However, a different rule is used for increasing the credit when receiving an innovative packet. The credit increments are determined based on the link loss probabilities in the entire network; for the exact expression, refer to [17].

7.2.2.4.1 Applications

Using network coding, the linear mixture packets can automatically weave their way to the destination, if one exists. Hence, network coding is as distributed, robust, and adaptive as flooding. As a result, network coding–based solutions are attractive candidates for multicasting protocols in multihop wireless networks.

Furthermore, the robustness offered by network coding makes it particularly attractive for networks with mobility (e.g., vehicular networks), sensor networks where nodes sleep most of the time to conserve energy, and other scenarios where connectivity is sparse. For these related applications of network coding–based multicasting, refer to, e.g., [25–27].

7.3 Network Coding in the Link Layer

In Figure 7.4 we see how network coding can be used to reduce the number of transmissions for packet exchanges between two nodes via an intermediate node, leveraging the broadcast medium. The gist of the packet exchange example in Figure 7.4 is as follows: At a certain moment, the left node has a, the right node has b, and the middle node has a and b. Thus, a mixture packet $a \oplus b$ can be decoded into a and b, respectively, at the right and left nodes.* This packet exchange example bears some similarities with the use of network coding for end-to-end multicasting. Indeed, we can interpret this packet exchange scenario as a virtual multicast session where initially a virtual source node sends packet a to the left node and packet b to the right node. Furthermore, the network coding-based information exchange constitutes the critical step in the minimum-energy multicast solution shown in Figure 7.3(b).

Despite the similarities, this example also has a distinctive feature: the mixture packet can be immediately decoded by the neighbors. The fact that mixing and demixing are both done locally brings in many advantages. For instance, a node can mix two packets passing through it, which may belong to different communication sessions; in contrast, in the random linear coding used for end-to-end multicasting, only packets belonging to the same end-to-end session can be mixed together. In fact, the mixing can be done in a way transparent to the communication sessions: they need not know that somewhere in the network their packets are mixed and then quickly demixed. In addition, a node need not buffer any mixture packet, resulting in a cleaner design that fits the existing networking framework better. To highlight this feature and differentiate it from the use of random linear coding for end-to-end multicasting, we use the name *local mixing* to refer to the use of network coding exemplified by Figure 7.4, where original packets are mixed locally at a node and then demixed immediately at the receivers.

Indeed, this technique has been generalized and developed into a link layer enhancement scheme for multihop wireless networks by Katti et al. [6]. As illustrated in Figure 7.9(a), the local mixing engine sits above the MAC layer (e.g., 802.11) and below the network layer. The network layer passes to the local mixing engine a list of packets

---

* In the following we use the name *source packet* to refer to a packet such as a, which was originally generated by a source node, and the name *mixture packet* to refer to a packet such as $a \oplus b$.

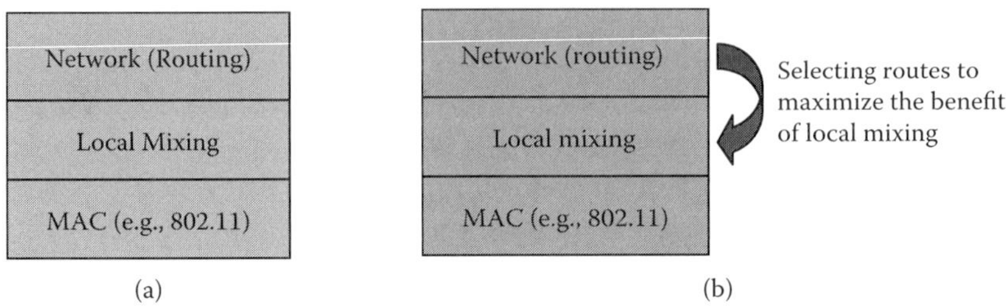

FIGURE 7.9 (a) The local mixing engine sits between the network layer and the MAC layer, and thus presents an enhanced link layer to the network layer. (b) Local mixing-aware routing schemes can better take advantage of the local mixing engine by generating traffic patterns that have more mixing opportunities. (From Wu et al., 2007. Copyright © 2007 IEEE. With permission.)

with their respective next-hops determined according to a certain routing scheme. The local mixing engine maintains information about the packets each neighbor has, and identifies opportunities to mix the outgoing packets to reduce the transmissions in the air. More specifically, each node snoops on the medium and buffers packets it hears. A node also informs its neighbors which packets it has overheard. This allows nodes to know roughly what packets are available at each neighbor (i.e., who has what). Knowing who has what in the neighborhood, a node examines its pending outgoing packets and forms output mixture packets if possible. In section 7.3.1, we explain the local mixing engine in detail.

The local mixing engine, on its own, can improve the link layer efficiency. The gain of this technique, however, critically depends on the traffic pattern in the network. For instance, if we have two flows traveling in opposite directions along a chain, then asymptotically the gain over conventional routing can approach 2:1 [3]. In addition, the results of [6] show that the throughput increase can be up to four times in a multihop wireless test bed for many UDP (user diagram protocol) flows among randomly chosen sources and destinations. The throughput gain is smaller in other traffic patterns (e.g., all traffic is to and from some Internet gateway). This motivates the following question: Can we make intelligent routing decisions that maximize the benefits offered by the local mixing engine? Figure 7.9(b) illustrates this concept. In section 7.3.2, we review a local mixing-aware routing scheme, proposed in [28].

7.3.1 Local Mixing

Consider the situation illustrated by Figure 7.10. A wireless router knows the source packets each neighbor has (i.e., who has what). It also knows who wants what because these are the packets in its output queue that it is supposed to forward to its neighbors. Then it can decide locally how to optimize the formation of mixture packets. A heuristic approach for generating the mixture packets is used in [6], which takes the packet at the head of the output packet queue and steps through the packet queue to greedily add packets to the mixture, while ensuring the neighbors can successfully demix. For example, in Figure 7.10, there are five packets in the output queue, $\boldsymbol{x}_1, \ldots, \boldsymbol{x}_5$; assume a

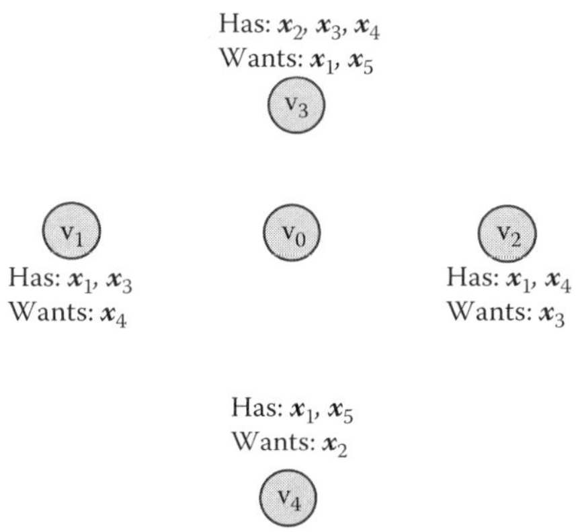

FIGURE 7.10 Knowing who has what and who wants what in a neighborhood, the local mixing engine identifies opportunities to mix the outgoing packets to reduce the resource consumption at v_0.

lower indexed packet is an earlier packet. Then the greedy procedure will use three transmissions: $\boldsymbol{x}_1 \oplus \boldsymbol{x}_2$, $\boldsymbol{x}_3 \oplus \boldsymbol{x}_4$, $\boldsymbol{x}_5$. However, there is a better solution: $\boldsymbol{x}_1 \oplus \boldsymbol{x}_3 \oplus \boldsymbol{x}_4$, $\boldsymbol{x}_2 \oplus \boldsymbol{x}_5$. A mathematical abstraction of the optimized formation of the mixture packets—the *local mixing problem*—is studied by Wu et al. [29] from an information theoretic point of view. Under the assumption that each neighbor discards the received packets that are polluted by sources it does not have or want, the optimal mixing is characterized.

Why would a node have packets meant for others? In Figure 7.4, node v_1 has packet $\boldsymbol{x}_1$ because it is the previous hop of $\boldsymbol{x}_1$. Another possibility is that neighboring nodes may overhear packets due to the broadcast nature of the wireless medium. For example, in Figure 7.10, packet $\boldsymbol{x}_1$ may follow a path $\ldots v_4 \rightarrow v_0 \rightarrow v_3 \ldots$; v_1 and v_2 may have overheard $\boldsymbol{x}_1$ when v_4 sent it to v_0.

How does a node get to know who has what? First, note that a node can obtain some partial information about its neighbors' data availability in a passive fashion. For example, node v_2 may infer that node v_1 holds packet $\boldsymbol{x}_1$ if v_1 recently received packet $\boldsymbol{x}_1$ or a mixture packet involving $\boldsymbol{x}_1$ from v_1, or if v_2 recently heard v_1 acknowledging the receipt of packet $\boldsymbol{x}_1$. This suffices for packet exchanges such as in Figure 7.4.

Passive inference does not incur any additional overhead. However, using passive inference alone, a node may only obtain a limited view of the neighbors' data availability. Katti et al. [6] extended this by proposing two techniques to obtain more information about local data availability: (1) Let each node explicitly announce to its neighbors the packets it currently has. (2) Let a node guess whether a neighbor has overheard a packet using information about the channel reception. In the former, each node can periodically compose *reception reports* to announce the packets it has overheard. The reception reports may also be piggybacked with ordinary packets. To implement guessing, nodes conduct measurements about the packet success probabilities to its neighbors and exchange the measurement results in the neighborhood. Such measurement and report functionality may already be needed by a routing protocol based on the expected transmission count (ETX) [30]. The guessing technique of [6] can be explained via Figure 7.10. Suppose v_4 sends a source packet $\boldsymbol{x}_1$ to v_0 without mixing; suppose v_0 knows that v_1 can receive a packet from v_4 with probability 0.8. When v_0 receives $\boldsymbol{x}_1$ sent by v_4, v_0 can infer that v_1 has overheard the packet with probability 0.8. Guessing may result in a more up-to-date knowledge about who has what; however, if the guess is wrong, the neighbor may fail to demix a packet intended for it. To compensate for the mistakes in guessing who has what, explicit ACKs can be used. Specifically, nodes can keep track of the packets that were sent but have not yet been acknowledged and retransmit packets after time-out.

7.3.1.1 Some Implementation Issues

In this section, we briefly review the key data structures and operations in a possible implementation.

Each packet has a variable length header that includes: (1) the IDs of the source packets being mixed and their respective receivers, (2) some piggybacked ACKs, and (3) some piggybacked reception reports. If no data packets were sent after a certain amount of time, then a dedicated control packet containing ACKs and reception reports is broadcast.

Each node maintains three separate buffers, `OverheardBuffer`, `ReceivedBuffer`, `SentBuffer`, holding the source packets that the node overheard, received, or sent, respectively. Upon receiving a packet, the packets in these three buffers are used for demixing. Reception reports describe new content in the `OverheardBuffer`. ACKs describe new content in the `ReceivedBuffer`.

Each node maintains a `WhoHasWhatTable` whose entries are of the form "node v_i has source packet $\boldsymbol{x}_j$ with probability p." Upon receiving a packet, the `WhoHasWhatTable` is updated according to the local mixing header. If the received packet is a source packet, guessing is also performed based on the measured channel reception probabilities.

A mixture packet may be intended for more than one receiver. However, the 802.11 protocol has a limited support for MAC layer broadcast (e.g., broadcast packets are not ACKed). To address this practical issue, Katti et al. [6] proposed to use "pseudo-broadcast." Specifically, the mixture packet is sent as a unicast packet addressed to one of the receivers. Nodes run in the promiscuous mode to overhear packets; upon receiving a packet, a node inspects the local mixing header to decide whether it is an intended receiver of the packet. A consequence of the pseudo-broadcast approach is that the sender cannot be sure whether the other intended receivers received the packet reliably. Such an issue can be addressed by using explicit ACKs, in addition to the ACK in 802.11 MAC [6].

After a packet is sent, the ingredient source packets are moved from the output queue into the `SentBuffer`. In addition, timer events are inserted so that the sent packets will be moved back to the output queue for retransmission if the ACKs do not arrive after a certain time threshold.

7.3.2 Local Mixing-Aware Routing

Consider the example setting illustrated in Figure 7.11. There is an existing long-term flow in the network, $C \to B \to A$. We want to find a good route for a flow from A to C. Due to the existence of the local mixing engine, the route $A \to B \to C$ is a good solution because the packets belonging to this new flow can be mixed with the packets belonging to the opposite flow $C \to B \to A$, resulting in improved resource efficiency. But how do we realize that $A \to B \to C$ is a better route than $A \to D \to C$?

Before explaining the local mixing-aware routing solution, we first review how routing is commonly done in wireless mesh networks. Conventionally, routing protocols in

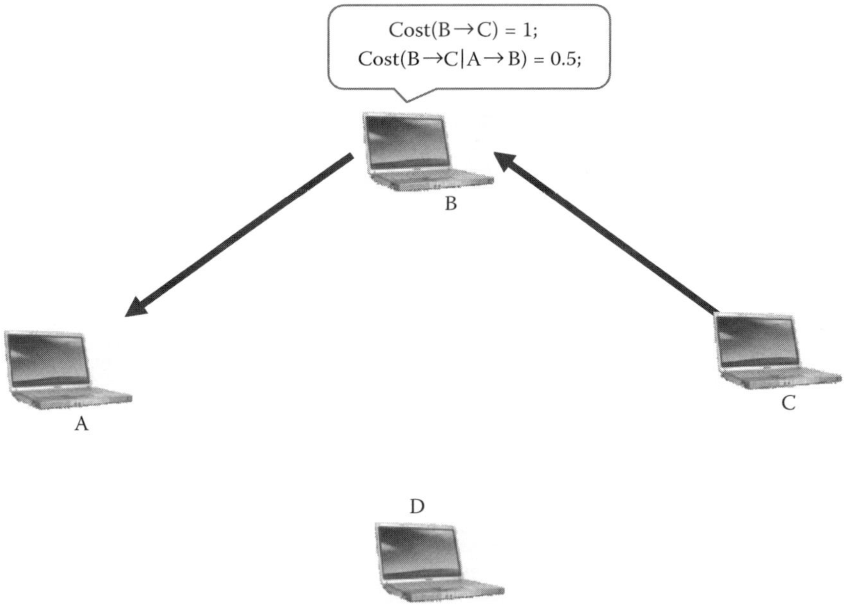

FIGURE 7.11 An example mesh networking scenario. Assume currently there is a long-term background flow, $C \to B \to A$. We want to find a good route for a flow from A to C.

wireless mesh networks have been based on finding shortest paths. Here the cost of a path is modeled as the sum of the costs on the constituting links, and the cost of a link typically reflects the link quality in one way or another. Hence, a natural first thought is to modify the link metrics to take into account the effect of the local mixing engine in reducing the transmissions in the air. For instance, can link $B \to C$ announce a lower cost? There are some issues in doing so, because a packet from D that traverses $B \to C$ may not share a ride with a packet from C that traverses $B \to A$, although a packet from A that traverses $B \to C$ can. We see from this example that in the presence of the local mixing engine, assessing the channel resource incurred by a packet transmission requires some context information about where the packet arrives from. For example, we can say that given the current traffic condition, the cost of sending a packet from B to C that previously arrived from A is smaller, because it can be mixed with the background flow. This observation can be modeled by a *conditional link cost*. Let cost $(B \to C|A \to B)$ denote the cost of sending a packet from B to C, conditioned on that the packet arrived from A.

Wu et al. [28] proposed to use a specific conditional link metric called the expected resource consumption (ERC) to model the resource saving due to local mixing. With the local mixing engine, several packets may share a ride in the air. Naturally, the passengers can share the airfare. In effect, each participating source packet is getting a discount. The ERC metric models the resource consumption while taking such discount into account. Consider a packet sent in the air. If it is a mixture of k source packets, then the ERC metric "charges" each ingredient source packet $1/k$ of the resource consumed by the packet transmission. Here the resource consumed by the transmission could be measured in terms of, e.g., air time, or consumed energy. Consider Figure 7.11. Suppose that the link $B \rightarrow C$ has a normal cost of 1. Node B may announce a lower conditional link cost, $\texttt{cost}(B \rightarrow C|A \rightarrow B)$, to reflect that a new packet from B can enjoy a 50% discount since it can be mixed with the background flow.

Using a set of conditional link costs, the cost of the path can be evaluated with a *Markovian metric* [28]. Consider a path $\mathcal{P} = v_0 \rightarrow v_1 \rightarrow \ldots \rightarrow v_k$. A Markovian metric models the cost of a path as the sum of the conditional costs on the links:

$$\texttt{cost}(\mathcal{P}) \triangleq \texttt{cost}(v_0 \rightarrow v_1) + \texttt{cost}(v_1 \rightarrow v_2 | v_0 \rightarrow v_1) + \ldots + \texttt{cost}(v_{k-1} \rightarrow v_k | v_{k-2} \rightarrow v_{k-1}). \tag{7.27}$$

The conventional routing metric can be viewed as a special case of the Markovian metric where all the conditional link costs are equal to their unconditional counterparts. The decomposition relation (7.27) is reminiscent of the decomposition of the joint probability distribution of random variables forming a Markov chain into a product of the conditional probabilities. Thus, a Markovian metric to an unconditional metric is like a Markov chain to a memoryless sequence of random variables. Due to this decomposition structure, the dynamic programming principle still applies, and thus finding the shortest path with a Markovian metric can still be done in polynomial time. In a practical network, support for the Markovian metric can be added easily into an existing routing framework that uses a conventional routing metric.

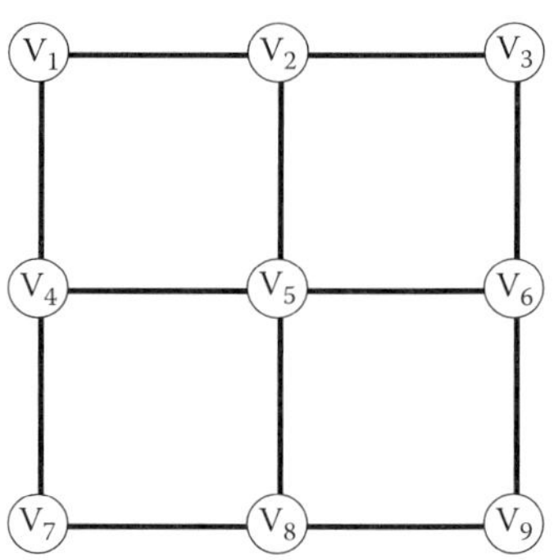

FIGURE 7.12 A nine-node grid network.

Let us now see some performance results from [28]. The simulation topology is the nine-node grid network scenario shown in Figure 7.12. Three UDP flows, $v_9 \rightsquigarrow v_1$, $v_1 \rightsquigarrow v_9$, and $v_3 \rightsquigarrow v_1$, are simulated. Each flow begins randomly between 50 and 60 seconds into the simulation. The results are depicted in Figure 7.13(a). Three systems, LQSR, LQSR+LM, and MMSR, are compared. LQSR [31] is a link state routing system using a link quality routing metric, which is used as a baseline. LQSR+LM is obtained by adding local mixing support in the link layer. MMSR uses local mixing in the link layer and a local mixing-aware routing system based on Markovian metric. The x-axis stands for the input traffic load. It is observed that LQSR cannot sustain the throughput imposed by the input flows to the network as the load increases. MMSR provides significant throughput gains compared to LQSR (up to 47%) and LQSR+LM (up to 15%). This is because not only does MMSR allow subsequent flows to mix with existing flows, but it explicitly tries to maximize mixing. In the example, the flow 1-2-3-6-9 is mixed with 9-6-3-2-1 with MMSR, due to the mutually beneficial discounts enjoyed by both flows. Figure 7.13(b) gives the amount of *resources* saved by

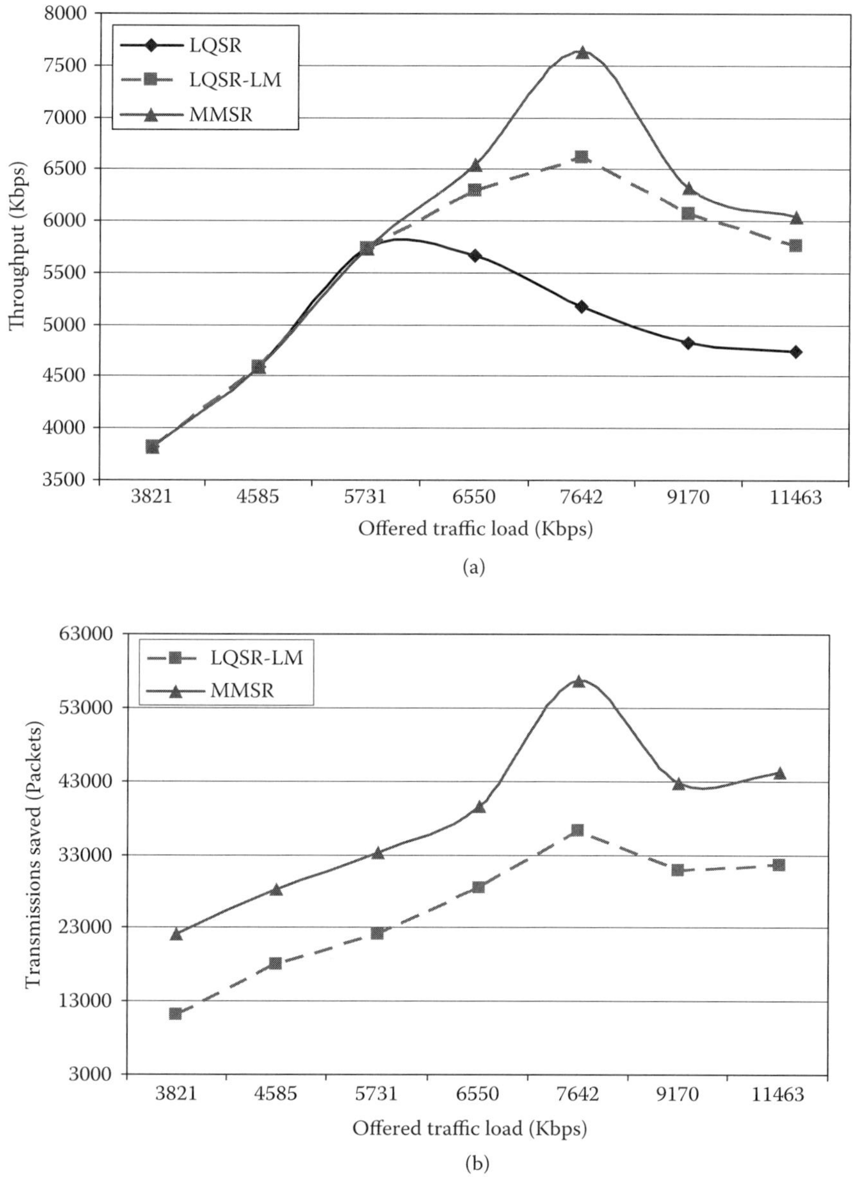

FIGURE 7.13 (a) Throughput comparison of MMSR, LQSR+LM, and LQSR. (b) Transmissions saved through mixing in MMSR and LQSR+LM. (From Wu et al., 2007. Copyright © 2007 IEEE. With permission.)

using MMSR; the y-axis is the number of original source packets minus the number of actual transmissions. MMSR consistently provides reduction of packet transmissions of over ten thousand packets across a wide variety of traffic demands.

In summary, the local mixing engine, on its own, can improve the link layer efficiency; it identifies mixing opportunities on the fly and takes advantage of them if they are present. Routing with a Markovian metric makes local mixing more useful as it creates more mixing opportunities. This can translate to notable resource saving and throughput gain, as confirmed by simulations. For a theoretical (flow-based) analysis of local mixing-aware routing, please see the recent work by Sengupta et al. [32].

7.4 Network Coding in the Physical Layer

In sections 7.2 and 7.3, the network coding techniques mix the information packets, and each packet is either lost or received free of error. In this section we discuss some physical layer mixing techniques that lead to further performance improvement. These techniques share some similarities with the use of mixing at the packet level; these techniques may also be used in conjunction with mixing at the packet level. We loosely classify these mixing techniques as network coding in the physical layer.

7.4.1 Mixing at the Modulator/Channel Coder

Consider the basic broadcasting scenario illustrated by Figure 7.14(a), where node v_0 wants to send a message W_2 to v_2 and a message W_1 to v_1. Suppose v_0 can communicate at a maximum rate of 3 Mbps with v_1 and at a maximum rate of 2 Mbps. Let R_1 and R_2 denote the long-term rates at which v_0 can send distinct messages to v_1 and v_2, respectively. Hence, the pairs $(R_1,R_2) = (3,0)$ and $(R_1,R_2) = (0,2)$ (Mbps) are achievable. In today's radio transceivers, node v_0 alternates between transmitting data to v_1 and transmitting data to v_2. Hence, as shown in Figure 7.14(b), the straight line connecting $(R_1,R_2) = (3,0)$ and $(R_1,R_2) = (0,2)$ is achievable by timesharing. Timesharing, however, is not the best possible scheme. For this scenario, it is known in information theory that a larger rate region is achievable. If the v_0 is related to v_1 and v_2 by additive white Gaussian noise (AWGN) channels, then the capacity region, i.e., the set of all achievable rate pairs, is [33–35]

$$\left\{\left(R_1,R_2\right)\middle| R_1 \le C\left(\frac{\alpha P}{N_1}\right),\ R_2 \le C\left(\frac{(1-\alpha)P}{\alpha P+N_2}\right)\right\}, \tag{7.28}$$

where P is the transmission power at v_0, N_1, N_2 are the noise levels at v_1 and v_2, respectively, $C(\mathsf{SNR}) = ½ \log2(1 + \mathsf{SNR})$, and α is an arbitrary constant in [0,1]. The boundary of the capacity region is illustrated by the upper curve in Figure 7.14(b). To achieve any rate pair in the capacity region, the *super-position coding* technique proposed by Cover [33] can be used. The transmitter generates two codebooks, one with 2^{nR_1} entries and power αP, and the other with 2^{nR_2} entries and power $(1 - \alpha)P$. To send messages $W_1 \in \{1, \ldots, 2^{nR_1}\}$ and $W_2 \in \{1, \ldots, 2^{nR_2}\}$ to v_1 and v_2, respectively, v_0 sends $\boldsymbol{x}_1(W_1) + \boldsymbol{x}_2(W_2)$, where $\boldsymbol{x}_1(W_1)$ is the codeword indexed by W_1 in the first codebook and $\boldsymbol{x}_2(W_2)$ is the codeword indexed by W_2 in the second codebook. Receiver v_2 decodes W_2 by treating

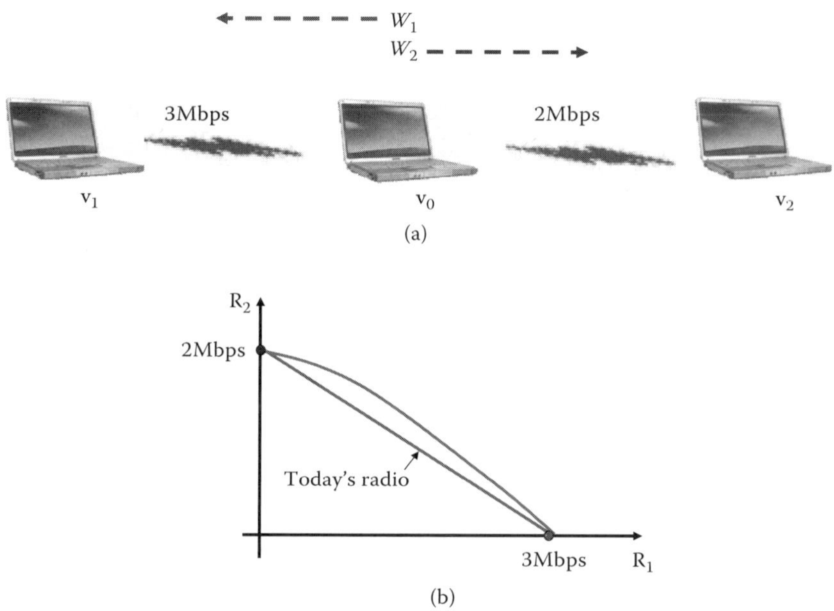

FIGURE 7.14 (a) The basic broadcasting scenario, where the node in the middle wants to send a packet a to v_2 and a packet b to v_1. (b) The capacity region of the additive white Gaussian broadcast channel is the upper curve.

$\boldsymbol{x}_1(W_1)$ as noise. Receiver v_1 decodes W_2 first, then subtracts $\boldsymbol{x}_2(W_2)$ and finally decodes W_1. Therefore, all points inside the region (7.28) can be achieved.

In a practical wireless network, sometimes a receiver may have prior knowledge about some messages destined to other nodes. For example, as illustrated by Figure 7.15(a), receiver v_2 may know the message W_1 *a priori*. Such a situation could arise because W_1 originates at v_2 and is then relayed by v_0 to v_1, or v_2 happens to overhear W_1 when some other nodes were sending W_1 to v_0. Interestingly, the presence of the prior knowledge can enlarge the rate region. As illustrated in Figure 7.15(b), if v_2 knows W_1 *a priori*, then the capacity region is given by the interior of the uppermost curve, i.e., $\{(R_1, R_2)|R_2 \leq 2, R_1 + R_2 \leq 3\}$ [7, 36, 37]. This means when v_0 is sending at the maximum rate (2 Mbps) to v_2, it can still pack information to v_1 at 1 Mbps for free!

The key constructive technique for achieving the enlarged rate regions in Figure 7.15 is *nested coding*. Let us first see an informal explanation of this technique. We show how to achieve the point $(R_1, R_2) = (1, 2)$. As illustrated by Figure 7.16, every microsecond, we encode 3 bits for every symbol via 8-PSK uncoded modulation, where the first two bits are from W_2 and the last bit is from W_1. In this case, receiver v_1 has a good channel with rate 3 Mbps; hence, it can differentiate the eight possible constellations and recover all three bits. Receiver v_2 only has a channel with rate 2 Mbps, but it knows the last bit of the symbol from its prior knowledge about W_1. Hence, to receiver v_2, the transmitted symbol appears to be from a four-point constellation. Thus, it can resolve the ambiguity and receive 2 bits about W_2 every microsecond.

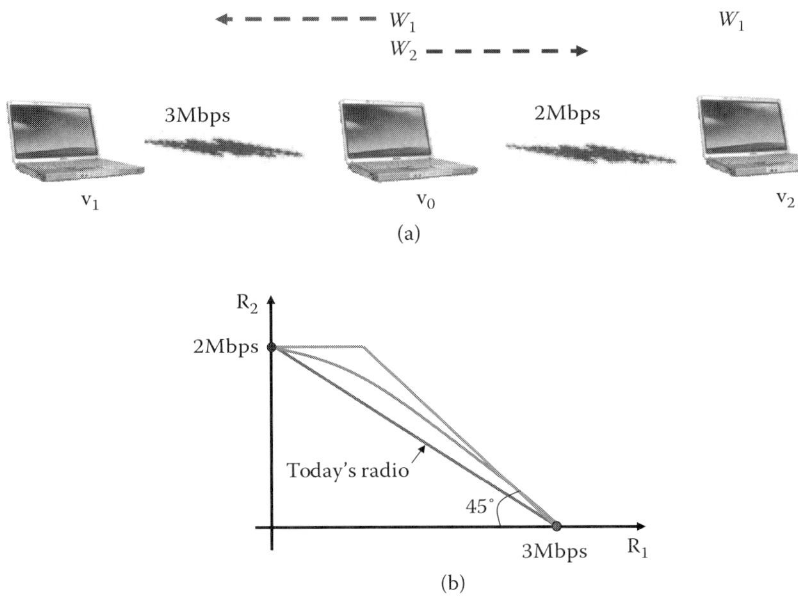

FIGURE 7.15 (a) Broadcasting when v_2 knows W_1 *a priori*. (b) The capacity region is enlarged with the side knowledge.

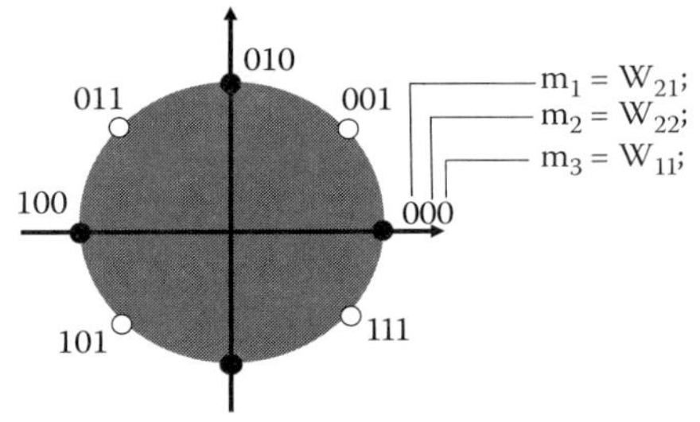

FIGURE 7.16 Illustration of the nested coding.

The above is an informal explanation of nested coding. More formally, consider coding over a block of n symbols; in nested coding, the transmitter uses a codebook that can be viewed as 2^{nR_1} sub-codebooks, each having 2^{nR_2} entries. The 2^{nR_1} sub-codebooks are indexed by W_1. To a receiver without prior knowledge, the transmitted codeword appears to be from a codebook of size $2^{n(R_1+R_2)}$. To a receiver that knows W_1 *a priori*, the transmitted codeword appears to be from a codebook of size 2^{nR_2}. Such nested coding has been used, for example, by Yang and Høst-Madsen [37], as an efficient scheme for cooperative relaying.

Regarding practical design, Xiao et al. [7] recently proposed a practical way of implementing nested coding. The scheme separately encodes multiple information packets via linear channel codes and then computes the XOR of the encoded packets at the physical layer prior to transmission. Specifically, consider mixing several information packets $\boldsymbol{i}_i, \ldots, \boldsymbol{i}_N$ together at the channel encoder. The resulting codeword is

$$\boldsymbol{c} = \boldsymbol{i}_1\boldsymbol{G}_1 \oplus \boldsymbol{i}_2\boldsymbol{G}_2 \oplus \ldots \oplus \boldsymbol{i}_N\boldsymbol{G}_N \tag{7.29}$$

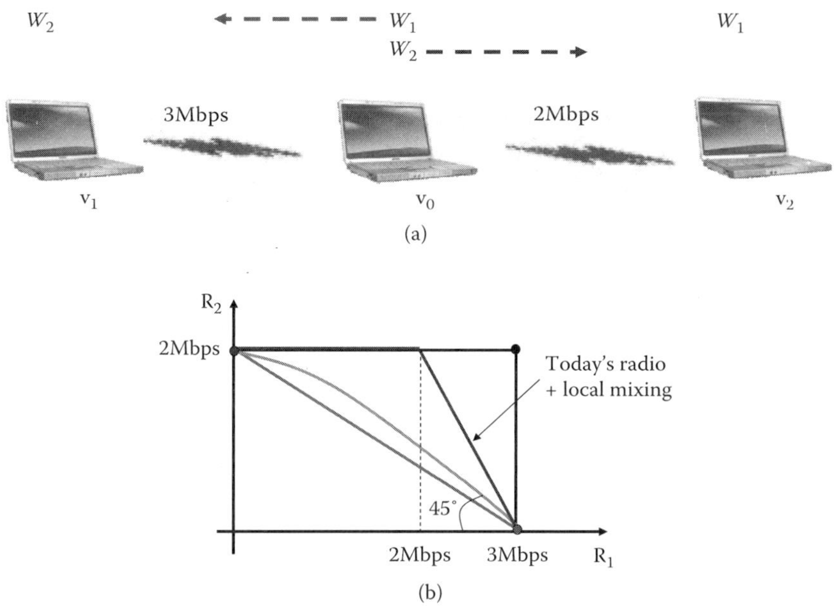

FIGURE 7.17 (a) Broadcasting when v_2 knows W_1 *a priori* and v_1 knows W_2 *a priori*. (b) The capacity region is the entire rectangle: $\{(R_1, R_2)|R_2 \leq 2, R_1 \leq 3\}$.

where $\boldsymbol{G}_i, \ldots, \boldsymbol{G}_N$ are different generator matrices. The mixture codeword is then modulated and sent for transmission. When decoding, a receiver that knows some packets can employ the *a priori* knowledge of some packets to reduce the effective code rate. The contribution of the known packets can be treated as a known scrambling bit pattern; channel decoding can then be modified to account for the scrambling bit pattern. For details about hard- and soft-decision decoding methods for such nested codes, refer to [7].

Note that the technique does not apply to the dual case where the better receiver v_1 knows W_2 *a priori*, but v_2 has no side knowledge. Indeed, it has been shown in [36] that in such case, the capacity region remains equal to that of the classical broadcast channel.

What if both receiver v_1 knows W_2 *a priori* and receiver v_2 knows W_1 *a priori*? Interestingly, in this case, as illustrated by Figure 7.17, the capacity region is the entire rectangle: $\{(R_1, R_2)|R_2 \leq 2, R_1 \leq 3\}$. So v_0 can simultaneously communicate at the maximum bit rates to both v_1 and v_2, without raising the power! Let us see how we can do this. As illustrated in Figure 7.18, we again use 8-PSK to send three bits, m_1, m_2, m_3. But now the first two bits are the XOR of the first two bits of W_1 with the first two bits of W_2. The last bit is the third bit of W_1. Again, node v_1 can recover all three transmitted bits. Then it can use its side knowledge about W_2 to recover W_1. Knowing W_1 *a priori*, node w_2 can also recover the transmitted bits by resolving the QPSK ambiguity; it can then recover W_2 from the received bits m_1, m_2.

Note that if we apply local mixing at the packet level, we can achieve the point $(R_1, R_2) = (2, 2)$ by XORing the two messages and sending them at 2 Mbps. Hence, by making some

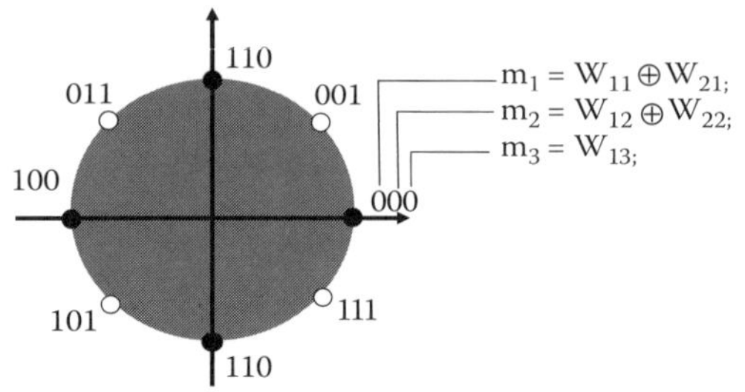

FIGURE 7.18 Joint network coding and nested coding.

Receiver \ Message	W_1	W_2	W_3	W_4	W_5
v_1	Wants		Wants	Has	Wants
v_2		Wants	Wants	Wants	Has

FIGURE 7.19 For two receivers, there are five types of messages: $W_1, \ldots, W_5$.

changes at the physical layer (e.g., using software-defined radios), the communication efficiency can be improved.

More generally, for two receivers, there could be five types of messages, as illustrated by Figure 7.19. The five-dimensional capacity region is characterized in [36]; the capacity region can be achieved by a combination of superposition coding, nested coding, and XORing of the message bits.

7.4.2 Mixing in the Air, Demixing by Cancelling Known Signals

Consider again the example in Figure 7.4. As shown in Figure 7.20(a), we can use three transmissions to exchange two packets between node 1 and node 3 via network coding. Recent works (see, e.g., [8–12]) noted that it is in fact possible to use three transmissions in two time slots to exchange two packets. As illustrated in Figure 7.20(b), in the first time slot, let nodes 1 and 3 transmit simultaneously. Therefore, node 2 will receive a noisy superposition of the two signals subject to channel distortion, e.g., delay, phase shift. Node 2 can amplify the received noisy superposition and broadcast it. Then the received signal at node 1 will be a different noisy superposition of the two signals subject to channel distortion. At a receiver, say node 1, since it knows its original signal a, it should be able to cancel out the contribution due to its own signal and then recover b.

Consider another scenario depicted in Figure 7.21 [10, 11]. Here we have a three-hop flow over a chain of four nodes. As illustrated in Figure 7.21(a), conventionally, the three links have to take turns in transmission. In particular, links $1 \rightarrow 2$ and $3 \rightarrow 4$ cannot be simultaneously active because node 1 and node 3's transmissions collide at node 2. Figure 7.21(b) shows that it is in fact possible to let node 1 and node 3 transmit concurrently. The critical observation is that node 2 already knows what node 3 is transmitting

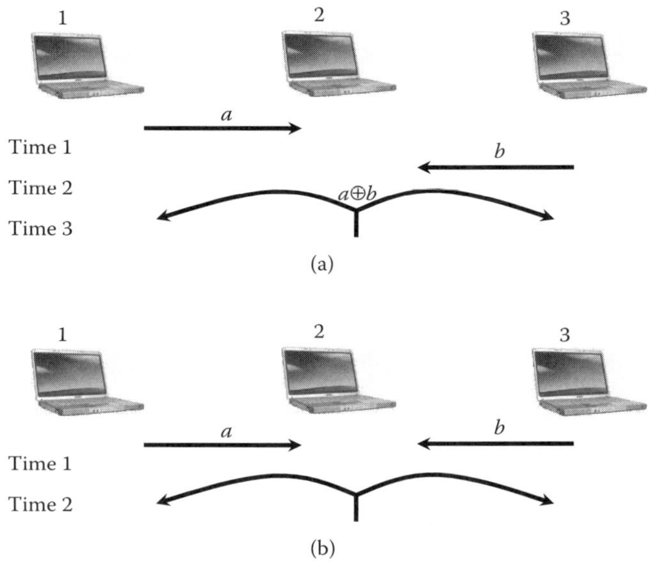

FIGURE 7.20 (a) With network coding, two packets can be exchanged in three steps. (b) With "mixing in the air," two packets can be exchanged in two steps.

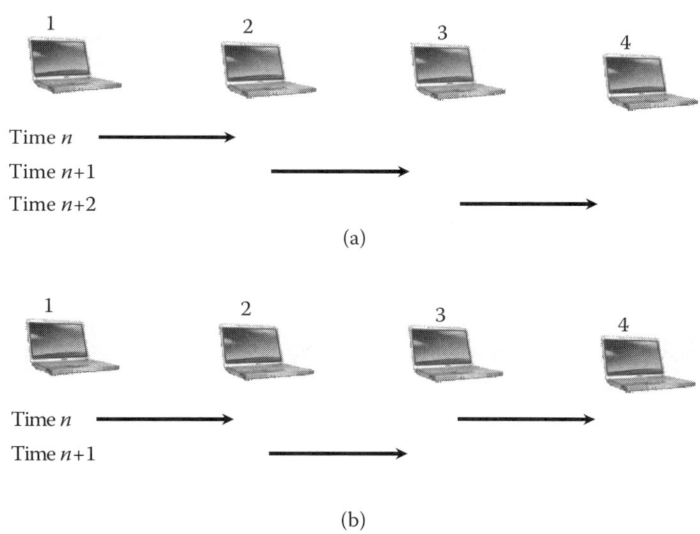

FIGURE 7.21 (a) Conventionally, the three links on a path have to take turns. (b) In fact, the first and third link can happen concurrently because node 3's transmission is known interference to node 2.

because it forwarded the packet earlier. Hence, it can cancel node 3's transmission out before decoding node 1's transmission.

The above discussion left out many practical issues, such as compensation of the channel distortion and the time shifts, and necessary changes in the higher layers to work in concert with the advanced physical layer. For more information about how to address these practical issues and for some theoretical analysis about the performance, refer to, e.g., [8–12].

7.5 Conclusion

An increasingly important application domain of network coding is wireless networks. This chapter provides an overview of the theory and practice of network coding in wireless networks.

In section 7.2, we started with network coding for multicasting. We showed how to obtain convex optimization formulations for network coding-based multicasting in wireless networks. We then discussed the core components of random linear mixing-based multicasting protocols. By having random mixture packets self-orchestrate multiple paths, network coding offers built-in error protection and adaptivity to topology changes due to joins, leaves, node or link failures, congestion, etc.; by employing a flooding-type delivery, network coding can be implemented in a distributed fashion easily. These properties render network coding particularly for unicasting and multicasting in mobile ad hoc networks.

In section 7.3, we discussed how network coding can be used as a generic technology that improves the link layer efficiency by exploiting the broadcast nature of the wireless medium. The local mixing engine buffers packets and opportunistically mixing packets passing by, to reduce the number of transmissions. The benefit is further increased with the recent development of local mixing-aware routing.

In section 7.4, we discussed some recent physical layer techniques that are based on the mixing and demixing of signals. By mixing at the modulator/channel coder and by exploiting the inherent mixing offered by the wireless medium, further performance gains can be achieved.

References

[1] R. Ahlswede, N. Cai, S.-Y. R. Li, and R. W. Yeung. 2000. Network information flow. *IEEE Trans. Inform. Theory* 46:1204–16.

[2] Y. Wu, P. A. Chou, and S.-Y. Kung. 2004. Minimum-energy multicast in mobile ad hoc networks using network coding. *IEEE Trans. Commun.* 53:1906–18.

[3] Y. Wu, P. A. Chou, and S.-Y. Kung. 2005. Information exchange in wireless networks with network coding and physical-layer broadcast. In *Proceedings of the 39th Annual Conference on Information Sciences and Systems (CISS)*, Baltimore. http://research.microsoft.com/~yunnanwu.

[4] P. Larsson, N. Johansson, and K. E. Sunnell. 2006. Coded bi-directional relaying. In *Proceedings of the Vehicular Technology Conference (VTC 2006)*, Melbourne, Australia, pp. 851–855

[5] M. Effros, T. Ho, and S. Kim. 2006. A tiling approach to network code design for wireless networks. In *Proceedings of the IEEE Information Theory Workshop (ITW'06)*, Punta del Este, Uruguay, pp. 62–66.

[6] S. Katti, H. Rahul, W. Hu, D. Katabi, M. Medard, and J. Crowcroft. 2006. XORs in the air: Practical wireless network coding. In *SIGCOMM*, Pisa, Italy, pp. 243–254.

[7] L. Xiao, T. E. Fuja, J. Kliewer, and D. J. Costello Jr. 2006. Nested codes with multiple interpretations. In *Proceedings of the Conference on Information Sciences and Systems (CISS)*, Princeton, NJ.

[8] B. Rankov and A. Wittneben. 2005. Spectral efficient protocols for nonregenerative half-duplex relaying. In *Proceedings of the 43rd Allerton Conference on Communications, Control and Computing*, Monticello, IL.

[9] P. Popovski and H. Yomo. 2006. Bi-directional amplification of throughput in a wireless multihop network. In *Proceedings of the Vehicular Technology Conference (VTC)*, pp. 588–93.

[10] S. Zhang, S.-C. Liew, and P. P. Lam. 2006. Hot topic: Physical-layer network coding. In *MobiCom*, Los Angeles, pp. 358–365.

[11] S. Katti, S. Gollakota, and D. Katabi. 2007. Embracing wireless interference: Analog network coding. In *ACM SIGCOMM*, Kyoto, Japan, pp. 397–408.

[12] S. J. Kim, P. Mitran, and V. Tarokh. 2007. Performance bounds for bi-directional coded cooperation protocols. In *Proceedings of the 27th International Conference on Distributed Computing Systems Workshops (ICDCSW '07).*

[13] Y. Wu, P. A. Chou, Q. Zhang, K. Jain, W. Zhu, and S.-Y. Kung. 2005. Network planning in wireless ad hoc networks: A cross-layer approach. *IEEE J. Select. Areas Commun.* 23:136–50.

[14] Y. Wu. 2005. Network coding for multicasting. PhD dissertation, Princeton University. http://research.microsoft.com/~yunnanwu/.

[15] J. Widmer, C. Fragouli, and J.-Y. L. Boudec. 2005. Low-complexity energy-efficient broadcasting in wireless ad-hoc networks using network coding. In *Proceedings of the 1st Workshop on Network Coding, Theory, and Applications (NetCod)*, Riva del Garda, Italy.

[16] C. Fragouli, J. Widmer, and J.-Y. L. Boudec. 2006. A network coding approach to energy efficient broadcasting: From theory to practice. In *Proceedings of INFOCOM*, Barcelona.

[17] S. Chachulski, M. Jennings, S. Katti, and D. Katabi. 2007. Trading structure for randomness in wireless opportunistic routing. In *ACM SIGCOMM*, Kyoto, Japan, pp. 169–180.

[18] J.-S. Park, M. Gerla, D. S. Lun, Y. Yi, and M. Médard. 2006. Codecast: A network-coding-based ad hoc multicasting protocol. *IEEE Wireless Commun.* 13:76–81.

[19] P. A. Chou, Y. Wu, and K. Jain. 2003. Practical network coding. In *Proceedings of the 41st Allerton Conference on Communication, Control, and Computing*, Monticello, IL.

[20] Y. Wu. 2006. A trellis connectivity analysis of random linear network coding with buffering. In *Proceedings of the International Symposium on Information Theory*, Seattle, WA.

[21] D. S. Lun, M. Medard, R. Koetter, and M. Effros. 2005. Further results on coding for reliable communication over packet networks. In *Proceedings of the International Symposium on Information Theory.*

[22] S.-Y. R. Li, R. W. Yeung, and N. Cai. 2003. Linear network coding. *IEEE Trans. Inform. Theory* 49:371–81.

[23] T. Ho, M. Médard, R. Koetter, D. R. Karger, M. Effros, J. Shi, and B. Leong. 2006. A random linear network coding approach to multicast. *IEEE Trans. Inform. Theory* 52:4413–30.

[24] P. Sander, S. Egner, and L. Tolhuizen. 2003. Polynomial time algorithms for network information flow. In *Symposium on Parallel Algorithms and Architectures (SPAA)*, San Diego, pp. 286–94.

[25] J. Widmer and J.-Y. L. Boudec. 2005. Network coding for efficient communication in extreme networks. In *Proceedings of ACM SIGCOMM'05 Workshops*, Philadelphia.

[26] G. Zhang, J. Neglia, J. Kurose, and D. Towsley. 2005. On the benefits of random linear coding for unicast applications in disruption tolerant networks. In *Proceedings of the 2nd Workshop on Network Coding, Theory, and Applications*, Boston.

[27] U. Lee, J.-S. Park, J. Yeh, G. Pau, and M. Gerla. 2006. Codetorrent: Content distribution using network coding in VANETs. In *1st International Workshop on Decentralized Resource Sharing in Mobile Computing and Networking (MobiShare'06)*, Los Angeles.

[28] Y. Wu, S. M. Das, and R. Chandra. 2007. Routing with a Markovian metric to promote local mixing. In *Proceedings of the IEEE INFOCOM 2007 Minisymposium*, Anchorage, Alaska.

[29] Y. Wu, J. Padhye, R. Chandra, V. Padmanabhan, and P. A. Chou. 2006. The local mixing problem. In *Proceedings of the Information Theory and Applications Workshop*, San Diego.

[30] D. S. J. D. Couto, D. Aguayo, J. Bicket, and R. Morris. 2003. High throughput path metric for multi-hop wireless routing. In *ACM MobiCom.*

[31] R. Draves, J. Pahdye, and B. Zill. 2004. Comparison of routing metrics for static multi-hop wireless networks. In *SIGCOMM.*

[32] S. Sengupta, S. Rayanchu, and S. Banerjee. 2007. An analysis of wireless network coding for unicast sessions: The case for coding-aware routing. In *Proceedings of INFOCOM*, Anchorage, AK.

[33] T. M. Cover. 1972. Broadcast channels. *IEEE Trans. Inform. Theory* 18:2–14.

[34] P. P. Bergmans. 1974. A simple converse for broadcast channels with additive white Gaussian noise. *IEEE Trans. Inform. Theory* 20:279–80.

[35] R. G. Gallager. 1974. Capacity and coding for degraded broadcast channels. *Probl. Inform. Transm.* 185–193.

[36] Y. Wu. 2007. Broadcasting when receivers know some messages *a priori*. In *Proceedings of the International Symposium on Information Theory.*

[37] Z. Yang and A. Høst-Madsen. 2005. Cooperation efficiency in the low power regime. In *Proceedings of the Asilomar Conference on Signals, Systems, and Computers*, Pacific Grove, CA.

8

A Survey of Wireless Sensor Networks: *Technologies, Challenges, and Future Trends*

Ali Alemdar
Queen's University

Mohamed Ibnkahla
Queen's University

8.1 Introduction

Wireless sensor networks are comprised of a variable number of autonomous electronic devices, with possible mechanic components, that have the capability of remote sensing, signal processing, and communication in an ad hoc fashion.

The basic principle is that these autonomous sensor nodes would be scattered over a certain geographic area; the nodes would have to be able to communicate via mesh networking to complete some objective goal. The geographic distribution of these nodes need not be predetermined; this facilitates the use of wireless sensor networks (WSNs) in almost any conceivable environment. However, the lack of infrastructure imposes various difficulties in the design of networking protocols for WSNs.

Possible methods of deployment for the nodes could be from moving ground/aerial units. The nodes would have to be self-contained and battery operated, able to gather data from the surrounding environment and forward it to some predetermined destination. The data gathering would be done via on-board sensors. Furthermore, they would also have to be fairly robust, as they were envisioned to operate in potentially hostile and extreme environments.

The number of nodes deployed, in general, is application specific; they can range anywhere from a handful of nodes to on the order of approximately 10^7 devices [27].

Once the data are processed, only germane information is forwarded. This alleviates the bandwidth constraints that are inherent in such power- and bandwidth-limited systems. Once the data have been partially processed, they are then forwarded to specific nodes in the network that are responsible for data fusion in the case of centralized data fusion. Another approach is to perform data fusion in a coordinated decentralized fashion that requires information sharing and cooperation among nodes in the network.

It is because of their fairly robust and autonomous nature that WSNs have been envisioned as an indispensable aid to the military, the health industry, industrial manufacturing, security applications, and precision agriculture.

8.2 The Development of the WSN

While the concept of wireless sensor networks has been around for some time, it is still a relatively nascent technology.

The earliest sensor network was the Sound Surveillance System (SOSUS) [7]. This system was deployed during the Cold War era to acoustically monitor Soviet submarines. However, this system is quite different from the modern vision of WSNs. While they were distributed, their processing capability was limited and the nodes in the detection array were wired together. A Distributed Sensor Network (DSN) program was started by a Defense Advanced Research Projects Agency (DARPA) initiative in 1978; key design paradigms and the components of a DSN were identified. Due to technological constraints, a physically realizable design was not feasible at the time.

WSNs took off in the late 1990s; with funding from DARPA, a solid foundation in the field was established. The original idea was to use COTS (commercial off-the-shelf) products to determine the feasibility of such a system. Initial designs for the sensor pods incorporated various sensing mechanisms (temperature, humidity, barometric pressure, light intensity, etc.). The idea was to use the sensing mechanisms in conjunction with on-board processing capabilities to greatly expand the capabilities of the nodes [8]. Furthermore, it was imperative that the nodes could communicate with each other in a wireless and ad hoc fashion.

The nodes, using COTS technology, were still relatively large for the applications for which they were intended. With advancements in electronics manufacturing, coupled with increased interest in the field from industry and academia, researchers were optimistic that nodes in the cubic millimeter range could be effectively designed using customized hardware.

By 2001 researchers at the University of California Berkeley were working on their third-generation node: Rene [5]. This matchbox-sized node was powered by a relatively large (in comparison to the node itself) battery pack. The performance of the sensor node was highly correlated to the quality of the battery pack. The research team realized that it was essential for its success that hardware be designed in conjunction with software. To fully utilize the potential of the hardware, the software would have to be designed in accordance with hardware constraints. Thus, the first comprehensive operating system for the wireless nodes was designed, and TinyOS was born.

8.3 Applications

There is a basic dichotomy in applications for WSNs: military and civilian. While the initial idea was to use WSNs for military applications, there has been a significant interest in WSNs from a civilian perspective.

For the military, one of the primary uses for WSNs would be remote sensing. Ideally, the nodes could be deployed en masse into hostile territory where they would establish an ad hoc network and sit in passive observation. One such project focused on pinpointing, through acoustic localization, the location of an enemy soldier [15]. The problem translated into how to effectively triangulate the position of the soldier given the arrival times of the muzzle blast at different locations on the battlefield. Acoustic localization in conjunction with knowledge of relative and absolute positioning of the nodes can also be used as an effective means of tracking mobile civilian and military conveyances. The Disposable Sensor Program is an initiative by the U.S. Army to deploy 10^4–10^7 disposable nodes with seismic, acoustic, radio frequency (RF), magnetic, chemical, biological, and infrared sensors [27]. This dense network of sensors would help to provide the army with a complete picture of almost any environment. These sensors would have to be relatively inexpensive and disposable.

The nodes themselves need not be homogeneous; they may contain differing capabilities/uses (static unmanned ground sensors vs. mobile robotic ground sensors) [5]. This gives the network designer tremendous amounts of flexibility, as he or she can retask or reconfigure nodes to serve specific interests.

Another possible application where WSNs could excel is as a replacement to minefields or so-called area denial technologies. The nodes would have actuation capabilities; through their ability to track mobile targets they could locate and incapacitate hostile entities.

There is also no dearth of applications of WSNs in the civilian domain. WSNs have been effectively used in habitat monitoring, telemedicine, inventory monitoring, traffic flow analysis, and security systems.

It has been shown that WSNs are an effective means of monitoring and observing physical phenomena. They often allow the user to gather data at resolutions and in geographic locations that are difficult to obtain otherwise [16, 17, 23]. Another salient feature of WSNs is that they allow users to gather long-term *in situ* data, thus providing a detailed macroscopic picture of the area of interest (AOI) [23]. The embedded computing and wireless connectivity of the nodes allow application-specific signal processing and on-demand reconfiguration [5]. Reconfiguration lends the sender network a tremendous amount of flexibility. As most WSNs are designed and deployed in an application-specific fashion, reconfiguration allows the user to observe initial data trends and the possible transient behavior of the system under study. If it is required, individual nodes can be queried to adjust various sensing parameters—the data acquisition rate being one specific example. However, as with any constrained system, there are certain important trade-offs. One major example that is particularly applicable to habitat monitoring is the trade-off between battery life and the data acquisition rate. The nodes only have a finite amount of energy to sense, communicate, and process data. Higher data acquisition rates rapidly deplete the available energy to power the node and its storage capacity. This and other issues will be discussed in section 8.4.

Habitat, environmental, and industrial process monitoring are applications that have received a significant amount of interest from both academia and commercial enterprises [5, 11, 14, 16, 17, 23, 26].

The use of WSNs in habitat monitoring allows the user to gather data that is untainted and unaltered by human interaction. It has been established that a human presence in many habitat-monitoring environments is detrimental to the collection of untainted data. *In situ* observation with WSNs avoids this unfavorable situation. The authors in [14] observed that even a negligible human presence at certain seabird colonies could potentially result in a tangible increase in the mortality rate of cormorant colonies over a given breeding year. Thus, WSNs provide an attractive and viable alternative to other human-based invasive data gathering methods.

Through the introduction of heterogeneous nodes into nesting petrels and other locations on Great Duck Island, Maine, researchers were able to monitor the dynamics of a seabird colony. The data gathered over this study were useful in creating a more complete picture of the habits of nesting cormorants.

Similar studies were conducted at the James San Jacinto Mountains Reserve (JMR) in Idyllwild, California [23]. Researchers were interested in getting a macroscopic picture of microclimate readings over an extensive geographic area. The study introduced sender nodes, located at different locations along the vertical and horizontal transects of a redwood tree. The nodes measured air temperature, relative humidity, and photosynthetically active radiation (PAR). Through this 44-day study, the researchers were able to effectively study and analyze the microclimate surrounding a redwood tree in coastal California.

Queen's University researchers, in collaboration with the Ontario Ministry of Natural Resources (MNR), Lunaris, Inc., and the Canadian Carbon Program [31] with the support of the Ontario Centres of Excellence (OCE), have recently initiated projects that aim to deploy large-scale wireless sensor networks for forest and environment monitoring. This sensor network would facilitate easy deployment, high-data-rate transmissions, high mobility, and controlled real-time sensing and communication. This network would be deployed along with MNR forest ecology stations to wirelessly gather factors such as heat, water and trace gas (mainly CO_2) fluxes above the canopy, air temperature, relative humidity, light, soil temperatures at different depths, soil moisture, stream

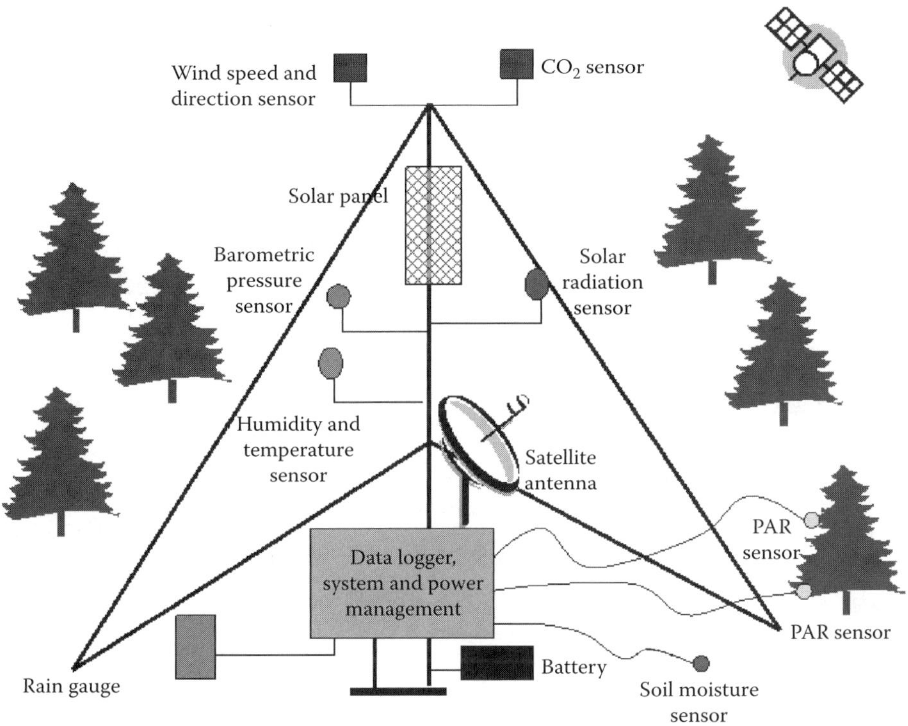

FIGURE 8.1 A classical ecology station equipped with a satellite transceiver.

flow parameters, and groundwater levels. Figure 8.1 displays a typical ecology station equipped with a satellite transmitter that sends collected data to researchers through a satellite link. Figure 8.2 displays the proposed WSN-based ecology station, where data are collected wirelessly and transmitted through the satellite link. Each physical sensing device is equipped with a wireless module that can transmit measured data to a relay node, which sends it to the station computer center (SCC). It is expected that the WSN technology will enable better spatial and temporal sampling, allowing ecologists and field biologists to obtain greater resolution in both time and space, and to unobtrusively collect new types of data, providing new insights on processes. Moreover, WSN facilitates the collection of diverse types of data at frequent intervals over large areas (e.g., temperature, image, PAR, etc.). Real-time data flows allow researchers to react rapidly to events, thus extending the laboratory to the field. Furthermore, WSNs allow remotely controlled sensing. For example, the numbers and locations of sensors can be chosen, optimized, changed, or activated and deactivated. Moreover, sensing parameters can be remotely throttled (e.g., frequency of sensing, data rate and type, sensing features, etc.).

Queen's University and the MNR are using WSNs for species-at-risk monitoring, with a special focus on endangered turtle species in Ontario (such as the Wood and

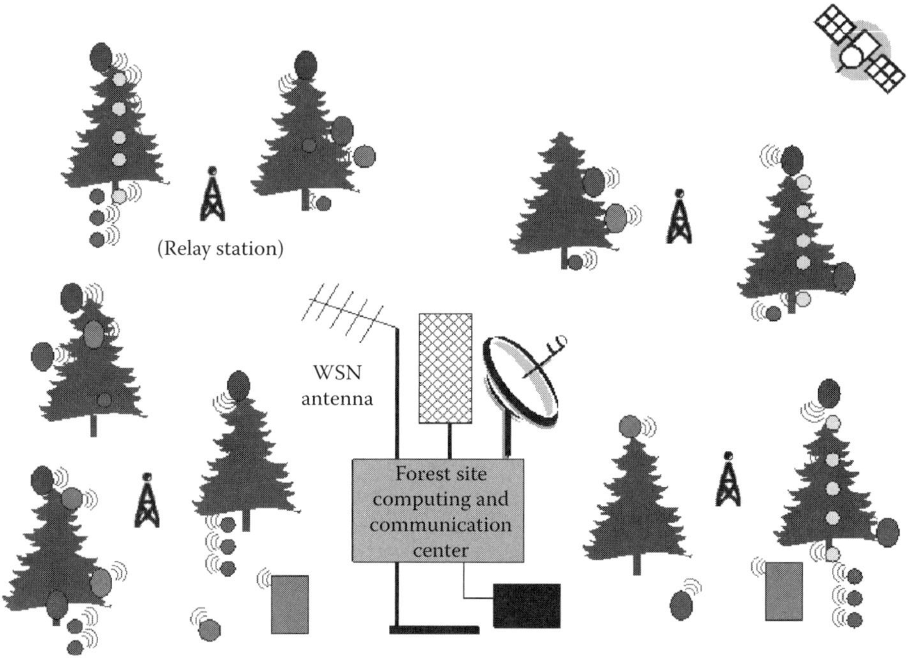

FIGURE 8.2 WSN-based ecology station.

Blanding's turtles). The technology can be designed to work on land and underwater, and would be able to provide information about turtle habitat use (location information) and relevant environmental factors, such as temperature and oxygen. Wireless sensor network technology would be more effective and less expensive than existing technologies for acquiring this information. Staff time would be greatly reduced, because data can be accessed directly via computer, which would considerably limit the need for site visits. Information would be available in real time, with no delays to download information from an on-site data storage unit. The researcher can remotely instruct the network to change parameters, such as what to measure and how often, without recapturing the animal or revisiting the site.

Environmental monitoring, in the hopes of disaster prevention, is a field of particular interest to many researchers. The authors in [18] studied how the introduction of WSNs in Bangladesh could be used as an effective means of disaster prevention. They propose that WSN could act as an early warning system that would allow authorities to effectively evacuate areas that would be subjected to floods, mudslides, or any conceivable natural disaster that could possibly endanger human lives. In this study, it was argued that the installation of nodes in various coastal areas, monitoring water and precipitation levels, could help to mitigate the effect of natural disasters on the human population.

Researchers at the University of Southampton have placed nodes in and around the Briksdalsbreen glacier in Norway [16, 17]. The main motivation for this project was to monitor the glacier morphology. The nodes act as an effective means to study glacial motion and their impact on global weather trends. There are various complications that arise when collecting data via nodes in such hostile environments: obvious mechanical problems are compounded by the glacier's location, and the remoteness of the glacier imposes distinct and discernible problems in communication among the nodes. Furthermore, not only do the nodes have to operate in such an extreme environment, but they also have to ensure the fidelity of the data that they measure.

As we can see, remote data sensing, for habitat and ecological monitoring, via WSNs provides researchers with tremendous amounts of flexibility in both data acquisition techniques and applications.

Advancements in micro-electro-mechanical systems (MEMS) technology coupled with improvements in medical image and information processing have led to rapid developments in Mobile Sensor Networks for Telemedicine (MSNT) [12]. MSNT aims to improve the quality of human life by coupling advancements in medicine and telemedicine with improvements in medical microsensor nodes. This is an interesting and emerging field with potential to appreciably improve our quality of life. To date, research in this field has primarily focused on either mobile networks for telemedicine or specific telemedicine systems. A significant body of research attempting to unify these two approaches has yet to materialize. It is this unification that promises to yield the most benefits for the health sector.

It should be fairly evident to the reader that there are a plethora of potential applications for WSNs. Despite current impediments, WSNs appear to be an extremely attractive technology that can improve the accuracy, quality, and efficacy of any study in which they are used.

Wireless sensor networks are also well suited to precision agriculture. Sensors placed in and around agricultural sites may monitor soil pH levels, humidity, chemical composition, or other pertinent parameters [2, 4]. After this *in situ* observation, the sensor network can inform an irrigation system so that localized areas in need of water may be attended to. The approach used by agriculturalists in their use of wireless sensor networks is different from that of, for example, habitat monitoring. The agriculturalists are not interested in sitting and interpreting data. They seek to conflate their in-depth knowledge with accurate *in situ* observations provided by the sensor network. By combining their technical expertise with ethnographic information, researchers were able to gain an accurate picture of an agriculturalist's work environment. This holistic approach enabled the researchers to tailor the salient features of a sensor network to suit the agriculturalist's needs. Precision agriculture is especially useful in environments in which water resources are scarce. By providing water to plants on a per-need basis, a significant reduction in water consumption can be achieved.

8.4 Challenges

In any constrained system there are certain inherent and unavoidable trade-offs. As nodes only have a finite amount of battery power that can be used for sensing, processing, and communication, understanding the trade-offs among battery life, signal processing, and wireless communication is of paramount importance.

8.4.1 Battery

To maximize the network lifetime, nodes must be designed with low power consumption in mind. This principle of low power consumption must be kept in mind when designing the sensor nodes; every aspect of their design must be geared toward minimizing the energy footprint of the sensing, processing, and communication equipment on the node. Ideally, WSNs would be deployed in the field and would have lifetimes measured in years. A typical AA battery can provide approximately 250 μA-years of charge; however, due to physical degradation and leakage currents, it is difficult to deploy AA-powered nodes for such extended periods [8].

Lithium batteries are a promising alternative to this problem. Their thin form factor suitably lends them to miniaturization, which is also an attractive feature [1].

Operating minute (centimeter- to millimeter-scale) nodes for long periods of time requires power consumption measured in tens to hundreds of microwatts per day. For this to become a reality, the nodes must be able to quickly and effectively enter and leave hibernating states. Duty cycling is an effective means of achieving this goal; however, in the hibernating state, leakage currents dominate the system energy budget. Leakage currents impose a significant restriction on the lifetime of the nodes. Current efforts to mitigate the effects of leakage currents using silicon-on-insulator (SOI) technology are a promising avenue of research; however, to date it has seen limited commercial use [8].

8.4.2 Integrated Circuits (IC)

Substituting commercial off the shelf (COTS) ICs with custom-designed ones for WSNs has significantly reduced their system energy footprint. Furthermore, with the ever-increasing trend toward miniaturization, custom-designed ICs have enabled the nodes to significantly shrink in size. However, integration of custom ICs to operate effectively with realistic node supply voltages is still a challenge and an open area of research [8].

8.4.3 Wireless Communication

In comparison to the energy expenditure of other components in a node, RF communication imposes a significantly greater energy footprint. As an example, 8-bit ADC requires two orders of magnitude less energy than does wirelessly transmitting and receiving a single 8-bit sample [8]. Thus, there is a significant impetus to develop energy-efficient communication strategies to help maximize the lifetime of the nodes. A thorough treatment of the energy requirements of wireless communication for WSNs can be found in [8].

Another factor that complicates the transmission of wireless data in WSNs is the effect of multipath fading. Investigation into the effect of radio irregularity on the performance of WSNs has shown that current radiowave propagation models do not accurately and sufficiently capture its characteristics. An accurate simulation model is an integral part of being able to fully characterize the behavior of WSNs. Research has established that the current spherical and disc [19, 28] propagation models are insufficient, and in their place, a radio irregularity model (RIM) has been proposed [28].

The heterogeneous nature of the wireless sensor nodes and the nonisotropic properties of the wireless channel are taken into consideration in the development of this model. It has shown that radio irregularity has an appreciable impact on the instituted routing protocols; however, the impact on the MAC sub-layer was relatively insignificant in comparison.

Furthermore, it has been often discussed that wideband radio is more effective at mitigating the effects of static multipath fading than its narrowband counterpart. While this is true for the most part in outdoor environments, the same cannot be said for rich scattering environments.

8.4.4 Multipath Routing

The dense scattering of nodes suitably lends itself to multipath routing. Many energy-efficient network protocols for WSNs use the concept of directed information diffusion as a means of disseminating pertinent data. Multipath routing in wireless sensor networks can be seen as a special case of mobile ad hoc networks [9]. While directed information diffusion addresses the issue of scalability, it can still be computationally intensive.

As the nodes are responsible for establishing an ad hoc network to forward and communicate data, they must then carry out the task of routing information. Networking protocols for wireless sensor networks must be robust, as link failure due to battery depletion, duty cycling, and multipath fading are evident impediments to reliable data delivery. Of the main routing protocols for WSNs, [28] has shown that radio irregularity has a significant impact on the efficacy of path-reversal (reverse-path forwarding is a technique used in multicast routing) and neighbor-discovery algorithms; however, the impact on multiround discovery is not as severe. While it may seem that multiround discovery is a natural candidate to succeed the aforementioned protocols, it must be noted that the associated overhead is also higher. Research has shown that symmetric geographic forwarding (SGF) and bounded distance forwarding (BDF) are suitable replacement routing protocols. Symmetric geographic forwarding is a location-based protocol where each node inserts the IDs of all of its discovered neighbors into a beacon message. When another node receives this beacon message and sees that its node ID is in the beacon message, it establishes a symmetric link between the nodes. If it does not see its ID in the beacon message, the link is then deemed to be asymmetric. In message forwarding the nodes would then only pass on data to nodes that are deemed symmetric. Bounded distance forwarding is simply a rule that sets limits on the single-hop distance of a forwarded message. If a node is outside of the allowed forward distance, the network protocol will not allow a message from a node to be forwarded to it. The allowed forward distance parameter is a function of the degree of radio irregularity evidenced in the network. The reader is encouraged to refer to [28] for an in-depth analysis of this topic.

Another approach to multipath routing utilizes a cross-layered design paradigm that successfully incorporates signal processing to optimize application-specific metrics. It has been shown [22] that traditional data metrics, such as throughput and delay, do not sufficiently capture the dynamic nature of wireless ad hoc networks. Instituting application-specific metrics such as miss detections, network lifetime, and energy efficiency enables the designer to have a more complete picture of the sensor network. A paradigm shift from user-centric to application- or data-centric protocols is necessary if we wish to fully integrate signal processing techniques into the network design. Instead of the classical resource-based allocation that is seen in many of today's MAC protocols (the Internet being an obvious example), a new data-type-based protocol would facilitate the integration of signal processing techniques into the protocol stack. One such approach is the Type-Based Multiple Access (TBMA) protocol. In TBMA, resources are allocated based on data type rather than among the sensor nodes themselves.

The authors in [22] discuss a novel event-driven, application-specific network design approach. In such applications, an event is used as a trigger for the sensor node to collect, process, and if necessary communicate information pertaining to this event with the rest of the network. To maximize the accuracy and efficacy of the network, it is obvious that a minimization of the probability of an undetected event is necessary. To this effect a cross-layered network protocol that incorporates performance bounds that include both network and signal parameters (the distance between pairs of nodes and their respective SNR being examples) was devised. This routing protocol is called Chernoff routing, as it incorporates Chernoff information in calculating the network metric. Chernoff routing aims to maximize the Chernoff information for a specific route subject to an energy constraint. Given data transmission over some route R and an energy constraint ε,

$$\max_{R:E(R)\leq\varepsilon} C(R). \tag{8.1}$$

Here, when transmitting over a route R, $C(R)$ represents the cumulative Chernoff information and $E(R)$ is the energy required to transmit data over that route. The Chernoff information is defined as [22]

$$C(R) \approx \sum_{\forall i \in R} C_i, \tag{8.2}$$

$$C_i = \frac{1}{2}\log\left(1 + \frac{P_{i|i-1}}{\sigma_w^2}\right), \tag{8.3}$$

where $P_{i|i-1}$ is the variance of signal innovation at node i with respect to all of the upstream nodes, and σ_w^2 represents the variance of measurement noise at each sensor. Hence, the link metric used in Chernoff routing is the mutual information between neighboring nodes. The optimal route is the one that maximizes the cumulative innovations entropy.

This application-specific approach is attractive in that it can be fine-tuned to parameters that are important and germane to the application, thus outperforming the traditional general design approach. By identifying parameters that are specific and pertinent to an application or a set of applications, network designers can tailor protocols to better suit the needs of the network under study. While there has been a substantial amount of work identifying and designing energy-efficient and robust transport protocols for wireless sensor networks, presenting the design in an application-specific framework is a largely unexplored research area.

8.4.5 Distributed Signal Processing and Time Synchronization

The problem of how to efficiently coordinate signal processing in a distributed framework has arisen as a significant challenge in the context of wireless sensor networks. Distributed signal processing has aspects that are not present in traditional centralized approaches. This shift in paradigm is due to the limited processing capabilities of the nodes and the effect of energy and bandwidth constraints [6]. It is undesirable for the nodes to share raw data, as this taxes the bandwidth requirements, and thus the battery of the nodes. This necessitates the ability for the nodes to perform basic local signal processing and data compression. The design of distributed signal processing algorithms is a challenging endeavor. As the network topology is not static; nodes may hibernate or die as a result of battery depletion. Thus, an effective distributed algorithm must be robust and impervious to network topology. (The problem of distributed signal processing is exacerbated by the hibernation of nodes; time synchronization, in effect knowing when to communicate, is a challenging problem for researchers.) Furthermore, in specific applications (e.g., target tracking), the need for accurate time synchronization data is paramount. To accurately localize a mobile signal, the network must ensure the fidelity of time-series data. Furthermore, in several security-related technologies, such as encryption, the need for accurate time-stamped data is very important [25].

There is a basic dichotomy in the approaches to distributed signal processing. One approach is to leave the data fusion to distinct fusion centers; the other is a distributed, cooperative effort among all the nodes in the sensor network. Figure 8.3 demonstrates the basic differences between these two approaches.

In the centralized data fusion approach, specific nodes act as data aggregation points. These fusion centers use distributed inference methods to gather and collate the data into a unified network view. These information sinks may have specialized hardware that enables them to efficiently and accurately perform data analysis. Once the data have been processed, if it is deemed necessary, the remote sensing nodes may then be notified and possibly retasked or reconfigured as per the requirements of the application.

These data aggregation nodes often employ graphical models to perform the data inference. A graphical model is a natural approach to data fusion, as it can effectively

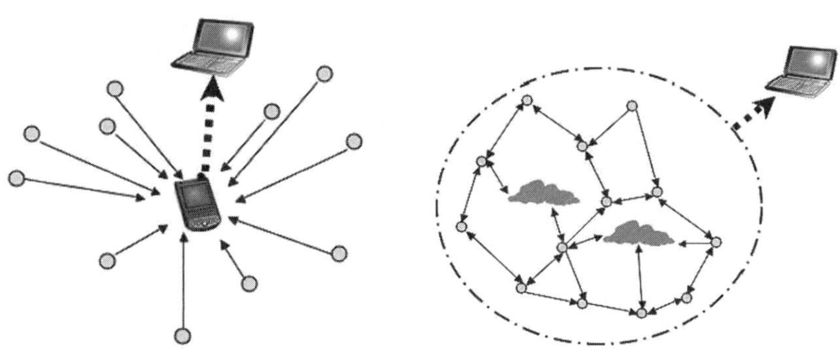

(a) (b)

FIGURE 8.3 (a) Centralized data fusion. (b) Distributed data fusion.

capture the topology of the network. Furthermore, a significant amount of research into the development of computationally efficient and accurate inference algorithms on graphs has been observed. Message-passing algorithms have been particularly effective in addressing this issue. The parallel nature of these algorithms is quite attractive, as they are scalable and well suited to physically distributed processors.

In many applications for WSNs researchers want to be able to quickly see how one scalar variable relates to another. For example, researchers might be interested in the roles that soil pH and relative humidity play in the growth of crops for agriculture. Thus, computationally efficient methods of performing such tasks are needed. The problem is compounded by the fact that often *in situ* observation of several variables for extended periods of time—applications in which wireless sensor networks excel—produces large volumes of data. Traditional data mining techniques are not feasible, as they are computationally expensive and require inordinate amounts of bandwidth and memory. The distributed index for multidimensional data (DIM) is an energy-efficient approach that promises to supersede traditional database approaches [13].

8.4.6 Security

Given the distributed nature of wireless sensor networks and their possible deployment in potentially hostile environments, the need for implementing stringent physical and network security measures is justified.

The term denial-of-service (DoS) attack encompasses any phenomenon that impedes the wireless sensor network from performing its objective goal [25]. These attacks may be attempts by malicious entities to subvert the network for some ulterior goal; they may also simply be a result of hardware or software failure. Regardless of the cause of the network disruption, WSNs must be designed with security in mind. It should be stressed that security is of paramount concern and must pervade all aspects of the design, from inception all the way to deployment. A relatively robust yet computationally lightweight intrusion detection system (IDS) may be sufficient to surmount a broad range of DoS attacks. Once again, design and security considerations are application specific; no general principle will sufficiently capture the necessities and complexities inherent in security for wireless sensor networks. For example, security measures for military applications of WSNs may require more sophisticated mechanisms to repel intelligent and coordinated attacks by determined adversaries. To effectively stave off DoS attacks, security measures must be implemented at all layers of the protocol stack. This is simplified in network protocols where each layer is disjoint; however, this comes at a performance penalty. It has been shown that a cross-layered protocol stack has tangible benefits for network performance, allowing global optimizations that previous protocols could not facilitate [3]. A survey of the potential benefits of cross-layering is addressed in section 8.5.2.

Attempts at instituting node authentication are complicated in the potentially amorphous and dynamic network topology that is often seen in wireless sensor networks. User authentication in geographically extensive distributed networks poses an almost insurmountable problem; this is exacerbated by the fact that nodes are often ID-less.

However, there are some rather simple mechanisms that can be implemented in the MAC protocol that add appreciable levels of security. Implementing random back-off periods before transmission, which affects the periodicity of the application—thus making the application less predictable, is a simple yet effective way of improving MAC layer performance. For a more thorough investigation of security concerns in WSNs, the reader is encouraged to refer to [25].

8.5 Future Trends

New advancements in technology have opened many potential applications for WSNs. In this section we define potential future applications and the technology that hopes to make this possible.

8.5.1 Batteries

The question of how to effectively power nodes, the workhorse behind WSNs, is an open problem for researchers. Currently, long-term deployment of sensor nodes is not feasible; a major contributing factor to this is the insufficiency of battery technology. Furthermore, for many potential applications, especially military, battery replacement is not an option. At the time of writing the majority of research focuses on how to operate efficiently, to complete some objective function, in the context of energy conservation. Researchers have identified that battery-driven design is an important paradigm that hardware designers must not overlook. New lithium-polymer batteries are an attractive alternative to traditional NiCad batteries [1]. The lithium-polymer batteries lend themselves suitably to miniaturization; they also can store energy at sufficient densities to enable their use in WSNs [8].

8.5.2 Cross Layer Adaptation

Cross layer design has emerged as a fascinating new area of research in WSNs. The basic idea behind cross-layer design is to make information available to multiple levels in the protocol stack. With this shared information, it is then possible to make more informed decisions and optimizations. The degree to which information is shared is an open area of research. A fully cross layered design shares information among all levels in the protocol; however, sacrificing the modularity also negatively affects the measure of security that the protocol can provide [3]. Thus, extra provisions must be made to ensure the safe and secure delivery of sensitive data. It has been shown that traditional legacy protocols, with disjoint layers, provide a measure of security; however, they are not suitable for the WSN environment, as they do not sufficiently capture its dynamic nature. By exploiting cross-layer interactions, researchers hope to develop a new framework that will help to solve some of the difficulties inherent in WSNs.

Recent research into cross layer design has looked to optimize different aspects of the sensor network; these approaches have been:

- Using cross layer design to improve the power efficiency of WSNs, thereby increasing the network lifetime
- Optimizing system throughput in hopes of making the problem of scalability of WSNs tractable
- Satisfying QoS requirements of WSNs
- Improving resource efficiency

In particular, improving the power efficiency of the system is of particular interest as it has a direct and tangible impact on the lifetime of the network. Instituting power-aware protocols that take into consideration the residual energy information (REI) of the sensor node and the channel state information (CSI) between nodes, and adjusting transmission range in response to dynamic network topology are a few of the many parameters that network designers must take into consideration when trying to maximize the efficacy of the sensor network [10, 20].

This problem is compounded by the fact that often there are inherent conflicts between optimization goals. An illustrative example is that of the accuracy of measured data: to increase resolution, it is necessary to sample at a higher rate; yet increasing this acquisition rate depletes the battery, and thus the network lifetime. Furthermore, optimization may require interaction among multiple layers in the protocol stack.

Figure 8.4 depicts an instance of a cross-layered design that implements a vertical module called network status (NeSt). The NeSt module allows nonadjacent layers in the protocol to communicate, thus enabling cross-layer optimization while keeping the layer separation principle intact. The vertical module is responsible for gathering and controlling all cross-layer interactions. To support cross-layering in this manner, the protocols must implement a vertical module; however, they are free to design the layers in the protocol stack as per the requirements of the network.

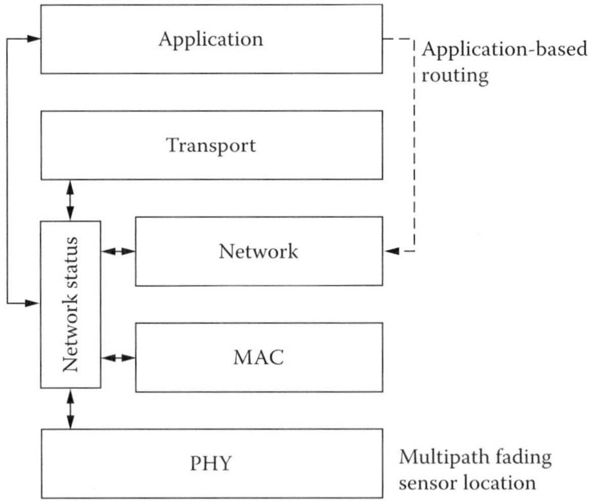

FIGURE 8.4 Protocol stack for an application-specific cross-layered architecture.

8.5.3 MEMS/CMOS/VLSI/Processing Power

With the ever-increasing drive toward miniaturization of components, a strong body of research has established that MEMS technology will likely enable researchers to integrate a wide selection of sensor equipment onto one chip [8]. A submillimeter node, with fully integrated components on a single chip, is not the most economically feasible implementation strategy. This is due in part to the fact that millimeter-scale on-chip antennae are not likely to perform as well as external antennae. However, an on-chip clock, using MEMS technology, is an attractive implementation consideration, as it is quite possible that it may outperform its off-chip counterparts. Thus, there are various factors to consider when designing nodes.

Research into low-power microprocessors has established a significant body of work, much of which has seen adaptation to the needs of microprocessors in the WSN environment. One avenue of research that has not garnered much attention in this field is the use of low-power field programmable grate arrays (FPGAs) as an alternative to microprocessors [1]. Other potential improvements could be the institution of frequency scaling for sensor nodes. During periods where the node is not hibernating and the microprocessor is not being heavily used, the clock frequency could be scaled down, thereby saving power and increasing the lifetime of the node.

8.5.4 Programming Abstraction/Software Tools

Many programmers that design software for WSNs are bogged down by implementation details in lower levels of the network protocol stack. To alleviate this problem and to be better able to deal with aggregate behavior of the WSN, the development of higher-level programming abstractions is necessary. This will enable programmers to deal with application functionality, thus increasing the productivity of the system as a whole [21]. With this in mind TinyOS, an embedded operating system for sensor networks, was developed. It allows programmers access to a large host of functions for sensor network design and deployment.

Often applications for WSNs generate large volumes of data. Developing a database view of the network-generated data would be very useful to both researchers and programmers. Being able to quickly query the network for specific data is imperative to the efficacy of the WSN.

8.5.5 Security/Privacy

Security and privacy concerns are issues that need to be addressed by any technology that is susceptible to tampering or deliberate malignant intervention by foreign entities. This is especially true in military and telemedicine applications, where the disclosure of personal or classified data is to be avoided at all costs. Current security mechanisms for WSNs are inadequate. Designing energy-efficient and computationally lightweight security protocols is critical to the development of WSNs. Security must be instituted at all levels in the protocol stack and must not be relegated to an afterthought [21, 25]. Ensuring network security in geographically extensive environments is a very challenging task; no single solution is likely to ensure the safety of the network.

8.5.6 Network Protocols/Multipath Routing

One avenue of research that has received a significant amount of interest from academia and industry is the institution of specially designed network protocols for WSNs. It has been established that the legacy Transmission Control Protocol/Internet Protocol (TCP/IP) protocol suite does not effectively operate in the WSN framework [3, 24]. This is due to the fact that TCP/IP maintains a table of network connectivity, giving each node a picture of the global topology. Maintaining such large tables requires large amounts of bandwidth and storage capacity; thus, it is not likely that TCP/IP will be a viable protocol for the WSN environment. Furthermore, IP-less nodes facilitate the deployment of large numbers of wireless sensors. Directed diffusion routing methods are an attractive alternative to TCP/IP in that they are more energy, bandwidth, and memory efficient. Furthermore, it is imperative for any networking protocol to exploit the knowledge of the CSI and REI of the nodes [20]. Herein lies a trade-off: although a node may have a good channel over which to transmit data, it may not have sufficient resources to do so. Thus, designing a network protocol that utilizes this information to help maximize the lifetime of the network is a challenging yet necessary endeavor.

Networking protocols that exploit mobility in the WSN framework have received attention from researchers. The protocols can be broken down into three disjoint categories:

8.5.6.1 Mobile Base Station (MBS)

In this protocol a mobile unit acts as a repository of information. Each node in the network forwards its local data to this information sink. Figure 8.5 represents the basic concepts behind the MBS approach. The nodes alternate between being sensor nodes and information sinks. The algorithm moves the position of the sink to maximize the lifetime of the network.

8.5.6.2 Mobile Data Collection (MDC)

In this protocol a mobile unit visits the nodes, where it then gathers the data from the nodes directly. The locally stored data are buffered until the MDC can arrive and the

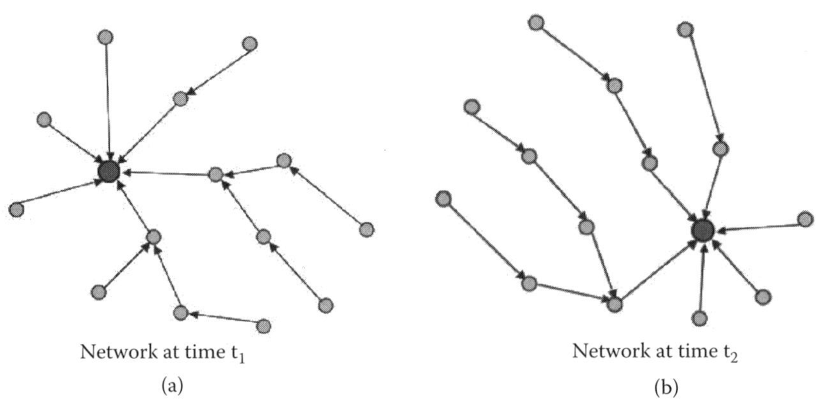

FIGURE 8.5 MBS approach: Move the sink in the network to distribute energy consumption.

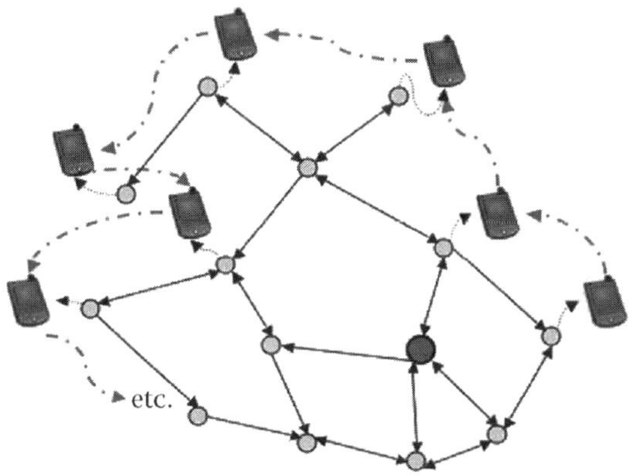

FIGURE 8.6 MDC approach: A mobile unit visits sensor nodes to gather data.

information can then be uploaded from the node. Figure 8.6 depicts the idea behind the MDC approach.

8.5.6.3 Rendezvous-Based Solutions

This protocol is simply a hybrid of the two schemes above. The locally stored node data are forwarded to points that are near the projected path of the mobile device.

There are a number of issues related with the listed protocols; particularly, the MDC proposal poses several problems for certain applications where gathering data from the nodes is not a feasible option. This might be due to the large number of nodes deployed or the possible environments in which they are situated.

In future applications for WSNs, the institution of tiered quality-of-service (QoS) requirements is a likely problem that researchers will face. Thus, designing provisions and protocols that will help facilitate the QoS requirements is necessary. The authors in [10] have proposed alternatives to legacy protocol implementations in which data, or urgent messages, can be delivered to nodes in the network with time guarantees.

8.5.7 Other Techniques for Extending the Network Lifetime

One major factor that could significantly extend the network lifetime is the institution of power scavenging techniques; certain MEMS devices have been able to convert mechanical vibrations into energy that can be stored or used to recharge a node's battery [8]. Other possible solutions include harvesting solar energy in the form of photovoltaic cells.

8.6 Hardware Implementation

In this section various hardware implementations of nodes and wireless sensor networks are detailed.

One of the most popular hardware implementations of the wireless sensor node is the MICA [29] line. Several MICA nodes with varied components, intended for different

TABLE 8.1 Comparison of the Salient Features of Different Nodes

Product	Frequency Band (MHz)	Transmit Data Rate (kbps)	Battery/ Voltage Supply	Size (mm)[a]	Tx Range (Indoor/Outdoor) (m)
MICA2 (MPR400CB)	868/916	38.4	2 × AA	58 × 32 × 7	—/150
MICA2DOT (MPR500CA)	868/916	38.4	3V Coin Cell	25 × 6	—/150
MICAz (MPR2400CA)	2400–2483.5	250	2 × AA	58 × 32 × 7	20–30/75–100
SmartMesh-XT (M1030)	902–928	76.8	2.7–3V	39.9 × 24.4 × 12.65	80/200
SmartMesh-XT (M2030)	2400–2483.5	250	2.7–3V	39.9 × 24.4 × 12.65	25/200
SmartMesh-XT (M2135)	2400–2483.5	250	2.7–3V	39.9 × 24.4 × 12.65	100/300

[a] These measurements do not include possible external battery packs.

end-user applications, have been designed and tested, enabling the user to custom tailor the nodes for his or her intended purpose. Another product similar to the MICA node is the SmartMesh-XT node [30]. Both product lines have nodes that can operate in the 900/2,400 MHz range. Table 8.1 gives a general overview of the salient features in some commercial node products.

The MICA2DOT's small form factor lends itself to a wide host of applications. Given its small size, the MICA2DOT is ideal for industrial process/inventory monitoring, and smart badges. The SmartMesh nodes use a proprietary Time-Synchronized Mesh Protocol (TSMP) to serve as a foundation for the wireless mesh network. The TSMP's salient features are time-synchronized communication, frequency hopping, dynamic network formation, redundant mesh routing, and secure message transfer.

8.7 Conclusion

The chapter has presented a survey of the applications, technologies, and challenges of WSNs. It is expected that this research area will remain very active in the following decades, with new applications and new challenges.

References

[1] M. Aboelaze and F. Aloul. 2005. Current and future trends in sensor networks: A survey. In *Second IFIP International Conference on Wireless and Communication Networks*, Dubai, UAE, pp. 551–55.

[2] A. Baggio. 2005. Wireless sensor networks in precision agriculture. In *Proceedings of the Workshop on Real-World Wireless Sensor Networks (REALWSN)*, Stockholm, Sweden.

[3] E. Borgie, M. Conti, and D. Delmastro. 2006. MobileMAN: Design, integration, and experimentation of cross-layer mobile multiphop ad-hoc networks. *IEEE Commun. Mag.* 44:80–85.

[4] J. Burrel, T. Brooke, and R. Beckwith. 2004. Vineyard computing: Sensor networks in agricultural production. *IEEE Pervasive Comput.* 3:38–45.

[5] Center for Embedded Sensing (CENS). 2006. *Annual progress report.*

[6] M. Cetin et al. 2006. Distributed fusion in sensor networks. *IEEE Signal Processing Mag.* 23:42–55.

[7] C. Chong and S. Kumar. 2003. Sensor networks: Evolution, opportunities, and challenges. *Proc. IEEE* 91:1247–56.

[8] B. Cook, S. Lanzisera, and K. Pister. 2006. SoC issues for RF smart dust. *Proc. IEEE* 94:1177–96.

[9] B. Deb, S.Bhatnagar, and Badri Nath. 2001. A *topology discovery algorithm for sensor networks with applications to network management.* Technical Report DCS-TR-441. Department of Computer Science, Rutgers University.

[10] E. Ekici, Y. Gu, and D. Bozdag. 2006. Mobility-based communication in wireless sensor networks. *IEEE Commun. Mag.* 44:56–62.

[11] J. J. Evans. 2005. Wireless sensor networks in electrical manufacturing. In *Proceedings of the Electrical Insulation Conference and Electrical Manufacturing Expo*, pp. 460–65.

[12] F. Hu and S. Kumar. 2003. QoS considerations in wireless sensor network for telemedicine. *Proc. SPIE* 5242:217–27.

[13] X. Li et al. 2003. Multi-dimensional range queries in sensor networks. In *Proceedings of ACM SenSys*, Los Angeles, pp. 63–75.

[14] A. Mainwaring et al. 2002. Wireless sensor networks for habitat monitoring. In *International Workshop on Wireless Sensor Networks and Applications*, Atlanta, GA, pp. 88–97.

[15] M. Maroti, G. Simon, A. Ledeczi, and J. Szipanovits. 2004. Shooter localization in urban terrain. *IEEE Comput.* 37:60–61.

[16] K. Martinez, J. Hart, and R. Ong. 2004. Environmental sensor networks. *IEEE Comput.* 37:50–56.

[17] K. Martinez et al. 2006. Deploying a sensor network in an extreme environment. In *Proceedings of the IEEE International Conference on Sensor Networks, Ubiquitous and Trustworthy Computing (SUTC)*, Taichung, Taiwan, pp. 186–93.

[18] A.-S. K. Pathan, C. S. Hong, and H.-W. Lee. 2006. Smartening the environment using wireless sensor networks in a developing country. In *International Conference on Advanced Communication Technology*, pp. 705–9.

[19] D. Puccinelli and M. Haenggi. 2006. Multipath fading in wireless sensor networks: Measurements and interpretation. In *International Wireless Communications and Mobile Computing Conference (IWCMC)*, Vancouver, Canada, pp. 1039–44.

[20] V. Raghunathan, S. Ganeriwal, and M. Srivastava. 2006. Emerging techniques for long lived wireless sensor networks. *IEEE Commun. Mag.* 44:108–14.

[21] J. Stankovic. 2004. Research challenges for wireless sensor networks. In *ACM Special Interest Group for Embedded Systems (SIGBED) Review*, vol. 1, pp. 9–12.

[22] Y. Sung et al. 2006. Signal processing for application specific ad hoc networks. *IEEE Signal Processing Mag.* 23:74–83.

[23] G. Tolle et al. 2005. A macroscope in the redwoods. In *Third ACM Conference on Embedded Networked Sensor Systems (SensSys)*, San Diego, pp. 51–63.

[24] C. Wang et al. 2006. A survey of transport protocols for wireless sensor networks. *IEEE Network* 20:34–40.

[25] A. Wood and J.Stankovic. 2002. Denial of service in sensor networks. *IEEE Comput.* 35:54–62.

[26] L.Yu, N. Wang, and X. Meng. 2005. Real-time forest fire detection with wireless sensor networks. In *Proceedings of the International. Conference on Wireless Communications, Networking, and Mobile Computing*, vol. 2, pp. 1214–17.

[27] Q. Zhao, A. Swami, and L. Tong. 2006. The interplay between signal processing and networking in sensor networks. *IEEE Signal Processing Mag.* 23:84–93.

[28] G. Zhou, T. He, S. Krishnamurthy, and J. Stankovic. 2004. Impact of radio irregularity on wireless sensor networks. In *The Second International Conference on Mobile Systems (MobiSys), Applications and Services*, Boston, pp. 125–38.

[29] Crossbow. Accessed April 21, 2008 from http://www.xbow.com.

[30] Dust Networks. Accessed April 21, 2008 from http://www.dustnetworks.com.

[31] Canadian Carbon Program. Accessed April 21, 2008 from http://www.fluxnet-canada.ca/.

9

Adaptive Routing in Wireless Sensor Networks

Hong Luo
Beijing University of Posts and Telecommunications

Guohua Zhang
University of Texas at Arlington

Yonghe Liu
University of Texas at Arlington

Sajal K. Das
University of Texas at Arlington

9.1 Introduction

By commanding a large number of sensor nodes capable of sensing, computing, and communication, wireless sensor networks (WSNs) are edging toward wide deployment in a plethora of applications [1, 2]. The communication architecture of a typical wireless

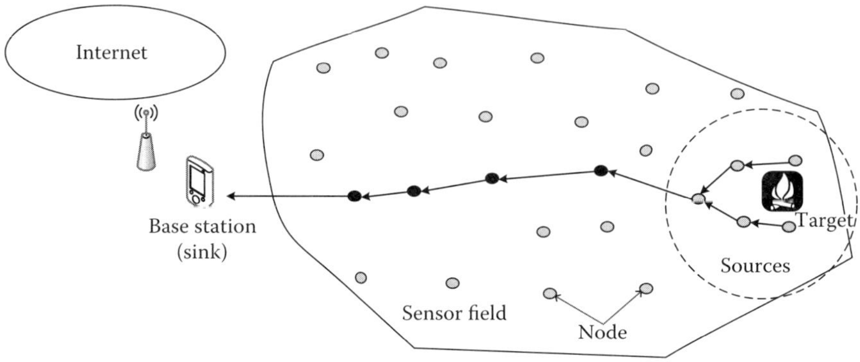

FIGURE 9.1 Communication architecture of a sensor network.

sensor network is illustrated in Figure 9.1. Usually, sensor nodes are scattered in the targeted field in either a random fashion or controlled manner. Each node senses its local environment, performs necessary computation and processing, and sends the sensory data through multihop or directly to the sink. The sink may be a fixed node or a mobile node capable of connecting the sensor network to an existing communication infrastructure where a user can access the reported data.

As WSNs are subject to strict resource constraints, particularly energy sources, energy-conserving techniques must be prioritized in designing any component of the network. For example, as a sensor node may need to operate over a long period of time on a battery with limited capacity, energy-aware communication protocols (including MAC protocols and routing protocols) are crucial to the long-term operation of WSNs. This can often be achieved through different approaches. For example, as a large number of sensor nodes are often densely deployed, multihop communication, as opposed to single-hop communication, is commonly adopted for its energy efficiency. At the same time, as data collected from densely deployed sensors are often redundant, data aggregation or data fusion can be employed as a useful strategy to curtail the network load and hence reduce energy consumption. Energy-efficient designs often penetrate into each component of the system, and cross-layer approaches are often employed.

9.1.1 Routing in Wireless Sensor Networks

Routing is a key component of a WSN for information gathering and critical in energy conservation. Generally, WSN can be considered a special case of mobile ad hoc networks (MANETs). Therefore, similar requirements and challenges often arise in MANETs as well, including the time-varying characteristics of wireless links, limited energy sources, possibility of link failures, bandwidth limitation, multihop communications, and ad hoc deployment. However, the unique characteristics of WSN introduce additional distinct requirements for data routing therein. These include:

- The destination in WSNs is often known and communication is normally from multiple data sources to the sink. As a result, the basic topology desired in data

gathering can often be abstracted as a spanning tree. In MANETs, on the contrary, communication is generally on a peer-to-peer basis (i.e., one-to-one).
- Data collected by sensor nodes in WSN are often highly correlated and subject to in-network processing.
- MANETs are usually characterized by highly dynamic topologies due to node mobility. In WSNs, nodes are often static, and thus the nature of the network dynamics is different.
- The energy constraints in WSN often are more severe than those in MANETs.

These unique requirements in a WSN often prevent it from directly adopting many end-to-end routing schemes proposed for MANETs.

9.1.2 Design Challenges in WSN Routing

Design of routing protocols in WSNs faces many challenges that must be overcome before efficient communication can be achieved. Some of these challenges and some design guidelines are listed below:

- *Ad hoc deployment*: Sensor nodes are often deployed randomly. This requires the system to be able to cope with the random node distribution and resulting network topology. The system should also be adaptive to changes in network connectivity as a result of node failure.
- *Communication range*: Node-to-node communication often is of short transmission ranges. Therefore, it is most likely that a data route will generally consist of multiple wireless hops.
- *Transmission media*: Traditional problems associated with wireless transmissions (e.g., channel fading and high error rate) will also affect the operation of sensor networks.
- *Fault tolerance*: Sensor nodes can fail due to the lack of power, physical damage, or environmental interference. The failure of certain sensor nodes should not affect the overall task of the sensor network. For example, MAC and routing protocols must be capable of forming new links and routes to the data collection sink. This may require actively adjusting transmission powers and data rates on the existing links, or rerouting packets through other regions of the network. Multiple levels of redundancy may be needed to provide fault tolerance.
- *Computation capabilities*: Sensor nodes have limited computing power and therefore may not be able to execute sophisticated network protocols. Lightweight routing protocols are demanded.
- *Quality of service*: While energy efficiency may be considered one quality of service metric for routing in WSNs, in certain applications, delay latency of data gathering is also critical.

Recently, routing protocols for WSNs have been the subject of extensive study. References [2, 3] serve as comprehensive surveys of classical routing protocols. In this chapter, however, we focus on energy-efficient routing designs that have self-adaptivity by jointly considering multiple factors, including node residual power, wireless link condition, data fusion, and sensory information. In particular, we will discuss energy-aware, channel-aware, fusion-aware, and information-aware adaptive routing protocols, with a focus on their design philosophy, advantages, drawbacks, and performance.

9.2 Energy-Aware Routing

Many energy-aware approaches have been proposed toward efficient routing for wireless sensor networks. To illustrate them, we use the example in Figure 9.2. Here, node T denotes the source node that produces the sensory data, PA is the available energy, and α_i denotes the energy required to transmit the data packet generated by node T over link i. From node T to the sink, there are four possible routes and corresponding metrics:

- Route 1: Sink-A-B-T, total $\alpha = 3$, min PA = 2, total PA = 4
- Route 2: Sink-A-B-C-T, total $\alpha = 6$, min PA = 2, total PA = 6
- Route 3: Sink-D-T, total $\alpha = 4$, min PA = 3, total PA = 3
- Route 4: Sink-E-F-T, total $\alpha = 6$, min PA = 1, total PA = 5

9.2.1 Minimum Total Energy Routing

Initially, energy-aware routing approaches focused on finding the minimum energy path from the source node to the sink.

- *Minimum total transmission energy routing (MTE)*: MTE minimizes the total energy consumed along a route. We denote the energy consumed when transmitting packets between node n_i and n_j as $p(n_i, n_j)$. Thus, the whole route requires energy

$$P_l = \sum_{i=0}^{D-1} p(n_i, n_{i+1}),$$

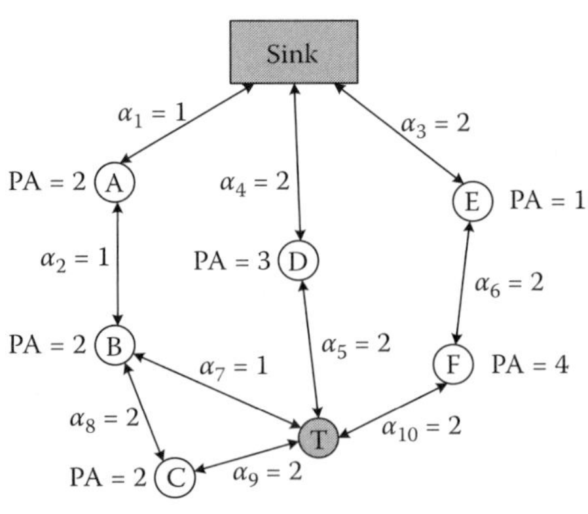

FIGURE 9.2 The power efficiency of the routes.

where n_0 and n_D are source and destination, respectively. So the scheme aims to determine the route with minimum energy consumption

$$P_o = \min_{l \in A} P_l,$$

where A is the set of all possible routes between the source and destination. In the example shown in Figure 9.2, route 1 is the MTE route.

- *Minimum hop routing (MH)*: In this type of routing protocol, the route with the minimum number of hops is preferred. If sensor nodes broadcast with the same power level without any power control, the same amount of energy is used on every link, i.e., all α_i are the same. In this case, MH is then equivalent to MTE. Route 3 in Figure 9.2 is the most efficient route based on this scheme.

MTE routing targets minimizing the total energy consumption for the data communication. However, a critical disadvantage here is that total energy consumption does not necessarily reflect the lifetime of each node. If all the traffic is routed through the minimum energy path to the destination, nodes on that path will run out of batteries quickly and may render other nodes useless due to network partition, even if those nodes do have available energy resources. Therefore, the remaining battery capacity of a node is also critical in routing design. Routing schemes considering maximum battery capacity indeed aim to schedule traffic along routes with more residual energy.

9.2.2 Maximum Battery Capacity Routing

In this type of routing scheme, the total available power (PA) of a route is calculated by summing the PA of each node along the route, where the route with the maximum PA is preferred. Based on this approach, for the example in Figure 9.2, route 2 will be selected. However, route 2 includes all nodes in route 1 and an extra node. Therefore, although it has a higher total PA, it is not a power-efficient one. As a result, it is important not to consider the routes derived by extending those that can connect the sensor to the sink as an alternative route. Eliminating route 2, route 4 is the preferred energy-efficient route under the maximum PA scheme.

As battery capacity is directly incorporated into the routing protocol, this metric can prevent a node's battery from being quickly exhausted and thereby increasing its lifetime. If all nodes have similar battery capacity, this metric will select a shorter-hop route. However, since only the summation of available power is considered, a route containing nodes with little remaining battery capacity may still be selected. For example, in Figure 9.2, although node E has much less available power than other nodes, the overall available power on route 4 is more than that on routes 1 and 2. Therefore, route 4 will be selected, which can reduce the lifetime of node E and hence is undesirable.

9.2.3 Max-Min Battery Capacity Routing

In order to consider individual node's remaining battery capacity and consequently extend network lifetime, several max-min battery capacity-based routing strategies were proposed:

- *Max-min PA node route (MMPA)*: To prevent a node's battery from being exhausted quickly, in these schemes, the PA of each route is not defined as the sum of the individual PA of each node, but as the minimum PA along that path. The route with maximum PA is preferred. In Figure 9.2, route 3 is the most efficient path and route 1 is the second most efficient path.
- *Min-max battery cost routing (MMBCR)*: This type of routing scheme also strives to avoid routing over the node with minimum residual energy. The cost of a route is not defined as the sum of individual battery cost of each node, but as the maximum battery cost along that route, and the routine that chooses a route with the minimum cost is preferred.

These schemes always try to avoid using the node with minimum residual energy, and thus improve the load balance among all nodes. But without consideration of the total transmission energy along a route, a route with large total transmission power may be chosen.

9.2.4 Combination of MTE and MMBCR

Message routing using paths only with nodes of high residual energy may be expensive compared with paths that have minimal power consumption. To solve this problem, schemes that combine the minimum total energy and residual battery capacity of each node are proposed to make routing decisions.

9.2.4.1 Conditional Max-Min Battery Capacity Routing

One scheme combining the advantages of both MTE and MMBCR was proposed in [4]. It chooses a route whose bottleneck residual energy is larger than a certain threshold. If there is more than one route satisfying this condition, it will select the one with minimum total transmission power, as in the case of MTE. When no route satisfies the condition, similar to MMBCR, it chooses a route with max-min residual energy. This scheme is termed conditional max-min battery capacity routing (CMMBCR). For the network shown in Figure 9.2, if the residual energy threshold is 2, route 1 is more energy efficient than route 2 and 3 since the total transmission power on route 1 is the lowest. When the residual energy threshold is 3 or 4, route 3 will be selected.

The basic idea behind CMMBCR is that when all nodes in possible routes between a source and a destination have sufficient remaining battery capacity (i.e., above a threshold), a route with the minimum total transmission power among these routes is chosen. Since less total power is required to forward packets for each connection, the relaying load for most nodes will be reduced, and their lifetime will be extended. However, if all routes have nodes with low battery capacity (i.e., below a threshold), routes including nodes with the lowest battery capacity should be avoided to extend the lifetime of these nodes. In CMMBCR, the battery capacity R_j^c for route j at time t is defined as

$$R_j^c = \min_{i \in route_j} PA_i^t. \tag{9.1}$$

Let A be a set containing all possible routes between any two nodes at time t and satisfying the following equation:

$$R_j^c \geq \gamma, \text{ for any route } j \in A. \tag{9.2}$$

Here, we assume the initial battery capacity is 100 and γ is a threshold ranging between 0 and 100. Let Q denote the set containing all possible paths between the specified source and the destination node at time t. If $A \cap Q \neq \Phi$, which implies that all nodes in certain paths have remaining battery capacity higher than γ, the scheme chooses a path in $A \cap Q$ by applying the MTE scheme. Otherwise, it will select route i with the maximum battery capacity: $R_i^c = \max\{R_j^c \mid j \in Q\}$.

If $\gamma = 0$, equation 9.2 is always true, and the scheme degrades to MTE. If $\gamma = 100$, equation 9.2 is always false, and the scheme is equivalent to MMBCR, as routes with less battery capacity will always be avoided. γ can be considered a protection margin: if a node's battery capacity falls below this value, it will be avoided to prolong the network's lifetime. The performance of CMMBCR is therefore dependent on the value of γ. By adjusting the value of γ, CMMBCR can maximize either the time when the first node runs out of battery or the lifetime of most nodes in the network.

9.2.4.2 Max-Min zP_{min}

Similarly, an approximation algorithm called *max-min* zP_{min} [5] was proposed that balances between the minimum transmission energy routing and the max-min residual energy routing. The basic idea of *max-min* zP_{min} is to find a max-min residual energy path by limiting its total power consumption. The *max-min* zP_{min} algorithm can be described as follows:

1. Find the path with the minimum transmission energy, denoted by P_{min}, through the Dijkstra algorithm.
2. Find the path with the minimum transmission energy in the graph. If the power consumption on that path is greater than zP_{min} or no path is found, the previous MTE path is the solution. The algorithm stops.
3. Otherwise, find the minimal residual energy on that path. Denote it by u_{min}.
4. Find all the edges whose residual energy fraction is smaller than or equal to u_{min}, and remove them from the potential set of edges.
5. Go to 2.

The resulting path is the max-min residual energy path with a limited total transmission energy, zP_{min}. An important factor in the *max-min* zP_{min} algorithm is the parameter z that measures the trade-off between the max-min path and the MTE path. When $z = 1$, the algorithm computes the minimal transmission energy path; when $z = \infty$, it computes the max-min path. In this approach, an adaptive computation approach of z is also provided. The performance of the adaptive *max-min* zP_{min} algorithm was shown to be close to the optimal solution obtained by linear programming.

Scalability of the algorithm was provided by a zone-based hierarchical routing approach. Zone-based routing is a hierarchical approach in which the area covered by the sensor network is divided into a small number of zones. Each zone has some sensors in geographic proximity, and each zone is treated as an entity. To perform routing, each zone is allowed to decide how it will route a message hierarchically across other zones. To send a message across the entire area, a global path from zone to zone is first determined. Nodes in a zone autonomously direct local routing and participate in estimating the zone power level. Each message is routed across the zones using information about their zone power estimates. A global controller for message routing, which may be the node with the highest power, is assigned the role of managing the zones. Therefore, there are three aspects in the zone-based routing scheme: (1) how the nodes in a zone collaborate to estimate the power of the zone, (2) how a message is routed within a zone, and (3) how a message is routed across zones. Here, (1) and (3) can use the *max-min* zP_{min} algorithm, which can be implemented in a distributed way. The max-min algorithm used in (2) is basically the Bellman-Ford algorithm, which can also be implemented as a distributed algorithm.

9.2.5 Hierarchical Routing

Low-energy adaptive clustering hierarchy (LEACH) [6] is a cluster-based protocol that minimizes energy dissipation in sensor networks. The purpose of LEACH is to randomly select sensor nodes as cluster heads so that high energy dissipation in communicating with the base station is spread to all sensor nodes in the network. The operation of LEACH is separated into two phases, the setup phase and the steady phase. The duration of the steady phase is longer than the duration of the setup phase in order to minimize the overhead. During the setup phase, a sensor node chooses a random number between 0 and 1. If this random number is less than the threshold $T(n)$, the sensor node is a cluster head. Here, $T(n)$ is calculated according to

$$T(n)=\begin{cases} \dfrac{P}{1-P\left[r \bmod (1/P)\right]} & \text{if } n\in G, \\ 0 & \text{otherwise} \end{cases} \tag{9.3}$$

where P is the desired percentage to become a cluster head, r is the current round, and G denotes the set of nodes that have not been selected as a cluster head in the last $1/P$ rounds. After the cluster heads are selected, the cluster heads advertise to all sensor nodes in the network their election. Upon receiving this advertisement, all the non-cluster-head nodes decide on the cluster to which they want to join, based on the signal strength of the advertisement. The non-cluster-head nodes inform the appropriate cluster heads that they will be members of the cluster. Afterwards, the cluster heads assign the time on which the sensor nodes can send data to the cluster heads based on a Time Division Multiple Access (TDMA) approach.

During the steady phase, the sensor nodes can begin sensing and transmitting data to the cluster heads. The cluster heads also aggregate data from the nodes in their cluster before sending these data to the base station. After a certain period of time is spent on the steady phase, the network goes into the setup phase again and enters into another round of selecting the cluster heads.

In the procedure of cluster-head election, LEACH considers the number of times the node has been a cluster head so far (i.e., up to the current round) but not the residual energy. But a cluster head located far away from the sink will spend more energy for communication than a cluster head closer to the base station, which will induce non-uniform energy distribution. To solve this problem, LEACH-C [7] and HEED [8], which incorporate residual energy into consideration when performing random cluster-head election, were proposed. LEACH-C provides a centralized cluster-head election scheme in which only the nodes with more residual energy than the average residual energy of the overall network can become cluster heads. HEED improves LEACH-C to a distributed version.

9.2.6 Energy-Aware Multipath Routing

The potential problem in the above protocols is that once the optimal route is determined, it will be used for every transmission. This may not be an ideal solution from the network's point of view. Using the optimal path frequently leads to energy depletion of the nodes along that path and, in the worst case, may lead to network partition. To counteract this problem, data forwarding could use different paths at different times; thus, any single path does not get energy depleted quickly. Compared with the single-path strategy, multipath routing can balance traffic loads among multiple nodes, and can respond to network dynamics where nodes can join and leave the network.

Chang and Tassiulas [9] proposed an algorithm to route data through a path whose nodes have the largest residual energy. The path is changed adaptively whenever a better path is discovered. The primary path will be used until its energy falls below the energy of the backup path, at which point the backup path is used. In this way, the nodes in the primary path will not deplete their energy resources due to continual use of the same route, thus achieving longer network life. However, the path-switching cost was not quantified in their paper.

Shah and Rabaey [10] have proposed an energy-aware routing protocol that uses a set of suboptimal paths occasionally to increase the lifetime of the network. These paths are chosen by means of a probability that depends on how low the energy consumption of each path is. The basic idea is that to increase the survivability of networks, it may be necessary to use suboptimal paths occasionally. This ensures that the optimal path's energy does not get depleted quickly, and the network degrades gracefully as a whole rather than getting partitioned. To achieve this, multiple paths are chosen between source and destinations, and each path is assigned a probability of being used, depending on the energy metric. Every time data are to be sent from the source to a destination, one of the paths is randomly chosen, depending on the probabilities. This means that none of the paths are used continuously. Additionally, different paths can improve tolerance to network dynamics.

Energy-aware routing is also a reactive routing protocol. It is a destination-initiated protocol where the consumer of data initiates the route request and maintains the route subsequently. Thus, it is similar to diffusion in certain ways. Multiple paths are maintained from source to destination. However, diffusion sends data along all the paths at regular intervals, while energy-aware routing uses only one path at all times. Due to the probabilistic choice of routes, it continuously evaluates different routes and chooses probabilities accordingly. The protocol has three phases: setup phase, data communication, and route maintenance.

- Setup phase
  1. The destination node initiates the connection by flooding the network in the direction to the source node. It also sets the cost field to zero before sending the request:

$$Cost(N_D)=0$$

  2. Every intermediate node forwards the request only to the neighbors that are closer to the source node than itself and farther away from the destination node. Thus, at a node N_i, the request is sent only to a neighbor N_j that satisfies

$$d(N_i,N_S)\geq d(N_j,N_S) \text{ and } d(N_i,N_D)\leq d(N_j,N_D),$$

     where $d(N_i, N_j)$ is the distance between N_i and N_j.
  3. On receiving the request, the energy metric for the neighbor that sent the request is computed and is added to the total cost of the path. Thus, if the request is sent from node N_i to node N_j, N_j calculates the cost of the path as

$$C_{N_j,N_i}=Cost(N_i)+Metric(N_j,N_i).$$

  4. Paths that have a very high cost are discarded and not added to the forwarding table. Only the neighbors N_i with paths of low cost are added to the forwarding table FT_j of N_j:

$$FT_j=\left\{i \,\middle|\, C_{N_j,N_i}\leq\alpha\cdot\left(\min_k C_{N_j,N_k}\right)\right\}$$

  5. Node N_j assigns a probability to each of the neighbors N_i in the forwarding table FT_j, with the probability inversely proportional to the cost:

$$P_{N_j,N_i}=\frac{\frac{1}{C_{N_j,N_i}}}{\sum_{k\in FT_j}\frac{1}{C_{N_j,N_k}}}$$

6. Thus, each node N_j has a number of neighbors through which it can route packets to the destination. N_j then calculates the average cost of reaching the destination using the neighbors in the forwarding table:

$$Cost(N_j) = \sum_{i \in FT_j} P_{N_j,N_i} C_{N_j,N_i}$$

7. This average cost, $Cost(N_j)$, is set in the cost field of the request packet and forwarded along toward the source node as in step 2.

- Data communication phase
  1. The source node sends the data packet to any of the neighbors in the forwarding table, with the probability of the neighbor being chosen equal to the probability in the forwarding table.
  2. Each of the intermediate nodes forwards the data packet to a randomly chosen neighbor in its forwarding table, with the probability of the neighbor being chosen equal to the probability in the forwarding table.
  3. This is continued until the data packet reaches the destination node.
- Routing maintenance: Localized flooding is performed infrequently from destination to source to keep all the paths alive.

The energy metric that is used to evaluate routes is a very important component of the protocol. Depending on the metric, the characteristics of the protocol can change substantially. The metric can include information about the cost of using the path, residual energy of the nodes along the path, topology of the network, etc. For energy-aware routing, a simple metric can be $Metric(N_j, N_i) = e_{ij}^{\alpha} R_i^{\beta}$. Here $Metric(N_j, N_i)$ is the energy metric between nodes N_i and N_j, e_{ij} is the energy used to transmit and receive on the link, and R_i is the residual energy at node N_i normalized to the initial energy of the node. The weight factors α and β can be chosen to find the minimum energy path or the path with nodes having the most energy, or a combination of both. Simulation results show that this energy-aware multipath routing scheme can increase network lifetime up to 40% over comparable schemes like directed diffusion routing. Furthermore, nodes also consume energy in a more balanced way across the network, ensuring a more graceful degradation of service with time.

9.3 Link-Aware Routing

In addition to resource constraints, time-varying wireless links are another challenge in routing design for wireless sensor networks. Coupled with potential interference from densely deployed nodes and multihop communications, link variations present more significant effects in data routing. Indeed, link variations present unique opportunities for routing algorithms to employ channel-adaptive techniques, which can even benefit from them. For example, channel-adaptive routing can use channel state information and cross-layer integration to route traffic along higher-capacity paths by consistently selecting channels with favorable conditions. This way, the overall system throughput can be increased. In this section, we provide an overview for channel-aware routing designs in wireless sensor networks.

9.3.1 Channel State Information

One way to facilitate routing based on channel state information (CSI) is to provide multiple next-hop alternatives to a transmitting node, whether it is a source node or an intermediate node along the path to the destination. Each time a packet is forwarded, one of these next-hop links would be selected based on current channel conditions. Two classes of routing protocols that can be used in practice to provide such alternatives on a per-hop basis are position-based routing and multipath routing.

Position-based routing algorithms belong to geographic routing that utilizes knowledge of the geographic locations of the final destination and neighboring nodes. In these schemes, packets are routed geographically toward the destination through a relay node that offers progress toward the final destination. A simple criterion to select the next-hop relay is to find a neighbor that is located closest to the destination. Without considering channel state information, a relay node with unfavorable channel condition can be chosen. If channel state information is available to a forwarding node, it is possible for it to combat channel variations by exploiting diversity among the channels to different neighboring nodes.

A second class of protocols that can be used adaptively is end-to-end route discovery algorithms modified to provide multiple routes between a source and destination. These algorithms do not require position information, but rely instead on neighbor information exchange to establish routes that minimize certain cost metrics such as the number of hops. Standard on-demand route discovery protocols only create a single route between a source and destination. However, modifications to the Ad hoc On-Demand Distance Vector (AODV) protocol, for example, have been proposed in [11] that would provide alternative next-hops along a route. These modified protocols can be used to establish a mesh of multiple paths between the source and destination. While originally intended to provide backup routes or to enable load balancing among different routes, multipath routing algorithms can be enhanced toward channel-adaptive routing. For example, a node with multiple next-hop alternatives can measure the channel state on the links, and then forward a packet based on the link quality and other metrics.

9.3.2 CSI-Based Routing

Over a wireless link, multipath fading and large-scale path loss can induce signal attenuation and affect signal-to-noise ratio (SNR). Channel measurements of SNR or simple signal strength can provide the channel state information (CSI). Channel-adaptive routing schemes are discussed in [12] and [13] based on measured CSI.

- *Maximum forward progress within radius R*: Maximum forward progress is the basic position-adaptive (PA) routing, which is based on position information only and not on CSI. In the PA scheme, the position of each node is described by its distance d from the source and the angle θ toward the final destination.

The progress of a successful transmission is defined as the distance covered by that link toward the final destination, i.e., $d\cos\theta$. When a transmitter has many candidate relays within radius R, it will:

1. Collect the position (d_i, θ_i) of each candidate relay.
2. Select the relay offering the greatest potential forward progress in the direction to the final destination:

$$\arg\max_i \{d_i \cos\theta_i\}.$$

PA routing neglects the impact of position on the probability of the transmission success. The effect is that relays offering large progress are favored even though those links tend to experience greater attenuation, and therefore lead to a lower probability of success.

- *Attenuation-adaptive routing (AA)*: This basic attenuation-adaptive routing selects the next-hop relay based on link attenuation only. No knowledge of the relay position is used. The algorithm for relay selection with AA routing is simply the following:
  1. Measure the link gain of each candidate relay.
  2. Select the relay with the maximum link gain.

  Since attenuation is proportional to the distance between two nodes, AA routing prefers a closer node as the next-hop relay. As a result, many hops will be required to reach the destination. This can potentially lead to a high total energy consumption and delivery delay.
- *Biggest progress with channel constraint*: This strategy chooses the next-hop relay offering the maximum forward progress under the constraint that the SNR of the selected link is not less than a certain SNR threshold. Thus, this strategy targets reducing the delivery delay by controlling the error probability.
- *Best channel with energy constraint*: This strategy chooses the node with the highest channel gain from all the nodes farther than r_{min} from the current node as the next-hop, where r_{min} is the required minimum hop length.
- *Attenuation and position adaptive (APA) routing*: In this scheme, performance is measured in terms of *information efficiency*, defined as the product of the progress (in distance) made by the transmission toward the final destination and the local throughput on the link. The throughput incorporates the spectral efficiency of the modulation-coding scheme conditioned on the fading/shadowing attenuation. Maximizing the information efficiency balances the need to minimize the number of hops along a route and the need to maximize the throughput on a given hop, a key trade-off inherent to multihop wireless networks. If all nodes employ the same and fixed modulation-coding scheme, information efficiency is equivalent to forward progress.

  APA routing selects the next-hop relay that maximizes the information efficiency toward the destination based on the knowledge of both the channel attenuation on each relay link and the position of each relay. Notice that channel attenuation is the combined effect of path loss due to transmitter-receiver distance and fading/shadowing on the link. Therefore, it is the most adaptive routing strategy compared to the above schemes. The algorithm for relay selection with APA routing at a forwarding node is as follows:

1. Collect the position (d, θ) and link gain of each candidate relay.
2. Select the relay that maximizes the information efficiency, which is a joint metric of relay's position and its fading attenuation.

9.3.3 Interference-Aware Routing

Communications among nodes can interfere with each other, and as a result, network capacity, connectivity, link quality, and bandwidth availability can vary dynamically on a variety of timescales. The problem is more severe in dense wireless sensor networks where interference can be heavy for the already limited bandwidth. Therefore, finding practical wireless interference-aware metrics for routing decisions is critical in WSNs.

An interference-aware metric named Network Allocation Vector Count (NAVC) for multihop wireless networks is proposed in [14], where IEEE 802.11 is used as the MAC protocol. Embedding the NAVC metric into the routing process can substantially increase the lifetime for densely deployed networks.

The scheme can be briefly outlined as follows. IEEE 802.11 MAC protocol uses a random back-off procedure to resolve medium contention conflicts and a virtual carrier-sense mechanism that exchanges request-to-send (RTS) and clear-to-send (CTS) frames to announce the impending use of the medium via setting the Network Allocation Vector (NAV) value. Once a node hears other nodes' transmissions, its NAV will be set to the busy state and this node has to remain silent for a duration equal to the Duration/ID field in the frame header. The higher the traffic rate is, the larger the accumulated NAV value will be, and vice versa. Therefore, the NAV in the 802.11 MAC protocol is a good indicator of the surrounding traffic.

The relationship between the NAV value and the available bandwidth and average delay in a time interval is captured by NAVC, which is the average NAV value in that time interval. Each node periodically computes its NAVC value by collecting the NAV value from the MAC layer. Thus, as long as the node is sensing the wireless medium, the delay and available bandwidth characteristics can be estimated by the NAVC. For example, if NAVC is less than 0.20, the delay is usually small (about 2 ms) and the number of transmitting nodes is small; when the NAVC is greater than 0.65, it indicates the network is overloaded.

NAVC, as a generic interference-aware metric, can be used in general on-demand routing protocols. With this metric, routes discovery can be done at the network layer using a route request–reply cycle with interference information obtained from the link layer. When the source node sends a route request (RREQ), two additional fields, heavy-node-number and navc-number, are included. Each intermediate node adds its NAVC value to navc-number and increases heavy-node-number if it has overloaded, and then forwards the RREQ. After receiving the route reply (RREP) from the destination, the source node selects the path with the least heavy-node-number and navc-number. Therefore, NAVC-driven routing can avoid selecting a route through neighborhoods with heavy interference.

9.3.4 Modulation-Adaptive Routing

Another channel-adaptive routing technique used in ad hoc networks and based on the adaptive physical layer design is called ABICM [15], in which a variable-throughput modulator and channel coding are used. In ABICM, when the CSI indicates that the channel is of a good quality, the transmitter employs a high-order modulation (e.g., 16 Quadrature Amplitude Modulation (16QAM)) and high-rate error correction code to boost the instantaneous throughput. On the other hand, when the channel quality is poor, the transmitter employs a lower-order modulation (e.g., Binary Phase-Shift-Keying (BPSK)) and low-rate error protection to protect the packet transmission at the expense of lower instantaneous throughput.

For example, based on the channel CSI, the quality of a channel can be classified into four classes: A, B, C, and D, with throughputs of 250, 150, 75, and 50 kbps, respectively (after adaptive channel coding and modulation). Then, a CSI-based "hop" can be defined in the following manner. If it is a link between two nodes with a channel quality of class A (with throughput of 250 kbps), then the distance between the two nodes is defined as one hop. We use this distance as a baseline metric. If it is a link between two nodes with a channel quality of class B (with a throughput of 150 kbps), the distance between the two nodes is considered 1.67 hops, because the transmission delay is now 1.67 times, compared with a link of class A. Thus, the distances of links with throughputs of 250, 150, 75, and 50 kbps are 1, 1.67, 3.33, and 5 hops, respectively.

According to this CSI-based hop count, a path with the minimum number of hops is selected. In [16], the Bandwidth-Guarded Channel Adaptive (BGCA) routing protocol is proposed for ad hoc networks. The main idea of BGCA is that when a link is in deep fading, the upstream node executes a local search to find a partial route to the destination. When a node has packets to send to a destination, it first generates a route request (RREQ) packet, which includes bandwidth requirement, hop count (CSI based) from the source, and intermediate node list.

When an intermediate node first receives the RREQ, it checks whether the available bandwidth of the link between it and its upstream node can satisfy the bandwidth requirement of the RREQ. If so, the intermediate node records this RREQ in its history hash table, appends its ID to the intermediate node list, and updates the hop count field by adding the hop distance between the upstream node and itself. Then, this intermediate node rebroadcasts the RREQ. Eventually, the destination node may receive several RREQs from all possible routes.

Figure 9.3(a) shows the broadcast of RREQ in the network. Each RREQ includes a full route from the source to the destination and related hop distance (CSI based). The destination node chooses the shortest route, S-A-B-C-D, and sends a route reply (RREP) along this route to the source, as shown in Figure 9.3(b) (three routes with hop distances of 9.33, 10.33, and 11, respectively).

When a downstream node notices channel quality degradation, it informs the upstream node of this CSI change. If the link deteriorates significantly and cannot satisfy the bandwidth requirement, the upstream node performs a local search to find a partial route to the destination. In Figure 9.3(c), the original route from source to destination is

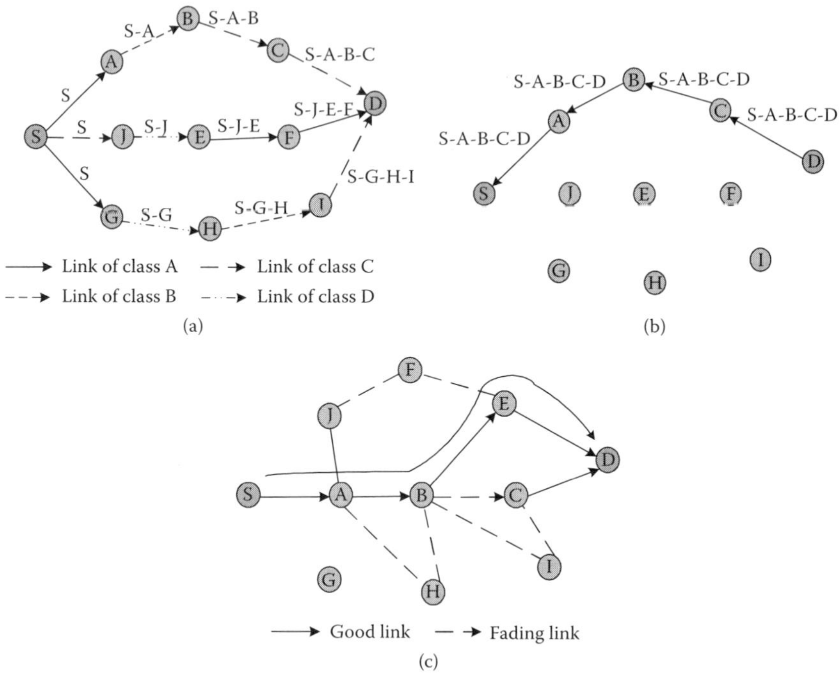

FIGURE 9.3 Illustration of the BGCA protocol: (a) broadcast of RREQ, (b) generation of RREP, and (c) change of links.

S-A-B-C-D. If link B-C is in deep fading and cannot satisfy the bandwidth requirement, node B performs the local search. The resultant new full route is S-A-B-E-D.

9.3.5 Joint-Adaptive Routing and Modulation

Since the CSI-based adaptive routing schemes assume that knowledge of the channel state information of wireless links is known, a natural extension would be to combine adaptive routing with adaptive modulation. A transmitting node would use knowledge of the channel information and (if available) position information to select the combination of relay and modulation scheme that maximizes the information efficiency (IE) of that hop.

The algorithm proposed in [13] for relay and modulation scheme selection with joint APA routing and adaptive modulation is described as follows:

1. Collect the position (d, θ) and link state information for each candidate relay.
2. Select the combination of relay and modulation scheme that maximizes the information efficiency, which is a joint metric of the relay's position and its channel state.

The algorithm for the simpler joint AA routing and adaptive modulation is as follows:

1. Measure the link gain of each candidate relay.
2. Select the relay with the maximum link gain.
3. Select the modulation scheme, for which the link gain of the selected relay is bounded by a given threshold.

TABLE 9.1 Summary of Adaptive Routing Schemes

Adaptive Scheme	Information Required	Remarks	IE Gain
Position adaptive (PA)	Positions of relays and final destination	Selects farther relays for greater progress per hop	1.4
Attenuation adaptive (AA)	Channel gains on source relay links	Selects closer relays for high SNR	1.5
Attenuation and position adaptive (APA)	Channel gains; position information; interference dispersion parameter (for conditional probability of success)	Combines CSI and position information to maximize the expected progress per hop	2.1
AA routing and adaptive modulation	Channel gains; channel gain thresholds	Achieves much of available gain in IE with a simple metric	3.2
APA routing and adaptive modulation	Channel gains; position information; interference dispersion parameter	Combines CSI and position information to maximize the IE per hop	4.2

Table 9.1 summarizes the key characteristics of some CSI-based adaptive routing schemes discussed above.

Even though the adaptive modulation technique mentioned above is proposed for ad hoc networks, routing schemes in sensor networks can readily adopt this technique. In [17], the authors study energy-aware and link-adaptive routing strategies for an ultra-wideband (UWB) sensor network. They use the ranging capabilities offered by UWB and employ jointly adaptive routing and adaptive modulation to benefit from favorable link conditions. Furthermore, they take the next-hop battery capacity information into account in the route decision for increased energy efficiency.

9.3.6 Cost-Aware Dynamic Routing

In WSNs, in addition to energy consumption, many other requirements, such as delay, reliability, or throughput, should be considered in different application-specific systems. To simultaneously satisfy different performance requirements, a cost-aware routing protocol is proposed in [18]. The protocol uses cost metrics to create gradients from a source to a destination node. The cost metrics consist of energy, node load, delay, and link reliability information that provide traffic differentiation by allowing choice among delay, reliability, and energy.

In this scheme, the sink advertises its presence by flooding routing advertisement (RADV) packets into the network. Nodes receiving the RADV packet calculate the cumulative cost to the sink and create forwarding gradients toward nodes that advertise the lowest costs. Once the gradients have been established, the sink requests data by sending an interest advertisement (IADV) on the reverse direction of gradients. The key in this protocol is the formulation of cost metric. Here, a cost function combines different cost metrics into a single value. Each cost metric is weighted with a separate scaling function to

- Maximize route reliability
- Emphasize energy usage by selecting either a route that uses least energy or a route that balances overall network energy usage
- Select a route with most available bandwidth
- Minimize end-to-end delay by choosing shortest-path and next-hop clusters that can be reached fastest

To establish the lowest cost gradient, a node N_i first selects the best next-hop by minimizing the cost C_{ij} that is required to reach the sink through node N_j. C_{ij} is calculated as

$$C_{ij} = C_j + \alpha \cdot e_{ij} + \beta \cdot r_{ij} + \gamma \cdot a_{ij},$$

where C_j is the cost from node j to sink, e_{ij} is the required transmission power to reach node j, r_{ij} is the the packet error rate on the link, a_{ij} is the access delay, and α, β, γ are scaling factors that determine the weight of each cost metric. The transmission power metric is used to minimize energy consumption along the route. Notice that packet error rate affects energy consumption and delays, since a packet sent via an unreliable link is likely to be retransmitted.

Then, node N_i calculates the cost C_i it advertises to its neighbors according to

$$C_i = C_{ij} + \delta \cdot L_i + \varepsilon(E_i),$$

where L_i is the node load, E_i is the remaining energy, δ is the scaling factor, and ε is an exponential scaling function. The remaining energy metric increases the lifetime of a node by increasing the cost for nodes with low energy. Node load is derived from average load (throughput or buffer states) and equalizes energy consumption in cluster heads. Furthermore, node load is used to select the route with the most available bandwidth.

9.4 Fusion-Aware Routing

Different from conventional networks with an ultimate goal of point-to-point (or point-to-multipoint) data forwarding, wireless sensor networks are often deployed to sense, process, and disseminate information of the targeted physical environments, for example, temperature, humidity, and sound. Due to the large number of sensor nodes and hence potentially immense amount of data, it is often impractical to gather all the sensory data from each individual sensor, in particular from the perspective of energy conservation. Therefore, data fusion is often employed as a key strategy to curtail the network load, and hence reduce energy consumption [19].

Performing data fusion in wireless sensor networks can be largely attributed to two reasons. On one hand, the user may be only interested in aggregated results of the sensory data. For example, only average measurement of the interested parameter may be needed. On the other hand, data from sensors in proximity may be highly correlated, and data fusion can effectively reduce the redundancy and hence network load.

Indeed, data fusion has been incorporated into a wide range of existing designs for wireless sensor networks. For example, for query processing in sensor networks, the TinyDB [20] and Cougar projects provide natural support for average, sum, count, and min/max operations in SQL (Structure Query Language)-like language. These supports are further generalized in [21] to include median, consensus value, histogram of the data distribution, and range queries. For routing schemes, directed diffusion [22] was proposed to achieve significant energy savings by allowing intermediate nodes to aggregate data streams. Subsequently, GIT [23] and other routing algorithms are proposed for performance improvement. At the same time, coding strategies targeted at data compression are also extensively explored in conjunction with data fusion for energy saving in gathering sensory data.

Despite the diverse work on data fusion in wireless sensor networks, a fundamental supporting mechanism is the underlying data routing scheme: the routing scheme will dictate when and where data streams will encounter, and hence fusion will be performed, and consequently, the effectiveness of the fusion itself. Joint study on fusion and routing using cross layer design will bring more energy-efficient solutions to wireless sensor network systems.

9.4.1 Routing Sensory Data with Fusion

A data gathering structure in a sensor network is usually a reversed multicast tree rooted at the sink. As sensory data are often correlated due to the physical proximity of corresponding sensor nodes, data fusion can effectively reduce the redundancy embedded therein. As a result, the network load can be curtailed, and hence energy can be saved. Performing data fusion while the packets are being routed in the network naturally is the most desirable strategy. Figure 9.4 illustrates the concept of en route in-network fusion. In this example, data packets from data sources *A* and *B* will be sent to the sink via another source node, *E*, and hence be fused there, while node *F* aggregates the data from sensor nodes *C* and *D*. Since sensory data from nodes *A* and *B* usually are correlated and contain redundant information, the fusion result in the form of an outgoing packet from *E* will be less than the summation of the incoming packets (including its own) in terms of data amount, and bigger than either incoming packet.

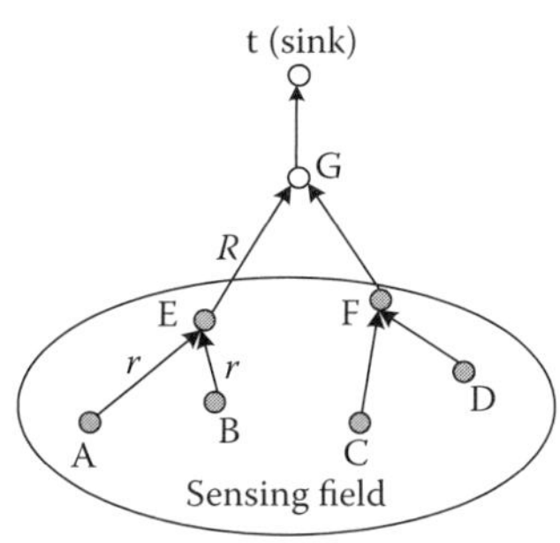

FIGURE 9.4 Illustration of data gathering with data fusion. Each node in the sensing field produces one packet of size r. After data fusion, the outgoing packet size from E satisfies $R < 3r$.

Reflected in the above example, a key factor in data fusion in wireless sensor networks is the routing scheme, as it will determine the incoming streams of a fusion node, the place fusion shall be performed, and ultimately the cost and benefit of the operation.

Different metrics can be used to measure the performance of a routing scheme supporting data fusion. The most important metric is perhaps *energy efficiency*, as the ultimate driving force for in-network data fusion is the reduction in energy consumption via reduced network load. Minimizing the network-wide energy consumption by optimizing the data routes and trading off the computational cost with the communication cost is often deemed the key problem therein. In addition, *scalability*, *delay*, and *adaptivity* should also be of concern. For example, owing to the potentially large number of sensor nodes, distributed routing decision and fusion operation are often preferred in order to scale well with an increasing number of nodes. At the same time, in-network fusion will inevitably introduce delay due to computation at intermediate nodes in the routing scheme. Therefore, a routing scheme with data fusion will need to provide mechanisms for handling time-sensitive information in order to satisfy the latency constraint. At the same time, time variations of the targeted environment and retasking of the network can often significantly affect the performance of the existing routing schemes. A routing scheme shall be capable of dynamically exploring these characteristics and adapt itself to new situations. Two key issues should be considered in aggregation routing design:

- *Aggregation ratio*: In certain special applications such as querying the average temperature, each node can form the outgoing packet by summing up the temperature and the number of reporting nodes. As a result, the outgoing packet size is the same as each incoming packet. We name this class of data fusion *full aggregation*. Similarly, we have two other classes, *partial aggregation* as the general case, and *zero aggregation* if the outgoing data amount equals the summation of all the incoming data.
- *Energy consumption*: Regardless of the amount of data reduction due to data fusion, the basic problem in routing sensory data is to find the best route to benefit both data transmission and data fusion. However, data fusion often is not free and will incur additional computation cost. Hence, the network will have to balance the trade-off in whether to execute data fusion, i.e., whether to trade computation cost with communication cost for network-wide energy conservation.

9.4.2 Classifying Routing Schemes with Data Fusion

Extensive research work has been devoted to providing energy-efficient routing algorithms for correlated data gathering. Depending on the design methodologies, they can be classified into three categories: routing driven, coding driven, and fusion driven.

- *Routing-driven algorithms*: This class of approaches assumes that data fusion can be done at any node without computation cost, and full aggregation is often implicitly assumed. The target of these algorithms is to minimize the total transmission cost for gathering the data. Aggregation occurs opportunistically when routes intersect. For example, in Figure 9.4, if both nodes A and B have chosen node E as their next-hop for routing sensory data toward the sink, their data will be fused at node E. Representative algorithms of this class include directed diffusion [22], LEACH [6], and PEGASIS [24].

- *Coding-driven algorithms*: This class of algorithms focuses on coding techniques at the source nodes and often only assumes partial aggregation is achievable. The premise of these algorithms is that data compression must rely on implicit or explicit side information and compressed packets cannot be fused again while they are being routed. Therefore, the target of these algorithms is efficient source coding strategy and energy-saving routes for raw data compression. Through the compression, it is expected that data amount can be minimized and further in-network fusion can be either completely avoided or dramatically minimized. In Figure 9.4, if the best side information to compress node *B*'s raw data is from node *E*, node *B* will send its raw data to node *E* and data compression will be performed on *E*. The resultant encoded data are then sent to the sink. The same can be done for node *C*'s raw data. However, the encoded data of nodes *B* and *C* will not be combined at node *G* even though they encounter there.
- *Fusion-driven algorithms*: In fusion-driven routing algorithms, routing paths are heavily dependent on data correlation in order to fully benefit from information reduction resulting from data aggregation. The target of these algorithms is efficient routing schemes that can maximally benefit from the fusion operation, and hence the total energy consumption. In contrast to coding-driven algorithms, fusion-driven algorithms often assume that data can be fused more than once along their path, as long as they are deemed beneficial. For example, in the above example, node *D*'s compressed data can be fused again with node *A*'s at *G* although they have been fused respectively at nodes *E* and *F*. Naturally, a different aggregation ratio must be accounted for in the algorithm.

9.4.3 Routing-Driven Routing Schemes

In WSNs, according to the different network architecture, different data aggregation model, and different application scenario, the data gathering structure supporting data aggregation is classified into three types: tree based, hierarchical (cluster based), and chain based. The tree-based structure is a reversed multicast tree rooted at the sink, in which sensors create information paths to the sink. When sensors have data matching the broadcasted query from the sink, they send their data along the path and data aggregation is performed at each intersection of points. In a cluster-based structure, sensors form clusters for energy saving. In each cluster, there is a cluster-head response for cluster management and data collection; all cluster members send their sensory data to their cluster head, and data aggregation is performed at that head. In the chain-based data gathering scheme, sensors form chains along which a node transmits and receives from a nearby neighbor. Aggregation is then performed while data move from node to node. Any data gathering structure has respective routing algorithms that can work together with data aggregation. Overall, routing-driven algorithms emphasize routing with minimum energy consumption and aggregation occurs opportunistically when routes intersect.

9.4.3.1 Directed Diffusion and Greedy Incremental Tree

Directed diffusion [22] is a data-centric and application-aware paradigm. It is data centric (DC) in the sense that all the data generated by sensor nodes are named by attribute-value pairs. DC performs in-network aggregation of data to yield energy-efficient data delivery. All sensor nodes in a directed diffusion-based network are application aware, which enables diffusion to achieve energy savings by selecting empirically good paths and by caching and processing data in the network. The main idea of the DC paradigm is to combine the data coming from different sources en route, for the purpose of eliminating redundancy, minimizing the number of transmissions, and thus saving network energy and prolonging its lifetime.

In the directed diffusion data dissemination paradigm, when the sink wants to query for data, it disseminates an interest, which is a task description, to all sensors, as shown in Figure 9.5(a). The task descriptors are named by assigning attribute-value pairs that describe the task. Each sensor node then stores the interest entry in its cache. The interest entry contains a timestamp field and several gradient fields. As the interest is propagated throughout the sensor network, the gradients from the source back to the sink are set up as shown in Figure 9.5(b). Hence, a gradient specifies an attribute value and a direction. Usually, the sink must refresh and reinforce the interest when it starts to receive data from the source. Then an empirically low delay path is selected to be reinforced. When the source has data for the interest, the source sends the data along the interest's low-delay gradient path, as shown in Figure 9.5(c). When many sources have similar data for the same interest, their data can be aggregated if encountered. The interest and data propagation and aggregation are determined locally.

From the viewpoint of data aggregation, directed diffusion establishes a low-latency aggregation tree rooted at the sink. Data from different sources (sensor nodes that detect phenomena) are opportunistically aggregated. Whenever similar data happen to meet at a branching node in the tree, the copies of similar data are replaced by a single message.

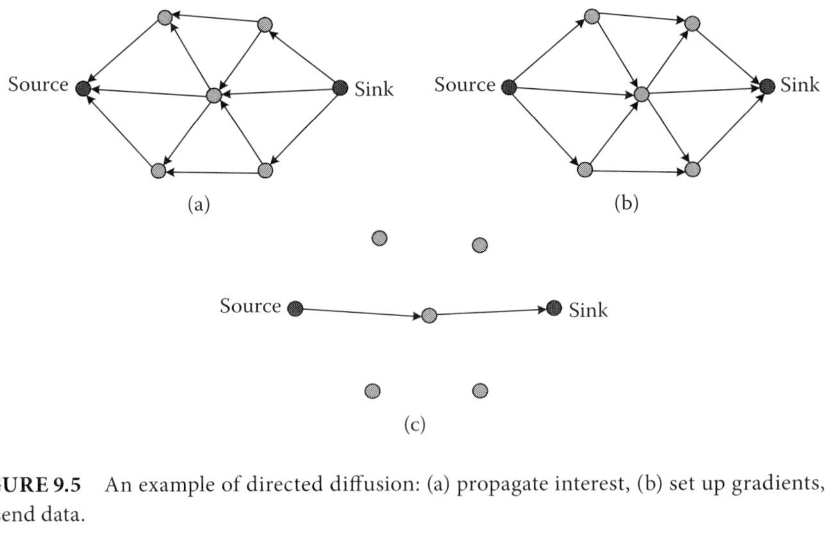

FIGURE 9.5 An example of directed diffusion: (a) propagate interest, (b) set up gradients, and (c) send data.

Directed diffusion establishes paths using path reinforcement, i.e., a node in the network decides to draw data from one or more neighbors in preference to other neighbors based on some metrics, such as the lowest delivery delay or transmission stability.

No matter what kind of metric is used in path reinforcement, opportunistic aggregation on such a tree, for example, a low-latency tree, often is not desirable, as data may not be aggregated or reduced near the places they are originated. This is observed in [23] and a strategy termed greedy incremental tree (GIT) for increased path sharing, and thus increased energy savings is proposed.

To construct a greedy incremental tree, a shortest path is established for only the first source node to the sink, whereas other source nodes are incrementally connected to the closest points on the current tree. Specifically, when a new source node needs to be added, each source node on the current tree generates an on-tree incremental cost message that contains the incremental energy cost to the existing tree required for delivering the new data. The preferred connecting point and path to be reinforced for the new source node is the node that can deliver the new data at the lowest energy cost. Owing to network dynamics, multiple connecting points and multiple paths can be reinforced. For energy efficiency, the unnecessary or inefficient paths will be pruned. The rule for pruning is to negatively reinforce neighbors from which no energy-efficient aggregates have been received.

The greedy-tree approach can achieve significant energy savings at high densities. The experiment results show that greedy aggregation can achieve up to 45% energy savings over opportunistic aggregation exemplified by directed diffusion [23].

9.4.3.2 LEACH

The low-energy adaptive clustering hierarchy (LEACH) described in section 9.2.5 is a cluster-based routing protocol that includes distributed cluster formation. It saves energy consumption not only by shrinking a communication range into a cluster, but by using localized coordination to enable scalability and robustness for dynamic networks and incorporating data fusion into the routing protocol in order to reduce the amount of information that must be transmitted to the base station.

Once clusters are set up, sensing data are then directly sent from cluster members to the cluster heads through direct communication. The data are then fused at the cluster head, and in turn, the aggregation result is sent to the the sink through direct or multihop communication. LEACH is widely used as a routing scheme for data fusion because of its good features, such as clustering architecture and localized coordination. To improve the scalability of LEACH to wide-range sensor networks, multilevel clustering can be used.

Because data collection is centralized and performed periodically, this protocol is most appropriate when constant monitoring by the sensor network is needed. Since cluster heads consume more energy than their members on cluster management and data fusion, to even the distribution of energy load among all sensors in the network, an adaptive clustering strategy is introduced, i.e., reclustering after a given interval with a randomized rotation of the energy-constrained cluster head so that energy dissipation in the sensor network is uniform. The authors also found, based on their simulation model, that only 5% of the nodes need to act as cluster heads.

But in this scheme, data are directly transmitted to cluster heads, and data fusion can only be performed at the cluster head; the data redundancy among neighbor nodes in each cluster is not explored, which indeed can reduce the data amount under transmission, and hence reduce the transmission energy through hop-by-hop transmission/aggregation in each cluster.

9.4.3.3 PEGASIS

PEGASIS is a chain-based protocol that allows only one node (the leader) to transmit to the sink in each round and requires other nodes instead to transmit only to respective nearby neighbors. The key ideas of PEGASIS are chaining and fusion. To construct a chain, each node finds its closest neighbor and forms a chain by employing the greedy algorithm. In each round, a node is chosen randomly to be the leader, and the leader initiates data transmission from the ends of the chain. Each node fuses its neighbor's data with its own to generate a single packet of the same length and then transmits it to its next neighbor. This is repeated until all the sensed data are collected at the leader node, which then transmits one data packet to the sink through direct or multihop communication. PEGASIS outperforms LEACH by eliminating the overhead of dynamic cluster formation, minimizing the sum of distances that nonleader nodes must transmit, and limiting the number of transmissions. To reduce the latency of data gathering, multilevel chaining can be used.

In routing-driven algorithms, it is usually assumed that all data packets can be aggregated to one packet with the same amount of data, i.e., the full-aggregation model is used implicitly here. Obviously, this assumption may not be applicable to a large number of wireless sensor networks.

9.4.4 Coding-Driven Routing Schemes

Coding-driven algorithms focus on compressing raw sensory data locally to reduce data redundancy, and hence reduce the total transmission energy. Different coding models will lead to different routing schemes. The model usually employed is joint entropy coding with implicit or explicit communication.

Theoretically, if complete knowledge of all data correlations is available in advance at each source, i.e., side information is known at each source, the best coding strategy is distributed source coding typified by Slepian-Wolf coding [25]. In this technique, compression is done at the original sources in a distributed manner to achieve the minimum entropy, and hence the need for data aggregation on the intermediate nodes can be completely avoided. A distributed approximation algorithm optimizing the rate allocation for nodes in the network is proposed in [26] and shortest path tree (SPT) is employed as the routing scheme. In [27], the reduction of transmission is achieved by having the sink track the correlation structure among nodes and then using this information to inform the sensor nodes of the number of bits they should use for encoding their measurements. The authors of [28] focus on source coding strategies with explicit side information, in which the raw data can only be compressed at the encoding node with side information from other nodes, and the encoded data are then transmitted to the sink without any further aggregation.

9.4.4.1 Distributed Source Coding: Slepian-Wolf Scheme

In [26], the authors employed distributed source coding, in particular Slepian-Wolf coding, as the model. Under this model, all sources can be coded with a total rate equal to the joint entropy without communication with each other, as long as their individual rates are at least equal to the conditional entropies. For example, if the nodes know in advance the correlation structure (which usually depends on the distance among the nodes), nodes $X_1, X_2, \ldots, X_N$ can code their data jointly with a rate $H(X_1, X_2, \ldots, X_N)$, where H denotes the entropy of the information from these nodes. In this case, SPT is optimal for any rate allocation, and therefore minimizing transmission energy consumption becomes a problem of rate allocation. This way, the optimization problem can be separated into two disjoint problems: a spanning tree construction and corresponding rate allocation.

Specifically, the algorithm for optimal Slepian-Wolf rate allocation can be summarized as follows:

1. Form the SPT from the source nodes to the sink.
2. On the SPT, let node X_1 denote the closest node to the sink and node X_N the furthest one from the sink on the SPT. Assign a rate to each source according to

$$R_1^* = H(X_1),$$

$$R_2^* = H(X_2 \mid X_1),$$

$$R_N^* = H(X_N \mid X_{N-1}, X_{N-2}, \ldots, X_1).$$

   In other words, the closest node to the sink is coded with a rate equal to its unconditioned entropy. Each of the other nodes is coded with a rate equal to its respective entropy conditioned on all other nodes that are closer to the sink than itself.

9.4.4.2 Approximate Distributed Source Coding

Evidently, assuming the availability of global correlation information is unrealistic. In order to use only local information to realize the distributed coding strategy, an approximation algorithm is proposed in [26]. In this algorithm, sensory data are coded locally at the node with a rate equal to the conditional entropy, where the conditioning is performed only on the subset formed by the neighbor nodes that are closer to the sink than the respective node. Therefore, only local information is required, and it can be implemented completely in a distributed manner. The rate allocation process is detailed as follows:

1. Form the SPT from the source nodes to the sink. Index each node according to its distance to the sink on the SPT tree. This will be used to determine which other nodes to condition when computing its rate assignment.
2. For each node i, in its neighborhood $\mathcal{N}(i)$, find the set C_i of nodes that are closer to the sink on the SPT than node i. Assign a rate to node i as $R_i^* = H(X_i|C_i)$.

Still, it provides a solution close to the optimum since the neglected conditioning (due to omitting of nodes farther away) is small in terms of rate for a correlation function that is sufficiently decaying with distance.

However, implementing the approximation algorithm in a practical setting is still an open problem since it demands additional communications at each source in order to obtain the knowledge of data correlations in order to perform the coding.

9.4.4.3 Source Coding with Explicit Side Information: MEGA

In [28], the authors studied the case where the reduction in rate by conditional coding is possible only when side information is explicitly available. The authors distinguished two classes of source coding with explicit side information: self-coding and foreign coding. Using self-coding, data are only allowed to be encoded at the producing node and only in the presence of side information from at least one different node. In contrast, with foreign coding, a node is only able to encode raw data originating from another node using its own data as it is routed toward the sink via itself.

In this approach, a sensor network is modeled as a graph $G = (V, E)$. The weight $w(e)$ is defined to be the cost of transmitting one bit of data on edge e. A raw data packet from node v_i is denoted by p_i with size s_i. If p_i is encoded with side information, the raw data packet p_j from node v_j, the encoded packet is denoted by p_i^j, and its corresponding size is s_i^j. The compression rate depends on the data correlation between the involved nodes v_i and v_j, denoted by the correlation coefficient $\rho_{ij} = 1 - s_i^j/s_i$.

The authors proposed a minimum-energy gathering algorithm (MEGA) for foreign coding. First, MEGA computes for each node v_i a corresponding encoding node v_j. To achieve this, MEGA builds a complete directed graph $\tilde{G} = (V, \tilde{E})$. The weight $\tilde{w}(e)$ for a directed edge $e = (v_i, v_j)$ in $\tilde{E}$ is defined as

$$\tilde{w}(e) = s_i(\sigma(v_i, v_j) + \sigma(v_j, t)(1 - \rho_{ij})), \tag{9.4}$$

where $\sigma(v_i, v_j)$ denotes the weight of a shortest path from v_i to v_j in G. The weight of an edge in $\tilde{G}$ therefore stands for the total energy consumption in order to route a data packet p_i to the sink using node v_j as an encoding relay. Then, a directed minimum spanning tree is constructed rooted at sink t, where each edge (v_i, v_j) represents that v_j is the best encoding node for v_i. Given this construction, raw data are delivered on the shortest path from each node v_i to its encoding relay v_j. After compression, the encoded data are then sent through the shortest path from v_j to t.

Therefore, the resulting topology of the algorithm is a superposition of two tree constructions. A directed minimum spanning tree is used to determine the encoding nodes for all data packets. Once a packet is encoded, it is routed on a shortest path toward the sink in order to save energy.

9.4.4.4 Source Coding with Explicit Side Information: LEGA

In self-coding, nodes can only encode their own raw data in the presence of other raw data routed through them. A low energy gathering algorithm (LEGA), an algorithm based on the shallow light tree (SLT), is proposed for this source coding model in [28]. SLT is a spanning tree that approximates both the minimum spanning tree (MST) and the SPT for a given node (e.g., the sink). The algorithm can be briefly summarized as follows.

Initially, the SLT spanning tree rooted at the sink node t is computed, which satisfies: (1) its total cost is at most $1+\sqrt{2}$ times the cost of the MST of the graph G, and (2) its distance between any node in V and the sink is at most $1+\sqrt{2}$ times the shortest path from that node to the sink. The sink t then broadcasts its raw data packet p_t to all its one-hop neighbor nodes in the SLT. When node v_i receives a raw data packet p_j from a neighboring node v_j, v_i encodes its locally measured data p_i using p_j, and transmits the packet p_i^j to the sink t via the path given by the SLT. Then, node v_i broadcasts its packet p_i to all its one-hop neighbors except v_j, in other words, to all its children but its parent in the SLT. The sink t has its own data p_t available locally (or it can use the data of one of its first-hop neighbors), and thus can perform recursive decoding of the gathered data, based on the encoded data that it has received from all other nodes in V. It has been proven that LEGA can achieve a $2(1+\sqrt{2})$ approximation of the optimal data gathering topology [28].

Although both MEGA and LEGA can achieve near-optimal performance under the source coding model with explicit side information, their performance is subject to high penalty in dense networks where adjacent data have high redundancy, as one source data can only be encoded once, and its data redundancy with other nodes cannot be eliminated.

9.4.5 Fusion-Driven Routing Schemes

Fusion-driven routing schemes usually are based on tree structures. In the process of data gathering, each leaf node transmits data to its parent. Once a parent receives data from all its children, it aggregates them and transmits the aggregated data to its own parent. This way, sensory data are aggregated along the routing path toward the sink.

In [26], the authors proved that the minimum-energy data gathering problem is NP-complete by applying a reduction set cover problem and claimed that the optimal result is between SPT and the traveling salesman path. To get the suboptimal solution, in [29], database queries are classified based on the level of aggregation, and different routing structures for different types of database queries were proposed. A cluster-based heuristic algorithm for gathering and aggregating correlated data is proposed in [30], where data are aggregated along the data gathering tree in each cluster. For prolonging the system lifetime, a class of data gathering trees is constructed and used for this purpose. In [31], a hierarchical matching algorithm is proposed, resulting in an aggregation tree with a logarithmic approximation ratio to the optimal for all concave aggregation functions. However, in this model, aggregation only depends on the number of nodes in the subtree rooted at the aggregation node, regardless of the correlation among the data. In [32], the authors took both transmission cost and aggregation cost into account and proposed a heuristic algorithm, Adaptive Fusion Steiner Tree (AFST). This algorithm can dynamically assign fusion decisions to routing nodes during the route construction process by evaluating whether fusion is beneficial to the network based on fusion/transmission costs and network/data structures.

9.4.5.1 Aggregation Model-Aware Routing Structure

In [29], the authors discussed the optimal routing schemes for data query. Since processing different types of query corresponds to different aggregation models—full aggregation, partial aggregation, or zero aggregation—the strategy is then to construct different routing trees for different aggregation models so as to maximize the system lifetime.

For full-aggregation queries, each node has the same transmission energy consumption, and its receiving energy is proportional to the amount of incoming data, which is dependent on the number of neighbors in the routing tree. To maximize the network lifetime, a node should have as few neighbors as possible to minimize power consumption. For this, minimum degree spanning tree (MDST) is the best solution. However, the specific numbers of neighbors can be determined according to their remaining energy in order to further extend the network lifetime. Therefore, the authors proposed to reduce the general routing problem to the MDST problem first, and then solve the MDST problem using approximation algorithms. The transformation process is described as follows:

1. Let T denote the system lifetime of an N nodes network; each node remains energy e_i. Then the maximum number of neighbors that a node i can have is

$$B_i = \left\lfloor 1 + \frac{1}{c_r}\left(\frac{e_i}{T} - 1\right)\right\rfloor. \tag{9.5}$$

   Here, it is assumed that transmitting one unit of data costs one unit of energy, while receiving one unit of data costs $c_r < 1$ amount of energy.
2. For every node i in the original graph, introduce $N - B_i$ auxiliary nodes and connect them to node i.
3. Construct the MDST on this graph and then delete the auxiliary nodes from the resultant MDST. The resulting spanning tree is the solution with a degree at most B_i for every node i.

It is proven that the approximation scheme is within a constant factor $(1 + c_r)$ of the optimal routing tree [29].

For partial-aggregation and zero-aggregation queries, the authors proposed a heuristic algorithm for computing an approximate routing tree. In this solution, the initial routing tree only contains the root, and the tree grows by adding edges until the tree spans all the vertices. The criterion of the newly added edge is to maximize the lifetime of the resulting new tree, which can be determined by calculating the amount of data each node will be required to forward. If multiple edges exist that will result in the same lifetime, the edge connected to the node with maximum energy will be selected. After getting this approximate routing tree, some improvement to the tree can be done by switching the parent of each node if the switching can prolong the system lifetime.

Using different routing structures for different queries can achieve better performance. However, the additional overhead incurred for route reconstruction when query type changes may be costly. Therefore, this scheme is in particular unsuitable in multitask sensor networks where the user interest, and hence query type, changes often.

9.4.5.2 Cluster-Based Maximum Lifetime Data Aggregation

In [30], the authors focus on maximizing the lifetime of data gathering for fully aggregated queries in large-scale sensor networks. Here, lifetime T of a system is defined as the number of data gathering rounds until the first sensor is drained of its energy.

In their model, each node i has energy ε_i and sends a data packet with k bits each round.

A *data gathering schedule* specifies, for each round, how the data packets from all the sensors are collected and transmitted to the sink. Observe that a schedule can be thought of as a collection of T directed trees, each rooted at the sink and spanning all the sensors, i.e., a schedule has one tree for each round. Each tree specifies how data packets are gathered and transmitted to the sink during a round of data gathering. Thus, these trees are called data aggregation trees. The lifetime of a schedule equals the lifetime of the system under that schedule. The heuristic Clustering-Based Maximum Lifetime Data Aggregation schedule (CMLDA) is detailed as follows.

Consider a sensor network having n nodes and one sink. Each sensor has initial energy ε. The energy consumed by a sensor i in receiving a k-bit data packet is given by RX_i, while the energy consumed in transmitting a data packet to sensor j is given by $TX_{i,j}$. Let the sensors be partitioned into m clusters $\phi_1,\ldots\phi_m$; each cluster is referred to as a super-sensor. The approach is to compute a maximum lifetime schedule for the super-sensors $\phi_1,\ldots\phi_m$ with the sink ϕ_{m+1}, and then use this schedule to construct aggregation trees for the sensors.

First, let the energy of each super-sensor ε_{ϕ_i} be the total initial energy of all sensors in ϕ_i, and let the distance between two super-sensors ϕ_i and ϕ_j be the maximum distance between any two nodes $u \in \phi_i$ and $v \in \phi_j$. Find an admissible flow network G for super-sensors $\phi_1,\ldots\phi_m$ with sink ϕ_{m+1}, and then compute a schedule consisting of a collection of aggregation super-trees $T_1, \ldots, T_k$, each rooted at ϕ_{m+1} and spanning over all the super-sensors. Next, a travel of the aggregation super-tree T_k is performed to construct the corresponding aggregation tree $\mathcal{A}$. For each visited super-sensor ϕ, add the edge (i, j), where $i \in \phi - \mathcal{A}$ and $j \in \mathcal{A}$, to $\mathcal{A}$ if pair (i, j) has maximum residual energy, defined as $min\{\varepsilon_i - TX_{i,j}, \varepsilon_j - RX_j\}$. The process is repeated until all sensors in ϕ are included in $\mathcal{A}$, upon which it continues with the next super-sensor in T_k.

CMLDA achieves a maximum lifetime schedule of the sensor networks, but it needs precomputation at the sink to get all aggregation trees and announce to all nodes.

9.4.5.3 Hierarchical Matching Algorithm

In [31], the authors used concave, nondecreasing cost functions to model the aggregation function applied at intermediate nodes. In a unified way, this model covers all three aggregation models, full aggregation, partial aggregation, and zero aggregation, discussed in section 9.4.2. To minimize the total energy consumption on data gathering, a hierarchical matching algorithm is proposed, resulting in an aggregation tree with a logarithmic approximation ratio to the optimal. The algorithm is detailed as follows:

1. Given k data sources, the initial source set S is constructed to include all k source nodes, and the initial aggregation tree is set to $T = \phi$.
2. The hierarchial matching algorithm runs in log$k$ phases. In each phase, initially a minimum-cost perfect matching in the subgraph induced by S is sought, and

the edges connecting each matched pair (u_i,v_i) are added into T. After this, for all matched pairs (u_i,v_i), one node out of u_i and v_i is chosen with equal probability, and it is removed from S. As a result, the size of S gets halved after each phase, and thus $|S| = 1$ after $\log k$ phases.

3. Add the edge connecting the single remaining element in S to the sink into T, and T is the resultant aggregation tree.

In this model, aggregation function only depends on the number of nodes providing data to the aggregation node regardless of the correlation among the available data. This assumption results in the disregard of the impact of data correlation in the solution.

9.4.6 Optimizing over Both Transmission and Fusion Costs

All algorithms discussed above have assumed the fusion operation function itself is free, and hence have neglected the fusion cost in the routing design procedure. Indeed, the cost for data aggregation may be negligible for certain types of networks. For example, sensor networks monitoring field temperature may use simple average, max, or min functions, which essentially cost nothing. However, other networks may require complex operations for data fusion. One example is hop-by-hop secure networks where encryption and decryption at intermediate nodes will significantly augment fusion cost even though the fusion function itself may be simple. It has been shown in [33] that energy consumption of a beam-forming algorithm for acoustic signal fusion is on the same order of that for data transmission. Moveover, in the experimental study described in [32], it is found that a typical aggregation function for vectorial data, such as image fusion, costs tens of nanojoules (nJ) per bit, which is on the same order as the communication cost reported in the literature [33]. In the case where the fusion cost is not negligible, it will significantly influence the routing structure and fusion decisions. This is illustrated in Figure 9.6.

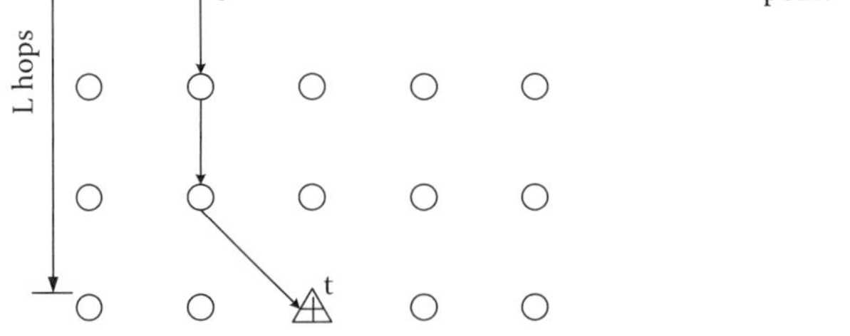

FIGURE 9.6 Illustration of fusion benefit, or disadvantage, in sensor networks.

In this example, sensor nodes are deployed on a grid, and sensed information of the source nodes is to be routed to sink t. Arrow lines form the aggregation tree in which nodes u and v initially aggregate data of areas A and B, respectively. As the sink is far away, u and v further aggregate their data at v and then send one fused data to the sink. Assume each hop has identical unit transmission cost c_0, the fusion cost is linear to the total amount of incoming data, and the unit fusion cost is q_0. Let $w(u)$ and $w(v)$ respectively denote the amount of data at u and v before the aggregation between u and v. The amount of resultant aggregated data at v can be expressed as $(w(u) + w(v))(1 - \sigma_{uv})$, where σ_{uv} represents the data reduction ratio owing to aggregation. In this scenario, if v performs data fusion, the total energy consumption of the route from v to t is given by

$$C_1 = L \cdot c_0(w(u)+w(v))(1-\sigma_{uv})+q_0(w(u)+w(v)). \tag{9.6}$$

On the contrary, if v does not perform data fusion, the total energy consumption of the network is simply the total relaying cost:

$$C_2 = L \cdot c_0(w(u)+w(v)). \tag{9.7}$$

Comparing equations (9.6) and (9.7), to minimize network-wide energy consumption, v should perform data fusion as long as $\sigma_{uv} \geq \frac{q_0}{Lc_0}$ to keep $C_1 < C_2$.

This simple example reveals that to minimize total network energy consumption, the fusion decision at an individual node has to be done based on the data reduction ratio due to aggregation, its related cost, and its effect on the communication costs at the succeeding nodes.

In [32], an algorithm optimized on both transmission and fusion costs was constructed. The proposed Adaptive Fusion Steiner Tree (AFST) also belongs to a fusion-driven routing scheme. However, the introduction of fusion cost additionally dictates that the algorithm also adaptively adjust its fusion decisions for sensor nodes along the route. Specifically, data fusion on intermediate nodes will automatically stop in AFST if the reduction of communication cost due to aggregation cannot overshadow the fusion cost. By evaluating whether fusion is beneficial to the network based on fusion/transmission costs and network/data structures, AFST dynamically assigns fusion decisions to routing nodes during the route construction process.

AFST is a hierarchical matching algorithm and also runs in phases. In each phase, three steps—matching, decision, and fusion—are performed. Comparing with the hierarchical algorithm discussed in the previous section, AFST defines a new metric employing data correlation explicitly for minimum-cost perfect matching, and introduces the fusion benefit calculation to determine whether fusion shall be performed.

To better present the algorithm, we first present a few notations. In a sensor network $G = (V, E)$, assume each source node v has weight w_v, denoting the data amount out going from it, and the data correlation between any two nodes, u and v, is denoted by data aggregation ratio σ_{uv}. Each link $e = (u, v)$ can be characterized by the unit transmission cost $c(e)$ and unit fusion cost ω. If node u uses node v as its fusion point, the fusion benefit is defined as the energy saving on transmitting u's additional data, $w(u)\sigma_{uv}$, on the shortest path from v to sink t, minus the energy consumption on fusing u and v's data. Let the initial source set S_i,($i = 0$) contain all the source nodes and the sink t. The algorithm is executed recursively on this source set in three steps:

- *The matching step*: In this step, the total cost on each edge $e = (u, v)$, denoted by $M(e)$, is calculated, which includes the cost of fusing u and v's data and the cost of transmitting data from u to v. Then, minimum-cost perfect matching is performed among the nodes in S_i according to the metric $M(e)$.
- *The decision step*: In this step, for each matched pair (u, v), if there is no fusion benefit, regardless of which node is selected as the fusion point, both nodes' data will be routed to the sink through SPT without any further fusion. Consequently, both nodes u and v will be removed from S_i. Otherwise, node pair (u, v) is deemed to be a fusion pair.
- *The fusion step*: In this step, for each fusion pair (u, v), one node is randomly chosen to be the fusion node. Subsequently, the nonfusion node will transfer its data to its corresponding fusion node. After that, all nonfusion nodes are removed from S_i and all remaining fusion nodes induce S_{i+1}.

The algorithm will then be reexecuted on S_{i+1}, until only the sink is left. In each run of the above algorithm, the size of S_i is reduced by at least half. More importantly, the resultant routing structure from AFST is partitioned into two parts: a lower part where aggregation is always performed, and an upper part where no aggregation occurs. This can be naturally employed to be the theoretical foundation of node clustering: within each cluster, fusion shall be performed; outside the clusters, data shall be directly transmitted to the sink without any further processing.

9.5 Information-Aware Routing

The information embedded in the data gathered from different parts of the sensing field may vary according to the monitored process and sensor deployment. According to the difference of information embedded in the sensing field, routing protocols can use different strategies when delivering different information to meet the system requirements and save energy simultaneously.

9.5.1 SPIN

A family of adaptive protocols, called Sensor Protocols for Information via Negotiation (SPIN), is proposed in [34]. These protocols disseminate all the information at each node to every node in the network, assuming that all nodes in the network are potential sinks. This enables a user to query any node and get the required information immediately. To reduce overhead, these protocols make use of the property that nearby nodes have similar data and thus distribute only data that other nodes do not have.

SPIN protocols are motivated by the observation that conventional protocols like flooding or gossiping waste energy and bandwidth by sending extra and unnecessary

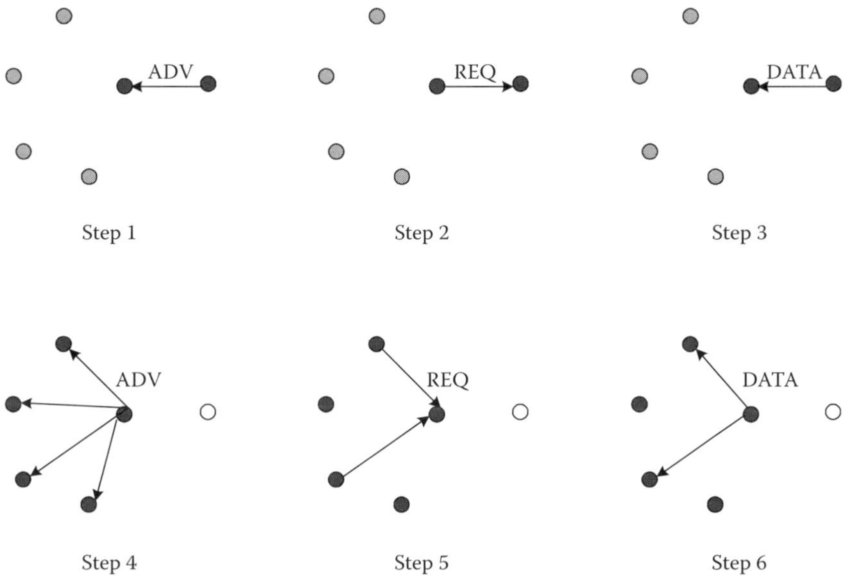

FIGURE 9.7 The SPIN protocols.

copies of data from sensors covering overlapping areas. Thus, nodes running SPIN assign metadata to completely describe their collected data. Since sensor nodes can operate more efficiently and conserve more energy by sending metadata instead of sending all the data, the SPIN family of protocols uses metadata negotiation before any data are transmitted. These protocols work in a time-driven fashion and distribute the information over the network, even when a user does not request any data.

SPIN has three types of messages: ADV, REQ, and DATA. ADV advertises new data, REQ requests data, and DATA is the actual message carrying the data. When a node obtains new data that it is willing to share, it broadcasts an ADV message containing metadata of the data, as shown in step 1 of Figure 9.7. If a neighbor is interested in the data, it sends a REQ message for the DATA and the DATA is sent to this neighbor node, as shown in steps 2 and 3 of Figure 9.7, respectively. The neighbor sensor node then repeats this process with its neighbors, as illustrated in steps 4–6 of Figure 9.7. As a result, sensor nodes in the entire sensor network, which are interested in the data, will get a copy.

SPIN protocols are well suited for an environment in which the sensors are mobile because they base their forwarding decisions on local neighborhood information. But the nodes around a sink could deplete their battery quickly if the sink is interested in too many events.

9.5.2 ReInForm

Consider a case where a temperature sensor in a forest senses 60°F on a normal spring day. On the other hand, another sensor senses a temperature of 1,000°F at the same time.

From the application's point of view, the packet containing the 1,000°F temperature is much more important, as it could indicate a fire and hence should reach the sink with high reliability and low latency. Thus, information-aware forwarding is demanded for such applications of sensor networks.

ReInForM, proposed in [35], provides a method of information-aware data delivery in sensor networks that adapts a multipath approach based on local channel error rates to achieve the data delivery at desired reliability. Desired transmission reliability is based on the information importance level. To attain desired reliability, ReInForM uses redundant copies of a packet to increase its end-to-end probability of data delivery. In other words, it sends multiple copies of each packet along multiple paths from the source to the sink. The degree of redundancy introduced is controlled using the desired reliability, the local channel error conditions, and neighborhood information available at each node. The main idea of ReInForm is described below.

In this scheme, the sink periodically broadcasts a routing update packet that is flooded throughout the entire network, and each node becomes aware of its neighbors and the hops to the sink. When a source generates a packet, it identifies the importance level of the packet and finds the corresponding reliability level r. Then the source will create N paths to the sink (N is a function of r). Afterwards, N copies are sent from the source to the selected N neighbors. The next-hop will determine the operation on the packet again, i.e., to discard it or create M multipaths to forward the received packet (M is a function of r and the channel error rate on the sender). The procedure will continue until the data packet reaches the sink.

9.6 Conclusions

Wireless sensor networks have revealed their vast potential in a plethora of applications. Due to the inherent resource constraints and uncertainty, wireless sensor networks demand adaptive techniques that can achieve high energy efficiency while accommodating the high dynamics of the system and environments. This chapter offers a comprehensive overview of the adaptive routing techniques proposed in the literature. To adapt to the system dynamics, these techniques have jointly considered information, including node residual energy, link condition and modulation, data fusion strategy, and application requirements in routing decisions. In particular, we have classified and discussed energy-aware, channel-aware, fusion-aware, and information-aware adaptive routing protocols, with a focus on their design philosophy, advantages, drawbacks, and performance.

Acknowledgments

The work was partially completed during the time Hong Luo was a visiting scholar at the University of Texas at Arlington. This work is partially supported by NSF grants CNS-0721951 and IIS-0326505, and Texas ARP grant 14-748779.

References

[1] C. Chong and S. Kumar. 2003. Sensor networks: Evolution, opportunities, and challenges. *Proc. IEEE* 91:1247–56.

[2] I. Akyildiz, W. Su, Y. Sankarasubramaniam, and E. Cayirci. 2002. A survey on sensor networks. *IEEE Commun. Mag.* 40:102–14.

[3] A. A. Ahmed, H. Shi, and Y. Shang. 2003. A survey on network protocols for wireless sensor networks. In *Proceedings of the IEEE International Conference on Information Technology: Research and Education*, Newark, NJ, pp. 301–305.

[4] C.-K. Toh. 2001. Maximum battery life routing to support ubiquitous mobile computing in wireless adhoc networks. *IEEE Commun. Mag. 39*:138–147.

[5] Q. Li, J. Aslam, and D. Rus. 2001. Online power-aware routing in wireless ad-hoc networks. In *Proceedings of the ACM MOBICOM*, Rome, pp. 97–107.

[6] W. R. Heinzelman, A. Chandrakasan, and H. Balakrishnan. 2000. Energy-efficient communication protocol for wireless microsensor networks. In *Proceedings of the 33rd Annual Hawaii International Conference on System Sciences*, Maui, pp. 1–10.

[7] W. Heinzelman, A. Chandrakasan, and H. Balakrishnan. 2002. An application-specific protocol architecture for wireless microsensor networks. *IEEE Trans. Wireless Commun.* 1:660–70.

[8] O. Younis and S. Fahmy. 2004. HEED: A hybrid energy efficient distributed clustering approach for ad hoc sensor networks. *IEEE Trans. Mobile Comput.* 3:366–79.

[9] J. H. Chang and L. Tassiulas. 2004. Maximum lifetime routing in wireless sensor networks. In *IEEE/ACM Trans. Networking* 12:609–619.

[10] R. C. Shah and J. M. Rabaey. 2002. Energy aware routing for low energy ad hoc sensor networks. In *Proceedings of the IEEE Wireless Communications and Networking Conference*, Orlando, FL, pp. 350–355.

[11] S. J. Lee and M. Gerla. 2000. AODV-BR: Backup routing in ad hoc networks. In *Proceedings of the IEEE International Conference on Wireless Communication Networks*, Chicago, pp. 1311–16.

[12] R. Veronesi, M. D. Pozzo, V. Tralli, and A. Giovanardi. 2005. Energy efficient forwarding strategies for wireless sensor networks in presence of fading and power control. In *Proceedings of the 16th IEEE International Symposium on Personal, Indoor and Mobile Radio Communications*, Berlin, Germany, pp. 1383–1388.

[13] M. R. Souryal, B. R. Vojcic, and R. L. Pickholtz. 2005. Information efficiency of multihop packet radio networks with channel-adaptive routing. *IEEE J. Select. Areas Commun.* 23:40–50.

[14] L. Ma, Q. Zhang, Y. Xiong, and W. Zhu. 2005. Interference aware metric for dense multi-hop wireless networks. In *Proceedings of the IEEE International Conference on Communications (ICC) 2005*, Seoul, Korea, pp. 1261–1265.

[15] V. Lau. 2002. Performance analysis of variable rate symbol-by-symbol adaptive bit-interleaved coded modulation for Rayleigh fading channels. *IEEE Trans. Vehicular Technol.* 51:537–50.

[16] X. Lin, Y. Kwok, and V. Lau. 2005. A quantitative comparison of ad hoc routing protocols with and without channel adaptation. *IEEE Trans. Mobile Comput.* 4:112–28.

[17] J. Xu, B. Peric, and B. R. Vojcic. 2005. Energy-aware and link-adaptive routing metrics for ultra wideband sensor networks. In *Proceedings of the 2nd International Workshop on Networking with UWB 2005*, Rome, pp. 1–8.

[18] J. Suhonen, M. Kuorilchto, M. Hannikainen, and T. D. Hamalainen. 2006. Cost-aware dynamic routing protocol for wireless sensor networks design and prototype experiments. In *Proceedings of the 17th Annual IEEE International Symposium on Personal, Indoor and Mobile Radio Communications*, Helsinki, Finland, pp. 1–5.

[19] B. Krishnamachari, D. Estrin, and S. Wicker. 2002. Impact of data aggregation in wireless sensor networks. In *Proceedings of the 22nd International Conference on Distributed Computing Systems*, Vienna, pp. 575–78.

[20] B. J. Bonfils and P. Bonnet. 2003. Adaptive and decentralized operator placement for in-network query processing. In *Proceedings of the 2nd International Workshop on Information Processing in Sensor Networks*, pp. 47–62.

[21] N. Shrivastava, C. Buragohain, S. Suri, and D. Agrawal. 2004. Medians and beyond: New aggregation techniques for sensor networks. In *Proceedings of ACM Sensys'04*, pp. 239–249.

[22] C. Intanagonwiwat, R. Govindan, D. Estrin, J. Heidemann, and F. Silva. 2003. Directed diffusion for wireless sensor networking. *IEEE/ACM Trans. Networking* 11:2–16.

[23] C. Intanagonwiwat, D. Estrin, R. Govindan, and J. Heidemann. 2002. Impact of network density on data aggregation in wireless sensor networks. In *Proceedings of the 22nd International Conference on Distributed Computing Systems*, Vienna, Austria, pp. 457–458.

[24] S. Lindsey and C. S. Raghavendra. 2002. Pegasis: Power-efficient gathering in sensor information systems. In *Proceedings of the IEEE Aerospace Conference*, Big Sky, MT, pp. 1125–1130.

[25] S. S. Pradhan and K. Ramchandran. 2003. Distributed source coding using syndromes (DISCUS): Design and construction. *IEEE Trans. Inform. Theory* 49:626–43.

[26] R. Cristescu, B. Beferull-Lozano, and M. Vetterli. 2004. On network correlated data gathering. In *Proceedings of IEEE INFOCOM*, Hong Kong, China, pp. 2571–2582.

[27] J. Chou, D. Petrovic, and K. Ramchandran. 2003. A distributed and adaptive signal processing approach to reducing energy consumption in sensor networks. In *Proceedings of IEEE INFOCOM*, San Franciso, CA, pp. 1054–1062.

[28] P. V. Rickenbach and R. Wattenhofer. 2004. Gathering correlated data in sensor networks. In *Proceedings of ACM Joint Workshop on Foundations of Mobile Computing*, Philadelphia, PA, pp. 60–66.

[29] C. Buragohain, D. Agrawal, and S. Suri. 2005. Power aware routing for sensor databases. In *Proceedings of IEEE INFOCOM*, Miami, FL, pp. 1747–1757.

[30] K. Dasgupta, K. Kalpakis, and P. Namjoshi. 2003. An efficient clustering-based heuristic for data gathering and aggregation in sensor networks. In *Proceedings of the IEEE Wireless Communications and Networking Conference*, New Orleans, LA, pp. 1948–1953.

[31] A. Goel and D. Estrin. 2003. Simultaneous optimization for concave costs: Single sink aggregation or single source buy-at-bulk. In *Proceedings of ACM-SIAM Symposium on Discrete Algorithms*, Baltimore, MD, pp. 499–505.

[32] H. Luo, J. Luo, Y. Liu, and S. K. Das. 2006. Adaptive data fusion for energy efficient routing in wireless sensor networks. *IEEE Trans. Comput.* 55:1286–99.

[33] A. Wang, W. B. Heinzelman, A. Sinha, and A. P. Chandrakasan. 2001. Energy-scalable protocols for battery-operated microsensor networks. *J. VLSI Signal Processing* 29:223–37.

[34] W. R. Heinzelman, J. Kulik, and H. Balakrishnan. 1999. Adaptive protocols for information dissemination in wireless sensor networks. In *Proceedings of the ACM MobiCom99*, Seattle, WA, pp. 174–185.

[35] B. Deb, S. Bhatnagar, and B. Nath. 2003. ReInForM: Reliable information forwarding using multiple paths in sensor networks. In *Proceedings of the 28th Annual IEEE International Conference on Local Computer Networks*, Bonn, Germany, pp. 406–415.

10

Coverage and Connectivity in Wireless Sensor Networks: *Lifetime Maximization*

Ananthram Swami
U.S. Army Research Laboratory

Qing Zhao
University of California

10.1 Introduction

A ubiquitous problem in wireless networks is that of providing connectivity and coverage. The problem of positioning cellular base stations has received considerable attention over the last two decades given its practical importance; see, e.g., [1–4]. Here, one wants to find the minimum number of base stations required to ensure continuity of coverage as a mobile moves within a given area. The emphasis is on maintenance of connectivity as the mobile moves from the domain of one base station to that of another. In a peer-to-peer network, the radios need to provide multihop connectivity. Here the emphasis is on finding radio deployments that would ensure that a given area is covered with high probability (whp). Interesting questions that have been posed and answered here include the following: What is the minimum density of nodes required to ensure connectivity given a fixed footprint covered by a node? Alternatively, if the density is fixed, what is the critical transmission power required to maintain connectivity?

Wireless sensor networks (WSNs) are increasingly finding applications in diverse areas such as battlefield surveillance, biohazard detection, habitat monitoring, area surveillance (e.g., to detect the onset of forest fires), and area denial (e.g., landmine replacement); e.g., see [5, 6]. A WSN provides specific services, and the performance of the network can be characterized in terms of the quality of service (QoS) that it provides. In a sniper detection network or in a forest fire detection network, QoS would be characterized by the time to detection and the accuracy of localization. Both depend upon the density and quality of sensors deployed, as well as on the communications network. In other target detection/tracking applications, we may be interested in optimal allocation of detection and tracking assets. Reporting from a WSN may be event driven, i.e., initiated by one or more of the sensors in response to a detection, or it could be query driven, i.e., initiated by the fusion center (FC) or monitoring station. In a third class of networks, it may be important for the FC to keep track of the environment all the time. In such a case, sensing and reporting take place continuously; we will discuss this in detail in the context of field (or two-dimensional signal) reconstruction. The sensing and communications ranges of nodes can be very different; for example, a cheap acoustic sensor may have a sensing range of 10 m and a magnetometer a range of 5 m; a Berkeley mote may have a communications range of 150–300 m. It is important to distinguish between the sensing and communications networks and the interplay between the two, in particular the spatial and temporal scales of correlation.

Wireless sensors in these scenarios are battery powered, and replacing or recharging batteries may be infeasible or impossible in remote or hostile terrain. Thus, minimizing the energy consumption is critical to maximizing the lifetime of the network. Lifetime can be defined as the length of time over which the network provides the desired QoS. If the sensors are cheap, the deployment could be dense, and since signals tend to be spatially correlated, we want to keep the active network as sparse as possible by duty cycling the nodes. Sensor management should take into account the possible redundancy in the deployed network versus the required QoS. Because of redundancy, multiple sensors might provide coverage of a given region; turning off all sensors but one would lead to energy savings and increase the sensing lifetime for that region. A further advantage of duty cycling is the recharging effect on batteries; taking this into account in a scheduling policy can further extend the WSN's lifetime [7, 8].

An interesting emerging application is in intelligent monitoring of people and baggage in airports [9]. The potentially powerful fixed infrastructure must cope with a high density of (low-mobility) RFID nodes and strict latency constraints. There is also an increasing interest in the passive monitoring of information flows at multiple locations, both in wired as well as in wireless networks. Here the issues are optimal placement of and optimal sampling at the monitors, subject to constraints on resource consumption, resilience to node and link failures, and dynamic changes in routing [10]. Passive monitoring may be useful for intrusion detection in a wireless network, as well as to monitor the health of the network, and to predict critical events such as congestion or resource depletion.

10.1.1 Sensing Models

Given the wide variety of sensing modalities available, the different environments in which these sensors might be deployed, and the diversity of potential applications, it is important to have some generic models of sensor coverage so as to provide general results. A typical model of sensor coverage is the disk model; any event that occurs within a distance R of the sensor is detected perfectly, anything beyond that range is not detected. This is a threshold model and often called the Boolean sensing model. More generally, one could associate a sensing footprint with a sensor; any event that occurs in the area $\mathcal{A}(s_i)$ can be detected by the i-th sensor. For example with a directional sensor, $\mathcal{A}$ may be a cone, whose radius and angular coverage depend upon sensor characteristics.

The disk model has also been used in the context of estimation. Consider a spatially stationary isotropic random process, $u(\boldsymbol{x})$, where $\boldsymbol{x}$ denotes the two-dimensional spatial variable, with mean μ, variance σ^2, and normalized correlation denoted by $R(d)$, where d denotes separation distance. We assume that the correlation is nonnegative and monotone nonincreasing; such assumptions hold for many spatial processes of interest. Suppose that we use the signal at the i-th sensor, say, $u(s_i)$, as an estimate, $\hat{u}(x)$, of the signal at any point x in the sensor's covering disk. How large can the radius of the disk be to ensure that the resulting distortion is below a threshold? For the mean-squared error (MSE) metric, this question was answered in [11]. The QoS requirement is characterized by the maximum distortion D in MSE:

$$E\left[\left(\hat{u}(\mathbf{x})-u(\mathbf{x})\right)^2\right]\le D\sigma^2, \quad \forall \mathbf{x}\in \mathcal{D}. \tag{10.1}$$

Let us define

$$r \triangleq \max\left(\left\{d : R(d)\ge 1-\frac{D}{2}, \quad d\in[0,d_{\max}]\right\}\right) \tag{10.2}$$

This value of r can be considered to be the sensing range of the node. To ensure signal reconstruction within the distortion constraint, every point in $\mathcal{D}$ must be no more than distance r from some sensor. As an example, if the correlation is exponentially decaying, $R(d) = \exp(-\lambda d)$, then $r = -\log(1 - D/2)/\lambda$, and for small distortion D, the desired sensing range $r \approx D/2\lambda$.

The FC could exploit the known correlation function to construct a linear predictor. It is easy to show that the corresponding sensing range would be

$$r_p \triangleq \max\left(\left\{d : R^2(d)\ge 1-D, \quad d\in[0,d_{\max}]\right\}\right). \tag{10.3}$$

As expected, $r_p \geq r$, i.e., the sensing range is larger if the FC exploits the signal correlation. This could be improved even further if the FC were to use the conditional expectation to estimate the signal. Note also that further reduction in distortion is possible if multiple sensors cover a point (we discuss *k*-connectivity later) so that interpolation/extrapolation techniques can be used. The two-dimensional spatial signal is band limited: one can also view this problem in the context of random/irregular sampling, an approach that may be particularly useful for asymptotic analysis. We stress that the sensing range thus depends not only upon the inherent properties of the sensor (e.g., noise figure), but also upon the sophistication of the algorithms implemented at the FC.

Typically, signals decay with distance; so a more realistic sensing model is to assume that the signal received from an event (or target) at a distance *d* from the sensor is of the form

$$u(d)=\alpha\Big/\left(d^{\beta}+\gamma\right), \tag{10.4}$$

where $\alpha > 0$, $\beta > 0$, and $\gamma > 0$ can represent factors due to technology and sensing modality. This is, of course, the typical RF path loss model.

The sensing range of a sensor depends upon the process being monitored, the modality (e.g., acoustic, seismic, infrared [IR], thermal), the quality (e.g., noise figure, processing capability, resolution) of the sensor, and the characteristics of the sensing environment (e.g., absorption, scattering, multipath fading). The sensing range is best characterized by a footprint that is typically irregular, and which may be expected to show statistical variability even within a suite of identically produced sensors. Characterization and utilization of these footprints is often difficult; disk models provide us with tractable models that can provide good insights.

10.2 Survey of Coverage and Connectivity

10.2.1 Definition of Coverage

A point in $\mathcal{D}$ is said to be covered if it lies within the sensing range of at least one sensor. Coverage could be deterministic or probabilistic. The simplest question is: Does a collection of sensing coverage footprints cover the domain $\mathcal{D}$? A *cover* is a subset of the sensors that provides complete coverage of the region,

$$\bigcup_i A(s_i)\supset \mathcal{D}.$$

A cover ensures that every point in the region of interest is within the sensing range of at least one sensor. We are interested in finding the *minimal* cover, i.e., the smallest subset of sensors that provides coverage. Given a deployment of sensors, the problem of finding the maximum number of disjoint covers is NP-complete [12]. This maximum number is an indication of the lifetime of the network (we will discuss this further in section 10.3). Given a desired QoS, how should we optimally position nodes? Given that both sensing and communications consume energy, how should one schedule these operations in a network to maximize lifetime while maintaining the desired QoS? We will cover these topics in this chapter.

There are even more interesting problems when the sensors can move, e.g., in a robotic network. How should a single robot or a group of robots move so as to obtain complete coverage of a building? How should robots equipped with acoustic sensors move so as to localize a sound source?

In this chapter, we provide a survey of results on the lifetime-maximizing coverage problem for static networks. We treat the problem of signal reconstruction in detail.

Many other concepts and definitions of coverage also exist. For example, *fractional area coverage* is the fraction of the region $\mathcal{D}$ that is covered by one or more sensors. *Node coverage fraction* is the largest fraction of nodes that can be removed without affecting coverage; this is a measure of redundancy. In the sequel we will define notions of k-coverage and barrier coverage as well.

10.2.2 Desirable Characteristics of Coverage Algorithms

A coverage algorithm must be *adaptive* (to node and link failures, duty cycling, changing environments; for example, if cued, one may want to allocate more resources to monitor an event). It must be *distributed* and localized without relying on a fixed communications graph. It must be *asynchronous* to accommodate events and agents that evolve at different timescales; this also enables the network to handle heterogeneity and duty cycling, and to account for propagation delays. Since global sync is not required, it could minimize overhead. The algorithm must be verifiably correct (given a postulated coverage, we should be able to verify that the desired QoS is satisfied).

10.2.3 Single-Hop Coverage

We first consider point and area coverage problems when the data exfiltration to the FC can be done in a single hop.

10.2.3.1 Point Coverage

In the point coverage problem, the objective is to cover a set of discrete points (targets) at known locations. Given finite energy constraints, the problem is to partition the set of N sensors into as many disjoint sets as possible, with each set providing a cover. Thus, the number of sets is the lifetime of the network, since the covers can be sequentially activated. This disjoint set covers (DSC) problem was shown to be NP-complete, and suboptimal heuristic designs were developed in [13]. A variation of the problem is to permit the sets to overlap, with each cover operating for a possibly different amount of time. This problem is also NP-complete [14], and suboptimal approaches based on LP and greedy algorithms are proposed.

In [15] the maximum set covers problem is extended to the case where sensors can choose one of M possible sensing ranges; associated with each range is a sensing cost. The problem is one of simultaneously finding covers and ranges. Centralized suboptimal solutions were proposed.

Considerations of energy consumption could also take into account the characteristics of the battery and the amplifier. In particular, cheap sensors are likely to have poor power amplifiers, thus limiting their linear dynamic range. Second, batteries exhibit a *charge recovery* phenomenon: terminal voltage can be higher at the end of the recovery period. Further, the capacity of a battery decreases as the discharge rate increases. Thus, the residual energy in a battery after n uses, each consuming E_i, depends upon the sequence of uses (see [7, 16]).

10.2.3.2 Barrier Coverage: Path Problems

Consider a scenario where sensors are deployed to detect intruders (perhaps as landmine replacement); coverage of the area will generally be nonuniform. Here barriers and sensors are metaphorically moats and alligators. Questions of interest here are: What is the path through the area that is worst (best) covered by the sensors? The worst covered path may need to be protected by other assets. If the sensors act as "guides," the best covered path may be the path of choice through an otherwise unknown territory. These problems are broadly referred to as barrier coverage problems.

The *maximal breach path* (MBP) and *maximal support path* (MSP) are formally defined as the paths on which the distance from any point to the closest sensor is maximized or minimized, respectively [17]. Based on the use of Voronoi diagrams and Delauney triangulation, centralized polynomial-time algorithms were developed in [17]. This approach was made more rigorous in [18], which also provides distributed solutions based on nearest-neighbor graphs and localized Delauney algorithms. The basic steps are a neighborhood discovery epoch (common to most algorithms), followed by a distributed Bellman-Ford algorithm to find the shortest-path routes. Since these routes are not unique, additional metrics such as shortest total distance or minimum energy consumption can be used.

Detectability is linked with passage time, or the time it takes an agent to traverse a given path. Minimum exposure time problems were introduced in [19], with a distributed solution proposed in [20]. Further complexity reductions are offered by the distributed algorithms of [21], which also considered the complementary maximal exposure path problem.

Consider an agent traversing from point A to point B in $\mathcal{D}$. If the sensors in $\mathcal{D}$ have constant sensing range r, then the probability of detection is simply the probability of sensors being present in a strip of width $2r$ centered along the path. Thus, the probability of detection can be written as $P_d(p) = 1 - \exp(-\lambda(\pi r^2 + 2r\|p\|))$, where $\|p\|$ is the length of the path [22]. More interestingly, a critical density exists; the probability of detection as the path length becomes unbounded goes to 1 (0) if the sensor density is above (below) a critical threshold.

For the signal model of (10.4), one can consider the case where the signals arriving at the sensors are transmitted to the FC, which sums them and compares them to a threshold. Although this detector does not have any particular sense of optimality, its asymptotic behavior yields insights. For this model, it has been shown in [22] that a critical density exists; detectability goes to 1 if this critical density is exceeded. An unexplored direction in these approaches is the consideration of noise at the sensors, and the inaccuracy of detection.

10.2.3.2.1 k-Barrier Coverage

In a relaxed setting, one may only require that it takes no more than k sensors to detect a target that completely traverses a "belt." Note that not every point in the region (or even path) has to be covered by any sensing disk. This notion was introduced in [23], which also considers the corresponding optimal sensor placement problem. A related idea is local barrier coverage [24], which guarantees the detection of any movement whose trajectory is confined to a slice of the belt region of deployment.

10.2.3.3 Area Coverage

A well-studied area coverage problem is the *art gallery problem* [25], in which one must determine the number and placement of cameras or guards necessary to cover an art gallery room such that *every* point in the gallery is observed by at least one camera. This problem can be solved optimally for the two-dimensional case, and is NP-hard in three-dimensional. The art gallery problem is a deterministic coverage problem, since we can control the position of each sensor. In WSN, the locations of the nodes cannot always be controlled due to the nature of their deployment (which might include being dropped from a helicopter, shot out of a cannon, or dispersed from a moving ground vehicle); hence, the notion of random coverage is more pertinent.

A disjoint set cover approach is considered in [12]; again, the idea is that these covers can be used sequentially, and maximizing the number of such covers would maximize the lifetime of the network. Crucial to this algorithm is the determination of areas that are least covered by sensors, a point that we will further take up in section 10.3. A different approach, based on the theory of dominating sets, is adopted in [26]. Distributed versions of these methods have been proposed. For example, in [27], nodes monitor local neighborhoods, and go to sleep if their area is already covered. To prevent blind spots and holes, a backoff-based scheme with a random timer is used. Every node starts the evaluation rule after a random time, and then broadcasts a status advertisement message to announce if it is available for turning off. A sleeping node will periodically wake up to check if it is still eligible to go back to sleep. This heuristic method often works well, but cannot guarantee a specific level of coverage (blind spots and holes), nor does it guarantee optimal energy usage (more sensors may be turned on than needed).

A variation of the method of [12] is presented in [28]. Here nodes are initially asleep, wake up periodically, and send out a probe. If the probe is acknowledged, i.e., one of its neighbors is awake, the node goes back to sleep for a random period of time. If it does not get an ACK, it goes into wake mode until its battery is exhausted. The coverage density is controlled by the radio range of the probe signal and also by the wake-up rate, both of which could be adjusted to maintain coverage and save energy.

Some fundamental limits on the coverage problem have been established under different assumptions. Reference [22] considers random uniform deployment of sensors in the two-dimensional plane with density λ and a common sensing range r. The fractional area coverage of such a network was shown to be $1 - \exp(-\lambda\alpha)$, where $\alpha = |\mathcal{A}|$ is the area of the sensing footprints of the nodes. Thus, given a desired fractional coverage rate (dictated by the application) and a sensing range (dictated by the sensors), one can readily compute the minimum required density of deployment.

10.2.3.3.1 k-Coverage

In a typical application such as monitoring, the sensor network is deployed to cover a given region. Often it suffices to have a single sensor cover a given point. By *cover* we mean that the signal received from an event or process occurring at that point can be reliably sensed by the sensor. Often it is desirable to have *k-coverage*, i.e., a minimum of k sensors must cover any given point in the region. For a typical detection application (e.g., trip wire), $k = 1$, but large values of k would be desired to confirm detection and for tracking. For localization, we need $k \geq 3$. k-Coverage is also important to cope with degraded links due to fading and target hiding due to local features (the pure geometric radius does not take terrain into account). With larger k, we can trade off inaccuracies in the quality of the sensors and be robust to environment-induced degradations as well. Of course, with larger k, the network expends more energy on sensing, processing, and communications. Thus, the trade-off ultimately is with the accuracy of monitoring (or the quality of service provided by the WSN) and the lifetime of the network. There are many interesting questions one can ask: How many sensors do we need to ensure k-coverage for different deployments (grid, random, random with some sensors at prespecified location) and for different QoS requirements (deterministic, probabilistic)? How should sensors be deployed? How should they be scheduled to maximize the lifetime of the network? How do we take into account duty cycling and account for probabilistic failure models (both in the sensing and in the communications back to a fusion center)?

10.2.3.4 Some Asymptotic Results

Several authors have established sufficient conditions on the density required to ensure k-connectivity assuming the disk model (but a general footprint model can be used). Recently, necessary and sufficient conditions on the required density for grid, uniform random, and Poisson deployments were established in [29], refining and extending the earlier work of [30–32]. Their results are for the asymptotic case when the area of the sensed region A grows but the sensing radius is kept fixed at $1/\sqrt{\pi}$. With p denoting the duty cycle factor (fractional on-time), they showed that asymptotically an area is k-connected whp if the density of active nodes in random deployments is of the form

$$\frac{np}{A} = \log(A) + 2k \log\log(A) + c(A),$$

where $c(A) \rightarrow \infty$. They also showed that for grid deployment, the corresponding required density is

$$-\frac{\log(1-p)}{A} = \log(A) + 2k \log\log(A) + 2\sqrt{-2\pi \log(A) \log(1-p)} + c(A).$$

Node density for grid placement is less than that for random deployment if $0 < \varepsilon \leq p \leq (1 - \varepsilon)$ for some constant $\varepsilon > 0$.

Some other problems related to area coverage have been addressed in [33–37]. In [33, 34], the following scenario is considered. The region $\mathcal{D}$ is a square of area $A = \ell^2$ that grows without bound; nodes are Poisson distributed in the plane with constant density λ and unit sensing range. A necessary and sufficient condition for k-coverage whp (asymptotic in ℓ) is the establishment of a critical density

$$\lambda = \log A + (k+1)\log\log A + c(\ell),$$

where $c(\ell) \to \infty$ [33]. For a fixed node density, upper bounds on the α-lifetime were also derived (only a fraction α of the region needs to be covered). Finally, duty cycling algorithms approaching the upper bound were also developed. For a fixed node density, and ignoring boundary conditions in a finite region, an order bound for the critical total power required to maintain k-connectivity was derived in [36]; nodes are allowed to vary their transmit powers, and the result is asymptotic in the node density. The verification problem, determining whether an area is k-covered, was considered in [37]; they derived an efficient polynomial-time algorithm that requires testing only the perimeter of the sensing regions of the nodes.

10.2.4 Multihop Coverage

We now consider point and area coverage problems when the data exfiltration to the FC must be done via a multihop route.

10.2.4.1 Point Coverage

With multihop connectivity, one must activate subsets of nodes that can ensure both sensing coverage and network connectivity. The problem is again one of partitioning the total set of sensors into such covering subsets. In [38], the partitions are allowed to overlap. Taking into account the energy consumed in the multihop transmissions, the problem is recast as a maximum cover tree problem and shown to be NP-complete. An upper bound is derived for the network lifetime, and suboptimal greedy algorithms are shown to perform close to the upper bound. The greedy algorithm takes into account both the residual energy at the sensors and the communication energy required to transmit over a link.

The same coverage and connectivity problem is considered in [39], but here the subsets of sensors are required to be disjoint. The authors consider a two-part approach. First, they consider the pure coverage problem for which they adopt the approach of [12]. To ensure connectivity, they then add sensors as needed. Finally, a pruning phase is considered wherein redundant nodes are removed. Multiple tree construction algorithms are considered. Because of the two-part approach, the solutions cannot guarantee optimality.

An additional degree of freedom can be obtained if the sensing ranges of the nodes are adjustable. Ideas used in the single-hop problems of [15] are now extended to the multihop case in [40]. Sensing energy is taken into account, but not the energy for transmissions. A two-phase approach is adopted. First, the connectivity problem is solved by creating a virtual backbone; the sensing ranges of the nodes on this backbone are set to the minimum value. Then the sensing ranges of the backbone nodes and the nonbackbone nodes are iteratively adjusted based on a utility function.

10.2.4.2 Area Connected Coverage

A fundamental result was established in [34]: a necessary and sufficient condition for coverage to imply connectivity is that the communication range be at least twice the sensing range. This result has subsequently been exploited in many approaches, essentially by ignoring the connectivity issue. The condition on the ranges is reasonable for many sensing modalities: as noted earlier, a cheap acoustic sensor may have a sensing range of 10 m and a magnetometer range of 5 m; a Berkeley mote may have a communications range of 150–300 m.

Geometric insights into the relationship between coverage and connectivity are also explored in [41]. They show that when the communications range is twice the sensing range, a set of nodes that k-cover a convex region form a k-connected communication graph. In such a graph, removing an arbitrary set of $k - 1$ nodes will still result in a connected network. They develop decentralized protocols. Dominating set theory approaches are adopted in [42] and [43] to develop distributed/localized algorithms for efficient routing.

The joint coverage and connectivity problem is also explored in [44], from the perspective of a query-driven report. The query requires a certain region of the domain to be sensed or covered; the selected set of sensors must provide sensing coverage and also form a communications link to the query node. They propose a centralized algorithm that relies on a greedy approach; a distributed version is also provided. These results are extended to k-coverage in [45]. These approaches, however, do not consider lifetime maximization.

10.2.5 Approaches, Objectives, and Characteristics

Table 10.1 provides a summary of different coverage algorithms that directly or indirectly seek to maximize lifetime.

10.3 Coverage-Based Information Retrieval

A sensor network must sense and report back to a gateway node, fusion center, or other designated node. Typically, the energy consumed in communications (transceiver energy consumption) is much larger than that consumed by the sensing device itself. As an example, the Rockwell WINS seismic sensor node consumes a mere 64 mW; with the processor on and with the radio unit it consumes 352 mW in sleep mode, 663 mW in idle, 687 mW in receive, and 1016 mW in transmit mode [46]. Sensors (sensing and signal processing algorithms) are getting more energy efficient, and although some parts of the transceiver operations can also be made more efficient, the dominant factor in energy consumption is the power required for transmission, and Maxwell's laws stay invariant and do not offer the promise of Moore's law!

Duty cycling wherein the transceiver is turned off can save considerable energy and has been the focus of concerted research efforts. Energy savings can be harvested at the PHY [47], MAC [48], networking [49], and application layers, and by judicious cross layer designs. Algorithms for maximizing the lifetime of a sensor network, as well as bounds on the lifetime, have been considered in [50], [51], and [53]. But these approaches do not

TABLE 10.1 Coverage Approaches: Characteristics and Assumptions: R_s Denotes Sensing Ranges

Ref.	Approach/Problem	Characteristics
	Single-Hop Coverage	
Point Coverage		
[13]	Disjoint cover sets	Centralized
[14]	Overlapping sets	Centralized
[15]	Overlapping sets	R_s adjustable, centralized
Barrier Coverage: Path Problems		
[17]	Voronoi and Delauney approaches	Centralized polynomial time
[18]	Maximal breach and support paths	Distributed solution
[19]	Minimum exposure time	Centralized and distributed
[21]	Minimum exposure time	Distributed solution
[21]	Maximal exposure problem	Distributed solution
[23]	*k-barrier coverage*	Distributed solution
[24]	Local barrier coverage	Localized solution
Area Coverage		
[12]	DSC-MCMC	Centralized
[26]	Disjoint dominating sets	Centralized
[27]	Sleep-wake scheduling	R_s adjustable, distributed
[28]	Sleep-wake scheduling with probes	Distributed
[34]	Density control of active nodes	Distributed
[33]	DSC	Centralized
[11]	DSC	R_s adjustable, centralized
	Multi-Hop Connectivity	
[39]	Tree designs, disjoint sets	Centralized
[38]	Maximum cover tree, with overlapping sets	Centralized
[40]	Tree designs, with adjustable sensing ranges	Centralized
[52]	Lifetime maximization with disjoint covers	Centralized
[42]	Disjoint dominating sets—lifetime	Distributed
[43]	Disjoint dominating sets—lifetime	Distributed
[41]	*k-coverage protocols*	Distributed

take into account coverage. We address this joint problem next; the results in this section are largely based on our work reported in [11].

10.3.1 Problem Formulation

N sensors $\{s_1, \ldots, s_N\}$ are randomly deployed to cover an area $\mathcal{D}$. Each sensor has an initial energy E_0 and covers a disk of radius r.* Sensors can communicate to the FC in one hop

---

* We assume that the sensors have equal initial energy and identical sensing ranges only for simplicity of exposition. This analysis extends to the general case.

and, if scheduled, consume energy $E_c^{(i)}$. We denote a sequence of cover sets by $\{C_j, j = 1, \ldots, K\}$. Recall that by definition if C_j consists of live sensors s_i, then

$$\bigcup_{s_i \in C_j} \mathcal{A}(s_i) \supset \mathcal{D} \quad \text{for} \quad j = 1, \ldots, K. \tag{10.5}$$

In the j-th data collection, measurements from the sensor nodes in the cover C_j are collected. Let b_{ij} be a Boolean variable that indicates whether the i-th sensor is used in the j-th cover. We want to find the best sequence of covers so as to maximize the lifetime of the network. Here, the network is considered to be dead if any part of the region is no longer covered by at least one sensor. A sensor is considered to be dead if its battery is depleted. The optimization problem is

$$\begin{aligned} &\text{Maximize Lifetime} && K \\ &\text{subject to Coverage Constraint:} && \bigcup\nolimits_{\{i|b_{ij}=1\}} \mathcal{A}(S_i) \supset \mathcal{D}, \text{ for } j = 1, \ldots, K, \\ &\text{and Energy Constraint:} && \sum\nolimits_{j=1}^{K} b_{ij} E_c^{(i)} \leq E_0, \text{ for } i = 1, \ldots, N, \\ &\text{where} && b_{ij} \in \{0,1\}. \end{aligned}$$

This problem is known to be NP-complete. We propose a suboptimal approach using a greedy algorithm.

10.3.2 Lifetime Upper Bound

We provide an upper bound on the network lifetime that provides full area coverage. This generalizes the results in [33] by allowing sensors to consume different transmission energies and thus have different lifetimes.

Definition A subregion is a set of points in $\mathcal{D}$ that are in the coverage range of a given set of sensors.

Thus, given N sensors, we have as many as 2^N possible subregions corresponding to all possible combinations of coverage/no coverage by each of the N sensors. By definition, the subregions are disjoint. By assumption, there is no subset of $\mathcal{D}$ that is not covered. We denote the L disjoint subregions by $\mathcal{F}_\ell$, $\ell = 1, \ldots, L$, $\mathcal{F}_i \cap \mathcal{F}_j = \phi$ for $i \neq j$, with

$$\bigcup\nolimits_{\ell=1}^{L} \mathcal{F}_\ell = D.$$

Each subregion is associated with a set of sensor nodes, $F_\ell = \{s_1^{(\ell)}, \ldots, s_{N_\ell}^{(\ell)}\}$, where N_ℓ is the number of sensors covering $\mathcal{F}_\ell$, and

$$\mathcal{F}_\ell \subset A\left(s_i^{(\ell)}\right) \quad \text{for} \quad i = 1, \ldots, N_\ell. \tag{10.6}$$

To ensure coverage, in each subregion $\mathcal{F}_\ell$, at least one of the sensors s_{n_ℓ} in F_ℓ must be selected for transmission. The network lifetime is therefore upper bounded by

$$\min_{F_\ell} \sum_{s_{n_i} \in F_\ell} \left\lfloor \frac{E_0}{E_c^{(n_i)}} \right\rfloor. \tag{10.7}$$

Note that the subregions that are least covered are the critical ones. To maximize the lifetime of the network, we must schedule sensors that cover these regions first. We consider a greedy approach for this next; see [11] for details.

10.3.3 A Greedy Approach to Lifetime Maximization

We propose a three-step approach. First, we identify the critical subregions and the sensors that cover them. If multiple sensors cover the critical subregion, we choose a covering sensor by taking into account its residual energy, the redundancy of coverage provided, and the energy consumed in transmission.

Step 1: Subregion creation. Given sensor locations and sensing range r, partition $\mathcal{D}$ into nonoverlapping regions, $\mathcal{F}_\ell$, $\ell = 1, \ldots, L$, and denote the associated sensor sets by F_ℓ. Each sensor is associated with a set of subregions that it covers, s_i: $S_i = \{\mathcal{F}_{n_1}, \ldots, \mathcal{F}_{n_i}\}$ such that

$$\bigcup_{n=n_1}^{n_i} \mathcal{F}_n = A(s_i).$$

Step 2: Cover set searching. The following two substeps are repeated until all the subregions are covered. The output is a cover set C_j for the j-th information retrieval operation.

1. *Critical subregions.* Let $E_r^{(i)}$ be the residual energy of the i-th sensor before the current data collection. The critical subregions are

$$F_c = \arg\min_{F_\ell} \sum_{s_i \in F_\ell} \left\lfloor \frac{E_r^{(i)}}{E_c^{(i)}} \right\rfloor. \tag{10.8}$$

2. *Sensor node selection.*
   - Compute the redundancy value v_i for the i-th sensor node where v_i = the number of $\mathcal{F}_c | \mathcal{F}_c \in S_i$. It is the selected critical subregions that are covered by the i-th sensor.
   - From the sensors covering the current critical subregion, choose the sensors satisfying $v_i = v_{\min}$, where $v_{\min} = \min\{v_i, i = 1, \ldots, N\}$.
   - Among the selected sensors, choose the sensor with the greatest value of a utility function $g(\cdot)$.
   - The subregions covered by the selected sensors are removed from the previous set of uncovered subregions.

Step 3: Residual energy update.

- For all the sensors in the current cover set, we update their residual energy $E_r^{(i)} \leftarrow E_r^{(i)} - E_c^{(i)}$.
- If $E_r^{(i)} < E_c^{(i)}$, sensor i is considered dead and dropped from the set of live sensors.
- If any sensors are removed from the set of live sensors, the subregions and their associated relationships with sensors are updated as in step 1.

The above steps are repeated until $\mathcal{D}$ is no longer fully covered by live sensors. This instant is the end of the network's life.

In the sensor selection step, we choose the sensor that has minimum value of v_i; thus, we prefer to use sensors that cover as few critical regions as possible so as to prolong the coverage of these regions. In [11], several other utility functions based on the residual energy and the consumed energy have been proposed and compared.

10.3.4 Performance Evaluation

In the following example sensor nodes are randomly deployed in a 20×20 area; sensors have identical sensing range r and also have identical initial energies. We consider the interesting case where the consumed energy is different from sensor to sensor, so that sensors have different lifetimes.

We compare the performance of the proposed method with the lifetime upper bound (LUB) and three other methods:

1. "GEOM": We choose the critical subregion as the subregion that is covered by the least number of sensors, and the utility function is the number of the uncovered subregions covered by a sensor.*
2. "MCLC": The most constrained, least constrained method of [12].
3. "RAND": We randomly choose a sensor from the sensor nodes that cover the critical subregion and add it to the cover set.

In the proposed method, we use the utility function $g_1(s_i) = E_r^{(i)}/E_c^{(i)}$. We also fix the number of sensors to $N = 100$ and the coverage range to $r = 5$ and vary the initial energy. Results are shown in Figure 10.1.

We note that the proposed method has near-optimal performance and is better than the other three methods. We have noticed that the improvement over MCLC is particularly pronounced when the sensors have different lifetimes, since MCLC does not explicitly take this into account.

10.3.5 Summary

We studied the problem of network lifetime maximization for QoS-specific information retrieval for the reconstruction of a spatially correlated signal field in a wireless sensor

* This method is appealing since it uses only geometry information to schedule the sensors.

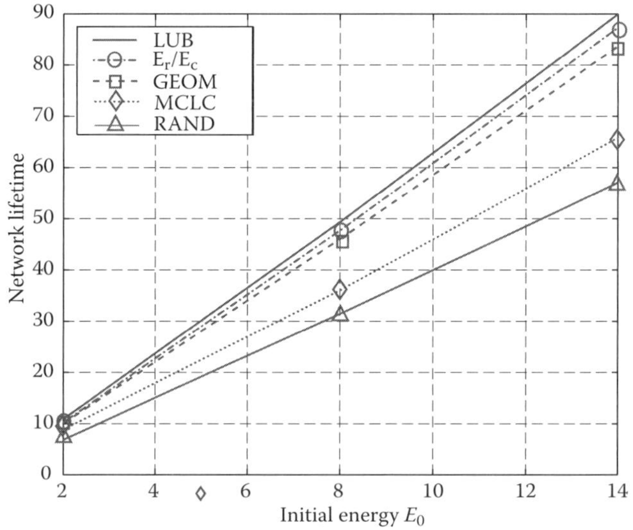

FIGURE 10.1 Lifetime versus initial energy.

network. We transformed the problem into a coverage-related problem, and proposed a greedy approach. The performance of the proposed approach was shown to be close to the upper bound on network lifetime.

10.4 Joint Design of Scheduling and Routing

We now extend the results of the previous section to the case where multihop routing to the FC is required.

10.4.1 Background

A survey of the literature on multihop connectivity and coverage was given earlier in section 10.2.4. We consider the joint problem of sensing coverage and connectivity for area coverage. Our optimality criterion is the network lifetime. To be realistic, our energy consumption model takes into account energy for sensing, reception, and transmission. Rather than considering cover trees, we formulate a more general flow assignment problem. The formulation leads to an integer programming problem that is NP-hard. We consider suboptimal approaches based on modified link criteria and show that performance is surprisingly close to an upper bound, which we establish. We will use the same notation as in the previous section, and introduce new symbols as needed. The results in this section are largely based on our work in [52].

We assume that the communication range is twice the sensing range, an assumption we argued earlier as realistic.

10.4.2 Problem Formulation

In any data collection, sensors that do not collect data may still be required to act as relays. We use the Boolean variable g_{ik} to denote whether data are generated at the i-th node during the k-th data collection. We let $q_k^{(i,j)}$ denote the flow on the edge connecting nodes i, j in the k-th collection; a positive value indicates flow into j; the q's are integers representing number of packets (reports).* We denote the gateway node or fusion center by s_D, and by V_k the set of nodes that are alive at the start of the k-th data collection. Let C_k denote the set of sensors that are used for sensing,

$$C_k = \{s_i \mid s_i \in V_k, \text{and } g_{ik} = 1\}. \tag{10.9}$$

C_k must satisfy the coverage constraint

$$\bigcup_{s_i \in C_k} \mathcal{A}(s_i) \supseteq \mathcal{D} \quad \text{for} \quad k = 1, \ldots, K.$$

The coverage condition can be written as

$$\sum_{i: s_i \in F_\ell} b_{ik} \geq 1, \quad \forall \ell, \quad \text{for} \quad k = 1, \ldots, K, \tag{10.10}$$

where the Boolean variable b_{ik} is 1 if node i is used for sensing during the k-th data collection. We have a flow conservation requirement to ensure that relays do not absorb packets:†

$$\sum_{j: s_j \in S^{(i)}} q_k^{(i,j)} = g_{ik}, \quad \forall i \in V_k - \{s_D\}, \quad \text{for} \quad k = 1, \ldots, K. \tag{10.11}$$

We also have energy constraints

$$\sum_{k=1}^{K} E_{u,k}^{(i)} \leq E_0^{(i)}, \quad \forall s_i \in V_k - \{s_D\}, \tag{10.12}$$

where $E_0^{(i)}$ is the initial energy of the i-th node and $E_{u,k}^{(i)}$ is the energy consumed by the i-th node in the k-th flow. This energy term is written as

---

* We do not consider the data aggregation problem here.

† We do not consider the problem of store-and-forward, which may be particularly important in the context of fading channels. We do not consider buffer overflows and dropped packets. Associated latency issues are also important.

$$E_{u,k}^{(i)} = e_s \cdot g_{ik} - e_r \cdot \sum_{j:j \in S_{\text{in},k}^{(i)}} q_k^{(i,j)} + e_t^{(i,j)} \cdot \sum_{j:j \in S_{\text{out},k}^{(i)}} q_k^{(i,j)}, \tag{10.13}$$

where e_s is the energy required for sensing, e_r is the energy required to receive one packet, and $e_t^{(i,j)}$ is the energy required to transmit one packet: $e_t^{(i,j)} = \varepsilon_t + \alpha \cdot d_{ij}^{\beta}$, where β is the path loss factor, d_{ij} the distance between nodes i and j, and ε and α constants associated with the transceiver. Thus, the three terms represent the energy consumed in sensing (if the sensor is allocated for sensing), the energy consumed in receiving all the incoming packets, and the energy consumed in transmitting all the outgoing packets. This formulation can be easily extended to other energy models to represent specific technology choices.

We want to maximize the lifetime K subject to the coverage, flow, and energy constraints (10.10) to (10.12), where b_{ik} and g_{ik} are Boolean variables and $q_k^{(i,j)}$ are integers.

This is an integer programming problem that is NP-hard. In the next section we derive a greedy algorithm, and show via simulations that its performance is close to an upper bound. To compute an upper bound, we need to consider energy consumed in sensing as well as that consumed in transmission. We can get a loose bound by finding the minimum energy that exfiltrates the packet to a node outside the region, not necessarily the sink node. Note that such a path might involve multiple nodes in the same subregion. The complexity of computing this upper bound lies in the complexity of finding the L subregions, and in finding the minimum energy paths for each of the N nodes. The upper bound itself is derived in [52] and will not be repeated here.

10.4.3 A Suboptimal Approach

We adopt the realistic assumption that the communications range is at least twice the sensing range. Thus, coverage implies connectivity as well. The set of live sensors therefore yields a communications graph where an edge exists between a pair of nodes only if their separation is less than the communications range.

Given the set of live sensors, we compute the subregions as in section 10.3.2. We define the residual lifetime of a subregion as the maximum number of data collections that can be used to reconstruct that subregion, based on the residual energy of the sensors covering it. Computing the exact lifetime is difficult. We could use the upper bound we just discussed, but computing that is also computationally demanding. Hence, we use a simpler metric: we approximate the lifetime of the subregion $\mathcal{F}_\ell$ as

$$L_r^{(\ell)} = \sum_{i \in F_\ell} \frac{E_r^{(i)}}{E_c^{(i)}}, \tag{10.14}$$

where $E_r^{(i)}$ is the residual energy of the i-th node and $E_c^{(i)}$ is the energy required to transmit to its closest node. The critical subregion is the one with the smallest $L_r^{(\ell)}$.

As in the one-hop case, our strategy is to first cover the critical subregions. If several sensors cover F_c, how should we select the sensor? We could use the sensor that has the minimum energy path to the sink. But this path might involve sensors covering the critical region. To take this into account, we define a metric called the critical value of a sensor, and define another metric that involves the residual sensor energy and the critical value. Each sensor s_i covers a set of subregions $S_i = \{\mathcal{F}_{n_1}, \ldots, \mathcal{F}_{n_i}\}$. We compute the residual lifetime $L_r^{(n_j)}$ for each of the subregions via (10.14). The critical value, $c_r^{(i)}$, of a sensor is then defined as the minimum of the residual lifetimes of the subregions covered by it. We then define a link cost:

$$\text{cost}\,(i,j)=\left(e_t^{(i,j)}\right)^{x_1}\cdot\left(E_r^{(i)}\right)^{-x_2}\cdot\left(c_r^{(i)}\right)^{-x_3}+\left(e_r\right)^{y_1}\cdot\left(E_r^{(j)}\right)^{-y_2}\cdot\left(c_r^{(j)}\right)^{-y_3}, \tag{10.15}$$

where e_r is the unit receiving energy, and x_k and y_k are integer constants that indicate the relative weights between the cost of receiving and sending a packet, the lifetime of the network, and the importance of the region covered by the sensor. Note that this link metric favors edges whose ends have either high critical values or large residual energies or low transmission/reception costs.

10.4.4 Numerical Performance Analysis

We assume that sensor nodes are randomly deployed in a 20×20 area following a uniform distribution. A sensing disk model with radius R_s is adopted; the maximum communication range is $R_c = 2R_s$. Initially all the nodes have the same energy. Our transmission energy model is $e_t^{(i,j)} = \alpha \cdot d_{ij}^{\beta}$, with $\alpha = 0.01$ and $\beta = 2$. Figures indicate results obtained by averaging over one hundred randomly generated network topologies.

We study the performance of the proposed suboptimal method and compare it with the lifetime upper bound. We also show the performance of a *separated design* method in which we separate the design of the sensor scheduling and information routing to maximize the network lifetime. We select the cover set according to the method in [11]. For routing, we find the minimum energy path using the method of [54]. We set $R_s = 4$ and $N = 100$. Figure 10.2 shows network lifetime versus initial energy. We see that the performance of the proposed suboptimal method is close to the upper bound. We also note that the proposed joint design provides substantial performance improvement over separated designs.

10.4.5 Summary

In this section, we considered the joint problem of scheduling sensors for coverage and information routing so as to maximize the network lifetime. The optimization problem is NP-complete, so we proposed a suboptimal design. We derived an upper bound and showed that the performance of the suboptimal design is close to a derived upper bound.

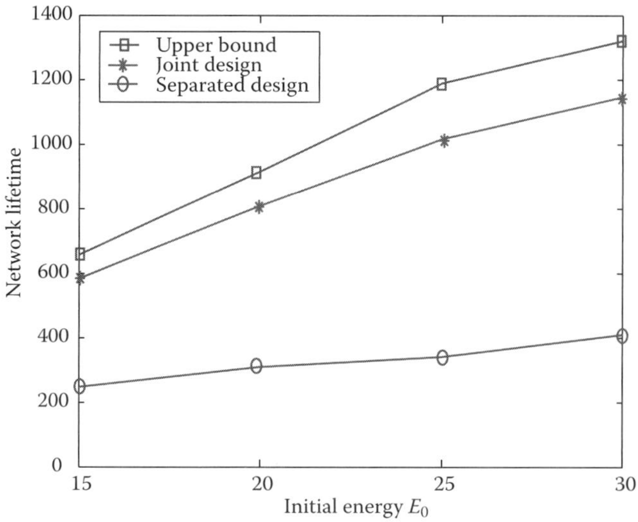

FIGURE 10.2 Lifetime versus initial energy.

10.5 Conclusions and Discussion

In this chapter, we have considered various concepts of coverage and connectivity and provided a survey of recent work on sensing and coverage and communication connectivity required to exfiltrate data to a gateway node in a static sensor network. We considered both the single-hop and the multiple hop cases. Taking into account energy consumed in sensing and communications, we presented an upper bound for the one-hop case. A greedy algorithm was presented and shown to have performance close to the upper bound. For the multi-hop scenario wherein sensors not required for coverage may be involved in relaying, we cast the problem as an integer optimization problem which is NP-hard. We developed a suboptimal greedy algorithm and showed that its performance is close to an upper bound that we derived.

In future work, we will consider the effect of channel fading and other characteristics on network lifetime. In this case, transmission energy is a random variable dictated by the channel. We will consider average and outage lifetimes. Sensing consumes energy, and sensing cost increases with sensing range (more sophisticated hardware, potentially more false alarms, more processing). Hence, sensing range should depend upon residual energy and could be optimized. Thus, as RE decreases, one may expect the sensing range to decrease, and more nodes to be activated. One could consider optimizing over the sensing range (with associated energy costs) as well as communications range (again with associated costs). We have not considered the multiple-access scheduling problem and latency issues. Both are important, as is the extension to store-and-forward (a sensor might store data and wait for a more favorable channel to conserve energy). Our discussion has been largely in the context of the disk model, which provides a tractable formulation. But other formulations wherein we can take into account multiple coverages provided by the sensors are also interesting. Another problem of interest is aggregation: if the relay nodes have some knowledge of the spatial correlation of the underlying field, they could optimally combine data rather than merely aggregating them. Extensions to k-coverage are of interest. Finally, distributed implementations are also of interest.

Acknowledgments

This work was supported in part by the Army Research Laboratory CTA on Communication and Networks under Grant DAAD19-01-2-0011 and by the National Science Foundation under Grants CNS-0627090 and ECS-0622200

References

[1] D. Stamatelos and A. Ephremides. 1996. Spectral efficiency and optimal base placement for indoor wireless networks. *IEEE J. Select. Areas Commun.* 14:651–61.

[2] K. S. Butterworth, K. W. Sowerby, and A. G. Williamson. 2000. Base station placement for in-building mobile communication systems to yield high capacity and efficiency. *IEEE Trans. Commun.* 48:658–69.

[3] M. Franceschetti, M. Cook, and J. Bruck. 2004. A geometric theorem for network design. *IEEE Trans. Comput.* 53:483–89.

[4] A. P. Azad and A. Chockalingam. 2006. Mobile base stations placement and energy aware routing in wireless sensor networks. In *Proceedings of IEEE WCNC 2006*, pp. 264–69.

[5] C. Chong and S. Kumar. 2003. Sensor networks: Evolution, opportunities, and challenges. *Proc. IEEE* 91:1247–56.

[6] Q. Zhao, A. Swami, and L. Tong. 2006. The interplay between signal processing and networking in sensor networks. *IEEE Signal Processing Mag.* 23:84–93.

[7] S. Dasika, S. Vrudhula, and K. Chopra. 2006. Algorithms for optimizing lifetime of battery powered wireless sensor networks. In *Sensor network operations*, ed. S. Phoha, T. La Porta, and C. Griffin. Piscataway, NJ: Wiley-IEEE Press, pp. 367–396.

[8] R. Tobin. 2003. US Army's BLUE radio. In *Proceedings of SPIE, Unattended Ground Sensor Technologies and Applications V*, Orlando, FL, vol. 5090.t

[9] J. Crowcroft, M. Scott, P. Brennan, J. Elmirghani, R. Penty, I. White, The Intelligent Networked Airport (TINA), white paper dated 2007, available at http://www.cl.cam.ac.uk/~jac22/out/tina-white-paper.pdf.

[10] K. Suh, Y. Guo, J. Kurose, and D. Towsley. 2006. Locating network monitors: Complexity, heuristics, and coverage. *J. Comput. Commun.* 29:1564–77.

[11] T. Zhao and Q. Zhao. 2007. Coverage-based information retrieval for lifetime maximization in sensor networks. In *Proceedings of the Conference on Information Sciences and Systems (CISS)*, Baltimore, pp. 220–225.

[12] S. Slijepcevic and M. Potkonjak. 2001. Power efficient organization of wireless sensor networks. In *Proceedings of the IEEE International Conference on Communications*, pp. 472–76.

[13] M. Cardei and D.-Z. Du. 2005. Improving wireless sensor network lifetime through power aware organization. *ACM Wireless Networks* 11:333–40.

[14] M. Cardei, M. T. Thai, Y. Li, and W. Wu. 2005. Energy-efficient target coverage in wireless sensor networks. In *Proceedings of IEEE INFOCOM 2005*, vol. 3, pp. 1976–84.

[15] M. Cardei, J. Wu, M. Lu, and M. O. Pervaiz. 2005. Maximum network lifetime in wireless sensor networks with adjustable sensing ranges. In *Proceedings of IEEE International Conference on Wireless and Mobile Computing, Networking and Communications*, vol. 3, pp. 438–45.

[16] D. Rakhmatov and S. Vrudhula. 2003. Energy management for battery-powered embedded systems. *ACM Trans. Embedded Comput. Syst.* 2:277–324.

[17] S. Meguerdichian, F. Koushanfar, M. Potkonjak, and M. B. Srivastava. 2001. Coverage problem in wireless ad-hoc sensor networks. In *Proceedings of IEEE International Conference on Computer Communications*, pp. 1380–87.

[18] X.-Y. Li, P.-J. Wan, and O. Frieder. 2003. Coverage in wireless ad hoc sensor networks. *IEEE Trans. Comput.* 52(6).

[19] S. Meguerdichian, F. Koushanfar, G. Qu, and M. Potkonjak. 2001. Exposure in wireless ad-hoc sensor networks. In *Proceedings of ACM International Conference on Mobile Computing and Networking*, pp. 139–50.

[20] S. Meguerdichian, S. Slijepcevic, V. Karayan, and M. Potkonjak. 2001. Localized algorithms in wireless ad-hoc networks: Location discovery and sensor exposure. In *Proceedings of ACM International Symposium on Mobile Ad-Hoc Networking and Computing*, pp. 106–16.

[21] G. Veltri, Q. Huang, G. Qu, and M. Potkonjak. 2003. Minimal and maximal exposure path algorithms for wireless embedded sensor networks. In *Proceedings of ACM International Conference on Embedded Networked Sensor Systems*, pp. 40–50.

[22] B. Liu and D. Towsley. 2004. A study of the coverage of large-scale networks. In *Proceedings of the IEEE International Conference on Mobile Ad Hoc and Sensor Systems (MASS)*, Ft. Lauderdale, FL, pp. 475–483.

[23] S. Kumar, T. H. Lai, and A. Arora. 2006. Barrier coverage with wireless sensors. *Springer Wireless Networks J. (WINET)*, pp. 817–834.

[24] Ai Chen, S. Kumar, and T. H. Lai. 2007. Designing localized algorithms for barrier coverage. In *MobiCom'07*, Montreal, pp. 63–74.

[25] J. O'Rourke. 1987. *Art gallery theorem and algorithms.* Oxford: Oxford University Press.

[26] M. Cardei, D. MacCallum, X. Cheng, M. Min, X. Jia, D. Li, and D.-Z. Du. 2002. Wireless sensor networks with energy efficient organizaton. *J. Interconnection Networks* 3:213–29.

[27] D. Tian and N. D. Georganas. 2002. A coverage-preserving node scheduling scheme for large wireless sensor networks. In *Proceedings of International Workshop on Wireless Sensor Networks and Applications*, pp. 32–41.

[28] F. Ye, G. Zhong, S. Lu, and L. Zhang. 2003. PEAS: A robust energy conserving protocol for long-lived sensor networks. In *Proceedings of International Conference on Distributed Computing Systems*, pp. 28–37.

[29] H. Zhang and J. C. Hou. 2006. Is deterministic deployment worse than random deployment for wireless sensor networks? In *Proceedings of IEEE INFOCOM'06*, pp. 1–13.

[30] T. K. Philips, S. S. Panwar, and A. N. Tantawi. 1989. Connectivity properties of a packet radio network model. *IEEE Trans. Inform. Theory* 35(5).

[31] S. Shakkotai, R. Srikant, and N. Shroff. 2003. Unreliable sensor grids: Coverage, connectivity and diameter. In *Proceedings of IEEE INFOCOM*, pp. 1073–1083.

[32] H. Zhang and J. Hou. 2004. On deriving the upper bound of α-lifetime for large sensor netwroks. In *Proceedings of ACM Mobihoc*, pp. 121–132.

[33] H. Zhang and J. C. Hou. 2005. On the upper bound of α-lifetime for large sensor networks. *ACM Trans. Sensor Networks* 1:272–300.

[34] H. Zhang and J. C. Hou. 2005. Maintaining sensing coverage and connectivity in large sensor networks. *Wireless Ad Hoc Sensor Networks Int. J.* 1:89–123.

[35] H. Zhang and J.C. Hou. 2005. Maxmizing α-lifetime for wireless sensor networks. In *International Workshop on Measurement, Modeling, and Performance Analysis of Wireless Sensor Networks*, San Diego, CA, pp. 70–77.

[36] H. Zhang and J. C. Hou. 2005. On the critical total power of asymptotic k-connectivity in wireless networks. In *Proceedings of IEEE INFOCOM'05*, pp. 13–17.

[37] C.-F. Huang and Y.-C. Tseng. 2003. The coverage problem in a wireless sensor network. In *ACM International Workshop on Wireless Sensor Networks and Applications*, pp. 115–21.

[38] Qun Zhao and M. Gurusamy. 2006. Maximizing network lifetime for connected target coverage in wireless sensor networks. In *IEEE International Conference on Wireless and Mobile Computing, Networking and Communications*, pp. 94–101.

[39] N. Jaggi and A. A. Abouzeid. 2006. Energy-efficient connected coverage in wireless sensor networks. In *Proceedings of the 4th Asian International Mobile Computing Conference*, Kolkata, India, pp. 77–86.

[40] M. Lu, J. Wu, M. Cardei, and M. Li. 2005. Energy-efficient connected coverage of discrete targets in wireless sensor networks. In *Proceedings of the International Conference on Computer Networks and Mobile Computing*, pp. 43–52, LNCS Vol. 3619.

[41] G. Xing, X. Wang, Y. Zhang, C. Lu, R. Pless, and C. Gill. 2005. Integrated coverage and connectivity configuration for energy conservation in wireless sensor networks. *ACM Trans. Sensor Networks* 1(1):36–72.

[42] J. Wu and H. Li. 1999. On calculating connected dominating set for efficient routing in ad hoc wireless networks. In *Proceedings of the III International Workshop on Discrete Algorithms and Methods for Mobile Computing and Communications*, pp. 7–14.

[43] J. Wu, F. Dai, M. Gao, and I. Stojmenovic. 2002. On calculating power-aware connected dominating sets for efficient routing in ad hoc wireless networks. *J. Commun. Networks* 4(1):50–70.

[44] H. Gupta, Z. Zhou, S. R. Das, and Q. Gu. 2006. Connected sensor cover: Self-organization of sensor networks for efficient query execution. *IEEE/ACM Trans. Networking* 14(1):55–67.

[45] Z. Zhou, S. Das, and H. Gupta. 2004. Connected k-coverage problem in sensor networks. In *Proceedings of the International Conference on. Computer Communications and Networks*, pp. 373–78.

[46] V. Raghunathan, C. Schurgers, S. Park, and M. B. Srivastava. 2002. Energy aware wireless microsensor networks. *IEEE Signal Processing Mag.* 19:40–50.

[47] E. Shih, S. H. Cho, N. Ickes, R. Min, A. Sinha, A. Wang, and A. Chandrakasan. 2001. Physical layer driven protocol and algorithm design for energy-efficient wireless sensor networks. *Mobile Computing and Networking*, pp. 272–87.

[48] W. Ye, J. Heidemann, and D. Estrin. 2002. An energy-efficient MAC protocol for wireless sensor networks. In *Proceedings of IEEE INFOCOM*, pp. 1567–76.

[49] J. H. Chang and L. Tassiulas. 2000. Energy conserving routing in wireless ad-hoc networks. *Proceedings of IEEE INFOCOM*, pp. 22–31.

[50] Y. Chen and Q. Zhao. 2005. On the lifetime of wireless sensor networks. *IEEE Commun. Lett.* 9(11):976–978.

[51] Y. Chen and Q. Zhao. 2007. An integrated approach to energy-aware medium access for wireless sensor networks. *IEEE Trans. Signal Processing* 55:3429–44.

[52] T. Zhao and Q. Zhao. 2007. Joint design of scheduling and routing based on connected coverage for optimal sensor network lifetime. In *Proceedings of SPIE'07*, San Diego.

[53] Y. Chen, Q. Zhao, V. Krishnamurthy, and D. Djonin. 2007. Transmission scheduling for optimizing sensor network lifetime: A stochastic shortest path approach. *IEEE Trans. Signal Processing* 55:2294–309.

[54] J.-H. Chang and L. Tassiulas. 2004. Maximum lifetime routing in wireless sensor networks. *IEEE/ACM Trans. Networking* 12:609–19.

11

Routing in Wireless Self-Organizing Networks

Marcelo Dias de Amorim
CNRS

Farid Benbadis
Université Pierre et Marie Curie

Mihail S. Sichitiu
North Carolina State University

Aline Carneiro Viana
INRIA

Yannis Viniotis
North Carolina State University

11.1 Introduction

Routing is a general term that finds applications in numerous areas, such as networking, transportation systems, electric circuits, postal systems, and neurobiology, to cite a few. By definition, routing is the mechanism responsible for determining the path across which objects (i.e., a packet, an electrical signal, a letter) are transported from a source to a destination.

In packet networks, routing is basically the way nodes fill their routing tables. A routing table is a data structure that stores descriptions of next-hops and possibly some cost toward a destination. From the routing table, a node builds a forwarding table that includes the transmit interface. We call the set of algorithms and procedures that build routing tables a *routing protocol*. As we will see in the rest of this chapter, routing may assume different forms depending on the characteristics of the underlying network. What should be clear from the start is that there is no routing approach that fits well all situations.

Routing is a difficult problem and fundamentally a major issue in any networking architecture. Let us first consider the case of the Internet, in which routing is a major investigation area, especially in the context of interdomain routing. The Internet is composed of autonomous systems (ASs), which are networks under a single administrative entity. Nodes are interconnected through two types of routing protocols: intradomain, which runs within an AS, and interdomain, which is used to interconnect ASs. Intradomain routing protocols (e.g., OSPF [37] and RIP [36]) rely on routing algorithms such as Bellman-Ford and Dijkstra that solve the single-source, shortest-path problem. In these algorithms, links are associated with a weight and nodes exchange knowledge they have on other nodes until a stable representation of the topology is achieved at each node of the network (algorithm convergence). Clearly, for these algorithms to converge, it is fundamental that all nodes cooperate and share information about the topology of the network to which they belong. This is not possible in the case of interdomain routing since AS administrators, for commercial reasons, do not agree to exchange information on their internal topologies by default. The *de facto* protocol for interdomain routing is the Border Gateway Protocol (BGP) [48]. The problem here is completely different from the intradomain case, since ASs must rely on policies to establish their routing tables.

Research on routing in wired networks has achieved a reasonable level of maturation. The same cannot be said about wireless networks, and in particular about *wireless self-organizing networks* (WSONs). The main characteristics of a WSON that set it apart from a traditional wired network are the lack of a management infrastructure (e.g., no centralized addressing scheme is present) and the dynamics of nodes. Additional peculiarities include the possible lack of geographical positioning infrastructure, the limited and variable capacity of wireless links, and the spontaneous nature of the topology (cf. section 11.2 for more details on the characteristics of WSONs).

The basic shift of paradigm between traditional networks and WSONs is related to the communication characteristics. In the traditional case, the network is in general

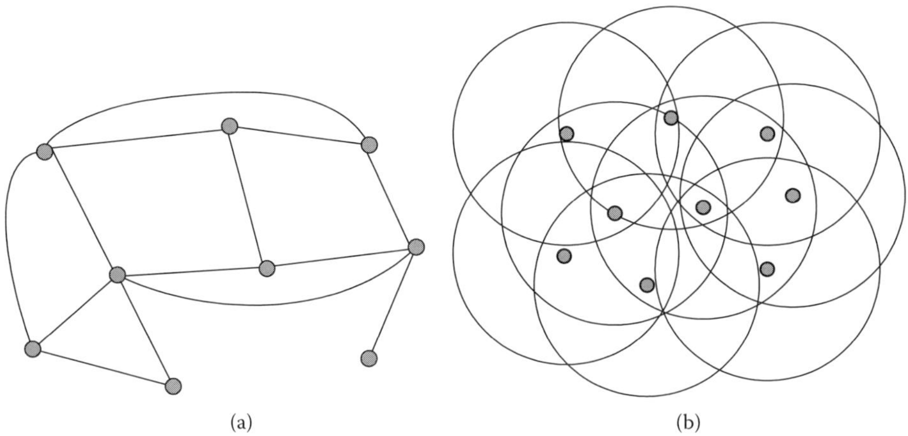

FIGURE 11.1 Representation of the communication geometry in wired versus WSON. (a) Wired network; (b) wireless self-organizing network.

represented by a connected graph, in which vertices are nodes of the network and arcs are connections between nodes. In wireless networks, the transmission medium is inherently broadcast at the physical layer, meaning that transmissions between neighbors interfere with transmissions of other nodes in the vicinity. Figure 11.1(a, b) illustrates the representation of wired and wireless self-organizing networks.

For these reasons, the definition of routing in self-organizing networks is an extension from the traditional statement, especially because, as we will see later in this chapter, sometimes there is even no notion of routing tables. It is important to underline again that there is no single routing solution that can be accommodated to any type of self-organizing network.

The goal of this chapter is to help the reader understanding the *foundations* of routing in WSONs and not only learning existing solutions.*

This chapter is structured as follows. In section 11.2 we present the characteristics of WSONs; in section 11.3 we present several categories of WSONs and their particularities. Section 11.4 focuses on additional requirements relevant to routing in WSONs. Section 11.5 decomposes the routing protocols in three generic modules corresponding to distinct functions of a WSON. We study the effect of various parameters of WSONs on the performance of the routing protocols in section 11.6. Sections 11.7 and 11.8 present, respectively, a classification and examples of routing protocols in WSONs. Section 11.9 concludes the chapter.

* Although this chapter is an introductory view of routing in wireless self-organizing networks, we assume that the reader has some basic notions on routing protocols.

11.2 Characteristics of Wireless Self-Organizing Networks

Wireless self-organizing networks also present a number of characteristics that set them apart from traditional "graph-based" networks. This chapter will focus on the particularities of these networks and on how routing must be rethought upon a radically different basis. In particular, routing protocols in WSONs must face the following characteristics (in section 11.4 we will present the architectural requirements imposed by WSONs):

- *Unpredictability of the wireless medium.* The conditions of the wireless medium are difficult to predict because of many natural phenomena, such as fading, multipath, attenuation, interference, and loss of line of sight. In addition, they make determining connectivity *a priori* difficult in WSONs. Contrary to the wired case, in which the channel can be characterized reasonably well, much more randomness is expected in the wireless case. This has a direct impact on the quality of the routes that may be highly variable in time.
- *Physical layer broadcast.* Unlike the Internet, where the point-to-point model of communication is dominant, communication in wireless networks is inherently broadcast. That is, when a node transmits some information, typically all nodes within its transmission range can receive it. Because of this broadcast nature of the medium, spectrum usage must be designed at a very fine-grained level. In this way, controlling access to the medium is of great importance. Contrary to the wired case, where collision can be detected, the wireless medium does not permit collision detection at the sender side. To this, we must also consider the problem of hidden terminals in which local communication impacts all the vicinity around the nodes involved. This is because the network model is no longer a graph but a collection of coverage surfaces.
- *Potentially high mobility.* Given that nodes communicate through the wireless medium, they can be mobile. In fact, humans are mobile by nature, and having the opportunity to move becomes a desired property; however, mobility implies a dynamic topology, which in turn implies more complex management algorithms for topology maintenance and routing.
- *Expensive routing.* Routing is an expensive operation in WSONs. The dynamic nature of these networks causes routes to be unstable and makes routing a resource-greedy operation. Furthermore, the more distant the corresponding node, the more expensive the communication. This invalidates several assumptions made for wired networks, such as: communication is carried out over an arbitrary overlay, which is not constrained by physical connectivity (there is a cooperative infrastructure); clique connectivity; and the cost of reaching a node is (almost) the same for all nodes.
- *Security vulnerability.* The architecture of wireless self-organized networks is inherently insecure. First, there is no central point or administrative entity that controls the access of nodes to the infrastructure. Second, transmissions are generally in broadcast mode, which opens the possibility for unauthorized

nodes to listen to the traffic. Malicious nodes can then perform attacks in a much easier way.

- *Energy constraints.* Nodes in WSONs are likely battery operated. It is then important to use routing protocols that are energy aware and energy efficient.
- *High heterogeneity.* In WSONs, it is likely that nodes are heterogeneous in their characteristics, such as memory availability, computation capacity, and transmitting power.
- *Data-centric routing.* Following the application requirements, some WSONs have stepped away from the Internet's point-to-point address-centric communication model and have adopted data-centric abstractions. In the data-centric paradigm, the nature of the data is more important than the identity of the source. In particular, routing decisions are taken primarily based on the type of data. This model of communication clearly favors in-network data processing and aggregation, in addition to imposing changes in the way routing and storage are performed in the network.

11.3 Classes of Wireless Self-Organizing Networks

In the previous paragraphs, we have discussed the general characteristics of wireless self-organizing networks. In this section, for the sake of clarity, we present the most significant classes of wireless self-organizing networks that have been receiving significant interest from the industry and academia in the last few years. The objective here is to help the reader understand that the domains of application of WSONs are quite large and that, although different, these networks show similar principles from the networking point of view.

11.3.1 Mobile Ad Hoc Networks

Having no need for preexisting fixed infrastructure, such as access points or base stations, the idea behind mobile ad hoc networks is that they can be established anytime and anywhere [2]. Such networks are intrinsically fault resilient because they do not operate under a fixed/centralized topology. Since all nodes are potentially mobile, the associated network composition is inherently time varying. Nodes join and leave by interacting with other nodes, without any centralized management. The absence of fixed infrastructure means that the nodes communicate directly with one another in a peer-to-peer fashion. Routing in such a context is thus challenging mainly because very little can be said *a priori* about the network.

Mobile ad hoc networks (MANETs) became a popular subject for research in the late 1990s with the advent of inexpensive IEEE 802.11 radio cards for personal computers. Minimal configuration and fast deployment make ad hoc networks suitable for emergency situations like natural or human-induced disasters, military conflicts, emergency medical situations, etc. Figure 11.2 illustrates a typical representation of a wireless mobile ad hoc network. In time, MANETs inspired other, more application-oriented types of WSONs.

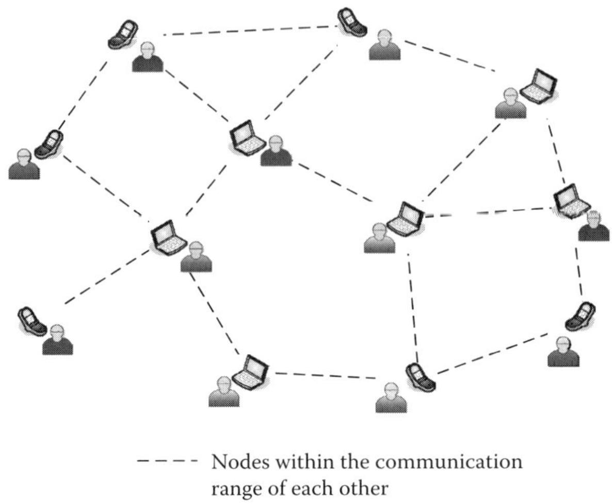

FIGURE 11.2 A mobile ad hoc network.

11.3.2 Wireless Sensor Networks

Wireless sensor networks (WSNs) consist of small nodes with sensing, computation, and wireless communications capabilities [4]. It is generally assumed that WSNs may contain hundreds or thousands of nodes (or more). This requires that the routing protocol be scalable; i.e., it must properly work not only in the case of huge topology sizes but also for extremely heterogeneous networks. Due to the absence of wired infrastructure and the limited characteristics of sensors (e.g., energy, computation capacity, memory limitations), WSNs introduce strong operation and management restrictions. In particular, mechanisms for energy optimization in WSNs constitute an important requirement. This optimization targets not only the reduction of energy consumption of a single sensor node, but also the extension of the entire network lifetime. At the network layer, for example, it is highly desirable to introduce approaches for energy-efficient route discovery and relaying of data from the sensor nodes to base stations, so that the lifetime of the network is maximized.

In addition to differences imposed by the wireless links, there are several other inherent particularities of a WSN. These include, for example, the fact that users of the network are data consumers, who access the network only to get answers for their queries. In addition, the inherent redundancy of sensor networks offers significant opportunities to network management approaches. One example is the paradigm of combining data from different sensors to eliminate redundant transmissions. This paradigm shifts the focus from traditional Internet-based address-centric approaches to a more data-centric approach [33].

Another particularity that makes routing in WSNs specific is the directionality of the traffic. As shown in Figure 11.3, information collected by sensors is transmitted toward a

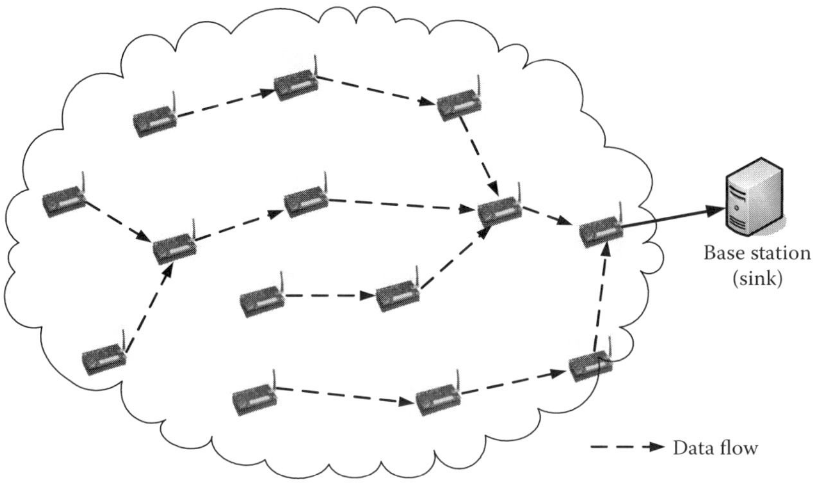

FIGURE 11.3 An example of a wireless sensor network.

sink (or base station), which is responsible for handling the sensed data. Finally, it is worth noticing that a wireless sensor network is cooperative by nature (common goal), contrary to wireless mobile ad hoc networks, where often nodes are assumed to be selfish.

Due to the inherent characteristics that distinguish these networks from other wireless self-organizing networks (e.g., mobile ad hoc networks), routing is also a challenge to the design and management of WSNs.

11.3.3 Wireless Mesh Networks

Wireless mesh networks (WMNs) are two-tiered networks composed of mesh routers and mesh clients, where mesh routers have minimal (or no) mobility and form the backbone of the network (Figure 11.4).* They provide network access for both mesh and conventional clients. A WMN is dynamically self-organized and self-configured, with the nodes in the network automatically establishing and maintaining mesh connectivity among themselves.

Since WMNs share common features with ad hoc networks, the routing protocols developed for ad hoc networks can be applied to the backbone of WMNs. Despite the availability of several routing protocols for ad hoc networks, the design of routing protocols for WMNs is still an active research area for several reasons. First, the requirements on power efficiency and mobility are much different between WMNs and ad hoc networks. In a WMN, nodes (mesh routers) in the backbone have limited mobility and no constraints on power consumption, while mesh client nodes usually require the support of mobility and a power-efficient routing protocol. Second, existing ad hoc routing

* There are many authors that consider the architecture of WMNs as three-tiered, because they separate routers that have Internet connectivity from the others.

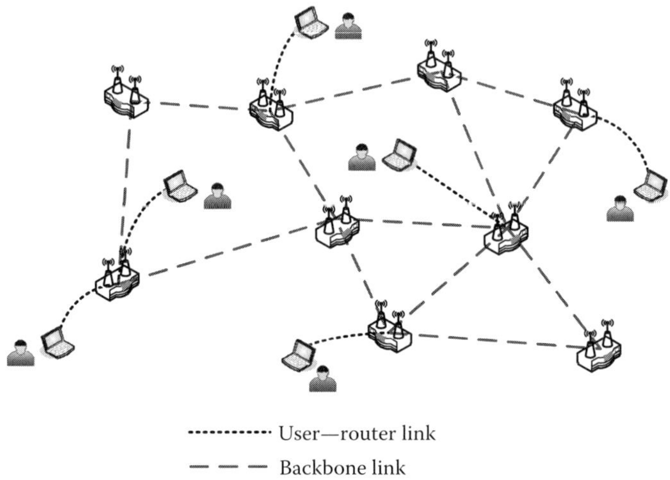

FIGURE 11.4 A representation of a wireless mesh network.

protocols still have limited scalability. Moreover, they treat the underlying MAC protocol as a transparent layer. The cross-layer interaction, for example, may be an interesting approach to improve the performance of the routing protocols in WMNs [25]. Those differences imply that the routing protocols designed for ad hoc networks, although functional, may not be optimized for WMNs.

Some routing strategies proposed in the specific case of WMNs are routing protocols with multiple performance metrics, multiradio routing, and multipath routing for load balancing and fault tolerance. The first class of routing protocols proposes to use performance metrics related to link quality to select routing paths [10, 11]. Multiradio technologies per node are also a candidate strategy in many situations, since the capacity of the network can be increased without modifying the MAC protocol [12]. Another possibility is to use multipath routing. The main objective of this method is to perform better load balancing and to provide high fault tolerance [38].

11.3.4 Vehicular Ad Hoc Networks

Vehicular ad hoc networks (VANETS) are one of the most promising special cases of ad hoc networks in which nodes are composed by vehicles in movement and eventually roadside base stations. Applications of VANETS are manifold, and can be generally classified into user and safety categories. Vehicles being able to communicate with each other and with their environment could not only help avoid crash situations and save lives on our streets, but also significantly improve the comfort of drivers and passengers. An example of a VANET is illustrated in Figure 11.5.

Vehicular communications have attracted much attention from both academia and industry in the last years. Although inspired from traditional wireless mobile ad

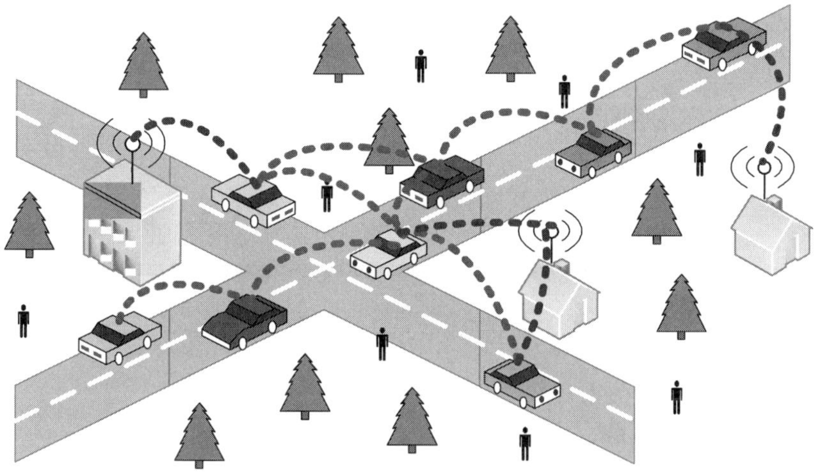

FIGURE 11.5 A vehicular ad hoc network.

hoc networks, VANETs introduce additional constraints regarding the high degree of mobility of the nodes, knowledge of geographical positions, and unique applications. In addition to vehicle-to-vehicle links, roadside-to-vehicle links can be used to access the Internet as well as updates on the traffic, maximum speed limit, nearby parking places, etc. All these characteristics have to be taken into account when designing an appropriate routing protocol for VANETs.

11.3.5 Disruption-Tolerant Networks

Disruption-tolerant networks (DTNs) are a specific class of networks in which connectivity between neighboring nodes is intermittent due either to mobility or to the varying conditions of the wireless medium.* In the literature, such a type of network is also referred to as *delay tolerant*, *intermittently connected*, or *highly partitioned*.

Most existing wireless routing protocols assume that there is (almost) always a connected path from a source to a destination. This assumption is clearly not valid in very fast moving networks and sparse mobile networks. This is the fundamental characteristic that makes the difference between DTNs and other WSONs. Intermittent connectivity introduces a supplementary dimension in the space of variables to be managed by the routing protocol.

Figure 11.6 shows a disruption-tolerant network at three different time instants. Observe that the protocol stack must support topology changes, as well as node departures and joins, while considering that disruptions might be short-lived or not.

* In general, in the literature, DTN refers to *delay*-tolerant network. Although these two networks share similar characteristics, in our case the "disruption" aspect is more important.

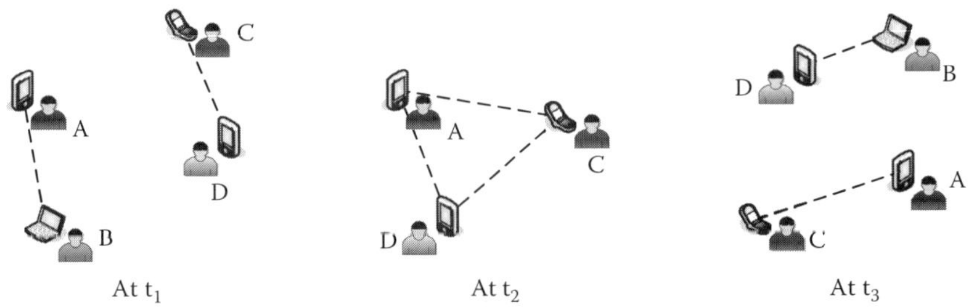

FIGURE 11.6 An example of a disruption-tolerant network at three different time instants.

11.4 Requirements of Wireless Self-Organizing Networks

When considering the application of routing protocols to wireless self-organizing networks, it is useful to consider the following main technical requirements:

- *Localized computation*: Decisions in the network must be taken in a local scope through simple neighborhood consensus.
- *Distributed nature*: Information and management responsibilities should be distributed among the nodes in the network.
- *Flexibility in route selection and dynamic-network management*: The addressing structure should offer flexibility in route selection. This issue has an impact on the dynamic-network management and affects performance in terms of path length, traffic concentration, and resilience to failures.
- *Scalability and low control message overhead*: Lookup operations should avoid heavy-overhead solutions like flooding the entire network to locate a node. Related requirements include simple forwarding decisions, low communication cost, and forwarding tables that are independent of the total number of nodes in the network.
- *Manageable complexity*: The addressing structure should be as flexible as possible when handling the addressing space allocation as nodes join, leave, or move.
- *Easy path computation*: Paths must be easily determined, independently of the complexity of the addressing structure. In other words, this means that the way nodes self-organize should have little influence on the complexity of the routing protocol.

Routing in such a context is challenging. In the Internet community, Mobile IP supports routing for mobile hosts [1, 43]. The Mobile IP solution works well if there exists a fixed routing infrastructure supporting the concept of "home agent," which stores location information of mobile nodes. Nevertheless, if all nodes, including home agents, are able to move, such a strategy cannot be directly applied.

In fixed networks, routing (or location) information is embedded into the location-dependent node address (e.g., IP addresses have been defined for both *identifying* and *locating* a node in the network). In wireless self-organizing networks, however, there is no static relationship between the node's location and the node's identifier as a consequence of the spontaneity and adaptability of the network. In this context, numerous routing protocols have been proposed in order to be adaptable to the network conditions imposed by each type of WSON.

11.5 Routing in Wireless Self-Organizing Networks

In practice, any routing protocol for WSONs can be defined as a composition of three fundamental architectural blocks: identification, location, and forwarding [9].* A routing protocol is composed of a combination of one, some, or all of the aforementioned blocks, and a particular composition determines the behavior of the routing architecture and its associated protocols.

11.5.1 Identification

The first step of any routing protocol is to identify nodes in the network. Identification can take two forms: naming and addressing.

There are in general two distinct uses for naming systems. First, a name may contain some semantics that represent the nature of the entity under consideration. For instance, in a DTN urban network, police-sgt3452 could be used to indicate that the entity is police sergeant number 3452. Similarly, sink-4-temperature could be an example of the name of sink number 4, which is used to collect temperature information in a sensor network. The second usage of names is for identifying nodes in a network. More precisely, names can be used (directly or indirectly) as inputs to the routing protocol. This means that nodes have to be uniquely recognized in the network.† A typical example is the use of distributed hash tables for routing. In this approach, the user-friendly name of the node is translated (through a hash function) into an identifier that belongs to a known space. This identifier is directly used by the routing protocol to locate nodes in the topology.

Names are particularly important in wireless self-organizing networks because the spontaneous nature of the network makes it difficult to establish any prior convention of node identification.

Addresses are responsible for giving a topological signification to node identifiers. By definition, an address is the basic information a node needs before communicating with another node, since it identifies the endpoints of a path. Consider, for example, the postal system. The hierarchical representation of postal addresses is the result of a consensus that allows the global system to be coherent. Because everyone respects this convention, routing letters between senders and receivers becomes possible. A special

* In [9], the authors propose a decomposition of the routing architecture into five building blocks: naming, addressing, disseminating, locating, forwarding. Without loss of generality, here we extend this definition by merging naming and addressing into the "identification" block, and dissemination and discovery into the "location service" block.

† It is generally required that identifiers be unique, unless the objective is to identify groups of nodes.

case of an addressing scheme is the one that considers the geographic coordinates of the nodes as their addresses. Such an approach has the advantage of uniquely representing nodes (since two nodes cannot occupy the same point in space) and easing the forwarding procedure (see section 11.5.3).

We can apply the same reasoning for the Internet. The term *IP address* is meaningful only if the hierarchy supposed to exist is respected and if paths are correctly associated to this hierarchy. Indeed, one of the main problems with the current architecture of the Internet is that IP numbers are used for both identifying and addressing nodes. Such a mixture of concepts is undesirable in wireless self-organizing networks because nodes are potentially mobile. If the same value were used for both identifying and addressing a node, then each time a node moved (i.e., its topological position changes) it would also be obliged to change the identifier.

11.5.2 Location Service

The core of any routing protocol is composed of the location service, which is responsible for establishing contacts between nodes that want to communicate with each other. When a source wants to communicate with a destination, it must request the location service to obtain the address of the destination by providing the destination's name. In this way, the location service can be seen as the mechanism through which nodes transform names into addresses.*

The result of a location service is either an address that explicitly indicates the topological position of a node (e.g., the geographic coordinates of the node) or some information on the path that packets will have to follow to reach the destination.

Any location service is in fact composed of two complementary steps: dissemination and discovery. Dissemination and discovery phases form a key part of the routing system. Moreover, the performance and efficiency of the routing scheme depend on their complementarity. Once a node knows its address, it has to make this information available to its potentially corresponding nodes, and eventually to the whole network. This is the dissemination phase. It can range from null (reactive protocols) to full (totally proactive protocols). Limited dissemination, where the management of location information is shared among nodes, is an intermediate case.

The next phase, which is closely related to the dissemination strategy, is the discovery (also known as localization) system, which is the act of obtaining a path to a particular node, its address, or its location. In fact, the degree of dissemination often determines the complexity of the discovery algorithm.

Loosely speaking, the location service can be performed in different ways according to how many nodes play the role of a location server. In *zero-dissemination* approaches, each node only knows its own position. This method requires that the source floods a route request in the network until the destination responds with its coordinates. Such an approach presents in general poor scalability properties because of the high traffic

* We focus here on location of nodes in a wireless self-organizing network. This terminology is not to be misinterpreted as services providing the location information on users or services at the application layer.

control overhead generated. There is also *full dissemination*, in which each node always knows the positions of all the other nodes in the topology in a proactive fashion. While discovering a node's position is fast and does not generate any traffic overhead, such a solution leads to high traffic overhead in order to keep databases constantly updated. *Partial-dissemination* methods distribute the whole topology information among a subset of nodes. When a node looks for a destination, it sends a route request to one of the servers. Although simple, this approach is not fully distributed because some nodes have more responsibilities than others.

For example, in location services that rely on distributed hash tables (DHTs), every node in the topology plays the role of a rendezvous point. A node n stores its location information in a rendezvous node r depending on n's identifier (cf. section 11.8.2). The problem in such a system is that, in order to keep an accurate location system, the location information must be updated every time a node moves. Depending on the dynamic nature of the topology, this may lead to high signaling overhead.

We can classify methods of location information dissemination based on when this information is updated. Periodic protocols disseminate routing information periodically, which allows rapid destination's location discovery and maintains network stability. The update period is important. When the update period is large, the information is not always up to date. When the period is short, network bandwidth, a precious resource in many WSONs, is consumed due to update packets. On the other hand, event-driven methods (where packets are sent only when an event occurs, a link failure, for example) may generate a large number of nonessential update packets if topology changes frequently. A good solution is a threshold, where update packets are not sent so frequently, but not only when events occur.

11.5.3 Forwarding

The ultimate goal of a routing scheme is the effective transmission of data, implemented by the forwarding scheme. It dictates the quality of the communication and is what users observe. The performance of the forwarding mechanism (e.g., path length, cost, delay, energy consumption) closely depends on the previous two components. For instance, messages can be forwarded following a hierarchy given by the addressing scheme or a geographic path computed through Euclidean distances between nodes. It can be predefined (source routing) or based on a hop-by-hop basis. The choice of what forwarding scheme to use also depends on desirable qualitative properties such as robustness, path length, power consumption, message delivery success, and scalability.

11.6 Parameters Having an Impact on the Routing Scheme

We are now able to derive a (nonexhaustive) list of useful parameters having an impact on the routing protocol. These observations can be used both in other analyses and to help designers proposing new solutions adapted to particular situations. In sections 11.7 and 11.8, we will use these parameters to better understand the suitability of the protocols to a particular situation. In other words, these parameters will be used to explain the issues that have an impact on the suitability of a routing solution to a particular set of characteristics of a network.

11.6.1 Mobility

Mobility is a parameter of major importance in the design of a routing protocol, and it requires a particularly careful design of the building blocks. Assigning to a node an address that is dependent on the node's neighborhood (e.g., in a tree-like fashion) is not appropriate since neighborhood information is more likely to change frequently as mobility increases; it is better to adopt an absolute addressing/positioning solution (e.g., GPS) and put more effort on the discovery algorithm. Indeed, since nodes move, their addresses are likely to have a short valid lifetime.* The higher the mobility, the less the usefulness of the dissemination phase, and the greater the need for an efficient discovery algorithm. Indeed, the effort of maintaining an accurate map of the network may be too costly compared to the level of link usage.

11.6.2 Intermittence of Connectivity

The WSON paradigm opens new ways for mobile users to get connected anywhere and anytime. An immediate consequence of this is the fact that connectivity may be intermittent. Due to user mobility, forwarding paths may be unstable and node reachability may be highly variable.

Existing wireless routing techniques, however, commonly assume that, for a connected topology, there is always a path from source to destination (although some works on MANETs consider splits/merges). This assumption is not always valid in realistic scenarios. The field of DTNs looks at enabling communication in the absence of end-to-end connectivity (due to frequent network partitions) or in the presence of links that are subject to long and variable delays. The idea is to explore the fact that users are nowadays more and more equipped with wireless devices, and that users that are physically close are potential data exchangers. It seems, then, interesting to exploit the resources of any available wireless communication possibility to deliver or store collected data in WSONs of intermittent connectivity.

11.6.3 Behavior of the Medium

In a WSON, the routing protocol is directly influenced by the characteristics of the underlying wireless medium, especially with respect to the dynamics of the perceived channel conditions. The wireless medium can present extreme variability. This is in opposition to the wired case, where the medium (a cable) can be characterized in a quite precise way.

---

* Note that in position-based routing protocols, the address of a node is given by the coordinates of the node in the geographical space.

11.6.4 Environment

In general, solutions for routing in WSONs only assume the existence of nodes and links. Some situations, however, require that the routing protocol be aware of the environment to which nodes belong. Examples are wireless sensor networks and vehicular mobile ad hoc networks. These networks are extremely sensitive to the conditions in which they operate. These conditions may include the presence of obstacles or limited space where nodes can be deployed (e.g., on top of buildings or spot lights in VANETs, or on trees and animals in WSNs).

11.6.5 Position Availability

To enable some applications, the nodes have to be equipped with GPS receivers: for example, in VANETs, the location of a traffic jam is essential for traffic monitoring and control applications; similarly, in military scenarios the nodes are likely equipped with GPS receivers to enable tactical situational awareness applications. One interesting property of WSONs is that the physical topology and the network topology are highly correlated (especially if omnidirectional antennas in a terrain without significant obstacles are used), i.e., if two nodes are physically close, there is likely a usable link between them. Thus, if the location of *significant* nodes is known (e.g., the location of the destination or of the immediate neighbors), several routing optimizations can be implemented. For example, in [32] the scope of a route request flood is restricted, based on knowledge of the possible location of the destination. Alternatively, a purely geographical forwarding algorithm (if the destination is known to be located in a particular direction, forwarding the packet to a neighbor in that direction) can be used. The main problem of geographical routing approaches is that, for some cases, they can reach dead ends and fail. Elaborate backtracking schemes can be employed to resolve this situation [51].

11.6.6 Reliability

Highly mobile WSONs are some of the most challenging types of networks for routing protocols. The hundreds of papers on routing in WSONs are a testament to the unique challenges posed by these networks. Years after classical routing protocols have been adopted and fine-tuned for the Internet, the "optimum" routing protocol for WSONs still eludes researchers. At its core, two phenomena are responsible for this situation: First, there is a very high rate of link changes associated with mobile large WSONs, and a corresponding high rate of route breakages. Second, compounding the problem, the wireless links themselves are inherently unreliable; thus, a loss of a packet is ambiguous—either the neighbor moved away (in which case the routing protocol should react and repair the path) or the transmission temporarily failed (in which case, the route should be maintained).

In addition to the two reasons above, imposing any type of structure or hierarchy in the network (e.g., electing cluster heads or forming a backbone) implies the introduction of new services that are not fully distributed. Changes in WSON topology or node membership may easily make these new services unavailable, thus further impacting the reliability of the routes in these networks.

Using (hot and disjoint) backup routes or fast reroute techniques can significantly improve the reliability of the routes in these networks at the cost of some communication and storage overhead. For example, each node can store two or more next-hop neighbors for each (active) destination.

11.6.7 Multicast and Broadcast Issues

Traditional multicast protocols (like PIM—Protocol Independent Multicast) have the same problem as traditional routing protocols—too much state has to be updated, resulting in a large overhead in large, dynamic WSONs. Therefore, several strategies to reduce the amount of updates have been devised [39]. Network broadcast, on the other hand, can be done relatively efficiently without using a significant amount of state at every node by taking advantage of the inherent broadcast nature of wireless communications at the physical layer. However, it was shown in [56] that the more state that is known (up to a two-hop neighborhood for each node), the more efficient the broadcast can be at low mobility. Of course, the performance degrades as the mobility increases.

11.7 Routing Protocols for WSONs: Classification

According to the building blocks described in section 11.5 routing protocols for WSONs can be classified into several categories. The overall view of this classification is illustrated in Figure 11.7.

11.7.1 Classification Based on the Addressing Scheme

Different addressing schemes can be applied to topologically identify nodes in a network. They fall in two categories, complete and incomplete, as described in the following:

Complete. An address is said to be complete when it embeds in some way the path to be followed by packets to reach the destination. An example is an address based on geographic coordinates. If a source knows its own coordinates and wants to send a packet to a destination, choosing the next-hop is trivial. The address of the destination (or its coordinates) is obtained through the location service that relies on the identifier of the destination. Upon obtaining the destination's address, it suffices to forward the packet to the neighbor that gets the message the closest to the destination, and this recursively until the latter receives the packet.*

Incomplete. We define an address as incomplete when it must be associated with an explicit path pointer to be topologically coherent. This means that the address of the destination is different for each potential source of data. The

* Of course, we assume here that the network is sufficiently dense so that there is always a neighbor that is closest to the destination. In the case where voids (dead ends) happen, nodes must apply alternative contouring solutions such as the right-hand rule [30].

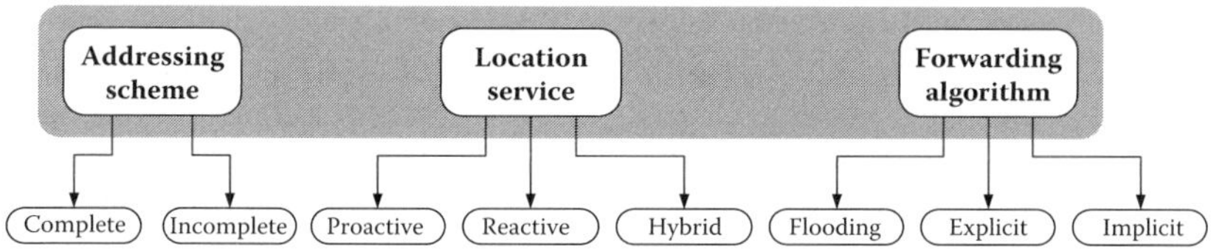

FIGURE 11.7 Classification.

solution used in the legacy Internet is a typical example of incomplete addressing scheme, where IP addresses alone are meaningless until they are associated with some next-hop information.

In WSONs, the same principle is used in protocols such as AODV and OLSR (cf. section 11.8 for more details on these protocols). Compared to the complete case, the main advantage of using incomplete addresses is that more elaborated routing protocols (e.g., routing with quality-of-service constraints) can be implemented in an easier way.

11.7.2 Classification Based on the Location Service

As previously described, the location service is composed of two functional blocks: dissemination and discovery. Based on this criterion, we can also classify routing protocols into three categories related to the type of location service:

Proactive. These protocols have been designed upon the same principles as routing protocols for wired networks. Proactive protocols [49], such as DSDV (Destination-Sequenced Distance-Vector routing protocol) [44], OLSR (Optimized Link State Routing protocol) [8], and HSR (Hierarchial State Routing protocol) [40], continuously maintain route entries for all destinations, including nodes to which no packets are sent. Routing tables are thus continuously updated, which allows nodes, at any moment, to communicate with each other without incurring additional delay and overhead for route establishment (in contrast with reactive protocolsm as described below).

Updating routing tables in this way allows fast communication establishment and low delay. However, a large amount of communication overhead is generated to keep all nodes' routing tables updated. Because of their ability to discover routes quickly, proactive protocols are appropriate for networks where quality of service is required and a high amount of communication exchanged.

Reactive. Reactive protocols [49] construct routes between sources and destinations as they are needed. Routes are thus discovered on demand. Usually, route discovery is based on flooding a request message, relayed by nodes, until it reaches the destination or a node aware of a way to reach the destination node. DSR (Dynamic Source Routing protocol) [28], TORA (Temporally-Ordered Algorithm protocol) [40], and AODV (Ad Hoc On-demand Distance Vector routing protocol) [42] are examples of reactive routing protocols.

Hybrid. Hybrid routing protocols are a mix of proactive and reactive protocols. HSLS (Hazy Sighted Link State routing protocol) [50], for example, uses a mathematical optimization to be able to use a link-state method with a reactive forwarding mechanism in order to optimize routing table updates. ZRP (Zone Routing Protocol) [19] is also a hybrid routing protocol that uses a proactive component within the zones and a reactive one between them. Distributed hash tables (DHTs), like Peernet [14] or Tribe [54], are also considered hybrid routing protocols. In these schemes, each node *n* is responsible for storing locations of some nodes in a distributed fashion. Each time a node moves, it must send its new position to the node that is responsible for storing its location information (note that this procedure is the dissemination part of the location service). We present in section 11.8.2 a complete example of a routing protocol based on DHTs.

11.7.3 Classification Based on the Forwarding Strategy

Routing protocols can also be classified according to their method of forwarding packets. We define three types of forwarding strategy: flooding, explicit, and implicit:

Data flooding. In this category of routing protocols, there is no location service at all. When a source node, *s*, needs to send a message to a destination node, *d*, it simply floods the network until the message reaches *d*. Although sending a message this way ensures that it will reach the destination (if *s* and *d* are connected), probably several times, it also reaches all the nodes in the topology. Because flooding is a simple routing algorithm, it is easy to implement. On the other hand, it requires a high amount of energy and bandwidth per message. Thus, flooding is not appropriate for large topologies with high communication rates.*

Explicit. Explicit forwarding is closely related to the incomplete addressing scheme. Indeed, nodes need an explicit definition of next-hop to forward packets (the next-hop is available in the forwarding table, computed in advance by the location service). In the case of DSR, for instance, the source node receives, as an answer to its request, the entire path the message will follow toward the destination. This route is then included in all the packets the source node generates.

Implicit. As in the previous case, implicit methods are also related to the addressing scheme adopted. Let us consider again the case where addresses are represented by the geographic coordinates of the nodes. When forwarding, a node must of course know its own coordinates, as well as the coordinates of its immediate neighbors and the ones of the destination (the latter is obtained through the location service). In the basic version, each node transmits the packet to the neighbor that is the closest to the destination. Since no specific

* This approach is quite expensive in terms of bandwidth consumption in large networks with large amounts of data to be exchanged. The problem exists when flooding a discovery message in reactive protocols. However, in reactive protocols, the discovery message is small and sent only once (while the path is valid). Data packets are then unicasted to the destination.

route is established in advance (instead, in an on-demand basis), this approach is classified as implicit.*

11.8 Routing Protocols for WSONs: Examples

In this section, we present a number of routing protocols for wireless self-organizing networks proposed in the literature. Of course, this list is far from comprehensive—there are probably thousands of routing protocols and variations for WSONs.

We do not present routing protocols for each class of WSON. Instead, we simply describe solutions that either are representative of the domain or have interesting properties. The reason for this choice is that, as it will become clear in the following, most of the solutions are quite generic. Furthermore, it is more natural to classify them in terms of achieved properties (e.g., robustness to mobility, resistance to density, energy consumed, etc.). However, when appropriate, we point out the type of network for which a particular solution has been designed.

11.8.1 Internet-Inspired Routing Protocols

The first wave of routing protocols for WSONs, mainly in the context of MANETs, have been largely inspired from fixed networks. Let us consider two known examples: AODV and OLSR.

The Ad-hoc On-Demand Distance Vector (AODV) is a reactive protocol for MANETs. In order to obtain a path to the destination, the source floods the network with route request (RREQ) packets until the destination is reached. As RREQ packets travel the network, they store the identifier of each node traversed. The destination responds with a route reply (RREP) packet to the source using the reverse path. Upon reception of an AODV RREQ packet, an intermediate node stores a route to the source (backward learning) and forwards the request. When the destination receives the RREQ packet, it responds using the path traced by the query.

To reduce the portion of the network flooded with broadcast messages, different routing protocols (e.g., AODV) use the expanding ring search (ERS) [3] technique during the route discovery phase. The idea is to limit the scope of the broadcast message with the utilization of the packet's TTL (Time To Live) field. ERS starts with a fixed value for the TTL, which is progressively incremented each time the search fails, until the destination is found or a maximum TTL is reached. An inherent problem of expanding ring search is the choice of the TTL values. Since the source does not know the exact location of the destination, it is difficult to determine the range of the search. On the one hand, if the range is short, further searches will be required until the destination is found. In this case, a potentially large number of links will be used unnecessarily and the route discovery latency will be higher. On the other hand, if the TTL is too large, network resources will be wasted in case of closely located destinations. AODV may perform ERS in two situations. The first one is when a new connection must be established and the source

* Note that the data flooding category could in fact be included in the implicit case. However, we preferred to set it apart because it does not rely on any location service.

does not have a route to the destination. The second one is a consequence of route loss, due to node mobility or failure. When the route is broken, a new broadcast is performed in order to find a new route. This frequent use of broadcast may degrade the overall system performance. Nevertheless, it is inherently difficult to determine the first route between two communicating nodes without performing a network-wide search. Thus, we argue that broadcast might be performed for the first search, but can be avoided for subsequent ones.

The Optimized Link-State Routing (OLSR) is a link-state proactive protocol for MANETs. Like all link-state protocols, each node in the network informs all other nodes in the network about the state of its links. Each time a new link is made or an old one breaks, all the nodes in the network have to be informed of this event. OLSR uses an optimized method of disseminating the link updates: instead of using pure flooding (where each node in the network broadcasts exactly once), each node chooses a set of neighbors (called multipoint relay [MPR]) to forward the updates to all the two-hop neighbors of the node. Thus, by using knowledge of the local (two-hop) topology, OLSR achieves a considerable reduction in the number of transmissions needed to update the information in the network. The nodes that are selected as an MPR by some neighbor nodes announce this information periodically in their control messages. Thereby, a node announces to the network that it has reachability to the nodes that have selected it as MPR. Once an MPR node has a complete knowledge of the topology of the network (through the received link-state updates), the nodes use a shortest-path algorithm like Dijkstra's to compute the routes in the network. Thus, in route calculation, the MPRs are used to form the route from a given node to any destination in the network.

11.8.2 DHT-Based Routing

Different from fixed networks, wireless self-organizing networks have no static relationship between the node's location and the node's identifier as a consequence of the spontaneity and adaptability of the network. This requires a dynamic association between identification and location of a node, and the specification of a mechanism to manage this association. In response to these requirements, distributed hash tables (DHTs) have been largely adopted as a scalable substrate to provide a number of properties, such as information distribution, location service, and location-independent identity, upon which a variety of self-organizing systems have been built. The functionality of decoupling identification from location, and of providing a general mapping between them, has made the DHT abstraction an interesting principle to be integrated in network-level routing protocols. The idea is to use a hash function to distribute nodes' location information among rendezvous points throughout the network. This hash function is also used by a source to identify the rendezvous point that stores a destination's location information. Distributed systems that make use of such a strategy inherit robustness, ease of operation, and scaling properties.

DHTs and rendezvous-based routing protocols have been the subject of a number of proposals in the literature [6, 14, 35, 55]. These protocols differ in the way they define the design issues and in the way DHTs are applied. The Grid project [16] combines the GLS (Geographic Location Service) [35] distributed location service—a DHT-based location system—with geographic forwarding. The characteristics of the Grid system are detailed in the following. GLS allows the creation of a location structure where the world is partitioned into a hierarchy of grids with squares of increasing size. A distributed location service is thus provided by replicating each node's current location at a small subset of location servers. A node selects location servers in each level of the grid hierarchy. PeerNet [14] proposes a network layer architecture with integrated support for routing between peers and explicit distinction between node identity and address. Addresses of l bits are organized as leaves of a binary tree of $l + 1$ levels, being physical nodes residing at the leaf level. Nonleaf vertices in the tree represent a range of addresses with a common prefix. An integrated distributed node location service maps the identifiers onto addresses using a known hash function. Finally, routing is a recursive procedure through the address tree, and can be seen as a hierarchical form of proactive distance-vector routing. Terminodes [6] is a long-term research project, aiming at deploying a new architecture for large-scale self-organized mobile ad hoc networks. The proposed framework encompasses all layers and explores interlayer interactions. Terminodes is built upon the concept of geographic forwarding and proposes a DHT-based location service to distribute and locate information throughout the topology. Each node advertises its current position to a geographical region called virtual home region (*VHR*) with a fixed center C_{VHR} and a variable radius r (adaptable to the density of the area containing the VHR). All nodes within the VHR store the node position information. Thus, the DHT is applied in order to do the mapping between the space of node identifiers and the geographic space of a terminode network.

11.8.3 Cross Layer Routing

Routing belongs squarely in layer 3 of the OSI networking stack [52]. Like any well-behaved protocol, routing processes should only communicate with peer routing processes in other devices through routing packets sent through the lower layer (the data link layer in this case). For the most part, this is the case with routing protocols in the Internet: although they use either IP packets or User Datagram Protocol (UDP) or Transmission Control Protocol (TPC) to exchange messages, the routing protocols do not interact in any other way with those layers.

Cross layer routing thus refers to the practice of designing a routing protocol that interacts with other layers in the networking stack. The main problem with this classical layering approach is that in WSONs, the resulting system can be rather inefficient if the routing protocol does not communicate with any other layer in the stack.

The exact type and degree of interaction with other layers (as well as the layer itself) vary widely in the literature, but the physical layer is most often used for increasing the performance of routing protocols. The MAC layer follows closely. Upper layer protocols, or even protocols that do not fit clearly in any one layer (e.g., energy management), may also interact with the routing protocol.

At the physical layer many routing protocols propose to use the received signal strength of a transmission from a neighbor to gauge the quality of that link. The primary goal is to avoid links with a low signal-to-noise ratio, as those links tend to have large transmission errors and low bandwidth. The routing protocol can embed this signal strength in the routing metric, and thus choose paths that avoid links with low signal strength. Other approaches consider the *decrease* in signal strength on a link as an indication that the link will soon be disconnected and that it is time to find an alternate route (the detection mechanism is similar to the handover procedure in cellular networks).

Some routing protocols consider the *channel* of the wireless card; it was shown that, especially for nodes with multiple interfaces, using different channels for those interfaces can significantly increase the capacity of the network [46]. The problem in this case is a joint routing and channel assignment.

Similarly, routing has often been coupled with transmission power control in an attempt to maximize the spatial reuse inherent in multihop networks while maintaining network connectivity [45]. In addition to maximizing the network capacity, controlling the transmission power of the transceiver can lead to lower power consumption, a problem especially important in wireless sensor networks, but also in MANETs if the devices are battery powered. Many research papers considered different strategies, although it turns out that the problem is far more complex than initially thought [20]. For the problem of minimizing power consumption, several optimization objectives can be considered. For example, one can optimize the power spent in transmitting a message from the source to the destination, to postpone as much as possible the death of a first node due to lack of battery power, to equalize power consumption in the network, etc. Those different optimization objectives may correspond to different routes in the network.

In other approaches, the MAC layer is tightly coupled with the routing protocol, both relying on each other to optimize their performance. For example, the MAC layer may help the routing layer discover new neighbors (and signal the disappearance of a neighbor), while the routing layer may expose its routing tables such that the MAC layer may find the next-hop of an ongoing packet reception. In fact, opportunistic routing often relies on a MAC-routing interaction for its performance.

In wireless sensor networks (and other data-driven networks), it is common (and natural) to combine routing and in-network processing (in an attempt to optimize one or another performance objective).

The examples above are just a small sample of the hundreds of approaches proposed in the literature. All these approaches rely on interactions between the routing protocol and other layers to optimize one or several parameters of the system. However, as it was convincingly argued [31], cross-layering has its own cost: the layered architecture has tremendous advantages that are not always apparent until we look at systems that passed the test of time—the Internet, computers, etc. One of the main advantages of the layered architecture is its *flexibility*, i.e., the capability of replacing one protocol with another, functionally equivalent protocol. Cross-layering destroys this flexibility for a boost in performance. However, it was shown that strictly layered protocol design is, up to a proportional constant, optimal [31].

11.8.4 Encounter-Based Routing

Mobility is one major factor in ad hoc network performance. While it brings some difficulties by frequently changing topology, it can provide opportunities that should be used to improve routing protocols. This aspect of mobility has recently been considered in FRESH [13] and EASE [18], where authors use the node mobility to facilitate routing. When two mobile nodes encounter each other (i.e., they are within the transmission range of each other), both record, in an encounter table, the time (and, in the case of EASE, the location) of this encounter. If two nodes have met in the past, the entry is just updated with the new time (and location) information; otherwise, a new entry is created. The time elapsed since the last update of an entry is the *age* of this information. This phase allows each node to remember its last encounter with other nodes that moved within its transmission range. The age and location information is then used to discover routes to other nodes. Each node is able to use the encounter table to compute a route toward the destination node, by iteratively discovering a set of intermediate nodes that have met the destination with decreasing encounter ages.

11.8.5 Epidemic Routing

Existing techniques in distributed systems, in particular in the Peer-to-Peer (P2P) communication paradigm, may be successfully adapted to explore new solutions in WSONs. It has been proved that systems that implement structured P2P strategies are inherently self-organizing, resilient to failures, and show scaling and reliability properties [5]. In this context, epidemic information dissemination algorithms have recently gained popularity as a robust and scalable way of propagating information in a peer-to-peer communication model [5]. Some of the epidemic dissemination-related properties that are of a WSON's interest constitute: resilience to failures in the infection process, reliable propagation over populations, simplicity and facility of deployment, stable behavior in the presence of a high rate of links or failures, and scalability potential.

From the epidemic algorithm point of view, each node of the network is potentially involved in the dissemination and forwards each message to a randomly selected set of nodes of limited size f, called the fan-out of the dissemination. More specifically, conventional gossiping-based routing protocols [3] use the gossip concept of a distributed system to select a random node to send a packet in order to avoid the blind broadcast of the packet. Vahdat and Becker [53] use only opportunistic contacts and an epidemic routing scheme, which consists of flooding to route data when nothing is known about the behavior of nodes.

11.8.6 Opportunistic Routing

In the case of DTNs, mobility, and in particular the knowledge of the mobility pattern of nodes, can be utilized to propagate messages in the system. That is, messages are forwarded to nodes that are likely to be moving toward a desired region of the network. During their mobility, these nodes offload the message opportunistically to other nodes they encounter. The decision regarding which nodes to offload a message to can be taken either by means of some heuristic or probabilistically. Hence, mobility becomes a substitute for dissemination along connected paths and an asset of data forwarding, rather than an obstacle.

In the literature, some interesting routing protocols explore the contact opportunity in the network to disseminate information. Jain et al. [27] try to improve the connectivity of a remote village to the Internet based on contacts who may be known in advance, and thus scheduled. Friedman et al. [29] try to predict contacts by taking advantage of knowledge concerning node mobility, and then route information. Grossglauser and Tse in [17] propose a two-hop forwarding algorithm, where data are forwarded from the source to a unique relay, and then, in a future contact, to the destination. However, that relay is chosen purely based on a first contact opportunity, and not considering any other information. In order to more carefully select potential relays during contact opportunities among wireless devices carried by humans, studies of human mobility performed by Chaintreau et al. [7] and Hui et al. [24] reveal interesting properties. Their results show that the distribution of intercontact time exhibits a heavy tail, such as one of a power law, over a large range of values. Nevertheless, they leave for a future work the design of forwarding algorithms that take into account human movement patterns.

11.8.7 Data-Centric Approaches

As stated before, the cooperative nature and the inherent redundancy of sensor networks offer significant opportunities to network management approaches. An example is the classical necessity in WSNs of sensor-to-sink communication. In this case, new network opportunities include, for example, the use of data aggregation for combining data coming from different sources toward a single sink. This points out the WSN requirement for data-centric approaches [33] instead of traditional address-centric approaches.

Data-centric approaches consist in finding routes from multiple sources to a destination that allows in-network consolidation of data, where aggregation of multiple input packets into a single output packet is performed by en route nodes. In this way, data aggregation may be performed to reduce data transmission by eliminating the redundancy. In the literature, this kind of approach can be classified as *reactive* and *proactive*.

Reactive data-centric approaches [23, 26, 33, 34] require flooding of data queries in the entire network. They are characterized by specific queries issued by sinks and by sensor answers returned directly to the issuer of the queries. General suboptimal schemes can be used to generate trees from sources to requesters of data. B. Krishnamachari et al. in [33] examine three schemes: center at nearest source (CNS), shortest-path tree (SPT), and greedy incremental tree (GIT). In addition, direct diffusion (DD) [15, 26] and Sensor Protocol for Information via Negotiation (SPIN) [21, 34] protocols are also reactive data-centric approaches and represent the two most commonly referenced wide-area sensor data dissemination techniques. Considering that event information can be named (by using single attribute types, including geographic location) and stored locally at a sensor node upon detection, direct diffusion involves sending requests for named data and allowing those nodes that have such data to respond. More specifically, sink nodes query for type information by disseminating interest (i.e., a range of values for one or more attributes). During interest propagation, a reverse data path is configured with associated gradients for data that match the interest. Following the configured gradients, while a datum is routed to the requester, it may be aggregated by intermediate nodes (those that provide reinforced data and positive gradients), resulting in a tree flowing from these nodes to the requester. Whenever similar data meet at a branching node in the tree, copies of similar data are replaced by a single message. This in-network aggregation represents a key feature of the direct diffusion method. DD does not require any packet forwarding methods other than flooding. The SPIN family proposes a resource-adaptive protocol and uses data negotiation in order to avoid waste of communication and energy resources of flooding-based approaches. Nodes running SPIN assign a high-level name to their data, called metadata, and perform metadata negotiation before transmission. SPIN makes use of the property that nodes in close proximity have correlated data, and hence there is a need to only distribute the data that other nodes do not possess. This ensures that there are no redundant data sent throughout the network, and enables a user to query any node and get the required information immediately. In addition, SPIN has access to the current energy level of the nodes and adapts the protocol behavior in function of the remaining energy.

On the other hand, proactive data-centric approaches [22, 47, 57] consist in storing relevant data by name at nodes within the network.* Thus, the goal here is to allow queries for data with a particular name to be sent directly to the node storing that named data, instead of flooding the entire network. It can also be envisaged to find good data aggregation trees from sources to destination, where each source is responsible for a specific type of collected data. In addition, proactive data-centric approaches can also provide location information about a given event, allowing subsequent queries to be directed toward the event's location. In the literature, the most referred to proactive approach is the Low-Energy Adaptive Clustering Hierarchy (LEACH) protocol [22]. LEACH proposes a clustering-based protocol for transmitting data from individual sensors to a sink node. LEACH uses localized coordination and incorporates data fusion into the routing protocol to reduce the amount of information that must be transmitted to the sink. The main features include local coordination for cluster formation among sensors, randomized rotation of cluster heads for improved energy utilization, and local data compression at cluster heads to reduce the amount of data sent from clusters to the sink. Depending on the amount of residual energy in the sensors, they elect themselves to be local cluster heads and broadcast their status to the other sensors in the network. Each sensor then chooses to associate itself with the cluster head that requires the minimum communication energy. At the end of this phase, cluster heads create a transmission schedule for the nodes in their clusters. This allows nodes to turn off their radio when they are not transmitting. Each cluster head aggregates the data received from the nodes in its cluster before transmitting the whole to the sink.

11.9 Conclusion

In this chapter, we studied the routing problem in wireless self-organizing networks. As we could see, routing in such networks involves a series of design issues (some of them quite orthogonal) that must be taken into account.

The main conclusion we can draw from the vast amount of knowledge accumulated in the area of routing in WSONs is that *there is no unique solution that fits all cases*. Nevertheless, invariants do exist that might help the designer conceive a protocol adapted to

* Names can be, for example, a class of the collected data.

her or his specific requirements. That is the reason we proposed to the reader a gradual understanding of the fundamental problems and principles, instead of a simple collection of existing solutions.

In fact, the deployment of wireless self-organizing networks raises a number of key design issues, such as scalability, spontaneity, ease of management, adaptability, and dynamically changing topologies. In this context, *routing* is particularly challenging and requires revisiting important components of the routing architecture/protocol suite. In this chapter, we have investigated the foundations of routing in WSONs with concern to their adaptability to mobility, large scale, heterogeneity, dynamics, and application requirements, to cite a few. Nevertheless, other important issues have great influence on the routing methodology and must receive more attention from the network community. Some of them are:

- Formal tools for designing routing protocols
- More realistic models for both the wireless medium and mobility
- Integrated solutions ranging from networking functionalities to hardware and software, as well as all the problems in between
- More experimental campaigns
- More applications running on top of these networks
- Benchmarks to help designers evaluate their solutions

References

[1] IP Routing for Wireless/Mobile Hosts (mobileip). IETF working group. Accessed March 2008 from http://www.ietf.org/html.charters/mobileip-charter.html.

[2] M. Abolhasan, T. Wysocki, and E. Dutkiewicz. 2004. A review of routing protocols for mobile ad hoc networks. *Ad Hoc Networks* 2:1–22.

[3] I. Akyildiz, W. Su, Y. Sankarasubramaniam, and E. Cayirci. 2002. A survey on sensor networks. *IEEE Communications Magazine* 40:102–14.

[4] I. F. Akyildiz, W. Su, Y. Sankarasubramaniam, and E. Cayirci. 2002. Wireless sensor networks: A survey. *Computer Networks* 38:393–422.

[5] K. P. Birman, M. Hayden, O. Ozkasap, Z. Xiao, M. Budiu, and Y. Minsky. 1999. Bimodal multicast. *ACM Transactions on Computer Systems (TOCS)* 17:41–88.

[6] L. Blazevic, S. Giordano, and J. Y. Boudec. 2002. Self organized terminode routing. *Journal of Cluster Computing* 5(2).

[7] A. Chaintreau, P. Hui, J. Scott, R. Gass, J. Crowcroft, and C. Diot. 2006. Impact of human mobility on the performance of opportunistic forwarding algorithms. IEEE Infocom 2006. Barcelona, Spain.

[8] T. Clausen and P. Jacquet. 2003. *Optimized link state routing protocol (OLSR)*. RFC 3626.

[9] M. Dias de Amorim, M. Sichitiu, F. Benbadis, Y. Viniotis, and S. Fdida. 2006. Dissecting the routing architecture of self-organizing networks. *IEEE Wireless Communications* 13(6):98–104.

[10] D. S. J. de Couto, D. Aguayo, J. Bicket, and R. Morris. 2003. A high-throughput path metric for multi-hop wireless routing. In *Proceedings of ACM Mobicom*, San Diego, pp. 134–46.

[11] R. Draves, J. Padhye, and B. Zill. 2004. Comparisons of routing metrics for static multi-hop wireless networks. In *Proceedings of ACM Sigcomm*, Portland, OR, pp. 133–144.

[12] R. Draves, J. Padhye, and B. Zill. 2004. Routing in multi-radio multi-hop wireless mesh networks. In *Proceedings of ACM Mobicom*, Philadelphia, PA, pp. 114–128.

[13] H. Dubois-Ferriere, M. Grossglauser, and M. Vetterli. 2003. Age matters: Efficient route discovery in mobile ad hoc networks using encounter ages. In *Proceedings of ACM Mobihoc*, Annapolis, MD, pp. 257–266.

[14] J. Eriksson, M. Faloutsos, and S. Krishnamurthy. 2004. Scalable ad hoc routing: The case for dynamic addressing. In *Proceedings of IEEE Infocom*, Hong Kong, pp. 1108–1119.

[15] D. Estrin, R. Govindan, J. Heidemann, and S. Kumar. 1999. Next century challenges: Scalable coordination in sensor networks. In *Proceedings of ACM Mobicom*, Seattle, WA, pp. 263–70.

[16] Grid Project. Accessed March 2008 from http://www.pdos.lcs.mit.edu/grid/.

[17] M. Grossglauser and D. Tse. 2002. Mobility increases the capacity of ad hoc wireless networks. *IEEE/ACM Transactions on Networking* 10(4).

[18] M. Grossglauser and M. Vetterli. 2003. Locating nodes with EASE: Last encounter routing in ad hoc networks through mobility diffusion. In *Proceedings of IEEE Infocom*, San Francisco, CA, pp. 1954–1964.

[19] Z. J. Haas. 1997. A new routing protocol for the reconfigurable networks. In *ICUPC'97*, San Diego, CA.

[20] M. Haenggi. 2004. Routing in ad hoc networks: A case for long hops. *IEEE Communications Magazine* 43(10):93–101.

[21] W. Heinzelman, J. Kulik, and H. Balakrishnan. 1999. Adaptive protocols for information dissemination in wireless sensor networks. In *Proceedings of ACM Mobicom*, Seattle, WA, pp. 174–85.

[22] W. Heinzelman, A. Chandrakasan, and H. Balakrishnan. 2000. Energy-efficient communication protocols for wireless microsensor networks. In *Proceedings of the Hawaiian International Conference on Systems Science*, Maui, Hawaii, pp. 8020–8029.

[23] J. M. Hellerstein, W. Hong, S. Madden, and K. Stanek. 2003. Beyond average: Toward sophisticated sensing with queries. In *Proceedings of the International Workshop on Information Processing in Sensor Networks.*

[24] P. Hui, A. Chaintreau, J. Scott, R. Gass, J. Crowcroft, and C. Diot. 2005. Pocket switched networks and the consequences of human mobility in conference environments. In *Proceedings of ACM SIGCOMM First Workshop on Delay Tolerant Networking and Related Topics*, Philadelphia, PA, pp. 244–251.

[25] L. Iannone, R. Khalili, K. Salamatian, and S. Fdida. 2004. Cross-layer routing in wireless mesh networks. In *1st International Symposium in Wireless Communication Systems*, Mauritius, pp. 319–323.

[26] C. Intanagonwiwat, R. Govindan, and D. Estrin. 2000. Directed diffusion: A scalable and robust communication paradigm for sensor networks. In *Proceedings of ACM Mobicom*, Boston, MA, pp. 56–67.
[27] S. Jain, K. Fall, and R. Patra. 2004. Routing in a delay tolerant network. In *Proceedings of ACM Sigcomm*, Portland, OR, pp. 145–58.
[28] D. Johnson, D. Maltz, and J. Broch. *DSR: The dynamic source routing protocol for multihop wireless ad hoc networks*, chap. 5, Addison-Wesley, pp. 139–172.
[29] T. Friedman, J. Leguay, and V. Conan. 2006. Evaluating mobility pattern space routing for DTNS. In *Proceedings of IEEE Infocom*, Barcelona, Spain, pp. 1–10.
[30] B. Karp and H. T. Kung. 2000. GPSR: Greedy perimeter stateless routing for wireless networks. In *Proceedings of ACM Mobicom*, Boston, MA, pp. 243–254.
[31] V. Kawadia and P. R. Kumar. 2005. A cautionary perspective on cross layer design. *IEEE Wireless Communications* 12:3–11.
[32] Y.-B. Ko and N. H. Vaidya. 1998. Location-aided routing (LAR) in mobile ad hoc networks. In *Proceedings of ACM Mobicom*, Dallas, TX, pp. 66–75.
[33] B. Krishnamachari, D. Estrin, and S. Wicker. 2002. The impact of data aggregation in wireless sensor networks. In *Proceedings of the IEEE International Conference on Distributed Computing Systems (ICDCSW)*, Washington, DC, pp. 575–78.
[34] J. Kulik, W. Heinzelman, and H. Balakrishnan. 2002. Negotiation-based protocols for disseminating information in wireless sensor networks. *Wireless Networks* 8:169–85.
[35] J. Li, J. Jannotti, D. S. J. De Couto, D. R. Karger, and R. Morris. 2000. A scalable location service for geographic ad hoc routing. In *Proceedings of ACM Mobicom*, Boston, MA.
[36] G. Malkin. 1998. *RIP version 2*. RFC 2453.
[37] J. Moy. 1998. *OSPF version 2*. RFC 2328.
[38] S. Mueller and D. Ghosal. 2004. Multipath routing in mobile ad hoc networks: Issues and challenges. In *Springer lecture notes in computer science*. 2965:209–234.
[39] K. Obraczka and G. Tsuduk. 1998. Multicast routing issues in ad hoc networks. In *Proceedings of the IEEE International Conference on Universal Personal Communications*, Washington, DC, pp. 751–56.
[40] V. D. Park and M. S. Corson. 1997. A highly adaptive distributed routing algorithm for mobile wireless networks. In *Proceedings of IEEE Infocom*, Kobe, Japan, pp. 1405–1413.
[41] G. Pei, M. Gerla, X. Hong, and C.-C. Chiang. 1999. Wireless hierarchical routing protocol with group mobility (WHIRL). In *IEEE Wireless Communications and Networking Conference*, New Orleans, LA, pp. 1538–1542.
[42] C. Perkins, E. Belding-Royer, and S. Das. 2003. *Ad hoc on-demand distance vector (AODV) routing*. RFC 3561.
[43] C. E. Perkins. 1997. *Mobile IP: Design principles and practices*. Reading, MA: Addison-Wesley.
[44] C. E. Perkins and P. Bhagwat. 1994. Highly dynamic destination sequenced distance-vector routing (DSDV) for mobile computers. In *Proceedings of ACM Sigcomm*, London, pp. 234–244.

[45] R. Ramanathan and R. Hain. 2000. Topology control of multihop wireless networks using transmit power adjustment. In *Proceedings of IEEE Infocom*, pp. 404–13.

[46] A. Raniwala and T. Chiueh. 2005. Architecture and algorithms for an IEEE 802.11-based multi-channel wireless mesh network. In *Proceedings of IEEE Infocom*, Miami, FL, pp. 2223–2234.

[47] S. Ratnasamy, D. Estrin, R. Govindan, B. Karp, and S. Shenker. 2002. Data-centric storage in sensornets. In *Proceedings of ACM Sigcomm*, pp. 137–142.

[48] Y. Rekhter, T. Li, and S. Hares. 2006. *A border gateway protocol 4 (BGP-4)*. RFC 4271.

[49] E. M. Royer and C.-K. Toh. 1999. A review of current routing protocols for ad hoc mobile wireless networks. *IEEE Personal Communications* 6:46–55.

[50] C. A. Santiváñez, R. Ramanathan, and I. Stavrakakis. 2001. Making link-state routing scale for ad hoc networks. In *ACM MobiHoc*, Long Beach, CA, pp. 22–32.

[51] I. Stojmenovic and X. Lin. 1999. Gedir: Loop-free location based routing in wireless networks. In *IASTED International Conference on Parallel and Distributed Computing and Systems*, Boston, MA, pp. 1025–1028.

[52] A. S. Tanenbaum. 2002. *Computer networks*. 4th ed. Upper Englewood Cliffs, NJ: Prentice-Hall.

[53] A. Vahdat and D. Becker. 2000. *Epidemic routing for partially-connected ad hoc networks*. Technical report. Department of Computer Science, Duke University.

[54] A. C. Viana, M. D. Amorim, S. Fdida, and J. F. Rezende. 2004. An underlay strategy for indirect routing. *ACM Wireless Networks* 10:747–58.

[55] A. C. Viana, M. D. Amorim, Y. Viniotis, S. Fdida, and J. F. Rezende. 2006. Twins: A dual addressing space representation for self-organizing networks. *IEEE Transactions on Parallel and Distributed Systems* 17:1468–81.

[56] B. Williams and T. Camp. 2002. Comparison of broadcasting techniques for mobile ad hoc networks. In *Proceedings of the 3rd ACM International Symposium on Mobile Ad Hoc Networking and Computing (MOBIHOC'02)*, pp. 194–205.

[57] Yuan Xue, Yi Cui, and Klara Nahrstedt. 2005. Maximizing lifetime for data aggregation in wireless sensor networks. *Mobile Networks and Applications* 10:853–64.

12

Selfishness in MANETs

Younghwan Yoo
Pusan National University

Dharma P. Agrawal
University of Cincinnati

12.1 What Is Selfishness?

A mobile ad hoc network (MANET) is basically a peer-to-peer, multihop mobile wireless network. In the absence of either a fixed communication infrastructure or any base stations (BSs) [1], mobile stations (MSs) constitute a network for themselves. Initially, MANETs have been developed for dangerous environments such as rescue and the battlefield so that emergency personnel or soldiers may be aware of the location of hazardous material or tactical situations. In these networks, all MSs are managed by a common authority (e.g., military or government agency); thus, they collaborate with each other for a common objective. On the other hand, recent proliferation of mobile communication devices and the portability of MANETs accelerate the growth of interest in commercialization of MANETs [2]. Various types of MSs used by different commercial entities have to form a MANET in a self-organizing manner and share their resources for global connectivity with achieving their own goals and objectives. In such type of MANET, it may be reasonable to assume that all MSs do not cooperate with each other all the time. MSs have many resource constraints: low battery energy, small memory capacity, and slow processing capability. They naturally want to use their restricted resources to maximize their own benefit. This benefit means what they can acquire by sending and receiving the packets related to their applications.

Selfishness in MANETs means that some MSs may refuse to serve others to save their own resources. Among various resources associated with MSs, energy is one of the most important, and so energy needs to be conserved as much as possible. In respect to energy consumption, data transmission is the most expensive operation. To transmit a bit over 10 or 100 m distance, MSs consume energy that can perform thousands to millions of arithmetic operations [3]. Thus, MSs may not give forwarding service to others and simply drop packets intentionally. They do not have any interests in supporting global connectivity, even though in the long run all the nodes benefit from such a commitment. MSs who just take advantage of other ones, not cooperating with others if the cooperation does not increase their own profit directly, are called *free riders*.

Selfishness also refers to security attacks on MANETs. These are called passive attacks, as distinguished from active attacks [4]. While active attacks* are performed by malicious nodes to intentionally harm the entire network, passive attacks† are done by selfish MSs whose goal is just to use their limited resources only for their own benefit. The terminology *passive* may give the impression that it is not as critical as the active attacks. However, the increase of the ratio of free riders degrades the packet delivery rate, ultimately resulting in network partitioning. Evaluation of existing schemes shows that even a small number of selfish MSs can impede entire communications and severely degrade overall network performance. Figure 12.1 compares the performance of cases with and without the selfishness prevention scheme Protocol-Independent Fairness Algorithm (PIFA) [6], which will be explained later. The figure illustrates the packet delivery rate against the ratio of selfish MSs among all MSs. In the case where any selfishness prevention is not installed, the packet delivery rates become seriously degraded, as the ratio of selfish MS increases. On the other hand, in the case where PIFA is utilized, the rate is maintained at an acceptable level, over 50%, even when half of all MSs are selfish. This result substantiates the necessity of having a selfishness prevention scheme.

The selfishness issue exists in P2P (peer-to-peer) networks, too. Some free riders just want to download information provided by others without uploading any useful information. However, the selfishness may be more critical in MANETs than in P2P networks because MSs in MANETs cannot get help from any infrastructure. The typical infrastructure network has dedicated nodes for important network operations such as

* They include denial of service (DoS), tunneling (wormhole attack), black hole, and impersonation [4, 5].

† Some researchers define the passive attack as the case where an attacker only illegally eavesdrops on communication to get important information.

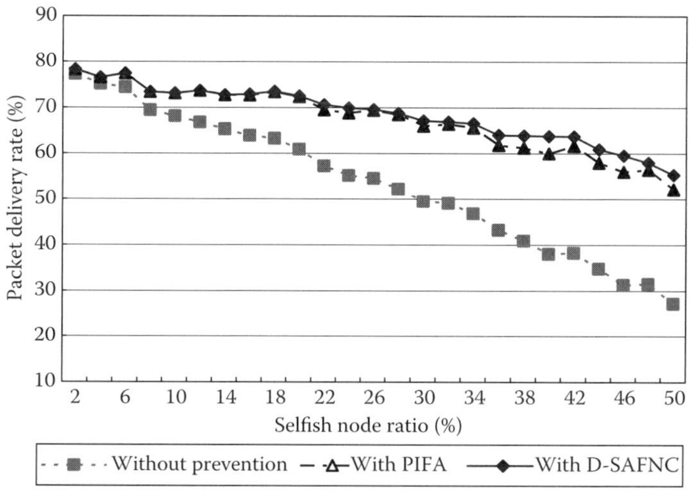

FIGURE 12.1 Packet delivery rate against selfish MS ratio [6]. From Yoo et al., 2005.

authorization, routing and forwarding, and network management. In MANETs, on the other hand, all these functions should be performed by all MSs themselves, although usual MSs cannot be trusted with some important network functions. Thus, security has become particularly important in MANETs. The conventional infrastructure network usually has a centralized authentication server and uses symmetric or asymmetric cryptographic mechanisms. However, MSs still have relatively small memory, low bandwidth, frequent communication errors, and limited battery power, in spite of recent remarkable advances in terms of power efficiency, flexibility, and robustness. This is why traditional security algorithms for infrastructure networks are difficult to be used for MANETs, and hence new approaches need to be developed for them.

MANETs become an easy target of passive attacks because most routing protocols, like DSDV (Destination Sequenced Distance Vector), TBRPF (Topology Broadcast based on Reverse Path Fowarding), AODV (Ad hoc On-demand Distance Vector), and DSR (Dynamic Source Routing) [1], operate on the assumption that all MSs honestly follow the protocol. Although several secure routing protocols have been proposed, such as SRP (Secure Routing Protocol), Ariadne, SEAD (Secure Efficient Ad hoc Distance Vector), ARAN (Authenticated Routing for Ad hoc Networks), and SPINS (Security Protocols for Sensor Networks) [4], they take care of only active attacks, not the selfishness problem. Actually, several selfishness prevention schemes have been reported since the year 2000. The proposed strategies can be divided into the following three groups: *reputation-based scheme*, *credit-payment scheme* (or *price-based scheme*), and *game theory scheme*. In reputation schemes, each MS observes other MSs' behavior and utilizes the information in the route discovery process. On the other hand, credit-payment schemes give credit (which is real or virtual currency) to MSs as a reward for packet forwarding. All MSs need the credit in order to generate their own packets. Finally, game theory–based schemes model the forwarding process as a game whereby all rational MSs gradually reach their own optimal strategies.

The following three sections introduce existing reputation-based methods, credit-based methods, and game theory methods. Then, after describing other work and summarizing, we forecast possible future directions of this research.

12.2 Reputation-Based Method

Secure routing protocols like SRP provide defense mechanisms against a variety of security attacks. Mainly taking into account active attacks, they can proactively prevent the selfishness problem as well. However, proactive schemes may not perfectly defend MANETs against node selfishness, because there always exists a way to bypass/fool the prevention scheme [7]. Hence, a reactive scheme to detect misbehaving MSs is absolutely essential.

12.2.1 Watchdog and Pathrater

Marti et al. [8] propose a method to mitigate bad effects of misbehaving MSs that may be selfish, malicious, broken, or overloaded. Each MS has two function modules on DSR: *watchdog* and *pathrater*. The watchdog *promiscuously* overhears neighbor MSs' transmission to check whether the neighbors forward packets correctly or not. In the promiscuous listening mode, MSs read all packets they can receive, not just packets addressed to them. If an MS finds out that a neighbor has dropped packets more times than a predefined threshold, it notifies the source node of the dropped packets by sending a message. This information is collected by the pathrater at each MS, which maintains a rating for every other MS in the entire MANET. This rating information is utilized to calculate reliability of paths, so as to avoid using a route with misbehaving MSs.

This scheme has several shortcomings. First, since the goal is not to enforce cooperation but to improve network throughput, detected selfish or malicious MSs are just circumvented during the route discovery process and remain unpunished. As a result, selfishness becomes a blessing to MSs themselves. Second, the promiscuous listening mode has critical limitations. For the promiscuous mode operation, wireless links must be bidirectional, while topology control techniques allow some unidirectional links between MSs. Moreover, the use of directional antenna makes it impossible for the watchdog to overhear neighbors' traffic. Third, if any collision is observed during overhearing (like Figure 12.2), the watchdog cannot determine if the collision is due to the neighbor's intentional misbehavior or simultaneous transmission of another neighbor. Fourth, packets may be dropped because of congestion and not because of selfishness. It will be unfair to punish the MSs that drop packets due to congestion. Lastly, each MS requires large memory to store transmitted packets, until forwarding by its neighbor has been confirmed. These buffered packets are compared with packets forwarded by its neighboring MS, to check if the neighbor transmits correct data.

12.2.2 Context-Aware Inference Method

Whereas watchdog and pathrater [8] focus on data forwarding, the context-aware inference mechanism [9] can detect misbehavior during the DSR route discovery as well.

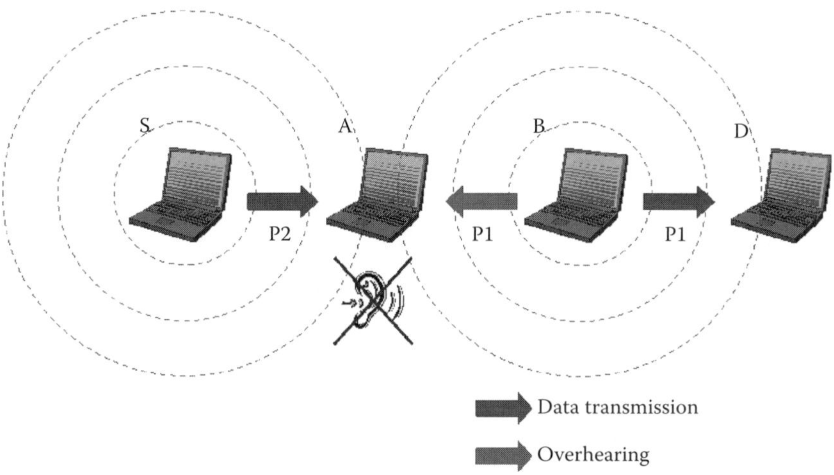

FIGURE 12.2 While MS *A* overhears MS *B* transmitting P1, P2 arrives at MS *A*.

Meanwhile, selfish MSs may falsely report link breakage errors to evade packet forwarding after the route discovery. Thus, the context-aware mechanism provides a way to prevent the false report, too. For misbehaving MSs, special *security messages* (SECMs) are sent to the source. In order not to unduly punish MSs by false accusation, source MSs do not convict the accused MSs until the same accusation arrives from more than three neighbors.

12.2.3 CONFIDANT

CONFIDANT (Cooperation of Nodes: Fairness in Dynamic Ad-hoc Networks) [10] requires each MS to have four components: *monitor, trust manager, reputation system*, and *path manager.* CONFIDANT is similar to the watchdog in [8], but the monitor not only promiscuously listens to the data transmission of neighboring MSs, but also observes how the neighbors treat route request (RREQ) messages. If any misbehavior is detected, the trust manager notifies all the friend MSs by sending *Alarm* messages. Trust managers at friend MSs determine the trustworthiness of the *Alarm*, based on the trust level of the sender. From this information, the reputation system computes the rate of each MS and creates a local rating list and a black list. These lists are exchanged with friend MSs. The local rating list maintains the trust level of each MS, and the black list contains MSs that should be avoided during route discovery. In addition, each MS discards RREQ originated by any MS in its black list. This is a notable difference from watchdog and pathrater [8], whereby selfish MSs are just detoured and not punished. Lastly, the path manager, similar to the pathrater in [8], sorts paths according to the reputation rate of MSs along the path and removes paths including malicious MSs. Like the watchdog [8], the monitor in CONFIDANT has the shortcoming that it cannot exactly tell selfish misbehavior from accidental packet drops.

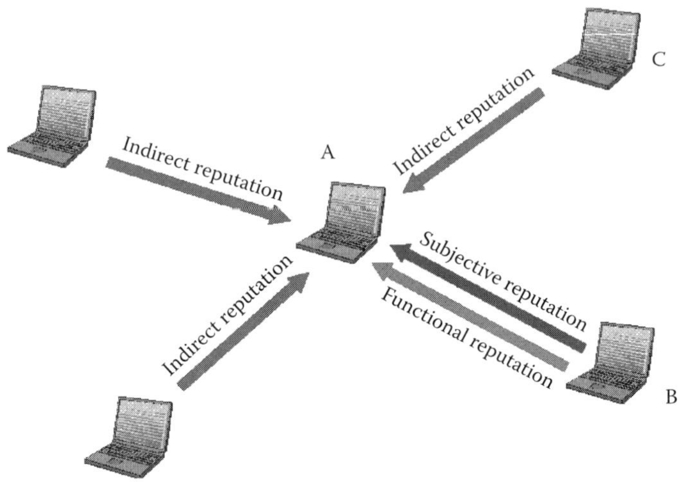

FIGURE 12.3 Reputation of MS *B* combined in MS *A* using CORE.

12.2.4 CORE

CORE (Collaborative Reputation Mechanism) [11] forces MSs to cooperate with each other through a collaborative monitoring technique; i.e., it utilizes positive reports from other MSs (indirect reputation) as well as its own observation on neighbors (subjective reputation), as shown in Figure 12.3. Each MS also monitors task-specific behavior of neighbors (functional reputation), for example, how to handle RREQ messages. These three types of reputation are integrated into a combined reputation, and MSs with reputation lower than a predefined threshold are excluded from the network. However, they can join MANET again if they increase their reputation over the threshold through good cooperation with others. It is noteworthy that MSs can send only positive reports for others, as indirect reputation; thus, malicious MSs cannot deliberately accuse other innocent MSs. Meanwhile, in order to mitigate the bad influence of possible wrong detection by each MS, an aging factor is introduced in computing the combined reputation. More weight is on past observations so that reputation of MSs may not be changed too frequently. This protects the MSs that cannot forward packets temporarily due to bad environmental conditions against undue punishment.

12.2.5 Robust Reputation System

Yau and Mitchell [12] came up with several problems that make CONFIDANT and CORE impractical in actual MANETs. First, the quantification of reputation is difficult since a single behavior can be thought of as a good or bad behavior, depending on the role of the MS. Second, it causes much overhead to reliably disseminate reputation in a MANET. Besides, the distributed reputation cannot be completely trusted due to false accusation of malicious MSs or possible collusion between them. Third, most reputation systems cannot tackle the *spoofing attack* whereby an MS continuously changes its

ID. Finally, as mentioned earlier, the promiscuous listening mode is never reliable in observing misbehavior of neighboring MSs.

To improve the robustness of a MANET, Yau and Mitchell propose a simple reputation system where MS utilizes only its own experience about neighboring MSs. Whenever requesting any service to neighbors, each MS updates the reputation of the neighbors based on their responses. Due to prudent reputation updates, malicious MSs cannot easily build up good reputation by cooperating for just a short period.

12.2.6 Friends and Foes

In Friends and Foes [13], MSs openly declare MS lists for which they will or will not provide forwarding service. Each MS periodically advertises three sets of MS IDs: (1) *friends* to which the MS is willing to provide service, (2) *foes* for which the MS will not provide any service, and (3) *selfish*, referring to the MS as a foe. While MSs forward packets from their friends, they drop packets coming from foes. They do not send packets to MSs in the *selfish* list. Although an MS is in the *foes* list of MS i, only if it is not included in the *selfish* list is there a chance for the MS to forward packets from MS i.

At the beginning, all other MSs are regarded as friends, but they are gradually moved into the *foes* or *selfish* set according to their observed behavior. For MS i to determine if any MS is its foe or not, it maintains *credits* of each MS, which is the difference between the numbers of packets MS i and the opponent have forwarded for each other. If the *credits* of any MS reach a threshold, i.e., if the MS just receives service from MS i unilaterally, then the MS is moved into the *foes* list of MS i. To determine which node is selfish, MS i promiscuously monitors how its neighbors deal with packets forwarded by it. If a neighbor does not handle the packets according to the protocol, MS i adds the neighbor into the *selfish* set. This method requires large computational and spatial complexity to collect and maintain various status variables such as *credits*, *friends* and *foes* of all other MSs, and recently forwarded packets for retransmission for the case where neighbors discard them. For this scheme to work correctly, every pair of MSs must have at least one path consisting of only well-behaving MSs.

12.2.7 SORI

The previous schemes do not suggest a clear formula to evaluate the reputation of each node. Thus, the goal of SORI (Secure and Objective Reputation-Based Incentive Scheme) [14] is to develop quantitative and objective ways to measure reputation. An MS N maintains a neighbor node list, NNL_N, through the promiscuous listening mode and keeps two counters, $RF_N(X)$ and $HF_N(X)$, for each neighbor node (denoted by X): $RF_N(X)$ is the total number of packets that MS N has requested MS X to forward, and $HF_N(X)$ is the number of packets that MS X has actually forwarded. Given $RF_N(X)$ and $HF_N(X)$, MS N creates a local evaluation record, $LER_N(X)$, which consists of $G_N(X)$ and $C_N(X)$. $G_N(X) = RF_N(X)/HF_N(X)$, and $C_N(X)$ denotes how MS N is confident in $G_N(X)$; the larger $RF_N(X)$ is, which means the more packets MS N has requested MS X to forward, the larger the confidence in $G_N(X)$ MS N has. These $G_N(X)$ and $C_N(X)$ are broadcast to other nodes and utilized to calculate the overall estimation record, $OER_M(X)$, in each MS, say M:

$$OER_M(X) = \frac{\sum_{i \in NNL_M \cup \{M\}, i \neq X} \lambda_M(i) \cdot C_i(X) \cdot G_i(X)}{\sum_{k \in NNL_M \cup \{M\}, k \neq X} \lambda_M(k) \cdot C_k(X)},$$

where $\lambda_M(i)$ is the credibility that MS M has on MS i.

SORI punishes the MSs with bad reputation. If the $OER_M(X)$ is lower than a threshold value, MS M drops packets originated from MS X with some probability. Since the probability is inversely proportional to $OER_M(X)$, an MS with the lower reputation would get the heavier penalty, but the MS is not totally isolated from others. This is a different point from CONFIDANT and CORE, where nodes with bad reputation are completely excluded from networking.

SORI has some weaknesses. First, it also uses promiscuous listening to measure the credibility of neighbor nodes; thus, it cannot work with recent networks using topology control and directional antenna. Second, although SORI provides a security mechanism using the one-way hash chain to prevent a node from changing its own reputation, it cannot keep a node from faking observations about neighbor nodes. An MS N may give a false impression on $OER(X)$ that other nodes have on MS X by adjusting its confidence value.

12.2.8 CineMA

CineMA (Cooperation Enhancements in MANETs) [15] focuses on the case where all MSs do not have the same selfishness prevention method, because it may be unrealistic that MSs operated by different entities use the same scheme. Moreover, depending on the type of device, the adoptable scheme may be different. CineMA is a group-based approach, which can work if just a portion* of entire MSs have it, while all the previous schemes require all MSs to use the same one.

Figure 12.4 illustrates how CineMA can detect selfish MSs. MSs with the CineMA module (CM1 and CM2) monitor neighbor MSs through the promiscuous listening mode, maintaining the number of packets a neighbor has received and forwarded in an *incoming list* and *forwarding list*. The ratio of the number of incoming packets and forwarded packets is referred to as the *cooperation level* of the MS, and CM1 and CM2 allocate their bandwidth to neighbor MSs according to the cooperation level. Whereas the complete isolation of selfish MSs is possible under the environment where all MSs use the same prevention scheme, it is not possible in the assumed network model. Instead, CineMA punishes selfish MSs by reducing bandwidth allocated to them.

CineMA is operable only on the source routing protocol like DSR, and the need for communication between CineMA MSs arises occasionally, for instance, for CM2 to know the information on the incoming packets to MS B, but secured communication between them is not suggested in this research. Moreover, CineMA MSs need large computing and memory resources to monitor neighbors all the time and to keep the incoming list and forwarding list for each neighbor.

* The portion is different depending on the network size and the transmission range of each MS.

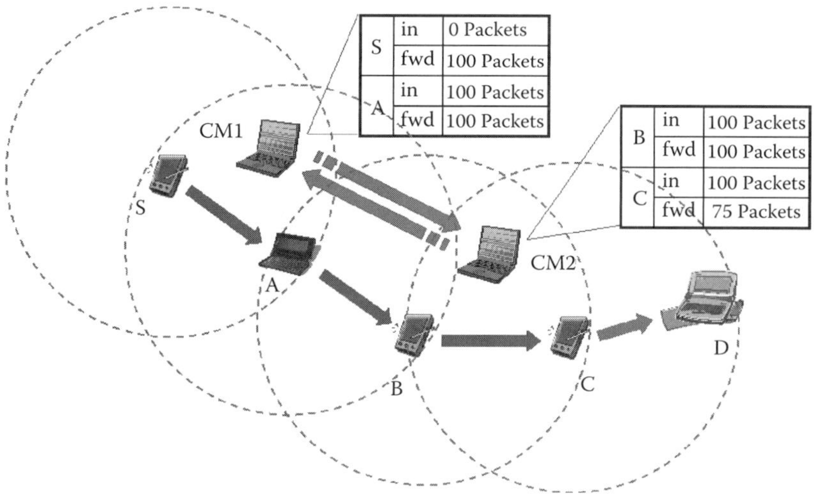

FIGURE 12.4 Operation of CineMA.

12.2.9 TWOACK

Most reputation-based methods monitor neighbor MSs' behavior through the promiscuous listening mode, which is called *passive acknowledgment*, as a contrary to *active acknowledgment*, using an ACK message. However, as stated before, the passive acknowledgment, or the promiscuous listening, may not work correctly when collisions occur or topology control is used. This is the reason that TWOACK and S-TWOACK [16] were developed. If an MS receives a packet, it sends an acknowledgment called TWOACK to the MS through which the data packet passed two hops before. Figure 12.5 illustrates this. With the TWOACK method, an MS must send an acknowledgment for every data packet, resulting in large message overheads. Thus, S-TWOACK allows one acknowledgment for several consecutive data packets to reduce the number of acknowledgments, similar to the go-back-N ARQ in TCP.

For these schemes, several issues need to be addressed. First, these methods can be used only with a source routing protocol, since an MS has to know the path of every data packet to send ACK to the MS two hops away. Moreover, as mentioned before, they cause large message overheads. Although S-TWOACK may reduce the message overheads to

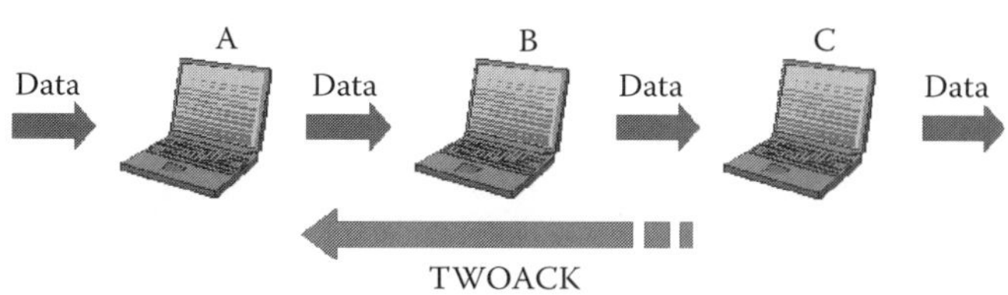

FIGURE 12.5 MS *C* sends MS *A* a TWOACK message.

some degree, each MS has to maintain the sequence number of every session that passes through it, for the correct operation of S-TWOACK. This means that all intermediate nodes of a connection have to perform a function in the transport layer during forwarding. Besides, if a TWOACK does not arrive before the timeout counter expires, an MS does not have any way to determine which is the misbehaving MS between the next-hop MS and the next-to-next MS. Thus, instead of the number of times of misbehavior for every MS, the proposed methods maintain the number of time of misbehavior for every link. For instance, in Figure 12.5, if MS *A* does not receive TWOACK for a packet it has forwarded, it increases the number of times of misbehavior for the links A-B and B-C. This results in larger storage overheads, compared to the case where the number is kept for each MS, since the number of links is generally larger than MSs. If the number of times of misbehavior of a link is over a threshold, two MSs at both ends of the link are convicted of misbehaving. However, they are just detoured in route discovery; moreover, they can enter the network again after a certain period. Thus, selfishness is a blessing to MSs, like in the watchdog and pathrater [8].

12.2.10 RIW

As stated earlier, CORE adopts an aging factor that places more weight on the past observation to protect MSs against undue punishment due to unintentional packet drops. This prudent and conservative punishment policy, however, may allow MSs to misbehave for a long time right after building up good reputation. Due to this concern, RIW (Reputation Indexing Window) [17] emphasizes the current feedback on neighbors' behavior rather than the old values by using an aging factor α:

$$Repu_{\text{new}} = \alpha\text{CurrFeedback} + (1 - \alpha)Repu_{\text{old}} \tag{12.1}$$

Ergo, isolated MSs may easily recover their reputation through just a short time period of cooperation. To mitigate this drawback, RIW proposes to assort feedback items (FIs) into three windows: RIW_1, RIW_2, and RIW_3. RIW_1 has the latest feedback and RIW_3 has the oldest feedback; the ratio of size of RIWs is RIW_1:RIW_2:RIW_3 = 10:30:60. Figure 12.6 depicts the three windows. If a new FI arrives at the leftmost position of the windows, all existing FIs are pushed one step in the right direction and the oldest one is discarded. The overall reputation is evaluated as

$$Repu = \lambda\text{RIW}_1 + \mu\text{RIW}_2 + \nu\text{RIW}_3 \tag{12.2}$$

It is claimed that this reputation reflects the latest change best when $\lambda = 0.66$, $\mu = 0.22$, and $\nu = 0.11$.

FIGURE 12.6 Reputation indexing window.

12.2.11 Light-weight Solution

The solution to the selfishness should be light-weight. Otherwise, it will consume too much energy, resulting in the short life span of MANETs. Too high cost for identifying selfish MSs is not rational at all from the perspective of economics. Thus, the light-weight solution [18] suggests on-demand monitoring and self-exemption considering battery status. An MS is not in the promiscuous listening mode all the time. Instead, it turns on the promiscuous listening module right after forwarding a packet, and if it detects its neighbor (say, *A*) dropping packets consecutively, MS *A* is suspected of being selfish. However, a single MS cannot determine exactly whether another one is selfish or not because it might not overhear neighbor MS's forwarding due to collisions or interference. Thus, it sends *A*'s neighbors an *Alert* message to wake up their promiscuous listening modules. This message means that help is needed to monitor MS *A*, and it includes the target node ID, the request node ID, and the observation period. Then, *A*'s neighbors observe MS *A* for the observation period, and if a neighbor MS thinks of MS *A* as a selfish node, it sends the request node an *Accuse* message. If more than half or more than a predetermined number of MSs bring in a verdict of "selfish" on MS *A*, all packets originated from MS *A* and destined to MS *A* are dropped at *A*'s neighbors.

As to self-exemption considering battery status, an MS can declare itself as a low-battery node so that it may get exemption from packet forwarding. In exchange for getting the exemption, it gets restriction on the number of packets it can generate.

This light-weight solution clearly has merits in terms of power consumption during the promiscuous listening mode. However, each MS has to maintain the neighbor node list of its neighbor MSs so that an *Alert* message can be sent. Much power and large memory space may be needed to collect and store that information. Moreover, a suspected MS also gets the *Alert* message, so it knows that its neighbors are observing it. As a result, it can evade the accusation by cooperating during the observation period only. Finally, an MS might declare that it has low battery when it does not have much to send, although it has high battery power in actual practice. Only when it needs to send much data does it participate in networking. If all MSs were to act like this, the overall communication of MANETs would be disrupted.

12.2.12 Reputation Index Table

As mentioned earlier, most reputation-based methods utilize promiscuous listening to monitor neighbor MSs' behavior, but this may not work well in any environment where packet collision exists or directional antenna is used. To overcome this weakness, Refaei et al. [19] proposed a method where each MS maintains a reputation index table for each of its neighbors, and the reputation is increased or decreased depending on whether packets have arrived at the final destination or not.

Figure 12.7 illustrates the operation of the protocol. At first, MS *S* sends packets to MS *D* through the path *S-A-B-D*. At any given time, MS *B* starts to behave selfishly, dropping packets coming from MS *A*. Then, MS *A* cannot receive delivery acknowledgment (e.g., Transmission Control Protocol [TCP] acknowledgment) from the final destination *D*, so MS *A* reduces the reputation of MS *B*. In turn, MS *S* reduces the reputation

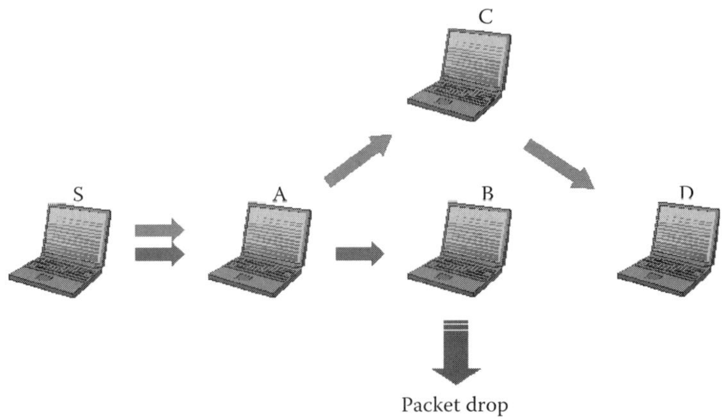

FIGURE 12.7 Operation of the reputation index table method.

of MS *A* because an MS in this method places all responsibility on the next-hop MS for packet delivery to the final destination. After some time, if MS *B*'s reputation falls below a threshold value, MS *A* isolates MS *B* and triggers a new route discovery process, resulting in the setup of a new path *S*-*A*-*C*-*D*. On the other hand, if MS *B* delivers packets to MS *D* properly, MS *B*'s reputation at MS *A* is increased, and in turn, MS *A*'s reputation at MS *S* is also increased.

This method has several drawbacks. First, this may not fit for MANETs with topology control, where the forward path for data and return path for ACK can be different. In that case, an MS may not receive an ACK from the next-hop MS through which an original data packet was forwarded, although it successfully arrived at the destination. Thus, the next-hop MS is deprived of the opportunity to increase its reputation. Second, consecutive false isolation can occur. In Figure 12.7, for instance, the reputation of MS *A* is also decreased due to MS *B*'s misbehavior, so MS *A* as well as MS *B* can be isolated at the same time. If there exist more MSs before MS *A*, they can be wrongly punished, too. Third, since the reputation index is maintained only for immediate neighbor MSs, if an MS is movable, it can continue to misbehave just by moving into a new position. Finally, a new cross-layer approach is needed, since intermediate MSs have to read delivery acknowledgments like TCP ACKs coming from the final destination MS.

12.2.13 Detection of Selfish Routing

Whereas most reputation-based methods focus on the detection of the selfish MSs that do not forward data packets, Wang et al. [20] introduce smart selfish MSs. They avoid being selected as an intermediate node by manipulating routing protocol, since various selfishness prevention methods make it dangerous to drop packets after being chosen as an intermediate node. This research first analyzes the AODV protocol and creates a finite state machine (FSM) for each MS to follow in the routing process from the perspective of its neighbor MS. Thereafter, if behavior of a neighbor MS is too much contrary to the FSM, the MS can be referred to as a smart selfish MS. However, either the propagation of this information to other MSs or the punishment of a smart selfish MS is not suggested in this research.

12.2.14 Integrated Detection

Besides the selfish routing, selfish nodes can misbehave through a variety of ways. Yokoyama et al. [21] summarize the types of selfish behaviors:

1. They do not relay RREQ messages to avoid being included in a path.
2. They do not send HELLO messages to hide themselves from others until they need to generate their own packets.
3. They intentionally delay RREQ relaying to reduce the probability of being included in a path.
4. They relay RREQ but not data packets.

The proposed detection method against these behaviors is very simple: promiscuous listening is used. Particularly for selfish behavior 3, the time taken for the next MS to relay RREQ is measured, and if is too long, the MS is regarded as a selfish node. However, it is not clear how large the "long" time is, because it depends on the status of the next MS's queue.

12.2.15 PCOM

All the previous schemes operate reactively in response to MS's misbehavior. This reactive operation, however, can cause excessive overheads for MSs with processing and battery constraints. Thus, Suzuki et al. [22] suggest Proactive Cooperation Mechanism (PCOM), which proactively prevents a selfish node from joining the network. An MS, say MS *A* in Figure 12.8, keeps a cooperation record (CR) for each of its neighbors through which at least one packet was forwarded to MS *A* before. When an MS is requested to forward data by a neighbor MS, it checks if it has the CR for the neighbor MS. If it does, it forwards the data; otherwise, the data is discarded. Therefore, a new MS cannot generate packets until at least one of its neighbors has CR for it. For its neighbor to do so, the new member has to show that it cooperates well with others by forwarding packets from others.

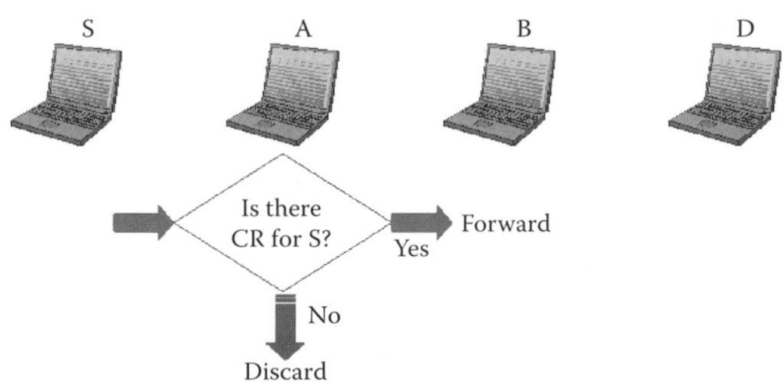

FIGURE 12.8 Forwarding process in PCOM.

The biggest shortcoming of PCOM is that once an MS has been able to record CR at the neighboring MSs, it is free to behave selfishly. That is, if an MS forwards some packets at first, and its neighbors come to have a CR for it, it can then discard all packets after that without getting any punishment. Also, all MSs may have no CR for any other MSs at the initialization stage of a network. In this case, communication between MSs cannot be performed at all, so PCOM provides Temporary CR Requests (TCRs) to resolve this deadlock. An MS can make its neighbor forward its data by sending a TCR, although the neighbor does not have a CR for it. However, free riders can abuse this TCR to send their own packets, since there is no way to distinguish them from pure TCRs to resolve the deadlock.

12.2.16 CI-DSR

Xu et al. [23] deal with two types of selfish MS: the type 1 model that refuses to forward data packets from others, and the type 2 model that refuses to join the route discovery process. In CI-DSR (Cooperation Inspirited Dynamic Source Routing), each node calculates the reliability of its neighbor with the ratio of the packets forwarded by the neighbor to the packets transmitted to the neighbor. This reliability is propagated to the neighboring MSs, and each MS, say K, creates joint evaluation reliability, $JER_K(X)$, for its neighbor MS, say X. The MS with the low reliability is punished by other ones by dropping its packets with some probability. This method also utilizes promiscuous listening and lacks appropriate proof for its correct operation.

12.2.17 Fellowship

Fellowship [24] defends MANET against both flooding and packet drop attack, because selfish MSs are regarded the same as malicious MSs in that they both disrupt the network service. Besides the promiscuous listening module to monitor neighbor MSs, each MS includes three important components: *rate limitation*, *enforcement*, and *restoration*. First, the rate limitation checks if a neighbor MS equally shares channel bandwidth with its neighbors, to prevent a flooding attack through the monopoly of a channel. Second, the enforcement forces each MS to contribute to network service as much as its obligation,* which is equally distributed among all MSs. Any MS that does not meet its obligation is expelled from the network. Finally, the restoration determines the reason of packet drops at the neighboring MSs. If the packets are dropped due to congestion, not because of any selfish reason, the MS should not be punished. However, it is not clear how to distinguish accidental packet drops from intentional ones.

The aforementioned methods in common assume that reputation evaluated by each MS is securely propagated to others, and that all MSs easily agree on the same overall reputation of a specific MS from the propagated information. However, this may be impractical in the error-prone wireless mobile environment, and the speed of reputation propagation has a great impact on the convergence speed of the reputation agreement.

---

* The authors call the Fellowship an obligation-based model.

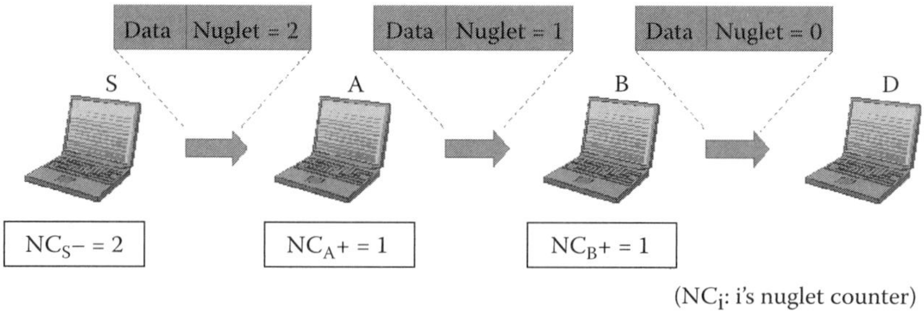

FIGURE 12.9 Operation of PPM.

Studying these issues, Liu and Yang [25] show that the reputation agreement can be attained through even localized propagation to neighbors only, if the reputation is propagated frequently enough and MSs utilize their own experience as a part of the update.

12.3 Credit-Payment Method

Credit-payment methods reward MSs participating in packet forwarding with credit, which can be virtual or real currency. This credit is used for all MSs in order to generate their own packets, because they have to pay forwarding cost to intermediate nodes.

12.3.1 PPM and PTM

Buttyan and Hubaux [26] first adopted the commercial transaction concept for selfishness prevention. MSs providing service to others are given credit, while MSs benefiting from the service pay for it. In the proposed two approaches, PPM (Packet Purse Model) and PTM (Packet Trade Model), MSs forwarding other MSs' packets are rewarded with virtual currency called *nuglet*,* which is necessary for those MSs to send and receive their own packets later. Which MS is charged nuglets for packet delivery is different depending on the model.

Figure 12.9 depicts the operation of PPM. The source MS loads nuglets into packets it generates, then intermediate MSs deduct them as a reward for packet forwarding. Since intermediate MSs will not forward but discard packets without sufficient nuglets, the source MS must know the amount of nuglets required for those packets to arrive at the destination. This means that source MSs should have the exact path information of every packet; thus, PPM can be used only with source routing protocols like DSR. In addition, forwarding MSs may take away more nuglets than they are supposed to, and may possibly drop packets right after taking out nuglets.

In the meantime, in PTM, as shown in Figure 12.10, intermediate MSs purchase packets from the previous ones and sell them to the next-hop MSs at higher prices. In the

* The name of *nugget* was originally used in their first paper; it was changed to *nuglet* afterwards.

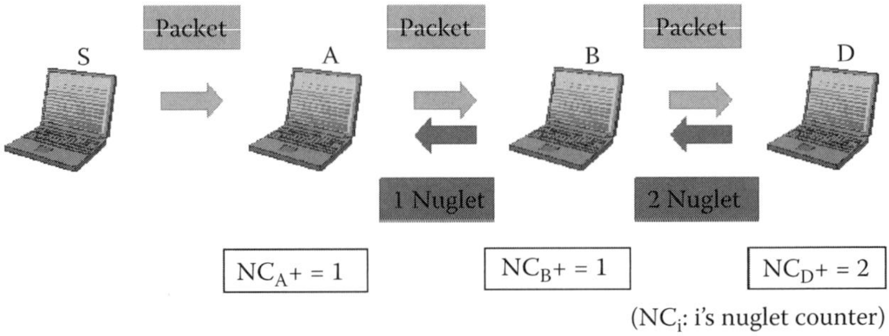

FIGURE 12.10 Operation of PTM.

end, the destination MS pays the final price to its predecessor. Although the use of PTM is not limited to source routing protocol, the MANET may be easily attacked by the denial of service (DoS) because MSs do not have to pay any credits to originate packets.

PPM and PTM have a common critical limitation: How can the validity of nuglets be trusted? Dishonest MSs may not only reuse nuglets already used once, but also freely increase their nuglets. To overcome this limitation, it is assumed that a tamper-resistant security module, like a special chip or a smart card, is installed in each MS, but this makes it hard for them to be widely accepted. A recent work [27] employs a public-key method and analyzes how each MS can maximize its benefits through the extension of PPM or PTM, which uses credit counters.

12.3.2 Ad hoc-VCG

The PPM method is modified a little in the Ad hoc-VCG scheme [28], which is a reactive routing protocol similar to DSR. It consists of two phases: *route discovery* and *data transmission*. During the route discovery, collecting RREQ messages, a destination MS determines how many payments are needed for intermediate MSs and notifies the actual payer. This payer may be either the source MS or the central bank, depending on the network architecture. The actual payment is performed during the data transmission phase.

The most significant issue in Ad hoc-VCG is that the correct operation completely depends on the destination MS. If a destination MS cannot exactly compute the required payment or does not report it to the payer, intermediate MSs may not receive their fair credit. Assuming that all MSs are possibly selfish, a destination MS in the central bank model has two reasons to evade the report on required payments. First, the energy to transmit the report to the central bank is wasted from the perspective of the destination, since the destination does not benefit from it at all. Second, the central bank periodically debits accounts of all MSs evenly to compensate for the credits it has paid to the intermediate MSs, which is similar to the premium of an insurance company. Thus, if a destination MS does not notify the central bank that it should pay credit to intermediate MSs, the amount of premium the bank has to collect from all MSs is reduced equally, resulting in the advantage of the destination MS itself.

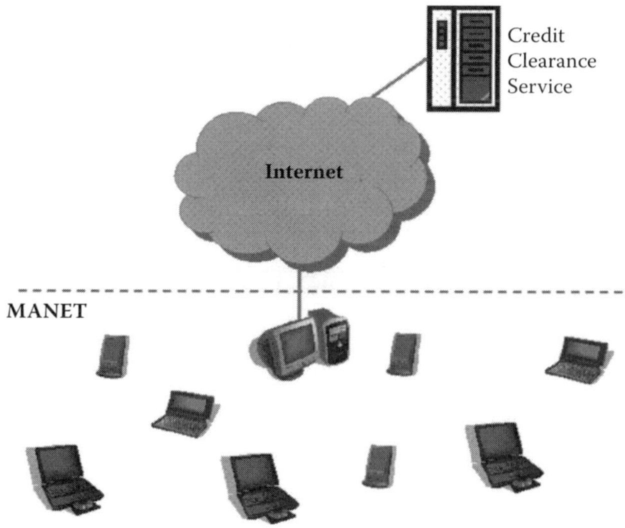

FIGURE 12.11 The architecture of Sprite.

12.3.3 Sprite

Sprite (a simple, cheat-proof, credit-based system) [29] has a central authority server called Credit Clearance Service (CCS), which maintains credit balance for all MSs, in the infrastructure network. Figure 12.11 illustrates the architecture of the Sprite method. As proof for the previous MS's contribution in data forwarding, each MS keeps a *receipt* whenever it receives a packet from the previous hop MS, and these receipts are reported to CCS whenever the MS can establish a fast connection to CCS. Collecting all the reports, CCS rewards both the sender of the receipts and intermediate MSs for the report transmission and packet forwarding, respectively. In return for the credits it has rewarded, CCS charges a premium to the source, but the amount of charged credit is not equal to that given to other MSs. In order to not only prevent cheating with false receipts but also encourage MSs to cooperate, CCS charges more credits than it is supposed to give to others and uses the surplus credits to induce MSs to send a true report. The Sprite algorithm was designed to make MSs sending a false report or colluding with another MS lose credits that they should get in actual.

CCS is an infrastructure server, which is not fit for the MANET architecture. Even though it is ignored, scalability may be a critical issue since MSs send a receipt for every packet to CCS. Moreover, each message is encrypted with a public/private key pair. Also, CCS needs the complete path information, so the use of Sprite is limited to source routing protocols like DSR.

12.3.4 Incentive Scheme for a Multihop Cellular Network

Salem et al. [30] adopted the credit-based incentive method into a multihop cellular network that combines MANET and the cellular network. Similar to a usual cellular

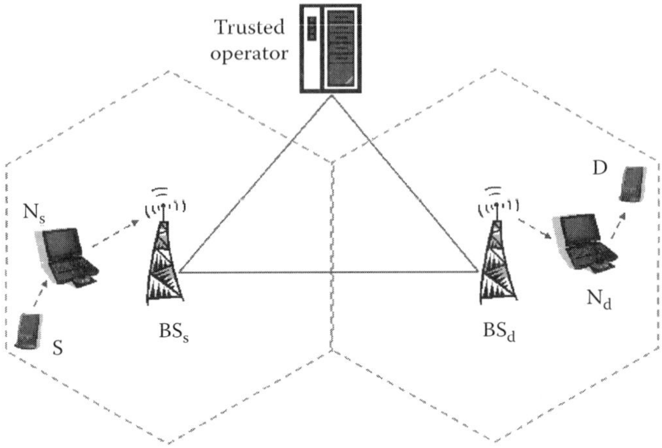

FIGURE 12.12 Incentive scheme in a multihop cellular network.

network, a cell is covered by a BS that is connected to other BSs via a backbone. However, unlike the cellular network, all MSs do not have to be directly connected to a BS for communication. Instead, an MS can communicate with a BS in the multihop manner through other MSs.

The proposed scheme forces MSs to communicate with each other through a BS, prohibiting direct communication between them. This results in the reduction of routing overheads, because an MS can reach any other MSs by maintaining just a single route to a BS. And a BS has a route to every MS existing in its own cell. It is assumed that a central trusted operator is present to manage the billing accounts of all MSs.

Figure 12.12 depicts the packet transmission between two MSs in different cells. Once a packet of MS S arrives at BS_s, the trusted operator deducts as many credits from the account of S as it should give to all forwarding MSs up to the destination D. The amount of required credits can be easily computed, since a BS has the route information to every MS in its cell.

This method prevents collusion between MSs as follows. A part of the credits is not practically given to N_s until a packet arrives at BS_s. Similarly, credits are not paid to N_d until the packet reaches the destination D and D sends the ACK to BS_d. However, even though a packet successfully arrives at the destination, MS D may not send the ACK to save its power. As provision against this type of selfish behavior, the trusted operator takes away some credits from the account of D before BS_d relays a packet to D. These credits are given back to D after the ACK arrives at BS_d. Communication between MS and BS is secured by the private key encryption.

12.3.5 Priority Forwarding

The aforementioned credit-based methods require all MSs to follow the same selfishness prevention protocol in order for the MANET to work correctly. In a practical MANET, however, a variety of heterogeneous MSs exist, and some of them may not even be aware of the presence of a credit-payment scheme, let alone how to handle the credits. Also, some MSs may not be able to make enough credits, not because they behave selfishly, but because they are just badly positioned in MANETs to be selected as intermediate forwarders.

To cope with this limitation, Raghavan and Snoeren [31] suggest two layered forwarding services: *priced priority forwarding* and *free best-effort forwarding*. The priced priority forwarding is processed before any free best-effort forwarding all the time. Instead, MSs that want to use the priority forwarding should pay some credits to intermediate MSs, whereas no credit is charged for the best-effort forwarding. In exchange for paying credits, MSs sending data through the priced forwarding want a guarantee that their packets will be properly handled; thus, the priced forwarding behavior is monitored by their neighbors in a promiscuous way. If it is found that some MSs do not support the priced priority forwarding fairly, the payment for the service is nullified.

This scheme borrows the model of CCS from Sprite [29] to manage the credits of all MSs. The difference is that the CCS in this scheme is just one of general MSs, whereas it is an infrastructure authorization server in the Sprite system. Although the CCS here accords with the MANET architecture better, it may be a burden for a single MS to maintain credits for all other MSs.

12.3.6 Willingness to Pay

The study in [32] is the first one that takes into account a set of MSs equipped with directional antennas. Route selection and flow allocation are based on the service price marked by relevant MSs, which combines current available bandwidth and power. This price is advertised during the route discovery. An MS s that wants to send data first determines a willingness to pay at time t, $\omega_s(t)$, which indicates the amount of credits it is willing to pay for its traffic. Afterwards, by summing up the price requested by intermediate MSs along each path, it determines the minimum cost path while satisfying $\omega_s(t)$. Using willingness to pay and cost of the found path, MSs adjust their resource usage accordingly. The price is said to be adapted depending on available bandwidth and power, but MSs cannot be restrained from declaring arbitrarily higher prices.

12.3.7 Truthful Multicast

Wang et al. [33] proposed a multicast routing protocol that encourages each MS to truthfully advertise its cost for relay in multicast routing structures. However, this method is based on some strong assumptions. First, in order that a pair of MSs always has a detour except for the current route between them, the network must be *biconnected*, which means the network should not be partitioned by the removal of one MS. Second, all receiver MSs should relay packets for other receivers in the multicast architecture for free whenever they are asked. The cost for the relay is compensated by the source MS later, but there is no guarantee for a relay MS to get fair credits. Finally, all MSs should try to maximize profits by forwarding packets all the time. Actually, however, some MSs that do not need credits any more may not have interest in packet forwarding.

24 Bytes

RID	NID	SEQ	I	O	S	T	OFN
4 Bytes	4 Bytes	6 Bytes	2 Bytes	2 Bytes	2 Bytes	2 Bytes	2 Bytes

FIGURE 12.13 Report message in PIFA.

12.3.8 PIFA

The use of most previous credit-based approaches is limited to the source routing protocols, as they need the complete route information from source to destination. Although PTM [26] and the incentive scheme for multihop cellular networks [30] are the exceptions, they depend on tamper-proof hardware or base station instead, which is usually not provided in a MANET environment. On the other hand, PIFA [6] is independent of the routing protocol type and implemented as a simple add-on to any routing protocol.

PIFA needs an authorization server named Credit Manager (CM) to keep track of the credits of all MSs. Either an MS with the largest power or an access point to the infrastructure network, if it is present, can be appointed as the CM node. It is assumed that the CM is specially managed by the administrator, and thus can be completely trusted.

All MSs periodically report to CM the number of packets they forwarded in each time interval. The message used for the report is shown in Figure 12.13. Collecting report messages from all MSs, CM verifies the credibility of them and rewards relay MSs accordingly. The purpose of the CM's credibility test is to check if reports from two neighboring MSs are consistent with each other. If the reports of two MSs are inconsistent, both of them receive one penalty point called Numbers of Alleged Manipulation (NAM). If the NAM of an MS exceeds a predefined threshold, the MS is excluded from the network. The NAM of an MS is decreased by half when its peer MS makes another inconsistent report with another MS. This is based on the intuition that a selfish MS will repeatedly attempt to cheat others. That is, if the same MS is found in multiple inconsistencies, the possibility of that MS being selfish is very high. Since the NAM is similar to reputation in reputation-based methods, PIFA can be said to have a feature of reputation-based approaches as well.

The weakness of PIFA is scalability, since a single CM has to handle all MS report messages.

12.3.9 D-SAFNC (D-PIFA)

To mitigate the problem caused by the centralization feature of PIFA, D-SAFNC (Distributed Self-Policing Architecture for Fostering Node Cooperation) [34] suggests the cluster architecture of PIFA. This method includes a cluster-head election algorithm called CB-DCA (Credit-Based Distributed Clustering Algorithm) in which the node with the minimum weight in its neighborhood is elected as the head. The weight is proportional to the node speed and inversely proportional to node credit, so that more reliable nodes may become cluster heads. It is similar to PIFA in the sense that each MS

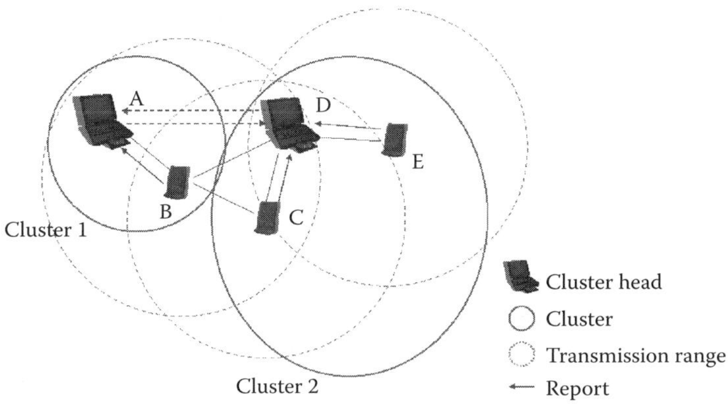

FIGURE 12.14 Operation of D-SAFNC.

periodically sends a report message about its contribution to network service, but its destination is not the CM, but the cluster head. As a result, burden on CM is distributed to many cluster heads. In some cases, communication between cluster heads is necessary, too. For instance, in Figure 12.14, MS *B* and MS *C* make a report to MS *A* and MS *D*, respectively. Thus, MS *A* and MS *D* should exchange their collected information to check the consistency between report messages from MS *B* and MS *C*. The rest of the algorithm is the same as in PIFA.

12.4 Game Theory Method

The game theory is a branch of economics for deriving the optimal strategy for every rational competitive player. The "rational" player tries to maximize her or his profits all the time. Although there exist a variety of subfields in the game theory, such as cooperative or noncooperative game and one-stage or repeated game, the common goal is to look for the *Nash equilibrium* point, where every player chooses a strategy to maximize his or her payoff regardless of other players' strategies. That is, in the *Nash equilibrium* point, any player cannot increase her or his payoff any more through strategy changes while other players' strategies remain fixed.

The game theory has been used in various fields of social economy, such as auction, tax policy, road design, and transportation networks, and recently, the use of it has been extended to various areas in communication networks, like packet forwarding and power control. Here, it is worth noting again that all players must be rational in the game theory. For instance, in a MANET, if some MSs have no interest in increasing their payoffs and just focus on power saving, the MANET may not be analyzed accurately using the game theory. Thus, assuming all rational MSs, all the following schemes model packet forwarding as a strategic game. Through the modeling, they find the optimal strategy for every MS. Although each method has a little different way to pursue the Nash equilibrium, there is a common notion behind them: assuming that all MSs are free to change their strategies, if a selfish MS is detected, all the other MSs change their strategies to punish it. Thus, the forwarding game will be gradually converged at one Nash equilibrium point where all MSs are satisfied.

12.4.1 GTFT

Srinivasan et al. [35] introduced GTFT (Generous TIT-FOR-TAT) and m-GTFT (multiple-GTFT), which are relay acceptance algorithms using the game theory in MANETs. GTFT is for the case where just one intermediate MS exists between source and destination, and m-GTFT is for when there are multiple relays. The objective of these algorithms is for an MS to balance the energy used for others with the energy serviced by others for it, and to find an optimal trade-off between the blocking probability and the power consumption. Since they pursue the fairness of energy consumption for each type of session between source and destination, spatial and computational complexity is remarkably reduced compared to the most previous algorithms pursuing fairness for each packet. Due to the self-organization property of practical MANETs, GTFT and m-GTFT take into account the presence of heterogeneous MSs: each MS has different resource constraints depending on its type (e.g., laptop, PDA, cell phone). In order for MS h to decide if it will accept relay requests, it maintains two variables, $\phi_h^j(k)$ and $\psi_h^j(k)$: $\phi_h^j(k)$ is the ratio of the requests successfully relayed by others to the total requests generated by MS h for session type j until time k, and $\psi_h^j(k)$ is the ratio of the requests relayed by MS h to the total requests arriving at MS h for session type j. The condition for MS h to reject a relay request of session type j is as follows:

$$\psi_h^j(k) > \tau_j \text{ or } \phi_h^j(k) < \psi_h^j(k) - \varepsilon, \tag{12.3}$$

where ε is a small positive number. In the former part, τ_j denotes the maximum relay ratio for session type j so as to fit with the amount of total traffic allowed for j. On the other hand, the later part means that the number of requests relayed by MS h should not be greater than the number of requests relayed by others for MS h by more than ε. The value ε indicates the generosity of each MS, which implies that an MS gives more services than it is supposed to do at some degree without any compensation. This research proves that the GTFT algorithm leads all MSs to a Nash equilibrium.

However, the proposed algorithms may suffer from large message overheads, because each MS should have sufficient information about the entire system, such as the total number of MSs, energy constraints of them, and the amount of requests for each session, in order to derive $\phi_h^j(k)$, $\psi_h^j(k)$, τ_j and an appropriate ε. Thus, although it is necessary to provide a mechanism to disseminate system information securely and efficiently, no algorithm is presented that can prevent MSs from sending false information to increase their own payoffs.

12.4.2 Presumption on Neighbor's Behavior

Urpi et al. [36] put constraints on not only the energy consumed to forward other MSs' packets, but also the energy to send their own packets. The proposed method utilizes the presumption on neighbor MSs' behavior to reduce as much wasted energy as possible in sending their packets. Although an MS has data to send, if it expects its neighbor not to relay its packets for some reason, it gives up sending them so as to save its energy. However, this mechanism has some strict assumptions: each MS knows not only all the past behaviors of its neighbors, but also their priorities in terms of resource consumption and throughput, which are used to determine how to act. Also, the naive definition of payoff for each MS is another drawback that is difficult to overcome.

12.4.3 CORE as a Game

In [37], two methods are suggested to model the CORE scheme [11]: from the perspective of a cooperative game approach and from that of a noncooperative game approach. In a cooperative game, players can reach a best agreement through communication among them, while players in a noncooperative game pursue their own profits independently. The well-known example of a noncooperative game is the prisoner's dilemma (PD). Two prisoners, P1 and P2, are interrogated in different rooms and have no way to communicate with each other. The prosecution suggests that they will be confined for the terms shown in Table 12.1 depending on whether they confess or not. If only one prisoner confesses, he or she will be discharged, whereas the other prisoner should serve 10 years in prison. Assume that P1 selects the strategy "not confess." If P2 also decides not to confess, both of them may get the best profit. However, if P2 confesses their crime, the payoff for P1 becomes significantly reduced, that is, his or her prison term is increased. On the other hand, if P1 selects the strategy "confess" from the beginning, her or his payoff can never be decreased depending on P2's strategy. Therefore, confession is the best strategy for himself or herself when P2's strategy is not known. As a result, "confess–confess" is the Nash equilibrium in this game. On the other hand, if they can cooperate, that is, if it is a cooperative game, both of them can get the best profit by selecting "not confess-not confess."

The proposed cooperative game approach extends this two-prisoner dilemma game to an N -prisoner dilemma. It is assumed that if one more MS cooperates with others, the increased payoff would be greater than the increased cost for the cooperation, and that the payoff would be shared by all MSs. Thus, MSs can maximize their payoffs all together by cooperating with all the others.

In the noncooperative game approach, the behavior of each MS is indicated with the ratio of energy E_{self}, which has been used for itself, to the sum of E_R and E_{PF}, which has been consumed for routing and forwarding for others. The utility function is based on the weighted difference between E_{self} and (E_R + E_{PF}), and it is adjusted depending on the

TABLE 12.1 Prisoner's Dilemma

P1 \ P2	Confess	Not Confess
Confess	(5, 5)	(0, 10)
Not confess	(10, 0)	(1, 1)

importance of the power for each MS. All MSs rationally try to maximize their profits while maintaining a good reputation.

12.4.4 Catch

Catch [38] tackles both the connectivity problem and the forwarding avoidance of selfish MSs. Catch utilizes anonymous messages where the sender identity is hidden, to keep connectivity between neighbor MSs. When an MS sends the anonymous message to its neighbors, if a neighbor MS does not reply an acknowledgment within a timeout period, then both the in- and out-connectivity to the neighbor are dropped. Although selfish MSs are desperate to hide themselves in order to avoid being selected as relay nodes, they should maintain at least one connectivity to send their own packets. However, since they cannot determine the sender of the anonymous message, they have no choice but to acknowledge all anonymous messages.

Meanwhile, the forwarding avoidance problem is solved by the watchdog systems in neighbor MSs. Cheaters are punished by being isolated from all other MSs. The authors adopt the game theory to prove that Catch induces cooperation to be an evolutionarily stable strategy (ESS) for all MSs. This proof is based on the following intuition: In a single-round game where strategies that MSs select once cannot be changed any more, payoff for each MS depends on the current strategies of all MSs, and the payoff is fixed forever. Thus, each MS selects the strategy to maximize its profit irrespective of other MSs' decisions, not considering other MSs' profits. However, if it is modeled as a repeated game whereby MSs can change their strategies every round and do not know which round is the final one, they have to consider the response of others in the next round when they determine current actions. If an MS misbehaves in this round, it will be punished by all other MSs in the next round. Similarly, prisoners in the PD game may be evolutionarily converged to the "not confess–not confess" strategy, if the game is designed as a repeated game. This research shows the possibility of the game theory being applied to develop a mechanism that enforces inter-Internet service provider (ISP) coordination as well.

12.4.5 SLAC

SLAC (Selfish Link and Behavior Adaptation to Produce Cooperation) [39] concentrates on cooperation between peers in P2P networks. Based on the concept of the repeated game, every MS can freely change its strategy and select its partner. Because every MS may well select another cooperative MS as its peer, selfish MSs are naturally isolated from others. However, it is not clearly stated how each MS fairly compares its performance against another MS and how to tell selfish MSs from cooperative ones exactly.

12.4.6 Incentive Scheduling

Game theory is also adopted to encourage cooperation between MSs in WWAN/WLAN [40]. The idea is simple: dividing all MSs into two groups, relay MS and nonrelay MS, BS allocates more time slots or power to relay MSs to encourage cooperative relay. If an MS wants to be a relay MS, it sends relay advertisement messages to other MSs and registers itself on BS. This scheme just focuses on improving overall throughput, ignoring energy consumption.

12.4.7 Multinode Attack-Resistant and Cheat-Proof Cooperation

Unlike the previous methods, Yu and Liu [41] consider the effect of malicious behavior as well as selfish misbehavior. Furthermore, they consider the error-prone characteristics of wireless channels in the analysis of cooperation strategies, while the previous methods assume the ideal channel condition.

For analysis, they model the secure routing and packet forwarding game where each player (MS) is either selfish or malicious. A player, say i, needs cost c_i to forward a packet for others, and it gets gain g_i when other ones forward a packet for it. Whereas the cost can be simply measured as the consumed energy, the gain may be an application-specific metric. The basic idea is that players may participate in a network only when the gain is not less than the cost. A player can choose its strategy in three stages: (1) route participation stage, where it can either accept or refuse RREQ from others; (2) route selection stage, where it can select one out of several discovered routes; and (3) packet forwarding stage, where it can either forward or drop a packet received from others. Each player determines its strategies to maximize its utility:

$$U_i(t) = \frac{S_i(t)g_i - F_i(t)c_i}{T_i(t)}, \tag{12.4}$$

where $T_i(t)$, $S_i(t)$, and $F_i(t)$ are the number of packets that a player i needs to send by time t, the number of packets successfully delivered to the destinations by time t, and the number of packets that it has forwarded by time t, respectively. Through analysis, the authors show that there exists at least one Nash equilibrium (NE) point.

Based on the analysis, the multinode attack-resistant and cheat-proof cooperation strategy is proposed. First, in the route participation stage, a player accepts RREQ if the following two conditions are satisfied: the originator is not marked as malicious, and the difference between the number of packets it has forwarded for the originator and the number of packets the originator has forwarded for it is less than a predefined threshold. In the route selection stage, a player selects the shortest path not containing any malicious MS on the condition that the expected gain is larger than the expected cost, i.e., $(1 - p_e)^n g_i > nc_i$, where p_e and n are the channel error ratio and hop length of the path. Finally, in the packet forwarding stage, a player i chooses its strategy according to $R_i(j,t)$ for nonmalicious player j, which is the number of packets that j has attempted to forward for i. The number of packets actually forwarded may be different due to communication error.

This method assumes that there exists a monitoring scheme where the source can detect who dropped its packet if the packet did not arrive at the destination. However, this may be a security hole that a selfish MS can attack by insisting on its innocence. Also, because each MS should periodically report private information such as its strategy, gain, and cost, control overhead may be too large. Furthermore, each player should maintain not only the number of packets it has forwarded for each other player, but also the number of packets the other players have forwarded for it.

Sprite [29] and the work done by Wang et al. [33] can be classified into this category also, as they determine the amount of charge and credit, so that truth-telling may be the optimal strategy for all MSs.

12.5 Other Methods

There exist other selfishness prevention schemes not included in the earlier three classes.

12.5.1 Token-Based Protocol

Yang et al. [42] introduce a token-based protocol whereby an MS should have a valid token to join the network. The token of each MS has a period of validity, and thus the token has to be refreshed before its expiration. In order to renew the token, the system secret is required, but each MS in the network has just a portion of the system secret. Therefore, it has to get other portions from enough neighboring MSs in order to refresh the token.

The proposed algorithm consists of four components: two proactive mechanisms, *Neighbor Verification* and *Security-Enhanced Routing Protocol*, for secure routing; and two reactive mechanisms, *Neighbor Monitoring* and *Intrusion Reaction*, for cooperative packet forwarding. First, Neighbor Verification determines the expiration time of each MS's token. The time is increased in proportion to the duration of an MS behaving well in the network; thus, a well-behaving MS can work for a long time without renewing its token. As the Security-Enhanced Routing Protocol, AODV-S is implemented in which MSs have the list of neighbors that are verified as MSs with a valid token. To provide more secure routing, each MS also maintains the route entries announced by its neighbors as well as its own routing table. Meanwhile, the Neighbor Monitoring module in each MS promiscuously monitors the neighbors and keeps the headers of the recently overheard packets. If selfish behavior is detected, the detector sends a single intrusion detection (SID) message to other MSs. Finally, the Intrusion Reaction module revokes the token of a cheater. If the number of issued SID messages reaches a predefined threshold, the accused MS is deprived of network access for good since it will not be able to renew the token again.

The proposed method has a lot of interesting features, but it also has some problems. First, each MS may suffer from large spatial and processing overhead, which are very limited resources for wireless MSs. Furthermore, sparse density of MSs and high mobility may be critical to this method since they make it difficult for an MS to find a sufficient number of neighbors that can provide a share of the system secret.

12.5.2 AD-MIX

AD-MIX [43] discourages selfish MSs from dropping packets by hiding destination addresses of the packets. Since some packets currently forwarded by them can be destined to return to themselves in the end, even selfish MSs do not rashly discard packets.

AD-MIX, however, makes an intentional loopback in path, resulting in a longer path and the waste of limited resources. This may shorten the life span of the MANET in the end.

12.5.3 SMT

SMT (Secure Message Transmission) [44] is a multipath routing protocol. Not tackling selfish MSs directly, it circumvents them by simultaneously sending redundant data via multiple paths. Destination can reconstruct the original message if only some parts of the redundant data arrive successfully. According to the probability of a packet arriving successfully, each path is given a rate of reliability. A path with a low rate is removed from the candidate path set. SMT has strength in supporting QoS requirements of real-time traffic.

12.5.4 PGP

Since Pretty Good Privacy (PGP) [45], [46] was first introduced, some literature has studied methods to evaluate the trust level of each user in a network. Although they are not directly related to selfishness prevention, it is worth introducing them since they help MSs select more reliable paths, resulting in the improvement of overall network performance. This is similar to watchdog [8] and TWOACK [16] in respect to the fact that all of them just detour unreliable MSs rather than enforce their cooperation with others. PGP gets rid of defects of a centralized architecture (e.g., the scalability issue) by making all entities play as certificate authorities, verifying and signing others' public keys independently. To do this, however, each entity is burdened with overheads to store certificates for all other entities. Each PGP entity classifies others into the four levels of trustworthiness of a certificate and its issuer.

Capkun et al. [47] utilize the small-world phenomenon [48] emerging in the PGP certificate graph to guarantee the trustworthiness of each MS. The small-world phenomenon indicates that any pair of people in the United States is connected with a chain of five or six acquaintances. Similarly, every MS can be reached through a chain of a few MSs. Unlike PGP, each MS in this method has certificates for a limited number of MSs. Instead, an MS depends on acquaintances' opinions to trust other MSs. If an MS needs the trust information for any MS, it requests the certificate on the target MS from its acquaintance MSs. If any of them do not have the certificate, they request it from their acquaintances sequentially. Sometimes, an MS that has the certificate for the target MS will appear and the certificate is relayed to the initial request issuer in a cascading way.

12.5.5 Trust Graph

Similar to PGP, Theodorakopoulos and Baras [49] suggest a pair of opinions on trust: the *trust* value, which indicates how much the corresponding MS can be trusted, and the *confidence* value, meaning how much the issuer is sure about the trust value. Through these opinions from current acquaintances, each MS accurately analogizes trustworthiness of MSs that it has not interacted with before, whereas entities in PGP utilize only the trust values directly assigned to themselves.

This research introduces a *trust graph* where nodes and directed edges are mapped to actual MSs and the trustworthiness each MS has for other ones, respectively. The collusion between MSs is taken into account in simulation by assuming that misbehaving MSs always have the best opinion for other misbehaving MSs.

12.5.6 Trust for a Specific Work

Pirzada and McDonald [50] propose to subdivide trust into several specific categories, instead of maintaining a single overall trust for each MS. In this representation, for instance, any MS can be trusted for packet forwarding but not for secure routing. It is also argued that discrete levels of trust, as in PGP systems, are not appropriate for reflecting various features in MANETs. Thus, they redesign the level of trust such that it has a value in a continuous range.

12.6 Summary

Table 12.2 lists important features and limitations of selfishness prevention schemes in this chapter. Generally speaking, reputation-based schemes are superior to the credit-based methods with respect to the scalability of the increased number of MSs, because reputation is usually managed in a distributed way, unlike credits. However, they have a critical limitation of the difficulty for use in future MANETs, since most of them adopt the promiscuous listening mode to observe neighbor MSs' behavior. Promiscuous listening cannot work in the presence of a unidirectional link.

For most credit-based schemes, scalability is the most critical issue, as expected, since current schemes usually have a centralized authority to manage credits for all MSs.

Game theory-based methods surpass both of the other schemes in many ways, but MSs have to collect a large amount of information on the entire network, which makes them difficult to be used in the practical MANET. Additionally, they can work correctly on the condition that all MSs in the game are rational, which means that they must try to maximize their profits continuously without exception.

While much effort is being concentrated on the development of robust and efficient selfish prevention, there are some studies with a different opinion, i.e., pessimistic prospects on this research. Particularly, Huang et al. [51] address a variety of drawbacks that may make both the reputation and the credit-payment incentive schemes impractical. Based on the adoption cycle of MANETs, they even insist that these methods are not needed at all, especially in the early stages of MANET. They enumerate the reasons as follows:

1. They cause large overhead in systems.
2. The credit-based methods may be critical to MSs on the outskirts of a network, because they may not earn sufficient credits due to their bad position as relays.
3. In the credit-based methods, real-time applications cannot send their packets until they have enough credits.
4. It is difficult to fairly manage credits.
5. It is not clear if the resources saved by selfish behavior can be motivation enough to tamper with MSs, even considering the cost to manipulate MSs.

TABLE 12.2 Selfishness Prevention Schemes

Name	Type	Manage	Feature	Limitation
Watchdog [8]	R	Distributed	Detouring selfish MSs, not punishing them	Dependence on promiscuous listening
Context aware [9]	R	Distributed	Misbehavior detection in the route discovery process as well	Offline agreement on a secret number
CONFIDANT [10]	R	Distributed	Isolation of selfish MSs	Dependence on promiscuous listening
CORE [11]	R	Distributed	Collaborative monitor on neighbor MSs	Slow reaction to MS's behavior
Robust reputation [12]	R	Distributed	Utilization of only local reputation	Ignorance of nonneighboring MSs
Friends and Foes [13]	R	Distributed	Individual relation between a pair of MSs	Large memory overhead
SORI [14]	R	Distributed	Quantitative and objective way to measure reputation	Possible for an MS to fake the observation about neighbors
CineMA [15]	R	Distributed	Group-based approach where a portion of MSs can run the algorithm	Large computing and memory overhead for CineMA MS
TWOACK [16]	R	Distributed	Acknowledgment for packets between MSs two hops away	Large message and memory overhead
RIW [17]	R	Centralized	Three-window weighted average for reputation to smooth MS status change	Arbitrary weight without a theoretical base
Light-weight [18]	R	Distributed	On-demand monitoring and self-exemption considering battery status	Need for information about neighbors of neighbors
RI table [19]	R	Distributed	Next-hop MS reputation according to the history of packet arrival at destination	Not fit for MANETs with topology control
Selfish routing [20]	R	Distributed	Smart selfish MSs not participating in the route discovery process	How to share smart selfish MS information
Integrated detection [21]	R	Distributed	Summary of various types of selfish behavior	Arbitrary time limit for delay RREQ relaying
PCOM [22]	R	Distributed	Proactive prevention of selfish MSs from joining the network	Absolute trust on the MSs with the cooperation record
CI-DSR [23]	R	Distributed	Detection of MSs not joining route discovery process as well	Limited to DSR
Fellowship [24]	R	Distributed	Rate limitation against flooding and cooperation enforcement	Hard to distinguish accidental packet drops from intentional ones

(continued on next page)

TABLE 12.2 Selfishness Prevention Schemes

Name	Type	Manage	Feature	Limitation
PPM/PTM [26, 27]	C	Distributed	First SRC and DSTN charge model for packet transmission	Tamper-proof hardware for security
Ad hoc-VCG [28]	C	Centralized	Two phases of cost calculation and payment for relays	Total dependence on destination's report
Sprite [29]	C, G	Centralized	Collusion prevention as well	Scalability issue with message overhead
Multihop cellular [30]	C	Centralized	Combined architecture of cellular network and MANET	Indirect communication between MSs
Priority forwarding [31]	C	Centralized	Two-layered service: free best-effort and priced priority forwarding	Dependence on an MS as a credit server
Willingness to pay [32]	C	Distributed	Adaptive price depending on the status of resource	Naive trust in each MS on the cost
Truthful multicast [33]	C, G	Centralized	Encouragement for truthful reporting in multicast routing tree	Limited to only biconnected networks
PIFA [6]	C, R	Centralized	Full compatibility to any type of routing	Dependence on an MS as credit server
D-SAFNC [34]	C, R	Distributed	Cluster architecture of PIFA	Not completely distributed
GTFT [35]	G	Distributed	A little generous for others' selfishness	Need for much system information
Presumption [36]	G	Distributed	Presumption on neighbor's behavior	Need for much information on neighbors
CORE game [37]	G	Distributed	Extension of the prisoner's dilemma	Too naive model for behavior of MS
Catch [38]	G	Distributed	Anonymous message: hidden sender ID	No proof of evolutionary stability
SLAC [39]	G	Distributed	Prisoner's dilemma in P2P	Lack of a way to fairly compare performance
Incentive scheduling[40]	G	Centralized	More time slots and power for relay MSs than nonrelay MSs	Relay MSs not relaying packets actually
Multinode cooperation[41]	G	Distributed	Attack-resistant and cheat-proof against malicious as well as selfish behavior	Need for a secure monitor that can detect who dropped packets
Token based [42]	—	Distributed	A partial of the system secret shared by MSs	Not on a sparse or high-mobility network
AD-MIX [43]	—	Distributed	Destination ID hidden in packets	Longer path by deliberate loopback
SMT [44]	—	Distributed	Redundant data via multiple paths	Increased amount of traffic

Note: R, reputation based; C, credit based; G, game theory.

Obviously, these limitations are overburdened in current MANETs, but it does not mean we should abandon the research. Considering that the operation of MANETs completely relies on voluntary participation of each MS, it is too naive to trust MSs unconditionally. Besides, we believe that all the technical problems, except for the last reason from the perspective of economics, could be resolved as time goes by. First, for credit-based methods, the large overhead is mainly due to the centralized architecture where a lot of messages should be concentrated on a central authority. For reputation-based or game theory-based methods, the large amount of messages to disseminate reputation or system information can be a problem, too. Hence, extensive research has been performed to develop self-organized algorithms without a central authority and ways to reduce the amount of messages. Actually, many efficient schemes have been proposed, although each of them still has a few weaknesses. Second, the problem of bad-positioned MSs will be naturally solved under the environment where all MSs can freely move. Besides, many algorithms try to solve the deadlock problems caused by the deficiency of credits either by distributing credits to all MSs periodically or by providing free best-effort forwarding separately from priced priority forwarding. Reputation-based methods do not have such a problem. Third, it is certain that the credit-based method may not support real-time traffic well, but users who want real-time service may purchase credits with real money. If any real-time service is really important to them, they might establish their own network. Fourth, for the fair management of credits, more and more robust schemes are emerging day by day.

Several stumbling blocks still remain, which need to be immediately cleared. First, we desperately need an alternative to the promiscuous listening in reputation-based algorithms. As mentioned earlier, the promiscuous mode cannot work in the presence of topology control or smart (directional) antenna. Second, most current algorithms do not have any countermeasure to the spoofing attack where MSs continuously change their IDs. The spoofing attack makes it useless either to build up reputation or to manage credits associated with a specific ID. These problems must be addressed carefully so that cooperative MSs may be fairly rewarded and selfish MSs may be surely punished.

References

[1] D. P. Agrawal and Q.-A. Zeng. 2003. Ad hoc and sensor networks. In *Introduction to wireless and mobile systems*, 2nd ed. chap. 13, pp. 297–348. Toronto, Canada: Brooks/Cole–Thomson Learning.

[2] N. Jain and D. P. Agrawal. 2005. Current trends in wireless sensor network design. *International Journal of Distributed Sensor Networks* 1:101–22.

[3] J. N. Al-Karaki and A. E. Kamal. 2004. Routing techniques in wireless sensor networks: A survey. *IEEE Wireless Communications* 11:6–28.

[4] R. Molva and P. Michiardi. 2003. Security in ad hoc networks. In *Proceedings of Personal Wireless Communications (PWC)*, pp. 756–775.

[5] H. Deng, W. Li, and D. P. Agrawal. 2002. Routing security in wireless ad hoc networks. *IEEE Communications Magazine* 40:70–75.

[6] Y. Yoo, S. Ahn, and D. P. Agrawal. 2005. A credit-payment scheme for packet forwarding fairness in mobile ad hoc networks. In *Proceedings of IEEE ICC*, pp. 3005–3009.

[7] S. Buchegger and J.-Y. L. Boudec. 2002. Cooperative routing in mobile ad-hoc networks: Current efforts against malice and selfishness. In *Proceedings of the Mobile Internet Workshop (Informatik)*, pp. 513–517.

[8] S. Marti, T. J. Giuli, K. Lai, and M. Baker. 2000. Mitigating routing misbehavior in mobile ad hoc networks. In *Proceedings of ACM MobiCom*, pp. 255–65.

[9] K. Paul and D. Westhoff. 2002. Context aware detection of selfish nodes in DSR based ad-hoc networks. In *Proceedings of IEEE Globecom*, pp. 178–82.

[10] S. Buchegger and J.-Y. L. Boudec. 2002. Performance analysis of the CONFIDANT protocol (cooperation of nodes: fairness in dynamic ad-hoc networks). In *Proceedings of ACM MobiHoc*, pp. 226–36.

[11] P. Michiardi and R. Molva. 2002. Core: A collaborative reputation mechanism to enforce node cooperation in mobile ad hoc networks. In *Proceedings of the Sixth IFIP Conference on Security Communications and Multimedia*, pp. 107–121.

[12] P.-W. Yau and C. J. Mitchell. 2003. Reputation methods for routing security for mobile ad hoc networks. In *Proceedings of the Mobile Future and Symposium on Trends in Communications*, pp. 130–37.

[13] H. Miranda and L. Rodrigues. 2003. Friends and Foes: Preventing selfishness in open mobile ad hoc networks. In *Proceedings of the International Conference on Distributed Computing Systems Workshops*, pp. 440–45.

[14] Q. He, D. Wu, and P. Khosla. 2004. SORI: A secure and objective reputation-based incentive scheme for ad-hoc networks. In *Proceedings of the IEEE Wireless Communications and Networking Conference (WCNC)*, pp. 825–30.

[15] M. Frank, P. Martini, and M. Plaggemeier. 2004. CineMA: Cooperation enhancements in MANETs. In *Proceedings of the International Conference on Local Computer Networks*, pp. 86–93.

[16] K. Balakrishnan, J. Deng, and P. K. Varshney. 2005. TWOACK: Preventing selfishness in mobile ad hoc networks. In *Proceedings of the IEEE Wireless Communications and Networking Conference (WCNC)*, vol. 4, pp. 2137–2142.

[17] W. J. Adams, G. C. Hadjichristofi, and N. J. Davis IV. 2005. Calculating a node's reputation in a mobile ad hoc network. In *Proceedings of the IEEE International Performance Computing and Communications Conference (IPCCC)*, pp. 303–7.

[18] Y. Wang and M. Singhal. 2005. A light-weight solution for selfish nodes problem considering battery status in wireless ad-hoc networks. In *Proceedings of WiMob*, pp. 299–306.

[19] M. T. Refaei, V. Srivastava, L. DaSilva, and M. Eltoweissy. 2005. A reputation-based mechanism for isolating selfish nodes in ad hoc networks. In *Proceedings of the International Conference on Mobile and Ubiquitous Systems: Networking and Services (MobiQuitous)*, pp. 3–11.

[20] B. Wang, S. Soltani, J. K. Shapiro, and P.-N. Tan. 2005. Local detection of selfish routing behavior in ad hoc networks. In *Proceedings of the IEEE International Symposium on Parallel Architectures, Algorithms and Networks*, pp. 392–99.

[21] S. Yokoyama, Y. Nakane, O. Takahashi, and E. Miyamoto. 2006. Evaluation of the impact of selfish nodes in ad hoc networks and detection and countermeasure methods. In *Proceedings of the International Conference on Mobile Data Management*, p. 95.

[22] T. Suzuki, M. Kobayashi, A. Khan, and M. Morita. 2006. Proactive cooperation mechanism based on cooperation records for mobile ad hoc networks. In *Proceedings of IEEE ICC*, pp. 3729–34.

[23] L. Xu, Z. Lin, and A. Ye. 2006. Analysis and countermeasure of selfish node problem in mobile ad hoc network. In *Proceedings of the International Conference on Computer Supported Cooperative Work in Design*, pp. 1027–1030.

[24] V. Balakrishnan, V. Varadharajan, and U. K. Tupakula. 2006 Fellowship: Defense against flooding and packet drop attacks in MANET. In *Proceedings of the IEEE/IFIP Network Operations and Management Symposium*.

[25] Y. Liu and Y. R. Yang. 2003. Reputation propagation and agreement in mobile ad hoc networks. In *Proceedings of the IEEE Wireless Communications and Networking Conference (WCNC)*, pp. 1510–15.

[26] L. Buttyan and J.-P. Hubaux. 2000. Enforcing service availability in mobile ad-hoc networks. In *Proceedings of ACM MobiHoc*, pp. 87–96.

[27] L. Buttyan and J.-P. Hubaux. 2003. Stimulating cooperation in self-organizing mobile ad hoc networks. *Mobile Networks and Applications* 8:579–92.

[28] L. Anderegg and S. Eidenbenz. 2003. Ad hoc-VCG: A truthful and cost-efficient routing protocol for mobile ad hoc networks with selfish agents. In *Proceedings of ACM MobiCom*, pp. 245–59.

[29] S. Zhong, J. Chen, and Y. R. Yang. 2003. Sprite: A simple, cheat-proof, credit-based system for mobile ad-hoc networks. In *Proceedings of IEEE Infocom*, pp. 1987–97.

[30] N. B. Salem, L. Buttyan, J.-P. Hubaux, and M. Jakobsson. 2003. A charging and rewarding scheme for packet forwarding in multi-hop cellular networks. In *Proceedings of ACM MobiHoc*, pp. 13–24.

[31] B. Raghavan and A. C. Snoeren. 2003. Priority forwarding in ad hoc networks with self-interested parties. In *Proceedings of the Workshop on Economics of Peer-to-Peer Systems*.

[32] J. Crowcroft, R. Gibbens, F. Kelly, and S. Östring. 2003. Modelling incentives for collaboration in mobile ad hoc networks. In *Proceedings of WiOpt*.

[33] W. Wang, X.-Y. Li, and Y. Wang. 2004. Truthful multicast routing in selfish wireless networks. In *Proceedings of ACM MobiCom*, pp. 245–59.

[34] L. Santhanam, N. Nandiraju, Y. Yoo, and D. P. Agrawal. 2006. Distributed self-policing architecture for packet forwarding fairness in wireless mesh networks. In *Proceedings of the IFIP International Conference on Personal Wireless Communications*, pp. 147–58.

[35] V. Srinivasan, P. Nuggehalli, C. F. Chiasserini, and R. R. Rao. 2003. Cooperation in wireless ad hoc networks. In *Proceedings of IEEE Infocom*, pp. 808–17.

[36] A. Urpi, M. Bonuccelli, and S. Giordano. 2003. Modelling cooperation in mobile ad hoc networks: A formal description of selfishness. In *Proceedings of WiOpt*.

[37] P. Michiardi and R. Molva. 2003. A game theoretical approach to evaluate cooperation enforcement mechanisms in mobile ad hoc networks. In *Proceedings of WiOpt*.

[38] R. Mahajan, M. Rodrig, D. Wetherall, and J. Zahorjan. 2004. Experiences applying game theory to system design. In *Proceedings of the ACM SIGCOMM Workshop on Practice and Theory of Incentives and Game Theory in Networked Systems (PINS)*, pp. 183–90.

[39] D. Hales. 2004. From selfish nodes to cooperative networks—Emergent link-based incentives in peer-to-peer networks. In *Proceedings of the IEEE International Conference on Peer-to-Peer Computing*, pp. 151–58.

[40] H.-Y. Wei and R. D. Gitlin. 2005. Incentive scheduling for cooperative relay in WWAN/WLAN two-hop-relay network. In *Proceedings of the IEEE Wireless Communications and Networking Conference (WCNC)*, pp. 1696–1701.

[41] W. Yu and K. J. R. Liu. 2007. Game theoretic analysis of cooperation stimulation and security in autonomous mobile ad hoc networks. *IEEE Transactions on Mobile Computing* 6:459–73.

[42] H. Yang, X. Meng, and S. Lu. 2002. Self-organized network-layer security in mobile ad hoc networks. In *Proceedings of the ACM Workshop on Wireless Security (WiSe)*, pp. 11–20.

[43] S. Sundaramurthy and E. M. Belding-Royer. 2003. The AD-MIX protocol for encouraging participation in mobile ad hoc networks. In *Proceedings of the IEEE International Conference on Network Protocols*, pp. 156–67.

[44] P. Papadimitratos and Z. J. Haas. 2003. Secure data transmission in mobile ad hoc networks. In *Proceedings of the ACM Workshop on Wireless Security (WiSe)*, pp. 41–50.

[45] P. R. Zimmermann. 1995. *The official PGP user's guide*. Cambridge, MA: MIT Press.

[46] The International PGP home page. http://www.pgpi.org.

[47] S. Capkun, L. Buttyan, and J.-P. Hubaux. 2002. Small worlds in security systems: An analysis of the PGP certificate graph. In *Proceedings of the ACM New Security Paradigms Workshop*, pp. 28–35.

[48] S. Milgram. 1967. The small world problem. *Psychology Today*, Vol. 61.

[49] G. Theodorakopoulos and J. S. Baras. 2004. Trust evaluation in ad hoc networks. In *Proceedings of ACM Workshop on Wireless Security (WiSe)*, pp. 1–10.

[50] A. A. Pirzada and C. McDonald. 2004. Establishing trust in pure ad-hoc networks. In *Proceedings of the 27th Conference of Australasian Computer Science*, pp. 47–54.

[51] E. Huang, J. Crowcroft, and I. Wassell. 2004. Rethinking incentives for mobile ad hoc networks. In *Proceedings of ACM SIGCOMM 2004*, pp. 191–196.

13

Mobile-Relay Forwarding in Opportunistic Networks

Giuseppe Anastasi
University of Pisa

Marco Conti
IIT-CNR, National Research Council

Andrea Passarella
IIT-CNR, National Research Council

Luciana Pelusi
IIT-CNR, National Research Council

13.1 Introduction

Opportunistic networks are one of the most interesting extensions of the legacy mobile ad hoc network (MANET) concept. Legacy MANETs are composed by mobile nodes that collaboratively set up a network plane by running a given routing protocol. Therefore, the—sometimes implicit—assumption behind MANETs is that the network is well connected, and nodes' disconnection is an exception to deal with. Most notably, if the destination of a given message is not connected to the network when the message is generated, then that message is dropped after a short time (i.e., the destination is assumed not to exist). Opportunistic networks are mobile wireless networks in which the presence of a continuous path between a sender and a destination is not assumed. Sender and destination nodes may never be connected to the network at the same time. The network is assumed to be highly dynamic, and the topology is thus extremely unstable and sometimes completely unpredictable. Nevertheless, the network must guarantee end-to-end delivery of messages despite frequent disconnections and partitions.

The opportunistic networking paradigm is particularly suitable to those environments characterized by frequent and persistent partitions. In section 13.7 we will survey the most relevant case studies that rely on this paradigm. However, we can anticipate here a couple of example scenarios where classical wireless networking approaches are not feasible and the opportunistic networking approach is the only viable solution. In the field of wildlife tracking, for example, some kinds of sensor nodes are used to monitor wild species. In these cases it is not easy (or possible sometimes) to have connectivity among a source sensor node and a destination data collector node. This is because the animals to be monitored move freely and there is no possibility to control them in such a way to favor connectivity. Opportunistic networks may also be exploited to bridge the digital divide. In fact, they can support intermittent connectivity to the Internet for underdeveloped or isolated regions. This can be obtained by exploiting mobile nodes that collect information to upload to the Internet as well as requests for Web pages or any kind of data that need to be downloaded from the Internet. Both data and requests are up- or downloaded to or from the Internet once the mobile data collector node reaches a location where connectivity is available.

It clearly emerges that routing and forwarding play a key role in opportunistic networks. However, given the intermittent connectivity, it is not always possible to define a complete route between the source and destination nodes at the moment the source is willing to deliver its message. Hence, routing is not intended in the classical way. Routes in opportunistic networks are usually computed "on the fly," while messages are forwarded. Routing is thus rather concerned with finding hop by hop a path to the destination. In fact, at each step the only decision that can be made is to whom the message is to be forwarded next. As a result, routing and forwarding are typically performed at the same time.*

In general, two main concepts are at the basis of routing/forwarding protocols for these networks. On the one hand, since topological information is unreliable, routing should exploit information pertaining to any layer of the stack to understand how to build routes. On the other hand, any communication opportunity should be exploited (at least, considered) for carrying messages closer to the eventual destination(s).

Different approaches to routing are possible, as discussed in section 13.2. Some (historical) routing approaches are based on a vast dissemination of data in all network directions. By spreading information throughout the network, the probability that messages eventually reach their destination is very high. However, these approaches cause severe resource consumption (e.g., bandwidth and memory at intermediate nodes) due to the

* In the following, we will use the terms *routing* and *forwarding* interchangeably.

frequent data exchanges involved. More recent approaches therefore tend to identify only one or a few preferential directions for data forwarding. These approaches are generally more computationally intensive than the previous ones, but consume smaller amounts of bandwidth and memory. Although we provide here a general classification and discussion of the possible routing approaches in opportunistic networks, we particularly focus on Mobile-Relay Forwarding (MRF). MRF assumes that there exist particular nodes (mobile relays) in the network that are exploited to collect messages from the source nodes, and to take messages (closer) to the destination. Routing approaches based on mobile relays (MRs) are very energy efficient because regular nodes are relieved of their routing workload, which is instead undertaken by MRs. Furthermore, this approach increases network scalability since the addition of extra nodes to the network does not imply an increment of routing complexity. This is particularly beneficial to scenarios that can potentially include a lot of (heterogeneous) devices like, for example, an urban environment. For example, MRs can be buses traveling in a city, while regular nodes can be pedestrians. Regular nodes wait for one of such MRs to pass nearby, and hand over messages to it. Usually, MRs have completely different mobility patterns with respect to regular nodes, cover larger distances, and are thus able to connect nodes that would not be able to communicate otherwise. Mobility of MRs can be either controllable or not, and MRs can be either already part of the system or deployed just for the sake of improving routing performance. Furthermore, MRs usually have fewer restrictions on the availability of resources with respect to the other nodes of the network. Therefore, they can greatly increase network connectivity and data delivery in opportunistic networks.

In this chapter we describe the different types of architecture that have been proposed to exploit the MRF concept (see section 13.3). Then, we identify the main issues to be addressed in the design of MR behavior (section 13.4), and focus on algorithms that control MR movements to optimize network performance (section 13.5). As opportunistic networks are composed by mobile devices, power conservation issues should be of primary concern. Therefore, we discuss power management techniques for MRF in section 13.6. Section 13.7 is devoted to the description of some relevant case studies highlighting how the MRF concept can be exploited in different scenarios. In particular, in section 13.7.4 we give special emphasis to a novel kind of network to which the MRF concepts can be applied, i.e., underwater sensor networks. Conclusions and open issues are discussed in section 13.8.

13.2 Routing Approaches in Opportunistic Networking

As highlighted above, routing is the most compelling challenge in opportunistic networking. The design of efficient routing strategies for opportunistic networks is generally a complicated task due to the absence of knowledge about the topological evolution of the network. Routing performance improves when more knowledge about the expected topology of the network can be exploited [Sush04]. Unfortunately, this kind of knowledge is not easily available, and a trade-off must be achieved between performance and knowledge requirement.

Depending on the particular routing approach followed, opportunistic networks may be classified at a very top level into two categories: *infrastructure-less* and *infrastructure-based* networks [Pelu06b]. Infrastructure-less networks are completely flat ad hoc networks (*without infrastructure*) where all the nodes equally take on routing/forwarding duties. In infrastructure-based networks (*with infrastructure*), instead, some form of infrastructure is exploited to forward messages opportunistically. The infrastructure is typically composed of special nodes that are in charge of messages forwarding, whereas the other nodes are generally relieved of the forwarding workload.

13.2.1 Infrastructure-less Opportunistic Networks

In infrastructure-less opportunistic networks two basic routing approaches are followed: *dissemination-based* and *context-based* routing. Dissemination-based algorithms are essentially forms of controlled flooding, and differentiate themselves for the policy used to limit flooding. Context-based approaches usually do not adopt flooding schemes, but use knowledge of the context that nodes are operating in to identify the best next-hop at each forwarding step. The following subsections offer an overview of both dissemination-based and context-based routing approaches, describing the most representative algorithms of each.

13.2.1.1 Dissemination-Based Routing

Routing techniques based on data dissemination perform delivery of a message to destination by simply diffusing it all over the network. The heuristic behind this policy is that since there is no knowledge of a possible path toward the destination, nor of an appropriate next-hop node, a message should be sent everywhere. It will eventually reach the destination by passing node to node. Dissemination-based techniques are very resource hungry. Moreover, due to the considerably high number of transmissions involved, dissemination-based techniques suffer from high contention and may potentially lead to network congestion. To increase the network capacity, the spreading radius of a message is typically limited by imposing a maximum number of relaying hops to each message or even by limiting the total number of message copies present in the network at the same time. When no relaying is further allowed, a node can only send directly to a destination when or in case it is met.

The first protocol exploiting dissemination techniques proposed in the literature is the *epidemic routing* protocol [Vahd00]. In epidemic routing messages diffuse in the network similarly to diseases or viruses, i.e., by means of pair-wise contacts between individuals/nodes. A node is *infected* by a message when it either generates that message or alternatively receives it from another node for forwarding. The infected node stores the message in a local buffer. A node is *susceptible* to infection when it has not yet received the message* but can potentially receive it in case it comes into contact with an infected node (i.e., a node that stores that message). The infected node becomes *recovered* (healed from the disease) once having delivered the message to the destination node, and as a result, it also becomes *immune* to the same disease and does not provide

* The message itself represents the infection/virus.

relaying to the same message any more. The dissemination process is somehow bounded because each message when generated is assigned a *hop count limit* giving the maximum number of hops that that message is allowed to traverse till the destination. When the hop count limit is 1, the message can only be sent directly to the destination node.

Further steps beyond epidemic routing are represented by *PROPHET* [Lind03] and the *MV routing* [Burn05] protocols. In both protocols, messages are exchanged during pair-wise contacts as in epidemic routing. However, a more sophisticated method to select the messages to forward to an encountered node is introduced. Basically, the choice depends on the *probability* of the encountered nodes to deliver the messages successfully to their eventual destinations. The delivery probability relies on observations on the meetings between nodes (in PROPHET), and on both the meetings between nodes and the visits of nodes to geographical locations that occurred in the recent past (in MV Routing).

Network-coding-based routing [Widm05] also falls in the category of dissemination-based algorithms, but takes an original approach to limit message flooding. Messages are combined together (encoded) at nodes before being forwarded. Then, the codes produced are sent out instead of the original messages. Codes are spread in different directions like in other dissemination-based routing protocols. The number of codes generated is higher than the number of original messages combined together. This is to achieve much more robustness against both packet and path loss. Encoding is performed at both source and intermediate nodes.

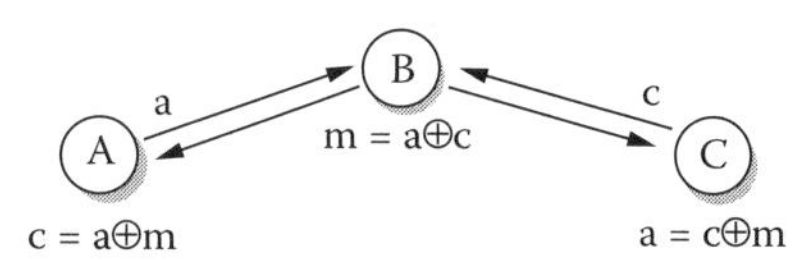

FIGURE 13.1 Example of network-coding efficiency.

Just to give a classical, and simplified, example of network coding, let A, B, and C, be the only three nodes of a small network (see Figure 13.1). Let node A generate the information *a* and node C generate the information *c*. Then, suppose the information produced needs to be known at all the nodes. Hence, nodes A and C send their information to node B, then node B, rather than sending two different packets for *a* and *c*, respectively, broadcasts a single packet containing *a xor c*. Once *a xor c* is received, both nodes A and C can finally infer the missing information (i.e., node A can infer *c* and node C can infer *a*). Network-coding-based routing can be generalized by recursively using erasure-coding techniques at intermediate nodes [Pelu06a]. It outperforms flooding, as it is able to deliver the same information with fewer messages injected into the network.

13.2.1.2 Context-Based Routing

Most of the dissemination-based techniques limit message flooding by exploiting knowledge about direct contact with destination nodes. Context-based routing exploits more information about the context nodes are operating in to identify suitable next-hops toward the eventual destinations. The usefulness of a host as next-hop for a message is hereafter referred to as *utility* of that host. Usually, such routing techniques are able to reduce significantly the message duplication and resource consumption (e.g., bandwidth, memory, energy) of dissemination-based techniques. Since they also reduce network congestion, it has been shown that they are able to reduce delays and message loss as well. The main cost paid for these advantages is the fact that context information must be kept at nodes and circulated among nodes. However, recent results show that resource consumption is far lower even when these additional costs are considered [Bold07a].

In the *context-aware routing* (CAR) protocol [Muso05] each node in the network is in charge of producing its own delivery probabilities toward each known destination host. Delivery probabilities are exchanged periodically so that, eventually, each node can compute the best carrier for each destination node. The best carriers are computed based on the nodes' context. Among the context attributes needed to elect the best carrier are, for example, the residual battery level, the rate of change of connectivity, the probability of being within reach of the destination, and the degree of mobility. When the best carrier receives a message for forwarding, it stores it in a local buffer and eventually forwards it to the destination node when met, or alternatively to another node with a higher delivery probability. Actually, CAR provides a framework for computing next-hops in opportunistic networks based on the *multiattribute utility theory* applied to generic context attributes. Simulation results show that CAR is more scalable than epidemic routing as the protocol overhead is approximately constant, regardless of the node buffer size.

In *MobySpace routing* [Legu06] the nodes' mobility pattern represents the context information used for routing. The protocol builds up a high-dimensional Euclidean space, named MobySpace, where each axis represents a possible contact between a couple of nodes and the distance along an axis measures the probability of that contact occurring. Two nodes that have similar sets of contacts and that experience those contacts with similar frequencies are close in MobySpace. The best forwarding node for a message is the node that is as close as possible to the destination in this space. This, in fact, improves the probability that the message will eventually reach the destination. Obviously, in this *virtual contact space* just described, the knowledge of all the axes of the space also requires the knowledge of all the nodes that are circulating in the space.*

Both CAR and MobySpace routing require full knowledge of possible destinations to enable forwarding. The History-Based Opportunistic Routing Protocol (HiBOP) [Bold07a, Bold07b] provides a framework for managing and exploiting context information that does not require all nodes to know each other. In HiBOP nodes exchange context information about the users when getting in touch. Each node remembers context information seen in the past (such information is enforced based on how often it is "seen" on encountered nodes). A node carrying a given message asks the encountered nodes to compute their delivery probability toward the destination(s). The delivery probability is computed based on the match between context information *about the destination* stored in the message itself, and context information stored by the encountered node itself. Messages are forwarded along a gradient defined by increasing match between the destination information and the context information of the carrying node. Hence, the algorithm dynamically selects as next-hops those nodes that share more and more context information with the destination(s). HiBOP exploits social relationships among users to identify good carriers for messages.

* [Legu06] also proposes an optimization that does not require knowledge of all contacts between nodes.

13.2.2 Infrastructure-Based Opportunistic Networks

Infrastructure-based opportunistic networks are characterized by the presence of special nodes that are used for collecting messages from source nodes and delivering them to their destinations. Such special nodes are generally more powerful than regular nodes, as they have a high energy budget and large storage capacity. They may act as a gateway toward a less challenged network (e.g., the Internet), or they can simply increase the connectivity between (regular) nodes in the network. Depending on the mobility of special nodes, we can distinguish opportunistic networks with *fixed infrastructure* and with *mobile infrastructure*, respectively. When using a fixed infrastructure, special nodes are stationary and are located at specific geographical points. On the other hand, in opportunistic networks with mobile infrastructure, special nodes move around in the network area following either predefined or completely random paths.

13.2.2.1 Routing Based on Fixed Infrastructure

A fixed infrastructure consists of special fixed nodes, i.e., base stations, which are sparsely deployed all over the network and act as message collectors. Base stations offer high capacity and robust data exchanges to the mobile nodes nearby. Moreover, they have high storage capacity to collect data from many nodes passing by. A source node wishing to deliver a message keeps it until it comes within reach of a base station, then forwards the message to the base station.

Base stations are generally gateways toward less challenged networks (e.g., they can provide Internet access or be connected to a local area network [LAN]). Hence, the goal of an opportunistic routing algorithm is to deliver messages to the gateways, which are supposed to be able to find the eventual destination more easily. Two variations of the protocol are possible. The first one works exactly as described above, and only node-to-base-station communications are allowed. As a result, messages experience fairly high delays. The classical example of this approach is the Infostation model [Good97]. A second version of the protocol allows both node-to-base-station and node-to-node communications. This means that a node wishing to send a message to a destination node delivers the message to the base station directly if it is within communication range; otherwise, it delivers the message opportunistically to a nearby node that will eventually forward it to the base station when encountered (routing schemes presented earlier can be used in this phase). Such a protocol has actually been proposed in the Shared Wireless Infostation Model (SWIM) [Smal03]. As the above examples show, historically, fixed base stations play a passive role in the opportunistic forwarding strategy because they simply act as information sinks (e.g., Infostations [Good97]). However, many benefits can be envisioned by running an opportunistic routing algorithm also at base stations. Base stations, for example, can simply collect the messages sent by the visiting nodes and then wait for the destination nodes to be within reach to forward the stored messages to them. Base stations of a mobile infrastructure (described in the next section) typically play such an active role.

Despite allowing energy saving at the mobile nodes (which are relieved of the forwarding workload, at least in the first version of the protocol), a routing approach relying on a fixed infrastructure is highly expensive due to the costs of the infrastructure. Moreover, it suffers from scalability issues since the addition of new nodes implies the expansion of the infrastructure. Using a mobile infrastructure instead of a fixed infrastructure is a valuable opportunity to realize a cheap and flexible solution. A mobile infrastructure is composed of mobile nodes that move around in the network place following either predetermined or arbitrary routes and gather messages from the nodes they approach. These special nodes may be referred to as *carriers*, *supports*, *forwarders*, *MULEs*, or even *ferries*. They can be the only entities responsible for the delivery of messages, when only node-to-carrier communications are allowed, or they can simply help increase connectivity in sparse networks and guarantee reachability of isolated nodes. In the latter case, delivery of messages is accomplished by both carriers and ordinary nodes and communications are allowed both from node to node and from node to carrier.

13.3 Forwarding Architectures for Opportunistic Networks with Mobile Relays

As mentioned in the introduction, the rest of this chapter is focused on opportunistic networks with MRs. Therefore, in this section we start the discussion by presenting the possible kinds of architectures for this approach.

Figure 13.2 shows the system architecture of opportunistic networking with MRs. We can distinguish the following three different components: regular nodes, MRs, and base stations.

Regular nodes (or simply nodes, for short) are the information sources and destinations. Depending on the specific application scenario, they may be fixed or mobile. For instance, in a sensor network nodes are typically stationary, while in a mobile ad hoc network (MANET) they are usually mobile.

Mobile relays (MRs) are specialized nodes that move throughout the network to collect data from source nodes and deliver it to the destination node or access point. They can follow a fixed or variable trajectory, at constant or variable speed. Therefore, the time interval between successive visits of an MR to the same node may be predictable, variable in a bounded range, or completely random. The number of MRs in a network

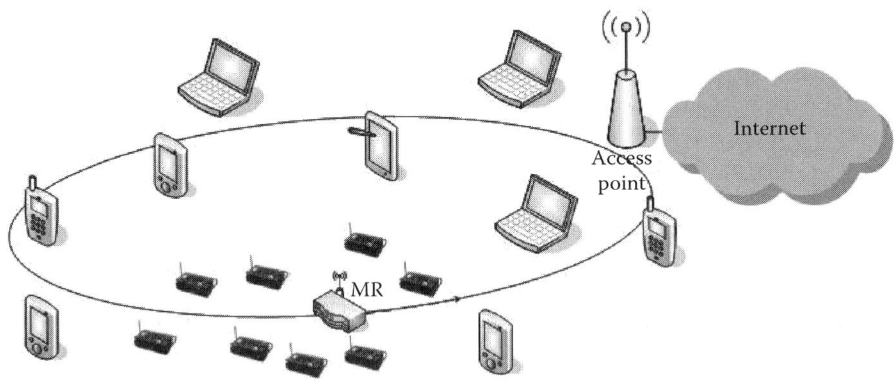

FIGURE 13.2 System architecture for opportunistic networking with mobile relays.

may vary depending on several factors, such as number of regular nodes, amount of traffic to manage, quality-of-service (QoS) requirements, and costs.

Access points (APs) are infrastructured nodes serving as gateways toward less challenged networks (e.g., they provide connectivity to the Internet or a LAN). Again, the number of APs can vary depending on the number of nodes, number of MRs, traffic load, QoS requirements, installation costs, and so on.

When designing an opportunistic networking system based on the above architecture, the following design issues need to be taken into account [Zhao03, Kans04]:

- *Node mobility.* Regular nodes may be stationary or mobile, depending on the application scenario. In case of mobile nodes, we can distinguish between *task-driven* and *message-driven* mobility. In task-driven mobility nodes move according to a path that is dictated by a specific task or goal (e.g., a person with a PDA moves to go to work). In message-driven mobility node movements are aimed at data transmission and reception in general (e.g., a person with a PDA moves toward an MR to exchange messages with it).
- *Coordination between nodes.* Typically there are many nodes in the network, densely or sparsely deployed, depending on the application requirements. Of course, nodes can communicate with an MR only when it is within their communication range. Therefore, either nodes are mobile or they must be deployed at a distance from the MR trajectory not greater than their communication range. Alternatively, nodes can organize themselves to form *clusters* [Soma06] or *regions* [Harr06]. Each cluster consists of a set of nodes that elect a specific node to act as a *gateway* node in charge of communication with the MRs. Nodes in the cluster send their messages to the gateway by *multihop communication*, and the gateway transmits such messages to the MRs (see Figure 13.3). The same approach is used for message reception.
- *MR mobility.* The mobility of MRs is a critical factor since it has direct impact on the success of message delivery as well as on the latency experienced by messages. To achieve good connectivity among nodes, MRs should be able to

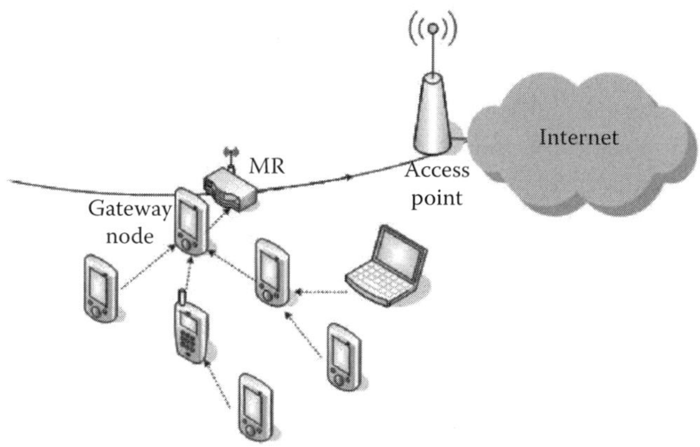

FIGURE 13.3 Opportunistic access through a gateway node via multihop communication.

approach as many nodes as possible and to visit them with an appropriate frequency. In practice, they may be either part of the environment (e.g., a bus or person) or intentionally included as part of the network infrastructure (e.g., a mobile robot). In the former case, there is typically no control on the MR mobility. In the latter case, it is usually possible to control the MR trajectory, speed, or both.

- *MR designation.* MRs may be either special nodes specifically designed to act as MRs or regular nodes that serve temporarily as MRs. In the former case, MRs are typically resource-rich devices. In the latter case, they have limited resources like all the other nodes in the system. In addition, an algorithm for designating MRs is required.
- *Number of MRs.* An opportunistic networking system may rely on one or more MRs, depending on performance, scalability, and reliability requirements. Obviously, the capacity of a single MR is limited by its movement capability. Increasing the number of MRs allows for increased scalability and geographic coverage. In addition, a system with multiple MRs is more resilient to MR failures. On the other hand, a larger number of MRs implies higher economical costs. Therefore, the optimal number of MRs must achieve a trade-off between performance and costs.
- *Coordination between MRs.* If there are multiple MRs in the system, they may have similar or different capabilities. Furthermore, they may operate independently of each other, or in cooperation. In the latter case, a message can be exchanged between several MRs before reaching the destination node or an access point.
- *MR trajectory.* The trajectory followed by an MR to visit nodes may be fixed or variable. In the latter case, it is adjusted dynamically depending on node requests, messages deadlines, etc. Obviously, this is possible only when an MR is part of the system infrastructure and can thus be controlled. In case of multiple MRs, a problem related to the MR trajectory design is the assignment of nodes to the MRs. The trajectory design should take into account not only routing but also load balancing among MRs.
- *MR speed.* MRs may move at a constant, variable, or controlled speed. In the last case, the speed can be controlled by the MR software and adjusted dynamically to improve communication performance. Again, this is possible only when MRs are part of the system infrastructure.
- *Power management and MR discovery.* As nodes have limited energetic resources, they should switch their radio to sleep (i.e., low-power) mode when they are not involved in communications with MRs. However, since MR arrivals are usually unpredictable, this may prevent nodes from discovering an incoming MR. Energy-efficient discovery schemes are thus required that minimize energy consumption while keeping the probability of missing contacts with MRs as low as possible.
- *Data collection and delivery.* A message generated at a source node requires several communications to reach the destination node or an access point. The message is first transmitted by the source (or gateway) node to an MR. In case of

multiple coordinated MRs, the message may be exchanged between several MRs before delivery to the destination node or access point. Protocols for efficient node-to-MR, MR-to-MR, and MR-to-AP communications are thus required.

In the following sections we will describe how the most relevant of the above issues have been addressed in practical opportunistic systems. Specifically, we will discuss the impact of different MR mobility patterns on forwarding, and also the related power management issues.

Before going on, we conclude this section with some comments about the data collection and delivery process. The most interesting aspect of this problem is managing communications between MRs and regular nodes (or the gateway node when a clustering approach is used). Communications between the nodes of a cluster and the related gateway usually borrow well-known techniques from the MANET literature (such as clustering), while communications between MRs and access points are not particularly challenging.

[Kans04] and [Soma06] use a *stop-and-wait* protocol for communication between the MR and a regular node (or the gateway node of a cluster). MR sends an acknowledgment to the sending node for each message correctly received. The node transmits the next message only after receiving acknowledgment from the MR. If the acknowledgment is not received within a predefined *timeout*, the node retransmits the message. The node starts transmitting data as soon as it discovers the MR in its proximity. No information about the location of the MR is exploited because such information may not be available in all systems. In [Anas07] it is analytically shown that using a window-based scheme with a window size greater than 1 provides a higher throughput, and for a fixed amount of data to transfer, it also lets the transfer time (and hence the energy consumption) decrease. However, increasing the window size beyond a given threshold may be impractical since the MR could move out of the communication range. This would result in useless message transmissions (and energy consumption).

13.4 Mobile Relays

As anticipated in section 13.3, MRs may be classified in two broad categories: they can be *part of the environment* or specifically designed as *part of the network infrastructure*. Depending on their nature, they may have different mobility patterns, as shown in Figure 13.4. When the MR is part of the environment its mobility is driven by the specific task the mobile element acting as MR is intended for, and cannot be controlled in any way. Conversely, when the MR is part of the system, its mobility can be controlled to improve the communication performance and extend the geographical coverage. However, even when the MR is not controllable, it may have different mobility patterns. If it follows a strict schedule, it has a completely predictable mobility (e.g., a shuttle for public transportation). On the opposite side, it may have a completely random behavior so that no reliable assumption can be made on its mobility. Finally, the MR may follow a mobility pattern that is neither predictable nor completely random. For example, this is the case of a public transportation bus, or a car, that moves in a city and whose speed is subject to large variation due to traffic conditions. In such a case, the MR mobility pattern, even if not predictable, can be learned based on successive observations and

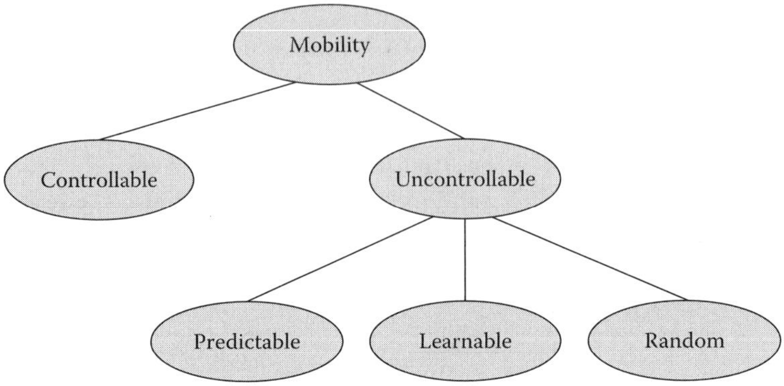

FIGURE 13.4 Classification of MR mobility.

estimated with some accuracy. Learning the MR mobility pattern and estimating times between successive MR visits to the same node is very important to save energy at the node, as will be shown in section 13.6.

A lot of examples of opportunistic networking systems have been proposed in the literature where MRs have different nature and mobility patterns.

[Shah03] proposes a three-tier architecture for energy-efficient data collection in sparse sensor networks based on data MULEs (Mobile Ubiquitous LAN Extensions). Data MULEs can be people, animals, cars, buses, etc., passing nearby sensor nodes and collecting data from them. Obviously, as they are part of the environment, applying the MR forwarding concept to this scenario comes for free. On the other hand, they typically move randomly and no control is possible on them. An approach similar to data MULEs is exploited in the Predictable Mobility Architecture described in [Chak03]. The authors rely on a public transportation shuttle for data collection in sensor networks inside a campus. Unlike data MULEs, the shuttle is assumed to have a strict schedule and predictable intervisit times. This greatly helps to optimize power consumption at nodes. Public transportation buses are also the mobile elements in the Ad Hoc City project [Jetc03]. This project is aimed at creating a city ad hoc network where the role of mobile routers is played by public buses.

Both the Zebranet [Juan02] and SWIM [Smal03] projects focus on tracking wild species and use MRs. In Zebranet, the animals to be tracked are zebras wearing special collars. The MR consists of a vehicle that periodically moves around in the savanna and collects data from the encountered zebras. Zebras collect data from other zebras and deliver them to the MR; thus, they also act as MRs. In the Shared Wireless Infostation Model (SWIM) [Smal03] special tags are applied to whales to perform periodic data monitoring. Data are diffused at each pair-wise contact between whales and finally arrive to special SWIM stations that can be fixed (on buoys) or mobile (on seabirds). Hence, both whale-to-whale and whale-to-SWIM station communications are allowed, and the MRs consist of both the mobile SWIM stations and the whales themselves. From the SWIM stations data are eventually forwarded to an access point on shore from where they will be finally delivered to their destination for processing and utilization.

Many opportunistic networking systems use *controllable* MRs. Among the most relevant ones we mention are, for example, message ferrying [Zhao03, Zhao04], interregional messenger [Harr06], and controllably mobile infrastructure [Kans04]. The main features of these systems are described in the next section.

Finally, there are several proposals (mainly targeted to sensor networks) that do not rely on MRs for transporting data mechanically from nodes to the access point. Instead, the access point itself is mobile and can change its position from time to time [Gand03, Akka05, LuoH05, Wang05]. This may be beneficial in terms of energy saving and decreased message latency. For example, moving the access point close to an area of heavy traffic or near loaded nodes helps reduce the total transmission power and extend the lifetime of nodes on the path of heavy traffic [Akka05]. However, this scenario is beyond the scope of this chapter.

13.5 Motion Control

MR motion control can be achieved only when using a controllable mobile element. Motion control can be performed along two orthogonal directions: space and time. In the space direction we can define, and adapt dynamically, the *trajectory* followed by the MR to visit nodes. In the time direction, we can control the MR *speed* and adjust it to improve the communication performance. Another valuable form of control in opportunistic environments is topology control. In this case, nodes may decide to increase (or decrease) the *transmit power* to increase (decrease) contact times. The three different forms of motion control are discussed in the next subsections.

13.5.1 Trajectory Control

Figure 13.5 shows a classification of trajectory control approaches. We can distinguish them into two broad categories depending on whether the path followed by the MR to visit nodes is *fixed* or *variable*.

When the MR trajectory is fixed there is actually no control. Therefore, the trajectory must be defined very carefully, especially when nodes are stationary. In particular, the

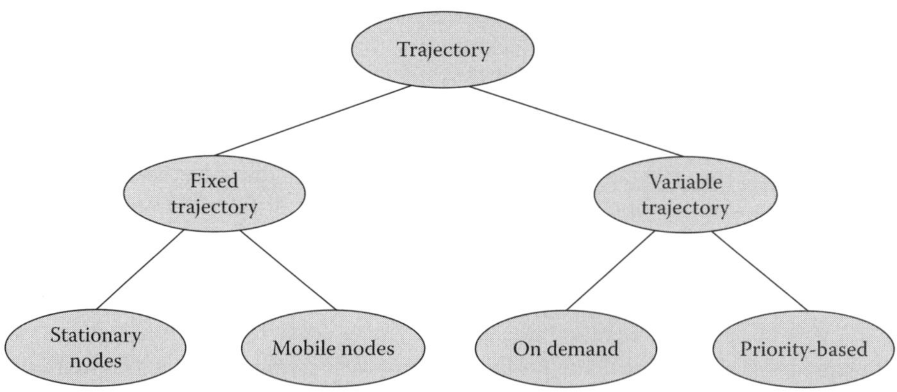

FIGURE 13.5 Classification of route control approaches.

following requirements need to be fulfilled. First, the MR trajectory should pass close to nodes, at a maximum distance less than the node communication range. Second, the MR should remain within the node communication range for enough time to permit a complete data exchange. Finally, the node trajectory must be feasible, i.e., compatible with geographical constraints. If the MR is a controllable mobile element (e.g., a mobile robot), its trajectory can be designed so as to address all the above requirements [Soma06, Zhao03]. If the MR is part of the environment (e.g., a bus), its trajectory depends on the specific task it carries out (e.g., public transportation) and cannot be changed. Hence, nodes must be deployed along the MR trajectory at a distance less than or equal to the node communication range [Jain06, Chak03, JeaS05]. The MR trajectory becomes less critical when nodes are mobile, as they can move toward the MR when they want to exchange data [Zhao04].

Trajectory control actually consists in adjusting the MR trajectory dynamically, based on node requirements As shown in Figure 13.5, trajectory control techniques can be classified into two main categories: *on-demand* and *priority-based* techniques. When using an on-demand approach, the MR adapts its movements to satisfy nodes' requests. Each time a node has data to exchange, it sends a service request to the MR by using a long-range radio, and the MR modifies its route to approach the requesting node. Nodes may be fixed or mobile [Zhao04]. Priority-based route control is used when nodes have different characteristics in terms of message generation rate or buffer size. Therefore, they need to be visited with different frequencies. In practice, each node is associated with a deadline, defined as the time when the node buffer will overflow. Nodes' deadlines are used to schedule the MR visits to nodes [Soma04, GuBo05, GuBo06].

13.5.2 Speed Control

The need for speed control comes from the evidence that network connectivity between nodes and MRs may be very different for nodes located at different places and, for each single node, may also vary over time. For example, connectivity may be difficult, or even impossible, due to the presence of physical obstacles. In addition, the wireless link quality may vary significantly from location to location due to the distance between nodes and MRs, the presence of multipath effects, and so on. For the same location, the wireless link quality may also vary from time to time, e.g., due to meteorological changes. Finally, the throughput experienced by a node depends on the density of nodes in its proximity.

The basic idea behind speed control is to improve the data communication performance by reducing the MR speed in places where the data communication is more difficult (i.e., the network throughput is lower), and increasing it where data communication performs better (i.e., the network throughput is higher). In practice, the speed control module at the MR monitors the throughput experienced in the communication, and adapts the MR speed to improve the data communication performance.

Speed control is also related to power management. As it will be shown in section 13.6, from the power management point of view, the time interval between two consecutive visits of the MR to the same node should be fixed or have very small variations. This allows nodes to sleep for the entire time between two successive visits, and save energy. Therefore, the speed control strategy should be designed in such a way to minimize variations in the time between consecutive visits.

Several strategies for speed control have been proposed in the literature. They are summarized below.

- *No control.* There is no control on the MR speed. This typically occurs when the MR is part of the environment, and hence its mobility is aimed at providing a service different from message collection/delivery (e.g., a public transportation service). In such a case, no control is possible, of course. However, even when the MR is a controllable mobile element (e.g., a mobile robot), dynamic speed control may not be implemented for reducing implementation complexity or costs. In such a case, the MR moves, for instance, at an approximately constant speed. The range of available speeds is obviously dictated by the mobile element acting as MR. Within this range, the optimal speed value can be chosen by taking into account several parameters, e.g., contact duration of each node, constraints on message latency, energy consumed by the mobile element for locomotion, and so on.
- *Stop and communicate.* This is the simplest form of speed control. As soon as the MR reaches a node, it stops for the time required by the node to (1) transfer all its data to the MR and (2) receive messages from the MR, if any. Then the MR moves toward the next node. Without any control on the time spent at different nodes, however, this approach may cause the total time taken for each path traversal to be variable. To avoid this drawback, [Kans04] proposes the *Stop to Collect Data* (SCD) algorithm, which is targeted at sensor networks. Let T be the maximum time the MR can take to complete a round across the network (T is imposed by constraints on message latency), and let s be the (constant) speed required to cover the entire path in a time less than or equal to T. In the SCD algorithm the MR moves at a constant speed of $2s$, and thus it requires a time $T/2$ to traverse the entire path. The remaining $T/2$ interval is used by the MR to stop at nodes to collect data. Specifically, if N is the number of nodes, the MR stops at each node for a time $T/(2N)$. A different distribution of extra time among network nodes would be possible and, perhaps, beneficial. Though simple, the SCD algorithm allows transferring a greater number of messages per visit with respect to the case without speed control, i.e., when the MR moves at a constant speed s [Kans04].
- *Communication-based speed control.* The stop-and-communicate approach described above does not rely on any data communication performance index to do speed control. It just stops for a fixed time when a new node is encountered. In addition, the MR can be either moving at a speed $2s$ or stopped. A finer control can be achieved by learning information about data communication of each single node, and varying the MR speed accordingly, like in the Adaptive Speed Control (ASC) algorithm [Kans04], where the speed is adjusted dynamically depending on the message loss rate experienced in the previous passage. The ASC algorithm is extended in [Soma06] to cope with scenarios where nodes are organized into clusters and transmit their messages to the MR through a cluster head. In the same paper it is also shown that adding more options to the MR motion actually does not produce any benefit. Again, the

ASC algorithm was originally proposed for wireless sensor networks where nodes are assumed to be static. However, it can easily be extended to other scenarios as well.

13.5.3 Topology Control

Topology control is another technique that can be used in combination with or as an alternative to trajectory and speed control. In the context of opportunistic networking based on MRs, the goal of topology control is to dynamically adjust the node's transmitting range to achieve the desired contact time with MR while reducing the energy consumed by the wireless interface (which is related to the transmission range). Besides reducing the energy consumption, in dense networks, topology control also reduces the probability of contention when accessing the wireless channel [Sant05].

As shown in Figure 13.6, given the trajectory and speed of the MR, the contact time duration depends on the node's transmission range. The basic idea of topology control can thus be exploited to derive the level of transmission power that allows the required contact time when the trajectory and speed of the MR are known. In addition, the duration of the contact time could also be adjusted dynamically by varying the node's transmission power and, consequently, its transmission range. This may be useful to cope with variations in the external conditions that affect the communication between the node and the MR (e.g., packet losses due to channel errors, collisions with neighboring nodes, etc.).

Topology control has been extensively studied in the context of traditional (i.e., multihop) ad hoc and sensor networks (a detailed survey can be found in [Sant05]). To the best of our knowledge, only a few proposals have been presented in the area of opportunistic networking based on MRs. In [Zhao04] the authors propose a trajectory control technique associated with a sort of topology control based on a dual radio. The MR follows

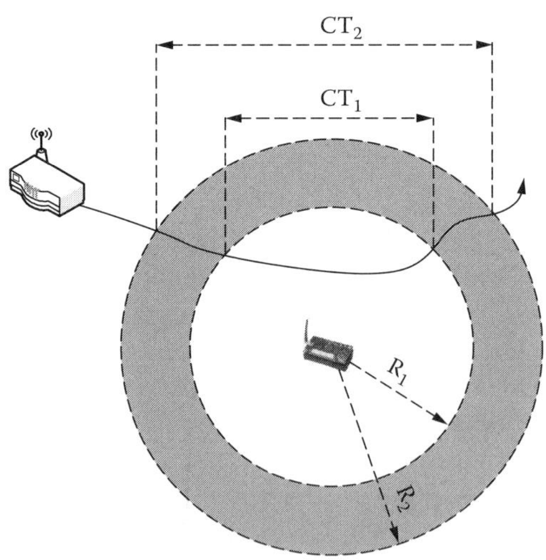

FIGURE 13.6 The contact time can be adjusted by varying the node's transmission range.

a default trajectory and periodically broadcasts its location using a long-range radio. When a node discovers that the MR is nearby, it sends a service request using its long-range radio. This message contains the node location as well. Upon receipt of a service request, the MR adjusts its trajectory to meet the node. When the two nodes are close enough, they start exchanging data using their short-range radio.

13.6 Power Management and MR Discovery

Since nodes are typically energy-constrained devices, a power management strategy is needed to save energy and increase nodes' lifetime. In the context of opportunistic networking, the objective of power management is to minimize energy consumption while missing as few contacts as possible to achieve an adequate performance level in terms of message latency and delivery ratio. Ideally, the node should sleep for most of the time and wake up only when the MR is within its communication range. In practice, this is infeasible because the node is not able to know exactly when the next contact will occur, unless the MR mobility pattern is known in advance (predictable mobility). Thus, the MR and the nodes agree on a discovery protocol that allows a timely MR discovery to the node with minimum energy consumption. Obviously, the discovery protocol can be optimized based on the knowledge available about the MR mobility.

The following scenarios have been identified in [JunA05].

- *Complete knowledge.* The time between two consecutive contacts with the MR is known in advance by the node (predictable mobility). This may happen when the MR is implemented on top of a controllable mobile element (e.g., a robot), or on a carrier with a fixed path and schedule (e.g., a shuttle bus with fixed schedule). Under the assumption of predictable MR mobility, a node can sleep for the time between two consecutive contacts, and wake up only for the time strictly needed to exchange messages with the MR. Obviously, in this scenario the power consumption is minimized [Chak03, JunA05]. Figure 13.7a shows a generic node transition between power modes during its activity. Upon departure of the MR (or when there are no more messages to exchange), the node calculates the time to the next contact, sets up a *sleeping timer*, and transitions to the sleeping mode. As soon as the sleeping timer expires, the node wakes up, enters the communication mode, and is ready for exchanging messages with the MR. Again, when the communication is over, or the MR exits the node's communication range, the node goes back to sleep.
- *Partial knowledge.* In practice, it is not very common to have complete knowledge of MR mobility. However, even if it is not known in advance, the MR behavior may be learned by observing successive MR passages. By exploiting learning techniques, the node can derive statistics about contact duration and time between contacts (e.g., mean, variance, distribution). Needless to say, the efficiency of power management depends on the degree of knowledge the node has about the MR mobility. Figure 13.7b shows the node transition diagram for this specific scenario. Let us assume the node is initially in communication mode. It remains in this mode until there are messages to exchange and the

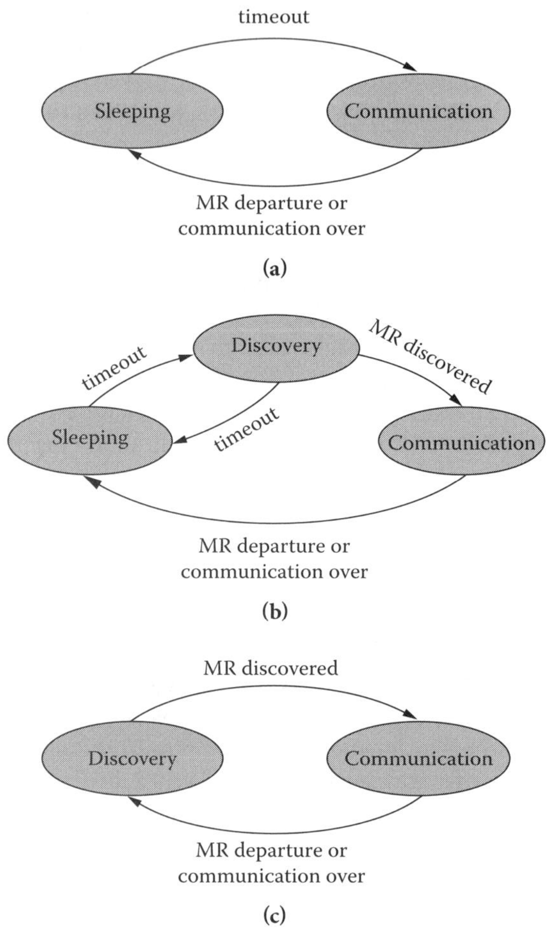

FIGURE 13.7 Transitions between different power modes under different degrees of knowledge about MR mobility: (a) complete knowledge, (b) partial knowledge, and (c) no knowledge.

MR is within communication range. Then, the node derives an estimate of the time to the next contact, and sets up a timer accordingly. Upon timer expiration the node enters the discovery mode. Unlike the previous scenario, now there is no guarantee that the MR is within the communication range when the node wakes up. In the discovery mode the node is waiting for the MR arrival. To this end, the node and the MR implement a distributed discovery algorithm to allow timely MR detection by the node (see below). To reduce energy consumption, nodes typically operate on a low duty cycle while in the discovery mode. In addition, the node remains in the discovery mode for a maximum discovery timeout. Then, it assumes that the contact was missed. Hence, it estimates the time to the next contact, sets up the sleeping timer, and switches back to the sleeping mode. Conversely, as soon as the node realizes that the

MR has entered its communication range, it switches to the fully operational mode (i.e., 100% duty cycle) and enters the communication mode to perform message exchange.

- *No knowledge.* The worst scenario is when there is no information available about MR mobility. For instance, this may occur when MRs move randomly through the network, and no assumption can be done about the times the MR will visit a node. In such a scenario, it is impossible to derive statistics (e.g., about the time between contacts). Therefore, each node must remain continuously active in looking for possible MR arrivals. The node transition diagram for this scenario is depicted in Figure 13.7c. Since the time a node passes in the discovery mode may be very large, the discovery algorithm must be very energy efficient, so as to allow a timely discovery of MR while keeping energy consumption low.

13.6.1 Discovery Algorithms

The discovery algorithm is a distributed algorithm used to allow a node to detect the presence of the MR as it enters the node's communication range. As the discovery phase may take a long time (especially in the no-knowledge scenario) to be carried out, energy efficiency should be of primary concern in the design of the discovery algorithm. Energy efficiency is typically achieved by putting nodes in a low duty cycle while in the discovery mode. Duty cycle reduces energy consumption but, at the same time, also increases the *discovery latency*, i.e., the time interval taken by a node to detect the MR presence inside its communication range. Obviously, the discovery latency should be as small as possible compared to the duration of the contact time to allow a larger amount of traffic to be exchanged between the node and the MR during the contact duration. The efficiency of a discovery scheme can be measured by means of the discovery ratio, defined as the average value of the discovered contact time (i.e., contact time less discovery latency) divided by the contact time [JunA06], i.e.,

$$\eta = E\left[\frac{contact\ time - discovery\ \ latency}{contact\,time}\right].$$

The design of a discovery algorithm must reach a trade-off between energy saving and discovery ratio. Of course, the discovery algorithm can be customized to the specific application scenario. Ideally, the most efficient discovery scheme consists in waking up a node exactly when the MR enters its communication range. This allows the maximum discovery ratio at the minimum energy cost. Unfortunately, this approach is difficult to implement in practice. In fact, its applicability is limited to the predictable mobility scenario where MR visit times are known in advance to each node, and the clocks of nodes and MR are synchronized. For all the other cases, a different approach must be used.

Figure 13.8 shows the main approaches proposed in the literature for discovery techniques. They can be broadly classified into two categories: *MR-triggered wake-up*

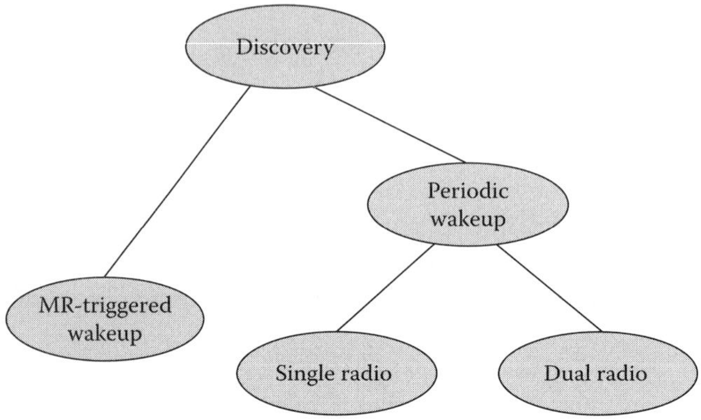

FIGURE 13.8 Classification of discovery approaches.

and *periodic wake-up*. In the former scheme, nodes are passive (i.e., their radio is kept in sleep mode during the discovery phase) and are awakened by the MR itself when it enters the node's communication range [GuSt05]. In the periodic wake-up scheme, nodes wake up periodically to look for possible MR arrivals [Shah03, Kans04, JunZ05, JunA05, JunA06, Jain06]. Finally, for more efficient energy consumption, MR discovery and data communication can be performed over two different radio channels (when available): a low-power channel for MR discovery and a high-power channel for data communication [Jain06, JunA06].

13.7 Relevant Case Studies

In this section we provide some relevant case studies on opportunistic systems based on mobile relays. Section 13.7.1 describes a scenario that may arise after a disaster when the existing infrastructure is unusable, and airplanes or terrestrial vehicles can be used as ferries to transport data between users in separated areas. Similarly, energy-efficient data collection in sensor networks can be performed by using mobile elements that can be either part of the external environment (section 13.7.2) or part of the network infrastrucuture (section 13.7.3). Finally, section 13.7.4 focuses on a special case of opportunistic sensor networks, i.e., underwater sensor networks with MRs.

13.7.1 Message Ferrying

A message ferrying scheme provides connectivity in sparse mobile ad hoc networks, which are characterized by sparse node deployment and network partitions that may last for extended periods of time. [Zhao04] assumes dealing with this kind of network and introduces extra mobile nodes named *Message Ferries* to offer a service of message relaying. *Message Ferries* move around in the network area and collect messages from the source nodes; then they provide forwarding of the collected messages. Message collection may happen in two ways:

- *Node-initiated message ferrying.* The ferry node moves around following a predefined and known path. Each node in the network has knowledge of the paths followed by active ferries. The node wishing to deliver a message moves toward the nearest ferry and, when sufficiently close, forwards its messages. Hence, the source node changes its trajectory to meet up with the ferry. This may obviously cause some degradation in the performance achieved by those applications that are currently running on the node during route deviation. Therefore, the source node controls its trajectory toward the ferry while striving to balance between performance degradation in the running tasks and performance gain in data delivery (i.e., minimizing message drops).
- *Ferry-initiated message ferrying.* The ferry node, again, moves around following a predefined default path. To let other regular nodes know its position with good approximation, the ferry periodically sends its position information via long-range broadcast signals. Any source node wishing to deliver messages sends a `ServiceRequest` to the ferry via a long-range radio signal. The source node also includes its current position in the `ServiceRequest`. After having received the request from the source node, the ferry changes its trajectory to meet up with the source node. The source node periodically communicates `LocationUpdates` to let the ferry adjust its trajectory in order to meet with it. The new trajectory of the ferry is computed with the aim to minimize message drops. When the ferry and the source node are close enough, message exchanges occur by means of short-range radio signals.

In both cases each node is expected to have location awareness, i.e., to know its position as well as the position of the ferries, for example, through GPS receivers.

A key issue of the message ferrying scheme is the design of the best trajectory that the ferries should follow to service the nodes. The trajectory design goal is to meet the traffic demand while minimizing the delay of data delivery. Obviously, better results can be met when multiple ferries are active in the network area, even though some effort should also be spent for their coordination or even synchronization in some cases. When a single ferry is available in the network area, good results are obtained by designing a route that consists of an ordered sequence of *way-points* and *waiting times* corresponding to these way-points [Tari06]. The ferry node traverses this so-determined route repeatedly and waits for the predefined waiting time at each way-point so that it can contact every node in the network with a certain probability. In fact, given for each node the probability to visit a particular place, the number and the places of way-points are determined such that the ferry can meet all the nodes with a given minimum probability. After having decided the particular set of way-points, the minimum path traversing them all is computed, and this is established to be the ferry trajectory.

By introducing multiple ferries into the network, the overall system becomes much more fault tolerant because even if a single ferry fails in collecting the data, other ferries can intervene in substitution. Moreover, the system becomes more scalable because a wider geographical area is covered and the traffic load is much more balanced over the entire network deployment. However, the presence of multiple ferries in the network also causes some extra costs, for example, to define the best trajectories that ferries should follow, and to assign to each ferry the best subset of nodes to serve. Indeed, sometimes some degree of synchronization among the ferries may also be needed, for example, when ferries are supposed to exchange messages with each other. Multiple ferries in the network can traverse the same trajectory starting at different moments and keeping fixed distance in between each other. Another possibility is that message ferries are assigned different trajectories, each one serving a specific subset of nodes, but that may also overlap, resulting in some nodes being visited for data collection more frequently than others. The ferry trajectories are computed in such a way to minimize, on average, the weighted delay between each pair of nodes. The well-known traveling salesmen problem is exploited for this purpose.

Ferries traveling throughout the network can be completely independent of each other such that they do not interact in any way; otherwise, they can exchange messages with each other to reduce the messages' delivery delay. In case ferries exchange messages with each other, they can do it directly when they meet each other, or by exploiting the static nodes they visit during the travel, the same ones that are data sources of the network, as relay nodes. This way a ferry can download the messages it carries to an intermediate stationary node, and another ferry later visiting the same node can upload these messages in order to carry them to the destination or to another intermediate node. As shown in [Zhao05], the best performance is achieved when multiple ferries travel through different trajectories and each one of them is assigned its own subset of nodes from which to upload data. Moreover, better performance is experienced when ferries do not interact with each other to exchange the messages they carry. In fact, performing ferry relaying is expensive since synchronization between ferries is necessary. The message ferry scheme scales well with the number of ferries in terms of throughput, delay, and resource requirements in both nodes and ferries.

13.7.2 Data MULEs

A data MULE system [Shah03, Jain06] is very similar to a message ferrying scheme. Data MULE systems are specifically designed for sparse sensor networks and focus on energy saving. They consist of a three-tier architecture:

- The lower level is occupied by sensor nodes that periodically perform data sampling from and about the surrounding environment.
- The middle level consists of mobile agents named *Mobile Ubiquitous LAN Extentions*, or MULEs for short. MULEs move around in the area covered by sensors to gather their data, which have previously been collected and temporarily stored in local buffers. Data MULEs can be, for example, people, animals, or vehicles. They move independently from each other and from sensor positions by following unpredictable trajectories. Whenever they get within reach of a sensor they gather information from it.
- The upper level consists of a set of wired access points (APs) and data repositories that receive information from the MULEs. They are connected to a central

data warehouse where the data received are synchronized and stored, multiple copies are identified, and acknowledgments are managed.

Sensor nodes are supposed to be immobile and continuously awake, waiting for a MULE to pass by for sending data to it. Sensor-to-MULE transmissions make use of short-range radio signals and hence do not consume too much energy. While moving around, when the MULE eventually passes by any AP deployed in the area, it transmits the collected sensors' data to it. MULEs are assumed to move independently of one another, each following a discrete random walk mobility model. No data exchange is assumed to occur among the MULEs, and finally, time synchronization is assumed to be present among sensors and MULEs.

Thanks to the short-range radio exchanges, the data MULE architecture is a very energy-efficient solution for data gathering in sparse sensor networks if compared to solutions based on the introduction of base stations to cover the entire area to monitor, and also to solutions based on the introduction of a high number of sensor nodes to form a dense, entirely connected sensor network. It also guarantees scalability and flexibility against the network size.

Unfortunately, this solution has a couple of limits, both depending on the randomness of the MULEs' motion. First, the latency for data arrival at the APs is considerable because some time elapses from the sampling instant to the moment the MULE takes the data, and then until the time the MULE actually reaches the AP and delivers the data to it. The second drawback is the fact that sensors have to continuously wait for any MULE to pass and cannot sleep. This leads to energy wastage.

When increasing the area to be monitored, the frequency of the visits to the sensors by MULEs naturally decreases, and an increase in the buffer size of the sensors is needed to prevent data loss. The latency experienced by the data monitored increases too. This effect can be alleviated by increasing the number of MULEs. When increasing the area to be monitored, the frequency of the visits of MULEs to the APs also decreases. This leads to a further increase in the latency of data and to the need to increase the buffer size at the MULEs to prevent data loss. An increase in the number of APs can help alleviate the above effects. In conclusion, the number of MULEs can be traded for the size of the sensors' buffers, whereas the number of APs can be traded for the size of the MULEs' buffers.

13.7.3 Mobile Controllable Infrastructure

[Soma06] addresses energy-efficient data collection from sparse wireless sensor networks through a mobile infrastructure consisting of a mobile base station. The primary purpose of this approach is to save part of the energy generally spent by sensor nodes in multihop transmissions toward a static sink node. In the framework developed, the mobile base station moves along a predetermined path that is fixed. Sensor nodes that are located in proximity of the mobile base station path send their data directly to the base station when in the communication range. Nodes that are far apart from the path followed by the base station send their data over a multihop path toward the base station when it passes by, or alternatively to one of the nodes that are positioned near the path of the base station. These nodes act as data repositories until the base station passes and finally collects all the pieces of data stored. Energy saving is addressed in that a large number of nodes are visited by the base station and can thus transmit data over a single-hop connection using short-range radio. The other nodes that are not in proximity of the path followed by the base station send their data over a multihop path that is shorter, and thus cheaper, with respect to the path established toward a fixed sink node in a classical dense wireless sensor network.

To manage this kind of data collection, nodes self-organize into clusters where cluster heads are the nodes that are nearer to the path of the base station, whereas the other nodes of the cluster send their data to the cluster head for storage until the next visit of the base station. Data from the sensor nodes of the cluster travel toward the cluster heads according to the directed diffusion protocol. Election of the cluster heads is done after the first traversal of the base station. During this first traversal the base station does not collect any data.

Transmissions from cluster heads to the base station occur only when the base station is in proximity to not wasting energy in useless transmissions. Hence, the base station periodically broadcasts POLL messages to inform of its approaching. Cluster heads that receive the POLL message from the base station start sending data to it. The base station acknowledges receipt of each message from a cluster head to inform it that the connection is still active and that the data are reliably delivered. Retransmissions are managed by cluster heads for the messages that are not acked. A cluster head stops transmitting when either it has sent out all the messages stored in cache or it realizes that the connection to the base station is lost, by not having received a POLL message for a certain time period.

The trajectory of the base station can be controlled in both space and time. However, changing the trajectory of the base station is not always possible in case of sensor networks because sensors may be deployed in places with obstacles, on rough terrain, or generally where unmanned vehicles can move only in certain directions. Hence, having a fixed path could often be a system requirement rather than a choice. Controlling the trajectory in time instead is considered to be a much more interesting possibility. The base station can move at a constant speed worked out, for example, depending on the buffer constraints of the cluster heads. Each cluster head is thus visited before its buffer runs out of space. However, better performance is experienced when the base station alternates between two states: moving at a certain constant speed or stopping. So base stations move fast in places with no, or only a few, sensors and stop in proximity of cluster heads where sensor deployment is denser. The determination of places where sensors deployment is denser (congested regions) is done at each traversal of the base station. The base station registers the identity of each node it has received a message from and the number of messages received from it. Given that each sensor node collects data at the same rate, and thus has the potential to send the same number of packets, the only reason why some nodes send fewer messages than others is that they are in a congested area with more sensors served by the same cluster head, which cannot succeed in sending all the data buffered during the limited visit of the base station. In the next traversal the base station stops for more time in regions that have previously been found congested.

13.7.4 Underwater Sensor Networks

Underwater sensor networks [Vasi05] have recently attracted a lot of attention in the research community. They are deployed to monitor and model the behavior of underwater ecosystems. They exploit the aforementioned data MULE communication system. These sensor networks gather physical variables such as water temperature, pressure, conductivity, turbidity, and pollutants' concentrations. Moreover, underwater sensors may collect images to measure visible changes in the deep underwater environment or even to classify species. The network consists of both static nodes and mobile nodes. Static nodes are sensor nodes that perform data collection and storage. They are extremely power efficient because they have little energy available and are not easily rechargeable. Mobile nodes are *autonomous underwater vehicles* (AUVs) that are responsible for data collection from the sensor nodes. They navigate the network to be within communication range of sensors to collect data from them. AUVs require much more power than sensor nodes because they navigate the sensor network; however, they are quite easily rechargeable. Static nodes are mostly in deep sleep mode and wake up every few seconds to determine if they are being signaled by mobile nodes nearby. Relieving static nodes from most of the communication and storage loads contributes to maximizing the network lifetime.

It has been found that the most efficient way for collecting data from an underwater sensor network is using a system capable of both optical and acoustic communications. Optical communications guarantee high data rate and high bandwidth but need line of sight to be established between the communication peers and can only cover short ranges. Acoustic communications, on the other hand, have the potential for higher transmission range but suffer from attenuation and reflections and allow lower bandwidth. Therefore, a trade-off needs to be met between communication range and data rate. Due to the broadcast nature of acoustic communication, when an acoustic transmission is holding, any other node is prevented from transmitting, even to signal an event. Nevertheless, an optical communication and an acoustic one may hold simultaneously. Hence, it has been established that the optical communication system is used for short-range line-of-sight data transfers between sensor nodes and AUVs (data mules). These transmissions are aimed at downloading the stored data from the sensor nodes and uploading commands to them. As they may involve much data, a faster communication system is more appropriate to use. The acoustic system is instead used to signal events over long distances and to transmit small amounts of data. Signaling an event allows the AUV to move to the area of interest, and may also trigger redeployment of the sensor network to concentrate on some important features in the environment. Acoustic communications are indeed particularly suitable for sensor node localization. In fact, the speed of sound in water is low enough to permit accurate timing of signals to determine the distance between nodes. Pair-wise node distances are then used to perform three-dimensional localization. The tasks of the mobile node are to establish a tour of the network, locate each node in the tour, one at a time, and hover above each node to download the data optically. During this period of communication the mobile node may also upload data to the static node, for example, to adjust its clock or to change the data sampling rate. The key challenges for underwater data muling are (1) locating the first node of the sequence to visit, (2) locating the next nodes of the sequence, (3) controlling the hover mode (for the mobile node), (4) accomplishing data transfer, and (5) synchronizing clocks so that the data collected by the sensor network are consistently time stamped. Localization of the first node of the data muling tour starts by positioning the robot in the general area of the network. Given that the general location is known in GPS coordinates, the AUV can perform surface navigation guided by GPS to move toward the node. Once close, the AUV descends to the optical communications range. At this point the AUV performs a spiral search to locate the node by making use of distributed localization algorithms built on top of acoustic ranging.

13.8 Conclusions

In this chapter we have provided a survey of routing approaches to opportunistic networks. This is a very hot topic, since opportunistic networks provide solutions for intrinsically disconnected ad hoc networks, which is one of the main points missing from the research on the legacy MANET paradigm. We have described in detail one of the most interesting cases of opportunistic networks, i.e., the Mobile-Relay Forwarding (MRF) approach. MRF assumes that a small subset of nodes have fewer restrictions in terms of resource constraints and follow completely different mobility patterns with respect to the vast majority of nodes in the network. These nodes, called mobile relays, can be, for example, mounted on buses roaming in a city, while regular nodes can be pedestrians' devices. Mobile relays are opportunistically exploited by the other users to bring messages to the destination, thus connecting nodes that would never be connected together, or anyway significantly improving the network performance. We have discussed a number of issues addressed in the literature with respect to MRF. Specifically, we have provided an extensive taxonomy of the system with respect to the type of mobile relay mobility, and we have discussed the case in which the mobility of mobile relays can be controlled. We have also described how power management can be achieved in such a scenario. Finally, we have presented some relevant case studies highlighting how the Mobile-Relay Forwarding concept can be exploited in different scenarios.

Despite the vast body of research in the field, there are still a number of open questions. Just to name a few, we can highlight MR motion control techniques, MR discovery under unpredictable (but learnable) mobility patterns, power management, and data communication protocols improving the performance of simple *stop-and-wait* protocols.

Acknowledgments

This work was funded partially by the European Commission under the FP6-2005-NEST-PATH MEMORY project and the FP6 IST-FET 027918 HAGGLE project, and partially by the Italian Ministry for Education and Scientific Research (MIUR) under the FIRB ArtDeco project.

References

[Akka05] K. Akkaya, M. Younis, and M. Bangad. 2005. Sink repositioning for enhanced performance in wireless sensor networks. *Comput. Networks* 49:512–34.

[Anas07] G. Anastasi, M. Conti, E. Monaldi, and A. Passarella. 2007. An adaptive data-transfer protocol for sensor networks with data mules. In *Proceedings of the IEEE International Symposium on a World of Wireless, Mobile, and Multimedia Networks (WoWMoM 2007)*, Helsinki. Available online.

[Bold07a] C. Boldrini, M. Conti, I. Iacopini, and A. Passarella. 2007. HiBOp: A history based routing protocol for opportunistic networks. In *Proceedings of IEEE WoWMoM 2007*, Helsinki. Available online.

[Bold07b] C. Boldrini, M. Conti, and A. Passarella. 2007. Impact of social mobility on routing protocols for opportunistic networks. In *Proceedings of the IEEE WoWMoM AOC Workshop 2007*, Helsinki. Available online.

[Burn05] B. Burns, O. Brock, and B. N. Levine. 2005. MV routing and capacity building in disruption tolerant networks. In *Proceedings of IEEE INFOCOM 2005*, Miami, Vol. 1, pp. 398–408.

[Chak03] A. Chakrabarti, A. Sabharwal, and B. Aazhang. 2003. Using predictable observer mobility for power efficient design of sensor networks. In *Proceedings of the International Workshop on Information Processing in Sensor Networks (IPSN 2003)*. Lecture Notes in Computer Science 2634. Palo Alto, CA: Springer.

[Gand03] S. Gandham, M. Dawande, R. Prakash, and S. Venkatesan. 2003. Energy-efficient schemes for wireless sensor networks with multiple mobile base stations. In *Proceedings of IEEE Globecom 2003*, San Francisco, Vol. 1, pp. 377–381.

[Good97] D. Goodman, J. Borras, N. Mandayam, and R. Yates. 1997. INFOSTATIONS: A new system model for data and messaging services. In *Proceedings of IEEE VTC'97*, vol. 2, pp. 969–973.

[GuBo05] Y. Gu, D. Bozdag, E. Ekici, F. Ozguner, and C. Lee. 2005. Partitioning based mobile element scheduling in wireless sensor networks. In *Proceedings of the IEEE Conference on Sensor, Mesh and Ad Hoc Communications and Networks (SECON 2005)*, Santa Clara, CA, pp. 386–95.

[GuBo06] Y. Gu, D. Bozdag, and E. Ekici. 2006. Mobile element based differentiated message delivery in wireless sensor networks. In *Proceedings of the IEEE Symposium on a World of Wireless, Mobile, and Multimedia Networks (WoWMoM 2006)*, Niagara Falls, NY, pp. 83–92.

[GuSt05] L. Gu and J. Stankovic. 2005. Radio-triggered wake-up for wireless sensor networks. *Real-Time Syst. J.* 29:157–82.

[Harr06] K. Harras and K. Almeroth. 2006. Inter-regional messengers scheduling in delay tolerant mobile networks. In *Proceedings of IEEE International Symposium on a World of Wireless, Mobile, and Multimedia Networks (WoWMoM 2006)*, Niagara Falls, NY, pp. 93–102.

[Jain06] S. Jain, R. Shah, W. Brunette, G. Borriello, and S. Roy. 2006. Exploiting mobility for energy efficient data collection in wireless sensor networks. *Mobile Networks Appl. (MONET)* 11:327–39.

[JeaS05] D. Jea, A. Somasundra, and M. Srivastava. 2005. Multiple controlled mobile elements (data mules) for data collection in sensor networks. In *Proceedings of IEEE International Conference on Distributed Computing in Sensor Systems (DCOSS 2005)*, Marina del Rey, CA, pp. 244–257.

[Jetc03] J. G. Jetcheva, Y.-C. Hu, S. PalChaudhuri, A. K. Saha, and D. B. Johnson. 2003. Design and evaluation of a metropolitan area multitier ad hoc network architecture. In *Proceedings of IEEE WMCSA 2003*, Monterey, CA, pp. 32–43.

[Juan02] P. Juang, H. Oki, Y. Wang, M. Martonosi, L. Peh, and D. Rubenstein. 2002. Energy-efficient computing for wildlife tracking: Design tradeoffs and early experiences with Zebranet. In *Proceedings of Architectural Support for Programming Languages and Operating Systems (ASPLOS)*, pp. 96–107.

[JunA05] H. Jun, M. Ammar, and E. Zegura. 2005. Power management in delay tolerant networks: A framework and knowledge-based mechanisms. In *Proceedings of IEEE Conference on Sensor and Ad Hoc Communication and Networks (SeCon 2005)*, pp. 418–429.

[JunA06] H. Jun, M. Ammar, M. Corner, and E. Zegura. 2006. Hierarchical power management in disruption tolerant networks with traffic-aware optimization. In *Proceedings of the ACM Workshop on Challenged Networks (CHANTS 2007)*, Pisa, Italy, pp. 245–252.

[JunZ05] H. Jun, W. Zhao, M. Ammar, E. Zegura, and C. Lee. 2005. Trading latency for energy in wireless ad hoc networks using message ferrying. In *Proceedings of IEEE PerCom Workshops, International Workshop on Pervasive Wireless Networking (PWN 2005)*, pp. 220–225.

[Kans04] A. Kansal, A. Somasundara, D. Jea, M. Srivastava, and D. Estrin. 2004. Intelligent fluid infrastructure for embedded networks. In *Proceedings of the Second International Conference on Mobile Systems, Applications, and Services (MobiSys)*, pp. 111–124.

[Legu06] J. Leguay, T. Friedman, and V. Conan. 2006. Evaluating mobility pattern space routing for DTNs. In *Proceedings of IEEE Infocom 2006*, Barcelona. Available online.

[Lind03] A. Lindgren, A. Doria, and O. Schelèn. 2003. Probabilistic routing in intermittently connected networks. *ACM Mobile Comput. Commun. Rev.* 7(3):19–20.

[LuoH05] J. Luo and J.-P. Hubaux. 2005. Joint mobility and routing for lifetime elongation in wireless sensor networks. *Proceedings of IEEE INFOCOM 2005*, Miami, Vol. 3, pp. 1735–1746.

[Muso05] M. Musolesi, S. Hailes, and C. Mascolo. 2005. Adaptive routing for intermittently connected mobile ad hoc networks. In *Proceedings of the 6th IEEE International Symposium on a World of Wireless, Mobile and Multimedia Networks (WoWMoM 2005)*, Taormina-Giardini Naxos, Italy, pp. 183–189.

[Pelu06a] L. Pelusi, A. Passarella, and M. Conti. To appear. Encoding for efficient data distribution in multi-hop ad hoc networks. In *Handbook of wireless ad hoc and sensor networks*, ed. A. Boukerche. New York: Wiley & Sons Publisher.

[Pelu06b] L. Pelusi, A. Passarella, and M. Conti. 2006. Opportunistic networking: Data forwarding in disconnected mobile ad hoc networks. *IEEE Commun. Mag.* 4:134–41.

[Sant05] P. Santi. 2005. Topology control in wireless ad hoc and sensor networks. *ACM Comp. Surv.* 37:164–94.

[Shah03] R. C. Shah, S. Roy, S. Jain, and W. Brunette. 2003. Data MULEs: Modeling a three-tier architecture for sparse sensor networks. In *Proceedings of the IEEE International Workshop on Sensor Network Protocols and Applications (SNPA 2003)*, pp. 30–41.

[Smal03] T. Small and Z. Haas. 2003. The shared wireless infostation model—A new ad hoc networking paradigm (or where there is a whale, there is a way). In *Proceedings of ACM MobiHoc*, 233–244.

[Soma04] A. Somasundara, A. Ramamoorthy, and M. Srivastava. 2004. Mobile element scheduling for efficient data collection in wireless sensor networks with dynamic deadlines. In *Proceedings of IEEE International Real-Time Systems Symposium (RTTS 2004)*, Lisbon, pp. 296–305.

[Soma06] A. Somasundara, A. Kansal, D. Jea, D. Estrin, and M. Srivastava. 2006. Controllably mobile infrastructure for low energy embedded networks. *IEEE Trans. Mobile Comput.* 5(8).

[Sush04] J. Sushant, K. Fall, and R. Patra. 2004. Routing in a delay tolerant network. In *Proceedings of SIGCOMM'04*, pp. 145–158.

[Tari06] M. M. B. Tariq, M. Ammar, and E. Zeruga. 2006. Message ferry route design for sparse ad hoc networks with mobile modes. In *Proceedings of the 7th ACM International Symposium on Mobile Ad Hoc Networking and Computing (MobiHoc 2006)*, Florence, Italy, pp. 37–48.

[Vahd00] A. Vahdat and D. Becker. 2000. *Epidemic routing for partially connected ad hoc networks*. Technical Report CS-2000-06. Department of Computer Science, Duke University, Durham, NC.

[Vasi05] I. Vasilescu, K. Kotay, D. Rus, M. Dunbabin, and P. Corke. 2005. Data collection, storage, and retrieval with an underwater sensor network. In *Proceedings of the 3rd ACM Conference on Embedded Networked Sensor Systems (SenSys)*, San Diego, pp. 154–165.

[Wang05] Z. Wang, S. Basagni, E. Melachrinoudis, and C. Petrioli. 2005. Exploiting sink mobility for maximizing sensor network lifetime. In *Proceedings of the Hawaii International Conference on System Science (HICSS-38)*, Big Island. Available online.

[Widm05] J. Widmer and J.-Y. Le Boudec. 2005. Network coding for efficient communication in extreme networks. In *Proceedings of ACM SIGCOMM 2005 Workshop on Delay Tolerant Networks*, Philadelphia, pp. 284–291.

[Zhao03] W. Zhao and M. Ammar. 2003. Message ferrying: Proactive routing in highly-partitioned wireless ad hoc networks. In *Proceedings of IEEE Workshop on Future Trends in Distributed Computing Systems (FTDCS 2003)*, Puerto Rico, pp. 308–314.

[Zhao04] W. Zhao, M. Ammar, and E. Zegura. 2004. A message ferrying approach for data delivery in sparse mobile ad hoc networks. In *Proceedings of ACM Mobihoc 2004*, Tokyo, pp. 187–198.

[Zhao05] W. Zhao, M. Ammar, and E. Zegura. 2005. Controlling the mobility of multiple data transport ferries in a delay-tolerant network. In *Proceedings of IEEE INFOCOM 2005*, Miami, Vol. 2, pp. 1407–1418.

14

Adaptive Techniques in Wireless Networks

Yuxia Lin
University of British Columbia

Vincent W.S. Wong
University of British Columbia

14.1 Introduction

IEEE 802.11-based wireless local area networks (LANs) have been widely deployed around the globe due to their low cost, robustness, and ease of deployment. The WLANs are set up in "hot spots" to provide high-data-rate network access for nomadic users. The original 802.11a/b/g standards [23–26] defined a physical layer (PHY) supporting data rates up to 54 Mb/s, and the corresponding medium access control (MAC) layer functionalities. With the increasing demand from real-time multimedia applications on WLANs, the 802.11e standard [27] was approved in 2005 to provide quality-of-service (QoS) support by differentiating different classes of traffic. The 802.11n working group [1] is in the process of finalizing a new PHY/MAC extension to increase the physical layer bit rate up to 600 Mbps. In order to extend the limited coverage area of WLAN, a new mesh networking mode is being defined by the 802.11s working group to allow significant network coverage extensions.

With the continued effort in the 802.11 standard groups, the 802.11-based WLAN has been evolving from a simple and inexpensive wireless data access scheme to a more sophisticated last-mile wireless broadband access technology, which provides high-throughput data access for a wide range of multimedia applications. End users would be able to access high-speed Internet, play online gaming, and use various voice-over Internet Protocol (VoIP) and video streaming applications. However, the original 802.11 MAC is designed with a rather simple carrier sense multiple access/collision avoidance (CSMA/CA) protocol. The major advantage of this MAC protocol is its simplicity, robustness, and fully distributed control. But one inherent drawback is that its efficiency is low and cannot provide hard quality-of-service (QoS) guarantees. For example, for a current 802.11a/g network with a physical data rate of 54 Mbps, sustainable throughput above the MAC layer is only around 30 Mbps due to protocol overhead at the MAC layer. To achieve the ultimate goal of providing high-throughput multimedia services with QoS support, different adaptive techniques need to be utilized in the system design in order to improve the network performance. For example, adaptive frame aggregation at the MAC layer can significantly reduce the protocol overhead and improve MAC efficiency. The auto-rate-selection function at the MAC layer can adapt the link data rate according to the received signal strength such that more reliable transmission may be achieved.

This chapter first provides a brief description of the IEEE 802.11-based WLANs. Some technical details about the 802.11 and 802.11e MAC layer design are summarized. The current MAC design sets the system parameters to some predefined values, which ignore the beneficial information about traffic and channel conditions. Several recent proposals of adaptive schemes for performance improvements in 802.11 WLANs are introduced and analyzed. Adaptive tuning of MAC parameters, such as the contention window (CW) size and carrier sensing range, adaptive scheduling, and cross layer design schemes are shown to improve the MAC performance under different traffic conditions. In the recently proposed 802.11n draft, the PHY layer provides bit-rate support of up to 600 Mbps. With this kind of high bit rate, the inefficiency in the MAC may severely reduce the achievable data rate above the MAC layer. One major method to improve MAC efficiency is to aggregate small frames into larger frames so that MAC overhead can be reduced. A frame size adaptation scheme for optimal aggregation is proposed and analyzed in this chapter.

14.2 Background

The IEEE 802 LAN/MAN standards committee develops and maintains several major local area and metropolitan area network (MAN) standards. The 802.11 working group designs WLAN standards to support wireless connectivity for fixed, portable, and moving stations within a local area. It defines the MAC and PHY functionalities for delivering MAC service data units (MSDUs) between two communicating stations' logical link control (LLC) layers. The main features provided by the current 802.11 family of standards include:

- Support of physical layer transmission rate from 1 Mbps up to 54 Mbps
- Contention-based and polling-based medium access control
- Network management services
- Network security

The 802.11 physical layer defines several wireless transmission techniques to achieve a raw transmission rate ranging from the basic 1 Mbps to 54 Mbps. An appropriate link rate is chosen according to channel signal conditions to maintain low transmission errors. Apart from providing the transmit and receive services for the MAC layer, it also provides clear channel assessment (CCA), which is essential for the MAC layer's random access scheme.

The 802.11 MAC layer initially defined two access mechanisms: the distributed coordination function (DCF) and point coordination function (PCF). DCF provides a contention-based access scheme where each station contends for channel access based on the carrier sense multiple access with collision avoidance (CSMA/CA) protocol. PCF can be used on a WLAN with an access point (AP) that acts as a point coordinator to schedule the transmission in the network and provide contention-free access. The IEEE 802.11e standard extends the basic DCF/PCF access schemes to include more sophisticated priority control such that traffic with different priorities may receive different QoS. In this chapter, we will focus on the adaptive issues mainly related to the MAC layer in 802.11 WLANs.

14.2.1 IEEE 802.11 MAC: DCF and PCF

The DCF is the mandatory access protocol for all 802.11a/b/g WLANs. It is based on the CSMA/CA protocol, which enables all wireless stations and the access point to share the wireless medium with asynchronous and fully distributed random contention. The fully distributed random access scheme has the advantages of being simpler and more robust than centralized access schemes such as time division multiple access (TDMA), frequency division multiple access (FDMA), and code division multiple access (CDMA) used in cellular networks. However, one major problem arising from this distributed control is that two stations may transmit at the same time, and the packets will be lost at the receiver due to collision. The CSMA/CA protocol mitigates the probability of collision by invoking a random backoff process after each successful transmission and each transmission collision. Several interframe spacing (IFS) intervals are defined to provide various levels of priority for different kinds of frames to access the channel. The different IFS intervals are illustrated in Figure 14.1. The short IFS (SIFS) is the shortest of the interframe spaces, which allows for transmission of the Acknowledgment (ACK) and Clear to Send (CTS) frames. The PCF IFS (PIFS) is longer than SIFS and is used by the AP operating under PCF to gain prioritized channel access. The DCF IFS (DIFS) is used before transmission of normal data and Request to Send (RTS) frames. The Extended IFS (EIFS) is used in DCF when a station receives a corrupted frame to allow enough time for another receiving station to possibly send an ACK frame.

The basic access scheme of the DCF involves sensing the channel to be idle for DIFS, finishing the necessary backoff, sending a data frame, and receiving the corresponding

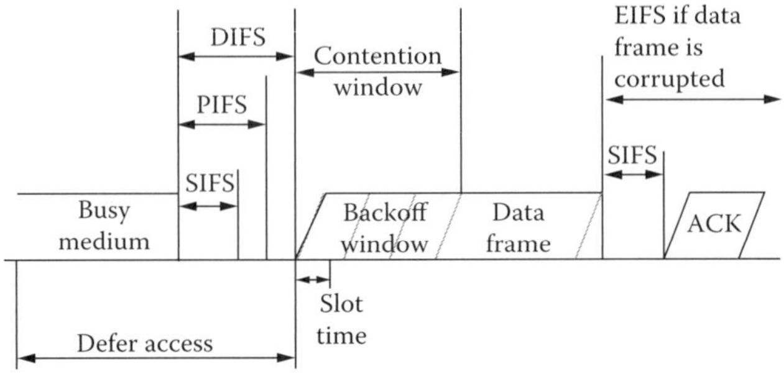

FIGURE 14.1 IEEE 802.11 DCF.

ACK, as shown in Figure 14.1. However, this basic access scheme suffers from the hidden terminal problem [10, 33, 69]. To mitigate the hidden terminal problem, the DCF protocol is extended with the optional Request to Send (RTS)/Clear to Send (CTS) control frames. Each station can perform a virtual carrier sensing and set a network allocation vector (NAV) based on reservation information received in the duration field of all frames.

The PCF is an optional operation mode available in an infrastructure WLAN where an AP is available to act as a point coordinator. It provides contention-free transfer of time-sensitive frames, such as VoIP data. At the beginning of the contention-free period, the point coordinator gains access to the wireless medium with the PIFS interframe spacing and then polls each wireless station. PCF may not function well when strong interferences exist, such as in overlapping WLANs [48, 59]. Due to its added complexity and reduced robustness compared to DCF, it is rarely deployed in commercial products.

14.2.2 IEEE 802.11e MAC for QoS Support: HCCA and EDCA

The DCF is currently the most widely deployed 802.11 MAC protocol in commercial products. However, it does not provide QoS guarantees. Real-time applications may degrade their performance when the network becomes congested [42, 66]. With the increasing demand for multimedia services (e.g., voice calls, video streaming) over wireless, the newly approved IEEE 802.11e standard [27] provides a QoS extension to the current 802.11a/b/g WLANs.

The IEEE 802.11e MAC architecture is shown in Figure 14.2. It defines a new hybrid coordination function (HCF) through the service of the DCF. Both HCF and DCF are present in all QoS-capable stations, while PCF is optional. The HCF includes a contention-based channel access method, Enhanced Distributed Channel Access (EDCA), for contention-based transfer and a controlled channel access, HCF Controlled Channel Access (HCCA), for contention-free transfer. A new concept of transmission opportunities (TXOPs) is introduced, which is defined with a starting time and a maximum duration. QoS stations (QSTAs) may obtain TXOPs using both EDCA and HCCA. Multiple frames can be transmitted without interruption during an obtained TXOP.

EDCA provides a prioritized QoS by enhancing the contention-based DCF. The EDCA model is shown in Figure 14.3. Higher-layer packets are classified into one of the

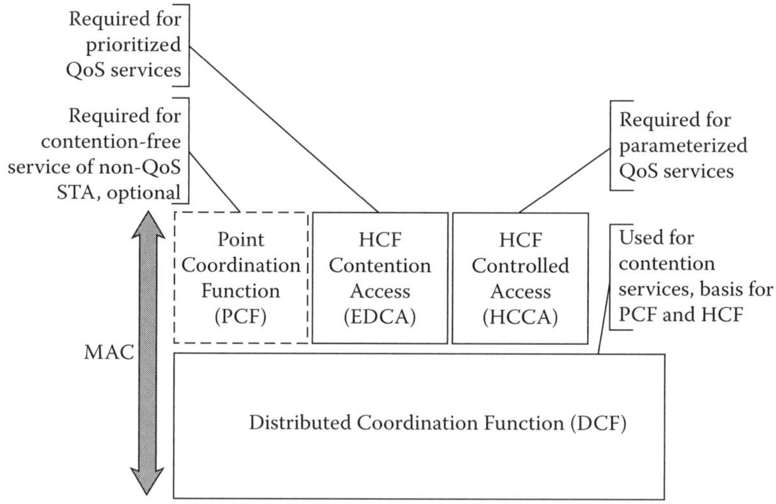

FIGURE 14.2 IEEE 802.11e MAC architecture.

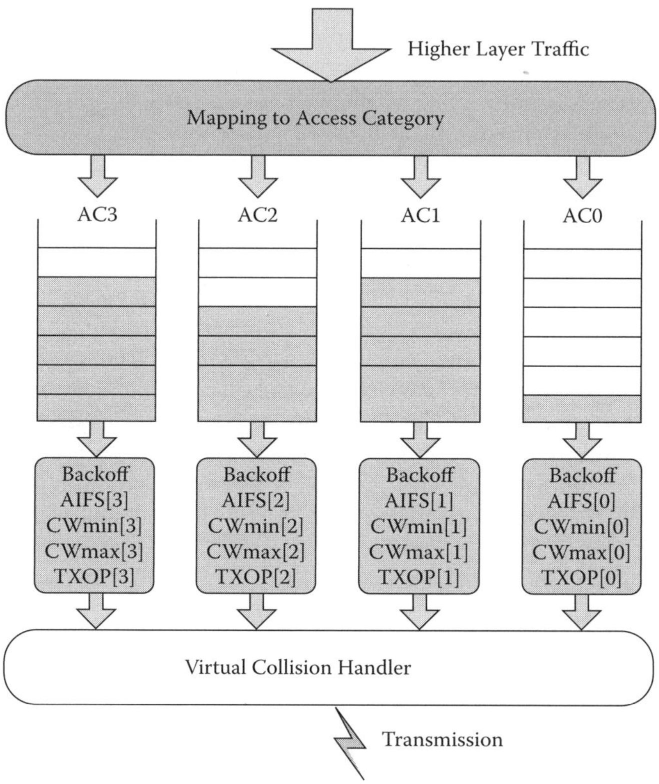

FIGURE 14.3 IEEE 802.11e EDCA: four access categories.

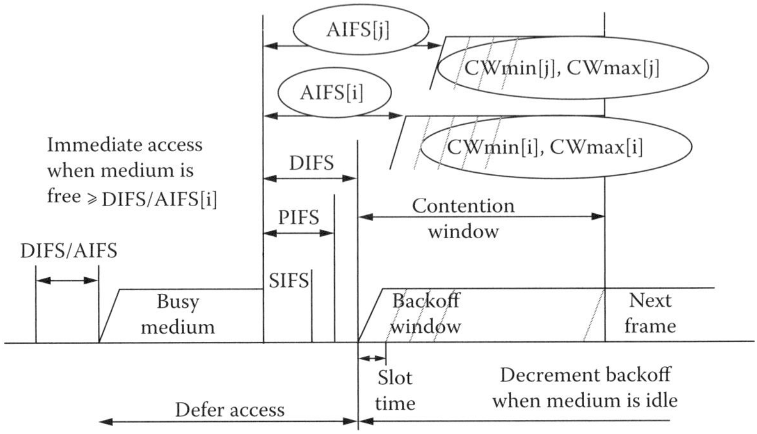

FIGURE 14.4 An Illustration of EDCA's timing sequence.

four access categories (ACs) at the MAC layer. Each AC has its own queue and channel access parameters, which include the arbitration interframe space (AIFS), minimum and maximum contention window sizes, and TXOP limits, as shown in Figure 14.4. The ACs gain different priority for channel access by differentiating these parameters. For example, the AC with a shorter AIFS and smaller contention window size will have a higher priority to access the wireless medium. However, due to its probabilistic nature of channel access, EDCA cannot provide hard QoS guarantees such as strict delay bound.

In order to provide parameterized QoS, HCCA has been proposed as a centralized polling scheme to allocate guaranteed channel access to traffic streams based on their QoS requirements. During a beacon interval, a hybrid coordinator (HC) that is collocated with the QoS access point (QAP) can access the wireless medium with its higher priority to initiate frame exchange sequences and allocate TXOPs to itself and other stations, so as to provide a limited-duration controlled access phase (CAP) for contention-free transfer of QoS data. The durations of TXOPs are calculated from the traffic specification (TSPEC), such as the mean/peak data rate and delay requirement announced at the beginning of the traffic flow. HCCA can be initiated during the contention-free period as well as the contention period. It is more flexible than PCF. With a good admission control and scheduling scheme, HCCA is able to provide guaranteed QoS to network flows. However, compared with EDCA, HCCA still presents much challenge for its actual implementation due to its higher complexity and cost concerns. With DCF-based 802.11a/b/g products dominating the market, EDCA is expected to be widely adopted as the multimedia solution to the 802.11-based WLANs.

14.3 Adaptive Techniques in the IEEE 802.11a/b/g Networks

In IEEE 802.11 WLANs, the MAC layer protocol is the main element that determines the efficiency of the wireless communications system. The CSMA/CA-based DCF has been shown to be inefficient in several communication scenarios [6, 21, 31]. This is partly due to the simple design of the standard, which fixes many operation parameters in the system. Many adaptive schemes have been shown to improve the network performance. Here, we first introduce some of the analytical models that strive to accurately describe the dynamic process of the DCF. Insights gained from these analyses of the protocol greatly help to derive efficient adaptive schemes.

14.3.1 Analytical Models for DCF

One major system performance figure for an IEEE 802.11 WLAN is its effective throughput above MAC layers. Even though the physical layer may provide a data rate up to 54 Mbps for an 802.11a/g network, the effective throughput above the MAC layer is significantly lower than the nominal physical data rate [21]. This is partly due to the relatively large PHY/MAC headers and partly due to the significant protocol overhead introduced by the CSMA/CA's binary exponential backoff mechanism.

Several factors contribute to the network's throughput, which include the network topology, traffic profile, and MAC protocol parameters. Figure 14.5 shows a simulation experiment by using the ns-2 simulator [2] for an 802.11b WLAN with ten wireless stations. Each station generates the same constant-bit-rate (CBR) traffic, and the data packet size is fixed at 1,500 bytes. Under very light traffic load, the network throughput is equal to the traffic load. It increases linearly with the offered load until it reaches the *maximum throughput*. Afterwards, the network enters the overload state and the throughput remains at the *saturation throughput* level. Although the saturation throughput is slightly lower than the maximum throughput due to excessive contentions, this stable state provides a good performance figure for a WLAN system.

Saturation throughput behavior of 802.11 WLANs has initially been studied by many simulation tests [7, 12, 60]. With the increasing success of the 802.11 WLANs, analytical models are constructed to mathematically compute the saturation throughput for a WLAN system. Among the works, Bianchi's model [6] has been one of the most widely known and accepted. It models a single station's backoff process with a two-dimensional

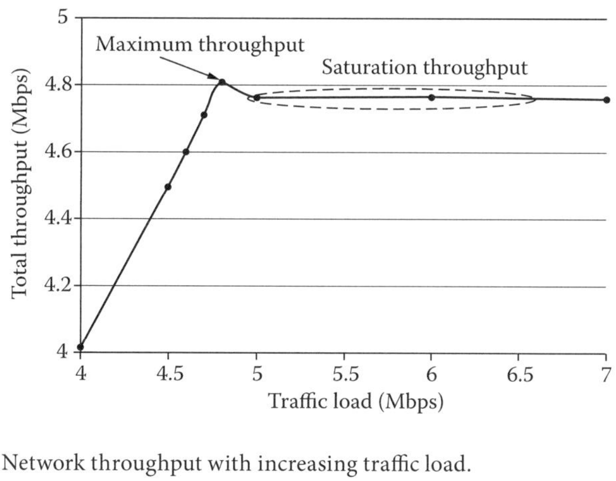

FIGURE 14.5 Network throughput with increasing traffic load.

Markov chain. Assuming that each transmission experiences identical and independent collision probability p, the stationary probability τ for a station to transmit in a generic time slot can be obtained from

$$\tau = \frac{2(1-2p)}{(1-2p)(W+1)+pW\left(1-(2p)^m\right)} \tag{14.1}$$

and

$$p = 1-(1-\tau)^{(n-1)}, \tag{14.2}$$

where W is the minimum contention window size, m is the number of backoff stages, and n is the number of competing stations in the network. With p and τ, the probability of successful transmission can be easily derived. The saturation throughput is obtained by dividing the expected payload information transmitted in a time slot by the expected length of a time slot. This method provides an accurate representation of DCF's binary exponential backoff process, and achieves good performance prediction. One major insight we can gain from this model is that, from (14.1), when m is small, $\tau \sim 2/(W+1)$. This shows that the transmission probability of a station, which is a key factor for controlling the network's contention level, is approximately inversely proportional to the contention window size. This crucial role of the contention window size makes it one of the most important adjustable parameters in the 802.11 DCF protocol. We will look at several examples for adaptive tuning of the CW in the next section.

Some other parallel works derive the saturation throughput without fully describing the detailed behavior of the binary exponential backoff. Cali et al. [10] modeled the backoff process with a p-persistent model, which we will discuss in some more detail in the next section. Tay and Chua [54] further simplified the 802.11 MAC modeling by using a mean value approach. Instead of studying the details of the stochastic process using Markov chain, their model tried to approximate the average value for a system variable wherever possible. This technique provides closed-form expressions for the collision probability and the saturation throughput. The model validation shows that even though this model omits many system details, it still achieves good accuracy, warranting its usefulness.

14.3.2 Examples of Adaptive Schemes

14.3.2.1 Dynamic Tuning of the Backoff Process

The main functionality provided by the DCF is the binary exponential backoff algorithm, which is a fully distributed mechanism for contention control. The key component in the backoff algorithm is the design of the contention window size. As evident from the analysis in the previous section, the CW size has a major impact on the contention level in the network. The 802.11 standard uses predefined minimum and maximum window sizes. To achieve better MAC efficiency, it is necessary to dynamically adjust the backoff process to maintain the network in an appropriate level of contention.

Cali et al. [10] modeled the 802.11 DCF protocol as a p-persistent random access protocol. A station transmits a packet in an idle time slot with probability $p = 1/(E[B] + 1)$, where $E[B]$ is the average value of the backoff slots. If the average contention window size is $E[CW]$, the average backoff slots can be expressed as $E[B] = (E[CW] - 1)/2$. As a result, the probability of packet transmission is $p = 2/(E[CW] + 1)$. Based on this assumption on the backoff process, the network saturated throughput can be derived as

$$S = \frac{\overline{m}}{\frac{1-p}{Np}t_{slot} + \frac{1-(1-p)^N - Np(1-p)^{N-1}}{Np(1-p)^{N-1}}\left[E[Coll] + t_{prop} + DIFS\right] + E[S]}, \tag{14.3}$$

where $\overline{m}$ is the average packet length, n is the number of transmitting stations in the network, t_{prop} is the maximum propagation delay, t_{slot} is the backoff slot time, $E[coll]$ and $E[S]$ are the average length of a collision and successful time slot, respectively. From this equation, if we know the number of contending stations N and the average packet length $\overline{m}$, we can determine the optimal p^* to maximize the network throughput. The optimal contention window size can be determined by $E[CW] = 2/p^* - 1$. However, solving for the optimal p^* from the nonlinear equation (14.3) is computationally intensive. As a result, a heuristic dynamic tuning algorithm is proposed to determine the p^* approximately. The heuristic algorithm is based on the following observation. The values of p lower than p^* correspond to the cases in which there is not sufficient contention such that most of the virtual time slot is wasted by the idle time slots $E[Idle]$, while p values greater than p^* correspond to excessive contention, where an average length of the virtual time slot is mainly caused by collisions. Hence, p^* can be approximated by choosing the p value that satisfies the following relationship:

$$E[Coll] \cdot E[N_c] = \left(E[N_c] + 1\right) \cdot E[Idle]. \tag{14.4}$$

When p^* is determined from the above heuristic, the contention window size is updated with the following moving average:

$$current\_cw \leftarrow \alpha \cdot current\_cw + (1-\alpha) \cdot (2/p^* - 1), \tag{14.5}$$

where $\alpha \in [0,1]$ is the smoothing factor.

Simulation tests show that the dynamic tuning algorithm significantly outperforms the original 802.11 standard in the aspect of saturation throughput.

However, the above dynamic backoff tuning algorithm depends on the accurate estimation of the number of actively transmitting stations in the network. There are several approximate methods for runtime estimating of the number of active stations, such as the Kalman filter-based algorithm proposed in [8]. To avoid the costly computation of the number of stations, further analysis shows that even though p^* is heavily influenced by the number of stations N, the product p^*N remains relatively constant for N greater than 2. With this observation, an asymptotically optimal backoff (AOB) algorithm is proposed in [9], where only slot utilization and the average size of the transmitted frames are needed for contention window size adaptation.

Some other schemes further modify the original 802.11 backoff process. Instead of only tuning the contention window size, the gentle DCF (GDCF) protocol proposed in [57] changes the collision resolution stage such that the contention window size CW is not reset to CW_{min} but is halved each time after c consecutive successful transmissions. This new design of the collision resolution stage makes use of the fact that always reducing CW to CW_{min} after a successful transmission may be too aggressive in a highly congested network. In order to determine the optimal c^*, a Markov chain model is extended from Bianchi's model in [6] to describe the new backoff behavior. The optimal value for c^* can be calculated, but it is also dependent on the number of stations in the network n. However, when $n \geq 10$, c^* is shown to be increasingly independent. So, this scheme may work without the exact knowledge of the number of active stations in a large-scale network.

14.3.2.2 Adaptive Carrier Sensing Range

The above dynamic tuning algorithms for the backoff process mainly deal with the contention control at the MAC layer. However, the performance of the DFC protocol is also heavily impacted by other aspects of the networking protocol stack. On the physical layer, the carrier sensing function provides the MAC layer with carrier sensing or CCA. This function is especially critical when the WLAN is operating under the ad hoc mode where a multihop topology may be formed, or when several infrastructure WLANs overlap and cause co-channel interferences. When two wireless stations within the transmission range of a receiver transmit simultaneously, the two transmissions may collide at the receiver and cause packet loss. Carrier sensing can reduce collisions by assessing the channel's energy level before transmitting. The IEEE 802.11 standard does not specify the carrier sensing range for implementation. Many studies, including the ns-2 simulator, use a static value for the carrier sensing range. But in fact, the carrier sensing range is a tunable parameter that may significantly affect the network performance in a multihop ad hoc scenario.

Figure 14.6 shows a segment in a typical ad hoc WLAN with a transmission from TX to RX and several other nodes (A, B, and C). There are several distances in this topology [67]:

- D: Separation distance between TX and RX
- R: Transmission range
- I: Interference range
- X: Physical carrier sensing range

It is generally assumed that $R < I < X$. The transmission range R is determined by the transmission power and the required signal-to-noise ratio (SNR) at the receiver. It is the largest distance with which the received SNR is equal to the SNR threshold for correct reception. The interference range is the largest distance with which another transmitter may disrupt the correct reception at RX by transmitting at the same time. The transmitter will deem the wireless channel busy if it senses an energy level equivalent to a transmitter within the carrier sensing range.

The carrier sensing range blocks any other station from transmitting within a radius of X, which reduces the probability of collisions. But it also decreases the channel spatial

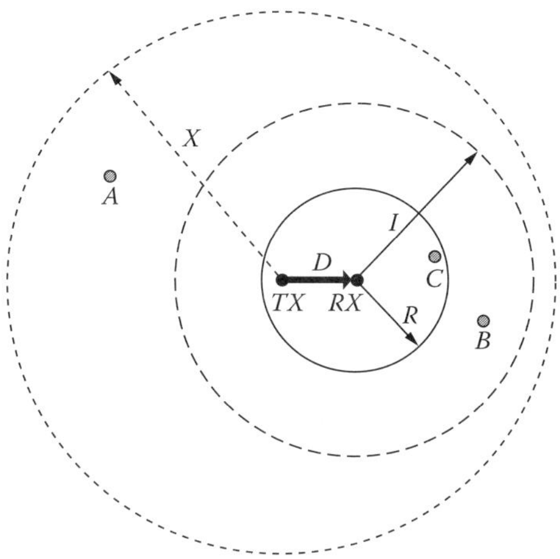

FIGURE 14.6 Communication ranges in an 802.11 ad hoc network.

reuse. The tuning of the carrier sensing range balances the trade-off between the channel spatial reuse and the probability of collision. Deng et al. [14] used extensive simulations to show that tuning of the carrier sensing range has a significant impact on the network's throughput and network reward function, which is a function of the network throughput and cost of packet collisions. Zhu et al. [67, 68] analytically derived the optimal carrier sensing threshold P_c in a homogeneous network with a regular topology where neighboring nodes are separated by equal distance D to be

$$P_c = P_D / S_0, \tag{14.6}$$

where P_D is the signal strength of the intended packet and S_0 is the minimum signal-to-interference-noise (SINR) ratio. Using this dynamic threshold, the network throughput is shown to approach 90% of the predicted theoretical upper bound in OPNET [3] simulations.

However, the Pc-based algorithm depends on periodic estimation and dissemination of physical channel SINR conditions, which creates extra protocol overhead. In [69], Zhu et al. proposed using the packet error rate (PER) as the metric for adaptively tuning the carrier sensing threshold. PER is readily available in each 802.11 WLAN card. A lower PER means less packet collisions, indicating a smaller number of hidden terminals. On the other hand, a higher PER means more packet collisions, which indicates that the carrier sensing range is too small. This will lead to more exposed terminals and less spatial reuse. As a result, PER can be used as a reasonable measurement for spatial reuse. The objective of the carrier sensing adaptation is to maintain the PER in a target range that reflects the optimal spatial reuse.

The proposed carrier sensing adaptation or CCA adaptation algorithm is shown in Figure 14.7. p_{max} and p_{min} are the targeted maximum and minimum PER. The CCA threshold P_c is periodically updated with the measured PER (P_m):

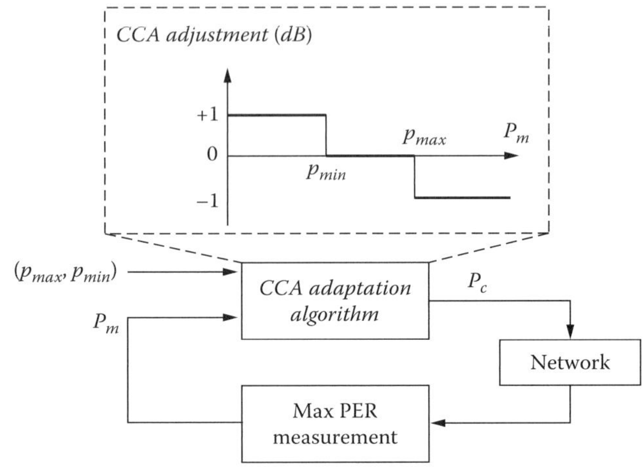

FIGURE 14.7 Carrier sensing adaptation system diagram.

$$P_c = \begin{cases} \max\left(P_{c_{min}}, P_c(dBm) - \delta(dB)\right), & \text{if } P_m > p_{max} \\ P_c, & \text{if } p_{min} \le P_m \le p_{max}, \\ \min(P_{c_{max}}, P_c + \delta), & \text{if } P_m < p_{min} \end{cases} \tag{14.7}$$

where δ is the step size for carrier sensing threshold adaptation, and $P_{c_{max}}$ and $P_{c_{min}}$ are the higher and lower bounds of P_c. The measured P_m is compared to p_{max} and p_{min} to decide whether to increase or decrease the CCA threshold of the transmitter. To determine the values for the parameters p_{max}, p_{min}, $P_{c_{max}}$ and $P_{c_{min}}$, a test bed experiment is conducted. Eight Cisco Aironet 1120 APs and eight IBM Thinkpad laptops with Intel Centrino™ radios form a test bed. The User Datagram Protocol (UDP) throughput shows an improvement of up to 300% when the adaptive carrier sensing mechanism is enabled. In [44], Ma and Roy also used the PER as a metric for adaptive carrier sensing, but they further considered jointly adapting the carrier sensing threshold and the multirate control by the MAC layer.

14.3.2.3 Cross Layer Design with Higher Layers

For a given wireless LAN environment, the solution to maximize network performance is usually a careful combination of several approaches addressing multiple aspects of the network behavior. Apart from MAC/PHY joint adaptation, as illustrated in the previous examples, joint adaptation with higher layers is critical for performance optimization. This approach is usually called the cross layer design in wireless networks [39, 56, 53].

For example, at the application layer, adaptive video streaming over 802.11 DCF or PCF has shown great performance gain [4, 19, 32, 55]. The video quality benefits from adapting the coding rate to the quality of the wireless channel. At the transport layer, wireless-aware congestion control is especially important because the high packet error rate on a wireless link undermines a TCP transport protocol's basic assumption that packet losses are mainly due to link congestions. Many adaptive protocols for TCP enhancement [13, 38] are proposed to adapt to the 802.11 WLAN channel.

14.4 Adaptation Techniques in the IEEE 802.11e Network

With the success of building the 802.11 DCF in millions of commercial products, there is great expectation that the DCF's QoS extension in 802.11e, EDCA, will dominate the next generation of QoS-enabled WLAN equipments. Experimental test beds [5] and numerous simulation tests [11, 20, 42, 48] have been carried out that show that EDCA is effective in traffic differentiation. In this section, we introduce some of the existing analytical models for EDCA and several adaptive EDCA schemes. Several adaptive scheduling schemes for HCCA are also described.

14.4.1 Analytical Models for EDCA

Most of the existing analytical models for 802.11e EDCA are extensions of the DCF models in section 14.3.1. It is more challenging to construct analytical models for EDCA than for DCF. Several models extend the DCF Markov chain model proposed by Bianchi. Xiao [61] modified the original two-dimensional Markov chain to three-dimensional, and analyzed the effects of the different contention window sizes on throughput, but the AIFS-based priority scheme is not included. Kong et al. [34] used a three-dimensional Markov chain to fully describe the backoff behavior of each AC in each station so that both AIFS and CW are considered. Robinson and Randhawa [51] kept the two-dimensional Markov chain, but divided the contention periods into different zones to account for the different collision probabilities during different contention periods. These multidimensional Markov chain analyses are highly computationally intensive, as they involve solving complex sets of nonlinear equations. Some simpler models involve extending Cali's p-persistent DCF model [16, 22], or extending Tay and Chua's mean value analysis [40].

14.4.2 Adaptive EDCA

EDCA is a QoS enhancement over DCF. As a result, most of the adaptive techniques available for DCF, as discussed in section 14.3.2, such as adaptive backoff and adaptive carrier sensing, can be extended to be used in EDCA [15, 47]. Furthermore, the EDCA function utilizes more system parameters, such as the AIFS and TXOP. AIFS is shown to have more significant impact on the traffic differentiation than CW. Large differences in different AC's AIFS values may tend to starve the lower-priority traffic [40]. Xu and Meng [64] proposed an adaptive adjustment algorithm for the AIFS values such that low-priority traffic may still gain some fair share of bandwidth when there is excessive contention from higher-priority traffic. Wang et al. [58] utilized neural networks to study the cross-layer interactions between the physical channel conditions, the EDCA

MAC, and the application layer QoS requirements. An adaptive scheme based on neural networks is proposed to adjust both the CW and AIFS values.

The TXOP is introduced in EDCA so that a station may transmit multiple frames separated by SIFS once it has gained access to the wireless medium. This improves system performance when wireless stations have many small packets for transmission, because each packet will not need to go through the time-consuming access contention. This operating mode is optional in 802.11e, and the standard does not specify the algorithm for obtaining the proper TXOPs. Majkowski and Palacio [45] propose using the multiplication of the average number of packets in each AC's queue and the average packet length as the TXOP. In [46], they further proposed an enhanced TXOP (ETXOP) scheme, where the TXOP time may be shared among different ACs to further increase system flexibility. In [49], Kim and Suh proposed to allocate more TXOPs for stations with weaker signal strength, which leads to lower data rate. This partly alleviates the unfairness problem faced in such networks.

Apart from the above schemes that adaptively tune the MAC parameters, some recent work is focusing on studying the best strategy of transmitting voice or video traffic over the 802.11e WLAN by taking into account the real-time traffic's characteristics. Ksentini et al. [36] propose a cross-layer architecture for transmitting H.264 video over 802.11e EDCA. The system's architecture is shown in Figure 14.8. The H.264 application consists of two sublayers. The video coding layer (VCL) provides the core video compression and coding. It sends coded video slices to the network abstraction layer (NAL). When the H.264's data partitioning technique is enabled, the coded slices can be classified into five types, which have different importance for video decoding. The most important slice is the one containing the Parameter Set Concept (PSC), which is essential for decoding all other slices. The other ones are the Instantaneous Decoding Refresh (IDR) slice and the Partition A, Partition B, and Partition C slices, with decreasing importance for decoding.

NAL is responsible for encapsulation of the coded slices into NAL units (NALU), which can be considered packets ready to be sent to the underlining network. Each NALU contains a header and the video data payload. Of particular interest to the cross-layer architecture is the NAL-Reference-Identification (NRI) field. The NRI contains two bits that indicate the priority of the NALU payload. An example of allocating an NRI value for the five different VCL slices is shown in Figure 14.8. This assignment may be adaptively adjusted if needed.

At the MAC layer, different NALUs are classified by their NRI values. As shown in Figure 14.8, the PSC slice is put into the highest AC[3]. IDR and Partition A slices are put into AC[2]. Partition B and C slices are put into AC[1]. Ksentini et al. proposes a simple marking algorithm to adaptively assign these AC priorities based on network QoS metrics. ns-2 simulations show much improvement in terms of delay and loss rate.

Other than the above cross-layer adaptive designs, it is also important to have proper admission control for voice or video flows. In [18], each mobile node measures the occupied bandwidth (or the average collision ratio) by each of its four ACs, and makes admission decisions with a simple threshold rule. In [35, 37, 50], the expected network throughput or delay performance is estimated based on an analytical model with measured parameters. Then, an admission control decision is made at the QAP based on the new flow's request. In [62], the QAP measures the medium utilization and announces

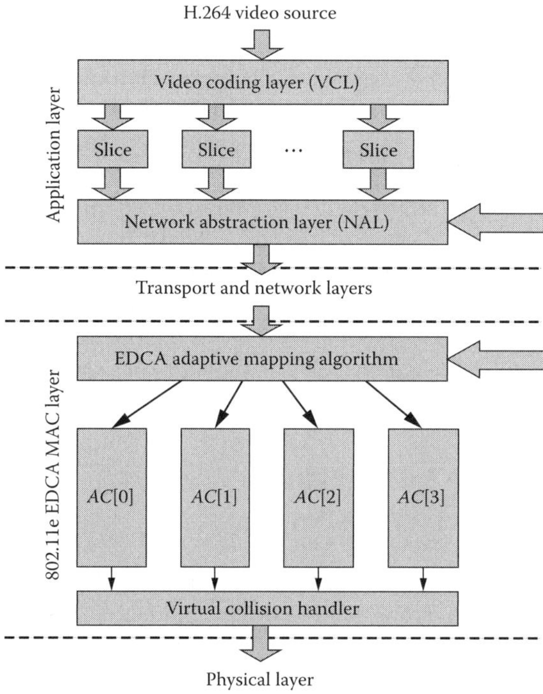

Slice type	*NRI value*
PSC	11
IDR	10
Partition A	10
Partition B	01
Partition C	01

NRI value	*AC*
11	3
10	2
01	1

FIGURE 14.8 H.264 over EDCA cross layer design architecture. (a) Frame format for A-MSDU; (b) frame format for MPDU.

the transmission opportunity budget (TXOPBudget) via periodic beacon signals for each AC (except AC[0] for best-effort traffic). When the TXOPBudget for one AC is depleted, new flows cannot gain transmission time and the existing flows cannot increase the transmission time either. Several enhancements are further proposed in [63] by adjusting the contention level from data traffic. In [41], an extension of the admission control to 802.11e multihop WLANs is proposed based on constructing contention cliques in the multihop topology.

14.4.3 Adaptive Scheduling for HCCA

The PCF defined in the legacy 802.11a/b/g standards did not succeed commercially due to its complexity and unsatisfactory performance. The newly proposed HCCA is an enhancement based on PCF, and still has some of its disadvantages compared to EDCA, such as its complexity, centralized control, and less robustness. As a result, it attracts less attention in the research community. But there is also some basic research work on HCCA, especially about its scheduling algorithm.

For a WLAN running HCCA, the hybrid coordinator (HC) is responsible for making polling schedules. In the 802.11e standard, a simple scheduling algorithm is defined. The HC schedules fixed batches of TXOPs at constant time intervals. Each batch contains one fixed length of TXOP per wireless station, based on the mean data rates declared in the respective traffic specifications (TSPECs). This scheduler accommodates the mean data rates of all traffic streams and performs well when the incoming traffic load does not deviate significantly from its declared mean data rate, such as constant-bit-rate traffic. However, for very bursty traffic, the performance degrades [17]. Grilo et al. [17] propose a more adaptive scheduler where the HC can poll QSTAs at variable intervals and assign variable-length TXOPs based on the earliest deadline of the waiting packets. The allocated TXOP is estimated from the mean data rate of traffic streams and the time interval between two successive transmissions. In [52], Skyrianoglou et al. further tune the TXOP duration based on the backlogged traffic in each station. Each station communicates the amount of its buffered data to the HC. The scheduled TXOP is set proportional to the time required to transmit the buffered data. Simulation tests are conducted to show the improvements in efficiency by this schedule algorithm.

14.5 Frame Aggregation Adaptation in High-Throughput WLANs

With the success of 802.11a/b/g WLANs, new applications are emerging for these wireless networks, such as video streaming, online gaming, and network-attached storage. Some of these new applications require extensive throughput support from the WLAN. The 802.11n proposal [30] aims to significantly improve the physical link data rate up to 600 Mbps by using multiple-input-multiple-output (MIMO) technology at the physical layer. Throughput performance at the MAC layer can be improved by aggregating several frames before transmission [43, 65]. Frame aggregation not only reduces the transmission time for preamble and frame headers, but also reduces the waiting time during the CSMA/CA random backoff period for successive frame transmissions. This section studies the frame aggregation techniques and introduces an optimal frame size adaptation algorithm.

14.5.1 The IEEE 802.11n High-Throughput PHY/MAC

There are several new MAC mechanisms in 802.11n for MAC efficiency enhancement, which includes the frame aggregation, block acknowledgment, bidirectional data transmission, etc.

Two frame aggregation methods are defined at the MAC layer. The first technique is by concatenating several MAC service data units (MSDUs) to form the data payload of a large MAC protocol data unit (MPDU). The PHY header and MAC header, along with the frame check sequence (FCS), are then appended to form the physical service data unit (PSDU). This technique is known as *MSDU aggregation* (A-MSDU). Figure 14.9(a) shows the frame format for A-MSDU.

The second technique is called *MPDU aggregation* (A-MPDU). It begins with each MSDU appending with its own MAC header and FCS to form a sub-MPDU. An MPDU delimiter is then inserted before each sub-MPDU. Padding bits are also inserted so that

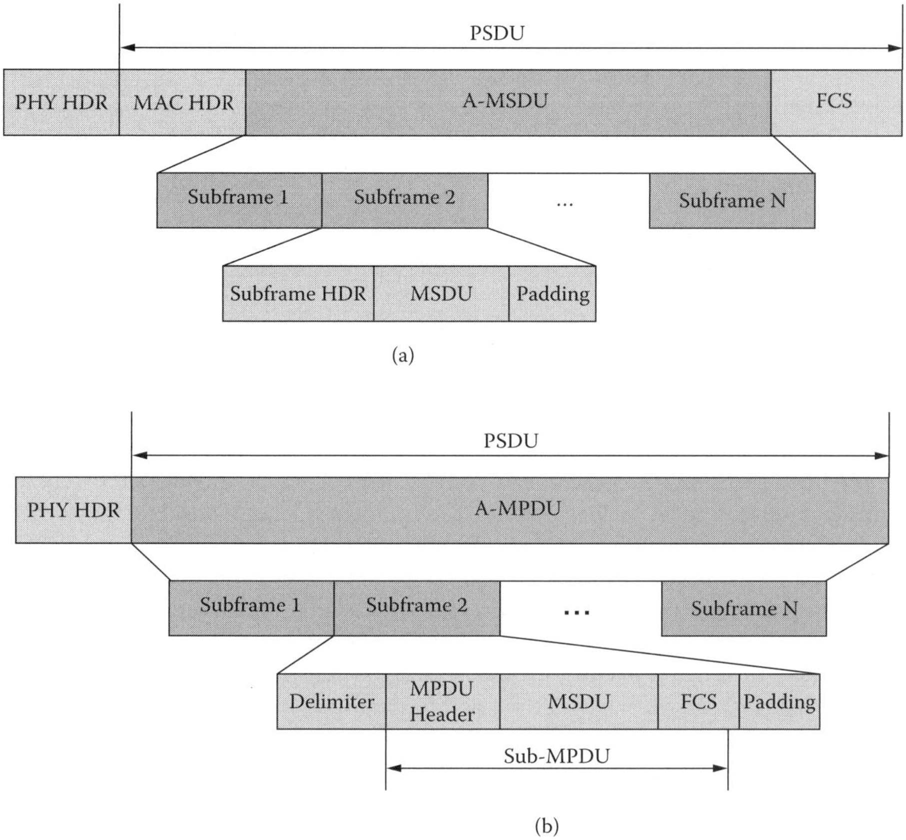

FIGURE 14.9 IEEE 802.11n frame aggregation format.

each sub-MPDU is a multiple of 4 bytes in length, which can facilitate subframe delineation at the receiver. Then, all the sub-MPDUs are concatenated to form a large PSDU. Figure 14.9(b) shows the frame format for A-MPDU.

The 802.11n standard also specifies a bidirectional data transfer method. If RTS/CTS is used, the current transmission sequence of *RTS-CTS-DATA-ACK* only allows the sender to transmit a single data frame. In the bidirectional data transfer method, the receiver may request a reverse data transmission in the CTS control frame. The sender can then grant a certain medium time for the receiver on the reverse link. The transmission sequence will then become *RTS-CTS-DATAf-DATAr-ACK*. This facilitates the transmission of some small feedback packets from the receiver and may also enhance the performance of TCP, which requires the transmission of TCP ACK segments.

In all of the above cases, Block Acknowledgment (BACK) can be used to replace the traditional ACK frame. The BACK can use a bit map to efficiently acknowledge each individual subframe within the aggregated frame. For the bidirectional data transfer, the reverse DATAr frame can contain a BACK to acknowledge the previous DATAf frame.

14.5.2 An Analytical Model of Frame Aggregation

Bianchi's model [6] can be extended to study the A-MPDU and A-MSDU frame aggregations under error-prone channels. In the analytical model, we assume that there are N mobile stations in the WLAN. Each mobile station has saturated traffic. The wireless channel has a bit error rate (BER) of P_b. The minimum contention window size is W and the maximum backoff stage is m. Since the size of an aggregated frame is large, the RTS/CTS access scheme is generally more efficient than the basic access scheme. As a result, only the access scheme with RTS/CTS will be discussed. In 802.11 WLANs, the control frames (RTS, CTS, BACK) are transmitted at the basic rate, which is much lower than the data rate and more robust in combating errors. Since the sizes of these control frames are much smaller than an aggregated data frame, they have a much lower frame error rate. In addition, the Physical Layer Convergence Procedure (PLCP) preamble and header are also transmitted at a lower rate. As a result, the frame error probabilities for control frames and preambles may be safely omitted for simplifying the analysis.

The possible timing sequences for A-MPDU and A-MSDU in the unidirectional transfer case are shown in Figure 14.10. The timing sequences for bidirectional data transfer are shown in Figure 14.11. In both figures, the DATA frame represents either an A-MPDU or an A-MSDU frame.

The system time can be broken down into virtual time slots where each slot is the time interval between two consecutive countdowns of backoff timers by nontransmitting stations. If we define p to be the unsuccessful transmission probability, conditioned on there being a transmission in a time slot, the transmission probability τ in a virtual slot is given in the same form as in (14.1). But when considering both collisions and transmission errors, p is changed from (14.2) to be

$$p = 1 - (1 - p_c)(1 - p_c), \tag{14.8}$$

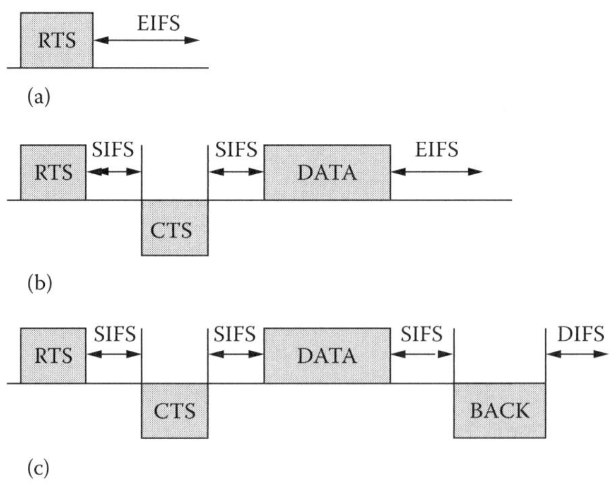

FIGURE 14.10 Unidirectional RTS/CTS access scheme. (a) RTS collision; (b) data frame corruption; (c) success.

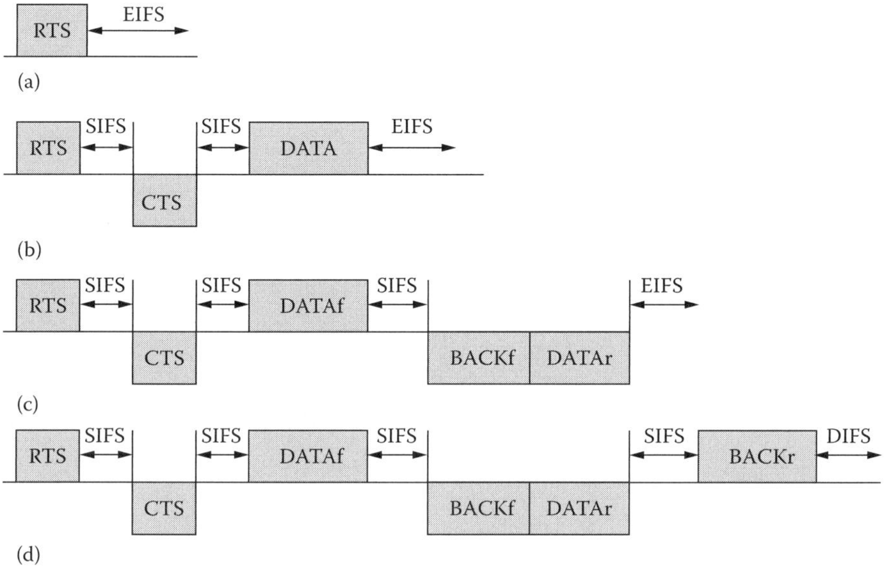

FIGURE 14.11 Bidirectional RTS/CTS access scheme. (a) RTS collision; (b) forward DATAf frame corruption; (c) reverse DATAr frame corruption; (d) success.

where $p_c = 1 - (1 - \tau)^{(N-1)}$ is the conditional collision probability and P_e is the error probability, on condition that there is a successful RTS/CTS transmission in the time slot. For unidirectional transfer, p_e is the error probability corresponding to the error case in Figure 14.10(b). For bidirectional transfer, we define p_e as a 2 × 1 vector $\mathbf{p_e} = [p_{e,1}, p_{e,2}]^T$ (where T is the transpose) corresponding to the two error cases in Figure 14.11(b, c). In the following, we will use the vector form for generality, and equation (14.8) for the bidirectional case is

$$p = 1 - (1 - p_c)(1 - p_{e,1}). \tag{14.9}$$

Note that only $p_{e,1}$ for Figure 14.10(b) contributes to p. This is because in the case of Figure 14.10(c), we follow our previous assumption that the $BACK_f$ control frame is error-free. Thus, $BACK_f$ for the forward frame is always successful and $DATA_f$'s sending station will not double its contention window in this case.

The probability of an idle slot P_{idle}, probability for transmission in a time slot P_{tr}, and probability for a noncollided transmission P_s can be calculated as

$$P_{idle} = (1 - \tau)^N, \tag{14.10}$$

$$P_{tr} = 1 - P_{idle} = 1 - (1 - \tau)^N, \tag{14.11}$$

$$P_s = \frac{N_\tau (1 - \tau)^{(N-1)}}{P_{tr}}. \tag{14.12}$$

The transmission failure probability due to error (no collisions but having transmission errors) is

$$\mathbf{p_{err}} = P_{tr}\, P_s\, \mathbf{p_e} = \left[p_{err,1},\ p_{err,2} \right]^T, \tag{14.13}$$

where $p_{err,1}$ and $p_{err,2}$ correspond to the two different error timing sequences for the bidirectional transfer in Figure 14.11. $\mathbf{p_{err}}$ reduces to a scalar for the unidirectional case.

The probability for a successful transmission (without collisions and transmission errors) is

$$P_{succ} = P_{tr} P_s \left(1 - p_{e,1} - p_{e,2}\right). \tag{14.14}$$

With the above probabilities, the network's saturation throughput can be calculated as

$$S = \frac{E_p}{E_t}, \tag{14.15}$$

where E_p is the number of payload information bits successfully transmitted in a virtual time slot, and E_t is the expected length of a virtual time slot:

$$E_t = T_{idle} P_{idle} + T_c P_{tr} \left(1 - P_s\right) + \mathbf{T_e}^T \mathbf{p_{err}} + T_{succ} P_{succ}, \tag{14.16}$$

where T_{idle}, T_c, and T_{succ} are the idle, collision, and successful virtual time slot's length, respectively. $\mathbf{T_e}$ is the virtual time slot length for an error transmission sequence. Similar to $\mathbf{p_e}$, it corresponds to a scalar for the unidirectional case, and a 2×1 vector for the bidirectional transfer timing sequence.

Apart from throughput, we study the average access delay experienced by each station in the unidirectional case. The access delay is defined as the delay between the time when an aggregated frame reaches the head of the MAC queue and the time that the frame is successfully received by the receiver's MAC. With the saturation throughput S, each frame takes an average of L_p/S to transmit (L_p is the aggregated frame's payload length). There are N stations competing for transmission. On average, the access delay is

$$d = N \frac{L_p}{S}. \tag{14.17}$$

To calculate S and d from equations (14.15) and (14.17), the parameters of E_p, T_{idle}, T_c, T_{succ}, $\mathbf{T_e}$, and $\mathbf{p_e}$ need to be determined. T_{idle} is equal to the system's empty slot time σ:

$$T_c = RTS + EIFS, \tag{14.18}$$

where RTS is the transmission time for an RTS frame. The other parameters are case dependent and will be discussed separately in the following subsections.

14.5.2.1 Unidirectional MAC

In the unidirectional case, the equations for T_{succ}, T_e, and E_p are as follows:

$$T_{succ} = RTS + CTS + DATA + BACK + 3SIFS + DIFS, \tag{14.19}$$

$$T_e = RTS + CTS + DATA\_EIFS + 2SIFS, \tag{14.20}$$

$$E_p = L_p P_{succ} = L_p P_{tr} P_s (1 - p_e), \tag{14.21}$$

where CTS, BACK, and DATA are the transmission times for CTS, BACK, and the aggregated data frame, respectively.

For A-MSDU, the equations for pe and Ep are

$$p_e = 1 - (1 - P_b)^L, \tag{14.22}$$

$$E_p = (L - L_{hdr})(1 - p_e), \tag{14.23}$$

where L is the aggregated MAC frame's size, and L_{hdr} is the total length of the MAC header and FCS.

For A-MPDU, error occurs when all the subframes become corrupted. The variables p_e and E_p can be expressed as

$$p_e = \prod_i \left(1 - (1 - P_b)^{L_i}\right), \tag{14.24}$$

$$E_p = \sum_i (L_i - L_{subhdr})(1 - P_b)^{L_i}, \tag{14.25}$$

where i is from 1 to the total number of aggregated sub-MPDUs, and L_i is the size for the ith sub-MPDU. L_{subhdr} is the total size of each sub-MPDU's delimiter, header, and FCS.

14.5.2.2 Bidirectional MAC

For the bidirectional MAC transfer function, there are also two aggregation methods: A-MPDU and A-MSDU. Due to the space limitation, we only present the results for A-MSDU aggregation here. The A-MPDU case can be derived in a similar way.

For error in the forward frame (see Figure 14.11(b)), we have:

$$T_{e,1} = RTS + CTS + DATAf + 2SIFS + EIFS, \quad (14.26)$$

$$p_{e,1} = 1-(1-P_b)^{DATAf}. \quad (14.27)$$

For error in the reverse frame (see Figure 14.11(c)), we have:

$$T_{e,2} = RTS + CTS + DATAf + BACKf + DATAr + 3SIFS + EIFS, \quad (14.28)$$

$$p_{e,2} = (1-P_b)^{DATAf}\left[1-(1-P_b)^{DATAr}\right]. \quad (14.29)$$

For a successful bidirectional frame transmission:

$$T_{succ} = RTS + CTS + DATAf + BACKf + DATAr + BACKr + 4SIFS + DIFS. \quad (14.30)$$

Since we assume that the BACK control frame is transmitted at the basic rate, $DATA_f$ will be successfully received in the case of Figure 14.11(c). Thus, the expected successful payload information transmitted, E_p, can be expressed as

$$\begin{aligned} E_p &= (L_f + L_r - 2L_{hdr})P_{succ} + (L_f - L_{hdr})p_{err,2} \\ &= (L_f + L_r - 2L_{hdr})P_{tr}P_s(1-p_{e,1}-p_{e,2}) + (L_f - L_{hdr})P_{tr}P_s p_{e,2}. \end{aligned} \quad (14.31)$$

To verify the accuracy of the above analytical model, simulations are carried out in the ns-2 simulator [2] for throughput and delay performance comparison with the analytical model. The parameters used in the simulation are from [28] and shown in Table 14.1.

TABLE 14.1 Simulation Parameters

Basic rate	54 Mbps
Data rate	144.4 Mbps
PLCP preamble	16 μs
PLCP header	48 bits
PLCP rate	6 Mbps
MAC header	192 bits
FCS (frame check sequence)	32 bits
Time slot	9 μs
SIFS	16 μs
Subframe header for A-MSDU	14 bytes
Delimiter for A-MPDU	4 bytes

14.5.2.3 Simulation Results for Unidirectional Data Transfer

In this simulation, there are ten wireless nodes and one access point in the network. All the wireless nodes have saturated CBR traffic directed to the access point. The BER varies from 0 to 10^{-3}. All the data packets passed down to the MAC layer are 100 bytes in length. The number of packets aggregated in one MAC frame varies from 1 to 80, which leads to an aggregated payload size from 100 bytes to 8 kbytes.

Figures 14.12 and 14.13 show the saturation throughput and access delay for the A-MSDU aggregation. Figures 14.14 and 14.15 show the saturation throughput and access delay for the A-MPDU aggregation. All the lines in the figures are the results obtained from the analytical model. The simulation results are shown in discrete marks. Comparison with the simulation results shows that the analytical model is accurate in predicting the network performance.

From these figures, we can observe that the saturation throughput decreases and the delay increases with increasing BER for both aggregation schemes. A-MSDU achieves a higher throughput than A-MPDU under ideal channel conditions (i.e., BER = 0). This is due to the fact that A-MSDU includes a lower overhead in the aggregation process than A-MPDU. However, under error-prone channels, the advantage of A-MSDU quickly diminishes. The curves in Figure 14.12 show that the throughput under A-MSDU first increases, and then decreases with increasing aggregated frame size in error-prone channels. This is because without the protection of FCS in individual subframes, a single bit error may corrupt the whole frame, which will waste lots of medium time usage and counteract the efficiency produced by an increased frame size. For A-MPDU, the throughput monotonically increases with increasing aggregated frame size. As a

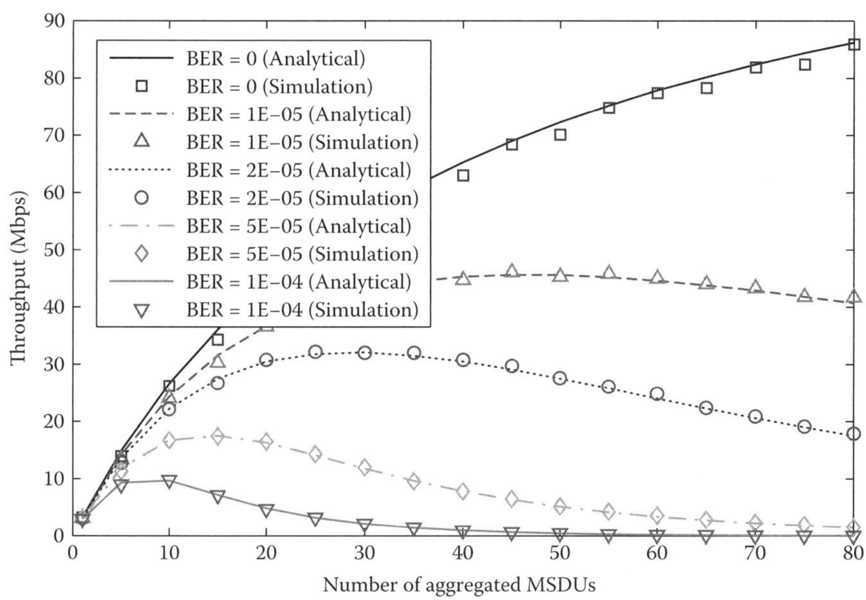

FIGURE 14.12 Saturation throughput for A-MSDU.

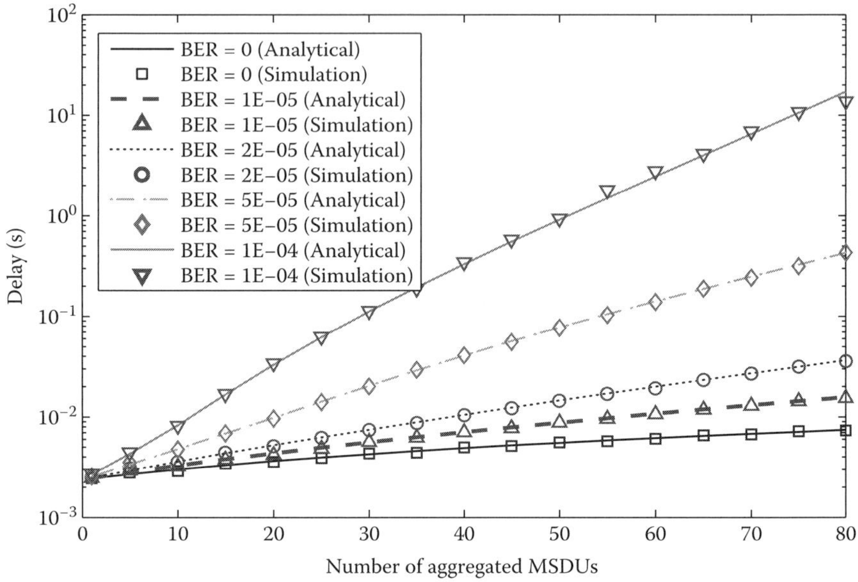

FIGURE 14.13 Access delay for A-MSDU.

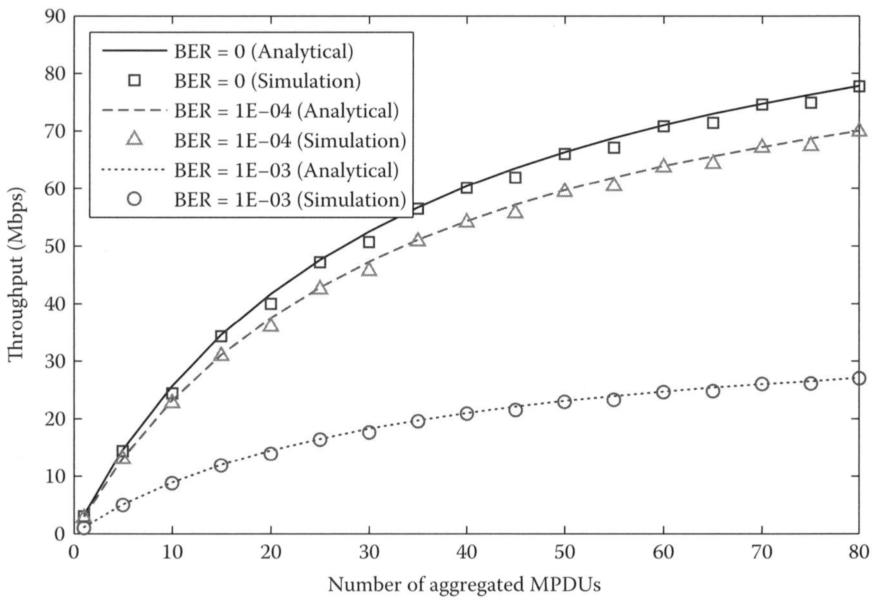

FIGURE 14.14 Saturation throughput for A-MPDU.

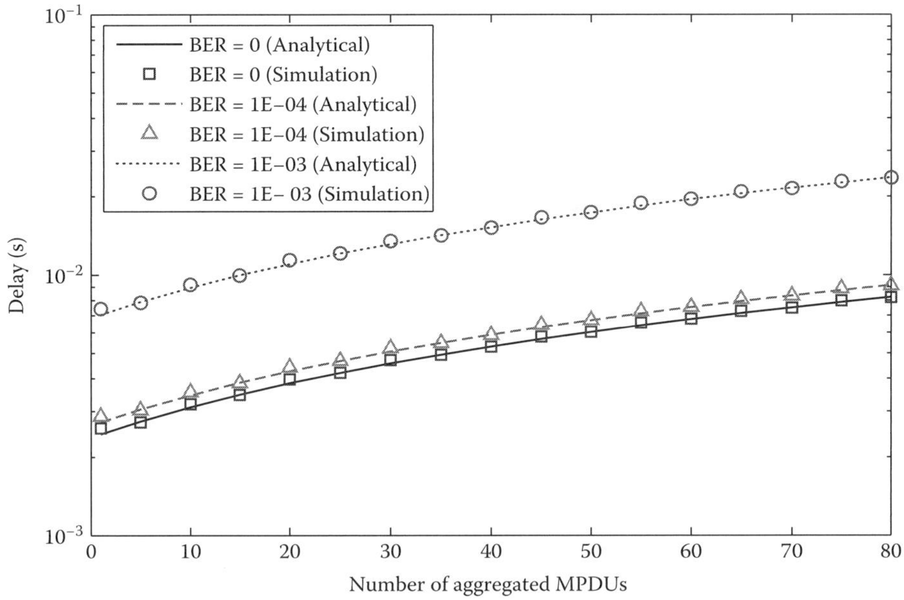

FIGURE 14.15 Access delay for A-MPDU.

result, it is more beneficial to use A-MSDU under good channel conditions and A-MPDU under bad channel conditions.

Although the throughput increases by increasing the aggregated frame size for A-MPDU, the frame size cannot be increased indefinitely due to the delay constraint by many applications. As a result, we need to choose the proper aggregation scheme and adapt parameters according to the different channel conditions and application requirements in order to achieve an optimal performance. In the next section, we will investigate the performance of a simple optimal frame size adaptation algorithm for A-MSDU under error-prone conditions.

14.5.2.4 Simulation Results for Bidirectional Data Transfer

The saturation throughput performance for A-MSDU with bidirectional data transfer under different BERs is shown in Figure 14.16. The numbers of aggregated MSDUs in the forward and reverse data aggregation are set to 20 and 1, respectively. The number of stations is varied from 5 to 30. The simulation results validate the accuracy of the analytical model in predicting the network performance. Comparing with Figure 14.12, the bidirectional transfer does not provide much gain from the aspect of saturation throughput performance. Its major contribution to system improvement is the interaction with the higher-layer protocols (e.g., TCP) for the transfer of acknowledgment segments in a timely manner.

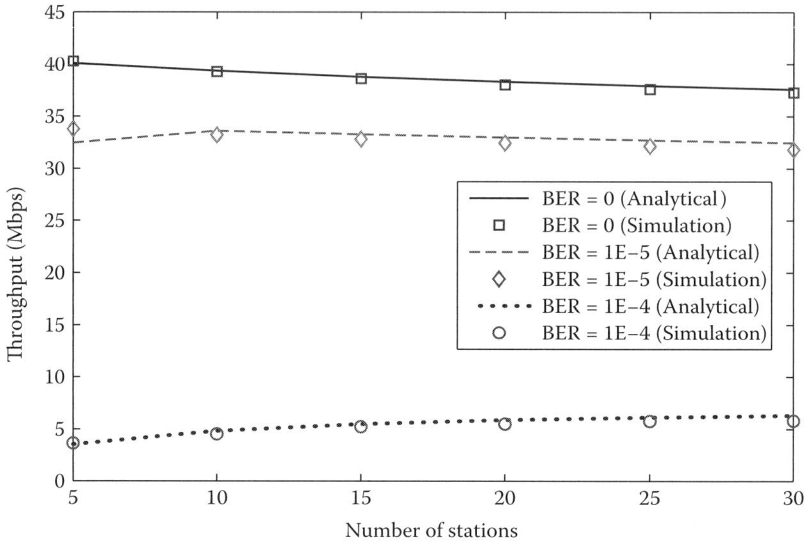

FIGURE 14.16 Saturation throughput under bidirectional data transfer.

14.5.3 Optimal Frame Size Adaptation

From Figure 14.12, we can observe that A-MSDU may reach a maximum throughput under different BER conditions. The optimal aggregated frame size L^* to achieve this maximum throughput varies with the channel's BER condition. To further determine the relationship between L^* and the number of contending stations, we conduct an experiment in which the number of stations changes from 10 to 30. The other parameters remain the same. The analytical and simulation results are shown in Figure 14.17.

From Figure 14.17, we can observe that the optimal aggregated frame size L^* is very sensitive to BER, but rather insensitive to the number of contending stations in the network. To this end, we propose a simple and efffective frame aggregation adaptation algorithm to be as follows: First, we determine the L^*-BER curve from the analytical model in section 14.5.2 by using an average number of stations N in the network. The L^*-BER curve gives the optimal aggregated frame size L^* under different channel bit error rates. Before transmitting an aggregated A-MSDU frame, the sending station will obtain an estimation of the channel BER, consult the L^*-BER curve for an optimal L^*, and then construct the aggregated frame with a size that is close to the optimal frame size.

The channel BER is a function of the modulation scheme and the signal-to-noise ratio (SNR). In general, for a given modulation and coding scheme, the BER can be determined from either a theoretical or an empirical BER-SNR curve. The SNR is measured at the receiver for each received frame. With the help of a closed-loop feedback mechanism, this SNR may be efficiently updated to the sender. For example, in the 802.11n proposals, there is a new MAC feature for channel management that is called the *receiver-assisted link adaptation* [29]. Channel conditions are fed back to the sender by control frames

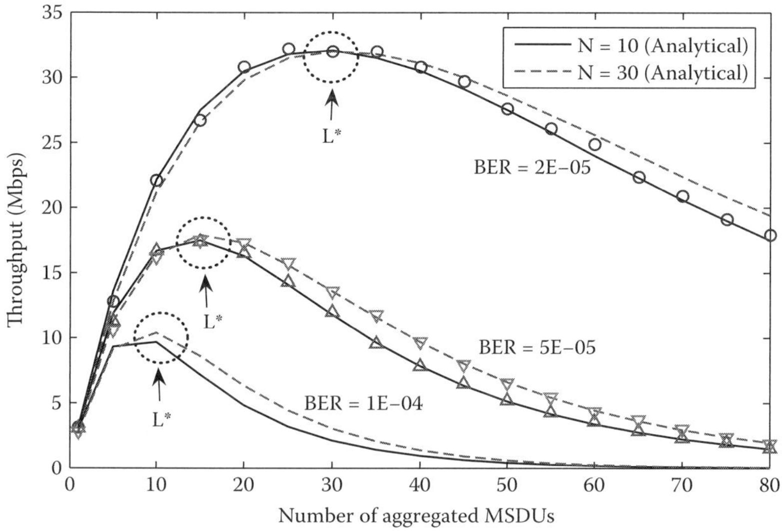

FIGURE 14.17 A-MSDU throughput under different numbers of stations.

in a closed-loop fashion. We assume that such a feedback mechanism is available in 802.11n to provide the sending station with the channel SNR information.

To determine the effectiveness of our proposed frame adaptation algorithm, we conduct an ns-2 simulation. We use the channel B error model from [28], which models a typical large open space and office environments that have non-line-of-sight conditions, and 100 ns rms delay spread. The 144.44 Mbps data rate is used, which leads to an effective transmission range of 25 m. The network topology consists of an open space of 50 × 50 m. The access point is fixed at the center of the area. There are *N* wireless nodes in the network, and they move according to a random waypoint mobility model. The maximum speed is 5 m/s and the pause time is 5 s. All the wireless terminals are saturated with CBR traffic. The number of stations *N* is varied from 5 to 20. The throughput performance of the optimal frame size adaptation algorithm is compared with a fixed frame aggregation model and a randomized frame aggregation model, where the aggregated frame sizes are randomly distributed between the minimum (100 bytes) and maximum (20 kbytes) frame size allowed. From the simulation results shown in Figure 14.18, we can observe that the adaptive frame aggregation algorithm achieves significant throughput gain over the other two algorithms.

14.6 Conclusions

This chapter introduced some of the common adaptive mechanisms available for performance improvement in the popular IEEE 802.11–based wireless LANs. Adaptive tuning of the system parameters has shown positive performance gains. Some other schemes, such as adaptive scheduling and frame aggregation, are also effective mechanisms available at the MAC layer. Cross-layer architectures, which jointly adapt the application, MAC, and

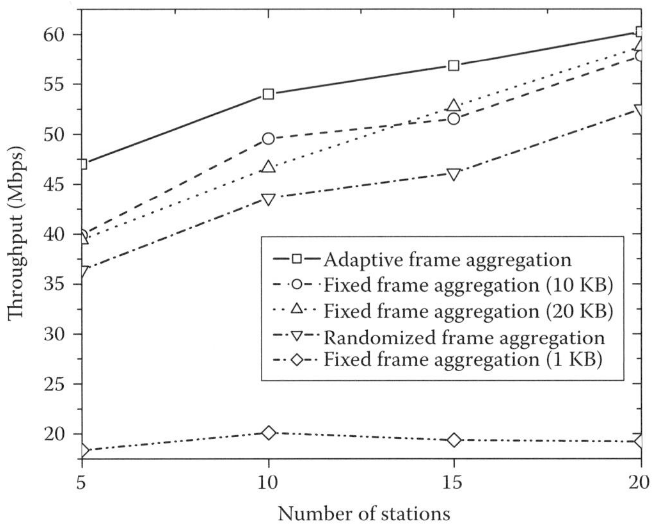

FIGURE 14.18 Throughput under different frame aggregation schemes.

physical operations, are promising techniques to further improve system performance by making adaptivity beyond the boundary of communication layers. All these emerging new techniques are good candidates for consideration when designing more efficient and more QoS-aware wireless LANs in the future. However, when trying to implement these adaptive schemes in real networks, we should also carefully consider the interaction of the adaptive algorithm on other aspects of the network, so that the wireless communication system can become more efficient as well as remain stable and robust.

References

[1] IEEE 802.11n Working Group. Accessed September 2007 from http://grouper.ieee.org/groups/802/11/Reports/tgn_update.htm.

[2] Ns-2 simulator. Accessed September 2007 from http://www.isi.edu/nsnam/ns/.

[3] OPNET. Accessed September 2007 from http://www.opnet.com/.

[4] Y. Andreopoulos, N. Mastronarde, and M. Van Der Schaar. 2006. Cross-layer optimized video streaming over wireless multihop mesh networks. *IEEE J. Select. Areas Commun.* 24:2104–15.

[5] A. Banchs, A. Azcorra, C. Garcia, and R. Cuevas. 2005. Applications and challenges of the 802.11e EDCA mechanism: An experimental study. *IEEE Network* 19:52–58.

[6] G. Bianchi. 2000. Performance analysis of the IEEE 802.11 distributed coordination function. *IEEE J. Select. Areas Commun.* 18:535–47.

[7] G. Bianchi, L. Fratta, and M. Oliveri. 1996. Performance evaluation and enhancement of the CSMA/CA MAC protocol for 802.11 wireless LANs. In *Proceedings of IEEE PIMRC*, Taipei, Taiwan, pp. 392–396.

[8] G. Bianchi and I. Tinnirello. 2003. Kalman filter estimation of the number of competing terminals in an IEEE 802.11 network. In *Proceedings of IEEE INFOCOM*, San Francisco, pp. 844–52.

[9] L. Bononi, M. Conti, and E. Gregori. 2004. Runtime optimization of IEEE 802.11 wireless LANs performance. *IEEE Trans. Parallel Distrib. Syst.* 15:66–80.

[10] F. Cali, M. Conti, and E. Gregori. 2000. Dynamic tuning of the IEEE 802.11 protocol to achieve a theoretical throughput limit. *IEEE/ACM Trans. Networking* 8:785–99.

[11] S. Choi, J. D. Prado, S. Shankar, and S. Mangold. 2003. IEEE 802.11e contention-based channel access (EDCF) performance evaluation. In *Proceedings of IEEE ICC*, Anchorage, AL, pp. 1151–56.

[12] B. P. Crow, I. Widjaja, J. G. Kim, and P. T. Sakai. 1997. IEEE 802.11 wireless local area networks. *IEEE Commun. Mag.* 35:116–26.

[13] R. de Oliveira and T. Braun. 2007. A smart TCP acknowledgment approach for multihop wireless networks. *IEEE Trans. Mobile Comput.* 6:192–205.

[14] J. Deng, B. Liang, and P. K. Varshney. 2004. Tuning the carrier sensing range of IEEE 802.11 MAC. In *Proceedings of IEEE Globecom*, Dallas, TX, pp. 2987–91.

[15] L. Gannoune. 2006. A comparative study of dynamic adaptation algorithms for enhanced service differentiation in IEEE 802.11 wireless ad hoc networks. In *Proceedings of IEEE AICT-ICIW*, Guadeloupe, French Caribbean, pp. 31–38.

[16] Y. Ge and J. Hou. 2003. An analytical model for service differentiation in IEEE 802.11. In *Proceedings of IEEE ICC*, Anchorage, AL, pp. 1157–62.

[17] A. Grilo, M. Macedo, and M. Nunes. 2003. A scheduling algorithm for QoS support in IEEE802.11e networks. *IEEE Wireless Commun. Mag.* 10:36–43.

[18] D. Gu and J. Zhang. 2003. A new measurement-based admission control method for IEEE 802.11 wireless local area networks. In *Proceedings of IEEE PIMRC*, Beijing, China, vol. 3, pp. 2009–13.

[19] L. Haratcherev, J. Taal, K. Langendoen, R. Lagendijk, and H. Sips. 2006. Optimized video streaming over 802.11 by cross-layer signaling. *IEEE Commun. Mag.* 44:115–21.

[20] D. He and C. Q. Shen. 2003. Simulation study of IEEE 802.11e EDCF. In *Proceedings of VTC*, Jeju, Korea, pp. 685–89.

[21] M. Ho, J. Wang, K. Shelby, and H. Haisch. 2003. IEEE 802.11g OFDM WLAN throughput performance. In *Proceedings of IEEE VTC-Fall*, Orlando, FL, pp. 2252–2256.

[22] J. Hui and M. Devetsikiotis. 2004. Performance analysis of IEEE 802.11e EDCA by a unified model. In *Proceedings of IEEE Globecom*, Dallas, TX, pp. 754–59.

[23] IEEE 802.11 WG. 1999. Wireless LAN medium access control (MAC) and physical layer (PHY) specifications. IEEE Standard 802.11.

[24] IEEE 802.11 WG. 1999. Wireless LAN medium access control (MAC) and physical layer (PHY) specifications: High-speed physical layer in the 5 GHz band. IEEE Standard 802.11a.

[25] IEEE 802.11 WG. 1999. Wireless LAN medium access control (MAC) and physical layer (PHY) specifications: Higher-speed physical layer extension in the 2.4 GHz band. IEEE Standard 802.11b.

[26] IEEE 802.11 WG. 2003. Wireless LAN medium access control (MAC) and physical layer (PHY) specifications: Further higher data rate extension in the 2.4 GHz band. IEEE Standard 802.11g.

[27] IEEE 802.11e WG. 2005. Wireless LAN MAC and PHY specifications amendment 8: MAC quality of service enhancements.

[28] IEEE 802.11n TGn Sync. 2005. TGn Sync proposal MAC simulation methodology.

[29] IEEE 802.11n TGn Sync. 2005. TGn Sync proposal technical specification.

[30] IEEE 802.11n WG. 2007. IEEE 802.11n draft 2.0.

[31] A. Kamerman and G. Aben. 2000. Throughput performance of wireless LANs operating at 2.4 and 5 GHz. In *Proceedings of IEEE PIMRC*, London, pp. 190–95.

[32] M. I. Kazantzidis. 2005. MAC intelligence for adaptive multimedia in 802.11 networks. *IEEE J. Select. Areas Commun.* 23:357–68.

[33] Y. Kim, J. Yu, S. Choi, and K. Jang. 2006, A novel hidden station detection mechanism in IEEE 802.11 WLAN. *IEEE Commun. Lett.* 10:608–10.

[34] Z. Kong, D. H. K Tsang, and B. Bensaou. 2004. Performance analysis of IEEE 802.11e contention-based channel access. *IEEE J. Select. Areas Commun.* 22:2095–106.

[35] Z. Kong, D. H. K. Tsang, and B. Bensaou. 2004. Measurement-assisted model-based call admission control for IEEE 802.11e WLAN contention-based channel access. In *Proceedings of the 13th IEEE Workshop on Local and Metropolitan Area Networks*, pp. 55–60.

[36] A. Ksentini, M. Naimi, and A. Gueroui. 2006. Toward an improvement of H.264 video transmission over IEEE 802.11e through a cross-layer architecture. *IEEE Commun. Mag.* 44:107–14.

[37] Y. Kuo, C. Lu, E. H. K. Wu, and G. Chen. 2003. An admission control strategy for differentiated service in IEEE 802.11. In *Proceedings of IEEE Globecom*, San Francisco, vol. 2, pp. 707–12.

[38] J. Lee, S. Kwon, and D. Cho. 2005. Adaptive beacon listening protocol for a TCP connection in slow-start phase in WLAN. *IEEE Commun. Lett.* 9:853–55.

[39] X. Lin, N. B. Shroff, and R. Srikant. 2006. A tutorial on cross-layer optimization in wireless networks. *IEEE J. Select. Areas Commun.* 24:1452–63.

[40] Y. Lin and V. W. S. Wong. 2006. Saturation throughput of IEEE 802.11e EDCA based on mean value analysis. In *Proceedings of IEEE WCNC*, Las Vegas, pp. 475–80.

[41] Y. Lin, V. W. S. Wong, and M. Cheung. 2006. An admission control algorithm for multi-hop 802.11e based WLANs. In *Proceedings of QShine*, Waterloo, Canada.

[42] A. Lindgren, A. Almquist, and O. Schelen. 2003. Quality of service schemes for IEEE 802.11 wireless LANs: An evaluation. *Mobile Networks Appl.* 8:223–35.

[43] C. Liu and A. P. Stephens. 2005. An analytic model for infrastructure WLAN capacity with bidirectional frame aggregation. In *Proceedings of IEEE WCNC*, pp. 113–19.

[44] H. Ma and S. Roy. 2006. Simple and effective carrier sensing adaptation for multi rate ad-hoc mesh networks. In *Proceedings of the 1st International Workshop on Wireless Mesh Networks and Application*, Vancouver, BC, pp. 795–800.

[45] J. Majkowski and F. C. Palacio. 2006. Dynamic TXOP configuration for QoS enhancement in IEEE 802.11e wireless LAN. In *Proceedings of IEEE SoftCOM*, Montreal, pp. 66–70.

[46] J. Majkowski and F. C. Palacio. 2006. Enhanced TXOP scheme for efficiency improvement of WLAN IEEE 802.11e. In *Proceedings of IEEE VTC*, Montreal, pp. 1–5.

[47] J. Naoum-Sawaya, B. Ghaddar, S. Khawam, H. Safa, H. Artail, and Z. Dawy. 2005. Adaptive approach for QoS support in IEEE 802.11e wireless LAN. In *Proceedings of IEEE WiMob*, Montreal, pp. 167–73.

[48] Q. Ni. 2005. Performance analysis and enhancements for IEEE 802.11e wireless networks. *IEEE Network* 19:21–27.

[49] E. Kim and Y. Suh, 2004. ATXOP: an adaptive TXOP based on the data rate to guarantee fairness for IEEE 802.11e wireless LANs. In *Proceedings of IEEE VTC-Fall*, Los Angeles, CA, pp. 2678-2682.

[50] D. Pong and T. Moors. 2003. Call admission control for IEEE 802.11 contention access mechanism. In *Proceedings of IEEE Globecom*, San Francisco, pp. 174–78.

[51] J. W. Robinson and T. S. Randhawa. 2004. Saturation throughput analysis of IEEE 802.11e enhanced distributed coordination function. *IEEE J. Select. Areas Commun.* 22:917–28.

[52] D. Skyrianoglou, N. Passas, and A. K. Salkintzis. 2006. ARROW: An efficient traffic scheduling algorithm for IEEE 802.11e HCCA. *IEEE Trans. Wireless Commun.* 5:3558–67.

[53] V. Srivastava and M. Motani. 2005. Cross-layer design: A survey and the road ahead. *IEEE Commun. Mag.* 43:112–19.

[54] Y. C. Tay and K. C. Chua. 2001. A capacity analysis for the IEEE 802.11 MAC protocol. *Wireless Networks* 7:159–71.

[55] M. van der Schaar, S. Krishnamachari, S. Choi, and X. Xu. 2003. Adaptive cross-layer protection strategies for robust scalable video transmission over 802.11 WLANs. *IEEE J. Select. Areas Commun.* 21:1752–63.

[56] M. van Der Schaar and S. N. Shankar. 2005. Cross-layer wireless multimedia transmission: Challenges, principles, and new paradigms. *IEEE Wireless Commun. Mag.* 12:50–58.

[57] C. Wang, B. Li, and L. Li. 2004. A new collision resolution mechanism to enhance the performance of IEEE 802.11 DCF. *IEEE Trans. Veh. Technol.* 53:1235–46.

[58] C. Wang, P. Lin, and T. Lin. 2006. A cross-layer adaptation scheme for improving IEEE 802.11e QoS by learning. *IEEE Trans. Neural Networks* 17:1661–65.

[59] W. Wang, S. C. Liew, and V. O. K. Li. 2005. Solutions to performance problems in VoIP over a 802.11 wireless LAN. *IEEE Trans. Veh. Technol.* 54:366–84.

[60] J. Weinmiller, M. Schlager, A. Festag, and A. Wolisz. 1997. Performance study of access control in wireless LANs-IEEE 802.11 DFWMAC and ETSI RES 10 HIPERLAN. *Mobile Networks* Appl. 2:55–67.

[61] Y. Xiao. 2004. Performance analysis of IEEE 802.11e EDCF under saturation condition. In *Proceedings of IEEE ICC*, Paris, pp. 170–174.
[62] Y. Xiao and H. Li. 2004. Local data control and admission control for QoS support in wireless ad hoc networks. *IEEE Trans. Veh. Technol.* 53:1558–72.
[63] Y. Xiao, H. Li, and S. Choi. 2004. Protection and guarantee for voice and video traffic in IEEE 802.11e wireless LANs. In *Proceedings of IEEE INFOCOM*, Hong Kong, vol. 3, pp. 2152–62.
[64] Z. Xu and L. Meng. 2006. A novel fair scheduling scheme based on dynamic AIFS in 802.11 wireless LANs. In *Proceedings of the International Conference on Communications, Circuits and Systems*, Guilin, China, pp. 1754–57.
[65] J. Yin, X. Wang, and D. P. Agrawal. 2004. Optimal packet size in error-prone channel for IEEE 802.11 distributed coordination function. In *Proceedings of IEEE WCNC*, pp. 1654–59.
[66] H. Zhai, X. Chen, and Y. Fang. 2005. How well can the IEEE 802.11 wireless LAN support quality of service? *IEEE Trans. Wireless Commun.* 4:3084–94.
[67] J. Zhu, X. Guo, L. L. Yang, and W. S. Conner. 2004. Leveraging spatial reuse in 802.11 mesh networks with enhanced physical carrier sensing. In *Proceedings of IEEE ICC*, Paris, pp. 4004–11.
[68] J. Zhu, X. Guo, L. L. Yang, W. S. Conner, S. Roay, and M. M. Hazra. Adapting physical carrier sensing to maximize spatial reuse in 802.11 mesh networks. In *ACM Wireless Commun. Mobile Comput.* 4(8):933–946.
[69] J. Zhu, B. Metzler, X. Guo, and Y. Liu. 2006. Adaptive CSMA for scalable network capacity in high-density WLAN: A hardware prototyping approach. In *Proceedings of IEEE INFOCOM*, Barcelona, Spain, pp. 1–10.

15

Tunable Security Services for Wireless Networks

Stefan Lindskog
Karlstad University
and
Norwegian University of Science and Technology

Anna Brunstrom
Karlstad University

Zoltán Faigl
Budapest University of Technology and Economics

15.1 Introduction

Global networking and mobile computing are two of the biggest trends in computing today. The popularity of the Internet is still growing, and Internet is now used for both private and commercial purposes. New networked applications are steadily emerging, and some of them put new demands on the underlying network. To meet such demands, the concept of tunable services has evolved. The basic idea with tunability is to provide mechanisms that can offer different service levels, which are expressed through well-defined parameters that are specified on the basis of need during operation. Bit rate, throughput, delay, and jitter are all examples of common performance parameters in packet networks. These performance parameters have been studied extensively and are all aimed to express (and guarantee) a certain service level with respect to reliability and performance. Another increasingly important issue in computer networks, and especially in wireless networks, is security.

Security has traditionally been thought of as a system or network attribute that was the result of the joint endeavors of the designer, maintainer, and user, among others. Even though security would never reach a 100% level, the aim was to provide as much security as possible, given the actual boundary conditions. With the advent of, e.g., many low-power computing and communication devices it has become desirable to trade security against other system attributes, such as performance and power consumption. Thus, in many situations, tunable rather than maximal security is desirable. This approach is also appropriate for multimedia applications that require tuning the security level in order to maintain performance at levels that are acceptable to users. Thus, in many situations, tunable security, rather than maximal security, is desirable.

The lack of mechanisms by which system owners and users can request a specific level of security as a service in the system, however, often makes it impossible to offer security based on need. All users are instead offered similar services, regardless of whether it is the desired level of security. All users are thus forced to bear the costs of either too much or too little protection. Furthermore, unnecessarily high levels of security can make systems more difficult to control. This could, for example, imply that network management becomes harder, processor loads on servers and clients increase, smaller handheld devices are not able to encrypt or decrypt data in real time, etc. This results in applications with unnecessarily high security and extra costs to users and system owners.

The objective of this chapter is to provide a survey of security services explicitly designed to offer tunability. The key characteristics of tunable security services are identified, allowing a structured and consistent description of the surveyed services. Besides introducing the general scope of each service, we specify the available security configurations, describe the tuning process, identify any external parameters that may influence the choice of a security configuration, and describe the decision process for selecting a security configuration. Collectively, the surveyed services provide a broad illustration of different methods for and issues involved in designing tunable, or adaptive, security services in wireless network environments.

The remainder of the chapter is organized as follows. In section 15.2, the concept of security services is introduced. Section 15.3 discusses relevant security and performance metrics related to this work. In section 15.4, the key characteristics of tunable security services are highlighted. Section 15.5 surveys existing tunable security services suitable for wireless environments. In total, eight different tunable security services are described in detail. Additional related services are also briefly introduced. In section 15.6, a discussion on the merits of the emphasized services is provided. Section 15.7 gives some concluding remarks. A list of literature references is provided at the end of the chapter.

15.2 Security Services

Security is typically provided through one or more security services. Both protective and detective security services are common today. A protective security service tries to prevent attacks from succeeding in the first place, while a detective service tries to detect attack attempts or attacks that have already succeeded in circumventing protection services. Diversity of defense is achieved by combining two or more security services, and is a general security principle used to enhance security [4]. For example, firewalls are used to block suspicious network traffic to and from internal networks. As a complement to firewalls, intrusion detection systems are used to detect insider and outsider intrusion attempts as well as successful intrusions.

Protective security services are typically good at preventing attacks that are difficult to detect, such as eavesdropping and traffic analysis. Detective security services, on the other hand, are important in case of attacks that are difficult to prevent, such as masquerading or spoofing, replay, modification (i.e., substitution, insertion, and destruction), and denial of service (DoS).

The type of security service necessary depends on what is to be protected. In a presumably insecure networking environment, security services are needed to protect data that are transferred over the network. There are data protection services to achieve data confidentiality to prevent unauthorized information disclosure, data integrity to detect modification and replay, data authenticity to prove assurance about the origin of the data, and nonrepudiation to provide protection against denial of origin or delivery by one entity to any third party [47]. Security services are also used to protect system resources, such as access points in wireless networks, routers, and servers, against unauthorized access.

Security services are especially important in wireless networks. Such networks are typically more vulnerable to some attack types than are fixed networks. This is of course due to the fact that wireless networks are different in nature from fixed networks. They use air as the medium for data transfers, which makes them more vulnerable to eavesdropping attacks because of lack of physical protection. They often share the medium; thus, DoS attacks are easier to perform [20].

Wireless networks are also more vulnerable to other attack types that exist in fixed networks, such as replay attacks, man-in-the-middle attacks, connection hijacking, and brute-force attacks. Characteristics that make wireless networks and their connected devices more vulnerable include the infrastructure-less operation of ad hoc networks; entities playing multiple roles, i.e., terminal and system functionalities in ad hoc and mesh networks; changing topologies; transient relationships due to mobility; resource constraints of devices running on batteries; and the importance of bandwidth consumption.

Although many security services exist, they are typically used as static services. In this chapter, however, we focus on tunable security services. The following definition of tunable security services is used:

> **Definition 1:** A tunable security service is a service that has been explicitly designed to offer various security configurations that can be selected during system operation.

Note that in some other contexts the terms *adaptable*, *adjustable*, *dynamic*, and *scalable* are used as synonyms for tunable. The reason for using *tunable* instead of the other terms is that *tunable* is still the term that seems to be most acceptable to people in the area.

According to the definition given above, trade-offs between security and performance can be specified through the choice of available security configurations. Tunable security services will thus be appropriate in both current and future wireless networks with heterogeneous devices with varying computing resources. Such services are also suitable for multimedia applications that require tuning the security level to maintain performance at levels that are acceptable to the users. In the rest of this chapter focus will be on data protection services that are designed to provide tunability.

15.3 Security and Performance Metrics

The services surveyed in this chapter illustrate how security can be traded off against different performance metrics, such as latency and energy consumption. This implies that the security for a given security service must be measurable requiring appropriate security metrics. The performance for a given security service must also be measurable requiring appropriate performance metrics. Another important aspect related to tunable security is usability. However, in this chapter we focus on security and performance issues.

15.3.1 Security Metrics

To be able to specify a security level, security must be quantifiable. At present, there exist no good quantitative measures [17, 30] of confidentiality or integrity that can be used to specify a tunable security level, with the possible exception of the security evaluation criteria [6, 14, 49] and the two uncertainty measures entropy [44] and guesswork [32]. Although availability has been treated by the dependability community in a measurable way, this is not applicable as a security measure, since all faults introduced are assumed to be unintentional and stochastic. A formal definition of availability as defined in the dependability community is given in [18, 35].

Although it is often hard to find quantitative measures for security, it may still be possible to order available security configurations. In fact, ordering of security configurations in many situations is good enough as a simple, but still very useful security measure. A simple example of such an ordering of encryption algorithms could be as follows:

$$AES > DES > RC4. \tag{15.1}$$

The interpretation of this relation is that a message encrypted with a data encryption standard (DES) algorithm is harder to break than a corresponding message encrypted with Rivest cipher 4 (RC4). Similarly, the advanced encryption standard (AES) algorithm offers better protection than both the DES algorithm and the RC4 algorithm.

When combining different types of algorithms, like in the transport layer security (TLS) ciphersuites [2, 7], the task of ranking them becomes even harder. In this case, the importance of the different security services must be weighted. A ranking of TLS ciphersuites with respect to security strengths has been performed by the National Institute of Standards and Technology (NIST) [5]. In its ranking, the most important aspect is the strength of the key exchange mechanism, followed by the server authentication mechanism, and then come data confidentiality and data integrity service during the sessions.

TABLE 15.1 Security Configurations in IEEE 802.15.4

Security Configuration	Description
Null	No security
AES-CBC-MAC-32	32-bit MAC only
AES-CBC-MAC-64	64-bit MAC only
AES-CBC-MAC-128	128-bit MAC only
AES-CTR	Encryption only
AES-CCM-32	Encryption and 32-bit MAC
AES-CCM-64	Encryption and 64-bit MAC
AES-CCM-128	Encryption and 128-bit MAC

Note: CBC, cipher block chaining; MAC, message authentication code; CTR, counter (encryption) mode; CCM, counter mode with CBC-MAC.

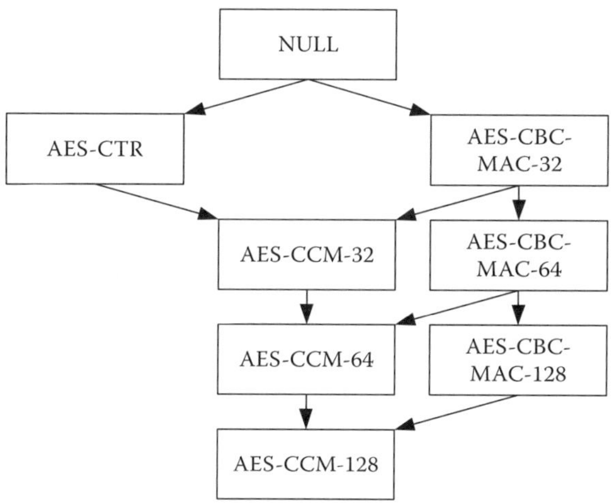

FIGURE 15.1 Partial order of the security configurations provided in IEEE 802.15.4. From Sastry and Wagner, 2004.

It may, however, not always be possible to fully order different security configurations with respect to the provided level of security. One such example was presented in [29] when describing the available security configurations in IEEE* 802.15.4, which is also referred to as Zigbee. The set of security configurations in IEEE 802.15.4 is summarized in Table 15.1 and described in [40]. The abbreviations used in the table are as follows: cipher block chaining (CBC), message authentication code (MAC), counter (encryption) mode (CTR), and counter mode with CBC-MAC (CCM). These are also used in the rest of the chapter. However, even though a total ordering is not possible, a partial ordering of the security configurations can still be defined. Such an ordering is depicted in Figure 15.1 for the eight different security configurations available in IEEE 802.15.4.

* IEEE is an abbreviation for Institute of Electrical and Electronics Engineers.

Based on the discussion above, it is evident that new ways of measuring security must be invented in order to be able to measure all aspects of security in communication networks. We also believe that new interpretations and definitions of security must be developed.

15.3.2 Performance Metrics

The implementation of security services adds an overhead to end-node devices and the network. The choice of which security configuration to use may have great impact on system performance. Latency, throughput, energy consumption, and utilization are all examples of performance metrics that can be relevant when selecting a set of security services, mechanisms, and algorithms for a particular communication session. These performance metrics are further discussed below.

Latency is a requirement that can be used as a guideline when selecting a particular security service and related algorithms. Given a communication protocol, end-node devices, and a specification of all needed supportive security services, e.g., certificate authority and public key server, an estimation of the expected latency can be calculated. Two types of latency can in fact be distinguished in such an environment: communication latency and computational latency. Communication latencies are those that are caused by the network, while computational latency is introduced by the nodes themselves. In both [28] and [51], latency is proposed as an explicit trade-off parameter to security. Both services are further described in section 15.5. Often there are very small differences in absolute terms between the latencies of a security service using different configurations. However, if the dominating part of the latency originates from specific parts of the system, and there are differences between the security configurations regarding the processing or space requirements for these bottlenecks, then it is meaningful to think about tuning to achieve different latency values. This was, for example, demonstrated in the performance analysis of Internet Protocol security (IPsec) in mobile IP version 6 (IPv6) scenarios in [9].

Similar to latency, throughput can also be used as a valuable parameter when selecting a security service with related parameters such as algorithms, modes, key lengths, etc. Throughput as an explicit trade-off parameter to security is proposed in [51], and it is also used as one of six trade-off parameters in [21]. These two services are further described in section 15.5. Throughput of a process in a system is the number of units served during a time interval. For the services described in section 15.5, throughput is measured by the number of frames or payload bits successfully transmitted on a network link, involving the processing and space overhead of the security services. This metric characterizes well the effect of security configurations for data link layer tunable security services. However, throughput of processes protected with different security configurations at a larger scale, or throughput of data streams protected at higher layers may also be an important metric in other service contexts. See [50] for a characterization of the general overheads imposed by IPsec regarding network throughput, or [12] for the throughput of Kerberos authentication transactions.

Energy consumption is yet another performance metric that is extremely valuable and important in today's mobile networks with resource-constrained, battery-driven handheld devices, or sensor network nodes. Security services that are based on cryptography typically require extensive processing power for security-related computations [22]. Such computations are highly energy consuming. By selecting the best combination of security service(s) and algorithms in a given situation, energy can be saved. Energy consumption as the dominating trade-off parameter to security is proposed in [23]. See section 15.5 for a detailed description of this service.

The utilization of a network node or a network link represents the fraction of time it is occupied by jobs, processes, or transmitted messages. An important aspect for service providers can be the utilization of servers by different security configurations. This may be important in case of node dimensioning, admission control, or the determination of possible performance gains in terms of CPU, memory, or network bandwidth utilization. A high utilization of a node or network link may also cause an increase in latency. Utilization is, for example, used as a performance metric in [9].

15.4 Categorization of Tunable Security Services

A scheme for categorizing tunable security services is introduced in this section. This characterization will later be used when analyzing and describing the surveyed services in order to allow a systematic analysis and comparison. The categorization includes the general scope of the service. It then highlights the core components that need to be considered in the design and analysis of a tunable security service: the available security configurations, the tuning process, external parameters that may influence the tuning, and the decision model used for the construction of the service. The characterization is partly based on the conceptual model for analyzing tunable security services proposed in [29].

15.4.1 General Scope

For each service described later in this chapter, the general scope of the service is identified. This includes a description of the type of security service offered as well as a description of the network type for which the service is designed. The trade-off provided by the service is also presented.

As described above, security is a complex concept, and many types of security services are available, depending on the security requirements that must be met. Most of the services covered in this chapter are related to the protection of data that are transferred over the network, focusing on data confidentiality, data integrity, and data authenticity.

Although all services described in the chapter are suitable for wireless networks, they differ in the assumptions made about the network. Some services are designed for very specific network types, such as IEEE 802.11. Other services make very few or no assumptions about the underlying network and can thus be applied over any network, wireless or fixed. Tunable services may utilize features and mechanisms at different layers within the communication stack (see Figure 15.2). A service that is independent of the communication stack is in this chapter referred to as an application layer tunable security service, while a service that makes use of features at the transport layer is referred to as a transport layer tunable security service. Similarly, services that utilize features at the

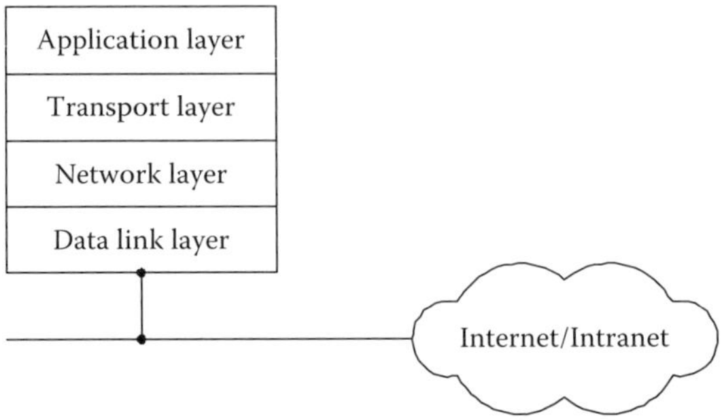

FIGURE 15.2 A typical communication stack used in IP-based networks.

network or data link layers are referred to as network and data link layer tunable security services, respectively.

The main idea behind a tunable security service is to offer a tunable trade-off between security and performance. Highlighting the offered trade-offs is therefore an important aspect in the general characterization of the surveyed services. As we will see later in the chapter, the trade-off offered by a service may be more or less well specified.

15.4.2 Security Configurations

As defined in definition 1, a tunable security service is a service that has been explicitly designed to offer various security configurations that can be selected during system operation. The possible security configurations that are offered by a service are thus a key characteristic in its description. In the remainder of the chapter, we will use S to denote the set of all possible security configurations provided by a service. Note that this set must contain at least two elements or configurations for any tunable security service. A service offering only one security configuration is by definition static.

The set of offered security configurations varies greatly between the surveyed services. However, the offered security configurations are typically constructed using two fundamentally different methods: algorithm selection and selective protection. Algorithm selection is when different security configurations are specified through the selection of a particular protection algorithm together with its related parameters. Selective protection means that different security configurations are created by varying the subset of data that is actually protected. A combination of algorithm selection and selective protection is also possible.

In order to design a meaningful tunable service, it must be possible to relate the strength of the different security configurations to each other. As discussed in section 15.3, the availability of quantifiable security metrics is limited and most of the surveyed services rely on an intuitive ordering of the strength of the different security configurations. In addition to an ordering based on the achieved security, an ordering of the available security configurations with respect to performance is also required.

15.4.3 Tuning Process

Another key component in the description of tunable security services is of course to identify the characteristics of the tuning process. This includes who the tuner is that has the possibility to influence the tuning, how the tuner preferences are expressed, and when the tuning may be performed.

The tuners provide the security- and performance-related preferences that influence the choice of a particular security configuration. Examples of possible tuners are end users, system administrators, and network operators, or any of their combinations. Applications may also act as tuners to guarantee prespecified security or performance requirements. Most services covered in this chapter allow the end users to act as tuners for the service.*

The tuner preferences may be expressed in many different ways and on different abstraction levels. Using a high abstraction level, a tuner may, for instance, make a selection between high, medium, and low security, as, for example, used in [23, 46]. An example of tuner preferences expressed at a low abstraction level is when the tuner identifies the exact data blocks that need protection; see, for example, [38]. The set of tuner preferences offered by a service is in this chapter represented by T. The elements in T express the tuner's preferences with regard to the trade-off between security and other performance parameters offered by the service. Note, however, that the trade-off is often expressed implicitly. A selection of high security in the example above says nothing about performance, but can be interpreted as a request to provide strong security even if this has negative effects on performance. A selection of low security indicates that it is important to select a security configuration that optimizes performance even if the security strength provided by the configuration is comparatively low.

According to our definition, a key characteristic of tunable security services is that they allow the security configurations to be selected at runtime. Still, the amount of tunability offered by a tunable security service may vary depending on when the tuning can be performed. We distinguish between two main classes of tunable services: per session and in session. In per-session tunable services, the security configuration is selected at the inception of a communication session and, once specified, remains fixed during the lifetime of the session. The highest degree of tunability is offered by in-session tunable services. In an in-session tunable service, the security configuration used can vary during the lifetime of a session.

15.4.4 Environment and Application Descriptors

The operation of a tunable security service may also be influenced by the current state of its environment or the state of the application. A tunable service designed to offer

* Handling selfishness, i.e., the overuse of common resources for one's own benefit by tuner entities, is an important aspect in tunable systems. The discussion here does not consider this topic, since the aim is to analyze and present their usefulness. However, in order to countermeasure selfish behavior, separate mechanisms that stimulate cooperation are necessary, such as reputation systems [3].

a trade-off between security and energy consumption, for instance, may use different security configurations depending on if it is operating in battery mode or on power. For example, Hager uses the remaining power of battery as an environmental parameter for context-aware and adaptive security services [11]. In this chapter, we let E denote the set of environment and application descriptors that may influence the selection of the security configuration for the service. Additional possible descriptors in E include type of equipment [28], network load [28], signal strength [11], application session length [11], and data size [23].

15.4.5 Decision Process

In a tunable security service, the security configuration to use is selected during system operation. The choice of a particular security configuration may be influenced by the tuner, through a set of tuner preferences, or by the current operating environment and application characteristics. A key question in the design of a tunable security service is thus to decide how to map a given combination of tuner preferences and environment and application descriptors to an appropriate security configuration. This decision process is captured by what we refer to as the tunable security (TS) function. The TS function was first introduced in [29]. Using the notation introduced above, the operation of a tunable security service can be formally expressed by its TS function as follows:

$$TS : T \times E \rightarrow S\,. \tag{15.2}$$

Recall from above that S is the set of available security configurations, T is the set specifying tuner preferences, and E is the set of relevant environment and application descriptors and constraints. Based on these three sets, the TS function illustrates the mapping from tuner preferences and environment and application characteristics to a particular security configuration.

As we will see later, the complexity of the TS function varies greatly between the surveyed services. For some services it provides a simple mapping from tuner preferences to security configurations, whereas other TS functions utilize more complex decision models that depend on a large number of input parameters. The last service surveyed in this chapter, i.e., adjustable authentication [21], provides an example of the latter. It uses the analytical hierarchy process (AHP) [39] as the decision model. In section 15.6, the complexity of the decision models used in all surveyed services is graded using a low, medium, and high scale.

To design an appropriate TS function, it is important to identify the security and performance implications of the different available security configurations under the various conditions in which the service operates. Since security and performance metrics are often imprecise, as was described above, the design of a TS function is more or less subjective. It provides one feasible security configuration for a given situation, but other feasible selections may also exist.

15.5 Survey of Tunable Security Services

This section contains a survey of existing tunable security services designed explicitly to offer various security levels. The characteristics of tunable security services introduced above are used in the description. For each service the scope of the service is defined, its security configurations are identified, the tuning process and any environmental or application descriptors that may influence the service are described, and the TS function is identified. The presentation below is organized into four subsections according to the dependence of the services on features at various communication layers, starting with the top-most layer first, i.e., application layer services. For each layer, two services are highlighted. In addition, some further tunable security services that are worth mentioning are shortly presented. Note that tunable security services have been proposed in other papers as well, e.g., [10, 36, 37, 48].

15.5.1 Application Layer Services

Most available tunable security services are implemented at the application layer. Two such services are described below using the criteria introduced in the previous section. The different services are introduced in complexity order, starting with the least complex. Some additional related services are also briefly mentioned.

15.5.1.1 Selective Encryption of MPEG Data

The first selected example of a tunable security service was proposed by Li et al. [26] and is aimed to protect video streams in Moving Picture Experts Group (MPEG) format. The suggested service provides different levels of data confidentiality. The level of protection provided can be traded off against the resulting encryption/decryption overhead, i.e., the computational latency. The overhead in turn influences the rate, and hence the quality, that can be supported in a real-time transfer of a protected MPEG video stream. The service makes no assumptions about the underlying network and can thus be applied in any networking environment. However, limiting the computational overhead of encryption/decryption is primarily useful for devices with limited computational capacity, making the service particularly well suited for wireless networks.

The service proposed by Li et al. is based on a selective encryption scheme. Their basic idea is to provide a protection hierarchy with three fixed encryption levels. The protection hierarchy is based on the three frame types used to encode MPEG video streams. In the proposed scheme, data encryption may be performed on intra (I) frames only, I and forward predicted (P) frames, or on all I, P, and bidirectional predicted (B) frames. Table 15.2 summaries the different alternatives in increasing order of security strengths. Based on the protection hierarchy, the set of possible security configurations is $\{S = I, IP, IPB\}$.

TABLE 15.2 Protection Hierarchy

Description	Abbreviation
Only I frames	*I*
I and P frames	*IP*
All I, P, and B frames	*IPB*

As described in [26], the security configuration to use is directly controlled by the user. Hence, the end user acts as the tuner. The set of tuner preferences is in this case equal to the set of possible security configurations: $T = \{I, IP, IPB\}$. Although not explicitly visible in T, the selection of a security configuration represents a trade-off between the level of protection achieved and the resulting encryption/decryption overhead. As mentioned above, the overhead may in turn influence the video quality that can be supported in real time. The security configuration to use is selected at the inception of a session and then remains fixed. Thus, tunability is provided on a per-session basis.

There is no explicit use of environment or application descriptors, since the security configuration is directly controlled by the user. We let E contain the empty set to represent this, thus $E = \{\emptyset\}$. As noted in the paper by Li et al., the level of protection or performance achieved will, however, be influenced by factors such as the size, content, and compression level of the video clip and the hardware used. These factors can, of course, be taken into account by the user when selecting the security configuration to use.

The TS function is in this case an obvious mapping from $T \times E$ to S as follows:

$$\begin{aligned} TS(T, E) &= S \\ TS(t, \emptyset) &= t, \end{aligned} \tag{15.3}$$

where $t \in T$. The simplicity of the TS function is an effect of the direct user control. It is up to the user to select an appropriate security configuration and to take the trade-off with performance into account for the given environment and application characteristics.

Based on a selective paradigm, many other application layer tunable encryption services have been proposed. A similar approach to the one described above was proposed by Meyer and Gadegast [33]. They proposed a tunable security solution for MPEG-1 multimedia data based on the layering structure of such video streams. Based on the layering structure, a protection hierarchy of five fixed encryption levels ranging from no encryption to full encryption is proposed. Similar ideas have also been used to protect compressed and uncompressed images [8] as well as compressed speech [43]. Support for tunable encryption has also been integrated directly into several multimedia applications, e.g., Nautilus [34] and Speak Freely [45].

15.5.1.2 Dynamic Authentication for High-Performance Network Applications

Schneck and Schwan [42] have designed and implemented a tunable authentication protocol for high-performance network applications such as teleconferencing or video streaming. It is called Authenticast. It provides data integrity, data origin authenticity, and nonrepudiation of origin in a tunable manner giving a trade-off between protection level, resource availability, and end-to-end delay (E2E). Authenticast provides three operation modes for data protection that are dynamically switched and tuned during sessions.

The first operation mode is called percentage-based authentication, referred to as $a(p)$ in the following. It enables the setting of the proportion, p, of data packets to be signed with a public key algorithm. The proportion p directly influences the security level, and provides a trade-off between security and CPU utilization:

$$p = \frac{\text{number of authenticated data packets}}{\text{total number of data packets}} \cdot 100. \quad (15.4)$$

The second operation mode is called delayed authentication, $b(n)$. It applies a buffer in order to collect a block of packets that are signed together once with a public key algorithm. The number of packets in one block, n, is variable. The trade-off is that by increasing n, the CPU utilization is decreased, but the E2E is increased. If only one packet in the group is incorrect, the whole group must be reprocessed. The security level is constant since every packet is always protected.

The third operation mode is called secret key connection, c. It applies a symmetric encryption scheme for per-packet data protection. The trade-off is that it reduces the computational overhead significantly, but does not provide a nonrepudiation service. It assumes preexisting shared keys between the client and the server.

The set of security configurations contains the above mentioned three operation modes with their input parameters. Thus, we have $S = \{a(p), b(n), c\}$.

The tuner is the user in Authenticast. Users specify a security level range by giving p_{min} and p_{max}, and an upper constraint for the E2E by specifying E2E_{max}. Thus, we get $T = \text{E2E}_{max} \times p_{min} \times p_{max}$.

The environmental and application characteristics that are taken into account in the decision process are the following: the packet drop events (PD), the current E2E between the client and the server, and the performance disparity* (DELTA) between the client and the server. From this it follows that $E = PD \times E2E \times DELTA$.

Exactly how to switch between different security levels is not clear from [42]. Authenticast provides a so-called adaptable authentication heuristic. It says that the initial security configuration of the protocol is full per-packet verification, with delayed authentication, $b(1)$. When the performance disparity overruns a LIMIT value between the client and the server, i.e., one or both cannot maintain performance with current security processing requests, they increase n to 10. This is performed only if the upper constraint on the E2E is not overrun. In case of an exceeded E2E_{max}, the protocol switches to percentage-based authentication with $p \leq 1/n$, i.e., $a(1/n)$. If packets are dropped due to overfilled receiver (or sender) buffers, Authenticast will decrease the security level using $a(p)$. A new level will, however, always be within the prespecified range of p. When the load has decreased, the initial security configuration may later be selected again. Either $b(n)$ with decreasing n or $a(p)$ with increasing p can be used to come back to the initial configuration $b(1)$.

A TS function similar to the adaptable authentication heuristic is illustrated in Figure 15.3. Note that the secret key connection mode is not included in the TS function above since it was not considered by the authors in the adaptive authentication heuristic. To consider this mode, the algorithm would have to be slightly revised.

* Performance disparity is not clearly defined in [41, 42]. It is related to the server and client relative processing speeds or their current utilization.

```
init configuration to b(n) where n=1
while inspecting E2E, delta and PD events

    # aborting transmission due to QoS violation
    if DELTA > LIMIT and E2E > E2E_max (there is no  possible 'b(n)'),
      if DELTA > LIMIT and p < p_min (there is no possible 'a(p)'),
         then ABORT transmission

    #  normal operation with 'b(n)'
    if DELTA > LIMIT, then use b(n) where n=10
    if DELTA < LIMIT, then use b(n) where n is decreased until it reaches 1

    # when E2E constraint is overrun then switch to 'a(p)'
    if E2E > E2E_max, then use a(p) where p=1/n
    if E2E < E2E_max, then use b(n) where n=1/p

    # in case of packet drops switch to 'a(p)'
    if PD event, then use a(p) where p=1/n
    if nomore PD event, then use a(p) where p is increased until it reaches 1
```

FIGURE 15.3 A possible TS function for adaptive authentication.

15.5.2 Transport Layer Services

Very few tunable security services take advantage of features at the transport layer. In fact, we are only aware of two such services. These are described below. We again describe the least complex service first.

15.5.2.1 Selective Security for TLS

The first tunable security service utilizing features at the transport layer was proposed by Portmann and Seneviratne [38]. The main idea proposed in the paper is to extend the TLS protocol [2, 7] with mechanisms for selective protection of data. The authors argue that only sensitive data parts need to be protected using encryption and message authentication codes, whereas nonsensitive parts are left unprotected. Leaving parts of the data unprotected is motivated by the need to integrate adaptation proxies with end-to-end security and also reduces computational overhead. Although content adaptation may be particularly useful in heterogeneous environments, such as wireless networks, the service can be used by any Transmission Control Protocol (TCP) or Stream Control Transfer Protocol (SCTP) application since it extends TLS. Data confidentiality, data integrity, and data authenticity are all provided, and the service could also be applied for other purposes.

To achieve selective protection, a new record layer type called `Cleartext Application Data` is proposed. With this, data can be transferred unprotected within a TLS connection. To utilize this new feature, a new application programming interface (API) method is needed, which is referred to as `write_clear()`. This new method is added to the existing API and is intended to be used when sending nonsensitive data. When sending sensitive data, the normal `write()` method is used. To be able to handle data sent with the two write methods accordingly on the receiver side, a new method `read_select()` is added to the TLS API. This method returns a value that informs the application whether or not the received data were protected.

The various security configurations that can be selected in this case can be described as a fraction as follows:

$$S = \frac{\text{number of encrypted TLS records}}{\text{total number of TLS records}}. \tag{15.5}$$

Note, however, that a higher value on S is not necessarily more secure than a lower one. This is due to the fact that some blocks may contain more sensitive information than others. Thus, the selection of the blocks also influences the level of security achieved. This is further discussed in [27]. Furthermore, Lundin et al. [31] present an extensive discussion on the achieved security level when selective encryption is used.

The selection of which data to protect can be made dynamically within a session, and it is controlled by the application. Hence, the application acts as the tuner, and the set of tuner preferences is in this case equal to the set of possible security configurations: $T = S$. Furthermore, since the security configuration is directly controlled by the user, we have $E = \{\emptyset\}$. This implies that the TS function is a direct mapping from $T \times E$ to S as follows:

$$TS(t,\emptyset) = t, \tag{15.6}$$

where $t \in T$. From its characteristics, this service is similar to the application layer service that provided selective encryption of MPEG data.

15.5.2.2 Tunable Network Access Control in IEEE 802.11i

The next presented tunable security service was presented in [28] and is based on the IEEE 802.11i standard [16] using open system authentication. This means that the Extensible Authentication Protocol (EAP) [1] is used for network access control. When EAP is the chosen protocol, an authentication method must be selected. Some examples of common EAP methods are: EAP-MD5,* EAP-TLS, EAP-TTLS,† and PEAP.‡ The assumption in this tunable security service is that EAP-TLS is used. This implies that TLS is used on top of EAP. With EAP-TLS a mutual authentication process can be performed before a station is allowed access to the wireless local area network (WLAN). The tunable service illustrates how latency can be traded against the strength of the authentication process in this context.

In TLS, a rich set of mutual authentication variants, which are based on different so-called ciphersuites, is provided. For simplicity, in this chapter we only consider the five ciphersuites listed in Table 15.3 (in descending order of security strength according to [5]). The notation in Table 15.3 is used in the TLS standard to specify combinations of security algorithms. For example, the first ciphersuite given above specifies that the ephemeral Diffie-Hellman (DHE) key exchange mechanism is used for key exchange, the digital signature standard (DSS) is used for signing and verification of certificates,

* MD5 is an abbreviation for Message Digest Algorithm 5.

† TTLS is an abbreviation for Tunneled TLS.

‡ PEAP is an abbreviation for Protected EAP.

TABLE 15.3 Considered TLS Ciphersuites

Ciphersuite	Abbreviation
TLS_DHE_DSS_WITH_AES_128_CBC_SHA	*TLS_A*
TLS_DHE_RSA_WITH_AES_128_CBC_SHA	*TLS_B*
TLS_RSA_WITH_AES_128_CBC_SHA	*TLS_C*
TLS_DH_DSS_WITH_AES_128_CBC_SHA	*TLS_D*
TLS_DH_RSA_WITH_AES_128_CBC_SHA	*TLS_E*

AES with a 128-bit key running in CBC mode is used for data encryption and decryption, and finally, the secure hash algorithm (SHA) is used as a MAC.

Note, however, that when TLS is used only for network access control, the selected data encryption and MAC algorithms have no effect during the authentication process. Based on the ciphersuites listed in Table 15.3, the set of provided security configurations is $S = \{TLS\_A, TLS\_B, TLS\_C, TLS\_D, TLS\_E\}$.

As mentioned above, latency is used as the performance indicator for the service, and five different tuner preferences have been defined: $T = \{SO, SM, BA, LM, LO\}$. *SO* denotes that only security is important. *SM* denotes that security is most important, but the importance of latency is not negligible. *BA*, which is an abbreviation for balanced, indicates that security and latency are equally important. *LM* specifies that latency is most important, but security must be taken into consideration when selecting a security configuration. Finally, *LO* implies that latency is the only important issue. The user acts as the tuner, and the tuning is performed on a per-session basis.

The set of environmental and application descriptors, *E*, is characterized by equipment type (*ET*) and network load (*NL*) as $E = ET \times NL$. *ET* is defined to have two elements: $ET = \{LE, HE\}$, where *LE* and *HE* denote low-end and high-end devices, respectively.* The network load considered in the paper corresponds to the channel load of an IEEE 802.11b access network. *NL* consists of three elements: $NL = \{MO, HI, SA\}$, where *MO*, *HI*, and *SA* are abbreviations for moderate, high, and saturated network load, respectively. The channel load is considered to be moderate at less than 45% utilization, high for 45–70% utilization, and saturated above 70% utilization.

The TS function defines the security configuration to select for each one of the pairs of possible user preferences and environmental descriptors. Its definition requires that the relation between the security configurations and the resulting latency is established and related to the achieved security and the user preferences. The latency estimation is presented in [28], where a more exact definition of the elements of *E* can also be found. In Tables 15.4 and 15.5, a summary of the total estimated latency, i.e., both network and computational latency, and the relative performance cost for the different ciphersuites considered here is presented.

Based on the data in Tables 15.4 and 15.5, the TS function is defined as follows, where | denotes a "logical or" and * denotes all possible alternatives. The input parameters of the

* For the measurements presented in paper [28], the low-end device was a 600 MHz Genuine Intel Celeron processor computer with 64 MB RAM, and the high-end device was a Pentium (R) 4 CPU, 3 GHz, 512 MB RAM computer.

TABLE 15.4 Total Estimated Latency in Seconds for the Low-End Device (and Relative Performance Cost) of the Considered Ciphersuites

Ciphersuite	Moderate (MO)	High (HI)	Saturated (SA)
TLS_A	5.69E-02 (100%)	7.90E-02 (100%)	1.89E+00 (98.8%)
TLS_B	4.41E-02 (77.5%)	6.63E-02 (83.9%)	1.89E+00 (98.8%)
TLS_C	5.63E-02 (99.0%)	7.87E-02 (99.6%)	1.92E+00 (100%)
TLS_D	4.28E-02 (75.3%)	6.34E-02 (80.2%)	1.74E+00 (90.7%)
TLS_E	3.04E-02 (53.4%)	5.07E-02 (64.2%)	1.74E+00 (90.8%)

TABLE 15.5 Total Estimated Latency in Seconds for the High-End Device (and Relative Performance Cost) of the Considered Ciphersuites

Ciphersuite	Moderate (MO)	High (HI)	Saturated (SA)
TLS_A	3.32E-02 (100%)	5.54E-02 (100%)	1.87E+00 (98.8%)
TLS_B	2.78E-02 (83.6%)	5.00E-02 (90.3%)	1.88E+00 (99.2%)
TLS_C	3.25E-02 (97.7%)	5.49E-02 (99.1%)	1.89E+00 (100%)
TLS_D	2.66E-02 (80.0%)	4.72E-02 (85.2%)	1.72E+00 (90.9%)
TLS_E	2.15E-02 (64.6%)	4.18E-02 (75.5%)	1.73E+00 (91.4%)

TS function are the tuner preference, the equipment type, and the current network load value (separated with commas):

$$\begin{aligned}
&TS(SO,(*,*)) = TLS_A \\
&TS(SM \mid BA,(*,MO \mid HI)) = TLS_B \\
&TS(SM \mid BA,(*,SA)) = TLS_A \\
&TS(LM,(*,*)) = TLS_D \\
&TS(LO,(*,MO \mid HI)) = TLS_E \\
&TS(LO,(*,SA)) = TLS_D.
\end{aligned} \tag{15.7}$$

Note that *TLS_C* is not the preferred security configuration in any of the cases above. This is due to the fact that this ciphersuite is neither especially secure nor efficient. Also note the TS function given above only represents one reasonable mapping. The definition of the TS function thus represents an important design decision in the construction of a tunable security service.

15.5.3 Network Layer Services

Network layer tunable security services are a little bit more common than services on the transport layer. Two network layer services are described below. In common for them are that they both make use of the IPsec protocol [24].

15.5.3.1 IPsec Modulation

The fifth service investigated was described by Spyropoulou et al. [46]. They propose a model for variant security based on user preferences. The main idea with the concept presented in the paper is to improve resource utilization, while at the same time maintaining security at an acceptable level. Different security levels can therefore be selected as a response to user requests. As a proof of concept, the proposed model is applied to create a tunable service at the network layer on top of IPsec. As the created service is based on IPsec, it provides data confidentiality, data integrity, and data authenticity and is applicable in any IP network.

A rich set of security configurations (S) is available in the service, since IPsec provides many different configurations. In the service, a user may select among three different security preferences: low (LO), medium (ME), and high (HI). Hence, $T = \{LO, ME, HI\}$ and the user is the tuner.

Their tunable service takes into account the application as well as the operational mode. Consequently, E is characterized by two components, application (AP) and operational mode (OM), such that $E = AP \times OM$. AP is the set of considered applications. In the paper, three different example applications are used: $AP = \{telnet, finger, ping\}$.

The system can operate in three modes. Normal operation (NO) mode is the default and initial mode. A system can enter impacted (IM) mode when, for example, it is overwhelmed with requests. However, not all services might be offered in this mode. Finally, in emergency (EM) mode strong security is always applied with very few configuration options. Hence, $OM = \{NO, IM, EM\}$. As a result of the system mode changing, the selected security configuration may change during a communication session.

Since the paper by Spyropoulou et al. focuses on the operation in IM mode, we only illustrate the TS function for this mode. A mapping to S for the different applications and user preferences when the system is operating in IM mode is presented in Table 15.6. From the table it is evident that five different security configurations are used in this particular mode: no IPsec processing, encapsulating security payload (ESP) processing with DES, ESP processing with triple DES (3DES), authentication header (AH) processing with hash function MAC (HMAC)-MD5, and AH processing with HMAC-SHA.

15.5.3.2 IPsec/IKE Adaptive Security

In [51], Yogender and Ali have developed a tunable security model based on IPsec and the Internet key exchange (IKE) [13] protocol. The proposed model allows IPsec to switch between various security levels based on defined quality-of-service (QoS) parameters, such as throughput and delay. Similarly to the previous example, the proposed model

TABLE 15.6 TS Function in Impacted Mode for IPsec Modulation

	Tuner Preferences (T)		
Application	Low (LO)	Medium (ME)	High (HI)
Telnet	No IPsec	ESP with DES	ESP with 3DES
Finger	No IPsec	AH with HMAC-MD5	AH with HMAC-SHA
Ping	No IPsec	No IPsec	No IPsec

TABLE 15.7 Defined Security Levels

Policies	Security Levels (S)						
	L1	L2	L3	L4	L5	L6	L7
Encryption algorithm (key length in bits)	DES (56)	Blowfish (8)	Blowfish (448)	IDEA[a] (128)	3DES (108)	3DES (168)	AES (256)
Integrity algorithm	MD5	MD5	SHA-1	SHA-1	SHA-1	SHA-1	SHA-1
IKE key refresh times[b]	NULL	3,000	2,400	1,600	1,400	800	300
IPsec key refresh times	NULL	2,400	1,600	1,000	800	300	100

[a] IDEA is an abbreviation for international data encryption algorithm.
[b] Refresh times are expressed in seconds.

provides all of data confidentiality, integrity, and authenticity and is applicable in any IP network, as it is based on IPsec.

Out of many possible IPsec configurations, seven different security levels are defined in the paper. These are summarized in Table 15.7. Based on the seven defined levels, the set of security configurations can be defined as $S = \{L1, L2, L3, L4, L5, L6, L7\}$, where L is an abbreviation for security level. With respect to security strength, the levels are in increasing order.

The tuning process for this service follows. The tuner is the user. The user preferences are expressed through a QoS requirement. The considered QoS parameters are average throughput, average session delay, and number of concurrent File Transfer Protocol (FTP) sessions that can be sustained. The QoS constraint may take into consideration one or a combination of the defined parameters, although the examples in the paper only consider one parameter at a time. For simplicity, we only consider one of the examples from the paper, the one using minimum average throughput (MAT) as the QoS constraint, since this example is the one most clearly explained. For this example, $T = \{MAT\}$. A value of 2,000 kbps is set for the MAT in the example.

The proposed adaptive switching model evaluates the achieved QoS at regular intervals and switches to the next, presumably more efficient, security level if the QoS constraint is not met. The achieved QoS varies over time and can be considered an environmental descriptor. Thus, for our example $E = \{AT\}$, where AT stands for the average achieved throughput. The TS function, which is invoked at regular intervals, can for this example then be described as follows:

$$TS(MAT, AT) = \begin{cases} next(L) & \text{if } AT < MAT \\ L & \text{otherwise} \end{cases}, \tag{15.8}$$

where L represents the current security level and $next(L)$ is a function that returns the next security level to switch to. How the next security level to switch to is determined, however, is not quite clear from the paper. In the example given, the initial security level is set to $L6$, which is the security level that has the least throughput performance of all regarded security levels. According to simulation results in the paper, the different security levels ($L1$, $L2$,…, $L7$) can be ordered, in increasing order of average throughput, as follows: $L6$, $L5$, $L1$, $L4$, $L3$, $L2$, $L7$.

Having made explicit the components of the tunable security service and the performance of the security configurations, we can see that a superior TS function for this case would be *TS(MAT,AT)* = *L7* since security level *L7* provides both the best security and the best throughput performance. This illustrates that a more secure configuration may not always have worse performance, and in this particular case the best choice would actually be a static configuration.

15.5.4 Data Link Layer Services

Two different tunable security services are described in this subsection. The first described service is specifically designed for IEEE 802.11–based wireless local area networks. The other service is a more generic service and can be used in conjunction with different data link protocols.

15.5.4.1 Tunable Packet Protection in IEEE 802.11

In [23], a tunable security model (TSM) that minimizes energy consumption while keeping a security level that satisfies the user's requirements is presented by Keeratiwintakorn and Krishnamurthy. The target environment is wireless networks with battery-driven devices. The main idea with TSM is to provide different protection mechanisms and configurations at a packet level. Proof of concept is illustrated through the IEEE 802.11 WLAN standard [15], applying the model at the data link layer for providing data confidentiality and integrity.

In IEEE 802.11, twenty-six packet* types are defined. For each packet type, the authors propose the type of protection service, i.e., message authentication or encryption, that should be used. In the paper, they propose that all packet types should be protected by a MAC. They also propose that the four different packet types used for data transfer should use data encryption in addition to the MAC.

An abstract representation of the security configurations, *S*, is given by the needed data protection time. From a desired protection time, an appropriate encryption key size (KS) and MAC size can be calculated. The formula for calculating KS (in bits) was inspired by Lenstra and Verheul [25] and is as follows:

$$KS = 56 + (y + y' - 1982)\left(\frac{12}{m} + \frac{1}{b}\right), \tag{15.9}$$

where y is the number of years needed for protection. The current year (e.g., 2008) is denoted y'. The average number of months that the CPU and memory performance are doubled is denoted by m, which is assumed to be 18 months. Finally, b is the number of years that the available attack budget is doubled. In the paper, $b = 10$ is used. A MAC size that is twice as long as the KS is used in order to provide a protection level of message authentication similar to that of encryption. In addition, the paper by Keeratiwintakorn and Krishnamurthy also discusses the number of operational rounds necessary for protecting data. However, the number of rounds is not taken into consideration here.

* Normally called frame, but here we use the authors' notation.

TABLE 15.8 Considered AES-Based Security Configurations

Description	Abbreviation
AES with a 192-bit key size and SHA with a 384-bit MAC size	AES_192_SHA_384
AES with a 128-bit key size and SHA with a 256-bit MAC size	AES_128_SHA_256
AES with a 128-bit key size and SHA with a 160-bit MAC size	AES_128_SHA_160
SHA with a 256-bit MAC	SHA_256
SHA with a 160-bit MAC	SHA_160

TABLE 15.9 Key and MAC Sizes in Bits When Using AES and SHA Year 2006

	Years of Protection								
Type	2	3	5	10	20	25	40	50	100
KS	128	128	128	128	128	128	128	128	192
MAC	160	160	160	256	256	256	256	256	384

At a low level, three different cipher schemes for providing security services are used. One is based on AES for encryption and SHA as MAC. The other is based on Rivest cipher 5 (RC5) for encryption and SHA as MAC. The third one uses both AES and RC5 with either CBC-MAC or SHA. In the latter case, AES is used for packets whose size is less than 100 bytes, and RC5 otherwise. Only the AES-based cipher schemes will be considered further here. The sets of considered low-level security configurations, expressed by algorithm type and key and MAC size, are summarized in Table 15.8.

A mapping from the abstract representation of security configurations, given in years of protection, to low-level key and MAC sizes is listed in Table 15.9 when using AES and SHA in year 2006 (i.e., $y' = 2006$). The KSs have been calculated using formula (15.9) and the MAC sizes, as argued above, should be twice as long as the KSs. We have furthermore assumed that only the standardized KSs for AES are used, i.e., 128, 192, and 256 bits. Additionally, SHA can produce variable MAC sizes of 160, 256, 384, and 512 bits. We always round up to the next available key or MAC size. This implies that if the formula gives a KS of 94 bits, for example, then 128 bits is used.

The tuning process is the following. The tuner is the user. Tuner preferences are expressed through a set of well-defined security levels. Security levels are defined based on the number of years the data must be protected. For example, a high security level may represent 100 years of data protection. In the paper, three different levels are defined: low (*LO*), medium (*ME*), and high (*HI*). Thus, $T = \{LO, ME, HI\}$.

The selection of which security configuration to use, i.e., which protection time is needed, is influenced by the packet type (*PT*). Hence, in our simplified scenario, $E = PT$. When the cipher scheme that uses both AES and RC5 is considered, the packet size also acts as an environmental descriptor influencing the selection.

The TS function can now be described. Table 15.10 shows the TS function using an abstract representation of the security configurations. From Table 15.9, it follows that

TABLE 15.10 TS Function for Tunable Packet Protection in IEEE 802.11

			Tuner Preferences (T)[a]		
Type	Packet	Service	LO	ME	HI
Management	Assoc Req	(1)[b]	5	10	20
	Assoc Resp	(1)	5	10	20
	Reassoc Req	(1)	5	10	20
	Reassoc Resp	(1)	5	10	20
	Probe Req	(1)	2	3	5
	Probe Resp	(1)	2	3	5
	Beacon	(1)	2	3	5
	ATIM	(1)	2	3	5
	Disassoc	(1)	2	3	5
	Authen	(1)	10	20	40
	Deauthen	(1)	2	3	5
Control	Action	(1)	2	3	5
	PS-Poll	(1)	2	3	5
	RTS	(1)	2	3	5
	CTS	(1)	2	3	5
	Ack	(1)	2	3	5
	CF-End	(1)	2	3	5
	CF-End+Ack	(1)	2	3	5
Data	Data	(1), (2)[c]	25	50	100
	Data+CF-Ack	(1), (2)	25	50	100
	Data+CF-Poll	(1), (2)	25	50	100
	Data+CF-Ack/Poll	(1), (2)	25	50	100
	Null	(1)	5	10	20
	Null+CF-Ack	(1)	5	10	20
	Null+CF-Poll	(1)	5	10	20
	Null+CF-Ack/Poll	(1)	5	10	20

[a] The columns indicate the number of years of protection applied for each packet type for a given tuner preference.

[b] (1) denotes a message authentication service.

[c] (2) denotes a message encryption service.

the corresponding TS function using the low-level security configurations is as follows, where | denotes a "logical or" and * denotes all possible alternatives:

$$
\begin{aligned}
&TS(LO, Assoc* \mid Reassoc*) = SHA\_160 \\
&TS(ME \mid HI, Assoc* \mid Reassoc*) = SHA\_256 \\
&TS(*, Probe* \mid Beacon \mid ATIM \mid Disassoc) = SHA\_160 \\
&TS(*, Authen) = SHA\_256 \\
&TS(*, Deauthen) = SHA\_160
\end{aligned}
\tag{15.10}
$$

$$TS(*, Action \mid PS-poll \mid RTS \mid CTS \mid Ack) = SHA\_160$$

$$TS(*, CF-End \mid CF-End+Ack) = SHA\_160$$

$$TS(LO \mid ME, Data*) = AES\_128\_SHA\_256$$

$$TS(HI, Data*) = AES\_192\_SHA\_384$$

$$TS(LO, Null*) = SHA\_160$$

$$TS(ME \mid HI, Null*) = SHA\_256.$$

As we can see, the TS function is for this service quite complex, even though we have considered a simplified scenario.

15.5.4.2 Adjustable Authentication

In [19], Johnson describes two light-weight authentication protocols suited for wireless devices with limited resources. In both schemes, authentication of data is performed on a per-packet basis at the data link layer. To each packet a light-weight authentication code (LAC) is added and is used as authenticator. The number of bits used for authentication is typically small. In the first protocol, i.e., the statistical 1-bit light-weight authentication (SOLA) protocol, only 1 bit per packet is used for authentication. The SOLA protocol was later extended to allow *k* bits for authentication. Both protocols are applicable to standards such as IEEE 802.11 and IEEE 802.15.1 (also referred to as Bluetooth).

Neither SOLA nor its extension is tunable according to the definition given in section 15.2. SOLA is, however, a possible alternative in a decision system for adequate authentication proposed in [21], and it is in this context that we consider it here. The decision system used in the paper is based on AHP [39]. In the case study, the authors use two authentication alternatives, which are referred to as light-weight authentication (LA), e.g., SOLA or its extension, and strong authentication (SA), e.g., HMAC using MD5 or SHA-1, respectively. This implies that $S = \{LA, SA\}$.

The selection of the most appropriate authentication method in *S* is derived from six different parameters collected from the operating environment and the end user. The parameters are as follows:

- Threat level (TL)
- Resources (RE)
- Position (PO)
- Content (CO)
- Throughput (TH)
- User assessment (UA)

For each parameter above, a value in the range 1–9 is assigned based on the value of the parameter for the given situation. Johnson et al. [21] do not discuss exactly how the collection and translation from measured values to the 1–9 scale is performed. However, higher-input values lead to higher values on the 1–9 scale. The UA parameter corresponds to the tuner preferences with the user acting as the tuner, and the rest of the parameters are environmental descriptors. Thus, $T = \{1, 2, \ldots, 9\}$. E, on the other hand, is much more complex and consists of all the other parameters. Thus, $E = TL \times RE \times PO \times CO \times TH$, where $\mathrm{TL} = RE = PO = CO = TH = \{1, 2, \ldots, 9\}$.

Since the decision model used in the paper is a so-called multicriteria decision-making model that is based on many parameters, a simple mapping to a preferred authentication method does not exist. The TS function is complex and is characterized by the AHP engine and all six input parameters.

An approach similar to the work presented above has also been proposed by Hager [11]. However, he has studied methods for selecting an appropriate encryption algorithm for specific wireless network applications in order to improve the efficiency of data encryption. A set of predefined operational parameters such as type of equipment, network topology, and communication characteristics is used to select an encryption algorithm. Three different decision models are investigated for the decision process. The first one is the classical AHP, as described above, the second is a deterministic model that makes decisions using predefined thresholds based on analytical results, and the third is a combination of the first two. The latter is a hybrid model referred to as the modified AHP model.

15.6 Discussion

Eight different tunable security services were presented in the previous section. The characteristics of the services are summarized in Tables 15.11 to 15.14, for the application layer services, transport layer services, network layer services, and data link layer services, respectively. The general description included in the tables provides the type of security service provided, the network type targeted by the service, and the performance parameter used in the trade-off with security. The method used to create different security configurations is also indicated. Furthermore, information on who the tuner is and the amount of tunability, as well as any environment and application descriptors for the service, is described. Finally, the complexity of the decision model is indicated. These aspects together give a fairly good picture of the features provided by the surveyed

TABLE 15.11 Summary of Application Layer Services

Aspect	Selective Encryption of MPEG Data	Authenticast
Security domain[a]	C	AI
Network type	General	General
Performance measures	Real-time video quality	End-to-end delay, resource utilization
Security configuration method	Selective protection	Combined
Tuner	End user	End user
Tunability	Per session	In session
Environment and application descriptors	—	End-to-end delay, packet loss client/server performance disparity
Decision model complexity	Low	Medium

[a] A, C, and I denote data authenticity, data confidentiality, and data integrity, respectively.

TABLE 15.12 Summary of Transport Layer Services

Aspect	Selective Security for TLS	Tunable Network Access Control in IEEE 802.11i
Security domain[a]	ACI	Network access control
Network type	TCP/SCTP-based network	IEEE 802.11i using EAP-TLS
Performance measures	Possibility for content adaptation, computational latency	Latency
Security configuration method	Selective protection	Algorithm selection
Tuner	Application	End user
Tunability	In session	Per session
Environment and application descriptors	—	Device type, network load
Decision model complexity	Low	Medium

[a] A, C, and I denote data authenticity, data confidentiality, and data integrity, respectively.

TABLE 15.13 Summary of Network Layer Services

Aspect	IPsec Modulation	IPsec/IKE Adaptive Security
Security domain[a]	ACI	ACI
Network type	IP networks	IP networks
Performance measures	Resource utilization	Throughput, session delay, number of concurrent FTP sessions
Security configuration method	Combined	Algorithm selection
Tuner	End user	End user
Tunability	In session	In session
Environmental and application descriptors	Application type, network mode	Throughput, session delay, number of concurrent FTP sessions
Decision model complexity	Medium	Medium

[a] A, C, and I denote data authenticity, data confidentiality, and data integrity, respectively.

TABLE 15.14 Summary of Data Link Layer Services

Aspect	Tunable Packet Protection in IEEE 802.11	Adjustable Authentication
Security domain[a]	CI	A
Network type	IEEE 802.11	Wireless networks
Performance measures	Energy consumption	Resource utilization
Security configuration method	Combined	Algorithm selection
Tuner	End user	End user
Tunability	In session	In session
Environmental and application descriptors	Packet type, packet size	Resources, position, content threat level, throughput
Decision model complexity	High	High

[a] A, C, and I denote data authenticity, data confidentiality, and data integrity, respectively.

services, even though it is not an exhaustive list of tunable security characteristics. As illustrated in the tables, a wide variety of tunable security services exist. The services target different security domains, aim to increase different performance metrics by tuning security, and take into account a range of different environment and application descriptors.

Existing tunable security services have successfully been implemented at the application, transport, network, and data link layers. A natural question then emerges: Where is it most suitable to implement tunable security services? As can be seen in Tables 15.11 through 15.14, it seems that the more lower-layer features are used, the more complex the service is. This is, of course, due to the fact that such services typically make use of multiple input parameters for selection of a certain security configuration. The use of multiple inputs strives to select the most appropriate configuration while taking all relevant aspects into account. In his PhD thesis, Johnson [20] coined the term *always best security* (ABS) to denote this. Note that with ABS it is the most suitable security solution that is selected, not the strongest one with respect to security. However, as the complexity increases, selecting the appropriate security configuration gets increasingly hard. It also gets increasingly difficult for the tuner to understand the implications of the choices available in the tuner preferences. Furthermore, the performance and security metrics available are often rough measures, making it difficult to use complex combinations of them. To fully provide the ABS concept, new and more fine-grained security and performance metrics are needed.

Eight different tunable security services are summarized in the tables. These services seem to be most adequate and valuable in resource-constrained environments such as wireless and mobile networks. When selecting a security solution for such an environment, a tunable security service should not be blindly selected. Instead, in all particular situations we argue for a careful evaluation of all possible solutions, i.e., both tunable and static. If a static service fulfills current as well as future requirements with respect to both security and performance, it is preferable to select that service. This is motivated by the fact that a tunable security service is always more costly to develop, verify for correctness, install and configure, and manage. As was also illustrated by the sixth service described, IPsec/IKE adaptive security, a more secure configuration may not always have worse performance. Hence, it is important to carefully evaluate the relation between security and performance for all considered solutions.

As mentioned, the level of security and performance for a service may not be fully quantifiable. In fact, many of the surveyed services have some subjective parts. The definition of security levels, the choice of weights for input parameters in the decision process, and the definition of their levels induce the existence of several possible solutions for the same situation.

15.7 Concluding Remarks

Adaptivity is an important tool to improve performance in wireless networks. In this chapter we have seen how this concept can be applied to security, through the provision of tunable security services. A tunable security service allows the security level to be dynamically configured and traded against other performance parameters. With an increasing demand for security and a trend toward increased heterogeneity in both network devices and access networks, it is reasonable to assume that the need for tunable security services will further increase in the future. Furthermore, tunable security services are useful in heterogeneous environments, as they make it sufficient to use the same service everywhere, since the service may self-adapt to the actual environment.

Eight different tunable security services have been analyzed in this chapter. The key characteristics of tunable security services were identified and used as a basis for the analysis. As was illustrated by the survey, tunable security can be implemented in many different ways, and the target environment and exact aim of the services differ. Still, the core building blocks needed to construct a tunable security service could be identified for all services. Collectively, the services also illustrate the need for appropriate performance and security metrics as a prerequisite for the construction of tunable security services. This is an area that still requires additional research.

An interesting continuation of this work would be to study tunable security services designed specifically for sensor networks. Devices in such environments typically have limited computational as well as battery power. Furthermore, the provided bandwidth is typically very limited. For these reasons, we believe that tunable security services would be even more suitable in sensor networks than in other wireless network settings.

Acknowledgment

This research was carried out as part of the work on the development of dynamic security services in heterogeneous wireless networks within the Network of Excellence in Wireless Communications (NEWCOM). The work performed by the first two authors was also supported by grants from the Knowledge Foundations of Sweden with TietoEnator and Ericsson as industrial partners.

References

[1] B. Aboba, L. Blunk, J. Vollbrecht, J. Carlson, and H. Levkowetz. 2004. *Extensible authentication protocol (EAP)*. RFC 3748.

[2] S. Blake-Wilson, M. Nystrom, D. Hopwood, J. Mikkelsen, and T. Wright. 2003. *Transport layer security (TLS) extensions*. RFC 3546.

[3] L. Buttyán and J.-P. Hubaux. 2007. *Security and cooperation in wireless networks: Thwarting malicious and selfish behavior in the age of ubiquitous computing*. Version 1.5.1. Cambridge University Press. Accessed October 4, 2007 from http://secowinet.epfl.ch.

[4] D. B. Chapman and E. D. Zwicky. 1995. *Building Internet firewalls*. Sebastopol, CA: O'Reilly & Associates.

[5] C. M. Chernick, C. Edington III, M. J. Fanto, and R. Rosenthal. 2005. *Guidelines for the selection and use of transport layer security (TLS) implementations.* Technical Report SP 800-52. National Institute of Standards and Technology (NIST).

[6] Common Criteria Implementation Board. 1999. *Common criteria for information technology security evaluation.* Version 2.1.

[7] T. Dierks and C. Allen. 1999. *The TLS protocol.* Version 1.0. RFC 2246.

[8] M. Van Droogenbroeck and R. Benedett. 2002. Techniques for a selective encryption of uncompressed and compressed images. In *Proceedings of Advanced Concepts for Intelligent Vision Systems (ACIVS'02)*, Ghent, Belgium, pp. 90–97.

[9] Z. Faigl, P. Fazekas, S. Lindskog, and A. Brunstrom. 2007. Performance analysis of IPsec in MIPv6 scenarios. In *Proceedings of the 16th IST Mobile Wireless Communication Summit*, Budapest, Hungary.

[10] J. Goodman and A. P. Chandrakasan. 1998. Low power scalable encryption for wireless systems. *Wireless Networks* 4:55–70.

[11] C. T. R. Hager. 2004. Context aware and adaptive security for wireless networks. PhD thesis, Virginia Polytechnic Institute and State University, Blacksburg, VA.

[12] A. Harbitter and D. A. Menascé. 2002. A methodology for analyzing the performance of authentication protocols. *ACM Transactions on Information System Security* 5:458–91.

[13] D. Harkins and D. Carrel. 1998. *The Internet key exchange (IKE).* RFC 2409.

[14] Information Technology Evaluation Criteria (ITSEC). 1993. *Provisional harmonized criteria.*

[15] Institute of Electrical and Electronic Engineers (IEEE). 1999. *Part 11: Wireless LAN medium access control (MAC) and physical layer (PHY) specifications.* IEEE Standard 802.11.

[16] Institute of Electrical and Electronic Engineers (IEEE). 2004. *Amendment 6: Medium access control (MAC) security enhancements.* IEEE Standard 802.11i.

[17] C. E. Irvine and T. E. Levin. 2000. Quality of security service. In *Proceedings of the 2000 New Security Paradigms Workshop (NSPW'00)*, Ballycotton, County Cork, Ireland, pp. 91–99.

[18] B. W. Johnson. 1989. *Design and analysis of fault tolerant digital systems.* Boston: Addison-Wesley.

[19] H. Johnson. 2004. Lightweight authentication in wireless networks. Licentiate thesis, Blekinge Institute of Technology, School of Engineering, Karlskrona, Sweden.

[20] H. Johnson. 2005. Toward adjustable lightweight authentication for network access control. PhD thesis, Blekinge Institute of Technology, Karlskrona, Sweden.

[21] H. Johnson, L. Isaksson, M. Fiedler, and S. F. Wu. 2006. A decision system for adequate authentication. In *Proceedings of the International Conference on Networking, International Conference on Systems and International Conference on Mobile Communications and Learning Technologies (ICNICONSMCL'06)*, Washington, DC.

[22] R. Karri and P. Mishra. 2003. Optimizing the energy consumed by secure wireless sessions: Wireless transport layer security case study. *Mobile Networks and Applications* 8:177–85.

[23] P. Keeratiwintakorn and P. Krishnamurthy. 2006. Energy efficient security services for limited wireless devices. In *Proceedings of the International Symposium on Wireless Pervasive Computing*, Phuket, Thailand.

[24] S. Kent and R. Atkinson. 1998. *Security architecture for the Internet protocol*. RFC 2401.

[25] A. K. Lenstra and E. R. Verheul. 2001. Selecting cryptographic key sizes. *Journal of Cryptology: The Journal of the International Association for Cryptologic Research* 14:255–93.

[26] Y. Li, Z. Chen, S. M. Tan, and R. H. Campbell. 1996. Security enhanced MPEG player. In *Proceedings of the 1996 International Workshop on Multimedia Software Development (MMSD'96)*, Berlin, Germany, pp. 169–76.

[27] S. Lindskog and A. Brunstrom. 2005. Design and implementation of a tunable encryption service for networked applications. In *Proceedings of the First IEEE/CREATE-NET Workshop on Security and QoS in Communications Networks (SecQoS 2005)*, Athens, Greece, pp. 258–66.

[28] S. Lindskog, A. Brunstrom, Z. Faigl, and K. Tóth. 2006. Providing tunable security services: An IEEE 802.11i example. In *Proceedings of the First Workshop on Enterprise Network Security (WENS 2006)*, Baltimore, MD.

[29] S. Lindskog, A. Brunstrom, R. Lundin, and Z. Faigl. 2006. A conceptual model of tunable security services. In *Proceedings of the 3rd International Symposium on Wireless Communication Systems (ISWCS 2006)*, Valencia, Spain, pp. 531–35.

[30] S. Lindskog and E. Jonsson. 2003. Adding security to QoS architectures. In *Perspectives on multimedia: Communication, media and information technology*, ed. R. Burnett, A. Brunstrom, and A. G. Nilsson, chap. 8, pp. 145–58. New York: John Wiley & Sons.

[31] R. Lundin, S. Lindskog, A. Brunstrom, and S. Fischer-Hübner. 2005. Using guesswork as a measure for confidentiality of selectively encrypted messages. In *Proceedings of the First Workshop on Quality of Protection (QoP 2005)*, Milano, Italy, pp. 173–184.

[32] J. Massey. 1994. Guessing and entropy. In *Proceedings of 1994 IEEE International Symposium on Information Theory*, Trondheim, Norway, p. 204.

[33] J. Meyer and F. Gadegast. 1995. Security mechanisms for multimedia data with the example MPEG-I video. Accessed October 4, 2007 from http://www.gadegast.de/frank/doc/secmeng.pdf.

[34] Nautilus secure phone home page. Accessed October 4, 2007 from http://nautilus.berlios.de/.

[35] D. Nicol, W. Sanders, and K. Trivedi. 2004. Model-based evaluation: From dependability to security. *IEEE Transactions on Dependable and Secure Computing* 1:48–65.

[36] C. S. Ong, K. Nahrstedt, and W. Yuan. 2003. Quality of protection for mobile multimedia applications. In *Proceedings of the 2003 IEEE International Conference on Multimedia & Expo (ICME'03)*, Baltimore, MD, vol. 2, pp. 137–140.

[37] M. Podesser, H. P. Schmidt, and A. Uhl. 2002. Selective bitplane encryption for secure transmission of image data in mobile environments. In *Proceedings of the 5th IEEE Nordic Signal Processing Symposium (NORSIG'02)*, Tromsø/Trondheim, Norway.

[38] M. Portmann and A. Seneviratne. 2001. Selective security for TLS. In *Proceedings of 9th International Conference on Networks (ICON'01)*, Bangkok, pp. 216–21.

[39] T. L. Saaty and M. S. Ozdemir. 2003. Why the magic number seven plus or minus two. *Mathematical and Computer Modelling* 38:233–44.

[40] N. Sastry and D. Wagner. 2004. Security considerations for IEEE 802.15.4 networks. In *Proceedings of the 2004 ACM Workshop on Wireless Security (WiSe'04)*, New York, pp. 32–42.

[41] P. Schneck and K. Schwan. 1997. *Authenticast: An adaptive protocol for high-performance, secure network applications.* Technical Report GIT-CC-97-22. Georgia Institute of Technology.

[42] P. A. Schneck and K. Schwan. 1998. Dynamic authentication for high-performance network applications. In *Proceedings of the Sixth IEEE/IFIP International Workshop on Quality of Service (IWQoS'98)*, Napa, CA, pp. 127–136.

[43] A. Servetti and J. C. De Martin. 2002. Perception-based selective encryption of G.729 speech. In *Proceedings of the 2002 IEEE International Conference on Acoustics, Speech, and Signal Processing*, Orlando, FL, vol.1, pp. 621–24.

[44] C. E. Shannon. 1949. Communication theory of secrecy systems. *Bell Systems Technical Journal* 28:656–715.

[45] Speak Freely homepage. http://www.speakfreely.org/ (accessed October 4, 2007).

[46] E. Spyropoulou, C. Ager, T. E. Levin, and C. E. Irvine. 2002. IPSec modulation for quality of security service. In *Proceedings of the Third Annual International Systems Security Engineering Association Conference (2002 ISSEA Conference)*, Orlando, FL.

[47] W. Stallings. 1998. *Cryptography and network security: Principles and practice.* 2nd ed. Upper Saddle River, NJ: Prentice Hall.

[48] S. W. Tak and E. K. Park. 2003. Adaptive secure software architecture for electronic commerce. *Software—Practice and Experience* 33:1343–57.

[49] U.S. Department of Defense. 1985. *Trusted computer system evaluation criteria (TCSEC).* DoD 5200.28-STD.

[50] C. Xenakis, N. Laoutaris, L. Merakos, and I. Stavrakakis. 2006. A generic characterization of the overheads imposed by IPsec and associated cryptographic algorithms. *Computer Networks* 50:3225–41.

[51] P. K. Yogender and H. H. Ali. 2006. Impacts of employing different security levels on QoS parameters in virtual private networks. In *Proceedings of the 24th IASTED International Multi-Conference on Parallel and Distributed Computing and Networks (PDCN)*, Innsbruck, Austria, pp. 142–149.

Index

B

D

E

F

G

H

I

J

K

L

M

N

O

P

S

T

U

V

W

Z

Adaptation in Wireless Communications

Edited by

Mohamed Ibnkahla

ADAPTIVE SIGNAL PROCESSING
in WIRELESS COMMUNICATIONS

ADAPTATION and CROSS LAYER DESIGN
in WIRELESS NETWORKS

THE ELECTRICAL ENGINEERING AND APPLIED SIGNAL PROCESSING SERIES

Edited by Alexander Poularikas

The Advanced Signal Processing Handbook: Theory and Implementation for Radar, Sonar, and Medical Imaging Real-Time Systems
Stergios Stergiopoulos

The Transform and Data Compression Handbook
K.R. Rao and P.C. Yip

Handbook of Multisensor Data Fusion
David Hall and James Llinas

Handbook of Neural Network Signal Processing
Yu Hen Hu and Jenq-Neng Hwang

Handbook of Antennas in Wireless Communications
Lal Chand Godara

Noise Reduction in Speech Applications
Gillian M. Davis

Signal Processing Noise
Vyacheslav P. Tuzlukov

Digital Signal Processing with Examples in MATLAB®
Samuel Stearns

Applications in Time-Frequency Signal Processing
Antonia Papandreou-Suppappola

The Digital Color Imaging Handbook
Gaurav Sharma

Pattern Recognition in Speech and Language Processing
Wu Chou and Biing-Hwang Juang

Propagation Handbook for Wireless Communication System Design
Robert K. Crane

Nonlinear Signal and Image Processing: Theory, Methods, and Applications
Kenneth E. Barner and Gonzalo R. Arce

Smart Antennas
Lal Chand Godara

Mobile Internet: Enabling Technologies and Services
Apostolis K. Salkintzis and Alexander Poularikas

Soft Computing with MATLAB®
Ali Zilouchian

Wireless Internet: Technologies and Applications
Apostolis K. Salkintzis and Alexander Poularikas

Signal and Image Processing in Navigational Systems
Vyacheslav P. Tuzlukov

Medical Image Analysis Methods
Lena Costaridou

MIMO System Technology for Wireless Communications
George Tsoulos

Signals and Systems Primer with MATLAB®
Alexander Poularikas

Adaptation in Wireless Communications - 2 volume set
Mohamed Ibnkahla

ADAPTIVE SIGNAL PROCESSING in WIRELESS COMMUNICATIONS

Edited by

Mohamed Ibnkahla

CRC Press
Taylor & Francis Group
Boca Raton London New York

CRC Press is an imprint of the
Taylor & Francis Group, an **informa** business

CRC Press
Taylor & Francis Group
6000 Broken Sound Parkway NW, Suite 300
Boca Raton, FL 33487-2742

Printed in the United States of America on acid-free paper
10 9 8 7 6 5 4 3 2 1

International Standard Book Number-13: 978-1-4200-4601-4 (Hardcover)

Library of Congress Cataloging-in-Publication Data

Adaptive signal processing in wireless communications / editor, Mohamed Ibnkahla.
p. cm. -- (Electrical engineering and applied signal processing series)
Includes bibliographical references and index.
ISBN 978-1-4200-4601-4 (alk. paper)
1. Adaptive signal processing. 2. Wireless communication systems. I. Ibnkahla, Mohamed. II. Title. III. Series.

TK5102.5.A296145 2008
621.382'2--dc22 2008025443

Visit the Taylor & Francis Web site at
http://www.taylorandfrancis.com

and the CRC Press Web site at
http://www.crcpress.com

Contents

Preface

Adaptive techniques play a key role in modern wireless communication systems. The concept of adaptation is emphasized in the *Adaptation in Wireless Communications Series* across all layers of the wireless protocol stack, ranging from the physical layer to the application layer.

This book is devoted to adaptation in the physical layer. It gives a tutorial survey of adaptive signal processing techniques used in wireless and mobile communication systems. The topics include adaptive channel modeling and identification, adaptive receiver design and equalization, adaptive modulation and coding, adaptive multiple-input-multiple-output (MIMO) systems, adaptive and opportunistic beam forming, and cooperative diversity. Moreover, the book addresses other important aspects of adaptation in wireless communications, such as software defined radio, reconfigurable devices, and cognitive radio. The book is supported by various new analytical, experimental, and simulation results and is illustrated by more than 160 figures, 20 tables, and 800 references.

I would like to thank all the contributing authors for their patience and excellent work. The process of editing started in June 2005. Each chapter has been blindly reviewed by at least two reviewers (more than 50% of the chapters received three reviews or more). I would like to thank the reviewers for their time and valuable contribution to the quality of the book.

Finally, a special thank you goes to my parents, my wife, my son, my daughter, and all my family. They all have been of great support for this project.

Mohamed Ibnkahla
Queen's University
Kingston, Ontario, Canada

Editor

Dr. Mohamed Ibnkahla earned an engineering degree in electronics in 1992, an M.Sc. degree in signal and image processing in 1992, a Ph.D. degree in signal processing in 1996, and the Habilitation à Diriger des Recherches degree in 1998, all from the National Polytechnic Institute of Toulouse (INPT), Toulouse, France.

Dr. Ibnkahla is currently an associate professor in the Department of Electrical and Computer Engineering, Queen's University, Kingston, Canada. He previously held an assistant professor position at INPT (1996–1999) and Queen's University (2000–2004).

Since 1996, Dr. Ibnkahla has been involved in several research programs, including the European Advanced Communications Technologies and Services (ACTS), and the Canadian Institute for Telecommunications Research (CITR). His current research is supported by industry and government agencies such as the Ontario Centers of Excellence (OCE), the Natural Sciences and Engineering Research Council of Canada (NSERC), the Ontario Ministry of Natural Resources, and the Ontario Ministry of Research and Innovation.

He is currently leading multidisciplinary projects designing, implementing and deploying wireless sensor networks for various applications in Canada. Among these applications are natural resources management, ecosystem and forest monitoring, species at risk tracking and protection, and precision agriculture.

Dr. Ibnkahla has published a significant number of journal papers, book chapters, technical reports, and conference papers in the areas of signal processing and wireless communications. He has supervised more than 40 graduate students and postdoctoral fellows. He has given tutorials in the area of signal processing and wireless communications in several conferences, including IEEE Global Communications Conference (GLOBECOM, 2007) and IEEE International Conference in Acoustics, Speech and Signal Processing (ICASSP, 2008).

Dr. Ibnkahla received the INPT Leopold Escande Medal for the year 1997, France, for his research contributions in signal processing; the Prime Minister's Research Excellence Award (PREA), Ontario, Canada in 2000, for his contributions in wireless mobile communications; and the Favorite Professor Award, Queen's University in 2004 for his excellence in teaching.

Contributors

Sofiène Affes
INRS-EMT
University of Quebec
Montreal, Quebec, Canada

Al-Mukhtar Al-Hinai
Department of Electrical and Computer Engineering
Queen's University
Kingston, Ontario, Canada

Mohamed-Slim Alouini
Department of Electrical and Computer Engineering
Texas A&M University (TAMU)-Qatar
Education City, Doha, Qatar

Moeness G. Amin
Center for Advanced Communications
College of Engineering
Villanova University
Villanova, Pennsylvania

Hüseyin Arslan
University of South Florida
Tampa, Florida

Bo-Yu Chang
National Tsing Hua University
Hsinchu, Taiwan

Alex B. Gershman
Darmstadt University of Technology
Darmstadt, Germany

Dennis L. Goeckel
Department of Electrical and Computer Engineering
University of Massachusetts
Amherst, Massachusetts

Philip K. F. Hölzenspies
University of Twente
Enschede, The Netherlands

Y.-W. Peter Hong
National Tsing Hua University
Hsinchu, Taiwan

Anders Høst-Madsen
Department of Electrical Engineering
University of Hawaii
Honolulu, Hawaii

Mohamed Ibnkahla
Department of Electrical and Computer Engineering
Queen's University
Kingston, Ontario, Canada

M. A. Khalighi
Institut Fresnel, UMR CNRS 6133
École Centrale Marseille
Marseille, France

Il-Min Kim
Department of Electrical and Computer Engineering
Queen's University
Kingston, Ontario, Canada

Young-Chai Ko
School of Electrical Engineering
Korea University
Seoul, Korea

André B. J. Kokkeler
University of Twente
Enschede, The Netherlands

Chun-Kuang Lin
National Tsing Hua University
Hsinchu, Taiwan

Richard K. Martin
Department of Electrical and Computer Engineering
Air Force Institute of Technology
Wright-Patterson AFB, Ohio

Haewoon Nam
Motorola, Inc.
Austin, Texas

Kyoung-Lae Noh
Department of Electrical and Computer Engineering
Texas A&M University
College Station, Texas

N. Prayongpun
GIPSA-Lab, UMR CNRS 5216
Département Images et Signal
ENSIEG, Domaine Universitaire
Saint Martin d'Hères, France

Khalid Qaraqe
Department of Electrical and Computer Engineering
Texas A&M University
College Station, Texas

K. Raoof
GIPSA-Lab, UMR CNRS 5216
Département Images et Signal
ENSIEG, Domaine Universitaire
Saint Martin d'Hères, France

Gerard K. Rauwerda
University of Twente
Enschede, The Netherlands

Erchin Serpedin
Department of Electrical and Computer Engineering
Texas A&M University
College Station, Texas

Besma Smida
Harvard University
Cambridge, Massachusetts

Gerard J. M. Smit
University of Twente
Enschede, The Netherlands

Vladimir Stanković
Department of Electronic and Electrical Engineering
University of Strathclyde
Glasgow, United Kingdom

Jitendra K. Tugnait
Department of Electrical and Computer Engineering
Auburn University
Auburn, Alabama

Shu-Hsien Wang
National Tsing Hua University
Hsinchu, Taiwan

Pascal T. Wolkotte
University of Twente
Enschede, The Netherlands

Yik-Chung Wu
Department of Electrical and Electronic Engineering
The University of Hong Kong
Hong Kong

Zixiang Xiong
Department of Electrical and Computer Engineering
Texas A&M University
College Station, Texas

Hong-Chuan Yang
Department of Electrical and Computer Engineering
University of Victoria
Greater Victoria, British Columbia, Canada

Zhihang Yi
Department of Electrical and Computer Engineering
Queen's University
Kingston, Ontario, Canada

Qiwei Zhang
University of Twente
Enschede, The Netherlands

1

Adaptation Techniques and Enabling Parameter Estimation Algorithms for Wireless Communications Systems

Hüseyin Arslan
University of South Florida

1.1 Introduction

Wireless communications systems have evolved substantially over the last two decades. The explosive growth of the wireless communication market is expected to continue in the future, as the demand for all types of wireless services is increasing. There is no doubt that the second generation of cellular wireless communications systems was a success. However, these systems were designed to provide good coverage for voice services so that a minimum required signal quality can be ensured over the coverage area. If the received signal quality is well above the minimum required level, the receivers do not exploit this. The speech quality does not improve much, as the quality is mostly dominated by the speech coder. On the other hand, if the signal quality is below the minimum required level, a call drop will be observed. Therefore, such a design requires the use of strong forward error correction (FEC) schemes, low-order modulations, and many other redundancies at the transmission and reception. In essence, the mobile receivers and transmitters are designed for the worst-case channel and received signal conditions. As a result, many users experience unnecessarily high signal quality from which they cannot benefit. While reliable communication is achieved, the system resources are not used efficiently.

New generations of wireless mobile radio systems aim to provide higher data rates and a wide variety of applications (like video, data, etc.) to mobile users while serving as many users as possible. However, this goal must be achieved under spectrum and power constraints. Given the high price of spectrum and its scarcity, the systems must provide higher system capacity and performance through better use of the available resources. Therefore, *adaptation techniques* have been becoming popular for optimizing mobile radio system transmission and reception at the physical layer as well as at the higher layers of the protocol stack.

Traditional system designs focus on allocating fixed resources to the user. Adaptive design methodologies typically identify the user's requirements and then allocate just enough resources, thus enabling more efficient utilization of system resources and consequently increasing capacity. Adaptive channel allocation and adaptive cell assignment algorithms have been studied since the early days of cellular systems. As the demand in wireless access for speech and data has increased, link and system adaptation algorithms have become more important.

For a given average transmit power, adaptation allows the users to experience better signal qualities. Adaptation reduces the average interference observed from other users, as they do not transmit extra power unnecessarily. As a result, the received signal quality will be improved over a large portion of the coverage area. These higher-quality signal levels can be exploited to provide increased data rates through rate adaptation. For a desired received signal quality, this might also translate into less transmit power, leading to improved power efficiency for longer battery life. On the other hand, for a desired minimum signal quality, this might lead to an increased coverage area or better frequency reuse. In addition, adaptive receiver designs allow the receiver to work with reduced signal quality values; i.e., a desired bit-error-rate (BER) or frame-error-rate (FER) performance can be achieved with a lower signal quality. Adaptive receivers can also enable reduced average computational complexities for the same quality of service, which again implies less power consumption. As can be seen, adaptation algorithms lead to improved performance, increased capacity, lower power consumption, increased radio coverage area, and eventually better overall wireless communications system design.

Many adaptation schemes require a form of measurement (or estimation) of various quantities (parameters) that might change over time. These estimates are then used to trigger or perform a multitude of functions, like the adaptation of the transmission and reception. For example, Doppler spread and delay spread estimations, signal-to-noise ratio (SNR) estimation, channel estimation, BER estimation, cyclic redundancy check (CRC) information, and received signal strength measurement are some of the commonly used measurements for adaptive algorithms. As the interest in the adaptation schemes increases, so does the research on improved (fast and accurate) parameter estimation techniques.

In this chapter, an overview of commonly used adaptation techniques and their applications for wireless mobile radio systems is given. Some of the commonly used parameters and their estimation using baseband signal processing techniques are explained in detail. Also, the current and future research issues regarding the improved parameter estimation and extensive use of adaptation techniques are discussed throughout the chapter. Note that there has been a significant amount of research on adaptation of wireless communications systems. This chapter is not intended to cover all these developments, but rather, it is intended to provide the readers an overview and conceptual understanding of adaptation techniques and related parameter estimation algorithms. More emphasis is given on signal processing perspectives of the adaptation of wireless communications systems.

1.2 Overview of Adaptation Schemes

In wireless mobile communications systems, information is transmitted through a radio channel. Unlike other guided media, the radio channel is highly dynamic. The transmitted signal reaches the receiver by undergoing many effects, corrupting the signal, and often placing limitations on the performance of the system.

Figure 1.1 illustrates a wireless communications system that includes some of the effects of the radio channel. The received signal strength varies depending on the distance relative to the transmitter, shadowing caused by large obstructions, and fading due to reflection, diffraction, and scattering. Mobility of the transmitter, receiver, or scattering objects causes the channel to change over time. Moreover, the interference conditions in the system change rapidly. Most important of all, the radio channel is highly random and the statistical characteristics of the channel are environment dependent. In addition to these changes, the traffic load, type of services, and mobile user characteristics and requirements might also vary in time. Adaptive techniques can be used to address all of these changing conditions.

The adaptation strategy can be different depending on the application and services. Constant BER constraint for a given fixed transmission bandwidth and constant throughput constraint are two of the most popular criteria for adaptation. In constant BER, a desired average or instantaneous BER is defined to satisfy the acceptable quality

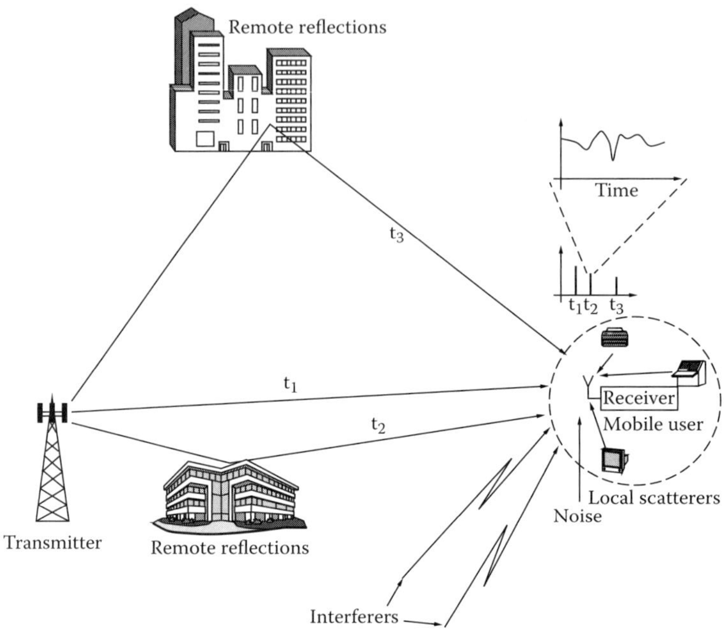

FIGURE 1.1 Illustration of some of the effects of a radio channel. Local scatterers cause fading; remote reflectors cause multipath and time dispersion, leading to ISI; mobility of the user or scatterers causes a time-varying channel; reuse of frequencies and adjacent carriers cause interference.

of service. Then the system is adapted to the varying channel and interference conditions so that the BER is maintained below the target value. In order to ensure this for all types of channel and interference conditions, the system changes power, modulation order, coding rate, spreading factor, etc. Note that this changes the throughput as the channel quality changes. On the other hand, for the constant throughput case, the adaptations are done to make sure that the effective throughput is constant, where the BER might change.

In general, it is possible to classify the adaptation algorithms as link and transmitter adaptation, adaptation of system resource allocation, and receiver adaptation. In the following sections, brief discussions of these adaptation techniques will be given.

1.2.1 Link and Transmitter Adaptation

A reliable link must ensure that the receiver is able to capture and reproduce the transmitted information bits. Therefore, the target link quality must be maintained all the time in spite of the changes in the channel and interference conditions. As mentioned earlier, one way to achieve this is to design the system for the worst-case scenario so that the target link quality can always be achieved.

If the transmitter sends more power for a specific user, the user benefits from it by having a better link quality, but the level of interference for the other users increases accordingly. On the other hand, if the user does not receive enough power, a reliable link cannot be established. In order to establish a reliable link while minimizing interference to other users, the transmitter should continuously control the transmitted power level. Power control is a simple form of adaptation that compensates for the variation of the received signal level due to path loss, shadowing, and sometimes fading. Numerous studies on power control schemes have been performed for various radio communications systems (see [1] and the references listed therein). In code division multiple-access (CDMA) systems, signals having widely different power levels at the receiver cause strong signals to swamp out weaker ones in a phenomenon known as the near–far effect. Power control mitigates the near–far problem by controlling the transmitted power.

It is possible to trade off power for bandwidth efficiency; i.e., a desired BER (or FER) can be achieved by increasing the power level or by reducing the bandwidth efficiency. One way of establishing a reliable link is to add redundancy to the information bits through FEC techniques. With no other changes, this would normally reduce the information rate (or bandwidth efficiency) of the communication. In the same way, high-quality links can be obtained by transmitting the signals with spectrally less efficient modulation schemes, like binary phase shift keying (BPSK) and quaternary PSK (QPSK). On the other hand, new-generation wireless systems aim for higher data rates made possible through spectrally efficient higher-order modulations. Therefore, a reliable link with higher information rates can be accomplished by continuously controlling the coding and modulation levels. Higher modulation orders with less powerful coding rates are assigned to users that experience good link qualities, so that the excess signal quality can be used to obtain higher data rates. Recent designs have exploited this with adaptive modulation techniques that change the order of the modulation [1, 2], as well as with adaptive coding schemes that change the coding rate [3, 4]. For example, the Enhanced General Packet Radio Service (EGPRS) standard introduces both Gaussian minimum shift keying (GMSK) and 8-PSK modulations with different coding rates through link adaptation and hybrid automatic repeat request (ARQ) [5]. The channel quality is estimated at the receiver, and the information is passed to the transmitter through appropriately defined messages. The transmitter adapts the coding and modulation based on this channel quality feedback. Similarly, variable spreading and coding techniques are present in third-generation CDMA-based systems [3], cdma2000 and wideband CDMA (WCDMA, or Universal Mobile Telecommunications System [UMTS]). Higher data rates can be achieved by changing the spreading factor and coding rate, depending on the perceived communication link qualities.

Adaptive antennas and adaptive beam-forming techniques have also been studied extensively to increase the capacity and to improve the performance of wireless communications systems [6]. The adaptive antenna systems shape the radiation pattern in such a way that the information is transmitted (for example, from a base station) directly to the mobile user in narrow beams. This reduces the probability of another user experiencing interference in the network, resulting in improved link quality, which can also be translated into increased network capacity. Although adaptive beam forming is an excellent way to utilize multiple-antenna systems to enhance the link quality, recently different flavors of the usage of multiantenna systems have gained significant interest. Space-time processing and multiple-input multiple-output (MIMO) antenna systems are some new developments that will allow further usage of multiple-antenna systems in wireless communications. Adaptive implementation of these technologies is important for successful and efficient integration of them into wireless communications systems.

1.2.2 Adaptive System Resource Allocation

In addition to physical link adaptation, system resources can also be allocated adaptively to reduce the interference and to improve the overall system quality. This includes adaptive power control, adaptive channel allocation, adaptive cell assignment, adaptive resource scheduling, adaptive spectrum management, congestion, handoff (mobility), admission, and load control strategies. Adaptive system resource allocation considers the current traffic load, as well as the channel and interference conditions. For example, the system could assign more resources to the mobiles that have better link quality to increase the throughput. Alternatively, the system could assign the resources to the user in such a way that the user experiences better quality for the current traffic condition.

Adaptive channel allocation and adaptive cell assignment in hierarchical cellular systems have been studied since the early days of cellular systems. Adaptive channel allocation increases the system capacity through efficient channel utilization and decreased probability of blocked calls [7]. Unlike fixed channel allocation, where the channels are assigned to the cells permanently and the assignment is done based on the worst-case scenario, in adaptive channel assignment, a common pool of channels is shared by many cells, and the channels are assigned with regard to the interference and traffic conditions.

Adaptive cell assignment can increase capacity without increasing the handoff rate. The cells can be assigned to the users depending on their mobility level. Fast-moving mobiles can be assigned to larger umbrella cells (to reduce the number of handoffs), while slow-moving mobiles are assigned to microcells (to increase capacity) [8].

Recently, research on increasing the average throughput of the system through water-filling-based resource allocation has gained significant interest [9–11]. The main idea is to allocate more resources to the users that experience better link quality, resulting in very efficient use of the available resources. The high-data-rate (HDR) system, which is based on a best-effort radio packet protocol, uses a water-filling-based approach in allocating system resources. Algorithms that deal with compromising the throughput to achieve fairness have also been studied [10, 11].

1.2.3 Receiver Adaptation

Digital wireless communication receiver performance is related to the required value of the signal-to-interference-plus-noise ratio (SINR) so that the BER (or FER) performance can be kept below a certain threshold for reliable communication. For a given complexity, if receiver A requires lower SINR than receiver B to satisfy the same error rate, receiver A is considered to perform better than receiver B.

Receiver adaptation techniques can increase the performance of the receiver, hence reducing the minimum required SINR. As mentioned before, this can be used to increase the coverage area for a fixed transmitted power, or it can be used to reduce the transmitted power requirement for a given coverage area. Moreover, receiver adaptation can reduce the average receiver complexity and the power drain from the battery for the same quality of service. In order to satisfy the desired BER performance, instead of running a computationally complex algorithm for all channel conditions, the receiver can choose the most appropriate algorithm given the system and channel conditions.

Advanced baseband signal processing techniques play a significant role in receiver adaptation. Baseband algorithms used for time and frequency synchronization, baseband filtering, channel estimation and tracking, demodulation and equalization, interference cancellation, soft information calculation, antenna selection and combining, decoding, etc., can be made adaptive depending on the channel and interference conditions.

Conventional receiver algorithms are designed for the worst-case channel and interferer conditions. For example, the channel estimation and tracking algorithms assume the worst-case mobile speed; the channel equalizers assume the worst-case channel dispersion; the interference cancellation algorithms assume that the interferer is always active and constant; and so on. Adaptive receiver design measures the current channel and interferer conditions and tunes the specific receiver function that is most appropriate for the current conditions. For example, a specific demodulation technique may work well in some channel conditions, but might not provide good performance in others. Hence, a receiver might include a variety of demodulators that are individually tuned to a set of channel classes. If the receiver could demodulate the data reliably with a simpler and less complex receiver algorithm under the given conditions, then it is desired to use that algorithm for demodulation.

1.3 Parameter Measurements

Many adaptation techniques require estimation of various quantities like channel selectivity, link quality, network load and congestion, etc. Here, we focus more on physical layer measurements from a digital signal processing perspective. As discussed earlier, link quality measures have many applications for various adaptation strategies. In addition, information on channel selectivity in time, frequency, and space is very useful for adaptation of wireless communications systems. In this section, these important parameters and their estimation techniques will be discussed.

1.3.1 Channel Selectivity Estimation

In wireless communications, the transmitted signal reaches the receiver through a number of different paths. Multipath propagation causes the signal to be spread in time, frequency, and angle. These spreads, which are related to the selectivity of the channel, have significant implications on the received signal. A channel is considered to be selective if it varies as a function of time, frequency, or space. The information on the variation of the channel in time, frequency, and space is very crucial in adaptation of wireless communications systems.

1.3.1.1 Time Selectivity Measure: Doppler Spread

Doppler shift is the frequency shift experienced by the radio signal when either the transmitter or receiver is in motion, and Doppler spread is a measure of the spectral

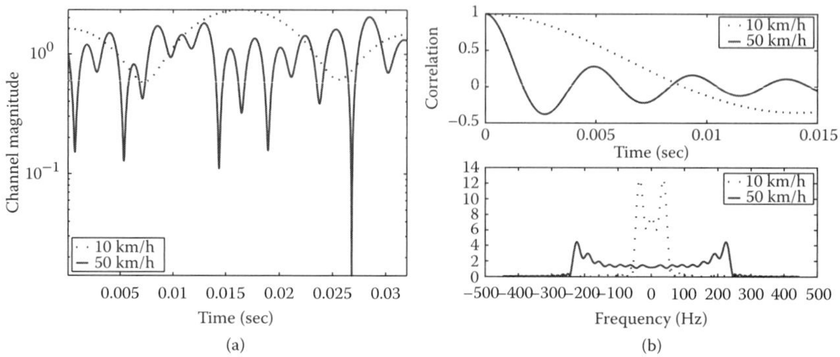

FIGURE 1.2 Illustration of the effect of mobile speed on time variation, time correlation, and Doppler spread of radio channel. (a) Channel time variation for different mobile speeds. (b) Time correlation of channel as a function of the time difference (separation in time) between the samples, and the corresponding Doppler spectrum in frequency.

broadening caused by the temporal rate of change of the mobile radio channel. Therefore, time-selective fading and Doppler spread are directly related. The coherence time of the channel can be used to characterize the time variation of the time-selective channel. It represents the statistical measure of the time window over which the two signal components have strong correlation, and it is inversely proportional to the Doppler spread. Figure 1.2 shows the effect of mobile speed on channel variation and channel correlation in time, as well as the corresponding Doppler spread values in frequency domain.

In an adaptive receiver, Doppler information can be used to improve performance or reduce complexity. For example, in channel estimation algorithms, whether using channel trackers or channel interpolators, instead of fixing the tracker or interpolation parameters for the worst-case Doppler spread value (as commonly done in practice), the parameters can be optimized adaptively based on Doppler spread information [12, 13]. Similarly, Doppler information could be used to control the receiver or transmitter adaptively for different mobile speeds, like variable coding and interleaving schemes [14]. Also, radio network control algorithms, such as handoff, cell assignment, and channel allocation in cellular systems, can utilize the Doppler information [8]. For example, as will be described later, in a hierarchical cell structure, the users are assigned to cells based on their speeds (mobility).

Doppler spread estimation has been studied for several applications in wireless mobile radio systems. Correlation and variation of channel estimates as well as correlation and variation of the signal envelope have been used for Doppler spread estimation [12]. One simple method for Doppler spread estimation is to use *differentials* of the complex channel estimates [15]. The differentials of the channel estimates are very noisy, which require low-pass filtering. The bandwidth of the low-pass filter is also a function of the Doppler estimate. Therefore, such approaches require adaptive receivers that continuously change the filter bandwidth depending on the previously obtained Doppler value. A Doppler estimation scheme based on the autocorrelation of complex channel estimates is described in [16]. Also, a maximum likelihood estimation-based approach, given the channel autocorrelation estimate, is utilized for Doppler spread estimation in

[17]. Channel autocorrelation is calculated using the channel estimates over the known field of the transmitted data.

Instead of using channel estimates, the received signal can also be used directly in estimating Doppler spread information. In [18], the Doppler frequency is extracted from the samples of the received signal envelope. Doppler information is calculated as a function of the squared deviation of the signal envelope. Similarly, in [19] the mobile speed is estimated as a function of the deviation of the averaged signal envelope in flat fading channels. For dispersive channels, pattern recognition, using the variation of pattern mean, can be used to quantify the deviation of signal envelope. In [20], the filtered received signal is used to calculate the channel autocorrelation values over each slot. Then, the autocorrelation estimate is used for identification of high- and low-speed mobiles. In [21], multiple antennas are exploited, where a linear relation between the switching rate of the antenna branches and Doppler frequency is given. Also, the level crossing rate of the average signal level has been used in estimating velocity [22, 23].

1.3.1.2 Frequency Selectivity Measure: Delay Spread

The multipath signals that reach the receiver have different delays as the paths that the signals travel through have different lengths. When the relative path delays are on the order of a symbol period or more, images of different transmitted symbols arrive at the same time, causing intersymbol interference (ISI). Delay spread is one of the most commonly used parameters that describes the time dispersiveness of the channel, and it is related to frequency selectivity of the channel. The frequency selectivity can be described in terms of coherence bandwidth, which is a measure of range of frequencies over which the two frequency components have a strong correlation. The coherence bandwidth is inversely proportional to the delay spread [24]. Figure 1.3 shows the effect of time dispersion on channel frequency variation and channel frequency correlation, as well as the corresponding power delay profiles.

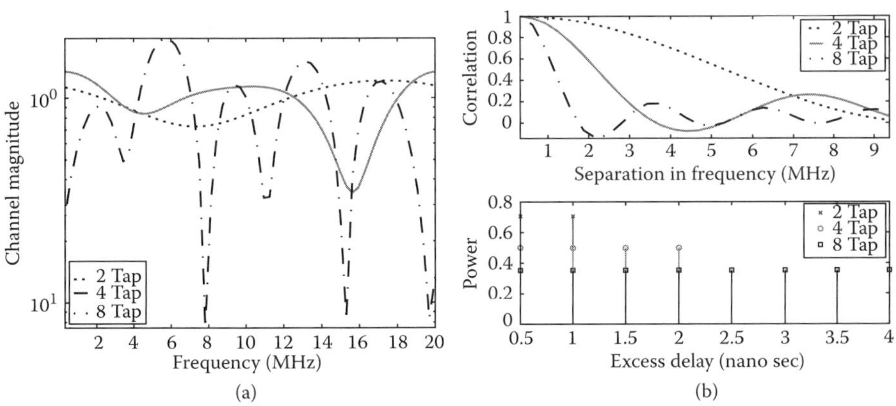

FIGURE 1.3 Illustration of the effect of time dispersion on channel frequency variation, channel frequency correlation, and delay spread. (a) Channel frequency variation for different delay spread values. (b) Channel frequency correlation as a function of separation in frequency and the corresponding power delay profiles.

Like time selectivity, the information about the frequency selectivity of the channel can be very useful for improving the performance of adaptive wireless radio systems. For example, in a time division multiple-access (TDMA)-based Global System for Mobile Communications (GSM), the number of channel taps needed for equalization might vary depending on channel dispersion. Instead of fixing the number of channel taps for the worst-case channel condition, we can change them adaptively [25], allowing simpler receivers with reduced battery consumption and improved performance. Similarly, in [26], a TDMA receiver with adaptive demodulator is proposed, using the measurement about the dispersiveness of the channel. Dispersion estimation can also be used for other parts of transmitters and receivers. For example, in frequency domain channel estimation using channel interpolators, instead of fixing the interpolation parameters for the worst expected channel dispersion, we can change the parameters adaptively depending on the dispersion information [27].

Although dispersion estimation can be very useful for many wireless communications systems, it is particularly crucial for orthogonal frequency division multiplexing (OFDM)-based wireless communications systems. OFDM, which is a multicarrier modulation technique, handles the ISI problem due to high-bit-rate communication by splitting the high-rate symbol stream into several lower-rate streams and transmitting them on different orthogonal carriers. The OFDM symbols with increased duration might still be affected by the previous OFDM symbols due to multipath dispersion. Cyclic prefix extension of the OFDM symbol avoids ISI from the previous OFDM symbols if the cyclic prefix length is greater than the maximum excess delay of the channel. Since the maximum excess delay depends on the radio environment, the cyclic prefix length needs to be designed for the worst-case channel condition. This makes the cyclic prefix a significant portion of the transmitted data, thereby reducing spectral efficiency. One way to increase spectral efficiency is to adapt the length of the cyclic prefix depending on the radio environment [28]. The adaptation requires estimation of maximum excess delay of the radio channel, which is also related to the frequency selectivity of the channel. In HiperLAN2, which is a wireless local area network (WLAN) standard, a cyclic prefix duration of 800 ns, which is sufficient to allow good performance for channels with delay spread up to 250 ns, is used. Optionally, a short cyclic prefix with 400 ns duration may be used for short-range indoor applications. Delay spread estimation allows adaptation of these various options to optimize the spectral efficiency. Other OFDM parameters that could be changed adaptively using the knowledge of the dispersion include OFDM symbol duration and OFDM subcarrier bandwidth.

Characterization of the frequency selectivity of the radio channel is studied in [29–31] using the level crossing rate (LCR) of the channel in frequency domain. Frequency domain LCR gives the average number of crossings per Hertz at which the measured amplitude crosses a threshold level. An analytical expression between LCR and the time domain parameters corresponding to a specific multipath power delay profile (PDP) is given. LCR is very sensitive to noise, which increases the number of level crossings and severely deteriorates the performance of the LCR measurement [31]. Filtering the channel frequency response reduces the noise effect, but finding the appropriate filter parameters is an issue. If the filter is not designed properly, one might end up smoothing the actual variation of frequency domain channel response. In [27], instantaneous root mean square (rms) delay spread, which provides information about local (small-scale) channel dispersion, is obtained by estimating the channel impulse response (CIR) in the time domain. The detected symbols in the frequency domain are used to regenerate the time domain signal through inverse fast Fourier transform (IFFT). This signal is then used to correlate the actual received signal to obtain CIR, which is then used for delay spread estimation. Since the detected symbols are random, they might not have good autocorrelation properties, which can be a problem, especially when the number of carriers is low. In addition, the use of detected symbols for correlating the received samples to obtain CIR provides poor results for low SNR values. In [28], the delay spread is also calculated from the instantaneous time domain CIR, wherein the CIR is obtained by taking IFFT of the frequency domain channel estimate. Channel frequency selectivity and delay spread information are calculated using the channel frequency correlation estimates in [24, 32]. An analytical expression between delay spread and coherence bandwidth is also given.

The level of time dispersion can be obtained by using known training sequences and a maximum likelihood-based algorithm. The channel can be modeled with different levels of dispersion. Using these various channel models, the corresponding channel estimates and the residual error can be calculated. From these residual error terms, a decision can be made about the level of dispersion. Note that when the channel is overmodeled, the residual error also becomes smaller. Hence, it is not necessarily true that the model that provides the smaller residual error is the most suitable one. The most appropriate model can be found by several information criteria algorithms, like Bayesian information criteria (BIC) or Akaike information criteria (AIC) [33].

1.3.1.3 Spatial Selectivity Measure: Angle Spread

Angle spread is a measure of how multipath signals are arriving (or departing) with respect to the mean arrival (departure) angle. Therefore, angle spread refers to the spread of angles of arrival (or departure) of the multipaths at the receiving (transmitting) antenna array [34]. Angle spread is related to the spatial selectivity of the channel, which is measured by coherence distance. Like coherence time and frequency, coherence distance provides the measure of the maximum spatial separation over which the signal amplitudes have strong correlation, and it is inversely proportional to angular spread, i.e., the larger the angle spread, the shorter the coherence distance. Figure 1.4 shows the effect of local scattering on angle of arrival. The local scattering in the vicinity of Receiver-2 results in larger angular spreads, as the received signals come from many different directions due to a richer local scattering environment. For a given receiver antenna spacing, this leads to less antenna correlations between the received antenna elements than in Receiver-1. Note that although the angular spread is described independent of the other channel selectivity values for the sake of simplicity, in reality the angle of arrival can be related to the path delay. The multipath components that arrive at the receiver earlier (with shorter delays) are expected to have similar angles of arrival (lower angle spread values).

Compared to time and frequency selectivity, spatial selectivity has not been studied widely in the past. However, recently there has been a significant amount of work in multiantenna systems. With the widespread application of multiantenna systems, it

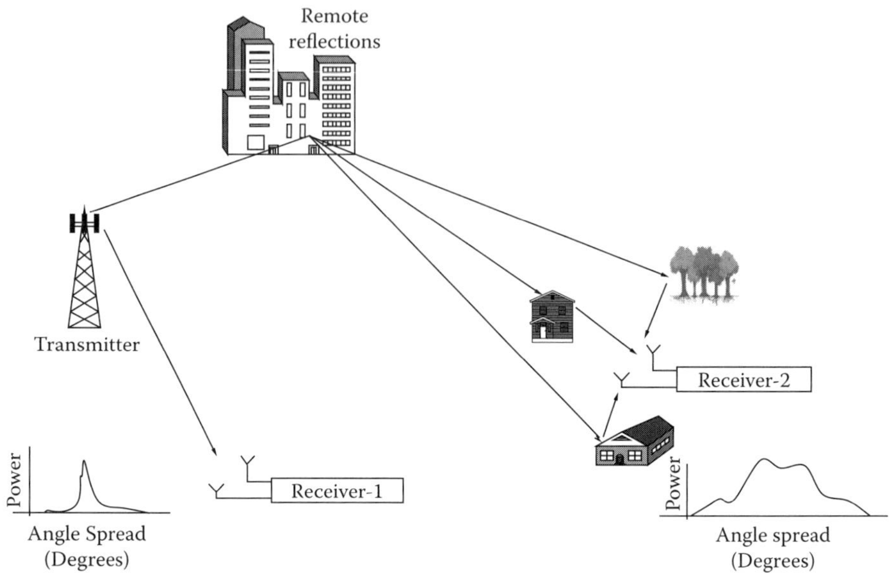

FIGURE 1.4 Illustration of the effect of different local scattering in angle of arrivals. Receiver-1 observes less angle spread than Receiver-2. Therefore, receiver antennas in Receiver-1 will have more correlations.

is expected that the need for understanding spatial selectivity and related parameter estimation techniques will gain momentum. Spatial selectivity will especially be useful when the requirement for placing antennas close to each other increases, as in the case of multiple antennas at the mobile units.

Spatial correlation between multiple-antenna elements is related to the spatial selectivity, antenna distance, mutual coupling between antenna elements, antenna patterns, etc. [35, 36]. Spatial correlation has significant effects on multiantenna systems. Full capacity and performance gains of multiantenna systems can only be achieved with low antenna correlation values. However, when this is not possible, maximum capacity can be achieved by employing efficient adaptation techniques. Adaptive power allocation is one way to exploit the knowledge of the spatial correlation to improve the performance of multiantenna systems [37]. Similarly, adaptive modulation and coding, which employs different modulation and coding schemes across multiantenna elements depending on the channel correlation, is possible [38, 39]. In MIMO systems, adaptive power allocation has been studied by using the knowledge of channel matrix estimate and eigenvalue analysis [40, 41].

1.3.2 Channel Quality Measurements

Channel quality estimation is by far the most important measurement that can be used in adaptive receivers and transmitters [3]. Different ways of measuring the quality of radio channel are possible, and many of these measurements are done in the physical layer using baseband signal processing techniques. In most of the adaptation algorithms,

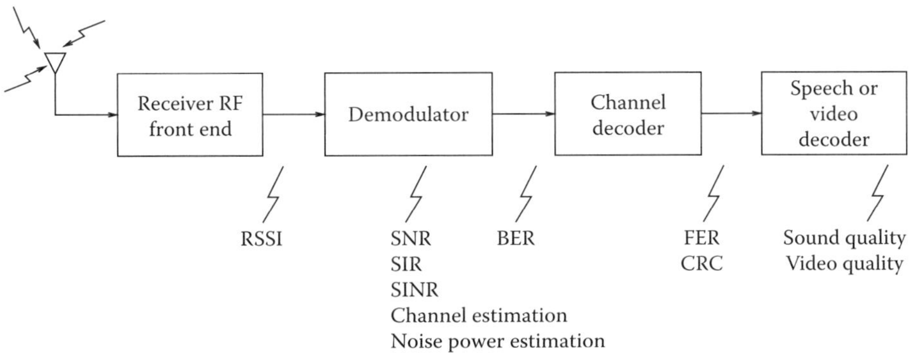

FIGURE 1.5 A simple wireless receiver that shows the estimation points of commonly used parameters.

the target quality measure is the FER or BER, as these are closely related to higher-level quality-of-service parameters like speech and video quality. However, reliable measurement of these qualities requires many measurements, and this causes delays in the adaptation as the process could be very long. Therefore, other types of channel quality measurements that are related to these might be preferred. When the received signal is impaired only by white Gaussian noise, analytical expressions can be found relating the BER to other measurements. For other impairment cases, like colored interferers, numerical calculations and computer simulations that relate these measurements to BER can be performed. Therefore, depending on the system, a channel quality is related to the BER. Then, for a target BER (or FER), a required signal quality threshold can be calculated to be used with the adaptation algorithm.

The measurements can be performed at various points of a receiver, depending on the complexity, reliability, and delay requirements. There are trade-offs in achieving these requirements at the same time. Figure 1.5 shows a simple example where some of these measurements can take place. In the following sections, these measurements will be discussed briefly.

1.3.2.1 Measures before Demodulation

Received signal strength (RSS) estimation provides a simple indication of the fading and path loss, and provides the information about how strong the signal is at the receiver front end. If the received signal strength is stronger than the threshold value, then the link is considered to be good. Measuring the signal strength of the available radio channels can be used as part of the scanning and intelligent roaming process in cellular systems. Also, other adaptation algorithms, like power control and handoff, can use this information. The RSS measurement is simply done by reading samples from a channel and averaging them [42]. Compared to other measurements, RSS estimation is simple and computationally less complex, as it does not require the processing and demodulation of the received samples. However, the received signal includes noise, interference, and other channel impairments. Therefore, receiving a good signal strength does not tell much about the channel and signal quality. Instead, it gives an indication of whether a strong signal is present in the channel of interest. For the measurement of RSS, the transmitter might send a pilot signal continuously, as in the WCDMA cellular system, or a link layer beacon can be transmitted at discrete time intervals, as in IEEE 802.11 WLANs.

Since the received signal power fluctuates rapidly due to fading, in order to obtain reliable estimates, the signal needs to be averaged over a time window to compensate for short-term fluctuations. The averaging window size depends on the system, application, variation of the channel, etc. For example, if multiple receiver antennas are involved at the receiver, the window can be shorter than that for a single-antenna receiver.

1.3.2.2 Measures during and after Demodulation

The signal-to-interference ratio (SIR), SNR, and SINR are the most common ways of measuring the channel quality during (or just after) the demodulation of the received signal. SIR (or SNR or SINR) provides information on how strong the desired signal is compared to the interferer (or noise or interference plus noise). Most wireless communications systems are interference limited; therefore, SIR and SINR are more commonly used. Compared to RSS, these measurements provide more accurate and reliable estimates at the expense of computational complexity and with additional delay.

There are many adaptation schemes where these measurements can be exploited. Link adaptation (adaptive modulation and coding, rate adaptation, etc.), adaptive channel assignment, power control, adaptive channel estimation, and adaptive demodulation are only a few of many applications.

SIR estimation can be employed by estimating signal power and interference power separately and then taking the ratio of these two. In many new-generation wireless communications systems, coherent detection, which requires estimation of channel parameters, is employed. These channel parameter estimates can also be used to calculate the signal power. The training (or pilot) sequences can be used to obtain the estimate of SIR. Instead of the training sequences, the data symbols can also be used for this purpose. For example, in [43], where SNR information is used as a channel quality indicator for rate adaptation, the cumulative Euclidean metric corresponding to the decoded trellis path is exploited for channel quality information. Another method for channel quality measurement is the use of the difference between the maximum likelihood decoder metrics for the best path and second-best path, as described in [44]. In a sense, in this technique, some sort of soft information is used for the channel quality indicator. However, this approach does not tell much about the strength of the interferer or the desired signal. There are several other ways of SNR measurement that are based on subspace projection techniques. These approaches can be found in [45] and in the references cited therein.

Often, in obtaining the estimates, the impairment (noise or interference) is assumed to be white and Gaussian distributed to simplify the estimation process. However, in wireless communications systems, the impairment might be caused by a strong interferer, which is colored. For example, in OFDM systems, where the channel bandwidth is wide and the interference is not constant over the whole band, it is very likely that some part of the spectrum is affected more by the interferer than the other parts. Figure 1.6 shows the OFDM frequency spectrum and two types of noise over this spectrum: colored and white. Hence, when the impairment is colored, estimates that take the color of the impairment into account might be needed [46].

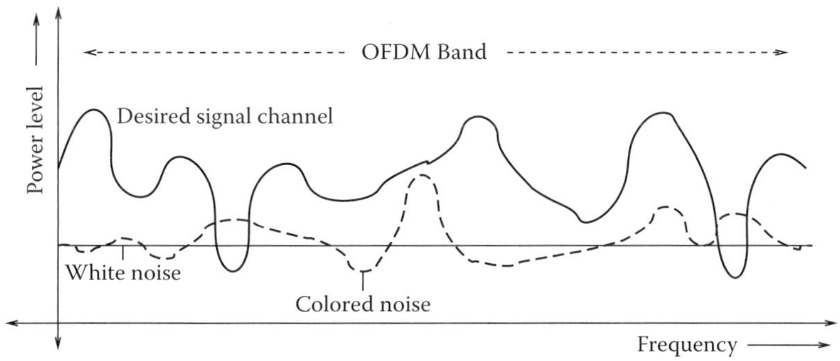

FIGURE 1.6 Representation of OFDM frequency channel response and noise spectrum. Spectrums for both white and colored noise are shown.

Note that since both the desired signal's channel and interferer conditions change rapidly, depending on the application, both short-term and long-term estimates are desirable. Long-term estimates provide information on long-term fading statistics due to shadowing and lognormal fading as well as average interference conditions. On the other hand, short-term estimates provide measurements of instantaneous channel and interference conditions. Applications like adaptive channel assignment and handoff prefer long-term statistics, whereas applications like adaptive demodulation, adaptive interference cancellation, etc., prefer short-term statistics.

For some applications, a direct measure of channel quality from channel estimates would be sufficient for adaptation. As mentioned above, channel estimates only provide information about the desired signal's power. It is a much more reliable estimate than RSS information, as it does not include the other impairments as part of the desired signal power. However, it is less reliable than SNR (or SINR) estimates, since it does not provide information about the noise or interference powers with respect to the desired signal's power.

Channel estimation for wireless communications systems has a very rich history. A significant amount of work has been done for various systems. In many systems, known information (like pilot symbols, pilot channels, pilot tones, training sequences, etc.) is transmitted along with the unknown data to help the channel estimation process. Blind channel estimation techniques that do not require known information transmission have also been studied extensively. For details on channel estimation for wireless communications systems, refer to [47, 48] and the references listed therein.

1.3.2.3 Measures after Channel Decoding

Channel quality measurements can also be based on postprocessing of the data (after demodulation and decoding). BER, symbol error rate (SER), FER, and CRC information are some of the examples of the measurements in this category. BER (or FER) is the ratio of the bits (or frames) that have errors relative to the total number of bits (or frames) received during the transmission. The CRC indicates the quality of a frame, which can be calculated using parity check bits through a known cyclic generator polynomial. FER can be obtained by averaging the CRC information over a number of frames. In order to calculate the BER, the receiver needs to know the actual transmitted bits, which is not possible in practice. Instead, BER can be calculated by comparing the bits before and after the decoder. Assuming that the decoder corrects the bit errors that appear before decoding, this difference can be related to BER. Note that the comparison makes sense only if the frame is error-free (good frame), which is obtained from the CRC information.

As mentioned earlier, although these estimates provide excellent link quality measures, reliable estimates of these parameters require observations over a large number of frames. Especially for low BER and FER measurements, extremely long transmission intervals will be needed. Therefore, for some applications these measures might not be appropriate. Note also that these measurements provide information about the actual operating condition of the receiver. For example, for a given RSS or SINR measure, two different receivers that have different performances will have different BER or FER measurements. Therefore, BER and FER measurements also provide information on the receiver capability as well as the link quality.

1.3.2.4 Measures after Speech or Video Decoding

The speech and video quality, the delays on data reception, and network congestion are some of the parameters that are related to the user's perception. Essentially, these are the ultimate quality measures that need to be used for adaptive algorithms. However, these parameters are not easy to measure, and in many cases, real-time measurement might not be possible. On the other hand, these measures are often related to the other measures mentioned above. For example, speech quality for a given speech coder can be related to FER of a specific system under certain assumptions [49]. However, as discussed in [49], some frame errors cause more audible damage than others. Therefore, it is still desired to find ways to measure the speech quality more reliably (and timely) and adapt the system parameters accordingly. Speech (or video) quality measures that take the human perception of the speech (or video) into account are highly desirable.

Perceptual speech quality measurements have been studied in the past. Both subjective and objective measurements are available [50]. Subjective measurements are obtained from a group of people who rate the quality of the speech after listening to the original and received speech. Then a mean opinion score (MOS) is obtained from their feedback. Although these measurements reflect the exact human perception that is desired for adaptation, they are not suitable for adaptation purposes because the measurements are not obtained in real time. On the other hand, the objective measurements can be implemented at the receiver in real time [51]. However, these measurements require a sample of the original speech at the receiver to compare the received voice with the original, undistorted voice. Therefore, they are also not applicable for many scenarios.

1.4 Applications of Adaptive Algorithms: Case Studies

1.4.1 Examples for Adaptive Receiver Algorithms

In this section, some representative examples for adaptive receiver algorithms will be discussed briefly. These algorithms can be employed in both base stations and mobile terminals, as well as in many other wireless receivers.

1.4.1.1 Channel Estimation with *A Priori* Information

Channel estimation is an integral part of standard adaptive receiver designs used in digital wireless communications systems. For conventional, coherent receivers, the effect of the channel on the transmitted signal must be estimated to recover the transmitted information. As long as the receiver estimates what the channel did to the transmitted signal, it can accurately recover the information sent.

The estimation of time-varying channel parameters is often based on an approximate underlying model of the radio channel. In fading environments, the coefficients of a channel model exhibit typical trends or quasi-periodic behavior in time, frequency, and space. The ability to track channel variation depends on how fast the channel changes in time, frequency, and space. As mentioned before, this is related to Doppler spread (time variation), delay spread (frequency variation), and angle spread (space variation). By utilizing *a priori* information about the channel variation, adaptive algorithms with larger memories can be designed without sacrificing tracking capability [15]. In contrast to the algorithms that do not exploit this information, adaptive algorithms provide a means of extrapolation of the channel coefficients in time, frequency, and space [13, 52]. For example, in [53], the step size of a simple least mean square (LMS) channel tracker is changed using the Doppler spread information. Similarly, the window size of a sliding window (moving average filtering)-based channel tracking algorithm can be adapted depending on Doppler spread and SNR information [54]. Wiener filtering, which is one of the most popular techniques for channel estimation using interpolation, is an excellent example in exploiting *a priori* information, as the optimal Wiener filter design requires knowledge of Doppler spread and noise power. In most conventional Wiener filtering designs, the worst-case expected Doppler spread values are used, degrading the performance of the algorithm for other Doppler spread values [55]. Recently, two-dimensional interpolation using Wiener filtering for OFDM-based wireless communications systems gained significant interest [28]. In this case, both Doppler spread and delay spread information, as well as noise variance estimates, can be used to optimize the channel tracker performance. Although we have mentioned a few examples, the usage of *a priori* information in channel estimation has been considered by many other authors. Further information can be found in [47, 48].

Figure 1.7 shows a simple coherent receiver structure with an adaptive channel tracker. The receiver includes a parameter measurement block that estimates the necessary parameters for the adaptation of the channel tracker. The necessary parameters can be estimated using the received signal and the output of the detector as described before. The detector requires the channel estimates that can be obtained from the channel tracker.

1.4.1.2 Adaptive Channel Length Truncation for Equalization

Time dispersion in wireless systems can cause ISI, which degrades the performance, often severely. Equalization is a technique used to counter the effects of ISI. In the Telecommunications Industry Association/Electronics Industry Association/Interim Standard 136 (TIA/EIA/IS-136, or simply IS-136) system, the channel can be assumed to be flat (nondispersive) with respect to the symbol duration most of the time. Equalization does

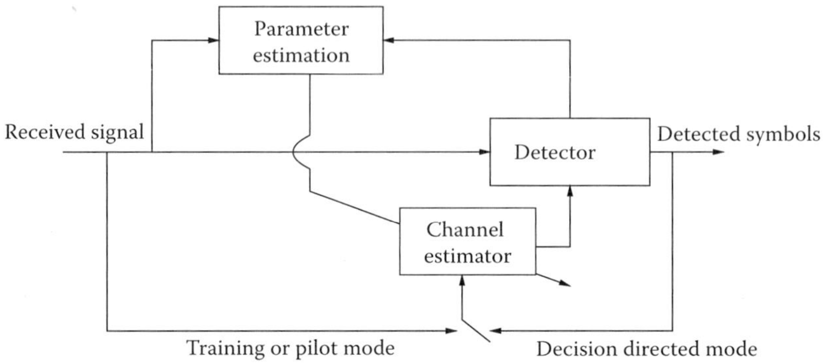

FIGURE 1.7 A simple adaptive channel estimation receiver.

not help much in nondispersive environments, and in fact hurts performance by trying to model dispersion that does not exist. However, in hilly terrain channel conditions, the channel is dispersive and requires equalization. Therefore, to design the receiver for the worst-case condition, equalization needs to be used for all the geographical conditions unnecessarily, resulting in a loss due to the mismatch of the implemented receiver to the fading scenario. An adaptive receiver, on the other hand, can have an algorithm that measures the dispersiveness of the channel and uses the appropriate demodulator based on the measurement [26]. This also results in conserving battery power.

In another cellular communications system, GSM, the symbol duration is relatively short compared to that in IS-136. Also, the pulse shaping itself introduces intentional ISI, so that equalization is required even in nondispersive channels. However, the number of channel taps needed for equalization might vary depending on the dispersion (the geographical area). Instead of fixing the number of channel taps for the worst-case condition, the number can be made adaptive [25], allowing simpler receivers with reduced battery consumption and improved performance. Again, the point emphasized here is to avoid overmodeling the signal. Figure 1.8 shows a simple example of an adaptive receiver that measures the level of dispersion and adapts the equalizer number of taps accordingly.

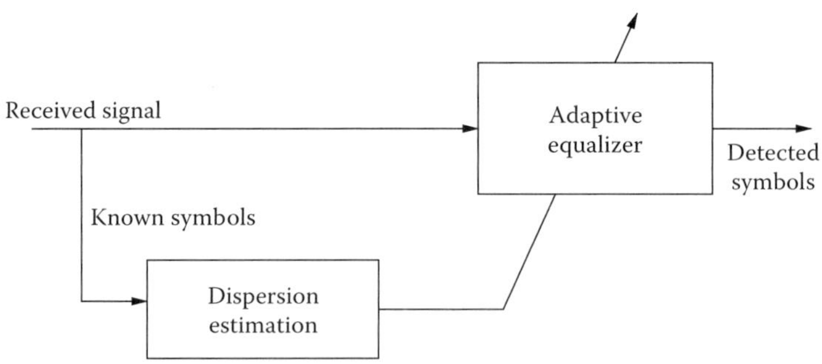

FIGURE 1.8 An adaptive receiver that uses the delay spread (time dispersion) estimate to adjust the equalizer.

1.4.1.3 Adaptive Interference Cancellation Receivers

The impairment sources in wireless mobile radio systems are numerous. Co-channel interference, which is caused by the reuse of carrier frequencies in nearby cells, is one of the major contributors. Another major interference source is adjacent channel interference, which is caused by the spectral overlap between adjacent channel users. Also, thermal noise and other impairment sources that are commonly modeled as additive white Gaussian noise (AWGN) degrade the performance of a receiver. The statistics of these disturbance sources are different. Conventional receivers commonly assume that the impairment at the receiver is white, which causes performance loss if the actual impairment is colored. By exploiting the statistics of the impairments, better receivers can be designed. For example, *interference whitening* is one such technique that partially suppresses the interference and optimizes the demodulator performance. However, at any given time, the kind of disturbance that is dominant at the receiver is not known before. In order to achieve the best possible performance in all situations, the receiver should estimate the possible disturbance source and adapt the receiver to the second-order statistics of the impairment. Such an adaptive receiver described in [56] improves the performance of the maximum likelihood–based receiver.

The interference can also be suppressed by employing interference cancellation techniques in the receivers. For example, joint demodulation (JD) of co-channel signals is a powerful technique for cancelling co-channel interference. In [57], it was shown that the capacity of the IS-136 system can be increased significantly by using a JD receiver. However, the JD receiver given in [57] works well only when there is a single dominant interferer, the mobile speed is low, and the channel is nondispersive. Otherwise, the conventional single-user demodulator (CD) works better than joint demodulation at the targeted operating SINR level. A simple and efficient solution to the above problem is an adaptive receiver that adapts the detector to the system conditions. Figure 1.9 illustrates

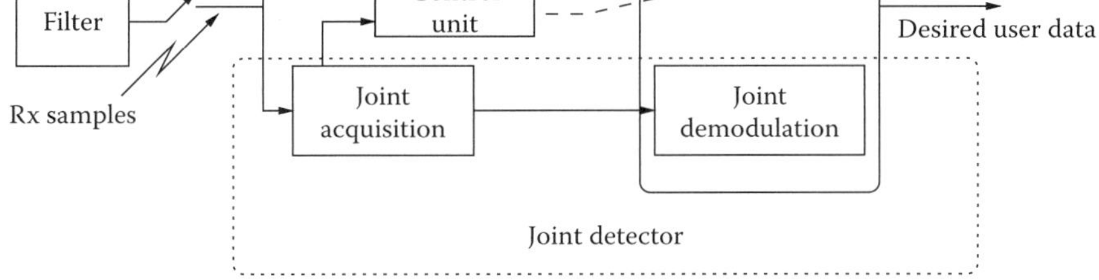

FIGURE 1.9 Example for adaptive interference cancellation receiver. A complex joint demodulation and a less complex single-user demodulation used adaptively based on the measured parameters.

the schematic of such an adaptive receiver. It contains the conventional detector, the joint detector, and a control unit to control the two detectors. For each slot, the control unit determines which of the two demodulators to use to recover the data symbols of the desired user. The control unit makes this decision on the basis of certain information obtained from conventional and joint acquisitions. The demodulator selected by the control unit outputs an estimate for the data symbols of the desired user. The details regarding the two demodulators can be found in [58].

The choice for the demodulator can be based on several criteria. Ideally, one would like to know the SNR, SIR, dominant interferer ratio ($I_1/I - I_1$, where I_1 is the dominant interferer and I is the total impairment, including the dominant interferer), and extent of ISI present in the system, among other parameters. Although these quantities are not generally available at the receiver, they can be estimated. For example, carrier and dominant interferer powers are estimated by averaging the corresponding channel tap strengths over multiple slots. The unmodeled impairment power is estimated from the accumulated Euclidean distance metric during the acquisition process (joint or conventional) over the training sequence of the desired signal.

1.4.1.4 Adaptive Soft Information Generation and Decoding

In digital wireless communications systems, forward error correction encoding is commonly used to provide a robust communication link. At the receiver, the decoder performance is optimized when the demodulator provides soft information for the encoded bits. The better soft information generation schemes require knowledge of the noise covariance, and often the noise covariance changes across the interleaving length. Therefore, a receiver should continuously measure the noise covariance and use these estimates for the improvement of soft bit values.

1.4.2 Examples for Link Adaptation and Adaptive Resource Allocation

In this section, some examples for adaptation of radio link and adaptive resource allocation will be discussed briefly. Examples in this area are numerous, and there has been a significant amount of research in this area.

1.4.2.1 Adaptive Power Control

Power control has a long and rich history in wireless communications systems [59–61]. Specifically, for CDMA-based cellular systems, adaptive power control has a significant role, as the performance and capacity of the CDMA systems are normally interference limited. Without power control, an interfering transmitter that is closer to the receiver than the desired signal's transmitter will cause a significant degradation, and this phenomenon is commonly referred to as the *near–far problem*. Power control handles this problem by adaptively controlling the user's power depending on the link quality and desired quality of service (QoS). As a result, the interference observed by other users due to this user will be less, which in turn reduces the average interference observed at the receivers. This results in a high-capacity system with improved battery life for the mobile terminals.

In voice-dominated cellular systems, the objective of the power control was mainly to maintain the minimal (target) link quality at a constant level for individual users. The data rates for all users are constant in this case, and each user experiences roughly the same quality of service. While this was appropriate for voice, recently, with the increased demand for multimedia services and high-speed data access, different objectives and cost functions to optimize the use of power resources have been developed. In mixed-traffic environments, the cost function for each service will be different, leading to different power allocation strategies [62]. Use of constant power along with variable coding, modulation, and spreading that adapts the data rate to the channel variations is one objective that some new-generation wireless systems have been adopting (rate control or rate adaptation) [43]. Also, water-filling types of power assignments, which assign more power to the users that have favorable channels, are being studied extensively [63].

In adaptive power control mechanisms, estimation of the link quality parameters is the key factor. Typical parameters used for adaptation include SIR, FER, and RSS. Doppler spread estimate can also be used to adjust the adaptation rate. Depending on the adaptation rate, power control can be classified as *fast power control* and *slow power control*. Fast power control compensates the changes in power level due to Rayleigh fading (small-scale fading), while slow power control is used for lognormal fading (shadowing) and path loss. The parameters that are used for them can also be different. For example, for fast power control, instantaneous SIR, SNR, SINR, and RSS can be more suitable than FER and BER, which might better suit slow power control. As mentioned in the parameter estimation section, parameter selection depends on the delay, complexity, and accuracy requirements. The estimation errors and delays, between measurements and adaptation of power, limit the efficient application of power control schemes. Therefore, more accurate and practical algorithms that estimate and predict the parameters to be used in adaptation are needed.

1.4.2.2 Adaptive Modulation and Channel Coding

Given the high price of spectrum and its scarcity, it is in the interest of operators to continue evolving their networks toward higher capacity and quality. Adaptive modulation and coding provide a framework to adjust modulation level and FEC coding rate depending on the link quality. Higher-order modulations (HOMs) allow more bits to be transmitted for a given symbol rate. On the other hand, HOM is less power efficient, requiring higher energy per bit for a given BER. Therefore, HOMs should be used only when the link quality is high, as they are less robust to channel impairments. Similarly, strong FEC and interleaving provide robustness against channel impairments at the expense of lower data rate and spectral efficiency, suggesting adaptation of coding rate based on the link quality. Figure 1.10 illustrates the capacity gain that can be achieved by employing adaptive modulation only. First, the BER performances of different modulations as a function of SNR are given in Figure 1.10(a). As can be seen, a desired BER can be achieved with low-order modulations for lower SNRs. Higher-order modulations need better link quality (higher SNRs) in order to obtain the same BER performance. Figure 1.10(b) shows the spectral efficiencies of different uncoded modulations, where an arbitrary packet size of 200 bits is used. Notice that the optimal spectral efficiency for

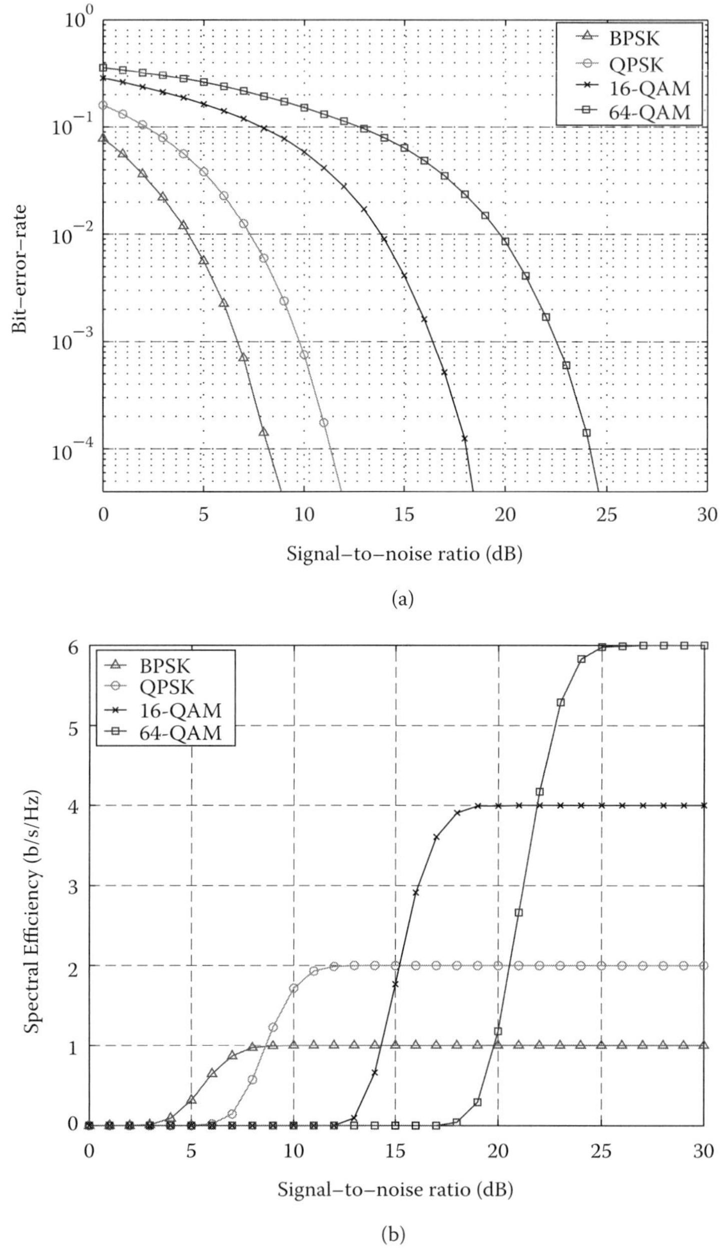

FIGURE 1.10 llustration of the BER and spectral efficiency of several modulation options. (a) BER plots of different modulations as a function of SNR. (b) Spectral efficiency of different modulation as a function of SNR.

TABLE 1.1 EGPRS Modulation and Coding Schemes and Peak Data Rates

Scheme	Modulation	Maximum Rate per Slot (kb/s)	Code Rate
MCS-1	GMSK	8.8	0.53
MCS-2	GMSK	11.2	0.66
MCS-3	GMSK	14.8	0.80
MCS-4	GMSK	17.6	1.00
MCS-5	8-PSK	22.4	0.37
MCS-7	8-PSK	44.8	0.76
MCS-6	8-PSK	29.6	0.49
MCS-8	8-PSK	54.5	0.92
MCS-9	8-PSK	59.2	1.00

different SNR regions can be obtained through the use of different modulations depending on the SNR.

Link adaptation using adaptive coding and modulation is deployed in some of the new-generation wireless communications systems. For example, EGPRS, which is the evolution of the second-generation GSM, employs two different modulation options (GMSK and 8-PSK) along with different coding rates, resulting in nine different modulation/coding options, as shown in Table 1.1 [5, 43]. In addition, EGPRS introduces the use of a type II hybrid ARQ system, commonly known within the specification as incremental redundancy. In link adaptation, the link quality is measured regularly and the most appropriate modulation and coding scheme is assigned for the next transmission interval. On the other hand, in the incremental redundancy scheme, information is first sent with low coding power (high coding rate). This results in a high bit rate if decoding is successful with this rate. However, if decoding fails with such a high rate, additional coded bits (redundancy) should be sent so that the transmitted bits can be decoded successfully. However, sending extra coded bits incrementally reduces the resulting bit rate and introduces undesired extra delay. Therefore, the initial code rate and modulation for the incremental redundancy scheme should be based on measurements of the link quality, instead of starting with any arbitrary rate [5]. As a result, by combining incremental redundancy with adaptive initial code rate, lower delays with lower memory requirements, and high data rates can be achieved. The different initial code rates are obtained by puncturing a different number of bits from a common convolution code (rate 1/3). Incremental redundancy operation is enabled by puncturing a different set of bits each time a block is retransmitted, whereby the code rate is gradually decreased toward one-third for every new transmission of the block.

Recent studies introduce new modulation and coding options together with other capacity enhancement techniques to further increase the data rate and throughput of EGPRS [64, 65]. Higher-order modulations like 16-QAM and 64-QAM are being proposed along with some more coding options to optimize the performance.*

* 16-QAM and64-QAM stand for 16-level and 64-level quadrature amplitude modulation, respectively.

Adaptive modulation and coding are also successfully employed for new-generation WLAN systems. HiperLAN2 and IEEE 802.11a, both of which use OFDM technology at the physical layer, allow four different modulation options (BPSK, QPSK, 16-QAM, and 64-QAM) with different coding rates. The coding rates are obtained with different puncturing patterns to a mother convolutional code, resulting in eight different modulation and coding options [66]. Similar to link adaptation in EGPRS, an appropriate modulation and coding scheme is used depending on the link quality. Therefore, a data rate ranging from 6 to 54 Mbit/s can be obtained by using various modes. BPSK, QPSK, and 16-QAM are used as mandatory modulation formats, whereas 64-QAM is applied as an optional mode.

Although only a couple of cases are given above, adaptive modulation and coding have attracted many new-generation wireless standards to consider them as options to increase the data rates, and there has been a significant amount of research in this area. Especially in conjunction with the advanced receiver algorithms that reduce the required SINR to lower values, the better link quality values can be exploited to increase the data rates further. Combining adaptive modulation with multiantenna transmitter and MIMO schemes based on the feedback-related channel estimates, channel quality, channel correlation, etc., is one of these interesting research areas. Based on the channel feedback information, the modulation type on multiantenna transmitters can be varied. Similarly, adapting the source coding with the channel coding or modulation is another interesting area of focus for link adaptation. For example, adaptive multirate (AMR) codec allows changing of the compression rate of speech depending on the link quality, as in GSM AMR. For weak link conditions, where heavy FEC is required, AMR has the ability to decrease the codec rate (more speech compression) to allocate more bits for FEC [49].

1.4.2.3 Adaptive Cell and Frequency Assignment

As mentioned before, radio spectrum is very expensive and limited. Efficient use of radio spectrum is very important to maximize the system capacity. The introduction of cellular technology was a major step toward efficient usage of finite spectrum through a concept called *frequency reuse*. The capacity of cellular systems is interference limited, dominated by co-channel interference (CCI) and adjacent channel interference (ACI). Early cellular systems aimed to avoid these major interference sources by designing systems for the worst-case interference conditions along with fixed channel allocation. This is often achieved by employing higher-frequency reuse and by allowing enough carrier spacing between adjacent channels. Both of these reduce the spectral efficiency. Later, more efficient spectrum usage strategies were developed that dynamically assign frequencies relative to current interference, propagation, and traffic conditions. In traditional cellular system designs, the allocation of frequency channels to cells is fixed, which means that each cell can use only a set of frequencies. Even if the other cells are not fully loaded, the cell that does not have any available frequency (fully loaded cell) cannot take advantage of it. In dynamic channel allocation, all the channels belong to a global pool and the channels are assigned according to a cost function that considers the

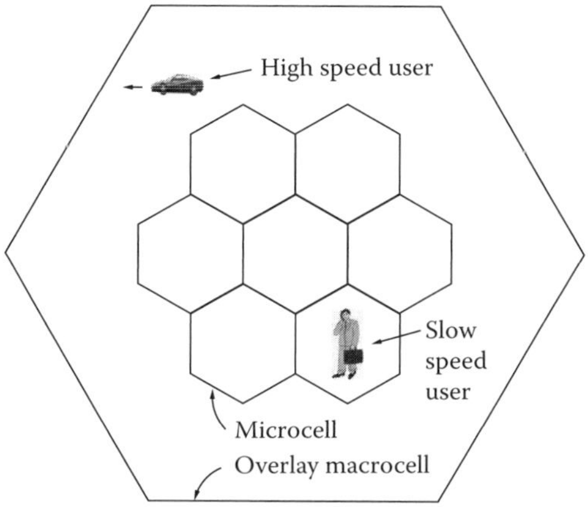

FIGURE 1.11 Illustration of two-layer hierarchical cell structure. High-speed mobiles are assigned to large cells, and low-speed mobiles are assigned to smaller cells.

CCI and ACI [67]. As a result, for nonuniform traffic conditions, the available channels can be used more efficiently.

Resource utilization has also evolved by employing a concept called hierarchical cellular structures (HCSs) [68]. The use of HCS has become a major component in third-generation mobile systems such as UMTS and IMT-2000. In an HCS, various cell sizes are deployed and small cell clusters are overlaid by larger cells. For example, Figure 1.11 shows a two-layer (e.g., microcell and macrocell) hierarchical system. Microcells increase capacity within a coverage area, but radio resource management becomes more difficult. The number of handoffs per cell is increased by an order of magnitude, and the time available to make a handoff is decreased. HCSs handle this by assigning cells to the mobiles depending on their speeds (Doppler spread estimate). For example, in the two-layer structure given in Figure 1.11, low-speed mobiles are assigned to microcells, whereas high-speed mobiles are assigned to macrocells. Hence, the macrocell-microcell overlay architecture provides a balance between maximizing the capacity per unit area and minimizing the number of handoffs [69]. As a result, the risk of call dropping is reduced, and there are other benefits, like lower handover delays, reduced switching load, and increased QoS. The HCS can be more than two layers (multilayer HCS). For example, picocellular layers can also be included in multilayer HCS. Similarly, communication satellite beams can overlay all the terrestrial layers at the highest hierarchical level.

Recently, dynamic allocation and multitiered design strategies are further generalized to take power control, cell handoff, traffic classes (like multimedia), and user priorities into account. Also, there are several studies toward combining link adaptation schemes with adaptive resource allocation. For example, adaptive modulation (and coding) can be combined with dynamic channel allocation. Similarly, adaptive modulation (and coding) can be combined with handover algorithms to introduce more intelligent handover strategies. All these developments require more sophisticated adaptation of the network, and they are based on many parameter measurements.

1.5 Future Research for Adaptation

Most of the techniques and issues that were described in the previous sections still need further research for the development of more efficient adaptation and parameter estimation algorithms. At the same time, there is a significant amount of effort in evolving current wireless communications systems to provide higher data rates, higher capacity, and better performance. New technologies are being introduced to accommodate these goals, like multicarrier wireless communications, MIMO, and ultrawideband (UWB). Adaptation techniques will be a significant factor in efficient and successful deployment of these technologies.

UWB is a promising technology for future data communications systems, high-accuracy (indoor) geolocation devices, sensor applications, etc. Any signal that occupies more than 500 MHz of bandwidth and meets the spectrum mask requirements enforced by spectrum regulation agencies is considered a UWB signal [70]. For example, in the United States, the Federal Communications Commission (FCC) has allocated 7.5 GHz of spectrum (between 3.1 and 10.6 GHz) for unlicensed use of UWB devices. One of the most popular UWB systems, which is based on impulse radio (IR), utilizes carrierless transmission with very low-power spectral density. IR-based UWB techniques are based on the transmission of nanosecond-level short pulses that generate extremely wide spectrums. This results in a covert noise-like signal in a radio channel. Note that within the transmission band of UWB, other technologies also coexist. For example, the OFDM-based WLAN technology at the 5 GHz U-NII band is a big concern for UWB signals, as it might create significant interference for UWB signals. In order to provide robustness against narrowband interference, adaptive implementation of UWB systems is very important. For this purpose, several strategies have been developed recently. Multiband UWB is one of the techniques proposed to reduce the effect of narrowband interference. In such techniques, the whole 7.5 GHz bandwidth is divided into several narrower bands that are still wider than 500 MHz. The information is transmitted in these bands depending on the narrowband interference situation. Several versions of multiband schemes are available, some of which can be found in [70]. Estimations of the existence and level of narrowband interference are interesting research topics that fall under the parameter estimation algorithms described before.

The wide bandwidth of UWB offers a capacity much higher than the current narrowband systems. Short-range data transmission rates of over 500 Mbps have been theoretically shown [71, 72]. However, these high data rates are only possible with excellent signal quality values and for short-range communications. High data rates can be traded off with longer ranges and for lower link quality values. Depending on the link quality and distance between transmitter and receiver, the rate can be changed through adaptation of the processing gain. UWB achieves processing gain due to pulse repetition, i.e., transmitting more than one pulse within a bit. For example, by transmitting 100 pulses per bit, a processing gain of 20 dB is obtained. Additional processing gain is obtained due to the low duty cycle, which is the ratio of the pulse repetition interval and the pulse width. Adaptation of processing gain is a research topic that needs to be explored for UWB systems. Similarly, multiple-access capability of UWB systems, which is primarily determined by the processing gain, needs to be explored. Adaptive multiaccess code design depending on delay spread of the channel is one of the interesting research areas. Also, adaptive multiuser detection techniques need to be studied for cancelling multiaccess interference. In summary, adaptation algorithms and related parameter estimation techniques for jointly optimizing the multiaccess capability, power consumption, data rate, and range of UWB systems are needed.

Adaptation of multicarrier systems has already gained some momentum. In multicarrier systems, the transmission bandwidth is much wider than the coherence bandwidth of the channel, resulting in frequency-selective fading channels. Therefore, different carriers experience different channel qualities. This leads to adaptation of each subcarrier individually. Adaptive bit/power loading can be used as an effective tool to get the highest capacity from a multicarrier system provided that the transmitter has the link quality information for each carrier. For example, adaptation of the modulation level for OFDM-based multicarrier systems has been studied recently [73]. The modulation level on different carriers can be changed depending on the link quality observed at the carriers. Transmitting different modulations on each carrier requires a large overhead for signaling. Therefore, approaches that group the neighboring carriers into subsets and use the same modulation in each subset are preferred. The signaling can also be avoided in time division duplexing (TDD) systems under certain assumptions. Unlike frequency division duplexing (FDD) systems, where the channels on the downlink and uplink are different, in TDD systems, using the assumption of the reciprocal and slowly varying channel, the transmitter and receiver can be assumed to experience the same channel response. Therefore, this might eliminate the need for signaling of the channel state information to the transmitter, if the channel estimates are used as the link quality measures. However, in this case, the receiver needs to know which modulation is used at the transmitter for each group. Blind modulation detection techniques can be used for this purpose [74]. Note that although the channels could be the same, the interferences observed in the transmitter and receiver are not necessarily the same. Therefore, the observed link qualities would be different at each end. Issues like these need to be studied further for successful implementation of adaptation techniques in multicarrier systems. Multicarrier CDMA (MC-CDMA), which combines OFDM modulation with CDMA-type multiple accessing, is a technology that is being pushed for fourth-generation cellular networks. The previous adaptation algorithms proposed for CDMA and OFDM technologies need to be revisited and optimized, and new adaptation algorithms need to be developed for MC-CDMA.

MIMO and multiantenna systems bring about a new dimension to wireless channels. The spatial dimension will be used in future communications systems for further improvement of the bandwidth and power efficiency. However, this dimension and the related parameter estimates need to be understood better. Research on parameter estimation for fast and accurate calculation of spatial selectivity, angular spread, antenna correlation, etc., is needed. Also, further research is required on the effect of mutual coupling between antenna elements, the effect of near-field scatterers on antenna patterns and antenna correlations, the exploitation of pattern selectivity when the spatial selectivity is not enough, and the generation of a desired pattern selectivity between antenna elements adaptively.

As described earlier, MIMO systems, which employ multiple-transmit and multiple-receive antennas, can provide huge capacity and improved performance gains by exploiting spatial selectivity of the channel. However, these gains, in reality, depend heavily on the statistical properties of the channel and the correlations between antenna elements. Among the factors that affect the antenna correlation are the characteristics of the scattering environment. Therefore, an optimal way of using multiple-antenna systems depends on the situation awareness. If the transmitter knows the instantaneous channel gains (the MIMO channel matrix), it can adapt the transmission to maximize the capacity of the MIMO system [40, 41]. Similarly, the instantaneous antenna correlation values can be exploited to adapt the transmission. In many cases, estimating the perfect instantaneous channel state and antenna correlation information and feeding this information back to the transmitter might not be possible. This is the case especially when the mobility is high. Instead, other parameter measures like partial (statistical) channel information, average channel selectivity, or angular spread would be useful for adapting the transmitter and receiver. Advanced signal processing techniques to calculate this partial channel and correlation information are needed.

Most of the previous adaptation techniques take place in physical and medium access control (MAC) layers. Future-generation wireless systems will also allow adaptive strategies at the higher networking layers. The higher layers will be more aware of the situation in lower layers. Cross-layer optimization algorithms and cost functions that involve many layers will be developed. This will also create the need for the development of new adaptation parameters and algorithms for the estimation of these parameters.

Current wireless communications systems are based on layered protocol design, and each layer is often designed and operated independently. For example, the channel variation is addressed by adapting the link in the physical or MAC layer using signal processing techniques as described before. The traffic load and delays are adapted by changing the routing tables, by adaptive channel and cell assignment techniques. The layered structure and adaptation of layers locally (and independently) simplify the network design. But the performance and capacity of the network is suboptimal, especially for addressing the requirements of future multimedia wireless services. The future applications will have different data rate, delay, power, and QoS requirements. Cross-layer adaptation could address these requirements by jointly optimizing multidimensional cost functions that involve all protocol layers [75–77]. As a result, networks with improved end-to-end performance subject to constraints in link quality and available network resources can be obtained, while being aware of application trade-offs.

1.6 Conclusion

Recently, the use of adaptation algorithms for better utilization of the available resources, like power and spectrum, has grown significantly. Several adaptation strategies to increase performance, data rate, capacity, and QoS of wireless communications systems have been introduced. Many of these adaptation techniques depend on accurate estimation of the various parameters. Therefore, further research on efficient parameter estimation techniques is still needed.

There is a significant amount of work needed for the evolution of the current wireless communications systems to accommodate the future demands. Ultrawideband, MIMO, and multicarrier wireless communications are some of the technologies that are being studied extensively. All of them have a common point: their capability to adapt the changing radio channel conditions. Adaptation of multicarrier communications and MIMO schemes has already gained some momentum. Adaptation algorithms for UWB still need to be explored. The flexibility of UWB makes it very attractive for employing successful adaptation schemes.

Acknowledgment

The author thanks T. Yücek for his help and the anonymous reviewers for their comments.

References

[1] S. Sampei. 1997. *Applications of digital wireless technologies to global wireless communications.* Englewood Cliffs, NJ: Prentice Hall.

[2] T. Ikeda, S. Sampei, and N. Morinaga. 2000. TDMA-based adaptive modulation with dynamic channel assignment for high-capacity communication systems. *IEEE Transactions on Communications* 49:404–12.

[3] S. Nanda, K. Balachandran, and S. Kumar. 2000. Adaptation techniques in wireless packet data services. *IEEE Transactions on Communications* 38:54–64.

[4] R. V. Nobelen, N. Seshadri, J. Whitehead, and S. Timiri. 1999. An adaptive radio link protocol with enhanced data rates for GSM evolution. *IEEE Communications Magazine* 54–63.

[5] A. Furuskar, S. Mazur, F. Muller, and H. Olofsson. 1999. EDGE: Enhanced data rates for GSM and TDMA/136 evolution. *IEEE Personal Communications Magazine* 6:56–66.

[6] S. Anderson, H. Dam, U. Forssen, J. Karlsson, F. Kronestedt, S. Mazur, and K. J. Molnar. 1999. Adaptive antennas for GSM and TDMA systems. *IEEE Personal Communications Magazine* 6:74–86.

[7] W. C. Lee. 1995. *Mobile cellular telecommunications.* New York: McGraw-Hill.

[8] G. Pollini. 1996. Trends in handover design. *IEEE Communications Magazine* 34:82–90.

[9] P. Bender, P. Black, M. Grob, R. Padavoni, N. Sindhushayana, and S. Viterbi. 2000. CDMA/HDR: A bandwidth efficient high speed wireless data service for nomadic users. *IEEE Communications Magazine* 38:70–77.

[10] A. Jalali, R. Padovani, and R. Pankaj. 2000. Data throughput of CDMA-HDR: A high efficiency-high data rate personal communication wireless system. In *Proceedings of the IEEE Vehicular Technology Conference*, Tokyo, vol. 3, pp. 1854–58.

[11] J. Holtzman. 2000. CDMA forward link waterfilling power control. In *Proceedings of the IEEE Vehicular Technology Conference*, Tokyo, vol. 3, pp. 1663–67.

[12] H. Arslan, L. Krasny, D. Koilpillai, and S. Channakeshu. 2000. Doppler spread estimation for wireless mobile radio systems. In *Proceedings of the IEEE WCNC Conference*, Chicago, vol. 3, pp. 1075–79.

[13] M. Sakamoto, J. Huoponen, and I. Niva. 2000. Adaptive channel estimation with velocity estimator for W-CDMA receiver. In *Proceedings of the IEEE Vehicular Technology Conference*, Tokyo, vol. 3, pp. 2024–28.

[14] D. Mottier and D. Castelain. 1999. A Doppler estimation for UMTS-FDD based on channel power statistics. In *Proceedings of the IEEE Vehicular Technology Conference*, Amsterdam, vol. 5, pp. 3052–56.

[15] L. Lindbom. 1992. Adaptive equalization for fading mobile radio channels. Licentiate thesis, Technology Department, Uppasala University, Uppasala, Sweden.

[16] M. Morelli, U. Mengali, and G. Vitetta. 1998. Further results in carrier frequency estimation for transmissions over flat fading channels. *IEEE Communications Letters* 2:327–30.

[17] L. Krasny, H. Arslan, D. Koilpillai, and S. Channakeshu. 2001. Doppler spread estimation in mobile radio systems. In *Proceedings of the IEEE WCNC Conference*, vol. 5, pp. 197–99.

[18] J. H. A. Sampath. 1993. Estimation of maximum Doppler frequency for handoff decisions. In *Proceedings of the IEEE Vehicular Technology Conference*, Secaucus, NJ, pp. 859–62.

[19] L. Wang, M. Silventoinen, and Z. Honkasalo. 1996. A new algorithm for estimating mobile speed at the TDMA-based cellular system. In *Proceedings of the IEEE Vehicular Technology Conference*, Atlanta, GA, vol. 2, pp. 1145–49.

[20] C. Xiao, K. Mann, and J. Olivier. 1999. Mobile speed estimation for TDMA-based hierarchical cellular systems. In *Proceedings of the IEEE Vehicular Technology Conference*, Amsterdam, vol. 2, pp. 2456–60.

[21] K. Kawabata, T. Nakamura, and E. Fukuda. 1994. Estimating velocity using diversity reception. In *Proceedings of the IEEE Vehicular Technology Conference*, Stockholm, vol. 1, pp. 371–74.

[22] W. Lee. 1998. *Mobile communications engineering.* New York: McGraw-Hill.

[23] M. Austin and G. Stüber. 1994. Eigen-based Doppler estimation for differentially coherent CPM. *IEEE Transactions on Vehicular Technology* 43:781–85.

[24] H. Arslan and T. Yücek. 2003. Delay spread estimation for wireless communication systems. In *Proceedings of the Eighth IEEE Symposium on Computers and Communications (ISCC 2003)*, Antalya, Turkey, pp. 282–87.

[25] J. T. Chen, J. Liang, H.-S. Tsai, and Y.-K. Chen. 1998. Joint MLSE receiver with dynamic channel description. *Journal on Selected Areas in Communications* 16:1604–15.

[26] L. Husson and J.-C. Dany. 1999. A new method for reducing the power consumption of portable handsets in TDMA mobile systems: Conditional equalization. *IEEE Transactions on Vehicular Technology* 48:1936–45.

[27] H. Schober and F. Jondral. 2002. Delay spread estimation for OFDM based mobile communication systems. In *Proceedings of the European Wireless Conference*, Florence, Italy, pp. 625–28.

[28] F. Sanzi and J. Speidel. 2000. An adaptive two-dimensional channel estimator for wireless OFDM with application to mobile DVB-T. *IEEE Transactions on Broadcasting* 46:128–33.
[29] K. Witrisal, Y.-H. Kim, and R. Prasad. 1998. RMS delay spread estimation technique using non-coherent channel measurements. *IEE Electronics Letters* 34:1918–19.
[30] K. Witrisal, Y.-H. Kim, and R. Prasad. 2001. A new method to measure parameters of frequency selective radio channel using power measurements. *IEEE Transactions on Communications* 49:1788–1800.
[31] K. Witrisal and A. Bohdanowicz. 2000. Influence of noise on a novel RMS delay spread estimation method. In *Proceedings of the IEEE PIMRC Conference*, London, vol. 1, pp. 560–66.
[32] H. Arslan and T. Yücek. 2003. Estimation of frequency selectivity for OFDM based new generation wireless communication systems. In *Proceedings of the 2003 World Wireless Congress*, San Francisco.
[33] T. Söderström and P. Stoica. 1997. *Applications of digital wireless technologies to global wireless communications*. Englewood Cliffs, NJ: Prentice Hall.
[34] A. Paulraj and B. Ng. 1998. Space-time modems for wireless personal communications. *IEEE Personal Communications Magazine* 5:36–48.
[35] M. K. Özdemir, H. Arslan, and E. Arvas. 2003. Mutual coupling effect in multi-antenna wireless communication systems. In *IEEE Globecom Conference*, San Francisco, vol. 2, pp. 829–833.
[36] M. K. Özdemir, H. Arslan, and E. Arvas. 2004. Dynamics of spatial correlation and implications on MIMO systems. *IEEE Communications Magazine*.
[37] D. S. Shiu, G. J. Foschini, M. J. Gans, and J. M. Kahn. 2000. Fading correlation and its effects on the capacity of multielement antenna systems. *IEEE Transactions on Communications* 48:502–13.
[38] M. Ivrlac, T. Kurpjuhn, C. Brunner, and W. Utschick. 2001. Efficient use of fading correlations in MIMO systems. In *Proceedings of the IEEE Vehicular Technology Conference*, Atlantic City, NJ, vol. 4, pp. 2763–67.
[39] S. Catreux, V. Erceg, D. Gesbert, and J. Heath. 2002. Adaptive modulation and MIMO coding for broadband wireless data networks. *IEEE Communications Magazine* 40:108–15.
[40] E. Telatar. 1995. *Capacity of multiantenna Gaussian channels*. Technical report. AT&T Bell Laboratories.
[41] G. J. Foschini and M. J. Gans. 1998. On limits of wireless communications in a fading environment when using multiple antennas. *Wireless Personal Communications* 6:311–35.
[42] TIA/EIA. 2000. *TDMA third generation wireless: Digital traffic channel layer 1*. TIA/EIA 136-131-B.
[43] K. Balachandran, S. Kabada, and S. Nanda. 1998. Rate adaptation over mobile radio channels using channel quality information. In *Proceedings of the IEEE Globecom'98 Communication Theory Mini Conference Record*, pp. 46–52.
[44] J. Jacobsmeyer. 1996. Adaptive data rate modem. U.S. Patent 5541955.

[45] M. Türkboylari and G. L. Stüber. 1998. An efficient algorithm for estimating the signal-to-interference ratio in TDMA cellular systems. *IEEE Transactions on Communications* 46:728–31.

[46] H. Arslan and S. Reddy. 2003. Noise variance and SNR estimation for OFDM based wireless communication systems. In *Proceedings of the 3rd International Conference on Wireless and Optical Communications*, Banff, Alberta, Canada.

[47] G. B. H. Arslan. 2001. Channel estimation in narrowband wireless communication systems. *Wireless Communications and Mobile Computing (WCMC) Journal* 1:201–19.

[48] G. Bottomley and H. Arslan. 2003. Channel estimation for time-varying channels in wireless communication systems. In *The Wiley encyclopedia of telecommunications*. New York: Wiley.

[49] K. Homayounfar. 2003. Rate adaptive speech coding for universal multimedia access. *IEEE Signal Processing Magazine* 20:30–39.

[50] S. Wolf, C. Dvorak, R. Kubichek, and C. South. 1991. How will we rate telecommunications system performance? *IEEE Communications Magazine* 29:23–29.

[51] S. Voran. 1999. Objective estimation of perceived speech quality: Part II. *IEEE Transactions on Speech and Audio Processing* 7:383–90.

[52] H. Arslan, R. Ramesh, and A. Mostafa. 1999. Interpolation and channel tracking based receivers for coherent Mary-PSK modulations. In *Proceedings of the IEEE Vehicular Technology Conference*, Houston, TX, vol. 3, pp. 2194–99.

[53] W. Liu. 1994. Performance of joint data and channel estimation using tap variable step size LMS for multipath fast fading channel. In *Proceedings of the IEEE Globecom Conference*, San Francisco, vol. 2, pp. 973–78.

[54] M. Benthin and K.-D. Kammeyer. 1997. Influence of channel estimation on the performance of a coherent DS-CDMA system. *IEEE Transactions on Vehicular Technology* 46:262–68.

[55] P. Schramm. 1999. Differentially coherent demodulation for differential BPSK in spread spectrum systems. *IEEE Transactions on Vehicular Technology* 48:1650–56.

[56] D. Hui and K. Zangi. 2001. An adaptive maximum-likelihood receiver for colored noise and interference. In *Proceedings of the IEEE Vehicular Technology Conference*, Atlantic City, NJ, vol. 4, pp. 2257–61.

[57] A. Hafeez, K. Molnar, and G. Bottomley. 1999. Co-channel interference cancellation for D-AMPS handsets. In *Proceedings of the IEEE Vehicular Technology Conference*, Houston, TX, vol. 2, pp. 1026–30.

[58] A. Hafeez, H. Arslan, and K. Molnar. 2001. Adaptive joint detection of co-channel signals for ANSI-136 handsets. In *Proceedings of the IEEE PIMRC Conference*, San Diego, vol. 2, pp. E-105–10.

[59] J. Zander. 1992. Performance of optimum transmitter power control in cellular radio systems. *IEEE Transactions on Vehicular Technology* 41:57–62.

[60] Z. Rosberg and J. Zander. 1998. Toward a framework for power control in cellular systems. *Wireless Networks* 4:215–22.

[61] S. Ulukus and R. D. Yates. 1998. Stochastic power control for cellular radio systems. *IEEE Transactions on Communications* 46:784–98.

[62] S. V. Hanly and D. N. Tse. 1999. Power control and capacity of spread-spectrum wireless networks. *Automatica* 35:1987–2012.

[63] A. Goldsmith. 1997. The capacity of downlink fading channels with variable rate and power. *IEEE Transactions on Vehicular Technology* 46:569–80.

[64] H. Arslan, T. J.-F. Cheng, and K. Balachandran. 2001. Physical layer evolution for GSM/EDGE. In *Proceedings of the IEEE Globecom Conference*, San Antonio, TX, vol. 5, pp. 8–27.

[65] H. Arslan, T. J.-F. Cheng, and K. Balachandran. 2001. Evolution of EDGE to higher data rates using QAM. In *Proceedings of the IEEE Vehicular Technology Conference*, Atlantic City, NJ, vol. 4, pp. 2267–71.

[66] A. Doufexi, S. Armour, M. Butler, A. Nix, D. Bull, J. McGeehan, and P. Karlsson. 2002. A comparison of the HIPERLAN/2 and IEEE 802.11a wireless LAN standards. *IEEE Communications Magazine* 40:172–80.

[67] I. Katzela and M. Naghshineh. 1996. Channel assignment schemes for cellular mobile telecommunication systems: A comprehensive survey. *IEEE Personal Communications Magazine* 3:10–31.

[68] N. D. Tripathi, N. J. H. Reed, and H. F. V. Landingham. 1998. Handoff in cellular systems. *IEEE Personal Communications Magazine* 5:26–37.

[69] G. Pollini. 1996. Trends in handover design. *IEEE Communications Magazine* 3:82–90.

[70] G. Aiello and G. Rogerson. 2003. Ultrawideband wireless systems. *IEEE Microwave Magazine* 4(2):36–47.

[71] J. Forrester, G. Evan, D. Leeper, and S. Srinivasa. 2001. Ultra-wideband technology for short or medium-range wireless communications. *Intel Technology Journal* 1–7.

[72] P. Mannion. 2002. Ultrawideband radio set to redefine wireless signaling. *EE Times*, pp. 71–84.

[73] L. Hanzo, C. H. Wong, and M. S. Yee. 2002. *Adaptive wireless transceivers: Turbo-coded, turbo-equalized and space-time coded TDMA, CDMA, and OFDM systems*. New York: John Wiley & Sons.

[74] S. Reddy, T. Yucek, and H. Arslan. 2003. An efficient blind modulation detection algorithm for adaptive OFDM systems. In *Proceedings of the IEEE Vehicular Technology Conference*, Orlando, FL, vol. 6, pp. 8–27.

[75] A. Goldsmith and S. Wicker. 2002. Design challenges for energy-constrained ad hoc wireless networks. *IEEE Wireless Communications* 9(4):8–27.

[76] T. Rappaport, A. Annamalai, R. Buehrer, and W. Tranter. 2002. Wireless communications: Past events and a future perspective. *IEEE Communications Magazine* 40:148–61.

[77] Z. Haas. 2001. Design methodologies for adaptive and multimedia networks. *IEEE Communications Magazine* 39:106–7.

2

Adaptive Channel Estimation in Wireless Communications

Jitendra K. Tugnait
Auburn University

2.1 Introduction

Propagation of signals through wireless channels (indoors or outdoors) results in the transmitted signal arriving at the receiver through multiple paths. These paths arise due to reflection, refraction, or diffraction in the channel. Multipath propagation results in a received signal that is a superposition of several delayed and scaled copies of the transmitted signal giving rise to frequency-selective fading. Frequency-selective fading (defined as changes in the received signal level in time) is caused by destructive interference among multiple propagation paths. The environment around the transmitter and the receiver can change over time, particularly in a mobile setting, leading to variations in the channel response with time. This gives rise to time-selective fading. Also, the channels may have a dominant path (direct path in line-of-sight channels) in addition to several secondary paths, or they may be characterized as having multiple "random" paths with no single dominant path.

Multipath propagation leads to intersymbol interference (ISI) at the receiver, which in turn may lead to high error rates in symbol detection. Equalizers are designed to compensate for these channel distortions. One may directly design an equalizer given the received signal, or one may first estimate the channel impulse response and then design an equalizer based on the estimated channel. After some processing (matched filtering, for instance), the continuous-time received signals are sampled at the baud (symbol) or higher (fractional) rate before processing them for channel estimation or equalization. It is therefore convenient to work with a baseband-equivalent discrete-time channel model. Depending upon the sampling rate, one has either a single-input single-output (SISO) (baud-rate sampling) or a single-input multiple-output (SIMO) (fractional sampling), complex discrete-time-equivalent baseband channel.

In this chapter, we present a review of various approaches to channel estimation for wireless mobile systems. Since approaches to channel estimation depend upon the underlying channel model, we also review various approaches to channel modeling. In section 2.2 we present the relevant channel models, including time-variant and time-invariant models. In section 2.3 various channel estimation methods suitable for block-by-block tracking are presented. In section 2.4 adaptive channel estimation approaches are reviewed. In section 2.5 we illustrate some of the reviewed approaches via two simulation examples to conclude the chapter.

Notation: Superscripts H, $*$, T, and $\dagger$ denote the complex conjugate transpose, complex conjugation, transpose, and Moore-Penrose pseudo-inverse operations, respectively. $\delta(\cdot)$ is the Kronecker delta function and $\mathbf{I}_N$ is the $N \times N$ identity matrix. The symbol $\otimes$ denotes the Kronecker product, and $tr(\mathbf{A})$ is the trace of a square matrix $\mathbf{A}$. The $(n,m)^{\text{th}}$ entry of a matrix $\mathbf{C}$ is denoted by $[\mathbf{C}]_{n,m}$.

2.2 Wireless Channel Models

2.2.1 Time-Variant (Doubly Selective) Channels

Consider a time-varying (e.g., mobile wireless) channel (linear system) with complex baseband, continuous-time, received signal $x(t)$ and transmitted complex baseband, continuous-time information signal $s(t)$ (with symbol interval T_s seconds) related by [12]

$$x(t) = \int_{-\infty}^{\infty} \tilde{h}(t;\tau)s(t-\tau)d\tau + w(t), \tag{2.1}$$

where $\tilde{h}(t;\tau)$ is the time-varying impulse response of the channel denoting the response of the channel at time t to a unit impulse input at time $t-\tau$, and $w(t)$ is the additive noise (typically white Gaussian). A delay Doppler spread function $H(f;\tau)$ is defined as the Fourier transform of $\tilde{h}(t;\tau)$ [2, 42]:

$$H(f;\tau)=\int_{-\infty}^{\infty}\tilde{h}(t;\tau)exp^{-j2\pi ft}\,dt\,. \tag{2.2}$$

If $|H(f;\tau)|\approx 0$ for $|\tau|>\tau_d$, then τ_d is called the (multipath) delay spread of the channel. If $|H(f;\tau)|\approx 0$ for $|f|>f_d$, then f_d is called the Doppler spread of the channel. Equation (2.1) is the most general form of a mobile channel discussed in this chapter.

In order to capture the complexity of the physical interactions characterizing the transmission through a real channel, $\tilde{h}(t;\tau)$ is typically modeled as a two-dimensional zero-mean random process. If $\tilde{h}(t;\tau)$ is wide-sense stationary in variable t, and $\tilde{h}(t;\tau_1)$ is uncorrelated with $\tilde{h}(t;\tau_2)$ for $\tau_1\neq\tau_2$ and any t, one obtains the well-known wide-sense stationary uncorrelated scattering (WSSUS) channel [2, 42, section 14].

In this chapter we will confine our attention to deterministic modeling of $\tilde{h}(t;\tau)$, which may be thought of as capturing realizations of the underlying random process.

2.2.1.1 Tapped Delay Line Model

We now consider a discrete-time channel model. If a linear modulation scheme is used, the baseband transmitted signal can be represented as

$$s(t)=\sum_{k=-\infty}^{\infty}s(k)g_T(t-kT_s), \tag{2.3}$$

where $\{s(k)\}$ is the information sequence and $g_T(t)$ is the transmit (low-pass) filter (typically a root raised cosine filter). Therefore, the baseband signal incident at the receiver is given by

$$x(t)=\sum_{k=-\infty}^{\infty}s(k)\int_{-\infty}^{\infty}\tilde{h}(t;\alpha)g_T(t-kT_s-\alpha)d\alpha+w(t). \tag{2.4}$$

After filtering with a receive filter with impulse response $g_R(t)$, the received baseband signal is given by

$$y(t)=\sum_{k=-\infty}^{\infty}s(k)\int_{-\infty}^{\infty}\int_{-\infty}^{\infty}g_R(t-\beta)\tilde{h}(\beta;\alpha)g_T(\beta-kT_s-\alpha)d\alpha d\beta+v(t), \tag{2.5}$$

where

$$v(t)=\int g_R(\tau)w(t-\tau)d\tau.$$

If the continuous-time signal $y(t)$ is sampled once every T_s seconds, we obtain the discrete-time sequence

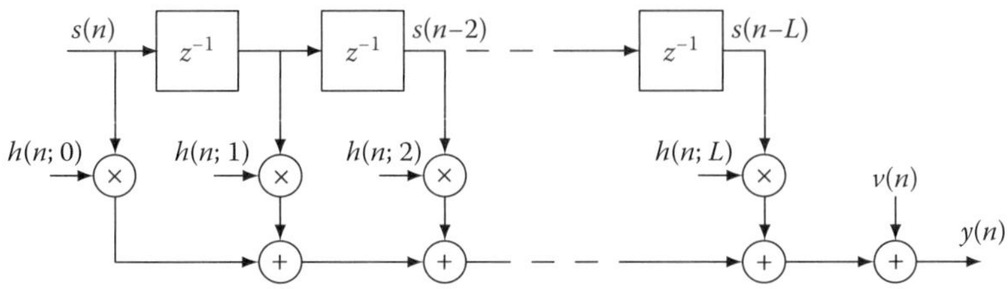

FIGURE 2.1 Tapped delay line model of frequency- and time-selective channel with finite impulse response. z^{-1} represents a unit (symbol duration) delay.

$$y(n) := y(t)|_{t=nT_s} = \sum_{k=-\infty}^{\infty} s(k)h(n;n-k) + v(n) = \sum_{l=-\infty}^{\infty} h(n;l)s(n-l) + v(n), \tag{2.6}$$

where $h(n;l)$ is the (effective) channel response at time n to a unit input at time $n - l$ and

$$h(n;n-k) := \int_{-\infty}^{\infty}\int_{-\infty}^{\infty} g_R(nT_s - \beta)\tilde{h}(\beta;\alpha)g_T(\beta - kT_s - \alpha)d\alpha d\beta. \tag{2.7}$$

Note that the noise sequence $\{v(n)\}$ in (2.6) is no longer necessarily white; it can be whitened by further time-invariant linear filtering (see [42]). Henceforth, we assume that a whitening filter has been applied to $y(n)$, but with an abuse of notation, we will still use (2.6). For a causal system, $h(n;l) = 0$ for $l < 0$ ($\forall n$), and for a finite-length channel of maximum length T_sL, $h(n;l) = 0$ for $l > L$ ($\forall n$). In this case we modify (2.6) as (recall the noise whitening filter)

$$y(n) = \sum_{l=0}^{L} h(n;l)s(n-l) + v(n). \tag{2.8}$$

The model (2.6) represents a time- and frequency-selective linear channel. A tapped delay line structure for this model is shown in Figure 2.1. For a slowly (compared to the baud-rate) time-varying system, one often simplifies (2.8) to a time-invariant system as

$$y(n) = \sum_{l=0}^{L} h(l)s(n-l) + v(n), \tag{2.9}$$

where $h(l) = h(0;l)$ is the time-invariant channel response to a unit input at time 0. The model (2.9) represents a frequency-selective linear channel with no time selectivity. It is the most commonly used model for receiver design.

Suppose that $h(n;l) = h(n)\delta(l,0)$ where $\delta(l,0)$ is the Kronecker delta located at 0, i.e., $\delta(l,0) = 1$ for $l = 0$ and $\delta(l,0) = 0$ for $l \neq 0$. Then we have the time-selective and frequency-non-selective channel whose output is given by

$$y(n) = h(n)s(n) + v(n). \tag{2.10}$$

Finally, a time-non-selective and frequency-non-selective channel is modeled as

$$y(n) = hs(n) + v(n), \tag{2.11}$$

where h is a random variable (or a constant).

2.2.1.2 Autoregressive (AR) Models

It is possible to accurately represent a WSSUS channel by a large-order AR model; see [34] and references therein. However, it is far more common to use a first-order AR model (*AR* (1)) given by [34, 35]

$$h(n;l) = \alpha_c h(n-1;l) + w_c(n), \tag{2.12}$$

where α_c is the AR coefficient, and the driving noise $w_c(n)$ is zero-mean complex Gaussian with variance σ_{wc}^2 and statistically independent of $h(n-1;l)$. Assume that $h(n;l)$ is also zero-mean complex Gaussian with variance σ_h^2. Then [35]

$$\alpha_c = \frac{1}{\sigma_h^2} E\{h(n;l)h^*(n-1;l)\}, \tag{2.13}$$

$$\sigma_{wc}^2 = \sigma_h^2 \left(1 - |\alpha_c|^2\right). \tag{2.14}$$

2.2.1.3 Basis Expansion Models

Recently, basis expansion models (BEMs) have been widely investigated to represent doubly selective channels in wireless applications [3, 13, 36, 46, 61], where the time-varying taps are expressed as superpositions of time-varying basis functions in modeling Doppler effects, weighted by time-invariant coefficients. Candidate basis functions include complex exponential (Fourier) functions [13, 36], polynomials [3], discrete prolate spheroidal sequences [61], etc. In contrast to AR models that describe temporal variation on a symbol-by-symbol update basis, a BEM depicts the evolution of the channel over a period (block) of time. Intuitively, the coefficients of the BEM approximation should evolve much more slowly in time than the channel, and hence are more convenient to track in a fast-fading environment.

Suppose that we include the effects of transmit and receive filters in the time-variant impulse response $h(t;\tau)$ in (2.1). Suppose that this channel has a delay spread τ_d and a Doppler spread f_d. Consider the kth block of data consisting of an observation window of T_B symbols where the baud-rate data samples in the block are indexed as $n = \bar{n}_k, \bar{n}_k + 1, \cdots, \bar{n}_k + T_B - 1, \bar{n}_k := (k-1)T_B$. If $2f_d\tau_d < 1$ (underspread channel), the complex exponential basis expansion model (CE-BEM) representation of $h(n;l)$ in (2.6) is given by

$$h(n;l)=\sum_{q=1}^{Q}h_q(l)e^{j\omega_q n},\ n=\bar{n}_k,\bar{n}_k+1,\cdots,\bar{n}_k+T_B-1, \tag{2.15}$$

where one chooses ($l = 0, 1, \cdots, L$, and K is an integer)

$$T:=KT_B,\quad K\geq 1, \tag{2.16}$$

$$Q\geq 2\lceil f_d T T_s\rceil+1, \tag{2.17}$$

$$\omega_q:=\frac{2\pi}{T}\left[q-(Q+1)/2\right],\quad q=1,2,\cdots,Q, \tag{2.18}$$

$$L:=\lfloor \tau_d/T_s\rfloor. \tag{2.19}$$

The BEM coefficients $h_q(l)$ remain invariant during this block, but are allowed to change at the next block, and the Fourier basis functions $\{e^{j\omega_q n}\}$ ($q = 1, 2, \cdots, Q$) are common for each block. If the delay spread τ_d and the Doppler spread f_d of the channel (or at least their upper bounds) are known, one can infer the basis functions of the CE-BEM [36]. Treating the basis functions as known, estimation of a time-varying process is reduced to estimating the invariant coefficients over a block of length T_B symbols. Note that the BEM period is $T = KT_B$, whereas the block size is T_B symbols. If $K > 1$ (e.g., $K = 2$ or $K = 3$), then the Doppler spectrum is said to be oversampled [32] compared to the case $K = 1$, where the Doppler spectrum is said to be critically sampled. In [13, 36] only $K = 1$ (henceforth called CE-BEM) is considered, whereas [32] considers $K \geq 2$ (henceforth called oversampled CE-BEM).

CE-BEM has a finite impulse response (FIR) structure in both time and frequency domains [46]. This unique time-frequency duality makes it a widely used model depicting the temporal variations of wireless channels. For $K = 1$, the rectangular window of this truncated discrete Fourier transform (DFT)-based model introduces spectral leakage [43]. The energy at each individual frequency leaks to the full-frequency range, resulting in significant amplitude and phase distortion at the beginning and the end of the observation window [61]. To mitigate this leakage, the oversampled CE-BEM with $K = 2$ or 3 has been considered in [32].

Equation (2.15) applies to single input single-output systems—one user and one receiver with symbol-rate sampling. It is easily modified to handle multiuser, multiple-transmit and -receive antennas, and higher-than-symbol-rate sampling—the basic representation remains essentially unchanged.

The representation $h(n;l)$ in (2.15) is a special case of a more general representation:

$$h(n;l)=\sum_{q=1}^{Q}h_q(l)\phi_q(n), \tag{2.20}$$

where $\{\phi_q(n)\}_{q=1}^{Q}$ are a set of orthogonal basis functions (over the time interval under consideration). Examples include wavelet-based expansions as in [37], polynomial bases as in [3], and other possibilities [40].

In discrete prolate spheroidal BEM (DPS-BEM), the ith DPS vector $\mathbf{u}_i := [u_i(0), \cdots, u_i(T_B - 1)]^T$ (called Slepian sequence in [61], which is a time-windowed [infinite] DPS sequence) is the ith eigenvector of a matrix **C** [48]: $\mathbf{CU}_i = \lambda_i \mathbf{u}_i$, where

$$\left[\mathbf{C}\right]_{n,m} = \frac{\sin\left[2\pi\left(n-m\right)f_d T_s\right]}{\left[\pi\left(n-m\right)\right]}$$

is the $(n,m)^{\text{th}}$ entry of **C** and $\lambda_1 \geq \lambda_2 \geq \cdots \geq \lambda_{T_B}$ are the eigenvalues of **C**. The Slepian sequences $\{u_q(n)\}$ are orthonormal over the finite time interval $[0,T_B - 1]$. The modeling error of the CE-BEM can result in a noticeable floor in BER curves [18]. The polynomial basis functions are neither time limited nor band limited, and their square bias varies heavily over the range of Doppler spread considered in [61]. DPS sequences are a good alternative as a basis set to approximate band-limited channels alleviating the spectral leakage of CE-BEM [61]. The (infinite) DPS sequences have their maximum energy concentration in an interval with length T while being band limited to $[-f_d T_s, f_d T_s]$, where $u_1(n)$ is the unique sequence that is band limited and most time concentrated, $u_2(n)$ is the next sequence having maximum energy concentration among the DPS sequences orthogonal to $u_1(n)$, and so on [48].

Figure 2.2 shows the channel modeling errors resulting from (critically sampled) CE-BEM, DPS-BEM, and oversampled CE-BEM ($K = 2$ or 3) when the underlying channel is a one-tap time-selective channel following Jakes' spectrum [22]. The results are

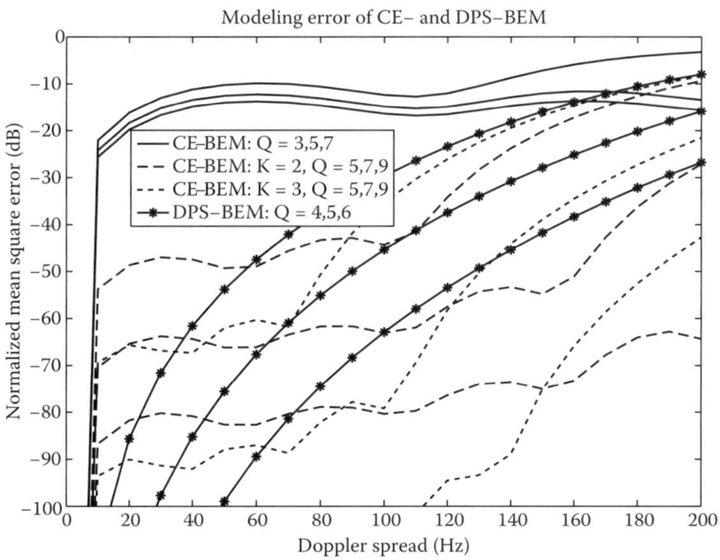

FIGURE 2.2 Channel modeling error for one-tap Jakes' channel, $T_B = 400$, $T_S = 25$ μs.

based on Monte Carlo averaging over one thousand runs with $T_B = 400$, $T_s = 25$ μs and varying Doppler spreads. (The results were obtained following the procedure in [61].) For a fixed value of Q, DPS-BEM provides the best fit, whereas CE-BEM (no oversampling) yields minor improvements with increasing Q. On the other hand, the basis functions in oversampled CE-BEM are not mutually orthogonal, leading to "analytical" difficulties. It is interesting to note that there exists a vast literature based on CE-BEM (no oversampling) where the large modeling errors are completely ignored and it is assumed that physical channel is accurately described by CE-BEM for both analysis and simulations; see, e.g., [1, 24–26, 31, 36, 45].

2.2.2 Time-Invariant Channels

After some processing (matched filtering, for instance), the continuous-time received signals are sampled at the baud (symbol) or higher (fractional) rate before being processed for channel estimation or equalization. It is therefore convenient to work with an equivalent baseband discrete-time white noise channel model [42, section 10.1]. For a baud-rate sampled system, the equivalent baseband channel model is given by

$$y(n)=\sum_{l=0}^{L} h(l)s(n-l)+v(n), \tag{2.21}$$

where $\{v(n)\}$ is a white Gaussian noise sequence with variance σ^2; $\{s(n)\}$ is the zero-mean, independent and identically distributed (i.i.d.), information (symbol) sequence, possibly complex, taking values from a finite set; $\{h(l)\}$ is an FIR linear filter (with possibly complex coefficients) that represents the equivalent channel, including the effects of the noise whitening filter; and $\{y(n)\}$ is the (possibly complex) equivalent baseband received signal. A tapped delay line structure for this model is shown in Figure 2.3.

The model (2.21) results in a single-input single-output (SISO) complex discrete-time baseband-equivalent channel model. The output sequence $\{y(n)\}$ in (2.21) is discrete-time stationary. When there is excess channel bandwidth [bandwidth > $\frac{1}{2}\times$ (baud rate)], baud-rate sampling is below the Nyquist rate, leading to aliasing and, depending upon the symbol timing phase, in certain cases, causing deep spectral notches in the sampled, aliased channel transfer function [15]. Linear equalizers designed on the basis of the

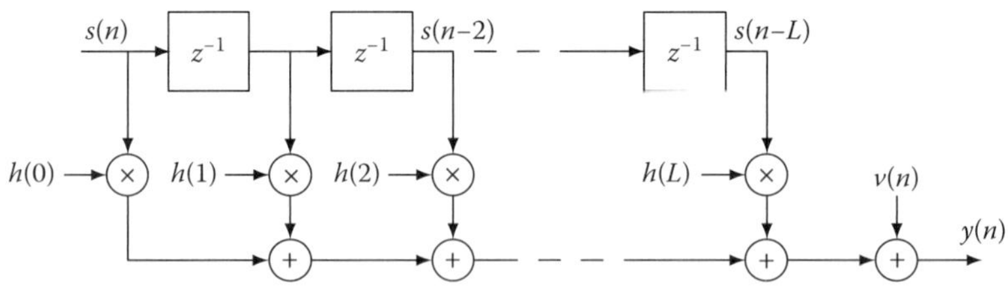

FIGURE 2.3 Tapped delay line model of the frequency-selective but time-non-selective baud-rate channel.

baud-rate sampled channel response are quite sensitive to symbol timing errors. Initially, in the trained case, fractional sampling was investigated to "robustify" the equalizer performance against timing errors. The model (2.21) does not apply to fractionally spaced samples, i.e., when the sampling interval is a fraction of the symbol duration. The fractionally sampled digital communications signal is a cyclostationary signal [9] that may be represented as a vector stationary sequence using a time series representation (TSR) [9, section 12.6]. Suppose that we sample at N times the baud rate with signal samples spaced T_s/N seconds apart where T_s is the symbol duration. Then a TSR for the sampled signal is given by

$$y_i(n)=\sum_{l=0}^{L} h_i(l)s(n-l)+v_i(n); \quad (i=1,2,\cdots,N), \tag{2.22}$$

where now we have N samples every symbol period, indexed by i. Notice, however, that the information sequence $s(n)$ is still one sample per symbol. It is assumed that the signal incident at the receiver is first passed through a receive filter whose transfer function equals the square root of a raised cosine pulse, and that the receive filter is matched to the transmit filter. The noise sequence in (2.22) is the result of the fractional-rate sampling of a continuous-time filtered white Gaussian noise process. Therefore, the sampled noise sequence is white at the symbol rate, but correlated at the fractional rate. Stack N consecutive received samples in the nth symbol duration to form an N vector $y(n)$ satisfying

$$\mathbf{y}(n)=\sum_{l=0}^{L} \mathbf{h}(l)s(n-l)+\mathbf{v}(n), \tag{2.23}$$

where $\mathbf{h}(n)$ is the vector impulse response of the SIMO-equivalent channel model given by

$$\mathbf{h}(n)=\begin{bmatrix} h_1(n) & h_2(n) & \cdots & h_N(n) \end{bmatrix}^T, \tag{2.24}$$

and $\mathbf{y}(n)$ and $\mathbf{v}(n)$ are defined similarly. A block diagram of model (2.22) is shown in Figure 2.4.

2.3 Channel Estimation

We first consider three types of channel estimators within the framework of maximizing the likelihood function. (Unless otherwise noted, the underlying channel model is given by the time-invariant model (2.23).) In general, one of the most effective and popular parameter estimation algorithms is the maximum likelihood (ML) method. The class of maximum likelihood estimators are optimal asymptotically.

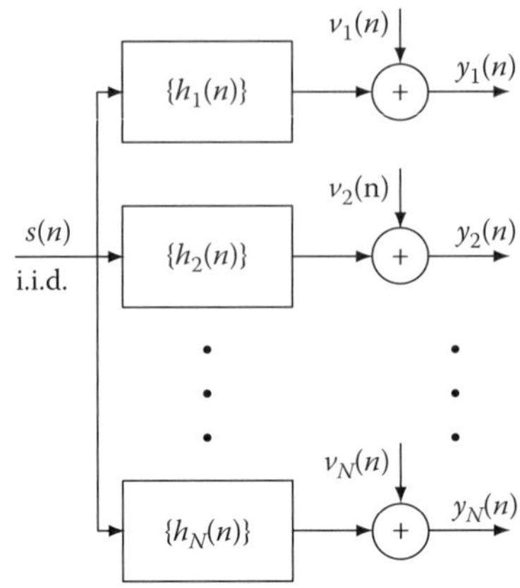

FIGURE 2.4 Block diagram of the fractionally sampled ($N\times$ baud-rate) frequency-selective but time-non-selective channel.

Let us consider the N vector channel model given in (2.23). Suppose that we have collected M samples of the observation $Y = [\mathbf{y}(T_B - 1), \cdots, \mathbf{y}^T(0)]^T$. We then have the following linear model:

$$Y = \begin{pmatrix} s(T_B-1)\mathbf{I}_N & s(T_B-2)\mathbf{I}_N & \cdots & s(T_B-L-1)\mathbf{I}_N \\ \vdots & \textit{Block} & \textit{Hankel} & \textit{Matrix} \\ s(0)\mathbf{I}_N & s(-1)\mathbf{I}_N & \cdots & s(-L)\mathbf{I}_N \end{pmatrix} \begin{pmatrix} \mathbf{h}(0) \\ \vdots \\ \mathbf{h}(L) \end{pmatrix} + \begin{pmatrix} \mathbf{v}(T_B-1) \\ \vdots \\ \mathbf{v}(0) \end{pmatrix}, \tag{2.25}$$

$$= \mathcal{T}(\mathbf{s})_{[T_B N]\times[N(L+1)]}\mathcal{H} + V$$

where $\mathbf{I}_N$ is a $N \times N$ identity matrix, s and V are vectors consisting of samples of the input sequence $\{s(n)\}$ and noise $\{\mathbf{v}(n)\}$, respectively, $\mathcal{H}$ is the vector of the channel parameters, and a block Hankel matrix has identical block entries on its block antidiagonals.

Let θ be the vector of unknown parameters that may include the channel parameters $\mathcal{H}$ and possibly the entire or part of the input vector s. Given the probability space that describes jointly the noise vector W and possibly the input data vector s, we can then obtain, in principle, the probability density function (pdf) of the observation Y. As a function of the unknown parameter θ, the pdf of the observation $f(Y|\theta)$ is referred to as the *likelihood function*. The maximum likelihood estimator is defined by the following optimization:

$$\hat{\theta} = \arg\max_{\theta\in\Theta} f(Y\,|\,\theta), \tag{2.26}$$

where Θ defines the domain of the optimization.

While the ML estimator is conceptually simple, and it usually has good performance when the sample size is sufficiently large, the implementation of the ML estimator is sometimes computationally intensive. Furthermore, the optimization of the likelihood function in (2.26) is often hampered by the existence of local maxima. Therefore, it is desirable that effective initialization techniques are used in conjunction with the ML estimation.

2.3.1 Training-Based Channel Estimation

The training-based channel estimation assumes the availability of the input vector s (as training symbols) and its corresponding observation vector Y. When the noise samples are zero mean, white Gaussian, i.e., v is a zero-mean Gaussian random vector with covariance $\sigma_v^2 \mathbf{I}_{TBN}$, the ML estimator defined in (2.26), with $\theta = \mathcal{H}$, is given by

$$\hat{\mathcal{H}} = \arg\min_{\mathcal{H}} \left\| Y - \mathcal{T}(\mathbf{s})\mathcal{H} \right\|^2 = \mathcal{T}^{\dagger}(\mathbf{s})Y, \tag{2.27}$$

where $\mathcal{T}^{\dagger}(\mathbf{s})$ is the Moore-Penrose pseudo-inverse of the $\mathcal{T}(\mathbf{s})$ defined in (2.25). This is also the classical linear least squares estimator, which can be implemented recursively, and it turns out to be the best (in terms of having minimum mean square error) among all unbiased estimators, and it is the most efficient in the sense that it achieves the Cramer-Rao lower bound. Various adaptive implementations can be found in [42].

2.3.1.1 Time-Variant Channels

In case of general time-varying channels represented by (2.8), a simple generalization of [4] (see also [36]) is to use a periodic Kronecker delta function sequence as training:

$$s(n) = \sum_j \delta(n - j\bar{P}). \tag{2.28}$$

With (2.28) as input to model (2.8), one obtains

$$y(n) = \sum_j h(n; n - j\bar{P}) + v(n), \tag{2.29}$$

so that if $\bar{P} > L$, we have for $0 \leq i \leq L$,

$$y(k\bar{P} + i) = \sum_j h(k\bar{P} + i; i) + v(k\bar{P} + i). \tag{2.30}$$

Therefore, one may take the estimate of $h(k\bar{P}; i)$ as

$$\hat{h}(k\bar{P}; i) = y(k\bar{P} + i) = h(k\bar{P} + i; i) + v(k\bar{P} + i). \tag{2.31}$$

For time samples between $k\bar{P}$ (k is an integer), linear interpolation may be used to obtain channel estimates.

If we use a CE-BEM representation (2.20), then one directly estimates the time-invariant parameters $h_q(l)$. From (2.8) and (2.20) we have

$$y(n)=\sum_{l=0}^{L}\sum_{q=1}^{Q}h_q(l)\phi_q(n)s(n-l)+v(n) \tag{2.32}$$

$$=\sum_{l=0}^{L}s(n-l)\underbrace{\left[\phi_q(n)\quad\cdots\quad\phi_q(n)\right]}_{E(n)}\underbrace{\begin{bmatrix}h_1(l)\\ \vdots\\ h_Q(l)\end{bmatrix}}_{H(l)}+v(n). \tag{2.33}$$

Collecting T_B samples of the observations $Y = [y(T_B - 1), \cdots, y(0)]^T$ we have the linear model

$$Y=\begin{pmatrix}s(T_B-1)E(T_B-1) & \cdots & s(T_B-L-1)E(T_B-1)\\ \vdots & & \vdots\\ I(0)E(0) & \cdots & I(-L)E(0)\end{pmatrix}\begin{pmatrix}H(0)\\ \vdots\\ H(L)\end{pmatrix}+\begin{pmatrix}v(T_B-1)\\ \vdots\\ v(0)\end{pmatrix}. \tag{2.34}$$

Now we have a model similar to (2.25) with a solution similar to (2.27).

2.3.2 Blind Channel Estimation

2.3.2.1 Combined Channel and Symbol Estimation

The simultaneous estimation of the input vector and the channel appears to be ill-posed; how is it possible that the channel and its input can be distinguished using only the observation? The key in blind channel estimation is the utilization of qualitative information about the channel and the input. To this end, we consider two different types of maximum likelihood techniques based on different models of the input sequence.

2.3.2.1.1 Stochastic Maximum Likelihood Estimation

While the input vector s is unknown, it may be modeled as a random vector with a known distribution. In such a case, the likelihood function of the unknown parameter $\theta = \mathcal{H}$ can be obtained by

$$f(Y\,|\,\mathcal{H})=\int f(Y\,|\,\mathbf{s},\mathcal{H})f(\mathbf{s})d\mathbf{s}, \tag{2.35}$$

where $f(\mathbf{s})$ is the marginal pdf of the input vector and $f(Y|\mathbf{s},\mathcal{H})$ is the likelihood function when the input is known. Assume, for example, that the input data symbol $s(k)$ takes, with equal probability, a finite number of values. Consequently, the input data vector s also takes values from the signal set $\{\mathbf{s}_1, \ldots, \mathbf{s}_K\}$. The likelihood function of the channel parameters is then given by

$$f(Y|\mathcal{H})=\sum_{i=1}^{K} f(Y|\mathbf{s}_i,\mathcal{H})\text{Prob}(\mathbf{s}=\mathbf{s}_i)=C\sum_{i=1}^{K}\exp\left\{-\frac{\left\|Y-\mathcal{T}(\mathbf{s}_i)\mathcal{H}\right\|^2}{2\sigma^2}\right\}, \tag{2.36}$$

where C is a constant, $\|Y\|^2 := Y^H Y$, Y^H is the complex conjugate transpose of the complex vector Y, and the stochastic maximum likelihood estimator is given by

$$\hat{\mathcal{H}}=\arg\min_{\mathcal{H}}\sum_{i=1}^{K}\exp\left\{-\frac{\left\|Y-\mathcal{T}(\mathbf{s}_i)\mathcal{H}\right\|^2}{2\sigma^2}\right\}. \tag{2.37}$$

The maximization of the likelihood function defined in (2.35) is in general difficult because $f(Y|\theta)$ is nonconvex. The expectation-maximization (EM) algorithm can be applied to transform the complicated optimization to a sequence of quadratic optimizations. Kaleh and Vallet [23] first applied the EM algorithm to the equalization of communication channels with input sequence having the finite alphabet property. By using a *hidden Markov model* (HMM), they developed a batch (off-line) procedure that includes the so-called forward and backward recursions. Unfortunately, the complexity of this algorithm increases exponentially with the channel memory.

To relax the memory requirements and facilitate channel tracking, on-line sequential approaches have been proposed in [28] for input with finite alphabet properties under an HMM formulation. Given the appropriate regularity conditions and a good initialization guess, it can be shown that these algorithms converge to the true channel value.

2.3.2.1.2 Deterministic Maximum Likelihood Estimation

The deterministic ML approach assumes no statistical model for the input sequence $\{s(k)\}$. In other words, both the channel vector $\mathcal{H}$ and the input source vector s are parameters to be estimated. When the noise is zero-mean Gaussian with covariance $\sigma_v^2\mathbf{I}_{T_BN}$, the ML estimates can be obtained by the nonlinear least squares optimization

$$\{\hat{\mathcal{H}},\hat{\mathbf{s}}\}=\text{argmin}\left\|Y-\mathcal{T}(\mathbf{s})\mathcal{H}\right\|^2. \tag{2.38}$$

The joint minimization of the likelihood function with respect to both the channel and the source parameter spaces is difficult. Fortunately, the observation vector Y is linear in both the channel and the input parameters individually. In particular, we have

$$Y = \mathcal{T}(\mathbf{s})\mathcal{H} + V = F(\mathcal{H})\mathbf{s} + V, \tag{2.39}$$

where

$$F(\mathcal{H}) = \begin{pmatrix} \mathbf{h}(0) & \cdots & \mathbf{h}(L) & & \\ & \ddots & & \ddots & \\ & & \mathbf{h}(0) & \cdots & \mathbf{h}(L) \end{pmatrix} \tag{2.40}$$

is the the so-called filtering matrix. We therefore have a separable nonlinear least squares problem that can be solved sequentially:

$$\{\hat{\mathcal{H}}, \hat{\mathbf{s}}\} = \arg\min_{\mathbf{s}} \left\{ \min_{\mathcal{H}} \left\| Y - \mathcal{T}(\mathbf{s})\mathcal{H} \right\|^2 \right\} \tag{2.41}$$

$$= \arg\min_{\mathcal{H}} \left\{ \min_{\mathbf{s}} \left\| Y - F(\mathcal{H})\mathbf{s} \right\|^2 \right\}. \tag{2.42}$$

If we are only interested in estimating the channel, the above minimization can be rewritten as

$$\hat{\mathcal{H}} = \arg\min_{H} \left\| \underbrace{(\mathbf{I} - F(\mathcal{H})\mathcal{F}^{\dagger}(\mathcal{H}))}_{P(\mathcal{H})} Y \right\|^2 = \arg\min_{\mathcal{H}} \left\| \mathcal{P}(\mathcal{H})Y \right\|^2, \tag{2.43}$$

where $\mathcal{P}(\mathcal{H})$ is a projection transform of Y into the orthogonal complement of the range space of $\mathcal{F}(\mathcal{H})$, or the noise subspace of the observation, and $\mathcal{F}^{\dagger}(\mathcal{H})$ denotes the pseudo-inverse of $\mathcal{F}(\mathcal{H})$. Discussions of algorithms of this type can be found in [50].

Similar to the HMM for the statistical maximum likelihood approach, the finite alphabet properties of the input sequence can also be incorporated into the deterministic maximum likelihood methods. These algorithms, first proposed by Seshadri [47] and Ghosh and Weber [12], iterate between estimates of the channel and the input. At iteration k, with an initial guess of the channel $\mathcal{H}^{(k)}$, the algorithm estimates the input sequence $\mathbf{s}^{(k)}$ and the channel $\mathcal{H}^{(k+1)}$ for the next iteration by

$$\mathbf{s}^{(k)} = \arg\min_{\mathbf{s} \in \mathcal{S}} \left\| Y - \mathcal{F}(\mathcal{H}^{(k)})\mathbf{s} \right\|^2 \tag{2.44}$$

$$\mathcal{H}^{(k+1)} = \arg\min_{H} \left\| Y - \mathcal{T}(\mathbf{s}^{(k)})\mathcal{H} \right\|^2, \tag{2.45}$$

where $\mathcal{S}$ is the (discrete) domain of $\mathbf{s}$. The optimization in (2.45) is a linear least squares problem, whereas the optimization in (2.44) can be achieved by using the Viterbi algorithm [42]. Seshadri [47] presented blind trellis search techniques. Reduced-state sequence estimation was proposed in [12]. Raheli et al. proposed a per-survivor processing technique in [44].

The convergence of such approaches is not guaranteed in general. Interesting examples have been provided in [5] where two different combinations of $\mathcal{H}$ and $\mathbf{s}$ lead to the same cost:

$$\left\| Y - \mathcal{T}(\mathbf{s})\mathcal{H} \right\|^2.$$

2.3.2.2 The Methods of Moments

Although the ML channel estimator discussed in section 2.3.2.1 usually provides better performance, the computation complexity and the existence of local optima are the two major difficulties. Therefore, simpler approaches have also been investigated.

2.3.2.2.1 SISO Channel Estimation

For baud-rate data, second-order statistics of the data do not carry enough information to allow estimation of the channel impulse response as a typical channel is non-minimum phase. On the other hand, higher-order statistics (in particular, fourth-order cumulants) of the baud-rate (or fractional-rate) data can be exploited to yield the channel estimates to within a scale factor.

Given the mathematical model (2.21), there are two broad classes of direct approaches to channel estimation, the distinguishing feature among them being the choice of the optimization criterion. All of the approaches involve (more or less) a least squares error measure. The error definition differs, however, as follows:

- *Fitting error*: Match the model-based higher-order (typically fourth-order) statistics to the estimated (data-based) statistics in a least squares sense to estimate the channel impulse response, as in [54] and [55], for example. This approach allows consideration of noisy observations. In general, it results in a nonlinear optimization problem. It requires availability of a good initial guess to prevent convergence to a local minimum. It yields estimates of the channel impulse response.
- *Equation error*: This is based on minimizing an "equation error" in some equation that is satisfied ideally. The approaches of [17] and [60] (among others) fall in this category. In general, this class of approaches results in a closed-form solution for the channel impulse response so that a global extremum is always guaranteed provided that the channel length (order) is known. These approaches may also provide good initial guesses for the nonlinear fitting error approaches. Quite a few of these approaches fail if the channel length is unknown.

Further details may be found in [14, 52, 56] and references therein.

2.3.2.2.2 SIMO Channel Estimation

Here we will concentrate upon second-order statistical methods, but first a few comments regarding indirect SIMO channel estimation. As noted in section 2.1, linear

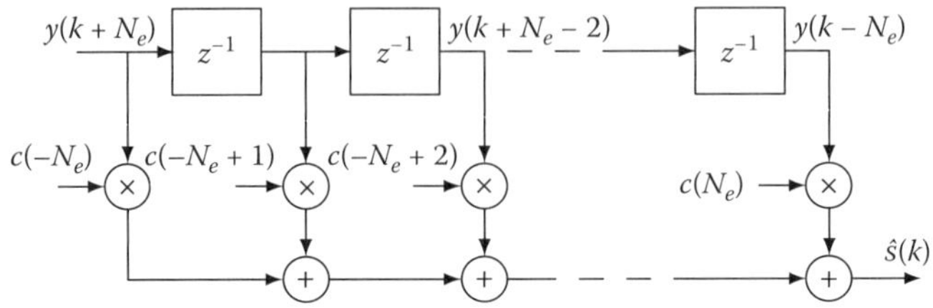

FIGURE 2.5 Structure of a baud-rate linear transversal equalizer.

equalizers designed on the basis of the baud-rate sampled received signal (see Figure 2.5 for a block diagram) are quite sensitive to symbol timing errors [15]. Therefore, fractionally spaced linear equalizers (typically with twice the baud-rate sampling: oversampling by a factor of two) are quite widely used to mitigate sensitivity to symbol timing errors. A fractionally spaced equalizer (FSE) in the linear transversal structure has the output

$$\hat{s}(k)=\sum_{n=-N_e}^{N_e}\left(\sum_{i=1}^{N}c_i(n)y_i(k-n)\right), \tag{2.46}$$

where $\{c_i(n)\}_{n=-N_e}^{n=N_e}$ are the $(2N_e + 1)$ tap weight coefficients of the ith subequalizer. Note that the FSE outputs data at the symbol rate. Similar to the SISO case, various criteria and cost functions exist to design the linear equalizers in both batch and recursive (adaptive) form.

Linear equalizers do not perform well when the underlying channels have deep spectral nulls in the passband. Several nonlinear equalizers have been developed to deal with such channels. Two effective approaches are:

- *Decision feedback equalizer* (DFE) is a nonlinear equalizer that employs previously detected symbols to eliminate the ISI due to the previously detected symbols on the current symbol to be detected. The use of the previously detected symbols makes the equalizer output a nonlinear function of the data. DFE can be symbol spaced or fractionally spaced.
- *Maximum likelihood sequence detector* estimates the information sequence to maximize the joint probability of the received sequence conditioned on the information sequence.

A detailed discussion may be found in [42].

Returning to the second-order statistical methods, for single-input multiple-output vector channels the autocorrelation function of the observation is sufficient for the identification of the channel impulse response up to an unknown constant [51, 53], provided that the various subchannels have no common zeros. This observation led to a number of techniques under both statistical and deterministic assumptions of the input sequence [50]. By exploiting the multichannel aspects of the channel, many of these techniques lead to a constrained quadratic optimization:

$$\hat{\mathcal{H}} = \arg\min_{\|\mathcal{H}\|=1} \mathcal{H}^H Q(Y)\mathcal{H}, \tag{2.47}$$

where $Q(Y)$ is a positive definite matrix constructed from the observation. Asymptotically (as either the sample size increases to infinity or the noise variance approaches zero), these estimates converge to true channel parameters.

2.3.3 Semiblind Approaches

Semiblind approaches utilize a combination of training-based and blind approaches. Here we present a brief discussion about the idea and refer the reader to a recent survey [7] for details. The objective of semiblind channel estimation (and equalization) is to exploit the information used by blind methods as well as the information exploited by the training-based methods. Semiblind channel estimation assumes additional knowledge of the input sequence. Specifically, part of the input data vector is known. Both the statistical and deterministic maximum likelihood estimators remain the same except that the likelihood function needs to be modified to incorporate the knowledge of the input. However, semiblind channel estimation may offer significant performance improvement over either the blind or the training-based methods, as demonstrated in the evaluation of the Cramer-Rao lower bound in [7].

There are many generalizations of blind channel estimation techniques to incorporate known symbols. In [6], Cirpan and Tsatsanis extended the approach of Kaleh and Vallet by restricting the transition of the hidden Markov model. In [30], knowledge of the known symbol is used to avoid the local maxima in the maximization of the likelihood function. A popular approach is to combine the objective function used to derive the blind channel estimator with the least squares cost in the training-based channel estimation. For example, a weighted linear combination of the cost for blind channel estimator and that for the training-based estimator can be used [16, 29, 33].

2.3.4 Superimposed Training-Based Approaches

In the superimposed training (hidden pilots)-based approach, one takes

$$s(n) = b(n) + c(n), \tag{2.48}$$

where $\{b(n)\}$ is the information sequence and $\{c(n)\}$ is a nonrandom periodic training (pilot) sequence. Exploitation of the periodicity of $\{c(n)\}$ allows identification of the channel without allocating any explicit time slots for training, unlike traditional training methods. There is no loss in data transmission rate. On the other hand, some useful power is wasted in superimposed training that could have otherwise been allocated to the information sequence. This lowers the effective signal-to-noise ratio (SNR) for the information sequence and affects the bit error rate (BER) at the receiver.

Superimposed training-based approaches have been discussed in [19, 20] and [38] for SISO systems. A block transmission method has been proposed in [10] and [11] where a data-dependent component is added to the superimposed training such that interference due to data (information sequence) is greatly reduced in channel estimation at the receiver. This method is applicable to time-invariant channels only, and it requires "data blocking" for block transmissions and insertion of a cyclic prefix in each data block. Its extension to a class of time-variant channels is given in [59]. The UTRA specification for third-generation (3G) systems [21] allows for a spread pilot (superimposed) sequence in the base station's common pilot channel, suitable for downlinks. Periodic superimposed training for channel estimation via first-order statistics for SISO systems has been discussed in [39, 41, 57, 58, 63]. In [8], performance bounds for training and superimposed training-based semiblind SISO channel estimation for time-varying flat fading channels have been discussed.

Suppose that the superimposed training sequence $c(n) = c(n + mP)$ $\forall m,n$ is a nonrandom periodic sequence with period P. Since $c(n)$ is P periodic, we have

$$c(n)=\sum_{m=0}^{P-1} c_m e^{j\alpha_m n}, \forall n, \;\; \alpha_m := 2\pi m/P , \tag{2.49}$$

where

$$c_m := \sum_{n=0}^{P-1} c(n) e^{-j\alpha_m n} / P.$$

Suppose that we use the Slepian (DPS) sequences $u_q(n)$ $(\equiv \phi_q(n))$ in the BEM. Then, by (2.20) and (2.49),

$$E\{\mathbf{y}(n)\} = \sum_{q=1}^{Q} \sum_{m=0}^{P-1} = \underbrace{\left[\sum_{l=0}^{L} c_m \mathbf{h}_q(l) e^{-j\alpha_m l} \right]}_{=:\mathbf{d}_{mq}} u_q(n) e^{j\alpha_m n}. \tag{2.50}$$

It follows that

$$\mathbf{y}(n) = \sum_{q=1}^{Q} \sum_{m=0}^{P-1} \mathbf{d}_{mq} u_q(n) e^{j\alpha_m n} + \mathbf{e}(n), \tag{2.51}$$

where $\{\mathbf{e}(n)\}$ is a zero-mean random sequence.

Define the cost function

$$J = \sum_{n=0}^{T-1} \left\| \mathbf{e}(n) \right\|^2. \tag{2.52}$$

Choose $\mathbf{d}_{mq}$'s to minimize J. We must have

$$\left.\frac{\partial J}{\partial \mathbf{d}_{mq}^*}\right|_{\mathbf{d}_{mq}=\hat{\mathbf{d}}_{mq}} = 0, \tag{2.53}$$

which leads to

$$\sum_{q'=1}^{Q}\sum_{m'=0}^{P-1} \hat{\mathbf{d}}_{m'q'} \left[\sum_{n=0}^{T-1} u_{q'}(n) u_q(n) e^{j(\alpha_{m'}-\alpha_m)n}\right] = \underbrace{\sum_{n=0}^{T-1} \mathbf{y}(n) u_q(n) e^{-j\alpha_m n}}_{=:\mathbf{g}_{mq}}. \tag{2.54}$$

Define

$$\mathbf{V} := \begin{bmatrix} 1 & 1 & \cdots & 1 \\ 1 & e^{-j\alpha_1} & \cdots & e^{-j\alpha_1 L} \\ \vdots & \vdots & \ddots & \vdots \\ 1 & e^{-j\alpha_{P-1}} & \cdots & e^{-j\alpha_{P-1} L} \end{bmatrix}, \tag{2.55}$$

$$\mathbf{D}_m := \left[\mathbf{d}_{m1}^T, \cdots, \mathbf{d}_{mQ}^T\right]^T, \quad \mathcal{D} := \left[\mathbf{D}_0^H, \cdots, \mathbf{D}_{P-1}^H\right]^H, \tag{2.56}$$

$$\mathbf{H}_l := \left[\mathbf{h}_1^T(l), \cdots, \ \mathbf{h}_Q^T(l)\right]^T, \mathcal{H} := \left[\mathbf{H}_0^H, \cdots, \mathbf{H}_L^H\right]^H, \tag{2.57}$$

$$\mathcal{C} := \left(diag\left\{c_0, \cdots, c_{P-1}\right\}\mathbf{V}\right) \otimes \mathbf{I}_{NQ}. \tag{2.58}$$

By the definition of $\mathbf{d}_{mq}$ in (2.56), it then follows that

$$\mathcal{C}\mathcal{H} = \mathcal{D}. \tag{2.59}$$

It is shown in [57] that if $P \geq L + 1$, $rank(C) = NQ(L + 1)$. Hence, we can determine the $\mathbf{h}_q(l)$'s uniquely by using the estimates of $\mathbf{d}_{mq}$'s.

Define $\hat{\mathcal{D}}$ as in (2.56) with $\mathbf{d}_{mq}$ replaced with $\hat{\mathbf{d}}_{mq}$, and similarly define $\mathcal{G}$ as in (2.56) with $\mathbf{d}_{mq}$ replaced with $\mathbf{g}_{mq}$. Then (2.54) leads to

$$\left(\Psi \otimes \mathbf{I}_N\right)\hat{\mathcal{D}} = \mathcal{G}, \tag{2.60}$$

where the entries of the $PQ \times PQ$ matrix Ψ are ($m,m' = 0, 1; \cdots, P - 1, q,q' = 1, 2, \cdots, Q$)

$$[\Psi]_{mQ+q,m'Q+q'} = \sum_{n=0}^{T-1} u_{q'}(n) u_q(n) e^{j(\alpha_{m'}-\alpha_m)n}. \tag{2.61}$$

The estimate of $\mathcal{D}$ is given by

$$\hat{\mathcal{D}} = (\Psi^{-1} \otimes \mathbf{I}_N)\mathcal{G}. \tag{2.62}$$

By (2.59) and (2.78) we have the channel coefficient estimate

$$\hat{\mathcal{H}} = \mathcal{C}^{\dagger}\hat{\mathcal{D}} = (\mathcal{C}^H\mathcal{C})^{-1}\mathcal{C}^H(\Psi^{-1} \otimes \mathbf{I}_N)\mathcal{G}. \tag{2.63}$$

The channel estimate is acquired by regenerating the DPS-BEM:

$$\hat{\mathbf{h}}(n;l) = \sum_{q=1}^{Q} \hat{\mathbf{h}}_q(l) u_q(n). \tag{2.64}$$

Remark 1: Using the fact that the (infinite) DPS sequences are band limited to the normalized frequency range $[-f_dT_s, f_dT_s]$, the time-limited DPS sequences, obtained by rectangular windowing over $0 \le n \le T-1$, approximately satisfy

$$\sum_{n=0}^{T-1} u_{q'}(n) u_q(n) e^{j(\alpha_{m'}-\alpha_m)n} \approx \delta(m'-m)\delta(q'-q), \tag{2.65}$$

if $f_dT_s \ll 1/P$ and T are multiples of P or if T is large, so that $\Psi \approx \mathbf{I}_{PQ}$. An estimate $\hat{\mathbf{d}}_{mq}$ of $\mathbf{d}_{mq}$, following (2.54) and (2.65), is given as

$$\hat{\mathbf{d}}_{mq} = \sum_{n=0}^{T-1} \mathbf{y}(n) u_q(n) e^{-j\alpha_m n}. \tag{2.66}$$

The estimation of the channel coefficients (2.63) is then given by

$$\hat{\mathcal{H}} = (\mathcal{C}^H\mathcal{C})^{-1}\mathcal{C}^H\hat{\mathcal{D}}. \tag{2.67}$$

Remark 2: As noted in [57], if the mean of the noise $\mathbf{v}(n)$ is unknown, say

$$E\{\mathbf{v}(n)\} = \mathbf{m}, \tag{2.68}$$

one should omit the first row (corresponding to α_0) of V in (2.55) (denote the resulting matrix by $\tilde{\mathbf{V}}$), and block $\mathbf{D}_0$ from $\mathcal{D}$ in (2.56) (denote the resulting matrix by $\tilde{\mathcal{D}}$). Define

$$\tilde{C} := \left(diag\left\{c_1 \cdots c_{P-1}\right\}\tilde{\mathbf{V}}\right) \otimes \mathbf{I}_{NQ}. \tag{2.69}$$

Then we have $\tilde{C}\mathcal{H} = \tilde{\mathcal{D}}$ and

$$\hat{\mathcal{H}} = (\tilde{C}^H \tilde{C})^{-1} \tilde{C}^H \hat{\tilde{\mathcal{D}}}, \tag{2.70}$$

where $\hat{\tilde{\mathcal{D}}}$ follows from $\tilde{\mathcal{D}}$ by using estimate (2.66). For identifiability, we now need $P \geq L + 2$. All our subsequent results hold true if appropriate substitutions are used.

2.4 Adaptive Channel Estimation

2.4.1 Block-Adaptive Channel Estimation Using CE-BEM

Here we summarize the time-multiplexed training approach of [36]. In [36] each transmitted block of symbols $\{s(n)\}_{n=0}^{T_B-1}$ is segmented into $\overline{P}$ subblocks of time-multiplexed training and information symbols. Each subblock is of equal-length l_b symbols with l_d information symbols and l_t training symbols ($l_b = l_d + l_t$). If s denotes a column vector composed of $\{s(n)\}_{n=0}^{T_B-1}$, then s is arranged as

$$\mathbf{s} := \begin{bmatrix} \mathbf{b}_0^T & \mathbf{c}_0^T & \mathbf{b}_1^T & \mathbf{c}_1^T & \cdots & \mathbf{b}_{\overline{P}-1}^T & \mathbf{c}_{\overline{P}-1}^T \end{bmatrix}^T, \tag{2.71}$$

where $\mathbf{b}_p$ ($p = 1, 0, \cdots \overline{P} - 1$) is a column of l_d information symbols and $\mathbf{c}_p$ is a column of l_t training symbols. We clearly have TB = $\overline{P}l_b$. Given (2.69) and CE-BEM (2.15), [36] has shown that (2.77) is an optimum structure for $K = 1$ with $l_t = 2L + 1$, $\overline{P} \geq Q$, and

$$\mathbf{c}_p := \begin{bmatrix} \mathbf{0}_L^T & \gamma & \mathbf{0}_L^T \end{bmatrix}^T, \gamma > 0. \tag{2.72}$$

Thus, given a transmission block of size T_B, $(2L + 1)\overline{P}$ symbols have to be devoted to training and the remaining $T_B - (2L + 1)\overline{P}$ are available for information symbols.

Let $n_p := pl_b + l_d + L$ ($p = 0, 1, \ldots \overline{P} - 1$) denote the location of (nonzero) γ's in the optimum $\mathbf{c}_p$'s in the P subblocks. Then by design, received signal (assuming timing synchronization)

$$y(n_p + l) = \gamma h(n_p + l; l) + v(n_p + l) \tag{2.73}$$

for $l = 0, 1, \cdots, L$. Using (2.15) in these $y(n_p + l)$'s, one can uniquely solve for $h_q(l)$'s via a least squares approach. The channel estimates are given by the CE-BEM (2.15) using the estimated BEM coefficients.

2.4.2 Adaptive Channel Estimation via Subblock Tracking

Suppose that we collect the received signal over a time interval of $\overline{T}$ symbols. We wish to estimate the time-variant channel using a channel model and time-multiplexed training (such as that discussed in section 2.4.1 and [36]), and subsequently using the estimated channel, estimate the information symbols. For CE-BEM, if we choose $\overline{T}$ as the block size, then in general the Q value will be very high, requiring estimation of a large number of parameters, thereby degrading the channel estimation performance. If we divide $\overline{T}$ into blocks of size T_B, and then fit CE-BEM block by block, we need smaller Q; however, estimation of $h_q(l)$'s is now based on a shorter observation size of T_B symbols, which might also degrade channel estimation performance. Thus, one has to strike a balance between estimation variance and block size. Such considerations do not apply to the AR channel model fitting. In the sequel, we propose a novel subblock tracking approach to CE-BEM channel estimation where we update estimates of $h_q(l)$'s every subblock based on all of the past training symbols.

By exploiting the invariance of the coefficients of CE-BEM over each block, hence each of the $\overline{P}$ subblocks per block of length T_B symbols, we seek subblock-wise tracking of the BEM coefficients of the doubly selective channel. Consider two overlapping blocks that differ by just one subblock: blocks with $n = m, m + 1, \cdots, m + T_B - 1$, where $m = m_0$ for the past block and $m = m_0 + l_b$ for the current block. If the two blocks overlap so significantly, one would expect the BEM coefficients to vary only a little from the past block to the current overlapping block. Therefore, rather than estimate $h_q(l)$'s anew with every non-overlapping block, as in section 2.4.1 and [36], we propose to track the BEM coefficients subblock by subblock using a first-order AR model for their variations.

Stack the channel coefficients in (2.15) into vectors

$$\mathbf{h}_l := \begin{bmatrix} h_1(l) & h_2(l) & \cdots & h_Q(l) \end{bmatrix}^T, \tag{2.74}$$

$$\mathbf{h} := \begin{bmatrix} \mathbf{h}_0^T & \mathbf{h}_1^T & \cdots & \mathbf{h}_L^T \end{bmatrix}^T \tag{2.75}$$

of size Q and $M := Q(L + 1)$, respectively. The coefficient vector in (2.75) for the pth subblock ($p = 0, 1, \cdots$) will be denoted by $\mathbf{h}(p)$. We assume that the BEM coefficients over each subblock are Markovian: a simplified model is given by the first-order AR process, i.e.,

$$\mathbf{h}(p) = \alpha\mathbf{h}(p-1) + \mathbf{w}(p), \tag{2.76}$$

where α is the AR coefficient, and the driving noise vector $\mathbf{w}(p)$ is zero-mean complex Gaussian with variance $\sigma_w^2\mathbf{I}_M$. If the channel is stationary and coefficients $h_q(l)$ are independent (as assumed in [36]), then by (2.76), $\sigma_w^2 = \sigma_h^2(1-|\alpha|^2)/Q$ with $\sigma_h^2 := E\{h(n;l)h^*(n;l)\}$. Since the coefficients evolve slowly, we have $\alpha \approx 1$ (but $\alpha < 1$ for tracking).

Under this formulation we do not have a strict definition of the block size T_B because, although we still use (2.15) for any n, we allow $h_q(l)$'s to change subblock by subblock based on the training symbols.

2.4.2.1 Subblock Tracking Using Kalman Filtering

Define $\varepsilon(n) := [e^{-j\omega_1 n}\ \ e^{-j\omega_2 n}\ \ \cdots\ \ e^{-j\omega_Q n}]^T$. If at time n the pth subblock is being received, by (2.63), (2.15)–(2.19), and (2.74) and (2.75), the received signal can be written as

$$y(n) = \mathbf{s}^T(n)\left[\mathbf{I}_{L+1} \otimes \varepsilon(n)\right]^H \mathbf{h}(p) + v(n), \tag{2.77}$$

where $\mathbf{s}(n) := [s(n)\ \ s(n-1)\ \ \cdots\ \ s(n-L)]^T$. Treating (2.76) and (2.77) as the state and the measurement equations, respectively, Kalman filtering can be applied to track the coefficient vector $\mathbf{h}(p)$ for each subblock.

We will employ the time-multiplexed training scheme proposed in [36] (see section 2.4.1), where each subblock (of equal-length l_b symbols) consists of a data session (of length l_d symbols) and a succeeding training session (of length $l_b = 2L + 1$ symbols). Using (2.73), at time $n_p + l$ ($p = 0, 1, \cdots$ and $l = 0, 1, \cdots, L$)

$$y(n_p + l) = \gamma E^H(n_p + l)\mathbf{h}_l(p) + v(n_p + l). \tag{2.78}$$

We intend to use only training sessions for subblock-wise channel tracking. Defining

$$\mathbf{y}(p) := \left[y(n_p)\quad y(n_p+1)\quad \cdots\quad y(n_p+L)\right]^T,$$

$$\mathbf{v}(p) := \left[v(n_p)\quad v(n_p+1)\quad \cdots\quad v(n_p+L)\right]^T,$$

$$\Psi(p) := \begin{bmatrix} E(n_p) & & & \\ & E(n_p+1) & & \\ & & \ddots & \\ & & & E(n_p+L) \end{bmatrix}^H,$$

by (2.78), we have

$$\mathbf{y}(p) = \gamma\Psi(p)\mathbf{h}(p) + \mathbf{v}(p). \tag{2.79}$$

Using this optimal training scheme, the measurement equation (2.77) is now simplified as (2.79). We have obtained a linear discrete-time system represented by (2.76) and (2.79). Kalman filtering is applied to track the channel BEM coefficients via the following steps [49]:

1. Initialization:

$$\hat{\mathbf{h}}(-1|-1)=0_M, \mathbf{R}_h(-1|-1)=\sigma_w^2\mathbf{I}_M;$$

2. Kalman recursion for $p = 0, 1, \cdots$
   - Time update:

$$\hat{\mathbf{h}}(p|p-1)=\alpha\hat{\mathbf{h}}(p-1|p-1),$$
$$\mathbf{R}_h(p|p-1)=|\alpha|^2\mathbf{R}_h(p-1|p-1)+\sigma_w^2\mathbf{I}_M;$$

   - Kalman gain:

$$\mathbf{R}_\eta(p)=|\gamma|^2\Psi(p)\mathbf{R}_h(p|p-1)\Psi^H(p)+\sigma_v^2\mathbf{I}_{L+1},$$
$$\mathbf{K}(p)=\mathbf{R}_h(p|p-1)\Psi^H(p)\mathbf{R}_\eta^{-1}(p);$$

   - Measurement update:

$$\hat{\mathbf{h}}(p|p)=\hat{\mathbf{h}}(p|p-1)+\mathbf{K}(p)\left[\mathbf{y}(p)-\gamma\Psi(p)\hat{\mathbf{h}}(p|p-1)\right],$$
$$\mathbf{R}_h(p|p)=\left[\mathbf{I}-\gamma\mathbf{K}(p)\Psi(p)\right]\mathbf{R}_h(p|p-1),$$

where $\hat{\mathbf{h}}(p|m)$ is the estimate of $\mathbf{h}(p)$ given the observations $\{\mathbf{y}(0), \mathbf{y}(1), \cdots, \mathbf{y}(m)\}$, and $\mathbf{R}_h(p|m)$ is the error covariance matrix of $\hat{\mathbf{h}}(p|m)$, defined as

$$\mathbf{R}_h(p|m):=E\left\{\left[\hat{\mathbf{h}}(p|m)-\mathbf{h}(p)\right]\left[\hat{\mathbf{h}}(p|m)-\mathbf{h}(p)\right]^H\right\}.$$

Now we generate the channel for the entire p^{th} subblock by the estimate $\hat{\mathbf{h}}(p|p)$ via the CE-BEM (2.15) as

$$\hat{h}(n;l)=\varepsilon^H(n)\hat{\mathbf{h}}_l(p|p) \tag{2.80}$$

for $n = pl_b, pl_b + 1, \cdots, (p + 1)l_b - 1$. The definition of $\hat{\mathbf{h}}_l(p|p)$ is similar to (2.74).

2.4.2.2 Kalman Detector for Equalization

The channel estimate given by (2.80) is fed into an equalizer for symbol detection. Another Kalman filter, together with a quantizer, acts as the symbol detector at the receiver end. The state and the measurement equations are now given by

$$\mathbf{s}_d(n) = \Phi \mathbf{s}_d(n-1) + \Gamma \bar{s}(n) + \Gamma \tilde{s}(n), \tag{2.81}$$

$$y(n) = \tilde{\mathbf{h}}_d^T(n) \mathbf{s}_d(n) + v(n), \tag{2.82}$$

where

$$\mathbf{s}_d(n) := \begin{bmatrix} s(n) & s(n-1) & \cdots & s(n-d) \end{bmatrix}^T,$$

$$\bar{s}(n) := E\{s(n)\}, \ \tilde{s}(n) := s(n) - \bar{s}(n),$$

$$\Phi := \begin{bmatrix} 0_d^T & 0 \\ \mathbf{I}_d & 0_d \end{bmatrix}, \ \Gamma := \begin{bmatrix} 1 & 0_d^T \end{bmatrix}^T,$$

$$\tilde{\mathbf{h}}_d(n) := \begin{bmatrix} \hat{h}(n;0) & \hat{h}(n;1) & \cdots & \hat{h}(n;L) & 0_{d-L} \end{bmatrix}^T,$$

where integer $d \geq L$; it will also be the equalization delay. Assume data symbols are zero mean and white. If $s(n)$ is a data symbol, we have $\bar{s}(n) = 0$, $\tilde{s}(n) = s(n)$; if $s(n)$ is a training symbol, $\bar{s}(n) = s(n)$, $\tilde{s}(n) = 0$. Details of Kalman filtering of the system described by (2.81) and (2.82) can be found in [34].

2.4.3 Symbol-Adaptive Joint Channel Estimation and Data Detection

Representative approaches in this category are [27, 34] and references therein. A Gauss-Markov model for channel variations (typically an autoregressive model) is coupled with a state-space model for received data to form an augmented state-space model with nonlinear measurement equation. This results in a nonlinear state estimation problem. In [27] a finite-length minimum mean square error (MMSE) DFE is used during non-data-aided periods to generate hard decisions. Reference [34] presents a low-complexity turbo equalization receiver for coded signals where a nonlinear Kalman filtering–based adaptive equalizer is coupled with a soft-in soft-out decoder. These approaches work well so long as the channel does not fade too fast.

2.5 Simulation Examples

In this section we present two simulation examples to illustrate some of the approaches to channel estimation. In both examples a random time- and frequency-selective Rayleigh fading channel is considered. We take $L = 2$ (three taps) in (2.63), and $h(n;l)$ are zero-mean complex Gaussian with variance $\sigma_h^2 = 1/(L+1)$. For different l's, $h(n;l)$'s are mutually independent and satisfy Jakes' model [22]. To this end, we simulated each single tap following [62] (with a correction in the appendix of [61]).

We consider a communication system with carrier frequency of 2 GHz, data rate of 40 kBd (kilo-Bauds), therefore $T_s = 25$ μs, and a varying Doppler spread f_d in the range of 0 to 400 HZ, or the normalized Doppler spread f_dT_s from 0 to 0.01 (corresponding to a maximum mobile velocity from 0 to 216 km/h). The additive noise was zero-mean complex white Gaussian. The (receiver) SNR refers to the average energy per symbol over one-sided noise spectral density. The time-multiplexed training scheme of [36] described in section 2.4.1 is adopted, where during data sessions the information sequence is modulated by binary phase-shift keying (BPSK) with unit power. The training session is described by (2.72) with $\gamma = \sqrt{2L+1}$ so that the average symbol power of training sessions is equal to that of data sessions.

We evaluate the performance of various approaches by considering the normalized channel mean square error (NCMSE) and the bit error rate (BER). The NCMSE is defined as

$$\text{NCMSE} := \frac{\sum_{i=1}^{M_r} \sum_{n=0}^{T-1} \sum_{l=0}^{L} \left\| \hat{\mathbf{h}}^{(i)}(n;l) - \mathbf{h}^{(i)}(n;l) \right\|^2}{\sum_{i=1}^{M_r} \sum_{n=0}^{T-1} \sum_{l=0}^{L} \left\| \mathbf{h}^{(i)}(n;l) \right\|^2},$$

where $\mathbf{h}^{(i)}(n;l)$ is the true channel and $\hat{\mathbf{h}}^{(i)}(n;l)$ is the estimated channel at the ith run, among total M_r runs.

2.5.1 Example 1

Here we consider block-adaptive channel estimation and restrict our attention to a single block. We compare superimposed training-based approaches with time-multiplexed training-based approaches. The superimposed training sequence was picked as a periodic repetition of a length 7 m-sequence (maximal length pseudo-random binary sequence) {1, 1,−1,1,1,1,−1}. In the simulations, the average transmitted power σ_c^2 in $c(n)$ was 0.3 of the power in $b(n)$, leading to a training-to-information power ratio TIR := $\sigma_c^2/\sigma_b^2 = \beta/(1-\beta) = 0.3$. We consider both critically sampled CE-BEM and DPS-BEM for channel modeling. For comparison, we consider a CE- or DPS-BEM-based periodically placed time-multiplexed training with zero padding, following the design of [36]. We took a training subblock of size $2L + 1 = 5$ symbols with the recommended structure

$$\left\{0,0,\sqrt{(2L+1)\left(\sigma_b^2+\sigma_c^2\right)},0,0\right\},$$

which follows an information data block of length 18 leading to a subblock of 23 symbols. This subblock was repeated over a record length of 418 symbols with a total of 16 subblocks. Thus, the training-to-information bit and power ratios are both 0.3 (the amplitude of the single nonzero training bit was picked to achieve this power ratio).

The results of our simulation averaged over five hundred runs are shown in Figures 2.6, 2.7, 2.8, and 2.9. For comparison, we plot the results of the CE- and DPS-BEM-based superimposed training schemes (denoted as SI in the figures), including the first-order

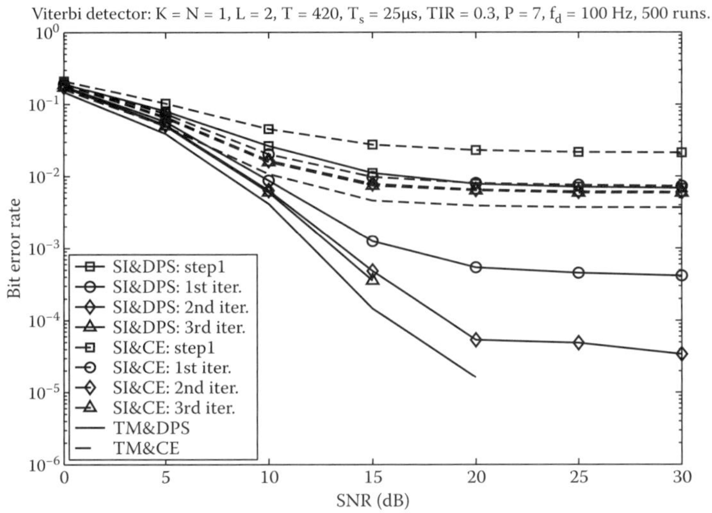

FIGURE 2.6 BER versus SNR for f_d = 100 Hz.

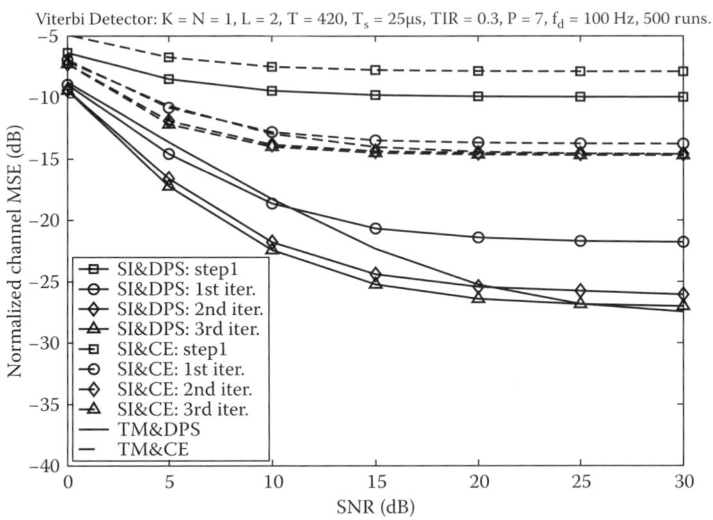

FIGURE 2.7 MSE versus SNR for f_d = 100 Hz.

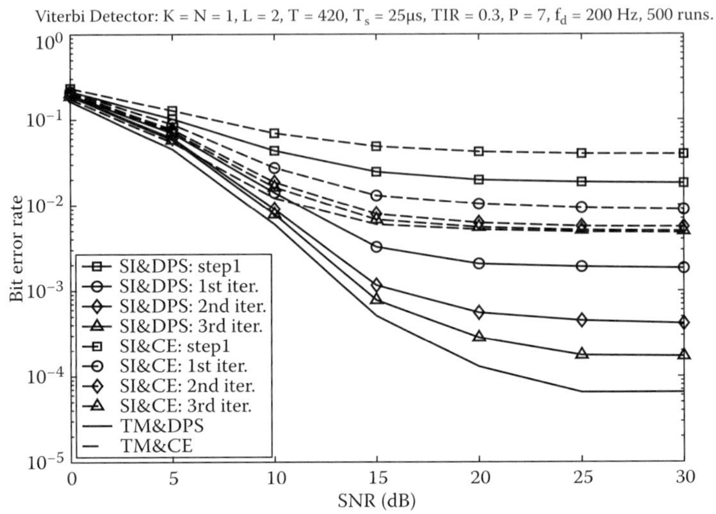

FIGURE 2.8 BER versus SNR for $f_d = 200$ Hz.

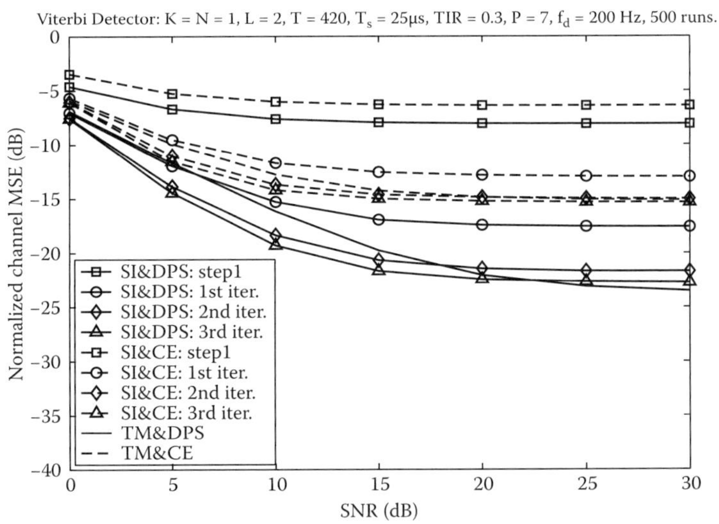

FIGURE 2.9 MSE versus SNR for $f_d = 200$ Hz.

statistics-based estimator (denoted as step 1 in the figures), and the DML approach after one, two, and three iterations (denoted as 1st iter., 2nd iter., and 3rd iter. in the figures), and TM training approaches (denoted as TM in the figures). From the four figures, we can see that after iterations, superimposed training-based estimation and detection performances improve a lot, because the information data that are viewed as interference by the first-order statistics-based estimator are now exploited to enhance the channel estimation for the next iteration. Therefore, the self-interference is effectively removed after iterations. The DML approach provides comparable error performance with TM training, but at a higher data transmission rate. The valuable bandwidth resources can thus be saved.

2.5.2 Example 2

Here we consider adaptive channel estimation at block-, subblock-, and symbol-adaptive levels using time-multiplexed training. Four estimation and tracking schemes are compared:

1. The block-adaptive channel estimation in [36] (see section 2.4.1). We consider two models corresponding to $T = 200$ and 400, respectively, so that $Q = 5$ and 9 by (2.17). In the figures, this scheme is denoted by BA.
2. The channel tracking scheme in [35] using the first-order AR model, where the time-varying channel is assumed to follow (2.12). Channel tracking is performed at training sessions only. For data sessions, the receiver updates the channel via $h(n;l) = \alpha_c h(n-1;l)$. We assume that only the upper bound of the Doppler spread is known. Then by (2.13) and (2.14) and Jakes' model, $\alpha_c = J_0(2\pi f_d T_s) = 0.999$ for $f_d T_s = 0.01$, where $J_0(\cdot)$ denotes the zero-th Bessel function of the first kind. This scheme is denoted by $AR(1)$-KF in the figures.
3. We also compare the approach of joint channel estimation and data detection via extended Kalman filtering in [34]. For fairness, the Turbo equalization procedure in [34] is omitted. The AR parameter of the channel also follows (2.13) and (2.14), as suggested by [34]. This scheme is denoted by Joint-KF.
4. Our proposed subblock-wise tracking using CE-BEM, which is denoted by SUBBLOCK. We also take $T = 200$ and 400 for two different settings of CE-BEM, and $Q = 5$ and 9 correspondingly.

The BERs for the schemes of BA, $AR(1)$-KF, and SUBBLOCK are evaluated by employing the Kalman detector described in section 2.4.2.2 with delay $d = 5$, using the channel estimates obtained by each scheme. In each run, a symbol sequence of length 5,000 is generated and fed into a random doubly selective channel. The first two hundred symbols are discarded in evaluations. All the simulation results are based on five hundred runs.

In Figures 2.10 and 2.11, the performances of the four schemes over different Doppler spreads f_d are compared. We set SNR = 20 dB, $l_t = 2L + 1 = 5$, and $l_d = 15$ symbols, so that 25% of the transmitted symbols are dedicated to training. For the BA scheme, we use the oversampled CE-BEM with $T_B = T/2$ and $K = 2$ in order to suppress spectral leakage. For our SUBBLOCK scheme, we take $\alpha = 0.99$ for $T = 200$, and $\alpha = 0.995$ for $T = 400$. Since more unknown parameters (BEM coefficients) are involved in SUBBLOCK, and therefore result in higher estimation variance, SUBBLOCK is slightly inferior to $AR(1)$-KF for slow-fading channels. As f_d increases, SUBBLOCK gradually outperforms the other three schemes, since the time variations of the channel have been well captured in CE-BEM. For the BA scheme, more basis functions do not necessarily translate into better performance since estimation variance also increases, whereas this strategy can well improve the performance of SUBBLOCK since all past data are implicitly utilized in Kalman filtering–based subblock tracking. Figures 2.12 and 2.13 compare the

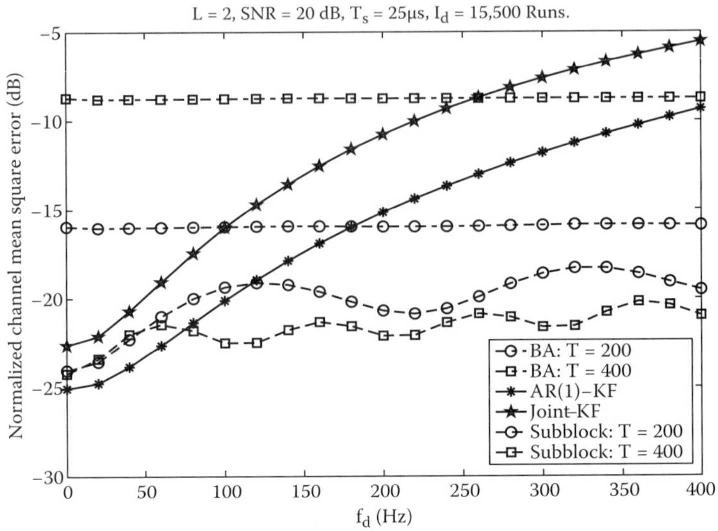

FIGURE 2.10 NCMSE versus f_d for SNR = 20 dB, l_d = 15, l_t = 5, l_c = 5, l_b = 5, $T_B = T/K$, and $K = 2$.

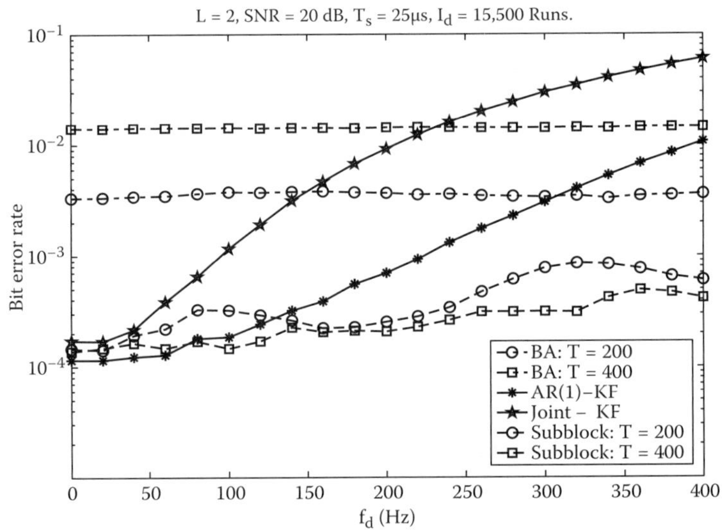

FIGURE 2.11 BER versus f_d for SNR = 20 dB, l_d = 15, l_t = 5, l_b = 20, $T_B = T/K$, and $K = 2$.

four schemes under different SNRs: results similar to those in Figures 2.10 and 2.11 are observed.

In Figures 2.14 and 2.15, longer data sessions are used to represent a more efficient transmission scenario, where we take l_d = 35 symbols (12.5% symbols are now devoted to training sessions). We now let $\alpha = 0.94$ for $T = 200$, and $\alpha = 0.97$ for $T = 400$ for

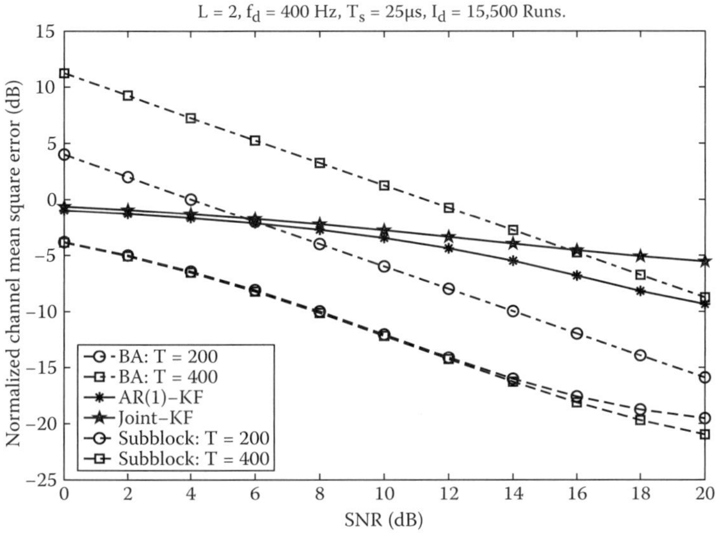

FIGURE 2.12 NCMSE versus SNR for $f_dT_s = 0.01$, $l_d = 15$, $l_t = 5$, $l_b = 20$, $T_B = T/K$, and $K = 2$.

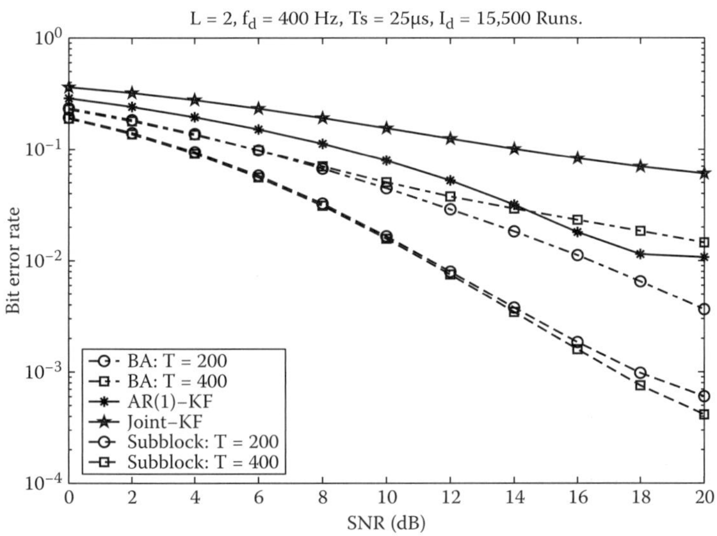

FIGURE 2.13 BER versus SNR for $f_dT_s = 0.01$, $l_d = 15$, $l_t = 5$, $l_b = 20$, $T_B = T/K$, and $K = 2$.

SUBBLOCK. For the BA scheme, an oversampled CE-BEM is not possible for $T = 200$, and we take $T_B = T$ for the two models. Comparing Figure 2.14 with Figure 2.12, we observe that the spectral leakage has worsened the channel estimation in BA (note the floors of the two BA curves in Figure 2.14), and thus the curves for symbol detection. Our SUBBLOCK maintains a satisfactory performance, while the performances of the schemes of $AR(1)$-KF and Joint-KF have severely deteriorated.

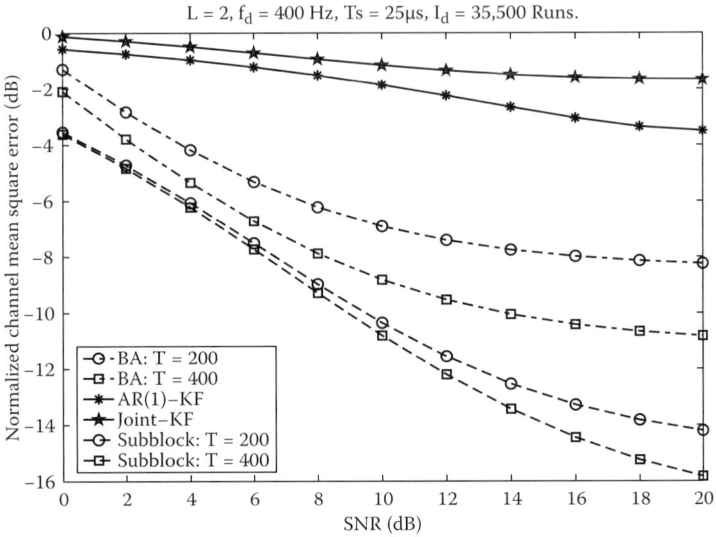

FIGURE 2.14 NCMSE versus SNR for $f_dT_s = 0.01$, $l_d = 35$, $l_c = 5$, $l_b = 40$, $T_B = T/K$, and $K = 1$.

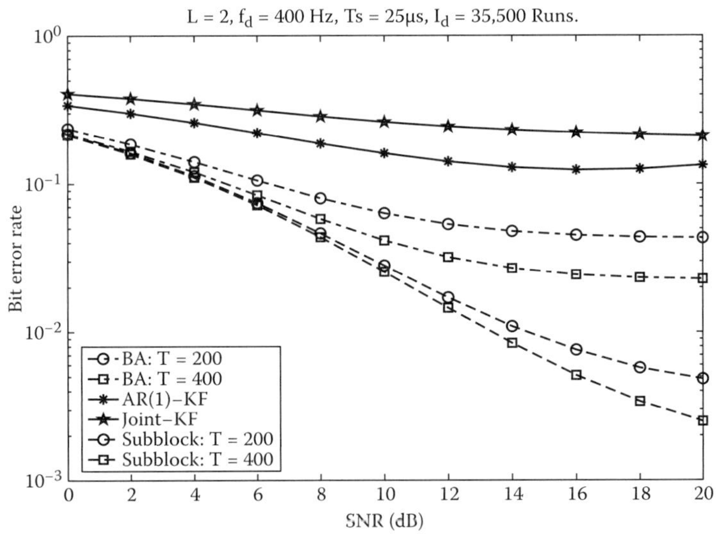

FIGURE 2.15 BER versus SNR for $f_dT_s = 0.01$, $l_d = 35$, $l_t = 5$, $l_b = 40$, $T_B = T/K$, and $K = 1$.

2.6 Conclusions

A review of various approaches to adaptive channel estimation for wireless mobile systems was presented. Emphasis was on linear baseband-equivalent models with a tapped delay line structure, and both time-invariant and time-variant (doubly-selective) models were discussed. Emphasis was on basis expansion modeling for time-variant channels where the basis functions are related to the physical parameters of the channel (such as Doppler and delay spreads). Channel modeling was followed by a discussion of various approaches to channel estimation, including training-based approaches, blind approaches, semiblind approaches and superimposed training-based approaches, and various approaches to channel adaptation. In the training-based approach a sequence known to the receiver is transmitted in the acquisition mode. In blind approaches no such sequence is available (or used), and the channel is estimated based solely on the noisy received signal exploiting the statistical and other properties of the information sequence. Semiblind approaches utilize a combination of training-based and blind approaches. In the superimposed training-based approaches a periodic (nonrandom) training sequence is arithmetically added (superimposed) at a low power to the information sequence at the transmitter before modulation and transmission. Channel adaptaion can be at the block level suitable for block transmissions, or the symbol-by-symbol level suitable for serial transmissions. Some of the approaches were illustrated via simulations.

Acknowledgment

This work was prepared in part under the support of the U.S. NSF under grant ECS-0424145.

References

[1] M.-A. R. Baissas and A. M. Sayeed. 2002. Pilot-based estimation of time-varying multipath channels for coherent CDMA receivers. *IEEE Trans. Signal Processing* 50:2037–49.

[2] P. A. Bello. 1963. Characterization of randomly time-variant channels. *IEEE Trans. Commun. Syst.* 11:360–93.

[3] D. K. Borah and B. Hart. 1999. Receiver structures for time-varying frequency-selective fading channels. *IEEE J. Selected Areas Commun.* 17:1863–75.

[4] J. K. Cavers. 1995. Pilot symbol assisted modulation and differential detection in fading and delay spread. *IEEE Trans. Commun.* 43:2206–12.

[5] K. M. Chugg. 1998. Blind acquisition characteristics of PSP-based sequence detectors. *IEEE J. Selected Areas Commun.* 16:1518–29.

[6] H. A. Cirpan and M. K. Tsatsanis. 1998. Stochastic maximum likelihood methods for semi-blind channel estimation. *IEEE Signal Processing Lett.* 5:21–24.

[7] E. de Carvalho and D. T. M. Slock. 2001. Semi-blind methods for FIR multichannel estimation. In *Signal processing advances in wireless and mobile communications*, ed. G. B. Giannakis, Y. Hua, P. Stoica, and L. Tong, chap. 7. Vol. I. Upper Saddle River, NJ: Prentice Hall.

[8] M. Dong, L. Tong, and B. M. Sadler. 2003. Optimal insertion of pilot symbols for transmissions over time-varying flat fading channels. In *Proceedings of the 2003 IEEE Workshop on Signal Processing Advances in Wireless Communications*, Rome, pp. 472–76.

[9] W. A. Gardner. 1989. *Introduction to random processes: With applications to signals and systems*. 2nd ed. New York: McGraw-Hill.

[10] M. Ghogho, D. McLernon, E. Alamdea-Hernandez, and A. Swami. 2005. Channel estimation and symbol detection for block transmission using data-dependent superimposed training. *IEEE Signal Processing Lett.* 12:226–29.

[11] M. Ghogho, D. McLernon, E. Alamdea-Hernandez, and A. Swami. 2005. SISO and MIMO channel estimation and symbol detection using data-dependent superimposed training. In *Proceedings of the 2005 IEEE ICASSP*, Philadelphia, pp. III-461–64.

[12] M. Ghosh and C. L. Weber. 1992. Maximum-likelihood blind equalization. *Optical Eng.* 31:1224–28.

[13] G. B. Giannakis and C. Tepedelenlioğlu. 1998. Basis expansion models and diversity techniques for blind identification and equalization of time-varying channels. *Proc. IEEE* 86:1969–86.

[14] G. B. Giannakis, Y. Hua, P. Stoica, and L. Tong, eds. 2001. *Signal processing advances in wireless and mobile communications: Trends in channel estimation and equalization*, ed. G. B. Giannakis, Y. Hua, P. Stoica, and L. Tong. Vol. 1. Upper Saddle River, NJ: Prentice Hall.

[15] R. D. Gitlin and S. B. Weinstein. 1981. Fractionally-spaced equalization: An improved digital transversal equalizer. *Bell Syst. Tech. J.* 60:275–96.

[16] A Gorokhov and P. Loubaton. 1997. Semi-blind second order identification of convolutive channels. In *Proceedings of the IEEE International Conference on Acoustics, Speech, Signal Processing*, Munich, Germany, pp. 3905–8.

[17] D. Hatzinakos and C. L. Nikias. 1991. Blind equalization using a tricepstrum based algorithm. *IEEE Trans. Commun.* 39:669–81.

[18] S. He and J. K. Tugnait. 2006. On bias-variance trade-off in superimposed training-based doubly selective channel estimation. In *Proceedings of the 2006 Conference on Information Systems and Sciences*, Princeton, NJ, pp. 1308–1313.

[19] P. Hoeher and F. Tufvesson. 1999. Channel estimation with superimposed pilot sequence. In *Proceedings of the IEEE GLOBECOM Conference*, Rio de Janeiro, pp. 2162–66.

[20] T. P. Holden and K. Feher. 1990. A spread spectrum based system technique for synchronization of digital mobile communication systems. *IEEE Trans. Broadcasting* 36:185–94.

[21] H. Holma and A. Toskala, eds. 2002. *WCDMA for UMTS: Radio access for third generation mobile communications*. 2nd ed. New York: Wiley.

[22] W. C. Jakes. 1974. *Microwave mobile communications*. New York: Wiley.

[23] G. K. Kaleh and R. Vallet. 1994. Joint parameter estimation and symbol detection for linear or nonlinear unknown dispersive channels. *IEEE Trans. Commun.* 42:2406–13.

[24] A. P. Kannu and P. Schniter. 2005. MSE-optimal training for linear time-varying channels. In *Proceedings of ICASSP*, vol. 3, pp. 789–92.

[25] A. P. Kannu and P. Schniter. 2005. Capacity analysis of MMSE pilot-aided transmission for doubly selective channels. In *Proceedings of the IEEE 6th Workshop on SPAWC*, pp. 801–5.

[26] A. P. Kannu and P. Schniter. 2006. Minimum mean-squared error pilot-aided transmission for MIMO doubly selective channels. In *Proceedings of the 2006 Conference on Information Systems and Sciences*, Princeton University, NJ.

[27] C. Komninakis, C. Fragouli, A. H. Sayed, and R. D. Wesel. 2002. Multi-input multi-output fading channel tracking and equalization using Kalman estimation. *IEEE Trans. Signal Processing* 50:1065–76.

[28] V. Krishnamurthy and J. B. Moore. 1993. On-line estimation of hidden Markov model parameters based on Kullback-Leibler information measure. *IEEE Trans. Signal Processing* 41:2557–73.

[29] S. Lasaulce, P. Loubaton, and E. Moulines. 2003. A semi-blind channel estimation technique based on second-order blind method for CDMA systems. *IEEE Trans. Signal Processing* 51:1894–904.

[30] J. Laurila, K. Kopsa, and E. Bonek. 1999. Semi-blind signal estimation for smart antennas using subspace tracking. In *Proceedings of the 1999 IEEE Workshop on Signal Processing: Advances in Wireless Communications*, Annapolis, MD, pp. 271–274.

[31] G. Leus, S. Zhou, and G. B. Giannakis. 2003. Orthogonal multiple access over time- and frequency-selective channels. *IEEE Trans. Inf. Theory* 49:1942–50.

[32] G. Leus. 2004. On the estimation of rapidly time-varying channels. In *Proceedings of the European Signal Processing Conference*, Vienna, pp. 2227–30.

[33] G. Leus. 2005. Semi-blind channel estimation for rapidly time-varying channels. In *Proceedings of the 2005 IEEE International Conference on Acoustics, Speech, Signal Processing*, Philadelphia, vol. III, pp. 773–76.

[34] X. Li and T. F. Wong. 2007. Turbo equalization with nonlinear Kalman filtering for time-varying frequency-selective fading channels. *IEEE Trans. Wireless Commun.* 6:691–700.

[35] Z. Liu, X. Ma, and G. B. Giannakis. 2002. Space-time coding and (Kalman) filtering for time-selective fading channels. *IEEE Trans. Commun.* 50:183–86.

[36] X. Ma, G. B. Giannakis, and S. Ohno. 2003. Optimal training for block transmissions over doubly selective wireless fading channels. *IEEE Trans. Signal Processing* 51:1351–66.

[37] M. Martone. 2000. Wavelet-based separating kernels for array processing of cellular DS/CDMA signals in fast fading. *IEEE Trans. Commun.* 48:979–95.

[38] F. Mazzenga. 2000. Channel estimation and equalization for M-QAM transmission with a hidden pilot sequence. *IEEE Trans. Broadcasting* 46:170–76.

[39] X. Meng and J. K. Tugnait. 2004. Superimposed training-based doubly-selective channel estimation using exponential and polynomial bases models. In *Proceedings of the 2004 Conference on Information Sciences and Systems*, Princeton University, NJ, pp. 621–626.

[40] M. Niedzwiecki. 2000. *Identification of time-varying processes.* New York: Wiley.

[41] A. G. Orozco-Lugo, M. M. Lara, and D. C. McLernon. 2004. Channel estimation using implicit training. *IEEE Trans. Signal Processing* 52:240–54.

[42] J. G. Proakis. 2001. *Digital communications.* 4th ed. New York: McGraw-Hill.

[43] J. G. Proakis and D. G. Manolaks. 1996. *Digital signal processing.* 3rd ed. Englewood Cliffs, NJ: Prentice-Hall.

[44] R. Raheli, A. Polydoros, and C. K. Tzou. 1995. Per-survivor processing: A general approach to MLSE in uncertain environments. *IEEE Trans. Commun.* 43:354–64.

[45] A. M. Sayeed, A. Sendonaris, and B. Aazhang. 1998. Multiuser detection in fast-fading multipath environment. *IEEE J. Selected Areas Commun.* 16:1691–701.

[46] A. M. Sayeed and B. Aazhang. 1999. Joint multipath-Doppler diversity in mobile wireless communications. *IEEE Trans. Commun.* 47:123–32.

[47] N. Seshadri. 1994. Joint data and channel estimation using blind trellis search techniques. *IEEE Trans. Commun.* 42:1000–11.

[48] D. Slepian. 1978. Prolate spheroidal wave functions, Fourier analysis, and uncertainty. V. The discrete case. *Bell Syst. Tech. J.* 57:1371–430.

[49] M. D. Srinath, P. K. Rajasekaran, and R. Viswanathen. 1996. *Introduction to statistical signal processing with applications.* Upper Saddle River, NJ: Prentice-Hall.

[50] L. Tong and S. Perreau. 1998. Multichannel blind channel estimation: From subspace to maximum likelihood methods. *Proc. IEEE* 86:1951–68.

[51] L. Tong, G. Xu, and T. Kailath. 1994. A new approach to blind identification and equalization of multipath channels. *IEEE Trans. Inform. Theory* 40:340–49.

[52] J. K. Tugnait. 1987. Identification of linear stochastic systems via second- and fourth-order cumulant matching. *IEEE Trans. Inform. Theory* 33:393–407.

[53] J. K. Tugnait. 1995. On blind identifiability of multipath channels using fractional sampling and second-order cyclostationary statistics. *IEEE Trans. Inform. Theory* 41:308–11.

[54] J. K. Tugnait. 1995. Blind estimation and equalization of digital communication FIR channels using cumulant matching. *IEEE Trans. Commun.* 43:1240–45.

[55] J. K. Tugnait. 1996. Blind equalization and estimation of FIR communications channels using fractional sampling. *IEEE Trans. Commun.* 44:324–36.

[56] J. K. Tugnait. 2001. Channel estimation and equalization using higher-order statistics. In *Signal processing advances in wireless and mobile communications: Trends in channel estimation and equalization*, ed. G. B. Giannakis, Y. Hua, P. Stoica, and L. Tong, chap. 1, pp. 1–39. Vol. 1. Upper Saddle River, NJ: Prentice Hall.

[57] J. K. Tugnait and W. Luo. 2003. On channel estimation using superimposed training and first-order statistics. In *Proceedings of the 2003 IEEE International Conference on Acoustics, Speech, Signal Processing*, Hong Kong, vol. 4, pp. 624–627.

[58] J. K. Tugnait and W. Luo. 2003. On channel estimation using superimposed training and first-order statistics. *IEEE Commun. Lett.* 8:413–15.

[59] J. K. Tugnait and S. He. 2006. Doubly-selective channel estimation using data-dependent superimposed training and exponential bases models. In *Proceedings of the 2006 Conference on Information Systems and Sciences*, Princeton University, NJ.

[60] J. Vidal and J. A. R. Fonollosa. 1996. Adaptive blind equalization using weighted cumulant slices. *Int. J. Adaptive Control Signal Processing* 10:213–38.

[61] T. Zemen and C. F. Mecklenbräuker. 2005. Time-variant channel estimation using discrete prolate spheroidal sequences. *IEEE Trans. Signal Processing* 53:3597–607.

[62] Y. R. Zheng and C. Xiao. 2003. Simulation models with correct statistical properties for Rayleigh fading channels. *IEEE Trans. Commun.* 51:920–28.

[63] G. T. Zhou, M. Viberg, and T. McKelvey. 2003. A first-order statistical method for channel estimation. *IEEE Signal Processing Lett.* 10:57–60.

3

Adaptive Coded Modulation for Transmission over Fading Channels

Dennis L. Goeckel
University of Massachusetts

3.1 Introduction

The main goal of modern communications systems is to make universal high-speed access to information a reality. Under nearly any realizable scenario, the end user in such a system is free of tethers, thus making robust high-speed data transfer across the wireless channel a topic of extreme importance. However, the wireless channel presents a number of challenges. In particular, the wireless signal experiences [55]:

1. Path loss: The signal attenuates with distance from the transmitter.
2. Shadowing: Large objects between the transmitter and receiver can obstruct the radio signal.
3. Multipath fading: Reflections from objects in the environment can add constructively or destructively at the receiver.

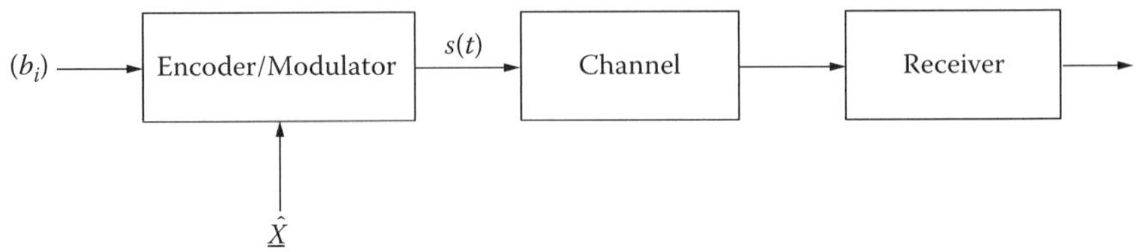

FIGURE 3.1 General adaptation framework: The transmitter sets parameters based on $\hat{\underline{X}}$, which contains information about the channel state, while forming the transmitted signal $s(t)$ from the information bit sequence (b_i). Note that $\hat{\underline{X}}$ can take many forms: path loss/shadowing estimates [47, 68, 70], number of errors corrected in previous packets [54, 55], explicit multipath fading estimates [31, 32], etc.

Since all three of these effects change with time or position, they cannot be known at the time that the system is designed. Thus, if no form of adaptation is performed while the system is operating, the system must be designed to in some sense deal with the worst case, which can be very expensive in terms of system resources. For example, if all users in a cellular system have to assume worst-case path loss and shadowing (e.g., behind a large building at the very edge of the cell) regardless of their location, they will transmit at maximum power, thus maximizing battery usage and the interference to other users. Therefore, wireless system adaptation has been a topic of critical interest in recent years.

Wireless link adaptation can be defined as shown in Figure 3.1: any altering of the parameters at the transmitter based on information about the current link state. Note that *link state* will be defined very generally; in particular, in addition to wireless channel conditions (path loss, shadowing, multipath fading), user data requirements will be included in the definition. Methods of adaptation can be classified by the type of adaptation performed (power, code rate, modulation, etc.) and the timescale of that adaptation.

The timescale of the adaptation depends on the link state phenomena for which measurements are provided to the adaptation algorithm. User data rate requirements, path loss, and shadowing change at a timescale that is long relative to the symbol rate. This makes measurements of such phenomena relatively robust [26, 64], particularly at high signal-to-noise ratios (SNRs), and has resulted in widespread penetration into current and pending systems of slow adaptations, such as power control [45, 66, 68] and data rate adaptation through variable spreading, code rate, or code aggregation [46]. Thus, throughout this chapter, it will be largely assumed that the measurements of the path loss and shadowing are accurate and known at both the transmitter and receiver, thus yielding a wireless system with a known given average received signal power but experiencing variable multipath fading.

Multipath fading is caused by the arrival at the receiver of many signal reflections, the superposition of which causes the instantaneous received signal power to vary widely [57, chapter 4] as described in section 3.2. Since significant nulls can occur, this signal fading is one of the most difficult problems to deal with in wireless communications systems. In particular, when the received power drops too low, a burst of bit errors can occur, and such bursts tend to dominate the error probability—even if the occurrence

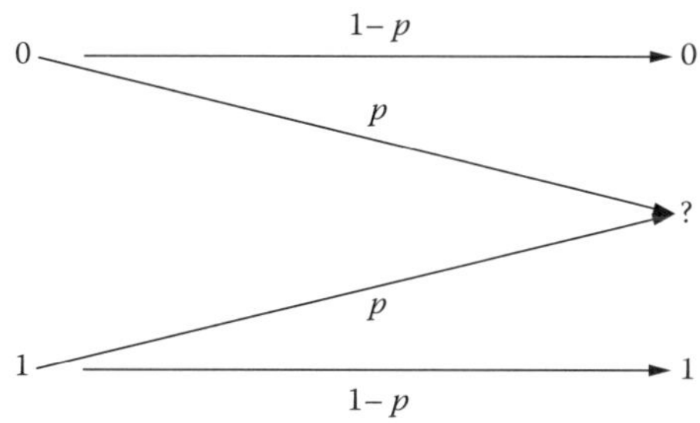

FIGURE 3.2 The binary erasure channel [16, p. 188] to represent a discrete-valued fading channel with channel state information available at the receiver.

of such system power drops is relatively unlikely. This results in a significantly higher required average received SNR for a given level of performance relative to systems operating over additive white Gaussian noise (AWGN) channels [53, p. 820]. However, unlike the effects of path loss and shadowing, which vary slowly and thus limit the ability of the system designer to average their effects over time, multipath fading varies relatively rapidly with time, position, and frequency. Thus, in many scenarios, well-designed systems achieve diversity, allowing them to average effectively over the effects of the multipath fading and thus significantly reduce its impact [53, p. 821]—even if there is no channel knowledge at the transmitter. For many systems, such nonadaptive solutions come at the cost of system latency or complexity, which motivates the consideration of transmission schemes that employ measurements of the multipath fading values.

Since this chapter will largely focus on the design of techniques for adaptation based on measurements of the multipath fading, it is important to understand the applicability of such adaptation. Thus, the gains from having knowledge of the channel at the transmitter will be a key topic discussed, and as motivated in the previous paragraph, it is often a question of the system complexity and latency allowable. To make this more concrete, consider the simple information theoretic example drawn from [16, p. 188], which is shown in Figure 3.2. First, to see how this represents a fading channel, consider a binary transmission system for which there are essentially no errors when the signal is transmitted over an AWGN channel (i.e., no fading); a simple example is coherently detected binary phase-shift keying (BPSK) with a relatively high SNR [53, p. 820]. Now assume that a BPSK system is operating over a discrete-valued fading channel described as follows. First, the state of the channel is independent for separate channel uses, which implies that a deep interleaver [53, p. 467] is employed. For a given channel use:

1. With probability p, the transmitted signal is multiplied by zero (hence disappears).
2. With probability $1 - p$, the transmitted signal is multiplied by $\alpha = 1/(1 - p)$ (hence amplified, which keeps the average received SNR identical to the AWGN case).

Assuming that channel state information (CSI), which throughout this paper will be the value of the multiplicative factor (in this case 0 or α), is available at the receiver, this yields the model shown in Figure 3.2. Consider signaling over the channel shown in

Figure 3.2 with and without channel state information at the transmitter. With CSI at the transmitter, transmission is halted when $\alpha = 0$ and a single bit is transmitted whenever $\alpha = 1/(1-p)$. Thus, with a very simple receiver identical to that for the BPSK system operating over the AWGN channel, the system reliably transmits $1-p$ information bits per channel use. Next, consider the case where there does not exist channel state information at the transmitter. Using information theoretic results [16, p. 188], the capacity of the channel without transmitter CSI is still $1-p$ information bits per channel use, but now it requires very long code words and the typical sequence decoding employed for the achievability statement of Shannon's capacity [59]. This simple example captures the key idea to adaptive signaling in response to the multipath fading in many cases—it will often not make sense from a Shannon capacity, but it can greatly simplify system design in practical systems [30].

Thus, adaptation in response to transmitter knowledge of the multipath fading has the promise of greatly simplifying the system design or, for a fixed system complexity, has the promise of greatly improving system performance (such as average data rate) [31, 32]. However, it is the very property that makes adaptation fruitful that also complicates its implementation; in particular, adaptation can be exploited because of the time-varying nature of user needs, path loss, shadowing, and multipath fading. But this time-varying nature makes that adaptation difficult; in particular, although changes in user needs and path loss generally happen over a long enough timescale that they can be reliably estimated, the time-varying nature of the shadowing [66] and the multipath fading [24, 27] make channel measurements outdated by the time they are ready to be used. In other words, the channel has changed since the measurements were performed, and thus the utility of such measurements in representing the current state of the channel can be questioned. This will be particularly exacerbated, of course, in systems that seek to adapt to the multipath fading [48].

The consideration of the design of signaling schemes that employ inherently outdated or noisy measurements is best done by carefully considering the channel characteristics *conditioned* on the measurements available. Naturally, if the support of the probability density function of the conditional channel given the measurements is very narrow, indicating that the system is fairly certain of the channel value, one can design coded modulation structures and rules for adapting those structures based on the assumption that the channel is fully known [31, 32] and suffer only mild degradations. However, such schemes can be very sensitive, even if the probability density function only shows a little spread around the estimated value [24, 27]. In such cases, not only must the rules of adaptation consider such spread, but the spread often will affect the types of coding and modulation structures that are effective, as demonstrated in section 3.3.3.

This chapter is organized as follows. In section 3.2, the system model that will be used throughout this work is presented. Section 3.3 provides a detailed derivation of the key issues in adaptive signaling using the simplest case of a system where there is only a single antenna employed at each the transmitter and receiver. Section 3.4 discusses recent extensions of these results to systems with multiple antennas at the transmitter and receiver, and section 3.5 presents conclusions and avenues for future work in the field.

3.2 Adaptive System Model

3.2.1 Model for a Wireless Link

The transmitted signal in a wireless communications system is affected by three factors: path loss, shadowing, and multipath fading. In complex baseband notation [53], the signal $r(t)$ that is received when the signal $s(t)$ is transmitted can be written as

$$r(t) = L(t)X(t)s(t) + n(t), \tag{3.1}$$

where $L(t)$ is a real-valued random process that represents the combined effect of the path loss and shadowing, $X(t)$ is a complex random process representing the effect of the multipath fading, and $n(t)$ is a stationary complex Gaussian random process with (two-sided) power spectral density $S_N(f) = N_0/2$, representing additive noise. In equation (3.1), the fading has been assumed to be frequency-non-selective [53, p. 816]; this is appropriate for a narrowband single-carrier system or a single subcarrier of a wideband orthogonal frequency division multiplexing (OFDM) system [6, 71]. Extensions of the concepts presented in this chapter to frequency-selective channels are conceptually straightforward, although such channels offer inherent natural diversity with little system latency, and hence often reduce the gain available through adaptive signaling.

Understanding the characteristics of the processes $L(t)$ and $X(t)$ in equation (3.1) is crucial in determining methods of adaptation based on such. The random process $L(t)$ is caused by path loss, which is determined by the distance the receiver is from the transmitter, and shadowing, which is determined by the existence of large objects between the transmitter and receiver. For a stationary user, the path loss and shadowing are generally modeled as constant, despite the fact that it could be argued that the movement of large objects can affect the shadowing. For a user in motion, the shadowing will be the more variable of the two effects, and the distance over which it is highly correlated can be roughly modeled as 100 m in a macrocellular suburban environment [35]. For a user at walking speed (say, 2 m/s), this implies that the shadowing correlation time is on the order of 50 s; for a user in a vehicle (say, 88 km/h), this implies that the shadowing correlation time is on the order of 4 s. This suggests that it is quite plausible to make estimates of the path loss and shadowing and to employ such in wireless communications systems. In fact, this is very often done in current and next-generation cellular system implementations [48]. Also, since $L(t)$ varies at a relatively long timescale, it will be assumed throughout the remainder of this chapter that it is measured accurately and known at the transmitter and receiver.

In contrast, consider the random process $X(t)$, which represents the multipath fading. In a wireless environment, the signal $s(t)$ is reflected to the receiver from many objects. Because the propagation distance is different for each of these reflections, the reflected signals will arrive at slightly different times at the receiver. For a narrowband system, which has a relatively long symbol interval, there will not be appreciable intersymbol interference (ISI) [53, p. 817]. However, because of the large carrier frequencies typically employed in modern wireless communications systems, even a small difference in arrival times for two paths can result in a large phase difference between those paths. For example, a path-length difference of only 1 ft results in the signal being delayed by 1 ns, which causes a full 2π rotation in phase when the carrier frequency is 1 GHz. Hence, the phase of any given arriving path is generally modeled as uniformly distributed. Since the process $X(t)$ is caused by the sum of very many roughly independent paths, projected onto each of the in-phase (real) and quadrature (imaginary) components, the central limit theorem [52, p. 214] motivates its modeling as a complex Gaussian random process [3].

By considering the genesis of the multipath fading as described above, it is easy to observe that the phase of a given path will change greatly for each movement of the reflecting object, the receiver, or the transmitter by one wavelength. Hence, even with only walking speed mobility (say, 2 m/s), a system with a 1 GHz carrier will yield a process $X(t)$ that changes independently six times per second (or, as commonly stated, with a 6 Hz Doppler frequency) [51, 56], which makes adaptation challenging, since feedback of the channel characteristics provided to the transmitter at some delay must accurately model the current channel fading for adaptation to be effective. Note that this problem will be exacerbated at higher mobilities and higher carrier frequencies.

Mathematically, $X(t) = X_R(t) + jX_I(t)$ will be assumed to be a zero-mean stationary Gaussian random process with an autocorrelation function of the real part $X_R(t)$ (or imaginary part $X_I(t)$) defined as

$$R_X(\tau) = E\left[X_R(t)X_R(t+\tau)\right] = E\left[X_I X_I(t+\tau)\right],$$

and the real part $X_R(t)$ and imaginary part $X_I(t)$ will be assumed to be independent of one another. The zero-mean assumption implies that a line-of-sight path is not present, which corresponds to the most pessimistic case—Rayleigh fading. Throughout this chapter, the popular Jakes model [41] will generally be adopted, which is characterized by $R_X(\tau) = J_0(2\tau f_d \tau)$, where $J_0(\cdot)$ is the zero-order Bessel function of the first kind and f_d is the Doppler frequency, which is defined as the number of wavelengths of motion of an object per second. It will be assumed that although the Doppler frequency might be large, it will not approach the symbol rate, and thus the channel $X(t)$ can be assumed to be constant over the support of a single signaling pulse $p(t)$, which is termed the *slowly fading* assumption in most digital communications texts [53, p. 816]. We hasten to emphasize, however, that the use of the word *slow* in this context is with reference to the symbol interval—not the amount of time between channel estimation and signal transmission, where even such "slow" multipath fading can have a significant effect.

3.2.2 Adaptation in Response to Path Loss/Shadowing

There are many forms of adaptation currently employed in response to path loss/shadowing in wireless communications systems. In fact, even the base station selection process, where a mobile generally decides to associate with the base station from which it sees the largest average received signal strength, can be viewed as a form of adaptation.

Such adaptations will be called "slow" adaptations throughout this work, and they will be characterized by schemes that adapt the transmitter at an interval on the order of (at least) many (hundreds of) symbols. A good tutorial on slow adaptations, particularly in current standards, is provided in [48].

First, consider wireless system adaptations that adapt depending on user needs. In particular, one of the key features of third-generation cellular systems is supporting users with high data rates. This is often done by simply allocating more of the time/bandwidth/code space to the users. For example, in Enhanced Data Rates for GSM Evolution (EDGE) systems, which are built on a time-division multiple-access (TDMA) framework, users with high-data-rate needs are allocated more time slots. In the code-division multiple-access (CDMA)-based IS-95 Revision B, high-data-rate users are allocated multiple spreading codes, which is termed code aggregation [48].

Next, consider adaptations based on the current channel conditions for a given user. In fixed-rate systems, such as first- and second-generation cellular telephone systems, where the rate of the vocoder is generally fixed, the key is to adapt the system such that acceptable performance is maintained at this fixed rate. The transmission technology and channel assumptions fix a minimum average received SNR γ_0 required for acceptable operation—the goal of adaptation is to maintain γ_0, which can be done by adapting the transmitted power in response to measurements of the path loss and shadowing. Methods of performing such adaptation include channel inversion, where the transmitted power is set proportional to the channel loss, and truncated channel inversion [e.g., 17, 66]. Truncated channel inversion is defined by a threshold L_0, which breaks the policy into two cases:

1. $L(t_0) \geq L_0$: The transmitted power is set to $\gamma_0/L(t_0)$, which results in an average received SNR of γ_0.
2. $L(t_0) < L_0$: The transmitted power is set to zero, which results in an outage.

Using this policy, the required average received SNR (and no more) is obtained whenever possible, but excessive power is not wasted by inverting the channel when there are large losses on the channel.

Outside of current standards, the setting of the rate (coding, modulation, spreading factor) of the system to the current average received SNR has clearly emerged as a critical topic. In particular, turbo codes [4] and low-density parity-check (LDPC) codes [23, 46] are approaching channel capacities on a variety of channels. Thus, assuming enough receiver complexity for the decoding of such codes and enough latency to allow perfect interleaving of the coded symbols, the rate of nearly error-free systems should approach the Shannon capacity [59] of the independent and identically distributed (IID) discrete-time Rayleigh fading channel, defined by:

$$Y_i = \alpha_i X_i + n_i,$$

where Y_i is the received sequence, α_i is the IID sequence of Rayleigh channel fading values, X_i is the transmitted sequence, and n_i is the noise sequence. The Shannon capacity for

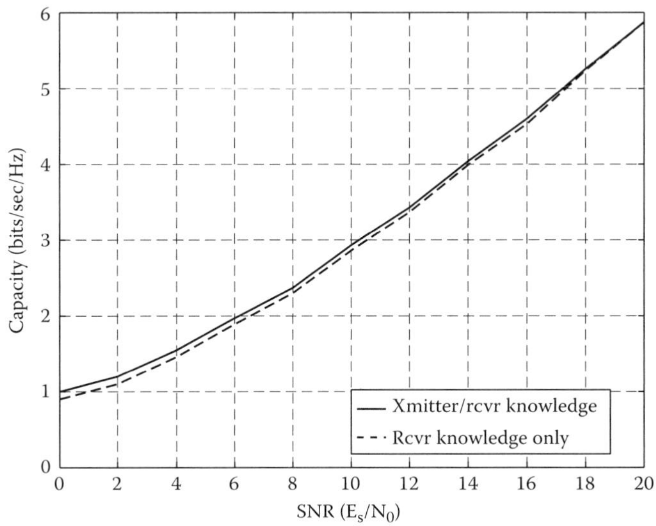

FIGURE 3.3 The Shannon capacity of an independent and identically distributed Rayleigh fading channel, assuming (1) perfect CSI available only at the receiver, and (2) perfect CSI available at both the transmitter and receiver [28]. Note that the gain in Shannon capacity resulting from having perfect CSI available only at the transmitter is only slight, as discussed in section 3.3.1.

such a channel without CSI at the transmitter is shown as the lower curve in Figure 3.3, where the SNR on the horizontal axis is the average received SNR (i.e., the SNR after the path loss and lognormal shadowing are considered). The only requirement for the highly efficient operation of such a system is the knowledge of this average received SNR at the transmitter so that the rate of the transmitter can be set appropriately, and this can be obtained by feedback of the path loss and shadowing. We emphasize that approaching the curves in Figure 3.3 still requires high decoding complexity and significant latency, which motivate whether adaptation with the additional knowledge of the values of the multipath fading can improve on the performance in Figure 3.3 in terms of performance versus system complexity.

3.2.3 Analytic Model for Fine-Scale Adaptation

The main portion of this chapter will be dedicated to the design and analysis of adaptive systems that use explicit measurements of the multipath fading to perform system adaptation. This is a topic that was considered in the 1970s [11, 37, 38] and then became popular again in the early 1990s [e.g., 1, 6, 14, 29, 30, 65].

A block diagram of the typically employed system is shown in Figure 3.4. Given the model shown in Figure 3.4 and channel model given in section 3.2.1, the key to designing adaptive signaling systems is considering signaling for the *conditional* channel for the symbol of interest (call it s_k) given the outdated measurement $\underline{\hat{X}} = (\hat{X}(t-\tau_1), \hat{X}(t-\tau_2), \ldots, \hat{X}(t-\tau_N))^T$. It will be assumed that a measurement $\hat{X}(t-\tau_i)$ is equal to the true value $X(t-\tau_i)$ plus additive Gaussian noise of variance

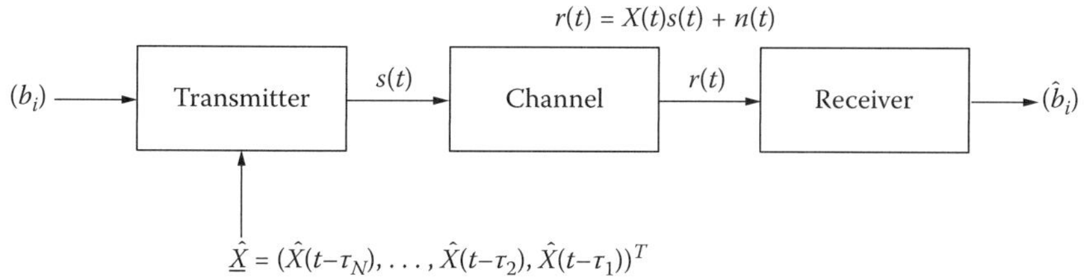

FIGURE 3.4 A block diagram of the system, where (b_i) is the sequence of information bits to be transmitted across the channel, $s(t) = \Sigma_{k=-\infty}^{\infty} s_k p(t - kT_s)$ is the transmitted signal, $r(t)$ is the received signal, $n(t)$ is additive white Gaussian noise, $\underline{\hat{X}}$ is the vector of outdated channel measurements, and $(\hat{b}_i)$ is the sequence of information bit estimates output from the receiver.

$$\sigma_\varepsilon^2 = \frac{1}{\frac{E_p}{N_0}}$$

in each of the in-phase and quadrature components. For example, such would be the case in an adaptive system employing pilot symbol–assisted modulation (PSAM) [12] with a pilot symbol energy of E_p [7, 24].

Note that the model in Figure 3.4 captures the critical issue of delay in the feedback path from the receiver to the transmitter, since the most recent estimate is assumed to have been made τ_1 seconds ago. In other words, the "outdated" nature of the estimates takes into account this key implementation issue in adaptive communications systems.

Denoting Y as the magnitude of the fading that multiplies s_k in the matched filter output for the kth symbol and using the fact that linear functionals of a Gaussian random process are jointly Gaussian, Y is Rician when conditioned on the vector $\underline{\hat{X}}$, with probability density function [25]

$$p_{Y|\underline{\hat{X}}}\left(y \mid \underline{x}\right) = \frac{y}{\sigma^2} e^{-\frac{y^2+s^2}{2\sigma^2}} I_0\left(\frac{ys}{\sigma^2}\right), \qquad y \geq 0, \tag{3.2}$$

where $I_0(\cdot)$ is the zero-order modified Bessel function. Using the assumption that $X(t)$ can be assumed constant over the support of $p(t)$ and normalizing the fading such that $E[(X_R(kTs))^2] = E[X_I(kT_s))^2] = 1$ (note that this simplification will make the average received energy twice that provided by simply the path loss and shadowing, which will be accounted for below), the noncentrality parameter in equation (3.2) is given by

$$s^2 = \left(\underline{\rho}^T \left(\Sigma_{\underline{X}} + \sigma_\epsilon^2 I_N\right)^{-1} \underline{x}_R\right)^2 + \left(\underline{\rho}^T \left(\Sigma_{\underline{X}} + \sigma_\epsilon^2 I_N\right)^{-1} \underline{x}_I\right)^2, \tag{3.3}$$

where I_N is an N by N identity matrix. The $(m, n)^{\text{th}}$ element of $\Sigma_{\underline{x}}$, the N by N autocorrelation matrix of the in-phase component of $\underline{\hat{X}}$ when the channel estimates are noiseless, is given by $R_x(\tau_{N-m+1} - \tau_{N-n+1})$, and the correlation vector of the in-phase component of $\underline{\hat{X}}$ with the in-phase component of the fading of interest is given by $\underline{\rho}$, where $\rho_i = R_x(\tau_{N-i+1})$. The parameter σ^2 in equation (3.2) is the mean square error of a minimum mean square error (MMSE) estimator [72, p. 54] of the in-phase (or quadrature) fading of interest, and is given by

$$\sigma^2 = 1 - \underline{\rho}^T \left(\Sigma_{\underline{x}} + \sigma_\epsilon^2 I_N\right)^{-1} \underline{\rho}\,. \tag{3.4}$$

Understanding the Rician density in equation (3.2) and the expression for the Rician noncentrality parameter s in equation (3.3) is key to designing effective adaptive coded modulation schemes. In particular, for $s = 0$, the Rician probability density function in equation (3.2) is equivalent to a Rayleigh probability density function, indicating that coded modulation structures designed for Rayleigh fading channels are pertinent for application when s is small; likewise, as $s \to \infty$, the (properly normalized) Rician density function approaches a delta function, thus indicating that the effective channel approaches an AWGN channel. Since coded modulation schemes for Rayleigh fading channels differ greatly from AWGN schemes, the interpretation of equation (3.3) is used extensively in the design of structures, as demonstrated in section 3.3.3.

There is one limitation to directly employing the result in equation (3.2). In particular, it presumes that the autocorrelation function $R_X(\tau)$ of the random process $X(t)$ is known at the transmitter; however, this autocorrelation function can vary greatly in wireless systems [51, p. 88–89]. Thus, it must generally be estimated [18, 19], either implicitly or explicitly, or uncertainties in it must be worked into system design [24, 27]. To address both possibilities in one framework, the autocorrelation function will be assumed to lie in some uncertainty class $\mathcal{R}$, which matches the approach taken in [27] directly. If it can be accurately estimated through techniques as described in [18, 19], this class can be shrunk accordingly (in the limit to a single autocorrelation function).

Thus, given the model for the system measurements and the measurements of the autocorrelation function, one should be able to ascertain (1) how much predictor error σ^2 will generally be in the system, and (2) what is the uncertainty class $\mathcal{R}$ over which some sort of robustness will be maintained. Understanding both of these for a given system configuration will be the key to understanding the design of the coded modulation. In particular, the former will allow the choice of the coded modulation structure, while the latter will allow one to design on that structure.

3.3 Adaptivity in Single-Input Single-Output Systems

3.3.1 Information Theoretic Bounds

Before considering the derivation of practical signaling schemes that adapt to the multipath fading, it is instructive to consider the improvement in Shannon capacity that is available when CSI is made available to the transmitter. Assume that the sequence of zero-mean complex Gaussian channel fading coefficients affecting the transmitted symbols forms an IID sequence; in other words, an IID Rayleigh fading channel is assumed.

If the criterion is to maximize the average data rate under an average power constraint, Goldsmith [30] has demonstrated that the information theoretic capacity when perfect CSI is available at both the transmitter and receiver is achieved with variable-power Gaussian codebooks, where the power depends on the current value of the channel fading. A comparison of the Shannon capacity when perfect CSI is available at both the transmitter and receiver with the capacity when CSI is available only at the receiver is shown in Figure 3.3. Note that, somewhat surprisingly, the gain in channel capacity is only slight. However, as with the example of the erasure channel in section 3.1, it should be remembered that this assumes very long code words and large decoding complexities. In particular, the minimum distance of the codes becomes very large, and since IID Rayleigh fading is assumed, the diversity achieved by a given code is very large—even for the case of CSI only at the receiver.

The analysis in the previous paragraph and Figure 3.3 applies to the Shannon capacity [59], which is generally appropriate if the average rate of a system is being considered. Recently, however, there has been significant interest in whether a system can transmit a fixed amount of information within a given time constraint—such approaches lead to measures such as the outage capacity [22] or the delay-limited capacity [36]. Such analyses are generally still done under the assumption of infinite-length code words, which are required for the random coding arguments invoked, but now under the assumption that a given code word will only see some small number of fades. Thus, in essence, this yields a view at system operation in a diversity-limited context. The metric is based on the probability that the system experiences a set of fading values for which it can communicate at the desired rate. In contrast to Figure 3.3, such analyses [8, 49] have demonstrated the significant gains possible when knowledge of the channel fading values is provided to the transmitter in addition to the receiver. This has motivated work in the design of practical adaptation schemes that focus on outage probability [43]. These results lead to the preliminary conclusion that the gain from having estimates of the channel fading at the transmitter is highly reliant on the decoder complexity and system latency allowed.

3.3.2 Design for Uncoded Systems

In this section, adaptive uncoded modulation will be designed. Unlike coded schemes, where the design is complicated by memory in the trellis and questions about the proper structure, uncoded schemes present a simple framework to demonstrate many of the key issues.

As described in section 3.2.3, there is a key issue of robustness to uncertainties in the autocorrelation function $R_X(\tau)$, which can be captured by designing for an uncertainty class $\mathcal{R}$ that shrinks to a single point when the autocorrelation is known or can be accurately estimated. If the class $\mathcal{R}$ is a single point as is often considered for prediction-based methods [18, 19, 44, 45], the application of equation (3.2) is identical regardless of the number of outdated estimates N employed. This is observed by noting that, for a vector of random variables drawn as samples from a stationary Gaussian random process, (1) the marginal probability density function is independent of the sampling time, and (2) the conditional probability density function for any one of the variables is

Gaussian when conditioned on the others, with variance given by σ^2 in equation (3.4). Thus, given the predicted value and σ^2, the choice of the signal set is independent of N. Hence, when the autocorrelation function $R_X(\tau)$ is known exactly, the design for $N = 1$ with the appropriate σ^2 is sufficient to characterize performance of a given scheme.

When $\mathcal{R}$ is made larger to capture uncertainties in the knowledge of $R_X(\tau)$, the design becomes greatly complicated [26]. Thus, robust design with general $\mathcal{R}$ with only a single outdated estimate ($N = 1$) will be considered; however, we hasten to note that, per the previous paragraph, designing adaptive coded modulation for known $R_X(\tau)$ and *any* N is a simplification of this case.

Since the case $N = 1$ will be considered, the quantity $\rho = R_X(\tau_1)$, which, since $X_R(t)$ and $X_I(t)$ are normalized to have unit energy, is the correlation coefficient of the in-phase (or quadrature) components of the multipath fading process between the time of channel estimation and the time of data transmission, will be important. Assuming that the estimates are noiseless ($\sigma_\epsilon^2 = 0$) implies that $\sigma^2 = 1 - \rho^2$, and it is observed that the mean square prediction error increases rapidly with decreasing correlation between the estimate and the current value, as expected. Throughout much of this chapter, systems will be designed for a given ρ, which captures the amount of information in the channel estimate about the current fading value. A conversion to mean square predictor error, if desired, can be obtained by the transformation $\rho = \sqrt{1-\sigma^2}$.

Designing robustly using a single outdate estimate requires performance to be guaranteed for all $\rho \in [\rho_{\min}, 1]$, where $\rho_{\min}$ is the minimum value of $R_X(\tau_1)$.

3.3.2.1 Design Rules

The design rules for uncoded systems have been well established by a number of authors [7, 27, 31]. The signal sets considered in this section will be 0-QAM (quadrature amplitude modulation) (no data transmitted), 2-QAM, 4-QAM, 16-QAM, and 64-QAM with two-dimensional Gray mapping, although the extension to any set of signal sets is immediate.

Following [27], let P_b be the target bit error probability for the system, which operates at the average received SNR E_s/N_0, where E_s is the average received energy per QAM symbol. For now, it will be assumed that the average energy E_s is not varied over time; energy adaptation will be discussed in detail below. Specification of the adaptive transmitter requires finding $\tilde{M}(h), \forall h$, where $\tilde{M}(h)$ is the number of signals in the QAM signal set employed when $|\hat{X}(kT_s - \tau_1)| = h$. If $\tilde{M}(h)$ is chosen such that P_b is maintained for each h,

$$\tilde{M}(h) = \max\left\{ M : \sup_{\rho_{\min} \le \rho \le 1} \tilde{P}_M\left(\frac{E_s}{N_0}, h, \rho\right) \le P_b \right\}, \tag{3.5}$$

where $\tilde{P}_M(E_s/N_0, h, \rho)$ is defined as the bit error probability of the M-QAM signal set at average received SNR E_s/N_0 when $R_X(\tau_1) = \rho$ and $|\hat{X}(kT_s - \tau_1)| = h$. Assume that maximum likelihood symbol detection, given the current channel fading amplitude, is employed on the samples of the matched filter output at the receiver. A tight approximation to the bit error rate of M-QAM modulations is given by [31]

$$P_M\left(y^2\frac{E_s}{N_0}\right)\approx 0.2\exp\left(-\frac{3}{4(M-1)}\frac{E_s}{N_0}y^2\right), \tag{3.6}$$

which will be employed for all M for much of the design work for uncoded systems. If errors in channel estimation *at the receiver* are considered, the right side of equation (3.6) will increase, of course, but it will often fit into the same functional form [7], which is convenient, since the same optimization will apply. Using equation (3.6) yields

$$\tilde{P}_M\left(\frac{E_s}{N_0},h,\rho\right)=E\left[P_M\left(Y^2\frac{E_s}{N_0}\right)\Big\|\hat{X}(kT_s-\tau_1)\Big|=h\right]$$
$$\approx\begin{cases}\dfrac{0.2\exp\left[-\dfrac{h^2\rho^2}{2(1-\rho^2)}\left(1-\dfrac{1}{1+\dfrac{3}{2}\dfrac{E_s}{N_0}\dfrac{(1-\rho^2)}{(M-1)}}\right)\right]}{1+\dfrac{3}{2}\dfrac{E_s}{N_0}\dfrac{(1-\rho^2)}{(M-1)}} & \rho<1\\ 0.2\exp\left(-\dfrac{3}{4}\dfrac{E_s}{N_0}\dfrac{h^2}{(M-1)}\right) & \rho=1\end{cases}, \tag{3.7}$$

where the second line is obtained by substituting equation (3.2) and equation (3.6) into the first line and evaluating the expectation over Y using [34, 6.614.3].

From equation (3.5), equation (3.7) must be evaluated at its supremum on $\rho\in[\rho_{\min},1]$. Since the right side of equation (3.7) is a continuous function on this closed interval, it achieves its maximum on this interval at a point that will be denoted ρ^*. The following solution is found by standard calculus techniques. Let

$$\tilde{\rho}=\begin{cases}0 & h\ge\sqrt{2}\\ \sqrt{\left(1+\dfrac{2(M-1)}{3}\dfrac{N_0}{E_s}\right)\dfrac{(2-h^2)}{2}} & 0\le h\le\sqrt{2}\end{cases}.$$

The worst-case autocorrelation is then given by

$$\rho^*=\begin{cases}\rho_{\min} & \tilde{\rho}\le\rho_{\min}\\ \tilde{\rho} & \rho_{\min}<\tilde{\rho}<1\\ 1 & 1\le\tilde{\rho}\end{cases}. \tag{3.8}$$

The signal set is specified using equations (3.7) and (3.8) in

$$\tilde{M}(h)=\max\left\{M:\tilde{P}_M\left(\frac{E_s}{N_0},h,\rho^*\right)\leq Pb\right\}.$$

Note that $\tilde{M}(h)$ is nondecreasing in h. Thus, the adaptive scheme can be specified by the values h_m, $m = 2, 4, 16, 64$, where h_m is defined as the threshold such that for $h \geq h_m$, m-QAM can be employed.

The discrete nature of the set of rates for any finite collection of signal sets hurts the performance of the system; in particular, for all h such that $h_m < h < h_{m+1}$, the estimate is better than that required to use m-QAM but not good enough to use $(m + 1)$-QAM. Energy adaptation provides a means to solve this problem [31]. Rather than employing the method of [31], an alternate method, which is analogous to truncated channel inversion and the power pruning of [20], is described here. The advantage of this method is that, with very little loss of optimality, it is easily extended to coded modulation structures, where the overall optimization problem of [31] is not easily framed when channel prediction is not perfect [27]. Once a signal set has been chosen, the system is essentially a fixed-rate system; thus, the goal changes from maximizing average rate to attempting to allow communication at this fixed rate with the least amount of power. Thus, after the signal set is chosen, equations (3.7) and (3.8) are used to decide the minimum energy required to maintain P_b given the channel estimate h, and this energy is employed rather than the average energy. Any excess energy is put into a "bank" on which successive symbols can draw.

3.3.2.2 Numerical Results

As discussed in section 3.3.1, systems with a significant amount of decoding complexity and allowable latency only have the potential for a small amount of improvement when CSI is provided to the transmitter. As might be expected, uncoded systems, which have the least decoder complexity and essentially no latency, benefit the most when transmitter CSI is available. In particular, uncoded systems operating over frequency-non-selective Rayleigh fading channels perform very poorly, because they do not achieve diversity. Because of this, coherently decoded quadrature phase-shift keying (QPSK) with only receiver CSI requires an SNR of 34 dB to achieve a bit error rate of 10^{-4} on a frequency-non-selective Rayleigh fading channel [53, p. 829], whereas the same technique requires an SNR of less than 10 dB to achieve the same bit error rate on an AWGN channel. The reason for this discrepancy is that the QPSK system operating over the Rayleigh fading channel is extremely susceptible to deep signal fades. Although the occurrence of such is relatively uncommon, the error rate during a bad fade can be orders of magnitude above that occurring when the average received SNR is observed, and thus these bad fades dominate the error rate.

In adaptive signaling, CSI is available at the transmitter. Arguably, the greatest utility of such information is that signaling can be avoided when bad fades are present. In particular, with perfect transmitter CSI [31], average rates in excess of 2 bits per symbol are possible at bit error rates of 10^{-5} for average received SNRs under 20 dB. Thus, there is a significant gain in system performance when transmitter CSI is available in uncoded systems.

However, as pointed out in [24, 27], the assumption of perfect CSI is dangerous when channel estimates are outdated or noisy, as would be the case with realistic delay in the feedback path from the receiver to the transmitter. In particular, the conditional density function given in equation (3.2) becomes Rician (rather than a delta function), and hence the conditional channel acts like a fading channel. For example, for the example described in section 3.3.2.1, adaptive signaling assuming perfect channel estimation can miss its target bit error rate by two orders of magnitude—even for the relatively high correlation coefficients of $\rho = 0.96$. In this case, bad predictions, which are relatively uncommon, lead to instantaneous error rates that are orders of magnitude above the target and thus dominate system performance. Using the design method of equation (3.2) reveals that there are still significant gains in adaptive signaling versus nonadaptive signaling when transmitter CSI is not perfect—even when the correlation coefficient drops as low as $\rho = 0.96$. We conclude from this section that adaptive signaling is particularly effective for simple, low-latency systems such as adaptive uncoded QAM systems [24, 27].

3.3.3 Coded Modulation Structures

As discussed in section 3.2.3, the conditional channel given an outdated measurement can vary from almost Rayleigh to almost AWGN—depending on both the channel estimate and the mean square prediction error σ^2. For small σ^2, the conditional channel is nearly always Rician with a large noncentrality component [53, p. 811] (hence approaching AWGN), whereas for large σ^2, the channel often approaches Rayleigh. It is well known that coded modulation structures optimized for AWGN channels [e.g., 63] are not well matched to Rayleigh channels [58]. Thus, the characterization of the mean square prediction error in a given system determines the types of coded modulation structures to be employed.

For systems where the mean square prediction error is anticipated to be nearly zero, coded modulation structures designed for the AWGN channel can be employed without interleaving [31]. For systems with a moderate amount of channel prediction error, structures designed for a Rayleigh fading channel can have aspects of structures designed for an AWGN channel embedded in them [27]. Finally, for adaptive systems where the mean square prediction error is expected to be large, adaptive bit-interleaved coded modulation (BICM) [50] is preferable. These three types of structures are presented below.

3.3.3.1 Coding Structures with (Nearly) Perfect Prediction

If the current channel fading $X(t)$ is known accurately at the transmitter (i.e., σ^2, the prediction error of an MMSE predictor, is small), the effective channel given the outdated estimate is roughly AWGN. Thus, coding structures designed for AWGN channels [63] should be employed [32]. In particular, a base trellis-coded modulation scheme [63] tuned to the average received SNR can be selected, and then uncoded bits can be added or deleted based on the channel estimate. This structure is shown in Figure 3.5. As

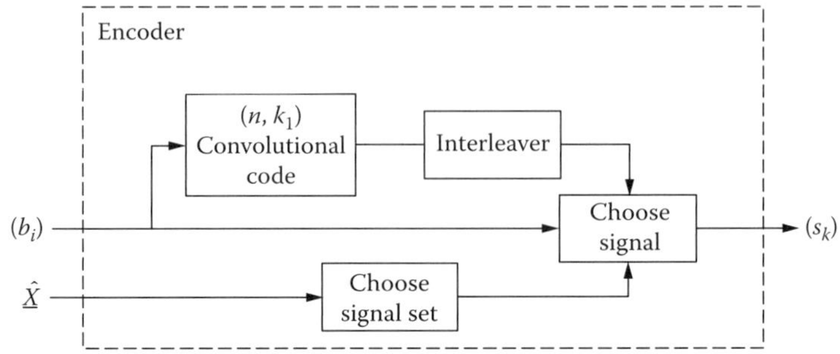

FIGURE 3.5 The adaptive trellis coding diagram—to be employed when performing adaptive signaling with low to moderate prediction errors. When the prediction error is low, the interleaver can be removed.

described in the legend to Figure 3.5, the interleaver can be removed in this case, since the Euclidean distance between two possible paths at the receiver can be precisely controlled. This structure allows symbol-by-symbol adaptation (unlike changing the rate of a convolutional encoder) and, because parallel branches are generally effective when communicating over AWGN channels, is an efficient coding structure over a wide range of instantaneous rates of the system.

3.3.3.2 Coding Structures with Moderate Prediction Error Statistics

When there is a moderate amount of predictor error power (i.e., moderate values of σ^2), the use of uncoded bits can be detrimental, since there will be a high number of channels that are not strongly Rician per equation (3.2), and it is well known that the use of uncoded bits on fading channels is problematic, per section 3.3.2.2. However, note that the channels become more Rician *as the predicted value increases.* This is fortuitous, because it allows the retention of the adaptive coded modulation structure shown in Figure 3.5, except that the base convolutional code is chosen to have no parallel branches [27]. Thus, when the estimate is small (and the channel nearly Rayleigh), a code appropriate for such a fading channel is employed [58]. When the channel estimate is large (and thus the channel strongly Rician), the structure in Figure 3.5 adds uncoded bits, which are appropriate in such a situation.

3.3.3.3 Coding Structures with Large Prediction Error Statistics

When there is a significant amount of prediction error (i.e., large values of σ^2), the use of parallel branches (i.e., uncoded bits) is not possible under almost any channel measurement, since from equation (3.2) it can be seen that the channel will be nearly Rayleigh with very high probability. Thus, symbol-by-symbol rate adaptation is desirable, but parallel branches are not allowable. A structure that allows such was presented in [50] and is shown in Figure 3.6. Note that the instantaneous rate is adapted, but all bits are coded, and thus the scheme retains diversity (in this case, against bad predictions) equal to the minimum Hamming distance of the convolutional code.

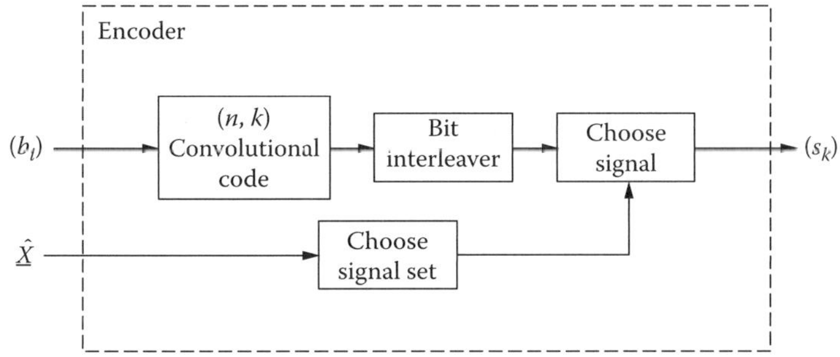

FIGURE 3.6 The adaptive bit-interleaved coded modulation scheme—to be employed when performing adaptive signaling with relatively frequent large prediction errors.

3.3.4 Designing with a Given Coded Modulation Structure

3.3.4.1 Design Rules

Unlike uncoded systems, which do not possess memory, and thus allow simple symbol-by-symbol adaptation, as demonstrated in section 3.3.2.1, the memory in coded modulation schemes complicates design. The main techniques that have been developed for coded modulation systems are described in [27] and [32]. Since the techniques in the later work of [27] include those in [32], the techniques of [27] will be briefly described. In particular, it is important to protect both the coded and uncoded information bits. This is done by maintaining the intersubset and intrasubset differences, which, roughly stated (see [27] for details), is the pairwise error probability between the two signal points in different and the same subsets [63], respectively. When the prediction error power is small, the intrasubset and intersubset differences for the coded modulation structures described by Figure 3.5 can be maintained by simply preserving the *received* Euclidean distance between adjacent points in the signal set [32]. When the prediction error power is moderate, the intersubset and intrasubset differences must be maintained separately [27]. For the adaptive BICM structure of Figure 3.6, there are only intersubset differences to be maintained [50].

3.3.4.2 Performance Results

As demonstrated in sections 3.3.1 and 3.3.2.2, the gains when CSI is available at the transmitter should decrease as system decoding complexity and latency are increased. Thus, it is anticipated that the gains described in this section will be smaller than those shown in section 3.3.2.2, and this is indeed observed. In particular, nonadaptive systems employing coded modulation and interleaving over frequency-non-selective Rayleigh fading channels have an enormous potential for gain, as evidenced by the vast difference between the performance of nonadaptive uncoded systems [53, p. 829] and the channel capacity shown in Figure 3.3. In contrast, adaptive signaling schemes with perfect prediction are signaling for a channel that is conditionally AWGN, which implies that the gains between uncoded systems [31] and channel capacity (see Figure 3.3) for systems with CSI at the transmitter are not so vast; in fact, they are similar to those attainable for AWGN channels [32].

The performance loss when perfect channel predictions are not available at the transmitter can be mitigated by employing the techniques of sections 3.3.3.2 and 3.3.3.3. Doing such with eight-state trellis codes provides gains over nonadaptive schemes that are on the order of 25 to 75% in data rate [27, 50], which, as expected, do not match the exorbitant gains seen in the uncoded case. One would expect the gains to decrease even further for more complicated codes and, in the complexity/latency limit, almost disappear as prescribed by Figure 3.3.

3.4 Adaptivity in Multiantenna Systems

Wireless systems employing multiple antennas at the transmitter or receiver have demonstrated both the theoretical [21, 62] and practical [21] ability to greatly increase system capacities far beyond those previously imagined. In particular, the disparate fading values between different pairs of antennas in multiple-input multiple-output (MIMO) systems lead to a large increase in the number of degrees of freedom of the system, and capacities can even dwarf those attainable for the AWGN channel at the same average transmitted SNR—even if channel state information is not available at the transmitter [21, 62].

Throughout this section, a multiple-antenna system will be referred to as an (M, N) system if it employs M transmit and N receive antennas. The system model generally employed for a narrowband MIMO system is given by

$$\underline{Y} = H\underline{X} + \underline{Z}, \tag{3.9}$$

where $\underline{X}$ is an $M \times 1$ vector whose j^{th} component represents the signal transmitted by the jth antenna. Similarly, the received signal and received noise are represented by $N \times 1$ complex vectors, $\underline{Y}$ and $\underline{Z}$, respectively. Generally, it is assumed that the entries of the $N \times M$ matrix H, whose entry (i, j) represents the fading from transmitter j to receiver i, are identically distributed zero-mean jointly complex Gaussian random variables. For many of the early results, the entries of H were considered to be independent [21, 62], although the impact of correlation of the entries has been widely considered in recent years [13, 60, 67].

First, a review of recent information theoretic results for systems that employ some form of channel information at the transmitter is considered. Although this topic is relatively new, single-user MIMO information theory has progressed very rapidly, and interesting results in multiuser MIMO information theory are starting to appear. Next, work concerned with the adaptation of practical structures built on equation (3.9) with knowledge of the values of H at the transmitter is considered.

3.4.1 Information Theoretic Considerations

An excellent recent tutorial of information theoretic considerations for MIMO systems, including adaptation-based ones on transmitter knowledge of the channel, can be found in [33].

3.4.1.1 MIMO Single-User Systems

It was established in early work on MIMO systems [62] that the Shannon capacity of an (N, N) MIMO system operating over a block-fading channel with CSI available at the transmitter and receiver is obtained by decomposing the channel into its eigenmodes, and then performing water filling [16, p. 349] on the eigenmodes based on the corresponding eigenvalues, where an eigenvalue indicates the SNR of the corresponding eigenmode. The system takes the convenient form of a single codebook designed for the AWGN channel followed by a beamformer that is adapted to each block [5].

As noted throughout this work, the assumption of perfect channel state information at the transmitter is problematic on wireless communication channels due to their time-varying nature. This is particularly true in the case of MIMO channels, since there are far more numerous coefficients to estimate than in the single-input single-output case. There has been a recent set of papers [40, 42, 61, 64] that consider the Shannon capacity of MIMO systems when there is mean and covariance feedback. The results in [40, 42, 61, 64] demonstrate the trade-offs in Shannon capacity associated with having channel state information available at the transmitter; in particular, they generalize the results in [62] and reveal when Shannon capacity can be obtained by beamforming—only a scalar codebook followed by a beamformer [33] rather than vector coding. In all cases, the Shannon capacity with knowledge of the channel at the transmitter only grows linearly with the number of antennas—only the leading constant and the simplicity of the system are possibly improved [2, 21], thus echoing the result of section 3.3 with respect to gains in ergodic capacity when CSI is available at the transmitter.

In [5], information theoretic measures based on the notion of outage are considered when there is perfect CSI available at both the transmitter and receiver. Recall that such measures (see section 3.3.1) attempt to capture the notion of system latency by limiting the number of fading blocks (and hence diversity) that a given code word experiences. In this case, a large number of antennas allows spatial diversity to be exploited, and in the limit, the effective channel can be made to look AWGN by employing a beamforming approach. Hence, an error control code designed for an AWGN channel concatenated with a beamformer is the optimal approach [5].

3.4.1.2 Multiuser MIMO Systems

The information theory of multiuser MIMO systems has only recently been explored. The most striking result in this context is the key gains that CSI at the transmitter [9, 10, 69] can provide, and this has generated a lot of interest in "dirty paper coding" methods [15]. In particular, unlike single-user MIMO systems, where generally only capacity-multiplying factors are improved or system complexity is decreased, CSI at the transmitter can allow the number of degrees of freedom to be increased [10, 33]. The facilitation of the penetration of such results into practical multiuser systems, where obtaining CSI at the transmitter can be complicated (and, at best, noisy and outdated), is an area that promises to be fruitful for research in the future.

3.4.2 Adaptive Coded Modulation for MIMO Systems

As demonstrated in section 3.4.1, there has been significant recent work establishing information theoretic bounds and methods for achieving those bounds for MIMO systems. It is now incumbent upon the communication theory community to translate those gains into practice. In particular, it will be important to consider the nature of obtaining channel state information at the transmitter. If the CSI provided to the transmitter is reliable and the channel is relatively constant over a long block, the path will be clear—standard scalar coding followed by beamforming matched to the current CSI. Code rate (and power) adaptation will only need to be done on a block-by-block basis. However, if the CSI is noisy or outdated, it will be interesting to consider whether the results of [40, 42, 61, 64] apply; that is, it will be interesting to consider the robustness of beamforming when practical coded modulation schemes are employed.

Recent work on adaptive coded modulation for multiple-antenna systems has followed the basic tenets above. In particular, there has been some consideration about how to perform adaptation in practical systems in the MIMO environment. In [2], consideration is given to a scheme that tracks the eigenspace of the system so that water-filling-type schemes can be employed—the parallel channel idea of [62]. Recent papers [e.g., 39] have considered different forms of adaptation—antenna selection, beamforming, and space-time coding. Because such papers generally rely on (nearly) perfect prediction assumptions, there has been little consideration of the effects of imperfect channel state information in these works. Recent work [73] has considered the impact of imperfect channel state information. In particular, robustness under imperfect channel state information is obtained by adding an Alamouti scheme over an inner beamformer (which would be ideal with perfect channel state information), and the optimality of such is shown for a system needing to transmit two information bits across the channel. These papers represent the start of research in a very important area that will bring the information theoretic gains of section 3.4.1 to application.

3.5 Conclusions

Adaptive signaling, where the transmitted signal in a wireless system is adjusted based on channel state information available to the transmitter, has been an area of significant research interest for over a decade. For single-user systems, the gain from having channel state information at the transmitter is generally a reduction in complexity and latency. Conversely, the gain of adaptive signaling is generally a function of the system complexity—large gains for very simple uncoded systems and almost no gain in the system Shannon capacity. However, for systems characterized by outage capacity or those employing multiple antennas, channel state information at the transmitter can have a significant role, particularly in systems with multiple users. This indicates the importance of the design of practical adaptive coded modulation for multiple-input multiple-output systems. As in single-antenna systems, variability of the channel coefficients between the time of channel estimation and the time of data transmission is a key concern that will need to be addressed.

References

[1] S. Alamouti and S. Kallel. 1994. Adaptive trellis-coded multiple-phase-shift keying for Rayleigh fading channels. *IEEE Transactions on Communications* 42:2305–14.

[2] B. Bannister and J. Zeidler. 2003. Feedback assisted transmission subspace tracking for MIMO systems. *Journal on Selected Areas in Communications*, 21:452–463.

[3] P. Bello. 1963. Characterization of randomly time-variant linear channels. *IEEE Transactions on Communications Systems* 11:360–93.

[4] C. Berrou, A. Glavieux, and P. Thitimajshima. 1993. Near Shannon limit error-correcting coding and decoding: Turbo-codes. In *Proceedings of the International Conference on Communications*, 1064–1070.

[5] E. Biglieri, G. Caire, and G. Taricco. 2001. Limiting performance of block-fading channels with multiple antennas. *IEEE Transactions on Information Theory* 47:1273–89.

[6] J. Bingham. 1990. Multicarrier modulation for data transmission: An idea whose time has come. *IEEE Communications Magazine* 28(5):5–14.

[7] X. Cai and G. Giannakis. 2003. Adaptive modulation with adaptive pilot symbol assisted estimation and prediction of rapidly fading channels. In *Proceedings of the Conference on Information Sciences and Systems.*

[8] G. Caire, G. Taricco, and E. Biglieri. 1999. Optimal power control for the fading channel. *IEEE Transactions on Information Theory* 45:1468–89.

[9] G. Caire and S. Shamai. 2000. On achievable rates in a multi-antenna broadcast downlink. In *Proceedings of the 38th Annual Allerton Conference on Communication, Control, and Computing*, pp. 1188–93.

[10] G. Caire and S. Shamai. On the achievable throughput of a multi-antenna Gaussian broadcast channel. *IEEE Transactions on Information Theory* 49(7):1691-1706.

[11] J. Cavers. 1972. Variable-rate transmission for Rayleigh fading channels. *IEEE Transactions on Communications* 20:15–22.

[12] J. Cavers. 1991. An analysis of pilot symbol assisted modulation for Rayleigh fading channels. *IEEE Transactions on Vehicular Technology* 40:686–93.

[13] C. Chuah, D. Tse, J. Kahn, and R. Valenzuela. 2002. Capacity scaling in MIMO wireless systems under correlated fading. *IEEE Transactions on Information Theory* 48(3):637–51.

[14] L. Cimini, Jr. 1995. Performance studies for high-speed indoor wireless communications. *Wireless Personal Communications* 2:67–85.

[15] M. Costa. 1983. Writing on dirty paper. *IEEE Transactions on Information Theory* 29(3):439-441.

[16] T. Cover and J. Thomas. 1991. *Elements of information theory.* New York: Wiley.

[17] L. Ding and J. S. Lehnert. 2000. Performance analysis of an uplink power control using truncated channel inversion for data traffic in a cellular CDMA system. In *Proceedings of the IEEE Vehicular Technology Conference*, 1673–1677.

[18] A. Duel-Hallen, S. Hu, and H. Hallen. 2000. Long-range prediction of fading signals: Enabling adaptive transmission for mobile radio channels. *IEEE Signal Processing Magazine* 17:62–75.

[19] T. Eyceoz, A. Duel-Hallen, and H. Hallen. 1998. Deterministic channel modeling and long range prediction of fast fading mobile radio channels. *IEEE Communication Letters* 2:254–56.

[20] E. Feig. 1990. Practical aspects of DFT-based frequency division multiplexing for data transmission. *IEEE Transactions on Communications* 38:929–32.

[21] G. Foschini. 1996. Layered space-time architecture for wireless communication in a fading environment when using multi-element antennas. *Bell Labs Technical Journal* 41–59.

[22] G. Foschini and M. Gans. 1998. On limits of wireless communications in a fading environment when using multiple antennas. *Wireless Personal Communications* 6:311–35.

[23] R. Gallager. 1963. *Low density parity-check codes.* Cambridge, MA: MIT Press.

[24] D. Goeckel. 1997. Robust adaptive coded modulation for time-varying channels with delayed feedback. In *Proceedings of the Thirty-Fifth Annual Allerton Conference on Communication, Control, and Computing*, pp. 370–79.

[25] D. Goeckel. 1998. Adaptive coding for fading channels using outdated fading estimates. In *Proceedings of the IEEE Vehicular Technology Conference*, pp. 1925–29.

[26] D. Goeckel. 1998. Strongly robust adaptive signaling for time-varying channels. In *Proceedings of the 1998 International Conference on Communications*, pp. 454–58.

[27] D. Goeckel. 1999. Adaptive coding for time-varying channels using outdated fading estimates. *IEEE Transactions on Communications* 47:844–55.

[28] A. Goldsmith, L. Greenstein, and G. Foschini. 1994. Error statistics of real-time power measurements in cellular channels with multipath and shadowing. *IEEE Transactions on Vehicular Technology* 43:439–46.

[29] A. Goldsmith. 1994. Variable-rate coded MQAM for fading channels. In *Proceedings of the IEEE Global Communications Conference: Communication Theory Miniconference*, pp. 186–90.

[30] A. Goldsmith. 1995. Capacity and dynamic resource allocation in broadcast fading channels. In *Proceedings of the Allerton Conference on Communications, Control, and Computing*, pp. 915–24.

[31] A. Goldsmith and S. Chua. 1997. Variable-rate variable-power MQAM for fading channels. *IEEE Transactions on Communications* 45:1218–30.

[32] A. Goldsmith and S. Chua. 1998. Adaptive coded modulation for fading channels. *IEEE Transactions on Communications* 46:595–602.

[33] A. Goldsmith, S. Jafar, N. Jindal, and S. Vishwanath. 2003. Capacity limits of MIMO channels. *IEEE Journal on Selected Areas in Communications* 21(5):684–702.

[34] I. Gradshteyn and I. Rhyzhik. 1980. *Table of integrals, series, and products.* New York: Academic Press.

[35] M. Gudmundson. 1991. Correlation model for shadow fading in mobile radio systems. *Electronics Letters* 27:2145–46.

[36] S. Hanly and D. Tse. 1998. Multi-access fading channels. Part II. Delay-limited capacities. *IEEE Transactions on Information Theory* 44:2816–31.

[37] J. Hayes. 1968. Adaptive feedback communications. *IEEE Transactions on Communications Technology* 16:29–34.

[38] V. Hentinen. 1974. Error performance for adaptive transmission on fading channels. *IEEE Transactions on Communications* 22:1331–37.

[39] S. Hu and A. Duel-Hallen. 2001. Combined adaptive modulation and transmit diversity using long range prediction for flat mobile radio channels. In *Proceedings of the Global Communications Conference*, 1257–1262.

[40] S. Jafar and A. Goldsmith. Transmitter optimization and optimality of beamforming for multiple antenna systems with imperfect feedback. *IEEE Transactions on Wireless Communications.*

[41] W. Jakes, Jr. 1974. *Microwave mobile communications.* New York: Wiley.

[42] E. Jorswieck and H. Boche. 2004. Channel capacity and capacity-range of beamforming in MIMO wireless systems under correlated fading with covariance feedback. *IEEE Transactions on Wireless Communications* 3(5): 1543-1553.

[43] K. Kamath and D. Goeckel. 2004. Adaptive modulation schemes for minimum outage probability in wireless systems. *IEEE Transactions on Communications 52(10): 1632–1635.*

[44] V. Lau and M. Macleod. 1998. Variable rate adaptive trellis coded QAM for high bandwidth efficiency applications in Rayleigh fading channels. In *Proceedings of the Vehicular Technology Conference*, pp. 348–52.

[45] V. Lau and M. Macleod. 2001. Variable-rate trellis coded QAM for flat-fading channels. *IEEE Transactions on Communications* 49:1550–60.

[46] D. MacKay. 1999. Good error-correcting codes based on very sparse matrices. *IEEE Transactions on Information Theory* 45:399–431.

[47] A. Monk and L. Milstein. 1995. Open-loop power control error in a land mobile satellite system. *Journal on Selected Areas in Communications* 13:205–12.

[48] S. Nanda, K. Balachandran, and S. Kumar. 2000. Adaptation techniques in wireless packet data services. *IEEE Communications Magazine 38(1):* 54–64.

[49] R. Negi and J. Cioffi. 2002. Delay-constrained capacity with causal feedback. *IEEE Transactions on Information Theory* 48:2478–94.

[50] P. Örmeci, X. Liu, D. Goeckel, and R. D. Wesel. 2001. Adaptive bit-interleaved coded modulation. *IEEE Transactions on Communications* 49:1572–81.

[51] K. Pahlavan and A. Levesque. 1995. *Wireless information networks.* New York: John Wiley & Sons.

[52] A. Papoulis. 1991. *Probability, random variables, and stochastic processes.* 3rd ed. New York: McGraw-Hill.

[53] J. Proakis. 2001. *Digital communications.* 4th ed. New York: McGraw-Hill.

[54] M. Pursley and C. Wilkins. 1999. Adaptive-rate coding for frequency-hop communications over Rayleigh fading channels. *Journal on Selected Areas in Communications* 17:1224–32.

[55] M. Pursley and J. Shea. 2000. Channel quality estimation with channel error counters for adaptive signaling in wireless communications. In *Proceedings of the International Symposium on Information Theory*, 109.

[56] T. Rappaport and C. McGillen. 1989. UHF fading in factories. *Journal on Selected Areas in Communications* 7:40–48.

[57] T. Rappaport. 1996. *Wireless communications.* New York: Prentice Hall.

[58] C. Schlegel and D. Costello, Jr. 1989. Bandwidth efficient coding for fading channels: Code construction and performance analysis. *Journal on Selected Areas in Communications* 7:1356–68.

[59] C. Shannon. 1948. A mathematical theory of communications. *Bell Systems Technical Journal* 27:379–423, 623–56.

[60] D. Shiu, G. Foschini, M. Gans, and J. Kahn. 2000. Fading correlation and its effect on the capacity of multielement antenna systems. *IEEE Transactions on Communications* 48:502–13.

[61] S. Simon and A. Moustakas. 2003. Optimizing MIMO antenna systems with channel covariance feedback. *Journal on Selected Areas in Communications 21(3):406-417.*

[62] I. Telatar. 1999. Capacity of multi-antenna Gaussian channels. *European Transactions on Telecommunications* 10:586–95.

[63] G. Ungerboeck. 1982. Channel coding with multilevel/phase signals. *IEEE Transactions on Information Theory* 28:55–67.

[64] E. Visotsky and U. Madhow. 2001. Space-time transmit precoding with imperfect feedback. *IEEE Transactions on Information Theory* 47:2632–39.

[65] B. Vucetic. 1991. An adaptive coding scheme for time-varying channels. *IEEE Transactions on Communications* 39:653–63.

[66] S. Wei and D. Goeckel. 2002. Error statistics for average power measurements in wireless communication systems. *IEEE Transactions on Communications* 50:1535–46.

[67] S. Wei, D. Goeckel, and R. Janaswamy. 2005. On the asymptotic capacity of MIMO systems with antenna arrays of fixed length. *IEEE Transactions on Wireless Communications 4(4):1608–1621.*

[68] R. Yates. 1995. A framework for uplink power control in cellular radio systems. *Journal on Selected Areas in Communications* 13:1341–47.

[69] W. Yu and J. Cioffi. 2001. Trellis precoding for the broadcast channel. In *Proceedings of the Global Communications Conference*, pp. 1344–48.

[70] J. Zander. 1992. Performance of optimum transmitter power control in cellular radio systems. *IEEE Transactions on Vehicular Technology* 41:57–62.

[71] R. van Nee and R. Prasad. 2000. *OFDM for wireless multimedia communications.* Norwood, MA: Artech House.

[72] H. Van Trees. 1968. *Detection, estimation, and modulation theory.* Vol. I. New York: Wiley.

[73] S. Zhou and G. Giannakis. 2002. Optimal transmitter eigen-beamforming and space-time block coding based on channel mean feedback. *IEEE Transactions on Signal Processing 50(10):2599–2613.*

4

MIMO Systems: *Principles, Iterative Techniques, and Advanced Polarization*

K. Raoof
Domaine Universitaire

M. A. Khalighi
École Centrale Marseille

N. Prayongpun
Domaine Universitaire

4.1 Introduction

This chapter considers the principles of multiple-input multiple-output (MIMO) wireless communication systems as well as some recent accomplishments concerning their implementation. By employing multiple antennas at both transmitter and receiver, very high data rates can be achieved under the condition of deployment in a rich-scattering propagation medium. This interesting property of MIMO systems suggests their use in the future high-rate and high-quality wireless communication systems. Several concepts in MIMO systems are reviewed in this chapter. We first consider MIMO channel models and recall the basic principles of MIMO structures and channel modeling. We next study the MIMO channel capacity and present the early developments in these systems concerning the information theory aspect. Iterative signal detection is considered next; it considers iterative techniques for space-time decoding. As the capacity is inversely proportional to the spatial channel correlation, MIMO antennas should be sufficiently separated, usually by several wavelengths. In order to minimize antennas' deployment, we present advanced polarization diversity techniques for MIMO systems and explain how they can help to reduce the spatial correlation in order to achieve high transmission rates. We end the chapter by considering the application of MIMO systems in local area networks, as well as their potential in enhancing range, localization, and power efficiency of sensor networks.

4.2 MIMO Systems and Channel Models

In this section, we present briefly a general MIMO communication structure. The definition of MIMO channel is then described with some related characteristics of wireless communication channels. Some recent MIMO channel models are also reported.

4.2.1 MIMO Communication Systems

Using multiple antennas [1] at both transmitter and receiver permits the increasing of the data rate by creating multiple spatial channels. Multiple receiving antennas can also be used to combat fading without expanding the bandwidth of the transmitted signal. In particular, with M_T transmitting and M_R receiving antennas, it is possible to achieve an M-time capacity of a single transmitting and single receiving antenna configuration where $M = \min\{M_T, M_R\}$. Figure 4.1 demonstrates a general system employing multiple transmitting and multiple receiving antennas to increase the data rate. A sequence of

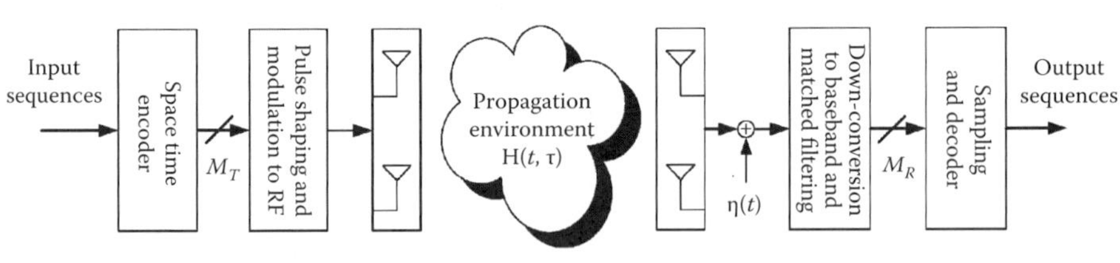

FIGURE 4.1 General configuration of a MIMO communication system.

input symbols is encoded by a space-time encoding function into an $M_T \times 1$ discrete-time complex baseband sequence $x[n]$ (n is a discrete-time index). The $x[n]$ sequence is subsequently transformed by a pulse-shaping filter into an $M_T \times 1$ continuous-time complex baseband sequence $x(t)$, and then the baseband signal is modulated with a transmission carrier. The transmission channel **H** superposes the transmitted signal due to the distortions of environment. At the reception side, under the assumption of synchronous sampling, the received signal $y(t)$ with additive noise is downconverted to baseband and sampled to produce a discrete-time signal sequence. Finally, the estimated symbols are decoded by the space-time decoding block.

If the channel is time invariant, the equivalent received signal at the receiving antenna with M_T elements at the transmitter (Tx) and M_R elements at the receiver (Rx) can be written as

$$\mathrm{y}(t) = \sum_{l=0}^{L-1} \mathbf{H}_l \mathrm{x}(t - \tau_l) + \eta(t). \tag{4.1}$$

If the channel is time variant, the overall MIMO relation can be formulated as

$$\mathrm{y}(t) = \int_{\tau} \mathbf{H}(t, \tau) \mathrm{x}(t - \tau) + \eta(t), \tag{4.2}$$

where $x(t)$ and $y(t)$ represent the transmitted and received signals,

$$\mathrm{x}(t) = [x_1(t), x_2(t), \ldots, x_{M_T}(t)]^T, \tag{4.3}$$

$$\mathrm{y}(t) = [y_1(t), y_2(t), \ldots, y_{M_R}(t)]^T, \tag{4.4}$$

and L denotes the number of resolvable multipaths, τ is the propagation delay, $\mathbf{H}(t, \tau)$ is the $M_R \times M_T$ time-variant channel matrix, $\mathbf{H}_l$ is the $M_R \times M_T$ channel matrix of resolvable path l, and finally, $\eta(t)$ is an additive noise.

4.2.2 MIMO Channel Stationarity Definition

In wireless communication, a stochastic time-variant linear channel usually employs wide-sense stationarity uncorrelated scattering (WSSUS) for stationary property [2, 3]. This WSSUS channel expresses uncorrelated attenuation in both time-delay and Doppler-shift domains. The quasi-WSSUS channel [2] is usually applied to real radio systems. It has the properties of a WSSUS channel for a limited bandwidth and for a limited time or within a limited environment. This kind of assumption is exceptionally useful in communication systems. For example, ergodic MIMO channel performance can be given by averaging the channel performances over many independent channel realizations considering that they have the same statistics.

In many practical communication systems, the WSSUS property is not truly satisfied because of path loss, shadowing, and varying propagation conditions such as delay-Doppler dispersion by band or time limitations at both Tx and Rx [4]. These effects cause the channel to be nonstationary, i.e., non-WSSUS.

4.2.3 Classification of MIMO Channel Models

Numerous MIMO channel models have already been proposed. They can be classified into physical and analytical models, as shown in Figure 4.2. On the one hand, the physical channel models focus on the characteristics of the environment and the electromagnetic wave propagation between the Tx and Rx, and consider the antenna configurations at both ends. On the other hand, the analytical models do not provide site-specific descriptions, as they do not take into account the wave propagation characteristics. The model impulse response is mathematically generated and related to the statistical properties of the propagation environment. However, due to its simplicity, an analytical channel model is very useful for producing a MIMO channel matrix for different kinds of communication systems.

In the literature, the physical models are decomposed into deterministic models [5–8] and geometry-based stochastic models [9–13]. Deterministic models, such as ray tracing and recording impulse response models, start by creating an artificial environment, and the channel response is then calculated for simulation purposes. However, these methods have high computational complexity. Geometry-based stochastic channel models (GSCMs), on the other hand, calculate the channel response by taking into account the characteristic of wave propagation, both site-specific Tx-Rx environments, and the scattering mechanisms. All parameters are statistically set to closely match the measured channel observation. By this approach, the channel response can be rapidly computed even for a multibounce scattering mechanism.

Analytical channel models can be further classified into correlation-based models, statistical cluster models, and propagation-based models. Correlation-based models

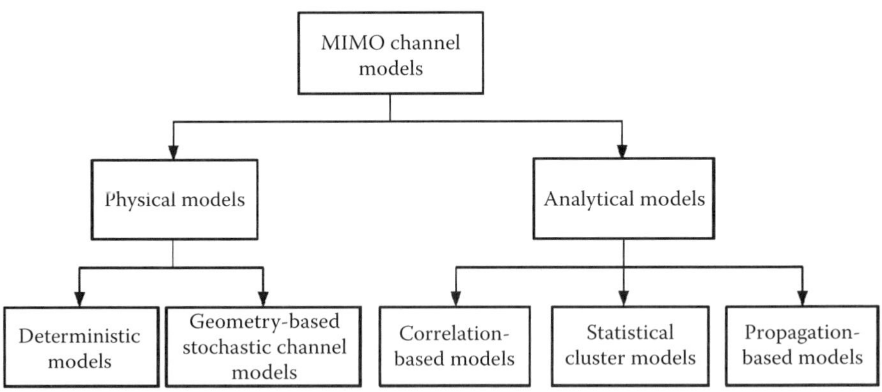

FIGURE 4.2 MIMO channel and propagation models.

contain the Tx-Rx channel correlation matrices. For example, the independent identically distributed (IID) model is proposed in the case of a rich-scattering environment with no spatial correlation [1, 14, 15]. The Kronecker model [16–18] assumes that the channel correlation is a product of the correlations at the Tx and Rx sides. Statistical cluster models determine physical parameters in a random manner without referring to the geometry of a physical medium. For example, the Saleh-Valenzuela model [19] uses two exponentially decaying amplitudes varying in time and distance of the clusters, while increasing delay time with the assumptions that the direction of departure (DOD) and the angle of arrival (AOA) are independent and identically distributed. The other models are propagation based, such as keyhole channel models [20], finite scattered model [21], maximum entropy model [22], virtual channel representation [23], etc.

In addition to the discussed models, there are several organizations that have proposed different MIMO channel models, such as COST 207, COST 231, COST 259 [24], COST 273 [25], 3GPP [26], and IEEE 802.16a, e.

4.3 MIMO Channel Capacity

At the end of the 1990s, pioneering works in Bell Laboratories showed for the first time that the use of multiple antennas at both sides of the transmission link can result in tremendous channel capacities, provided that the propagation medium is rich scattering [1, 14, 27–29]. This increase in capacity is obtained without any need for extra bandwidth or extra transmission power. Multipath propagation, previously regarded as an impediment to reliable communication, was shown to be exploitable for increasing the data throughput. In this section, we explain briefly how exploiting the spatial dimension can lead to an increase in the system spectral efficiency. Only single-user applications are considered. For more discussions and details, the reader is referred to [15] and the references therein.

4.3.1 Capacity of a Fading Channel

Let us first recall the definition of the capacity for a fading channel. For a time-varying channel, the capacity C becomes a random variable whose instantaneous value depends on the channel realization [30]. In such a case, the Shannon capacity of the channel may even be zero. Indeed, if we choose a transmission rate for communication, there may be a nonzero probability that the channel realization is incapable of supporting it. The two most used definitions for the channel capacity are *ergodic* capacity and *outage* capacity. The ergodic capacity, C_{erg}, which is the expected value of C, is suitable for fast varying channels. The outage capacity, C_{out}, is usually used when considering packet-based transmission systems where the block-fading model properly describes the channel. If the preassumed channel capacity is too optimistic, i.e., larger than the instantaneous capacity, a channel outage may occur. The outage capacity, or more correctly, the capacity versus outage, as seen from its name, is the channel capacity conditioned to an outage probability P_{out}. Obviously, there is a trade-off between the expected data throughput and the outage probability.

In the expressions that we will provide in this section, C will denote the instantaneous channel capacity.

4.3.2 MIMO Capacity

We present the MIMO channel capacity for different cases of channel state information (CSI) at the transmitter (Tx) and receiver (Rx).

4.3.2.1 General Assumptions

The global scheme of the transmission link is shown in Figure 4.3. We denote by M_T and M_R the number of antennas at Tx and Rx, respectively. The communication channel includes the effect of transmit/receive antennas and the propagation medium. We neglect the effect of the antenna patterns and assume the *far-field* conditions; that is, dominant reflectors are assumed to be sufficiently far from the Tx and Rx. We consider the simple case of a frequency-non-selective (flat) fading channel that is true for narrowband communications. This assumption is mostly valid in indoor applications [31, 32]. In the equivalent baseband representation, each subchannel is characterized by a complex, circularly symmetric random variable that is assumed to be normalized in power. The entire MIMO channel is described by a channel matrix $\boldsymbol{H}$ of dimension $(M_R \times M_T)$. For instance, the entry H_{ij} of $\boldsymbol{H}$, characterizes the subchannel between the ith receive and the jth transmit antenna. Finally, we assume that the total transmit power at each sample time (corresponding to each channel use) is constrained to P_T.

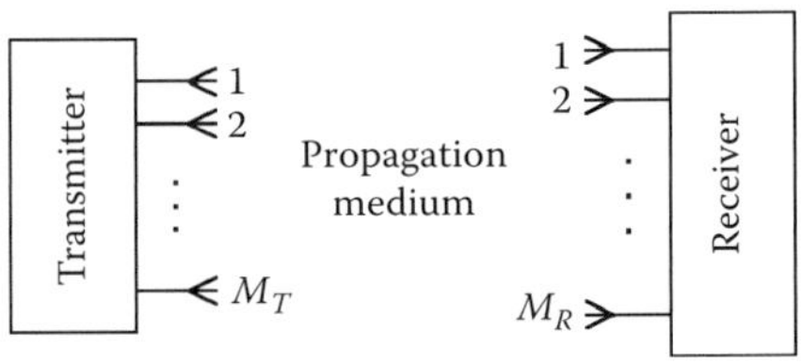

FIGURE 4.3 Global scheme of a MIMO communication structure.

4.3.2.2 CSI Known to Rx but Unknown to Tx

This is the classical case that is usually considered in the literature. The channel is assumed to be known (e.g., perfectly estimated) at the Rx but unknown to the Tx. Since the Tx does not know the channel, it is logical to distribute the available power uniformly on the transmit antennas. In fact, this is the optimal way for power allotment over the M_T antennas in this case. We denote by ρ_T the total average received signal-to-noise ratio (SNR) at the receiver array $\rho_T = P_T/\sigma_n^2$, where σ_n^2 is the variance of the additive white Gaussian complex noise. The MIMO channel capacity in units of bps/Hz is

$$C = \log_2 \det\left[\boldsymbol{I}_{M_R} + \frac{\rho_T}{M_T} \boldsymbol{H}\boldsymbol{H}^\dagger \right]. \tag{4.5}$$

Here $\boldsymbol{I}_{M_R}$ is the $(M_R \times M_R)$ identity matrix. Equation (4.5) can also be written in a different form if we consider the singular value decomposition of $\boldsymbol{H}$:

$$\boldsymbol{H} = \boldsymbol{U}_H\, \boldsymbol{\Lambda}_H\, \boldsymbol{V}_H^\dagger \tag{4.6}$$

where † denotes complex conjugate transpose, $\boldsymbol{U}_H$ and $\boldsymbol{V}_H$ are unitary matrices of dimensions $(M_R \times M_R)$ and $(M_T \times M_T)$, respectively, and Λ_H is an $(M_R \times M_T)$ matrix containing the singular values of $\boldsymbol{H}$. Let us define $M = \min\{M_T, M_R\}$. We denote these singular values by $\lambda_{H,i}$, $i = 1, \ldots, M$. The MIMO capacity can be written in the following form:

$$C = \log_2 \det\left[\boldsymbol{I}_{M_R} + \frac{\rho_T}{M_T}\Lambda_H \Lambda_H^\dagger\right] = \sum_{i=1}^{M} \log_2\left(1 + \frac{\rho_T}{M_T}\lambda_{H,i}^2\right). \tag{4.7}$$

4.3.2.3 CSI Known to Both Tx and Rx

This is the case when the estimated CSI at the Rx is provided for the Tx, which is practically feasible when the channel varies slowly in time. Providing the CSI for the Tx can be done using a (hopefully low-bandwidth) feedback channel or via the reverse link when the communication takes place in a duplex mode. We assume that CSI is provided perfectly and without any delay to the Tx. In this case, the Tx can allot the available power on the antennas in an optimal manner in order to achieve the maximum capacity [33]. This capacity is often called *known CSI* capacity or *water-filling* (WF) capacity. For this purpose, we should weight the transmitted symbols vector $\boldsymbol{x}$ by the matrix $\boldsymbol{V}_H$, and the received signal vector by the matrix $\boldsymbol{U}^\dagger_H$ [15, 34]. This can be regarded as an *optimal beamforming* solution. By this weighting, we can in fact consider an equivalent channel $\boldsymbol{H}_{eq}$ between $\boldsymbol{x}$ and $\boldsymbol{y}$:

$$\boldsymbol{H}_{eq} = \boldsymbol{U}_H^\dagger \, \boldsymbol{H} \, \boldsymbol{V}_H = \Lambda_H. \tag{4.8}$$

In other words, the MIMO channel matrix $\boldsymbol{H}$ is decomposed into several parallel independent single-input single-output subchannels. The number of these subchannels is equal to rank($\boldsymbol{H}$), and their gain is given by the singular values of $\boldsymbol{H}$. Let $\boldsymbol{R}_X$ be the autocorrelation matrix of $\boldsymbol{x}$, with eigenvalues $\lambda_{X,i}$, $i = 1, \cdots, M$. The optimal WF solution consists in distributing the available power P_T, optimally over the equivalent parallel subchannels, which results in [15, 35]

$$\lambda_{X,i} = \left(\psi - \frac{\sigma_n^2}{\lambda_{H,i}^2}\right)^+, \quad i = 1, \cdots, M \tag{4.9}$$

where $(s)^+ = s$ if $s > 0$, and 0 otherwise. Also, ψ is a constant that is determined so as to satisfy the constraint on the total transmit power,

$$\sum_{i=1}^{M} \lambda_{X,i} = P_T.$$

The WF solution imposes that we allocate more power to best subchannels and lower (or perhaps no) power to worse ones. Now, the WF capacity C_{WF} is given by [15, 35]

$$C_{\mathrm{WF}} = \sum_{i=1}^{M} \log_2\left(1 + \frac{\lambda_{X,i}\lambda_{H,i}^2}{\sigma_n^2}\right). \tag{4.10}$$

4.3.2.4 CSI Unknown to Both Tx and Rx

If the channel is not known to the Tx and the Rx, we can consider (4.5) as an upper bound on capacity. Let Δ be the coherence interval of the channel in units of channel uses. As Δ tends to infinity, the channel capacity approaches this upper bound, because with a greater Δ, tracking the channel variations becomes more possible for the Rx [36]. For the same reason, there is less difference between the capacity and the upper bound for higher SNR values. This difference becomes more considerable, however, for larger M_T or M_R [36]. For a fast varying channel, the capacity is far less than the Rx-perfect-knowledge upper bound, because practically there is no possibility to estimate the channel at the Rx. For M > Δ, no increase is achieved in the MIMO capacity by an increase in M [36]. It is shown in [37] that at high SNR and for a rich-scattering propagation medium, the MIMO unknown CSI capacity increases linearly with $M^*(1 - M^*/\Delta)$, where

$$M^* = \min\left\{M_T, M_R, \left\lfloor \frac{\Delta}{2} \right\rfloor\right\}.$$

4.3.3 Some Numerical Results

We present some numerical results, excluding the case where CSI is unknown to both the Tx and the Rx. We consider the outage capacity C_{out} for an outage probability of $P_{out} = 0.01$ and will refer to it simply as capacity. Let us first consider the case where CSI is known only to the Rx. We will call this case *no-WF*. Figure 4.4 contrasts the outage capacity of MIMO, single-input multiple-output (SIMO), and multiple-input single-output (MISO) systems under the conditions of uncorrelated Rayleigh flat fading [38] and $P_{out} = 0.01$. SNR stands for ρ_T. For the MIMO system, we have $M_T = M_R = M$. Uncorrelated fading necessitates enough antenna spacings at the Tx and the Rx that depend, in turn, on the propagation conditions [39, 16]. The MIMO capacity increases linearly with M and is much more considerable than that of MISO and SIMO systems.

Now consider the case of known CSI at both the Tx and the Rx. For $M_R \geq M_T$, the improvement in capacity by performing WF, which we call the *WF gain*, is considerable for low SNR and a large number of transmit antennas [33]. But this gain is much more considerable when $M_T > M_R$. For example, Figure 4.5 shows curves of no-WF and WF capacities for two cases of $M_R > M_T = 4$ and $M_T > M_R = 4$. Notice that the WF capacity of an (M_T, M_R) system is equal to that of an (M_R, M_T) system. Also, the WF capacity of a MISO system is equal to that of the equivalent SIMO system.

The gain in capacity by WF is especially interesting for the case of correlated channels. For instance, for the case of Ricean fading [40, 41], curves of capacity versus the Ricean factor (RF) are presented in Figure 4.6, for a MIMO system with $M_T = M_R = M$ and two cases of $M = 2$ and $M = 4$. RF represents the percentage of power received from

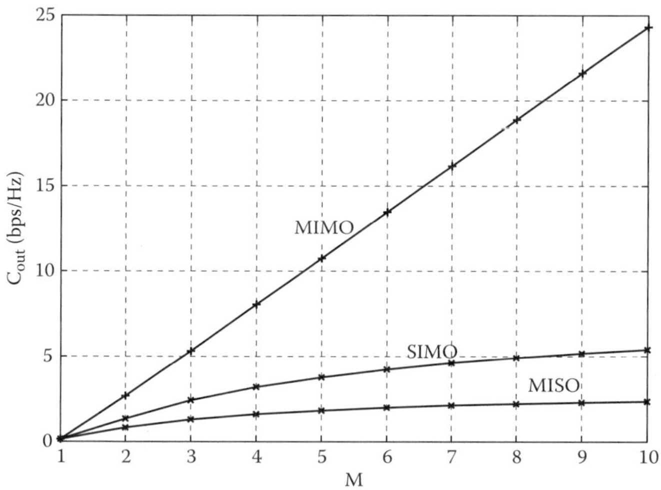

FIGURE 4.4 Capacity for MIMO, SIMO, and MISO structures; uncorrelated Rayleigh flat fading; SNR = 10 dB, $P_{out} = 0.01$.

the line of sight (LOS) to the total received average power [41]. For relatively high RF that can be regarded as a more correlated channel, the WF gain is quite considerable.

4.4 Iterative Signal Detection

Signal detection is a crucial part of the transmission system. Among the various detection techniques proposed for the case of MIMO systems, there are iterative (also called *turbo*) detectors. This is what we are going to focus on in this section. In effect, since the invention of turbo-codes by Berrou and Glavieux [42], who proposed iterative decoding of parallel concatenated convolutional codes, the *turbo principle* has been applied to several problems in communications, such as channel equalization [43], channel estimation [44], synchronization [45], multiuser detection [46, 47], and, of course, MIMO signal detection [48]. The turbo principle consists of the exchange of *soft* information between two different stages of the Rx, mostly including the soft channel decoder. In MIMO systems too, iterative processing has attracted special attention as it makes a good compromise between complexity and performance. Before presenting the basics of iterative detection, we have to present a brief introduction on space-time coding that is performed at the Tx. We mostly consider frequency-non-selective (flat) fading conditions and single-carrier modulation.

4.4.1 Space-Time Coding and Decoding

An important aspect in the implementation of MIMO systems is to appropriately distribute redundancy in space and in time at the Tx, what is called space-time (ST) coding [49]. To date, there has been considerable work on this subject, and a variety

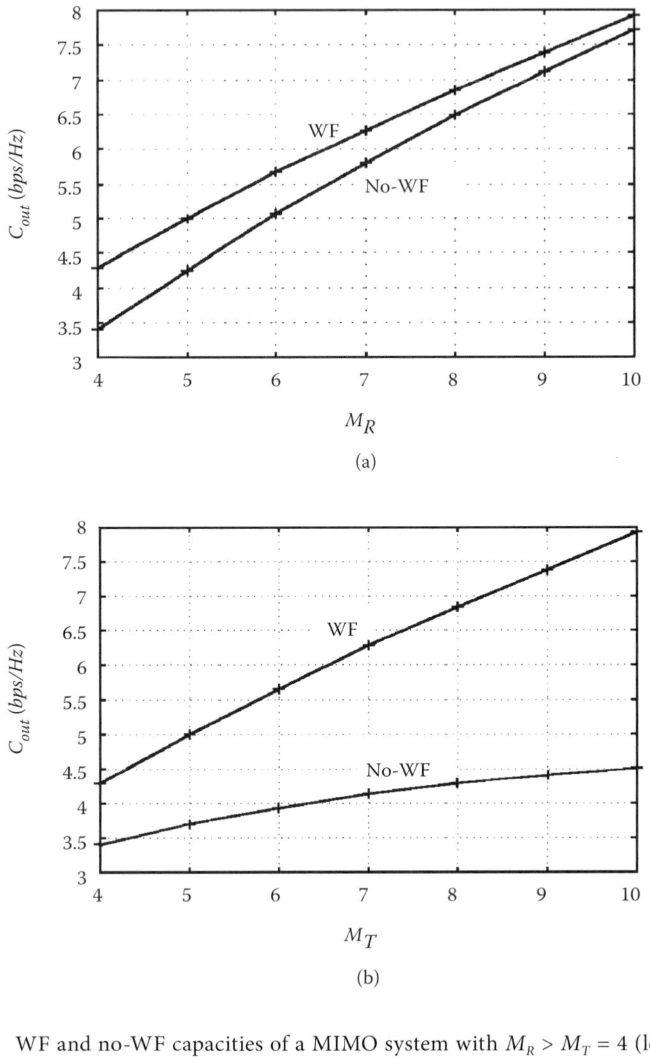

FIGURE 4.5 WF and no-WF capacities of a MIMO system with $M_R > M_T = 4$ (left) and $M_T > M_R = 4$ (right), Rayleigh flat fading, SNR = 3 dB, $P_{out} = 0.01$.

of ST schemes have been proposed for MIMO systems. The key criteria in the design of ST codes are the coding gain and the diversity gain. The first one aims at achieving high rate by capitalizing on the MIMO capacity, whereas the latter aims at profiting from the space diversity to reduce fading at the Rx. The two extreme schemes corresponding to these criteria are respectively spatial multiplexing and transmit diversity. For instance, orthogonal space-time block codes (OSTBCs) [50, 51] aim at diversity gain and some coding rate; space-time trellis codes (STTCs) [52] and linear constellation precoding (LCP) [53] aim at both coding and diversity gain; spatial multiplexing or the V-BLAST architecture maximizes the coding rate [28]; and the more general family of

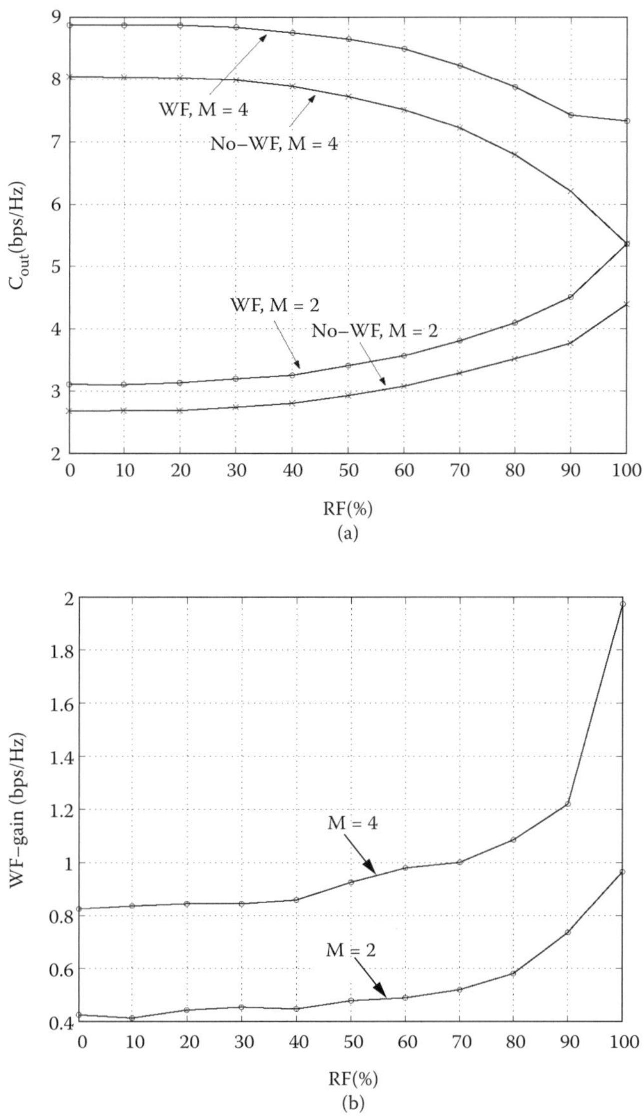

FIGURE 4.6 WF and no-WF capacities (left) and WF gain in capacity (right) for a MIMO system with $M_T = M_R = M$, Ricean flat fading, SNR = 10 dB, $P_{out} = 0.01$.

linear dispersion (LD) codes [54], which maximize the mutual information between the Tx and the Rx, allow flexible rate-diversity trade-off.

Apart from the problem of code construction, one important criterion in the choice of the appropriate ST scheme could be its decoding complexity. ST orthogonal designs like OSTBCs offer full diversity and can be decoded using an optimal decoder with linear complexity. However, these schemes suffer from low rate, especially for a large number

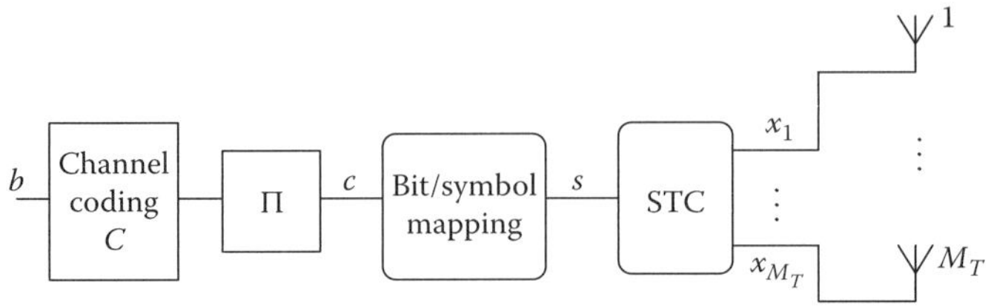

FIGURE 4.7 Block diagram of the BICM transmission scheme.

of transmit antennas. Moreover, full-rate OSTBCs exist only for a restricted number of transmit antennas and modulations [50]. Nonorthogonal schemes, on the other hand, offer higher coding rates, but their optimal decoder becomes prohibitively complex for a large number of transmit antennas and large signal constellation sets. This is especially the case for STTC schemes that, although offering high rates and good diversity gains, are complex to decode and, moreover, suffer from long decoding delays.

For nonorthogonal schemes, instead of performing complex optimal decoding, we may use suboptimal decoding based on simple linear-algebraic techniques such as sphere decoding [55] or interference-cancelling-based decoding [28, 48]. For either solution, the Rx performance can be improved considerably by performing iterative detection.

4.4.1.1 ST Coding, Tx Scheme

In addition to using a special ST scheme, we usually perform channel coding at the Tx. Let us consider bit-interleaved coded modulation (BICM) [56] for which a typical scheme is shown in Figure 4.7. The advantage of BICM is its flexibility regarding the choice of the code and the bit-symbol mapping, as well as its conformity to iterative detection. In Figure 4.7, the binary data b are encoded by a channel code C, before being interleaved (the block Π). The output bits c are then mapped to symbols according to a given constellation set. We will mostly consider QAM modulation with B bits per symbol. Power-normalized symbols s are next combined according to a given ST scheme and then transmitted on M_T antennas.

4.4.2 General Formulation of LD Codes

Before talking about ST decoding, let us present the general formulation of the LD codes from [54] that can be equally used for other ST schemes as well. Let $\boldsymbol{S}$ of dimension $(Q \times 1)$ be the vector of data symbols prior to ST coding:

$$\boldsymbol{S} = \left[s_1, s_2, \cdots, s_Q\right]^t, \tag{4.11}$$

where $.^t$ denotes transposition. By ST coding, these symbols are mapped into a $(M_T \times T)$ matrix $\boldsymbol{X}$, where T is the number of channel uses. We define the ST coding rate as $R_{\text{STC}} = Q/T$. Corresponding to an encoded matrix $\boldsymbol{X}$, we receive the $(M_R \times T)$ matrix $\boldsymbol{Y}$. We separate the $\Re$ and $\Im$ parts of the entries of $\boldsymbol{S}$ and $\boldsymbol{X}$ and stack them row-wise in vectors $\mathcal{S}$ of dimension $(2Q \times 1)$ and $\mathcal{X}$ of dimension $(2M_T T \times 1)$, respectively. For instance,

$$\mathcal{S} = \left[\Re\{s_1\} \ \Im\{s_1\}, \ \cdots, \ \Re\{s_Q\} \ \Im\{s_Q\} \right]^t. \tag{4.12}$$

We have then $\mathcal{X} = \mathcal{F}\,\mathcal{S}$, where the $(2M_T T \times 2Q)$ matrix $\mathcal{F}$ depends on the actual ST scheme (see [54] for more details). Let the $(M_R \times M_T)$ matrix $\boldsymbol{H}$ represent our flat channel. Similar to $\mathcal{X}$, we construct the $(2M_R T \times 1)$ vector $\mathcal{Y}$ from $\boldsymbol{Y}$. Vectors $\mathcal{X}$ and $\mathcal{Y}$ are then related through a $(2M_R T \times 2M_T T)$ matrix $\mathcal{H}$:

$$\mathcal{Y} = \mathcal{H}\,\mathcal{X} + \mathcal{N} \tag{4.13}$$

where $\mathcal{N}$ is the vector of real AWGN of zero mean and variance σ_n^2. Matrix $\mathcal{H}$ is composed of $(2T \times 2T)$ segments $\mathcal{H}_{ij}$, $i = 1, \cdots, M_R$, $j = 1, \cdots, M_T$, described below:

$$\mathcal{H}_{ij} = \begin{bmatrix} \boldsymbol{H}_{ij} & 0 & \cdots & 0 \\ 0 & \boldsymbol{H}_{ij} & \cdots & 0 \\ \vdots & & \ddots & \vdots \\ 0 & \cdots & 0 & \boldsymbol{H}_{ij} \end{bmatrix} \tag{4.14}$$

The (2×2) elements $\boldsymbol{H}_{ij}$ are obtained from each entry H_{ij} of the initial matrix $\boldsymbol{H}$:

$$\boldsymbol{H}_{ij} = \begin{bmatrix} \Re\{H_{ij}\} & -\Im\{H_{ij}\} \\ \Im\{H_{ij}\} & \Re\{H_{ij}\} \end{bmatrix}. \tag{4.15}$$

Now, we can describe the ST encoder and channel input/output relationship by considering an equivalent channel matrix $\mathcal{H}_{eq}$ of dimension $(2M_R T \times 2Q)$:

$$\mathcal{Y} = \mathcal{H}\,\mathcal{F}\,\mathcal{S} + \mathcal{N} = \mathcal{H}_{eq}\,\mathcal{S} + \mathcal{N}. \tag{4.16}$$

4.4.3 Iterative Detection for Nonorthogonal ST Schemes

We assume that $\mathcal{H}_{eq}$ and σ_n^2 are known at the Rx. Having received the vector $\mathcal{Y}$, we should extract from it the transmitted data $\mathcal{S}$. As we perform channel coding together with ST coding, the idea of iterative detection comes to mind. Indeed, by profiting in this way from the channel coding gain, we can obtain a good performance after only a few iterations and approach the optimal ST decoder + channel decoder performance. This is, of course, the case for nonorthogonal ST schemes. In what follows, we explain the principle of iterative detection and explain in detail the ST decoding part.

The block diagram of such an Rx is shown in Figure 4.8. Soft-input soft-output signal detection and channel decoding are performed. For MIMO signal detection or ST

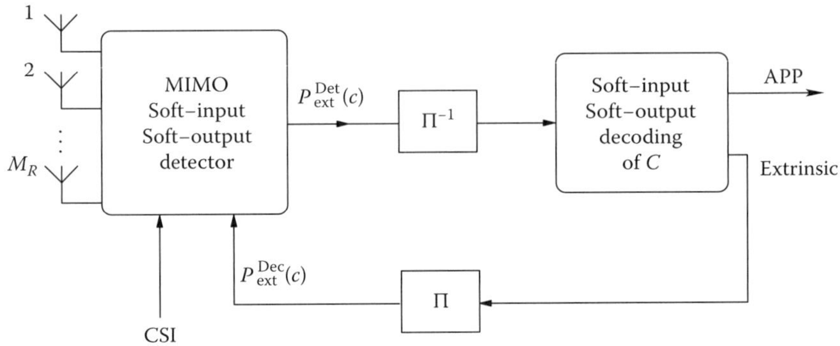

FIGURE 4.8 Block diagram of the receiver.

decoding, we can use the optimal maximum *a posteriori* (MAP) algorithm, or a sub-optimal solution based on sphere decoding or interference cancelling, for example. In fact, the optimal MAP detector becomes too complex to implement in practice, especially for large Q or large signal constellations.

Soft channel decoding, on the other hand, can be done using the well-known forward-backward algorithm [57], the soft-output Viterbi algorithm (SOVA) [58], or a simplification of them [59]. For the final decision making on the transmitted data bits, we use the *a posteriori* probabilities at the decoder output. Like in any other turbo processing case, *extrinsic* information is exchanged between the two blocks of MIMO detector and channel decoder. In what follows, we explain the principle of MIMO detection based on MAP, sphere decoding, and soft interference cancelling, while focusing on the third approach.

4.4.3.1 MAP Signal Detection

We present here the formulation of the MAP detector based on probabilities. It can also be implemented using logarithmic likelihood ratios (LLRs). Remember the expression of $\mathcal{Y}$ from (4.16). The MIMO detector provides at its output extrinsic probabilities on the coded bits c. Let $\mathcal{Q}$ be the cardinality of $\mathcal{S}$ of size $q \triangleq |\mathcal{Q}| = 2^{BQ}$. Let also c_i, $i = 1, \cdots, BQ$ be the bits corresponding to a vector of symbols $\mathcal{S} \in \mathcal{Q}$. The extrinsic probability on the bit c_j at the MIMO detector output, $P_{\text{ext}}^{\text{Det}}(c_j)$, is calculated as follows [60]:

$$P_{\text{ext}}^{\text{Det}}\left(c_j = 1\right) = K \sum_{\substack{\mathcal{S} \in \mathcal{Q} \\ c_j = 1}} \exp\left(-\frac{\left\|\mathcal{Y} - \mathcal{H}_{eq}\mathcal{S}\right\|^2}{\sigma_n^2}\right) \prod_{\substack{i=1 \\ i \neq j}}^{BQ} P_{\text{ext}}^{\text{Dec}}\left(c_i\right), \tag{4.17}$$

where K is the normalization factor satisfying $P_{\text{ext}}^{\text{Det}}(c_j = 1) + P_{\text{ext}}^{\text{Det}}(c_j = 0) = 1$. The probability $P_{\text{ext}}^{\text{Dec}}(c_i)$ is in fact the *a priori* information on bit c_i, fed back from the channel decoder. At the first iteration, where no *a priori* information is available on bits c_i, $P_{\text{ext}}^{\text{Dec}}$ are set to 1/2. The summation in (4.17) is taken over the product of the conditional channel likelihood (the exponential term) given a vector $\mathcal{S}$, and the *a priori* probability on this symbol, i.e., the term $\Pi\, P_{\text{ext}}^{\text{Dec}}$. In this latter term, we exclude the *a priori* probability corresponding to the bit c_j itself, so as to respect the exchange of *extrinsic* information between the channel decoder and the MIMO detector. Also, this term assumes independent coded bits c_i, which is true for random interleaving of large size.

4.4.3.2 Sphere Decoding

As it was seen, in the calculation of the extrinsic probabilities, the MAP detector considers the exhaustive list of all possibly transmitted symbol vectors. Hence, the complexity of the MAP detector grows exponentially with the number of transmit antennas M_T and the number of bits per modulation symbol B. By sphere decoding, these probabilities are calculated based on a nonexhaustive list [55]. Corresponding to a vector $\mathcal{Y}$, we only take into account those lattice points that are in a hypersphere of radius R around $\mathcal{Y}$. The main detection tasks are then setting the radius R as well as determining which lattice points are within the sphere. In this way, the average complexity of the detector under flat fading conditions becomes polynomial (often subcubic) under high SNR [61]. At low SNR, however, the detector complexity can be very high. A drawback of this method is the time variability of the detector complexity, as it depends on SNR.

4.4.3.3 Parallel Interference Cancelling-Based Detection

The block diagram of the detector based on soft-parallel interference cancellation (soft-PIC) is shown in Figure 4.9, where the detector soft outputs are considered in the form of LLR. Reformulation is trivial if we want to consider these soft outputs in the form of probability (like in Figure 4.8). We will refer to the corresponding receiver scheme as turbo-PIC. Soft-PIC is essentially composed of the three blocks of PIC detector, conversion to LLR, and soft estimation of transmit symbols.

4.4.3.3.1 PIC Detector

In order to present general expressions for the detector, let us denote by $\hat{\gamma}_p$, $p = 1, \cdots, 2Q$, the detected data at the PIC detector output, corresponding to the real or imaginary parts of s_q, $q = 1, \cdots, Q$. At the first iteration, $\hat{\gamma}_p$ are obtained via minimum mean square error (MMSE) filtering [48, 62, 63]:

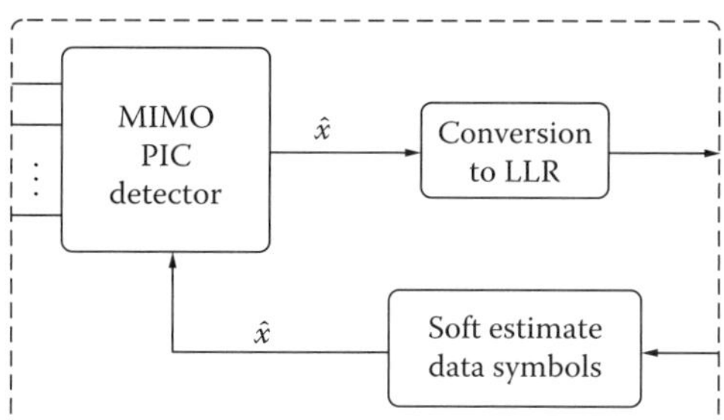

FIGURE 4.9 Block diagram of the soft-PIC detector.

$$\hat{\gamma}_p = \mathcal{W}_p^\dagger \mathcal{Y} = h_p^\dagger \left(\mathcal{H}_{eq}\mathcal{H}_{eq}^\dagger + \sigma_n^2 \boldsymbol{I}\right)^{-1} \mathcal{Y}, \tag{4.18}$$

where $\mathcal{W}_p$ denotes the filter, h_p is the pth column of $\mathcal{H}_{eq}$, and $(.)^\dagger$ stands for transpose complex conjugate. Note that the entries of $\mathcal{H}_{eq}$ are real values, and hence $(.)^\dagger$ is equivalent to transposition. From the second iteration, we can calculate soft estimates $\tilde{\mathcal{S}}$ of the transmitted data using the soft decoder outputs. Using these estimates, we perform interference cancelling followed by a simple zero forcing (ZF) or MMSE filtering:

$$\hat{\mathcal{Y}}_p = \mathcal{Y} - \mathcal{H}_p \tilde{\mathcal{S}}_p, \quad \hat{\gamma}_p = \mathcal{W}_p^\dagger \hat{\mathcal{Y}}_p, \tag{4.19}$$

$$\text{ZF: } \mathcal{W}_p = \frac{1}{h_p^\dagger h_p} h_p, \quad \text{MMSE: } \mathcal{W}_p = \frac{1}{\left(h_p^\dagger h_p + \sigma_n^2\right)} h_p, \tag{4.20}$$

where $\tilde{\mathcal{S}}_p$ of dimension $((2Q - 1) \times 1)$ is $\tilde{\mathcal{S}}$ with its pth entry removed, and $\mathcal{H}_p$ of dimension $(2M_R T \times (2Q - 1))$ is the matrix $\mathcal{H}$ with its pth column removed. Notice that, compared to the exact MMSE filtering proposed in [48], (4.20) are simplified solutions that assume almost perfect estimation of data symbols and permit a considerable reduction of the computational complexity. Thanks to iterative processing, the performance loss due to this simplification would be negligible. In the results that we present later, we will consider the simplified ZF solution.

4.4.3.3.2 Conversion to LLR

For QAM modulation with B (an even number) bits per symbol, we can attribute $m = B/2$ bits to the real and imaginary parts of each symbol. Let, for instance, the bit c_i correspond to the real (imaginary) part of the symbol s_q. Let also $a_{1,j}$ and $a_{0,j}$, $j = 1, \cdots, B/2$ denote the real (imaginary) part of the signal constellation points, corresponding to $c_i = 1$ and $c_i = 0$, respectively. Remember that the signal constellation points have normalized average power. The LLR corresponding to c_i is calculated as follows [62]:

$$\text{LLR}_i = \log_{10} \frac{\sum_{j=1}^{2^{m-1}} \exp\left(-\frac{1}{2\sigma_p^2}\left(\hat{\gamma}_p - \alpha_{1,j}\right)^2\right)}{\sum_{j=1}^{2^{m-1}} \exp\left(-\frac{1}{2\sigma_p^2}\left(\hat{\gamma}_p - \alpha_{0,j}\right)^2\right)}, \quad i = 1, \cdots, m, \tag{4.21}$$

where σ_p^2 is the variance of noise plus the residual interference (RI) that intervenes in the detection of $\hat{\gamma}_p$, and is assumed to be Gaussian. Note that as the detection is performed on blocks of Q complex symbols, or in other words, on blocks of $2Q$ *real symbols* in our model, the RI comes in fact from $(2Q - 1)$ other real symbols in the corresponding channel use [64]. In LLR calculation, we need the variances σ_p^2. These variances can be calculated analytically as shown in [46], or estimated at each iteration and for each one of

2Q real symbols, as done in [64]. To simplify the detector further, we may neglect the RI and take into account only the noise variance. For not too large signal constellation sizes, this simplification causes a negligible performance loss [65, 66].

4.4.3.3.3 *Soft Estimation of Transmit Symbols*

Each element $\tilde{\gamma}_p$ of the vector $\tilde{S}$ is obtained by taking a summation over all the possible values of the real part (or imaginary part) of the signal constellation, multiplied by the corresponding probability calculated using the soft decoder output [62, 63]. It is preferable to use the *a posteriori* information from the decoder output rather than extrinsic information in the calculation of $\tilde{\gamma}_p$. This has the advantage of permitting a better and faster convergence of the Rx.

4.4.3.4 Case Study

As we focus here on the turbo-PIC detector as the suboptimal solution, we just compare the performance of this detector with that of turbo-MAP. We consider the simplified implementation of soft-PIC based on ZF filtering given in (4.20). For this comparison, we consider the simple spatial multiplexing (V-BLAST) ST scheme and the case of four transmit antennas, $M_T = 4$, while we take M_R between 1 and 4. The Tx and Rx schemes correspond to Figures 4.7 and 4.8, respectively. We consider Gray bit/symbol mapping and random interleaving, as well as the Rayleigh flat quasi-static channel model. The nonrecursive and nonsystematic convolutional (NRNSC) channel code $(5, 7)_8$ (in octal representation) is considered with rate $R_c = 1/2$. SNR is considered in the form of E_b/N_0, where E_b is the average received energy per information bit and N_0 is the unilateral noise power spectral density; E_b/N_0 includes the Rx array gain, M_R.

Curves of bit error rate (BER) versus E_b/N_0 are given in Figure 4.10. In fact, the performances of turbo-PIC and turbo-MAP are relatively close to each other for $M_R \geq M_T$. Turbo-PIC can still be used for certain values of $M_R < M_T$, mostly for $M_R > M_T/2$ [63]. So, for these M_R values where turbo-PIC converges properly, it would be preferred to turbo-MAP due to its considerably lower complexity. Better performances are obtained for turbo-PIC if the variance of the RI is taken into account in LLR calculation [63].

4.4.4 Orthogonal versus Nonorthogonal ST Schemes

In practice, to attain a desired spectral efficiency, we should adopt the most appropriate scheme by fixing the degrees of freedom of the system, that is, the signal constellation, the channel coding rate, and the ST coding scheme. The answer to the question "What is the most suitable combination?" is not obvious for moderate to high spectral efficiencies. In effect, if a low spectral efficiency is required, an OSTBC scheme together with a powerful turbo-code would be a suitable solution, as the reduction in the overall coding rate is best invested in turbo channel codes [67]. To attain high spectral efficiencies with OSTBC schemes, however, we have to use large signal constellations and increase the channel coding rate by puncturing the encoder output bits. Use of larger signal constellations complicates the tasks of synchronization and detection at the Rx and also results in a higher SNR required to attain a desired BER. On the other hand, puncturing results in a reduced channel code robustness against noise. Higher ST coding rates are offered

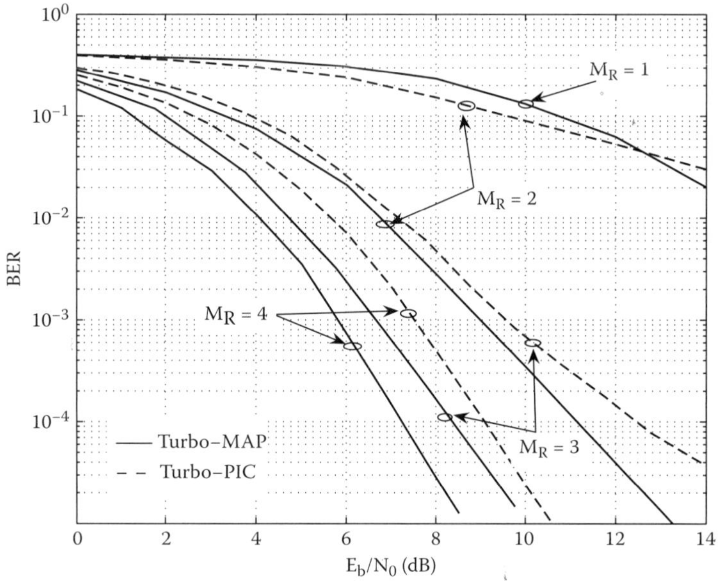

FIGURE 4.10 Comparison of turbo-PIC and turbo-MAP detectors, $M_T = 4$, $(5, 7)_8$ channel code, QPSK modulation, 64 channel uses per frame.

by nonorthogonal schemes, hence relaxing the conditions on signal constellation and channel coding. Here, a simple (suboptimal) iterative detector can be used for ST decoding, as explained in the previous subsection, and we may approach the optimal detection performance after few iterations. Nevertheless, the detector remains more complex, as compared to the OSTBC case. However, this increased Rx complexity is quite justified; using such an appropriate nonorthogonal ST scheme and iterative detection, we obtain a considerable gain in performance with respect to OSTBC choice [64–66]. Results in [65, 66] have also confirmed that the gain obtained by using nonorthogonal with respect to orthogonal schemes is still considerable, and even more important when channel estimation errors are taken into account.

4.4.4.1 Case Study

We consider the case of a (2×2) MIMO system, Gray bit/symbol mapping and random interleaving, as well as the Rayleigh flat block-fading channel model with $N_c = 32$ independent fades per frame. The number of channel uses corresponding to a frame is 768. The NRNSC channel code $(133, 171)_8$ is considered with rate $R_c = 1/2$. The ST schemes we consider are shown in Table 4.1, where η is the spectral efficiency in units of bps/Hz. As the OSTBC scheme, we consider the Alamouti code [51]. Using the formulation of LD codes that we presented in section 4.4.2, we have $Q = T = 2$, $R_{STC} = 1$, and

TABLE 4.1 Different ST Schemes for a (2×2) MIMO System with $\eta = 2$ bps/Hz

ST Scheme	R_{STC}	Modulation	R_c
Alamouti	1	16-QAM	1/2
V-BLAST	2	QPSK	1/2
GLD	2	QPSK	1/2

$$X = \begin{bmatrix} s_1 & -s_2^* \\ s_2 & s_1^* \end{bmatrix}. \tag{4.22}$$

For the OSTBC case, the decoding is performed once using (4.18). We also consider two nonorthogonal schemes. The first one is the simple V-BLAST scheme described by $X = [s_1\ s_2]^t$, for which $Q = 2$, $T = 1$, and $R_{STC} = 2$. The second one is the optimized scheme proposed in [68] and called *Golden code*, which we denote by GLD. For this code that offers full rate and full diversity with the property of nonvanishing determinant, we have $Q = 2$, $T = 2$, and $R_{STC} = 2$:

$$X = \frac{1}{\sqrt{5}} \begin{bmatrix} \alpha(s_1 + \theta\, s_2) & \gamma\bar{\alpha}(s_3 + \bar{\theta}\, s_4) \\ \alpha(s_3 + \theta\, s_4) & \bar{\alpha}(s_1 + \bar{\theta}\, s_2) \end{bmatrix}, \tag{4.23}$$

where

$$\theta = \frac{1+\sqrt{5}}{2}, \alpha = 1 + j(1-\theta), \bar{\theta} = 1 - \theta, \bar{\alpha} = 1 + j(1-\bar{\theta}), \gamma = j = \sqrt{-1}.$$

The factor $1/\sqrt{5}$ ensures normalized transmit power per channel use. For η = 2 bps/Hz, performance curves are shown in Figure 4.11, where again perfect channel knowledge is assumed at the Rx. For V-BLAST and GLD schemes, BER curves are shown for the second and fourth iterations, where almost full Rx convergence is attained. We see that, by using the V-BLAST scheme, we gain about 3.3 and 3.75 dB in SNR at BER = 10^{-4} after two and four iterations, respectively, compared to Alamouti coding. The corresponding gains by using GLD code are about 3.5 and 4.3 dB, respectively. We note that even when, for the reasons of complexity and latency, only two iterations are to be performed, the gain in SNR compared to the Alamouti scheme is still considerable.

4.5 Advanced Polarization Diversity Techniques for MIMO Systems

The initial research demonstrates that the MIMO channel capacity based on the uncorrelated channel model can be proportionally increased by increasing the number of antennas. However, in practice, the performance of the MIMO communication channel is affected by spatial correlation, which is dependent on antenna array configurations (such as radiation pattern, antenna spacing, and array geometry), and propagation channel characteristics, which are dependent on the environment (such as number of channel paths, distribution and properties of scatterers, angle spread, and cross-polarization discrimination). Thus, the antenna arrays at Tx and Rx should be properly designed to reduce the spatial correlation effects and to improve the communication performance.

However, it is possible to reduce these effects by increasing antenna array spacing, but this solution is not always suitable in some wireless applications where the array size is

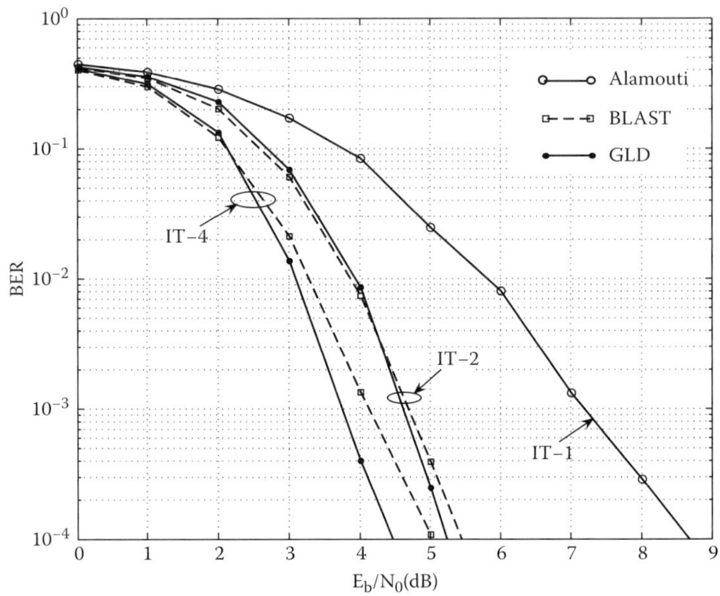

FIGURE 4.11 (2×2) MIMO system, turbo-PIC detection, $(133, 171)_8$ channel code, $\eta = 2$ bps/Hz, each frame contains 768 channel uses corresponding to 32 independent fades.

limited. Therefore, in order to eliminate the spatial correlation effects and remain with high transmission performance, there are essentially two diversity techniques, such as angular [69] and polarization diversity techniques [70]. For the pattern diversity technique, the radiation of antennas should be generated in a manner to isolate the radiation pattern. For polarization diversity techniques [71], the antennas are designed to radiate with orthogonal radiation polarizations to create uncorrelated channels across different array elements. The polarization diversity techniques could be applied in point-to-point communication systems such as intermobile base station communications, mobile satellite communications, high-resolution localization systems, military communications, etc.

Finally, there are also other diversity techniques, such as multimode diversity [72], that exploit the difference of high-order modes to obtain low correlated channels across the modes and a combination of pattern and polarization diversity techniques [73, 74] that take together the advantages of orthogonal radiation patterns and polarizations.

4.5.1 Antennas

In practice, not only the propagation environment for the multiple antenna systems has an important role in determining the transmission performance, but also the proper implementation of the antennas plays a dominant role. For example, uniform linear arrays, uniform circular arrays, and cube antenna arrays give different performances

TABLE 4.2 Patterns of Different Electric Dipoles

	E_x	E_y	E_z	E_ϕ	E_θ
$E_\theta(\theta, \phi)$	$-\cos\theta\cos\phi$	$-\cos\theta\cos\phi$	$-\sin\theta$	0	1
$E_\phi(\theta, \phi)$	$\sin\phi$	$-\cos\phi$	0	1	0

in terms of channel capacity. Moreover, different array configurations produce different correlation effects. In this part, we will analyze five types of antennas [75], *x*-, *y*-, and *z*-oriented dipole antennas, azimuth, and elevation isotropic antennas, applied to a uniform linear array.

The radiating patterns of the antennas are considered in the far-field case, and are also simplified by neglecting path loss and distance phase. Hence, these radiating patterns are simply dependent on the azimuth and elevation angle direction as shown in Table 4.2. A general expression of radiation patterns is given by [75]

$$E = E_\theta(\theta,\phi)\vec{\theta} + E_\phi(\theta,\phi)\vec{\phi}, \tag{4.24}$$

where $E_\theta(\theta,\phi)$ and $E_\phi(\theta,\phi)$ are the amplitudes of polarization vector at the $\vec{\theta}$- and $\vec{\phi}$-directions, and *x*, *y*, and *z* are the antenna orientations. As different types of antennas are employed in this chapter, it is necessary to normalize the radiation pattern when comparing all channel performances. Thus, all radiation patterns of an antenna are normalized by that of an isotropic antenna, and they can be written as

$$G = \sqrt{\frac{\iint\limits_{\Delta\phi,\Delta\theta} \sin\theta d\theta d\phi}{\iint\limits_{\Delta\phi,\Delta\theta}\left[\left|E_\theta(\theta,\phi)\right|^2 + \left|E_\phi(\theta,\phi)\right|^2\right]\sin\theta d\theta d\phi}}\left(E_\theta(\theta,\phi)\vec{\theta} + E_\phi(\theta,\phi)\vec{\phi}\right) \tag{4.25}$$

where G is the antenna gain that is used for the computation of the channel matrix.

4.5.2 Cross-Polarization Discrimination

In wireless communications, due to the interactions of environment, such as diffractions, reflection, and refraction, the transmitted signals are generally not only attenuated but also depolarized. Depolarization is the change of the original state of the polarization of the electromagnetic wave propagated from the Tx.

Cross-polarization discrimination (XPD) is defined as the power ratio of the co-polarization and cross-polarization components of the mean incident wave. The higher the XPD, the less energy that is coupled in the cross-polarized channel. Therefore, there are two transmission cases, azimuth transmission (χ_θ) and elevation transmission (χ_ϕ), as follows:

$$\chi_\theta = \frac{E\left\{\left|E_{\theta\theta}\right|^2\right\}}{E\left\{\left|E_{\theta\phi}\right|^2\right\}}, \quad \chi_\phi = \frac{E\left\{\left|E_{\phi\phi}\right|^2\right\}}{E\left\{\left|E_{\phi\theta}\right|^2\right\}}, \tag{4.26}$$

where $E_{\theta\phi}$ denotes the θ-polarized electric field, which is propagated from a Tx and received in the ϕ polarization. It has the same explanation for $E_{\theta\theta}$, $E_{\phi\phi}$, and $E_{\phi\theta}$. In the literature, some measurement campaigns have been carried out, and they concluded that the XPD depended on the physical obstacles, the distance between Tx and Rx, and the delay of multipath components of each environment [76]. Therefore, for simplicity, XPD can be approximated by a Gaussian statistical distribution with average μ and variance σ^2 [77, 78]. For urban environments, the mean value of XPD can vary from 0 to 16 dB and the standard deviation can change from 3 to 9 dB.

4.5.3 Geometry-Based Stochastic Channel Models

We now focus on a useful model for simulation purposes, geometry-based stochastic modeling or geometric scattering modeling [10, 79, 80], which can be easily exploited to examine the performance of different antenna patterns and polarizations. This model is based on the assumption that scatterers around the Tx and Rx influence the direction of departure (DOD) and the direction of arrival (DOA), respectively, within transmit and receive scattering areas. Scatterers are randomly located according to a certain probability distribution. In particular, the scatterers are used to represent the depolarization and attenuation mechanism of incident waves traveling from the transmitters.

In Figure 4.12, a scattering geometry is shown. A uniform linear array of z-oriented dipole antennas at both Tx and Rx is employed. The heights of Tx and Rx are the same level. Moreover, transmit and receive scatterers are uniformly distributed within an angular region defined by $|\phi + \pi/2| \leq \Delta\phi/2$ in elevation area and $|\theta + \pi/2| \leq \Delta\theta/2$ in azimuth area. In order to determine one propagation path, from one transmit scatterer to one receive scatterer, we consider that there is a double depolarization mechanism

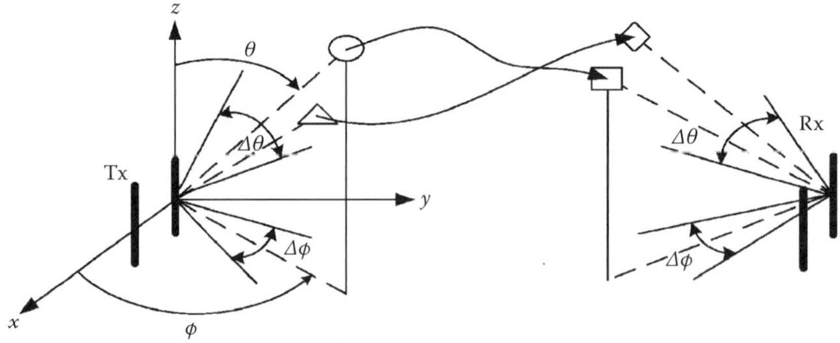

FIGURE 4.12 Geometries of MIMO channel.

replaced by one scattering matrix. One of the propagation path channels occurs when one of the transmit and one of the receive scatterers are randomly linked. To reduce the computational cost, a scatterer is used to generate only one propagation channel. Then the actual channel impulse response is established by a sum of propagation channels. We also assume that the channel coherence bandwidth is larger than the transmitted bandwidth of the signal. This channel is usually called frequency-non-selective or flat fading channel.

In case of far-field transmission without the line-of-sight channel, the flat fading transmission channel between antenna p at the Tx and antenna m at the Rx can be expressed as

$$h_{mp}(t,f)=\frac{1}{\sqrt{N_S}}\sum_{i=1}^{N_S}a_m^{(i)}a_p^{(i)}\exp\left\{-j\vec{k}^{(i)}\cdot\vec{v}_{Rx}t-j\vec{k}'^{(i)}\cdot\vec{v}_{Tx}t+\varphi_{mp}\right\}$$
$$\left[G_\theta^m(\theta_i,\phi_i)\quad G_\phi^m(\theta_i,\phi_i)\right]\mathbf{S}_{mp}^{(i)}\begin{bmatrix}G_\theta^p(\theta_i,\phi_i)\\ G_\phi^p(\theta_i,\phi_i)\end{bmatrix},\tag{4.27}$$

where t is time; f is frequency; N_s is the number of scatterers at the Rx and Tx; $\vec{v}_{Rx}$ and $\vec{v}_{Rx}$ are the velocity vectors of the Tx and Rx; $\vec{k}'^{(i)}$ and $\vec{k}^{(i)}$ are the vectors of wave number in the direction of the ith transmit scatterer and the ith receive scatterer, where $|\vec{k}^{(i)}| = |\vec{k}'^{(i)}| = 2\pi/\lambda$; $G_\theta^p(\theta_i,\phi_i)$ and $G_\phi^p(\theta_i,\phi_i)$ are the gain in the $\vec{\theta}$ and $\vec{\phi}$ directions of the pth transmit antenna in the direction of the ith transmit scatterer; $G_\theta^m(\theta_i,\phi_i)$ and $G_\phi^m(\theta_i,\phi_i)$ are the gain in the $\vec{\theta}$ and $\vec{\phi}$ directions of the mth receive antenna in the direction of the ith receive scatterer; $a_m^{(i)}$ is the mth element of the local vector of the receive antenna, so that the local receive vector can be expressed as $\mathbf{a}_{\text{Rx}}^{(i)} = [1\ \ \exp\{-j\vec{k}^{(i)}\cdot\vec{r}_1\}\ \cdots\ \exp\{-j\vec{k}^{(i)}\cdot\vec{r}_{M_R-1}\}]$; $a_p^{(i)}$ is the pth element of the local vector of the transmit antenna, so that the local transmit vector can be expressed as $\mathbf{a}_{\text{Tx}}^{(i)} = [1\ \ \exp\{-j\vec{k}'^{(i)}\cdot\vec{r}'_1\}\ \cdots\ \exp\{-j\vec{k}'^{(i)}\cdot\vec{r}'_{M_T-1}\}]$; and $\mathbf{S}_{mp}^{(i)}$ are a 2×2 scattering matrix for the ith transmit scatterer and the ith receive scatterer for $i = 1 \ldots N_S$ wave components. A scattering matrix contains the polarization mechanism as defined by

$$\mathbf{S}_{mp}^{(i)}=\begin{bmatrix}\sqrt{\dfrac{\chi_\theta}{1+\chi_\theta}}\exp\left\{j\beta_{mp}^{(\theta\theta)}\right\} & \sqrt{\dfrac{1}{1+\chi_\phi}}\exp\left\{j\beta_{mp}^{(\phi\theta)}\right\}\\ \sqrt{\dfrac{1}{1+\chi_\theta}}\exp\left\{j\beta_{mp}^{(\theta\phi)}\right\} & \sqrt{\dfrac{\chi_\phi}{1+\chi_\phi}}\exp\left\{j\beta_{mp}^{(\phi\phi)}\right\}\end{bmatrix},\tag{4.28}$$

where $\beta_{mp}^{(\phi\theta)}$ denotes phase offset of the ith incident wave, which changes from $\vec{\phi}$ directions to $\vec{\theta}$ directions superposing on the m-p channel, and χ_θ and χ_ϕ denote the ratio of the co-polarized average received power to the cross-polarized average received power. In [77], after sufficiently reflecting the propagation signal between transmitters and receivers, the polarization state of the signal will be independent of the transmitted signals.

4.5.4 Spatial Correlation and Angle Spread Effects

The consecutive MIMO systems based on the spatial diversity technique are directly influenced by the spatial correlation effect (or antenna correlation effect) [16]. This effect is drastically dependent on array configurations and environment characteristics. Therefore, the antenna arrays at both Tx and Rx should be properly designed or adapted to decrease spatial correlation effects.

As illustrated in Figure 4.13, the spatial correlation of a 2×2 uniform linear antenna array depends on the antenna spacing and the angle spread (AS). The general expression of spatial correlation [81] between two antenna elements can be written as

$$\rho_{i,j} = \frac{\iint\limits_{\Delta\theta,\Delta\phi} a_i \cdot a_j^* \sin(\theta)\, p(\theta,\phi)\, d\theta d\phi}{\sqrt{\iint\limits_{\Delta\theta,\Delta\phi} |a_i|^2 \sin(\theta)\, p(\theta,\phi)\, d\theta d\phi \cdot \iint\limits_{\Delta\theta,\Delta\phi} |a_j|^2 \sin(\theta)\, p(\theta,\phi)\, d\theta d\phi}}, \tag{4.29}$$

where a_i is the local value of the ith transmit or receive antenna and a_j is the local value of the jth transmit or receive antenna in the local vector. The scalar $p(\theta,\phi)$ is the joint probability density function (pdf) of the angles of arrival for the receiving spatial correlations or of the angles of departure for the transmitting spatial correlations.

When the narrow angle spread of incident fields occurs in the transmitting or receiving side, the separation between antennas should be expanded in order to reduce the spatial correlation problem, as shown in Figure 4.13. This shows the spatial correlation of 2×2 MIMO z-oriented antennas in the case of a uniform distribution of the angles

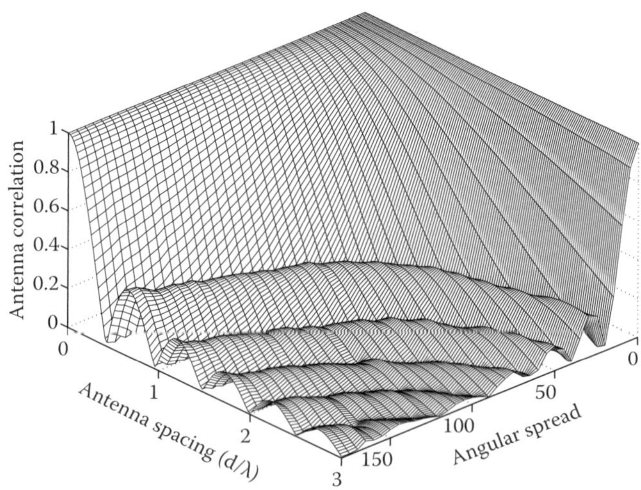

FIGURE 4.13 Spatial correlation of uniform linear antenna array for single-polarized configuration.

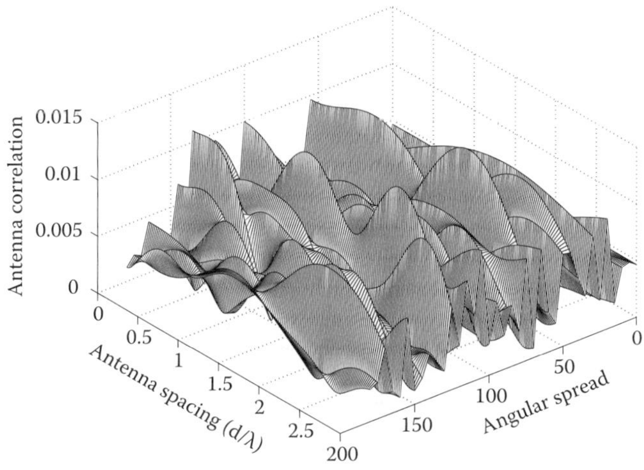

FIGURE 4.14 Spatial correlation of uniform linear antenna array for dual-polarized configuration.

of arrival within a region. It demonstrates that the received signals become uncorrelated when the antenna spacing is sufficiently increased and the angle spread is quite wide. However, wide antenna spacing may preclude implementation in some applications where size is a limitation. The use of polarized antennas is promising low spatial correlation. The spatial correlation is demonstrated in Figure 4.14, while the 2×2 polarized MIMO configuration employs a pair of y- and z-oriented dipole antennas at both Tx and Rx. The spatial correlation of dual-polarized MIMO is much lower than that of single-polarized MIMO in all simulation scenarios.

Although there are only two diversity branches, the use of polarized antennas does allow the antenna elements to be collocated without the correlation effect. However, there is considerable interest in many diversity branches by applying a combination of the pattern and polarization diversity [74, 82].

4.5.5 Capacity of Polarized Channels

Antenna polarization diversity is very useful in MIMO systems for enhancing channel capacity. Indeed, employing polarization diversity can reduce the antenna array size and also the spatial correlation; then we can obtain a better capacity. That is why the multi-polarized antennas become more and more interesting in MIMO transmission systems. In this section, MIMO systems are investigated to show the potential of using multi-polarized antennas for differently oriented dipole antennas.

We assume that the channel state information (CSI) is perfectly known to the Rx but unknown to the Tx. This is in theory what happens to signals propagating through an urban and an indoor environment. In the case of a random channel model, the channel matrix (4.27) is stochastic, and then the capacity given by (4.5) is also random. In this situation, the ergodic capacity can be obtained by taking the expectation of capacity over all possible channel realizations.

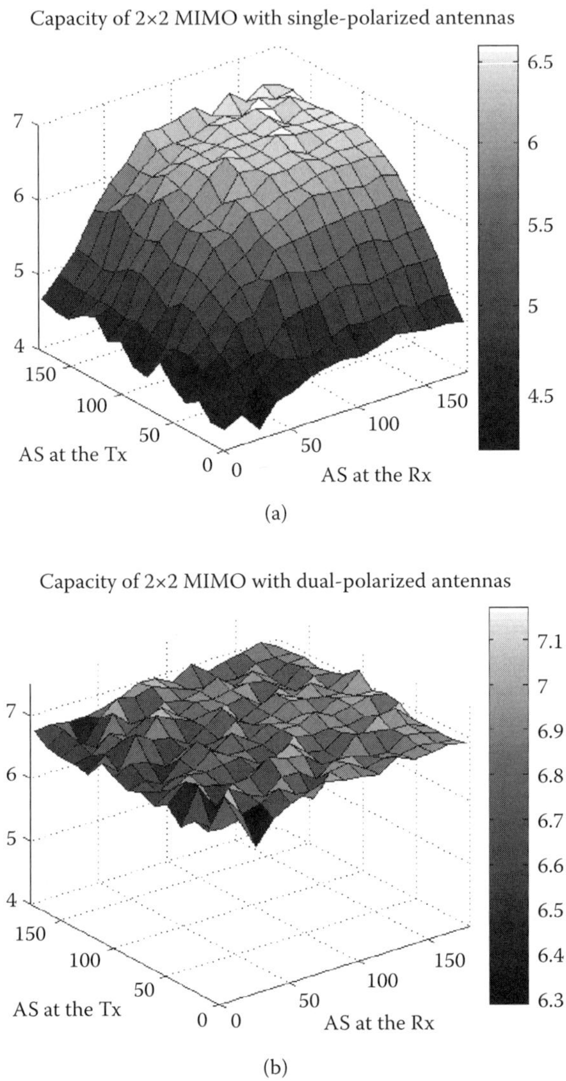

FIGURE 4.15 (2×2) MIMO channel capacity of isotropic antennas: (a) single-polarization configuration and (b) dual-polarization configuration.

Figure 4.15 demonstrates the 2×2 MIMO channel capacity of isotropic antennas with single polarization (a) and dual polarization (b). In this case, the single-polarization configuration exploits only the azimuth isotropic antenna, and the dual polarization configuration applies the azimuth and elevation isotropic antennas. The channel capacity is examined in function of angular spread (AS) with twenty scatterers distributed around the Tx and Rx. As mentioned in the previous section, the XPD is defined for

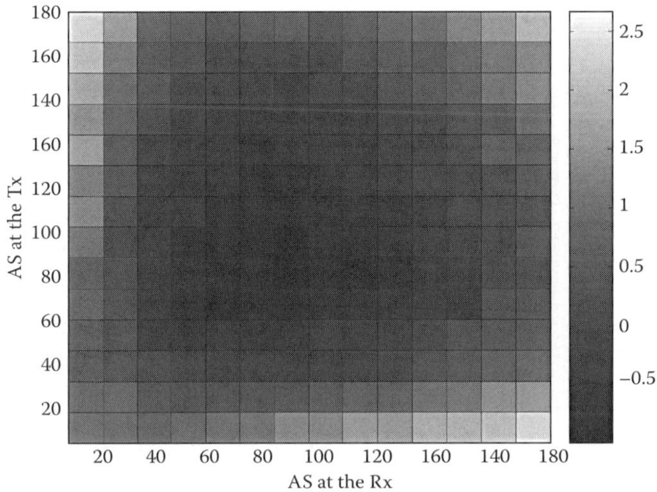

FIGURE 4.16 Difference between the triple-polarized and single-polarized channel capacities of 3 × 3 MIMO systems employing dipole antennas.

the urban case by χ_θ and $\chi_\phi \sim \mathcal{N}(0, 9)$. As shown in Figure 4.15a, the MIMO channel capacity increases as the angle spread increases at Tx and Rx for the same polarization antennas. In contrast, dual polarization improves the channel capacity due to the lower antenna correlation, as shown in Figure 4.15b. However, when the spatial correlation of the single-polarized antennas is lower, the channel capacity is proportional until 6.7 dB for AS > 80°. It should be noted that the MIMO channel capacity is significantly dependent on the antenna correlation.

Figure 4.16 shows the difference between the triple-polarized and single-polarized channel capacities ($\Delta C = C_{\text{triple-polar}} - C_{\text{single-polar}}$) of the 3 × 3 MIMO system versus the average XPD and AS. The propagation environment has the same conditions as the previous section. For the triple-polarization configuration, *x*-, *y*-, and *z*-oriented dipole antennas are employed. This can represent a combination of angular and polarization techniques. We found that for the triple-polarization case, the average power of the subchannel can be unfortunately lost when the angular spread is not large enough until covering all antennas. As seen in Figure 4.16, the single-polarized channel capacity can be superior to the triple-polarized channel capacity because single polarization has a low spatial correlation and triple polarization loses the subchannel power due to insufficient angular spread. However, generally speaking for most scenarios, triple polarizations can maintain a higher capacity with respect to the single-polarization case.

4.5.6 Impact of Depolarization Effect on MIMO Configurations

In this section, we investigate the impact of the depolarization effect on 4 × 4 MIMO systems with single- and dual-polarization configurations. While pattern and polarization

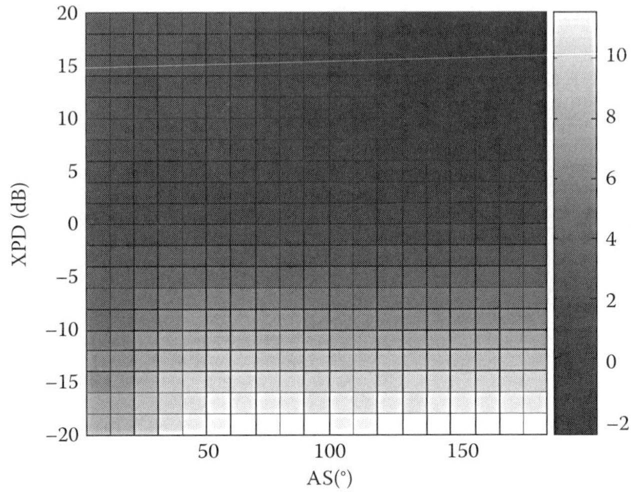

FIGURE 4.17 Difference between the dual-polarized and single-polarized channel capacities of 4 × 4 MIMO systems in the functions of XPD and AS.

diversity techniques are employed, the special correlation effect can be reduced or eliminated when there is no pattern interference. Nevertheless, the cross-polarization discrimination (XPD) becomes the most important parameter because XPD represents the ratio of the co-polarized average received power to the cross-polarized average received power. Then, for a high XPD value, less energy is coupled between the cross-polarized wireless channels. At lower XPD and higher *K*-factor values [83], multipolarized antenna arrays can give high capacity. However, at higher XPD and lower spatial correlation, a single-polarized antenna array can provide even better results.

Figure 4.17 explains the difference between the dual-polarized and single-polarized channel capacity ($\Delta C = C_{\text{dual-polar}} - C_{\text{single-polar}}$) of 4 × 4 MIMO systems versus XPD and AS. Same angle spreads at Tx and Rx are considered in the simulation. For high XPD and sufficiently large angle spread, we can note that the MIMO channel capacity of the single-polarized antenna is superior to that of the dual-polarized antenna because of the subchannel power loss. The Frobenius norm of the MIMO channel is used to investigate the total channel power. It confirms that with a high XPD and low spatial correlation the average transmission power of single-polarized isotropic antenna arrays is $\|\boldsymbol{H}\|_F \approx MN$, but that of dual-polarized isotropic antenna arrays is always $\|\boldsymbol{H}\|_F \approx M^{\theta}N^{\theta} + M^{\phi}N^{\phi}$ for high XPDs and $\|\boldsymbol{H}\|_F \approx M^{\theta}N^{\phi} + M^{\phi}N^{\theta}$ for low XPDs.

Figure 4.18 shows the XPD impact on 4 × 4 MIMO systems with single- and dual-polarization configurations. We employ four azimuth isotropic antennas for the single-polarization configuration and two elevation and two azimuth isotropic antennas for the dual-polarization configurations, which are applied to a uniform linear antenna array with a $\lambda/2$ antenna separation. It shows that the channel power of single-polarization configuration augments significantly with respect to XPD until $\|\boldsymbol{H}\|_F = 4 \times 4 = 16$, while that of dual-polarization configuration keeps around $\|\boldsymbol{H}\|_F = 2 \times 2 + 2 \times 2 = 8$.

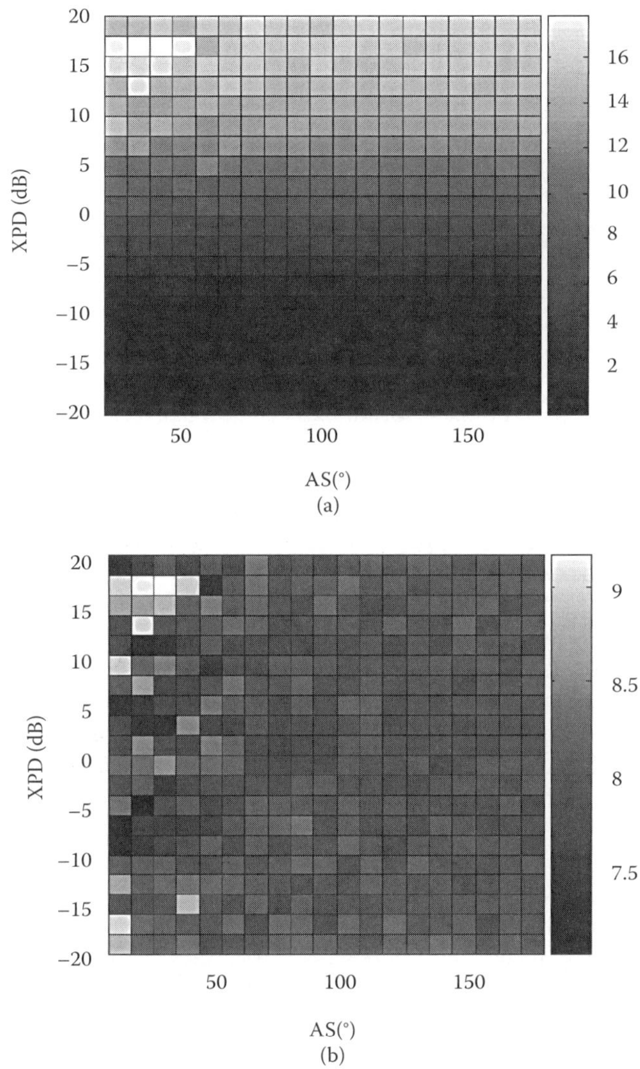

FIGURE 4.18 Frobenius norm of 4 × 4 MIMO systems in terms of XPD and AS with (a) a single-polarization configuration and (b) a dual-polarization configuration.

4.5.7 Adaptive MIMO Polarized-Antenna Selection Technique (AMPAS)

In the previous sections, we defined the scattering mechanisms that are used to represent not only the attenuation of traveling waves but also the polarization of the electromagnetic wave. The achieved performance in capacity is calculated under the assumption

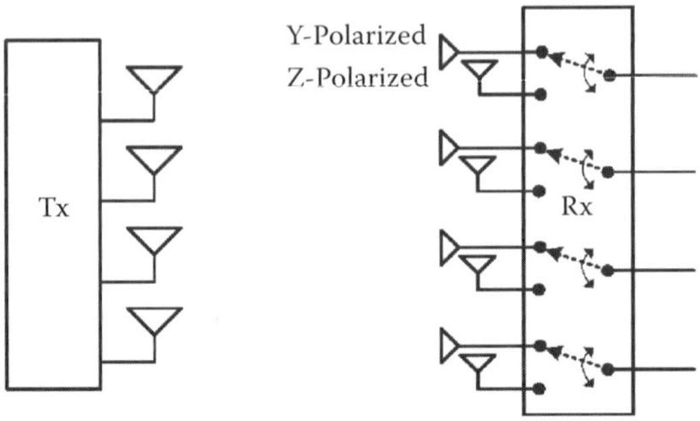

FIGURE 4.19 Adaptive polarization technique (AMPAS) at Rx.

that the average received power is normalized and the channel attenuation is neglected. Subsequently, the variation of polarization is characterized only by the XPD effects.

As shown in Figure 4.17, this phenomenon affects directly the performance of non-polarized MIMO systems. In the single-polarization communications case, low XPD causes higher losses in channel power; in other words, there is some sort of mismatch in polarization. That is why we will apply adaptive techniques to reduce this mismatch in the polarization of MIMO systems. This technique is called adaptive MIMO polarized-antenna selection technique (AMPAS). The principle of this method is to choose properly the antenna polarizations that optimize the receiving signal power while minimizing fading correlation antenna effects. In Figure 4.19, an example of an adaptive polarization system employing four z-oriented dipoles at Tx and four pairs of y- and z-oriented dipoles at Rx is illustrated. Simulation results based on three-dimensional ray-tracing techniques show that the channel capacity obtained by an adaptive polarization increases 7–13% in comparison to the single-polarization channel capacity.

Another example of an adaptive polarization system is based on the rotation of the antenna elements according to the polarization of traveling waves at Rx. The proposed MIMO system consists of P half-wavelength dipole antennas that are rotated against one another by the rotation angle $\gamma = 180°/P$ with phase centers at the same point at both Tx and Rx.

Figure 4.20 demonstrates the obtained performances of 1 × 1 SISO, 1 × 2 SIMO, and 2 × 2 MIMO communication systems while the receiving antenna is rotating on the y-z plane with χ_θ and $\chi_\phi \sim \mathcal{N}(0,5)$. Performance can be better enhanced if polarization at the Rx is properly matched to that of incident waves. In contrast, it can be worse if they are not well matched, as shown in Figure 4.20a (at rotational angle (γ) ≈ 140°). Thus, for improving the MIMO channel capacity, the receiving antenna elements should be rotated to find maximum receiving signals while minimizing fading correlation antenna effects.

While the other systems employ only z-oriented dipoles at Tx and y- and z-oriented dipoles at Rx in the case of the same antenna position (b and c) and the $\lambda/2$-separated antenna elements (d), the rotations of two dipoles on plane y-z can provide different performances, as shown in Figure 4.20b–d. The channel performance is maximized in Figure 4.20b when the polarizations of receiving antennas are well matched, and

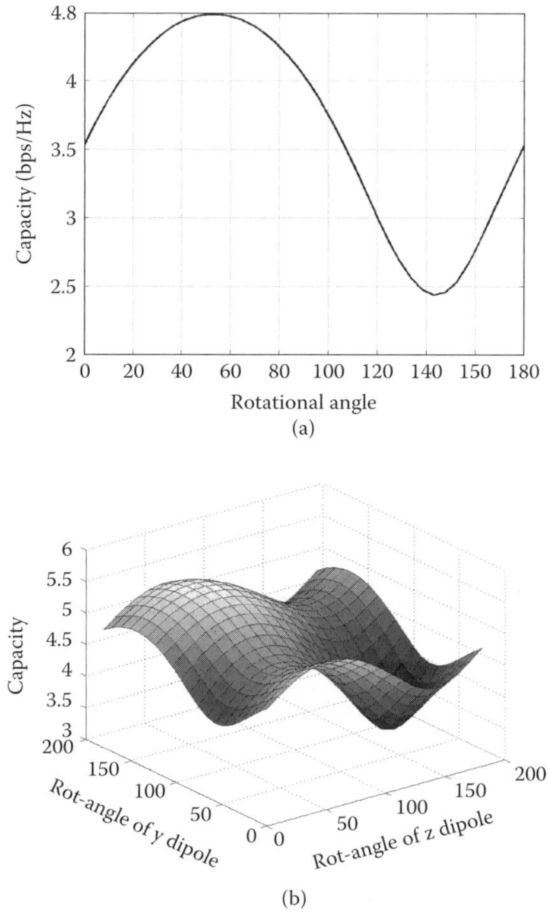

FIGURE 4.20 Performances of (a) 1 × 1 SISO, (b) 1 × 2 SIMO, (c) 2 × 2 MIMO as the same antenna position, and (d) 2 × 2 MIMO as the separated antenna position while the receiving antenna is rotating on plane *y-z*.

is minimized if they are mismatched. In Figure 4.20c, when the antennas have been rotated to nearly the same position, high correlation is produced. Subsequently, the channel capacity is reduced even if the polarizations are correctly matched. However, it has better performance in the situation of polarization diversity when the antenna rotation has the difference of 90°. Moreover, we observe the MIMO capacity with spatial diversity as illustrated in Figure 4.20d, where the antenna correlation is reduced and the channel capacity is improved.

In order to achieve a better transmission performance, the polarized antenna selection can exploit together all diversity techniques, such as pattern, spatial, and polarization diversity. Pattern diversity should be employed when a large angle spread is detected. Spatial diversity could be exploited when a high antenna correlation is observed.

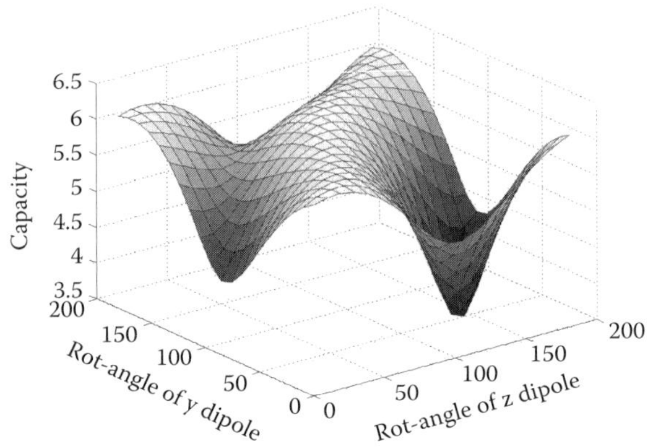

(c)

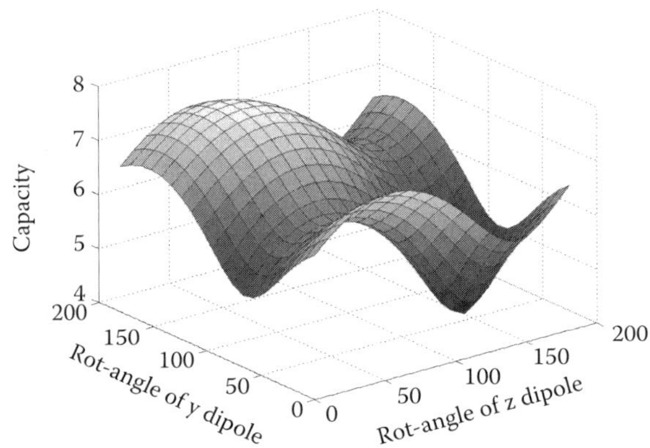

(d)

FIGURE 4.20 (continued)

Polarization diversity should be used when a low XPD occurs. Therefore, the employed diversity should be properly selected according to the propagation environment.

4.6 MIMO Applications

4.6.1 Wireless LAN-Based MIMO

In this chapter it was shown that employing MIMO systems could achieve higher performance in data transmission. MIMO signaling can increase network bandwidth, range, and reliability. Recently, many communication systems began to take advantage of this channel capacity enhancement, such as wireless local area network (WLAN) and wireless metropolitan area network (WMAN).

The IEEE 802.11 WLAN and IEEE 802.16 WMAN standards are based on orthogonal frequency division multiplexing (OFDM). OFDM is a multicarrier modulation system, reducing the required bandwidth but keeping the modulated signals orthogonal so they do not interfere with each other. An important high-data-transmission-rate extension of these standards could be based on MIMO. An advantage of these systems is that they are principally deployed in indoor environments and suburban environments that are characterized by a rich multipath.

There are also some motivations in order to improve the performance and transmission rate in MIMO-OFDM systems when the CSI is available at the transmitter. As mentioned in previous sections, the water-filling technique is used to optimize the distribution of the total transmit power over transmit antennas. Therefore, information symbols and power could be optimally allocated over space and frequency in MIMO-OFDM communication.

The number of antennas utilized in a MIMO (WLAN) router, for example, can vary; a typical MIMO router contains three or four antennas. These systems are driving the need for the next broadband revolution focused on home networking. Such systems will cover next-generation game consoles, video on demand, HDTV, and other new products. These services are creating a more sophisticated home entertainment environment, together with a high level of quality of service (QoS) to facilitate multimedia connectivity.

The intelligence behind the antenna polarization is described as adaptive polarization. The router receives feedback from the client adapter and has the ability to focus the polarization of the signals. As signal travels between an access point and wireless card, it will bounce off of walls, ceilings, and any other obstacle, resulting in multiple reflections of the original signal arriving by different paths and different polarizations. By applying adaptive antenna polarization algorithms (AMPAS), these reflections can be used to improve the signal-to-noise ratio, instead of having just one copy of the original signal.

This adaptive transmitting feature provides a more reliable signal at extreme ranges. Moreover, MIMO can eliminate dead spots, delivering reliable whole-home coverage with all the speed you need for application in the future. Today, one can say, wireless is faster than leased wire systems.

Finally, MIMO networking has the potential to increase communications data rates by 10–20 times above those of current systems. Such systems will use multipath reflections to create parallel channels in the same frequency bandwidth, thereby increasing spectral efficiency. In the next section sensor networks will be discussed as an application of a MIMO-based system.

4.6.2 MIMO for Cooperative Sensor Networks

A sensor network can be considered a self-contained circuit with its sensor and RF interface, as shown in Figure 4.21. Recent hardware advances allow more signal processing functionality to be integrated into a single chip. For example, it is possible to integrate an RF transceiver, sensor interface, and baseband processors into one device that is as small as a piece of coin and can be used as a fully functional wireless sensor node. Such wireless nodes typically operate with small batteries; that is why these sensor nodes have limited power capabilities. In many scenarios, the wireless nodes must operate without battery replacement for many years. Consequently, minimizing the energy consumption

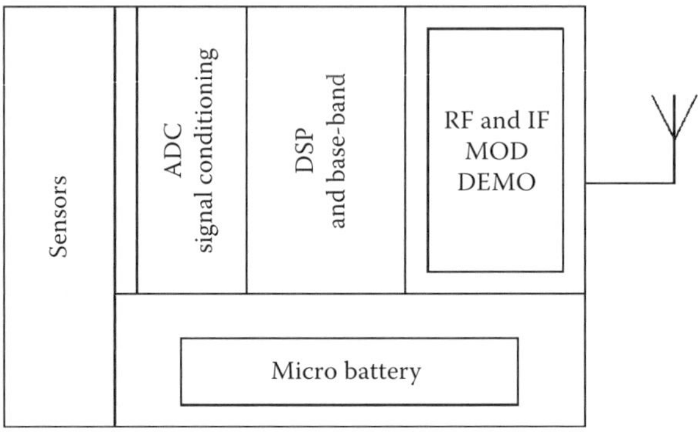

FIGURE 4.21 Typical wireless sensor node.

is a very important design consideration, and energy-efficient transmission schemes must be used for the data transfer in sensor networks.

In addition, because sensor nodes will be deployed in remote and oftentimes dangerous locations, their maintenance (in particular, battery replacement) will be unlikely [84].

Sensor networks are a new attraction for many potential applications, such as industrial, military, geolocalization, surveillance, intrusion detection, and environmental monitoring [84, 85].

Robust communications between sensor nodes are highly demanded at low power. As was shown, MIMO communication promises performance enhancements over conventional single-input single-output (SISO) technology without increasing the bandwidth consumed by the system or the total power radiated from a transmitter. MIMO technology has promising characteristics that make it a serious candidate for sensor network communication technology. Signal processing techniques that use multiple transmit and receive antennas, such as space-time coding (ST coding), have been shown to increase transmission reliability.

In a surveillance application, the ability of sensor nodes to relay data is critical to the utility and effectiveness of the sensor network.

For a given node density, nodes are more likely to be out of range, thus inhibiting communication. In a situation such as this, the extended range of MIMO is of greater importance because it enables cohesion (the ability of the sensor nodes to form a completely connected network), which guarantees the success of the final application [86].

New protocols for target reporting and a procedure for target localization that conserve energy have recently been developed [86, 87]. In [88], the authors summarize and compare several routing MIMO technologies.

Mean path length provides a measurement of the impact of MIMO communications on a wireless sensor network. Mean path length provides a rough estimate of the amount of time and energy expended in a data transmission from one node to another in the network.

Most significant mean path length reduction is provided by MIMO in the low or mid range of node densities because the internode spacing is such that MIMO can reliably form some links that SISO cannot. In the elongated region scenario this trend holds, though it is less apparent [89].

There is also an increasing need for mobile networks with distributed transmitters and receivers, typically referred to as mobile ad hoc networks (MANETs). There, transmitters and receivers do not pool their information together, either due to geographical dispersiveness, the bandwidth and resource limitation, or due to security/privacy concerns.

Recognizing that multiple antennas at the transceivers provide inherent multiplexing capability due to their spatial selectivity, it is attractive to study MIMO communication in ad hoc networks with interference transmission.

Energy-efficient communication techniques typically focus on minimizing the transmission energy only, which is reasonable in long-range applications where the transmission energy is dominant in the total energy consumption.

In cooperative sensor networks, we allow the cooperation among sensors for information transmission or reception, so that energy consumption as well as transmission delays over some distance ranges can be reduced.

In conclusion, for the same throughput requirement, MIMO systems require less transmission energy than SISO systems. However, direct application of multiantenna techniques to sensor networks is impractical due to the limited physical size of a sensor node, which typically can only support a single antenna. If individual single-antenna nodes allowed cooperating on information transmission or reception, a cooperative MIMO system can be constructed such that energy-efficient MIMO schemes can be deployed [90]. Finally, MIMO can provide significant network performance improvements in power consumption, latency, and network robustness.

References

[1] E. Telatar. 1999. Capacity of multi-antenna Gaussian channels. *Eur. Trans. Telecommun.* 10:585–95.

[2] P. Bello. 1963. Characterization of randomly time-variant linear channels. *IEEE Trans. Commun. Syst.* 11:360–93.

[3] J. D. Parsons. 2000. *The mobile radio propagation channel.* New York: Wiley.

[4] G. Matz. 2005. On non-WSSUS wireless fading channels. *IEEE Trans. Wireless Commun.* 4:2465–78.

[5] J. W. McJown and R. L. Hamilton. 1991. Ray tracing as a design tool for radio networks. *IEEE Network Mag. 5(6):*27–30.

[6] T. Kurner, D. J. Cichon, and W. Wiesbeck. 1993. Concepts and results for 3D digital terrain-based wave propagation models: An overview. *IEEE J. Selected Areas Commun.* 11:1002–12.

[7] R. Kouyoumjian and P. Pathak. 1974. A uniform geometrical theory of diffraction for an edge in a perfectly conducting surface. *Proc. IEEE* 62:1448–61.

[8] A. R. N. G. E. Athanasiadou and J. P. McGeehan. 2000. A microcellular ray-tracing propagation model and evaluation of its narrowband and wideband. *IEEE J. Selected Areas Commun.* 18:322–35.

[9] P. Petrus, J. Reed, and T. Rappaport. 2002. Geometrical-based statistical macrocell channel model for mobile environments. *IEEE Trans. Commun.* 50:495–502.

[10] J. C. Liberti and T. Rappaport. 1996. A geometrically based model for line-of-sight multipath radio channels. In *Proceedings of VTC*, pp. 844–48.

[11] C. Oestges, V. Erceg, and A. Paulraj. 2003. A physical scattering model for MIMO macro-cellular broadband wireless channels. *IEEE J. Selected Areas Commun.* 21:721–29.

[12] A. F. Molisch, A. Kuchar, J. Laurila, K. Hugl, and R. Schmalenberger. 2003. Geometry-based directional model for mobile radio channels—Principles and implementation. *Eur. Trans. Telecommun.* 14:351–59.

[13] K. Raoof and N. Prayongpun. 2005. Channel capacity performance for MIMO polarized diversity systems. In *Proceedings of WCNM*, pp. 1–4.

[14] G. J. Foschini and M. J. Gans. 1998. On limits of wireless communications in a fading environment when using multiple antennas. *Wireless Personal Commun.* 6:311–35.

[15] M. A. Khalighi, K. Raoof, and G. Jourdain. 2002. Capacity of wireless communication systems employing antenna arrays, a tutorial study. *Wireless Personal Commun.* 23:321–52.

[16] D. Shiu, G. J. Foschini, M. J. Gans, and J. M. Kahn. 2000. Fading correlation and its effect on the capacity of multi-element antenna systems. *IEEE Trans. Commun.* 48:502–13.

[17] J. Kermoal et al. 2002. A stochastic MIMO radio channel model with experimental validation. *IEEE J. Selected Areas Commun.* 20:1211–26.

[18] H. Özcelik et al. 2003. Deficiencies of the "Kronecker" MIMO radio channel model. In *Elec. Lett.* 39(16):1209–1210.

[19] A. Saleh and R. Valenzuela. 1987. A statistical model for indoor multipath propagation. *IEEE J. Selected Areas Commun.* 5:128–37.

[20] D. Gesbert et al. 2002. Outdoor MIMO wireless channels: Models and performance prediction. *IEEE Trans. Commun.* 50:1926–34.

[21] A. Burr. 2003. Capacity bounds and estimates for the finite scatterers MIMO wireless channel. *IEEE J. Selected Areas Commun.* 21:812–18.

[22] M. Debbah and R. Muller. 2005. MIMO channel modeling and the principle of maximum entropy. *IEEE Trans. Inf. Theory* 51(5).

[23] A. Sayeed. 2002. Deconstructing multiantenna fading channels. *IEEE Trans. Signal Processing* 50:2563–79.

[24] L. M. Correia. 2001. *Wireless flexible personalised communication*. COST 259 Final Report. New York: Wiley.

[25] L. M. Correia. 2006. *Mobile broadband multimedia networks*. COST 273 Final Report. London: Elsevier.

[26] 3GPP. 2003. *Spatial channel model for multiple input multiple output (MIMO) simulations*. 3GPP2 TR 25.996.

[27] G. J. Foschini. 1996. Layered space-time architecture for wireless communication in a fading environment when using multi-element antennas. *Bell Labs Tech. J.* 1:41–59.

[28] G. D. Golden, G. J. Foschini, R. A. Valenzuela, and P. W. Wolniansky. 1999. Detection algorithm and initial laboratory results using V-BLAST space-time communication architecture. *Electronic Lett.* 35:14–16.
[29] G. G. Raleigh and V. K. Jones. 1999. Multivariate modulation and coding for wireless communication. *IEEE J. Selected Areas Commun.* 17:851–66.
[30] E. Biglieri, J. Proakis, and S. S. Shitz. 1998. Fading channels, information-theoretic and communications aspects. *IEEE Trans. Inf. Theory* 44:2619–92.
[31] W. C. Jakes. 1998. *Microwave mobile communications.* New York: John Wiley & Sons.
[32] H. Hashemi. 1993. The indoor radio propagation channel. *Proc. IEEE* 81:943–68.
[33] M. A. Khalighi, J. M. Brossier, G. Jourdain, and K. Raoof. 2001. Water filling capacity of Rayleigh MIMO channels. In *Proceedings of PIMRC*, San Diego, vol. A, pp. 155–58.
[34] J. B. Anderson. 2000. Array gain and capacity for known random channels with multiple element arrays at both ends. *IEEE J. Selected Areas Commun.* 18:2172–78.
[35] G. G. Raleigh and J. M. Cioffi. 1998. Spatio-temporal coding for wireless communication. *IEEE Trans. Commun.* 46:357–66.
[36] T. L. Marzetta and B. M. Hochwald. 1999. Capacity of a mobile multiple-antenna communication link in Rayleigh flat fading. *IEEE Trans. Inf. Theory* 45:139–57.
[37] L. Zheng and D. N. C. Tse. 2002. Communication on the Grassmann manifold: A geometric approach to the noncoherent multiple-antenna channel. *IEEE Trans. Inf. Theory* 48:359–83.
[38] B. Sklar. 1997. Rayleigh fading channels in mobile digital communication systems. Part I. Characterization. Part II. Mitigation. *IEEE Commun. Mag.* 35:90–109.
[39] J. Salz and J. H. Winters. 1994. Effect of fading correlation on adaptive arrays in digital mobile radio. *IEEE Trans. Veh. Technol.* 43:1049–57.
[40] P. F. Driessen and G. J. Foschini. 1999. On the capacity formula for multiple input-multiple output wireless channels: A geometric interpretation. *IEEE Trans. Commun.* 47:173–76.
[41] M. A. Khalighi, J. M. Brossier, G. Jourdain, and K. Raoof. 2001. On capacity of Ricean MIMO channels. In *Proceedings of PIMRC*, San Diego, vol. A, pp. 150–54.
[42] C. Berrou and A. Glavieux. 1996. Near optimum error correcting coding and decoding: Turbo-codes. *IEEE Trans. Commun.* 44:1261–71.
[43] C. Douillard, M. Jézéquel, C. Berrou, A. Picart, P. Didier, and A. Glavieux. 1995. Iterative correction of intersymbol interference: Turbo-equalization. *Eur. Trans. Telecommun.* 6:507–11.
[44] M. A. Khalighi and J. J. Boutros. 2006. Semi-blind channel estimation using EM algorithm in iterative MIMO APP detectors. *IEEE Trans. Wireless Commun.* 5:3165–73.
[45] N. Noels, C. Herzet, A. Dejonghe, V. Lottici, H. Steendam, M. Moeneclaey, M. Luise, and L. Vandendorpe. 2003. Turbo synchronization: An EM algorithm interpretation. In *Proceedings of ICC*, Anchorage, AK, vol. 4, pp. 2933–37.
[46] X. Wang and H. V. Poor. 1999. Iterative (turbo) soft interference cancellation and decoding for coded CDMA. *IEEE Trans. Commun.* 47:1046–61.

[47] H. El Gamal and E. Geraniotis. 2000. Iterative multiuser detection for coded CDMA signals in AWGN and fading channels. *IEEE J. Selected Areas Commun.* 18:30–41.

[48] M. Sellathurai and S. Haykin. 2002. Turbo-BLAST for wireless communications: Theory and experiments. *IEEE Trans. Signal Processing* 50:2538–46.

[49] B. Vucetic and J. Yuan. 2003. *Space-time coding.* Chichester, England: John Wiley & Sons Ltd.

[50] V. Tarokh, H. Jafarkhani, and A. R. Calderbank. 1999. Space-time block codes from orthogonal designs. *IEEE Trans. Inf. Theory* 45:1456–67.

[51] S. M. Alamouti. 1998. A simple transmit diversity technique for wireless communications. *IEEE J. Selected Areas Commun.* 16:1451–58.

[52] V. Tarokh, N. Seshadri, and A. R. Calderbank. 1998. Space-time codes for high data rate wireless communication: Performance criterion and code construction. *IEEE Trans. Inf. Theory* 44:744–65.

[53] Y. Xin, Z. Wang, and G. B. Giannakis. 2003. Space-time diversity systems based on linear constellation precoding. *IEEE Trans. Wireless Commun.* 2:294–309.

[54] B. Hassibi and B. M. Hochwald. 2002. High-rate codes that are linear in space and time. *IEEE Trans. Inf. Theory* 48:1804–24.

[55] E. Viterbo and J. J. Boutros. 1999. A universal lattice code decoder for fading channels. *IEEE Trans. Inf. Theory* 45:1639–42.

[56] G. Caire, G. Taricco, and E. Biglieri. 1998. Bit-interleaved coded modulation. *IEEE Trans. Inf. Theory* 44:927–46.

[57] L. Bahl, J. Cocke, F. Jelinek, and J. Raviv. 1974. Optimal decoding of linear codes for minimizing symbol error rate. *IEEE Trans. Inf. Theory* 20:284–87.

[58] J. Hagenauer, E. Offer, and L. Papke. 1996. Iterative decoding of binary block and convolutional codes. *IEEE Trans. Inf. Theory* 42:429–45.

[59] P. Robertson, P. Hoeher, and E. Villebrun. 1997. Optimal and sub-optimal maximum a posteriori algorithms suitable for turbo decoding. *Eur. Trans. Telecommun.* 8:119–25.

[60] J. J. Boutros, N. Gresset, and L. Brunel. 2003. Turbo coding and decoding for multiple antenna channels. In *Proceedings of the International Symposium on Turbo Codes and Related Topics*, Brest, France, pp. 1–8..

[61] B. Hassibi and H. Vikalo. 2005. On the sphere-decoding algorithm I. Expected complexity. *IEEE Trans. Signal Processing* 53:2806–18.

[62] M. A. Khalighi and J. J. Boutros. 2005. Channel estimation in turbo-BLAST detectors using EM algorithm. In *Proceedings of ICASSP*, Philadelphia, vol. III, pp. 1037–40.

[63] M. A. Khalighi, J. J. Boutros, and J.-F. Hélard. 2006. Data-aided channel estimation for turbo-PIC MIMO detectors. *IEEE Commun. Lett.* 10:350–52.

[64] M. A. Khalighi and J.-F. Hélard. 2005. Should MIMO orthogonal space-time coding be preferred to non-orthogonal coding with iterative detection? In *Proceedings of ISSPIT*, Athens, pp. 340–45.

[65] M. A. Khalighi, J.-F. Hélard, and S. Bourennane. 2006. Choice of appropriate space-time coding scheme for MIMO systems employing channel coding under BICM. In *Proceedings of SPAWC*, Cannes, France, pp. 1–5.

[66] M. A. Khalighi, J.-F. Hélard, and S. Bourennane. 2006. Contrasting orthogonal and non-orthogonal space-time schemes for perfectly-known and estimated MIMO channels. In *Proceedings of ICCS*, Singapore, pp. 1–5.

[67] V. Le Nir, J.-M. Auffray, M. Hélard, J.-F. Hélard, and R. Le Gouable. 2003. Combination of space-time block coding with MC-CDMA technique for MIMO systems with two, three and four transmit antennas. In *Proceedings of the IST Mobile Communications Summit Conference*, Aveiro, Portugal.

[68] J.-C. Belfiore, G. Rekaya, and E. Viterbo. 2005. The golden code: A 2×2 full-rate space-time code with nonvanishing determinants. *IEEE Trans. Inf. Theory* 51:1432–36.

[69] R. G. Vaughan. 1998. Beam spacing for angle diversity. In *Proceedings of Globecom*, vol. 2, pp. 928–33.

[70] P. Kyritsi et al. 2002. Effect of antenna polarization on the capacity of a multiple element system in an indoor environment. *IEEE J. Selected Areas Commun.* 20:1227–39.

[71] V. Eiceg, H. Sampath, and Catreux-Erceg, S. 2006. Dual-polarization versus single-polarization MIMO channel measurement results and modeling. *IEEE Trans. Wireless Commun.* 5:28–33.

[72] T. Svantesson. 2000. On the potential of multimode antenna diversity. In *Proceedings of VTC*, pp. 2368–2372.

[73] M. A. J. T. Svantesson and J. W. Wallace. 2004. Analysis of electromagnetic field polarizations in multiantenna systems. *IEEE Trans. Wireless Commun.* 3:641–46.

[74] J. B. Andersen and B. N. Getu. 2002. The MIMO cube—A compact MIMO antenna. *Wireless Personal Multimedia Commun.* 1:112–14.

[75] C. A. Balanis. 1997. *Antenna theory.* New York: John Wiley & Sons.

[76] P. Soma, D. S. Baum, V. Ercegl, R. Krishnamoorthyl, and A. J. Paulraj. 2002. Analysis and modeling of multiple-input multiple-output (MIMO) radio channel based on outdoor measurements conducted at 2.5 GHz for fixed BWA applications. In *Proceedings of ICC*, pp. 272–76.

[77] R. G. Vaughan. 1990. Polarization diversity in mobile communications. *IEEE Trans. Veh. Technol.* 39(3).

[78] J. Lempiainen, J. K. Laiho-Steffens, and A. F. Wacker. 1997. Experimental results of cross polarization discrimination and signal correlation values for a polarization diversity scheme. In *Proceedings of VTC*, pp. 1498–502.

[79] T. Svantesson. 2002. Correlation and channel capacity of MIMO systems employing multi-mode antennas. *IEEE Trans. Veh. Technol.* 51(6).

[80] A. F. Molisch. 2004. A generic model for the MIMO wireless propagation channels in macro- and microcells. *IEEE Trans. Signal Processing* 52:61–71.

[81] S. K. Yong and J. S. Thompson. 2003. A three-dimensional spatial fading correlation model for uniform rectangular arrays. *IEEE Antennas Wireless Propagation Lett.* 2.

[82] M. R. Andrews, P. P. Mitra, and R. de Carvalho. 2001. Tripling the capacity of wireless communication using electromagnetic polarization. *Nature* 409:316–18.

[83] N. Prayongpun and K. Raoof. 2006. MIMO channel capacities in presence of polarization diversity with and without line-of-sight path. *J. WSEAS Trans. Commun.* 5:1744–50.

[84] I. E. Akyildiz et al. 2002. A survey on sensor networks. *IEEE Commun. Mag. 40(8):*102–14.

[85] C. Y. Chong and S. P. Kumar. 2003. Sensor networks: Evolution, opportunities, and challenges. *Proc. IEEE* 91:1247–56.

[86] J. Burdin and J. Dunyak. 2006. *Cohesion of wireless sensor networks with MIMO communications.* Technical report. MITRE Corp.

[87] L. Godard, J. Lienard, and K. Raoof. 2006. *Localisation des objets communicants.* Technical report. Laboratoire LIS.

[88] Q. Jiang and D. Manivarman. 2004. Routing protocols for sensor networks. In *Consumer Communications and Networking Conference*, pp. 93–98.

[89] J. Burdin and J. Dunyak. 2006. *Enhancing the performance of wireless sensor networks with MIMO communications.* Technical report. MITRE Corp.

[90] S. Cui, A. J. Goldsmith, and A. Bahai. 2004. Energy-efficiency of MIMO and cooperative MIMO techniques in sensor networks. *IEEE J. Selected Areas Commun 22(6):1089–1098.*

5

Adaptive Modeling and Identification of Nonlinear MIMO Channels Using Neural Networks

Mohamed Ibnkahla
Queen's University

Al-Mukhtar Al-Hinai
Queen's University

5.1 Introduction

Multiple-input multiple-output (MIMO) systems have been well looked at and investigated as a means of increasing channel capacity. The MIMO concept can offer significantly increased high data rate and capacity, with no additional bandwidth, when the channel exhibits rich scattering and its variations can be accurately tracked [1].

In order to achieve high data rates and fulfill the power requirement at the same time, MIMO communication systems may be equipped with high-power amplifiers (HPAs) [11, 14, 15, 18]. However, HPAs introduce nonlinearity to the system when operating near their nonlinear saturation regions [12]. This nonlinearity may restrict the communication

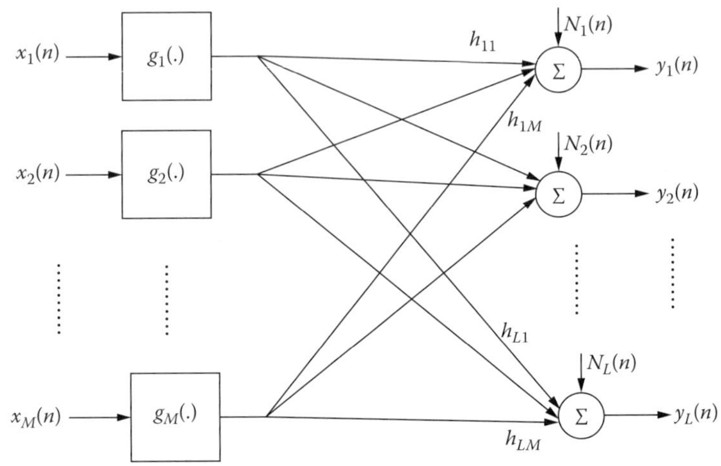

FIGURE 5.1 Nonlinear MIMO system.

system to low-order modulation schemes. In addition to the nonlinear behavior of the HPAs, a further challenge lies in the variations of the wireless fading channel [5, 17, 18].

Therefore, to reach maximum throughput, knowledge of accurate and timely channel state information (CSI) is needed by the receiver for accurate detection and demodulation.

Modeling the channel time-varying parameters and the HPA nonlinearity [5, 12] is a highly challenging task, especially when both the nonlinearity and fading parameters are unknown.

The nonlinear MIMO channel studied in this chapter is depicted in Figure 5.1. It is composed of M inputs, M memoryless nonlinearities (representing the HPAs), a linear combiner H (representing the propagation channel), and L outputs (representing the receiving antennas).

The chapter proposes a block-oriented neural network (NN) approach [4, 6] to adaptively identify the overall MIMO input-output transfer function and characterize each component of the system (i.e., the memoryless nonlinearities and the linear combiner). The proposed NN model is composed of a set of memoryless NN blocks followed by an adaptive linear combiner. Each block in the adaptive system aims at identifying the corresponding block in the unknown MIMO system.

The chapter is organized as follows. Section 5.2 presents the system to be identified and the NN algorithm. Section 5.3 presents some applications and simulation results.

5.2 System Model and Neural Network Algorithm

5.2.1 System Model

The nonlinear MIMO channel is presented in Figure 5.1. The system is assumed to be composed of M zero-mean uncorrelated inputs $x_i(n)$, $i = 1, \ldots, M$. Each input is nonlinearly transformed by a memoryless nonlinearity $g_i(.)$. The outputs of these nonlinearities are then linearly combined by an $L \times M$ matrix $H = [h_{ij}]$.

The jth output of the MIMO channel can be expressed as

$$y_j(n) = \sum_{i=1}^{M} h_{ji}(n) g_i(x_i(n)) + N_j(n), \tag{5.1}$$

where N_j is a white noise.

The system input-output relationship can be expressed in a matrix form as

$$\begin{bmatrix} y_1(n) \\ y_2(n) \\ \dots \\ y_L(n) \end{bmatrix} = H \times \begin{bmatrix} g_1(x_1(n)) \\ g_2(x_2(n)) \\ \dots \\ g_M(x_M(n)) \end{bmatrix} + \begin{bmatrix} N_1(n) \\ N_2(n) \\ \dots \\ N_L(n) \end{bmatrix}. \tag{5.2}$$

Matrix H is a propagation matrix [1] that may be time varying.

In our modeling approach, only the structure of the MIMO system is assumed known. That is, we know that the MIMO system is composed of linear memoryless blocks and a linear combining matrix, but we do not know what their values and behaviors are.

5.2.2 Neural Network Scheme

The neural network [2, 3, 5] used for modeling the nonlinear MIMO system is represented in Figure 5.2. It is composed of M neural network blocks. Each block k has a scalar input $x_k(n)$ ($k = 1, \dots, M$), N neurons, and a scalar output:

$$NN_k(n) = \sum_{i=1}^{N} c_{ki} f\left(a_{ki} x_k(n) + b_{ki}\right), \quad k = 1, \dots, M, \tag{5.3}$$

where f is the NN activation function (a sigmoid transform). a_{ki}, c_{ki}, b_{ki} represent, respectively, the input weight, bias term, and output weight of the ith neuron of the kth block. The output NN_k of the kth block is connected to the jth output of the system through weight w_{jk}. The system j^{th} output is then expressed as

$$s_j(n) = \sum_{k=1}^{M} w_{jk} NN_k(n), \quad j = 1, ,L. \tag{5.4}$$

Weights w_{jk} will be put in a matrix form: $W = [w_{jk}]$; $j = 1, \dots, L$; $k = 1, \dots, M$.

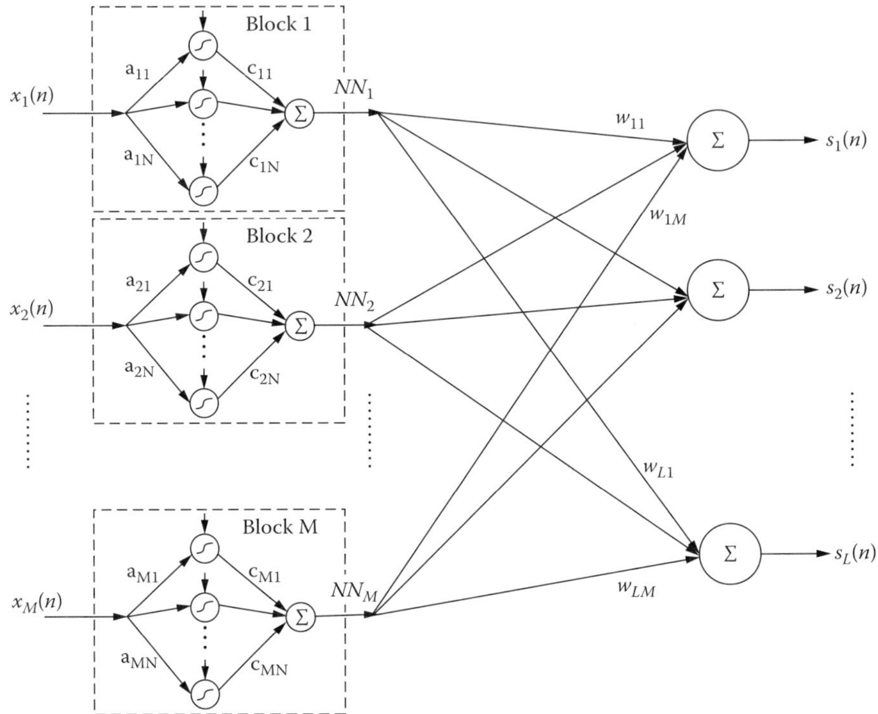

FIGURE 5.2 NN identification structure.

This can be expressed in a matrix form as

$$\begin{bmatrix} s_1(n) \\ s_2(n) \\ \dots \\ s_L(n) \end{bmatrix} = W \times \begin{bmatrix} NN_1(x_1(n)) \\ NN_2(x_2(n)) \\ \dots \\ NN_M(x_M(n)) \end{bmatrix}. \tag{5.5}$$

By choosing this structure, our goal is to characterize each part of the unknown system by the corresponding part in the modeling structure [4, 6]: each block $NN_k(.)$ would identify the corresponding nonlinearity $g_k(.)$, and matrix W would identify matrix H, to within a scaling factor for each transmitting antenna. The scaling factors can be determined by knowing the transmission antennas' average powers.

5.2.3 Learning Algorithm

In supervised learning [2, 3, 8, 9, 13], the unknown MIMO system and the proposed NN model are fed with the same input vector (Figure 5.3). At each iteration, the NN parameters are updated so that to minimize a cost function, which is defined in this

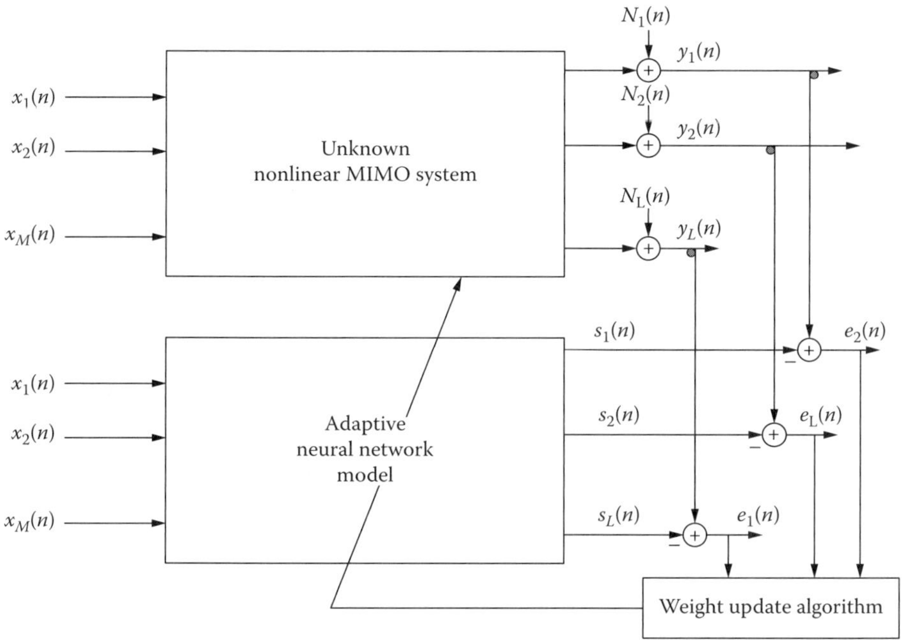

FIGURE 5.3 Adaptive system diagram.

chapter as the sum of the squared errors between the unknown system outputs and the corresponding outputs of the model:

$$J(n)=\sum_{j=1}^{L}\left(e_j(n)\right)^2, \tag{5.6}$$

where $e_j(n) = y_j(n) - s_j(n)$.

A gradient descent approach is used in this chapter for the updating equations:

$$w_{jk}(n+1)=w_{jk}(n)-\mu\frac{\partial J(n)}{\partial w_{jk}(n)}=w_{jk}(n)+2\mu e_j(n)NN_k(n), \tag{5.7}$$

$$c_{ki}(n+1)=c_{ki}(n)-\mu\frac{\partial J(n)}{\partial c_{ki}(n)}=c_{ki}(n)+2\mu f\left(a_{ki}x_k(n)+b_{ki}\right)\sum_{l=1}^{L}w_{lk}e_l(n), \tag{5.8}$$

$$a_{ki}(n+1)=a_{ki}(n)-\mu\frac{\partial J(n)}{\partial a_{ki}(n)}=a_{ki}(n)+2\mu c_{ki}x_k(n)f'\left(a_{ki}x_k(n)+b_{ki}\right)\sum_{l=1}^{L}w_{lk}e_l(n), \tag{5.9}$$

$$b_{ki}(n+1)=b_{ki}(n)-\mu\frac{\partial J(n)}{\partial b_{ki}(n)}=b_{ki}(n)+2\mu c_{ki}f'\left(a_{ki}x_k(n)+b_{ki}\right)\sum_{l=1}^{L}w_{lk}e_l(n), \quad (5.10)$$

where μ is a small positive constant.

The adaptive system is initialized with a set of small random-weight values. The learning curve is defined as the evolution of the mean squared error (MSE) during the learning process. In modeling fixed (i.e., static) MIMO systems, the MSE error starts to decrease until the system reaches a steady state and only slight changes in the weights are observed. In that case, the learning process may be stopped. However, if the MIMO system is time varying, then the learning process should be maintained in order to keep tracking the time-varying parameters.

We have chosen to compare our results to the classical multilayer perceptron (MLP) [2]. The MLP is composed of M inputs, a number of nonlinear neurons in the first layer, followed by L linear combiners (Figure 5.4).

The MLP is trained using the backpropagation algorithm [2, 3]. It allows a black-box modeling [8, 9, 13] of the unknown MIMO system, i.e., the modeling of the overall nonlinear MIMO input-output transfer function, without being able to characterize the different parts of the unknown system (i.e., the memoryless nonlinearities and the combining matrix H).

For a fair comparison between the MLP structure and the proposed NN block structure, we will take the same total number of neurons in each scheme. Therefore, for the block structure, if we denote the number of neurons in each block by N, then the total number of neurons will be equal to MN. This structure will be compared to an MLP with MN neurons.

Table 5.1 compares the computational complexity of each algorithm:

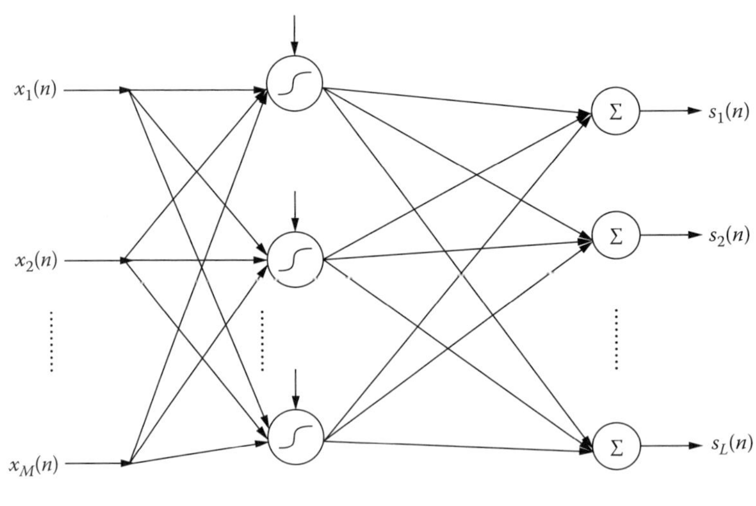

FIGURE 5.4 MLP structure.

TABLE 5.1

	Parameters	Sigmoid Transforms	Multiplications	Additions
Proposed NN	M(3N + L)	MN	M(2N + L)	M(2N + L + 1)
MLP	MN(M + L + 1)	MN	MN(M + L)	MN(M + L + 1)

The number of sigmoid transforms is the same, since we have the same number of neurons in the two structures. However, the MLP requires much more multiplications and additions. This is because in the MLP structure, all inputs are connected to all first-layer neurons (*MN* neurons in total), whereas in the block structure, each input is connected to one block that is composed of *N* neurons only.

5.3 Applications and Simulation Results

5.3.1 Modeling and Identification of MIMO Transmitters with Nonlinear Amplifiers and RF Coupling Interference

In this section, we apply the proposed NN for modeling and identification of MIMO RF transmitters that are equipped with nonlinear HPAs. Figure 5.5 shows the coupling terms generated by RF interference [10, 16]. In this application, the unknown nonlinear HPA transfer functions to be identified are taken from a family of nonlinear functions of the form

$$g_i(x) = \alpha_i x \exp\left(\frac{-\beta_i x^2}{2}\right),$$

where α_i and β_i are positive constants.

For the simulations presented in this section, we have taken the following parameters: $M = L = 2$, $\alpha_1 = \alpha_2 = 1$, $\beta_1 = 1$, $\beta_2 = 2$. For each output, the noise is taken as white Gaussian with variance σ_i^2. For the simulation, the inputs were zero-mean white Gaussian processes with unit variance.

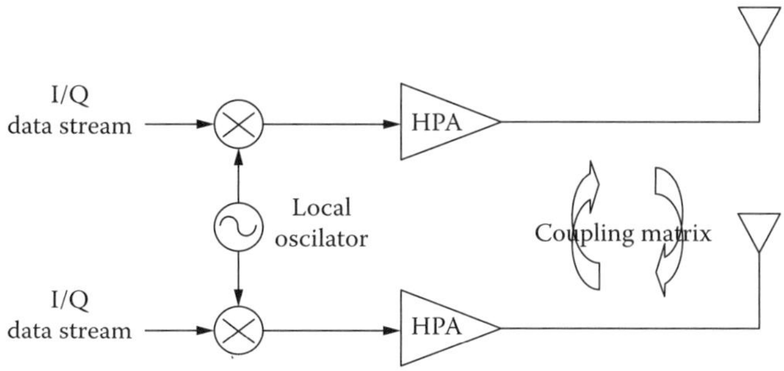

FIGURE 5.5 Example of a two-dimensional MIMO RF transmitter, including the coupling matrix between antennas.

TABLE 5.2

	Parameters	Sigmoid Transforms	Multiplications	Additions
Proposed NN	34	10	24	26
MLP	50	10	40	50

The coupling matrix simulated in this example was taken as

$$H = \begin{bmatrix} 1 & 0.3 \\ 0.3 & 1 \end{bmatrix}.$$

In this example, the proposed block-oriented NN scheme was composed of two blocks of $N = 5$ neurons each. The total number of neurons is then equal to 10. We have compared this scheme to an MLP structure composed of ten neurons. The *erf* function has been taken for the activation function.

The complexity of the algorithms is displayed in Table 5.2.

As can be seen in the table, the number of multiplications and additions in the MLP structure is almost twice that of the proposed block structure.

We have tested both algorithms for a range of μ values belonging to the interval $[10^{-5}\ 10^{-2}]$ under various initial conditions. The proposed block structure has always outperformed the MLP.

Figure 5.6 shows the learning curves for the block structure and the MLP for $\mu = 0.005$. The proposed approach shows lower MSE and faster convergence speed.

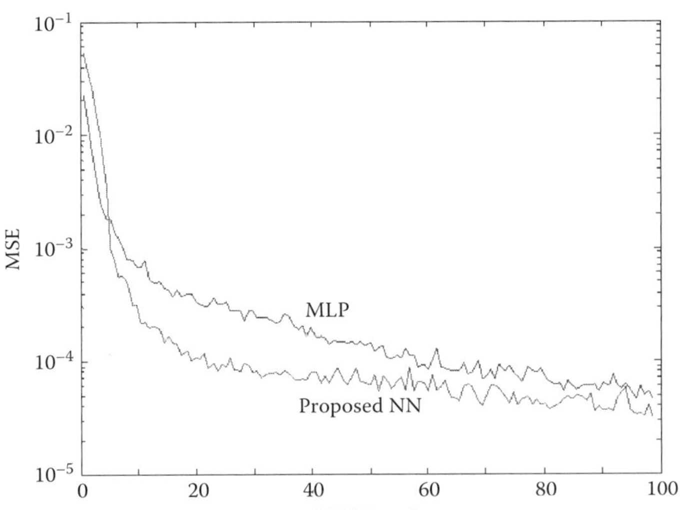

FIGURE 5.6 Smoothed MSE curves (an averaging window of 1,000 samples has been used). $\mu = 0.005$.

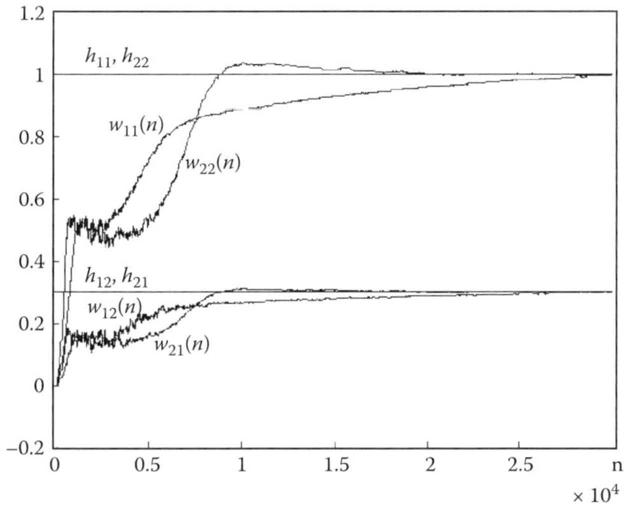

FIGURE 5.7 Evolution of the weights (normalized) during the learning process.

Moreover, the block structure allows excellent characterization of each of the nonlinearities as well as the coupling matrix H, to within scaling factors. That is, $NN_1(x)$ has converged to $\gamma_1 g_1(x)$, $NN_2(x)$ has converged to $\gamma_2 g_2(x)$, and W has converged to

$$\begin{bmatrix} \frac{1}{\gamma_1} & 0 \\ 0 & \frac{1}{\gamma_2} \end{bmatrix} \times H,$$

where γ_1 and γ_2 are real constants.

It is easy to show that if the output of block k (i.e., $NN_k(x_k)$) is multiplied by a given factor γ_k, and if the corresponding weights $\{w_{ki}\}$, $i = 1, \ldots, L$ are multiplied by the inverse of that factor, i.e., $1/\gamma_k$, then the overall NN system transfer function remains unchanged (equation (5.5)). This reveals that the adaptive system has several equivalent stationary points. The values of γ_k can be determined if the amplifiers' output average powers or operating points are given.

Figure 5.7 shows the transient behavior of matrix W weights (after normalization with the scaling factors). The scaling factors here were $\gamma_1 = -1.1246$ and $\gamma_2 = -0.9319$. Figures 5.8 and 5.9 show that the memoryless nonlinearities have been successfully modeled by our NN approach.

5.3.2 Application to Fault Detection

In this section, the proposed scheme is applied to fault detection in nonlinear MIMO systems.

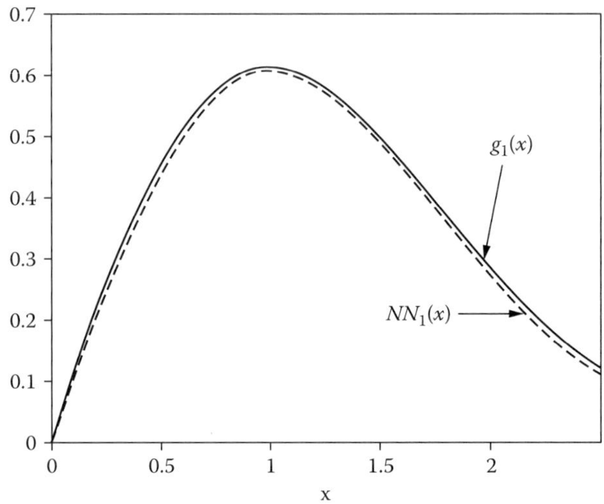

FIGURE 5.8 Unknown function $g_1(x)$ and normalized $NN_1(x)$.

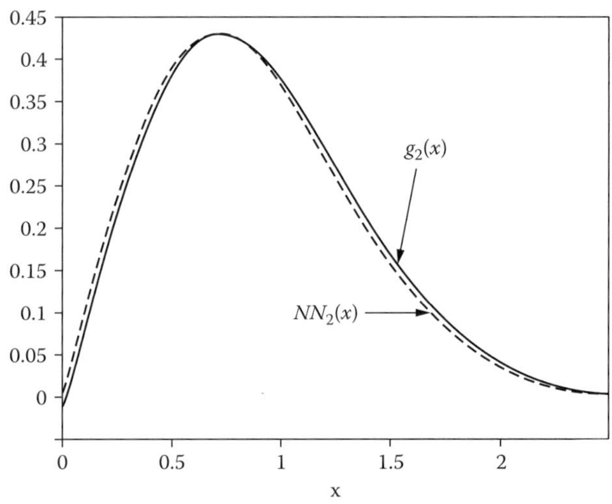

FIGURE 5.9 Unknown function $g_2(x)$ and normalized $NN_2(x)$.

Here we consider the case studied in section 5.3.1 (modeling MIMO RF transmitters). We have simulated an abrupt change in $g_1(x)$. The change occurred during the learning process at the 12,500th iteration. $g_2(x)$ and the other parameters of the MIMO system were unchanged.

The learning curves are shown in Figure 5.10 for the block NN structure and MLP. As expected, the MSE error increases at the time of change, then decreases. Here again, the block NN structure performs better than the MLP. Figure 5.11 shows that the block structure has correctly detected and characterized the change in $g_1(x)$. $g_2(x)$ and H (not

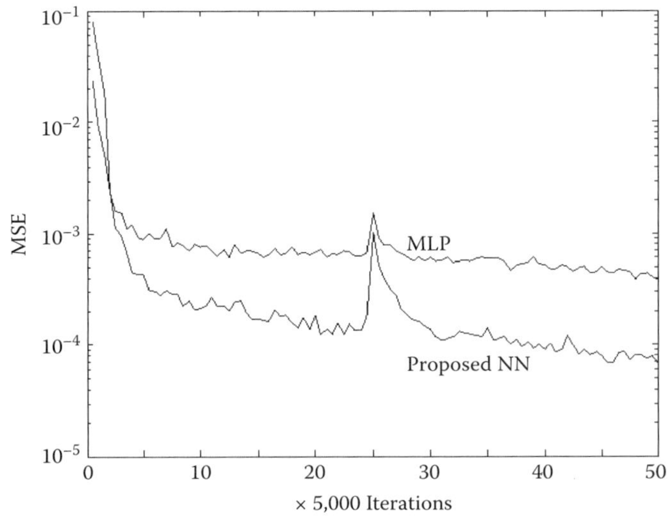

FIGURE 5.10 Fault detection: learning curves.

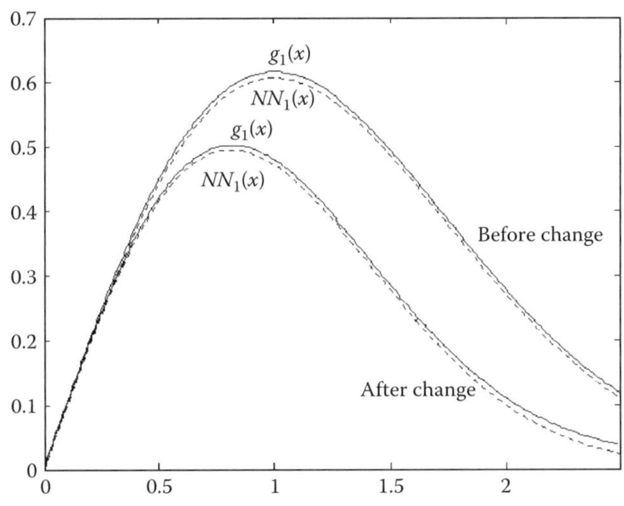

FIGURE 5.11 $g_1(x)$ and $NN_1(x)$ (normalized) before and after the change.

shown in the figure) were correctly identified as well. We have studied other MIMO systems with higher dimensions (3 × 3 and 4 × 4 systems) in which we simulated changes in more than one nonlinearity; the proposed NN model was always capable of characterizing the changes.

On the other hand, the black-box MLP structure could not characterize the change. By looking at the MLP learning curve (Figure 5.10), one could only predict that "something unusual" happened to the unknown system (or to the backpropagation algorithm) at iteration 12,500, without any further indication of what exactly happened during the learning process.

5.3.3 Tracking of Slowly Time-Varying Propagation Channels

This section deals with another application in which matrix H is slowly time varying.

In this application, we consider a multiple antenna transmitter/receiver system (Figure 5.1), with $M = L = 2$. The propagation channel H coefficients are modeled as slowly time-varying Ricean fading gains with a Ricean factor $K = 5$, and a normalized Doppler frequency of 0.0001.

Figure 5.12 shows the learning curves of the block structure and MLP. It can be noticed that the MSE errors here are higher than those of the static case studied in section 5.3.1 (Figure 5.6). This is because of the fact that gradient descent algorithms do not provide excellent tracking capabilities in time-varying environments.

As shown in Figure 5.12, our approach is faster and yields lower MSE than the MLP. Moreover, the proposed structure allows good identification of the unknown nonlinearities (Figures 5.13 and 5.14). The time-varying coefficients were correctly tracked by matrix W. An illustration is made for $w_{11}(n)$ versus $h_{11}(n)$ (Figure 5.15).

5.3.4 Nonlinear MIMO Channel Receiver Design

This section applies the NN identification scheme to MIMO receiver design [17, 18]. A V-BLAST (Vertical Bell Laboratories Layered Space-Time) receiver is proposed. It

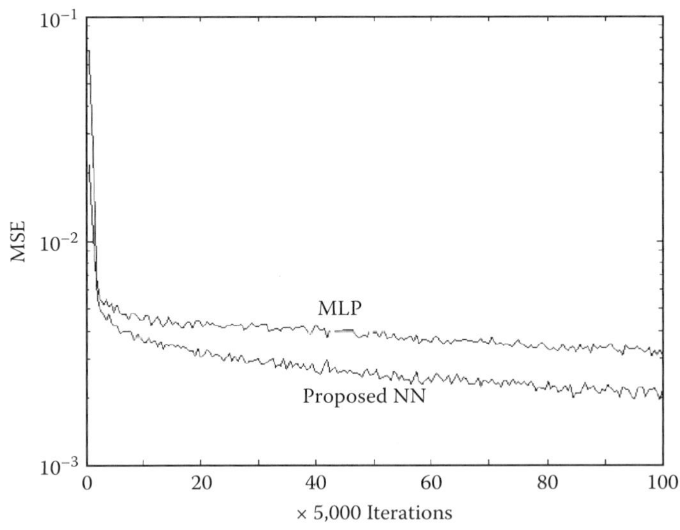

FIGURE 5.12 Learning curves: time-varying case.

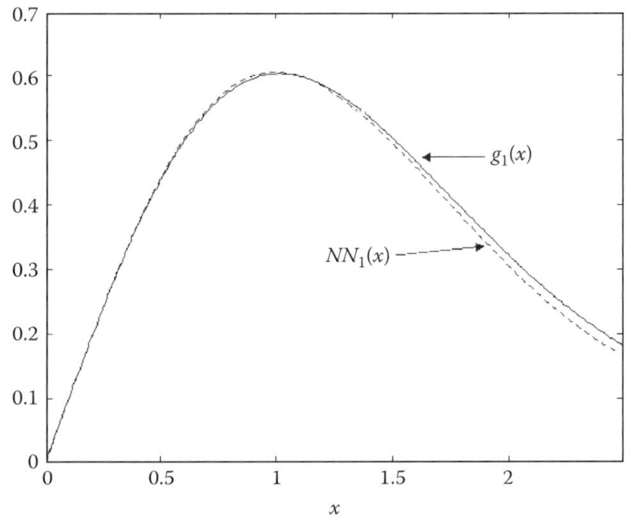

FIGURE 5.13 $g_1(x)$ and $NN_1(x)$ (normalized): time-varying case.

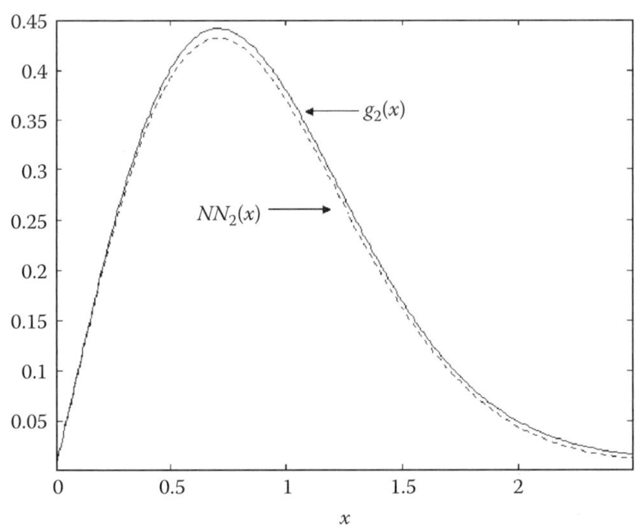

FIGURE 5.14 $g_2(x)$ and $NN_2(x)$ (normalized): time-varying case.

consists of a NN MIMO channel estimator (NNCE) and a ZF V-BLAST (zero-forcing V-BLAST) detection algorithm [1, 7, 11]. The NNCE performs an online estimation of the nonlinear MIMO channel. The estimated channel state information (CSI) is provided to the ZF V-BLAST detection algorithm, which gives an estimation of the transmitted symbols (Figure 5.16).

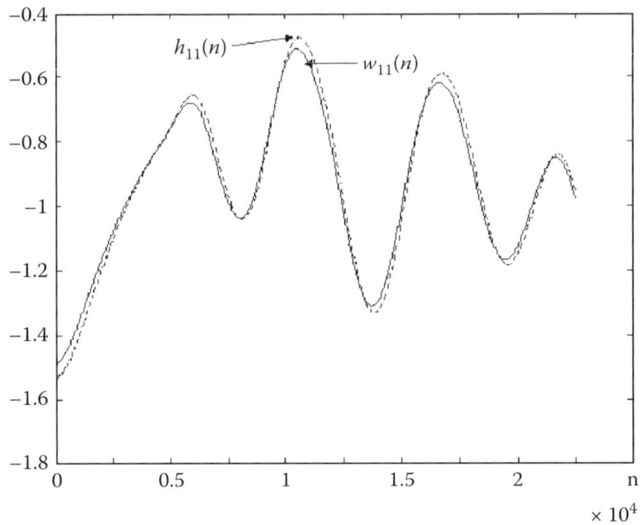

FIGURE 5.15 Identification of the time-varying channel: $h_{11}(n)$ and normalized $w_{11}(n)$.

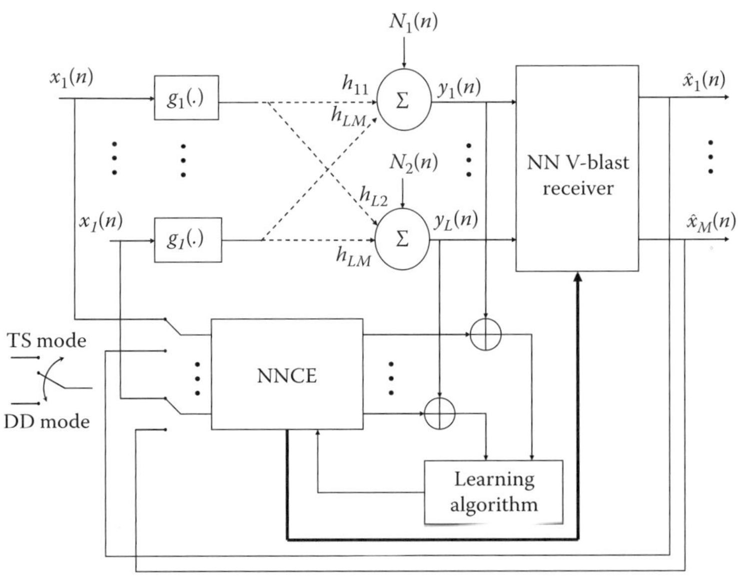

FIGURE 5.16 NN-based V-BLAST receiver.

The NNCE operation consists of two modes: training sequence (TS) mode and decision-directed (DD) mode. In the TS mode the amplitudes of the transmitted signals are used to train the NNCE. The TS mode continues until the NNCE is fully trained, then the NNCE switches to the DD mode. In the DD, the amplitudes of the detected signals ($\hat{x}_k(n)$, $k = 1, \ldots, M$) are used to train the NNCE.

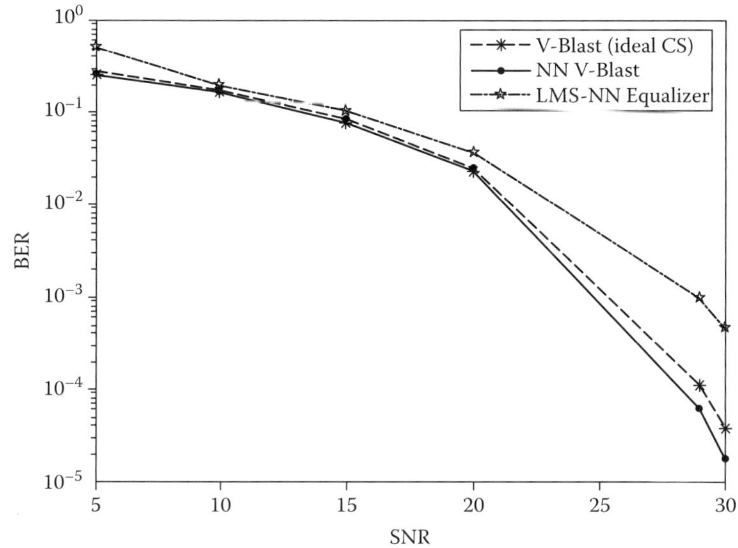

FIGURE 5.17 BER performance versus SNR.

Here the transmitted signals are 16-QAM modulated, and $M = L = 2$. The unknown propagation channel is modeled as time-varying Rician fading gains with a Rician factor $K = 5$ and normalized Doppler frequency of $f_D = 10^{-4}$. For the modeling and identification part, the noise is taken as white Gaussian with variance 0.001. However, for the detection part, the noise is varied depending on the channel signal-to-noise ratio (SNR). Each subnetwork in each block is composed of $N = 5$ neurons. It is assumed in the simulations that the HPA transmission powers are known to the receiver. The receiver is trained using a training sequence of fifty thousand transmitted symbols, after which the DD mode is activated.

Figure 5.17 shows the BER performance of our NN V-BLAST receiver. The proposed receiver achieves a bit error rate (BER) of 10^{-4} at an SNR of 29.2 dB. The NN V-BLAST receiver is compared to a least mean squares (LMS) tracker following a memoryless neural network (LMS-NN) equalizer [6]. It is composed of a linear combiner, W^{-1}, acting as an inverter to the channel matrix H, and a NN structure, acting as an inverter to the channel nonlinearities. The general structure of the LMS-NN equalizer is shown in Figure 5.18. BER simulation results show that our NN V-BLAST receiver outperforms the LMS-NN equalizer in terms of BER. We can also see that the performance of the NN V-BLAST receiver is close to that of the ideal V-BLAST receiver, which assumes perfect channel knowledge. This should not be surprising, since the different parts of the channel have been well identified.

5.4 Conclusion

The chapter proposed an NN approach for modeling nonlinear MIMO channels composed of a set of memoryless nonlinearities followed by a linear combining channel. We

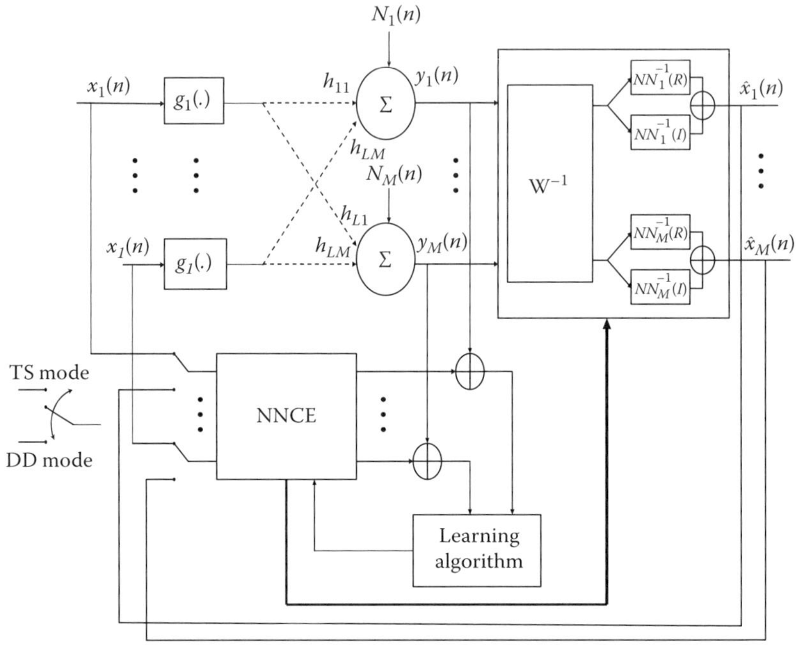

FIGURE 5.18 LMS-NN equalizer.

followed a block-oriented NN approach for the MIMO system modeling. The NN model allows excellent modeling and identification of the different parts of the unknown MIMO system; the learning process is being made by using the MIMO system input-output data. The proposed approach outperforms the black-box MLP method in terms of computational complexity, convergence speed, and MSE performance. The proposed scheme has been successfully applied to tracking of slowly time-varying MIMO channels, detection and characterization of changes in MIMO channels, and receiver design for nonlinear Rician MIMO channels.

References

[1] H. Bolcskei. 2004. MIMO systems: Principles and trends. In *Signal processing for mobile communications handbook*, ed. M. Ibnkahla, chap. 12. Boca Raton, FL: CRC Press.

[2] S. Haykin. 1999. *Neural networks: A comprehensive foundation*. Englewood Cliffs, NJ: Prentice Hall.

[3] S. Haykin. 1996. *Adaptive filter theory*. Englewood Cliffs, NJ: Prentice Hall.

[4] M. Ibnkahla, N. J. Bershad, J. Sombrin, and F. Castanié. 1998. Neural network modeling and identification of non-linear channels with memory: Algorithms, applications and analytic models. *IEEE Trans. Signal Processing* 46(5):1208–1220.

[5] M. Ibnkahla, Q. Rahman, A. Sulyman, H. Al-Asady, Y. Jun, and A. Safwat. 2004. High speed satellite mobile communications: Technologies and challenges. *Proc. IEEE* 92(2):312–39.

[6] M. Ibnkahla. 2000. Applications of neural networks to digital communications: A survey. *Signal Processing* 80:1185–215.

[7] T. Javornik, G. Kandus, S. Plevel, G. White, and A. Burr. 2003. V-BLAST algorithm performance in non-linear channel. In *IEEE Computer as a Tool Conference (EUROCON)*, vol. 1, pp. 183–87.

[8] L. Ljung. 1999. *System identification: Theory for the user.* Englewood Cliffs, NJ: Prentice Hall.

[9] K. S. Narendra and K. Parthasarathy. 1990. Identification and control of dynamical systems using neural networks. *IEEE Trans. Neural Networks* 1:4–27.

[10] J. Pedro and S. Maas. 2005. A comparative overview of microwave and wireless power amplifier behavioral modeling approaches. *IEEE Trans. Microwave Theory Tech.* 53:1150–63.

[11] G. Poitau and A. Kouki. 2004. Impact of realistic amplification models on dynamic VBLAST optimization. In *Proceedings of the IEEE Vehicular Technology Conference*, pp. 894–97.

[12] A. Saleh. 1981. Frequency-independent and frequency-dependent nonlinear models of TWT amplifiers. *IEEE Trans. Commun.* 29(11).

[13] J. Sjoberg et al. 1995. Nonlinear black box modeling in system identification: A unified overview. *Automatica* 31:1691–724.

[14] A. I. Sulyman and M. Ibnkahla. 2004. Performance analysis of nonlinearly amplified M-QAM signals in MIMO channels. In *Proceedings of IEEE ICASSP'04*, Montreal, Canada.

[15] A. I. Sulyman and M. Ibnkahla. 2005. Performance of space-time codes over nonlinear MIMO channels. In *Proceedings of the International Symposium on Signal Processing and Its Applications (ISSPA'05)*, Sydney, Australia.

[16] S. Woo, D. Lee, K. Kim, H. Hur, C. Lee, and J. Laskar. 2005. Combined effects of RF impairments in the future IEEE 802.11n WLAN systems. In *Proceedings of the IEEE Vehicular Technology Conference*, vol. 2, pp. 1346–49.

[17] S. Yang, J. Xi, and X. Mu. 2005. Decision aided joint compensation of clipping noise and nonlinearity for MIMO-OFDM systems. In *Proceedings of the IEEE International Symposium on Communications and Information Technology ISCIT 2005*, vol. 1, pp. 725–28.

[18] S. Yang, J. Xi, F. Wang, X. Mu, and H. Kobayashi. 2005. Decision aided compensation of residual frequency offset for MIMO-OFDM systems with nonlinear channel. In *Proceedings of the International Symposium on Intelligent Signal Processing and Communication Systems ISPACS 2005*, pp. 113–16.

6

Joint Adaptive Transmission and Switched Diversity Reception

Hong-Chuan Yang
University of Victoria

Young-Chai Ko
Korea University

Haewoon Nam
Motorola, Inc.

Mohamed-Slim Alouini
Texas A&M University (TAMU)-Qatar

6.1 Introduction

Future wireless communication systems have to provide multimedia services to battery-operated portable terminals. These systems must efficiently utilize the limited bandwidth and power resources to enable high-data-rate high-reliability transmission over wireless fading channels. Adaptive modulation and diversity combining are two of the most important enabling techniques for future wireless communication systems. Adaptive modulation can achieve high spectral efficiency over wireless channels [1–3]. The basic idea of adaptive modulation is to match the modulation parameters, such as constellation size and coding rate, to the prevailing fading channel conditions while maintaining the instantaneous error rate below a target value. Usually, the modulation mode is chosen based on the comparison results of received signal strength with several predetermined thresholds. Diversity combining, on the other hand, can improve the reliability of wireless fading channels by appropriately combining differently faded information-bearing signals [4]. While adaptive modulation can certainly benefit from diversity combining through the improved channel quality, the design and analysis of these two techniques were usually carried out separately.

From a performance perspective, the optimal combining solution is the well-known maximum ratio combining (MRC) [5, chapter 12]. With MRC, the combiner output is obtained by coherently summing up the optimally weighted signal replicas from all diversity paths. Meanwhile, the implementation complexity of MRC is high, especially when the number of diversity paths is large. Indeed, MRC not only requires the same number of RF chains as the number of available diversity paths, but also mandates the complete and simultaneous knowledge of the channel condition for each diversity path. To reduce the implementation complexity of the MRC diversity combiner, generalized selection combining (GSC) was proposed and extensively studied (see, for example, [6–11] and references therein). The receiver with GSC combines a fixed number of best diversity paths as per the rules of the optimal MRC scheme. As one can intuitively expect, diversity combining techniques such as MRC and GSC improve the spectral efficiency of adaptive transmission systems by allowing a higher-order transmission mode. On the other hand, these combining schemes operate rather independently of the adaptive transmission system.

Recently, there has been a growing interest in adaptive combining schemes for processing power-saving purposes [12–17]. The basic idea of these power-saving schemes is to use the diversity combiner resource adaptively in a way that the output signal satisfies a certain quality requirement. For example, minimum selection GSC (MS-GSC) was proposed by Kim et al. [12] as a power-saving implementation of the GSC scheme. With MS-GSC, the receiver combines the least number of best branches such that the combined SNR exceeds a certain predetermined threshold. It has been shown [13–15] that MS-GSC can save a considerable amount of processing power by keeping fewer branches active on average, while still providing nearly the same performance as MRC. Output-threshold MRC (OT-MRC) is another adaptive combining scheme [17], which can further reduce the number of path estimations by sequentially combining additional paths if necessary. It is interesting to note that both adaptive modulation and adaptive combining concepts utilize some predetermined threshold in their operation. Very recently, based on this observation, two joint adaptive modulation and diversity combining designs, where both the transmission mode and combiner structure are adaptively determined based on the fading channel condition, have been proposed and analyzed [18, 19].

We note at this point that both MRC- and GSC-based diversity combining schemes require the implementation of multiple RF chains. While this may not be an issue for conventional cellular or wireless LAN systems, it certainly becomes very challenging or even prohibitive for future generations of wireless personal area network (PAN) systems operating over 60 GHz bands with millimeter (MM)-wave technology, which are being currently standardized by the IEEE 802.15.3c group. Only until very recently, RF-CMOS technology was able to produce chips operating over the 60 GHz frequency range, but still with considerable high cost. On the other hand, implementing multiple MM-wave antennas at the receiver is achievable at a reasonable price. In the resulting multiple antennas with a single RF chain receiver scenario, only traditional selection or switched combining schemes can be employed to explore the diversity benefit. In this chapter, we complement the previous work on joint adaptive transmission and diversity combining by extending the design and analysis in [18, 19] to the lower-complexity selection and switched combining schemes. The proposed joint design can also apply to a transmit diversity system to achieve high efficiency with low feedback load and no power spreading loss.

The selection combining (SC) scheme [4, section 6.2] has much lower complexity than MRC. The receiver with SC uses only the best diversity paths of all available ones. This nevertheless requires the estimation and comparison of the quality indicators, such as the signal-to-noise ratio (SNR), of all diversity paths. With switched combining, where switch-and-stay combining (SSC) is the popular example [20–27], the complexity is further reduced by eliminating the need for simultaneously checking the path quality of all diversity paths. In fact, SSC, where the combiner switches to the other branch only when the output SNR of the current branch is below a certain threshold, can be viewed as the origin of aforementioned adaptive combining schemes such as MS-GSC and OT-MRC. Recent applications (such as transmit diversity, for example) have also motivated studies of multibranch switch-and-examine combining (SEC) [28, 29]. In particular, with SEC, if the current path is not of acceptable quality, then the combiner switches and examines the quality of the next available path. This switching-and-examining process is repeated until either an acceptable path is found or all available diversity paths have been examined. In the latter case, the combiner either settles on the last examined path [28] or connects to the receiver the path with the best quality among all examined paths, and this SEC variant is known as SEC with post-examine selection (SECps) [29].

In this chapter, we study joint adaptive transmission and diversity combining with SC, SEC, and SECps schemes. We first examine the spectral efficiency benefit provided by the SC scheme for adaptive transmission systems. Then, capitalizing on the adaptive nature of switched combining, we develop more integrated joint designs based on the SEC and SECps schemes. In particular, the resulting system jointly selects the most appropriate transmission mode and diversity paths based on the current channel conditions and the desired bit error rate (BER) requirement. Depending on the primary objective of the joint design, we arrive at a scheme that requires a minimum number of path estimations (termed as minimum estimation scheme) and a scheme with a high bandwidth efficiency (termed as bandwidth-efficient scheme), both of which satisfy the desired BER requirement. For both schemes under consideration, we quantify through accurate analysis their processing complexity (quantified in terms of average number of paths estimated), spectral efficiency (quantified in terms of average number of transmitted bits/s/Hz), and performance (quantified in terms of average BER). Finally, some selected numerical examples are presented to illustrate the mathematical formalism.

The rest of this chapter is organized as follows. Section 6.2 contains the general description of the system and channel model under consideration. While section 6.3 addresses the design and analysis of SC-based joint adaptive modulation and diversity combining schemes, section 6.4 is dedicated to the operation and performance of switched combining–based schemes. Finally, section 6.5 provides some concluding remarks. In all sections, selected numerical examples are provided together with some related discussions and interpretations.

6.2 System and Channel Models

In this section, we first present the adaptive modulation scheme adopted in this work. After discussing the discrete-time implementation model for the proposed transmission system, we introduce the block-fading channel model under consideration.

6.2.1 Adaptive Modulation

We adopt the constant-power variable-rate uncoded M-ary quadrature amplitude modulation (M-QAM) scheme studied in [2].* With this adaptive modulation scheme, the mode selection is solely based on the fading channel condition. In particular, the SNR range is divided into N regions and the constellation size $M = 2^n$ is used during the subsequent data burst transmission if the output SNR of diversity combiner ends up being in the nth region $[\gamma_{T_n}, \gamma_{T_{n+1}})$, where $n = 2, 3, \cdots, N$. Note that in this case, n bits are carried by each symbol. The region boundaries, denoted by γ_{T_n}, are determined such that the instantaneous BER for the chosen constellation is below a certain required value, denoted by BER_0. More specifically, to meet a BER requirement of 0.1%, the thresholds can be set by solving the following equations:

$$\gamma_{T_n} = \text{BER}_n^{-1}(0.001), \tag{6.1}$$

where $\text{BER}_n^{-1}(\cdot)$ is the inverse BER expression. After applying a numerical method to solve (6.1) with the exact BER expression for square M-QAM given in [30], we obtain the exact E_s/N_0 values for the thresholds, γ_{T_n}, for the instantaneous BER requirement of 10^{-2}, 10^{-3}, and 10^{-4} cases, and these values are summarized in Table 6.1. Note that γ_{T_1} is equal to 0 or $-\infty$ dB, while $\gamma_{T_{n+1}} = +\infty$.

TABLE 6.1 SNR Thresholds (E_s/N_0) at Each Modulation Level to Satisfy 1, 0.1, and 0.01% Bit Error Rate for 2_n-ary QAM, Respectively

n	γ_{T_n} (dB) for $\text{BER}_0 = 10^{-2}$	γ_{T_n} (dB) for $\text{BER}_0 = 10^{-3}$	γ_{T_n} (dB) for $\text{BER}_0 = 10^{-4}$
2	7.33	9.64	11.35
3	11.84	13.32	16.07
4	13.90	16.63	18.23
5	17.83	19.79	22.29
6	19.73	22.86	24.30
7	23.85	25.91	28.24
8	25.43	28.94	30.23

* Note that the design can be easily extended to include the coding scheme, in which case we should use the proper SNR thresholds for different modulation and coding scheme combinations.

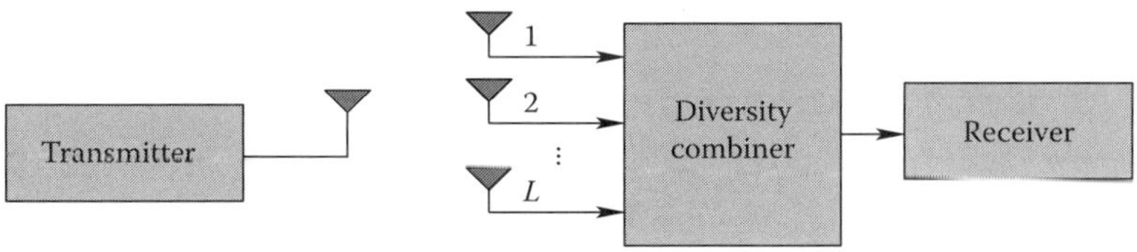

FIGURE 6.1 Block diagram of a receiver space diversity system.

6.2.2 System Model

In this work, we consider a receiver space diversity system as shown in Figure 6.1. In particular, L antennas are used at the receiver to create differently faded replicas of the transmitted signal. Because of the hardware complexity constraint that is imposed by MM-wave systems, we assume that the receiver can only process a single selected diversity branch. We adopt a discrete-time implementation for the proposed transmission system. More specifically, short guard periods are periodically inserted into the transmitted signal. During these guard periods, the receiver performs a series of operations, including diversity path estimations and their comparisons, in order to select the appropriate diversity branch and the suitable adaptive modulation mode to be used during the subsequent data burst reception. Once these decisions are made at the receiver, the adaptive modulation mode is fed back to the transmitter via an error-free reverse channel before the guard period ends. After that, the transmitter and receiver are configured accordingly throughout the subsequent data burst transmission.

6.2.3 Channel Model

We assume that the length of the guard period plus data burst is of the order of the channel coherence time, and therefore, the faded signal amplitude remains constant during each guard period–data burst pair and de-correlates after that. We also assume that the received signal on each antenna branch experiences an independent and identically distributed (i.i.d.) fading process. As such, the faded SNR, denoted by γ_i, $i = 1, \cdots, L$, on each diversity branch shares a common probability density function (PDF) and cumulative distribution function (CDF). In Table 6.2, we summarize the PDF, $p_\gamma(x)$, and the CDF, $F_\gamma(x)$, of the received SNRs under three popular fading models: Rayleigh, Rice, and Nakagami-m. In Table 6.2, $\bar{\gamma}$ is the average SNR, $\Gamma(\cdot)$ is the Gamma function [31, section 8.31], $I_0(\cdot)$ is the modified Bessel function of the first kind with zero order [31, section 8.43], $\Gamma(\cdot,\cdot)$ is the incomplete Gamma function [31, section 8.35], and $Q_1(\cdot,\cdot)$ is the first-order Marcum Q-function [32].

6.3 Joint Adaptive Modulation and Selection Combining

In this section, we consider the analysis of a joint adaptive modulation and selection combining system. We first present the mode of operation of the transmission system

TABLE 6.2 Statistics of the Faded SNR for the Three Fading Models under Consideration

Model	Rayleigh	Rice/Nakagami-n	Nakagami-m
Parameter	.	$K = n^2 \geq 0$	$m \geq \frac{1}{2}$
PDF ($p_\gamma(x)$)	$\frac{1}{\bar{\gamma}} e^{-\frac{x}{\bar{\gamma}}}$	$\frac{(1+n^2)e^{-n^2}}{\bar{\gamma}} e^{-\frac{1+n^2}{\bar{\gamma}}x} I_0\left(2n\sqrt{\frac{1+n^2}{\bar{\gamma}}x}\right)$	$\left(\frac{m}{\bar{\gamma}}\right)^m \frac{x^{m-1}}{\Gamma(m)} e^{-\frac{mx}{\bar{\gamma}}}$
CDF ($F_\gamma(x)$)	$1-e^{-\frac{x}{\bar{\gamma}}}$	$1-Q_1\left(n\sqrt{2}, \sqrt{\frac{2(1+n^2)}{\bar{\gamma}}x}\right)$	$1-\frac{\Gamma\left(m, \frac{mx}{\bar{\gamma}}\right)}{\Gamma(m)}$

during guard periods. The statistics of the received SNR for the SC-based joint design are then derived before this result is applied to the performance and spectral efficiency analysis.

6.3.1 Mode of Operation and Statistics

During each guard period, the receiver performs the following operations. First, the receiver estimates the faded SNR of each antenna branch and selects the best branch after proper comparisons. The combiner output SNR is then given by $\gamma_c = \max[\gamma_1, \gamma_2, \ldots, \gamma_L]$. After that, the receiver determines the appropriate modulation scheme by sequentially comparing the output SNR γ_c with the threshold γ_{T_n}, $n = N, N-1, \cdots, 2$. Whenever the receiver finds that the output SNR is smaller than $\gamma_{T_{n+1}}$ but greater than γ_{T_n}, it selects the modulation mode n for the subsequent data burst and feeds back that mode index to the transmitter. If the combined SNR of all L available branches is below γ_{T_2}, the receiver may ask the transmitter to either (1) transmit using the lowest modulation mode, i.e., QPSK, in violation of the target instantaneous BER requirement (option 1), or (2) buffer the data and wait until the next guard period for more favorable channel conditions (option 2). Note that option 1 is useful in supporting delay-sensitive traffic with hard delay constraint, especially when coupled with a powerful upper-layer error correction mechanism.

Based on the mode of operation described above, we can see that the received SNR, γ_c, of the SC-based scheme is the same as the output SNR of a traditional L branch selection combiner. In other words, the CDF of the received SNR, $F_{\gamma_c}(\cdot)$, is given by

$$F_{\gamma_c}(\gamma) = \begin{cases} F_{\gamma_c}^{SC}(\gamma), & \text{for option 1;} \\ \begin{cases} F_{\gamma_c}^{SC}(\gamma), & \gamma > \gamma_{T_2} \\ F_{\gamma_c}^{SC}(\gamma_{T_2}), & 0 < \gamma \leq \gamma_{T_2} \end{cases} & \text{for option 2,} \end{cases} \tag{6.2}$$

where $F_{\gamma_c}^{SC}(\cdot)$ denotes the CDF of the combined SNR with L branch SC, which is given for i.i.d. fading environment by

$$F_{\gamma_c}^{SC}(\gamma)=\left(F_\gamma(\gamma)\right)^L. \tag{6.3}$$

Correspondingly, the PDF of the received SNR, $p_{\gamma_c}(\cdot)$, is given by

$$p_{\gamma_c}(\gamma)=\begin{cases} p_{\gamma_c}^{SC}(\gamma), & \text{for option 1;} \\ p_{\gamma_c}^{SC}(\gamma)\mathcal{U}\left(\gamma-\gamma_{T_2}\right)+F_{\gamma_c}^{SC}(\gamma_{T_2})\delta(\gamma), & \text{for option 2,} \end{cases} \tag{6.4}$$

where $\mathcal{U}(\cdot)$ is the Heaviside unit step function, $\delta(\cdot)$ is the Dirac Delta function and $p_{\gamma_c}^{SC}(\cdot)$ denotes the PDF of the combined SNR with L branch SC, which is given for the i.i.d. fading environment by

$$p_{\gamma_c}^{SC}(\gamma)=L\left(F_\gamma(\gamma)\right)^{L-1}p_\gamma(\gamma). \tag{6.5}$$

6.3.2 Performance and Efficiency Analysis

6.3.2.1 Average Spectral Efficiency

The average spectral efficiency of an adaptive modulation system can be calculated as [2, equation (33)]

$$\eta=\sum_{n=2}^{N} n\, p_n,$$

where p_n is the probability that the nth constellation is used. For the SC-based system under consideration, it can be shown that p_n is given by

$$p_n=\begin{cases} \begin{cases} F_{\gamma_c}^{SC}\left(\gamma_{T_{n+1}}\right)-F_{\gamma_c}^{SC}\left(\gamma_{T_n}\right), & n\geq 3; \\ F_{\gamma_c}^{SC}\left(\gamma_{T_3}\right), & n=2 \end{cases} & \text{for option 1;} \\ F_{\gamma_c}^{SC}\left(\gamma_{T_{n+1}}\right)-F_{\gamma_c}^{SC}\left(\gamma_{T_n}\right), & \text{for option 2,} \end{cases} \tag{6.6}$$

Therefore, the average spectral efficiency of the SC-based scheme is given by

$$\eta=\begin{cases} N-\sum_{n=3}^{N} F_{\gamma_c}^{SC}\left(\gamma_{T_n}\right), & \text{for option 1;} \\ N-\sum_{n=2}^{N} F_{\gamma_c}^{SC}\left(\gamma_{T_n}\right), & \text{for option 2.} \end{cases} \tag{6.7}$$

6.3.2.2 Average Error Rate

The average BER for an adaptive modulation system can be calculated as [2, equation (35)]

$$\langle BER \rangle = \frac{1}{\eta} \sum_{n=2}^{N} n\, \overline{BER_n}, \tag{6.8}$$

where $\overline{BER_n}$ is the average error rate for constellation n, and which is given by

$$\overline{BER_n} = \int_{\gamma_{T_n}}^{\gamma_{T_{n+1}}} BER_n(\gamma) p_{\gamma_c}^{SC}(\gamma) d\gamma, \tag{6.9}$$

where $BER_n(\gamma)$ is the conditional BER of the constellation n over an additive white Gaussian noise (AWGN) channel given that its SNR is equal to γ, which is given in [30] for uncoded M-QAM. Therefore, the average BER of the SC-based scheme under consideration can be calculated as

$$\langle BER \rangle = \begin{cases} \dfrac{\int_{0}^{\gamma_{T_3}} BER_2(\gamma)\; p_{\gamma_c}^{SC}(\gamma)\; d\gamma + \sum_{n=3}^{N} n \int_{\gamma_{T_n}}^{\gamma_{T_{n+1}}} BER_n(\gamma)\; p_{\gamma_c}^{SC}(\gamma)\; d\gamma}{N - \sum_{n=3}^{N} F_{\gamma_c}^{SC}\left(\gamma_{T_n}\right)}, & \text{for option 1;} \\[2ex] \dfrac{\sum_{n=2}^{N} n \int_{\gamma_{T_n}}^{\gamma_{T_{n+1}}} BER_n(\gamma)\; p_{\gamma_c}^{SC}(\gamma)\; d\gamma}{N - \sum_{n=2}^{N} F_{\gamma_c}^{SC}\left(\gamma_{T_n}\right)}, & \text{for option 2.} \end{cases} \tag{6.10}$$

6.3.3 Numerical Examples

Figure 6.2 plots the average spectral efficiency of the SC-based scheme with both options and for different numbers of receive antennas L. As we can see, as the number of receive antennas increases, the average spectral efficiency of the system improves, but with diminishing gain. If we compare the average spectral efficiency of the two options with the same number of antennas, we can see that for a high-SNR region, both options have nearly the same spectral efficiency, whereas for a low-SNR region, option 1 offers higher efficiency than option 2. This is because when the channel condition is poor, option 2 will buffer the data while option 1 continues transmission at the cost of violating the BER constraint.

The BER constraint violation can be immediately observed from Figure 6.3, where we plot the average error rate of the SC-based scheme with both options and different numbers of receive antenna L. The target BER is set to be 10^{-3} in this figure. We also observe that when the average SNR is greater than a certain value (10 dB for the L = 4 case), the average BER of the SC-based scheme with option 1 will always be smaller than the target BER of 10^{-3}. When the number of receive antennas increases, the SNR region where the BER constraint is satisfied with option 1 also increases. For option 2, however, more receive antennas will only lead to a better BER performance over the very high-SNR region.

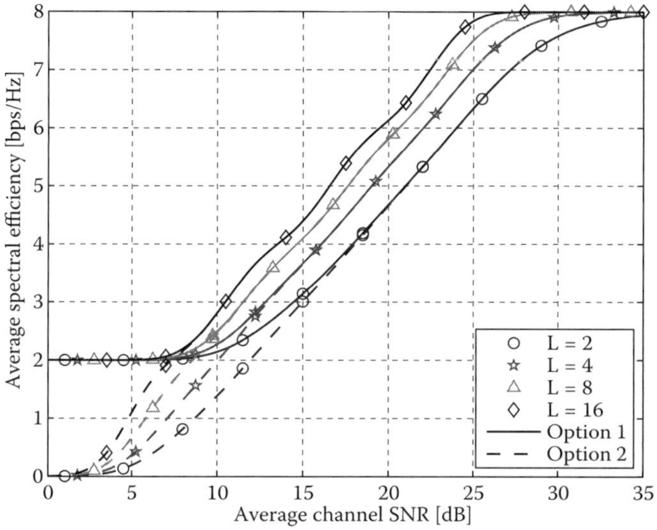

FIGURE 6.2 Average spectral efficiency of the SC-based scheme for both options and different numbers of antennas L.

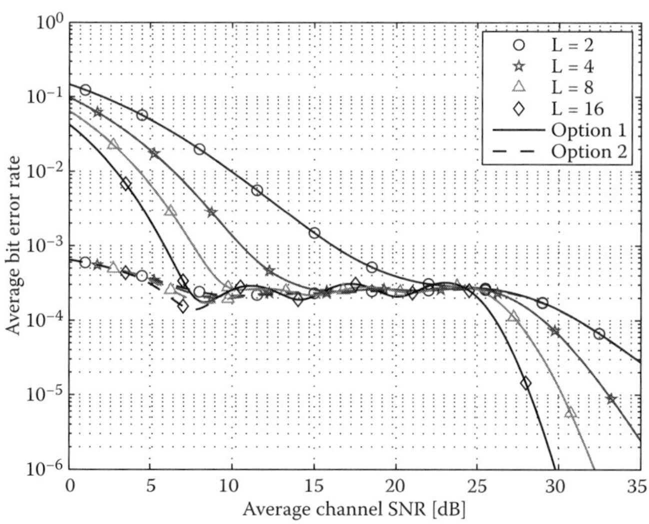

FIGURE 6.3 Average error rate of the SC-based scheme for both options and different numbers of antennas L.

6.4 Joint Adaptive Modulation and Switched Combining

In this section, we design and analyze joint adaptive modulation and switched combining schemes. Note that switched combining schemes offer an even lower-complexity combining solution than SC since they involve less channel estimation and fewer and simpler comparisons. In what follows, we first present several switched combining schemes and then study two variants of the joint adaptive modulation and diversity combining designs based on switched combining, namely, minimum estimation schemes and bandwidth-efficient schemes.

6.4.1 Switched Combining

6.4.1.1 SSC/SEC

Dual-branch SSC is one of the most widely studied switched combining schemes. With SSC, the receiver uses the current branch until it becomes unacceptable, and then it switches and stays on the other branch regardless of its quality. The branch acceptance is determined by comparing the SNR of the current branch with a preselected fixed threshold, denoted by γ_T, which can be implemented with a threshold detector. Note that the major complexity saving of SSC over SC is that SSC needs only the SNR estimate of the current path and a simpler comparison to reach the combining decision. Because the receiver simply stays on the switch-to branch irrespective of its quality, SSC cannot benefit from additional diversity paths.

In multibranch scenario, i.e., $L > 2$, we can generalize SSC in order to take advantage of the additional diversity paths, and this leads to the SEC scheme [28]. In this case, the receiver with SEC cyclically switches between the L antenna branches. Branch switching occurs only when the received SNR of the currently used branch is below the threshold γ_T, and as such, this branch becomes unacceptable. Unlike SSC, the receiver examines the received SNR of the switch-to branch and switches again if it is found unacceptable. The receiver will repeat this process until either it finds an acceptable branch or all L antenna branches have been examined. In the latter case, it uses the last examined antenna branch for the data reception. It can be seen that the SEC scheme retains the complexity advantage of less path estimation and simple comparison.

Based on the mode of operation of SSC/SEC, it can be shown that the CDF and PDF of the output SNR with L-SEC are given by [28]

$$F_{\gamma_c}^{SEC}(\gamma)=\begin{cases}\left[F_{\gamma}(\gamma_T)\right]^{L-1}F_{\gamma}(\gamma), & \gamma<\gamma_T;\\ \sum_{j=0}^{L-1}\left[F_{\gamma}(\gamma)-F_{\gamma}(\gamma_T)\right]\left[F_{\gamma}(\gamma_T)\right]^{j} & \\ \quad +\left[F_{\gamma}(\gamma_T)\right]^{L}, & x\geq\gamma_T,\end{cases} \tag{6.11}$$

and

$$p_{\gamma_c}^{SEC}(\gamma)=\begin{cases}\left[F_\gamma(\gamma_T)\right]^{L-1} p_\gamma(\gamma), & \gamma<\gamma_T;\\ \sum_{j=0}^{L-1}\left[F_\gamma(\gamma_T)\right]^j p_\gamma(\gamma), & \gamma\geq\gamma_T,\end{cases} \tag{6.12}$$

respectively, where $F_\gamma(\cdot)$ and $p_\gamma(\cdot)$ are the common CDF and PDF of received SNR per path. Note that when $L = 2$, (6.11) and (6.12) reduce to the CDF and PDF for dual-branch SSC, respectively.

6.4.1.2 SECps

We can further improve the performance of SSC/SEC by better exploiting the available channel estimates in the worst-case scenario for which no acceptable diversity path is found. Since all available diversity paths have been examined in this case, the receiver may use the strongest one among all these unacceptable paths, instead of randomly choosing the last unacceptable path, for data reception. We term the resulting scheme SECps [29]. In particular, the receiver with SECps tries to use an acceptable path by examining as many diversity paths as necessary, as is the case with SEC. When no acceptable path is found after examining all diversity paths, unlike conventional SEC, the receiver with SECps selects the best unacceptable path, i.e., the one with the highest SNR, for data reception.

More specifically, the receiver first estimates the SNR of the currently used diversity path, denoted by γ_1, without loss of generality, and compares it with the threshold γ_T. If the current diversity path is acceptable (i.e., $\gamma_1 \geq \gamma_T$), then the receiver continues to use it for data reception (i.e., $\gamma_c = \gamma_1$). Otherwise (i.e., $\gamma_1 < \gamma_T$), the receiver tries to find an acceptable path by sequentially examining the other $L - 1$ diversity paths. This process is continued until either an acceptable path is found or all available diversity paths have been estimated. In the latter case, since all path SNRs are known, the receiver compares the estimate path SNRs and selects the one with the largest SNR for data reception. Compared to conventional SEC, SECps requires the additional complexity for the occasional selection of the best path among all unacceptable paths. However, SECps takes full advantage of the available path estimates. In particular, by using the best unacceptable path instead of a randomly chosen one (usually the last one examined) as in SEC, SECps can deliver better performance than conventional SEC.

Based on the mode of operation of SECps, it can be shown that the CDF and PDF of the output SNR with L-SECps are given by [29]

$$F_{\gamma_c}^{SEC}(\gamma)=\begin{cases}1-\sum_{i=0}^{L-1}\left[F_\gamma(\gamma_T)\right]^i\left[1-F_\gamma(\gamma)\right], & \gamma\geq\gamma_T;\\ \left[F_\gamma(\gamma)\right]^L, & \gamma<\gamma_T.\end{cases} \tag{6.13}$$

and

$$p_{\gamma_c}^{SEC}(\gamma)=\begin{cases}\sum_{i=0}^{L-1}\left[F_\gamma(\gamma_T)\right]^i\, p_\gamma(\gamma), & \gamma\geq\gamma_T;\\ L\left[F_\gamma(\gamma)\right]^{L-1}\, p_\gamma(\gamma), & \gamma<\gamma_T,\end{cases} \tag{6.14}$$

respectively.

6.4.2 Minimum Estimation Schemes

6.4.2.1 Mode of Operation

The primary objective of the minimum estimation schemes based on switched combining is to minimize the number of path estimations for path selection at the receiver. Once this primary objective is met, this scheme tries to afford the largest possible spectral efficiency while meeting the required target BER. Based on these objectives, the diversity combiner will perform just enough combining operations such that at least the lowest adaptive modulation mode, i.e., QPSK ($M = 4$), will exhibit an instantaneous BER smaller than the predetermined target value. In particular, the receiver tries to increase the output SNR Γ above the threshold for QPSK, i.e., γ_{T_2}, by performing SSC/SEC or SECps diversity. The receiver will sequentially estimate the received SNR of each diversity path and compare it with γ_{T_2}. The first acceptable path, i.e., the one with SNR greater than γ_{T_2}, will be used for data burst reception. Then, the receiver starts to determine the modulation mode to be used by checking in which interval the resulting output SNR falls. In particular, the receiver sequentially compares the output SNR with respect to the thresholds, $\gamma_{T3}, \gamma_{T4}, \cdots, \gamma_{TN}$. Whenever the receiver finds that the output SNR is smaller than $\gamma_{T_{n+1}}$ but greater than γ_{T_n}, it selects the modulation mode n for the subsequent data burst and feeds back that particular modulation mode to the transmitter. If the received SNRs of all L available branches are below γ_{T_2}, the receiver may ask the transmitter to either (1) transmit using the lowest modulation mode in violation of the target instantaneous BER requirement (option 1) or (2) buffer the data and wait until the next guard period for more favorable channel conditions (option 2). For option 1, the receiver will use the last unacceptable path for data reception for the case of SSC/SEC and the strongest unacceptable path instead for the case of SECps.

6.4.2.2 Statistics of the Received SNR

Based on the mode of operation described above, we can see that the received SNR, Γ, of the minimum estimation scheme based on SSC/SEC is the same as the combined SNR of the SSC/SEC scheme with γ_{T_2} as the output threshold. In other words, the CDF of the received SNR, $F_\Gamma(\cdot)$, of the SSC/SEC-based minimum estimation scheme is given by

$$F_\Gamma(\gamma)=\begin{cases}F_{\gamma_c}^{SEC(\gamma_{T2})}(\gamma), & \text{for option 1;}\\ \begin{cases}F_{\gamma_c}^{SEC(\gamma_{T2})}(\gamma), & \gamma>\gamma_{T_2};\\ F_{\gamma_c}^{SEC(\gamma_{T2})}(\gamma_{T_2}), & 0<\gamma\leq\gamma_{T_2}\end{cases} & \text{for option 2,}\end{cases} \tag{6.15}$$

where $F_{\gamma_c}^{SEC(\gamma_{T_2})}(\cdot)$ denotes the CDF of the combined SNR with L branch SEC and using γ_{T_2} as an output threshold, which is given for the i.i.d. fading environment in (6.11). Correspondingly, the PDF of the received SNR, $p_\Gamma(\cdot)$, is given by

$$p_\Gamma(\gamma) = \begin{cases} p_{\gamma_c}^{SEC(\gamma_{T_2})}(\gamma), & \text{for option 1;} \\ p_{\gamma_c}^{SEC(\gamma_{T_2})}(\gamma)\mathcal{U}(\gamma-\gamma_{T_2}) + F_{\gamma_c}^{SEC(\gamma_{T_2})}(\gamma_{T_2})\delta(\gamma), & \text{for option 2,} \end{cases} \tag{6.16}$$

where $p_{\gamma_c}^{SEC(\gamma_{T_2})}(\cdot)$ denotes the PDF of the combined SNR with L branch SSC/SEC and using γ_{T_2} as an output threshold, which is given for the i.i.d. fading environment by (6.12).

Similarly, the CDF and PDF of the received SNR, Γ, of the minimum estimation scheme based on SECps can be obtained as

$$F_\Gamma(\gamma) = \begin{cases} F_{\gamma_c}^{SECps(\gamma_{T_2})}(\gamma), & \text{for option 1;} \\ \begin{cases} F_{\gamma_c}^{SECps(\gamma_{T_2})}(\gamma), & \gamma > \gamma_{T_2}; \\ F_{\gamma_c}^{SECps(\gamma_{T_2})}(\gamma_{T_2}), & 0 < \gamma \leq \gamma_{T_2} \end{cases} & \text{for option 2,} \end{cases} \tag{6.17}$$

and

$$p_\Gamma(\gamma) = \begin{cases} p_{\gamma_c}^{SECps(\gamma_{T_2})}(\gamma), & \text{for option 1;} \\ p_{\gamma_c}^{SECps(\gamma_{T_2})}(\gamma)\mathcal{U}(\gamma-\gamma_{T_2}) + F_{\gamma_c}^{SECps(\gamma_{T_2})}(\gamma_{T_2})\delta(\gamma), & \text{for option 2,} \end{cases} \tag{6.18}$$

respectively, where $F_{\gamma_c}^{SECps(\gamma_{T_2})}(\cdot)$ and $p_{\gamma_c}^{SECps(\gamma_{T_2})}(\cdot)$ denote the CDF and PDF of the combined SNR with L branch SECps and γ_{T_2} as an output threshold, which is given for the i.i.d. fading environment by (6.13) and (6.14), respectively.

6.4.2.3 Performance and Efficiency Analysis

With the statistics of the received SNR Γ derived for minimum estimation schemes based on both SSC/SEC and SECps, we evaluate the performance and efficiency of the system. Following an approach similar to the one used for the SC-based schemes in the previous section, we obtain the average spectral efficiency of the SSC/SEC-based minimum estimation scheme as

$$\eta = \begin{cases} N - \sum_{n=3}^{N} F_{\gamma_c}^{SEC(\gamma_{T_2})}(\gamma_{T_n}), & \text{for option 1;} \\ N - \sum_{n=2}^{N} F_{\gamma_c}^{SEC(\gamma_{T_2})}(\gamma_{T_n}), & \text{for option 2.} \end{cases} \tag{6.19}$$

The average BER of the SSC/SEC-based minimum estimation scheme can be similarly obtained as

$$\langle BER\rangle=\begin{cases}\dfrac{\int_0^{\gamma_{T_3}} BER_2(\gamma)\, p_{\gamma_c}^{SEC(\gamma_{T_2})}(\gamma)\, d\gamma+\sum_{n=3}^{N} n\int_{\gamma_{T_n}}^{\gamma_{T_{n+1}}} BER_n(\gamma)\, p_{\gamma_c}^{SEC(\gamma_{T_2})}(\gamma)\, d\gamma}{N-\sum_{n=3}^{N} F_{\gamma_c}^{SEC(\gamma_{T_2})}(\gamma_{T_n})}, & \text{for option 1;}\\[2ex] \dfrac{\sum_{n=2}^{N} n\int_{\gamma_{T_n}}^{\gamma_{T_{n+1}}} BER_n(\gamma)\, p_{\gamma_c}^{SEC(\gamma_{T_2})}(\gamma)\, d\gamma}{N-\sum_{n=2}^{N} F_{\gamma_c}^{SEC(\gamma_{T_2})}(\gamma_{T_n})}, & \text{for option 2.}\end{cases} \tag{6.20}$$

Similar results can be obtained for SECps-based schemas by using the CDF and PDF of the combined SNR with L branch SECps given in (6.13) and (6.14) instead, but these results are omitted here for conciseness.

The complexity saving of switched combining schemes over selection combining schemes manifests in the average number of path estimations. Note that SC always needs L path estimations, i.e., $N_E^{SC} = L$. On the other hand, the receiver with SSC/SEC and SECps needs to estimate additional paths only if necessary. Whenever an acceptable path is found, the receiver stops path estimation. Therefore, the average number of path estimations needed with the SSC/SEC- or SECps-based minimum selection scheme is given by

$$N_E^{SEC/SECps}=1+\sum_{i=1}^{L-1}\left[P_\gamma(\gamma_{T_2})\right]^i=\frac{1-\left[P_\gamma(\gamma_{T_2})\right]^L}{1-P_\gamma(\gamma_{T_2})}. \tag{6.21}$$

Therefore, the switched combining–based scheme requires less path estimations than the SC-based scheme on average, i.e.,

$$N_E^{SEC}=N_E^{SECps}\le N_E^{SC}.$$

6.4.3 Bandwidth-Efficient Schemes

6.4.3.1 Mode of Operation

The primary objective of the switched combining-based bandwidth-efficient scheme is to maximize the spectral efficiency. As such, the receiver with this scheme performs as many combining operations as needed so that the highest achievable modulation mode can be used while satisfying the instantaneous BER requirement. More specifically, the receiver tries first to increase the output SNR Γ above the threshold of highest modulation mode 2^N-QAM, i.e., γ_{T_N}, by performing SSC/SEC or SECps. The receiver will sequentially estimate the received SNR of each diversity path and compare it with γ_{T_N}. Whenever a path with SNR greater than γ_{T_N} is found, the receiver stops path estimation, uses that path for data burst reception, and informs the transmitter to use 2^N-QAM as the modulation mode for the subsequent data burst. If the received SNRs of all the available paths are below γ_{T_N}, the receiver selects the modulation mode corresponding to the SNR interval in which the combined SNR falls. With SSC/SEC, the combined SNR is the SNR of the last unacceptable path, whereas with SECps, it is the SNR of the best unacceptable path. In particular, the receiver sequentially compares the output SNR with the thresholds, $\gamma_{T_{N-1}}, \gamma_{T_{N-2}}, \cdots, \gamma_{T_2}$. Whenever the receiver finds that the output SNR is smaller than $\gamma_{T_{N+1}}$ but greater than γ_{T_n}, it selects the modulation mode n for the subsequent data burst and feeds back this selected mode to the transmitter. If, in the worst case, the combined SNRs of all the available paths ends up being below γ_{T_2}, the receiver has the same two termination options as for the minimum estimation scheme (i.e., to transmit using the lowest modulation mode [option 1] or to wait until the next guard period [option 2]).

6.4.3.2 Statistics of the Received SNR

Based on the mode of operation described above, we can see that the received SNR, Γ, of the bandwidth-efficient scheme is the same as the combined SNRs of SSC/SEC or SECps diversity with γ_{T_N} as the output threshold. Therefore, the CDF of the received SNR of this bandwidth-efficient scheme based on SEC is given by

$$F_\Gamma(\gamma)=\begin{cases} F_{\gamma_c}^{SEC(\gamma_{T_N})}(\gamma), & \text{for option 1;} \\ \begin{cases} F_{\gamma_c}^{SEC(\gamma_{T_N})}(\gamma), & \gamma>\gamma_{T_2}; \\ F_{\gamma_c}^{SEC(\gamma_{T_N})}(\gamma_{T_2}), & 0<\gamma\le\gamma_{T_2} \end{cases} & \text{for option 2,} \end{cases} \tag{6.22}$$

where $F_{\gamma_c}^{SEC(\gamma_{T_N})}(\cdot)$ denotes the CDF of the combined SNR with L branch SEC and γ_{T_N} as an output threshold. Correspondingly, the PDF of the received SNR is given by

$$p_\Gamma(\gamma)=\begin{cases} p_{\gamma_c}^{SEC(\gamma_{T_N})}(\gamma), & \text{for option 1;} \\ p_{\gamma_c}^{SEC(\gamma_{T_N})}(\gamma)\mathcal{U}(\gamma-\gamma_{T_2})+F_{\gamma_c}^{SEC(\gamma_{T_N})}(\gamma_{T_2})\delta(\gamma), & \text{for option 2,} \end{cases} \tag{6.23}$$

where $p_{\gamma_c}^{SEC(\gamma_{T_N})}(\cdot)$ denotes the PDF of the combined SNR with L branch SEC and γ_{T_N} as output threshold. The CDF and PDF of received SNR with the SECps-based bandwidth-efficient scheme can be similarly obtained, but are omitted here for conciseness.

6.4.3.3 Performance and Efficiency Analysis

With the statistics of the received SNR at hand, we study the performance, efficiency, and complexity of the bandwidth-efficient scheme as we did for the minimum estimation. For conciseness, we just list the analytical results in what follows for the SSC/SEC-based scheme, while noting that those for the SECps-based scheme can be easily obtained by using the appropriate CDF and PDF instead.

Average spectral efficiency:

$$\eta = \begin{cases} N - \sum_{n=3}^{N} F_{\gamma_c}^{SEC(\gamma_{T_N})}\left(\gamma_{T_n}\right), & \text{for option 1;} \\ N - \sum_{n=2}^{N} F_{\gamma_c}^{SEC(\gamma_{T_N})}\left(\gamma_{T_n}\right), & \text{for option 2.} \end{cases} \tag{6.24}$$

Average BER:

$$\langle BER \rangle = \begin{cases} \dfrac{\int_0^{\gamma_{T_3}} BER_2(\gamma)\, p_{\gamma_c}^{SEC(\gamma_{T_N})}(\gamma)\, d\gamma + \sum_{n=3}^{N} n \int_{\gamma_{T_n}}^{\gamma_{T_{n+1}}} BER_n(\gamma)\, p_{\gamma_c}^{SEC(\gamma_{T_N})}(\gamma)\, d\gamma}{N - \sum_{n=3}^{N} F_{\gamma_c}^{SEC(\gamma_{T_N})}\left(\gamma_{T_n}\right)}, & \text{for option 1;} \\ \dfrac{\sum_{n=2}^{N} n \int_{\gamma_{T_n}}^{\gamma_{T_{n+1}}} BER_n(\gamma)\, p_{\gamma_c}^{SEC(\gamma_{T_2})}(\gamma)\, d\gamma}{N - \sum_{n=2}^{N} F_{\gamma_c}^{SEC(\gamma_{T_N})}\left(\gamma_{T_n}\right)}, & \text{for option 2.} \end{cases} \tag{6.25}$$

Average number of path estimation:

$$N_E^{SEC/SECps} = 1 + \sum_{i=1}^{L-1} \left[P_\gamma\left(\gamma_{T_N}\right)\right]^i = \frac{1 - \left[P_\gamma\left(\gamma_{T_N}\right)\right]^L}{1 - P_\gamma\left(\gamma_{T_N}\right)}. \tag{6.26}$$

6.4.4 Numerical Examples

In this section, we examine the different design trade-offs involved in the joint adaptive modulation and switched combining schemes through several selected numerical examples.

Figure 6.4 plots the average spectral efficiency of the SEC-based minimum estimation schemes. It is interesting to see that the average spectral efficiency of this scheme for both options and different numbers of receive antennas overlaps for the high-SNR region. This is because when the channel condition is favorable, the first path examined will always be acceptable and will be used for modulation mode selection. As a result, the system becomes equivalent to the no-diversity case. We also observe from Figure 6.4 that over the low- to medium-SNR region, the system can benefit from an increasing number of receive antennas, while this benefit is more significant for option 2 than for option 1. Note that the probability of no transmission for option 2 is reduced when the number of receive antennas increases. Figure 6.4 also shows that option 1 has a considerable spectral efficiency advantage over option 2 in the low-SNR region. This advantage

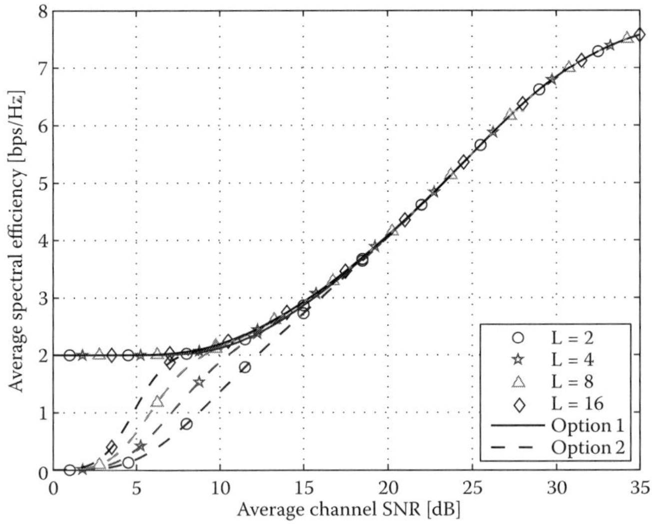

FIGURE 6.4 Average spectral efficiency of SEC-based minimum estimation schemes for both options and different numbers of antennas *L*.

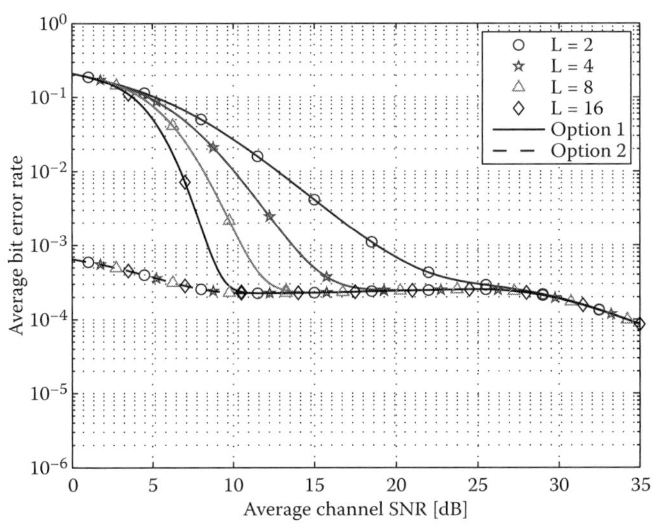

FIGURE 6.5 Average error rate of SEC-based minimum estimation schemes for both options and different numbers of antennas *L*.

again comes at the cost of the violation of the BER constraint, as shown in Figure 6.5. In Figure 6.5, we plot the average error rate of the SEC-based minimum estimation scheme with both options and different numbers of receive antenna *L*. It is clear that option 1 constantly violates the target BER requirement over the low- to medium-SNR region,

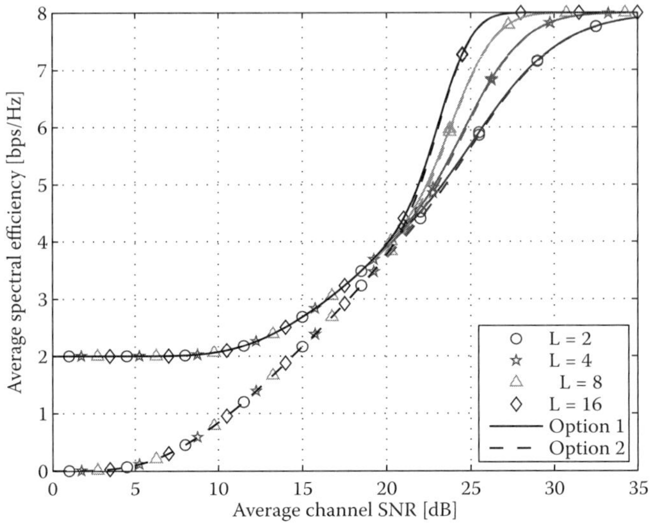

FIGURE 6.6 Average spectral efficiency of SEC-based bandwidth-efficient schemes for both options and different numbers of antennas L.

whereas option 2 always satisfies the requirement. We also notice from Figure 6.5 that while increasing the number of receive antennas can considerably alleviate the BER violation for option 1, it has nearly no effect on the BER performance of option 2.

We plot the average spectral efficiency of the SEC-based bandwidth-efficient schemes for both options and different numbers of antennas L in Figure 6.6. Unlike the minimum estimation schemes, increasing the number of receive antennas has little effect on the spectral efficiency for either option over the low- to medium-SNR region, whereas option 1 has certain spectral advantage over option 2 for the same reason given earlier. For the high-SNR region, the spectral efficiency of both options converges and benefits considerably from the additional number of receive antennas. This is because when the channel condition is favorable, having more antennas will increase the chance of using the largest possible constellation size and, as such, increase spectral efficiency. On the other hand, when the channel condition is poor, the largest constellation is not feasible, and based on the operation of the bandwidth-efficient scheme, the receiver will always choose constellation size based on the last examined path. As a result, the system becomes equivalent to the no-diversity case. Figure 6.7 shows the average BER of the SEC-based bandwidth-efficient schemes. As expected, option 1 violates the BER requirement over a large SNR range, while option 2 always satisfies the requirement. Again, increasing the number of antennas helps improve the BER performance of option 1 over the high-SNR region but affects little that of option 2.

It can be expected that the spectral efficiency of the SECps-based minimum estimation scheme is the same as that of the SEC-based scheme, and that the SECps-based scheme leads to an improved error performance for option 1 over the low-SNR region due to the occasional selection of the best unacceptable paths in the worst case. We omit the numerical examples of the SECps-based minimum estimation scheme for conciseness and focus

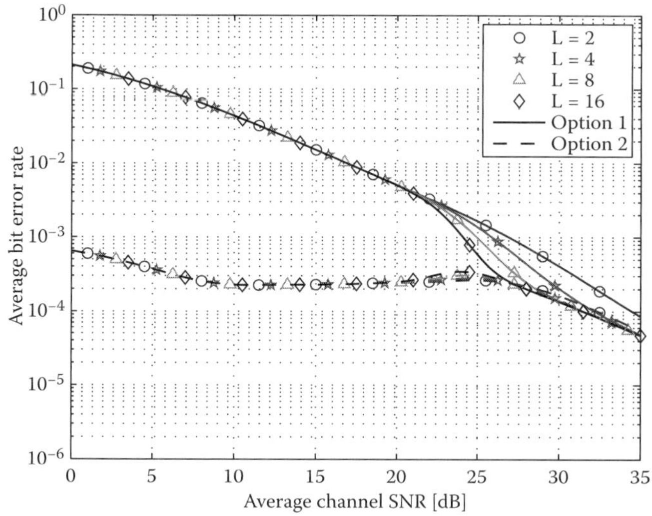

FIGURE 6.7 Average error rate of SEC-based bandwidth-efficient schemes for both options and different numbers of antennas L.

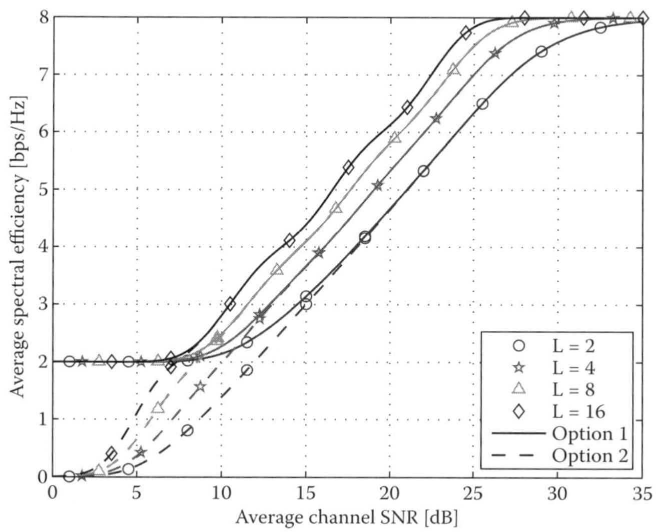

FIGURE 6.8 Average spectral efficiency of SECps-based bandwidth-efficient schemes for both options.

on those of the SECps-based bandwidth-efficient scheme here. From Figure 6.8, unlike the SEC-based bandwidth-efficient scheme, the spectral efficiency of the SECps-based scheme benefits from the increasing number of antennas over all SNR regions. Basically, when the largest constellation is not feasible, the SECps-based bandwidth-efficient

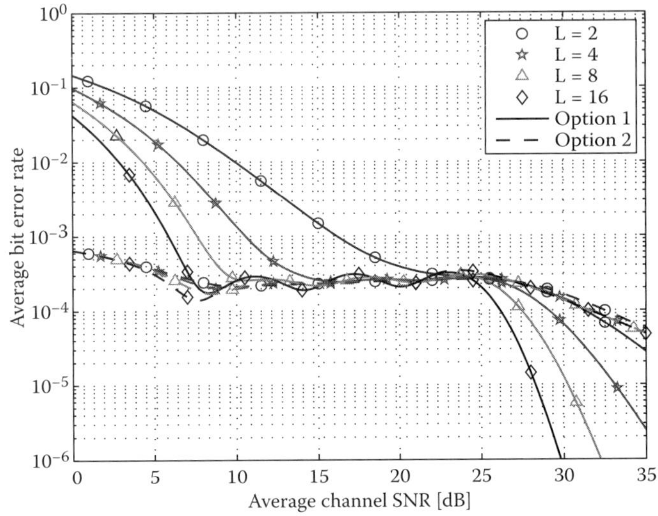

FIGURE 6.9 Average error rate of SECps-based bandwidth-efficient schemes for both options.

scheme can use a larger constellation than the SEC-based scheme since the best path is used in this case. Comparing Figures 6.2 and 6.8, we can conclude that the SECps-based bandwidth-efficient scheme achieves nearly the same spectral efficiency as the SC-based scheme. From Figure 6.9, we observe that similar to the SEC-based scheme, the BER performance of the SECps-based bandwidth-efficient scheme with option 1 shows a more significant improvement than with option 2 when the number of antennas increases. But the difference is that now the BER improvement manifests over all the SNR region. In fact, the average BER performance of the SECps-based bandwidth-efficient scheme with option 1 is very similar to that of the SC-based scheme shown in Figure 6.3.

We now compare the performances and spectral efficiencies of different schemes in Figures 6.10 and 6.11, respectively, for a fixed number of receive antennas (L = 4). The target BER is set to 10^{-4} in these figures. In particular, the average BER performance of different schemes with option 1 is shown in Figure 6.10. Note that we use subscript 1 to denote the minimum estimation schemes based on SEC or SECps, and subscript N to denote the bandwidth-efficient schemes. It can be seen that the SECps-based bandwidth-efficient scheme offers nearly the same error performance as the SC-based scheme. The complexity advantage of the SEC-based scheme comes at the cost of certain error performance degradation. It is worth pointing out that for the high-SNR region, the average BER of the SEC-based bandwidth-efficient scheme becomes better than that of the SECps-based minimum estimation scheme. Figure 6.11 compares the average spectral efficiencies of different schemes with option 2. Again, the SECps-based bandwidth-efficient scheme offers the same high spectral efficiency as the SC-based scheme. The SEC-based minimum estimation scheme shows a certain spectral advantage over the no-diversity case in the low-SNR region, whereas this advantage is more pronounced for

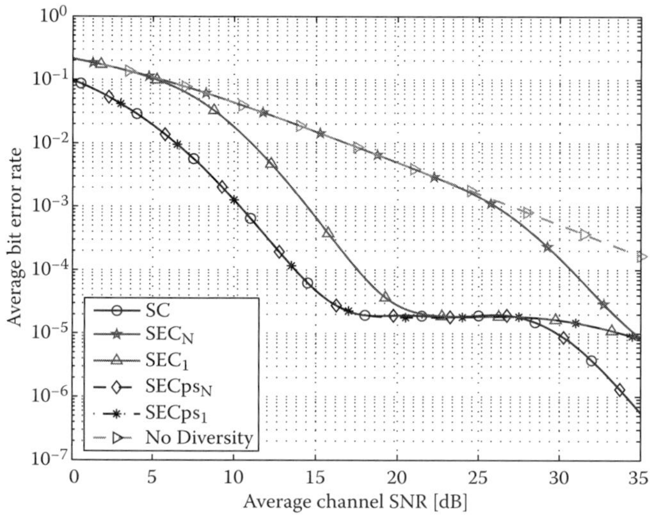

FIGURE 6.10 Comparison of the average error rate of SC-, SEC-, and SECps-based schemes (all for option 1).

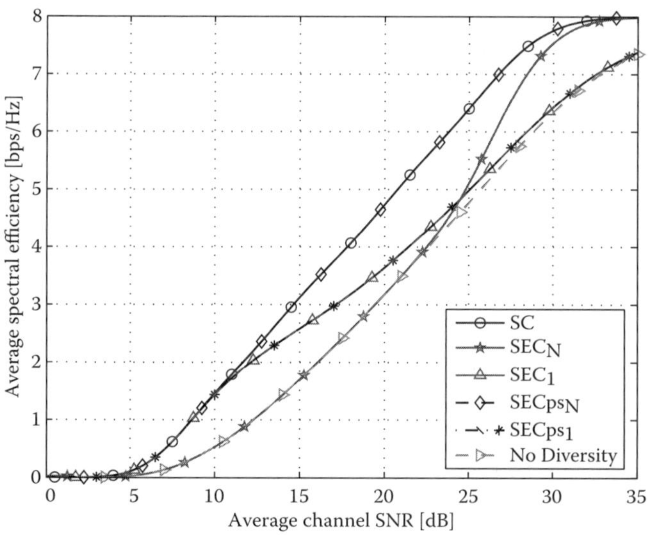

FIGURE 6.11 Comparison of the average spectral efficiency of SC-, SEC-, and SECps-based schemes (all for option 2).

the SEC-based bandwidth-efficient scheme in the high-SNR region. Finally, it is interesting to note that the spectral efficiencies of the SEC-based and SECps-based minimum estimation schemes for option 2 are exactly the same as one would intuit.

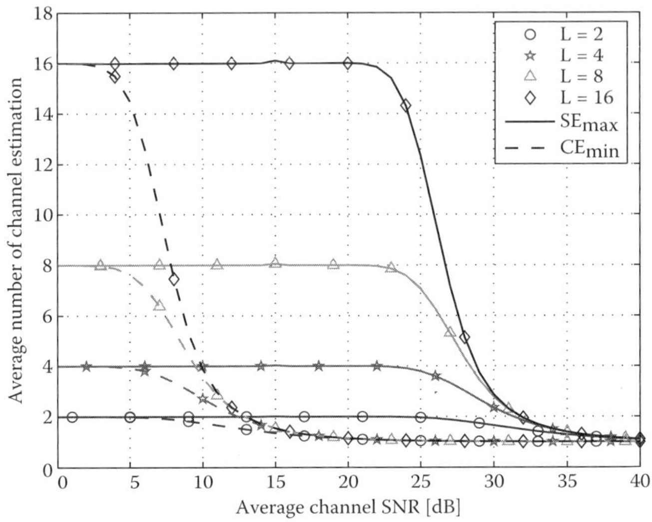

FIGURE 6.12 Comparison of the average number of channel estimation for the bandwidth-efficient and minimum estimation schemes.

Finally, we examine the average number of channel estimations required by switched combining–based joint designs in Figure 6.12. We use SE_{max} to denote the bandwidth-efficient schemes and CE_{min} the minimum estimation schemes. As we can see, as the average SNR increases, the number of channel estimations for all cases decreases from L to 1. For the low- to medium-SNR range, the minimum estimation schemes always require much less channel estimation than the bandwidth-efficient scheme, especially when L is large. Both types of schemes require less than L channel estimations, which is required by the SC-based scheme, in the high-SNR region. Therefore, the complexity advantage of switched combining-based schemes is that on average, the system needs to perform less channel estimation for diversity path and modulation mode selection.

6.5 Concluding Remarks

In this chapter, we investigated the joint design of adaptive modulation and switched diversity combining. The proposed system selects the diversity path and modulation mode jointly based on the fading channel condition and target BER requirement. Both SC and different variants of switched combining schemes, including SSC, SEC, and SECps, were considered. For the switched combining-based systems, we took into account both minimum channel estimation and maximum spectral efficiency design objectives. For the resulting schemes, we accurately quantified the performance, spectral efficiency, and complexity. It is observed from the selected numerical examples that the SECps-based bandwidth-efficient scheme can achieve nearly the same spectral efficiency and link reliability as SC–based schemes while requiring less path estimation on average. We also noticed that the SEC-based schemes offer different design trade-offs. With its low implementation complexity, we believe that the proposed joint design can be readily applied to the emerging MM-wave-based WPAN systems.

Acknowledgments

This work is supported in part by a discovery grant from NSERC Canada, in part by the Ministry of Information and Communications (MIC), Korea, under the Information Technology Research Communication (ITRC) support program supervised by the Institute of Information Technology Advancement (IITA) (IITA-2008-C1090-0801-0037), and in part by the Qatar Foundation for Education, Sciences, and Community Development, Doha, Qatar.

References

[1] A. J. Goldsmith and S.-G. Chua. 1998. Adaptive coded modulation for fading channels. *IEEE Trans. Commun.* 46:595–602.

[2] M.-S. Alouini and A. J. Goldsmith. 2000. Adaptive modulation over Nakagami fading channels. *Kluwer J. Wireless Commun.* 13:119–43.

[3] K. J. Hole, H. Holm, and G. E. Øien. 2000. Adaptive multidimensional coded modulation over flat fading channels. *IEEE J. Select. Areas Commun.* 18:1153–58.

[4] G. L. Stüber. 2000. *Principles of mobile communications.* 2nd ed. Norwell, MA: Kluwer Academic Publishers.

[5] R. Prasad. 1998. *Universal wireless personal communications.* Boston: Artech House.

[6] N. Kong and L. B. Milstein. 1999. Average SNR of a generalized diversity selection combining scheme. *IEEE Commun. Lett.* 3:57–59.

[7] M. Z. Win and J. H. Winters. 1999. Analysis of hybrid selection/maximal-ratio combining in Rayleigh fading. *IEEE Trans. Commun.* 47:1773–76.

[8] M.-S. Alouini and M. K. Simon. 2000. An MGF-based performance analysis of generalized selective combining over Rayleigh fading channels. *IEEE Trans. Commun.* 48:401–15.

[9] A. Annamalai and C. Tellambura. 2002. Analysis of hybrid selection/maximal-ratio diversity combiner with Gaussian errors. *IEEE Trans. Wireless Commun.* 1:498–512.

[10] Y. Ma and S. Pasupathy. 2004. Efficient performance evaluation for generalized selection combining on generalized fading channels. *IEEE Trans. Wireless Commun.* 3:29–34.

[11] S. W. Kim, Y. G. Kim, and M. K. Simon. 2004. Generalized selection combining based on the log-likelihood ratio. *IEEE Trans. Commun.* 52:521–24.

[12] S. W. Kim, D. S. Ha, and J. H. Reed. 2003. Minimum selection GSC and adaptive low-power RAKE combining scheme. In *Proceedings of the IEEE International Symposium on Circuits and Systems (ISCAS'03)*, Bangkok, Thailand, vol. 4, pp. 357–60.

[13] P. Gupta, N. Bansal, and R. K. Mallik. 2005. Analysis of minimum selection H-S/MRC in Rayleigh fading. *IEEE Trans. Commun.* 53:780–84.

[14] H.-C. Yang. 2006. New results on ordered statistics and analysis of minimum-selection generalized selection combining (GSC). *IEEE Trans. Wireless Commun.* 5:1876–85.

[15] R. K. Mallik, P. Gupta, and Q. T. Zhang. 2005. Minimum selection GSC in independent Rayleigh fading. *IEEE Trans. Veh. Technol.* 54:1013–21.

[16] M.-S. Alouini and H.-C. Yang. 2005. Minimum estimation and combining generalized selection combining (MEC-GSC). In *Proceedings of the IEEE International Symposium on Information Theory (ISIT'05)*, Adelaide, Australia, pp. 578–82.

[17] H.-C. Yang and M.-S. Alouini. 2005. MRC and GSC diversity combining with an output threshold. *IEEE Trans. Veh. Technol.* 54:1081–90.

[18] H.-C. Yang, N. Belhaj, and M.-S. Alouini. 2006. Bandwidth-efficient-power-greedy joint adaptive modulation and diversity combining. In *Proceedings of the International Symposium on Information Theory (ISIT '2006)*, Seattle, WA, pp. 937–941.

[19] Y.-C. Ko, H.-C. Yang, and M.-S. Alouini. 2006. Adaptive modulation and diversity combining based on output-threshold MRC. In *Proceedings of the 63rd IEEE Semiannual Vehicular Technology Conference (VTC'2006)*, Melbourne, Australia, pp. 1693–97.

[20] W. C. Jakes. 1994. *Microwave mobile communication*. 2nd ed. Piscataway, NJ: IEEE Press.

[21] M. A. Blanco and K. J. Zdunek. 1979. Performance and optimization of switched diversity systems for the detection of signals with Rayleigh fading. *IEEE Trans. Commun.* 27:1887–95.

[22] M. A. Blanco. 1983. Diversity receiver performance in Nakagami fading. In *Proceedings of the IEEE Southeastern Conference*, Orlando, FL, pp. 529–32.

[23] A. A. Abu-Dayya and N. C. Beaulieu. 1994. Analysis of switched diversity systems on generalized-fading channels. *IEEE Trans. Commun.* 42:2959–66.

[24] A. A. Abu-Dayya and N. C. Beaulieu. 1994. Switched diversity on microcellular Ricean channels. *IEEE Trans. Veh. Technol.* 43:970–76.

[25] G. Femenias and I. Furió. 1997. Analysis of switched diversity TCM-MPSK systems on Nakagami fading channels. *IEEE Trans. Veh. Technol.* 46:102–7.

[26] Y.-C. Ko, M.-S. Alouini, and M. K. Simon. 2000. Analysis and optimization of switched diversity systems. *IEEE Trans. Veh. Technol.* 49:1569–74.

[27] C. Tellambura, A. Annamalai, and V. K. Bhargava. 2001. Unified analysis of switched diversity systems in independent and correlated fading channel. *IEEE Trans. Commun.* 49:1955–65.

[28] H.-C. Yang and M.-S. Alouini. 2003. Performance analysis of multibranch switched diversity systems. *IEEE Trans. Commun.* 51:782–94.

[29] H.-C. Yang and M.-S. Alouini. 2006. Improving the performance of switched diversity with post-examining selection. *IEEE Trans. Wireless Commun.* 5:67–71.

[30] K. Cho and D. Yoon. 2002. On the general BER expression of one- and two-dimensional amplitude modulation. *IEEE Trans. Commun.* 50:1074–80.

[31] I. S. Gradshteyn and I. M. Ryzhik. 1994. *Table of integrals, series, and products.* 5th ed. San Diego: Academic Press.

[32] A. H. Nuttall. 1975. Some integrals involving the Q_M function. *IEEE Trans. Information Theory* 1:95–96.

7

Adaptive Opportunistic Beamforming in Ricean Fading Channels

Il-Min Kim
Queen's University

Zhihang Yi
Queen's University

7.1 Introduction

Fading is one of the major impairments of wireless channels, and it has traditionally been seen as an obstacle to reliable data transmission. Considerable efforts have been devoted to combat fading and to send data more reliably over wireless channels. One of the key methods to mitigate fading is to implement *diversity* techniques in wireless communication systems. There are several different ways to achieve the diversity: frequency diversity [1], time diversity [2], and spatial diversity [3–6].

Recently, a new type of diversity, so-called *multiuser diversity*, has received a lot of attention. Multiuser diversity exploits the fact that in a multiuser system with independently varying channels, there is likely to be at least a single user whose channel is near its peak at any time, if the number of users is sufficiently large. By allowing only the user having the best channel to use the system resource at a given time, the total throughput of the entire system can be maximized. In [7], Knopp and Humblet focused on the uplink channel of a multiuser system. In order to maximize the total information-theoretic capacity of the system, they showed that the optimum strategy was to transmit the signals of the user with the best channel at any time. Similar results were also obtained by studying the downlink channel from the base station to the mobile users [8]. In [9], the space-time coding was combined with multiuser diversity. In [10] and [11], the interaction between multiple-antenna diversity and multiuser diversity was discussed in detail. Furthermore, the multiuser diversity technique has been implemented in practical communication systems, such as the downlink of the IS-856 system [12].

In a multiuser system exploiting multiuser diversity, channel fading can be considered a source of randomization providing multiuser diversity, as opposed to an impairment. Therefore, large multiuser diversity gain is achieved when the dynamic range of the channel fluctuation is large or the variation rate of the channel is fast. In practice, however, the channel fluctuation may not be large enough to provide satisfactory multiuser diversity. Furthermore, when the channel fading is slower than the delay constraint of a system or an application, the user cannot wait until its channel reaches the peak, and thus, the multiuser diversity gain may get smaller. Addressing these problems, in [13], Viswanath et al. proposed *opportunistic beamforming*, which artificially induced channel fluctuation when the fluctuation of the underlying physical channel was small or the fading was slow. They studied a system where the base station was equipped with multiple antennas and the same signal was transmitted from the antennas after being multiplied by pseudorandom weight coefficients. The phase and magnitude of each weight coefficient were changing in a controlled but pseudorandom fashion. By using a single pilot signal, the signal-to-noise ratio (SNR) of the overall *equivalent* channel was measured at every user and was fed back to the base station. Based on the SNR feedback, the base station picked the user with the best equivalent channel and the data only for this user were transmitted.

Over the past few years, many works have been devoted to extend the opportunistic beamforming technique. In [14], multiple weight coefficients were used at every time slot and the one producing the highest SNR was chosen. By doing this, better performance could be achieved at the expense of the increased feedback overhead. Opportunistic beamforming was combined with the water-filling method and extended to multiple-input multiple-output systems in [15]. Furthermore, several new opportunistic schemes, including opportunistic cophasing and antenna selection, were proposed and their performance was analyzed in [16].

Very recently, Kim et al. proposed an adaptive version of opportunistic beamforming in Ricean fading channels [17]. This new scheme improved the performance substantially over Ricean fading channels without introducing multiple weight coefficients or increasing the feedback overhead. Unlike the opportunistic beamforming in [13], which generated the weight coefficients in a pseudorandom fashion, the improved opportunistic beamforming generated the weight coefficients more intelligently by estimating the directions of arrival (DOAs) of the users.

The outline of the rest of this chapter is as follows. In section 7.2, the fundamental idea and theories behind multiuser diversity are reviewed. In section 7.3, opportunistic beamforming [13] is presented in detail. In section 7.4, the adaptive version of opportunistic beamforming [17] is discussed. Finally, some conclusions are drawn in section 7.5.

7.2 Multiuser Diversity

In this section we present fundamental ideas and theories of multiuser diversity. We consider the downlink of a wireless communication system where the base station with a single antenna is communicating with M users. Let $h_k(t)$ denote the channel coefficient from the base station to the kth user at time slot t, $\eta_k(t)$ the additive noise at the kth user at time slot t, and $s(t)$ the transmitted signal from the base station at time slot t. At a given time slot t, the power of $s(t)$ is P and $\eta_k(t)$ are modeled by circularly symmetric complex Gaussian random variables with zero mean and variance ρ_k^2, i.e., $\eta_k(t) \sim \mathcal{CN}(0, \rho_k^2)$. The received signal $r_k(t)$ at the kth user at time slot t is given by

$$r_k(t) = h_k(t)s(t) + \eta_k(t). \tag{7.1}$$

The instantaneous SNR $\gamma_k(t)$ of the kth user is $\gamma_k(t) = P|h_k(t)|^2/\rho_k^2$. We define the sum capacity of this downlink channel as the maximum achievable sum of the long-term average data rate transmitted to all users. It has been shown that the sum capacity is achieved by transmitting to the user with the highest $\gamma_k(t)$ at every time slot t [8].

In Figure 7.1, we compare the sum capacity of the downlink channel under Rayleigh fading and additive white Gaussian noise (AWGN) environments, which have the same average SNR. We note that, with more than two users, the sum capacity under the

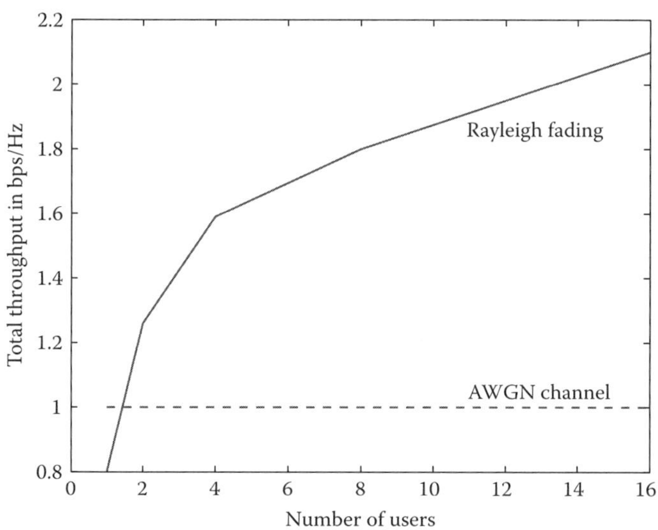

FIGURE 7.1 Sum capacity of Rayleigh fading and AWGN environments, average SNR = 0 dB. (Reproduced from Viswanath et al., 2002. © 2002, IEEE. With permission.)

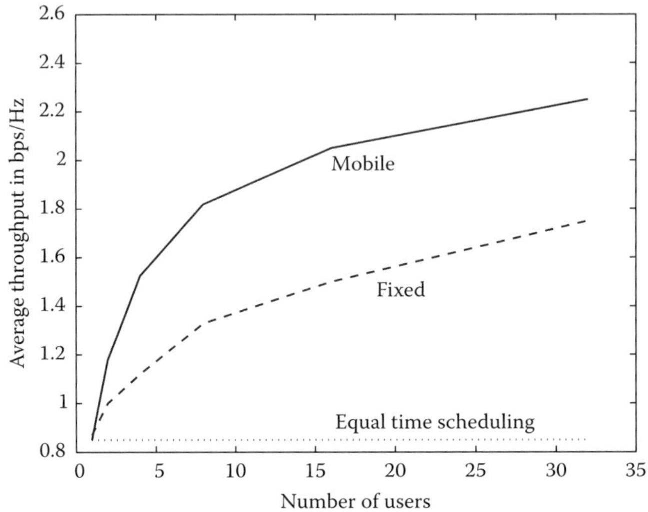

FIGURE 7.2 Multiuser diversity gain in fixed and mobile environments. (Reproduced from Viswanath et al., 2002. © 2002, IEEE. With permission.)

Rayleigh fading environment is larger than that under the AWGN environment. This is due to multiuser diversity. Figure 7.2 gives more insight into how the dynamic range and the variation rate of the channel fluctuation affect the benefits of multiuser diversity. The figure demonstrates the total throughput of the downlink channel of the IS-856 system under the following two environments: (1) fixed—users are stationary but there are objects moving around them, and (2) mobile—users move at a walking speed (3 km/h). Under both environments, the total throughput increases with the number of users, but the increase is more impressive in the mobile case. This is because the dynamic range and the variation rate of the channel fluctuation for the mobile case are larger than those for the fixed case.

In order to exploit multiuser diversity, every user needs to estimate its own instantaneous SNR $\gamma_k(t)$ by a pilot signal and feed it back to the base station. On the other hand, the base station needs to properly schedule the transmission to all the users and adapt the data rate based on the instantaneous channel condition. This technique has been adopted in some communications systems, such as IS-856 [18].

In practical communication systems, two important problems must be solved before multiuser diversity is implemented: fairness and delay. If the fading statistics of the users' channels are the same, transmitting data only to the user with the highest $\gamma_k(t)$ not only maximizes the throughput of the system, but also maximizes the throughput of every user. In reality, however, the fading statistics are not symmetrical. For example, some users may have better average SNRs, because they are closer to the base station. Furthermore, practical systems usually have a delay constraint on every user. Thus, one user cannot wait for too long until the user's instantaneous SNR becomes the highest among all the users.

In order to solve the fairness and delay problem, the proportional fair scheduling algorithm has been developed in [12]. In this algorithm, the kth user feeds back the requested data rate $R_k(t)$ to the base station, where $R_k(t)$ is the data rate the user can support at time slot t. The proportional fair scheduling algorithm also stores the average throughput $T_k(t)$ of every user in a past window of length t_c. At any time slot t, the proportional fair scheduling algorithm transmits data to the user with the largest

$$\frac{R_k(t)}{T_k(t)}. \tag{7.2}$$

In order to see how the proportional fair scheduling algorithm works, we consider a system with only two users. If the two users have identical fading statistics, the average throughput $T_k(t)$ of each user will converge to the same value. Thus, the proportional fair scheduling algorithm just picks the user with the highest $R_k(t)$, and it is fair for every user in the long term. It is possible that the first user's channel is better than that of the second user on average. Always transmitting to the user with the highest $R_k(t)$ implies that the first user will be served for most of the time and the second user will not be served in a resource fair manner. This problem is solved by using the proportional fair scheduling algorithm. Because the proportional fair scheduling algorithm selects the user based on $R_k(t)/T_k(t)$, a user is selected when its instantaneous channel condition is high relative to its own average channel condition over the time scale t_c. To an extreme case, if the second user is not served in the past window of length t_c, $T_2(t)$ becomes zero and the base station will transmit data to the second user immediately.

7.3 Opportunistic Beamforming

As we have seen in section 7.2, the dynamic range and the variation rate of the channel fluctuation determine the performance gain of multiuser diversity. Thus, if one can induce larger and faster channel fluctuation, higher gain will be achieved. This can be realized by opportunistic beamforming [13]. In this scheme, N antennas are deployed at the base station and there are M users in the system. Let $h_{n,k}(t)$ denote the channel coefficient from the nth antenna to the kth user at time slot t. The transmitted signal $s(t)$ is first multiplied by a complex weight coefficient

$$w_n(t)=\sqrt{\alpha_n(t)}e^{j\phi_n(t)}$$

and then transmitted by the nth antenna, for $n = 1, \cdots, N$. In order to satisfy the power constraint, it is assumed that

$$\sum_{n=1}^{N} \alpha_n(t)=1.$$

Thus, the received signal $r_k(t)$ at the kth user is given by

$$r_k(t)=\left(\sum_{n=1}^{N}\sqrt{\alpha_n(t)}e^{j\phi_n(t)}h_{n,k}(t)\right)s(t)+\eta_k(t). \tag{7.3}$$

The overall equivalent channel gain $\tilde{h}_k(t)$ is

$$\tilde{h}_k(t)=\sum_{n=1}^{N}\sqrt{\alpha_n(t)}e^{j\phi_n(t)}h_{n,k}(t), \tag{7.4}$$

and the overall equivalent instantaneous SNR is

$$\tilde{\gamma}_k(t)=P\frac{|\tilde{h}_k(t)|^2}{\rho_k^2}. \tag{7.5}$$

The power fraction $\alpha_n(t)$ is modeled as samples of a random variable varying from 0 to 1, and the artificial phase shift $\phi_n(t)$ is modeled as samples of a random variable uniformly distributed over $[0,2\pi)$. At different transmit antennas, $\alpha_n(t)$ and $\phi_n(t)$ are varying independently.

The kth user feeds the equivalent instantaneous SNR $\tilde{\gamma}_k(t)$ back to the base station, and the base station selects the user by using the proportional fair scheduling algorithm. Note that the kth user does not need to estimate $h_{n,k}(t)$. Actually, the existence of multiple antennas at the base station can be completely transparent to the users. Note that the dynamic range and the variation rate of the equivalent channel $\tilde{h}_k(t)$ can be controlled by $w_n(t)$.

The performance of opportunistic beamforming is demonstrated in Figure 7.3 under the same simulation environment as in Figure 7.2. Two antennas are deployed at the base stations and $\alpha_n(t) = 1/\sqrt{2}$ at any time slot t. It can be seen that the performance of the fixed case is considerably improved by using opportunistic beamforming. This is because the channel is changing faster and the dynamic range of the channel fluctuation is larger after the use of $\alpha_n(t)$ and $\phi_n(t)$. In the following, the slow fading and fast fading environments are considered separately.

7.3.1 Slow Fading

In the slow fading environment, we assume that $h_{n,k}(t) = h_{n,k}$ in a time slot t. If only one antenna is used at the base station, the SNR $\gamma_n(t)$ of every user will remain constant at every time slot t, and hence, no multiuser diversity can be exploited. However, by using opportunistic beamforming, the equivalent channel $\tilde{h}_k(t)$ will vary in time and provide the opportunity for achieving multiuser diversity.

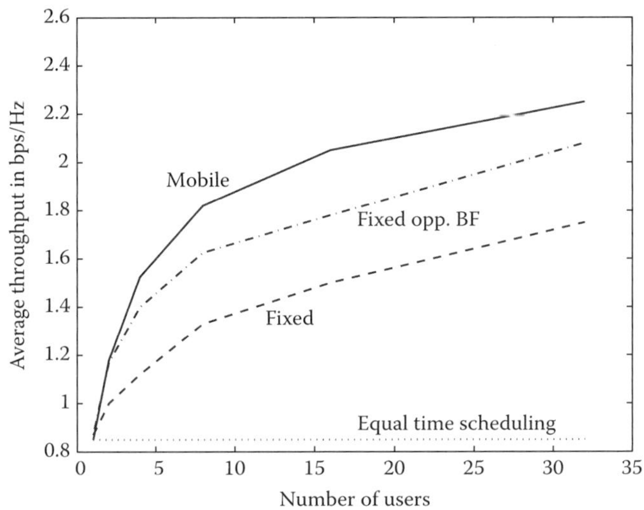

FIGURE 7.3 Amplification in multiuser diversity gain with opportunistic beamforming in a fixed environment. (Reproduced from Viswanath et al., 2002. © 2002, IEEE. With permission.)

For a particular user k, it is well known that coherent transmit beamforming can maximize $\tilde{\gamma}_n(t)$ by setting [13]

$$\alpha_n = \frac{|h_{n,k}|^2}{\sum_{n=1}^{N} |h_{n,k}|^2}, \tag{7.6}$$

$$\phi_n = -\arg(h_{n,k}), \tag{7.7}$$

which is called the beamforming configuration. In this case, however, the base station needs to know $h_{n,k}$, for $n = 1, \cdots, N$, and hence, this amount of feedback information might be prohibitive. On the other hand, opportunistic beamforming selects the user when its equivalent SNR is at the peak. Therefore, opportunistic beamforming may achieve the performance of coherent transmit beamforming, while only requiring the feedback of $\tilde{\gamma}_k(t)$. This is confirmed by the following theorem.

Theorem 7.1

[13] "Suppose the slow fading states of the users are i.i.d and are discrete, and the joint stationary distribution of $(\alpha_1(t), \ldots, \alpha_N(t), \phi_1(t), \ldots, \phi_2(t))$ *is the same as that of*

$$\left(\frac{|h_{1k}|^2}{\sum_{n=1}^{N} |h_{n,k}|^2}, \cdots, \frac{|h_{n,k}|^2}{\sum_{n=1}^{N} |h_{n,k}|^2}, -\arg(h_{1k}), \cdots, -\arg(h_{n,k}) \right) \tag{7.8}$$

for the slow fading state of any individual user k. Then, almost surely, we have

$$\lim_{M\to\infty} KT_k^{(M)} = R_k^{bf} \tag{7.9}$$

for all k. $T_k^{(M)}$ *is the average throughput of user k in a system with M users and* R_k^{bf} *is the instantaneous data rate that user k achieves when it is in the beamforming configuration, i.e. when its instantaneous SNR is*

$$P\sum_{n=1}^{N}\left|h_{n,k}\right|^2 / \rho_k.\text{"}$$

Proof: See appendix A in [13].

When the number of users is sufficiently large, theorem 7.1 implies that, with very high probability, one user is selected when it is in its beamforming configuration and every user is allocated an equal amount of time. The stationary distributions of $\alpha_n(t)$ and $\phi_n(t)$ required by this theorem are given in closed form in [13]. In a slow Rayleigh fading environment with ten antennas at the base station, the throughput of one specific user is given in Figure 7.4. The proportional fair scheduling algorithm is employed. One can see that the throughput converges to that of coherent transmit beamforming asymptotically with the number of users. Figure 7.5 demonstrates the total throughput of all users.

7.3.2 Fast Fading

We have seen that opportunistic beamforming can considerably improve the performance when the underlying channels are slow fading. But if the underlying channels are already fast fading, can opportunistic beamforming improve the performance? The performance gain of opportunistic beamforming is from the randomization of $\alpha_n(t)$ and $\phi_n(t)$, which makes the dynamic range of the channel fluctuation larger and the variation rate of the channel faster. Therefore, if the dynamic range of the equivalent channel $\tilde{h}_k(t)$ becomes larger after using opportunistic beamforming, the system will achieve better performance.

We first consider the independent Rayleigh fading environment, where $h_{n,k}(t)$ are i.i.d circularly symmetric complex Gaussian random variables with zero mean. In this case, it is easy to see from (7.4) that the distribution of $\tilde{h}_k(t)$ is exactly the same as that of the $h_{n,k}(t)$. The use of $\alpha_n(t)$ and $\phi_n(t)$ neither makes the dynamic range of the channel fluctuation larger nor makes the variation rate of the channel faster. Therefore, in an independent fast Rayleigh fading environment, opportunistic beamforming is not able to improve the performance.

In contrast, when the underlying channels are Ricean fading channels, opportunistic beamforming can bring considerable performance gain. The Ricean channel can be modeled as [19]

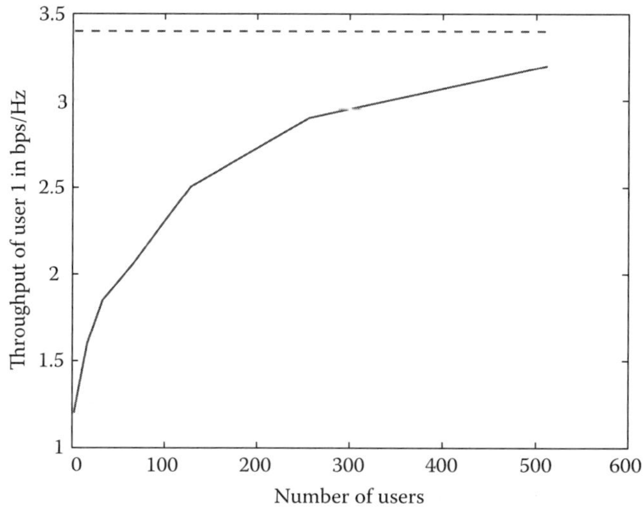

FIGURE 7.4 Throughput in bps/Hz for user 1 multiplied by number of users scheduled for slow Rayleigh fading at average SNR = 0 dB with the proportional fair scheduling algorithm. Performance of coherent transmit beamforming for user 1 and scheduled at all times is plotted as a dotted line. There are ten antennas in this experiment. (Reproduced from Viswanath et al., 2002. © 2002, IEEE. With permission.)

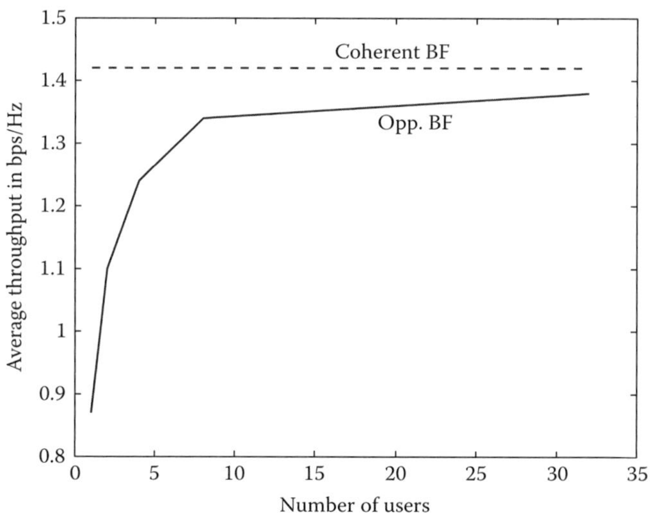

FIGURE 7.5 Total throughput in bps/Hz averaged over slow Rayleigh fading at average SNR = 0 dB with the proportional fair scheduling algorithm. Performance of coherent transmit beamforming is also plotted. (Reproduced from Viswanath et al., 2002. © 2002, IEEE. With permission.)

$$h_{n,k}(t) = \sqrt{\frac{K_k}{1+K_k}} \exp(jn\theta_k) + \sqrt{\frac{1}{1+K_k}} b_{nk}(t), \tag{7.10}$$

where K_k is the Ricean K-factor and θ_k is related to the DOA of the user. The first term in (7.10) denotes the line-of-sight (LOS) component of the channel. The second term is the diffused component and $b_{nk}(t) \sim \mathcal{CN}(0,1)$. Then the equivalent channel gain becomes

$$\tilde{h}_k(t) = \sum_{n=1}^{N} \sqrt{\frac{\alpha_k(t)K_k}{1+K_k}} \exp\left(j(n\theta_k + \phi_n(t))\right) + \sum_{n=1}^{N} \sqrt{\frac{\alpha_n(t)}{1+K_k}} \exp(j\phi_n(t)) b_{nk}(t). \tag{7.11}$$

It can be seen that the randomization of $\alpha_n(t)$ and $\phi_n(t)$ will not change the dynamic range of the diffused term, which is the same as in the Rayleigh fading environment. However, the randomization of $\alpha_n(t)$ and $\phi_n(t)$ induces the fluctuation of the LOS component significantly, which leads to performance improvement. Therefore, if the LOS component is more dominant than the diffused path, larger fluctuation will be created by opportunistic beamforming and more performance gain can be achieved. To an extreme case that $K_k \to \infty$, the channel reduces to the slow fading case and opportunistic beamforming can improve the performance considerably, as we have seen in section 7.3.1. Figure 7.6 shows the total throughput for Ricean fading channels with $K_k = 10$. There is an impressive improvement in performance after using opportunistic beamforming. For comparison, the performance curve for the Rayleigh fading channels is plotted.

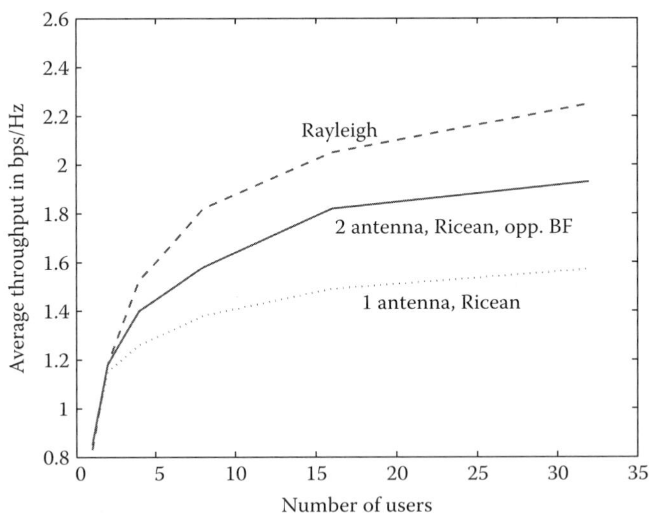

FIGURE 7.6 Total throughput as a function of the number of users under Ricean fading, with and without opportunistic beamforming. The power allocations $\alpha_n(t)$ are uniformly distributed in [0, 1] and the phases $\theta_n(t)$ are uniform in [0,2π]. (Reproduced from Viswanath et al., 2002. © 2002, IEEE. With permission.)

7.4 Adaptive Opportunistic Beamforming in Ricean Channels

In section 7.3, it was indicated that the throughput of opportunistic beamforming con verged to that of coherent transmit beamforming, when the number of users was sufficiently large. However, the number of users required grows exponentially with the number of antennas. For example, in Figure 7.4, the throughput is approximately 65% of coherent transmit beamforming even with one hundred users. In order to improve the throughput up to 90% of coherent transmit beamforming, the number of users must be increased up to about four hundred, which may not be a practical number of users in a cell. With a realistic number of users, the performance of opportunistic beamforming falls much lower than that of coherent transmit beamforming. This problem can be solved by the adaptive opportunistic beamforming proposed in [17] over Ricean channels. This section is devoted to presenting the adaptive opportunistic beamforming technique and its performance. In this section, we will refer to opportunistic beamforming in [13] as conventional opportunistic beamforming in order to distinguish it from adaptive opportunistic beamforming.

7.4.1 Adaptive Opportunistic Beamforming

The downlink of a single cell is considered, where the number of users is M and each user has a single antenna. The base station is equipped with a *linear* antenna array with N elements. The fading channel is a Ricean channel modeled by (7.10). The parameter θ_k is related to the DOA Θ_k of the kth user as follows:

$$\theta_k = \frac{2\pi d f_0 \cos\Theta_k}{c}, \tag{7.12}$$

where c is the speed of propagation of the plane wave, f_0 is the carrier frequency of the transmitted signal, and d is the spacing between two antenna elements [4].

It is assumed that a mini-slot exists at the beginning of each time slot. Through the mini-slot, a known pilot signal $s_k(t)$ is transmitted from every antenna after being multiplied by a weight coefficient $w_n(t)$. When the transmission power is uniformly allocated to the N antenna elements, $w_n(t)$ is given by

$$w_n(t) = \frac{1}{\sqrt{N}} \exp(jn\phi(t)). \tag{7.13}$$

The signal $r_k(t)$ received by the kth user during the mini-slot of time slot t is given by

$$r_k(t) = \tilde{h}_k(t) s_k(t) + \eta_k(t), \tag{7.14}$$

where $\eta_k(t) \sim \mathcal{CN}(0, \rho_k^2)$. The overall equivalent channel $\tilde{h}_k(t)$ for user k is given by

$$\tilde{h}_k(t)=\sqrt{a_k}\sum_{n=1}^{N}\exp\left(jn\left(\theta_k+\phi(t)\right)\right)+B_k(t), \tag{7.15}$$

where

$$\sqrt{a_k}=\sqrt{\frac{K_k}{N(1+K_k)}}, \tag{7.16}$$

$$B_k(t)=\sqrt{\frac{1}{N(1+K_k)}}\sum_{n=1}^{N}b_{nk}(t)\exp\left(jn\phi(t)\right). \tag{7.17}$$

It is easy to show that $B_k(t) \sim \mathcal{CN}(0,\sigma_k^2)$, where $\sigma_k^2 = 1/(1+K_k)$.

Let $H_k(t)$ denote the magnitude of $\tilde{h}_k(t)$:

$$H_k(t)=\left|\tilde{h}_k(t)\right|=\left|\sqrt{a_k}\sum_{n=1}^{N}\exp(jn(\theta_k+\phi(t)))+B_k(t)\right|. \tag{7.18}$$

Each user measures only $H_k(t)$ and then determines the data rate $R_k(t)$ at which the data can be reliably transmitted from the base station to the user with a predetermined SNR threshold. The determined rates $R_k(t)$ for all the users are fed back to the base station. The requested rate $R_k(t)$ can be expressed as $R_k(t) = f(H_k(t))$, where $f(\cdot)$ is a nondecreasing function, which can be assumed to be known at the base station. Hence, the values of $H_k(t)$ are also assumed to be known at the base station in this section.

On the right-hand side of (7.18), the first term

$$\sqrt{a_k}\sum_{n=1}^{N}\exp\left(jn\left(\theta_k+\phi(t)\right)\right)$$

is related to the LOS component and the second term $B_k(t)$ to the diffused component. Let $G_k(t)$ denote the magnitude of the LOS component:

$$G_k(t)\overset{def}{=}\left|\sqrt{a_k}\sum_{n=1}^{N}\exp\left(jn\left(\theta_k+\phi(t)\right)\right)\right|. \tag{7.19}$$

Then one can easily see that $G_k(t)$ is maximized with $\phi(t) = -\theta_k$ and the maximum value is $\sqrt{a_k}N$. Motivated by this observation, the following adaptive opportunistic beamforming algorithm is proposed to improve the performance [17]:

1. *As in the conventional opportunistic beamforming, the information on $H_k(t)$, or equivalently $R_k(t)$, is fed back to the base station.*
2. *The base station estimates the DOAs* of the users using the values of $H_k(t)$.† The estimated DOA of user k is denoted by $\hat{\theta}_k$, $k = 1, \cdots, M$.*
3. *The base station conducts the proportional fair scheduling to choose a user. Let us assume that user k^* is chosen.*
4. *When the base station transmits the data to user k^* during a time slot excluding the mini-slot, the artificial phase shift $\phi(t)$ of the weight coefficient is set to $-\hat{\theta}_{k^*}$ and this weight coefficient is multiplied to the transmitted data.*

Note that this adaptive scheme forms the beams only in the directions where users really exist, as opposed to conventional opportunistic beamforming, which forms the beams blindly over the omnidirectional space. Therefore, the adaptive scheme can improve the performance without wasting resources such as time and power. In adaptive opportunistic beamforming, the LOS-related component of $H_k(t)$, which is $G_k(t)$, is maximized. Therefore, the performance heavily depends on the K-factors. Specifically, larger K-factors $\{K_k\}$ result in larger performance improvement; smaller K-factors result in smaller improvement. In an extreme case where the K-factors are zero, which is the Rayleigh case, the adaptive scheme reduces to the conventional scheme because $\phi(t)$ is chosen as a sample of a random variable uniformly distributed over $[0,2\pi)$.

The vital step of adaptive opportunistic beamforming is to estimate the DOA as accurately as possible with the information available. In the previous publications, a number of algorithms for DOA estimation have been studied, such as the multiple signal classification (MUSIC) method [20], Root-MUSIC [21], and the estimation of signal parameters via the rotational invariance technique [22] (also see [4] and the references therein). Those previous methods, however, are not applicable to the system considered, because the base station has only very limited channel information: the magnitude values $H_k(t)$ of the equivalent channels. In the next section, a new and efficient DOA estimation algorithm requiring only $H_k(t)$ values is proposed for use in adaptive opportunistic beamforming.

7.4.2 Estimation of Users' DOAs

In this section, a maximum-likelihood (ML) estimator of $\{\theta_k\}_{k=1}^{M}$ is developed.‡ To this end, the probability density function (PDF) and cumulative density function (CDF) of $H_k(t)$ are first derived in the following theorem:

* To be precise, Θ_k is the DOA of user k. However, since Θ_k is uniquely determined by θ_k, we simply refer to θ_k as the DOA of user k in this section.

† The DOAs must be estimated based only on $H_k(t)$ values to ensure that the channel estimation algorithm of the receivers and the feedback overhead remain the same as in conventional opportunistic beamforming.

‡ Note that ML estimators can be asymptotically considered as minimum variance unbiased estimators [23].

Theorem 7.2

[17] "Let $A_k(t)$ be defined as follows:

$$A_k(t) \overset{def}{=} \begin{cases} \sqrt{a_k} + 2\sqrt{a_k} \sum_{n=1}^{(N-1)/2} \cos\left(n\left(\theta_k + \phi(t)\right)\right), & \text{if } N \text{ is odd} \\ 2\sqrt{a_k} \sum_{n=1}^{N/2} \cos\left(\left(n - \frac{1}{2}\right)\left(\theta_k + \phi(t)\right)\right), & \text{if } N \text{ is even.} \end{cases} \tag{7.20}$$

Then the PDF $p_{H_k(t)}(\cdot)$ and CDF $F_{H_k(t)}(\cdot)$ of $H_k(t)$ are given by

$$p_{H_k(t)}(x) = \frac{2x}{\sigma_k^2} \exp\left(-\frac{x^2 + A_k^2(t)}{\sigma_k^2}\right) I_0\left(\frac{2xA_k(t)}{\sigma_k^2}\right), \tag{7.21}$$

$$F_{H_k(t)}(x) = 1 - Q_1\left(\frac{\sqrt{2}A_k(t)}{\sigma_k^2}, \frac{\sqrt{2}x}{\sigma_k^2}\right), \tag{7.22}$$

where $I_0(\cdot)$ is the modified Bessel function of order zero and $Q_m(\cdot,\cdot)$ is the Marcum Q-function."

Proof: See appendix A in [17].

It is very interesting to see that (7.21) is exactly the same form as a Ricean distribution, except that $A_k(t)$ is a summation of multiple cosine functions. When $N = 1$, (7.21) reduces to the well-known Ricean PDF. As in the conventional Ricean case, we define $\Omega_k(t)$ to be the total power of (7.21) as follows:

$$\Omega_k(t) \overset{def}{=} E_{H_k(t)}\left[H_k^2(t)\right] = A_k^2(t) + \sigma_k^2. \tag{7.23}$$

Also, the K-factor of (7.21) can be defined as follows:

$$K_k(t) = \frac{A_k^2(t)}{\sigma_k^2}. \tag{7.24}$$

It follows that (7.21) can be rewritten as

$$p_{H_k(t)}(x) = \frac{2x\left(1 + K_k(t)\right)}{\Omega_k(t)} \exp\left(-K_k(t) - \frac{\left(1 + K_k(t)\right)x^2}{\Omega_k(t)}\right) I_0\left(2x\sqrt{\frac{K_k(t)\left(K_k(t) + 1\right)}{\Omega_k(t)}}\right). \tag{7.25}$$

The DOAs are estimated using the $H_k(t)$ values during an estimation period $\mathcal{P}$, which is defined to be a set of L mini-slots. Each estimation period is divided into Q subestimation periods $\mathcal{P}_i$, $i = 1, \cdots, Q$, such that $\mathcal{P} = \mathcal{P}_1 \cup \mathcal{P}_2 \cup \cdots \cup P_Q$ and $\mathcal{P}_i \cap \mathcal{P}_j = 0$, $i \neq j$. Each $\mathcal{P}_q$ is composed of L_q mini-slots with

$$L = \sum_{q=1}^{Q} L_q.$$

In subestimation period $\mathcal{P}_q$, the artificial phase shift $\phi(t)$ of adaptive opportunistic beamforming is set to ϕ_q, i.e., $\phi(t) = \phi_q \in [0,2\pi)$, for $t \in \mathcal{P}q$, $q = 1, 2, \cdots, Q$.* For example, we can set $\phi_q = (q/Q)2\pi$, for $q = 1, 2, \cdots, Q$. We define $A_{k,q}$, $\Omega_{k,q}$, and $\mathcal{K}_{k,q}$ as follows: $A_{k,q} = A_k(t)|_{t\in\mathcal{P}_q}$, $\Omega_{k,q} = \Omega_k(t)|_{t\in\mathcal{P}_q} = A_{k,q}^2 + \sigma_k^2$, and $\mathcal{K}_{k,q} = \mathcal{K}_k(t)|_{t\in\mathcal{P}_q}$.

Given σ_k and $A_{k,q}$ at the base station, the ML estimation of θ_k is given by

$$\hat{\theta}_k^{ML} = \arg \max_{\theta_k \in [0,2\pi)} \prod_{q=1}^{Q} \left(\frac{2}{\sigma_k^2}\right)^{L_q} \left\{ \prod_{t\in\mathcal{P}_q} H_k(t) \exp\left(-\frac{H_k^2(t) + A_{k,q}^2}{\sigma_k^2}\right) I_0\left(\frac{2H_k(t)A_{k,q}}{\sigma_k^2}\right)\right\} \quad (7.26)$$

for $k = 1, \cdots, M$. Taking logarithm, a simpler estimator is given by

$$\hat{\theta}_k^{ML} = \arg \max_{\theta_k \in [0,2\pi)} \sum_{q=1}^{Q} \sum_{t\in P_q} \left[\ln I_0\left(\frac{2H_k(t)A_{k,q}}{\sigma_k^2}\right) - A_{k,q}^2 \right] \quad (7.27)$$

for $k = 1, \cdots, M$. In order to reduce the computational complexity further at the loss of the ML optimality, suboptimum estimators may be considered. In particular, noting that $G_k(t)$ is maximized when $\phi_q = -\theta_k$, an efficient and very simple suboptimum estimator can be given by

$$\hat{\theta}_k^{SUB} = -\left(q^*/Q\right)2\pi, \quad (7.28)$$

$$q^* = \arg \max_{q=1,\cdots,Q} \frac{1}{L_q} \sum_{t\in P_q} H_k^2(t). \quad (7.29)$$

7.4.3 Estimation of *K*-Factors of the Physical Channels

In the proposed DOA estimation algorithms, it was assumed that the exact $\{K_k\}_{k=1}^{M}$ values were known at the base station. In practical systems, however, each user needs

* Recall that in conventional opportunistic beamforming, $\phi(t)$ is randomly chosen from $[0,2\pi)$. Also, note that the pilot overhead is the same for the adaptive and conventional schemes, because only one mini-slot of fixed length is used for pilot signaling in each time slot for both schemes.

to estimate the K-factor and feed it back to the base station. Unfortunately, this may raise some problems, such as the change of the receiver of every user and the increased feedback overhead. These problems can be completely avoided if the base station can directly estimate $\{K_k\}$ based on only $H_k(t)$ values, which are already available at the base station. To this end, we first have the following theorem, which gives the relation between the K-factors $\{K_k\}$ of the physical channels $h_{n,k}(t)$ and the K-factors $\{\mathcal{K}_{k,q}\}$ of (7.21).

Theorem 7.3

[17] "When $\phi_q = (q/Q)2\pi$, $q = 1, \cdots, Q$, the K-factors $\{K_k\}$ of the physical channels $h_{n,k}(t)$ can be expressed by the K-factors $(\mathcal{K}_{k,q}\}$ of (7.21) as follows:

$$K_k = \lim_{Q\to\infty} \frac{1}{Q}\sum_{q=1}^{Q} \mathcal{K}_{k,q}, \qquad k=1,\cdots,M." \tag{7.30}$$

Proof: See appendix D in [17].

Then the following theorem shows that $\{\mathcal{K}_{k,q}\}$ can be estimated only by using $\{\Omega_{k,q}\}$ and $\{H_k(t)\}$:

Theorem 7.4

[17] "The K-factors $\{\mathcal{K}_{k,q}\}$ of (7.21) can be expressed in terms of $\Omega_{k,q} = E_q[H_k^2(t)]$ and $E_q[H_k^4(t)]$:

$$\mathcal{K}_{k,q} = \frac{\sqrt{1-\gamma_{k,q}}}{1-\sqrt{1-\gamma_{k,q}}}, \tag{7.31}$$

where

$$\gamma_{k,q} = \frac{-\Omega_{k,q}^2 + E_q\left[H_k^4(t)\right]}{\Omega_{k,q}^2} \tag{7.32}$$

and E_q denotes the expectation with respect to $H_k(t)$ for $t \in \mathcal{P}_q$."

Proof: See proof of theorem 6 in [17].

The final step is to estimate $\Omega_{k,q}^2$ and $E_q[H_k^4(t)]$ values in (7.32). They can be estimated as in the classical single-input single-output (SISO) systems. Let $\hbar(t)$ and $\bar{L}$ denote the received signal envelope at time t and the total number of samples obtained in the SISO system, respectively. Then it is well known that

$$\left(1/\bar{L}\right)\sum_{t=1}^{\bar{L}} \hbar^2(t)$$

is the ML estimate of $E_\hbar(t)[\hbar^2(t)]$ [24], and $E_\hbar(t)[\hbar^4(t)]$ can be estimated simply as

$$\left(1/\bar{L}\right)\sum_{t=1}^{\bar{L}} \hbar^4(t)$$

[25]. Exactly in the same manner, one can show that $\Omega_{k,q}$ and $E_q[H_k^4(t)]$, and thus $\gamma_{k,q}$, can be estimated as follows:

$$\hat{\gamma}_{k,q} = \frac{-\left(\hat{\Omega}_{k,q}^{ML}\right)^2 + \frac{1}{L_q}\sum_{t\in P_q} H_k^4(t)}{\left(\hat{\Omega}_{k,q}^{ML}\right)^2}, \tag{7.33}$$

where $\hat{\Omega}_{k,q}^{ML}$ denotes the ML estimate of $\Omega_{k,q}$, which is given by

$$\hat{\Omega}_{k,q}^{ML} = \frac{1}{L_q}\sum_{t\in P_q} H_k^2(t). \tag{7.34}$$

It follows that the *K*-factors are estimated as follows:

$$\hat{K}_k = \frac{1}{Q}\sum_{q=1}^{Q}\hat{K}_{k,q} = \frac{1}{Q}\sum_{q=1}^{Q}\frac{\sqrt{1-\hat{\gamma}_{k,q}}}{1-\sqrt{1-\hat{\gamma}_{k,q}}}. \tag{7.35}$$

By analytically deriving the Cramer-Rao lower bound (CRLB) of the *K*-factors $\{K_k\}_{k=1}^{M}$, it has been demonstrated that this is a good estimator [17].

7.4 Performance Evaluation

In this section, the performance of the ML estimator of DOAs, the Ricean *K*-factor estimation algorithm, and the adaptive opportunistic beamforming scheme based on DOAs and the *K*-factor estimation are numerically evaluated.

7.4.1 ML Estimation of Users' DOAs

In order to concentrate on the performance evaluation of the DOA estimation, the true values of the Ricean *K*-factor are assumed to be known at the base station. Figure 7.7

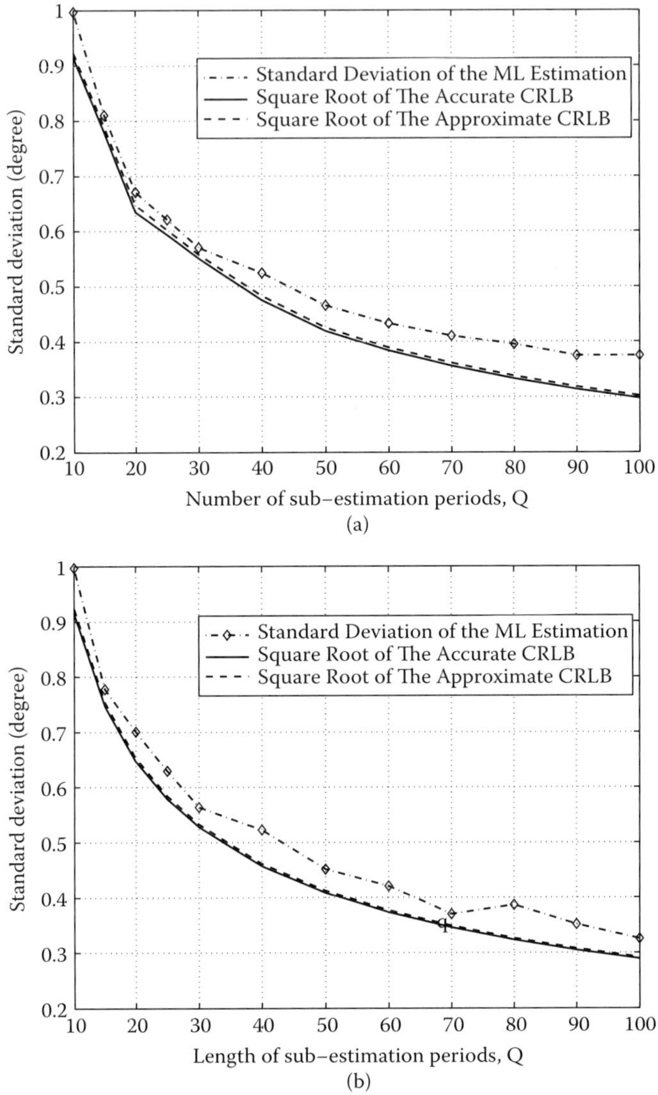

FIGURE 7.7 Comparison of CRLB and the estimation variance of the ML estimator of $\{\theta_k\}_{k=1}^{M}$ (a) $N = 8$, $K_k = 5$, $M = 1$, $L_q = 10$; (b) $N = 8$, $K_k = 5$, $M = 1$, $Q = 10$. (Reproduced from Kim et al., 2006. © 2006, IEEE. With permission.)

shows the square root of the CRLB and the standard deviation of the ML estimator of $\{\theta_k\}_{k=1}^{M}$, when $\{\theta_k\}_{k=1}^{M}$ are generated randomly from $[0,2\pi)$ and $\phi_q = (q/Q)2\pi$. One can see that the standard deviation of the DOA estimation algorithm is very close to the CRLB. This agrees with the theory that the proposed ML estimator of DOAs is asymptotically a minimum variance estimator.

7.4.2 PDF of $\max_{k=1,\cdots,M} H_k(t)$

We define $\mathcal{H}_M(t)$ as follows: $\mathcal{H}_M(t) \stackrel{def}{=} \max_{k=1,\ldots,M} H_k(t)$. The PDFs of $\mathcal{H}_M(t)$ of conventional and adaptive opportunistic beamforming algorithms are compared in Figure 7.8. It can be seen that adaptive opportunistic beamforming considerably shifts the range of $\mathcal{H}_M(t)$

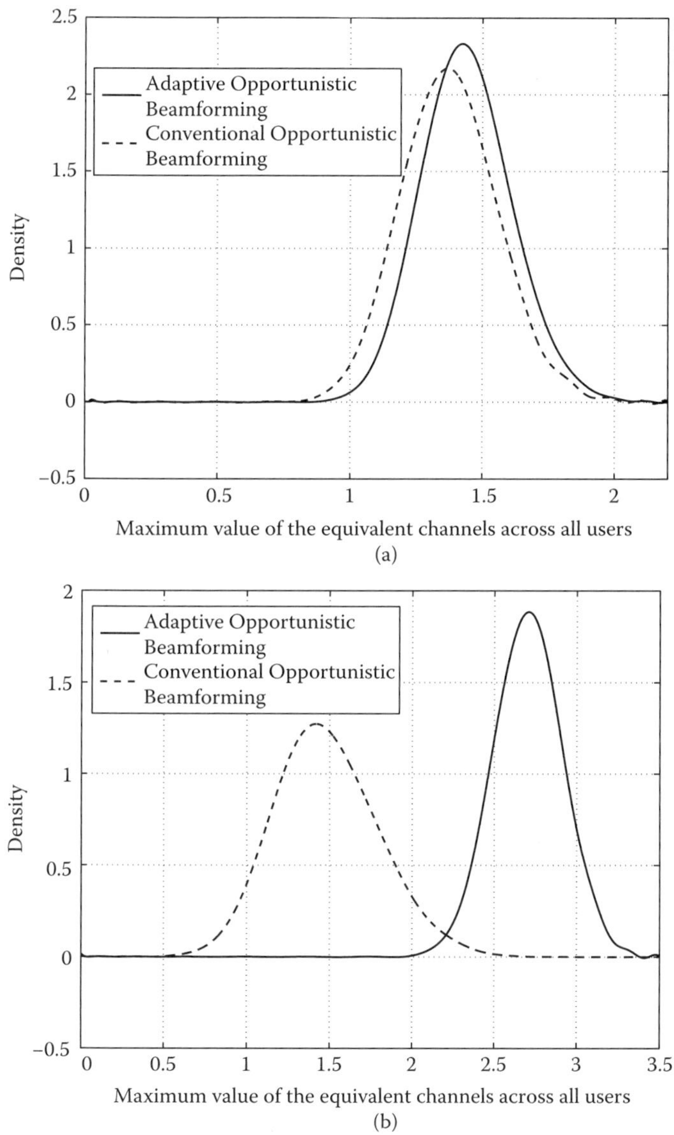

FIGURE 7.8 Comparison of the PDF of the maximum value $\mathcal{H}_k(t)$ of the equivalent channels in conventional and adaptive opportunistic beamforming: (a) $K_k = 10$, $M = 5$, $N = 2$; (b) $K_k = 10$, $M = 5$, $N = 8$. (Reproduced from Kim et al., 2006. © 2006, IEEE. With permission.)

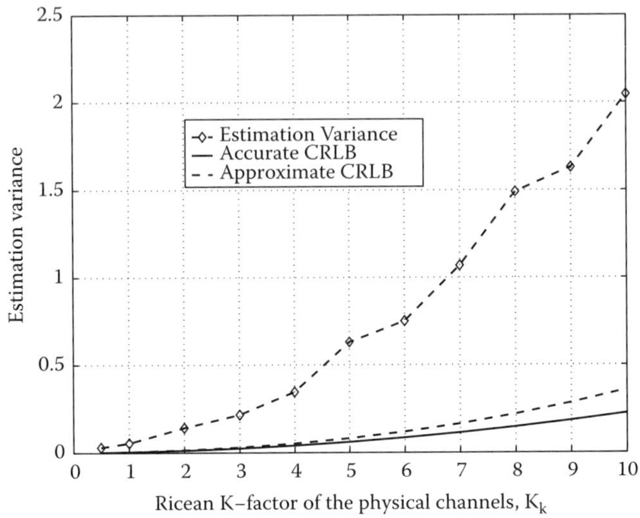

FIGURE 7.9 Comparison of the CRLB and the estimation variance of the *K*-factor estimator of the physical channels $h_{nk}(t)$, $N = 8$, $M = 1$, $Q = 50$, $L_q = 20$. (Reproduced from Kim et al., 2006. © 2006, IEEE. With permission.)

toward $+\infty$. This implies that the maximum channel gain exploited by multiuser diversity has increased significantly, and thus, the multiuser diversity gain can be notable. As can be seen from (7.18), adaptive opportunistic beamforming achieves more improvement when the number N of transmission antennas grows. This is because the value of $H_k(t)$ increases with N when $\phi(t) \simeq -\theta_k$, and in turn, the value of $\mathcal{H}_M(t)$ increases with N.

7.4.3 Estimation of *K*-Factors of the Physical Channels

In order to evaluate the performance of the *K*-factor estimator, the estimation variance and the CRLB are compared in Figure 7.9 when $\phi_q = (q/Q)2\pi$. One can see that the difference between the CRLB and the actual variance grows with the *K*-factor.* However, the difference itself is not very large. For example, when $K_k = 10$, the difference is approximately 2, and thus the standard deviation is approximately 1.4, which is 14% of the true K_k value. From the numerical results in the next section, it turns out that the effect of the *K*-factor estimation errors on the throughput performance is negligible.

7.4.4 Throughput

The final performance measure is the throughput obtained by different schemes. In slow fading channels, the length t_c of the past window is $t_c = +\infty$, and hence, the proportional fair scheduling algorithm converges to select the user with the highest $R_k(t)$. The

* In many previous *K*-factor estimators for SISO systems, the estimation variance also increases with the *K*-factor. For example, see [25].

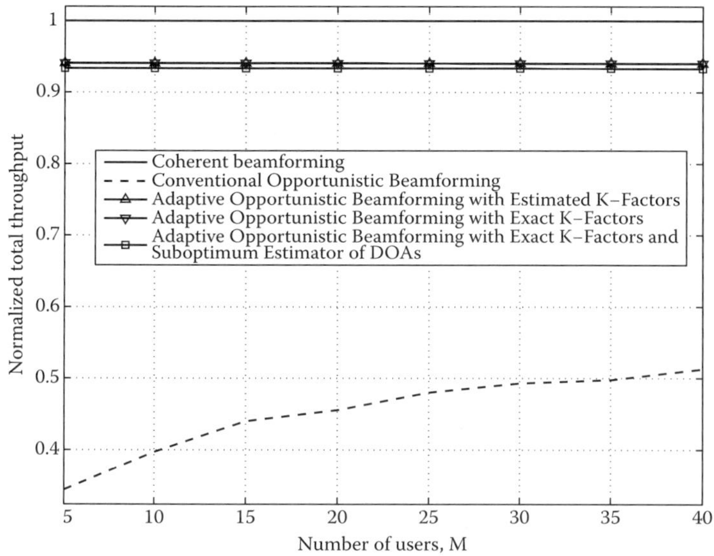

FIGURE 7.10 Normalized total throughput in the slow Ricean fading channels, $K_k = 10$, $N = 8$, $Q = 50$, $L_q = 20$. (Reproduced from Kim et al., 2006. © 2006, IEEE. With permission.)

throughput values of coherent beamforming, conventional opportunistic beamforming, and adaptive opportunistic beamforming are compared. Particularly, to see the impact of the *K*-factor estimation errors on the performance of adaptive opportunistic beamforming, the throughput is evaluated with exact *K*-factors and estimated *K*-factors. All throughput values obtained are normalized by the throughput of coherent transmit beamforming, which is the performance limit. The average SNR $PH_k^2(t)/\rho_k^2$ of each user is set to 0 dB, where P is the transmission power.

Figure 7.10 shows that adaptive opportunistic beamforming considerably outperforms conventional opportunistic beamforming, and actually the performance is very close to that of coherent beamforming. In particular, even with the small number of users, the adaptive scheme performs very well. Also, the throughput degradation due to the *K*-factor estimation errors and the use of the suboptimum DOA estimator is negligible. This implies that the estimation of users' DOAs is still very accurate with the errors of *K*-factor estimation, and the suboptimum DOA estimator is working very well at a very low complexity.

Figure 7.11 shows the impact of the number of antennas on the performance. As discussed before, the performance of conventional opportunistic beamforming heavily depends on the number of antennas, as it forms beams randomly and blindly. On the other hand, the adaptive scheme works much better than the conventional one with few antennas.

Finally, Figure 7.12 shows the effect of the *K*-factor on the performance. With smaller *K*-factor, the performance improvement diminishes. In the extreme case of $K = 0$, the conventional scheme and the adaptive scheme have the same performance, as expected.

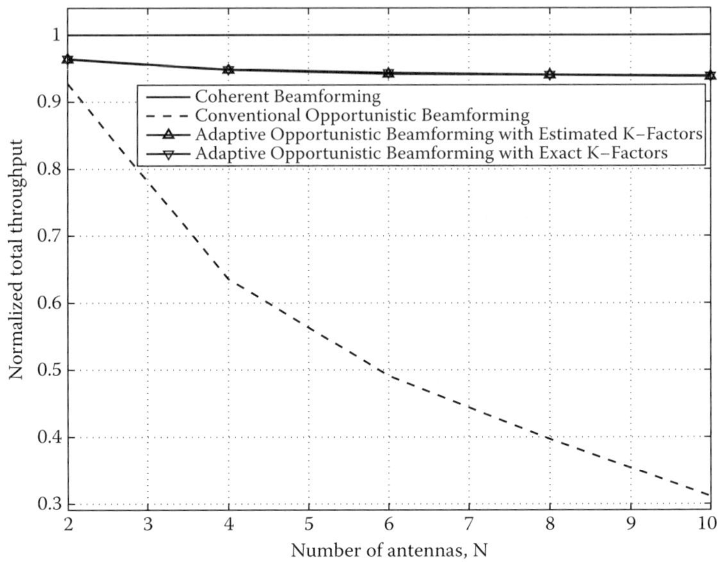

FIGURE 7.11 Normalized total throughput in the slow Ricean fading channels, $K_k = 10$, $M = 10$, $Q = 50$, $L_q = 20$. (Reproduced from Kim et al., 2006. © 2006, IEEE. With permission.)

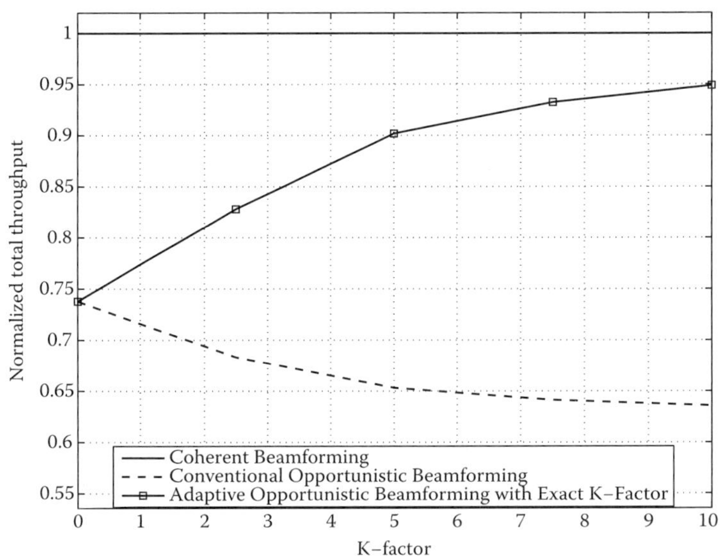

FIGURE 7.12 Normalized total throughput in the slow Ricean fading channels, $M = 10$, $N = 4$, $Q = 50$, $L_q = 20$. (Reproduced from Kim et al., 2006. © 2006, IEEE. With permission.)

On the other hand, as K grows, the performance improvement by the adaptive scheme becomes significant, because the diffused component $B_k(t)$ in (7.15) can be neglected and the LOS component becomes more dominant.

7.5 Conclusions

In the context of multiuser diversity, fading is seen as a resource that can be exploited, not an impediment that must be combated. In a multiuser system where every user has an independent fading channel, transmitting data only to the user with the best channel condition not only increases the throughput of the system but also increases the throughput of every user in the long term on average. The multiuser diversity technique has been adopted in practical systems, such as the downlink of the IS-856 system.

In order to achieve the higher multiuser diversity gain, the dynamic range of the channel fluctuation should be larger and the variation rate of the channel should be faster. When the underlying physical channels have small fluctuation and change slowly, opportunistic beamforming artificially induces the channel fluctuation by using some pseudorandom weight coefficients at the base station. By doing this, higher multiuser diversity can be achieved. However, this opportunistic beamforming requires a large number of users, especially with many transmit antennas at the base station.

Adaptive opportunistic beamforming can solve this problem very well in Ricean fading channels. The adaptive opportunistic beamforming algorithm generates the weight coefficients not randomly, but intelligently by estimating the DOAs of the users. That is, each beam is generated only in the direction where users really exist. This enables adaptive opportunistic beamforming to achieve excellent performance even with a small number of users.

References

[1] G. K. Kaleh. 1996. Frequency-diversity spread-spectrum communication system to counter bandlimited Gaussian interference. *IEEE Trans. Commun.* 44:886–93.

[2] Y. E. Dallal and S. Shamai. 1992. Time diversity in DPSK noisy phase channels. *IEEE Trans. Commun.* 40:1703–15.

[3] L. C. Godara. 1997. Applications of antenna arrays to mobile communications. Part I. Performance improvement, feasibility, and system considerations. *Proc. IEEE* 85:1031–60.

[4] L. C. Godara. 1997. Applications of antenna arrays to mobile communications. Part II. Beamforming and direction-of-arrival considerations. *Proc. IEEE* 85:1195–245.

[5] V. Tarokh, N. Seshadri, and A. R. Calderbank. 1998. Space-time codes for high data rate wireless communication: Performance criterion and code construction. *IEEE Trans. Inform. Theory* 44:744–65.

[6] J. H. Winters. 1998. Smart antennas for wireless systems. *IEEE Personal Commun.* 5:23–27.

[7] R. Knopp and P. Humblet. 1995. Information capacity and power control in single cell multiuser communications. In *Proceedings of the IEEE International Computer Conference (ICC'95)*, Seattle, WA, pp. 331–335.

[8] D. N. C. Tse. 1997. Optimal power allocation over parallel Gaussian channels. In *Proceedings of the International Symposium on Information Theory*, Ulm, Germany, p. 27.

[9] R. Gozali, R. M. Buehrer, and B. D. Woerner. 2003. The impact of multiuser diversity on space-time block coding. *IEEE Commun. Lett.* 7:213–15.

[10] J. Jiang, R. M. Buehrer, and W. H. Tranter. 2004. Antenna diversity in multiuser data networks. *IEEE Trans. Commun.* 52:490–97.

[11] E. G. Larsson. 2004. On the combination of spatial diversity and multiuser diversity. *IEEE Commun. Lett.* 8:517–19.

[12] D. N. C. Tse. 2002. Transmitter directed, multiple receiver system using path diversity to equitably maximize throughput. U.S. Patent 6449490.

[13] P. Viswanath, D. N. C. Tse, and R. Laroia. 2002. Opportunistic beamforming using dumb antennas. *IEEE Trans. Inform. Theory* 48:1277–94.

[14] I.-M. Kim, S.-C. Hong, S. S. Chassemzadeh, and V. Tarokh. 2005. Opportunistic beamforming based on multiple weighting vectors. *IEEE Trans. Wireless Commun.* 4:2683–87.

[15] J. Chung, C.-S. Hwang, K. Kim, and Y. K. Kim. 2003. A random beamforming technique in MIMO systems exploiting multiuser diversity. *IEEE J. Selected Areas Commun.* 21:848–55.

[16] N. Sharma and L. H. Qzarow. 2005. A study of opportunism for multiple-antenna systems. *IEEE Trans. Inform. Theory* 51:1808–14.

[17] I.-M. Kim, Z. Yi, D. Kim, and W. Chung. 2006. Improved opportunistic beamforming in Ricean fading channels. *IEEE Trans. Commun.* 54:2199–211.

[18] P. Bender, P. Black, M. Grob, R. Padovani, N. Sindhushayana, and A. Viterbi. 2000. CDMA/HDR: A bandwidth efficient high speed wireless data service for nomadic users. *IEEE Commun. Mag.* 38:70–78.

[19] D. Tse and P. Viswanath. 2005. *Fundamentals of wireless communication.* Cambridge: Cambridge University Press.

[20] R. O. Schmidt. 1986. Multiple emitter location and signal parameter estimation. *IEEE Trans. Antennas Propagation* 34:276–80.

[21] A. Barabell. 1983. Improving the resolution of eigenstructured based direction finding algorithms. In *Proceedings of IEEE ICASSP' 83*, pp. 336–39.

[22] R. Roy and T. Kailath. 1989. ESPRIT—Estimation of signal parameters via rotational invariance techniques. In *IEEE Trans. Acoustics Speech Signal Processing* 37:984–95.

[23] S. M. Kay. 1993. *Fundamentals of statistical signal processing.* Upper Saddle River, NJ: Prentice-Hall PTR.

[24] A. Abdi, C. Tepedelenlioglu, M. Kaveh, and G. Giannakis. 2001. On the estimation of the *K* parameters for the Rice fading distribution. *IEEE Commun. Lett.* 5:92–94.

[25] C. Tepedelenlioglu, A. Abdi, and G. B. Giannakis. 2003. The Ricean *K* factor: Estimation and performance analysis. *IEEE Trans. Wireless Commun.* 2:799–810.

8

Adaptive Beamforming for Multiantenna Communications

Alex B. Gershman
Darmstadt University of Technology

8.1 Introduction

Beamforming is a versatile approach to signal spatial filtering that has found numerous applications in diverse areas, including radar, sonar, wireless communications, geophysics, speech and audio processing, ultrasonic imaging, biomedicine, radio astronomy, and other fields [1]. Early attempts of applying beamforming to wireless communications go back to the late 1970s and early 1980s [2–4]. During the last two decades, there has been a major trend to use multiantenna transceivers in wireless communication systems to facilitate the explosive growth of the number of users and meet their rapidly increasing demands for new high-data-rate services [5–20]. As a result, spatial division multiple access (SDMA) technology recently became one of the key concepts in third and higher generations of mobile communication systems. In particular, the receive (uplink) and transmit (downlink) beamforming techniques used at multiantenna base stations (BSs) have been shown to enable efficient mitigation of multiuser interference and offer substantial improvements in system capacity and performance [6, 7, 14, 19].

In this chapter, we provide an overview of fundamentals and recent advances in the field of receive and transmit beamforming for multiantenna communication systems. The chapter is organized as follows. In section 8.2, the basic receive and transmit signal models are introduced. Section 8.3 is devoted to the receive beamforming problem and methods. In the same section, applications of receive adaptive beamforming to space-time multiuser multiple-input multiple-output (MIMO) receivers are highlighted. The transmit beamforming problem and methods are overviewed in section 8.4, and conclusions are given in section 8.5.

8.2 Basic Signal Models

Let us consider a receive (transmit) M-element BS array depicted in Figure 8.1, whose sensors are weighted by the weight vector $\mathbf{w} = [w_1, w_2, \ldots, w_M]^T$, where $(\cdot)^T$ denotes the transpose. Assume that there are L single-antenna users with flat fading channels. The $M \times 1$ uplink user channel vectors (hereafter referred to as user *spatial signatures*) are denoted by $\mathbf{a}_l$, $l = 1, \ldots, L$, while the $M \times 1$ downlink channel vectors are denoted by $\mathbf{h}_l$, $l = 1, \ldots, L$. Note that for each user, the uplink and downlink channel vectors may differ from each other because the difference in the uplink and downlink frequencies in the frequency division duplex (FDD) mode and channel variability in the time division duplex (TDD) mode may violate the uplink-downlink reciprocity property [14]. In the sequel, we assume without any loss of generality that the first user is the user of interest.

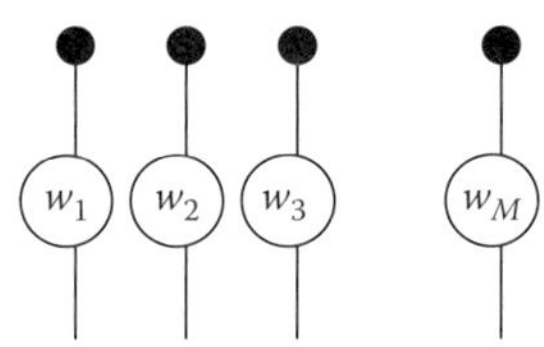

FIGURE 8.1 Receive/transmit beamformer.

8.2.1 The Uplink Case

In the uplink case, the baseband $M \times 1$ complex signal vector received at the BS array can be written as

$$\mathbf{x}(t) = \sum_{l=1}^{L} s_l(t)\mathbf{a}_l + \mathbf{n}(t) = \mathbf{A}\mathbf{s}(t) + \mathbf{n}(t), \tag{8.1}$$

where $s_l(t)$ is the baseband receive signal waveform of the lth user, $\mathbf{n}(t)$ is the $M \times 1$ vector of additive sensor noise, $\mathbf{A} \triangleq [\mathbf{a}_1, \mathbf{a}_2, \ldots, \mathbf{a}_L]$, $\mathbf{s}(t) \triangleq [s_1(t), s_2(t), \ldots, s_L(t)]^T$, and t is the time index. The additive noise is assumed to be zero mean and spatially white, that is, its correlation matrix is given by

$$\mathrm{E}\{\mathbf{n}(t)\mathbf{n}^H(t)\} = \sigma^2\mathbf{I}, \tag{8.2}$$

where $\mathbf{I}$ is the identity matrix, E{·} denotes the statistical expectation, and $(\cdot)^H$ stands for the Hermitian transpose.

The baseband complex signal at the output of the receive beamformer can be written as

$$y(t) = \mathbf{w}^H\mathbf{x}(t) = \sum_{l=l}^{L} s_l(t)\mathbf{w}^H\mathbf{a}_l + \mathbf{w}^H\mathbf{n}(t). \tag{8.3}$$

8.2.2 The Downlink Case

Now, let us consider the transmit beamforming mode with single BS and the same signal sent to all users. The baseband signal received by the lth user can be expressed as

$$z_l(t) = s(t)\mathbf{h}_l^H\mathbf{w} + n_l(t), \tag{8.4}$$

where $s(t)$ is the transmit baseband signal, $\mathbf{w}$ is the BS weight vector, and $n_l(t)$ is additive noise at the lth user.

The latter model can be further extended to the case of K different BSs and L mobile users. Let $\mathbf{w}_l$ be the weight vector used at the BS assigned to the lth user to transmit the baseband signal $s_l(t)$ to this user. Let us also define the BS cell site index $c(l)$ as the index of the particular BS that is assigned to the lth user. Note that $c(l) = c(m)$ if both the lth and m^{th} users are assigned to the same BS, and $c(l) \neq c(m)$ if these users are assigned to different BSs. Using these notations, the vector of signals transmitted from the kth BS can be expressed as [14]

$$\mathbf{x}_k(t) = \sum_{i \in \mathcal{G}(k)} s_i(t)\mathbf{w}_i, \tag{8.5}$$

where

$$\mathcal{G}(k) \triangleq \{i : c(i) = k\} \tag{8.6}$$

is the set of indices of all weight vectors that are used at the kth BS (or equivalently, the set of indices of all users that are assigned to this BS). Equation (8.5) implies that the kth BS transmits only to the users that are assigned to it rather than to all the users in the cellular network.

Using (8.5), the baseband signal received by the lth user can be modeled as [14]

$$z_l(t) = \sum_{k=1}^{K} \mathbf{h}_{l,k}^H\mathbf{x}_k(t) + n_l(t), \tag{8.7}$$

where $\mathbf{h}_{l,k}$ is the downlink channel vector between the kth BS and the lth user.

8.3 Receive Beamforming

The goal of receive beamforming is, given the knowledge of the spatial signature of the user of interest, to receive the signal of this user (hereafter referred to as the *desired signal*) with the maximal possible gain while suppressing the interfering users and noise as much as possible. The spatial signature of the user of interest has to be estimated in advance using, for example, traditional training-based techniques or blind approaches; see [21–25] and references therein.

8.3.1 Maximal Ratio Combining

The most traditional and simplest beamforming strategy, called *maximal ratio combining* (MRC), is to maximize the signal-to-noise ratio (SNR):

$$\mathrm{SNR} = \frac{\mathrm{E}\left\{\left|s_1(t)\mathbf{w}^H\mathbf{a}_1\right|^2\right\}}{\mathrm{E}\left\{\left|\mathbf{w}^H\mathbf{n}(t)\right|^2\right\}} = \frac{\sigma_1^2\left|\mathbf{w}^H\mathbf{a}_1\right|^2}{\sigma^2\mathbf{w}^H\mathbf{w}} \tag{8.8}$$

ignoring the effect of multiuser interference. Here, σ_l^2 is the power of the lth user, and as already mentioned before, the first user is assumed to be the user of interest.

From the Cauchy-Schwartz inequality it follows that the SNR in (8.8) is maximized with

$$\mathbf{w}_{\mathrm{MRC}} = \alpha\mathbf{a}_1, \tag{8.9}$$

where α is an arbitrary constant that does not affect the SNR value. It should be stressed that the MRC approach belongs to the class of nonadaptive (conventional) beamforming techniques because its weight vector does not depend on the received data.

It is well known that the MRC approach works acceptably well if

$$\sigma_1\left\|\mathbf{a}_1\right\|^2 \gg \sigma_i\left|\mathbf{a}_1^H\mathbf{a}_i\right|, \quad i = 2, \ldots, L, \tag{8.10}$$

where $\|\cdot\|$ hereafter denotes the 2-norm of a vector or the Frobenius norm of a matrix. From (8.10), it follows that the MRC method can only be used when the users are spatially well separated (so that $\|\mathbf{a}_1\|^2 \gg |\mathbf{a}_1^H\mathbf{a}_i|$), and when the powers of interfering users, σ_i^2 ($i = 2, \ldots, L$), do not substantially exceed the power of the user of interest, σ_1^2. However, if the power of one or more interfering users is substantially higher than that of the user of interest, then the MRC technique is not a proper approach anymore, and we have to resort to *adaptive* beamforming methods.

8.3.2 Minimum Variance Beamforming

The main idea of adaptive beamforming is to maximize the signal-to-interference-plus-noise ratio (SINR) rather than the SNR. Therefore, in contrast to the MRC approach, the effect of multiuser interference is no longer ignored. The output SINR is given by

$$\mathrm{SINR} = \frac{\mathbf{w}^H \mathbf{R}_s \mathbf{w}}{\mathbf{w}^H \mathbf{R}_{i+n} \mathbf{w}}, \tag{8.11}$$

where $\mathbf{R}_s$ is the correlation matrix of the desired signal component in (8.1), whereas $\mathbf{R}_{i+n}$ is the correlation matrix of the interference and noise components. From (8.1) it follows that

$$\mathbf{R}_s = \sigma_1^2 \mathbf{a}_1 \mathbf{a}_1^H, \tag{8.12}$$

and, therefore, (8.11) can be rewritten as

$$\mathrm{SINR} = \frac{\sigma_1^2 \left| \mathbf{w}^H \mathbf{a}_1 \right|^2}{\mathbf{w}^H \mathbf{R}_{i+n} \mathbf{w}}. \tag{8.13}$$

To obtain the optimal weight vector that maximizes the SINR in (8.11), one can minimize the output interference-plus-noise power while maintaining the distortionless array response to the desired signal [26, 27]:

$$\min_{\mathbf{w}} \mathbf{w}^H \mathbf{R}_{i+n} \mathbf{w} \quad \text{subject to} \quad \mathbf{w}^H \mathbf{R}_s \mathbf{w} = 1. \tag{8.14}$$

Taking into account (8.12), and therefore using (8.13) instead of (8.11), the problem of maximizing SINR under the distortionless response constraint can be rewritten in a simpler form [26]:

$$\min_{\mathbf{w}} \mathbf{w}^H \mathbf{R}_{i+n} \mathbf{w} \quad \text{subject to} \quad \mathbf{w}^H \mathbf{a}_1 = 1. \tag{8.15}$$

The beamforming problems (8.14) and (8.15) are commonly referred to as the *minimum variance* (MV) problems. The solution to the general MV beamforming problem (8.14) is given by [27]

$$\mathbf{w}_{\mathrm{opt}} = \mathcal{P}\left\{ \mathbf{R}_{i+n}^{-1} \mathbf{R}_s \right\}, \tag{8.16}$$

where $\mathcal{P}\{\cdot\}$ stands for the *principal eigenvector* of a matrix. Note that this eigenvector should be normalized to enable the resulting weight vector to satisfy the distortionless response constraint in (8.14). However, it is clear from (8.11) that any rescaling of the weight vector by a complex nonzero constant does not alter the output SINR (8.11). Hence, the aforementioned normalization is immaterial.

The solution to the simplified problem (8.15) is given by

$$\mathbf{w}_{\mathrm{opt}} = \alpha \mathbf{R}_{i+n}^{-1} \mathbf{a}_1, \tag{8.17}$$

where α can be obtained from the distortionless response constraint in (8.15) and is equal to $\alpha = 1/\mathbf{a}_1^H \mathbf{R}_{i+n}^{-1} \mathbf{a}_1$ [26]. Again, this constant is immaterial and therefore will be dropped in the sequel.

8.3.2.1 The SMI Algorithm

In practical applications, the true matrix $\mathbf{R}_{\mathrm{i+n}}$ is unavailable but can be estimated from the receiver data* as [6, 26, 28]

$$\hat{\mathbf{R}} = \frac{1}{N}\sum_{t=1}^{N}\mathbf{x}(t)\mathbf{x}^{H}(t), \tag{8.18}$$

where N is the number of snapshots available. The key idea of the sample matrix inversion (SMI)-based version of the MV beamformer is to replace $\mathbf{R}_{\mathrm{i+n}}$ by (8.18) [28]. For the MV beamforming problems in (8.14) and (8.15), the SMI weight vectors are given by [26, 28]

$$\mathbf{w}_{\mathrm{MV}} = \mathcal{P}\{\hat{\mathbf{R}}^{-1}\mathbf{R}_{s}\}, \tag{8.19}$$

$$\mathbf{w}_{\mathrm{MV}} = \hat{\mathbf{R}}^{-1}\mathbf{a}_{1}, \tag{8.20}$$

respectively.

The use of $\hat{\mathbf{R}}$ instead of $\mathbf{R}_{\mathrm{i+n}}$ in (8.19) or (8.20) is known to cause severe performance degradation in the case when the desired signal component is present in the beamformer snapshots used to estimate the sample correlation matrix (in the sequel, the latter case is referred to as the *signal-present* case as opposed to the *signal-free* case). Although in the signal-free case the output SINR of the SMI beamformer (8.20) rapidly converges to the optimal SINR value with increasing N [28], in the signal-present case this convergence becomes much slower [29], and the beamformer performance degrades severely even in the presence of small errors between the presumed and actual spatial signatures of the user of interest [27, 30]. Such signature errors can be caused by high user mobility and wireless channel variability, as well as by a limited amount of training symbols and the effect of multiuser interference that may prevent obtaining spatial signature estimates of acceptable quality. The effect of the beamformer performance degradation due to spatial signature errors is commonly known as *signal self-nulling*.

Another typical cause of the beamformer performance degradation in mobile communications is a highly nonstationary behavior of the propagation channel and, in particular, high mobility of interfering users [31]. The effect of such nonstationarity on the performance of receive adaptive beamformers is that the array weights may not be able to adapt fast enough to compensate for the interfering user motion [30, 31]. This phenomenon is usually referred to as *interference undernulling*.

* These data can be either information-bearing or training symbols. Therefore, there is no need to increase the amount of training data when using the SMI-based MV beamformers in multi-antenna wireless communication systems.

8.3.3 Robust Minimum Variance Beamforming

8.3.3.1 Linearly Constrained Minimum Variance Beamformer

A classic approach to prevent signal self-nulling and improve the robustness of MV beamforming against spatial signature errors is to use additional point or derivative mainlobe constraints [26, 32, 33].

In the case of point constraints, the sample version of the MV beamforming problem (8.15) is modified as [26]

$$\min_{\mathbf{w}} \mathbf{w}^H \hat{\mathbf{R}} \mathbf{w} \quad \text{subject to} \quad \mathbf{C}^H \mathbf{w} = \mathbf{f}, \tag{8.21}$$

where $\mathbf{C}$ is the $M \times P$ matrix of P constrained steering vectors in the neighborhood of the presumed spatial signature of the user of interest, and the $P \times 1$ vector $\mathbf{f} = [1, 1, \ldots, 1]^T$. In the case of derivative constraints, the problem can be formulated in the same way as in (8.21), but the matrix $\mathbf{C}$ and the vector $\mathbf{f}$ should be defined in a different way [26]. The multiple constraints given by the equation $\mathbf{C}^H \mathbf{w} = \mathbf{f}$ in (8.21) help to stabilize the mainlobe area of the array beampattern, and therefore to prevent signal self-nulling effects. This approach is commonly referred to as linearly constrained minimum variance (LCMV) beamforming.

The solution to (8.21) is given by [26]

$$\mathbf{w}_{\text{LCMV}} = \hat{\mathbf{R}}^{-1}\mathbf{C}\left(\mathbf{C}^H \hat{\mathbf{R}}^{-1}\mathbf{C}\right)^{-1}\mathbf{f}. \tag{8.22}$$

Although the LCMV beamformer is a popular technique in radar and sonar, its major shortcoming in wireless communications is that it requires exact knowledge of the array manifold, that is, it assumes the structure of the user-of-interest spatial signature to be known up to the signal direction-of-arrival parameters. In traditional array processing–based source localization techniques, the simplest free-space propagation model is typically adopted, that is, the source spatial signatures are assumed to be plane waves. However, in wireless communications, the plane wave structure of the user spatial signatures may be distorted by multipath scattering effects [34], and it is rather difficult to precisely parameterize the spatial signature vectors in terms of directions-of-arrival (DOAs). This may limit the use of the LCMV beamforming approach in wireless communications.

8.3.3.2 Diagonally Loaded Minimum Variance Beamformer

A popular approach to improve the robustness of MV beamformers is the so-called diagonal loading technique, whose idea is to *regularize* the adaptive array weights by adding a quadratic penalty term to the objective function in (8.14) or (8.15) [35–37]. The sample version of such a modified objective function can be written as

$$\mathbf{w}^H \hat{\mathbf{R}} \mathbf{w} + \gamma \mathbf{w}^H \mathbf{w} = \mathbf{w}^H (\hat{\mathbf{R}} + \gamma \mathbf{I}) \mathbf{w}, \tag{8.23}$$

where γ is a fixed loading factor that penalizes high-norm realizations of the weight vector. Minimizing this objective function under the distortionless response constraints used in (8.14) and (8.15), the following SMI-based diagonally loaded minimum variance (DLMV) beamformers can be obtained [26, 27, 35, 36]:

$$\mathbf{w}_{\text{DLMV}} = \mathcal{P}\left\{(\hat{\mathbf{R}}+\gamma\mathbf{I})^{-1}\mathbf{R}_s\right\}, \tag{8.24}$$

$$\mathbf{w}_{\text{DLMV}} = (\hat{\mathbf{R}}+\gamma\mathbf{I})^{-1}\mathbf{a}_1. \tag{8.25}$$

From (8.24) and (8.25), it can be seen that the diagonal loading operation can be viewed as injecting an artificial amount of white noise into the main diagonal of the sample correlation matrix. This warrants that, even though in the small sample case $\hat{\mathbf{R}}$ may be singular, the diagonally loaded matrix $\hat{\mathbf{R}} + \gamma\mathbf{I}$ is always positive definite, and therefore invertible.

In addition to ensuring that the correlation matrix is invertible, the diagonal loading approach is known to substantially improve the performance of the SMI-based MV beamformers (8.19) and (8.20) in the signal-present case. This improvement is especially pronounced in scenarios with mismatched spatial signature of the user of interest [30, 36, 38].

Despite the popularity of the DLMV beamforming technique, its applications may be limited by the fact that the optimal choice of the loading factor γ is scenario-dependent, and therefore, *ad hoc* choices of γ proposed in the literature [35–37] can lead to a drastic degradation of the beamformer performance.

8.3.3.3 Eigenspace-Based Beamformer

Another useful technique that helps to prevent signal self-nulling is the eigenspace-based beamformer [29, 39]. The essence of this beamformer is to reduce the spatial signature errors by projecting the presumed signature onto the estimated signal-plus-interference subspace obtained by means of the eigendecomposition of the sample correlation matrix

$$\hat{\mathbf{R}} = \hat{\mathbf{E}}\hat{\mathbf{\Lambda}}\hat{\mathbf{E}}^H + \hat{\mathbf{G}}\hat{\mathbf{\Gamma}}\hat{\mathbf{G}}^H ,$$

where the $M \times L$ matrix $\hat{\mathbf{E}}$ contains the L signal-plus-interference subspace eigenvectors of $\hat{\mathbf{R}}$, and the $L \times L$ diagonal matrix $\hat{\mathbf{\Lambda}}$ contains the corresponding eigenvalues of this matrix. Similarly, the $M \times (M - L)$ matrix $\hat{\mathbf{G}}$ contains the $(M - L)$ noise subspace eigenvectors of $\hat{\mathbf{R}}$, and the $(M - L) \times (M - L)$ diagonal matrix $\hat{\mathbf{\Gamma}}$ contains the corresponding eigenvalues. The total number of users is assumed to be known at the receiver and less than M. The weight vector of the eigenspace-based beamformer can be written as

$$\mathbf{w}_{\text{eig}} = \hat{\mathbf{R}}^{-1}\mathbf{P}_{\hat{\mathbf{E}}}\mathbf{a}_1, \tag{8.26}$$

where $\mathbf{P}_{\hat{\mathbf{E}}} = \hat{\mathbf{E}}(\hat{\mathbf{E}}^H\hat{\mathbf{E}})^{-1}\hat{\mathbf{E}}^H = \hat{\mathbf{E}}\hat{\mathbf{E}}^H$ is the orthogonal projection matrix onto the estimated signal-plus-interference subspace.

If the number of users is low and their SNRs are high, the eigenspace-based beamformer is known to provide a significant improvement of the robustness against spatial signature errors compared to the SMI-based MV beamformer [39]. However, the eigenspace-based approach may degrade severely if the number of users is comparable with the number of BS antennas, or if the SNRs of some of the users are low. In the latter case, subspace swap effects may cause a severe performance degradation due to the errors in estimating the user spatial signature subspace [27]. This makes it very difficult to use the eigenspace-based beamformer in mobile communications where the number of users served may be comparable to the number of BS array sensors, and some user powers may be low due to near–far effects.

8.3.3.4 MV Beamformers with Data-Dependent Beampattern Null Constraints

In mobile communications, user spatial signatures can rapidly change in time because of the user mobility and channel variability effects. As has been mentioned before, this may cause interference undernulling and, as a result, lead to a substantial performance degradation of BS receive beamformers. To preserve the beamformer performance in nonstationary scenarios, the idea of artificially broadening the adaptive beampattern nulls in interferer directions can be used [31, 40–42], where either point or derivative data-dependent constraints can be exploited to specify the array beampattern in the adaptive null areas.

One such technique developed in [41] uses derivative data-driven constraints (DDCs). The essence of this approach is to replace the sample correlation matrix in the MV or DLMV beamformers by the following modified sample correlation matrix:

$$\hat{\mathbf{R}}_{\text{mod}} = \sum_{p=0}^{P} \zeta_p \mathbf{B}^p \hat{\mathbf{R}} \mathbf{B}^p, \tag{8.27}$$

where $\mathbf{B}$ is an $M \times M$ diagonal matrix whose entries are determined by the known array geometry, P is the highest order of the data-dependent constraints, and ζ_p ($p = 0, \ldots, P$, $\zeta_0 = 1$) determine a proper trade-off between the constraints of different orders. In practice, $P = 1$ has been demonstrated to suffice for providing a substantial robustness against interferer motion [41]. In the case $P = 1$, (8.27) can be rewritten in a simpler form:

$$\hat{\mathbf{R}}_{\text{mod}} = \hat{\mathbf{R}} + \zeta \mathbf{B} \hat{\mathbf{R}} \mathbf{B}, \tag{8.28}$$

where ζ determines the required trade-off between the null depth and width. Under several mild conditions, the optimal value of ζ was shown to remain independent of the user parameters and can be straightforwardly obtained from the known BS array parameters [41].

Another approach to widen the adaptive beampattern nulls is based on data-driven point constraints [31, 40] and is commonly called the correlation matrix tapering (CMT) technique [42]. The essence of this technique is to replace the sample correlation matrix in the MV or DLMV beamformer by the so-called *tapered* sample correlation matrix:

$$\hat{\mathbf{R}}_{\text{tap}} = \hat{\mathbf{R}} \odot \mathbf{T}, \tag{8.29}$$

where $\mathbf{T}$ is the $M \times M$ matrix taper and $\odot$ denotes the Schur-Hadamard (element-wise) matrix product. Particular designs of matrix tapers are discussed in [31], [40], and [42].

An interesting link between the DDC and CMT approaches was found in [43], where it has been proven that the matrix (8.27) can be viewed as a tapered correlation matrix (8.29) with some particular choice of $\mathbf{T}$. Hence, the DDC approach can be interpreted in terms of the CMT approach.

Although both the DDC and CMT approaches have been shown to provide excellent robustness against moving interferers with plane wavefronts, they may severely degrade when the wavefronts of the interferers deviate from the plane wavefront form [27]. This may essentially limit the application of these two approaches in wireless communications.

8.3.3.5 Worst-Case Minimum Variance Beamformers

All the previously discussed robust beamforming techniques use rather *ad hoc* ways to incorporate the robustness feature into the SMI-based MV beamforming scheme. Recently, more theoretically rigorous techniques have been proposed that improve the robustness of MV beamforming by means of *worst-case* performance optimization [38, 44–46]. Such worst-case MV beamformers are discussed in this section.

The first worst-case robust MV (RMV) beamformer was developed in [38], where the mismatch

$$\delta \triangleq \tilde{\mathbf{a}}_1 - \mathbf{a}_1 \tag{8.30}$$

between the actual spatial signature $\tilde{\mathbf{a}}_1$ of the user of interest and its presumed version $\mathbf{a}_1$ was assumed to be norm-bounded by some known constant ε, that is,

$$\|\boldsymbol{\delta}\| \leq \varepsilon. \tag{8.31}$$

Equation (8.31) corresponds to the case when all possible mismatched spatial signatures of the user of interest belong to a *spherical uncertainty set*. The essence of the approach proposed in [38] is to add robustness to the SMI-based MV beamforming problem by maximizing the output SINR for the worst-case spatial signature mismatch that satisfies (8.31). This is equivalent to minimizing the output beamformer power under the worst-case distortionless response constraint, which should be satisfied for all mismatched spatial signature vectors that belong to the spherical uncertainty set of (8.31). The latter problem can be written as the following modification of the sample MV problem [38]:

$$\min_{\mathbf{w}} \mathbf{w}^H \hat{\mathbf{R}} \mathbf{w} \qquad \text{subject to} \qquad \left| \mathbf{w}^H (\mathbf{a}_1 + \delta) \right| \geq 1 \qquad \text{for all} \quad \left\| \delta \right\| \leq \varepsilon. \tag{8.32}$$

At first glance, the problem in (8.32) appears to be computationally intractable because it involves minimization of a quadratic function subject to infinitely many nonconvex quadratic constraints. However, it has been found in [38] that (8.32) can be converted to a much simpler form:

$$\min_{\mathbf{w}} \mathbf{w}^H \hat{\mathbf{R}} \mathbf{w} \quad \text{subject to} \quad \mathbf{w}^H \mathbf{a}_1 \geq \varepsilon \|\mathbf{w}\| + 1, \tag{8.33}$$

where the constraint in (8.33) can be shown to hold with equality at the optimal point of (8.33). The problem in (8.33) belongs to the class of convex second-order cone programming (SOCP) problems [47–49], which can be easily solved using efficient interior point methods [50] at complexity $O(M^3)$.

An alternative way to solve the problem in (8.33) is to resort to Newton-type algorithms. Several approaches of that type (all with complexities $O(M^3)$) are described in [46–52] to solve problem (8.33) and its extensions to the ellipsoidal uncertainty case [46, 52], and to a more general class of beamformers or multiuser detectors [44, 51, 52].

In [45], the RMV beamformer of (8.33) has been generalized to account for interferer nonstationarity in addition to the spatial signature errors. The essence of the approach of [45] is, in addition to modeling the uncertainty in the spatial signature vector, to model data nonstationarity by considering an uncertainty in the data matrix

$$\mathbf{X} \triangleq [\mathbf{x}(1), \ldots \mathbf{x}(N)]. \tag{8.34}$$

The sample correlation matrix can be expressed through the data matrix (8.34) as

$$\hat{\mathbf{R}} = \frac{1}{N} \mathbf{X}\mathbf{X}^H. \tag{8.35}$$

To take into account the nonstationarity of the array data, the so-called *mismatch matrix*

$$\mathbf{\Delta}_x \triangleq \tilde{\mathbf{X}} - \mathbf{X} \tag{8.36}$$

was introduced in [45], where $\tilde{\mathbf{X}}$ and $\mathbf{X}$ stand for the actual and presumed data matrices, respectively. Here, $\mathbf{X}$ is the data matrix acquired by the beamformer, while the actual data matrix $\tilde{\mathbf{X}}$ may differ from $\mathbf{X}$ because of a nonstationary character of sensor array snapshots. That is, the data samples in $\mathbf{X}$ can become irrelevant at the time when the beamformer is used (and when the actual, yet unknown, data matrix is $\tilde{\mathbf{X}}$ rather than $\mathbf{X}$). In such nonstationary scenarios, the actual sample correlation matrix is given by

$$\hat{\tilde{\mathbf{R}}} = \frac{1}{N} \tilde{\mathbf{X}}\tilde{\mathbf{X}}^H = \frac{1}{N} (\mathbf{X} + \mathbf{\Delta}_x)(\mathbf{X} + \mathbf{\Delta}_x)^H. \tag{8.37}$$

To make the SMI-based MV beamformer jointly robust against both the interference nonstationarity and spatial signature error effects, the authors of [45] have exploited an idea similar to that in [38]. In particular, it has been assumed that the norms of both the spatial signature mismatch $\boldsymbol{\delta}$ and the data matrix mismatch $\boldsymbol{\Delta}_x$ are norm-bounded by some known constants ε and η as

$$\|\boldsymbol{\delta}\| \leq \varepsilon, \tag{8.38}$$

$$\|\boldsymbol{\Delta}_x\| \leq \eta. \tag{8.39}$$

Then, the problem in (8.32) can be extended as [45]

$$\min_{\mathbf{w}} \max_{\|\Delta_x\|\leq\eta} \left\|(\mathbf{X} + \boldsymbol{\Delta}_x)^H \mathbf{w}\right\| \quad \text{subject to} \quad \left|\mathbf{w}^H(\mathbf{a}_1 + \delta)\right| \geq 1 \quad \text{for all} \quad \left\|\delta\right\| \leq \varepsilon \tag{8.40}$$

This problem has been converted in [45] to an equivalent form:

$$\min_{\mathbf{w}} \left\|\mathbf{X}^H \mathbf{w}\right\| + \eta\left\|\mathbf{w}\right\| \qquad \text{subject to} \qquad \mathbf{w}^H \mathbf{a}_1 \geq \varepsilon\left\|\mathbf{w}\right\| + 1. \tag{8.41}$$

The problem in (8.41) belongs to the class of convex SOCP problems and, similar to (8.33), can be efficiently solved using standard interior point methods [50].

Another interesting approach to extend problem (8.32) to a more general class of problems and, at the same time, to turn the solution for the weight vector into a closed form has been developed in [44]. The authors of [44] have considered the so-called general-rank signal case where the array response to the user of interest is characterized by the signal correlation matrix $\mathbf{R}_s$ rather than the spatial signature $\mathbf{a}_1$, and then have relaxed the so-obtained problem to come up with a closed-form beamformer.

Following [44], let us consider the errors between the presumed and actual interference-plus-noise and signal correlation matrices. These unknown error matrices can be expressed as

$$\boldsymbol{\Delta}_s \triangleq \tilde{\mathbf{R}}_s - \mathbf{R}_s, \tag{8.42}$$

$$\boldsymbol{\Delta}_{i+n} \triangleq \tilde{\mathbf{R}}_{i+n} - \mathbf{R}_{i+n}, \tag{8.43}$$

where $\mathbf{R}_s$ and $\mathbf{R}_{i+n}$ are the presumed signal and interference-plus-noise correlation matrices, respectively, whereas $\tilde{\mathbf{R}}_s$ and $\tilde{\mathbf{R}}_{i+n}$ are the actual values of these matrices. It should be stressed that the presumed value of the interference-plus-noise correlation matrix is given by the sample array correlation matrix, that is, $\mathbf{R}_{i+n} = \hat{\mathbf{R}}$ in (8.43).

In the case of nonzero mismatches $\boldsymbol{\Delta}_s$ and $\boldsymbol{\Delta}_{i+n}$, equation (8.11) for the output SINR has to be rewritten as

$$\text{SINR} = \frac{\mathbf{w}^H \tilde{\mathbf{R}}_{\text{s}} \mathbf{w}}{\mathbf{w}^H \tilde{\mathbf{R}}_{\text{i+n}} \mathbf{w}}. \tag{8.44}$$

In the spirit of [38], the approach of [44] assumes that the error matrices $\boldsymbol{\Delta}_{\text{s}}$ and $\boldsymbol{\Delta}_{\text{i+n}}$ are norm-bounded by known constants ε and γ, that is,

$$\left\|\boldsymbol{\Delta}_{\text{s}}\right\| \leq \varepsilon, \qquad \left\|\boldsymbol{\Delta}_{\text{i+n}}\right\| \leq \gamma. \tag{8.45}$$

To provide robustness against errors in the signal and interference-plus-noise correlation matrices, it has been suggested in [44] to obtain the beamformer weight vector by means of maximizing the worst-case output SINR:

$$\max_{\mathbf{w}} \min_{\boldsymbol{\Delta}_{\text{s}},\boldsymbol{\Delta}_{\text{i+n}}} \frac{\mathbf{w}^H (\mathbf{R}_{\text{s}} + \boldsymbol{\Delta}_{\text{s}})\mathbf{w}}{\mathbf{w}^H (\hat{\mathbf{R}} + \boldsymbol{\Delta}_{\text{i+n}})\mathbf{w}} \qquad \text{for all } \left\|\boldsymbol{\Delta}_{\text{s}}\right\| \leq \varepsilon, \ \left\|\boldsymbol{\Delta}_{\text{i+n}}\right\| \leq \gamma. \tag{8.46}$$

Clearly, (8.46) may be only an approximate (relaxed) version of the worst-case SINR optimization problem in the case when the worst-case values of the matrices $\boldsymbol{\Delta}_{\text{s}}$ and $\boldsymbol{\Delta}_{\text{i+n}}$ lead to the matrices $\mathbf{R}_{\text{s}} + \boldsymbol{\Delta}_{\text{s}}$ and $\hat{\mathbf{R}} + \boldsymbol{\Delta}_{\text{i+n}}$ that are not positive semidefinite.* In other words, to warrant the strict equivalence of these two problems, the constraints

$$\mathbf{R}_{\text{s}} + \boldsymbol{\Delta}_{\text{s}} \succeq 0, \tag{8.47}$$

$$\hat{\mathbf{R}} + \boldsymbol{\Delta}_{\text{i+n}} \succeq 0 \tag{8.48}$$

should be added to (8.46), where $\mathbf{Q} \succeq 0$ for any matrix $\mathbf{Q}$ means that this matrix is positive semidefinite. The authors of [44], however, drop these two constraints with the motivation that they make the optimization problem much more complicated, and that ignoring them can only make the problem more conservative.

It has been shown in [44] that the worst-case error matrices $\boldsymbol{\Delta}_{\text{s}}$ and $\boldsymbol{\Delta}_{\text{i+n}}$ satisfying (8.45) are given by

$$\boldsymbol{\Delta}_{\text{s}}^{\text{WC}} = -\varepsilon \frac{\mathbf{w}\mathbf{w}^H}{\left\|\mathbf{w}\right\|^2}, \tag{8.49}$$

$$\boldsymbol{\Delta}_{\text{i+n}}^{\text{WC}} = \gamma \frac{\mathbf{w}\mathbf{w}^H}{\left\|\mathbf{w}\right\|^2}, \tag{8.50}$$

* Note that any correlation matrix must by definition be positive semidefinite.

and therefore, the solution to the RMV beamforming problem (8.46) can be expressed as

$$\mathbf{w}_{\text{RMV}} = \mathcal{P}\left\{(\hat{\mathbf{R}}+\gamma\mathbf{I})^{-1}(\mathbf{R}_s - \varepsilon\mathbf{I})\right\}. \tag{8.51}$$

From (8.49) and (8.51) it follows that, although $\boldsymbol{\Delta}_{\text{i+n}}^{\text{WC}}$ satisfies (8.48), the constraint (8.47) is always violated with $\boldsymbol{\Delta}_{\text{s}}^{\text{WC}}$. Therefore, as expected, problem (8.46) is an approximate (conservative) version of the worst-case SINR optimization problem.

Taking into account that $\mathbf{R}_s = \sigma_1^2\mathbf{a}_1\mathbf{a}_1^H$ and absorbing σ_1^2 in ε (that is, assuming without any loss of generality that $\sigma_1^2 = 1$), equation (8.51) can be rewritten as

$$\mathbf{w}_{\text{RMV}} = \mathcal{P}\left\{(\hat{\mathbf{R}}+\gamma\mathbf{I})^{-1}(\mathbf{a}_1\mathbf{a}_1^H - \varepsilon\mathbf{I})\right\}. \tag{8.52}$$

From (8.51) and (8.52) it can be seen that the approach of [44] is equivalent to a combination of negative and positive diagonal loading. In the case of $\varepsilon = 0$, (8.51) and (8.52) simplify to the DLMV beamformer weight vectors (8.24) and (8.25), respectively.

An interesting interpretation of these two types of diagonal loading has been established in [44], where it has been obtained that (8.51) can be interpreted as a generalization of the DLMV beamformer whose amount of positive diagonal loading is *scenario-adaptive* rather than fixed. Similar adaptive diagonal loading interpretation of the RMV beamformer (8.33) has been found in [38], where it has been shown that the amount of diagonal loading in the latter RMV beamformer is optimally matched to the uncertainty in the spatial signature of the user of interest.

Although the worst-case RMV beamformers are known to be very robust techniques, they might be overly conservative because the actual worst operational conditions may occur in practice with a very low probability. Thus, obtaining less conservative robust alternatives to the worst-case techniques is of significant interest. This motivated the authors of [53–55] to develop an alternative, more flexible approach to RMV beamforming that can provide the robustness against spatial signature errors with a certain selected probability (i.e., using soft probabilistic constraints rather than deterministic worst-case constraints). In particular, the probabilistically constrained counterpart of problem (8.32) can be written as [53]

$$\min_{\mathbf{w}} \mathbf{w}^H\hat{\mathbf{R}}\mathbf{w} \text{ subject to } \Pr\left\{\left|\mathbf{w}^H(\mathbf{a}_1+\boldsymbol{\delta})\right| \geq 1\right\} > p, \tag{8.53}$$

where $\boldsymbol{\delta}$ is assumed to be a random vector drawn from some known distribution, $\Pr\{\cdot\}$ is the probability operator whose explicit form can be obtained from the statistical assumptions on the steering vector error, and p is some preselected probability threshold value. It can be seen that, in contrast to the *hard* constraint used in (8.32) (that requires the distortionless response to be maintained for all norm-bounded error vectors in the uncertainty set), the *soft* constraint in (8.53) maintains the distortionless response only for the error vectors $\boldsymbol{\delta}$ whose probability is high enough, while skipping the distortionless response condition for those values of $\boldsymbol{\delta}$ that are unlikely to occur. Therefore, the constraint in (8.53) can be interpreted as an *outage probability* constraint that maintains this probability acceptably low.* This makes the beamformer design in (8.53) more flexible than that in (8.32) because using soft constraints, more freedom is left for minimizing the objective function. Furthermore, the choice of the design parameter p in (8.53) is dictated by the quality-of-service (QoS) requirements (outage probability), and therefore is easier to specify than the robustness parameter ε in (8.32).

Although the problem in (8.53) describes a potentially more attractive beamformer design than (8.32), the main challenge of using (8.53) is that this problem belongs to the class of *chance programming* problems [56] that are rather difficult to solve. To enable a practical solution, several approximations of (8.53) by deterministic convex optimization problems have been considered in [53–55].

8.3.4 Extensions of Minimum Variance Beamforming to MIMO Communications

In this section, a useful extension of the theory of MV beamforming to the problem of designing multiantenna multiuser receivers for space-time block-coded multiple-input multiple-output (MIMO) communications is considered. The problem of MIMO receiver design is generally more difficult that the receive beamformer design problem because in MIMO communications, both the receive BS antenna and mobile transmitters have multiple antennas, and the information symbols may be sent in different time slots through different antennas of each transmitter using some space-time code. The problem is to design a linear receiver that enables the decoding of each symbol of the user of interest while mitigating multiaccess interference (MAI) caused by the remaining users [57–60]. To illustrate the problem, a typical multiuser MIMO uplink scenario is shown in Figure 8.2.

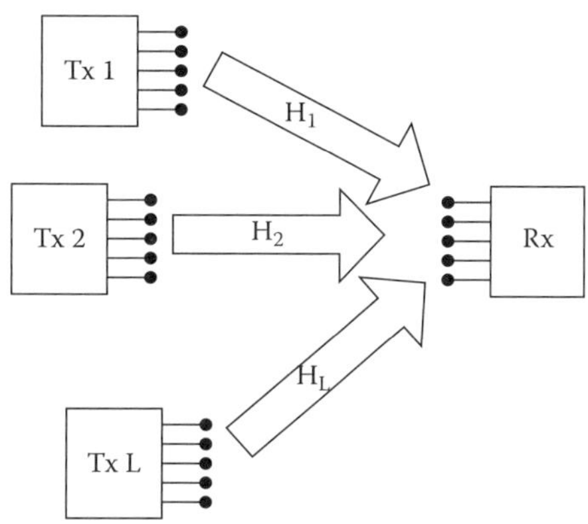

FIGURE 8.2 Uplink multiuser MIMO scenario.

Let us first consider a point-to-point flat block-fading MIMO system with M_t transmit and M_r receive antennas. Assuming that the channel is used at the times 1, 2, ..., T with the block length T, the input-output relationship for such a system can be written as [9, 10, 20]

$$\mathbf{X} = \mathbf{G}\mathbf{H} + \mathbf{N}, \tag{8.54}$$

* For the RMV beamforming problem (8.53), the outage probability is equal to $1 - p$.

where $\mathbf{H}$ is the $M_t \times M_r$ complex channel matrix, $\mathbf{G}$ is the $T \times M_t$ complex matrix of the transmitted signals, $\mathbf{X}$ is the $T \times M_r$ complex matrix of the received signals, and $\mathbf{N}$ is the $T \times M_r$ matrix of noise.

Let us denote complex information symbols prior to space-time encoding as $s_1, \ldots, s_J$ and define the $J \times 1$ symbol vector $\mathbf{s} \triangleq [s_1, \ldots, s_J]^T$. The $T \times M_t$ matrix $\mathbf{G} = \mathbf{G}(\mathbf{s})$ is called an orthogonal space-time block code (OSTBC) if [16–18]:

- All elements of $\mathbf{G}(\mathbf{s})$ are linear functions of the J complex variables $s_1, \ldots, s_J$ and their complex conjugates.
- For any $\mathbf{s}$, it satisfies $\mathbf{G}^H(\mathbf{s})\,\mathbf{G}(\mathbf{s}) = \|\mathbf{s}\|^2\,\mathbf{I}$.

Assuming the OSTBC signaling, the matrix $\mathbf{G}(\mathbf{s})$ can be written as [16, 61, 62]

$$\mathbf{G}(\mathbf{s}) = \sum_{l=1}^{J} \left(\mathbf{C}_l \mathrm{Re}\{s_l\} + \mathbf{D}_l \mathrm{Im}\{s_l\} \right), \tag{8.55}$$

where $\mathbf{C}_l \triangleq \mathbf{G}(\mathbf{e}_l)$, $\mathbf{D}_l \triangleq \mathbf{G}(j\mathbf{e}_l)$, $\mathbf{e}_l$ is the $J \times 1$ vector having one in the lth position and zeros elsewhere, and $j \triangleq \sqrt{-1}$. Using (8.55), the MIMO model (8.54) can be rewritten as [59, 62]

$$\underline{\mathbf{X}} = \mathcal{A}\underline{\mathbf{s}} + \underline{\mathbf{N}}, \tag{8.56}$$

where the "underline" operator for any matrix $\mathbf{P}$ is defined as

$$\underline{\mathbf{P}} \triangleq \begin{bmatrix} \mathrm{vec}\{\mathrm{Re}(\mathbf{P})\} \\ \mathrm{vec}\{\mathrm{Im}(\mathbf{P})\} \end{bmatrix}, \tag{8.57}$$

vec{·} is the vectorization operator stacking all columns of a matrix on top of each other, and the $2\,M_r T \times 2J$ real matrix $\mathcal{A} = \mathcal{A}(\mathbf{H})$ is given by

$$\mathcal{A} = \left[\underline{\mathbf{C}_1\mathbf{H}}, \ldots, \underline{\mathbf{C}_J\mathbf{H}}, \underline{\mathbf{D}_1\mathbf{H}}, \ldots, \underline{\mathbf{D}_J\mathbf{H}} \right]. \tag{8.58}$$

This matrix can be shown to satisfy the following orthogonality property (that, in turn, is due to the orthogonality property of the space-time code):

$$\mathcal{A}^T\mathcal{A} = \|\mathbf{H}\|^2\,\mathbf{I}. \tag{8.59}$$

From (8.58), it can be seen that the matrix $\mathcal{A}$ in (8.56) captures both the effects of the OSTBC and the channel, while the vector $\underline{\mathbf{s}}$ depends on the information symbols only. The columns of $\mathcal{A}$ can be viewed as spatio-temporal signatures that describe the receiver response to each entry of $\underline{\mathbf{s}}$ (i.e., to real or imaginary parts of each information symbol).

Obviously, there is a strong similarity between the vectorized MIMO model in (8.56) and the beamforming snapshot model in (8.1).

The established similarity between the models (8.1) and (8.56) opens an avenue for extending MV beamforming techniques to MIMO communications. A linear MIMO space-time receiver can be expressed as [57, 59]

$$\hat{\underline{\mathbf{s}}} = \mathbf{W}^T \underline{\mathbf{X}}, \tag{8.60}$$

where $\hat{\underline{\mathbf{s}}}$ is the receiver estimate of the vector $\underline{\mathbf{s}}$, and $\mathbf{W} \triangleq [\mathbf{w}_1, \mathbf{w}_2, \ldots, \mathbf{w}_{2J}]$ is the $2M_rT \times 2J$ matrix of the receiver weights. Note that each entry of $\underline{\mathbf{s}}$ requires a separate weight vector for estimation and subsequent decoding. For example, the vector $\mathbf{w}_l$ can be interpreted as the space-time receiver weight vector for the lth entry of $\underline{\mathbf{s}}$.

The MIMO receiver counterpart of the MRC technique is commonly referred to as the matched-filter (MF) receiver, and can be written as [59, 62]

$$\hat{\underline{\mathbf{s}}} = \frac{1}{\|\mathbf{H}\|^2} \mathcal{A}^T \underline{\mathbf{X}}. \tag{8.61}$$

The MF receiver (8.61) corresponds to the following weight matrix:

$$\mathbf{W}_{\text{MF}} = \frac{1}{\|\mathbf{H}\|^2} \mathcal{A}. \tag{8.62}$$

When followed by the simple nearest-neighbor symbol-by-symbol decoder, this receiver is known to be equivalent to the maximum likelihood (ML) space-time decoder for the single-user (point-to-point) case [62].

In the multiuser case (when multiple multiantenna transmitters simultaneously communicate with a multiantenna receiver; see Figure 8.2), the models in (8.54) and (8.56) can be straightforwardly modified to account for multiple users. For example, if all the transmitters use the same OSTBC, (8.56) can be rewritten as [59]

$$\underline{\mathbf{X}} = \sum_{l=1}^{L} \mathcal{A}_l \underline{\mathbf{s}_l} + \underline{\mathbf{N}}, \tag{8.63}$$

where $\mathbf{s}_l$ is the $J \times 1$ vector of information symbols of the lth user,

$$\begin{aligned} \mathcal{A}_l &= \left[\underline{\mathbf{C}_1\mathbf{H}_l}, \ldots, \underline{\mathbf{C}_J\mathbf{H}_l}, \underline{\mathbf{D}_1\mathbf{H}_l}, \ldots, \underline{\mathbf{D}_J\mathbf{H}_l}\right] \\ &\triangleq [\mathbf{a}_{l,1}\ \mathbf{a}_{l,2} \cdots \mathbf{a}_{l,2J}], \end{aligned} \tag{8.64}$$

and $\mathbf{H}_l$ is the matrix channel between the lth user and the receiver.

In the multiuser case, the MF receiver (8.62) is no longer the optimal ML decoder, and the complexity of the optimal ML decoding technique grows exponentially with the number of users. Therefore, in this case suboptimal but simple linear receivers are of particular interest [57, 59, 60].

As before, we assume without any loss of generality that the first transmitter is the user of interest. Then, using (8.60), we can express the output vector of a linear receiver as

$$\hat{\underline{\mathbf{s}}}_1 = \mathbf{W}^T \underline{\mathbf{X}}, \tag{8.65}$$

where for each user of interest, a separate matrix $\mathbf{W}$ should be used.

Given the matrix $\mathbf{W}$, the information symbols of the user of interest can be estimated as

$$\hat{\mathbf{s}}_1 = [\mathbf{I}, j\mathbf{I}]\hat{\underline{\mathbf{s}}}_1, \tag{8.66}$$

where the dimension of the identity matrices in (8.66) is $J \times J$. Using the linear estimate (8.66), the lth information symbol can be detected as the nearest to the lth entry of the $\hat{\mathbf{s}}_1$ signal constellation point.

Using the concept of MV beamforming, the receiver weight matrix $\mathbf{W}$ can be designed to maximally suppress interference while preserving a distortionless response toward the signal of the transmitter-of-interest. Specifically, for each entry of $\underline{\mathbf{s}}_1$, the receiver output power has to be minimized while preserving the distortionless response for that particular entry of $\underline{\mathbf{s}}_1$. This is equivalent to solving the following optimization problem [59]:

$$\min_{\mathbf{w}_l} \mathbf{w}_l^T \hat{\mathcal{R}} \mathbf{w}_l \quad \text{subject to} \quad \mathbf{a}_{1,l}^T \mathbf{w}_l = 1 \quad \text{for all} \quad l = 1, 2, \ldots, 2J, \tag{8.67}$$

where

$$\hat{\mathcal{R}} = \frac{1}{N} \sum_{t=1}^{N} \underline{\mathbf{X}(t)}\,\underline{\mathbf{X}(t)}^T \tag{8.68}$$

is the sample estimate of the $2M_rT \times 2M_rT$ correlation matrix $\mathcal{R} \triangleq \mathrm{E}\{\underline{\mathbf{X}}\,\underline{\mathbf{X}}^T\}$ of the vectorized data, and $\mathbf{X}(t)$ denotes the tth data block.

Taking into account that problem (8.67) can be solved independently for each l, the solution to (8.67) can be written as [59]

$$\mathbf{w}_{\mathrm{MV},l} = \frac{1}{\mathbf{a}_{1,l}^T \hat{\mathcal{R}}^{-1} \mathbf{a}_{1,l}} \hat{\mathcal{R}}^{-1} \mathbf{a}_{1,l}, \quad l = 1, 2, \ldots, 2J. \tag{8.69}$$

Although the MV receiver (8.69) is able to suppress MAI, it does not completely cancel *self-interference* that, for the lth entry of $\underline{\mathbf{s}}_1$, is caused by entries of $\underline{\mathbf{s}}_1$ other than the lth one. Clearly, self-interference is treated in (8.67) in the same way as MAI. Thus, when the MAI component is strong, self-interference may not be sufficiently rejected. Note, however, that complete cancellation of self-interference is a strongly desirable property. Otherwise, the symbol-by-symbol detector loses its optimality [59].

To ensure that the self-interference component is completely cancelled, the following additional zero-forcing constraints can be added to (8.67):

$$\mathbf{a}_{1,l}^T \mathbf{w}_m = 0 \quad \text{for all} \quad m \neq l, \quad m, l = 1, 2, \ldots, 2J. \tag{8.70}$$

Problem (8.67) with the additional constraints (8.70) can be expressed as

$$\min_{\mathbf{W}} \text{tr}\{\mathbf{W}^T \hat{\boldsymbol{\mathcal{R}}} \mathbf{W}\} \quad \text{subject to} \quad \boldsymbol{\mathcal{A}}^T \mathbf{W} = \mathbf{I}, \tag{8.71}$$

where tr{·} denotes the trace of a matrix.

The solution to (8.71) is given by [59]

$$\mathbf{W}_{\text{MV}} = \hat{\boldsymbol{\mathcal{R}}}^{-1} \boldsymbol{\mathcal{A}} (\boldsymbol{\mathcal{A}}^T \hat{\boldsymbol{\mathcal{R}}}^{-1} \boldsymbol{\mathcal{A}})^{-1}. \tag{8.72}$$

Clearly, this MV receiver can be interpreted as a combination of the prewhitener $\hat{\boldsymbol{\mathcal{R}}}^{-1/2}$ and decorrelator receiver $\hat{\boldsymbol{\mathcal{R}}}^{-1/2} \boldsymbol{\mathcal{A}} (\boldsymbol{\mathcal{A}}^T \hat{\boldsymbol{\mathcal{R}}}^{-1} \boldsymbol{\mathcal{A}})^{-1}$. Furthermore, using the property (8.59), it can be seen that in the specific case of $\hat{\boldsymbol{\mathcal{R}}} \propto \mathbf{I}$, the MV receiver (8.72) simplifies to the MF receiver (8.62). The latter property is in agreement with the well-known fact that the MF receiver ignores the effect of MAI treating it as a white noise.

Several robust modifications of the MV receivers (8.69) and (8.72) have been considered in the literature to improve their robustness in the case of imperfect receive channel state information (CSI). In [59], diagonally loaded modifications of (8.69) and (8.72) have been developed. In [63], the approach of [38] has been used to design worst-case optimization-based RMV receivers that explicitly account for norm-bounded CSI errors. For example, the following worst-case modification of (8.67) has been proposed in [63]:

$$\min_{\mathbf{w}_l} \mathbf{w}_l^T \hat{\boldsymbol{\mathcal{R}}} \mathbf{w}_l \quad \text{subject to} \quad \min_{\|\boldsymbol{\Delta}\| \leq \varepsilon} \mathbf{w}_k^T \mathbf{a}_l(\hat{\mathbf{H}}_1 + \boldsymbol{\Delta}) \geq 1 \quad \text{for all} \quad l = 1, \ldots, 2J, \tag{8.73}$$

where, for the sake of clarity, the user-of-interest spatio-temporal signature $\mathbf{a}_{l,1}$ is explicitly denoted as $\mathbf{a}_l(\mathbf{H}_1)$ to stress that it is a function of the user-of-interest channel $\mathbf{H}_1$. In (8.73), another version of this signature, $\mathbf{a}_l(\hat{\mathbf{H}}_1 + \boldsymbol{\Delta})$, is used, where $\boldsymbol{\Delta}$ stands for the receive CSI error of the user-of-interest:

$$\boldsymbol{\Delta} \triangleq \mathbf{H}_1 - \hat{\mathbf{H}}_1, \tag{8.74}$$

and $\mathbf{H}_1$ and $\hat{\mathbf{H}}_1$ denote the actual and presumed (estimated) channel matrices of the user of interest, respectively. Similar to [38], it is assumed in (8.73) that the CSI error is norm-bounded by some known constant ε, that is,

$$\|\Delta\| \leq \varepsilon. \tag{8.75}$$

The essence of the RMV receiver design problem in (8.73) is to minimize the receiver output power while maintaining the distortionless response for the worst-case CSI errors.

It has been proved in [63] that for any OSTBC,

$$\|\Delta\| = \|\delta_l\| \quad \text{for all} \quad l = 1, \ldots, 2J, \tag{8.76}$$

where

$$\delta_l \triangleq \mathbf{a}_l(\mathbf{H}_1) - \mathbf{a}_l(\hat{\mathbf{H}}_1) \tag{8.77}$$

is the error between the actual spatiotemporal signature $\mathbf{a}_l(\mathbf{H}_1)$ and its mismatched (presumed) value $\mathbf{a}_l(\hat{\mathbf{H}}_1)$. Therefore, problem (8.73) can be rewritten as

$$\min_{\mathbf{w}_l} \mathbf{w}_l^T \hat{\mathcal{R}} \mathbf{w}_l \quad \text{subject to} \quad \min_{\|\delta_l\| \leq \varepsilon} \mathbf{w}_l^T (\mathbf{a}_l(\hat{\mathbf{H}}_1) + \delta_l) \geq 1 \quad \text{for all} \quad l = 1, \ldots, 2J. \tag{8.78}$$

This problem is mathematically equivalent to $2J$ decoupled RMV beamforming problems (8.32). Hence, using the results of [38], problem (8.78) can be reformulated as [63]

$$\min_{\mathbf{w}_l} \mathbf{w}_l^T \hat{\mathcal{R}} \mathbf{w}_l \quad \text{subject to} \quad \mathbf{w}_l^T \mathbf{a}_l(\hat{\mathbf{H}}_1) \geq \varepsilon \|\mathbf{w}_l\| + 1 \quad \text{for all} \quad l = 1, \ldots, 2J, \tag{8.79}$$

where the constraint can be shown to be satisfied with equality. Similar to (8.33), the problem in (8.79) belongs to the class of standard SOCP problems and can be efficiently solved using either modern convex optimization tools [50] or Newton-type algorithms [46, 51, 52].

In addition to (8.79), one more RMV receiver has been developed in [63] as a robust extension of the MV receiver (8.72). The later RMV receiver enables better cancellation of self-interference and, as a result, offers an improved performance compared to (8.79).

Using the concept of probabilistically constrained RMV beamforming, several promising extensions of multiuser RMV MIMO receivers of [63] have been developed in [55, 64–67] using soft (outage probability-based) rather that deterministic worst-case distortionless response constraints. An interesting open problem is how to extend these receiver techniques to other than orthogonal space-time codes with higher transmission rates.

8.4 Transmit Beamforming

In wireless communications, there are two basic ways to use transmit antenna arrays: for space-time data encoding [8, 9, 16] and spatial data multiplexing (beamforming) [11, 14]. Although space-time coding techniques do not require any CSI at the transmitter, beamforming methods typically require an accurate knowledge of the transmit CSI.

Certain combinations of space-time coding and beamforming at the transmitter are also possible, that may require only partial CSI [68, 69].

In the sequel, we will consider the transmit beamforming problem that includes the following (related) cases:

1. *Unicasting.* Different data streams should be delivered to different users.
2. *Broadcasting.* The same data content should be delivered to all users.
3. *Multicasting.* The same data content should be broadcasted to a selected group of users, but different data streams should be transmitted to different groups of users.

Clearly, the first two cases represent two extremes of the most general multicasting case. In particular, broadcasting and unicasting can be viewed as single-group multicasting and multicasting with single-user groups, respectively.

8.4.1 Unicast Transmit Beamforming

8.4.1.1 Traditional Techniques

Early formulations of the transmit beamforming problem were mostly developed in the context of voice services in a cellular mobile radio network, where from the operator perspective, the system should provide an acceptable QoS for each user and serve as many users as possible, while radiating as low power as possible [11, 14].

The QoS requirements can be set up in the form of the lowest admissible value of the received SINR at each mobile. Using (8.7), the receive SNR of the ith user can be expressed as [14]

$$\text{SINR}_i = \frac{\mathbf{w}_i^H \mathbf{R}_{i,c(i)} \mathbf{w}_i}{\sigma_i^2 + \sum_{m=1;m\neq i}^{L} \mathbf{w}_m^H \mathbf{R}_{i,c(m)} \mathbf{w}_m}, \tag{8.80}$$

where σ_i^2 is the noise power of the ith user, $\mathbf{R}_{i,c(m)}$ is the correlation matrix of the downlink channel between the BS serving the mth mobile and the ith mobile, and the other parameters used in (8.80) have been defined in section 8.2.

The numerator of (8.80) represents the receive signal power at the ith mobile, whereas the denominator of (8.80) contains the noise and interference powers at the same mobile. The interference terms are given by the sum of powers of transmissions that are intended for other than the ith mobile but which are received by the ith mobile. Clearly, this is a *crosstalk* type of interference that should be avoided to guarantee an acceptable quality of the voice message.

Equation (8.80) views all channels as stochastic random vectors. This representation is suitable for the case of fast fading where the downlink channel vectors themselves are unavailable at the transmitter, and only their correlation matrices are known. In the opposite case of slow channel fading, it is more natural to view the downlink channels as deterministic vectors. In the latter case, the downlink channel correlation matrix is rank one and is given by $\mathbf{R}_{i,c(m)} = \mathbf{h}_{i,c(m)}\mathbf{h}_{i,c(m)}^H$. Then, (8.80) can be rewritten in a simpler form:

$$\mathrm{SINR}_i = \frac{\left|\mathbf{w}_i^H \mathbf{h}_{i,c(i)}\right|^2}{\sigma_i^2 + \sum_{m=1;m\neq i}^{L} \left|\mathbf{w}_m^H \mathbf{h}_{i,c(m)}\right|^2}. \tag{8.81}$$

The traditional approach to optimizing the beamformer weight vectors is to minimize the total transmit power under the constraints guaranteeing that an acceptable QoS is provided for each user. Mathematically, this problem can be expressed as [11, 14]

$$\min_{\{\mathbf{w}_l\}_{l=1}^L} \sum_{l=1}^{L} \mathbf{w}_l^H \mathbf{w}_l \quad \text{subject to} \quad \mathrm{SINR}_i \geq \gamma_i \quad \text{for all} \quad i=1,\ldots,L, \tag{8.82}$$

where γ_i is the minimum acceptable QoS for the ith user. This is a quadratic optimization problem with quadratic nonconvex constraints. Different algorithms have been proposed in the literature to solve (8.82). For example, the algorithms of [11] and [70] separate the problem into the following two subproblems: downlink power control and downlink beamforming with norm-one weight vectors. Using this representation of the problem at hand, the approaches of [11] and [70] propose to find the solution to (8.82) in an iterative way, by finding the optimal norm-one weight vectors for given power levels, and then updating the power levels based on these weight vectors.

Another powerful approach to solve (8.82) has been proposed in [71] (see also [14]) using convex optimization. The key idea of this approach is to reformulate problem (8.82) in a convex form as follows. Using the notation $\mathbf{W}_i \triangleq \mathbf{w}_i\mathbf{w}_i^H$, problem (8.82) with the user SINRs (8.80) can be transformed to [14, 71]

$$\begin{aligned} \min_{\{\mathbf{W}_l\}_{l=1}^L} \sum_{l=1}^{L} \mathrm{tr}\{\mathbf{W}_l\} \quad \text{subject to} \quad & \mathrm{tr}\{\mathbf{R}_{i,c(i)}\mathbf{W}_i\} - \gamma_i \sum_{m\neq i} \mathrm{tr}\{\mathbf{R}_{i,c(m)}\mathbf{W}_m\} \geq \gamma_i\sigma_i^2, \\ & \mathbf{W}_i = \mathbf{W}_i^H, \ \mathbf{W}_i \succeq 0 \quad \text{for all} \ i=1,\ldots,L, \end{aligned} \tag{8.83}$$

where the last two constraints guarantee that the matrices $\mathbf{W}_i$ are Hermitian and positive semidefinite for all $i = 1, \ldots, L$. Note that the last constraint can be viewed as a convex relaxation of the nonconvex rank-one constraint rank $\{\mathbf{W}_i\} = 1$.

The problem in (8.83) belongs to the class of convex semidefinite programming (SDP) problems, and therefore, it can be efficiently solved using modern convex optimization tools [50]. Moreover, it has been proved in [14] that for this problem, $\mathbf{W}_i \succeq 0$ is exactly equivalent to rank $\{\mathbf{W}_i\} = 1$, and therefore, the replacement of the latter constraint by the former one is not a relaxation but actually an equivalent reformulation of the problem. After obtaining the optimal values for $\mathbf{W}_i$, $i = 1, \ldots, L$, this property enables recovery of the optimal weight vectors $\mathbf{w}_i$, $i = 1, \ldots, L$ in a simple way, from the principal eigenvectors of $\mathbf{W}_i$, $i = 1, \ldots, L$.

In the slow fading case, where the downlink channel vectors are available at the transmitter and (8.81) can be used instead of (8.80), problem (8.82) can be rewritten in a SOCP form [14]:

$$\min_{\{\mathbf{w}_l\}_{l=1}^{L}} \sum_{l=1}^{L} \mathbf{w}_l^H \mathbf{w}_l \quad \text{subject to} \quad (\mathbf{w}_i^H \mathbf{h}_{i,c(i)})^2 - \gamma_i \sum_{m \neq i} \left| \mathbf{w}_m \mathbf{h}_{i,c(m)} \right|^2 \geq \gamma_i \sigma_i^2, \qquad (8.84)$$

$$\mathbf{w}_i^H \mathbf{h}_{i,c(i)} \geq 0, \ \mathrm{Im}\{\mathbf{w}_i^H \mathbf{h}_{i,c(i)}\} = 0 \quad \text{for all} \quad i = 1, \ldots, L.$$

The SOCP problem (8.84) has much less optimization variables than the SDP problem (8.83) and can be solved easier than the latter problem; see [14].

An interesting and practically relevant extension of the problems (8.81) and (8.82) is, in addition to optimizing the beamformer weight vectors, to maximize the number of admitted users under the condition that the problem remains feasible. Such an extension of the transmit beamforming approach of [11] has been proposed in [72–74], where two computationally efficient joint beamforming and user admission control techniques have been developed using SDP and SOCP convex optimization approaches.

Another recent trend in transmit beamforming is to maximize the system throughput under certain power constraints rather than to minimize the total transmit power [75–79]. A growing interest in such types of transmit strategy is motivated by a higher emphasis of delay-tolerant high-rate packet data services for the third- and fourth-generation (3G and 4G) communication systems [79].

To illustrate the concept of sum-capacity maximization, let us consider the single-cell multiuser model with the downlink user channels $\mathbf{h}_l$, $l = 1, \ldots, L$ (see section 8.2). The baseband signal to be sent to the lth user at time t is denoted as $s_l(t)$, while the signal received at the lth user is denoted as $z_l(t)$. Introducing the notations

$$\mathbf{s}(t) \triangleq [s_1(t), s_2(t), \ldots, s_L(t)]^T, \qquad (8.85)$$

$$\mathbf{z}(t) \triangleq [z_1(t), z_2(t), \ldots, z_L(t)]^T, \qquad (8.86)$$

$$\mathbf{n}(t) \triangleq [n_1(t), n_2(t), \ldots, n_L(t)]^T, \qquad (8.87)$$

$$\mathcal{H} \triangleq [\mathbf{h}_1, \mathbf{h}_2, \ldots, \mathbf{h}_L]^H, \qquad (8.88)$$

$$\mathcal{W} \triangleq [\mathbf{w}_1, \mathbf{w}_2, \ldots, \mathbf{w}_L], \qquad (8.89)$$

the receive signal model can be written as

$$\mathbf{z}(t) = \mathcal{H}\mathcal{W}\mathbf{s}(t) + \mathbf{n}(t). \qquad (8.90)$$

The sum capacity of such a vector Gaussian channel $\mathbf{z} = \mathcal{H}\mathbf{y} + \mathbf{n}(t)$ with the total transmit power constrained to $\mathcal{P}$ is given by [79]

$$C = \max_{\mathbf{R}_y} \log\det\left\{\mathbf{I} + \mathcal{H}\mathbf{R}_y\mathcal{H}^H\right\}, \tag{8.91}$$

where $\mathbf{R}_y \triangleq \mathrm{E}\{\mathbf{y}\mathbf{y}^H\}$ and the maximization in (8.91) is performed over positive semi-definite matrices that satisfy the power constraint

$$\mathrm{tr}\{\mathbf{R}_y\} \leq \mathcal{P}. \tag{8.92}$$

The SINR of the lth user can be written as

$$\mathrm{SINR}_l = \frac{\left|\mathbf{h}_l^H \mathbf{w}_l\right|^2}{\sigma_l^2 + \sum_{m \neq l} \left|\mathbf{h}_l^H \mathbf{w}_m\right|^2}. \tag{8.93}$$

Using (8.90)–(8.93), the transmit beamforming problem can be formulated as [79]

$$\max_{\mathcal{W}} \sum_{l=1}^{L} \log(1 + \mathrm{SINR}_l) \quad \text{subject to} \quad \left\|\mathcal{W}\right\|^2 \leq \mathcal{P}. \tag{8.94}$$

The essence of (8.94) is to maximize the total system throughput under the transmit power constraint. In simple words, the maximum in (8.94) is achieved by means of exploiting *multiuser diversity* that suggests always to transmit to the strongest-channel users. Several computationally efficient algorithms to solve (8.94) have been proposed in the literature; see [77–79] and references therein.

8.4.1.2 Robust Extensions

The idea of incorporating robustness against transmitter CSI errors in the downlink beamforming problem (8.82) has been discussed in [14] and [80]. This approach uses an idea related to that used in [44] for receive RMV beamforming. More specifically, the following upper and lower bounds on downlink channel correlation matrices are considered in [80]:

$$\mathbf{R}_{i,c(m)}^{\text{lower}} \preceq \mathbf{R}_{i,c(m)} \preceq \mathbf{R}_{i,c(m)}^{\text{upper}}, \tag{8.95}$$

where

$$\mathbf{R}_{i,c(m)}^{\text{lower}} = \mathbf{R}_{i,c(m)} - \xi\mathbf{I}, \qquad \mathbf{R}_{i,c(m)}^{\text{upper}} = \mathbf{R}_{i,c(m)} + \xi\mathbf{I}, \tag{8.96}$$

and ξ determines the bound on the norm of downlink correlation matrix errors. To account for the CSI mismatch, it is proposed in [14] and [80] to enforce the QoS constraints for all possible values of the downlink channel correlation matrices that satisfy (8.95). Using this approach, the robust modification of (8.83) can be expressed as

$$\min_{\{\mathbf{W}_l\}_{l=1}^L} \sum_{l=1}^{L} \mathrm{tr}\{\mathbf{W}_l\} \quad \text{subject to } \mathrm{tr}\left\{(\mathbf{R}_{i,c(i)} - \xi\mathbf{I})\mathbf{W}_i\right\} - \gamma_i \sum_{m \neq i} \mathrm{tr}\left\{(\mathbf{R}_{i,c(m)} + \xi\mathbf{I})\mathbf{W}_m\right\} \geq \gamma_i \sigma_i^2,$$

$$\mathbf{W}_i = \mathbf{W}_i^H, \ \mathbf{W}_i \succeq 0 \quad \text{for all} \quad i = 1, \ldots, L. \tag{8.97}$$

Problem (8.97) is also a convex SDP problem that can be solved at the same complexity as (8.83). It can be interpreted as the worst-case modification of the nonrobust problem (8.83).

A much simpler yet suboptimal approach to robust unicast downlink beamforming has been recently proposed in [81]. The essence of this technique is, following the ideas of the algorithms of [11], [70], and [82], to split the transmit beamforming problem into the power control and weight vector optimization subproblems, and then to utilize simple decentralized algorithms for the weight vector optimization, while using a centralized worst-case optimization-based technique for adjusting the transmit powers. This approach has been shown in [81] to have a performance comparable to that of [80], while enjoying both a lower computational cost and a substantially reduced degree of the required cooperation between the network BSs compared to the algorithm of [80].

8.4.2 Broadcast Transmit Beamforming

The traditional broadcasting strategy is to radiate the transmitted power isotropically or using a fixed transmit beampattern. However, in such a case the transmit CSI about the user channels is ignored, and as a result, the users with weak channels can experience a severe QoS degradation. In future data broadcasting/multicasting applications, the CSI for all the intended users is likely to be available at the transmitter. Therefore, this CSI can be exploited to improve the performance with respect to the traditional broadcasting scheme [83].

Let a single multiantenna transmitter with the weight vector $\mathbf{w}$ broadcast the signal $s(t)$ to L single-antenna users whose channels $\mathbf{h}_l$, $l = 1, \ldots, L$, are known at the transmitter. The signal received at the lth user can be modeled as (8.4), and the received SNR at the lth user is given by

$$\mathrm{SNR}_l = \frac{\left|\mathbf{w}^H \mathbf{h}_l\right|^2}{\sigma_l^2}. \tag{8.98}$$

The SNRs in (8.98) can be viewed as the user QoS values. Let γ_l be the minimum acceptable QoS value for the lth user. Then, defining the normalized channel vectors

$$\mathbf{g}_l \triangleq \frac{\mathbf{h}_l}{\sqrt{\gamma_l \sigma_l^2}}, \quad l=1,\ldots,L, \tag{8.99}$$

the optimal broadcasting problem can be written as [83]

$$\min_{\mathbf{w}} \mathbf{w}^H \mathbf{w} \quad \text{subject to} \quad \left|\mathbf{w}^H \mathbf{g}_l\right|^2 \geq 1 \qquad \text{for all} \qquad l=1,\ldots L. \tag{8.100}$$

According to (8.100), the optimal broadcast weight vector is designed by minimizing the total transmit power under the constraints that the minimum acceptable QoS is guaranteed for each user.

Unfortunately, the problem in (8.100) is NP-hard [83], and hence it cannot be solved in polynomial time. To obtain a reasonably simple approximate solution to (8.100), it has been proposed in [83] to reformulate this problem in terms of $\mathbf{Z} \triangleq \mathbf{w}\mathbf{w}^H$ and $\mathbf{Q}_l \triangleq \mathbf{g}_l\mathbf{g}_l^H$ as

$$\begin{aligned} &\min_{\mathbf{Z}} \operatorname{tr}\{\mathbf{Z}\} \quad \text{subject to} \quad \operatorname{tr}\{\mathbf{Z}\mathbf{Q}_l\} \geq 1 \quad \text{for all} \quad l=1,\ldots,L, \\ &\qquad\qquad\qquad\qquad \mathbf{Z}^H = \mathbf{Z}, \ \operatorname{rank}\{\mathbf{Z}\} = 1, \end{aligned} \tag{8.101}$$

and then to replace the nonconvex constraint rank{**Z**} = 1 by its relaxed convex version $\mathbf{Z} \succeq 0$.

The so-obtained problem is a convex SDP problem. However, in contrast to the problem in (8.83), the rank and semidefinite constraints on **Z** are not equivalent to each other; that is, the matrix $\mathbf{Z}_{\text{opt}}$ obtained by solving the relaxed SDP problem is not rank one in general [83]. Therefore, to recover the optimal value of weight vector from $\mathbf{Z}_{\text{opt}}$, the so-called *randomization* approach was adopted in [83].

Another alternative broadcast beamforming problem setting in the case of fixed transmit power constraint is to maximize the minimum receiver SNR subject to this constraint. This problem can be written as [83]

$$\max_{\mathbf{w}} \min_{l} \left\{\left|\mathbf{w}^H \mathbf{h}_l\right|^2 \big/ \sigma_l^2\right\}_{l=1}^{L} \quad \text{subject to} \quad \mathbf{w}^H \mathbf{w} \leq \mathcal{P}. \tag{8.102}$$

This problem can also be shown to be NP-hard, but it can be relaxed to a convex SDP form in nearly the same way as problem (8.100); see [83] for more details.

A useful worst-case design-based robust modification of problem (8.100) has been discussed in [84]. The authors of [84] have assumed the norm-bounded channel errors $\boldsymbol{\nu}_l \triangleq \tilde{\mathbf{g}}_l - \mathbf{g}_l$; $\|\boldsymbol{\nu}_l\| \leq \kappa$ (with $\tilde{\mathbf{g}}_l$ and $\mathbf{g}_l$ being the actual and presumed values of the normalized channel vector, respectively), and modified problem (8.100) as

$$\min_{\mathbf{w}} \mathbf{w}^H \mathbf{w} \quad \text{subject to} \quad \left|\mathbf{w}^H (\mathbf{g}_l + \boldsymbol{\nu}_l)\right|^2 \geq 1 \qquad \text{for all} \qquad \|\boldsymbol{\nu}_l\| \leq \kappa, \ l=1,\ldots,L. \tag{8.103}$$

A simple and elegant algorithm has been proposed in [84] to find an approximate solution to (8.103) using a properly rescaled weight vector of the original nonrobust problem (8.100) or its convex relaxed reformulation presented in [83]. Another robust technique that solves (8.103) has been developed in [85] using the convex SDP approach.

8.4.3 Multicast Transmit Beamforming

The multicast transmit beamforming problem is a natural, yet nontrivial, extension of the broadcast problem to the case when different data streams have to be transmitted to different groups of users.

Let us modify the broadcast scenario by assuming a total of I ($1 \le I \le L$) multicast groups $\{\mathcal{D}_1, \ldots, \mathcal{D}_I\}$, where D_i is the index set of receivers participating in the ith group [86, 87]. It will be assumed that no user can be shared by more than one group, that is, $\mathcal{D}_i \cap \mathcal{D}_k = \varnothing$ for any $i \neq k$ and $\mathcal{D}_1 \cup \mathcal{D}_2 \cup \cdots \cup \mathcal{D}_I = \{1, 2, \ldots, L\}$, which also implies that

$$\sum_{i=1}^{I} |\mathcal{D}_i| = L.$$

Using these conventions, the joint design of the transmit beamformers for multiple user groups amounts to minimizing the total transmit power subject to the QoS constraints [86]:

$$\min_{\{\mathbf{w}_i\}_{i=1}^{I}} \mathbf{w}_i^H \mathbf{w}_i \quad \text{subject to} \quad \frac{|\mathbf{w}_k^H \mathbf{h}_m|^2}{\sigma_k^2 + \sum_{l \neq k} |\mathbf{w}_l^H \mathbf{h}_m|^2} \geq \gamma_m \quad \text{for all} \quad m \in \mathcal{D}_k, \quad k, l = 1, \ldots, I, \tag{8.104}$$

where $\mathbf{w}_i$ is the weight vector used for the ith group and, as before, γ_k and $\mathbf{h}_k$ are the minimal acceptable QoS and the downlink channel vector of the kth user.

Clearly, as (8.104) represents a generalization of (8.100), it it also NP-hard. Moreover, in contrast to (8.100), problem (8.104) can be infeasible due to crosstalk-type interference between the user groups. Using an approach similar to that used in [83] for relaxing broadcast problem (8.100) to a convex form, the authors of [86] have derived a relaxation of (8.104) to the following convex SDP problem:

$$\begin{aligned} \min_{\{\mathbf{Z}_i\}_{i=1}^{I}} \sum_{i=1}^{I} \mathrm{tr}\{\mathbf{Z}_i\} \quad \text{subject to} \quad & \mathrm{tr}\{\mathbf{Q}_m \mathbf{Z}_k\} \geq \gamma_m \sum_{l \neq k} \mathrm{tr}\{\mathbf{Q}_m \mathbf{Z}_l\} + \gamma_m \sigma_m^2 \\ & \text{for all} \quad m \in \mathcal{D}_k, \quad k, l = 1, \ldots, I, \\ & \mathbf{Z}_i^H = \mathbf{Z}_i, \ \mathbf{Z}_i \succeq 0 \quad \text{for all} \quad i = 1, \ldots, I. \end{aligned} \tag{8.105}$$

Similar to (8.101), problem (8.105) can be directly solved using available convex optimization tools [50].

The authors of [86] also derived a convex optimization-based algorithm to solve the multicast generalization of the max-min fair beamforming problem (8.102). The design of robust worst-case multicast beamformers is currently an open problem.

8.5 Conclusions

An overview of adaptive beamforming methods for wireless multiantenna communications has been presented. Both the receive and transmit beamforming problems and techniques have been discussed in detail. As many popular recent approaches to receive and transmit beamforming are based on modern optimization theory, our chapter has a strong emphasis on convex optimization-based techniques. More details about using convex optimization methods in wireless communications can be found in the recent tutorial papers [48] and [49].

Acknowledgment

This work was supported by the German Research Foundation (DFG) under Grant GE 1881/1-1.

References

[1] B. D. Van Veen and K. M. Buckley. 1988. Beamforming: A versatile approach to spatial filtering. *IEEE Acoust. Speech Signal Processing Mag.* 5:4–24.

[2] J. Mayhan. 1976. Nulling limitations for a multiple-beam antenna. *IEEE Trans. Antennas Propagation* 24:769–79.

[3] R. T. Compton. 1978. An adaptive array in a spread spectrum communication system. *Proc. IEEE* 66:289–98.

[4] J. Winters. 1982. Spread spectrum in a four-phase communication system employing adaptive antennas. *IEEE Trans. Commun.* 30:929–36.

[5] G. J. Foschini. 1996. Layered space-time architecture for wireless communication in a fading environment when using multielement antennas. *Bell Labs Tech. J.* 1:41–59.

[6] L. C. Godara. 1997. Application of antenna arrays to mobile communications. II. Beam-forming and direction-of-arrival considerations. *Proc. IEEE* 85:1195–245.

[7] T. S. Rapapport, ed. 1998. *Smart antennas: Adaptive arrays, algorithms, and wireless position location*. Piscataway, NJ: IEEE Press.

[8] S. M. Alamouti. 1998. A simple transmit diversity technique for wireless communications. *IEEE J. Selected Areas Commun.* 45:1451–58.

[9] V. Tarokh, N. Seshadri, and A. R. Calderbank. 1998. Space-time codes for high data rate wireless communication: Performance criterion and code construction. *IEEE Trans. Inform. Theory* 44:744–65.

[10] I. E. Telatar. 1999. Capacity of multi-antenna Gaussian channels. *Eur. Trans. Telecommun.* 10:585–96.

[11] F. Rashid-Farrokhi, K. J. R. Liu, and L. Tassiulas. 1998. Transmit beamforming and power control for cellular wireless systems. *IEEE J. Selected Areas Commun.* 16:1437–50.

[12] S. Kapoor, D. J. Marchok, and Y.-F. Huang. 1999. Adaptive interference suppression in multiuser wireless OFDM systems using antenna arrays. *IEEE Trans. Signal Processing* 47:3381–91.

[13] J. Razavilar, F. Rashid-Farrokhi, and K. J. R. Liu. 2000. Traffic improvements in wireless communication networks using antenna arrays. *IEEE J. Selected Areas Commun.* 18:458–71.

[14] M. Bengtsson and B. Ottersten. 2001. Optimal and suboptimal transmit beamforming. In *Handbook of antennas in wireless communications*, ed. L. Godara. Boca Raton, FL: CRC Press, chapter 18.

[15] C. Farsakh and J. A. Nossek. 1998. Spatial covariance based downlink beamforming in an SDMA mobile radio system. *IEEE Trans. Commun.* 46:1497–506.

[16] E. G. Larsson and P. Stoica. 2003. *Space-time block coding for wireless communications*. Cambridge, UK: Cambridge University Press.

[17] A. Paulraj, R. Nabar, and D. Gore. 2003. *Introduction to space-time wireless communications*. Cambridge, UK: Cambridge University Press.

[18] A. B. Gershman and N. D. Sidiropoulos, eds. 2005. *Space-time processing for MIMO communications*. New York: John Wiley & Sons.

[19] T. Kaiser, A. Boudroux, H. Boche, J. R. Fonollosa, J. B. Andersen, and W. Utschick, eds. 2005. *Smart antennas—State-of-the-art*. EURASIP Book Series on Signal Processing and Communications. New York: Hindawi.

[20] H. Bölcskei, D. Gesbert, C. B. Papadias, and A. J. van der Veen, eds. 2006. *Space-time wireless systems—From array processing to MIMO communications*. Cambridge, UK: Cambridge University Press.

[21] S. S. Jeng, H. P. Lin, G. Xu, and W. J. Vogel. 1995. Measurements of spatial signature of an antenna array. In *Proceedings of PIMRC'95*, Toronto, vol. 2, pp. 669–72.

[22] A. J. Weiss and B. Friedlander. 1996. Almost blind steering vector estimation using second-order moments. *IEEE Trans. Signal Processing* 44:1024–27.

[23] A. L. Swindlehurst. 1998. Time delay and spatial signature estimation using known asynchronous signals. *IEEE Trans. Signal Processing* 46:449–62.

[24] D. Astely, A. L. Swindlehurst, and B. Ottersten. 1999. Spatial signature estimation for uniform linear arrays with unknown receiver gains and phases. *IEEE Trans. Signal Processing* 47:2128–38.

[25] Y. Rong, S. A. Vorobyov, A. B. Gershman, and N. D. Sidiropoulos. 2005. Blind spatial signature estimation via time-varying user power loading and parallel factor analysis. *IEEE Trans. Signal Processing* 53:1697–710.

[26] H. L. Van Trees. 2002. *Optimum array processing*. New York: Wiley.

[27] A. B. Gershman. 2003. Robustness issues in adaptive beamforming and high-resolution direction finding. In *High-resolution and robust signal processing*, ed. Y. Hua, A. B. Gershman, and Q. Cheng. New York: Marcel Dekker.

[28] I. S. Reed, J. D. Mallett, and L. E. Brennan. 1974. Rapid convergence rate in adaptive arrays. *IEEE Trans. Aerospace Electronics Syst.* 10:853–63.

[29] D. D. Feldman and L. J. Griffiths. 1994. A projection approach to robust adaptive beamforming. *IEEE Trans. Signal Processing* 42:867–76.

[30] A. B. Gershman. 1999. Robust adaptive beamforming in sensor arrays. *AEU Int. J. Electronics Commun.* 53:305–14.

[31] J. Riba, J. Goldberg, and G. Vazquez. 1997. Robust beamforming for interference rejection in mobile communications. *IEEE Trans. Signal Processing* 45:271–75.

[32] S. P. Applebaum and D. J. Chapman. 1976. Adaptive arrays with main beam constraints. *IEEE Trans. Antennas Propagation* 24:650–62.

[33] K. Takao, M. Fujita, and T. Nishi. 1976. An adaptive antenna array under directional constraint. *IEEE Trans. Antennas Propagation* 24:662–69.

[34] K. I. Pedersen, P. E. Mogensen, and B. H. Fleury. 2000. A stochastic model of the temporal and azimuthal dispersion seen at the base station in outdoor propagation environments. *IEEE Trans. Vehicular Technol.* 49:437–47.

[35] Y. I. Abramovich. 1981. Controlled method for adaptive optimization of filters using the criterion of maximum SNR. *Radio Eng. Electronic Physics* 26:87–95.

[36] H. Cox, R. M. Zeskind, and M. H. Owen. 1987. Robust adaptive beamforming. *IEEE Trans. Acoust. Speech Signal Processing* 35:1365–76.

[37] B. D. Carlson. 1988. Covariance matrix estimation errors and diagonal loading in adaptive arrays. *IEEE Trans. Aerospace Electronic Syst.* 24:397–401.

[38] S. Vorobyov, A. B. Gershman, and Z.-Q. Luo. 2003. Robust adaptive beamforming using worst-case performance optimization: A solution to the signal mismatch problem. *IEEE Trans. Signal Processing* 51:313–24.

[39] L. Chang and C. C. Yeh. 1992. Performance of DMI and eigenspace-based beamformers. *IEEE Trans. Antennas Propagation* 40:1336–47.

[40] R. J. Mailloux. 1995. Covariance matrix augmentation to produce adaptive array pattern troughs. *IEE Electronics Lett.* 31:771–72.

[41] A. B. Gershman, U. Nickel, and J. F. Böhme. 1997. Adaptive beamforming algorithms with robustness against jammer motion. *IEEE Trans. Signal Processing* 45:1878–85.

[42] J. R. Guerci. 2000. Theory and application of covariance matrix tapers to robust adaptive beamforming. *IEEE Trans. Signal Processing* 47:977–85.

[43] M. A. Zatman. 2000. Comment on "Theory and application of covariance matrix tapers for robust adaptive beamforming." *IEEE Trans. Signal Processing* 48:1796–800.

[44] S. Shahbazpanahi, A. B. Gershman, Z.-Q. Luo, and K. M. Wong. 2003. Robust adaptive beamforming for general-rank signal models. *IEEE Trans. Signal Processing* 51:2257–69.

[45] S. Vorobyov, A. B. Gershman, Z.-Q. Luo, and N. Ma. 2004. Adaptive beamforming with joint robustness against mismatched signal steering vector and interference nonstationarity. *IEEE Signal Processing Lett.* 11:108–11.

[46] R. G. Lorenz and S. P. Boyd. 2005. Robust minimum variance beamforming. *IEEE Trans. Signal Processing* 53:1684–96.

[47] M. Lobo, L. Vandenberghe, S. P. Boyd, and H. Lebret. 1998. Applications of second-order cone programming. *Lin. Algebra Appl.* 284:193–228.

[48] Z.-Q. Luo. 2003. Applications of convex optimization in signal processing and digital communication. *Math. Programming* 97B:177–207.

[49] Z.-Q. Luo and W. Yu. 2006. An introduction to convex optimization for communications and signal processing. *IEEE J. Selected Areas Commun.* 24:1426–38.

[50] J. F. Sturm. 1999. Using SeDuMi 1.02, a MATLAB toolbox for optimization over symmetric cones. *Optim. Meth. Software* 11/12:625–53.

[51] K. Zarifi, S. Shahbazpanahi, A. B. Gershman, and Z.-Q. Luo. 2005. Robust blind multiuser detection based on the worst-case performance optimization of the MMSE receiver. *IEEE Trans. Signal Processing* 53:295–305.

[52] P. Stoica and J. Li, eds. 2006. *Robust adaptive beamforming.* Hoboken, NJ: John Wiley & Sons.

[53] Y. Rong, S. A. Vorobyov, and A. B. Gershman. 2005. Robust adaptive beamforming using probability-constrained optimization. In *Proceedings of the IEEE Workshop on Statistical Signal Processing*, Bordeaux, France, pp. 934–39.

[54] S. A. Vorobyov, Y. Rong, and A. B. Gershman. 2006. Robust minimum variance adaptive beamformers and multiuser MIMO receivers: From worst-case to probabilistically constrained designs. In *Proceedings of ICASSP'06*, Toulouse, France, vol. 5, pp. 977–80.

[55] S. A. Vorobyov, A. B. Gershman, and Y. Rong. 2007. On the relationship between the worst-case optimization-based and probability-constrained approaches to robust adaptive beamforming. In *Proceedings of ICASSP'07*, Honolulu, HI, vol. 2, pp. 977–980.

[56] A. Prékopa. 1995. *Stochastic programming.* Dordrecht, Netherlands: Kluwer Academic Publishers.

[57] H. Li, X. Lu, and G. B. Giannakis. 2002. Capon multiuser receiver for CDMA systems with space-time coding. *IEEE Trans. Signal Processing* 50:1193–204.

[58] D. Reynolds, X. Wang, and H. V. Poor. 2002. Blind adaptive space-time multiuser detection with multiple transmitter and receiver antennas. *IEEE Trans. Signal Processing* 50:1261–76.

[59] S. Shahbazpanahi, M. Beheshti, A. B. Gershman, M. Gharavi-Alkhansari, and K. M. Wong. 2004. Minimum variance linear receivers for multiaccess MIMO wireless systems with space-time block coding. *IEEE Trans. Signal Processing* 52:3306–13.

[60] A. Nordio and G. Taricco. 2006. Linear receivers for the multiple-input multiple-output multiple-access channel. *IEEE Trans. Commun.* 54:1446–56.

[61] B. Hassibi and B. M. Hochwald. 2002. High-rate codes that are linear in space and time. *IEEE Trans. Inform. Theory* 48:1804–24.

[62] M. Gharavi-Alkhansari and A. B. Gershman. 2005. Constellation space invariance of orthogonal space-time block codes. *IEEE Trans. Inform. Theory* 51:331–34.

[63] Y. Rong, S. Shahbazpanahi, and A. B. Gershman. 2005. Robust linear receivers for space-time block coded multi-access MIMO systems with imperfect channel state information. *IEEE Trans. Signal Processing* 53:3081–90.

[64] Y. Rong, S. A. Vorobyov, and A. B. Gershman. 2004. A robust linear receiver for multi-access space-time block coded MIMO systems based on probability-constrained optimization. In *Proceedings of IEEE VTC'04*, Milan, Italy, vol. 1, pp. 118–22.

[65] Y. Rong, S. A. Vorobyov, and A. B. Gershman. 2005. Robust linear receiver design for multi-access space-time block coded MIMO systems using stochastic optimization. In *Proceedings of the IEEE Workshop on Statistical Signal Processing*, Bordeaux, France, pp. 65–70.

[66] R. Wang, H. Li, and T. Li. 2006. Robust multiuser detection for multicarrier CDMA systems. *IEEE J. Selected Areas Commun.* 24:673–83.

[67] Y. Rong, S. A. Vorobyov, and A. B. Gershman. 2006. Robust linear receivers for multi-access space-time block coded MIMO systems: A probabilistically constrained approach. *IEEE J. Selected Areas Commun.* 24:1560–70.

[68] G. Jöngren, M. Skoglund, and B. Ottersten. 2002. Combining beamforming and space-time block coding. *IEEE Trans. Inform. Theory* 48:611–27.

[69] S. Zhou and G. B. Giannakis. 2002. Optimal transmitter eigen-beamforming and space-time block coding based on channel mean feedback. *IEEE Trans. Signal Processing* 50:2599–613.

[70] E. Visotsky and U. Madhow. 1999. Optimum beamforming using transmit antenna arrays. In *Proceedings of IEEE VTC'99*, Houston, TX, vol. 1, pp. 851–56.

[71] M. Bengtsson and B. Ottersten. 1999. Optimal downlink beamforming using semidefinite optimization. In *Proceedings of the 37th Annual Allerton Conference on Communications, Control, and Computing*, pp. 987–96.

[72] E. Matskani, N. D. Sidiropoulos, Z.-Q. Luo, and L. Tassiulas. Submitted. Convex approximation techniques for joint multiuser downlink beamforming and admission control. *IEEE Trans. Wireless Commun.*

[73] E. Matskani, N. D. Sidiropoulos, Z.-Q. Luo, and L. Tassiulas. 2007. Joint multiuser downlink beamforming and admission control: A semidefinite relaxation approach. In *Proceedings of ICASSP'07*, Honolulu, HI, vol. 3, pp. 585–588.

[74] E. Matskani, N. D. Sidiropoulos, Z.-Q. Luo, and L. Tassiulas. 2007. A second-order cone deflation approach to joint multiuser downlink beamforming and admission control. In *Proceedings of the IEEE Workshop on Signal Processing Advances in Wireless Communications*, Helsinki, Finland, pp. 1–5.

[75] G. Caire and S. Shamai. 2003. On the achievable throughput of a multi-antenna Gaussian broadcast channel. *IEEE Trans. Inform. Theory* 49:1691–706.

[76] S. Vishwanath, N. Jindal, and A. Goldsmith. 2003. Duality, achievable rates, and sum-rate capacity of Gaussian MIMO broadcast channels. *IEEE Trans. Inform. Theory* 49:2658–68.

[77] Z. Tu and R. S. Blum. 2003. Multiuser diversity for a dirty paper approach. *IEEE Commun. Lett.* 7:370–72.

[78] Q. H. Spencer, A. L. Swindlehurst, and M. Haardt. 2004. Zero-forcing methods for downlink spatial multiplexing in multi-user MIMO channels. *IEEE Trans. Signal Processing* 52:461–71.

[79] G. Dimić and N. D. Sidiropoulos. 2005. On downlink beamforming with greedy user selection: Performance analysis and a simple new algorithm. *IEEE Trans. Signal Processing* 53:3857–68.

[80] M. Bengtsson. 2000. Robust and constrained downlink beamforming. In *Proceedings of EUSIPCO'00*, Tampere, Finland, pp. 1433–36.

[81] M. Biguesh, S. Shahbazpanahi, and A. B. Gershman. 2004. Robust downlink power control in wireless cellular systems. *EURASIP J. Wireless Commun. Networking* 2:261–72.

[82] H. Boche and M. Schubert. 2001. A new approach to power adjustment for spatial covariance based downlink beamforming. In *Proceedings of ICASSP'01*, Salt Lake City, UT, vol. 5, pp. 2957–60.

[83] N. D. Sidiropoulos, T. N. Davidson, and Z.-Q. Luo. 2006. Transmit beamforming for physical-layer multicasting. *IEEE Trans. Signal Processing* 54:2239–51.

[84] E. Karipidis, N. D. Sidiropoulos, and Z.-Q. Luo. 2006. Convex transmit beamforming for downlink multicasting to multiple co-channel groups. In *Proceedings of ICASSP'06*, Toulouse, France, vol. 5, pp. 973–76.

[85] I. Wajid, A. B. Gershman, S. A. Vorobyov, and Y. A. Karanouh. 2007. Robust multiantenna broadcasting with imperfect channel state information. In *Proceedings of the IEEE Workshop on Computer Advances in Multi-Sensor Adaptive Processing*, U.S. Virgin Islands.

[86] E. Karipidis, N. D. Sidiropoulos, and Z.-Q. Luo. 2008. Submitted. Quality of service and max-min fair transmit beamforming to multiple co-channel multicast groups. *IEEE Trans. Signal Processing* 56:1268–1279.

[87] E. Karipidis, N. D. Sidiropoulos, and Z.-Q. Luo. 2005. Transmit beamforming to multiple co-channel multicast groups. In *Proceedings of the IEEE Workshop on Computer Advances in Multi-Sensor Adaptive Processing*, Puerto Vallarta, Mexico, pp. 109–12.

9

Adaptive Equalization for Wireless Channels*

Richard K. Martin
Air Force Institute of Technology

Adaptive equalizers have been in use for about four decades. Since then, the applications and allowable complexity have changed dramatically, but the basic design approaches are largely unchanged. This chapter reviews common methodologies for designing trained and blind adaptive equalizers, including fast algorithms and slower, low-complexity algorithms. Application of these principles to modern wireless communication systems is demonstrated via discussion of popular equalizers for two newly popular modulation formats, multicarrier and ultrawideband communications. The chapter concludes with a discussion of the interaction of the adaptive equalizer with other adaptive blocks, such

* The views expressed in this chapter are those of the author and do not reflect the official policy or position of the U.S. Air Force, Department of Defense, or U.S. government. This document has been approved for public release, distribution unlimited.

as an adaptive carrier frequency offset estimator, an adaptive gain controller, and adaptive frequency-domain equalizers (for multicarrier systems).

9.1 Introduction and Historical Perspective

In this section we discuss the motivation for adaptive equalizers, including a discussion of notation. We then give a brief historical perspective, following the proposal of adaptive equalizers in the 1960s, to their heyday of research in the 1980s, through current research.

9.1.1 Equalizer Structure and Notation

Consider a single-input multiple-output (SIMO) wireless communication channel. This can be obtained via multiple sensors or oversampling by a factor of P at the receiver. The channel at sensor p or sampling subinterval p is often modeled as

$$y^{(p)}[n]=\sum_{k=0}^{L} h^{(p)}[n,k]x[n-k]+w^{(p)}[n], \quad p\in\{1,\ldots,P\}, \tag{9.1}$$

where $w^{(p)}[n]$ is additive noise (typically white and Gaussian), the channel coefficient $h^{(p)}[n,k]$ gives the dependence of output n on input n-k, and L is the channel order. This is sometimes called the time-lag representation of the channel [1]. Throughout this chapter, we will use n to index the time value (or sample number or iteration) and k to index the lag, i.e., the dependence of the current output on the input that occurred k samples ago, and p will index the receive antennas (or oversampling subsequence). When abundance of indices permits, discrete time indices will be indicated in square brackets, element indices of a matrix or vector will be indicated as subscripts, and all other indices will be in superscripts with parentheses. Matrices and vectors will be in boldface. The superscripts $(\cdot)^*$, $(\cdot)^T$, and $(\cdot)^H$ denote complex conjugate, matrix transpose, and Hermitian (conjugate transpose), respectively. All other superscripts are powers.

When the channel is time invariant, it is not dependent on n and the model simplifies to a discrete convolution,

$$\begin{aligned} y^{(p)}[n] &= \underbrace{\sum_{k=0}^{L} h^{(p)}[k]x[n-k]}_{\mathbf{x}^T[n]\mathbf{h}^{(p)}}+w^{(p)}[n], \quad p\in\{1,\ldots,P\}. \\ \mathbf{h}^{(p)} &= \left[h^{(p)}[0],\ldots,h^{(p)}[L]\right]^T, \\ \mathbf{x}[n] &= \left[x[n],\ldots,x[n-L]\right]^T. \end{aligned} \tag{9.2}$$

Unless otherwise noted, we will use the channel model of (9.2), although we will discuss some recent work that features the more challenging model of (9.1). Ideally, the channel is unity for a single value of the delay and zero otherwise, indicating a single propagation path from the transmitter to the receiver. However, realistically, there are multiple delayed and attenuated reflections, called multipath, leading to many other nonzero channel coefficients.

The goal of the receiver is to estimate the sequence $x[n]$ from the sequence $y[n]$. There are various ways to do this. Optimally, one would estimate the entire sequence $x[n]$ from the entire sequence $y[n]$, and the maximum likelihood solution is called maximum likelihood sequence estimation (MLSE), which is typically implemented via the Viterbi algorithm [2]. However, MLSE is typically extremely complex (of the order A^L, where A is the alphabet size of the transmitted signal). Thus, it is impractical for most wireless channels, which can span hundreds of samples, with A typically ranging from 4 to 64.

Equalization is a relatively low-complexity alternative to MLSE, although its performance is suboptimal. Equalization is the act of processing the output of the communication channel with the goal of recovering the input, often in the form of a filtering operation. This can be a linear finite impulse response (FIR) filter, a linear infinite impulse response (IIR) filter, or a nonlinear decision-feedback equalizer (DFE). Other equalization techniques exist, but due to considerations of brevity, this chapter focuses on equalization via linear filtering, as it is the most popular approach, although we briefly discuss the DFE and its variants.

The linear equalizer of order N is modeled as

$$\begin{aligned} \hat{x}[n] &= \sum_{p=1}^{P}\left(\sum_{k=0}^{N} f^{(p)}[k]\,y^{(p)}[n-k]\right) = \mathbf{f}^T\mathbf{y}[n], \\ \mathbf{f} &= \left[f^{(1)}[0],\ldots,f^{(1)}[N],f^{(2)}[0],\ldots,f^{(P)}[N]\right]^T, \\ \mathbf{y}[n] &= \left[y^{(1)}[n],\ldots,y^{(1)}[n-N],y^{(2)}[n],\ldots,y^{(P)}[n-N]\right]^T. \end{aligned} \tag{9.3}$$

Later in this chapter, we will discuss methods of designing $\mathbf{f}$ for various communication standards that are currently enjoying widespread popularity. First, however, we discuss the reasons for making $\mathbf{f}$ adaptive, and give a brief history of equalizer design.

9.1.2 Motivation for Adaptation

Although the entirety of this book deals with adaptive processing, there is need to specifically motivate adaptive equalization. In a wireless environment, the channel model is highly dependent on the physical location of the reflectors that lead to multipath. As objects move, the channel model must change accordingly. In particular, if either the transmitter or receiver is mobile, every channel coefficient will gradually change, and the equalizer must change accordingly.

There are two ways to deal with this changing environment. One could assume the channel is static over a short time window, and then compute the optimal equalizer for that window via a batch processing algorithm [e.g., 3]. For each new block, a new equalizer could be computed from scratch, and that equalizer could be used to equalize the data in that particular block. Alternatively, an adaptive algorithm can recursively compute the equalizer by tweaking the coefficient values from the previous time step. The former approach has the advantage of optimality when the channel truly is static and the window is large enough to average out the noise. However, the complexity can be very high, since optimal solutions generally require matrix inversions, computation of generalized eigenvectors [3], or singular value decompositions [4], depending on the problem, and these must be repeated every block. The latter approach has the advantage of making use of the solution from the previous time step, and the complexity is usually limited to computation of a matrix-vector multiply, or often only vector-vector and vector-scalar multiplies, and moreover, the complexity is evenly spread out over time (rather than as a lump sum at initialization). However, convergence to a good value often requires many more data samples than optimal batch processes. Thus, adaptive equalizers are by no means universally superior, but they are often preferred when computational power is at a premium and mobility is high, e.g., in a mobile handset.

9.1.3 History of Adaptive Equalizers

Adaptive equalizers, sometimes called automatic equalizers, have been in use since the 1960s [5, 6]. Of particular note is the introduction of the least mean square (LMS) algorithm, sometimes called the Widrow-Hoff algorithm [5, 7]. LMS is still used today as a benchmark for comparison of adaptive equalizers, due to its low complexity and its convergence to the minimum mean squared error (MMSE) equalizer for a static channel. Research on adaptive equalization became more widespread in the 1970s, motivated by the need to equalize the impulse responses of telephone lines [8–10]. This research focused on the comparison of different cost functions, and on hybrid equalizer structures, such as the combination of a partial equalizer and a reduced-complexity MLSE. However, this research typically assumed the availability of a sufficiently long training signal, which reduces the channel throughput, and is not even available in surveillance environments.

A "blind" (or "self-recovering") equalizer is one that relies on known statistical properties of the transmitted signal, rather than on a training signal. A blind, adaptive equalizer was first introduced in 1975 in [11], which replaced the training signal in the LMS algorithm with the output of a decision device at the receiver. This idea was later termed decision direction (DD). However, it is dependent on the ability to make good decisions at initialization, which is not always the case. A more sophisticated blind equalizer, the constant modulus algorithm (CMA), was introduced in the early 1980s [12–14]. CMA assumes the transmitted data has a constant modulus, and the equalizer attempts to restore this property. However, CMA can be extended to non-constant modulus sources [15], in which case it may be viewed as a dispersion (or effective noise power)-minimizing algorithm. Despite the age of DD and CMA, most blind adaptive equalizers proposed more recently are rooted in these two algorithms. Exceptions often involve finding alternate signal properties to restore.

Since the mid-1980s, adaptive equalizer research has focused less on development of new algorithms and more on either characterizing popular algorithms or tweaking them for performance improvement or complexity reduction. Notable extensions for trained equalizers include application of nonlinearities to the data and error functions in the LMS algorithm [17], modifications to exploit sparsity in the equalizer [18], and consideration of alternate cost functions, such as the fourth power of the error instead of the square [19, 20], or hybrids of the two. Notable extensions for blind equalizers include the use of multiple moduli in CMA for higher-order constellations [21], and hybrid cost functions that retain the benefits of CMA but provide improved convergence speed [22].

In the past few years, there has been a surge of interest in algorithms that exploit the idea of CMA, but with a different contour in the constellation space. CMA tries to force the constellation points back onto a circular contour in the complex plane. Similarly, the square contour algorithm (SCA) [23], also called the constant square algorithm (CQA) [24], uses a square contour, which has a constant infinity norm (as opposed to the constant 2-norm of a circle). This idea was generalized to include all possible norms on the complex plane in [24] and [25], leading to a plethora of possible algorithms, including extended CMA (ECMA) [25] and the constant norm algorithm with an L6 norm (CNA-6) [24]. However, not all constellations are built on square or circle patterns; hence, [26] proposed the use of a more complicated cross-shaped contour, leading to the constant cross algorithm (CXA).

Further information on equalization of slowly time-varying channels can be obtained from a variety of survey papers. Qureshi's paper [16] surveys early work in trained adaptive equalizers (pre-1985). Reviews of blind adaptive equalization algorithms can be found in [27] and [28]. One of the more popular "encyclopedias" of adaptive filter algorithms is Haykin's book [29].

Up until around 2000, almost all adaptive equalizers were designed for the case in which the channel was frequency selective but quasi-static. That is, the channel impulse response was assumed to be relatively constant and to approximately obey the model of (9.2), but it could drift over time. This allows a gradient-descent type of equalizer to keep up. In the past few years, interest has risen in equalization of channels that vary rapidly with respect to the symbol period [1, 30–33]. In such a case, the channel is both time and frequency selective, or doubly selective, and is modeled more appropriately by (9.1) than by (9.2). Such channels cannot be equalized by gradient-descent equalizers (trained or blind) since the update rule cannot keep up with the speed of the channel variations, and more complex methods of equalization are required. One possibility is to take the two-dimensional Fourier transform of the channel convolution matrix, and then perform the equalization in this frequency domain. This is most appropriate for systems such as orthogonal frequency division multiplexing (OFDM), in which the data are already encoded in the frequency domain. For small amounts of time variation, MMSE symbol estimation can be performed by considering the interference from adjacent frequency bins due to the Doppler spread. The complexity can be kept manageable by ignoring the small interference coefficients at large Doppler spreads [31] or by windowing in the time domain to restrict the Doppler spread [1]. However, [1] and [31] were derived specifically in the context of OFDM, for which the frequency-domain input signal is discrete, so this approach is not applicable to other modulation schemes. A more general approach is to model both the channel and the equalizer in the time domain as matrices rather than as vectors [30]. This explicitly accounts for the time variations in the model, and allows for far more degrees of freedom in the equalizer. The challenge then becomes accurately determining all of the channel coefficients from limited observable data, and of computing the equalizer with a moderate amount of complexity. These are difficult goals, and are the primary drawbacks of this approach [1], although this can be mitigated somewhat by explicitly exploiting the sparse structure of the channel matrix [33]. Another approach that requires less degrees of freedom is to use a basis expansion model (BEM) [32]. A BEM models the doubly selective channel using basis signals that allow for both time and frequency selectivity, by substituting

$$h^{(p)}\left[n,k\right]=\sum_{q=-Q/2}^{Q/2} e^{j2\pi qn/N}h^{(p,q,k)},\quad k\in\left\{0,\ldots,L\right\} \tag{9.4}$$

into (9.1), where q indexes the Doppler spread and $h^{(p,q,k)}$ is the strength of Doppler component q at lag k for receive antenna p. For each time instant, the channel is modeled as a sum of complex exponentials with different Doppler spreads; hence, the amount of Doppler that is modeled is controlled (as opposed to the matrix model of [30], which can in principle allow arbitrarily large Doppler). The equalizer can also be modeled using the BEM, and hence it is also effectively a time-varying filter.

This chapter covers adaptive equalizers, and the equalizers discussed above for doubly selective channels are not adaptive in the classical sense of a recursive update rule. However, they do perform equalization by explicitly accounting for the time variations in the channel model, which is the essence of adaptive equalization. As mobility becomes more pervasive in communications and as data rates are pushed higher, many adaptive equalizers may need to be recast in more complex forms, such as BEMs.

9.2 Adaptive Equalizer Algorithm Formulation

This section covers adaptive algorithm design approaches, in the context of traditional digital communication systems (later sections discuss emerging digital modulation formats). We begin with a discussion of methods of trained and blind equalizer design, with a focus on the design philosophy rather than a specific algorithm. We then cover methods of improving algorithm performance, by either accelerating convergence or exploiting the sparse structure in wireless channels.

9.2.1 Trained Adaptive Algorithm Design Methodologies

Most trained adaptive equalizers take the form of a stochastic gradient descent of a cost function. By far, the most popular choice of cost function is the mean squared error (MSE), where the error is the difference of a desired signal $d[n]$ and the filter output. However, the point at which the error is measured can vary from application to application, leading to different algorithms.

In a traditional single-carrier communication system, the input-output relation is adequately modeled by (9.2) with no additional preprocessing at the transmitter. The data sequence $x[n]$ is in general complex valued and from a discrete signal constellation. When training (knowledge of $x[n]$ for short, intermittent time intervals) is available, the desired signal in wireless communications is almost always a delayed version of the input, $d[n] = x[n - \Delta]$. Then the error signal is

$$e[n] = \underbrace{x[n-\Delta]}_{d[n]} - \underbrace{\mathbf{f}^T \mathbf{y}[n]}_{\hat{x}[n-\Delta]}. \tag{9.5}$$

The cost function to be minimized is

$$J = E\left[\left|e[n]\right|^q\right] \tag{9.6}$$

with $q = 2$ leading to the MSE cost function, although $q = 1$ and $q = 4$ are sometimes encountered in the literature, leading to the mean absolute error (MAE) and mean fourth error (MFE) cost functions, respectively.

Computing a stochastic gradient descent of (9.6) requires computing the gradient with respect to $\mathbf{f}$ and removing the expectation. Then the new equalizer is additively adjusted in the direction of the negative gradient by a small step size μ,

$$\mathbf{f}[n+1] = \mathbf{f}[n] - \mu \nabla_{\mathbf{f}} J. \tag{9.7}$$

For $q = 2$, this leads to the LMS algorithm,

$$\mathbf{f}[n+1] = \mathbf{f}[n] + \mu\, e[n] \mathbf{y}^*[n \tag{9.8}$$

Due to the vector structure of $\mathbf{f}$ and $\mathbf{y}[n]$, (9.8) accounts for the SIMO channel model. Similarly, $q = 1$ leads to the error sign LMS algorithm [34], and $q = 4$ leads to the least mean fourth (LMF) algorithm [19].

Almost all adaptive equalizers in the literature follow this same structure: a cost function is proposed (usually in the form of (9.6) or a hybrid of several such cost functions), and a stochastic gradient descent update is computed. In sections 9.3 and 9.4, we will talk about modifications that can be applied to most gradient descent algorithms to improve performance.

9.2.2 Blind Adaptive Algorithm Design Methodologies

The biggest challenge in adaptive equalization comes when training is unavailable. Even in standards that include training, there is typically a long interval of data between training symbols. If this is the case, the error signal of (9.5) cannot be computed for use in the cost function.

Blind algorithms often turn to known statistical properties of the source signal in lieu of training. For example, when the source data comes from a finite constellation, then a pseudo-training signal can be formed from the nearest constellation point. Denoting $Q\{\cdot\}$ as the decision or quantization function that selects the nearest constellation point, an error signal and associated cost function can be computed as

$$\begin{aligned} e[n] &= Q\{\hat{x}[n]\} - \hat{x}[n], \\ J &= E\left[\left|e[n]\right|^2\right]. \end{aligned} \tag{9.9}$$

The associated stochastic gradient descent algorithm is called decision-directed LMS (DD-LMS). It is identical to (9.8) except for the new error signal. DD-LMS is commonly used due to its simplicity, but it is actually a member of a far broader class of algorithms, called Bussgang algorithms [35], which use the error signal

$$e[n] = g\{\hat{x}[n]\} - \hat{x}[n], \tag{9.10}$$

and MSE cost function (as in (9.9)), where $g\{\cdot\}$ is a nonlinear function that estimates the source signal from the equalizer output using Bayesian techniques.

If reliable decisions cannot be made due to excessive noise, an alternative is CMA. CMA uses the error signal and cost function

$$e[n] = \left|\hat{x}[n]\right|^2 - \gamma, \tag{9.11}$$

and MSE cost function (as in (9.9)). If the transmitted signal is not constant modulus, CMA can still be used with a performance penalty, or the multiple modulus algorithm (MMA) can be used [21]. Table 9.1 compares the error signals, cost functions, and parameters of the most common trained and blind algorithms, from which most algorithm variants are derived. Note that it is possible to manipulate the form of the CMA error term to show that it is a special case of the Bussgang algorithm, and hence some sources list it as such [29], but it is not really in the same spirit as other Bussgang algorithms (i.e., it does not make a nonlinear estimation of the source symbol), and hence it is listed separately in Table 9.1.

In emerging communications standards, the properties exploited in this section are often not readily available. In this case, the algorithm designer must search for other properties. For example, known zero padding or redundancy (such as repetition) in the transmitted signal may be exploited. Sections 9.3 and 9.4 discuss specific examples in more detail.

9.2.3 Algorithm Acceleration Techniques

The primary drawback of LMS and particularly CMA is that they take a very large number of symbols to converge. Thus, much recent research has focused on accelerating

TABLE 9.1 A Comparison of Common Trained (T) and Blind (B) Adaptive Equalizer Errors, Costs, and Parameters, When the Algorithm Can Be Written as a Stochastic Gradient Descent

	Error Signal	Cost Function	Parameters	Cost Name	Algorithm Name	Reference
T	$e[n] = x[n] - \hat{x}[n]$	$E\left[\lvert e[n]\rvert^q\right]$	$q = 1$	MAE	Error sign LMS	[34]
T			$q = 2$	MSE	LMS	[5]
T			$q = 4$	MFE	LMF	[19]
B	$e[n] = g\{\hat{x}[n]\} - \hat{x}[n]$					
B	$g\{x\} = \text{sign}(x)$	$E\left[\lvert e[n]\rvert^q\right]$	$q = 2$		Sato	[11]
B	$g\{x\} = Q\{x\}$	$E\left[\lvert e[n]\rvert^q\right]$	$q = 2$	DD	DD-LMS	[29]
B	$g\{x\} =$ Bayesian estimator of source	$E\left[\lvert e[n]\rvert^q\right]$	$q = 2$		Bussgang	[35]
B	$e[n] = \lvert\hat{x}[n]\rvert_l^p - \gamma$	$E\left[\lvert e[n]\rvert^q\right]$	$(p,q,l) = (1,1,2)$			[13]
B			$(p,q,l) = (2,2,2)$	CM	CMA	[13]
B			$(p,q,l) = (\text{var},2,2)$	Godard	Godard	[12]
B			$(p,q,l) = (1,2,\infty)$		SCA or CQA	[25]
B			$(p,q,l) = (2,2,4)$		ECMA	[25]
B			$(p,q,l) = (2,2,6)$		CNA-6	[24]

Note: The delay Δ is omitted for simplicity. In each section of the table, the general form of the algorithm is given in the first line, and the next few lines give more specific cases. The variables $x[n], \hat{x}[n], e[n]$ are the source signal, its estimate, and the error signal used by the equalizer, respectively.

their convergence speed. There are techniques such as recursive least squares (RLS) that improve the speed at the cost of an increase in the number of computations (e.g., $O\{N^2\}$ per iteration for RLS, as opposed to $O\{N\}$ per iteration for LMS). However, herein we focus on acceleration techniques that have only a minimal increase in computational overhead, as is appropriate for a mobile handset or any other wireless device with power and computational restrictions.

One type of acceleration technique is to exploit the fact that wireless channels are often very sparse. This is reflected in the fact that most wireless standards define "test channels" via ray models with only about four rays over the span of many symbol intervals.

One difficulty in exploiting sparsity comes from the fact that even when the channel impulse response is sparse, the corresponding equalizer is not necessarily sparse. This is shown in Figure 9.1, which plots the magnitudes of the impulse response coefficients for (a) an example sparse channel, (b) the corresponding equalizer, which is only

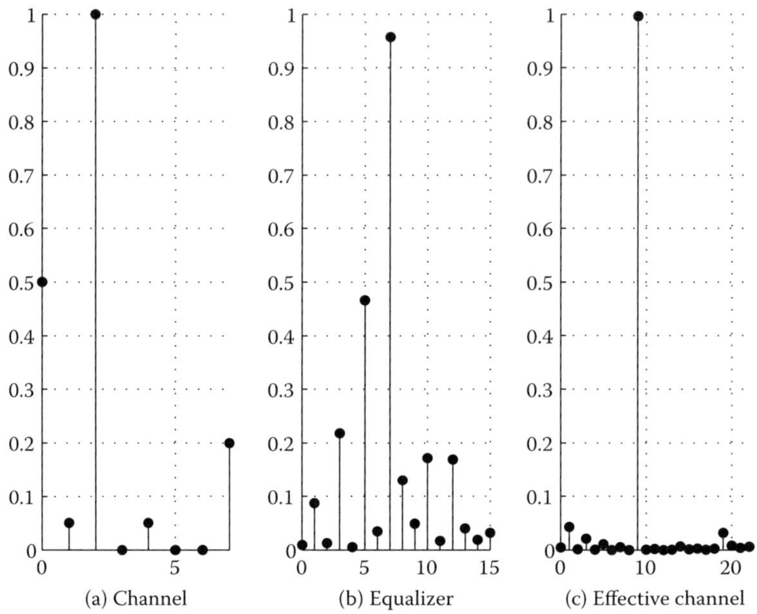

FIGURE 9.1 Even when a multipath channel is sparse, the corresponding linear equalizer need not be sparse. (a) Magnitudes of channel coefficients. (b) Magnitudes of zero-forcing equalizer coefficients. (c) Magnitudes of coefficients of resulting effective channel.

somewhat sparse, and (c) the resulting effective channel. This means there are several ways to exploit sparsity in adaptive equalization:

- Form an adaptive equalizer that is only somewhat sparse, as in Figure 9.1(b).
- Form an adaptive channel identifier that exploits sparsity, then periodically use the current estimate to compute the equalizer.
- Use an alternate equalizer structure that exploits sparsity, such as a partial feedback equalizer (PFE) [36], as shown in Figure 9.2, and adapt it via an algorithm that exploits sparsity.

We now discuss these adaptive algorithms for sparse linear filters (identifiers or equalizers).

An algorithmic paradigm that has received much attention recently due to its fast convergence for sparse adaptive filters is the use of proportionate adaptation [37], as proposed by Duttweiler [18]. The idea is to update large taps more quickly than small taps, since they are more important. The many small taps do little to reduce MSE, and if the filter is known *a priori* to be sparse, many of the small taps may actually be zero, and their updates are noise driven. Although recent advances in proportionate adaptation have been motivated by echo cancellation, the algorithms are equally applicable to wireless systems in which the channels or equalizers are known to be sparse, for example, digital television channels. As shown in [38], a histogram of measured channel coefficients for digital television (U.S. standard) follows an inverse power law distribution,

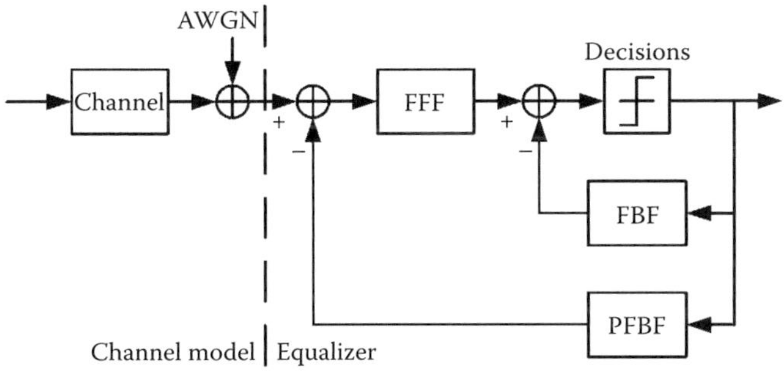

FIGURE 9.2 Block diagram of a partial feedback equalizer (PFE). The partial feedback filter (PFBF) cancels the sparse channel coefficients before the feedforward filter (FFF) smears them out, which cannot be done in a conventional DFE. Then the FFF and the feedback filter (FBF) operate as a normal DFE would. The PFBF can be long and sparse, allowing the FFF and FBF to be short.

whereas histograms of the coefficients of the feedforward and feedback filters in the optimal DFE follow decaying exponential distributions. Thus, the probability of a tap being large is very small.

The proportionate normalized LMS (PNLMS) algorithm [18] is the most widely referenced proportionate adaptation algorithm. The update rule, generalized from [18] to complex-valued signals, is

$$\begin{aligned}
\gamma_{\min}[n] &= \rho \max\{\delta_0, |\mathbf{f}_0[n]|, \ldots, |\mathbf{f}_N[n]|\} \\
\boldsymbol{\gamma}_k[n] &= \max\{\gamma_{\min}[n], |\mathbf{f}_k[n]|\}, \quad k \in \{0, \ldots, N\} \\
\mathbf{g}_k[n] &= \frac{\boldsymbol{\gamma}_k[n]}{\left(\frac{1}{N+1}\right)\sum_{l=0}^{N} \boldsymbol{\gamma}_l[n]}, \quad k \in \{0, \ldots, N\} \\
\mathbf{G}[n] &= \mathrm{diag}[\mathbf{g}_0[n], \ldots, \mathbf{g}_N[n]] \\
\mathbf{f}[n+1] &= \mathbf{f}[n] + \frac{\mu\,\mathbf{G}[n]\mathbf{y}^*[n]e[n]}{\mathbf{y}^T[n]\mathbf{G}[n]\mathbf{y}^*[n]+\delta},
\end{aligned} \tag{9.12}$$

where the error signal $e(n)$ is as in (9.5), and the set of constants $\{\rho, \delta, \delta_0\}$ are called regularizers, mollifiers, or biases, and serve to keep the numerator and the denominator of the update term from going to zero. Ignoring the mollifiers, the kth diagonal element of the matrix $\mathbf{G}[n]$ is the fraction of the 1-norm of the equalizer that occurs in filter tap k. Thus, the product $\mu\,\mathbf{G}[n]$ can be thought of as a vector step size with an average value of μ, but whose elements are proportionate to the corresponding equalizer tap magnitudes.

The denominator $\mathbf{y}^T[n]\mathbf{G}[n]\mathbf{y}^*[n] + \delta$ normalizes by the (tap-weighted) power of the filter input, which essentially makes the scale of the update insensitive to the scale of the filter input signal, and this aspect of PNLMS is taken from the normalized LMS (NLMS) algorithm [29]. Note that PNLMS and its successors were originally developed for echo cancellation, for which this normalization term was needed. However, in this chapter, we discuss equalization; hence, the power of the input signal is less variable and the normalization term is not necessary. Thus, henceforth we omit it.

By weighting the update terms by the current tap magnitudes, PNLMS forces the large taps to converge quickly at the expense of the small taps. Since many of the small taps should be zero anyway, this mitigates the amount of misadjustment (jitter-induced excess MSE) [7] while still allowing the large (and thus more important) taps to converge quickly. Thus, NLMS is an adaptation rule that distributes the update strength equally across all taps whether they need it or not, whereas PNLMS is a "capitalist" adaptation rule that allows the rich (moderate, partially converged values) to get richer (large, converged values). The sparser the optimal filter is, the more the large taps can benefit from the minimal adaptation of the small taps; hence, PNLMS is ideally suited to sparse channels.

Much recent literature has focused on analyzing and improving PNLMS. Deng and Doroslovački [39] have extended the idea of proportionate adaptation to find the optimal weighting function. That is, instead of populating the matrix $\mathbf{G}[n]$ with the normalized filter tap magnitudes, which is essentially a linear weight, they considered applying an arbitrary function to the filter taps before normalizing. They then computed the function that led to the fastest convergence, in terms of having *all* filter taps reach values within some small ε of their optimal values. They concluded that the optimal function was logarithmic rather than linear, so that the first three lines of (9.12) can be replaced by

$$\mathbf{g}_k[n] = \frac{\ln\left(\left|\mathbf{f}_k[n]\right|/\varepsilon\right)}{\left(\frac{1}{N+1}\right)\sum_{l=0}^{N}\ln\left(\left|\mathbf{f}_l[n]\right|/\varepsilon\right)}, \tag{9.13}$$

where ε serves as a mollifier when filter taps are small. The resulting algorithm is called μ-law PNLMS (MPNLMS). The natural log can be replaced by a piecewise-linear function with a minimal effect on performance [39].

Another approach is to combine the advantages of NLMS and PNLMS. The simplest approach is to alternate between the two [40, 41]. A more sophisticated approach was presented in [42], in which an improved PNLMS (IPNLMS) algorithm was created that combines NLMS and PNLMS in every update, and does not require the true weight vectors to be sparse. This is especially important for wireless equalization, since as stated above, sparse channels do not always lead to sparse equalizers. The motivation for IPNLMS is that the max function in (9.12) is very harsh, and that if the parameter estimates are inaccurate, the proportionality can amplify the error and have a detrimental effect on performance. Thus, they modify the weights in (9.12) to

$$\gamma_k[n] = \frac{(1-\alpha)}{2}\left(\frac{1}{N+1}\right)\left\| \mathbf{f}[n] \right\|_1 + \frac{(1+\alpha)}{2}\left| \mathbf{f}_k[n] \right|,$$
$$\mathbf{g}_k[n] = \frac{\gamma_k[n]}{\left(\frac{1}{N+1}\right)\sum_{l=0}^{N} \gamma_l[n]}, \quad k \in \{0, \ldots, N\}, \tag{9.14}$$

where $\|\cdot\|_1$ is the 1-norm of a vector and $\alpha \in [-1,1]$ is a user-defined weighting parameter. Observe that $\|\gamma[n]\|_1 = \|\mathbf{f}[n]\|_1$; for $\alpha = -1$ we get the NLMS algorithm, and for $\alpha = 1$ we get the PNLMS algorithm.

Proportionate adaptation rules are compared in Figures 9.3 to 9.5. In Figure 9.3, NLMS, PNLMS, MPNLMS, and IPNLMS are compared for channel identification, in which just over half of the channel coefficients are zero, and most of the rest are small. In Figures 9.4 and 9.5, a similar comparison is made for equalization. The magnitudes of the optimal equalizer taps are presented in Figure 9.4, sorted for easy analysis of the level of sparsity, and the learning curves are presented in Figure 9.5. Observe that for both the identification problem and the equalization problem, all proportionate adaptation algorithms see a performance gain. However, PNLMS and MPNLMS were originally developed for echo cancellation, in which as many as 90% of the channel taps may be essentially zero. In wireless identification or equalization, the optimal filters are not

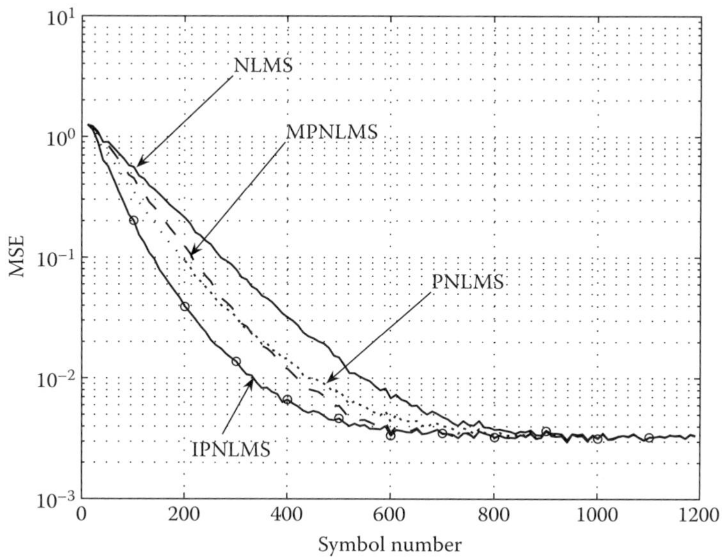

FIGURE 9.3 A demonstration of the convergence improvement of adaptive identification of wireless channels due to the use of proportionate adaptation. The algorithms considered are all in the normalized LMS family: NLMS [29], PNLMS [18], MPNLMS [39], and IPNLMS [42]. The channel had five of eleven nonzero taps, and thus was only somewhat sparse.

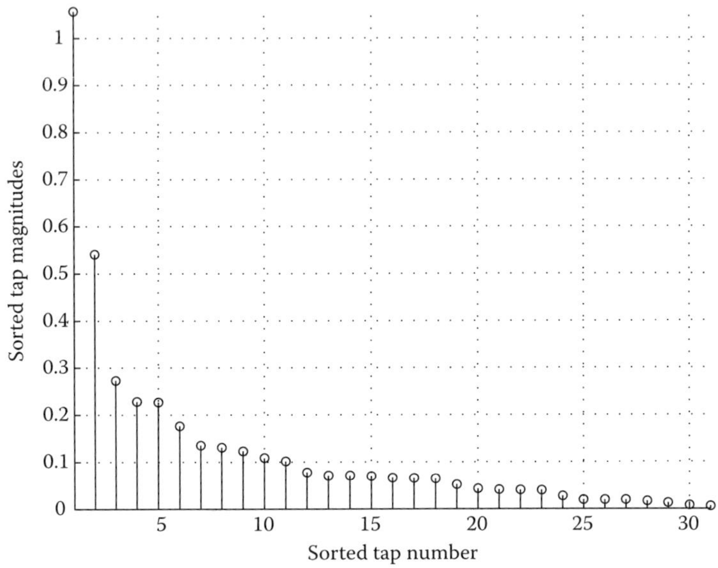

FIGURE 9.4 The tap magnitudes of an optimal equalizer, sorted in descending order. Note that although there are no zero values, the equalizer is still somewhat sparse, since two-thirds of the values are below 10% of the peak value.

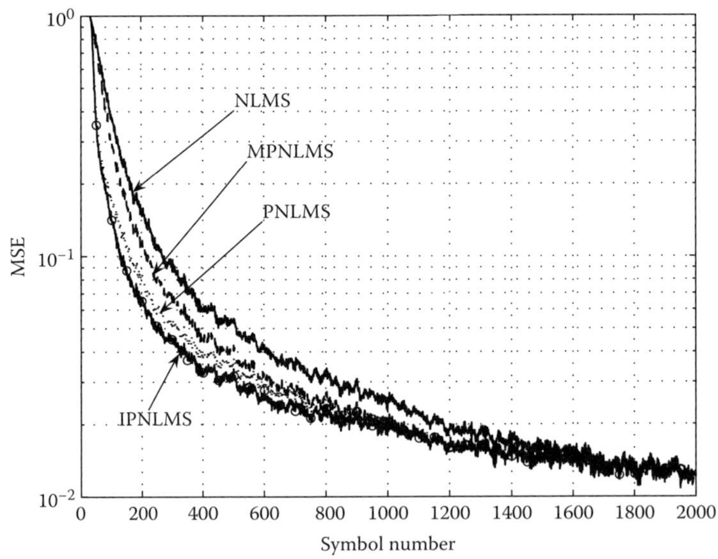

FIGURE 9.5 A demonstration of the convergence improvement of adaptive equalization of wireless channels due to the use of proportionate adaptation. The algorithms considered are all in the normalized LMS family: NLMS [29], PNLMS [18], MPNLMS [39], and IPNLMS [42]. The optimal equalizer tap magnitudes are shown in Figure 9.4.

nearly so sparse; hence, the gains are only on the order of a factor of 2 in convergence time. IPNLMS was designed to work for filters that are both sparse and nonsparse, and thus it outperforms the other algorithms in this case.

9.2.4 Complexity Reduction Techniques for Sparse Adaptive Equalizers

There are two benefits to *a priori* knowledge that the ideal filter is sparse. First, as discussed in the previous section, knowledge of sparsity can be incorporated into the adaptation rule to accelerate convergence of the important taps while suppressing (often advantageously) the adaptation of the less important taps. Second, the adaptation rule itself can be implemented sparsely, saving computations. Adaptation rules that are themselves sparse are generally referred to as partial-update algorithms. Sometimes they may incur a performance penalty, but if the system truly is sparse, they may actually improve performance. Whether or not the true system is actually sparse, they will always reduce the computational load of the equalizer by a fixed, known amount. Hence, they are very attractive for implementation in adaptive equalizers in low-power battery-operated devices.

As the name implies, partial-update algorithms only update a subset of M of the equalizer coefficients (out of the total of $N + 1$ coefficients) at each iteration. The distinctions between different partial-update algorithms are primarily the number of taps updated per iteration and the method by which the taps to be updated are chosen. Computational complexity is reduced by roughly a factor of M/N, but the exact amount depends upon the algorithm that is being partially updated.

One approach is to simply cycle through the taps to be updated at each iteration, as in sequential or periodic partial-update LMS [43]. For this baseline approach, one would expect the algorithm to take N/M times longer to converge. More sophisticated partial-update algorithms attempt to choose the M taps to be updated such that the best update possible can be formed under this constraint. However, care must be taken to ensure that the computational burden of the tap selection mechanism does not begin to balance out the computational savings of the partial update. Max-NLMS [44] chooses a single tap to update at each iteration, chosen as the tap with largest corresponding input. M-max-NLMS [45] generalizes this to update the M taps with the M largest inputs. The rationale is that the taps with the largest inputs will have the largest innovations, since the updates in NLMS algorithms are proportionate to the tap inputs. Heuristically, the largest innovations should lead to the largest improvement in the cost function. By selecting the M taps that yield the best improvement to the cost function, the update term removes the least useful steps and focuses on the most useful steps. This mitigates the misadjustment due to adapting the unnecessary taps, which can improve the performance of the algorithm, depending on how necessary or unnecessary each tap is. Since choosing the largest inputs only requires a small number of comparisons (essentially a subtraction operation each) and no multiplies, this selection criterion has very little complexity overhead.

All of these partial-update algorithms can be cast into the generic gradient descent form of

$$\mathbf{g}_j[n] = \begin{cases} 1, & (\text{select } M \text{ values of } j \text{ by some method}) \\ 0, & \text{else} \end{cases}$$
$$\mathbf{G}[n] = \text{diag}[\mathbf{g}[n]] \tag{9.15}$$
$$\mathbf{f}[n+1] = \mathbf{f}[n] + \mu\,\mathbf{G}[n]\mathbf{y}^*[n]e[n],$$

where $\mathbf{G}[n]$ is a diagonal matrix of ones and zeros that turn on or off the updates to each coefficient. (For applications such as echo cancellation, an NLMS normalization term as in the denominator of (9.12) can be explicitly included, or it can be included implicitly by dividing the step size μ by the normalization term.) With this structure, the partial update of (9.15) has the same form as the proportionate update of (9.12), and in either case, $\mu\,\mathbf{G}[n]$ effectively forms a vector step size.

Specific tap selection rules for (9.15) have the mathematical form

$$\mathbf{g}_j^{\text{sequential}}[n] = \mathbf{g}^{\text{sequential}}_{(j-1)_{\text{mod}(N+1)}}[n-1]$$
$$\mathbf{g}_j^{\text{maxLMS}}[n] = \begin{cases} 1, & j = \arg\max_{0 \le k \le N} |\mathbf{y}_k[n]| \\ 0, & \text{else} \end{cases} \tag{9.16}$$
$$\mathbf{g}_j^{\text{MmaxLMS}}[n] = \begin{cases} 1, & j \in \{\text{indices of } M \text{ largest } |\mathbf{y}_k[n]|\} \\ 0, & \text{else} \end{cases}.$$

Observe that max-NLMS and M-max-NLMS are better suited for equalization than channel identification when the application is digital communications. This is because digital communication channels often have inputs drawn from a constant modulus source (e.g., $\pm 1 \pm j$), in which case all of the inputs to the channel model would have the same magnitude, and no selection could be made, whereas the equalizer inputs contain intersymbol interference and noise, with a wide range of magnitudes.

Other recent work has investigated extensions to M-max-NLMS that further reduce the complexity. Selective-block NLMS [46] is like M-max NLMS, but in terms of blocks of taps rather than individual taps. Similarly, the short-sort algorithm [47] recognizes that in many cases, the significant taps are grouped together. The algorithm identifies a contiguous block of significant taps and always updates those, and performs M-max NLMS on the remaining taps, which reduces the sorting overhead. A slightly more sophisticated approach is to monitor an activity measure for each tap [48] (instead of simply comparing input magnitudes), and then favor those taps when updating. Table 9.2 lists popular proportionate adaptation and partial update algorithms.

Proportionate adaptation rules and partial-update rules were explicitly designed to exploit sparsity in the filter impulse response. PNLMS and its variants and M-max-LMS

TABLE 9.2 Proportionate Adaptation (PA) and Partial-Update (PU) Algorithms that Can Be Written in the Form $\mathbf{f}[n] = \mathbf{f}[n] + \mu e[n]\mathbf{G}[n]\mathbf{y}^*[n]$, with $\mathbf{G}[n] = \text{diag}[\mathbf{g}[n]]$

Type	Name	Value of $\mathbf{g}[n]$	Reference
	NLMS	$\mathbf{g}[n]=[\underbrace{1,1,\ldots,1}_{N+1 \text{ of them}}]^T$	[29]
PA	PNLMS	$\gamma_k[n]=\max\{\lvert\mathbf{f}_k[n]\rvert,\rho\max\{\delta_0,\lVert\mathbf{f}[n]\rVert_\infty\}\}$ $\mathbf{g}_k[n]=\dfrac{\gamma_k[n]}{\left(\frac{1}{N+1}\right)\sum_{l=0}^{N}\gamma_l[n]},\quad k\in\{0,\ldots,N\}$	[18]
PA	IPNLMS	$\gamma_k[n]=\frac{(1-\alpha)}{2}\left(\frac{1}{N+1}\right)\lVert\mathbf{f}[n]\rVert_1+\frac{(1+\alpha)}{2}\lvert\mathbf{f}_k[n]\rvert,\quad -1\le\alpha\le 1$ $\mathbf{g}_k[n]=\dfrac{\gamma_k[n]}{\left(\frac{1}{N+1}\right)\sum_{l=0}^{N}\gamma_l[n]},\quad k\in\{0,\ldots,N\},$	[42]
PA	MPNLMS	$\mathbf{g}_k[n]=\dfrac{\ln(\lvert\mathbf{f}_k[n]\rvert/\varepsilon)}{\left(\frac{1}{N+1}\right)\sum_{l=0}^{N}\ln(\lvert\mathbf{f}_l[n]\rvert/\varepsilon)},\quad k\in\{0,\ldots,N\}$	[39]
PU	Sequential PU NLMS	$\mathbf{g}_j^{\text{sequential}}[n]=\mathbf{g}^{\text{sequential}}_{(j-1)_{\text{mod}(N+1)}}[n-1]$ $\mathbf{g}[0]=\{\text{user-defined vector of } M \text{ ones, } N\text{-}M+1 \text{ zeros}\}$	[43]
PU	Max-LMS	$\mathbf{g}_j^{\text{maxLMS}}[n]=\begin{cases}1, & j=\arg\max_{0\le k\le N}\lvert\mathbf{y}_k[n]\rvert\\ 0, & \text{else}\end{cases}$	[44]
PU	M-max-LMS	$\mathbf{g}_j^{\text{MmaxLMS}}[n]=\begin{cases}1, & j\in\{\text{indices of } M \text{ largest } \lvert\mathbf{y}_k[n]\rvert\}\\ 0, & \text{else}\end{cases}$	[45]

Note: The table lists the composition of the vector $\mathbf{g}[n]$ for each algorithm.

and its variants were primarily motivated by channel identification for acoustic echo cancellation, although they have been applied to wireless channels as well. If we wish to apply them to equalization of wireless channels, it makes sense to use an equalizer structure that preserves the sparsity of the channel in its own impulse response. Motivated primarily by complexity reduction, the PFE and several related equalizer structures have been recently proposed that satisfy this goal [36]. In the PFE, as shown in Figure 9.2, a partial feedback filter (PFBF) cancels the sparse channel coefficients before the feedforward filter (FFF) smears them out, which cannot be done in a conventional DFE. Then the FFF and the feedback filter (FBF) operate as a normal DFE would. The PFBF can be long and sparse, allowing the other filters to be short.

9.3 Cyclic-Prefixed Communication Systems

Cyclic-prefixed systems come in two varieties: multicarrier and single-carrier cyclic prefixed. Multicarrier modulation is arguably the most popular choice of modulation format for communication standards that have emerged in the past decade. Examples include wireless local area networks (IEEE802.11a, HIPERLAN/2 in Europe, and MMAC in Japan), European Digital Video/Audio Broadcast (DVB/DAB), digital subscriber loops (DSL), the terrestrial repeaters in Sirius and XM satellite radio, power line communications, and the proposed "multiband" standard for ultrawideband communications, among others.

Given that the signal structure of cyclic-prefixed modulation differs markedly from that of single-carrier modulation, it provides a good framework for demonstrating the principles of adaptive equalizer design. Moreover, adaptive equalizers for cyclic-prefixed systems are fairly recent research products, and are of interest in and of themselves. In this section, we first review the signal structure, and then discuss trained and blind adaptive equalizers. Emphasis is placed on design methodology rather than on the algorithms themselves, since communications standards fall in and out of favor, but the design principles are equally applicable to other modulation formats.

The idea of cyclic-prefixed communications is to perform the equalization in the frequency domain. By considering a frequency-selective channel as a bank of parallel narrowband flat fading channels, equalization can be performed via a one-tap complex scalar on each tone. The collection of these scalars is called a frequency-domain equalizer (FEQ). However, in order to use the fast Fourier transform (FFT) to convert between time and frequency domains, the signal is necessarily discrete in time and block processed. Thus, some additional mathematical tricks are required in order to equalize in the frequency domain. Specifically, element-wise multiplication of the channel and the input data in the frequency domain is equivalent to their *circular* convolution in the time domain, but multipath propagation as modeled by (9.2) is a *linear* convolution. In order to make the convolution appear periodic (equivalently, circular), the input signal is made periodic over a short duration. A cyclic prefix (CP) is inserted as an extension to the start of each transmitted block by making a copy of the last ν samples. Thus, for a block initially of length N_{FFT}, the cyclically extended block of length $\bar{N} = N_{FFT} + \nu$ is periodic with period N_{FFT}. Provided that the channel order satisfies $L \leq \nu$, then the last N_{FFT} outputs of each block will appear to be a periodic (circular) convolution of the channel and the original N_{FFT} inputs of that block. Under this condition, equalization can be accomplished by taking an FFT, multiplying element-wise by the FEQ, and then taking an inverse FFT (IFFT) of the result. Since the FFT inherently induces block-based processing, this section uses the time index b to denote the block number, in addition to the previously introduced time index n that denotes the sample number.

Block diagrams of the two existing cyclic-prefixed systems are shown in Figure 9.6, namely, single-carrier cyclic-prefixed (SCCP) modulation (sometimes called single-carrier frequency-domain equalization [SC-FDE]) and OFDM. The distinction is that SCCP systems store the data in the time domain, so they follow the format discussed above, whereas multicarrier systems store the data in the frequency domain, so an additional

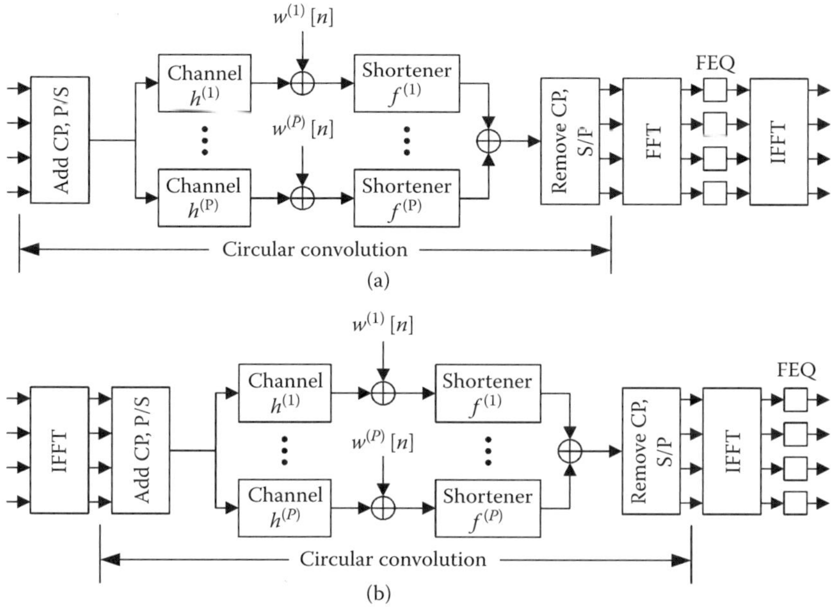

FIGURE 9.6 Block diagrams of (a) single-carrier cyclic-prefixed (SCCP) modulation and (b) multicarrier modulation. (I)FFT denotes (inverse) fast Fourier transform, P/S denotes parallel-to-serial conversion, and S/P the reverse.

IFFT is needed before transmission and the final IFFT discussed above is redundant and thus omitted. Multicarrier systems are older, and thus far more common in both the technical literature and industry standards. However, their transmitted IFFT outputs follow a Gaussian distribution (via the central limit theorem). This means that multicarrier systems tend to have a high peak-to-average power ratio, whereas SCCP systems do not.

In general, the channel order L is unknown, which often causes problems for adaptive equalizers. In the case of cyclic-prefixed systems, the CP length ν must be set without knowledge of L, in which case the condition $L \leq \nu$ may not be satisfied. Moreover, the CP introduces redundancy, reducing the data rate by a factor of $N_{FFT}/(N_{FFT} + \nu)$, so ν may be set small to keep the redundancy small even if the channel order may be larger than the maximum tolerable ν. Either way, the excess channel length induces interference and should be mitigated. This is often done via a channel-shortening equalizer (CSE). (Most literature refers to this as a time-domain equalizer [TEQ], but we avoid that term here since all the equalizers discussed in the previous sections also operate in the time domain, and the term *CSE* avoids confusion.) The CSE shortens the effective channel to the CP length, and then the FEQ can perform equalization. Like a standard equalizer, the CSE is modeled as in (9.3), and it is applied to the channel output as shown in Figure 9.6.

The next two sections discuss adaptive algorithms for the FEQ and CSE, in turn.

9.3.1 Adaptive FEQ Algorithms

Adaptation rules for the FEQ are easily formed as analogous to or generalizations of the commonly accepted adaptation rules for traditional equalizers, although there are some nuances due to modulation format. For the time being, we assume that the CSE is operating perfectly. The model of the remaining receiver processing of block b is

$$\begin{aligned}
\mathbf{y}[b] &= \left[y\left[(b-1)\bar{N}+\nu+1+\Delta\right],\ldots,y\left[b\bar{N}+\Delta\right]\right]^T \\
\tilde{\mathbf{y}}[b] &= \mathbf{F}\mathbf{y}[b] \\
\tilde{\mathbf{x}}[b] &= \mathbf{d}\odot\tilde{\mathbf{y}}[b] \\
\hat{\mathbf{x}}_{mc}[b] &= Q\{\tilde{\mathbf{x}}[b]\} \\
\hat{\mathbf{x}}_{sccp}[b] &= Q\{\mathbf{F}^{-1}\tilde{\mathbf{x}}[b]\},
\end{aligned} \tag{9.17}$$

where Δ accounts for the propagation delay as in (9.5), $\mathbf{F}$ denotes the Fourier transform of size N_{FFT} with inverse transform $\mathbf{F}^{-1} = \mathbf{F}^H$, $\mathbf{d}$ is the FEQ as a length N_{FFT} vector, $\odot$ denotes Hadamard (element-wise) multiplication, and $Q\{\cdot\}$ quantizes the output to the nearest constellation point.

First, consider trained adaptation rules. In many multicarrier standards, training is not available on every tone. For example, in DVB operating in 2K mode, the FFT size and number of subcarriers is 2,048. Of these, 1,705 are used and the rest are left null as a guard band, to mitigate adjacent channel interference. Of the used subcarriers, 45 are continual pilot tones (transmitting a pseudo-random binary sequence as training). Also, of these 1,705, every third tone is used as a pilot, but only once every four symbols, in alternating fashion [49]. This arrangement of pilots across time and frequency is shown in Figure 9.7. Thus, at best, every third coefficient of the channel in the frequency domain has training (and not all of them at once), which complicates trained adaptation. This can be dealt with by assuming the channel coherence is high enough that the remaining coefficients can be interpolated.

On a subcarrier that has training, a simple one-tap LMS or RLS adaptation rule can be created. For time or frequency indices when training is unavailable, a simple one-tap DD-LMS or CMA adaptation rule can be implemented, or the FEQ value can be interpolated across time and frequency from indices where training is available.

Adaptation of the FEQ for SCCP systems is slightly more complicated, since the outputs of the FEQ are not expected to be finite alphabet until after the final IFFT. Also, since SCCP has not yet been deployed in an industry standard, there are no set formats for training availability. Instead of indexing the transmitted signal by block number and subcarrier number, an SCCP transmission can be indexed by block number and sample number within that block, and training could be staggered in some fashion. What is critical for a simple FEQ adaptation rule is that the output of the FEQ be comparable to some known data. Since the FEQ is in the frequency domain, this requires that an

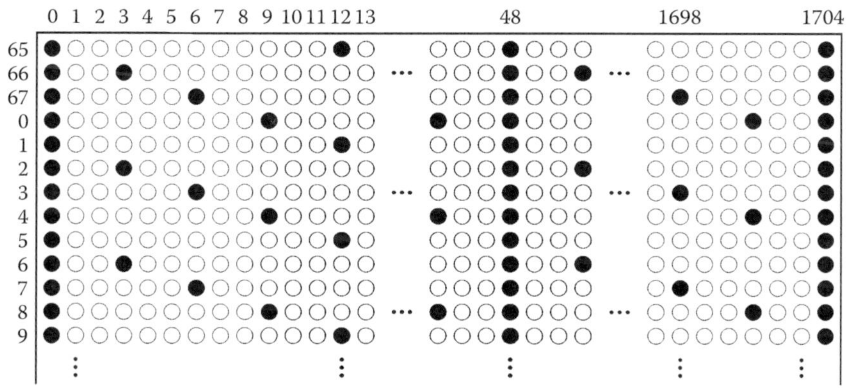

FIGURE 9.7 Diagram of the continual and staggered pilot tones (training) in the European standard for Digital Video Broadcast. The column index k is the frequency or subcarrier index, and the row index b is the block or symbol index. Shaded circles represent training data. The continual pilots are not regularly spaced in frequency; the full list is available in [49].

entire time-domain block contain training for any given frequency coefficient to be considered known. In this case, a simple one-tap LMS or RLS adaptation rule can be implemented for each FEQ coefficient. Similarly, one-tap DD-LMS and CMA rules can be formed when training is not available. In vector form, the LMS, DD-LMS, and CMA FEQ updates for an SCCP system are [50]

$$\begin{aligned}
\mathbf{d}_{\mathrm{LMS}}[b+1] &= \mathbf{d}_{\mathrm{LMS}}[b] + \mu\,\tilde{\mathbf{y}}^*[b] \odot \left(\mathbf{F}\mathbf{x}[b] - \tilde{\mathbf{x}}[b]\right) \\
\mathbf{d}_{\mathrm{DD}}[b+1] &= \mathbf{d}_{\mathrm{DD}}[b] + \mu\,\tilde{\mathbf{y}}^*[b] \odot \left(\mathbf{F}Q\{\mathbf{F}^{-1}\tilde{\mathbf{x}}[b]\} - \tilde{\mathbf{x}}[b]\right) \\
\mathbf{d}_{\mathrm{CMA}}[b+1] &= \mathbf{d}_{\mathrm{CMA}}[b] - \mu\,\tilde{\mathbf{y}}^*[b] \odot \left(\mathbf{F}\mathbf{e}_{\mathrm{CMA}}[b]\right) \\
\mathbf{e}_{\mathrm{CMA}}[b] &= \left(\left(\mathbf{F}^{-1}\tilde{\mathbf{x}}[b]\right)^* \odot \left(\mathbf{F}^{-1}\tilde{\mathbf{x}}[b]\right) - \gamma\right) \odot \left(\mathbf{F}^{-1}\tilde{\mathbf{x}}[b]\right),
\end{aligned} \tag{9.18}$$

and the RLS FEQ update with forgetting factor ρ is

$$\begin{aligned}
\mathbf{R}[b] &= \rho\left(\mathbf{R}[b-1] + \operatorname{diag}\left[\tilde{\mathbf{y}}^*[b] \odot \tilde{\mathbf{y}}[b]\right]\right) \\
\mathbf{P}[b] &= \rho\left(\mathbf{P}[b-1] + \tilde{\mathbf{y}}^*[b] \odot \left(\mathbf{F}\mathbf{x}[b]\right)\right) \\
\mathbf{d}_{\mathrm{RLS}}[b] &= \mathbf{R}^{-1}[b]\,\mathbf{P}[b].
\end{aligned} \tag{9.19}$$

Due to the diagonal structure of $\mathbf{R}$, the RLS update requires only marginally more computations and converges much faster than LMS [50]; hence it is preferable of the two.

9.3.2 Adaptive CSE Algorithms

Now we return to the problem of CSE adaptation in cyclic-prefixed systems. As mentioned in the discussion of FEQ adaptation, training in multicarrier systems may be spread across time and frequency, and it may be allocated in the frequency domain. In order to train based on the equalizer output, we first assume that time-domain training is available, which is equivalent to assuming that training is available on all subcarriers in a given symbol. In this case, the obvious choice is to use an LMS-like equalization rule. However, this is complicated by the fact that the goal of the CSE is not to make the effective channel simply a delay, so we cannot form an error signal by comparing to the delayed training data as in (9.5). Instead, the goal of the CSE is to make the effective channel short, but (to a first-order approximation) we do not care about what shape the effective channel takes so long as it is short. Falconer and Magee [10] solved this problem in the context of CSE design in conjunction with MLSE by forming an additional filter, $\mathbf{b}$, of length $\nu + 1$, called the target impulse response (TIR). Then the CSE and TIR can each be adapted in turn by an LMS algorithm.

Mathematically, the error function and cost are

$$\begin{aligned} e[n] &= \mathbf{b}^T\mathbf{x}[n-\Delta] - \mathbf{f}^T\mathbf{y}[n], \\ J &= E\left[\left|e[n]\right|^q\right], \end{aligned} \tag{9.20}$$

where Falconer and Magee chose $q = 2$, corresponding to an LMS algorithm, yet others have considered $q = 4$, corresponding to an LMF algorithm [51]. Since the trivial setting $\mathbf{b} = \mathbf{0}, \mathbf{f} = \mathbf{0}$ permits a zero-cost solution, a constraint must be enforced, such as a unit-tap or unit-norm filter. Iterating through the adaptation of each filter and maintenance of the constraint, the MMSE adaptive CSE algorithm is

$$\begin{aligned} \mathbf{f}[n+1] &= \mathbf{f}[n] + \mu e[n]\mathbf{y}^*[n] \\ \hat{\mathbf{b}}[n+1] &= \mathbf{b}[n] - \mu e[n]\mathbf{x}^*[n-\Delta] \\ \mathbf{b}[n+1] &= \hat{\mathbf{b}}[n+1] \Big/ \left\|\hat{\mathbf{b}}[n+1]\right\|_2 . \end{aligned} \tag{9.21}$$

The choice of a larger q will add factors in the first two lines, and the choice of an alternate constraint will require a different projection in the third line of (9.21). For $\nu = 0$, $\mathbf{b}$ becomes a scalar, and (9.21) reduces to the LMS algorithm for $\mathbf{f}$ alone.

The coupling of $\mathbf{b}$ and $\mathbf{f}$, along with the constraint on $\mathbf{b}$, makes formulation of an RLS adaptation rule more complicated than for a traditional equalization problem. Use of the unit-norm constraint leads to an optimal solution for $\mathbf{b}$ as an eigenvector rather than a least squares solution, and hence RLS does not apply. However, if a unit-tap constraint is used, the optimal $\mathbf{b}$ can be written as the solution to either a generalized eigen problem [52] or a least squares problem [50]. Although this constraint does lead to a suboptimal

MSE compared to the unit-norm constraint [52], it does permit the use of an RLS algorithm operating on the concatenated parameter vector

$$\mathbf{v} = \left[\mathbf{w}^T, \mathbf{b}_0, \ldots, \mathbf{b}_{i_0-1}, \mathbf{b}_{i_0+1}, \ldots, \mathbf{b}_\nu\right]^T, \tag{9.22}$$

where $\mathbf{b}_{i_0}$ is the tap that is constrained to unity. The full RLS algorithm is given in [50].

We will return to the issue of trained CSE adaptation in the case of training on intermittent subcarriers momentarily. However, this requires a discussion of blind adaptation first. Since the CSE output is not expected to be finite alphabet, DD and constant modulus methods cannot immediately be used. However, the FEQ output is expected to be finite alphabet. Thus, we have two options: form a blind cost at the FEQ output, and propagate its dependence on the CSE back through the FEQ and FFT; or apply the "property restoral" concept for development of alternate blind cost functions and algorithms. We withhold the former discussion until section 9.5, since the FEQ is adaptive as well, which leads to a coupling of adaptive blocks. For now, we focus on properties of cyclic-prefixed systems that can be restored by the CSE.

A property common to both multicarrier and SCCP systems is the redundancy induced by the CP. This redundancy has been exploited for adaptive carrier frequency offset estimation [53, 54] as well as CSE adaptation [50]. To make use of this property, we form an error signal and cost function as

$$\begin{aligned} e[b] &= \mathbf{f}^T \mathbf{y}\left[(b-1)\bar{N} + \nu + \Delta\right] - \mathbf{f}^T \mathbf{y}\left[b\bar{N} + \Delta\right], \\ J &= E\left[\left|e[b]\right|^q\right], \end{aligned} \tag{9.23}$$

along with a constraint (again, unit norm or unit tap) to avoid the trivial solution $\mathbf{f} = \mathbf{0}$. Due to the linearity of the error with respect to $\mathbf{f}$, when $q = 2$ we get a simple LMS-like update rule, although a periodic projection is required to enforce the constraint, as in (9.21). The resulting algorithm is called multicarrier equalization by restoration of redundancy (MERRY), and is given by

$$\begin{aligned} \tilde{\mathbf{y}}[b] &= \mathbf{y}\left[(b-1)\bar{N} + \nu + \Delta\right] - \mathbf{y}\left[b\bar{N} + \Delta\right] \\ e[b] &= \mathbf{f}^T[b]\tilde{\mathbf{y}}[b], \\ \hat{\mathbf{f}}[b+1] &= \mathbf{f}[b] - \mu\, e[b]\tilde{\mathbf{y}}^*[b] \\ \mathbf{f}[b+1] &= \hat{\mathbf{f}}[b+1] \Big/ \left\|\hat{\mathbf{f}}[b+1]\right\|_2 . \end{aligned} \tag{9.24}$$

More details and an RLS-like implementation are available in [50].

Another property that can be exploited in multicarrier (but not SCCP) systems is the guard band in the frequency domain, which was mentioned in section 9.3.1. Often, the subcarriers on the band edges are left as zeros at the transmitter in order to limit adjacent channel interference. These are often referred to as null tones. If the channel is benign, then at the receiver, the null tones should also be zero. Thus, we can form a cost function as

$$J = \sum_{j \in \text{null tones}} \beta_j E\left[\left|\tilde{y}[b]\right|^2\right], \tag{9.25}$$

where the weights β_j would typically be all ones. This is very similar to a decision-directed algorithm, with only one valid constellation point (i.e., zero). The difference is that (9.25) is invariant to magnitude and phase distortion, since $0 \cdot Ae^{j\theta} = 0$. Thus, the effects of the FEQ can be ignored, which would not be the case for a DD-LMS algorithm operating on the non-null tones. A gradient descent of (9.25) was proposed in [55] and further analyzed in [56], and it is sometimes referred to as the carrier nulling algorithm (CNA; not to be confused with the constant norm algorithm of the same acronym, discussed in the previous section). It also has an LMS-like structure, although with part of an FFT involved to relate the values on the null tones to the CSE output:

$$\begin{aligned} \mathbf{Y}_{\text{null}} &= \mathbf{F}_{null}\mathbf{R}\mathbf{f}[b], \\ \hat{\mathbf{f}}[b+1] &= \mathbf{f}[b] - \mu\, \mathbf{R}^H \mathbf{F}_{\text{null}}^H \mathbf{Y}_{\text{null}} \\ \mathbf{f}[b+1] &= \hat{\mathbf{f}}[b+1] \big/ \left\|\hat{\mathbf{f}}[b+1]\right\|_2, \end{aligned} \tag{9.26}$$

where $\mathbf{R}$ is the $N \times L$ Toeplitz matrix:

$$\mathbf{R} = \begin{bmatrix} r\left[(b-1)\bar{N}+\nu+1+\Delta\right], & \ldots, & r\left[(b-1)\bar{N}+\nu+1+\Delta-L\right] \\ \vdots & & \vdots \\ r\left[b\bar{N}+\Delta\right], & \ldots, & r\left[b\bar{N}+\Delta-L\right] \end{bmatrix}, \tag{9.27}$$

and where $\mathbf{F}_{null}$ is obtained by taking the rows of the FFT matrix indexed by the null tones, and $\mathbf{Y}_{null}$ is similarly obtained from the FFT output vector.

Now we can reconsider the case in which training is intermittently provided on some subcarriers. Let S_{null}, $S_{pilot}[b]$, and $S_{data}[b]$ be the sets of subcarriers containing null tones, pilot tones, or unknown data (respectively) during block b, with $S_{null} \cap S_{pilot}[b] \cap S_{data}[b] = \{0, \ldots, N-1\}$. Then we can form the semiblind cost function

$$J = \sum_{\substack{1\text{ period}\\ \text{of } b}} \left(\underbrace{\sum_{j \in S_{null}} \beta_j E\left[\left|\tilde{\mathbf{y}}_j[b]\right|^2\right]}_{\text{CNA term}} + \underbrace{\sum_{j \in S_{pilot}[b]} \beta_j E\left[\left|\mathbf{x}_j[b] - \tilde{\mathbf{x}}_j[b]\right|^2\right]}_{\text{MSE term}} + \underbrace{\sum_{j \in S_{data}[b]} \beta_j E\left[\left(\left|\tilde{\mathbf{x}}_j[b]\right|^2 - \gamma\right)^2\right]}_{\text{CMA (or DD) term}} \right). \tag{9.28}$$

The weights β_j control the relative use of the trained MSE term and the blind CNA and CMA terms. Note that a DD cost could be used instead of or in addition to the CMA cost.

A comparison of the learning curves of trained and blind adaptive CSE algorithms is shown in Figure 9.8. The simulated system had parameters consistent with IEEE 802.11a: an FFT size of 64 with 12 null tones, a CP of length 16, and complex baseband channels with Rayleigh-distributed taps and an approximately exponential delay profile. The SIMO channel model had one transmit antenna and two receive antennas. The performance metric is the BER, and a combination of 4-ary quadrature amplitude modulation (QAM) per tone and differential encoding was used to remove the need for and effects of an adaptive FEQ.

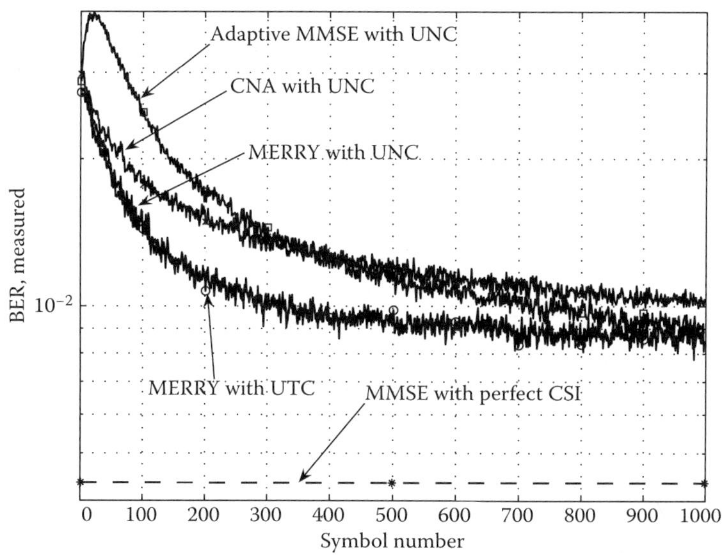

FIGURE 9.8 A performance comparison of trained and blind adaptive channel-shortening equalizers. The algorithms are MMSE [10], MERRY [50], and CNA [55]. MERRY is shown with both a unit-norm and a unit-tap constraint (UNC, UTC), and since the difference is negligible, MMSE and CNA only use a UNC.

The algorithms compared in Figure 9.8 are the adaptive MMSE algorithm [10], MERRY [50], and CNA [55, 56]. The optimal nonadaptive MMSE design with perfect channel state information (CSI) is also shown. In principle, all of these could use either a unit-norm or unit-tap constraint. Both are shown for the case of MERRY; however, since there is no appreciable difference, only a single constraint is shown for the remainder. Also, the blind algorithms can only update once per block (OFDM symbol), but the MMSE design can update as often as once per sample. For a fair comparison, the MMSE design was only updated once per block, so that the update complexity per symbol of the various algorithms would be comparable.

Observe that the various algorithms considered all had comparable learning rates for a given allowable complexity. However, the trained MMSE algorithm could in principle update $\bar{N}$ times more often (with a corresponding increase in complexity).

To recap, trained and blind adaptive equalizers for multicarrier systems can be designed using the same principles as more conventional communication systems. Trained algorithms are generally based on an MMSE framework, and blind algorithms are generally based on a property restoral framework. The unique aspects of adaptive algorithms in multicarrier systems include the need for an adaptive target impulse response for trained MMSE adaptation, the need for a constraint for most trained and blind adaptive algorithms, and the trade-offs associated with operating at the block rate versus the sample rate.

9.4 Ultrawideband Communication Systems

Like multicarrier systems, ultrawideband (UWB) systems have received much attention in recent literature due to the fact that their modulation format is very distinct from single-carrier amplitude modulation formats, although it is not in as widespread implementation as multicarrier modulation. This section reviews UWB modulation and discusses the application of adaptive equalizer design principles to this modulation format.

A UWB system is often defined as one in which the bandwidth divided by the carrier frequency exceeds ¼. The idea is that a very large bandwidth is used, but the total power is moderate—hence the power per Hertz is on the order of the noise floor. This is akin to spread-spectrum modulation in its motivation, but the process of modulation differs. Presently, there are two popular formats of UWB in the literature: pulse position modulation (PPM) and a multiband format, much like OFDM. In this chapter, we focus on PPM, due to its implications for adaptive equalizer design.

In the PPM version of UWB, information is transmitted by very short, infrequent pulses. By transmitting an impulse-like pulse, the bandwidth is very large. In some cases, the need for RF hardware can be avoided completely, since an impulsive transmission can spread the data up to very high frequencies without upconversion. Thus, PPM UWB is sometimes called impulse radio.

Consider a PPM system that transmits one of M possible symbols each time slot. The time slot is divided into $K = M/2$ chips. A symbol consists of transmitting a +1 in one chip and zeros in the other $K - 1$ chips (so there are K of these to choose from), or transmitting a –1 in one chip and zeros in the other $K-1$ chips (so there are K of these as well). One bit of information is contained in the sign of the transmission, and the other $\log_2(K)$ bits of information are contained in the position of the transmission. This means that like multicarrier modulation, the receiver must process the data in block format. After equalization and symbol (block) synchronization, blocks of K samples are parallelized and fed into a decision device, then a single decision is made to produce the estimate of the transmitted symbol.

9.4.1 Trained Adaptation

Since PPM operates on a block basis, the MSE cost function should be block based as well. Defining the actual and desired equalizer output for block b as

$$\begin{aligned}\mathbf{z}[b]&=\left[z[Kb],\ldots,z[Kb-K+1]\right]^T\\ \mathbf{x}[b]&=\left[x[Kb-\Delta],\ldots,x[Kb-K+1-\Delta]\right]^T,\end{aligned} \tag{9.29}$$

the MSE cost function defined in [57] is

$$J=E\left[\left\|\mathbf{z}[b]-\mathbf{x}[b]\right\|_2^2\right]. \tag{9.30}$$

Because $\mathbf{z}[b]$ is a vector of samples, it is formed by the multiplication of a *matrix* of equalizer inputs,

$$\mathbf{Y}[b]=\begin{bmatrix} y[Kb] & \cdots & y[Kb-K+1] \\ \vdots & \iddots & \vdots \\ y[Kb-L+1] & \cdots & y[Kb-L-K+2]\end{bmatrix}, \tag{9.31}$$

and the equalizer, i.e., $\mathbf{z}[b]=\mathbf{Y}^T[b]\mathbf{f}$. This means that an LMS-style gradient descent of (9.30) is not simply a vector update rule; rather, there is a matrix-vector product involved. Specifically, a trained, adaptive LMS equalizer for PPM systems takes the form [57, 58]

$$\mathbf{f}[b+1]=\mathbf{f}[b]-\mu\mathbf{Y}[b]\left(\mathbf{z}[b]-\mathbf{x}[b]\right). \tag{9.32}$$

Thus, the approach is similar to trained, adaptive equalization in traditional communication systems, but the inherent block structure introduces a slightly more complicated update rule, in terms of structure and computations.

An alternate approach would be to use a DFE. Trained zero-forcing DFE algorithms are discussed in [59, 60].

9.4.2 Blind Adaptation

The simplest blind equalizer is DD-LMS. However, even DD-LMS has a slight twist in the case of PPM. The DD cost function and update rule are

$$J = E\left[\left\|\mathbf{z}[b] - \mathbf{Q}\{\mathbf{z}[b]\}\right\|_2^2\right]$$
$$\mathbf{f}[b+1] = \mathbf{f}[b] - \mu\mathbf{Y}[b]\left(\mathbf{z}[b] - \mathbf{Q}\{\mathbf{z}[b]\}\right). \tag{9.33}$$

The twist is that the decision function $Q\{\cdot\} : \mathbb{R}^K \mapsto \mathbb{R}^K$ operates in vector form, rather than element-wise. The vector argument is considered in its entirety and is mapped to the nearest vector in the signal constellation [57, 58].

A variant of CMA can also be derived for PPM, which will be called linear transversal equalizer adaptation for biorthogonal modulation, blindly (LTBOMB). Like DD-LMS for PPM, the distinction from traditional adaptive algorithms lies in the block structure of PPM. The cost function and algorithm are [57, 58]

$$J = E\left[\left(\left\|\mathbf{z}[b]\right\|_2^2 - 1\right)^2\right]$$
$$\mathbf{f}[b+1] = \mathbf{f}[b] - \mu\left(\mathbf{z}^T[b]\mathbf{z}[b] - 1\right)\mathbf{Y}[b]\mathbf{z}[b]. \tag{9.34}$$

However, as implied by the time indices, the algorithm only updates once per block, i.e., once every K chips. Again, note the matrix-vector structure of the update rule, as opposed to standard CMA, which uses a scalar times vector update rule.

An alternative to CMA for blind, adaptive equalization of traditional communication systems is the Shalvi-Weinstein algorithm (SWA) [61]. SWA is similar to CMA insofar as it looks at higher-order statistics of the equalizer output, but SWA attempts to maximize the magnitude of the kurtosis. In [58] and [62], the SWA philosophy is used to create a blind, adaptive equalizer for PPM, called the recovery of M-ary biorthogonal signals via p-norm equivalence (TROMBONE). The cost function and algorithm are

$$J = E\left[\left\|\mathbf{z}[b]\right\|_2^4 - \left\|\mathbf{z}[b]\right\|_4^4\right], \quad \text{such that } \left\|\mathbf{z}[b]\right\|_2^2 = 1$$
$$\hat{\mathbf{f}}[b+1] = \mathbf{f}[b] - \mu\mathbf{Y}[b]\left(\mathbf{z}^T[b]\mathbf{z}[b]\mathbf{I}_K - \left(\mathrm{diag}\left[\mathbf{z}[b]\right]\right)^2\right)\mathbf{z}[b] \tag{9.35}$$
$$\mathbf{f}[b+1] = \frac{\hat{\mathbf{f}}[b+1]}{\left\|\hat{\mathbf{f}}[b+1]\right\|_2},$$

where $\mathbf{I}_K$ is the $K \times K$ identity matrix. The last line implements the constraint (similar to the MMSE, MERRY, and CNA algorithms for multicarrier systems) to constrain the equalizer away from the trivial all-zero solution. As with the DD-LMS and CMA variants of (9.33) and (9.34), the SWA variant of (9.35) only updates once per block.

To recap, traditional methods of trained and blind adaptive equalizer design can be modified for use in PPM-based UWB systems. However, the cost functions and algorithms must be modified to account for the block structure. The resulting algorithms only update once per block, they generally require matrix-vector products in the update term, as opposed to the more traditional scalar-vector products, and the decision function is a vector-to-vector mapping rather than a scalar-to-scalar mapping.

9.5 Interaction of Equalizer with Other Adaptive Blocks

Many equalizer designs are produced in isolation: the equalizer explicitly or implicitly assumes that all other adaptive blocks in the system are working perfectly. However, even in a simple single-carrier receiver, the signal amplitude, timing frequency/phase, equalizer, and carrier frequency/phase are all interdependent [15, Table 9.2). In newer modulation formats with more or different blocks, there can be additional interdependence. We conclude this chapter with a discussion of the merits of joint analysis and design of the equalizer and other blocks in an adaptive receiver, though a full treatment of this subject is beyond the current state of the art.

One of the classical examples of this dependence is the dependence of an adaptive equalizer on accurate carrier frequency offset (CFO) estimation. For a trained, LMS algorithm, the equalizer inputs and outputs will not match their model if there is a residual CFO. At the same time, a trained CFO estimator needs ISI-free received data in order to form a good estimate. This can be a chicken-and-egg problem. However, the CMA equalizer uses a cost function that depends only on the magnitude of the equalizer output, and not the phase. In the presence of a residual CFO, the received data will have a linearly increasing additive phase, causing the signal constellation to spin about the origin. Since CMA does not care about the phase, it can remove the ISI even in the presence of the CFO-induced spinning. Then the CFO estimator can operate on the ISI-free equalizer output, and remove the CFO [15].

This method of delaying the need for CFO correction cannot be as easily incorporated into multicarrier systems, since as discussed in section 9.3, blind multicarrier equalizers do make use of a constant modulus cost function in the time domain. An alternate method of dealing with the coupling of an adaptive equalizer and another adaptive block is to jointly adapt the two. In the case of multicarrier equalization and CFO correction, the CNA, MERRY, and DD cost functions used for equalization can also be used to adjust the CFO [53, 54, 63, 64]. By forming a single cost function at the output of the two adaptive blocks in series, both algorithms may converge to their optimal setting [63].

As mentioned in section 9.3, in multicarrier systems, the CSE and FEQ must both be adapted simultaneously. Algorithms such as DD-LMS and CMA cannot directly be used for the CSE since the CSE output is not expected to be finite alphabet. However, the final FEQ output is. By forming a decision-directed or constant modulus cost at the FEQ output, both the CSE and FEQ can adapt based on this cost. If the FEQ uses a relatively higher step size than the CSE, then the CSE will adapt slowly, and the FEQ will track it and quickly reach its best value for the current CSE setting. This concatenation of adaptive equalizers allows the use of traditional cost functions. The penalty is that the FEQ step size has an upper bound for stability, and the CSE step size must be much lower, so adaptation cannot be very swift. This need for step-size-based timescale separation is present for most coupled adaptive systems adapting over a single cost function, for example, the adaptive CSE and target response in the MMSE filter [10, 65].

9.6 Summary

This chapter has discussed the design of adaptive equalizers. We began with a historical perspective and a discussion of the need for an adaptive equalizer. We then discussed popular methods of creating adaptation rules, making them converge quickly, and reducing their computational load, all of which are necessary if the equalizer is part of a small wireless device. As specific examples of creating adaptation rules, we discussed recent literature on adaptive equalization in two currently popular communication standards, multicarrier and ultrawideband. We concluded with a discussion of the interaction of an adaptive equalizer and other adaptive blocks within the receiver.

References

[1] Schniter, P. 2004. Low-complexity equalization of OFDM in doubly-selective channels. *IEEE Trans. Signal Processing* 52:1002–11.

[2] Viterbi, A. J. 1967. Error bounds for convolutional codes and an asymptotically optimal decoding algorithm. *IEEE Trans. Inf. Theory* 13:260–69.

[3] Tong, L., Xu, G., and Kailath, T. 1994. Blind identification and equalization based on second-order statistics: A time domain approach. *IEEE Trans. Inf. Theory* 40:340–49.

[4] Moulines, E., Duhamel, P., Cardoso, J., and Mayrargue, S. 1995. Subspace methods for the blind identification of multichannel FIR filters. *IEEE Trans. Signal Processing* 43:516–25.

[5] Widrow, B., and Hoff, Jr., M. E. 1960. Adaptive switching circuits. In *Proceedings of the WESCON Conference Recordings*, Part 4, pp. 96–104.

[6] Lucky, R. W. 1965. Automatic equalization for digital communication. *Bell Syst. Tech. J.* 44:547–88.

[7] Widrow, B., et al. 1976. Stationary and nonstationary learning characteristics of the LMS adaptive filter. *Proc. IEEE* 64:1151–62.

[8] Gitlin, R. D., and Mazo, J. E. 1973. Comparison of some cost functions for automatic equalization. *IEEE Trans. Commun.* 21:233–37.

[9] Qureshi, S., and Newhall, E. 1973. An adaptive receiver for data transmission over time-dispersive channels. *IEEE Trans. Inf. Theory* 19:448–57.

[10] Falconer, D. D., and Magee, F. R. 1973. Adaptive channel memory truncation for maximum likelihood sequence estimation. *Bell Syst. Tech. J.* 1541–62.

[11] Sato, Y. 1975. A method of self-recovering equalization for multilevel amplitude-modulation systems. *IEEE Trans. Commun.* 23:679–82.
[12] Godard, D. N. 1980. Self-recovering equalization and carrier tracking in two-dimensional data communication systems. *IEEE Trans. Commun.* 28:1867–75.
[13] Treichler, J. R., and Agee, B. G. 1983. A new approach to multipath correction of constant modulus signals. *IEEE Trans. Acoustics Speech Signal Processing* 31:459–72.
[14] Foschini, G. J. 1985. Equalizing without altering or detecting data. *AT&T Tech. J.* 64:1885–911.
[15] Treichler, J. R., Larimore, M. G., and Harp, J. C. 1998. Practical blind demodulators for high-order QAM signals. *Proc. IEEE* 86:1907–26.
[16] Qureshi, S. 1985. Adaptive equalization. *Proc. IEEE* 73:1349–87.
[17] Sethares, W. A. 1992. Adaptive algorithms with nonlinear data and error functions. *IEEE Trans. Signal Processing* 40:2199–206.
[18] Duttweiler, D. L. 2000. Proportionate normalized least-mean-squares adaptation in echo cancellers. *IEEE Trans. Speech Audio Processing* 8:508–18.
[19] Walach, E., and Widrow, B. 1984. The least mean fourth (LMF) adaptive algorithm and its family. *IEEE Trans. Inf. Theory* 30:275–83.
[20] Hübscher, P. I., Nascimento, V. H., and Bermudez, J. C. M. 2003. New results on the stability analysis of the LMF (least mean fourth) adaptive algorithm. In *Proceedings of the International Conference on Acoustics, Speech, and Signal Processing*, vol. VI, pp. 369–72.
[21] Sethares, W. A., Rey, G. A., and Johnson, Jr., C. R. 1989. Approaches to blind equalization of signals with multiple modulus. In *Proceedings of the International Conference on Acoustics, Speech, and Signal Processing*, vol. 2, pp. 972–75.
[22] Banovic, K., Abdel-Raheem, E., and Khalid, M. A. S. 2005. Hybrid methods for blind adaptive equalization: New results and comparisons. In *Proceedings of the IEEE Symposium on Computers and Communications*, pp. 275–80.
[23] Thaiupathump, T., and Kassam, S. A. 2003. Square contour algorithm: A new algorithm for blind equalization and carrier phase recovery. In *Proceedings of the 37th Asilomar Conference on Signals, Systems, and Computers*, vol. 1, pp. 647–51.
[24] Goupil, A., and Palicot, J. 2007. New algorithms for blind equalization: The constant norm algorithm family. *IEEE Trans. Signal Processing* 55:1436–44.
[25] Li, X.-L., and Zhang, X.-D. 2006. A family of generalized constant modulus algorithms for blind equalization. *IEEE Trans. Commun.* 54:1913–17.
[26] Abrar, S., and Qureshi, I. M. 2006. Blind equalization of cross-QAM signals. *IEEE Signal Processing Lett.* 13:745–48.
[27] Johnson, Jr., C. R., et al. 1998. Blind equalization using the constant modulus criterion: A review. *Proc. IEEE* 86:1927–50.
[28] Johnson, Jr., C. R., et al. 2000. The core of FSE-CMA behavior theory. In *Unsupervised adaptive filtering: Blind deconvolution*, ed. S. Haykin, pp. 13–112. Vol. II. New York: Wiley.
[29] Haykin, S. 2002. *Adaptive filter theory.* 4th ed. Upper Saddle River, NJ: Prentice-Hall.
[30] Stamoulis, A., Diggavi, S. N., and Al-Dhahir, N. 2002. Intercarrier interference in MIMO OFDM. *IEEE Trans. Signal Processing* 50:2451–64.

[31] Cai, X., and Giannakis, G. B. 2002. Low-complexity ICI suppression for OFDM over time- and frequency-selective Rayleigh fading channels. In *Proceedings of the Asilomar Conference on Signals, Systems, and Computers*, pp. 1822–1826.

[32] Barhumi, I., Leus, G., and Moonen, M. 2005. Time-varying FIR equalization for doubly selective channels. *IEEE Trans. Wireless Commun.* 4:202–14.

[33] Ahmed, S., Sellathurai, M., Lambothoran, S., and Chambers, J. 2006. Low complexity iterative method of equalization for single carrier with cyclic prefix in doubly selective channels. *IEEE Signal Processing Lett.* 13:5–8.

[34] Verhoeckx, N. A. M., Van dem Elzer, H., Snijders, F. A. M., and Van Gerwer, P. J. 1979. Digital echo cancellation for baseband data transmission. *IEEE Trans. Acoust. Speech Signal Processing* 27:768–81.

[35] Bellini, S. 1986. Bussgang techniques for blind equalization. In *Proceedings of the IEEE Global Telecommunications Conference*, pp. 1634–40.

[36] De, P., Bao, J., and Poon, T. 1999. A calculation-efficient algorithm for decision feedback equalizers. *IEEE Trans. Consumer Electronics* 45:526–632.

[37] Chen, Z., Haykin, S., and Gay, S. L. 2003. Proportionate adaptation: New paradigms in adaptive filters. In *Least-mean-square adaptive filters*, ed. S. Haykin and B. Widrow. New York: John Wiley & Sons, pp. 293–334.

[38] Martin, R. K., Sethares, W. A., Williamson, R. C., and Johnson, Jr., C. R. 2002. Exploiting sparsity in adaptive filters. *IEEE Trans. Signal Processing* 50:1883–94.

[39] Deng, H., and Doroslovački, M. 2006. Proportionate adaptive algorithms for network echo cancellation. *IEEE Trans. Signal Processing* 54:1794–803.

[40] Gay, S. L. 1998. An efficient, fast converging adaptive filter for network echo cancellation. In *Proceedings of the Asilomar Conference on Signals, Systems, and Computers*, pp. 394–98.

[41] Nekuii, M., and Atarodi, M. 2004. A fast converging algorithm for network echo cancellation. *IEEE Signal Process. Lett.* 11:427–30.

[42] Benesty, J., and Gay, S. L. 2002. An improved PNLMS algorithm. In *Proceedings of the IEEE International Conference on Acoustics, Speech, and Signal Processing*, vol. 2, pp. 1881–84.

[43] Douglas, S. C. 1997. Adaptive filters employing partial updates. *IEEE Trans. Circuits Syst. II* 44:209–16.

[44] Douglas, S. C. 1995. Analysis and implementation of the max-NLMS adaptive filter. In *Proceedings of the 29th Asilomar Conference on Signals, Systems, and Computers*, Pacific Grove, CA, vol. 1, pp. 659–63.

[45] Aboulnasr, T., and Mayyas, K. 1999. Complexity reduction of the NLMS algorithm via selective coefficient update. *IEEE Trans. Signal Processing* 47:1421–24.

[46] Schertler, T. 1998. Selective block update of NLMS type algorithms. In *Proceedings of the IEEE International Conference on Acoustics, Speech, and Signal Processing*, Seattle, vol. 3, pp. 1717–20.

[47] Naylor, P., and Sherliker, W. 2003. A short-sort M-max NLMS partial update adaptive filter with applications to echo cancellation. In *Proceedings of the IEEE International Conference on Acoustics, Speech, and Signal Processing*, Hong Kong, vol. 5, pp. 373–76.

[48] Homer, J. 1998. Detection guided LMS estimation of sparse channels. In *Proceedings of the IEEE Global Communications Conference*, Sydney, vol. 6, pp. 3704–9.

[49] European Telecommunications Standards Institute. 2001. *Digital video broadcasting (DVB); Framing structure, channel coding and modulation for digital terrestrial television*. ETSI EN 300 744 V1.4.1.

[50] Martin, R. K. 2007. Fast-converging blind adaptive channel shortening and frequency-domain equalization. *IEEE Trans. Signal Processing* 54:102–10.

[51] Yap, K. S., and McCanny, J. V. 2002. Improved time-domain equalizer initialization algorithm for ADSL modems. In *Proceedings of the 6th International Symposium on DSP for Communication Systems*, Sydney-Manly, Australia, pp. 253–58.

[52] Al-Dhahir, N., and Cioffi, J. M. 1996. Efficiently computed reduced-parameter input-aided MMSE equalizers for ML detection: A unified approach. *IEEE Trans. Inf. Theory* 42:903–15.

[53] van de Beek, J.-J., Sandell, M., and Borjesson, P. O. 1997. ML estimation of time and frequency offset in OFDM systems. *IEEE Trans. Signal Processing* 45:1800–5.

[54] Choi, Y., Voltz, P. J., and Cassara, F. A. 2001. ML estimation of carrier frequency offset for multicarrier signals in Rayleigh fading channels. *IEEE Trans. Veh. Technol.* 50:644–54.

[55] de Courville, M., Duhamel, P., Madec, P., and Palicot, J. 1996. Blind equalization of OFDM systems based on the minimization of a quadratic criterion. In *Proceedings of the IEEE International Conference on Communications*, Dallas, TX, pp. 1318–21.

[56] Romano, F., and Barbarossa, S. 2003. Non-data aided adaptive channel shortening for efficient multi-carrier systems. In *Proceedings of the IEEE International Conference on Acoustics, Speech, and Signal Processing*, Hong Kong, vol. 4, pp. 233–36.

[57] Klein, A. G., Johnson, Jr., C. R., and Duhamel, P. 2005. On blind equalization of M-ary bi-orthogonal signaling. In *Proceedings of the IEEE International Conference on Acoustics, Speech, and Signal Processing.*

[58] Klein, A. G., Johnson, Jr., C. R., and Duhamel, P. 2007. On blind equalization of biorthogonal signaling. *IEEE Trans. Signal Processing* 55:1421–35.

[59] Barry, J. R. 1994. Sequence detection and equalization for pulse-position modulation. In *Proceedings of the IEEE International Conference on Communications*, vol. 3, pp. 1561–656.

[60] Varanasi, M. K. 1997. Equalization for multipulse modulation. In *Proceedings of the IEEE International Conference on Personal Wireless Communications*, pp. 48–51.

[61] Shalvi, O., and Weinstein, E. 1990. New criteria for blind deconvolution of nonminimum phase systems (channels). *IEEE Trans. Inf. Theory* 36:312–21.

[62] Klein, A. G., Johnson, Jr., C. R., and Duhamel, P. 2006. A blind algorithm based on difference of norms for equalization of biorthogonal signals. In *Proceedings of the IEEE International Conference on Acoustics, Speech, and Signal Processing.*

[63] Martin, R. K. 2006. Joint blind adaptive synchronization and channel shortening. *IEEE Trans. Signal Processing* 54:4194–203.

[64] Liu, Z., and Weng, B. 2003. Finite-alphabet based blind carrier frequency offset estimation for differentially coded OFDM. In *Proceedings of the IEEE Global Telecommunications Conference*, vol. 1, pp. 327–31.

[65] Chow, J. S., Cioffi, J. M., and Bingham, J. A. C. 1993. Equalizer training algorithms for multicarrier modulation systems. In *Proceedings of the IEEE International Conference on Communications*, pp. 761–65.

10

Adaptive Multicarrier CDMA Space-Time Receivers

Besma Smida
Harvard University

Sofiène Affes
University of Quebec

10.1 Motivation of the Chapter

One important challenge for future wireless networks is the design of appropriate transceivers that can reliably transmit high data rates at a high bandwidth efficiency. Multicarrier code division multiple access (MC-CDMA) systems in particular have received considerable attention, because they have the attractive feature of high spectral efficiency and because they can be easily implemented using fast Fourier transform (FFT) without significantly increasing the transmitter and receiver complexities [1, 2].

The multicarrier systems include different combinations of multicarrier modulation (accomplished by orthogonal frequency division multiplex [OFDM]) and direct-sequence code division multiple access (DS-CDMA). This combination provides both high-data-rate transmission and multiple access capabilities. An excellent overview of the different multicarrier CDMA systems is found in [2] and [3]. They can be divided into two categories of multicarrier CDMA: one combines multicarrier modulation with frequency-domain spreading, and the other transmits several DS-CDMA waveforms in parallel with the spreading operation performed in time. The transmitter proposed here belongs to the second group, and it can be divided into MC-DS-CDMA and multitone (MT) CDMA, the difference between the two being the subcarrier frequency separation.

However promising, challenges remain before multicarrier CDMA can achieve its full potential. One open area is the design of transceivers that will enable the future upgrade of current wireless networks beyond the third generation (3G). Transceivers selected for early implementation need to achieve high spectrum efficiency in realistic propagation channels while being robust to imperfections such as time and frequency mismatch. Multicarrier CDMA, similar to other multicarrier schemes, is sensitive to the signal distortion generated by the imperfect frequency downconversion at the receiver due to local oscillator frequency offset. It has been found that carrier frequency offset (CFO) gives rise to a reduction of the useful signal power and to the intercarrier interference (ICI) [5]. Furthermore, one of the major obstacles in detecting multicarrier CDMA signals is interference. The multiple access interference (MAI) and the intersymbol interference (ISI), which are inherited from conventional DS-CDMA, affect likewise the performance of multicarrier CDMA systems.

The subject of this work is the design of a new adaptive multicarrier CDMA space-time receiver that provides solutions to these problems, with particular emphasis on the comparison of the MC-DS-CDMA– and MT-CDMA–air interface configurations. The proposed receiver, named MC-ISR* (multicarrier interference subspace rejection), will hence (1) perform blind channel identification and equalization as well as fast and accurate joint synchronization in time and frequency, and (2) mitigate the full interference effect. In addition, the assessment of this new receiver is oriented toward an implementation in a future, real-world wireless system.

* This work was presented in part at the IEEE SPAWC 2005 and IEEE ISSPA 2005 conferences and accepted for publication in *IEEE Transactions on Vehicular Technology* [6].

10.2 Overview of Multicarrier CDMA for Wireless Communications

In this section, we review concisely the class of multicarrier CDMA schemes, which have been discussed in the literature. Specifically, we discuss their parameters, spectral characteristics, advantages, and disadvantages in terms of design and structure. For more detailed information on the different multicarrier CDMA schemes, the reader is referred to the excellent monographs by Hanzo et al. [3].

The multicarrier CDMA schemes are categorized mainly in two groups. One spreads the original data stream using a given spreading code, and then modulates a different subcarrier with each chip (i.e., the spreading is in the frequency domain), and the other spreads the serial-to-parallel (S/P) converted data streams using a given spreading code, and then modulates a different subcarrier with each of the data streams (spreading is in the time domain).

10.2.1 Frequency-Domain Spreading Multicarrier CDMA: FD-MC-CDMA

Figure 10.1a shows the spectrum of the multicarrier CDMA scheme associated with frequency-domain spreading [7–10]. We refer to this scheme as FD-MC-CDMA. The FD-MC-CDMA transmitter spreads the original data stream over N_c subcarriers using a given spreading code in the frequency domain. This scheme does not include serial-to-parallel data conversion, and there exists no spreading modulation on each subcarrier. Therefore, the data rate on each of the N_c subcarriers is the same as the input data rate. However, by spreading each data bit across all of the N_c subcarriers, the fading effects of multipath channels are mitigated. In this FD-MC-CDMA system, the subcarrier frequencies are chosen to be orthogonal to each other, i.e., the subcarrier frequencies satisfy the following condition:

$$f_i - f_j = \frac{n|i-j|}{T},$$

where $n \in \mathbb{N}$ and T is the symbol duration. Therefore, the minimum spacing Δ between two adjacent subcarriers satisfies $1/T$, which is a widely used assumption [7–10] and is also the case employed in Figure 10.1a. If no overlap is assumed, then the minimum spacing Δ between two adjacent subcarriers is $2/T$.

Yee et al. [7] have considered an FD-MC-CDMA system, in which the subcarriers' frequency separation is higher than the coherence bandwidth of the channel, and therefore the individual subcarriers experience independent fading. As a result, the frequency diversity is maximized. This is the main advantage of the FD-MC-CDMA scheme over other multicarrier CDMA schemes [2]. However, this system may require a considerable transmission bandwidth. Besides that, a large delay spread per subcarrier would lessen this bandwidth requirement. But in a frequency-selective fading channel, different

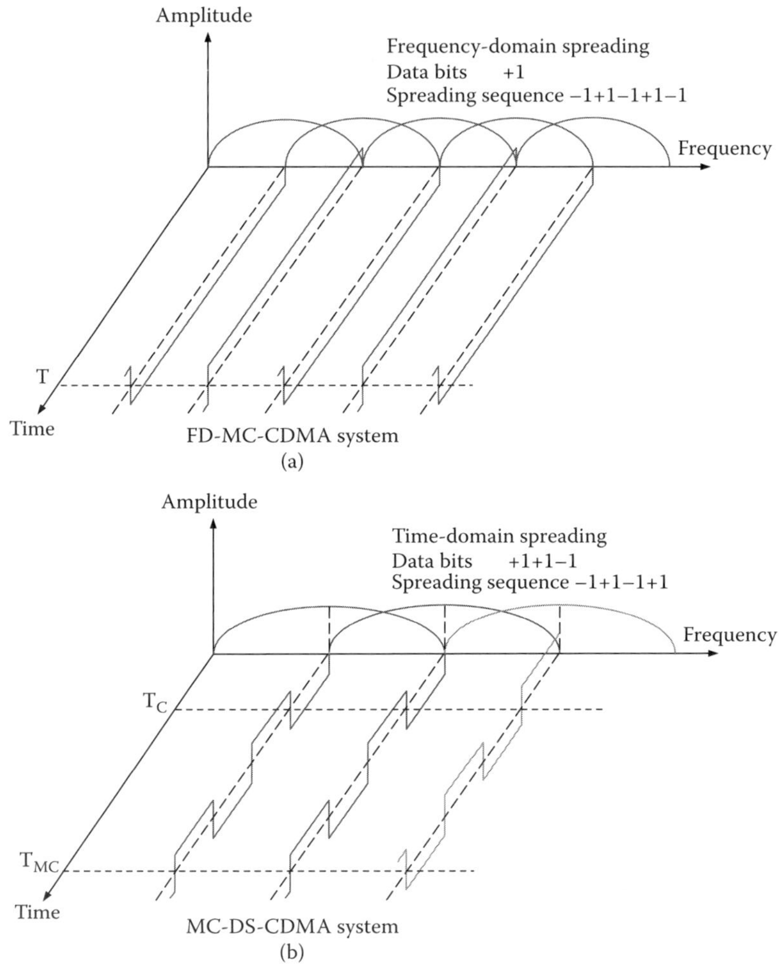

FIGURE 10.1 Spectra of different multicarrier CDMA schemes: (a) FD-MC-CDMA, (b) MC-DS-CDMA, (c) MT-CDMA.

subcarriers may encounter different amplitude attenuations and phase shifts, which can consequently destroy the orthogonality of the subcarriers.

10.2.2 Time-Domain Spreading Multicarrier CDMA: MC-DS-CDMA

In [11], a multicarrier DS-CDMA system named MC-DS-CDMA has been proposed. The MC-DS-CDMA transmitter spreads the serial-to-parallel converted data streams using a given spreading code in the time domain so that the resulting spectrum of each subcarrier can satisfy the orthogonality condition with the minimum frequency separation [11]. This scheme was originally proposed for an uplink communication system, because

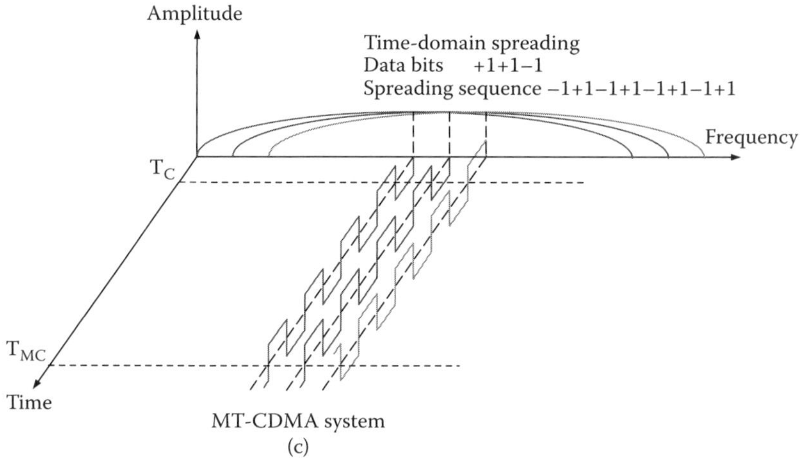

FIGURE 10.1 (continued).

this characteristic is effective for establishing a quasi-synchronous channel. In [12], Kondo and Milstein proposed a similar transmitter, except that band-limited subcarrier signals and larger subcarrier separation are employed. This scheme yields both frequency diversity improvement and narrowband interference suppression. The spectrum of the MC-DS-CDMA signal with 50% overlap having three subcarriers is shown in Figure 10.1b. In the MC-DS-CDMA system, the subcarrier frequencies are usually chosen to be orthogonal to each other after spreading, which can be formulated as

$$f_i - f_j = \frac{n|i-j|}{T_c},$$

where $n \in \mathbb{N}$ and T_c is the chip duration. Therefore, the minimum spacing Δ between two adjacent subcarriers satisfies $1/T_c$.

Frequency diversity in MC-DS-CDMA systems can be achieved by repeating the transmitted signal in the frequency (F) domain with the aid of several subcarriers [12–14]. Alternatively, in MC-DS-CDMA systems the F-domain repetition can be replaced by F-domain spreading [15] using a spreading code. One of the advantages of using F-domain spreading instead of F-domain repetition in MC-DS-CDMA systems is that frequency diversity can be achieved without reducing the maximum number of users supported by the system [15, 16]. The MC-DS-CDMA scheme can provide the following advantages [13]. First, the spreading processing gain is increased compared to the corresponding single-carrier DS-CDMA scheme. Second, the effect of multipath interference is mitigated because of DS spreading. Third, frequency/time diversity can be achieved. Finally, a longer chip duration may lead to more relaxed synchronization schemes.

10.2.3 Time-Domain Spreading Multicarrier CDMA: MT-CDMA

The multitone CDMA (MT-CDMA) scheme was proposed by Vandendorpe in [17, 18]. The MT-CDMA transmitter spreads the serial-to-parallel converted data streams using a given spreading code in the time domain, so that the spectrum of each subcarrier prior to the spreading operation can satisfy the orthogonality condition with the minimum frequency separation [18]. Therefore, there exists a strong spectral overlap among the different subcarrier signals after DS spreading. The spectra associated with three subcarriers for an MT-CDMA signal are shown in Figure 10.1c. In an MT-CDMA system, the subcarrier frequencies are chosen to be orthogonal to each other with the minimum frequency separation before spreading, which can be formulated as

$$f_i - f_j = \frac{n|i-j|}{T_{MC}},$$

where $n \in \mathbb{N}$ and T_{MC} is the symbol duration after S/P. It can be shown that the minimum spacing of the subcarrier frequencies is $1/T_{MC}$.

Unfortunately, the MT-CDMA scheme suffers from intercarrier interference because of the strong spectral overlap among the different subcarriers. However, the capability to use longer spreading codes results in the reduction of self-interference and multiple access interference, compared to the spreading codes assigned to a corresponding single-carrier DS-CDMA scheme. The MT-CDMA scheme uses longer spreading codes than the corresponding single-carrier DS-CDMA scheme [2], where the relative code-length extension is in proportion to the number of subcarriers. Therefore, the MT-CDMA system can accommodate more users. Simulation results will later show the advantages of MT-CDMA in increasing throughput and bandwidth efficiency.

10.2.4 Multicarrier CDMA Transmitter Selection

We had briefly outlined the features of a number of multicarrier CDMA systems, which have been studied in the literature. So far, many reports are dedicated to the BER performance comparisons of DS-CDMA with multicarrier CDMA systems. These works show that all multicarrier CDMA schemes—MC-CDMA [19–21], MC-DS-CDMA [13, 14], and MT-CDMA [22]—outperform DS-CDMA.

In addition, it can be shown that there are trade-offs associated with each multicarrier CDMA scheme considered. Each technique has different benefits and drawbacks, depending on the intended applications [2, 14, 23]. Yang and Hanzo showed in [24] that MC-DS-CDMA has the highest degree of freedom in the family of CDMA schemes that can be beneficially exploited during the system design and reconfiguration procedures. The MC-DS-CDMA constitutes a trade-off between DS-CDMA and MC-CDMA in the context of the system's architecture and performance. By employing multiple subcarriers, MC-DS-CDMA typically requires lower-chip-rate spreading codes than DS-CDMA. It necessitates a lower number of subcarriers than MC-CDMA due to imposing DS spreading on each subcarrier's signal. Consequently, MC-DS-CDMA typically requires lower-rate signal processing than DS-CDMA and has lower worst-case peak-to-average power fluctuation than MC-CDMA.

Therefore, we will study the MC-DS-CDMA–air interface in this chapter. In addition, we broaden our analysis by considering the family of generalized MC-DS-CDMA transceivers, defined in [4]. This generalized scheme includes the subclasses of MT-CDMA and MC-DS-CDMA as special cases. Simulation results will later show the advantages of MT-CDMA in increasing throughput and bandwidth efficiency.

In the following sections, we will adopt this general view and simply refer to it as MC-CDMA in the remainder of the chapter unless otherwise required. We also assume the uplink of an asynchronous multicellular multicarrier CDMA system with C in-cell active users. For the sake of simplicity, we assume that all users use the same subcarriers and transmit with the same modulation at the same rate.

10.2.5 MC-CDMA Transmission Model

This section explains in more detail the MC-CDMA scheme adopted in this chapter. The block diagram of the MC-CDMA transmitter is shown in Figure 10.2. The input information sequence of the uth user is first converted into $N_c = 2K + 1$ parallel* data sequences $b^u_{-K,n}, \ldots, b^u_{0,n}, \ldots, b^u_{K,n}$, where n is the time index. The datum $b^u_{K,n} \in C_{\mathcal{M}}$ is $\mathcal{M}$-PSK modulated and differentially† encoded at rate $1/T_{MC}$, where $T_{MC} = N_c \times T$ is the symbol duration after S/P conversion, T is the symbol duration before S/P, and $C_{\mathcal{M}} = \{\ldots, e^{j2\pi m/\mathcal{M}}, \ldots\}$, $m \in \{0, \ldots, \mathcal{M} - 1\}$. The resulting S/P converter output is then spread with a random spreading code $c_u(t)$ at a rate $1/T_c$. The spreading factor, defined as the ratio between the chip rate and the symbol rate, is $L = T_{MC}/T_c$. We write the spreading-code segment over the nth period T_{MC} as

$$c^u_n\left(t\right) = \sum_{l=0}^{L-1} c^u_{1,n} \phi\left(t - 1T_c - nT_{MC}\right), \tag{10.1}$$

where $c^u_{l,n} = \pm 1$ for $l = 0, \ldots, L - 1$ is a random sequence of length L and $\phi(t)$ is the chip pulse. We consider square-root raised cosine chip pulses with roll-off factor β (see appendix). Closed-loop power control is taken into account at the transmitter by the amplification factor $a_u(t)$. All the data are then modulated in baseband by the inverse discrete Fourier transform (IDFT) and summed to obtain the multicarrier signal. No guard interval is inserted. Indeed, the channel identification and equalization are achieved by MC-STAR (multicarrier spatiotemporal array receiver) [32], and simulation results have shown that the guard interval length does not affect the link-level performance. MC-STAR exploits the intrinsic channel diversity by combining and equalizing the multipath signals. We

* We selected an odd number of subcarriers to have a central frequency, but the model can easily be rearranged to operate with an even number of subcarriers.

† We can also use pilot symbols for coherent modulation and detection [25], but that is beyond the scope of this chapter.

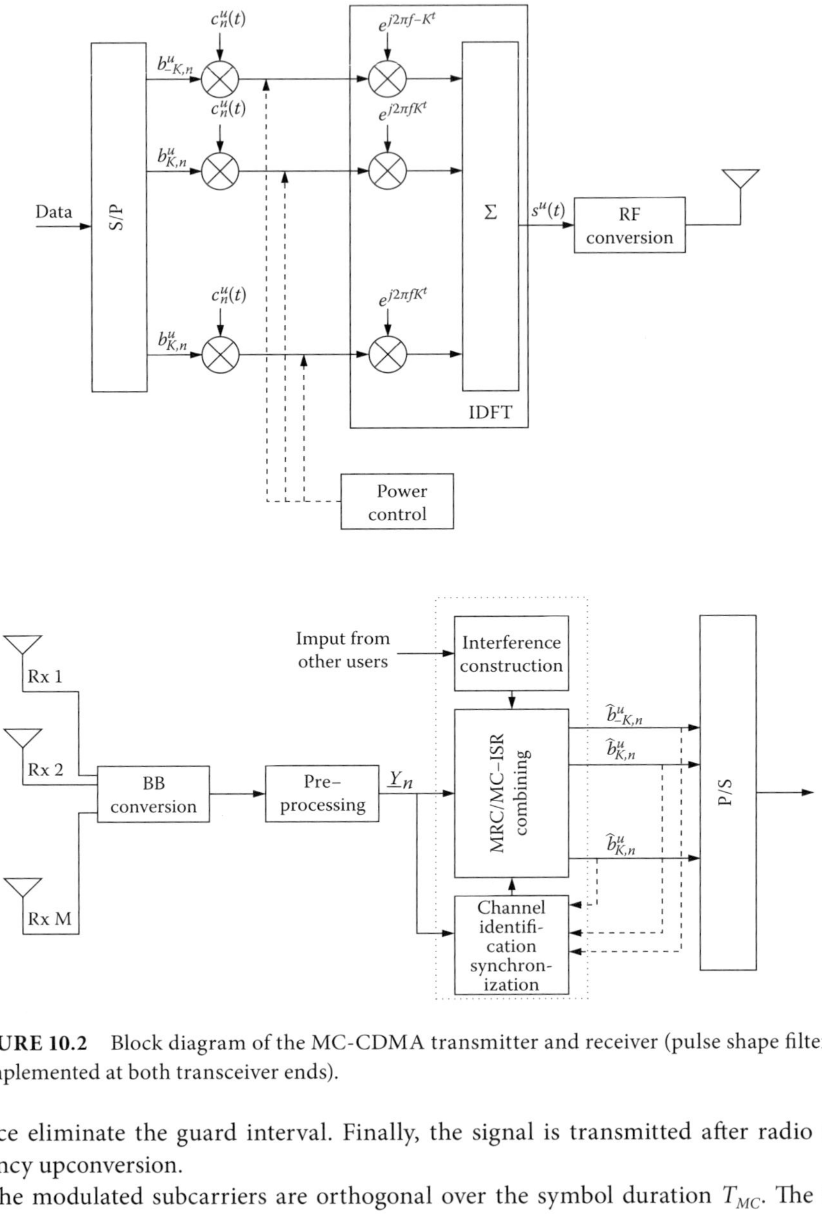

FIGURE 10.2 Block diagram of the MC-CDMA transmitter and receiver (pulse shape filtering is implemented at both transceiver ends).

hence eliminate the guard interval. Finally, the signal is transmitted after radio frequency upconversion.

The modulated subcarriers are orthogonal over the symbol duration T_{MC}. The frequency corresponding to the kth subcarrier is $f_k = \lambda \times k/T_{MC}$. The transmitter belongs to the family of MT-CDMA if λ is set to 1, and to the class of MC-DS-CDMA if λ is set to L (see resulting signal spectra in Figure 10.3). Indeed, in an MT-CDMA system, the subcarrier frequencies are chosen to be orthogonal harmonics with minimum frequency separation before spreading. By contrast, in MC-DS-CDMA, the subcarrier frequencies are chosen to satisfy the orthogonality condition with minimum possible frequency after spreading. The transmitted signal of the uth user is given by

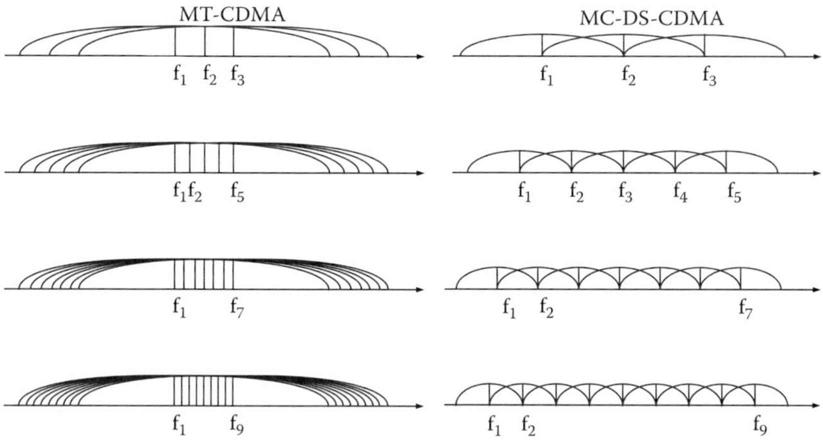

FIGURE 10.3 Different configurations of MT-CDMA and MC-DS-CDMA spectra within the same bandwidth.

$$s^u(t) = \sum_{k=-K}^{K} \sum_{n=-\infty}^{\infty} a^u(t) b_{k,n}^u c_n^u(t) e^{j2\pi f_k t}. \tag{10.2}$$

The transmitted signal's bandwidth is

$$BW = \frac{(N_c - 1)\lambda}{T_{MC}} + \frac{(1+\beta)}{T_c}. \tag{10.3}$$

10.2.6 Channel Model for Multicarrier Transmission

We consider an uplink transmission to M receiving antennas at the base station. The channel is assumed to be a slowly varying frequency-selective Rayleigh channel with delay spread $\Delta\tau$. For each kth subcarrier of user u, the key channel parameter is the number of resolvable paths, P_k^u, which is given by

$$P_k^u = \left\lfloor \frac{\Delta\tau}{T_c} \right\rfloor + 1, \tag{10.4}$$

where T_c is the chip duration. In practice, the number of multipaths depends also on the choice of the noise threshold used to differentiate between the received multipath components and the thermal noise. Typical delay-spread values are in the range of 0.4–4 μs in outdoor mobile radio channels, and the number of multipaths P_k^u varies between 2 and 5 with 3.84 megachips per second (Mcps) resolution [26]. The M-dimensional complex low-pass equivalent vector representation of the impulse response experienced by subcarrier k of the uth user, for a receiver equipped with M antennas, is

$$H_k^u(t) = \frac{\rho_k^u(t)}{(r^u)^e(t)} \sum_{p=1}^{P_k^u} \mathcal{G}_{k,p}^u(t)\delta\left(t-\tau_{k,p}^u(t)\right). \tag{10.5}$$

where $\rho_k^u(t)$ and $(r^u)^e(t)$ model the effects of shadowing and path loss, respectively, $r^u(t)$ is the distance from the uth user to the base station, and e is the path loss exponent. We assume their variations in time to be very slow, and hence nearly constant over several symbol durations. The M-dimensional complex vector $\mathcal{G}_{k,p}^u(t)$* denotes the fading and the array response from the user to the antenna elements of the receiver, and $\tau_{k,p}^u(t)$ represents the propagation time delay along the pth path. We note here that the large-scale path loss that includes free-space path loss and shadowing is the same for all subcarriers of the same user. Moreover, the number of resolvable paths and their propagation time delays depend on the reflecting objects and scatterers and can be assumed equal for all subcarriers [28]. Therefore, we omit the index k from ρ, P, and τ ($\rho_k^u = \rho^u$, $P_k^u = P^u$, and $\tau_{k,p}^u = \tau_p^u$) and reformulate equation (10.5) as

$$H_k^u(t) = \frac{\rho^u(t)}{(r^u)^e(t)} \sum_{p=1}^{P^u} \mathcal{G}_{k,p}^u(t)\delta\left(t-\tau_p^u(t)\right). \tag{10.6}$$

A frequency-domain channel model for a multicarrier system can be characterized by the coherence bandwidth† [28]:

$$B_c \simeq \frac{1}{2\pi\Delta\tau}. \tag{10.7}$$

When the frequency separation λ/T_{MC} is less than B_C, the MC-CDMA system is subject to correlated fading over different subcarriers. Fades across taps (multipaths) are mutually independent for the same carrier. However, fading for the same tap across different carriers is correlated. The envelope correlation coefficient between subcarrier k and subcarrier k' for user u is [28]

$$\rho_{k,k'}^u = E\left[\left|\mathcal{G}_{k,p,m}^u(t)\right|\left|\mathcal{G}_{k',p,m}^u(t+\tau)\right|\right] = \frac{(1+\lambda_{k,k'})\tilde{E}\left(\frac{2\sqrt{\lambda_{k,k'}}}{1+\lambda_{k,k'}}\right) - \frac{\pi}{2}}{2-\frac{\pi}{2}}, \tag{10.8}$$

* We may characterize $\mathcal{G}_{k,p}^u(t)$ in a space manifold parameterized by angles of arrivals [27]. However, a space characterization requires perfect antenna calibration and adequate sensor positioning.

† The coherence bandwidth is defined as the bandwidth over which the envelope correlation is above 0.5 [28].

with

$$\lambda_{k,k'} = \frac{J_0\left(2\pi f_D \tau\right)}{\sqrt{1+\left[2\pi\left(f_k - f_{k'}\right)\Delta\tau\right]^2}}, \tag{10.9}$$

where E is the expectation function, $\mathcal{G}^u_{k,p,m}(t)$ is the fading and the antenna response from user u to antenna m of the receiver along the pth path, $\tilde{E}$ is the complete elliptic integral of the second kind, J_0 is the zeroth-order Bessel function, and f_D is the maximum Doppler frequency. We adopt the approach proposed in [29, 30] to generate correlated Rayleigh channels across subcarriers. We also assume that the received channel multipath components across the M antennas are independent.

10.2.7 Received Signal

For a multicellular MC-CDMA system with C in-cell users and $N_c = 2K + 1$ carriers, the received signal is the superposition of signals from all users and all subcarriers. Hence, the M-dimensional observation vector received, after downconversion, by the antenna array can be expressed as follows:

$$\begin{aligned} X(t) &= \sum_{u=1}^{C} \sum_{k=-K}^{K} \sum_{n=-\infty}^{\infty} H^u_k(t) \otimes a^u(t) b^u_{k,n} c^u_n(t) e^{j2\pi(f_k+\Delta f^u)t} + N(t), \\ &= \sum_{u=1}^{C} \sum_{k=-K}^{K} \sum_{n=-\infty}^{\infty} X^u_{k,n}(t) + N(t), \end{aligned} \tag{10.10}$$

where $\otimes$ denotes time convolution and Δf^u models the carrier frequency offset (CFO), which is assumed equal for all subcarriers. This is a realistic assumption since there is only one oscillator per transmitter (see Figure 10.2). On the downlink, the CFO is even equal for all in-cell users (i.e., $\Delta f^u = \Delta f \ \forall \ u \in \{1, \ldots, C\}$). The noise term $N(t)$ includes the thermal noise received at the antennas as well as the out-cell interference. The contribution $X^u_{k,n}(t)$ of the nth data symbol over the kth carrier of user u to the received vector $X(t)$ is given by

$$\begin{aligned} X^u_{k,n}(t) &= H^u_k(t) \otimes a^u(t) b^u_{k,n} c^u_n(t) e^{j2\pi(f_k+\Delta f^u)t} \\ &= \psi^u_k(t) b_{k,n} \sum_{p=1}^{P^u} G^u_{k,p}(t) \varepsilon^u_{k,p}(t) c^u_n\left(t-\tau^u_p\right) e^{j2\pi(f_k+\Delta f^u)(t-\tau^u_p)}. \end{aligned} \tag{10.11}$$

Along the pth path, $G^u_{k,p}(t) = (\sqrt{M}/\|\mathcal{G}^u_{k,p}(t)\|)\mathcal{G}^u_{k,p}(t)$ is the propagation vector over the kth subcarrier of the uth user with norm $\sqrt{M}$, and $(\varepsilon^u_{k,p})^2(t) = \|\mathcal{G}^u_{k,p}(t)\|^2/\sum_{p=1}^{P_u}\|\mathcal{G}^u_{k,p}(t)\|$ is the fraction of the total received power on the kth subcarrier of user u:

$$\left(\psi_k^u\right)^2(t)=\left(\frac{\rho^u(t)}{\left(r^u\right)^e(t)}\right)^2\left(a^u\right)^2(t)\sum_{p=1}^{p^u}\frac{\left\|\mathcal{G}_{k,p}^u(t)\right\|^2}{M}. \tag{10.12}$$

10.2.8 Interference Analysis

We define the matched-filtered observation vector of frame number n over a time interval $[0, T_{MC})$ as

$$Y_n(t)=\frac{1}{T_c}\int_{D_\phi}X\left(nT_{MC}+t+t'\right)\phi\left(t'\right)dt', \tag{10.13}$$

where D_ϕ denotes the temporal support* of $\phi(t)$. After sampling at a multiple of the chip rate, we frame the observation into overlapping blocks of constant length N_p. The oversampling ratio k_s is defined as the number of samples per chip. In DS-CDMA and MT-CDMA systems we need no more than one sample per chip ($k_s = 1$). In contrast, in an MC-DS-CDMA system, a higher sampling frequency is necessary for the receiver. Indeed, the sampling frequency has to satisfy the Nyquist sampling theorem, which states that the sampling interval must be smaller than the inverse of the double-sided bandwidth of the sampled signals. Hence, the smallest number greater than the number of subcarriers N_c is an adequate oversampling ratio for MC-DS-CDMA. The resulting processing block duration $T_P = N_P(T_c/k_s)$ is equal to $T_{max} + \overline{\Delta\tau}$. The processing period $T_{max} = LT_c$ contains N_c carrier symbols targeted for detection. The frame overlap $\overline{\Delta\tau} < T_{max}$, which is larger than the delay spread, allows multipath tracking [31]. Hence, we obtain the $M \times N_P$ matched-filtered observation matrix:

$$\mathbf{Y}_n=\left[Y_n(0), Y_n\left(T_c/k_s\right),\ldots, Y_n\left(\left(N_p-1\right)T_c/k_s\right)\right]. \tag{10.14}$$

It can be expressed as

$$\mathbf{Y}_n=\sum_{u=1}^{C}\sum_{k=-K}^{K}\sum_{n'=-\infty}^{\infty}\mathbf{Y}_{n',k,n}^u+N_n, \tag{10.15}$$

where the baseband preprocessed thermal noise and the out-cell interference contribute $\mathbf{N}_n$, and where symbol n' of carrier k of user u contributes its observation matrix $\mathbf{Y}_{n',k,n}^u$, obtained by

* For a rectangular pulse, D_ϕ is $[0, T_c]$. In practice, and as assumed in this chapter, it is the temporal support of a truncated square-root raised cosine, $D_\phi = [-N_{src}T_c, N_{src}T_c]$, where N_{src} stands for the truncation span of the shaping pulse in chip samples around 0.

$$\mathbf{Y}^u_{n',k,n} = \left[Y^u_{n',k,n}(0), Y^u_{n',k,n}\left(T_c/k_s\right), \ldots, Y^u_{n',k,n}\left(\left(N_p - 1\right)T_c/k_s\right)\right], \tag{10.16}$$

where

$$Y^u_{n',k,n}\left(t\right) = \frac{1}{T_c}\int_{D_\phi} X^u_{k,n'}\left(nT_{MC} + t + t'\right)\phi\left(t'\right)dt'. \tag{10.17}$$

As a result of the stationarity assumptions stated in section 10.2.6, $Y^u_{n',k,n}(t)$ can be developed into

$$\begin{aligned} Y^u_{n',k,n}\left(t\right) &\simeq e^{j2\pi\Delta f^u nT_{MC}}\psi^u_k\left(nT_{MC}\right)b^u_{k,n'}\frac{1}{T_c}\sum_{p=1}^{P}\int_{D_\phi} G^u_{k,p}\left(nT_{MC} + t + t'\right) \\ &\quad .\varepsilon^u_{k,p}\left(nT_{MC} + t + t'\right)e^{j2\pi f_k\left(nT_{MC} + t + t' + \tau^u_p\right)}c^u_{n'}\left(nT_{MC} + t + t' - \tau^u_p\right)\phi\left(t'\right)dt' \\ &\simeq e^{j2\pi\Delta f^u nT_{MC}}\psi^u_k\left(nT_{MC}\right)b^u_{k,n'}V^u_{n',k,n}\left(t\right) \\ &\simeq \psi^u_k\left(nT_{MC}\right)b^u_{k,n'}U^u_{n',k,n}\left(t\right), \end{aligned} \tag{10.18}$$

where the spread channel vector without CFO $V^u_{n',k,n}(t)$ is obtained by

$$V^u_{n',k,n}\left(t\right) = \frac{1}{T_c}\int_{D_\phi} H^u_k\left(nT_{MC} + t + t'\right) \otimes c^u_{n'}\left(nT_{MC} + t + t'\right)e^{j2\pi f_k\left(nT_{MC} + t + t'\right)}\phi\left(t'\right)dt', \tag{10.19}$$

and the spread channel vector is

$$U^u_{n',k,n}\left(t\right) = e^{j2\pi\Delta f^u nT_{MC}}V^u_{n',k,n}\left(t\right). \tag{10.20}$$

We assumed in the development of equation (10.18) that $\psi_k(nT_G + t + t')$ is constant during the interval $t' \in D_\phi$. We also considered that the frequency offset is small compared to the symbol rate ($\Delta f^u T_{MC} \ll 1$); thus, $e^{j2\pi\Delta f^u(nT_{MC}+t+t')} \simeq e^{j2\pi\Delta f^u nT_{MC}}$ for $t' \in D_\phi$ and $t \in [0, T_{MC})$. Substituting equation (10.18) in equation (10.15) gives

$$\begin{aligned} \mathbf{Y}_n &= \sum_{u=1}^{C}\sum_{k=-K}^{K}\sum_{n'=-\infty}^{\infty} b^u_{k,n'}\psi^u_{k,n}e^{j2\pi\Delta f^u nT_{MC}}\mathbf{V}^u_{n',k,n} + \mathbf{N}_n \\ &= \sum_{u=1}^{C}\sum_{k=-K}^{K}\sum_{n'=-\infty}^{\infty} b^u_{k,n'}\psi^u_{k,n}\mathbf{U}^u_{n',k,n} + \mathbf{N}_n, \end{aligned} \tag{10.21}$$

where $\psi^u_{k,n} = \psi^u_k(nT_{MC})$. Due to asynchronism and multipath propagation, each user's carrier observation matrix carries information from the current as well as the previous and future symbols of the corresponding user's carrier. We therefore have

$$\mathbf{Y}_n = \sum_{u=1}^{C}\sum_{k=-K}^{K}\sum_{n'=n-1}^{n+1} b^u_{k,n'}\psi^u_{k,n}\mathbf{U}^u_{n',k,n} + \mathbf{N}_n. \tag{10.22}$$

Without loss of generality, let us focus on the detection of the nth symbol carried by the kth carrier of the desired user assigned index $d \in \{1, \dots C\}$, i.e., $b^d_{k^d,n^d}$. Using equation (10.22) and defining a vector $\underline{V}$ as a matrix $\mathbf{V}$ reshaped column-wise, we can rewrite the observation matrix for a desired user d with respect to its nth symbol of carrier k targeted for detection in the following simpler vector form:

$$\begin{aligned}\underline{Y}_n &= \underbrace{s^d_{k,n}\underline{U}^d_{n,k,n}}_{\text{desired signal}} + \underbrace{\sum_{\substack{u=1\\u\neq d}}^{C}\sum_{k'=-K}^{K}\sum_{n'=n-1}^{n+1} s^u_{k',n'}\underline{U}^u_{n',k',n}}_{\underline{I}^d_{MAI,k,n}} \\ &\quad + \underbrace{\sum_{\substack{k'=-K\\k'\neq k}}^{K}\sum_{n'=n-1}^{n+1} s^d_{k',n'}\underline{U}^d_{n',k',n}}_{\underline{I}^d_{\text{ICI},k,n}} + \underbrace{\sum_{\substack{n'=n-1\\n'\neq n}}^{n+1} s^d_{k,n'}\underline{U}^d_{n',k,n}}_{\underline{I}^d_{\text{ISI},k,n}} + \underline{N}_n, \\ &= s^d_{k,n}\underline{U}^d_{n,k,n} + \underline{I}^d_{k,n} + \underline{N}_n,\end{aligned} \tag{10.23}$$

where $s^u_{k',n'} = \psi^u_{k',n}b^u_{k',n'}$ and $(\psi^u_{k,n})^2$ are the n'^{th} signal component of the k'^{th} carrier of user u, and the received power of user u over carrier k, respectively. The total interference $\underline{I}^d_{k,n}$ includes three types of interference: (1) the multiple access interference $\underline{I}^d_{\text{MAI},k,n}$ is the interference due to the N_c carriers from the other in-cell users $u \neq d$; (2) the inter-carrier interference $\underline{I}^d_{\text{ICI},k,n}$ is the interference due to the other carriers, $k' \neq k$, from the same user d; and (3) the intersymbol interference $\underline{I}^d_{\text{ISI},k,n}$ is the interference due to the same carrier k from the same user d. The noise vector $\underline{N}_n$, which comprises the preprocessed thermal noise and the interference due to out-of-cell users, is assumed to be uncorrelated both in space and time with variance σ^2_N.

In previous work [32], we proposed a receiver named MC-STAR. MC-STAR assumes the interference $\underline{I}^d_{k,n}$ as another contribution to the noise $\underline{N}_n$. Hence, the signal component of the desired user's carrier is extracted by spatiotemporal maximum ratio combining (MRC) as follows:

$$\hat{s}^d_{k,n} = \underline{W}^{d^H}_{MRC,k,n}\underline{Y}_n = \frac{\underline{\hat{U}}^{d^H}_{n,k,n}\underline{Y}_n}{\left\|\underline{\hat{U}}^d_{n,k,n}\right\|^2}, \tag{10.24}$$

where anywhere in the chapter the notation $\hat{\alpha}$stands for an estimate of a given variable α, $(.)^H$ is the Hermitian operator, and $\underline{W}^d_{MRC,k,n}$ is the MRC beamformer. Equation (10.23) shows that the net interference increases with the number of interferers and subcarriers, which severely limits the capacity of the MC-CDMA system with simple MRC receivers. Therefore, in the next section, we shall use the data decomposition of equation (10.23) to formulate the interference suppression problem and propose a new MC-CDMA receiver with full interference suppression capabilities.

10.2.9 Multiuser Detection Techniques for MC-CDMA Systems

Conventional multicarrier CDMA detectors—such as the matched filter, the RAKE combiner, and the MC-STAR receiver—are optimized for detecting the signal of a single desired user. RAKE combiners exploit the inherent multipath diversity in CDMA, since they essentially consist of matched filters for each resolvable path of the multipath channel. The outputs of these matched filters are then coherently combined according to a diversity combining technique, such as maximum ratio combining, equal gain combining, or selection diversity combining [13, 54]. Unlike RAKE-type receivers, which assume perfect knowledge of the channel [3], we proposed in previous work a full space-time receiver solution, named MC-STAR [43], that jointly implements adaptive channel identification and synchronization in both time and frequency.* These conventional single-user detectors are inefficient, because the interference is treated as noise and there is no utilization of the available knowledge about the mobile channel or the spreading sequences of the interferers.

In order to mitigate the problem of MAI, [33] proposed and analyzed the optimum multiuser detector for asynchronous Gaussian multiple access channels. This optimum detector significantly outperforms the conventional detector, and it is near–far resistant, but unfortunately its complexity grows exponentially with the number of interfering users. Following this work, numerous suboptimum multiuser detectors have been proposed for a variety of channels, data modulation schemes, and transmission formats. Since MC-CDMA systems also contain a DS-CDMA component, traditional suboptimum multiuser detection techniques can be performed on each carrier with some form of adaptation. A variety of linear multiuser receivers have been investigated for MC-CDMA systems such as the minimum mean square error (MMSE) detector [34] and the combination of MMSE and the decorrelator detector [35]. Interference cancellation (IC) schemes constitute another variant of multiuser detection that has been applied to MC-CDMA systems. They can be broadly divided into two categories: parallel cancellation (PIC) [37, 38] and successive cancellation (SIC) [36]. At each stage in the detector, the estimates of all the other users from the previous stage were used for reconstructing an estimate of the MAI, and this estimate was then subtracted from the interfered signal representing the wanted bit. A novel class of multicarrier multiuser detectors, referred to as subspace blind detectors, was proposed by [39] and [40], where only the spreading sequence and the delay of the desired user were known at the receiver. Based on this knowledge, a blind subspace tracking algorithm was developed for estimating the data of the desired user.

* MC-STAR is our starting receiver; hence, we will provide a short overview of this receiver in section 10.3.

Most of these multiuser receivers have focused on multiple access interference while ignoring the ICI. In addition, important system design issues such as carrier frequency offset recovery (CFOR) have often been neglected. In multiuser detection, the CFO of one user not only degrades the detection of that user itself, but also makes the receiver based on the ideal carrier frequency acquisition no longer optimal, thus degrading the detection of the other users [41]. An alternative multiuser detection technique, denoted interference subspace rejection (ISR), has been proposed for DS-CDMA [42]. This technique offers different modes. Each mode characterizes the interference vector in a different way and accordingly suppresses it. The flexibility and robustness inherent to ISR make its exploitation in multicarrier systems of great interest.

10.3 Proposed Adaptive Multicarrier CDMA Receiver: MC-ISR

This section is dedicated to the description, performance analysis, and implementation of the proposed MC-ISR receiver [6]. After a short overview of MC-STAR [43], which is our starting single-user receiver, we will describe and evaluate the adaptive interference rejection procedure that characterizes the proposed MC-ISR receiver.

10.3.1 The General Concept of MC-STAR

The adaptive receiver MC-STAR implements joint space-time-frequency processing over the despread data to improve the spectrum efficiency of the MC-CDMA system. The adaptive blind channel identification and equalization as well as the acquisition and tracking of multipaths and CFO are carried out on each subcarrier. However, their modules are interconnected to ensure proper information exchange and joint processing over carriers.* Mathematical details of the different adaptive procedures and their connections are provided in [43]. In this section, we explain the advantages of MC-STAR by describing the intermediate stages in our development that led to this receiver.

At the beginning, we extend original STAR, proposed for DS-CDMA [31], to a multicarrier system by placing STAR on each subcarrier. This extension requires a modification of the time-delay tracking procedure. Indeed, we introduce an intermediate transformation of the time response to reallow estimation of the multipath delays by simple linear regression.

Multi-carrier CDMA systems are very sensitive to the CFO. Therefore, we further introduce joint time-delay and frequency synchronization. The effect of the CFO on the performance of the spatiotemporal array receiver was not addressed in [31]. The space-time separation of the channel enables us to decouple time and carrier frequency synchronization. We can hence estimate the CFO by linear regression (LR) of the phase variation of each fading coefficient. Once an estimate of the carrier frequency offset

---

* The complexity of MC-STAR, which is approximately the complexity of STAR multiplied by the number of subcarriers, can be assessed using the results established in [44]. The latter suggests that MC-STAR can be implemented today on a single FPGA.

estimate $\widehat{\Delta f}$ is available by exploiting diversity in space, time, and frequency, we implement carrier frequency offset recovery (CFOR) in an adaptive closed-loop structure, where we feed back the estimate of the frequency offset to the input of the receiver. The CFOR reduces the time variations in the spatiotemporal propagation channel due to Δf to much weaker fluctuations due to the residual $\delta f = \Delta f - \widehat{\Delta f}$. It results in much weaker identification errors and enables further reduction of the carrier frequency estimation error δf.

At this stage, the receiver still consists of independent modules on each subcarrier. The purpose of the last step is to improve the performance of the overall receiver by interconnecting these modules and performing joint multicarrier processing. We exploit the intercarrier correlation, intrinsic to a multicarrier system, as a type of *frequency gain* to improve the performance by joint multicarrier channel identification and synchronization operations. Indeed, in the context of a multicarrier system, the adjacent subcarriers are exposed to correlated fading, especially if the delay spread of the channel is relatively low, resulting in relatively large coherence bandwidth. Hence, averaging the adjacent subcarrier channel parameters should improve the BER performance when transmitting over such low-dispersive fading channels. Along this perspective, the parameters common to all subcarriers can be estimated more accurately by averaging their estimates over all subcarriers. These parameters include the number of multipaths, their corresponding time delays, and the frequency offset. Other channel parameters, such as the channel fading coefficients, are correlated but not identical over all subcarriers. Therefore, combining them may not achieve the expected performance enhancement. We thus introduce a moving average technique over subcarriers with high correlation. The fact that subcarriers are highly correlated implies similar or identical channel parameters over subcarriers. Yet, the noise is uncorrelated across subcarriers, and hence the similar/common parameters can be estimated more accurately by averaging their estimates over all subcarriers, yielding the so-called frequency gain. The variance of the resulting estimation error is lower than the variance of the estimation error without frequency gain. Please bear in mind that we used the term *frequency gain* and not *frequency diversity*, which relies on the fact that the fading is different over different subcarriers.

10.3.2 Multicarrier Interference Subspace Rejection (MC-ISR)

Provided that an instantaneous estimate of the total interference $\hat{\underline{I}}^d_{k,n} = \hat{\underline{I}}^d_{\mathrm{MAI},k,n} + \hat{\underline{I}}^d_{\mathrm{ICI},k,n} + \hat{\underline{I}}^d_{\mathrm{ISI},k,n}$ is made available at the receiver (see section 10.3.4), we can eliminate it and yet achieve distortionless response to the desired signal by imposing the following simple constraints to the combiner $\underline{W}^d_{k,n}$:

$$\begin{cases} \underline{W}^{d^H}_{k,n}\hat{\underline{U}}^d_{n,k,n} & =1, \\ \underline{W}^{d^H}_{k,n}\hat{\underline{I}}^d_{k,n} & =0, \end{cases} \Rightarrow \begin{cases} \underline{W}^{d^H}_{k,n}\hat{\underline{U}}^d_{n,k,n} & =1, \\ \underline{W}^{d^H}_{k,n}\left(\hat{\underline{I}}^d_{\mathrm{MAI},k,n}+\hat{\underline{I}}^d_{\mathrm{ICI},k,n}+\hat{\underline{I}}^d_{\mathrm{ISI},k,n}\right) & =0. \end{cases} \tag{10.25}$$

The first constraint guarantees a distortionless response to the desired signal, while the second directs a null to the total interference realization and thereby cancels it.

Exploiting the general framework developed in [42], the solution to the specific optimization problem in equation (10.25) is the MC-ISR combiner $\underline{W}^d_{k,n}$ given as follows:

$$\underline{W}^d_{k,n} = \frac{\Pi^d_{k,n}\hat{\underline{U}}^d_{n,k,n}}{\hat{\underline{U}}^{d^H}_{n,k,n}\Pi^d_{k,n}\hat{\underline{U}}^d_{n,k,n}}, \tag{10.26}$$

$$Q_n = 1\Big/\left(\hat{\underline{I}}^{d^H}_{k,n}\hat{\underline{I}}^d_{k,n}\right) = 1\Big/\left\|\hat{\underline{I}}^d_{k,n}\right\|^2, \tag{10.27}$$

$$\mathbf{\Pi}^d_{k,n} = \mathbf{I}_{N_T} - \hat{\underline{I}}^d_{k,n}\hat{\underline{I}}^{d^H}_{k,n} \times Q_n, \tag{10.28}$$

where $N_T = M \times N_P$ is the total space dimension and $\mathbf{I}_{N_T}$ denotes an $N_T \times N_T$ identity matrix. First, we form the projector $\mathbf{\Pi}^d_{k,n}$ orthogonal to the total interference realization. Second, we project the estimated response vector $\hat{\underline{U}}^d_{n,k,n}$ and normalize it to derive the combiner. We use this combiner instead of MRC (used by MC-STAR) to extract the nth signal component of the kth carrier of the desired user as

$$\hat{s}^d_{k,n} = \underline{W}^{d^H}_{k,n}\underline{Y}_n. \tag{10.29}$$

Unlike most of the multiuser receivers proposed for MC-CDMA, which focus on multiple access interference while ignoring the intercarrier interference, MC-ISR fully suppresses the total interference resulting from MAI, ISI, and ICI by simple yet efficient nulling.* In addition, the interference suppression is made adaptive to track the current situation of the wireless channel and the interference. Simulation results will later show that ICI is not negligible and that full adaptive interference suppression is required to improve the MC-CDMA system performance.

10.3.3 Link/System-Level Performance Analysis

This section is dedicated to the performance analysis of the MC-ISR receiver based on the Gaussian assumption (GA). We exploit the analysis results of DS-CDMA ISR recently developed in [45] at the link level and extend them to MC-ISR. Additionally, we broaden the scope of the analysis to the system level.

10.3.3.1 Link-Level Performance

For the sake of simplicity, we assume temporarily perfect channel identification and perfect CFO estimation and recovery. Later in the simulations, we will use the channel

* The formulation of MC-ISR can be extended to MMSE-type criteria [42].

and CFO estimates provided by MC-STAR* [32]. The postcombined signal can be formulated as

$$\hat{s}^d_{k,n} = \underline{W}^{d^H}_{k,n}\underline{Y}_n = s^d_{k,n} + \delta^d_{MAI,k,n} + \delta^d_{ICI,k,n} + \delta^d_{ISI,k,n} + \underline{W}^{d^H}_{k,n}\underline{N}_n, \tag{10.30}$$

where $\delta^d_{MAI,k,n}$, $\delta^d_{ICI,k,n}$, and $\delta^d_{ISI,k,n}$ are the combining residuals of $\underline{I}^d_{\mathrm{MAI},k,n}$, $\underline{I}^d_{\mathrm{ICI},k,n}$, and $\underline{I}^d_{\mathrm{ISI},k,n}$, respectively. We assume here that the interference rejection residuals $\delta^d_{MAI,k,n}$, $\delta^d_{ICI,k,n}$, and $\delta^d_{ISI,k,n}$ are Gaussian random variables with zero mean. Hence, we only need to evaluate their variances. Note that the residuals would be null (i.e., $\delta^d_{MAI,k,n} = \delta^d_{ICI,k,n} = \delta^d_{ISI,k,n} = 0$) if the reconstruction of the interference were perfect (i.e., $\hat{\underline{I}}^d_{k,n} = \underline{I}^d_{k,n}$), and hence $\hat{s}^d_{k,n} = s^d_{k,n} + \underline{W}^{d^H}_{k,n}\underline{N}_n$ would be corrupted only by the residual noise, which is Gaussian with zero mean and variance:

$$\mathrm{Var}\left[\underline{W}^{d^H}_{k,n}\underline{N}_n\right] = \bar{\kappa}\sigma^2_N, \tag{10.31}$$

where

$$\bar{\kappa} = \mathrm{E}\left[\left\|\underline{W}^d_{k,n}\right\|^2\right] = \frac{ML-1}{ML-2},$$

is a measure of the enhancement of the white noise compared to MRC ($\bar{\kappa} = 1$ for MRC) [45]. However, in practice the interference vector is reconstructed erroneously due to wrong tentative data decisions and power control errors, and hence $\hat{s}^d_{k,n}$ is further corrupted by non-null residual interference rejection components. Therefore, we introduce the error indicating variables

$$\xi^u_{k,n} = \hat{b}^{u*}_{k,n} b^u_{k,n}$$

and

$$\lambda^u_{k,n} = \hat{\psi}^{u*}_{k,n}\,\psi^u_{k,n} \Big/ \left\|\hat{\psi}^u_{k,n}\right\|^2$$

where $(.)^*$ means complex conjugate. $\xi^u_{k,n}$ models the symbol estimation error provided by MRC at the initial stage. $\lambda^u_{k,n}$ characterizes the power control error. $\xi^u_{k,n}$ and $\lambda^u_{k,n}$ equal 1 when the estimated data symbol and the power control are perfect; otherwise, they are complex numbers. Since

$$\underline{Y}^u_{n',k',n} = s^u_{k',n'}\underline{U}^u_{n',k',n} = b^u_{k',n'}\psi^u_{k',n}\underline{U}^u_{n',k',n} = \xi^u_{k',n'}\lambda^u_{k',n}\hat{b}^u_{k',n'}\hat{\psi}^u_{k',n}\underline{U}^u_{n',k'n} = \xi^u_{k',n'}\lambda^u_{k',n}\hat{\underline{Y}}^u_{n',k',n},^{\dagger}$$

* Simulations will show little deviation from analysis in the operating BER region.

† Here we assume perfect time and frequency synchronization.

we can rewrite equation (10.23) as

$$\underline{Y}_n = \underline{Y}^d_{n,k,n} + \sum_{\substack{u=1\\u-d}}^{C} \sum_{k'=-K}^{K} \sum_{n'=n-1}^{n+1} \xi^u_{k',n'} \lambda^u_{k',n} \hat{\underline{Y}}^u_{n',k',n}$$
$$+ \sum_{\substack{k'=-K\\k'\neq k}}^{K} \sum_{n'=n-1}^{n+1} \xi^d_{k',n'} \lambda^d_{k',n} \hat{\underline{Y}}^d_{n',k',n} + \sum_{\substack{n'=n-1\\n'\neq n}}^{n+1} \xi^d_{k,n'} \lambda^d_{k,n} \hat{\underline{Y}}^d_{n',k,n} + \underline{N}_n. \tag{10.32}$$

The signal after MC-ISR combining is then

$$\underline{W}^{d^H}_{k,n} \underline{Y}_n = s^d_{k,n} + \sum_{\substack{u=1\\u\neq d}}^{C} \sum_{k'=-K}^{K} \sum_{n'=n-1}^{n+1} \xi^u_{k',n'} \lambda^u_{k',n} \underline{W}^{d^H}_{k,n} \hat{\underline{Y}}^u_{n',k',n}$$
$$+ \sum_{\substack{k'=-K\\k'\neq k}}^{K} \sum_{n'=n-1}^{n+1} \xi^d_{k',n'} \lambda^d_{k',n} \underline{W}^{d^H}_{k,n} \hat{\underline{Y}}^d_{n',k',n} + \sum_{\substack{n'=n-1\\n'\neq n}}^{n+1} \xi^d_{k,n'} \lambda^d_{k,n} \underline{W}^{d^H}_{k,n} \hat{\underline{Y}}^d_{n',k,n} \tag{10.33}$$
$$+ \underline{W}^{d^H}_{k,n} \underline{N}_n.$$

The MC-ISR combiner $\underline{W}^d_{k,n}$ satisfies the optimization property in equation (10.25), and thus

$$\underline{W}^{d^H}_{k,n} \hat{\underline{I}}^d_{k,n} = 0 \Rightarrow \text{Var}\left[\underline{W}^{d^H}_{k,n}\left(\hat{\underline{I}}^d_{\text{MAI},k,n} + \hat{\underline{I}}^d_{\text{ICI},k,n} + \hat{\underline{I}}^d_{\text{ISI},k,n}\right)\right] = 0. \tag{10.34}$$

This result allows the derivation of the variance of the interference rejection residuals, as shown in the appendix. Let $\bar{\psi}^2_D = E[(\bar{\psi}^d_k)^2]$ be the average power of the kth carrier of the desired user and $\bar{\psi}^2_I$ be the average interference power on each interfering carrier. The variances of the residual $\underline{I}^d_{\text{MAI},k,n}$ can be written as

$$\text{Var}\left[\delta^d_{MAI,k,n}\right] = (C-1)\frac{\bar{\psi}^2_I}{L}\left[\varsigma(\beta) + \chi_k(\beta)\right](1 + \rho_\lambda - \rho_\xi)\bar{\kappa}, \tag{10.35}$$

where $\varsigma(\beta) = 1 - \beta/4$ and

$$\chi_k(\beta) = \begin{cases} \dfrac{\beta}{8}, & \text{if } k = -K \text{ or } K, \\ \dfrac{\beta}{4}, & \text{if } k = -K+1, \ldots, K-1, \end{cases} \tag{10.36}$$

for MC-DS-CDMA ($\lambda = L$ and $f_k = k/T_c$) and

$$\chi_k(\beta) = \sum_{\substack{k'=-K \\ k' \neq k}}^{K} \vartheta\left(\left|k - k'\right|\right), \tag{10.37}$$

where

$$\vartheta(x) = \begin{cases} 1 - \frac{\beta}{2} - \frac{x}{2L} + \frac{3\beta}{4\pi}\sin\left(\frac{\pi x}{\beta L}\right) + \left(\frac{\beta}{4} - \frac{x}{4L}\right)\cos\left(\frac{\pi x}{\beta L}\right), & \text{if } 0 \le x/L \le \min(\beta, 1-\beta) \\ 1 - \frac{x}{L} & \text{if } \beta \le x/L \le 1-\beta \text{ and } \beta < 0.5 \\ \frac{3}{4} - \frac{\beta}{4} - \frac{x}{4L} + \frac{3\beta}{4\pi}\sin\left(\frac{\pi x}{\beta L}\right) + \left(\frac{\beta}{4} - \frac{x}{4L}\right)\cos\left(\frac{\pi x}{\beta L}\right) + & \\ \quad \frac{3\beta}{8\pi}\sin\left(\frac{\pi x}{\beta L} - \frac{\pi}{\beta}\right) - \left(\frac{x}{8L} - \frac{1-\beta}{8}\right)\cos\left(\frac{\pi x}{\beta L} - \frac{\pi}{\beta}\right) & \text{if } 1-\beta \le x/L \le \beta \text{ and } \beta > 0.5 \\ \frac{3}{4} + \frac{\beta}{4} - \frac{3x}{4L} + \frac{3\beta}{8L}\sin\left(\frac{\pi x}{\beta L} - \frac{\pi}{\beta}\right) - \left(\frac{x}{8L} - \frac{1-\beta}{8}\right)\cos\left(\frac{\pi x}{\beta L} - \frac{\pi}{\beta}\right) & \text{if } \max(\beta, 1-\beta) \le x/L \le 1 \end{cases} \tag{10.38}$$

for MT-CDMA ($\lambda = L$ and $f_k = k/T_{Mc}$). The expressions of

$$\rho_\xi = \mathrm{E}\left[\xi^{u}_{k',n'}\lambda^{u}_{k',n}\xi^{u'*}_{k',n'}\lambda^{u'*}_{k',n}\right]$$

and

$$\rho_\lambda = \mathrm{E}\left[\left(\lambda^{u}_{k,n}\right)^2\right] - 1$$

are derived for a Rayleigh fading channel with P paths to yield

$$\begin{aligned} \rho_\xi &= \left(1 - \left(1 - \cos\left(2\pi/\mathcal{M}\right)\right)S_{rec}\right)^2, \\ \rho_\lambda &= \frac{4\pi^2\left(f_D \times \tau_{PC}\right)^2}{P-1}, \end{aligned} \tag{10.39}$$

where S_{rec} is the symbol error rate in the previous MC-ISR stage, f_D is the maximum Doppler frequency, and τ_{PC} is the power control feedback delay. The variances of the residuals $\underline{I}^{d}_{\mathrm{ICI},k,n}$ and $\underline{I}^{d}_{\mathrm{ISI},k,n}$ can be written as

$$\mathrm{Var}\left[\delta^{d}_{ICI,k,n}\right]=\frac{\overline{\psi}_D^2}{L}\,\delta_{is}\chi_k(\beta)\left(1+\rho_\lambda-\rho_\xi\right)\overline{\kappa},$$
$$\mathrm{Var}\left[\delta^{d}_{ISI,k,n}\right]=\frac{\overline{\psi}_D^2}{L}\,\delta_{is}\varsigma(\beta)\left(1+\rho_\lambda-\rho_\xi\right)\overline{\kappa}, \tag{10.40}$$

where $\delta_{is} = (P - 1)/P$ is a measure of the relative impact of the interference generated by the other paths on a given path of the desired user (for a Rayleigh fading channel with P equal paths). The SINR on the kth carrier can be estimated as

$$\mathrm{SINR}_{ISR,k}=\frac{\overline{\psi}_D^2}{\mathrm{Var}\left[\delta^{d}_{\mathrm{MAI},k,n}\right]+\mathrm{Var}\left[\delta^{d}_{\mathrm{ICI},k,n}\right]+\mathrm{Var}\left[\delta^{d}_{\mathrm{ISI},k,n}\right]+\overline{\kappa}\sigma_N^2}. \tag{10.41}$$

Note that the SINR expression above applies to MRC as well by setting $\overline{\kappa} = 1$ and $\rho_\lambda = \rho_\xi = 0$ in equations (10.35) and (10.40). Note also that in [46], we provide the variance of the interference for an MC-CDMA system with a rectangular pulse. In this chapter, we improve the analytical performance evaluation by deriving the variance of the interference with a more practical band-limited square-root raised cosine waveform. The BER performance on the kth carrier is then given as follows:

$$P_e^k=\Omega\left(\mathrm{SINR}_{ISR,k}\right), \tag{10.42}$$

where Ω represents the single-user bound (SUB), which is classically defined as a conditional Gaussian Q-function over ψ_D and ψ_I. When using this classical representation, the average BER is derived by first finding the probability density functions (pdfs) of ψ_D and ψ_I and then averaging over those pdfs. Since it is difficult to find a simple expression for the pdfs of ψ_D and ψ_I that takes into consideration antenna diversity, imperfect power control, and imperfect channel identification, we may consider an approximative pdf. In this analysis, we choose to simulate Ω without imposing any pdf approximation. For each multicarrier configuration, we run single-user and single-carrier link-level simulations. We reproduced as much as possible most of the real-world operating conditions: time and frequency synchronization, imperfect power control, channel identification errors, antenna diversity, etc. These link-level simulations gave a realistic Ω : $BER = \Omega(SNR)$. The simulations will later consider a multiuser and multicarrier environment. The average BER performance of the MC-ISR receiver is given by

$$P_e=\frac{1}{2K+1}\sum_{k=-K}^{K}P_e^k. \tag{10.43}$$

10.3.3.2 System-Level Performance

In order to compare the different MC-ISR configurations, the link-level curves provide a good picture of the performance of each system. But limiting comparisons to the

BER performance is not sufficient because the data rate is not equal for all configurations. Hence, we translate the link-level results into system-level results in terms of total throughput under the following three assumptions: (1) all users are received with an equal average power (i.e., $\bar{\psi}_D^2$ and $\bar{\psi}_I^2$) [4]; (2) all the cells have the same average load of C users per cell; and (3) the out-cell to in-cell interference ratio f is set to 0.6 [47]. Given these assumptions in an interference-limited system (thermal noise is low compared to interference), the link-level SIR at the base station antennas (ignoring ISI for simplicity) is

$$SIR_{ISR} = \frac{1}{(C-1)\alpha + \frac{1}{L}\delta_{is}\chi(\beta)(1+\rho_\lambda-\rho_\xi)\bar{\kappa} + Cf\lambda}, \tag{10.44}$$

where

$$\begin{aligned} \chi(\beta) &= \max_k\left[\chi_k(\beta)\right], \\ \alpha &= \frac{1}{L}\left(\varsigma(\beta)+\chi(\beta)\right)\left(1+\rho_\lambda-\rho_\xi\right)\bar{\kappa}, \\ \gamma &= \frac{1}{L}\left(\varsigma(\beta)+\chi(\beta)\right)\bar{\kappa}. \end{aligned} \tag{10.45}$$

$(C-1)\alpha$ is the normalized variance of the residual MAI ($u \neq d$), $\frac{1}{L}\delta_{is}\chi(\beta)(1+\rho_\lambda-\rho_\xi)\bar{\kappa}$ is the normalized variance of the residual ICI ($k' \neq k$ and $u = d$), and $Cf\gamma$ is the normalized variance of the out-cell interference. Note that equation (10.44) is derived from equation (10.41) and the assumption of negligible thermal noise and ISI.

The maximum number of users that can access the system C_{max} can be hence calculated by the simple procedure illustrated in Table 10.1. After initialization, this procedure increments the capacity C, until the SIR_{ISR} given by equation (10.44) no longer exceeds the required SNR_{req}. The SRN_{req} is the required SNR, derived from link-level simulations, to meet a BER of 5% in order to achieve a QoS of 10^{-6} after channel decoding. In step 2.2 of Table 10.1, we use the fact that the SIR expression applies to MRC by setting $\bar{\kappa} = 1$ and $\rho_\lambda = \rho_\xi = 0$ in equation (10.44). In step 2.3, we evaluate the symbol error rate S_{MRC} after the MRC stage as follows:

$$S_{MRC} = \Omega(\mathrm{SIR}_{MRC}), \tag{10.46}$$

where Ω represents the single-user bound (SUB). Note that multistage MC-ISR is considered in step 2.5. The total throughput is hence $\mathcal{T}_{max} = C_{max} \times R_b = C_{max} \times R_s \times \log_2(\mathcal{M})$ where R_b and R_s are the bit rate and symbol rate over all subcarriers, respectively.

10.3.4 MC-ISR Receiver Implementation

As mentioned in section 10.3.2, the proposed MC-ISR receiver requires accurate channel parameter estimates and data decisions to reconstruct the total interference $\underline{\hat{I}}^{d}_{k,n}$ and

TABLE 10.1 Capacity Computation Procedure

1. Initialize capacity $C = 0$.
2. Start computation loop:

2.1 Increment capacity $C = C + 1$.

2.2 Compute the SIR with MRC:

$$SIR_{MRC} = \frac{L}{(C-1)\left(\varsigma(\beta)+\chi(\beta)\right)+\chi(\beta)\delta_{is}+fC\left(\varsigma(\beta)+\chi(\beta)\right)}.$$

2.3 Compute the symbol error rate (SER) after MRC stage

$$S_{MRC} = \Omega\left(\mathrm{SIR}_{MRC}\right).$$

2.4 Compute ρ_λ and ρ_ξ.

2.5 Compute the SIR:

$$SIR_{ISR_1} = \frac{1}{(C-1)\alpha+\frac{1}{L}\delta_{is}\chi(\beta)\left(1+\rho_\lambda-\rho_\xi\right)\bar{\kappa}+Cf\gamma}.$$

If number of stages $S > 1$, start the loop; else go to 2.6.

For s = 2 : S.

Compute the symbol error rate (SER) after the $s - 1$ stage:

$$S_{ISR_{s-1}} = \Omega\left(\mathrm{SIR}ISR_{s-1}\right)$$

Compute ρ_ξ.

Compute the SIR:

$$SIR_{ISR_s} = \frac{1}{(C-1)\alpha+\frac{1}{L}\delta_{is}\chi(\beta)\left(1+\rho_\lambda-\rho_\xi\right)\bar{\kappa}+Cf\gamma}.$$

End.

2.6 If $SIR_{ISR_s} > SNR_{req}$ go to 2.1; else exit.

3. Decrement capacity $C = C - 1$.

null it reliably. Unlike previous works on interference suppression or multiuser detection [34, 38] that assume perfect knowledge of the channel, we propose here an adaptive receiver that uses runtime information about the channel and interference. Indeed, MC-ISR jointly implements channel identification and synchronization in both time and frequency, using MC-STAR [32], as well as signal combining with full interference suppression capabilities. Figure 10.2 shows the block diagram of the proposed receiver implementation, divided in four main modules. The first module is a preprocessor that downconverts the received signal to baseband, then passes it through the chip-matched filter before sampling and data block framing. The second module is a signal combiner that provides symbol estimates from the data observation, first by MRC in an initial iteration, then by MC-ISR in one or more iterative stages. The third module is an adaptive channel identifier and synchronizer from MC-STAR that implements closed-loop CFOR and estimates all the channel parameters (multipath time delays and their phases and amplitudes, received power, CFO). The fourth module is a null-constraint generator common to all in-cell users. It gathers the data decisions and channel parameter estimates from the second and third modules dedicated to each in-cell user-carrier pair in order to reconstruct the total in-cell signal vector $\underline{I}_n$. Then to each combiner, say of the desired user-carrier pair as illustrated in Figure 10.1b, it passes on the associated null constraint (i.e., $\underline{I}^d_{k,n} = \underline{I}_n - s^d_{k,n}\underline{U}^d_{n,k,n}$) calculated with the least computations by simple subtraction from $\underline{I}_n$ of the desired signal contribution from the corresponding user-carrier pair.

The implementation of an adaptive closed-loop CFOR* jointly with multicarrier and multiuser detection (here by MC-ISR) requires careful attention regarding the order in which these two tasks should be processed. Indeed, conventional operation of CFOR at an early processing stage† prior to interference suppression would require (on the uplink only) as many independently CFO-compensated observations and interference null constraints as received in-cell users, thereby resulting in a tremendous complexity increase. Here we develop an efficient post-interference-suppression CFOR scheme by splitting the MC-ISR combining operation of equation (10.29) into two steps, an observation-cleaning projection and an MRC combining, and by inserting CFO compensation in between as follows:

$$\underline{Y}^d_{\Pi,k,n} = \Pi^d_{k,n}\underline{Y}_n, \tag{10.47}$$

$$\widehat{\Delta f}^d_n = \widehat{\Delta f}^d_{n-1} + \widehat{\delta f}^d_n, \tag{10.48}$$

$$\dot{\underline{Y}}^d_{\Pi,k,n} = \underline{Y}^d_{\Pi,k,n} e^{-j2\pi\widehat{\Delta f}^d_n nT}, \tag{10.49}$$

$$\hat{\underline{V}}^d_{\Pi,k,n} = \Pi^d_{k,n}\hat{\underline{V}}^d_{\Pi,k,n}, \tag{10.50}$$

$$\hat{s}^d_{k,n} = \frac{\hat{\underline{V}}^{d^H}_{\Pi,k,n}\dot{\underline{Y}}^d_{\Pi,k,n}}{\left\|\hat{\underline{V}}^d_{\Pi,k,n}\right\|^2}. \tag{10.51}$$

The cleaning projection of equation (10.47) results in an almost interference-free observation $\underline{Y}^d_{\Pi,k,n}$ and allows for CFO estimation and compensation in equations (10.48) and (10.49), respectively, using the CFOR module of the single-user MC-STAR (refer to [32] and [48] for details on how to estimate the CFO adjustment term in equation (10.49)), and for MRC combining in equation (10.51) using the projected estimate of the

* In contrast to open-loop structures, closed-loop CFOR reduces the channel time variations and greatly improves their tracking [32].

† Usually CFOR is embedded in the RF chain or plugged to the preprocessor output.

spread channel vector without CFO $\underline{\hat{V}}^d_{\Pi,k,n}$. To the best of our knowledge, we are the first to report on and address this issue and to propose an efficient scheme for closed-loop CFOR in a multiuser detection context. It is important to mention here that if $\Delta f^u = \Delta f \ \forall \ u \in \{1, \ldots, C\}$ (i.e., downlink), then there is no need to estimate the CFO for the MC-ISR to null the in-cell interference. Indeed, the MC-ISR combiner $\underline{W}^d_{k,n}$ satisfies the optimization property in equation (10.25). Thus, it is not affected by the CFO of other users, i.e.,

$$\underline{W}^{d^H}_{k,n} \underline{\hat{I}}^d_{k,n} = 0 \ \Rightarrow \ e^{j2\pi\Delta f n T_{MC}} \left[\underline{W}^{d^H}_{k,n} \underline{\hat{I}}^d_{k,n} \right] = 0. \tag{10.52}$$

Once the MC-ISR projection is performed in equation (10.47) after reconstruction of $\underline{\hat{I}}^d_{k,n}$ without CFO, we implement the same CFOR scheme implemented in part by equations (10.48) and (10.49). Hence, like the near–far resistant detector proposed in [41], the multiuser CFOR problem can be transformed on the downlink into a single-user CFOR problem, and conventional single-user methods can therefore be used to estimate the frequency offset.*

To validate the efficiency of the proposed CFOR strategy in a multicarrier and multiuser detection scheme on the uplink, we consider a multiuser DBPSK MT-CDMA system with seven subcarriers, a spreading factor of 96, and five in-cell users ($N_c = 7$, $L = 96$, $C = 5$). We select the setup that will be introduced in section 10.4.1. The frequency offset normalized by the subcarrier separation ($\Delta f \times T_{MC}$) is set to 0.005 (i.e., $\Delta f = 200$ Hz).[15] Figure 10.4 shows the link-level results of MC-ISR with and without CFOR. Results suggest that a CFO of 200 Hz has a serious impact on the performance of MC-CDMA, and that the link-level gain with the proposed CFOR is in the range of 1 dB at a BER of 5% before channel decoding. By comparing the link-level curves of MC-ISR with CFOR and MC-ISR without a frequency offset (i.e., CFO=0 Hz), we notice that CFOR compensates almost completely the performance loss due to the frequency offset. These results confirm the need for and the efficiency of the proposed CFOR in a multicarrier and multiuser detection context.

10.4 Simulation Results

10.4.1 Simulation Setup

We consider an MC-CDMA system operating at a carrier of 1.9 GHz with maximum bandwidth of 5 MHz. We select a frequency offset Δf of 200 Hz, the maximum error tolerated by 3G standards† (≡ 0.1 ppm) for the frequency mismatch between the mobile and the base station [49]. We assume a frequency-selective Rayleigh fading channel with $P^u = P$ propagation paths with exponentially decreasing powers. The channel is

* The study of the CFOR performance is provided in [43].

† We select $\Delta f = 200$ Hz to show that even CFO residuals below the maximum value tolerated by 3G standards result in significant losses in performance.

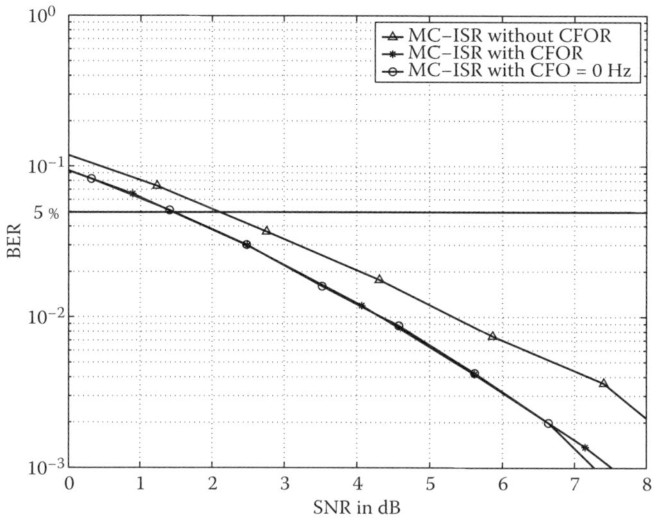

FIGURE 10.4 BER versus SNR for MT-CDMA MC-ISR, $L = 96$, $N_c = 7$, $C = 5$, with and without CFOR.

correlated across subcarriers and varying in time with Doppler shift f_D. We suppose a low Doppler situation $f_D = 8.8$ Hz unless otherwise mentioned. We consider that time delays vary linearly in time with a delay drift of 0.049 ppm. The receiver has $M = 2$ antennas. We implement closed-loop power control operating at 1,600 Hz and adjusting the power in steps of ±0.25 dB. An error rate on the power control bit of 5% and a feedback delay of 0.625 ms are simulated. The simulation parameters common to all multicarrier system configurations are listed in Table 10.2.

Table 10.3 shows the parameters specific to each multicarrier CDMA configuration. We choose as a reference the 3G DS-CDMA ($N_c = 1$) system with spreading factor $L = 32$ and a chip rate of 3.84 Mcps. We assume frequency-selective fading with $P = 3$ propagation paths. One of the features of MT-CDMA is that for a constant bandwidth the ratio between the spreading factor L and $2K = N_c - 1$ is constant. We hence maintain the same chip rate (3.840 Mcps) by changing the spreading factor and the number of subcarriers, as shown in Figure 10.3. We consider four MT-CDMA configurations. Since they use the same chip rate, there are three paths in each MT-CDMA subcarrier. For a fair comparison among different configurations of MC-DS-CDMA, the bandwidth should be the same. By reducing the chip rate, we varied the number of subcarriers while maintaining the orthogonality between them, as illustrated in Figure 10.3. Due to the reduction in bandwidth, each subcarrier in MC-DS-CDMA has either two paths (i.e., $P = 2$) or one path (i.e., $P = 1$, frequency-non-selective fading) for $N_c = 3$ and $N_c \geq 5$, respectively. The main performance criterion is the link-level SNR required per carrier to meet a BER of 5% in order to achieve a QoS of BER = 10^{-6} after channel decoding and the resulting system-level throughput. The user's data rate is calculated by summing the data rates over all subcarriers.

TABLE 10.2 Simulation Parameters

Parameter	Value	Comment
BW_{max}	5 MHz	Maximum bandwidth
M	2	Number of antennas
f_c	1.9 GHz	Central carrier frequency
f_D	8.8 Hz	Doppier frequency (5 kmph)
Δf	200 Hz	Frequency offset
f_{PC}	1,600 Hz	Frequency of PC updating
Δ_{PC}	±0.25 dB	Power control adjustment
PC_{min}^{max}	±30 dB	Power control stage
BER_{PC}	5%	Simulated PC bit error rate
$\delta\tau/\delta t$	0.049 ppm	Time-delay drift
$\Delta\tau$	4 chips	Delay spread
L_g	0	Guard interval length
N_s	2	Number of multistages
β	0.22	Roll-off factor
$2N_{src} + 1$	9	Number of pulse samples

10.4.2 Validation of the Performance Analysis

In this section, we investigate the accuracy of the analytical performance analysis in section 10.3.3 under realistic channel conditions. Indeed, we do not assume perfect channel identification; instead, we use the channel estimate provided by MC-STAR [32]. We validate the Gaussian approximation (GA) of the residual interference by comparison with simulation results. Since the SUB (Ω) is not known explicitly in this case, it has been obtained from extensive simulations. We consider two configurations: DBPSK MT-CDMA (L=64, N_c = 3) and DBPSK MC-DS-CDMA (L=32, N_c = 3). Figure 10.5 shows the link-level performances. It is seen, not surprisingly, that the GA is accurate in the presence of moderate background noise. The accuracy of GA increases at larger loads or at lower Doppler situations (speed of V = 5 kmph). Despite the realistic channel model employed and the channel estimate errors, there is a very good match between the analytical and simulation results for both MT-CDMA and MC-DS-CDMA in the target BER region (5%). This suggests that the analytical evaluation is accurate in a low Doppler situation.

10.4.3 Advantage of Full Interference Suppression

The imperfect frequency downconversion due to the instability of local oscillators combined with the multipath effect destructs the subcarriers' orthogonality and hence causes ICI. In this section we evaluate the advantage of full interference suppression on the link-level performance of MC-CDMA. We plot the link-level performances of MT CDMA (L = 64, N_c = 3, C = 8) and MC-DS-CDMA (L = 32, N_c = 3, C = 8) with MC-MRC (i.e., multicarrier receiver with MRC combining) and MC-ISR in Figure 10.6. It is clear that MC-ISR performs better than MC-MRC. Indeed, in low Doppler (speed of

TABLE 10.3 Parameters of Each Multicarrier System Configuration

Parameter	DS-CDMA	MT-CDMA					MC-DS-CDMA					Comment
λ	—	1					L					Subcarrier spacing parameter
N_c	1	3	5	7	9	11	3	5	7	9	11	Number of subcarriers
L	32	64	128	192	256	320	32	32	32	32	32	Spreading factor
R_c in Mcps	3.840	3.840	1.4549	0.8975	0.6488	0.5081	0.4175					Chip rate
P	3	3	2	1	1	1	1					Number of paths per subcarrier
R_s in kbaud	120	180	150	140	135	132	136.4	140.2	141.9	142.9	143.5	Symbol rate over all subcarriers
R_b for DBPSK in kbps	120	180	150	140	135	132	136.4	140.2	141.9	142.9	143.5	Peak rate of DBPSK
R_b for D8PSK in kbps	360	540	450	420	405	396	409.2	420.6	425.7	428.7	430.5	Peak rate for D8PSK
BW_{nor}	1	1.026	1									Bandwidth normalized vs. DS-CDMA

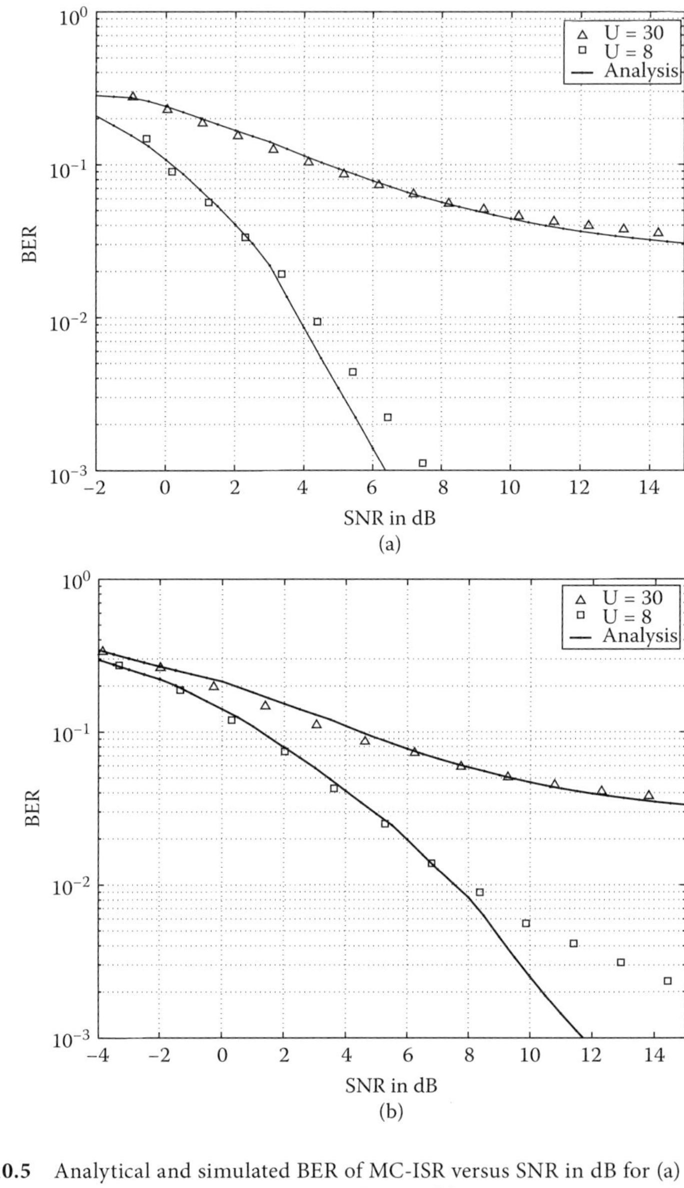

FIGURE 10.5 Analytical and simulated BER of MC-ISR versus SNR in dB for (a) MT-CDMA, $L = 64$, $N_c = 3$, DBPSK, and (b) MC-DS-CDMA $L = 32$, $N_c = 3$, DBPSK.

$V = 5$ kmph) we report 1.85 and 2 dB gains in SNR for MT-CDMA and MC-DS-CDMA, respectively. Note that the SNR gains are more important in high Doppler situations (speed of $V = 50$ kmph). At a bit error rate of 5%, MC-ISR performs 5.5 and 3 dB better than MC-MRC for MT-CDMA and MC-DS-CDMA, respectively.

In order to evaluate the specific impact of ICI on the link-level performance, we compare the BER curves of MT-CDMA ($L = 64$, $N_c = 3$, $C = 8$) MC-ISR with and without full

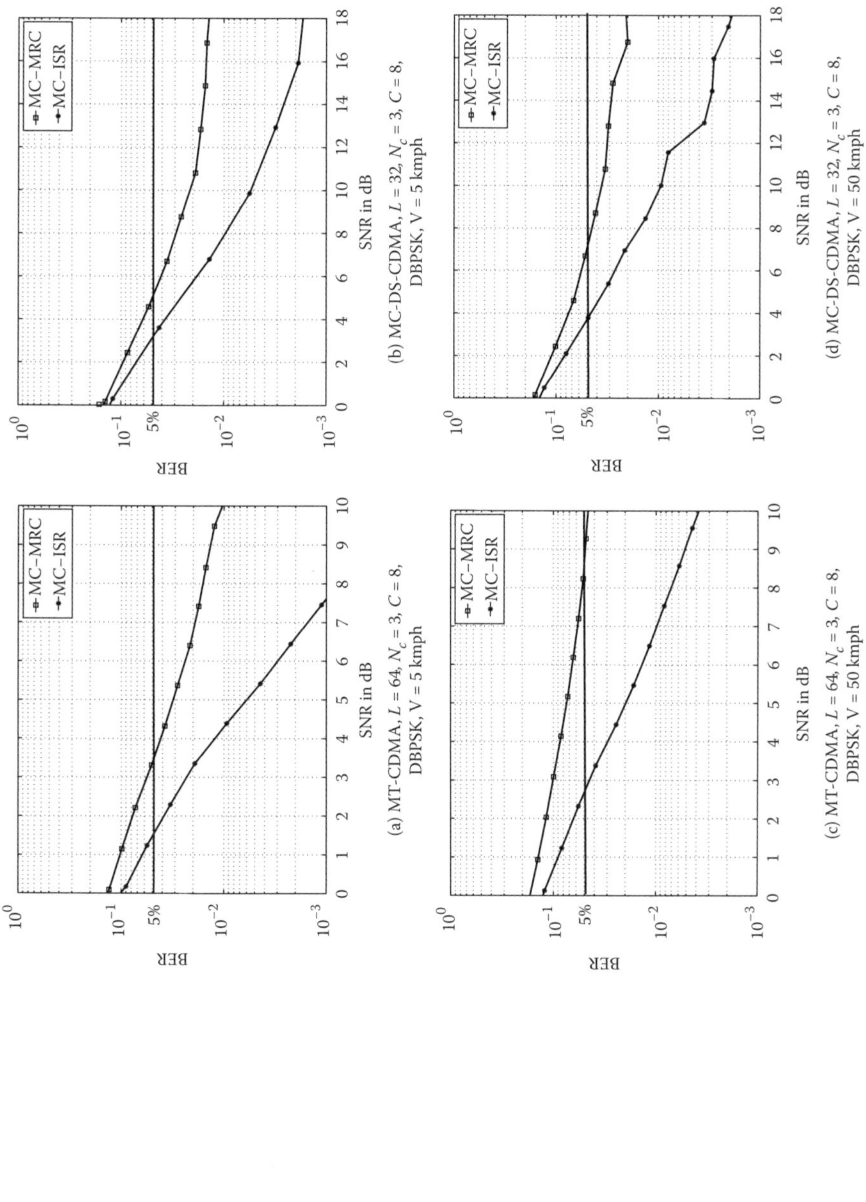

Figure 10.6 BER versus SNR in dB of MC-MRC and MC-ISR.

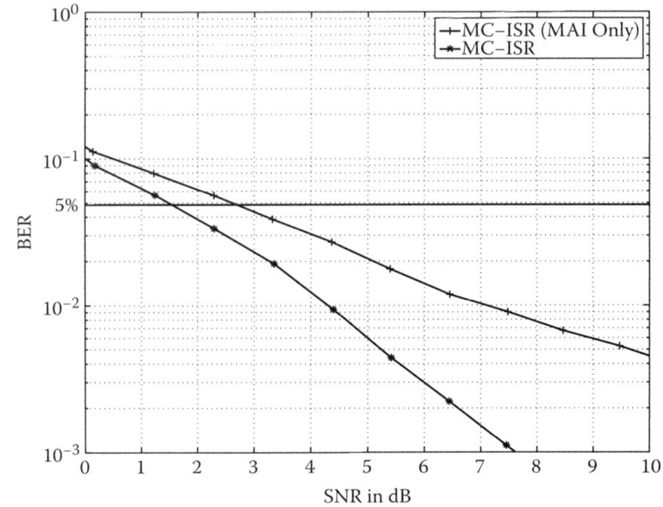

FIGURE 10.7 BER versus SNR in dB of MT-CDMA MC-ISR, $L = 64$, $N_c = 3$, $C = 8$, with and without full interference suppression (i.e., with and without ICI suppression).

interference suppression (i.e., with and without ICI suppression). Figure 10.7 shows that MC-ISR with full interference rejection is required to improve the system performance. Indeed, at a bit error rate of 5%, MC-ISR with full interference suppression performs 1.2 dB better than MC-ISR with MAI suppression only. In order to capture in more detail the gains achieved by ICI suppression in MC-CDMA, we proceed in Figure 10.8 to additional comparisons between the link-level BER performances of MC-ISR and MC-MRC in a single-user context (i.e., $C = 1$, no MAI, only ICI, and negligible ISI). Starting from the reference situation of Figure 10.8a with $L = 64$, $N_c = 7$, and DBSPK where the reported SNR gain due to ICI suppression is about 0.5 dB, the results suggest that ICI suppression is even more advantageous at higher-rate transmissions, and increasingly so when we move to the scenarios of Figures 10.8b–d; i.e., when we increase the number of carriers to $N_c = 11$ (SNR gain is about 1 dB), reduce the processing gain to $L = 32$ (SNR gain is about 3 dB), or increase the modulation order to D8PSK (SNR gain far exceeds 5 dB if not infinite), respectively. These results further confirm the benefits of ICI rejection in a full interference suppression scheme using MC-ISR.*

10.4.4 MT-CDMA, MC-DS-CDMA, and DS-CDMA Performance Comparison

This section is dedicated to the performance comparison of the proposed MC-ISR receiver with two potential next-generation multicarrier CDMA–air interface configurations: MT-CDMA and MC-DS-CDMA. Single-carrier ISR [42] for 3G DS-CDMA–air

* Simulation results performed in the framework of this contribution and reported in [50] show that the ICI rejection is even more beneficial in the case of one receiving antenna ($M = 1$).

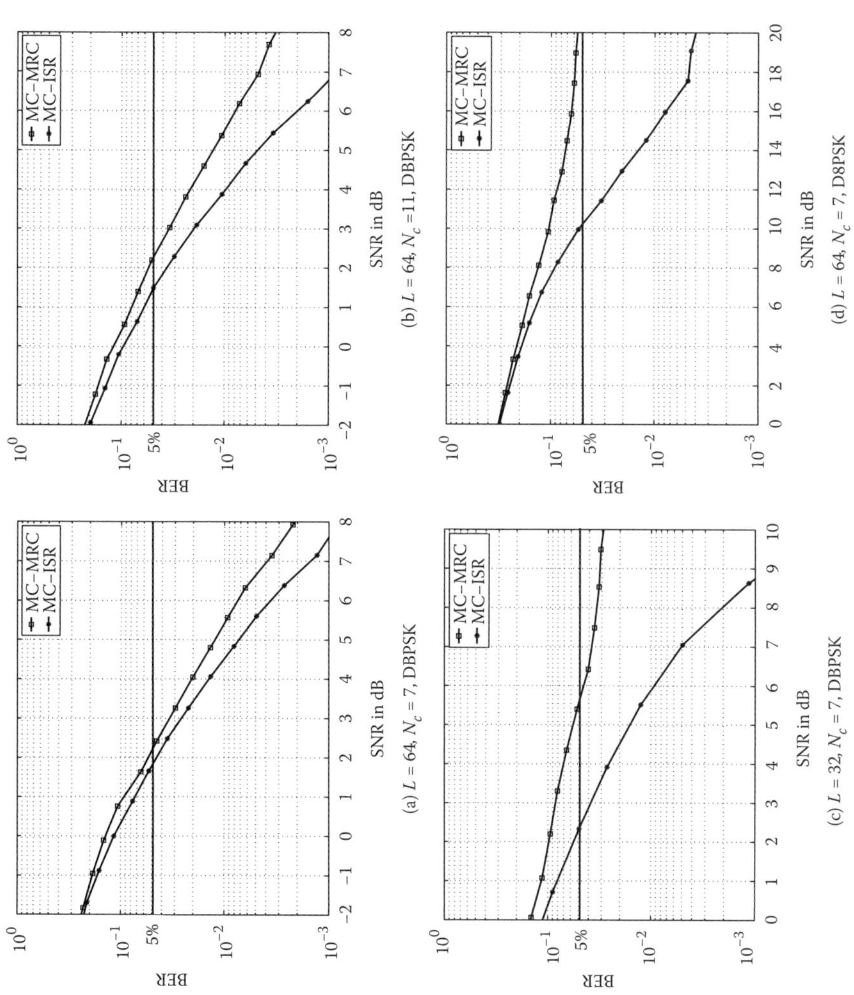

(a) $L = 64, N_c = 7$, DBPSK

(b) $L = 64, N_c = 11$, DBPSK

(c) $L = 32, N_c = 7$, DBPSK

(d) $L = 64, N_c = 7$, D8PSK

FIGURE 10.8 BER versus SNR in dB of MC-ISR and MC-MRC with single-user MT-CDMA.

interface is also considered as a reference. In addition, in order to provide a more detailed picture of the aggregate gain of the proposed MC-ISR receiver, we also compared its performance with that of MC-MRC over the same two multicarrier CDMA–air interface configurations, as well as with that of single-carrier MRC over current 3G DS-CDMA. First, we derive the SNR_{req} from link-level simulations. Then, we translate the link-level results into system-level results using the procedure in Table 10.1. In Table 10.4, we provide the required SNR and the total throughput of DBPSK and D8PSK modulated data for DS-CDMA, MT-CDMA, and MC-DS-CDMA. For DBPSK modulation, we observe that we can improve the system performance by increasing the number of subcarriers. Indeed, the total throughput continues to increase despite the increase in the number of carriers. But a gain saturation is encountered as the number of subcarriers increases. Note, however, that the throughput increase is more important with MC-ISR due to ICI suppression. Table 10.4 also shows that MT-CDMA outperforms MC-DS-CDMA with DBPSK modulation because it uses longer spreading sequences and exploits the subcarrier correlation. Moreover, due to the reduced subcarrier bandwidth, MC-DS-CDMA has less frequency diversity, while MT-CDMA is better able to exploit path diversity, and hence achieves better performance. Note also that MC-DS-CDMA is more robust against ICI, but in applying MC-ISR, this advantage over MT-CDMA becomes obsolete and the performance gap between MT-CDMA and MC-DS-CDMA increases.

Next, we compare different configurations with D8PSK modulation. We notice a link-level deterioration for MT-CDMA as the number of subcarriers increases. Indeed, higher-order modulation is more sensitive to the residual ICI. MC-DS-CDMA is much less affected by this phenomenon because it is much more robust to ICI thanks to the higher subcarrier spacing. Therefore, with high-order modulation, MC-DS-CDMA outperforms MT-CDMA when the number of subcarriers is high enough. We notice also that with the MC-MRC combiner, D8PSK MC-DS-CDMA outperforms D8PSK MT-CDMA even with a small number of subcarriers. It is clear, however, that D8PSK is less efficient than DBPSK modulation for all air interface configurations. In Table 10.4 we highlight the most spectrum-efficient MC-ISR–air interface configuration for each modulation. For both modulations MT-CDMA has the best link-level performance and the highest throughput (for a tested number of carriers less than or equal to 11). MT-CDMA with nine subcarriers and DBPSK modulation outperforms all other configurations and provides a throughput about 115% higher than that achievable with single-carrier MRC over a 3G DS-CDMA–air interface. The net benefits due to the proposed MC-ISR combiner and to the potential migration to a next-generation MT-CDMA–air interface are about 80 and 15%, respectively.

10.5 Conclusions

In this chapter we propose an adaptive multicarrier CDMA space-time receiver with full interference suppression capabilities named MC-ISR. First, we derived a complete model of the interference that takes into account MAI, ISI, and ICI in a multipath fading

TABLE 10.4 Required SNR and Maximum Throughput of DS-CDMA, MT-CDMA, and MC-CDMA for DBPSK and D8PSK (best performance values for each modulation are in bold)

MC-STAR Configuration	DS-CDMA	MT-CDMA					MC-DS-CDMA				
N_c	1	3	5	7	9	11	3	5	7	9	11
Modulation	DBPSK										
SNR_{req} in dB with MC-MRC	0.76	0.74	0.76	0.59	0.57	0.69	3.7	2.8	2.7	2.55	3
τ_{max} in kbps with MC-MRC	2,040	2,160	2,100	2,240	2,295	2,244	1091.2	1402	1419	1,571.9	1,435
SNR_{req} in dB with MC-ISR	0.76	0.74	0.75	0.55	0.5	0.62	3.62	2.8	2.6	2.48	3
τ_{max} in kbps with MC-ISR	3,600	3,960	4050	4,200	4,320	4,224	2,182.4	2,523.6	2,554.2	2,572.2	2,439.5
Modulation	DSPSK										
SNR_{req} in dB with MC-MRC	8.57	8	9.28	9	10	11	11	10.25	10.34	10.05	10.8
τ_{max} in kbps with MC-MRC	1,080	1,080	900	840	810	792	818.4	841.2	851.4	857.4	361
SNR_{req} in dB with MC-ISR	8.57	7.86	8.8	8.9	9.4	10.5	10.94	10.25	10.2	10.02	10.8
τ_{max} in kbps with MC-ISR	1,800	2,160	1,350	1,260	1,215	792	1,227.6	1,261.8	1,277.1	1,286.1	861

channel with timing and frequency mismatch. Based on this model, we proposed a new adaptive multicarrier interference subspace rejection receiver. We incorporated the least complex and more practical ISR interference rejection mode to simultaneously suppress MAI, ISI, and ICI at the signal combining step. We also proposed a realistic implementation of the new MC-ISR receiver that includes an efficient strategy for carrier offset recovery in a multicarrier and multiuser detection scheme. MC-ISR supports both the MT-CDMA- and MC-DS-CDMA–air interfaces. Furthermore, the assessment of the new MC-ISR receiver was oriented toward an implementation in a future, real-world wireless system. Indeed, we analyzed the performance of MC-ISR in an unknown time-varying Rayleigh channel with multipath, carrier offset, and cross-correlation between subcarrier channels and took into account all channel estimation errors. As another contribution in this work, we derived a link/system-level performance analysis of MC-ISR based on the Gaussian assumption (GA) and validated it by simulations. Under realistic propagation conditions and in the presence of channel estimation errors, simulation results validated the performance analysis and confirmed the net advantage of the full interference suppression capabilities of MC-ISR. With two receiving antennas and nine MT-CDMA subcarriers in 5 MHz bandwidth, MC-ISR provides about 4,320 kbps at low mobility for DBPSK, i.e., an increase of 115% in throughput over current 3G DS-CDMA with MRC.

Appendix

Derivation of the Interference Variance after MC-ISR Combining

Our goal is to estimate the variances:

$$\begin{aligned}\mathrm{Var}\left[\delta^{d}_{MAI,k,n}+\delta^{d}_{ICI,k,n}+\delta^{d}_{ISI,k,n}\right]=\mathrm{Var}\Bigg[&\sum_{\substack{u=1\\u\neq d}}^{C}\sum_{k'=-K}^{K}\sum_{n'=n-1}^{n+1}\xi^{u}_{k',n'}\lambda^{u}_{k',n}\underline{W}^{d^H}_{k,n}\hat{\underline{Y}}^{u}_{n',k',n}\\&+\sum_{\substack{k'=-K\\k'\neq k}}^{K}\sum_{n'=n-1}^{n+1}\xi^{d}_{k',n'}\lambda^{d}_{k',n}\underline{W}^{d^H}_{k,n}\hat{\underline{Y}}^{d}_{n',k',n}\\&+\sum_{\substack{n'=n-1\\n'\neq n}}^{n+1}\xi^{d}_{k,n'}\lambda^{d}_{k,n}\underline{W}^{d^H}_{k,n}\hat{\underline{Y}}^{d}_{n',k,n}\Bigg].\end{aligned}\tag{10.53}$$

Let us consider the general problem of deriving the variance of the sum of random complex variables. We first introduce the variables x_α, $\alpha\in\{1,\ldots,N_t\}$ and ξ_α, $\alpha\in\{1,\ldots,N_t\}$, with the following properties: $\mathrm{E}[\xi_\alpha\xi^*_{\alpha'}] = M_\xi$, $\forall\alpha\neq\alpha'$, $\mathrm{E}[\xi_\alpha\xi^*_\alpha] = V_\xi$, $\mathrm{E}[x_\alpha] = 0$, and $\mathrm{Var}[\sum_{\alpha=1}^{N_t} x_\alpha] = 0$. Then we assume that ξ_α and x_α are independent. Thus, we derive the variance as follows:

$$\begin{aligned}\mathrm{Var}\left[\sum_{\alpha=1}^{N_t}\xi_\alpha x_\alpha\right]&=\sum_{\alpha=1}^{N_t}\mathrm{Var}\left[\xi_\alpha x_\alpha\right]+\sum_{\alpha=1}^{N_t}\sum_{\alpha'=1\alpha'\neq\alpha}^{N_t}\mathrm{E}\left[\xi_\alpha\xi_{\alpha'}x_\alpha x_{\alpha'}\right]\\&=\sum_{\alpha=1}^{N_t}V_\xi\mathrm{Var}\left[x_\alpha\right]+\sum_{\alpha=1}^{N_t}\sum_{\alpha'=1\alpha'\neq\alpha}^{N_t}\mathrm{E}\left[\xi_\alpha\xi_{\alpha'}\right]\mathrm{E}\left[x_\alpha x_{\alpha'}\right]\\&=\sum_{\alpha=1}^{N_t}V_\xi\mathrm{Var}\left[x_\alpha\right]+\sum_{\alpha=1}^{N_t}\sum_{\alpha'=1\alpha'\neq\alpha}^{N_t}\rho_\xi\mathrm{E}\left[x_\alpha x_{\alpha'}\right].\end{aligned}\quad(10.54)$$

From $\mathrm{Var}[\sum_{\alpha=1}^{N_t} x_\alpha] = 0$ we have

$$\begin{aligned}&\mathrm{Var}\left[\sum_{\alpha=1}^{N_t}x_\alpha\right]=\sum_{\alpha=1}^{N_t}\mathrm{Var}\left[x_\alpha\right]+\sum_{\alpha=1}^{N_t}\sum_{\alpha'=1\alpha'\neq\alpha}^{N_t}\mathrm{E}\left[x_\alpha x_{\alpha'}\right]=0\\&\Rightarrow\sum_{\alpha=1}^{N_t}\sum_{\alpha'=1\alpha'\neq\alpha}^{N_t}\mathrm{E}\left[x_\alpha x_{\alpha'}\right]=-\sum_{\alpha=1}^{N_t}\mathrm{Var}\left[x_\alpha\right].\end{aligned}\quad(10.55)$$

Then, by substituting equation (10.55) in equation (10.54) we obtain

$$\mathrm{Var}\left[\sum_{\alpha=1}^{N_t}\xi_\alpha x_\alpha\right]=\left(V_\xi-\rho_\xi\right)\sum_{\alpha=1}^{N_t}\mathrm{Var}\left[x_\alpha\right].\quad(10.56)$$

Now we apply the same procedure to derive the variance of $\delta^d_{MAI,k,n} + \delta^d_{ICI,k,n} + \delta^d_{ISI,k,n}$. We substitute ξ_α with $\xi^u_{k',n'}\lambda^u_{k',n}$ and x_α with $\underline{W}^{d^H}_{k,n}\underline{\hat{Y}}^u_{n',k'n}$. The MC-ISR combiner $\underline{W}^d_{k,n}$ satisfies the optimization property in equation (10.25); thus,

$$\underline{W}^{d^H}_{k,n}\underline{\hat{I}}^d_{k,n}=0\;\Rightarrow\;\mathrm{Var}\left[\underline{W}^{d^H}_{k,n}\left(\underline{\hat{I}}^d_{\mathrm{MAI},,k,n}+\underline{\hat{I}}^d_{\mathrm{ICI},,k,n}+\underline{\hat{I}}^d_{\mathrm{ISI},,k,n}\right)\right]=0.\quad(10.57)$$

Then,

$$\begin{aligned}\mathrm{Var}\left[\delta^d_{MAI,k,n}+\delta^d_{ICI,k,n}+\delta^d_{ISI,k,n}\right]&=\left(V_\xi-\rho_\xi\right)\sum_{\substack{u=1\\u\neq d}}^{C}\sum_{k'=-K}^{K}\sum_{n'=n-1}^{n+1}\mathrm{Var}\left[\underline{W}^{d^H}_{k,n}\underline{\hat{Y}}^u_{n',k',n}\right]\\&+\left(V_\xi-\rho_\xi\right)\sum_{\substack{k'=-K\\k'\neq k}}^{K}\sum_{n'=n-1}^{n+1}\mathrm{Var}\left[\underline{W}^{d^H}_{k,n}\underline{\hat{Y}}^d_{n',k',n}\right]\\&+\left(V_\xi-\rho_\xi\right)\sum_{\substack{n'=n-1\\n'\neq n}}^{n+1}\mathrm{Var}\left[\underline{W}^{d^H}_{k,n}\underline{\hat{Y}}^d_{n',k,n}\right].\end{aligned}\quad(10.58)$$

We consider $E[\|\underline{W}_{k,n}^d\|^2] = \bar{\kappa}$, which is a measure of the enhancement of the white noise compared to the MRC combiner [45]. We also assume that the combiners $\underline{W}_{k,n}^d$ and $\hat{\underline{Y}}_{n',k'n}^u$ Thus, we derive the variance of the residual interference as follows:

$$\begin{aligned}\mathrm{Var}\left[\delta_{MAI,k,n}^d+\delta_{ICI,k,n}^d+\delta_{ISI,k,n}^d\right]&=\left(V_\xi-\rho_\xi\right)\bar{\kappa}\sum_{\substack{u=1\\u\neq d}}^{C}\sum_{k'=-K}^{K}\sum_{n'=n-1}^{n+1}\mathrm{Var}\left[\hat{\underline{Y}}_{n',k',n}^u\right]\\&+\left(V_\xi-\rho_\xi\right)\bar{\kappa}\sum_{\substack{k'=-K\\k'\neq k}}^{K}\sum_{n'=n-1}^{n+1}\mathrm{Var}\left[\hat{\underline{Y}}_{n',k',n}^d\right]\\&+\left(V_\xi-\rho_\xi\right)\bar{\kappa}\sum_{\substack{n'=n-1\\n'\neq n}}^{n+1}\mathrm{Var}\left[\hat{\underline{Y}}_{n',k,n}^d\right],\\&=\left(V_\xi-\rho_\xi\right)\bar{\kappa}\mathrm{Var}\left[\hat{\underline{I}}_{k,n}^d\right].\end{aligned}\tag{10.59}$$

In the developments of equation (10.59), we exploited the fact that we transmit different data sequences over distinct subcarriers for a given user, and hence assumed that the cross-correlation terms from different subcarriers are zero.

In the following, we will derive the values of V_ξ and ρ_ξ under the following three assumptions: (1) the error indicating variables $\xi_{k',n'}^u$ and $\lambda_{k',n}^u$ are independent; (2) all the random sequence variables $(\xi_{k',n'}^u)$ and $(\lambda_{k',n}^u)$ are independent and identically distributed; and (3) $E[\lambda_{k',n}^u]$. Given these assumptions we derive V_ξ as follows:

$$\begin{aligned}V_\xi&=E\left[\xi_{k',n'}^u\lambda_{k',n}^u\xi_{k',n'}^{u*}\lambda_{k',n}^{u*}\right]=E\left[\xi_{k',n'}^{u*}\xi_{k',n'}^{u*}\right]E\left[\lambda_{k',n}^u\lambda_{k',n}^{u*}\right]\\&=E\left[\lambda_{k',n}^u\lambda_{k',n}^{u*}\right]=\left[1+\rho_\lambda\right].\end{aligned}\tag{10.60}$$

In order to evaluate V_ξ, we exploit the expression of the variance of the power control error in [51]. Hence, ρ_λ varies with the maximum Doppler frequency f_D (equation (10.51) in [51]), yielding:

$$\rho_\lambda=\frac{4\pi^2\left(F_D\times\tau_{PC}\right)^2}{P-1},\tag{10.61}$$

where τ_{PC} is the power control feedback delay. Below we derive the expectation

$$\rho_\xi=E\left[\xi_{k',n'}^u\lambda_{k',n}^u\xi_{k',n'}^{u'*}\lambda_{k',n}^{u'*}\right]=E\left[\xi_{k',n'}^u\right]E\left[\xi_{k',n'}^{u'*}\right]E\left[\lambda_{k',n}^u\right]E\left[\lambda_{k',n}^{u'*}\right]=E\left[\xi_{k',n'}^u\right]^2.$$

If $S_{rec}\ll 1$, the value of ρ_ξ can be derived as follows [45]:

$$\rho_\xi\approx\left(1-\left(1-\cos\left(2\pi/\mathcal{M}_i\right)\right)S_{rec}\right)^2.\tag{10.62}$$

The interference $\underline{I}_{k,n}^{d}$ is approximated as a Gaussian distributed random variable with zero mean. Only its variance needs to be evaluated to derive the variance of the residual interference in equation (10.59).

Derivation of the Interference Variance for Band-Limited MC-CDMA

The chip waveform has been noted to be an important system parameter for DS-CDMA and MC-DS-CDMA. Hence, the performances of DS-CDMA and MC-DS-CDMA with various time-limited and band-limited chip waveforms have been investigated. However, for all the MT-CDMA systems found in the literature, a time-limited waveform is generally employed [4, 18, 52, 53]. Since we consider a practical square-root raised cosine pulse, the focus of this appendix is to derive the variance of the interference of MC-CDMA (including MC-DS-CDMA and MT-CDMA) with a band-limited square-root raised cosine waveform. Let $G(f)$ be the Fourier transform of the raised cosine filter:

$$G(f)=\begin{cases} T_c, & 0\le |f| \le \dfrac{1-\beta}{2T_c} \\ \dfrac{T_c}{2}\left\{1+\cos\left[\dfrac{\pi T_c}{\beta}\left(|f|-\dfrac{1-\beta}{2T_c}\right)\right]\right\}, & \dfrac{1-\beta}{2T_c}\le |f| \le \dfrac{1+\beta}{2T_c}. \\ 0, & |f| > \dfrac{1+\beta}{2T_c} \end{cases} \tag{10.63}$$

Let $\bar{\psi}_D^2 = E[(\psi_k^d)^2]$ be the average power of the kth carrier of the desired user and $\bar{\psi}_I^2$ be the average power on each interfering carrier (assumed equal for all u and all k). Using the general results in [54], one has

$$\text{Var}\left[\underline{I}_{MAI,k,n}^{d}\right]=(C-1)\frac{\bar{\psi}_I^2}{L}\left[\varsigma(\beta)+\chi_k(\beta)\right], \tag{10.64}$$

where

$$\varsigma(\beta)=\frac{1}{T_c}\int_{-\infty}^{\infty} G^2(f)\,df \tag{10.65}$$

and

$$\chi_k(\beta)=\sum_{\substack{k'=-K \\ k'\neq k}}^{K}\frac{1}{T_c}\int_{-\infty}^{\infty} G(f)G\big(f-(f_k-f_{k'})\big)\,df\,. \tag{10.66}$$

It is easy to obtain $\varsigma(\beta) = 1 - \beta/4$). To obtain $\chi_k(\beta)$, we need to consider the MC-DS-CDMA and MT-CDMA systems separately. After mathematical evaluations of the integral, we obtain:

$$\chi_k(\beta) = \begin{cases} \dfrac{\beta}{8}, & \text{if } k = -K \text{ or } K, \\[2ex] \dfrac{\beta}{4}, & \text{if } k = -K+1, \ldots, K-1, \end{cases} \tag{10.67}$$

for MC-DS-CDMA ($\lambda = L$ and $f_k = k/T_c$) and

$$\chi_k(\beta) = \sum_{\substack{k'=-K \\ k' \neq k}}^{K} \vartheta\left(\left|k - k'\right|\right), \tag{10.68}$$

where

$$\vartheta(x) = \begin{cases} 1 - \dfrac{\beta}{2} - \dfrac{x}{2L} + \dfrac{3\beta}{4\pi}\sin\left(\dfrac{\pi x}{\beta L}\right) + \left(\dfrac{\beta}{4} - \dfrac{x}{4L}\right)\cos\left(\dfrac{\pi x}{\beta L}\right), & \text{if } 0 \le x/L \le \min(\beta, 1-\beta) \\[2ex] 1 - \dfrac{x}{L} & \text{if } \beta \le x/L \le 1-\beta \text{ and } \beta < 0.5 \\[2ex] \dfrac{3}{4} - \dfrac{\beta}{4} - \dfrac{x}{4L} + \dfrac{3\beta}{4\pi}\sin\left(\dfrac{\pi x}{\beta L}\right) + \left(\dfrac{\beta}{4} - \dfrac{x}{4L}\right)\cos\left(\dfrac{\pi x}{\beta L}\right) + & \\ \quad \dfrac{3\beta}{8\pi}\sin\left(\dfrac{\pi x}{\beta L} - \dfrac{\pi}{\beta}\right) - \left(\dfrac{x}{8L} - \dfrac{1-\beta}{8}\right)\cos\left(\dfrac{\pi x}{\beta L} - \dfrac{\pi}{\beta}\right) & \text{if } 1-\beta \le x/L \le \beta \text{ and } \beta > 0.5 \\[2ex] \dfrac{3}{4} + \dfrac{\beta}{4} - \dfrac{3x}{4L} + \dfrac{3\beta}{8L}\sin\left(\dfrac{\pi x}{\beta L} - \dfrac{\pi}{\beta}\right) - \left(\dfrac{x}{8L} - \dfrac{1-\beta}{8}\right)\cos\left(\dfrac{\pi x}{\beta L} - \dfrac{\pi}{\beta}\right) & \text{if } \max(\beta, 1-\beta) \le x/L \le 1 \end{cases} \tag{10.69}$$

for MT-CDMA ($\lambda = L$ and $f_k = k/T_{Mc}$). The variances of the residual ICI and ISI interferences from the same user can be written as

$$\begin{aligned} \text{Var}\left[\underline{I}_{ICI,k,n}^{d}\right] &= \frac{\bar{\Psi}_D^2}{L}\delta_{is}\chi_k(\beta), \\ \text{Var}\left[\underline{I}_{ISI,k,n}^{d}\right] &= \frac{\bar{\Psi}_D^2}{L}\delta_{is}\varsigma(\beta), \end{aligned} \tag{10.70}$$

where $\delta_{is} = (P-1)/P$ is a measure of the relative impact of the interference generated by the other paths on a given path of the desired user.

References

[1] L. Hanzo, T. Killer, M. S. Munster, and B. J. Choi. 2003. *OFDM and MC-CDMA for broadband multiuser communications, WLANs and broadcasting.* New York: John Wiley & Sons.

[2] S. Hata and R. Prasad. 1997. Overview of multicarrier CDMA. *IEEE Commun. Mag.* 35:126–33.

[3] L. Hanzo, L.-L. Yang, E.-L. Kuan, and K. Yen. 2003. *Single- and multi-carrier CDMA multi-user detection, space-time spreading, synchronisation and standards.* New York: John Wiley & Sons.

[4] Y. Lie-Liang and L. Hanzo. 2002. Performance of generalized multicarrier DS-CDMA over Nakagami-m fading channels. *IEEE Trans. Commun.* 50:956–66.

[5] H. Steendam and M. Moeneclaey. 2001. The effect of carrier frequency offsets on downlink and uplink MC-DS-CDMA. *IEEE J. Selected Areas Commun.* 19:2528–36.

[6] B. Smida, S. Affes, K. Jamaoui, and P. Mermelstein. 2008. A multicarrier-CDMA space-time receiver with full interference suppression capabilities. *IEEE Trans. Veh. Technol.* 57(1):363–379.

[7] N. Yee, J. P. Linnarz, and G. Fettweis. 1993. Multi-carrier CDMA in indoor wireless radio networks. In *Proceedings of the IEEE International Symposium on Personal, Indoor and Mobile Radio Communications (PIMRC 1993)*, pp. 109–13.

[8] A. Chouly, A. Brajal, and S. Jourdan. 1993. Orthogonal multicarrier techniques applied to direct sequence spread spectrum CDMA systems. In *Proceedings of the IEEE Global Telecommunications Conference (GLOBECOM 1993)*, vol. 3, pp. 1723–28.

[9] K. Fazel and L. Papke. 1993. On the performance of convolutionally-coded CDMA/OFDM for mobile communication system. In *Proceedings of the IEEE Personal, Indoor and Mobile Radio Communications Symposium (PIMRC 1993)*, pp. 468–72.

[10] A. Chouly, A. Brajal, and S. Jourdan. 1993. Orthogonal multicarrier technique applied to direct sequence spread spectrum CDMA systems. In *Proceedings of the IEEE GLOBECOM '93*, pp. 1723–28.

[11] V. DaSilva and E. S. Sousa. 1993. Performance of orthogonal CDMA codes for quasi-synchronous communication systems. In *Proceedings of the IEEE International Conference on Universal Personal Communications (ICUPC 1993)*, vol. 2, pp. 995–99.

[12] S. Kondo and L. B. Milstein. 1993. On the use of multicarrier direct sequence spread spectrum systems. In *Proceedings of the IEEE Military Communinations Conference (MILCOM 1993)*, vol. 1, pp. 52–56.

[13] E. A. Sourour and M. Nakagawa. 1996. Performance of orthogonal multicarrier CDMA in a multipath fading channel. *IEEE Trans. Commun.* 44:356–67.

[14] X. Gui and T. S. Ng. 1999. Performance of asynchronous orthogonal multicarrier CDMA system in frequency selective fading channel. *IEEE Trans. Commun.* 47:1084–91.

[15] L. L. Yang and L. Hanzo. 2002. Broadband MC DS-CDMA using space-time and frequency-domain spreading. In *Proceedings of the IEEE VTC'2002*, pp. 1632–36.

[16] Z. Luo, J. Liu, M. Zhao, Y. Liu, and J. Gao. 2006. Double-orthogonal coded space-time-frequency spreading CDMA scheme. *IEEE J. Selected Areas Commun.* 24:1244–55.

[17] L. Vandendorpe. 1993. Multitone direct sequence CDMA system in an indoor wireless environment. In *Proceedings of the IEEE First Symposium on Communications and Vehicular Technology in the Beneluz*, pp. 4.1.1–4.1.8.

[18] L. Vandendorpe. 1995. Multitone spread spectrum multiple access communications system in a multipath Rician fading channel. *IEEE Trans. Veh. Technol.* 44:327–37.

[19] S. Kaiser. 1995. OFDM-CDMA versus DS-CDMA: Performance evaluation for fading channels. In *Proceedings of the IEEE International Conference on Communications (ICC 1995)*, vol. 3, pp. 1722–26.

[20] J.-Y. Oh and M.-S. Lim. 1999. The bandwidth efficiency increasing method of multi-carrier CDMA and its performance evaluation in comparison with DS-CDMA with RAKE receivers. In *Proceedings of the IEEE Vehicular Technology Conference (VTC 1999)*, vol. 1, pp. 561–65.

[21] S. B. Slimane. 1999. Bandwidth efficiency of MC-CDMA signals. *IEEE Electronics Lett.* 35:1797–98.

[22] X. Zhang, T.-S. Ng, and J. Wang. 1999. Capacity comparison of single-tone and multitone CDMA systems. In *Proceedings of the IEEE Vehicular Technology Conference (VTC 1999)*, vol. 1, pp. 243–47.

[23] S. Hata and R. Prasad. 1996. An overview of multi-carrier CDMA. In *Proceedings of the IEEE International Symposium on Spread Spectrum Techniques and Applications*, vol. 1, pp. 107–14.

[24] L. L. Yang and L. Hanzo. 2003. Multicarrier DS-CDMA: A multiple access scheme for ubiquitous broadband wireless communications. *IEEE Commun. Mag.* 41:116–24.

[25] S. Affes and P. Mermelstein. 2003. Adaptive space-time processing for wireless CDMA. In *Adaptive signal processing: Application to real-world problems*, ed. J. Benesty and A. H. Huang. Berlin: Springer, chapter 10.

[26] 3GPP. 1997. *Universal Mobile Telecommunications System (UMTS); Selection procedures for the choice of radio transmission technologies of the UMTS*. 3GPP TR 101 112 V3.1.0.

[27] S. Affes, D. Feng, L. Ge, and P. Mermelstein. 2002. Does direction-of-arrival estimation help channel identification in multi-antenna CDMA receivers? In *IEEE ISWC'2002*, pp. 167–68.

[28] W. C. Jakes, ed. 1974. *Microwave mobile communications*. New York: Wiley.

[29] B. Natarajan, C. R. Nassar, and V. Chandrasekhar. 2000. Generation of correlated Rayleigh fading envelopes for spread spectrum applications. *IEEE Commun. Lett.* 4:9–11.

[30] R. B. Ertel and J. H. Reed. 1998. Generation of two equal power correlated Rayleigh fading envelopes. *IEEE Commun. Lett.* 2:276–78.

[31] S. Affes and P. Mermelstein. 1998. A new receiver structure for asynchronous CDMA: STAR the spatio-temporal array-receiver. *IEEE J. Selected Areas Commun.* 16:1411–22.

[32] B. Smida, S. Affes, J. Li, and P. Mermelstein. 2005. Multicarrier-CDMA STAR with time and frequency synchronization. In *IEEE ICC 2005*, vol. 4, pp. 2493–99.

[33] S. Verdu. 1986. Minimum probability of error for asynchronous Gaussian multiple access channels. *IEEE Trans. Information Theory* 32:85–96.

[34] X. Weiping and L. B. Milstein. 1998. MMSE interference suppression for multi-carrier DS-CDMA in frequency selective channels. In *IEEE GLOBECOM 1998*, vol. 1, pp. 259–64.

[35] H. Lie-Liang, H. Wei, and L. Hanzo. 2003. Multiuser detection in multicarrier CDMA systems employing both time-domain and frequency-domain spreading. In *IEEE PIMRC 2003*, vol. 2, pp. 1840–44.

[36] F. Lin and L. B. Milstein. 2000. Successive interference cancellation in multicarrier DS/CDMA. *IEEE Trans. Commun.* 48:1530–40.

[37] W. Huahui, K. W. Ang, K. Yen, and Y. H. Chew. 2003. An adaptive PIC receiver with diversity combining for MC-DS-CDMA system. In *IEEE VTC 2003*, vol. 2, pp. 1055–59.

[38] W. Huahui, K. W. Ang, K. Yen, and Y. H. Chew. 2000. Performance analysis of an adaptive PIC receiver for asynchronous multicarrier DS-CDMA system. In *IEEE PIMRC 2003*, vol. 2, pp. 1835–39.

[39] W. Nabhane and H. V. Poor. Blind joint equalization and multiuser detection in dispersive MCCDMA/MC-DS-CDMA/MT-CDMA channels. In *Proceedings of the 2002 IEEE Military Communications Conference*, Anaheim, CA, vol. 2, pp. 814–819.

[40] J. Namgoong, T. F. Wong, and J. S. Lehnert. 2002. Subspace multiuser detection for multicarrier DS-CDMA. *IEEE Trans. Commun.* 48:1892–908.

[41] M. Peng, Y. J. Guo, and S. K. Barton. 2001. Multiuser detection of asynchronous CDMA with frequency offset. *IEEE Trans. Commun.* 49:952–60.

[42] S. Affes, H. Hansen, and P. Mermelstein. 2002. Interference subspace rejection: A framework for multiuser detection in wideband CDMA. *IEEE J. Selected Areas Commun.* 20:287–302.

[43] B. Smida, S. Affes, J. Li, and P. Mermelstein. 2007. A spectrum-efficient multi-carrier CDMA array-receiver with diversity-based enhanced time and frequency synchronization. *IEEE Trans. Wireless Commun.* 6:2315–27.

[44] S. Jomphe, K. Cheikhrouhou, J. Belzile, S. Affes, and J.-C. Thibault. 2005. Area-efficient advanced multiuser WCDMA receiver. In *IEEE Northeast Workshop on Circuits and Systems NEWCAS'05*, pp. 296–99.

[45] H. Hansen, S. Affes, and P. Mermelstein. 2005. *Mathematical derivation of interference subspace rejection*. Technical Report INRS-EMT, EMT-014-0105.

[46] B. Smida and S. Affes. 2005. On the performance of interference subspace rejection for next generation multicarrier CDMA. In *IEEE ISSPA'2005*.

[47] T. S. Rappaport. 1999. *Wireless commmunications: Principles and practice*. Upper Saddle River, NJ: Prentice Hall PTR, pp. 417–435.

[48] B. Smida, S. Affes, and P. Mermelstein. 2003. Joint time-delay and frequency offset synchronization for CDMA array-receivers. In *IEEE SPAWC 2003*, pp. 499–504.

[49] 3rd Generation Partnership Project (3GPP). 2002. *Technical specification group (TSG), radio access network (RAN), UE radio transmission and reception (FDD).* 3GPP TS 25.101 V5.4.0.

[50] B. Smida, S. Affes, K. Jamaoui, and P. Mermelstein. 2005. A multicarrier-CDMA receiver with full interference suppression and carrier frequency offset recovery. In *IEEE SPAWC'05*, pp. 435–439.

[51] A. Abrardo and D. Sennati. 2000. On the analytical evaluation of closed-loop power-control error statistics in DS-CDMA cellular systems. *IEEE Trans. Veh. Technol.* 49:2071–80.

[52] Y. Lie-Liang and L. Hanzo. 2003. Performance of generalized multicarrier DS-CDMA using various chip waveforms. *IEEE Trans. Commun.* 51:748–52.

[53] Q. M. Rahman and A. B. Sesay. 2001. Performance analysis of MT-CDMA system with diversity combining. *In IEEE MILCOM 2001*, vol. 2, pp. 1360–1364.

[54] S. Kondo and B. Milstein. 1996. Performance of multicarrier DS CDMA systems. *IEEE Trans. Commun.* 44:238–246.

11

Cooperative Communications in Random Access Networks

Y.-W. Peter Hong
National Tsing Hua University

Shu-Hsien Wang
National Tsing Hua University

Chun-Kuang Lin
National Tsing Hua University

Bo-Yu Chang
National Tsing Hua University

11.1 Introduction

Multiple-input multiple-output (MIMO) systems [1–4] have been studied extensively in the literature to combat fading in wireless communication systems. By mounting multiple antennas on each communication terminal and by having them placed sufficiently apart on the device, each user is able to gain access to multiple independent fading paths, which can be exploited to improve communication efficiency. With access to the independent fading paths, diversity and multiplexing gains [5] can be achieved by transmitting the same message on different paths [2] or by exploiting the additional degrees of freedom to construct multiple parallel transmission channels [3]. The gains increase with the number of transmit and receive antennas. However, in practice, it is difficult to place multiple antennas on a single terminal while trying to maintain independence among the different fading paths, not to mention the cost of the devices. This is especially true as the dimension of the devices decreases. For this reason, cooperative communications have been proposed to achieve MIMO gains without requiring multiple antennas on each terminal.

Cooperative communications [e.g., 6–12] allow users in the system to cooperate by relaying each other's messages to the destination. With only one antenna on each device, the users may cooperatively form a distributed antenna array to emulate the spatial diversity and multiplexing gains in conventional MIMO systems. Most works in the literature on cooperative communications focus on the physical layer designs, such as the coding, modulation, and diversity-combining techniques. However, it is equally important to study the subject from a medium access control (MAC) layer perspective, in which case the system throughput and the efficiency of resource allocation are considered instead of the outage or error probabilities. In fact, the major drawback of cooperative systems is the loss of bandwidth efficiency due to the need for interuser coordination since the antennas are deployed at distributed locations. That is, the messages exchanged for coordination occupy the bandwidth resources that otherwise could have been used for noncooperative transmissions. Therefore, MAC protocols designed for cooperative systems, i.e., protocols that allocate the resources for cooperative transmissions or interuser coordination, may have a notable impact on the system performance. In this chapter, we provide a comprehensive survey of MAC protocols for cooperative random access networks. In contrast to most works in the literature, where a perfect centralized scheduling is usually assumed, we focus on the random access scenario where users decide locally when to transmit their own packets and when to cooperate.

Random access protocols allow users in the system to access the network without complex interactions with a centralized control. Essentially, users access the channel based only on some simple local rule whenever they have a packet to transmit and devise strategies to deal with interference or collisions after they occur. This is in contrast to centralized scheduling where collision and interference are prevented with the additional cost of communication overhead. Random access protocols are particularly useful for cooperative systems that span over a large network, such as that of sensor and ad hoc networks, where centralized coordination may be costly. However, the advantages of cooperation are not evident in random access networks since no centralized control is available to coordinate their cooperative operations.

In this chapter, we shall discuss the use of cooperative transmissions in four different random access systems: (1) the slotted ALOHA system [13, 14], (2) CSMA/CA in IEEE 802 legacy systems [15, 16], (3) the random access system with collision resolution [17, 18], and (4) the distributed ad hoc network [9, 19, 20]. Specifically, for the slotted ALOHA system, cooperative transmission and queuing schemes have been proposed in [13, 14] where each user is allowed to record the packet transmitted by its partners if the partner failed in its previous transmission attempt. The user recording the packet may then relay the message to the destination in subsequent time slots. The queue stability is used as the performance measure in these works. In [15, 16], similar cooperative relaying mechanisms have also been considered for the IEEE 802 legacy system. These schemes exploit the benefits of cooperative relaying, but do not consider the receiver's ability to perform signal combining. In [18], the authors proposed advanced signal processing methods to resolve collision resolution at the destination by modeling the problem as a multiuser detection or source separation problem. The performance of this scheme is further enhanced with cooperative relaying as proposed in [17]. Furthermore, in [9, 19, 20], cooperation is used to facilitate the broadcasting operations in large-scale distributed ad hoc networks. Specifically, with users cooperatively relaying the same signal (i.e., broadcasting), the aggregate signal arriving at each receiver can be simply treated as a multipath signal and exploited through diversity-combining techniques. Through these discussions, we demonstrate that, in random access networks, cooperation can be performed opportunistically rather than being fully coordinated.

11.2 Fundamentals of Cooperative Communications

The basic operations of a cooperative communication system can be described with the canonical two-user example illustrated in Figure 11.1. Suppose we have two users, user 1 and user 2, that are transmitting to the common destination *D*. At any instant in time, one of the users will serve as the source, while the other user cooperates by relaying the message from the source to the destination. As a result, the destination receives two packets encoded from the same message, one directly from the source and the other from the relay. The destination fails to decode the message only if the channels on both paths are unreliable at the time of transmission.

Based on relaying, many cooperation schemes have been proposed in the literature, such as amplify-and-forward (AF), decode-and-forward (DF) [7], coded cooperation [10, 21], quantize-and-forward, compress-and-forward [22], etc. The most intuitive and widely applied methods are the AF and DF schemes. In the AF scheme, the users, which play the role of the relays, simply amplify the signal that they receive without explicitly decoding the messages. On the other hand, in the DF scheme, the relays first decode and retransmit either a repetition of the original message or a re-encoded message to improve the coding efficiency. If the message is re-encoded (with forward error correction capabilities), this scheme is then referred to as the coded cooperation scheme. All of these methods can be employed with the addition of distributed space-time coding or

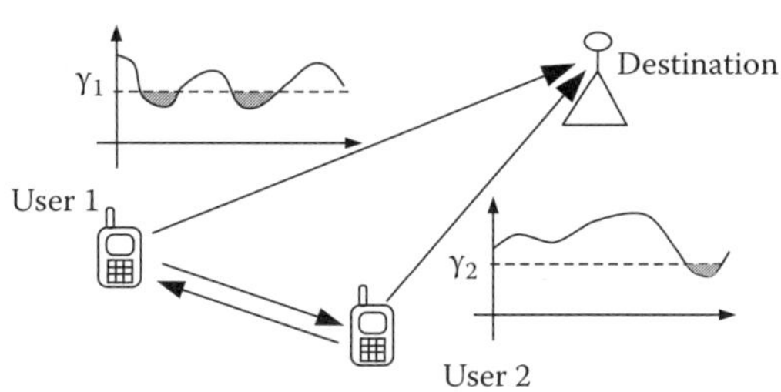

FIGURE 11.1 Cooperative communications with two users and one destination.

antenna selection techniques. Moreover, when the channel state information (CSI) can be further exploited at the transmitters, methods such as power allocation, precoding, or transmit beamforming can also be applied [11]. Details of these methods can be found in [6–11] and the references therein.

These cooperation methods can be easily extended to a network with $N > 2$ number of users. One approach is to divide the network into cooperating pairs that cooperate according to the mechanics shown in Figure 11.1. Another approach is to have, at each instant in time, one user serve as the source while the remaining users relay cooperatively the messages to the destination. In the first case, the cooperating pairs can be viewed as a super-user contributing a sum traffic to the network, while in the latter case, the relaying users effectively form a distributed antenna array possibly providing diversity and multiplexing gains for the source [12], similar to that in conventional multiple-antenna systems. Any combination of these two approaches can be taken as well.

Before each cooperative transmission, a coordination phase is required to synchronize the participating users. Inevitably, additional bandwidth and power resources are consumed in this process, as compared to the conventional MIMO system. In fact, this is the main cause of inefficiency in cooperative systems, but is often overcome by the significant gains in reliability and throughput. Two coordination methods are considered in the literature: coordination through direct interuser communication and coordination using feedback from the destination. In the first case, the users simply transmit their messages to each other and indicate the coding or modulation methods that should be used [7–9, 19, 23]. In the latter case, the feedback from destination is used to control and coordinate the messages transmitted by the users [15, 18, 24, 25].

In the past, most work on cooperative communications focused largely on the physical layer signal processing, such as the coding, modulation, diversity-combining schemes, power control, etc. A tutorial on these cooperative communication schemes can be found in [10, 11, 26]. However, most of the systems proposed in the literature assume strict synchronization between users or require a high degree of coordination, which is hard to achieve in practice. In the following, we provide examples of how cooperative communications can be applied to random access networks with reduced synchronization requirements.

11.3 Cooperation in Slotted ALOHA Random Access Networks

In random access networks (RAN), the users' transmissions are based only on simple local rules, with minimal interference from the central control. The lack of centralized control may result in collision between users and, in turn, reduce the system throughput. However, the simplicity, scalability, and robustness of random access protocols make RAN a favorable choice for low-traffic local area networks or multihop ad hoc networks. While cooperation is shown to be advantageous with perfect synchronization, it is not obvious whether cooperation can be helpful in random access networks and, if so, by how much. In fact, since the users behave in a decentralized manner, cooperation can only be achieved opportunistically in these scenarios. In this section, we study the effect of cooperation on one of the most fundamental random access protocols—the slotted ALOHA system [27].

11.3.1 Slotted ALOHA in a Wireless Network

In this section, let us start by reviewing the conventional slotted ALOHA system before incorporating the concept of cooperation in later sections.

Consider a network with N users, denoted by the set $\mathcal{U} = \{1, 2, \cdots, N\}$, transmitting to an access point (AP) through the wireless fading channel, as shown in Figure 11.2. In the slotted ALOHA system, the time is divided into equal-length time slots with the duration equal to the transmission time of one packet. The beginning and ending of each time slot are synchronized for all users, and a transmission can only occur at the beginning of a time slot. Let us assume, without loss of generality, that the length of each time slot is equal to 1 such that the transmission in the mth time slot occurs during the time $t \in [m, m+1)$. If a user, say user i, has a packet to transmit, it will transmit in the current time slot with probability p_i, which is independent from all users and over all time slots. If more than one user is transmitting in the same time slot, a collision will occur and no packet will be successfully received at the AP.

In contrast to conventional slotted ALOHA systems, where a transmission is assumed to be successful as long as there is no collision, the transmission of a packet in a wireless system may fail due to channel fading. Suppose that the channel experienced by user i in the mth time slot is parameterized by the variable $\gamma_i[m]$. For example, $\gamma_i[m]$ may be the signal-to-noise ratio (SNR) of the signal received at the AP corresponding to the transmission by user i in the mth time slot. Assume that $\gamma_i[m]$ is independent and identically distributed (i.i.d.) over time with the distribution function F_{γ_i}. Given that user i is the only user transmitting in the current time slot, the probability that the destination receives the packet correctly can be modeled by the probability $\Psi(\gamma_i[m])$, which is a function of the channel state. Let $\{H_i[m]\}_{m=0}^{\infty}$ be a sequence of i.i.d. Bernoulli random variables with probability $\psi_i = \mathbf{E}_{\gamma_i}[\Psi(\gamma_i[m])]$, where the event $H_i[m] = 1$ indicates the event of a successful transmission. After each time slot, the access point will feedback $\{\mathbf{0,1,}\boldsymbol{e}\}$ information to the users, where $\mathbf{0}$ indicates an idle slot, $\mathbf{1}$ indicates a success, and $\boldsymbol{e}$ indicates an error, due to either collision or channel fading.

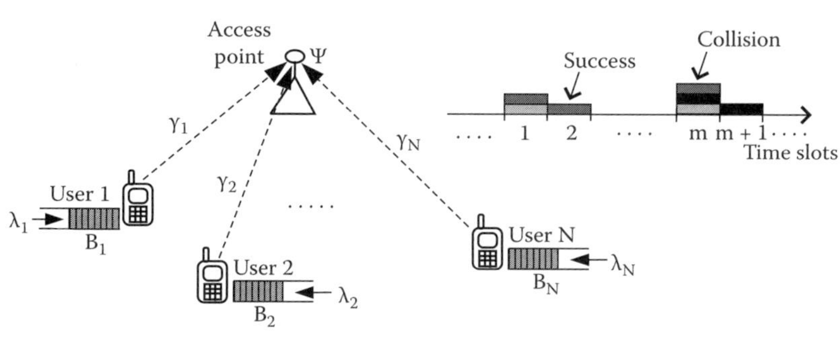

FIGURE 11.2 Illustration of the slotted ALOHA system.

Suppose that the arrival of messages at each user, say user i, is modeled as a sequence of i.i.d. Bernoulli random variables, $\{A_i[m]\}_{m=0}^{\infty}$, with rate λ_i packets per time slot. Let B_i, $i = 1, \cdots, N$, be the buffer at user i that stores the packets awaiting transmission and assume that there is no limit on the occupancy of the buffer. The number of packets in the buffer B_i at the beginning of the mth time slot is denoted by $Q_i[m]$, which is referred to as the *queue state*. Let $\{\tilde{V}_i[m]\}_{m=0}^{\infty}$ be a sequence of Bernoulli random variables with $\tilde{V}_i[m] \in \{0,1\}$ and probability $p_i = \Pr(\tilde{V}_i[m] = 1)$, where $\tilde{V}_i[m] = 1$ indicates the event that user i attempts a transmission in the mth time slot regardless of whether it has a packet to transmit. The event that user i actually transmits is then denoted by $V_i[m] = \tilde{V}_i[m] \cdot \chi_N(Q_i[m])$, where $\mathbb{N}$ is the set of natural numbers 1, 2, 3, …, and $\chi_{\mathbb{E}}(x)$ is the indicator function, defined as

$$\chi_{\mathbb{E}}(x) = \begin{cases} 1, & x \in \mathbb{E} \\ 0, & x \notin \mathbb{E}. \end{cases}$$

The evolution of the queue states can then be written as

$$Q_i[m+1] = (Q_i[m] - D_i[m])^+ + A_i[m], \ \forall i = 1, \cdots, N, \tag{11.1}$$

where

$$D_i[m] = H_i[m] V_i[m] \prod_{j \neq i} \left(1 - V_j[m]\right)$$

is the departure process and $(x)^+ = \max\{x,0\}$. It follows that $Q_i[m+1]$ is independent of $Q_i[m-1]$ if given the values of $\{Q_i[m], \forall i\}$, and thus $\mathbf{Q}[m] = [Q_1[m], \cdots, Q_N[m]]$ for $m = 0$, 1, 2, …, forms a discrete-time Markov chain.

In the discussions on MAC protocols, we are often concerned with the maximal stable throughput of the system, which is defined as the maximum arrival rate that each user can accommodate without causing the queues to become unstable.

Definition 11.1

The system is stable under the arrival rates $(\lambda_1, \lambda_2, \cdots, \lambda_N)$ if there exists a set of transmission probabilities $p_1, p_2, \cdots, p_N$ such that

$$\lim_{m \to \infty} \Pr\{Q_i[m] < x\} = G_i(x) \text{ and } \lim_{x \to \infty} G_i(x) = 1, \ \forall i.$$

Our goal is then to characterize the possible values of arrival rates for which the system remains stable, which is defined as the *stability region* of the system. However, finding the stability region of the slotted ALOHA system has been a difficult task. Over the past 30 years, only the stability region of a two-user [27] or three-user [28] system has been completely characterized, although inner and outer bounds have been derived for the case where $N > 3$. In the following, we shall compare the two-user stability region of the noncooperative and cooperative cases to illustrate the cooperative advantages in the slotted ALOHA system.

11.3.2 Stability Region of the Noncooperative Slotted ALOHA System

To derive the stability region of the slotted ALOHA system, Rao and Ephremides [27] proposed the use of an auxiliary hypothetical system that has the same arrival process and follows the same transmission policy as the original system, but is less likely to be stable. This system is called a *dominant system*. The stability region of the dominant system may be easier to derive if it is set up properly, but is only guaranteed to be an inner bound of the true stability region. However, in certain cases, the stability region of both systems may actually coincide, as is the case in the following example. Even if they do not coincide, the dominant system still provides useful insights on the behavior of the original system.

A dominant system can be constructed by assuming that a subset of users in the network are *fully loaded*, i.e., they always have a packet to transmit regardless of the actual queue state in the original system. Under this assumption, the departure rate of each user will be no larger than that in the original system since contention between users increases.

Interestingly, following the same procedures as in [27], we can show that the systems considered in the following satisfy the conditions of Loynes' formulation, and thus, by applying Loynes' theorem [29], we can say that the system is stable when the arrival rates $\lambda_1, \cdots, \lambda_N$ are smaller than the service rates $\mu_1, \ldots, \mu_N$, for a given set of transmission probabilities $p_1, \cdots, p_N$, i.e., $\lambda_i < \mu_i$ for all i. The service rate of user i refers to the average number of user i's packets that can be served in each time slot. On the other hand, the system is *unstable* if there exists i such that $\lambda_i > \mu_i$. In this case, our task reduces to finding the departure rate of each system.

To derive the stability region of the two-user system, let us first consider the case where user 1 is fully loaded. In this case, a packet from user 2 can be successfully received at the base station if and only if the reception is successful, which occurs with probability ψ_2, and user 1 does not transmit. Therefore, the service rate of user 2, which is a function of p_1 and p_2, is equal to $\mu_2^{1*}(p_1, p_2) = \psi_2 p_2(1 - p_1)$. If user 2 transmits at a rate λ_2 strictly less than the service rate $\mu_2^{1*}(p_1, p_2)$, it follows from Little's theorem [30] that the probability the buffer of user 2 is non-empty is equal to

$$\frac{\lambda_2}{\psi_2 p_2(1-p_1)},$$

and the service rate of user 1 is equal to

$$\mu_1^{1^*}(p_1, p_2) = \psi_1 p_1\left[1 - \frac{\lambda_2}{\psi_2(1-p_1)}\right].$$

Hence, given p_1 and p_2, the stability region obtained by assuming that user 1 is fully loaded is given by

$$\mathcal{R}^{1^*}(p_1,p_2)=\left\{(\lambda_1,\lambda_2):\lambda_1<\mu_1^{1^*}(p_1,p_2),\lambda_2<\mu_2^{1^*}(p_1,p_2)\right\}. \tag{11.2}$$

For $\lambda_2 < \psi_2 p_2(1 - p_1)$, the bounds for λ_1 are not only inner bounds, but in fact coincide with that of the true system. More specifically, when λ_1 is greater than $\mu_1^{1^*}(p_1, p_2)$, user 1 will be unstable in the dominant system and the queue length will go to infinity without emptying with finite probability, which means that there are sample paths that do not return to zero infinitely often. For such sample paths, the queue state will not return to zero after a certain time, in which case the dominant system and the original system will be identical. This shows that the original system will be unstable for these values of λ_1 as well.

Similarly, by assuming that user 2 is fully loaded, the stability region can be expressed as

$$\mathcal{R}^{2^*}(p_1,p_2)=\left\{(\lambda_1,\lambda_2):\lambda_1<\mu_1^{2^*}(p_1,p_2),\lambda_2<\mu_2^{2^*}(p_1,p_2)\right\}, \tag{11.3}$$

where $\mu_1^{2^*}(p_1, p_2) = \psi_1 p_1(1 - p_2)$ and

$$\mu_2^{2^*}(p_1,p_2)=\psi_2 p_2\left[1-\frac{\lambda_1}{\psi_1(1-p_2)}\right].$$

By taking the union of $\mathcal{R}^{1^*}(p_1, p_2)$ and $\mathcal{R}^{2^*}(p_1, p_2)$ for all possible transmission probabilities, the stability region of the two-user slotted ALOHA system is given by

$$\mathcal{R}=\bigcup_{(p_1,p_2)\in[0,1]^2}\left[\mathcal{R}^{1^*}(p_1,p_2)\bigcup\mathcal{R}^{2^*}(p_1,p_2)\right]=\left\{(\lambda_1,\lambda_2):\sqrt{\frac{\lambda_1}{\psi_1}}+\sqrt{\frac{\lambda_2}{\psi_2}}<1\right\}. \tag{11.4}$$

The boundary of the stability region is plotted in Figure 11.3 for the cases where $\psi_1 = 0.75$ and $\psi_2 = 0.5$ (solid curve) and $\psi_1 = \psi_2 = 0.25$ (dash-dotted curve). The case with no channel errors (dotted curve), $\psi_1 = \psi_2 = 1$, is also plotted as a reference.

11.3.3 Cooperation in a Two-User Slotted ALOHA System

In the conventional slotted ALOHA system, each user transmits independently of others, and thus, the stable throughput of each user is limited by its local channel quality. It is possible that the users experiencing bad channels on average will be seriously backlogged, while the others, i.e., the ones that experience good channels, have nothing to transmit. Intuitively, if the users are willing to cooperate, those that are more capable can help by relaying the messages from the less capable users and thereby increase their stable throughput.

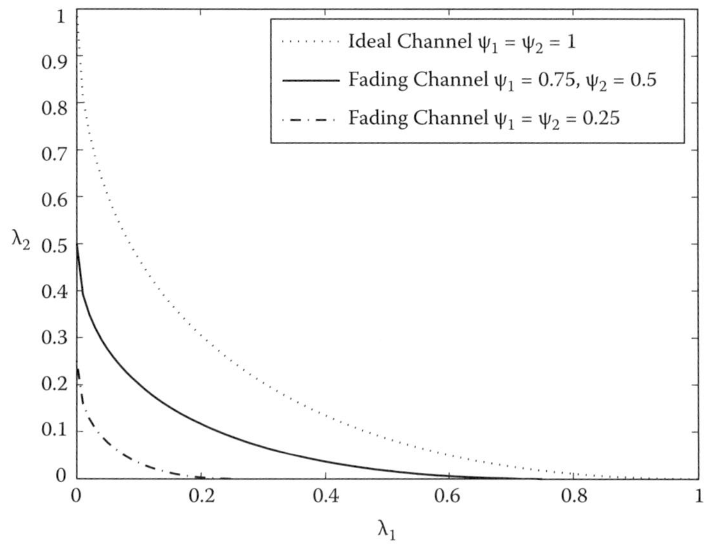

FIGURE 11.3 The two-user slotted ALOHA system without cooperation.

Let us consider a two-user slotted ALOHA system as shown in Figure 11.4. Following the same mechanics as the slotted ALOHA system, we consider the simplest class of cooperation based on decode-and-forward relaying. In the cooperative system, each idle user (i.e., users that have no packet to transmit) will continuously listen for a transmission from the other user and help by relaying the packet to the AP if the original transmission failed in its attempt to reach the AP. In order to do this, we assume that each user is equipped with an additional buffer, which will be used to store the messages transmitted by the other user. To distinguish from buffer B_i, we denote this buffer by CB_i for user i and refer to it as the *cooperative buffer*. In this case, B_i will be called the *source buffer*. Under the half-duplex assumption, a user is able to receive the message from the other user only if it is not transmitting in the same time slot. Moreover, we assume that

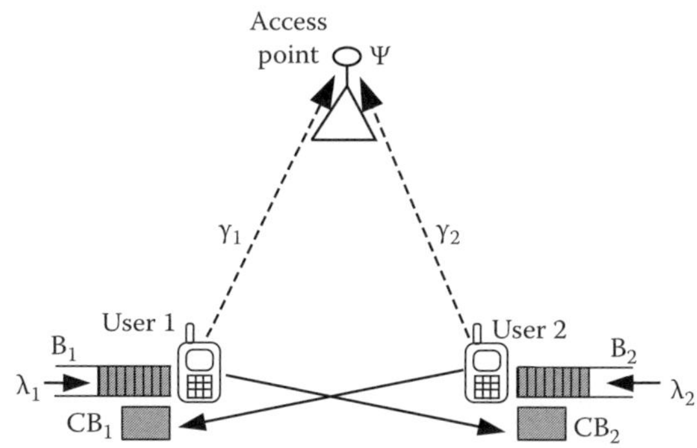

FIGURE 11.4 The two-user slotted ALOHA system with cooperation.

the message will always be correctly received by the cooperating users as long as the transmission is collision-free. This basically assumes that the users are sufficiently close to each other.

Suppose that user 1 is transmitting a certain packet for the first time from its source buffer while user 2 remains silent. During this time slot, user 2 will be able to overhear the message transmitted by user 1 and record the message in its cooperative buffer (i.e., CB_2). If an ***e*** feedback is received from the access point, the packet will remain at the head of line (HOL) and both users will be allowed to retransmit the packet in later time slots until one of them succeeds. When a transmission is eventually successful, i.e., a **1** feedback is received from the access point, the packet will be dropped from both users. Since no new packet is transmitted by the source user until the HOL packet is successfully transmitted, the length of the cooperative queue cannot exceed 1 and, thus, the stability of this queue will not be a concern. Let $CQ_1[m]$, $CQ_2[m] \in \{0, 1\}$ and $Q_1[m]$, $Q_2[m] \in \{0, 1, 2, \ldots\}$ represent the number of packets stored in the buffers CB_1, CB_2, B_1, and B_2, respectively, at the beginning of the mth time slot.

Let p_i be the transmission probability of user i and let $A_i[m]$, $V_i[m]$, $\tilde{V}_i[m]$, and $H_i[m]$ be as defined in the previous subsection. If user i decides to transmit in the current slot, and both B_i and CB_i are nonempty, user i will choose the packet from B_i with probability q_i, and CB_i with probability $\bar{q}_i = 1 - q_i$. To incorporate this into the queue evolution equation, we introduce the new Bernoulli random process $\{U_i[m]\}_{m=0}^{\infty}$, where $\Pr\{U_i[m] = 1\} = q_i$ and $U_i[m] = 1$ indicate the event that user i transmits a packet from B_i instead of CB_i when both the buffers are nonempty. A packet from the source buffer B_i will be transmitted either if (1) CB_i is empty and B_i is not or if (2) B_i and CB_i are both nonempty but user i chooses to transmit from B_i, i.e., $U_i[m] = 1$. Therefore, the event that a transmission attempt is made by a packet in B_i is represented with

$$V_i^{coop}[m] = V_i[m] \cdot \chi_{\{0\}}(CQ_i[m]) + V_i[m] \cdot U_i[m] \cdot \chi_{\mathbb{N}}(CQ_i[m]), \tag{11.5}$$

where $V_i[m] = \tilde{V}_i[m] \cdot \chi_{\mathbb{N}}(Q_i[m])$. Notice that $V_i^{coop}[m] = 1$ if a packet is transmitted from B_i, and $V_i^{coop}[m] = 0$ otherwise. On the other hand, a packet from the cooperative buffer CB_i is transmitted if user i attempts a transmission and the packet transmitted is not from B_i. Therefore, we have

$$CV_i^{coop}[m] = \tilde{V}_i[m][1 - \chi_{\{0\}}(Q_i[m]) \cdot \chi_{\{0\}}(CQ_i[m])] - V_i^{coop}[m]. \tag{11.6}$$

More specifically, a packet from the cooperative buffer is transmitted if $CV_i^{coop}[m] = 1$; otherwise, $CV_i^{coop}[m] = 0$. Since the successful transmission of a packet can be achieved either by the source itself or by the cooperating user, the departure process of user i is

$$\begin{aligned} D_i[m] = {} & V_i^{coop}[m] H_i[m] (1 - V_j^{coop}[m] - CV_j^{coop}[m]) \\ & + CV_j^{coop}[m] H_j[m] (1 - V_i^{coop}[m] - CV_i^{coop}[m]), \quad \text{for } i = 1,2 \text{ and } i \neq j. \end{aligned} \tag{11.7}$$

The evolution of the source queues is given in (11.1), while that of the cooperative queue is

$$CQ_i[m+1]=(CQ_i[m]-D_j[m])^+ + V_j^{coop}[m](1-H_j[m])(1-V_i^{coop}[m])\chi_{\{0\}}(CQ_i[m]). \quad (11.8)$$

The queue states $\{CQ_1[m],CQ_2[m],Q_1[m],Q_2[m]\}_{m=0}^{\infty}$ form a discrete-time Markov chain.

11.3.4 Stability Region of the Two-User Cooperative Slotted ALOHA System

Due to the complex interactions among the four queues $Q_1[m]$, $Q_2[m]$, $CQ_1[m]$, and $CQ_2[m]$, the stability region of the cooperative system is difficult to obtain. However, we are able to derive an inner bound by analyzing a dominant system where we let both users be fully loaded, i.e., they always have a packet available for transmission. The stability region of this dominant system will be referred to as the *fully loaded region* in the following.

When the users are fully loaded, the packet in CB_i will always be transmitted with probability $\bar{q}_i$ as long as it is nonempty, i.e., $CQ_i[m] = 1$. This is to say that when the users are fully loaded, the states of the cooperative buffers are independent of the states of the source buffers. In this case, $\{CQ_1[m],CQ_2[m]\}_{m=0}^{\infty}$ forms a finite-state Markov chain with four states: $S_0 = (0,0)$, $S_1 = (0,1)$, $S_2 = (1,0)$, and $S_3 = (1,1)$. For a given set of transmission probabilities p_1, p_2, q_1, and q_2, the Markov chain yields a steady-state distribution denoted by the probabilities π_0, π_1, π_2, and π_3, where π_i is the steady-state probability of state S_i. The service rates for users 1 and 2, respectively, are given by

$$\mu_1(p_1,p_2,q_1,q_2)=(\pi_1+\pi_3)\bar{q}_2 p_2(1-p_1)\psi_2+[1-(\pi_2+\pi_3)q_1]p_1(1-p_2)\psi_1,$$

$$\mu_2(p_1,p_2,q_1,q_2)=(\pi_2+\pi_3)q_1 p_1(1-p_2)\psi_1+[1-(\pi_1+\pi_3)\bar{q}_2]p_2(1-p_1)\psi_2.$$

The stability region of the fully loaded system is

$$\mathcal{R}_{coop}=\bigcup_{\substack{(p_1,p_2)\in[0,1]^2\\(q_1,q_2)\in[0,1]^2}}\mathcal{R}_{coop}(p_1,p_2,q_1,q_2), \quad (11.9)$$

where $R_{coop}(p_1, p_2, q_1, q_2) = \{(\lambda_1,\lambda_2):\lambda_1 < \mu_1(p_1, p_2, q_1, q_2), \lambda_2 < \mu_2(p_1, p_2, q_1, q_2)\}$ is the fully loaded region given fixed values of p_1, p_2, q_1, and q_2. The region in (11.9) serves as an inner bound to the stability region of the cooperative slotted ALOHA system.

11.3.5 Cooperative Slotted ALOHA with Channel Awareness

The cooperation considered up to this point exploits the diversity gains to improve the throughput of the system. Since the same packet may be transmitted by both users in the network, the probability of success will not be limited by the local channel quality of any single source. However, the performance can be further improved by exploiting knowledge of the local channel state information (CSI) at the users [13]. For example, if a user experiences bad channels, it should reduce its transmission probability to avoid collision with the other users. This is the subject of this section.

In a multiuser system, it is well known that, with knowledge of the CSI, the throughput of the system can be increased by scheduling users with the best channel to transmit in each time slot. This advantage is referred to as multiuser diversity in the literature [31]. To exploit this in a random access system, one can adopt a channel-aware transmission control function, as proposed in [32, 33], that determines the transmission probability of each user based on its local channel state in each time slot. Specifically, instead of transmitting with a fixed transmission probability, we assume that each user, say user i, adjusts its transmission probability based on its local channel state $\gamma_i[m]$ according to the function $s_i(\gamma_i[m])$. The average transmission probability is defined as

$$p_i = \int s_i(\gamma_i)dF_{\gamma_i}(\gamma_i),$$

where $F_{\gamma_i}(\gamma_i)$ is the distribution function of the channel state γ_i.

Consider a cooperative network that consists of two users and assume that both users are fully loaded. Similar to the case without channel awareness, the cooperative queue states $\{CQ_1[m], CQ_2[m]\}_{m=0}^{\infty}$ form a four-state Markov chain where the stationary distribution can also be derived. The service rates under the fully loaded assumption are given by

$$\begin{aligned}\mu_{1,CSI}(s_1,s_2,q_1,q_2) &= (\pi_1+\pi_3)\bar{q}_2\beta_2(1-p_1)+[1-(\pi_2+\pi_3)\bar{q}_1]\beta_1(1-p_2),\\ \mu_{2,CSI}(s_1,s_2,q_1,q_2) &= (\pi_2+\pi_3)\bar{q}_1\beta_1(1-p_2)+[1-(\pi_1+\pi_3)\bar{q}_2]\beta_2(1-p_1),\end{aligned} \tag{11.10}$$

for users 1 and 2, respectively, where $\beta_i = \mathbf{E}_{\gamma_i}[s_i(\gamma_i)\Psi(\gamma_i)]$. By Loynes' theorem, the fully loaded region for fixed s_1, s_2, q_1, and q_2 is

$$\mathcal{R}_{coop,CSI}(s_1,s_2,q_1,q_2)=\{(\lambda_1,\lambda_2):\lambda_1<\mu_{1,CSI}(s_1,s_2,q_1,q_2),\lambda_2<\mu_{2,CSI}(s_1,s_2,q_1,q_2)\}. \tag{11.11}$$

For fixed $s_1(\gamma_1)$, $s_2(\gamma_2)$, we take the union of the regions in (11.11) over q_1, q_2 to obtain

$$\begin{aligned}\mathcal{R}_{coop,CSI}(s_1,s_2)=\{(\lambda_1,\lambda_2):\lambda_1 &< \frac{\beta_2(1-p_1)+\beta_1(1-p_2)}{\beta_2(1-p_1)+p_1(1-p_2)}p_1(1-p_2)\equiv V_{1,x}(s_1,s_2),\\ \lambda_2 &< \frac{\beta_1(1-p_2)+\beta_2(1-p_1)}{\beta_1(1-p_2)+p_2(1-p_1)}p_2(1-p_1)\equiv V_{2,y}(s_1,s_2),\\ \lambda_1+\lambda_2 &< \beta_1(1-p_2)+\beta_2(1-p_1)\equiv k(s_1,s_2)\}.\end{aligned} \tag{11.12}$$

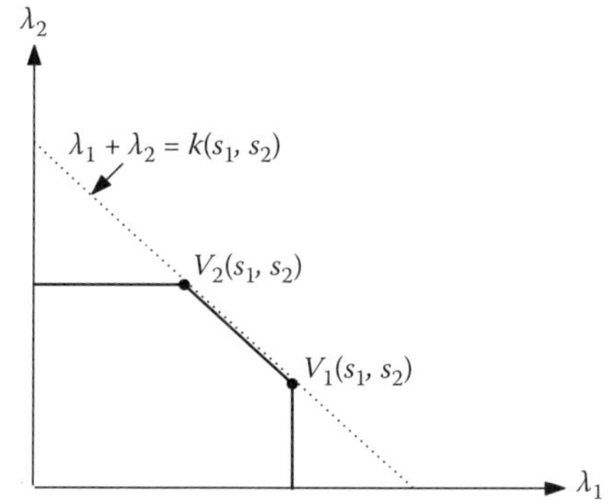

FIGURE 11.5 Stability region for given $s_1(\gamma_1)$ and $s_2(\gamma_2)$.

The region given above takes on the shape illustrated in Figure 11.5, where $V_1(s_1,s_2) = (V_{1,x}(s_1,s_2), k(s_1,s_2) - V_{1,x}(s_1,s_2))$ and $V_2(s_1,s_2) = (k(s_1,s_2) - V_{2,y}(s_1,s_2), V_{2,y}(s_1,s_2))$. The maximum service rate of user 1 is achieved by setting $q_1 = 1$ and $q_2 = 0$ (i.e., user 1 always transmits from the source buffer, and user 2 always transmits from the cooperative buffer), and similarly for user 2. By taking the union of the pentagons in Figure 11.5 over all s_1, s_2, we obtain the fully loaded region for the cooperative channel-aware system.

11.3.6 Performance Comparisons

In Figure 11.6, we plot the boundaries of three regions: (1) the stability region of the conventional slotted ALOHA system, (2) the inner bound of the cooperative system, and (3) the inner bound of the cooperative system with channel-aware transmission control. We consider, as an example, the reception model where

$$\Psi(\gamma) = \begin{cases} 1, & \text{if } \gamma \geq \gamma_{th} \\ 0, & \text{otherwise,} \end{cases}$$

which says that the packet is perfectly received if the channel state exceeds a certain threshold γ_{th} and fails otherwise. For the case with channel awareness, we assume that the channel-aware transmission control functions $s_i(\cdot)$, for all i, are step functions such that the transmission probability is 1 if the local channel state exceeds a certain value, and 0 otherwise. This was shown to be optimal in [13, 32].

Figure 11.6(a) is plotted for the case where $\psi_1 = \mathbf{E}_{\gamma_1}[\Psi(\gamma_1)] = 0.1$ and $\psi_2 = \mathbf{E}_{\gamma_2}[\Psi(\gamma_2)] = 0.1$, and Figure 11.6(b) is plotted for the case where $\psi_1 = 0.2$ and $\psi_2 = 0.8$. As shown in the figures, cooperation enlarges the stability region, especially when one of the users has a much better channel than the other. With cooperation, the user that on average experiences a bad channel can utilize help from its partner to increase the throughput of its own packets. When both users experience bad channels, little advantage can be gained through pure cooperation, as shown in Figure 11.6(a). However, a significant gain is

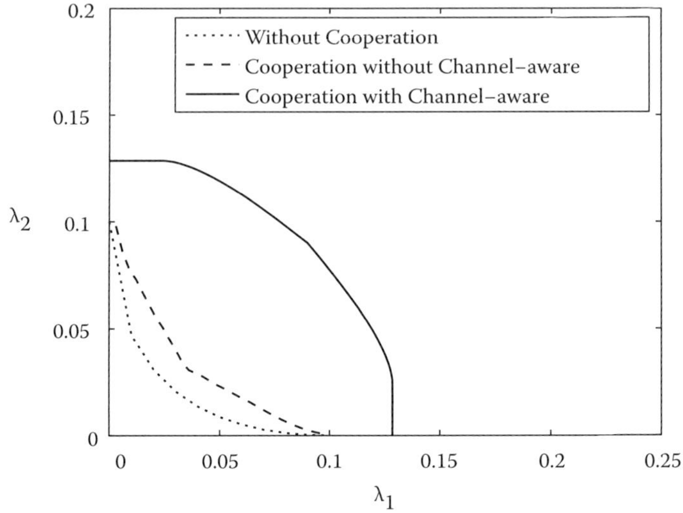

(a) $\psi_1 = 0.1$ and $\psi_2 = 0.1$

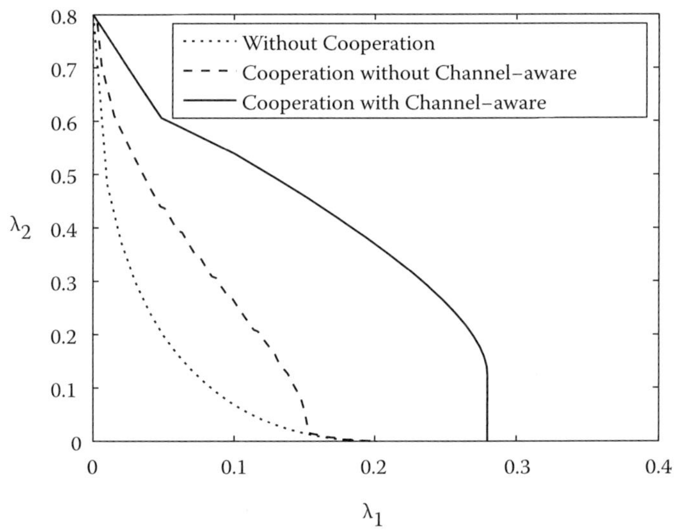

(b) $\psi_1 = 0.2$ and $\psi_2 = 0.8$

FIGURE 11.6 The stability regions and fully loaded regions for the cases with and without cooperation and channel awareness.

achieved when the users are able to adjust their transmission probabilities according to their local channel state. The channel-aware transmission control allows the users to exploit the multiuser diversity and increase the throughput of both cooperative users.

Although significant gains are already demonstrated in the above examples, it is worthwhile to notice that the form of cooperation discussed above does not utilize diversity-combining techniques at the AP. However, these techniques can be easily incorporated into our model to further improve the system performance. The analysis of these systems is subject to future investigation. The discussions for the case of N users can be found in [34]. Furthermore, we can also consider the use of multiple-packet reception in the cooperative slotted ALOHA system. Readers are referred to [35] for detailed discussions.

11.4 Cooperation with CSMA/CA in IEEE 802 Legacy Systems

Cooperation in its simplest form can be easily incorporated into the IEEE legacy systems. In fact, Liu, Tao, and Panwar [15] proposed a cooperative MAC protocol for wireless local area networks (WLANs) by making small changes to the IEEE 802.11b MAC. The idea is to use, in addition to the *request-to-send* and *clear-to-send* (RTS/CTS) messages, the *helper request-to-send* (HTS) message to help reserve the channel for the duration of the cooperative transmission and to coordinate the transmission between cooperating users.

There are two MAC operations in IEEE 802.11b: the distributed coordination function (DCF) and the point coordination function (PCF). The DCF is a contention-based method that provides asynchronous data transfers, and the PCF is a schedule-based method used to provide connection-oriented file transfers. Cooperation is less problematic in the latter case since a centralized control is used to schedule the users' transmissions. The cooperative MAC protocol proposed in [15] is based on the contention-based method, where no centralized control from the access point is required, and therefore is suitable for the ad hoc scenario. The DCF is designed based on the *Carrier Sensing Multiple Access with Collision Avoidance* (CSMA/CA) protocol. In DCF, a user must first sense whether a transmission by a neighboring user is in progress before transmitting its own data, i.e., *physical carrier sensing*. However, this does not avoid the hidden terminal problem, that is, the event that a user in the vicinity of the destination is transmitting simultaneously with the source and interfering with the reception at the destination. To avoid this problem, RTS and CTS messages are broadcast by the source and the destination before the data packet is actually transmitted to inform the neighboring users that a transmission with a certain duration is to follow. This is referred to as *virtual carrier sensing*.

For example, suppose that user 1 intends to transmit a data packet to user 2, as shown in Figure 11.7. Before transmission of the packet, user 1 first senses (physically) whether the channel is idle. If it is, user 1 will send an RTS message to the destination (i.e., user 2), which will reserve the channel from its neighboring users for a duration indicated in the network allocation vector (NAV) of the RTS message. If the destination is ready to receive the packet, that is, no other user is transmitting in its vicinity, it will reply with a CTS message to confirm the transmission and to reserve the channel from users in its vicinity. With multi-rate modulation and coding schemes available in the IEEE 802.11 standard, the users are able to adapt their transmission rates based on their local

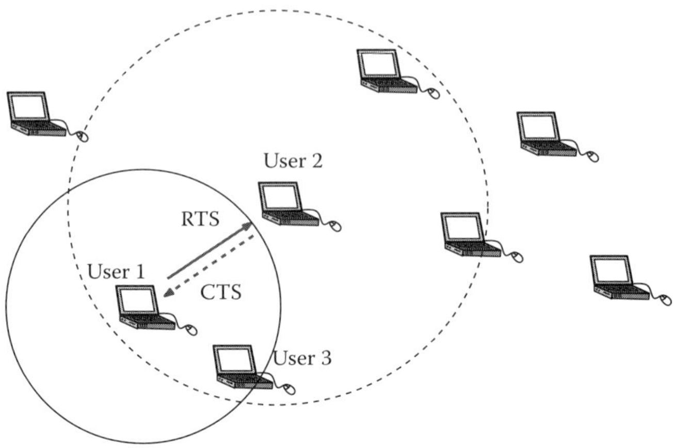

FIGURE 11.7 Carrier Sensing Multiple Access with Collision Avoidance (CSMA/CA).

channel quality. The time duration reserved by the RTS/CTS messages is then adjusted accordingly. When only a low data rate is achievable between two distant users, the transmission channel may be occupied for a long period of time but with low channel utilization, which significantly reduces the system throughput. To improve upon this, the source can ask a close-by user to relay the message to the destination, assuming that the source-relay and the relay-destination channels have sufficiently high transmission rates to reduce the total transmission time.

Cooperation can also be considered with slight modifications to the CSMA/CA protocol, as illustrated in Figure 11.8. In this case, each user is required to maintain a helper table that records the information about the potential helpers in its vicinity. The table contains four fields (e.g., see Figure 11.8(a)): the helper ID (i.e., MAC address), the time that a packet from the helper is last received, the rate between the helper and the destination (i.e., R_{hd}), and the rate between the source and the helper (i.e., R_{sh}). Before each transmission, the user searches this table for the best helper and determines whether cooperation is actually advantageous in its case.

For example, suppose that the source node (i.e., user 1) wants to transmit an N-bit packet to the destination node (i.e., user 2), as illustrated in Figure 11.8(b). With R_{sd} being the rate of the source-destination link and R_{sr}, R_{rd} the rate of the source-relay and relay-destination links, each user can compute the time that it takes for both direct and cooperative transmissions. Specifically, for direct transmission, the time that it takes is equal to N/R_{sd}, and for cooperative transmissions with relay r, the time is equal to

$$\left(\frac{N}{R_{sr}}+\frac{N}{R_{rd}}\right).$$

Before each transmission, the source first computes the cooperative transmission time corresponding to each helper and selects the user with the shortest time as the potential helper. The source then compares this with the time required for direct transmission

	Helper MAC address	Received time of the last packet from helper	Rate between helper and destination	Rate between helper and source
♯1	00-07-5E-30-44-0D	T_1	300 Kb/s	100 Kb/s
♯2	00-13-D4-4F-4C-0F	T_2	500 Kb/s	600 Kb/s
♯3	⋮	⋮	⋮	⋮
♯4	⋮	⋮	⋮	⋮

(a)

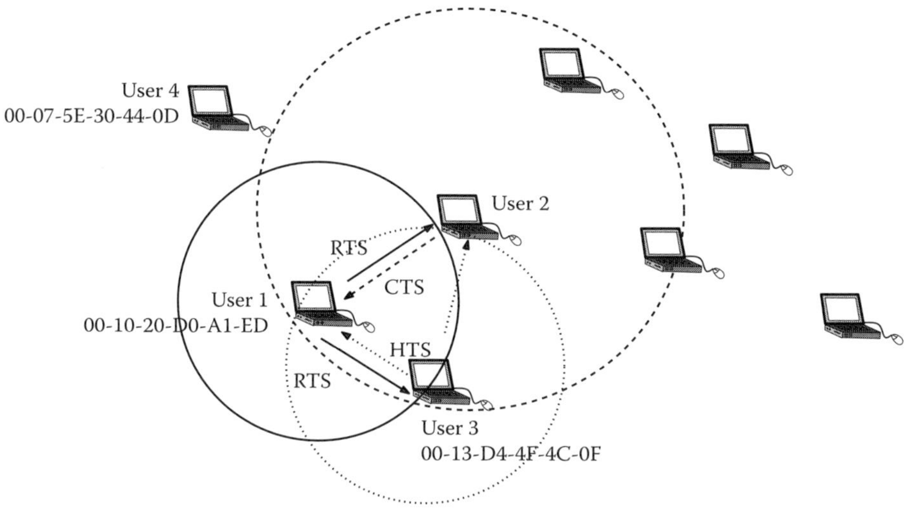

(b)

FIGURE 11.8 Illustration of the cooperative MAC protocol.

and requests help from another user only if relaying can speed up the transmission. In other words, user 1 selects user i to be the potential helper, where

$$i = \arg\min_r \left(\frac{N}{R_{sr}} + \frac{N}{R_{rd}} \right),$$

and requests cooperation only if

$$\left(\frac{N}{R_{si}} + \frac{N}{R_{id}} \right) > \frac{N}{R_{sd}}.$$

To inform the destination about the availability of the relay node and reserve the transmission time for the relay node, an additional HTS control message is exchanged among the source, relay, and destination. Specifically, when the source node decides to transmit cooperatively, it will include in the RTS packet the ID of the potential helper and the corresponding rate indicated in the helper's table. When the helper receives the message, it will check whether the rate indicated by the source is achievable. Inconsistencies between the helper table and the actual achievable rate occur when the environment varies and the entries in the table become out of date. If the helper is able to cooperate, it will respond with an HTS message to notify the source and the destination that cooperation is available and reserve the channel from its neighboring users. On the other hand, if the relay node is busy (due to its own transmission or because of the transmission of a neighbor), or it cannot achieve the rate requested by the source node, it will simply remain silent, and after a certain time-out period expires, the destination node will proceed and reply with a CTS message to reserve the channel for direct transmission.

As shown above, the concept of cooperation can be easily incorporated into the IEEE 802.11 MAC with little changes to the original design. Several enhancement methods are also given in [15] to reduce the transmission overhead and increase the system throughput.

11.5 Cooperation-Enhanced Collision Resolution Methods

The essence of random access protocols is to enable users to access the channel in a distributed fashion and deal with the collision afterwards. In the cooperative MAC protocols introduced above, the advantages of relaying are exploited to combat fading in the wireless channel, allowing the users that experience bad channels to be helped by the users that experience good channels. However, these methods do not exploit the cooperative advantages in resolving the interference or collision at the destination. In this and the following sections, we show that, with cooperation among users and appropriate signal processing techniques at the destination, the mixture of signals that is received during collision can be utilized to enhance the detection performance. In particular, we describe in this section the system, proposed first by Tsatsanis et al. in [18] and then by Lin and Petropulu in [17], that utilizes the variations in channel gains of the multiple relaying paths to perform signal separation when collision occurs.

Consider a network of N nodes transmitting to an access point, similar to that shown in Figure 11.2. Conventionally, when more than one user is transmitting in the same time slot, a mixture of signals will be received at the destination, causing strong interference among each other. In this case, it is likely that no packet will be received successfully, resulting in the so-called *collision*. The mixture of signals is usually discarded at the destination, and no information is extracted from the signals, which is clearly a waste of energy and bandwidth.

Due to this reason, Tsatsanis et al. [18] proposed a network-assisted collision resolution method. In this method, the destination records the mixture of signals whenever a collision occurs and enables the same set of colliding users to continue transmitting simultaneously in a certain number of subsequent time slots. Specifically, let $\mathbf{x}_i[n] = [x_{i,1}[n], \cdots, x_{i,M}[n]]^T$ be the M-bit message transmitted by user i in the nth time slot, and let $\mathcal{I}[n] \subset \{1, 2, \ldots, N\}$ be the set of users that are transmitting in the nth time slot. A collision occurs when the cardinality of $\mathcal{I}[n]$ is greater than 1, i.e., $|\mathcal{I}[n]| > 1$. The received signal at the destination is given by

$$\mathbf{y}[n]=[y_1[n],\cdots,y_M[n]]^T = \sum_{i\in\mathcal{I}[n]} h_{i,d}[n]\mathbf{x}_i[n]+\mathbf{v}[n]$$

where $h_{i,d}[n]$ is the channel between user i and the destination in the nth time slot, and $\mathbf{v}[n] = [v_n(1), \cdots, v_n(M)]^T$ is the additive noise vector with i.i.d. elements. The channel is assumed to be static over each time slot, but varying rapidly over different time slots.

Suppose that the destination or access point is able to estimate accurately the number of users transmitting in the same time slot and, through the cyclic redundancy check (CRC), is also able to detect an error whenever it occurs. After each time slot, the destination sends a **0**, **1**, or ***e*** feedback, with **0** indicating that the channel was idle, **1** indicating that all packets transmitted in the time slot were correctly received, and ***e*** indicating that at least one packet was not correctly decoded. When an *e* feedback is received, the users that were originally transmitting in the time slot will retransmit their messages again in the next time slot. Therefore, a sequence of collisions will occur, as shown in Figure 11.9. Suppose that the retransmissions (and thus collisions) occur over $\hat{K}$ consecutive time slots. Then, with $\mathcal{I}[n] = \{i_1, \ldots, i_K\}$, the destination receives the messages

$$\begin{aligned}\mathbf{Y}[n]&=[\mathbf{y}[n],\mathbf{y}[n+1],\cdots,\mathbf{y}[n+K-1]]\\ &=[\mathbf{x}_{i_1}[n],\cdots,\mathbf{x}_{i_K}[n]]\cdot\begin{bmatrix} h_{i_1,d}[n] & \cdots & h_{i_1,d}[n+k-1] \\ \vdots & \ddots & \vdots \\ h_{i_K,d}[n] & \cdots & h_{i_K,d}[n+K-1]\end{bmatrix}+[\mathbf{v}[n],\cdots,\mathbf{v}[n+K-1]] \quad (11.13)\\ &=\mathbf{X}[n]\mathbf{H}[n]+\mathbf{V}[n].\end{aligned}$$

In the absence of noise, the messages $\mathbf{X}[n]$ can be decoded by solving the set of linear equations $\mathbf{Y}[n] = \mathbf{X}[n]\mathbf{H}[n]$, given that the rank of $\mathbf{H}[n]$ is greater than K. Assuming that the channel coefficients for all users and all time slots are continuous random variables, e.g., Rayleigh fading, the matrix $\mathbf{H}[n]$ will have full rank with probability 1. Therefore, in the noiseless case, we need only $\hat{K}= K$ time slots in order to resolve the message of K users. Hence, there is no loss of throughput due to collision.

When noise is present, the maximum likelihood estimate of the message is given by

$$\hat{\mathbf{X}}[n]=\arg\min_{\mathbf{X}}\left\| \mathbf{Y}[n]-\mathbf{X}\cdot\mathbf{H}[n] \right\|_F^2, \quad (11.14)$$

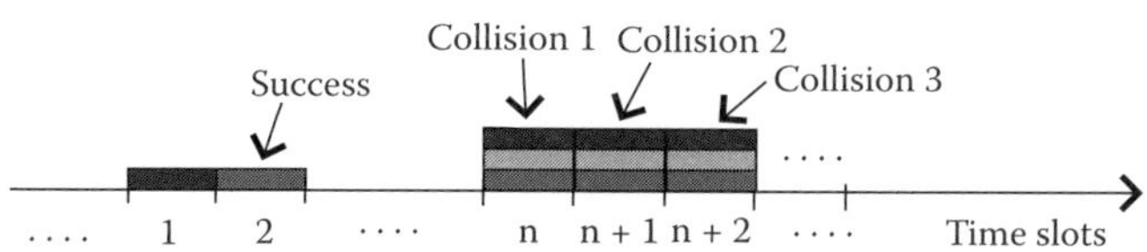

FIGURE 11.9 Network-assisted diversity for collision resolution.

where $\|\cdot\|^2$ represents the Frobenius norm. A suboptimal linear solution,

$$\hat{\mathbf{X}}[n] = \mathbf{Y}[n] \cdot \mathbf{H}^{-1}[n], \tag{11.15}$$

can also be used provided that $\mathbf{H}[n]$ is full rank.

However, when the channel varies slowly with respect to the time slot index n, one may need a large number of transmissions before the set of linear equations yield a reliable solution to the channel messages. To improve upon this, Lin and Petropulu [17] proposed the use of cooperative relaying to increase the variations of the channel gains in each transmission. Namely, upon collision, instead of asking the same users to retransmit continuously in consecutive time slots, the new strategy assigns a single node to transmit based on some predetermined order. The order can be determined, for example, by giving a unique index from 1 to N to each user upon deployment and allocating the retransmission time slot to the user that has an index equal to the time index m modulo N. The node that is assigned to transmit can be either one of the nodes participating in the collision, in which case it will simply retransmit its own message, or any other node in the network that amplifies and retransmits the mixture of signals that it overheard in the previous time slot. This period of repeated retransmissions is referred to as the *cooperative transmission epoch* (CTE). The CTE terminates when the receiver gains sufficient knowledge to recover the packets from all users. An illustration of this process is given in Figure 11.10.

Specifically, when a collision occurs in slot n, we assume that all users that are not involved in the collision will receive a mixture of signals similar to that at the destination. The signal received at user r is given by

$$\mathbf{z}_r[n] = \sum_{i \in \mathcal{I}[n]} h_{i,r}[n]\mathbf{x}_i[n] + \mathbf{v}_r[n],$$

where $h_{i,r}[n]$ is the channel coefficient between user i and user r during the nth time slot and $\mathbf{v}_r[n]$ is the noise at user r. Upon collision, the system will enter the CTE, during which $\hat{K} - 1$ users, denoted by $r_{n+1}, \cdots, r_{n+\hat{K}-1}$, will be assigned to retransmit. Suppose that the retransmitting node r_{n+k} belongs to the set $\mathcal{I}[n]$, then the user will simply transmit its own message and the signal received at the destination will be

$$\mathbf{y}[n+k] = h_{rd}[n+k]\mathbf{x}_r[n] + \mathbf{v}[n+k].$$

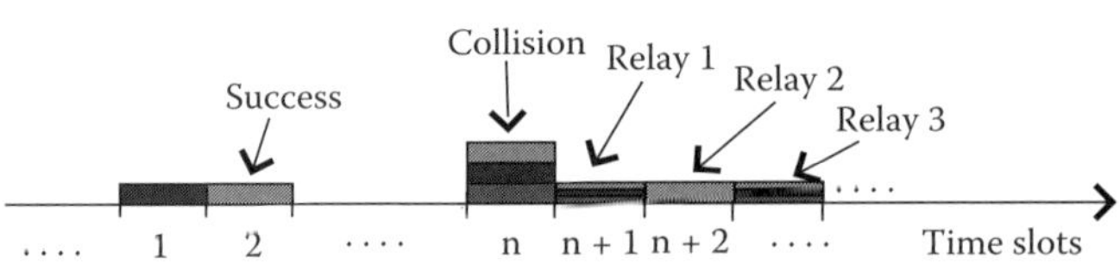

FIGURE 11.10 Cooperative diversity for collision resolution.

On the other hand, if $r_{n+k} \notin \mathcal{I}[n]$, then it will amplify and forward the signal with amplifying gain equal to $c[n + k]$, which is determined by the energy constraint at each node. The signal received at the destination is given by

$$\mathbf{y}[n+k] = h_{rd}[n+k] \cdot c[n+k] \cdot \mathbf{z}_r[n] + \mathbf{v}[n+k].$$

With amplify-and-forward relaying, the signal can also be decomposed into a signal component and a noise component, and the channel coefficient of the signal will be equal to the product of the source-to-relay and the relay-to-destination channels. Therefore, the detection methods given in (14.14) and (14.15) are equally applicable for this case.

By fully exploiting the signals that were received during collision, the system throughput may increase dramatically due to significant improvements in bandwidth inefficiency. Significant gains in energy efficiency may also be observed since all signals are combined for detection at the receiver. This concept can be exploited in ad hoc networks as described in the following section.

11.6 Asynchronous Cooperation in Multihop Ad Hoc Networks

In ad hoc wireless networks, users communicate over multihop relaying paths with no predetermined network infrastructure. In fact, multihop relaying is basically a form of cooperative communications but does not exploit diversity-combining techniques at the destinations. That is, as the messages are transmitted over a multihop route, each user in the route utilizes for detection only the signal coming from the single closest transmitter, even though the messages transmitted by all users in the same route are identical. The signals received from other users are essentially treated as interference. This can be improved upon by considering diversity combining at the destination, but must be done without the strict synchronization requirements that are prohibitive in large networks.

For network broadcasting applications, the *opportunistic large arrays* (OLA) system was proposed in [9, 19] to exploit the advantages of signal combining in cooperative multihop networks. The key idea is to treat the mixture of signals that arrive at unsynchronized time instants as an artificial multipath signal and utilize existing techniques such as the RAKE receiver or equalizers to resolve the multipath signal [20].

Consider a network broadcasting scenario where we have one source node transmitting a message to all other nodes in the network through multihop transmissions. At the beginning of the process, the source node first broadcasts its message through the wireless medium to all other users. Each user that is able to reliably decode the message will then retransmit the same message, which is again heard by all other nodes in the network. The users that have not yet transmitted will then take the mixture of signals received from their upstream nodes and decode the message using standard RAKE receivers or equalizers [9, 20]. Given that the transmission energy is sufficiently high, these operations will trigger an avalanche of signals propagating through the network. The intermediate nodes can either transmit a repetition of the decoded message or simply amplify and forward the original signal. To reduce the effect of error or noise propagation, the reliability of the signal received at each node can be defined based on different criteria, such as the received SNR or the probability of error. The mechanics of this system emulate the chanting of *Olé!* in a football stadium, where the signal first starts from a single source and propagates to the entire stadium as more and more people follow.

Suppose that there are N nodes in the network. Let us start by considering a symbol-by-symbol relaying scheme described as follows. At the beginning of the network broadcast operation, the source node first transmits an M-ary symbol represented by one of the waveforms $p_m(t)$, $m \in \{1, \ldots, M\}$, which has average energy normalized to 1,

$$\frac{1}{M}\sum_{m=1}^{M}\int \left|p_m(t)\right|^2 dt = 1.$$

In the case where all the intermediate nodes are eventually able to correctly decode the message, but at different instants in time, the signal received at user i can be written as

$$y_i(t) = \sum_{n=1}^{N} h_{i,n}(t)\cdot\sqrt{\varepsilon_n}\cdot p_m(t-\tau_{i,n}(t)) + n_i(t) = s_{i,m}(t) + n_i(t), \tag{11.16}$$

where $n_i(t)$ is the additive white Gaussian noise (AWGN) at the ith node with variance N_0, $\tau_{i,n}(t)$ is the time that user n transmits plus the propagation delay between the ith node and the nth node, ε_n is the energy of the pulse transmitted by user n, and $h_{i,n}(t)$ is the channel fading coefficient between user i and user n. The relaying nodes act as active scatters that form the multipath signal $s_{i,m}(t)$. Interestingly, this signal is unique for each user i due to the different channel gains and propagation delays experienced by the incoming signals. We refer to this as the *network signature* of user i given that the mth symbol was transmitted.

Assume that $h_{i,n}(t)$ and $\tau_{i,n}(t)$ are constant over the symbol duration T_s. Notice that T_s is proportional to the maximum delay spread of the signals $s_{i,m}(t)$, for all i, which is denoted by $\Delta\tau$. Specifically, we have

$$\Delta\tau = \max_i \left\{ \sqrt{\frac{\int_{-\infty}^{\infty}(t-\bar{\tau}_i)^2 |s_{i,m}(t)|^2\, dt}{\int_{-\infty}^{\infty}|s_{i,m}(t)|^2\, dt}} \right\}, \tag{11.17}$$

where

$$\bar{\tau}_i = \frac{\int_{-\infty}^{\infty} t\left|s_{i,m}(t)\right|^2 dt}{\int_{-\infty}^{\infty}\left|s_{i,m}(t)\right|^2 dt} \tag{11.18}$$

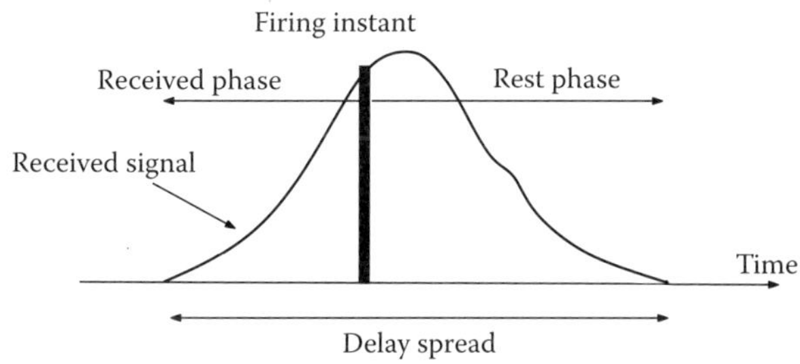

FIGURE 11.11 Signal received by active nodes.

is the average delay of the signals received at user i. With the symbol duration $T_s \approx \Delta\tau$, the rate $R_s \approx 1/\Delta\tau$ is achieved without intersymbol interference. Nevertheless, a higher rate is achievable with the use of equalizers at the destination, which is shown in [20]. An example of the signature waveform is depicted in Figure 11.11.

The signal received at each user can be divided into two phases: the receive phase and the rest phase. In the *receive phase*, the user first accumulates the signal transmitted by its upstream nodes until the signal energy is sufficient to perform reliable detection. At this instant, the symbol is detected and retransmitted to other users in the network. This instant is called the *firing instant*. Once a user has transmitted, it will then shut down until the downstream signals fade away. This period of time is called the *rest phase* and is used to prevent an infinite feedback, which may cause the system to become unstable.

The operations described above can be demonstrated with a simple four-node example, as shown in Figure 11.12. Assume node 1 is the source node and nodes 2–4 are the relays. We can see from the figure that node 2 will be the first to receive the signal from node 1. By the time user 2 accumulates sufficient signal energy for reliable detection, it will immediately retransmit the packet to nodes 3 and 4. By shutting down the receivers in the rest phase, the nodes that transmitted earlier in the symbol period will not receive signals from their downstream nodes.

Two main properties of OLA allow it to improve over the conventional multihop broadcasting scheme. The first property is its ability to combine the signals from all relays, including those that were transmitted from nodes that are located far away and were originally omitted in conventional networks. This enables a tremendous amount of energy savings, as detailed in [19]. Second, the simultaneously transmitting signals, which were originally treated as interference, are also combined for detection. Not only does this increase the energy efficiency, but it also reduces the network congestion and significantly decreases the broadcasting delay [see 9]. A specific equalizer design that performs the signal combining is proposed by Wei, Gofeckel, and Valenti in [20], where a random delay at the users is also proposed to increase the delay diversity of the system.

OLA provides a simple and efficient way to achieve broadcast communications in wireless networks. The system scales naturally to a large network since users are only required to follow a simple local rule. The integrate-and-fire mechanism at each user emulates the behavior of many biological networks, such as the flashing of fireflies, the firing of neurons, or even highway traffic patterns. These examples show that complex large-scale behaviors can result from the interaction between simple local mechanisms.

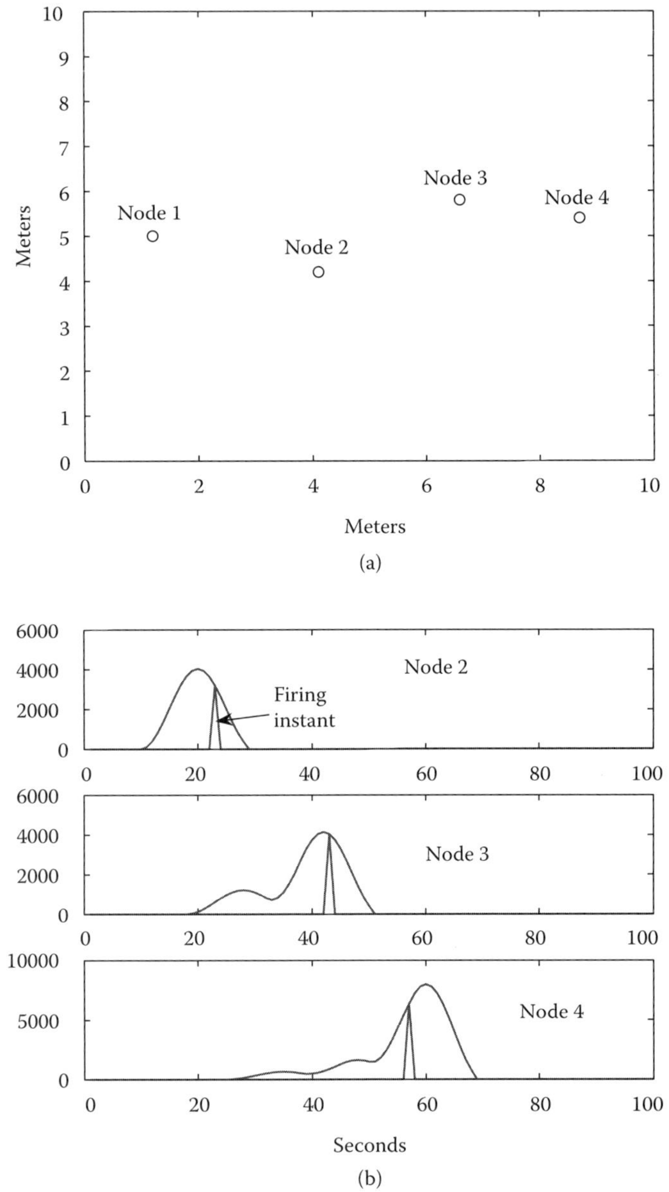

FIGURE 11.12 Demonstration of the signal at nodes 2–4. (a) Node location; (b) OLA signals received at nodes 2–4.

Biological networks are widely believed to be near optimal as a result of elimination and evolution. It is interesting to see how the efficiency of communication systems can be improved by emulating these biological networks. Discussions on these issues can be found in [36, 37] and the references therein.

11.7 Conclusion

Cooperative communications is a promising technique that has been proposed to combat fading without requiring multiple antennas on each device. Most studies in the literature focus on the signal processing aspects in the physical layer while assuming strict synchronization among the cooperating users. This strict synchronization requirement is perhaps the most difficult problem to overcome when realizing a cooperative network.

In this chapter, we studied the advantage of cooperative communications in a random access scenario where the users transmit based only on simple local rules. Some efficiency will be lost inevitably due to the lack of coordination, but significant improvements can be attained with cooperation. Two concepts have been repeatedly explored in this chapter: the user's ability to relay the information from the other users and the diversity-combining technique at the destination. With proper design of the transmission protocol, these added functionalities may significantly improve the throughput of the system. In this chapter, these advantages have been exploited in (1) the slotted ALOHA system, (2) the CSMA/CA system, (3) the collision resolution, and (4) multihop ad hoc networks. These discussions highlight the importance of cross-layered considerations when designing cooperative systems.

References

[1] J. H. Winters. 1987. On the capacity of radio-communication systems with diversity in a Rayleigh fading environment. *IEEE J. Selected Areas Commun.* 5:871–78.

[2] V. Tarokh, N. Seshadri, and A. R. Calderbank. 1998. Space-time codes for high data rate wireless communication: Performance analysis and code construction. *IEEE Trans. Inf. Theory* 44:744–65.

[3] G. J. Foschini. 1996. Layered space-time architecture for wireless communication in fading environments when using multi-element antennas. *Bell Syst. Tech. J.* 1:44–59.

[4] A. Goldsmith. 2005. *Wireless communications.* Cambridge: Cambridge University Press.

[5] L. Zheng and D. N. C. Tse. 2003. Diversity and multiplexing: A fundamental tradeoff in multiple-antenna channels. *IEEE Trans. Inf. Theory* 49:1073–96.

[6] T. Cover and A. El Gamal. 1979. Capacity theorems for the relay channel. *IEEE Trans. Inf. Theory* 25:572–84.

[7] J. N. Laneman, D. N. C. Tse, and G. W. Wornell. 2004. Cooperative diversity in wireless networks: Efficient protocols and outage behavior. *IEEE Trans. Inf. Theory* 50:3062–80.

[8] A. Sendonaris, E. Erkip, and B. Aazhang. 2003. User cooperation diversity. Part I. System description. Part II. Implementation aspects and performance analysis. *IEEE Trans. Commun.* 51(11).

[9] A. Scaglione and Y.-W. Hong. 2003. Opportunistic large arrays: Cooperative transmission in wireless multi-hop ad hoc networks for the reach back channel. *IEEE Trans. Signal Processing* 51(8).

[10] A. Nosratinia, T. E. Hunter, and A. Hedayat. 2004. Cooperative communication in wireless networks. *IEEE Commun. Mag.* 42:74–80.
[11] Y.-W. Hong, W.-J. Huang, F.-H. Chiu, and C.-C. J. Kuo. 2007. Cooperative communications in resource-constrained wireless networks. *IEEE Signal Processing Mag.* 24:47–57.
[12] K. Azarian, H. El Gamal, and P. Schniter. 2005. On the achievable diversity-multiplexing tradeoff in half-duplex cooperative channels. *IEEE Trans. Inf. Theory* 51:4152–72.
[13] Y.-W. Hong, C.-K. Lin, and S.-H. Wang. 2007. On the stability of two-user slotted ALOHA with channel-aware and cooperative users. In *Proceedings of the IEEE Workshop on Wireless Networks: Communication, Cooperation and Competition (WNC3).*
[14] Y.-W. Hong, C.-K. Lin, and S. H. Wang. 2007. On the stability region of finite-user slotted ALOHA with cooperative relays. In *Proceedings of the International Symposium on Information Theory (ISIT).*
[15] P. Liu, Z. Tao, and S. Panwar. 2005. A cooperative MAC protocol for wireless local area networks. In *Proceedings of the IEEE International Conference on Communications (ICC)*, vol. 5, pp. 2962–2968.
[16] P. Liu, Z. Tao, Z. Lin, E. Erkip, and S. Panwar. 2006. Cooperative wireless communications: A cross-layer approach. *IEEE Wireless Commun. Mag.* 13:84–92.
[17] R. Lin and A. P. Petropulu. 2005. A new wireless network medium access protocol based on cooperation. *IEEE Trans. Signal Processing* 53:4675–84.
[18] M. K. Tsatsanis, R. Zhang, and S. Banerjee. 2000. Network-assisted diversity for random access wireless networks. *IEEE Trans. Commun.* 48:702–11.
[19] Y.-W. Hong and A. Scaglione. 2006. Energy-efficient broadcasting with cooperative transmission in wireless ad hoc networks. *IEEE Trans. Wireless Commun.* 5(10).
[20] S. Wei, D. L. Goeckel, and M. Valenti. 2006. Asynchronous cooperative diversity. *IEEE Trans. Wireless Commun.* 5:1547–57.
[21] M. Janani, A. Hedayat, T. E. Hunter, and A. Nosratinia. 2004. Coded cooperation in wireless communications: Space-time transmission and iterative decoding. *IEEE Trans. Signal Processing* 52:362–71.
[22] G. Kramer, M. Gastpar, and P. Gupta. 2005. Cooperative strategies and capacity theorems for relay networks. *IEEE Trans. Inf. Theory* 51:3037–63.
[23] J. N. Laneman and G. W. Wornell. 2003. Distributed space-time-coded protocols for exploiting cooperative diversity in wireless networks. *IEEE Trans. Inf. Theory* 49:2415–25.
[24] Y.-W. P. Hong and A. Scaglione. 2008. Group testing for binary Markov sources: Data-driven group queries for cooperative sensor networks. *IEEE Trans. Inf. Theory.*
[25] Y.-W. Hong and A. Scaglione. 2004. On multiple access for distributed dependent sources: A content-based group testing approach. In *Proceedings of the IEEE Information Theory Workshop*, San Antonio, TX, pp. 298–303.
[26] F. H. P. Fitzek and M. D. Katz, eds. 2006. *Cooperation in wireless networks: Principles and applications: Real egoistic behavior is to cooperate.* The Netherlands: Springer.
[27] R. Rao and A. Ephremides. 1988. On a stability of interacting queues in multiaccess system. *IEEE Trans. Inf. Theory* 34:918–30.

[28] W. Szpankowski. 1994. Stability conditions for some multiqueue distributed systems: Buffered random access systems. *Adv. Appl. Probability* 26:498–515.
[29] R. Loynes. 1962. The stability of a queue with non-independent inter-arrival and service times. *Proc. Cambridge Philos. Soc.* 58:497 520.
[30] D. Gross and C. M. Harris. 1998. *Fundamentals of queueing theory.* 3rd ed. New York: John Wiley & Sons.
[31] R. Knopp and P. A. Humblet. 1995. Information capacity and power control in single-cell multiuser communications. In *IEEE International Conference on Communications*, vol. 1, pp. 331–335.
[32] S. Adireddy and L. Tong. 2005. Exploiting decentralized channel state information for random access. *IEEE Trans. Inf. Theory* 51:537–61.
[33] X. Qin and R. A. Berry. 2006. Distributed approaches for exploiting multiuser diversity in wireless networks. *IEEE Trans. Inf. Theory* 52:392–413.
[34] C.-K. Lin and Y.-W. P. Hong. 2008. On the finite-user stability region of slotted ALOHA with cooperative users. In *International Conference on Communications (ICC).*
[35] Y. E. Sagduyu and A. Ephremides. 2006. A game-theoretic look at throughput and stability in random access. In *Proceedings of the Military Communications Conference (MILCOM)*, pp. 1–7.
[36] E. Bonabeau, M. Dorigo, and G. Theraulaz. 1999. *Swarm intelligence: From natural to artificial systems.* Oxford: Oxford University Press.
[37] Y.-W. Hong and A. Scaglione. 2003. Swarming activities to distribute information in large sensor networks. In *Proceedings of the Military Communications Conference (MILCOM)*, Boston, pp. 682–687.

12

Cooperative Diversity: *Capacity Bounds and Code Designs*

V. Stanković
University of Strathclyde

Anders Høst-Madsen
University of Hawaii

Zixiang Xiong
Texas A&M University

12.1 Introduction

In this chapter we discuss methods for conveying information in a wireless network based on cooperation among network nodes. The methods are particulary suitable for decentralized wireless ad hoc networks where information is transmitted in multihops.

A wireless ad hoc network is a network without infrastructure and centralized control. It consists of a large number of possibly mobile nodes that communicate with each other over wireless links. Low cost and simple reconfiguration make ad hoc networks attractive in both commercial and military applications such as wireless local area network (LAN) (e.g., IEEE 802.11x) and metropolitan area network (MAN) (e.g., DARPA's GLOMO), home networks (e.g., HomeRF), device networks (e.g., based on Bluetooth or ZigBee), and sensor networks (e.g., SmartDust, WINS).

In contrast to a traditional infrastructure wireless network (e.g., a cellular network), where information is transmitted from one user to another via a control base station, an ad hoc network allows peer-to-peer communication from a sending node to a destination node. That is, the information can travel directly from a sending to receiving node in a single hop.

However, since wireless channels are often poor, single-hop routing requires either high transmission power, and consequently causes increased interference, or complex multiple access schemes. To achieve significant power savings and keep complexity low, information should be conveyed to a destination through multiple intermediate nodes. Whereas transmission over a single-hop channel has already been intensively studied and is well understood, cooperative communication in multinode networks is still an open research problem, which recently has received considerable attention, inspired by the papers [1, 2].

Communication over a wireless channel is limited by interference, fading, multipath, path loss, and shadowing. The main design challenge in ad hoc networks lies in devising communication methodologies in a decentralized manner, based on the knowledge of local conditions only, to overcome these limitations. An additional design issue has to do with the high dynamics of an ad hoc network, where nodes frequently join and leave the network.

One way of achieving high performance is to employ multiple transmitter and receiver antennas at nodes. This multiple antenna system increases the capacity and improves robustness to fading and interference by means of *spatial diversity* and *data rate multiplexing* [3, 4]. However, building multiple antennas at each node can be expensive, impractical, and often infeasible, especially for small and simple nodes such as those used in sensor networks.

Another recently proposed solution for achieving spatial diversity without requiring multiple antennas at any node is *cooperative diversity* [1, 2]. It is based on grouping several nodes (each with only one antenna) together into a cluster to form a large transmit or receive antenna array. Collaborative clusters are formed in an ad hoc fashion by negotiations among neighboring nodes without centralized control (see Figure 12.1). Cooperative diversity naturally arises in ad hoc networks as it enables great power savings with cheap, simple, and mobile nodes, while supporting decentralized routing and control algorithms. However, it is not limited to ad hoc networks, as it can be useful in infrastructure networks as well.

The simplest nontrivial setup is when the nodes form pairs, i.e., clusters of two. In a two-transmitter two-receiver cooperative channel, the two single-antenna transmitters want to communicate messages to the two remote single-antenna receivers over the same wireless radio channel. In *transmitter cooperation*, the two transmitters first exchange their messages, and then start to act as a single two-antenna broadcast transmitter. On the other hand, in *receiver cooperation*, the two receivers exchange their received signals and act as a single two-antenna multiple access receiver. In general, the two transmitters as well as the two receivers can collaborate among each other to form a virtual multiple-input multiple-output (MIMO) channel with two transmitter and two receiver antennas. The main goal of node cooperation is to achieve spatial diversity and

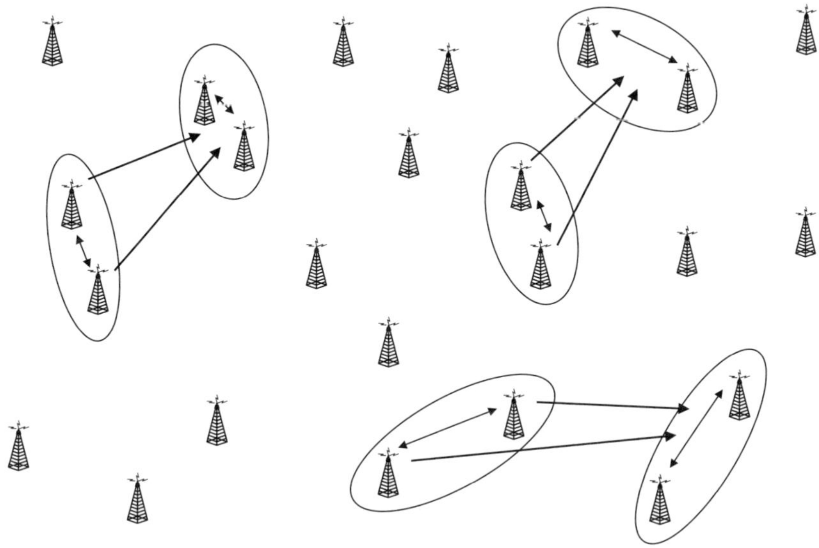

FIGURE 12.1 A wireless network with cooperative diversity. Closely located transmitters form a transmitter cluster, whereas closely located receivers form a receiver cluster.

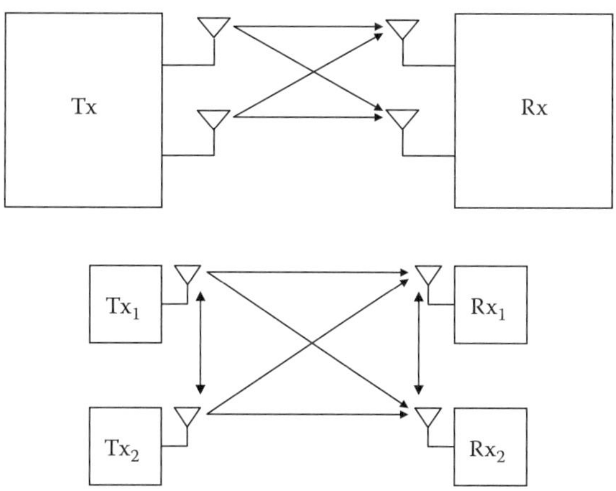

FIGURE 12.2 MIMO channel (up) versus cooperative diversity (down).

rate multiplexing (that is, to mimic a MIMO transceiver) without increasing the number of antennas at a single node.

The difference between a MIMO channel and a cooperative diversity channel is highlighted in Figure 12.2. In a two-antenna MIMO channel, a transmitter/receiver is equipped with two transmitting antennas. In a two-transmitter two-receiver cooperative diversity channel, two antennas at the transmitter/receiver side are physically separated; hence, a somewhat lower performance is expected. However, if the two transmitters/receivers are closely located, then the performance of a well-designed cooperative diversity system should be close to that of a MIMO system.

Note that the concept of cooperative diversity can be extended to the case where multiple-antenna transceivers cooperate to build a virtual transceiver with a higher number of antennas. However, in this chapter we focus on the simplest scenario where each node is equipped with a single transceiver/receiver antenna.

This chapter is organized as follows. In the next section we give an information-theoretical model of two-transceiver two-receiver cooperative channels. In section 12.3 we present the capacity bounds of the classic three-node relay channel [5], which is a building block of cooperative diversity, and cooperative channels. The next section focuses on recent advances made in practical code designs for the relay, transmitter, and receiver cooperative channels. The final section points to open challenges and opportunities for future research.

12.2 Information-Theoretical Model

Consider the channel model depicted in Figure 12.3 (upper left), where the transmitter in node 1 wants to send message $\omega_1 \in \{1, \ldots, M_1\}$ to the receiver in node 3; likewise, the transmitter in node 2 intends to send message $\omega_2 \in \{1, \ldots, M_2\}$ to the receiver in node 4.

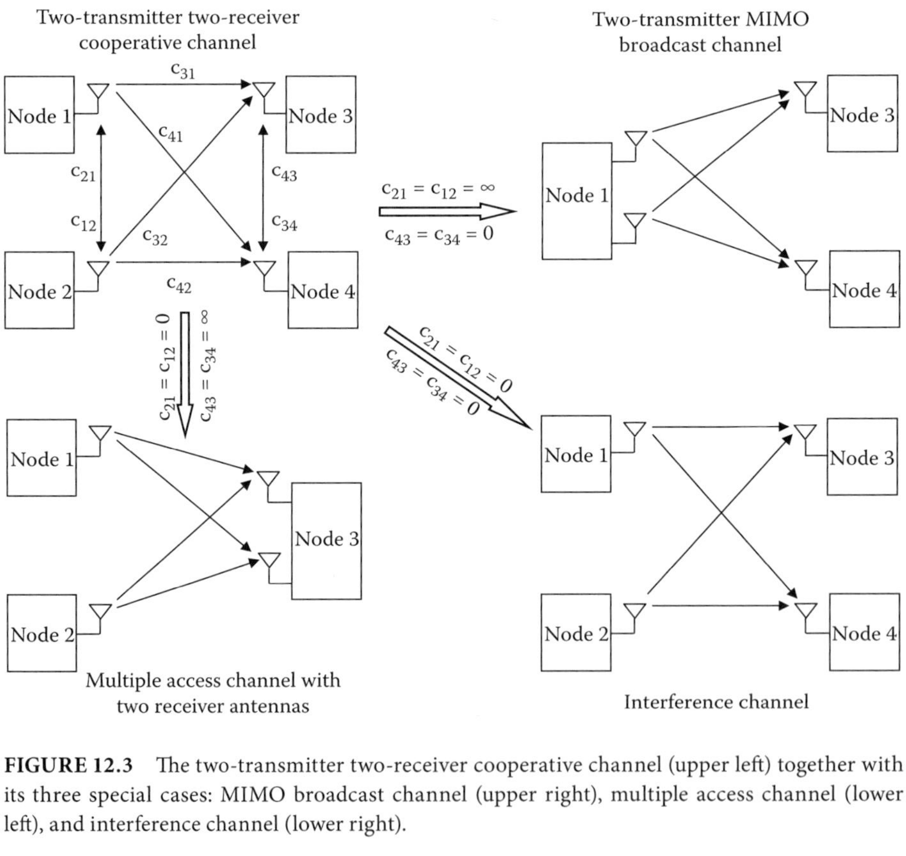

FIGURE 12.3 The two-transmitter two-receiver cooperative channel (upper left) together with its three special cases: MIMO broadcast channel (upper right), multiple access channel (lower left), and interference channel (lower right).

Specifically, node i (i = 1, 2, 3, 4) transmits a block $x_i[n]$ of N symbols at a time with n = 1, …, N, while being subject to an average power constraint

$$\frac{1}{N}\sum_{n=1}^{N}|x_i[n]|^2 \le P_i.$$

The rate of the transmission from node i is then

$$R_i = \frac{\log M_i}{N}.$$

We assume that the channel between node i and node j is a Rayleigh flat-fading channel [6] with channel coefficient c_{ij}, which is an independent identically distributed (i.i.d.), complex, zero-mean, Gaussian random variable. The nodes can either be able to transmit and receive simultaneously in the same time slot or at the same frequency (full-duplex) or not, in which case the nodes transmit during one time slot, or at one frequency, and receive during the next time slot, or at another frequency (half-duplex).

At the symbol level, the received signals at nodes 1 and 2 are given by

$$y_1[n] = c_{12}x_2[n] + z_1[n] \tag{12.1}$$

and

$$y_2[n] = c_{21}x_1[n] + z_2[n], \tag{12.2}$$

respectively, and the signals received by nodes 3 and 4:

$$y_3[n] = c_{31}x_1[n] + c_{32}x_2[n] + c_{34}x_4[n] + z_3[n] \tag{12.3}$$

and

$$y_4[n] = c_{41}x_1[n] + c_{42}x_2[n] + c_{43}x_3[n] + z_4[n], \tag{12.4}$$

respectively, where z_i, i = 1, …, 4, are i.i.d., circular, complex, zero-mean, additive, Gaussian noises. Without loss of generality, we assume that the noises are of unit power and $c_{31} = c_{42} = 1$. In the transmitter cooperative channel, we set $c_{43} = c_{34} = 0$, and for receiver cooperation, $c_{21} = c_{12} = 0$.

If cooperation is perfect, then transmitter cooperation leads to a two-antenna MIMO broadcast channel [7] ($c_{21}, c_{12} \to \infty$ in the sense that the channel between the two transmitters becomes an ideal noiseless channel with zero delay), receiver cooperation reduces to a two-user multiple access channel (MAC) [8] with two receiver antennas ($c_{43}, c_{34} \to \infty$), and the general setup with both transmitter and receiver cooperation becomes a single MIMO channel [3] with two transmitter and two receiver antennas ($c_{21}, c_{12}, c_{43}, c_{34} \to \infty$).

On the other hand, when cooperation is not allowed, i.e., $c_{21} = c_{12} = c_{43} = c_{34} = 0$, the channel degenerates to the interference channel [9]. Figure 12.3 depicts three of these four simplifications.

When we restrict the channels to be quasi-static, then all channel coefficients are constant during transmission of each block of N symbols. In the *synchronous model* of (12.1)–(12.4), we assume that the nodes are perfectly synchronized and have full channel state information (CSI), i.e., each node knows instantaneous values of all channel coefficients and their statistics. While it is relatively simple to achieve symbol/time synchronization between nodes, carrier synchronization, which requires phase-locking separated microwave oscillators, is challenging in practice. Therefore, we also consider the *asynchronous model*, where random phase offsets due to oscillator fluctuations are added to the transmitted signals. We include these random phases in the channel coefficients, so that the model stays the same as (12.1)–(12.4). Under the asynchronous model for receiver cooperation, the transmitters do not have any CSI, whereas the receivers need to know only the magnitudes of all channel coefficients, not their phases. Thus, receiver cooperation is suitable in the systems with simple transmitters. On the other hand, under the asynchronous model for transmitter cooperation, the transmitters must know the magnitudes of all channel coefficients.

We discuss the diversity and data rate gains achievable by node cooperation, while focusing on the high-signal-to-noise ratio (SNR) regime, where the data rates are mainly limited by interference. (In the low-SNR regime, the influence of channel noise prevails, and hence the gain from cooperation is reduced.) The *diversity gain* [10], defined as

$$d=-\lim_{\text{SNR}\to\infty}\frac{\log P_e(\text{SNR})}{\log \text{SNR}},$$

shows how fast the probability of decoding error P_e decays with SNR. A higher d means lower P_e at the same SNR, and thus a more reliable system. The data rate gain is usually decoupled into a *multiplexing gain* and an *additive gain*. The multiplexing gain (or degree of freedom) [10] shows how fast the rate increases with SNR and is given by

$$r=\lim_{\text{SNR}\to\infty}\frac{R(\text{SNR})}{\log \text{SNR}},$$

where R(SNR) denotes the sum of data rates of transmitting nodes for a given SNR. The additive gain (or the high-SNR power offset) [11, 12] is a shift of the R(SNR) function from the origin at high SNRs, i.e.,

$$a=\lim_{\text{SNR}\to\infty} R(\text{SNR})-r\log(\text{SNR}).$$

If all the limits exist, then R(SNR) in the high-SNR regime can be approximated by a line of slope r and SNR offset a, i.e.,

$$R(\text{SNR})\approx r\log(\text{SNR})+a.$$

It is well known that perfect cooperation (a two-antenna MIMO transceiver) achieves a diversity gain of $d = 2$ and a multiplexing gain of $r = 2$ [3]. On the other hand, the interference channel (without node cooperation) in Figure 12.3 provides no diversity or multiplexing gain (i.e., $d = 1$ and $r = 1$).

12.3 Capacity Bounds

In this section, we first present capacity bounds for cooperative diversity, indicating a multiplexing gain of only 1 at high SNRs, which is a somewhat negative result. However, the main message is that node cooperation *can* provide a large additive gain and a diversity gain of 2.

While the capacities of most point-to-point channels are known, this is not the case for wireless multinode channels. Indeed, we only know the capacities of the Gaussian MAC and the broadcast channel. For all other multinode channels, e.g., the relay and interference channels, capacities are known only in special cases. However, it is possible to obtain upper and lower bounds on the capacity, which are often very close, thereby practically indicating the capacity. A lower bound is the rate that can be attained by some coding scheme and is therefore an achievable rate. All rates higher than the upper bound cannot be achieved. If the lower and upper bounds overlap entirely, the complete rate region is known. However, if the two bounds do not overlap, the gap between them characterizes the unknown region.

There are two main ideas in obtaining achievable rates for cooperative channels. The first idea is based on nodes decoding messages from other nodes and re-encoding them. The second lies in exploiting the joint statistics between the data at cooperating nodes by means of *coding with side information*, i.e., Wyner-Ziv coding [13] or dirty-paper coding [14]. Specifically, it turns out that Wyner-Ziv coding achieves the capacity of receiver cooperation (asymptotically as the interference and SNR approach infinity), while dirty-paper coding plays a major role in transmitter cooperation. Below we give a brief summary of coding with side information.

12.3.1 Coding with Side Information

Distributed source coding addresses separate compression and joint decompression of correlated sources [8]. Its foundation was laid by Slepian and Wolf [15], who defined the rate region for lossless compression of two correlated discrete sources showing a surprising result: separate encoding and joint decoding suffer no rate loss compared to the case when the sources are compressed jointly. The framework was extended and generalized in [16], where the problem of lossy compression under distortion constraints, called multiterminal source coding, was posed and the bounds given.

A special case of multiterminal source coding is source coding with side information at the decoder, or Wyner-Ziv coding (WZC). The WZC problem considers lossy compression of source X under the distortion constraint when a correlated source S—called side information—is available at the decoder but not at the encoder (see Figure 12.4). This rate-distortion problem was first considered by Wyner and Ziv in [13], where the

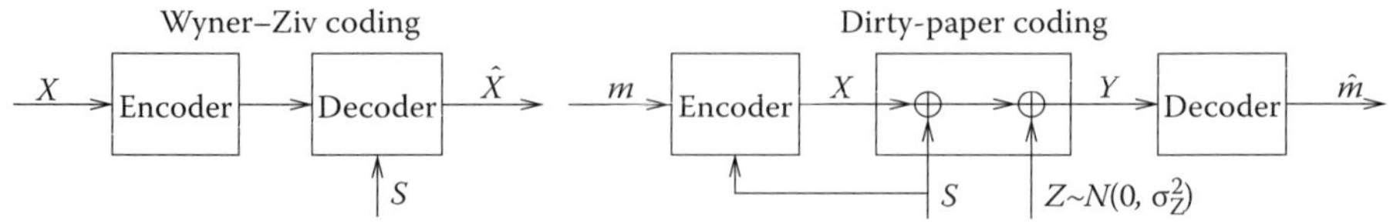

FIGURE 12.4 Coding with side information. WZC refers to lossy source coding of X with decoder side information S, whereas DPC considers channel encoding of message m with encoder side information S over an AWGN channel.

minimum rate for compressing X was derived. In general, WZC incurs a rate loss when compared to the case with S, also available at the encoder. However, if the correlation between X and S is modeled as $X = S + Z$, with Z being an i.i.d., memoryless Gaussian random variable, independent of S, then there is no rate loss with WZC under the mean-squared error (MSE) distortion measure.

The information-theoretical dual [17] of WZC is channel coding with side information at the encoder, or Gelfand-Pinsker coding [18], where the encoder has perfect (noncausal) knowledge of the side information or CSI. The limits on the rate at which messages can be transmitted to a receiver are given in [18]. In general, there is a rate loss compared to the case when the receiver also knows noncausally the CSI, i.e., the encoder side information. However, when the channel is additive white Gaussian noise (AWGN), Gelfand-Pinsker coding does not suffer any rate loss. In this case we have the celebrated dirty-paper coding (DPC) problem [14], shown in Figure 12.4 (right), where the decoder can completely cancel out the effect of the interference caused by the side information.

Practical WZC and DPC both involve source-channel coding. WZC can be implemented by first quantizing the source X, followed by Slepian-Wolf coding of the quantized X with side information S at the decoder [19]. Using syndrome-based channel coding for compression, Slepian-Wolf coding here plays the role of conditional entropy coding. For DPC, source coding is needed to quantize the side information to satisfy the power constraint. In the meantime, the quantizer induces a constrained channel, for which practical channel codes can be designed to approach its capacity. Indeed, limit-approaching code designs [20–22] have appeared for both WZC and DPC recently.

12.3.2 The Relay Channel

Since cooperative diversity is largely based on relaying messages, its information-theoretical foundation is built upon the landmark 1979 paper of Cover and El Gamal [23] on capacity bounds for relay channels. We thus start with the relay channel, give the theoretical bounds on its capacity, and describe proposed coding strategies in the Gaussian and Rayleigh flat-fading environments. Then we proceed with extensions to two-transmitter two-receiver cooperative channels.

The relay channel, introduced by van der Meulen in [5], is a three-node channel where the source communicates to the destination with the help of an intermediate relay node. It is shown in Figure 12.5. The source broadcasts encoded messages to the relay and destination. The relay processes the received information and forwards the resulting signal to the destination. The destination collects signals from both the source and relay

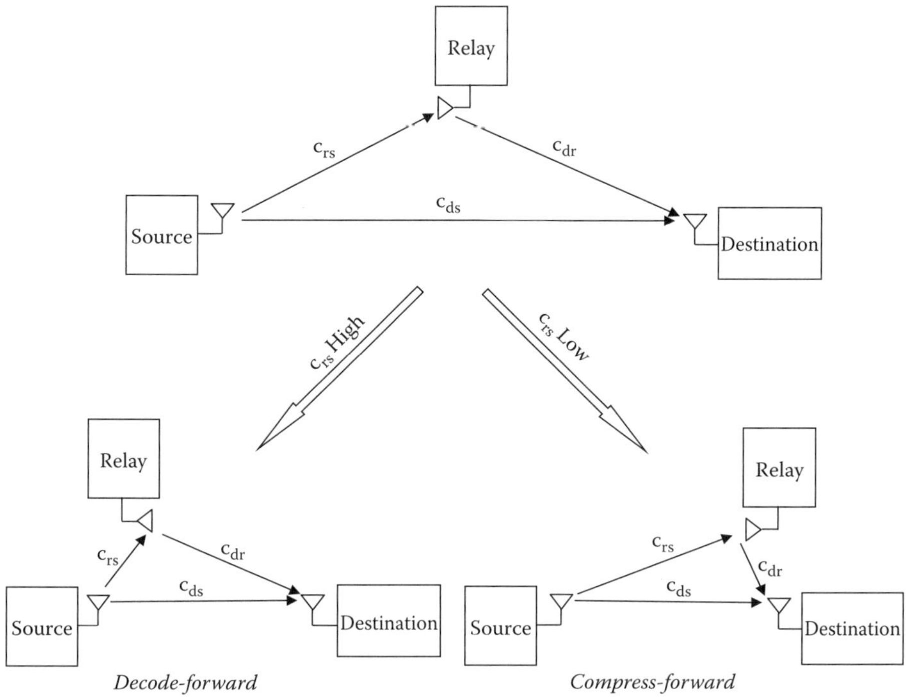

FIGURE 12.5 The wireless relay channel. Decode-forward works better when the relay is close to the source, but compress-forward is preferred when the relay is close to the destination.

node before attempting to recover the information. The relay's task is thus to facilitate decoding at the destination by means of spatial/temporal diversity; that is, it enables the message to be sent via different paths and in different time slots, so that the destination can improve decoding by exploiting two different independently corrupted looks of the same message.

The capacity for the general relay channel is still unknown, and we can only speak of the upper bound, above which reliable communication is not possible, and the lower bound, which can asymptotically be achieved with developed coding strategies; it is still unknown whether there exists a coding scheme that can operate in the uncertainty region between the two bounds.

Cover and El Gamal [23] derived upper and lower bounds on the capacity of the general relay channel using random coding and converse arguments. These two bounds coincide only in a few special cases [24, 25] (e.g., the degraded Gaussian case [23]).

The wireless relay channel is shown in Figure 12.5, where c_{rs}, c_{ds}, and c_{dr} denote channel coefficients. There are two setups in relaying: full-duplex and half-duplex. In the full-duplex setup, the relay is able to transmit and receive simultaneously on the same frequency. The capacity bounds are given by Cover and El Gamal [23], and efficient practical designs performed by Zhang and Duman [26]. Implementing full-duplex relaying, however, is a microwave design challenge (e.g., due to the large difference in the transmitting and receiving signal power levels). A simpler setup is half-duplex relaying, in which the relay does not simultaneously receive and transmit. Half-duplex relaying [24, 27–30] can be implemented with lower complexity by using either time division, frequency division, or code division, which are equivalent from an information theory point of view [8]. In the following, we will focus on time-division half-duplex relaying and discuss both capacity bounds and practical designs.

In time-division relaying, a frame of length n is divided into two parts: a relay-receive period of length $n\alpha$, $0 \le \alpha \le 1$, and a relay-transmit period of length $n(1-\alpha)$. In the relay-receive period, the source transmits a codeword x_{s1}. The relay overhears this transmission, processes its received signal y_r in some way, and transmits a codeword $x_r = f_r(y_r)$ in the relay-transmit period. While the relay transmits, the source simultaneously transmits another codeword x_{s2}. The codeword x_{s2} is not heard by the relay, as it is in transmit mode, and is therefore transmitted directly to the destination. One way to accomplish this is to split the message $m \in \{1, \ldots, M\}$ into two parts, m_1 and m_2, at the source. Then, m_1 is encoded into the $n\alpha$-length codeword $x_{s1}(m_1)$, and the remaining m_2 is encoded into an $n(1-\alpha)$-length codeword $x_{s2}(m_2)$. At the symbol level, the received signals at the relay and destination during the relay-receive period are

$$y_r[n]=c_{rs}x_{s1}(m_1)[n]+z_{rs}[n] \tag{12.5}$$

and

$$y_{d1}[n]=c_{ds}x_{s1}(m_1)[n]+z_{ds}[n], \tag{12.6}$$

respectively, where z_{rs} and z_{ds} are independent white Gaussian noises with unit power. During the relay-transmit period in the asynchronous case, the relay sends an $n(1-\alpha)$-length codeword $x_r(m_1)$ to the destination, which receives

$$y_{d2}[n]=c_{ds}x_{s2}(m_2)[n]+c_{dr}x_r(m_1)[n]+z[n], \tag{12.7}$$

where z is again a white Gaussian noise with unit power.

In the synchronous case, the system can additionally use the antennas at the source and the relay as a two-antenna transmit array. Suppose that the source is able to completely predict what the relay will send in the relay-transmit period; then the source can transmit the same signal with a phase shift calibrated so that the two signals add up coherently at the destination. The received signal is then

$$y_{d2}[n]=c_{ds}x_{s2}(m_2)[n]+(c_{dr}+c_{ds}A)x_r(m_1)[n]+z[n], \tag{12.8}$$

where A is a complex constant subject to a power constraint and with such a phase that $|c_{dr} + c_{ds}A|$ is maximized. If the source can only partially predict what the relay will transmit, it is still possible to take advantage of this partial coherency.

The optimum operation at the relay is not known, but several coding schemes have been proposed [2, 23, 25, 27] to obtain achievable bounds on the rate region. These schemes can be classified into *decode-forward* and *observe-forward* [23], although hybrid schemes are also possible [23].

The main operation of decode-forward (DF) is full decoding at the relay node. Upon receiving y_r, the relay node first decodes m_1 and then re encodes it before forwarding the resulting codeword $x_r(m_1)$ to the destination during the relay-transmit period. It should be emphasized that the relay might use a different codebook than the source. In any case, the source can completely predict what the relay will transmit, and full coherency is therefore possible. The destination attempts to reconstruct message m by combining the signals received during the relay-receive and relay-transmit periods using either successive list decoding [8, 23], backward decoding [31], or decoding based on parallel Gaussian channel arguments [28], which all result in the same achievable rate region. Although DF can be very efficient in some scenarios [24, 25], since the relay must perfectly decode the source message, the achievable rates are bounded by the capacity of the channel between the source and the relay. That is, the capacity of the DF relay channel cannot be higher than that of the source-relay channel; hence, if the channel between the source and relay is poor, the relay is useless, and direct transmission from the source to the destination is a better option.

To alleviate this problem of DF, a class of observe-forward schemes has been proposed, where the relay does not attempt to decode the signal from the source; it merely forwards a processed version of its received signal to the destination.

The simplest observe-forward scheme is *amplify-forward* (AF) [32], in which the relay, sticking to its rudimentary role, just amplifies the received signal before forwarding. A more sophisticated scheme is *compress-forward* (CF), which is rooted in the original work of Cover and El Gamal [23], where the relay compresses the signal it has received from the source within certain distortion. Since y_r (received by the relay node) and y_{d1} (received by the destination node) are independently corrupted versions of the same encoded message $x_{s1}(m_1)$, they are correlated. Thus, the relay node can employ WZC [13] when compressing y_r by treating y_{d1} as the decoder side information. The Wyner-Ziv compressed signal is then channel encoded to $x_r(m_1)$ before being forwarded to the destination, which recovers m_2 and m_1 using successive cancellation decoding that involves several steps. First, $\hat{x}_r(m_1)$ is reconstructed by assuming $x_{s2}(m_2)$ as the noise, and then it is subtracted from y_{d2} before m_2 is decoded. Second, to reconstruct m_1, y_r is estimated from $\hat{x}_r(m_1)$ using Wyner-Ziv decoding with y_{d1} as the decoder side information; maximum ratio combining on y_{d1} and the obtained estimate $\hat{y}_r$ is then invoked to recover m_1. CF based on WZC has higher computational complexity than DF, but it gives many rate points that are not achievable with any other coding strategies. It provides the best solution [23–25] when the relay is close to the destination node.

12.3.3 Capacity Bounds of the Gaussian Half-Duplex Relay Channel

In the flat-fading environment, channel coefficients vary in time. The upper and lower bounds on capacity can be computed by averaging over all channel realizations (with optimally allocated power). This average capacity is called *ergodic capacity*. The bounds on ergodic capacity for the full-duplex and half-duplex flat-fading relay channel are given in [24]. However, in practice, it is more convenient to use *outage capacity* [33, 34]. Since in the fading channel case, the source does not know the exact value of channel coefficients ($\mathbf{c} = [\mathbf{c}_{sr}, \mathbf{c}_{sd}, \mathbf{c}_{rd}]$) for a particular channel realization, it cannot determine the maximum achievable rate; consequently, the sending rate R can be higher than the capacity of the instantaneous channel realization $C(\mathbf{c})$, in which case the destination cannot decode. The probability of this event, $P(R > C(\mathbf{c}))$, is called the outage probability. The outage capacity is then the maximum achievable rate with the outage probability less than a certain level p; it can be computed as the $(1 - p)$ percentile of the rate for the specific value of $\mathbf{c}$.

In the following, we summarize the outage capacity bounds of the Gaussian half-duplex relay channel, which can also be used for computing the outage capacity of a wireless quasi-static flat-fading channel (as done in [24]). The bounds for the full-duplex Gaussian channel can be found in [8, 23, 24].

Under the assumption that the nodes are synchronized and have perfect CSI, i.e., each node knows instantaneous values of all channel coefficients and their statistics, an upper bound on the capacity of the Gaussian half-duplex relay channel (although channel coefficients are in general assumed to be complex, in this case they are positive real constants) is derived in [24, 27] and given by

$$C_{ub} = \max_{0 \le \rho \le 1, 0 \le \alpha \le 1, 0 \le k \le 1} \min\{C_{ub1}, C_{ub2}\}, \tag{12.9}$$

where

$$C_{ub1} = \frac{\alpha}{2}\log\left(1 + \left(|c_{rs}|^2 + |c_{ds}|^2\right)\frac{kP_s}{\alpha}\right) + \frac{1-\alpha}{2}\log\left(1 + (1-\rho^2)|c_{ds}|^2\frac{(1-k)P_s}{1-\alpha}\right),$$

$$C_{ub2} = \frac{\alpha}{2}\log\left(1 + |c_{ds}|^2\frac{kP_s}{\alpha}\right)$$
$$+\frac{1-\alpha}{2}\log\left(1 + |c_{ds}|^2\frac{(1-k)P_s}{1-\alpha} + |c_{dr}|^2 P_r + \frac{2\sqrt{\rho^2|c_{ds}|^2|c_{dr}|^2(1-k)P_sP_r}}{1-\alpha}\right),$$

and P_s and P_r are the average source and relay power constraints, respectively. The parameter ρ reflects the correlation between the source and relay signals, and it can be written in closed form [24, 27]. It is clear from the bound above that the highest multiplexing gain r is 1. However, the full-diversity gain of 2 can be achieved with a simple AF scheme [2, 35].

The rate bound of DF is [24, 27]:

$$R_{DF} \le \max_{0 \le \rho \le 1, 0 \le \alpha \le 1, 0 \le k \le 1} \min\{R_{DF1}, R_{DF2}\}, \tag{12.10}$$

where

$$R_{DF1} = \frac{\alpha}{2}\log\left(1+\left|c_{ds}\right|^2 \frac{kP_s}{\alpha}\right)+\frac{1-\alpha}{2}\log\left(1+(1-\rho^2)\left|c_{ds}\right|^2 \frac{(1-k)P_s}{1-\alpha}\right)$$

and $R_{DF2} = C_{ub2}$. The achievable rate with CF is [24]

$$R_{CF} \leq \max_{0\leq\alpha\leq 1, 0\leq k\leq 1} \{R_{CF1}(\alpha,k)+R_{CF2}(\alpha,k)\}, \tag{12.11}$$

where

$$R_{CF1}(\alpha,k)=\alpha\frac{1}{2}\log\left(1+\left|c_{ds}\right|^2 \frac{kP_s}{\alpha}+\frac{\left|c_{rs}\right|^2 kP_s}{\alpha(1+\sigma_\omega^2)}\right)$$

and

$$R_{CF2}(\alpha,k)=(1-\alpha)\frac{1}{2}\log\left(1+\left|c_{ds}\right|^2 \frac{(1-k)P_s}{1-\alpha}\right),$$

with σ_ω^2 being the WZC noise [13] given by

$$\sigma_\omega^2=\frac{\alpha+\left(\left|c_{rs}\right|^2+\left|c_{ds}\right|^2\right)kP_s}{\left\{\left[1+\left|c_{dr}\right|^2 P_r\Big/\left(1-\alpha+\left|c_{ds}\right|^2(1-k)P_s\right)\right]^{\frac{1-\alpha}{\alpha}}-1\right\}\left(\alpha+\left|c_{ds}\right|^2 kP_s\right)}.$$

DF and CF give the best-known results on the achievable rates for the half-duplex relay channel (however, a hybrid approach may give a higher rate). Depending on the parameters, either DF or CF can be superior. Indeed, DF outperforms CF when the link between the source and relay is better than that between the relay and destination (e.g., when the relay is close to the source). On the other hand, CF provides higher rates when the link between the relay and destination is clean (e.g., when the relay is close to the destination). See Figure 12.5. We show in Figure 12.6 for one setup the rate bounds in (12.10) and (12.11), achievable with DF and CF, respectively, together with the upper bound given by (12.9) and the rate bound with multihop transmission, which is given by the minimum between the capacity at the source-relay link, $\frac{1}{2}\log(1 + |c_{rs}|^2 P_s)$, and the capacity at the relay-destination link, $\frac{1}{2}\log(1 + |c_{dr}|^2 P_r)$. We plot the rate gain relative to direct transmission (i.e., no relaying) as a function of $|c_{rs}|^2$. The increase in $|c_{rs}|^2$ can be construed as the result of decreased distance between the source and the relay. It is seen from Figure 12.6 that CF outperforms DF for low $|c_{rs}|^2$. When $|c_{rs}|^2 < |c_{ds}|^2 = 0$ dB, DF is worse than direct transmission. On the other hand, CF always outperforms direct

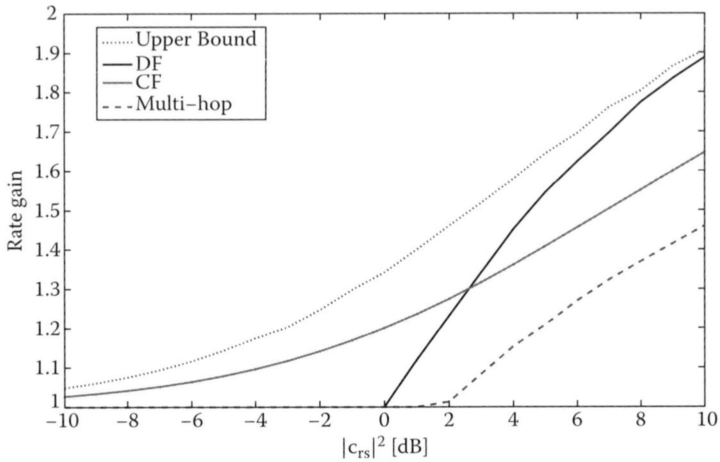

FIGURE 12.6 The multihop bound and the upper bound on the capacity together with the achievable bounds of DF and CF for the Gaussian half-duplex relay channel, assuming $|c_{ds}|^2$ = 0 dB, $|c_{dr}|^2$ = 10 dB, and $P_s = P_r$ = 5 dB. The rate gain over direct transmission is shown as a function of $|c_{rs}|^2$.

transmission. Thus, even if the link between the source and relay is poor, the relay can still help somewhat by using CF.

It is instructive to compare relay channel signaling with a traditional multihop ad hoc network, where physical layer communication and networking are typically separated. Such a comparison will show how cooperative diversity can help increase the performance over traditional networking. In a traditional multihop network, the source transmits a packet either directly to the destination or to the relay, which would decode it, re-encode it, and transmit it to the destination. Relay channel signaling improves upon this in several ways:

1. The destination uses the signals from both the source and relay for decoding, as opposed to only one of them.
2. The relay uses a different codebook for encoding in DF than the source, which is similar to using error-correcting codes rather than repetition coding.
3. The relay can use soft information, as in CF, which resembles using soft decisions in decoding error-correcting codes rather than hard decisions.
4. The source is allowed to transmit new information simultaneously with the relay's transmission, which at high SNR brings a large increase in rate.
5. In the synchronous case, the relay can use coherency to combine signals constructively, achieving a gain similar to that in MIMO systems.

12.3.4 Receiver Cooperation

In receiver cooperation, two (closely located) single-antenna receivers cooperate to facilitate decoding messages from two remote single-antenna transmitters.

The channel model is shown in Figure 12.3 (upper left) with $c_{21} = c_{12} = 0$, where node 1 (node 2) wants to communicate to node 3 (node 4). We consider only the full-duplex case, i.e., nodes 3 and 4 can simultaneously transmit and receive. Since the cooperative channel can be viewed as a combination of the interference channel and the relay channel, its best achievable rate regions are obtained by combining DF or CF coding techniques for the relay channel with coding for the interference channel [9]. In receiver cooperation, a receiver node processes the received information and forwards the result to the other receiver node to help decoding. Because the distance between the two receivers is expected to be much smaller than that between a transmitter and a receiver, CF with WZC provides the highest achievable rates. Indeed, from (12.2), the received signals in nodes 3 and 4 at time instants i and $i+1$ are

$$\begin{aligned} y_3[i] &= x_1[i] + c_{32}x_2[i] + c_{34}x_4[i] + z_3[i], \\ y_3[i+1] &= x_1[i+1] + c_{32}x_2[i+1] + c_{34}x_4[i+1] + z_3[i+1], \\ y_4[i] &= c_{41}x_1[i] + x_2[i] + c_{43}x_3[i] + z_4[i], \\ y_4[i+1] &= c_{41}x_1[i+1] + x_2[i+1] + c_{43}x_3[i+1] + z_4[i+1]. \end{aligned} \tag{12.12}$$

In CF [36], the receiver in node 3 (or node 4) employs WZC to compress the signal $y_3[i]$ (or $y_4[i]$) it has received, while assuming $y_4[i]$ (or $y_3[i]$) as the decoder side information, before passing the resulting codeword $x_3[i+1]$ (or $x_4[i+1]$) to the collaborating receiver in node 4 (or node 3). Node 3 starts by decoding $x_4[i+1]$ from $y_3[i+1]$ while treating $x_1[i+1] + c_{32}x_2[i+1]$ as part of the Gaussian noise. In *forward decoding*, $x_4[i]$, recovered in the previous time instant, is Wyner-Ziv decoded using $y_3[i]$ as the decoder side information, resulting in an estimate of $y_4[i]$. Next, the joint or individual decoding technique [9] proposed for the interference channel is employed to reconstruct $x_1[i]$ (and $x_2[i]$) from the obtained estimates

$$y_3[i] - c_{34}\hat{x}_4[i] = x_1[i] + c_{32}x_2[i] + z_3[i]$$

and

$$\hat{y}_4[i] - c_{43}x_3[i] = c_{41}x_1[i] + x_2[i] + z_4[i].$$

A similar procedure can be performed at node 4. Besides forward decoding, it is also feasible to employ *backward decoding*, where the decoder starts by decoding the previously received block of symbols and proceeding backwards. In [36], forward decoding is combined with either joint or individual decoding [9], and backward decoding is used with joint decoding, giving three different decoding choices. Since nodes 3 and 4 can use three different decoding methods each, there are nine possibilities, each providing a different rate bound. To obtain the best achievable CF rate bound, the maximum of all nine rate bounds should be taken.

In a similar manner, DF can be extended to receiver cooperation. The DF strategy for receiver cooperation can be found in [36].

12.3.5 Capacity Bounds of the Gaussian Full-Duplex Receiver Cooperative Channel

In this section we summarize the results of [36], where the upper and lower bounds on capacity of receiver cooperation are derived for both CF and DF. The derived upper bounds are tighter than the standard max-flow-min-cut bound. It is given by:

$$R_1 + R_2 \le \log\left(1+|c_{41}|^2 P_1 + P_2 + |c_{43}|^2 P_3\right) + \log\frac{1+\left(1+|c_{41}|^2\right)P_1}{1+|c_{41}|^2 P_1}$$

in the asynchronous case, and

$$R_1 + R_2 \le \log\left(1+|c_{41}|^2 P_1 + P_2 + |c_{43}|^2 P_3 + 2\sqrt{|c_{43}|^2 P_2 P_3 + |c_{43}|^2 |c_{41}|^2 P_1 P_3}\right) + \log\frac{1+\left(1+|c_{41}|^2\right)P_1}{1+|c_{41}|^2 P_1}$$

in the synchronous case. Note that we get a symmetric set of rate bounds if nodes 1 and 2 are exchanged with nodes 3 and 4. Achievable (lower) rate bounds for CF and DF can be found in [36]. Figure 12.7 shows the sum-rate $R_1 + R_2$ as a function of the received SNR on the direct link between nodes 1 and 3. The received SNR at the link between nodes 3 and 4 is 30 dB higher than that from the direct link—an indication that the cooperating nodes are close together. The average powers of all four nodes are the same. All channels are independent Rayleigh flat fading, meaning that each c_{ji} is an i.i.d. Gaussian random variable, and the results are averaged over simulated ensembles of channel realizations. For comparison purposes, the rate limits of the two-user MAC with two antennas at the receiver (with perfect cooperation) and the interference channel (without cooperation) are included.

It is seen from Figure 12.7 that receiver cooperation with CF gives an additive gain that can be up to 20 dB higher than no cooperation, CF always performs close to the upper bound, and there is no gain from synchronization. Interestingly, receiver cooperation performs close to using two receiver antennas at low and medium SNRs, thus providing a multiplexing gain of 2. However, at the high SNRs, the multiplexing gain drops to 1, and the rate gain over the noncooperative case boils down to a high additive gain.

We can see that, in the high-SNR regime, receiver cooperation gives a multiplexing gain of only $r = 1$, as opposed to the two-user MAC with two receiver antennas, which results in $r = 2$. However, the additive gain with receiver cooperation, which is upper bounded by

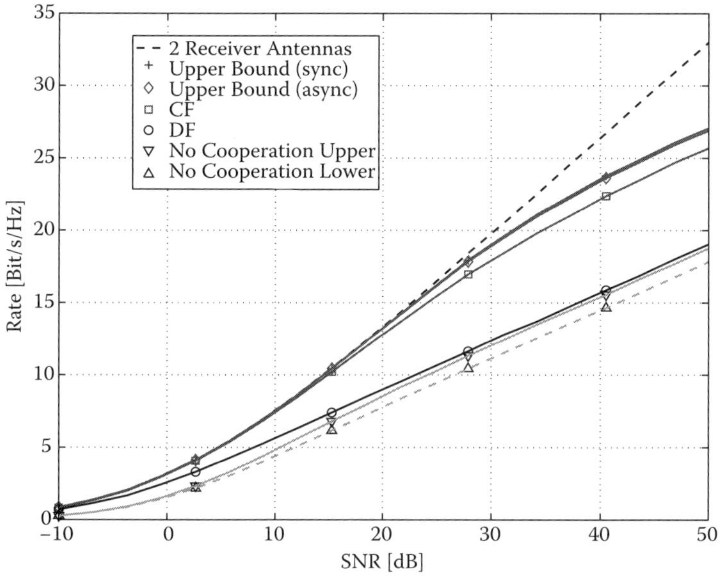

FIGURE 12.7 Bounds on the sum-rate $R_1 + R_2$ as a function of the received SNR from the direct link between nodes 1 and 3 for receiver cooperation. The received SNR at the link between nodes 3 and 4 is 30 dB higher than that at the direct link. The average powers of all four nodes are the same.

$$a \le \min\left\{\log\left(|c_{41}|^2 P_1 + P_2 + |c_{43}|^2 P_3\right) + \log\left(1 + |c_{41}|^{-2}\right),\right.$$
$$\left.\log\left(P_1 + |c_{32}|^2 P_2 + |c_{34}|^2 P_4\right) + \log\left(1 + |c_{32}|^{-2}\right)\right\}, \tag{12.13}$$

can be very high. On the other hand, for $|c_{41}|,|c_{32}| > 1$ it is shown in [36] that CF with forward joint decoding gives an additive gain of

$$a = \min\left\{\log\left(|c_{41}|^2 P_1 + P_2 + |c_{43}|^2 P_3\right), \log\left(P_1 + |c_{32}|^2 P_2 + |c_{34}|^2 P_4\right)\right\}. \tag{12.14}$$

Note that the gain in (12.14) is identical to that in (12.13) except for the $\log(1 + |c_{41}|^{-2})$ and $\log(1 + |c_{32}|^{-2})$ terms, which are small for large $|c_{41}|$ and $|c_{32}|$. Thus, CF with WZC achieves capacity asymptotically as $|c_{41}|$, $|c_{32}|$, and the SNRs go to infinity. All other cooperative strategies (including DF) give no additional gain over that of no cooperation, whose additive gain is

$$a = \min\left\{\log\left(|c_{41}|^2 P_1 + P_2\right), \log\left(P_1 + |c_{32}|^2 P_2\right)\right\}. \tag{12.15}$$

However, these strategies are useful in the medium- or low-SNR regimes. From (12.14) and (12.15) we see that the gain of receiver cooperation comes from the terms $|c_{43}|^2 P_3$ and $|c_{34}|^2 P_4$, which depend on the channel between the two receivers. Since this channel is expected to be good (a node should cooperate with its "best neighbor"), this gain can be very high. An interesting conclusion from [36] is that the gain from exploiting full synchronization in receiver cooperation is very limited. Thus, in practice it is enough to resort to the asynchronous cooperation, which significantly saves the hardware cost. However, as pointed out in [37], optimal power allocation is essential in achieving the full additive gain.

12.3.6 Transmitter Cooperation

In transmitter cooperation, two (close) single-antenna transmitters collaborate in communicating to two (remote) single-antenna receivers. The channel model is depicted in Figure 12.3 (upper left) with $c_{43} = c_{34} = 0$. As in receiver cooperation, we restrict to the full-duplex case, where nodes 1 and 2 can simultaneously transmit and receive.

It is shown in [36, 37] that, in contrast to receiver cooperation, synchronization helps a lot when the transmitters cooperate. That is, if the two transmitters are synchronized, they can completely cancel out the interference using DPC.

The DPC technique was exploited in [7, 38] to find the capacity of the Gaussian MIMO broadcast channel. For the two-antenna broadcast channel with two receivers [7], the main idea is to decompose the MIMO channel into two interference channels and perform successive encoding, in which the message for the second receiver is dirty-paper encoded while assuming the previously encoded message for the first receiver is known interference (the side information). In this way, the second receiver can completely cancel out the interference from the signal for the first receiver. However, to achieve full capacity, the transmitter has to perform optimal channel decomposition using precoding with the output vector $\mathbf{x} = \mathbf{B}[u_1 \;\; u_2]^T$, where $\mathbf{B}$ is a 2×2 precoding matrix that has to satisfy the power constraint, and u_1 and u_2 are the encoded codewords (with unit power) intended for the first and second receivers, respectively, and obtained via successive dirty-paper encoding. Assuming a 2×2 channel matrix $\mathbf{H}$ and unit-power Gaussian noise, the achievable rates for the first and second receivers are

$$R_1 = \log\left(1 + \frac{|s_{11}|^2}{1 + |s_{12}|^2}\right)$$

and $R_2 = \log(1 + |s_{22}|^2)$, respectively, where s_{ij} are the entries of matrix $\mathbf{S} = \mathbf{HB}$.

The coding strategy of [7] is extended to transmitter cooperation in [36, 39]. In [39], it is assumed that the channel between the two transmitters is orthogonal to the channels between the transmitters and receivers (which can be achieved by means of multiple access techniques). Thus, collaboration between the transmitters does not cause interference at the receivers, which simplifies the code design. This orthogonality assumption is removed in [36], where a coding scheme that in each time instant exploits *three* DPCs is proposed.

We outline here a simplified (hence more practical) solution proposed in [40] based on *one* DPC (in conjunction with backward decoding). During the ith time instant, the transmitter in node j, $j = 1, 2$, sends

$$x_j[i] = A_j U_j\left(\omega_j[i]\right) + t_{j1} U_1^0\left(\omega_1[i-1], U_2^0\left(\omega_2[i-1]\right)\right) + t_{j2} U_2^0\left(\omega_2[i-1]\right).$$

Here U_2^0, U_1, and U_2 are Gaussian codebooks (e.g., standard channel codes in practice) of unit power that encode $\omega_2[i-1]$, $\omega_1[i]$, and $\omega_2[i]$, respectively. U_1 and U_2 are used for exchanging messages between the transmitters and appear as part of the background noise at the receivers. Assuming correct decoding of $U_1(\omega_1[i-1])$ and $U_2(\omega_2[i-1])$ in time instant $i-1$, the two transmitters can now act as a single two-antenna broadcast transmitter, and the coding strategy of [7] described above can be applied. Thus, the unit-power codebook U_1^0 can encode $\omega_1[i-1]$ using DPC with $U_2^0(\omega_2[i-1])$ as the side information. The scaling factors A_i and t_{ij} are selected to maximize the rate while satisfying the input power constraints.

In the asynchronous case, DPC cannot be exploited, and the resulting known achievable rates are strictly below those in the synchronous case. However, so far there exist no upper bounds that actually prove that the gains cannot be obtained without synchronization. Although it is possible to use CF based on WZC in transmitter cooperation, since the two transmitters are closely located, DPC always dominates. This is why WZC is not considered in this setup.

12.3.7 Capacity Bounds of the Gaussian Full-Duplex Transmitter Cooperative Channel

In this section we summarize the capacity bounds of the Gaussian full-duplex transmitter cooperative channel. As in the case of receiver cooperation, we give only final results without derivations. All proofs can be found in [36].

Based on the argument exploited in [7] that the capacity region depends only on the marginal distribution of the noises at the receivers and not on their correlation, the following upper bounds on capacity are derived in [36]. For $|c_{41}| < 1$, the upper bound is

$$R_2 \le \log \frac{1+|c_{41}|^2 P_1 + P_2}{|c_{41}|^2 2^{R_1} \dfrac{1+P_1}{1+\left(1+|c_{21}|^2\right)P_1} + 1 - |c_{41}|^2},$$

in the asynchronous case, and

$$R_2 \le \log \frac{1+\left(|c_{41}|\sqrt{P_1}+\sqrt{P_2}\right)^2}{|c_{41}|^2 2^{R_1} \dfrac{1+P_1}{1+\left(1+|c_{21}|^2\right)P_1}+1-|c_{41}|^2}$$

in the synchronous case, and the sum-rate bound $R_1 + R_2$ is an increasing function of R_1. If $|c_{41}| > 1$, the upper bound is

$$R_1+R_2 \le \log\left(1+|c_{41}|^2 P_1+P_2\right)+\log\frac{1+\left(|c_{41}|^2+|c_{21}|^2\right)P_1}{1+|c_{41}|^2 P_1}$$

in the asynchronous case, and

$$R_1+R_2 \le \log\left(1+\left(|c_{41}|\sqrt{P_1}+\sqrt{P_2}\right)^2\right)+\log\frac{1+\left(|c_{41}|^2+|c_{21}|^2\right)P_1}{1+|c_{41}|^2 P_1} \tag{12.16}$$

in the synchronous case. There is also a symmetric set of rate bounds by exchanging nodes 1 and 2 with nodes 3 and 4.

Achievable bounds in asynchronous (without DPC) and synchronous systems (with three DPCs) can be found in [36]. The achievable bound for the scheme that used one DPC is given in [40]. Figure 12.8 shows the sum-rate bounds $R_1 + R_2$ as functions of the received SNR on the direct link between nodes 1 and 3. The simulation setup is similar to that for receiver cooperation with the received SNR at the cooperative link (between nodes 1 and 2) being 30 dB higher than that at the direct link, again indicating that the cooperating transmitters are close together.

The achievable bounds of the synchronous system with DPC are usually close to the upper bound, although noticeable gaps exist in certain SNR ranges. There is only a small performance loss if only one DPC is used instead of three. The additive gain compared to the noncooperative case is up to 15 dB in the high-SNR regime. Transmitter cooperation with DPC performs close to using two transmitter antennas at low and medium SNRs, giving a multiplexing gain of 2. However, at high SNRs, the multiplexing gain is only 1.

Similar to receiver cooperation, in the high-SNR regime, transmitter cooperation only gives a multiplexing gain of $r = 1$ (in contrast to the two-antenna broadcast channel, which results in $r = 2$). The additive gain can be high. For example, when $|c_{41}| < 1$, in the synchronous case it is bounded by

$$a \le \log\left(\left(|c_{41}|\sqrt{P_1}+\sqrt{P_2}\right)^2\right)+\log\left(\frac{1+|c_{21}|^2}{|c_{41}|^2}\right). \tag{12.17}$$

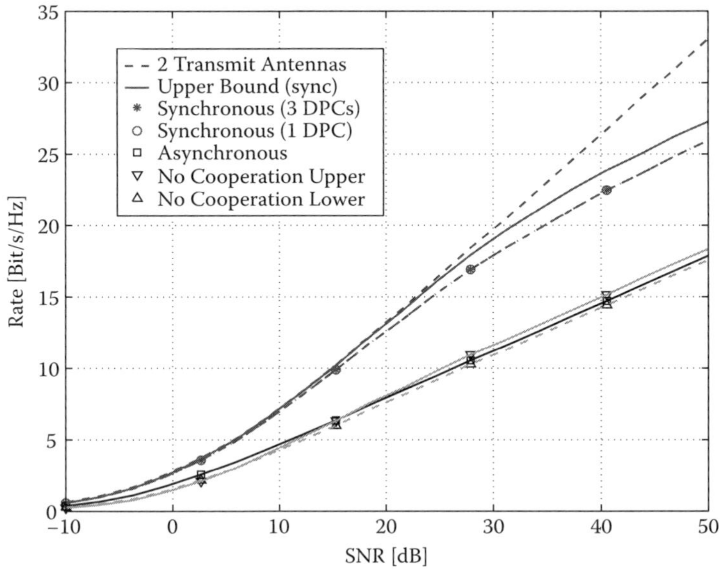

FIGURE 12.8 The bounds on the sum-rate $R_1 + R_2$ as a function of the received SNR at the direct link between nodes 1 and 3 for transmitter cooperation. The received SNR at the link between nodes 1 and 2 is 30 dB higher than that at the direct link. The average powers of all four nodes are the same. The upper bound is for the synchronous system.

By exchanging nodes 1 and 2 with nodes 3 and 4, we get another symmetric rate bound. Besides the multiplexing gain of $r = 1$, DPC achieves a high-SNR additive gain of

$$a = \log\left(\left(|c_{41}|\,|t_{12}| + |t_{22}|\right)^2\right) + \log\frac{|c_{21}|^2}{|c_{41}|^2}. \tag{12.18}$$

Comparing (12.17) and (12.18), we see that the additive gain of using DPC is approximately equal to that from the upper bound when $|t_{12}|^2 \approx P_1$ and $|t_{22}|^2 \approx P_2$, which corresponds to the scenario with weak interference, i.e., $|c_{41}|,|c_{32}| \ll 1$. In this case, the gain compared to no cooperation in (12.15) is $\min\{\log(|c_{12}|),\log(|c_{21}|)\}$, which can be significant, because the channel between the two transmitters is expected to be good. This is illustrated in Figure 12.9, which shows the high-SNR additive gain for a symmetric cooperative channel ($|c_{41}| = |c_{32}|$, $|c_{21}| = |c_{12}|$). Note that under strong interference, i.e., $|c_{32}|,|c_{41}| \gg 1$, there is no gain from cooperation, which might seem somewhat surprising and is in contrast to receiver cooperation. For weak interference, on the other hand, there is a high gain from cooperation. This is true even when the link between the transmitters is weak ($|c_{21}|^2 = -6$ dB in Figure 12.9), and it can be explained by the fact that there is no known signaling for the interference channel with weak interference, while cooperation can help in this scenario.

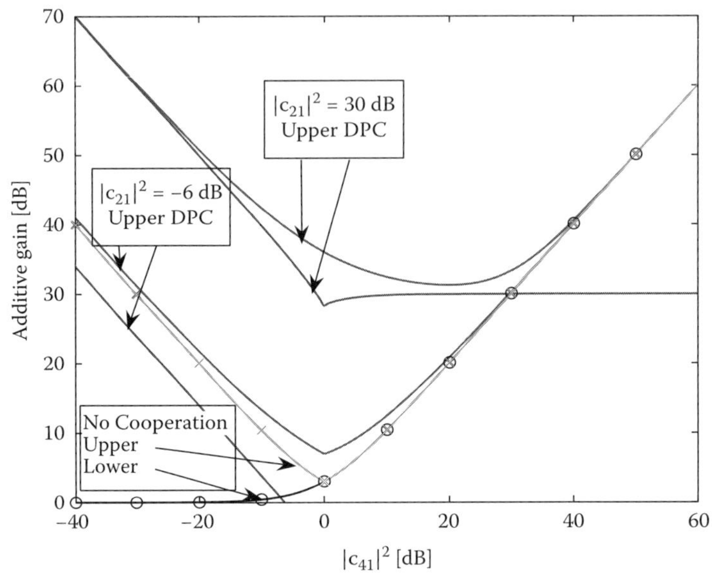

FIGURE 12.9 The bounds on the additive gain for a symmetric transmitter cooperative channel ($|c_{41}| = |c_{32}|$, $|c_{21}| = |c_{12}|$), averaged over the relative phases of c_{41} and c_{32}. The upper bound is for a synchronous system.

12.4 Practical Designs

In the previous section, we described coding methods that could lead to the highest rates while assuming ideal coding, signaling, and infinite block length. In this section we describe practical systems based on AF, DF, and CF for the relay and cooperative channels, with the focus on capacity-approaching code designs.

Efficient practical coding protocols for wireless relay channels appeared in [41], where, for AF and DF schemes, receivers based on maximum-likelihood and maximum SNR criteria were developed. The work was extended to wireless multirelay channels in [32], where space-time codes [3] with AF and DF are designed to enable simultaneous transmission from all relays on the same channel without receive collision. It is further shown that the proposed schemes achieve the full-diversity gain. An AF space-time code for a single relay is proposed in [35]. Note that the works of [32, 35] only outline space-time code designs without practical implementation of channel codes.

Practical DF schemes for a half-duplex flat-fading relay channel based on distributed convolutional and turbo coding are proposed in [28]. The best scheme of [28] exploits a recursive systematic convolutional code at both the source and the relay. It results in a powerful distributed turbo code, which besides a spatial diversity gain of DF, achieves extra coding gain due to interleaving.

Extending their work on practical full-duplex relaying [26], Zhang and Duman [29] recently provided a DF design for half-duplex fading relay channels, where in a given time slot, the source and relay both transmit over the same channel, resulting in a high rate gain at the price of receive collision. The design in [29] exploits turbo coding with binary phase shift keying (BPSK) modulation [6]. It is consistent with the optimal DF scheme for the half-duplex relay channel described in section 12.3.2, with the simplification that coefficient A in (12.8) is always set to zero. Decoding is based on parallel Gaussian channel arguments [28]. Similar to the MAC setting, the destination exploits a MAP detector to extract information from the received mixture signal. Recent work [42] on low-density parity-check (LDPC) code [43] design for the half-duplex relay channel based on DF also reported a similar loss of 1.2 dB to the theoretical limit.

The systems in [28, 29, 32, 35] demonstrate the great advantage of relaying as compared to direct and multihop transmission. However, because these systems exploit AF or DF, they can approach the lower bound of DF at best, which is far away from the CF limit in many cases. Indeed, as shown in section 12.3.2, when the relay is close to the destination, CF gives rate points that are not achievable with any other coding strategies. Moreover, it was shown in [44] that CF achieves optimal diversity-multiplexing trade-off in half-duplex relay systems.

Practical CF code designs for the half-duplex relay channel recently appeared in [45–47]. The design of [45] is a quantize-forward scheme that does not exploit WZC at the relay. The design of [46] is based on WZC at the relay and uses scalar quantization and convolutional codes, but it does not exploit the limit-approaching CF scheme of [24], and no theoretical bounds or performance comparisons (to the bounds) were given.

A practical CF code design based on WZC for the half-duplex Gaussian relay channel, which closely follows the CF scheme outlined in section 12.3.2, is proposed in [47]. The scheme of [47] exploits BPSK modulation; hence, the signal to be compressed by WZC at the relay and the side information at the destination are not jointly Gaussian as assumed in section 12.3.2 and [36]. Instead, the source and the side information are Gaussian mixture generated from the BPSK modulation. Although the theoretical achievable rate of WZC for this model is yet unknown, a lower bound and an upper bound are derived in [47], which in the case when the relay is close to the destination are close to each other, indicating practically system capacity. The code design relies on practical WZC based on nested lattice quantization [48] followed by Slepian-Wolf coding [15] of the nested quantization index as a second stage of binning for further compression [49]. Thus, Slepian-Wolf coding here plays the role of conditional entropy coding of the nested quantization indices (given the decoder side information).

Practical Slepian-Wolf coding is implemented via channel coding (see [19] for a review of channel code designs for Slepian-Wolf coding). Since the Slepian-Wolf compressed bitstream is to be transmitted over a noisy channel from the relay to the destination, channel coding is needed to protect them. This calls for distributed joint source-channel coding (DJSCC) [50], i.e., joint Slepian-Wolf compression and channel protection. In the practical implementation of [47], irregular repeat-accumulate (IRA) codes [51] were used by designing one multilayer code to take care of two channels: one is the physical noisy channel between the relay and the destination, and another is the "virtual" correlation channel [19], which characterizes the correlation between the quantized source at the relay and the decoder side information at the destination.

In particular, the message m is split at the source into two parts, m_1 and m_2, which are protected independently by two different LDPC codes [43] and BPSK modulated before being transmitted in two separate fractions of a time slot. The relay compresses its received signal using WZC and adds error protection against the noise (and interference) in the link between the relay and the destination. WZC and error protection are performed jointly using DJSCC; the resulting codeword $x_r(m_1)$ is sent during the relay-transmit period. The destination starts by recovering m_2 using successive cancellation decoding: $x_r(m_1)$ is first reconstructed using distributed joint source-channel decoding, and is then subtracted from y_{d2} (interference cancellation) so that m_2 can be recovered with the first LDPC decoder having z as the only noise in the channel. The main idea behind distributed joint source-channel decoding is to view the system as transmitting the symbols over two channels—the first being the actual transmission channel (i.e., the MAC) with noise $z + x_{s2}(m_2)$, which describes the distortion experienced by the parity-check symbols of the IRA code, and the second being the virtual correlation channel [19] between y_r and the side information y_{d1}. The destination estimates $\hat{y}_r$ by employing a conventional IRA decoder for these two parallel channels. Finally, m_1 is recovered using maximum ratio combining (of $\hat{y}_r$ and $y_{d1} = c_{ds}x_{s1}(m_1) + z_{ds}$) and a second LDPC decoder for the direct transmission channel.

Figure 12.10 compares this CF design with the best practical DF design of [29] for the half-duplex Gaussian relay channel. Different theoretical bounds are also depicted (deri-

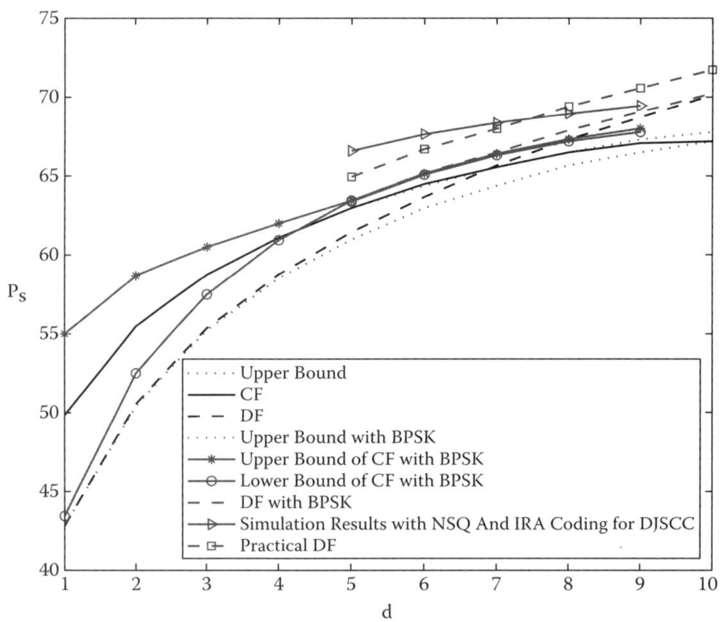

FIGURE 12.10 Half-duplex Gaussian relay channel. The average transmitting power from the source P_s is shown as a function of the distance d from the source to relay. The transmitting rate is 0.5 and the average transmitting power from the relay is $P_r = 70$ dB. The relay is moving along the line from the source to destination. BPSK signaling is assumed.

vations of the bounds in the BPSK case are not included here due to space limitations and interested readers are referred to [47] for details). The transmission rate is fixed at 0.5 bit per channel use, and the average relay power is $P_r = 70$ dB.

It is assumed that the relay is located along a straight line from the source to the destination, which are 10 m apart. The channel coefficient of the link from sender i to receiver j (sender i could be the source or relay, and receiver j could be the relay or destination) is $c_{ij}^2 = K_o d_{ij}^{-n}$ [28], where d_{ij} is the distance from sender i to receiver j, n is the path loss coefficient, $K_o = (c/4\pi d_o f_c)^2$, c is the light speed, d_o is the free-space reference distance, and f_c is the transmission frequency. The experimental setup is fixed with $f_c =$ 2.4 GHz carrier frequency, path loss coefficient $n = 3$, and free-space reference distance $d_o = 1$ m.

Figure 12.10 shows the average source power P_s as a function of the distance d between the source and relay. The additive white Gaussian noises over the transmitting channels are all set to be with unit variance. It is seen that, when $d > 8$ m, CF outperforms DF theoretically, and the practical CF code of [47] is preferable to the practical DF code [29]. Note that when 7.5 m $< d <$ 8 m, where DF is superior in theory, the CF scheme still performs better than the practical DF code.

The above practical WZC-based design can be extended to the case of the two-receiver cooperative channel by following the CF coding strategy with forward individual decoding described in section 12.3.4. Each transmitter is equipped with one LDPC channel encoder, and each receiver performs one distributed joint source-channel encoding step and two channel decoding steps.

In contrast to the scarcity of WZC-based CF designs for receiver cooperation, there have been more code designs for transmitter cooperation. For example, since the publication of the work on user cooperation [1], several research groups [32, 52–55] have developed practical designs based on AF and DF for the wireless two-transmitter cooperative channel. A common characteristic of these designs is the avoidance of receive collision by transmitting signals over orthogonal channels, which simplifies the code design. Specifically, the two transmitters (or cooperative partners) send encoded messages over orthogonal channels during the first fraction of a time slot. Each transmitter decodes the signal it has received from its partner and, in the case of successful decoding, either re-encodes the recovered message using the partner's codebook (repetition-based DF) or generates additional parity symbols out of a rate-compatible code (DF based on incremental redundancy). The resulting codewords are then forwarded over orthogonal channels to the receiver during the second fraction of a time slot. If a transmitter cannot successfully decode its partner's message, it switches to either the noncooperative mode or AF. Orthogonal signaling is achieved by using time-, frequency-, or code-division or space-time coding.

A DF design based on incremental redundancy for a flat-fading transmitter cooperative channel, dubbed coded cooperation, is proposed in [52, 53]. Two users, partners, first exchange their messages; then each user sends only additional redundancy bits for its partner message. That is, after successful decoding, the user transmits only additional protection bits of a rate-compatible code for the partner's message, so that its overall code rate is decreased. In addition, whenever the partner's message cannot be decoded,

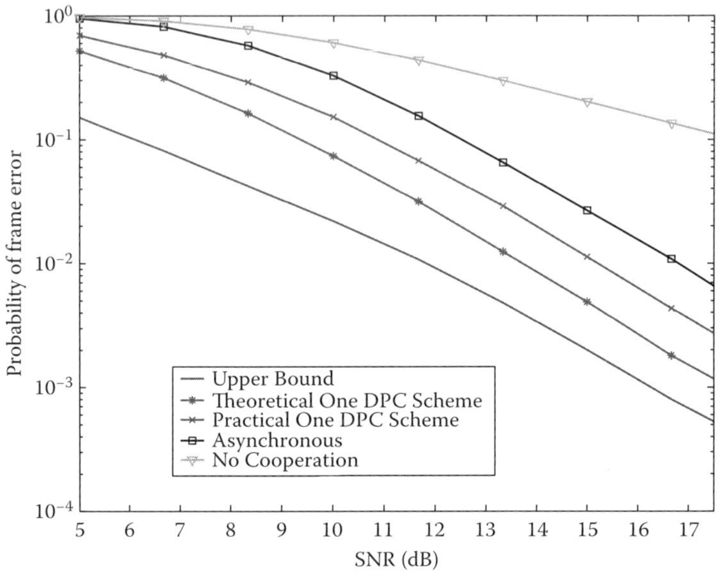

FIGURE 12.11 Simulation results obtained with the DPC-based scheme of [40] for transmitter cooperation, together with various theoretical bounds. The probability of frame error is shown versus the received SNR at the direct link. The rate constraints are fixed at $R_1^o = R_2^o = 1$ bit/sample, and the received SNR at the link between nodes 1 and 2 is 30 dB higher than that at the direct link between nodes 1 and 3.

the user switches to a noncooperative mode. A low-complexity implementation with rate-compatible punctured convolutional (RCPC) codes and a more complex design with space-time turbo coding (efficient on both slow and fast flat-fading channels) are given in [52] and [53], respectively.

A similar DF scheme, based on incremental redundancy, reported in [54] exploits convolutional codes optimized for two-transmitter cooperation in a Rayleigh flat-fading environment. The scheme is shown to be able to achieve the full-diversity gain.

Although the above DF schemes can provide the full-diversity gain, they do so at the expense of decreased rate gain. On the other hand, we know from section 12.3.6 that DPC provides the highest achievable rate over a transmitter cooperative channel. Practical DPC is exploited in [40] to design codes for a wireless two-transmitter cooperative channel. The scheme follows closely coding strategy outlined in section 12.3.6 based on one DPC only. The DPC scheme of [22] is used at the transmitters. It is based on trellis-coded quantization and turbo trellis-coded modulation. Punctured turbo trellis-coded modulation is used to facilitate message exchanges between the two transmitters. Practical design results are shown in Figure 12.11, where the frame error rate against the SNR at the direct link between nodes 1 and 3 is shown. The figure indicates a loss of 1.5 dB from the achievable bound at 2% frame error rate.

12.5 Conclusions

Due to its low complexity and decentralized nature, cooperative diversity arises as a strong candidate for conveying information in emerging wireless ad hoc networks. This motivates research in determining its ultimate performance limit. For a two-transmitter two-receiver cooperative channel, the theory shows that, at least in the high-SNR regime, in contrast to a two-antenna MIMO system, cooperative diversity cannot achieve the full multiplexing gain of 2. Thus, there is a cost paid for having only one antenna at each node. This result indicates that in achieving the full multiplexing gain of 2, tight coordination among the transmit/receive antennas is necessary, which is possible only if they are placed together. However, cooperative diversity does offer high additive rate gains when compared to the noncooperation case, and the key in achieving these gains lies in coding with side information (e.g., DPC and WZC). In transmitter cooperation, synchronization between the two transmitters is essential in obtaining high data rates; DF with DPC is the best coding strategy, coming very close to the upper limits if the interference is weak. Unfortunately, when the interference is strong, transmitter cooperation does not help (in the high-SNR regime). In contrast, receiver cooperation is beneficial in both weak and strong interference scenarios, and CF with WZC is the dominant coding strategy that asymptotically achieves the capacity as the interference and SNR approach infinity. More importantly, full synchronization between the nodes is not necessary for receiver cooperation. Interestingly, as pointed out in [37], optimal power allocation is crucial to realize the full performance gain of receiver cooperation, whereas it provides only a marginal gain in transmitter cooperation.

Owing to its promising application in wireless ad hoc networks, cooperative diversity has been studied intensively recently. Many problems are still open. For example, posed more than 30 years ago, the capacity of the simplest Gaussian relay channel—a building block of cooperative diversity—is still unknown. Recent achievements [38] in providing the full-capacity region for a Gaussian MIMO broadcast channel using DPC might inspire new ideas for solving this problem. The theoretical bounds reported so far are mainly for the full-duplex setups with up to four nodes, where either two transmitters or two receivers cooperate. Providing results for half-duplex cooperative channels should be a research priority. Treating an ad hoc network where the two transmitters and two receivers simultaneously cooperate is another possible research direction. Combining DPC and WZC could lead to the largest achievable rate region, but such a theoretical treatment is not straightforward. In addition, extensions to larger networks with more than four nodes that require cross-layer designs are very challenging because of the additional problem associated with selecting the best partner for cooperation. Also, considering cases where the relay has side information in the form of correlated source [56] or with multiantenna relaying [57] is of interest. Finally, exploiting network coding [58] for user cooperation [59] is a promising approach.

The reported practical designs still suffer performance loss compared to the theoretical limits. Closing this gap with better code designs while staying at acceptable complexity is an urgent research task. The practical designs proposed so far are only for wireless relay and two-transmitter two-receiver cooperative channels. Substantial research efforts are needed to construct practical systems based on cooperative diversity for larger ad hoc networks.

References

[1] A. Sendonaris, E. Erkip, and B. Aazhang. 2003. User cooperation diversity part I and part II. *IEEE Trans. Commun.* 51:1927–48.

[2] J. Laneman, D. Tse, and G. Wornell. 2004. Cooperative diversity in wireless networks: Efficient protocols and outage behavior. *IEEE Trans. Inform. Theory* 50:3062–80.

[3] D. Tse and P. Viswanath. 2005. *Fundamentals of wireless communications.* Cambridge: Cambridge University Press.

[4] A. Goldsmith. 2005. *Wireless communications.* Cambridge: Cambridge University Press.

[5] E. van der Meulen. 1971. Three-terminal communication channels. *Adv. Appl. Probability* 3:120–54.

[6] J. Proakis. 2000. *Digital communications.* New York: McGraw-Hill.

[7] G. Caire and S. Shamai (Shitz). 2003. On the achievable throughput of a multi-antenna Gaussian broadcast channel. *IEEE Trans. Inform. Theory* 49:1691–706.

[8] T. Cover and J. Thomas. 1991. *Elements of information theory.* New York: Wiley.

[9] A. Carleial. 1978. Interference channels. *IEEE Trans. Inform. Theory* 24:60–70.

[10] L. Zheng and D. Tse. 2003. Diversity and multiplexing: A fundamental tradeoff in multiple antenna channels. *IEEE Trans. Inform. Theory* 49:1073–96.

[11] S. Shamai (Shitz) and S. Verdu. 2001. The impact of frequency-flat fading on the spectral efficiency of CDMA. *IEEE Trans. Inform. Theory* 47:1302–27.

[12] A. Lozano, A. Tulino, and S. Verdu. 2005. High-SNR power offset in multi-antenna communication. *IEEE Trans. Inform. Theory* 51:4134–51.

[13] A. Wyner and J. Ziv. 1976. The rate-distortion function for source coding with side information at the decoder. *IEEE Trans. Inform. Theory* 22:1–10.

[14] M. Costa. 1983. Writing on dirty paper. *IEEE Trans. Inform. Theory* 29:439–41.

[15] D. Slepian and J. Wolf. 1973. Noiseless coding of correlated information sources. *IEEE Trans. Inform. Theory* 19:471–80.

[16] T. Berger. 1977. Multiterminal source coding. In *The information theory approach to communications*, ed. G. Longo. New York: Springer-Verlag, pp. 173–231.

[17] S. Pradhan, J. Chou, and K. Ramchandran. 2003. Duality between source coding and channel coding and its extension to the side information case. *IEEE Trans. Inform. Theory* 49:1181–203.

[18] S. Gelfand and M. Pinsker. 1980. Coding for channel with random parameters. *Probl. Contr. Inform. Theory* 9:19–31.

[19] Z. Xiong, A. Liveris, and S. Cheng. 2004. Distributed source coding for sensor networks. *IEEE Signal Processing Mag.* 21:80–94.

[20] U. Erez and S. ten Brink. 2005. A close to capacity dirty paper coding scheme. *IEEE Trans. Inform. Theory* 51:3417–32.

[21] A. Bennatan, D. Burshtein, G. Caire, and S. Shamai. 2006. Superposition coding for side-information channels. *IEEE Trans. Inform. Theory* 52:1872–89.

[22] Y. Sun, A. Liveris, V. Stanković, and Z. Xiong. 2005. Near-capacity dirty-paper code designs based on TCQ and IRA codes. In *Proceedings of the ISIT-2005 IEEE International Symposium on Information Theory*, Adelaide, Australia, pp. 184–188.

[23] T. Cover and A. El Gamal. 1979. Capacity theorems for the relay channel. *IEEE Trans. Inform. Theory* 25:572–84.

[24] A. Høst-Madsen and J. Zhang. 2005. Capacity bounds and power allocation for the wireless relay channel. *IEEE Trans. Inform. Theory* 51:2020–40.

[25] G. Kramer, M. Gastpar, and P. Gupta. 2005. Cooperative strategies and capacity theorems for relay networks. *IEEE Trans. Inform. Theory* 51:3037–63.

[26] Z. Zhang and T. Duman. 2005. Capacity approaching turbo coding and iterative decoding for relay channels. *IEEE Trans. Commun.* 53:1895–905.

[27] M. Khojastepour, A. Sabharwal, and B. Aazhang. 2003. On capacity of Gaussian "cheap" relay channel. In *Proceedings of IEEE Globecom-2003*, San Francisco, vol. 3, pp. 1776–1780.

[28] M. Valenti and B. Zhao. 2003. Distributed turbo codes: Towards the capacity of the relay channel. In *Proceedings of the VTC-2003 IEEE Vehicular Technology Conference*, Orlando, FL, pp. 322–326.

[29] Z. Zhang and T. Duman. 2005. Capacity approaching turbo coding for half-duplex relaying. In *Proceedings of the ISIT-2005 IEEE International Symposium on Information Theory*, Adelaide, Australia, pp. 1888–1892.

[30] Y. Liang and V. Veeravalli. 2005. Gaussian orthogonal relay channel: Optimal resource allocation and capacity. *IEEE Trans. Inform. Theory* 51:3284–89.

[31] C. Zeng, F. Kuhlmann, and A. Buzo. 1989. Achievability proof of some multiuser channel coding theorems using backward decoding. *IEEE Trans. Inform. Theory* 35:1160–65.

[32] J. Laneman and G. Wornell. 2003. Distributed space-time coding protocols for exploiting cooperative diversity in wireless networks. *IEEE Trans. Inform. Theory* 49:2415–25.

[33] L. Ozarow, S. Shamai, and A. Wyner. 1994. Information theoretic consideration for cellular mobile radio. *IEEE Trans. Veh. Technol.* 43:359–77.

[34] I. Telater. 1999. Capacity of multi-antenna Gaussian channels. *Eur. Trans. Telecommun.* 10:585–95.

[35] R. Nabar, H. Bölcskei, and F. Kneubühler. 2004. Fading relay channels: Performance limits and space-time signal design. *IEEE J. Select. Areas Commun.* 22:1099–109.

[36] A. Høst-Madsen. 2006. Capacity bounds for cooperative diversity. *IEEE Trans. Inform. Theory* 52:1522–44.

[37] C. Ng and A. Goldsmith. 2005. Capacity gains from transmitter and receiver cooperation. In *Proceedings of the ISIT-2005 IEEE International Symposium on Information Theory*, Adelaide, Australia, pp. 397–401.

[38] H. Weingarten, Y. Steinberg, and S. Shamai. 2006. The capacity region of the Gaussian multiple-input multiple-output broadcast channel. *IEEE Trans. Inform. Theory* 52:3936–64.

[39] N. Jindal, U. Mitra, and A. Goldsmith. 2004. Capacity of ad-hoc networks with node cooperation. In *Proceedings of the ISIT-2004 IEEE International Symposium on Information Theory*, Chicago, pp. 271–275.

[40] M. Uppal, V. Stanković, A. Høst-Madsen, and Z. Xiong. 2006. Code design for transmitter cooperation. In *Proceedings of the MSRI Workshop on Mathematics of Relaying and Cooperation in Communication Networks*, Berkeley, CA.

[41] J. Laneman and G. W. Wornell. 2000. Energy-efficient antenna sharing and relaying for wireless networks. In *Proceedings of the WCNC-2000 IEEE Wireless Communications and Networking Conference*, vol. 1, pp. 7–12.

[42] A. Chakrabarti, A. de Baynast, A. Sabharwal, and B. Aazhang. 2007. Low density parity check codes for the relay channel. *IEEE J. Selected Areas Commun.* 25:280–91.

[43] R. G. Gallager. 1963. *Low density parity check codes*. Cambridge, MA: MIT Press.

[44] M. Yuksel and E. Erkip. 2007. Diversity-multiplexing tradeoff in half-duplex relay systems. In *Proceedings of the ICC-2007 IEEE International Conference on Communications*, Glasgow, UK, pp. 689–694.

[45] A. Chakrabarti, A. de Baynast, A. Sabharwal, and B. Aazhang. 2006. Half-duplex estimate-and-forward relaying: Bounds and code design. In *Proceedings of the ISIT-2006 IEEE International Symposium on Information Theory*, Seattle, WA, pp. 1239–1243.

[46] R. Hu and J. Li. 2006. Practical compress-forward in user cooperation: Wyner-Ziv cooperation. In *Proceedings of the ISIT-2006 IEEE International Symposium on Information Theory*, Seattle, WA, pp. 489–493.

[47] Z. Liu. 2007. Slepian-Wolf coded nested lattice quantization for Wyner-Ziv coding: High-rate performance analysis, code designs, and application to cooperative networks. PhD thesis, Texas A&M University.

[48] R. Zamir, S. Shamai, and U. Erez. 2002. Nested linear/lattice codes for structured multiterminal binning. *IEEE Trans. Inform. Theory* 48:1250–76.

[49] Z. Liu, S. Cheng, A. Liveris, and Z. Xiong. 2006. Slepian-Wolf coded nested quantization for Wyner-Ziv coding: High-rate performance analysis and code design. *IEEE Trans. Inform. Theory* 52:4358–4379.

[50] Q. Xu, V. Stanković, and Z. Xiong. 2006. Layered Wyner-Ziv video coding for transmission over unreliable channels. *Signal Processing* 86:3112–25.

[51] H. Jin, A. Khandekar, and R. McEliece. 2000. Irregular repeat-accumulate codes. In *Proceedings of the 2nd International Symposium on Turbo Codes and Related Topics*, 1–8.

[52] T. Hunter and A. Nosratinia. 2006. Diversity through coded cooperation. *IEEE Trans. Wireless Commun.* 5:283–89.

[53] M. Janani, A. Hedayat, T. Hunter, and A. Nosratinia. 2004. Coded cooperative in wireless communications: Space-time transmission and iterative decoding. *IEEE Trans. Signal Processing* 52:362–71.

[54] A. Stefanov and E. Erkip. 2004. Cooperative coding for wireless networks. *IEEE Trans. Commun.* 52:1603–12.

[55] M. Souryal and B. Vojcic. 2004. Cooperative turbo coding with time-varying Rayleigh channels. In *Proceedings of the ICC-2004 IEEE International Conference on Communications*, Paris, pp. 356–60.

[56] D. Gunduz, C. T. K. Ng, E. Erkip, and A. J. Goldsmith. 2007. Source transmission over relay channels with correlated side information. In *Proceedings of the ISIT-2007 IEEE International Symposium on Information Theory*, Nice, France.

[57] Y. Fan, H. V. Poor, and J. Thompson. 2007. Cooperative multiplexing in a half-duplex relay network: Performance and constraints. In *Proceedings of the 45th Annual Allerton Conference on Communications, Control and Computing*, Monticello, IL.

[58] R. Ahlswede, N. Cai, S.-Y.R. Li, and R. W. Yeung. 2000. Network information flow. *IEEE Transactions on Information Theory* 46:1204–16.

[59] M. Yu, J. Li, and R. Blum. 2007. User cooperation through network coding. In *Proceedings of the ICC-2007 IEEE International Conference on Communications*, Glasgow, UK, pp. 4064–4069.

13

Time Synchronization for Wireless Sensor Networks

Kyoung-Lae Noh
Texas A&M University

Yik-Chung Wu
The University of Hong Kong

Khalid Qaraqe
Texas A&M University

Erchin Serpedin
Texas A&M University

13.1 Introduction

With the help of recent technological advances in micro-electro-mechanical systems (MEMS) and wireless communications, low-cost, low-power, and multifunctional wireless sensing devices have been developed. When these devices are deployed over a wide geographical region, they can collect information about the environment and efficiently collaborate to process such information by forming a distributed communication network, called the *wireless sensor network* (WSN). WSN is a special case of wireless ad hoc network, and assumes a multihop communication framework with no common infrastructure, where the sensors spontaneously cooperate to deliver information by forwarding packets from a source to a destination. The feasibility of WSNs keeps growing rapidly, and WSNs have been regarded as fundamental infrastructures for future ubiquitous communications due to a variety of promising potential applications: monitoring the health status of humans, animals, plants, and environment; control and instrumentation of industrial machines and home appliances; homeland security; detection of chemical and biological threats and leaks; etc. [1, 2].

Time synchronization is a procedure for providing a common notion of time across a distributed system. It is crucial for WSNs in performing a number of fundamental operations, such as:

- Data fusion: Data fusion is a main operation in all distributed networks for processing and integrating in a meaningful way the collected data, and it requires some or all nodes in the network to share a common timescale.
- Power management: Energy efficiency is a key designing factor for WSNs since sensors are usually left unattended without any maintenance and battery replacement for their lifetimes after deployment. Most energy-saving operations strongly depend on time synchronization. For instance, the duty cycling (sleep and wake-up modes control) helps the nodes to save huge energy resources by spending minimal power during the sleep mode. Thus, network-wide synchronization is essential for efficient duty cycling, and its performance is proportional to the synchronization accuracy.
- Transmission scheduling: Many scheduling protocols require time synchronization. For example, the Time Division Multiple Access (TDMA) scheme, one of the most popular communications schemes for distributed networks, is only applicable to a synchronized network.

Moreover, many localization, security, and tracking protocols also demand the nodes to time stamp their messages and sensing events. Therefore, time synchronization appears as one of the most important research challenges in the design of energy-efficient WSNs.

In general, synchronization is considered a critical problem for distributed wireless ad hoc networks due to its decentralized nature and the timing uncertainties introduced by the imperfections in hardware oscillators and message delays in physical and Medium Access Control (MAC) layers. All these uncertainties cause the local clocks of different nodes to drift away from each other over the course of a time interval. In the context of the Internet (a kind of distributed network), time synchronization has been thoroughly studied and investigated. In the Internet, the Network Time Protocol (NTP) [3] is employed ubiquitously due to its diverse advantages, such as scalability, robustness, and self-configurability. Besides, NTP does not rely on GPS and is a software-based protocol. However, NTP presents a number of challenges when applied to WSNs due to the unique nature of sensor networks: limited power resources, wireless channel conditions, and dynamic topology caused by mobility and failure. Therefore, different types of synchronization schemes have to be explicitly designed for WSN applications to cope with these challenges (see also the surveys in [4–9] for additional motivations in this direction).

Research works on time synchronization in the context of WSN roughly began to appear in 2002, where [5], for the first time, pointed out NTP cannot be directly applied to WSN and described some important characteristics and design principles of time synchronization in WSNs. In the same year and the next, two important time synchronization protocols for WSNs were reported: Timing-sync Protocol for Sensor Networks (TPSN) [17] and Reference Broadcasting Synchronization (RBS) [18]. These two protocols set the stage for two fundamental approaches of time synchronization in WSNs. After that, many extensions and generalizations of TPSN and RBS, and many different synchronization schemes based on other ideas, have been proposed in the literature. Notice that symbol timing synchronization in the physical layer, which is critical for accurate symbol detection at the receiver, is a different problem and is out of the scope of this chapter.

The purpose of this chapter is threefold. First, this chapter summarizes the fundamental features and theoretical results encountered in time synchronization of WSNs. Second, it represents the survey of existing time synchronization protocols for WSNs, focusing mainly on the signal processing aspects, the most recent developments in this field. Finally, this chapter discusses the need for adaptive time synchronization schemes for WSNs, analyzes the features of the recently reported adaptive time synchronization protocols, and proposes several research directions for improving their performance.

The rest of this chapter is organized as follows. In section 13.2, the general clock model for time synchronization is first introduced and analyzed. Some important features that have to be considered when designing time synchronization protocols for WSNs are presented. Additionally, various delay components in timing message delivery are categorized. Section 13.3 presents three general and fundamentally different time synchronization approaches: sender-receiver, receiver-receiver, and receiver-only synchronization. These basic approaches are analyzed and compared to illustrate the common and different characteristics in clock synchronization of WSNs. Section 13.4 categorizes and surveys the existing synchronization protocols and relates them to the results presented in section 13.3. In section 13.5, results concerning the importance and effectiveness of adaptive time synchronization schemes are presented, and the most important adaptive synchronization protocols are introduced as well. Finally, section 13.6 summarizes and concludes this chapter.

13.2 Signal Models for Time Synchronization

13.2.1 Definition of Clock

Every individual sensor in a network has its own clock. The counter in a sensor is increased in accordance with the zero-crossings or the edges of the periodic output signal of the local oscillator. When the counter reaches a certain threshold value, an interrupt is created and delivered to the memory. The frequency of the oscillator and the threshold value determine the resolution of the clock. Ideally, the clock of a sensor node should be configured such that $C(t) = t$, where t stands for the ideal or reference time. However, due to the imperfections of the clock oscillator, the clock function of the ith node is modeled as

$$C_i(t)=\theta_o+\theta_s \cdot t+e, \tag{13.1}$$

where the parameters θ_o and θ_s are called clock offset (phase difference) and clock skew (frequency difference), respectively, and e stands for random noise.

Assuming the effect of random noise e is negligible, from (13.1), the clock relationship between two nodes, say node 1 and node 2, can be represented by

$$C_1(t)=\theta_o^{(12)}+\theta_s^{(12)} \cdot C_2(t),$$

where $\theta_o^{(12)}$ and $\theta_s^{(12)}$ are the relative clock offset and skew between node 1 and node 2, respectively. Thus, $\theta_o^{(12)} = 0$ and $\theta_s^{(12)} = 1$ when the two clocks are perfectly synchronized. Suppose there are L nodes in the network, then the global network-wide synchronization is achieved when $C_i(t) = C_j(t)$ for all $i, j = 1, \cdots, L$.

Time synchronization in wireless sensor networks is a complicated problem due to the following reasons. First, every single oscillator has unique clock parameters regardless of its type. For instance, according to the data sheet of a typical crystal-quartz oscillator commonly used in sensor networks, the frequency of a clock varies up to 40 ppm, which means clocks of different nodes can lose as much as 40 ms in a second. In other words, every single oscillator might assume a different skew parameter ranging from −20 to 20 ppm.

Notice that in general, the clock skew θ_s is a time-dependent random variable (RV) and there are two concepts used often in clock terminology regarding the nature of time-dependent randomness present in clock parameters. These concepts are referred to as *short-term* and *long-term* stabilities, respectively. For the oscillators currently used in sensor networks, all these parameters are almost constant for short-term time intervals [10]. Besides, the total power of the noise process is too small to be effective in short time spans [11]. Therefore, the parameters of a clock are assumed to be constants for the time period of interest.

As far as the long-term stability is concerned, the clock parameters are subject to change due to environmental or other external effects such as temperature, atmospheric pressure, voltage changes, and hardware aging [10]. Hence, in general, the relative clock offset keeps changing with time, which means that the network has to perform periodic time resynchronization to adjust the clock parameters.

13.2.2 Design Considerations

Time synchronization for conventional wired networks has been thoroughly studied and a plethora of synchronization protocols have been developed as surveyed in [1]. For wireless sensor networks, there are a number of unique and important factors to be considered when designing time synchronization protocols as listed below.

- *Energy consumption*: Energy consumption is momentous in wireless sensor networks due to their limited and generally nonrechargeable power resources. Hence, the design of wireless sensor networks should be subjected to maintaining minimal energy expenditure in each sensor node. Various types of power control procedures, such as sleep/wake-up modes and dynamic routing controls, are commonly considered in this regard. Time synchronization is one of the critical components contributing to energy consumption due to the highly energy consuming radio transmissions for achieving clock synchronization. Indeed, the energy consumption required for time synchronization of a node is approximately 17% of the total energy spent by a node [12]. Pottie and Kaiser showed in [13] that the radio frequency (RF) energy required to transmit 1 bit over 100 m (i.e., 3 J) is equivalent to the energy required to execute 3 million instructions. Therefore, developing efficient synchronization algorithms represents an ideal mechanism for trading computational energy for reduced (RF) communication energy. In the sequel, energy efficiency is the main concern in designing time synchronization protocols.

- *Latency*: Latency in message delivery is a fundamental factor when designing communications networks. For networks based on multihop transmissions like wireless sensor networks, this is even more critical because the uncertainty in message delivery significantly increases as the number of hops increases. Besides, the effects of channel variations, mobility, and the ad hoc nature of wireless sensor networks make this problem more complex. Efficient localization and time synchronization protocols are necessary for reducing the latency error and jitter.
- *Security and reliability*: Network security has gained huge attention in recent years as the networks become more accessible and vulnerable due to the development of sophisticated spying techniques and devices. Besides, unlike wired networks, far more frequent message losses occur in wireless networks because of the time-varying nature of wireless channels. Therefore, a mechanism to cope with message losses and malicious attacks in time synchronization will be necessary for wireless sensor networks.
- *Network topology changes*: The performance of a time synchronization protocol is closely related to the network topology, i.e., it varies with the density and distribution of sensors in the network. Therefore, any shift in the location or scale of sensors incurs a network topology change, which requires at its turn a new self-configuration. Mobility of the sensors and battery timeouts are the main reasons for this change. Hence, for dynamic sensor networks, time synchronization protocols should be able to adapt well to frequent network topology changes.
- *Scalability*: Scalability is another important factor in the design of synchronization protocols. The computational complexity of synchronization algorithms becomes a critical problem as the number of sensors becomes very large. Besides, many other crucial MAC operations, such as multihop routing and network configuration, highly depend on the network scalability as well.

13.2.3 Delay Components in Timing Message Delivery

The main role of time synchronization in a distributed network is to ensure a common timescale for all the network nodes, and to provide the right temporal coordination among all the nodes engaged in a collaborative and distributed interaction with the physical environment. Timing mismatch arises mainly from different setup times of nodes and time variations introduced by local oscillators running at different frequencies. Environmental variations, such as temperature and aging, also drive local clock oscillators to run unpredictably. All these uncertainties cause the local clocks of different nodes to drift away from each other over the course of a time interval.

Assume two nodes need to be synchronized. One of the nodes sends its current time to a neighboring node; if there is absolutely no delay in the message delivery, that neighboring node can immediately know the difference between its clock and its neighbor's clock. Unfortunately, in a real wireless network, various delays affect the message delivery, making time synchronization much more difficult than it seems to be. In general, a series of timing message transmissions is required to estimate the relative time offsets among nodes. In some sense, time synchronization in wireless sensor networks can be regarded as a process of removing the nondeterministic delays during timing message transmission over wireless channels.

There are a number of nondeterministic delays while transferring messages between nodes. Kopetz and Ochsenreiter for the first time analyzed the structure of message delays and characterized the delay components according to the process of message delivery [14]. The delay components in message delivery can be categorized as follows:

1. Send time: The time spent in building the message at the application layer, including other delays introduced by the operating system when processing the send request. The send time is nondeterministic and can be up to hundreds of milliseconds depending on the workload of the system.
2. Access time: The time waiting for accessing the channel after reaching the MAC layer. This is the most significant factor and highly variable according to the specific MAC protocol. The access time is nondeterministic and varies from milliseconds up to seconds depending on the current network traffic.
3. Transmission time: The time for transmitting a message at the physical layer. This delay can be estimated by the length of a message and the speed of radio in the medium and is in the order of tens of milliseconds.
4. Propagation time: The actual time for a message to transmit from the sender to the receiver in a wireless channel. The propagation time is deterministic and less than 1 μs, which is almost negligible compared with the other delay components.
5. Reception time: The time required for receiving a message at the physical layer, which is the same as the transmission time. In some cases, this delay has been categorized as a part of the receive time, to be presented next.
6. Receive time: Time to construct and send the received message to the application layer at the receiver. It is a corresponding component of the send time on the transmitter side and can be varied due to the variable delays introduced by the operating system.

Note that the time delay in message transmission is also dependent on other factors, such as hardware platform, error correction code, and modulation scheme. The estimated time delay discussed above in each component is based on the Mica2 platform [15]. More detailed analysis can be found in [16].

13.3 Fundamental Approaches to Time Synchronization

Time synchronization in wireless sensor networks can be achieved by transferring a group of timing messages to the target sensors. The timing messages contain the information about the time stamps measured by the transmitting sensors. There exist two well-known approaches for time synchronization in wireless sensor networks, which are categorized as sender-receiver synchronization (SRS) and receiver-receiver synchronization (RRS). SRS is based on the traditional model of two-way message exchanges between a pair of nodes. For RRS, the nodes to be synchronized first receive a beacon packet from a common sender, then compare their receiving time readings of the beacon packet to compute the relative clock offset. Most of the existing time synchronization protocols rely on one of these two approaches. For instance, the Network Time Protocol (NTP) [3] and the Timing-sync Protocol for Sensor Networks (TPSN) [17] adopt SRS since they depend on a series of pairwise synchronizations that assume two-way timing message exchanges. Notice also that the Reference Broadcast Synchronization (RBS) protocol [18] relies on RRS since it requires pairs of message exchanges among children nodes (except the reference) to compensate their relative clock offsets.

Recently, a new approach for time synchronization, called receiver-only synchronization (ROS), was proposed. It aims at minimizing the number of required timing messages and energy consumption during synchronization while preserving a high level of accuracy [19]. This approach can be used to achieve network-wide synchronization with much less timing messages than other well-known existing protocols such as TPSN and RBS.

Next we will present and analyze each of these synchronization approaches and illustrate how the general design considerations can be resolved in them. For all these approaches, we only present the underlying signaling mechanisms for performing pairwise synchronization, i.e., synchronizing a pair of nodes, since network-wide synchronization can be simply achieved by performing a group of pairwise synchronizations.

13.3.1 Sender-Receiver Synchronization

This approach is based on the classical two-way timing message exchange mechanism between two adjacent nodes. Consider a parent node P and one of its children nodes, node A, in Figure 13.1. The clock model for the two-way message exchange is depicted in Figure 13.2, where $\theta_o^{(AP)}$ denotes the clock offset between node A and node P and timing messages are assumed to be exchanged multiple (N) times [4, 17]. Here, the time stamps made during the ith message exchange $T_{1,i}^{(A)}$ and $T_{4,i}^{(A)}$ are measured by the local clock of node A, and $T_{2,i}^{(P)}$ and $T_{3,i}^{(P)}$ are measured by the local clock of node P, respectively. Node A transmits a synchronization packet, containing the value of time stamp $T_{1,i}^{(A)}$ to node P. Node P receives it at time $T_{2,i}^{(P)}$ and transmits an acknowledgment packet to node A at $T_{3,i}^{(P)}$. This packet contains the value of time stamps $T_{1,i}^{(A)}$, $T_{2,i}^{(P)}$, and $T_{3,i}^{(P)}$. Then, node A finally receives the packet at $T_{4,i}^{(A)}$.

As discussed before, packet delays can be characterized into several distinct components: send, access, transmission, propagation, and receive times. These delay components

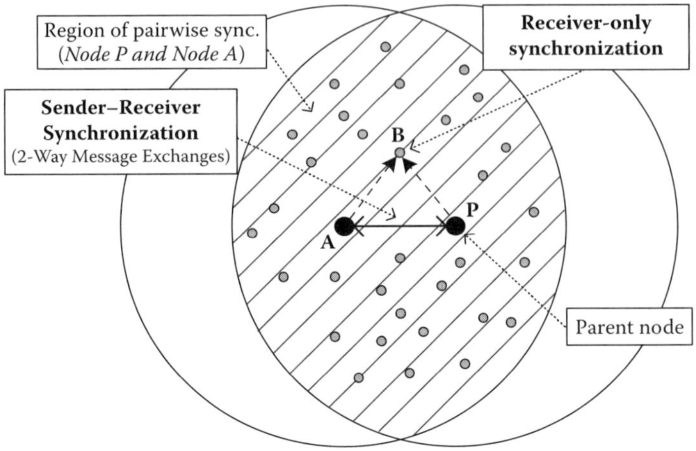

FIGURE 13.1 Sender-receiver synchronization and receiver-only synchronization.

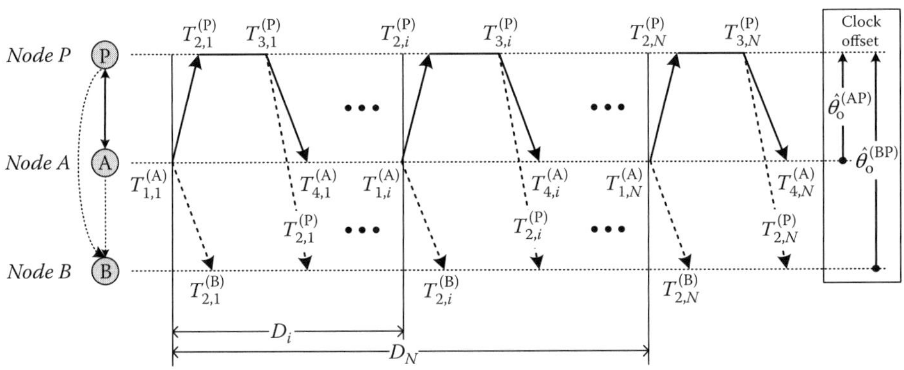

FIGURE 13.2 Clock synchronization model of SRS (node P and node A) and ROS (node B).

are divided into two parts: the fixed portion d and the variable portion X_i. The variable portion of delays depends on various network parameters (e.g., network status, traffic, etc.) and setup, and therefore no single delay model can be found to fit for every case. Thus far, several probability density function (PDF) models have been proposed for modeling random delays, the most widely deployed ones being Gaussian, Gamma, exponential, and Weibull PDFs [20, 21]. The Gaussian delay model is appropriate if the delays are thought to be the addition of numerous independent random processes. In [18], the chi-squared test showed that the variable portion of delays can be modeled as Gaussian distributed random variables (RVs) with 99.8% confidence. On the other hand, a single-server M/M/1 queue can fittingly represent the cumulative link delay for point-to-point hypothetical reference connection, where the random delays are independently modeled as exponential RVs [22]. Thus, we assume the random portions of delays are either normal or exponentially distributed RVs.

13.3.1.1 Clock Offset Estimation

Suppose that the clock frequencies of two nodes remain equal during the synchronization period, and both $X_i^{(AP)}$ and $X_i^{(PA)}$ are normal distributed RVs with mean μ and variance $\sigma^2/2$. From Figure 13.2, $T_{2,i}^{(P)}$ and $T_{4,i}^{(A)}$ can be expressed as

$$T_{2,i}^{(P)} = T_{1,i}^{(A)} + \theta_o^{(AP)} + d^{(AP)} + X_i^{(AP)}, \tag{13.2}$$

$$T_{4,i}^{(A)} = T_{3,i}^{(P)} + \theta_o^{(PA)} + d^{(PA)} + X_i^{(PA)}, \tag{13.3}$$

where $\theta_o^{(PA)} = -\theta_o^{(AP)}$, $d^{(AP)}$, and $X_i^{(AP)}$ denote the fixed and random portions of timing delays in the message transmissions from node A to node P, respectively. By defining the delays in uplink $U_i \triangleq T_{2,i}^{(P)} - T_{1,i}^{(A)}$ and downlink $V_i \triangleq T_{4,i}^{(A)} - T_{3,i}^{(P)}$, the ith delay observations corresponding to the ith timing message exchange are given by $U_i = \theta_o^{(AP)} + d^{(AP)} + X_i^{(AP)}$ and $V_i = \theta_o^{(PA)} + d^{(PA)} + X_i^{(PA)}$, respectively. Then, the likelihood function based on the observations $\{U_i\}_{i=1}^N$ and $\{V_i\}_{i=1}^N$ is given by

$$L\left(\theta_o^{(AP)}, \mu, \sigma^2\right) = (\pi\sigma^2)^{-\frac{N}{2}} e^{-\frac{1}{\sigma^2}\left[\sum_{i=1}^{N}(U_i - d^{(AP)} - \theta_o^{(AP)} - \mu)^2 + \sum_{i=1}^{N}(V_i - d^{(PA)} + \theta_o^{(AP)} - \mu)^2\right]},$$

where N is the number of message exchanges. Differentiating the log-likelihood function leads to

$$\frac{\partial \ln L\left(\theta_o^{(AP)}\right)}{\partial \theta_o^{(AP)}} = -\frac{2}{\sigma^2}\sum_{i=1}^{N}\left[\theta_o^{(AP)} + d^{(AP)} - d^{(PA)} - (U_i - V_i)\right].$$

The fixed portions of delays are mainly determined by the propagation delays, and both up- and downlink channels have the same distance. Thus, the fixed portions of delays $d^{(AP)}$ and $d^{(PA)}$ are assumed to be equal, and are denoted by d for the rest of this chapter. Indeed, the propagation delay is less than 1 μs for ranges under 300 m, hence almost negligible when compared to other dominant delay components whose ranges are about hundreds of milliseconds [16]. The maximum likelihood estimate (MLE) of clock offset is given by [25]

$$\hat{\theta}_o^{(AP)} = \arg\max_{\theta_o^{(AP)}}\left[\ln L\left(\theta_o^{(AP)}\right)\right] = \frac{\bar{U} - \bar{V}}{2}. \tag{13.4}$$

Thus, node A can be synchronized to the parent node P by simply taking the difference of the average delay observations $\bar{U}$ and $\bar{V}$.

For exponential random delays $X_i^{(PA)}$ and $X_i^{(AP)}$ with the same mean λ, the likelihood function based on the observations $\{U_i\}_{i=1}^N$ and $\{V_i\}_{i=1}^N$ becomes

$$L(\theta_o^{(AP)},\lambda)=\lambda^{-2N}e^{-\frac{1}{\lambda}\sum_{i=1}^{N}\left[U_i+V_i-2d\right]}\cdot\prod_{i=1}^{N}I\left[U_i-\theta_o^{(AP)}-d\geq 0, V_i+\theta_o^{(AP)}-d\geq 0\right],$$

where $I(\cdot)$ stands for the indicator function (i.e., $I(\cdot)$ is 1 whenever its inner condition holds, otherwise it is equal to 0). In [23], Jeske proved that the maximum likelihood estimator of $\theta_o^{(AP)}$ exists when d is unknown and exhibits the same form as the estimator proposed in [24], namely,

$$\hat{\theta}_o^{(AP)}=\frac{\min\limits_{1\leq i\leq N} U_i-\min\limits_{1\leq i\leq N} V_i}{2}. \tag{13.5}$$

Notice from (13.4) and (13.5), it is clear that if only one round of message exchange is performed (N = 1), the MLE of clock offset for both exponential and Gaussian delay models becomes $\hat{\theta}_o^{(AP)}=(U-V)/2$, which is exactly the same clock offset estimator adopted in [17].

13.3.1.2 Joint Clock Offset and Skew Estimation

The clock offset between two nodes generally keeps increasing due to the difference of clock parameters of each oscillator. Thus, a model with the same clock frequency is not sufficient for long-term synchronization. Indeed, applying the clock skew correction mechanism increases the synchronization accuracy and guarantees the long-term reliability of synchronization.

Figure 13.3 shows the effect of clock offset (θ_o) and skew (θ_s) on timing message exchanges between two nodes. Without loss of generality, the reference time $T_{1,1}^{(A)}$ is set to be zero. Here, the time stamp at node P in the ith uplink message $T_{2,i}^{(B)}$ is given by

$$\begin{aligned}T_{2,i}^{(P)}&=T_{1,i}^{(A)}+\theta_o^{(AP)}+\theta_s^{(AP)}(T_{1,i}^{(A)}+d+X_i^{(AP)})+d+X_i^{(AP)}\\&=(1+\theta_s^{(AP)})(T_{1,i}^{(A)}+d+X_i^{(AP)})+\theta_o^{(AP)},\end{aligned} \tag{13.6}$$

where the term $\theta_s^{(AP)}(T_{1,i}^{(A)}+d+X_i^{(AP)})$ is due to the effect of clock skew. Similarly, the time stamp at node P in the ith downlink message $T_{3,i}^{(P)}$ takes the equations

$$\begin{aligned}T_{3,i}^{(P)}&=T_{4,i}^{(A)}+\theta_o^{(AP)}+\theta_s^{(AP)}(T_{4,i}^{(A)}-d-X_i^{(PA)})-d-X_i^{(PA)}\\&=(1+\theta_s^{(AP)})(T_{4,i}^{(A)}-d-X_i^{(PA)})+\theta_o^{(AP)},\end{aligned} \tag{13.7}$$

where the term $\theta_s^{(AP)}(T_{4,i}^{(A)}-d-X_i^{(PA)})$ is again due to the effect of clock skew. For an easier illustration, we introduce the simplified notations $\theta_s\triangleq\theta_s^{(AP)}$, $\theta_o\triangleq\theta_o^{(AP)}$, $T_{1,i}\triangleq T_{1,i}^{(A)}$, $T_{4,i}\triangleq T_{4,i}^{(A)}$, $T_{2,i}\triangleq T_{2,i}^{(P)}$, $T_{3,i}\triangleq T_{3,i}^{(P)}$, $X_i\triangleq X_i^{(AP)}$, and $Y_i\triangleq X_i^{(PA)}$ in this section, respectively.

Assuming $\{X_i\}_{i=1}^N$ and $\{Y_i\}_{i=1}^N$ are zero-mean independent Gaussian distributed RVs with variance $\sigma^2/2$, then the joint PDF of $\mathbf{X}\triangleq\{X_i\}_{i=1}^N$ and $\mathbf{Y}\triangleq\{Y_i\}_{i=1}^N$ is given by

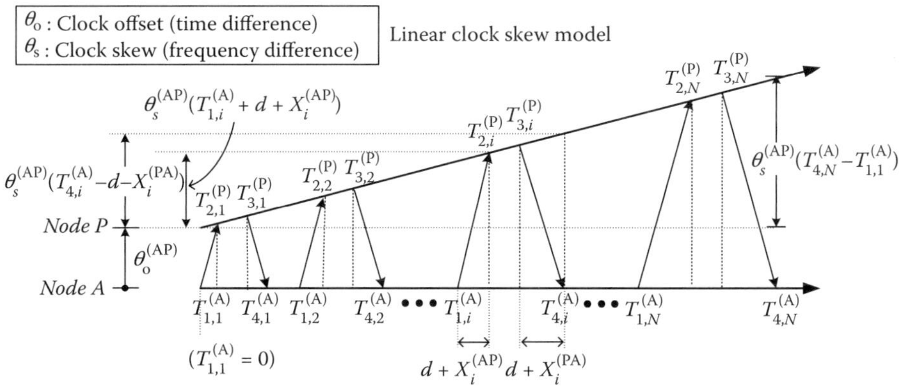

FIGURE 13.3 Two-way timing message exchange model that assumes clock offset and skew.

$$f_{\mathbf{X},\mathbf{Y}}(\mathbf{x},\mathbf{y}) = (\pi\sigma^2)^{-\frac{N}{2}} e^{-\frac{1}{\sigma^2}\sum_{i=1}^{N}\left[\left(\frac{\theta_o - T_{2,i} + (T_{1,i}+d)(1+\theta_s)}{1+\theta_s}\right)^2 + \left(\frac{\theta_o - T_{3,i} + (T_{4,i}-d)(1+\theta_s)}{1+\theta_s}\right)^2\right]}.$$

Further assuming that the fixed portion of delay d is known and $\theta'_s \triangleq 1/(1+\theta_s)$, then the log-likelihood function (ignoring irrelevant additive and multiplicative constants) for (θ_o, θ'_s), based on observations $\{T_{1,i}\}_{i=1}^N$, $\{T_{2,i}\}_{i=1}^N$, $\{T_{3,i}\}_{i=1}^N$, and $\{T_{4,i}\}_{i=1}^N$, is given by

$$\ln L(\theta_o, \theta'_s) = -\sum_{i=1}^{N}\left\{\left[\theta'_s(\theta_o - T_{2,i}) + (T_{1,i}+d)\right]^2 + \left[\theta'_s(\theta_o - T_{3,i}) + (T_{4,i}-d)\right]^2\right\}. \quad (13.8)$$

It has been shown in [25] that the values of θ_o and θ_s that maximize the above log-likelihood function are given, respectively, by

$$\hat{\theta}_o^{ML} = \frac{\sum_{i=1}^{N}(T_{1,i}+T_{4,i})\sum_{i=1}^{N}(T_{2,i}^2+T_{3,i}^2) - \sum_{i=1}^{N}(T_{2,i}+T_{3,i})Q}{\sum_{i=1}^{N}(T_{2,i}+T_{3,i})\sum_{i=1}^{N}(T_{1,i}+T_{4,i}) - 2NQ}, \quad (13.9)$$

$$\hat{\theta}_s^{ML} = \frac{-2N\left[\sum_{i=1}^{N}(T_{1,i}+T_{4,i})\sum_{i=1}^{N}(T_{2,i}^2+T_{3,i}^2) - Q\sum_{i=1}^{N}(T_{2,i}+T_{3,i})\right]}{\sum_{i=1}^{N}(T_{1,i}+T_{4,i})\left[\sum_{i=1}^{N}(T_{2,i}+T_{3,i})\sum_{i=1}^{N}(T_{1,i}+T_{4,i}) - 2NQ\right]} + \frac{\sum_{i=1}^{N}(T_{2,i}+T_{3,i})}{\sum_{i=1}^{N}(T_{1,i}+T_{4,i})} - 1, \quad (13.10)$$

where

$$Q \triangleq \sum_{i=1}^{N} \left(T_{1,i}T_{2,i} + T_{3,i}T_{4,i} + (T_{2,i} - T_{3,i})d\right).$$

Note that the joint MLE depends on the value of the fixed portion of delays d, which is assumed to be known in this section. Although estimating d is an achievable task, we do not consider d another unknown (nuisance) parameter due to the inherent highly nonlinear and complex operations required for estimating d.

The Cramer-Rao lower bound (CRB) for the vector parameter $\theta = [\theta_o, \theta_s]^T$ can be derived from the 2×2 Fisher information matrix $\mathbf{I}(\theta)$ by taking its inverse. From (13.6), the second-order derivatives of the log-likelihood function with respect to θ_o and θ'_s are found as

$$\frac{\partial^2 \ln L\left(\theta_o, \theta'_s, \sigma^2\right)}{\partial \theta_o^2} = -\frac{4N\theta'^2_s}{\sigma^2},$$

$$\frac{\partial^2 \ln L\left(\theta_o, \theta'_s, \sigma^2\right)}{\partial \theta'^2_s} = -\frac{2}{\sigma^2}\sum_{i=1}^{N}\left[(T_{2,i} - \theta_o)^2 + (T_{3,i} - \theta_o)^2\right],$$

$$\frac{\partial^2 \ln L\left(\theta_o, \theta'_s, \sigma^2\right)}{\partial \theta_o \theta'_s} = -\frac{2}{\sigma^2}\sum_{i=1}^{N}\left(2\theta'_s\theta_o - \theta'_s T_{2,i} + T_{1,i} - \theta'_s T_{3,i} - T_{4,i}\right).$$

Taking the negative expectations yields

$$-E\left[\frac{\partial^2 \ln L\left(\theta_o, \theta'_s, \sigma^2\right)}{\partial \theta_o^2}\right] = \frac{4N\theta'^2_s}{\sigma^2},$$

$$-E\left[\frac{\partial^2 \ln L\left(\theta_o, \theta'_s, \sigma^2\right)}{\partial \theta'^2_s}\right] = \frac{2}{\sigma^2}\sum_{i=1}^{N} E_{X_i,Y_i}\left[\frac{(X_i + T_{1,i} + d)^2 + (Y_i - T_{4,i} + d)^2}{\theta'^2_s}\right]$$

$$\overset{(a)}{=} \frac{2\sum_{i=1}^{N}\left((T_{1,i} + d)^2 + (T_{4,i} - d)^2 + 2\sigma^2\right)}{\sigma^2 \theta'^2_s},$$

$$-E\left[\frac{\partial^2 \ln L\left(\theta_o, \theta'_s, \sigma^2\right)}{\partial \theta_o \theta'_s}\right] = -\frac{2}{\sigma^2}\sum_{i=1}^{N} E_{X_i,Y_i}\left[2\theta'_s\left(2\theta_o - T_{2,i} - T_{3,i}\right) + T_{1,i} + T_{4,i}\right]$$

$$\overset{(b)}{=} \frac{2N}{\sigma^2}\left(\overline{T_1} + \overline{T_4}\right),$$

where (a) and (b) are due to $X_i = \theta'_s(T_{2,i} - \theta_o) - (T_{1,i} + d)$ and $Y_i = \theta'_s(\theta_o - T_{3,i}) + (T_{4,i} - d)$, and

$$\bar{T}_1 = \sum_{i=1}^{N} T_{1,i}/N$$

and

$$\bar{T}_4 = \sum_{i=1}^{N} T_{4,i}/N$$

stand for the average of time observations. Therefore,

$$\mathbf{I}(\theta) = \begin{bmatrix} -E\left[\frac{\partial^2 \ln L(\theta_o, \theta'_s, \sigma^2)}{\partial \theta_o^2}\right] & -E\left[\frac{\partial^2 \ln L(\theta_o, \theta'_s, \sigma^2)}{\partial \theta_o \theta'_s}\right] \\ -E\left[\frac{\partial^2 \ln L(\theta_o, \theta'_s, \sigma^2)}{\partial \theta'_s \theta_o}\right] & -E\left[\frac{\partial^2 \ln L(\theta_o, \theta'_s, \sigma^2)}{\partial \theta'^2_s}\right] \end{bmatrix},$$

$$= \frac{2}{\sigma^2}\begin{bmatrix} 2N\theta'^2_s & N(\bar{T}_1 + \bar{T}_4) \\ N(\bar{T}_1 + \bar{T}_4) & \frac{1}{\theta'^2_s}\sum_{i=1}^{N}\left[(T_{1,i}+d)^2 + (T_{4,i}-d)^2 + \sigma^2\right] \end{bmatrix}. \tag{13.11}$$

The CRB can be obtained by taking the $[i,i]$th element of the inverse of the Fisher information matrix (i.e., var $(\hat{\theta}_i) \geq [\mathbf{I}^{-1}(\theta)]_{ii}$), and the inverse $\mathbf{I}^{-1}(\theta)$ is given by

$$\mathbf{I}^{-1}(\theta) = \frac{\sigma^2}{2}\begin{bmatrix} \frac{P}{\theta'^2_s N\left[2P - N(\bar{T}_1+\bar{T}_4)^2\right]} & \frac{-(\bar{T}_1+\bar{T}_4)}{2P - N(\bar{T}_1+\bar{T}_4)^2} \\ \frac{-(\bar{T}_1+\bar{T}_4)}{2P - N(\bar{T}_1+\bar{T}_4)^2} & \frac{2\theta'^2_s}{2P - N(\bar{T}_1+\bar{T}_4)^2} \end{bmatrix}, \tag{13.12}$$

where

$$P \triangleq \sum_{i=1}^{N}\left[(T_{1,i}+d)^2 + (T_{4,i}-d)^2 + 2\sigma^2\right].$$

Consequently, the CRBs of the joint clock offset and skew estimator are given by

$$\text{var}(\hat{\theta}_o^{ML}) \geq \frac{\sigma^2(1+\theta_s)^2 P}{2N\left[2P - N(\bar{T}_1+\bar{T}_4)^2\right]}, \tag{13.13}$$

$$\mathrm{var}(\hat{\theta}_s^{ML}) \geq \left(\frac{\partial\theta_s}{\partial\theta'_s}\right)^2 \cdot \frac{\sigma^2\theta_s'^2}{2P - N\left(\bar{T}_1 + \bar{T}_4\right)^2} = \frac{\sigma^2\left(1+\theta_s\right)^2}{2P - N\left(\bar{T}_1 + \bar{T}_4\right)^2}. \tag{13.14}$$

In fact, finding the joint MLE of clock skew requires quite a number of computations as in (13.10), and the fixed portion of delays d must be known (or estimated), which might not be applicable for wireless sensor networks consisting of low-end terminals. In practice, it requires an additional estimation procedure, which might deteriorate the robustness of the joint MLE. To overcome this limitation, a family of robust and simple clock offset and skew estimators that do not require prior knowledge of d have been proposed in [25].

13.3.2 Receiver-Only Synchronization

Due to the power constraint, the communication range of a sensor is strictly limited to a (radio-geometrical) circle whose radius depends on the transmission power (see Figure 13.1). In this figure, every node within the checked area (e.g., node B) can receive messages from both node P and node A. Suppose that node P is a parent (or reference) node, and node P and node A perform a pairwise synchronization using two-way timing message exchanges [17]. Then, all the nodes in the common coverage region of node P and node A (checked region) can receive a series of synchronization messages containing the information about the time stamps of the pairwise synchronization. Using this information, node B can also be synchronized to the parent node, node P, with no extra timing message transmissions. This approach is called receiver-only synchronization (ROS). In general, all the sensor nodes lying within the checked area can be synchronized by only receiving timing messages using ROS. Here, node P and node A can be regarded as super nodes since they provide synchronization beacons for all the nodes located in their vicinity.

In Figure 13.1, consider an arbitrary node, say node B, in the checked region. While node P and node A exchange time messages, node B can overhear these time messages. Hence, node B is capable of observing a set of time readings ($\{T_{2,i}^{(B)}\}_{i=1}^N$) at its local clock when it receives packets from node A, as depicted in Figure 13.2. Besides, node B can also receive the information about a set of time stamps $\{T_{2,i}^{(P)}\}_{i=1}^N$ obtained by receiving the packets transmitted by node P. Considering the effects of both clock offset and skew, the reception time at node P in the ith uplink message $T_{2,i}^{(P)}$ is given by

$$T_{2,i}^{(\mathrm{P})} = T_{1,i}^{(\mathrm{A})} + \theta_{\mathrm{o}}^{(\mathrm{AP})} + \theta_{\mathrm{s}}^{(\mathrm{AP})} \cdot (T_{1,i}^{(\mathrm{A})} - T_{1,1}^{(\mathrm{A})}) + d^{(\mathrm{AP})} + X_i^{(\mathrm{AP})}, \tag{13.15}$$

where $\theta_{\mathrm{s}}^{(\mathrm{AP})}$ stands for the relative clock skew between node A and node P. Likewise, the reception time at node B in the ith uplink message $T_{2,i}^{(B)}$ can be represented by

$$T_{2,i}^{(\mathrm{B})} = T_{1,i}^{(\mathrm{A})} + \theta_{\mathrm{o}}^{(\mathrm{AB})} + \theta_{\mathrm{s}}^{(\mathrm{AB})} \cdot (T_{1,i}^{(\mathrm{A})} - T_{1,1}^{(\mathrm{A})}) + d^{(\mathrm{AB})} + X_i^{(\mathrm{AB})}, \tag{13.16}$$

where $\theta_o^{(AB)}$ and $\theta_s^{(AB)}$ stand for the relative clock offset and skew between node A and node B, and $d^{(AB)}$ and $X_i^{(AB)}$ denote the fixed and random portions of timing delays in the message transmission from node A to node B, respectively. Here, $X_i^{(AB)}$ is assumed to be a normal distributed RV with mean μ and variance $\sigma^2/2$.

The linear regression technique can be applied to synchronize node B and compensate the effects of the relative clock skew between node P and node B. Subtracting (13.16) from (13.15) gives

$$T_{2,i}^{(P)} - T_{2,i}^{(B)} = \theta_o^{(BP)} + \theta_s^{(BP)} \cdot (T_{1,i}^{(A)} - T_{1,1}^{(A)}) + d^{(AP)} - d^{(AB)} + X_i^{(AP)} - X_i^{(AB)}. \quad (13.17)$$

Since $d^{(AB)}$ and $d^{(AP)}$ are fixed values and $X_i^{(AB)}$ and $X_i^{(AP)}$ are normal distributed RVs, the noise component can be defined by $z[i] \triangleq \mu' + X_i^{(AP)} - X_i^{(AB)}$, where $\mu' \triangleq d^{(AP)} - d^{(AB)}$ and $z[i] \sim \mathcal{N}(\mu',\sigma^2)$. Let $x[i] \triangleq T_{2,i}^{(P)} - T_{2,i}^{(B)} - \mu'$ and $w[i] \triangleq z[i] - \mu'$, then the set of observed data can be written in matrix notation as follows:

$$\mathbf{x} = \mathbf{H}\theta + \mathbf{w},$$

where $\mathbf{x} = [x[1]\ x[2]\ \cdots\ x[N]]^T$, $\mathbf{w} = [w[1]\ w[2]\ \cdots\ w[N]]^T$, $\theta = [\theta_o^{(BP)}\ \theta_s^{(BP)}]^T$, and

$$\mathbf{H} = \begin{bmatrix} 1 & 1 & \cdots & 1 \\ 0 & T_{1,2}^{(A)} - T_{1,1}^{(A)} & \cdots & T_{1,N}^{(A)} - T_{1,1}^{(A)} \end{bmatrix}^T.$$

Note that the noise vector $\mathbf{w} \sim \mathcal{N}(0,\sigma^2\mathbf{I})$ and the matrix **H** is the observation matrix whose dimension is $N \times 2$. From [26, theorem 3.2, p. 44], the minimum variance unbiased (MVU) estimator for the relative clock offset and skew is given by $\hat{\theta} = \mathbf{g}(\mathbf{x})$, where $\mathbf{g}(\mathbf{x})$ satisfies

$$\frac{\partial \ln p(\mathbf{x};\theta)}{\partial \theta} = \mathbf{I}(\theta)(\mathbf{g}(\mathbf{x}) - \theta). \quad (13.18)$$

Since the noise vector **w** is zero mean and Gaussian distributed, from the results in [26, p. 85], the derivative of the log-likelihood function can be written as

$$\frac{\partial \ln p(\mathbf{x};\theta)}{\partial \theta} = \frac{\mathbf{H}^T\mathbf{H}}{\sigma^2}[(\mathbf{H}^T\mathbf{H})^{-1}\mathbf{H}^T\mathbf{x} - \theta], \quad (13.19)$$

where $\mathbf{H}^T\mathbf{H}$ is assumed to be invertible. Therefore, comparing (13.18) with (13.9) yields

$$\hat{\theta} = (\mathbf{H}^T\mathbf{H})^{-1}\mathbf{H}^T\mathbf{x}, \quad (13.20)$$

$$\mathbf{I}(\theta) = \frac{\mathbf{H}^T\mathbf{H}}{\sigma^2}, \quad (13.21)$$

where $\mathbf{I}(\theta)$ is the Fisher information matrix. After some mathematical manipulations, the joint clock offset and skew estimator can be expressed as [19]

$$\begin{bmatrix} \hat{\theta}_o^{(BP)} \\ \hat{\theta}_s^{(BP)} \end{bmatrix} = \frac{1}{N\sum_{i=1}^{N} D_i^2 - \left[\sum_{i=1}^{N} D_i\right]^2} \begin{bmatrix} \sum_{i=1}^{N} D_i^2 \sum_{i=1}^{N} x[i] - \sum_{i=1}^{N} D_i \sum_{i=1}^{N} [D_i \cdot x[i]] \\ N\sum_{i=1}^{N} [D_i \cdot x[i]] - \sum_{i=1}^{N} D_i \sum_{i=1}^{N} x[i] \end{bmatrix}, \tag{13.22}$$

where Di $\triangleq T_{1,i}^{(A)} - T_{1,1}^{(A)}$. The Cramer-Rao lower bound (CRB) can be obtained by inverting the Fisher information matrix $\mathbf{I}(\theta)$. From (13.21), the Fisher information matrix is given by

$$\mathbf{I}(\theta) = \frac{1}{\sigma^2} \begin{bmatrix} N & \sum_{i=1}^{N} D_i \\ \sum_{i=1}^{N} D_i & \sum_{i=1}^{N} D_i^2 \end{bmatrix}.$$

Then, inverting $\mathbf{I}(\theta)$ yields

$$\mathbf{I}^{-1}(\theta) = \frac{\sigma^2}{N\sum_{i=1}^{N} D_i^2 - \left[\sum_{i=1}^{N} D_i\right]^2} \begin{bmatrix} \sum_{i=1}^{N} D_i^2 & -\sum_{i=1}^{N} D_i \\ -\sum_{i=1}^{N} D_i & N \end{bmatrix}. \tag{13.23}$$

Hence, from (13.23), the CRBs for the relative clock offset and skew become

$$\text{var}(\hat{\theta}_o^{(BP)}) \geq \frac{\sigma^2 \sum_{i=1}^{N} D_i^2}{N\sum_{i=1}^{N} D_i^2 - \left[\sum_{i=1}^{N} D_i\right]^2} \tag{13.24}$$

and

$$\text{var}(\hat{\theta}_s^{(BP)}) \geq \frac{\sigma^2 N}{N\sum_{i=1}^{N} D_i^2 - \left[\sum_{i=1}^{N} D_i\right]^2}. \tag{13.25}$$

Notice further that the regularity conditions for the CRBs hold:

$$E\left[\frac{\partial \ln p(\mathbf{x};\theta)}{\partial \theta}\right]=\begin{bmatrix} E\left[\frac{\partial \ln p(\mathbf{x};\theta)}{\partial \theta_o^{(BP)}}\right] \\ E\left[\frac{\partial \ln p(\mathbf{x};\theta)}{\partial \theta_s^{(BP)}}\right] \end{bmatrix}$$

$$=\begin{bmatrix} E\left[\frac{1}{\sigma^2}\sum_{i=1}^{N}\left[x[n]-\theta_o^{(BP)}-\theta_s^{(BP)}\cdot D_i\right]\right] \\ E\left[\frac{1}{\sigma^2}\sum_{i=1}^{N}\left\{\left[x[n]-\theta_o^{(BP)}-\theta_s^{(BP)}\cdot D_i\right]\cdot D_i\right\}\right] \end{bmatrix}=0.$$

Consequently, using the results in (13.22), node B can be synchronized to node P. Likewise, all the other nodes in the checked region in Figure 13.1 can be simultaneously synchronized to the parent node, node P, without any additional timing message transmissions, thus saving a significant amount of energy. Besides, there is no loss of synchronization accuracy when compared with other approaches [19].

13.3.3 Receiver-Receiver Synchronization

Receiver-receiver synchronization is an approach to synchronize a set of children nodes who receive the beacon messages from a common sender (a reference or parent node). Consider a parent (reference) node P and arbitrary nodes A and B, which locate within the communication range of the parent node in Figure 13.4. Suppose, in Figure 13.5 both node A and node B receive the ith beacon from node P at time instants $T_{2,i}^{(A)}$ and $T_{2,i}^{(B)}$ of their local clocks, respectively. Nodes A and B record the arrival time of the broadcast packet according to their own timescales and then exchange their time stamps. Suppose

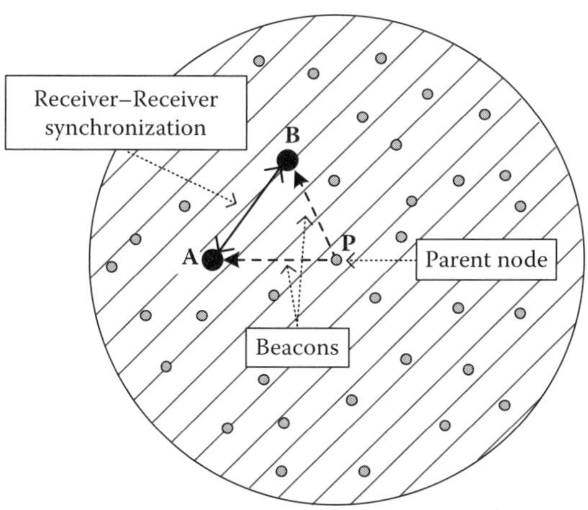

FIGURE 13.4 Receiver-receiver synchronization.

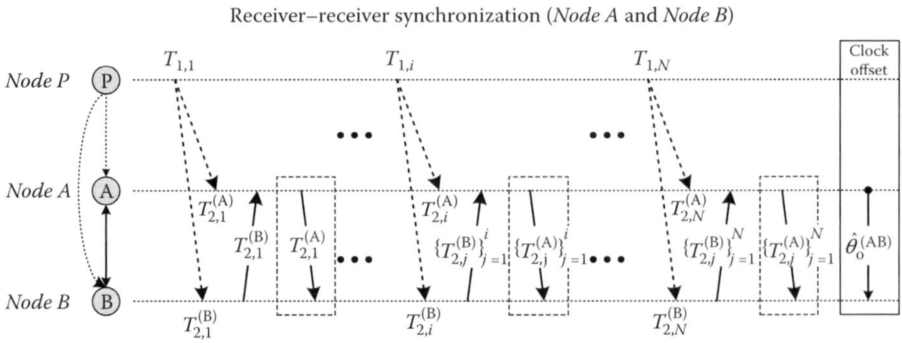

FIGURE 13.5 Clock synchronization model of RRS.

$X_i^{(PA)}$ denotes the nondeterministic delay components (random portion of delays) and $d^{(PA)}$ denotes the deterministic delay component (propagation delay) from node P to node A; then $T_{2,i}^{(A)}$ can be written as

$$T_{2,i}^{(A)} = T_{1,i} + d^{(PA)} + X_i^{(PA)} + \theta_o^{(PA)} + \theta_s^{(PA)} \cdot (T_{1,i} - T_{1,1}), \tag{13.26}$$

where $T_{1,i}$ is the transmission time at the reference node, and $\theta_o^{(PA)}$ and $\theta_s^{(PA)}$ are the clock offset and skew of node A with respect to the reference node, respectively. Similarly, we can decompose the arrival time at node B as

$$T_{2,i}^{(B)} = T_{1,i} + d^{(PB)} + X_i^{(PB)} + \theta_o^{(PB)} + \theta_s^{(PB)} \cdot (T_{1,i} - T_{1,1}), \tag{13.27}$$

where $d^{(PB)}$, $X_i^{(PB)}$, $\theta_o^{(PB)}$, and $\theta_s^{(PB)}$ stand for the propagation (fixed) delay, random portion of delays, clock offset, and skew of node B with respect to the reference node, respectively.

Subtracting (13.27) from (13.26), we obtain

$$T_{2,i}^{(A)} - T_{2,i}^{(B)} = \theta_o^{(BA)} + \theta_s^{(BA)} \cdot (T_{1,i} - T_{1,1}) + d^{(PA)} - d^{(PB)} + X_i^{(PA)} - X_i^{(PB)}, \tag{13.28}$$

where $\theta_o^{(BA)} \triangleq \theta_o^{(PA)} - \theta_o^{(PB)}$ and $\theta_s^{(BA)} \triangleq \theta_s^{(PA)} - \theta_s^{(PB)}$ are the relative clock offset and skew between node A and node B at the time they receive the ith broadcast packet from the reference node, respectively. Here, we assume these random portions of delays $X_i^{(PA)}$ and $X_i^{(PB)}$ are normal distributed RVs with mean μ and variance $\sigma^2/2$. Indeed, (13.28) assumes exactly the same form as (13.17). Hence, the same steps can be applied to derive the joint clock offset and skew estimator for ROS. More specifically, let the noise component $z[i] \triangleq \mu' + X_i^{(BA)}$, where $\mu' \triangleq d^{(PA)} - d^{(PB)}$ and $z[i] \sim \mathcal{N}(\mu',\sigma^2)$. Let us also define $x[i] \triangleq T_{2,i}^{(A)} - T_{2,i}^{(B)} - \mu'$ and $w[i] \triangleq z[i] - \mu'$. Using similar steps as in ROS, it is straightforward to show that the same form of the joint clock offset and skew estimator (13.22) can also be applied to RRS. Consequently, there is no difference between ROS and RRS with regard to the accuracy of synchronization since the effects of random delays are the same. Likewise, the CRB for RRS can also be obtained using a similar procedure as in ROS. When there is no relative clock skew ($\theta_s^{(BA)} = 0$), it is straightforward to show that the maximum likelihood estimator of the relative clock offset $\hat{\theta}_o^{(BA)}$ becomes

$$\hat{\theta}_o^{(BA)} = \frac{1}{N}\sum_{i=1}^{N}\left[T_{2,i}^{(A)} - T_{2,i}^{(B)}\right], \tag{13.29}$$

which is the equivalent to the estimator presented in [18].

The main benefit of this approach is that all nondeterministic delay components on the transmitter side (send time and access time) are eliminated. Thus, a high degree of synchronization accuracy can be achieved using this approach.

13.4 Existing Time Synchronization Protocols

Thus far, a number of protocols have been suggested to solve the problem of time synchronization in distributed networks. For general computer networks, NTP has been adopted as the standard time synchronization scheme of the Internet [3]. Although NTP was shown to perform well in computer networks, it is not directly applicable to wireless sensor networks due to the unique challenges sensor networks face: limited power resources, wireless channel conditions, dynamic topology changes, etc. (recall also the design considerations presented in section 13.2.2). NTP enjoys unlimited (or rechargeable) energy resources and a relatively static topology in computer networks. However, these are not available in sensor networks. Therefore, different types of time synchronization protocols have been proposed to meet the design requirements of wireless sensor networks [2].

Ideally, a time synchronization protocol should be able to work optimally in terms of all design requirements of time synchronization, which are energy efficiency, scalability, precision, security, reliability, and robustness to network dynamics. However, the complex nature of wireless sensor networks makes it very difficult to optimize the protocol with respect to all these requirements simultaneously. Due to the trade-offs in satisfying these requirements, each protocol is designed to put distinct emphases on different requirements.

Assuming various criteria, time synchronization protocols can be categorized into different classes:

- **Master-slave versus peer-to-peer**
  - *Master-slave*: Where first a tree-like network hierarchy is arranged, and upon the completion of this arrangement only the connected nodes in the hierarchy synchronize with each other.
  - *Peer-to-peer*: Where any pair of nodes in the network can synchronize with each other.

- **Clock correcting versus untethered clock**
    - *Clock correcting*: Where the clock function in memory is modified after each run of the time synchronization process.
    - *Untethered clock*: Where every node maintains its own clock as it is, and keeps a time-translation table relating its clock to other nodes' clocks. Thus, instead of updating its clock constantly, each node translates the time information in the data packets coming from other nodes to its own clock by using the time-translation table.
- **Synchronization approach**
    - *Sender-receiver*: Where one of two nodes, which are synchronizing with each other, sends a time-stamp message while the other one receives it.
    - *Receiver-receiver*: Where a reference node transmits synchronization signals and two synchronizing nodes receive these signals and record the time of receptions (time stamps).
    - *Receiver-only*: Where a group of nodes can be simultaneously synchronized by only listening to the message exchanges of a pair of nodes.
- **Pairwise synchronization versus network-wide synchronization**
    - *Pairwise synchronization*: Where the protocols are primarily designed to synchronize two nodes, although they usually can be extended to handle synchronization of a group of nodes.
    - *Network-wide synchronization*: Where the protocols are primarily designed to synchronize a large number of nodes in the network.

Additional classifications can be found in [4]. In the following, we will summarize the existing time synchronization protocols based on the last category.

13.4.1 Pairwise Synchronization

13.4.1.1 Timing-Sync Protocol for Sensor Networks (TPSN)

TPSN [17] uses the two-way message exchange mechanism, as discussed in the sender-receiver synchronization approach described in section 13.3.1, to achieve the synchronization between two nodes. With only one round of message exchanges, and without any statistical model on the variable delay components $X_i^{(AP)}$ and $X_i^{(PA)}$ in (13.2) and (13.3), a simple estimate for $\theta_o^{(AP)}$ is proposed in [17] as

$$\hat{\theta}_o^{(AP)} = \frac{U_i - V_i}{2}, \tag{13.30}$$

where $U_i \triangleq T_{2,i}^{(P)} - T_{1,i}^{(A)}$ and $V_i \triangleq T_{4,i}^{(A)} - T_{3,i}^{(P)}$. Notice that in the original form of TPSN, it does not estimate clock skew; therefore, frequent application of TPSN is needed to keep the clock offset between two nodes under a certain limit.

13.4.1.2 Maximum Likelihood Estimation for Clock Offset Based on Two-Way Message Exchanges

Assume the clock offset $\theta_o^{(AP)}$ is constant for N rounds of message exchanges. If $X_i^{(AP)}$ and $X_i^{(PA)}$ in (13.2) and (13.3) are exponentially distributed with the same unknown mean λ, and when $d \triangleq d^{(AP)} = d^{(PA)}$ is unknown, it is proved in [23] that the ML estimator of $\theta_o^{(AP)}$ is given by

$$\hat{\theta}_o^{(AP)} = \frac{\min_{1\le i\le N} U_i - \min_{1\le i\le N} V_i}{2}. \tag{13.31}$$

On the other hand, with $X_i^{(AP)}$ and $X_i^{(PA)}$ in (13.2) and (13.3) being modeled as independent and normally distributed RVs with the same mean μ and variance $\sigma^2/2$, the maximum likelihood (ML) estimate for $\theta_o^{(AP)}$ takes the equation (derived in section 13.3.1)

$$\hat{\theta}_o^{(AP)} = \frac{\frac{1}{N}\sum_{i=1}^{N} U_i - \frac{1}{N}\sum_{i=1}^{N} V_i}{2}. \tag{13.32}$$

Notice from (13.30)–(13.32), it is clear that if only one round of message exchange is performed, the TPSN presented in (13.30) is the ML estimator under both exponential and Gaussian delay models.

13.4.1.3 Joint Clock Offset and Skew Estimation Based on Two-Way Message Exchanges

When clock skew exists between two nodes, the clock offset between them will increase linearly, as shown in Figure 13.3. In order to establish long-term synchronization, it is more efficient to estimate jointly the clock offset and skew. In section 13.3.1, we derived the joint offset and skew ML estimators (see equations (13.9) and (13.10)), when the variable delays $X_i^{(PA)}$ and $X_i^{(AP)}$ are modeled as independent Gaussian distributed RVs. When $X_i^{(PA)}$ and $X_i^{(AP)}$ are exponentially distributed RVs, the likelihood function for joint estimation of the clock offset and skew is very complicated. However, a solution to this problem has been recently reported in [27].

Notice that the joint offset and skew ML estimators (equations (13.9) and (13.10)) under Gaussian delay assumption are quite complicated. Besides, there is no simple closed-form solution for the ML joint offset and skew estimation when the delays are exponentially distributed. For these reasons, a family of robust and simple clock offset and skew estimators, named maximum likelihood like estimators (MLLEs), has been proposed in [25].

13.4.1.4 Tiny-Sync and Mini-Sync

Tiny-sync and Mini-sync [28] are two lightweight clock synchronization protocols that also use the two-way message exchanges. Node A and node P exchange messages just like in Figure 13.3. The only difference here is that node P replies to node A immediately after receiving the message, i.e., $T_{2,i}^{(P)} = T_{3,i}^{(P)}$. Assuming the clocks between node A and node P are linearly related, from (13.6) and (13.7) we have

$$\frac{T_{2,i}^{(P)} - \theta_o^{(AP)}}{1+\theta_s^{(AP)}} = T_{1,i}^{(A)} + d + X_i^{(AP)},$$

$$\frac{T_{2,i}^{(P)} - \theta_o^{(AP)}}{1+\theta_s^{(AP)}} = T_{4,i}^{(A)} - d - X_i^{(PA)}.$$

Since d, $X_i^{(AP)}$, and $X_i^{(PA)}$ are all nonnegative, defining $\theta'_s \triangleq 1/(1 + \theta_s^{(AP)})$ and $\theta'_o \triangleq \theta_o^{(AP)}/(1 + \theta_s^{(AP)})$, we obtain

$$T_{1,i}^{(A)} \leq \theta'_s T_{2,i}^{(P)} + \theta'_o \leq T_{4,i}^{(A)}. \tag{13.33}$$

The 3-tuple of time stamp ($T_{1,i}^{(A)}$, $T_{2,i}^{(P)}$, and $T_{3,i}^{(A)}$) is called a data point. With N message exchanges, the goal is to find θ'_o and θ'_s such that they satisfy (13.33) for $1 \leq i \leq N$. In general, this is a linear programming problem and there are an infinite number of solutions for this problem [29]. Although more time stamps would generate tighter bounds on θ'_o and θ'_s, unfortunately, at the same time, the computational and storage requirements of the linear programming approach also increase. Thus, such an approach appears to be not suitable for implementation in wireless sensor nodes, which have strictly limited memory and computing resources.

Tiny-sync and Mini-sync tackle the problem as finding the best-fit line that lies between the bound sets defined by the data points. Based on the observation that not all data points are useful, Tiny-sync preserves only four constraints (the ones that yield the best bounds on the estimate) out of all data points. This results in a very efficient algorithm. However, it is shown by a counterexample [28] that this scheme does not always produce the optimal solution since some data points are considered useless and discarded at a certain time, a step that actually might provide a better bound if it is properly considered with another data point that is yet to come.

Mini-sync is an improved version of Tiny-sync in the sense that it finds the optimal solution with increased complexity (but still with lesser complexity than the linear programming approach). Mini-sync basically uses an additional criterion to determine whether the data point can be safely discarded.

13.4.1.5 Reference Broadcast Synchronization (RBS)

RBS [18] is based on the RRS approach discussed in section 13.3.3. Let the time stamps recorded at node A and node B for receiving the ith common packet be denoted as $T_{2,i}^{(A)}$ and $T_{2,i}^{(B)}$, respectively. The estimate of the clock offset between node A and node B is proposed in [18] as

$$\hat{\theta}_o^{(BA)} = \frac{1}{N}\sum_{i=1}^{N}\left[T_{2,i}^{(A)} - T_{2,i}^{(B)}\right], \tag{13.34}$$

where N stands for the total number of common packets received by node A and node B. We have shown in section 13.3.3 that the above estimator is actually the ML estimator for the clock offset, assuming the random portions of the delays in message deliveries are Gaussian distributed RVs, and there is no clock skew. When there is a clock skew between node A and node B, least-squares linear regression is proposed in [18] to estimate the clock skew.

The main advantage of RBS is that by comparing the time stamps of a common packet at two different nodes, it removes the largest sources of nondeterministic error (send time and access time) from the transmission path. Thus, RBS provides a high degree of synchronization accuracy. Note also that RBS can be applied to commodity hardware and existing software in sensor networks as it does not need access to the low levels of the operating system.

13.4.1.6 Clock Offset and Skew Estimation Based on Broadcast Clock

Under the setting that a sensor node observes and synchronizes to a broadcast clock, [30] derives the ML estimator for clock offset and skew with the broadcast message delay being modeled as uniformly distributed RVs. It is shown that the ML estimate in this case is generally not unique. Furthermore, the support of the likelihood function is not convex, which leaves out the possibility of taking the mean of all equally likely solutions. This motivated [30] to consider the linear estimator for the clock offset and skew. Under the same setting, [31] derives the joint ML clock offset and skew estimator with the assumption that the broadcast message delays are modeled as exponentially distributed RVs. It is shown in [31] that a unique joint ML clock offset and skew estimate exists under certain conditions, as opposed to the case of uniformly distributed delay. Furthermore, the Gibbs sampler was introduced in [31] to further enhance the performance of the joint ML estimator.

13.4.1.7 Flooding Time Synchronization Protocol (FTSP)

In [16], it is argued that if one can time-stamp the message at the MAC layer, this immediately eliminates three sources of delay uncertainties: transmit, access, and receive times. In this case, the main delivery delay comes from transmission and reception times at the radio chips (see section 13.2.3). These delays can be further decomposed into:

1. Interrupt handling time, which is the delay between the radio chip raising and the microcontroller responding to an interrupt
2. Encoding time, which is the time it takes for the radio chip to encode and transform the message into a radio wave
3. Decoding time, which is the time for the radio chip at the receiver to transform the radio wave back into binary data
4. Byte alignment time, which is the delay at the receiver to synchronize with the byte boundary at the physical layer

FTSP [16] uses a single broadcasted message to establish synchronization points between sender and receivers, while eliminating the jitter of interrupt handling and encoding/decoding times by utilizing multiple MAC layer time stamps on both the sender and receiver sides. Furthermore, the skew of the clock between sender and receiver is estimated using multiple messages and linear regression.

13.4.2 Network-Wide Synchronization

Until this point, we have only described the time synchronization between two neighboring sensor nodes. In this section, we will discuss protocols for network-wide synchronization.

13.4.2.1 Extension of TPSN

In order to establish a global timescale for all the nodes in the sensor field based on TPSN, [17] proposes to create a hierarchical structure (spanning tree) in the network (named level discovery phase) before pairwise synchronization is performed between adjacent levels (named synchronization phase). The level discovery phase consists of the following steps:

1. Select a root node using an appropriate leader election algorithm and assign a 0 level to the root node.
2. The root node broadcasts a level discovery packet (LDP) containing the identity and level of the packet.
3. Every node that receives an LDP assigns its level to a level greater (by one) than that of the received packet and sends a new level discovery packet attaching its own level (once being assigned a level, a node neglects future packets requesting level discovery to avoid flooding congestion).
4. Repeat step 3 until every node in the network successfully assigns a level.

After the spanning tree is formed, the root node initiates the synchronization phase by synchronizing all the nodes in level 1. Next, the nodes in level 1 synchronize with the nodes in level 2, and so on, until all the nodes have been synchronized. Notice that the synchronization error of a node with respect to the root node is a nondecreasing function of the hop distance because the random signal errors over each hop add up. A number of different searching algorithms can be considered in the construction of the spanning tree. For instance, Van Greunen and Rabaey suggested some preliminary ideas on constructing spanning trees with low depth in order to improve the accuracy of synchronization [12].

13.4.2.2 Lightweight Time Synchronization (LTS)

Also based on two-way message exchanges, [12] proposes two network-wide synchronization protocols. The first one is called centralized multihop LTS, which is basically the same protocol as the extension of TPSN discussed above. The other one is called distributed multihop LTS. This distributed LTS algorithm moves the resynchronization from the root node to the nodes that need resynchronization. When a node A determines that it needs to be resynchronized, it will send a resynchronization request to the root node. In order for node A to resynchronize, all nodes along the routing path from the root node to node A will be synchronized in a pairwise fashion.

13.4.2.3 Extension of RBS

The RBS protocol discussed in the above subsection can only synchronize a set of nodes that lie within a single broadcast domain. In order to synchronize a large sensor network,

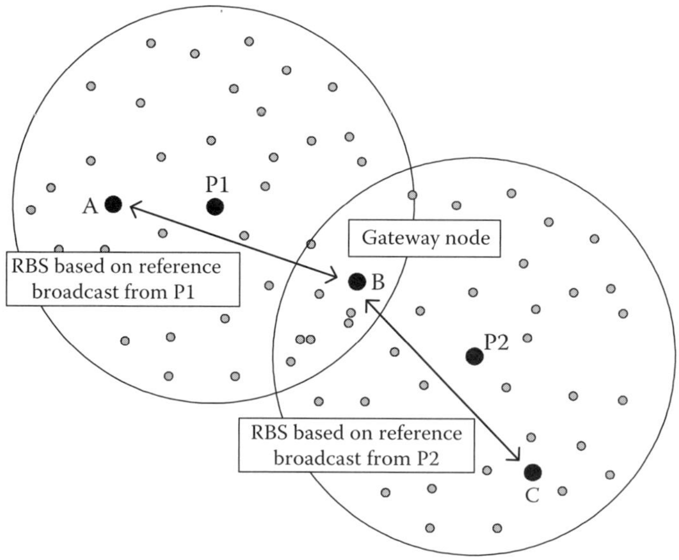

FIGURE 13.6 Extension of RBS to multihop.

[18] proposes to use *gateway* nodes for converting time stamps from one neighborhood's time base to another. The idea is illustrated in Figure 13.6. Nodes P1 and P2 send out synchronization beacons, and they create two overlapping neighborhoods, where node B lies in the overlapping area. Since node A and node B lie within the same neighborhood, their clock relationship (i.e., clock offset and skew) can be established from node P1's reference broadcast. Similarly, the clock relationship between node B and node C can be established from node P2's reference broadcast. Therefore, the clock relationship between node A and node C can be computed with node B acting as a gateway.

13.4.2.4 Extension of FTSP

FTSP can be extended to network-wide synchronization in a straightforward manner. First, a root node, to which the whole network is being synchronized, is elected by the network. Nodes that are within the broadcast radius of the root node can receive time-stamped messages from the root node. They then estimate the offset and skew of their own local clocks, thus synchronizing with the root node. The newly synchronized nodes can then broadcast synchronization messages to other nodes in the network. The advantage of this flooding process is that it begins with the root node, and there is no need to have a level hierarchy, as opposed to TPSN.

13.4.2.5 Pairwise Broadcast Synchronization

Pairwise broadcast synchronization (PBS) employs both sender-receiver and receiver-only synchronization approaches to achieve network-wide synchronization with high energy efficiency [19]. As discussed in section 13.3.2, in PBS a number of sensor nodes can be synchronized by only overhearing timing messages being exchanged between pairs of nodes, which significantly reduces the overall energy consumption by decreasing the number of required timing messages in synchronization. PBS requires a much smaller number of timing messages than other existing protocols, such as RBS, TPSN, and FTSP, and its benefits remarkably increase as the sensors are more densely deployed.

Let N_{RBS}, N_{TPSN}, N_{FTSP}, and N_{PBS} denote the numbers of required timing messages for synchronization in RBS, TPSN, FTSP, and PBS, respectively. In TPSN, since every node in the network is connected to its parent node (except the root node), there are $L - 1$ branches (edges) in a hierarchical tree, where L is the overall number of sensor nodes [17]. For TPSN, $2N$ timing messages are required in every pairwise synchronization, hence $N_{TPSN} = 2N(L - 1)$. This result can be applied to other level-based SRS protocols without loss of generality. For RBS, the reference node must broadcast the beacon packet N times. Besides, every sensor node must send time readings upon receiving the broadcast beacons of all the other nodes in the network to compensate relative clock offsets among each other [18]. Thus, $N_{RBS} = N + L(L - 1)/2$, since the number of unique pairs in the network is $L(L - 1)/2$. In FTSP, each sensor node must send its timing messages once upon receiving the timing messages from another sensor due to its flood-based communication procedure [16]. Hence, the number of required timing messages in FTSP becomes $N_{FTSP} = NL$.

It is remarkable that the required numbers of timing messages for all the above-mentioned protocols are proportional to the number of sensors in the network L or its square L^2. However, PBS needs only $2N$ timing messages in every synchronization period, i.e., $N_{PBS} = 2N$, assuming all the nodes lie within a single broadcast neighborhood. Hence, N_{PBS} does not depend on the number of sensors in the network, which incurs an enormous amount of energy savings. Moreover, this gain proportionally increases with respect to the scale of the network. Consequently, the benefit of PBS over RBS, TPSN, and FTSP is clear and huge in terms of energy savings with the cost of allocating two super nodes in the network. In case there exist other nodes located outside of the checked region in Figure 13.1, likewise RBS, the network could be divided into a number of separated groups (clusters), and they could be synchronized by additional pairwise synchronizations among the super nodes in different groups, i.e., global synchronization can be achieved by a sequence of pairwise synchronizations among the super nodes. Here, diverse grouping and pair selection algorithms can be considered according to the type of the network. For instance, assuming the level hierarchy of the network is established, there are groups of parents and children nodes, where a group consists of a parent and its children nodes. Here, every parent node can search the connectivity among its children nodes to select the best synchronization pairs that maximize the number of nodes performing ROS, i.e., minimizing the number of pairwise synchronizations. In fact, no network-wide connection search is required in this case because of its limited and known set of scanning nodes.

13.4.2.6 Time-Diffusion Synchronization Protocol (TDP)

TDP [32] is a protocol enabling the sensor network to reach an equilibrium time with the clocks of individual sensors within a small time deviation from the equilibrium time. The protocol can be understood as periodically applying three phases: (1) election of master/diffused leader nodes, (2) time-diffusion procedure, and (3) peer evaluation

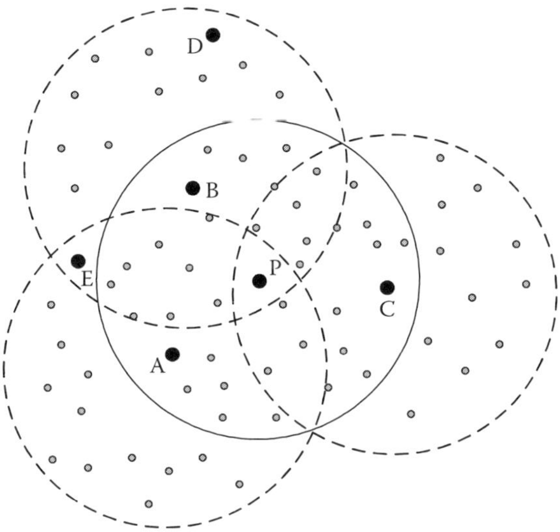

FIGURE 13.7 Time-diffusion synchronization protocol.

procedure. It is shown analytically in [32] that the TDP enables the clocks in the whole network to converge to a unique value.

In the first phase, master nodes are elected in the sensor field. The election criteria include the quality of clock and the energy resources of a particular node. Referring to Figure 13.7, assume that node P is elected to be the master node (here we illustrate the concept with one master node, while in more complicated networks, more than one master node might be possible). Node P then sends a number of time-stamped messages to its neighbors. Once the neighbors receive the messages, they self-determine if they would become diffused leader nodes, based on the results of the last round peer evaluation procedure (the third phase). In Figure 13.7, nodes A, B, and C are the elected diffused leader nodes. The elected diffused leader nodes respond to the master node, thus enabling the master node to measure the average and the standard deviation of the round-trip delay from its neighbors. At the same time, the diffused leader nodes start sending messages to their own neighbors to measure the mean and standard deviation of round-trip delay to their neighbors. The process is repeated until all the nodes have been covered.

In the second phase, the time information from the master node will be diffused (with the help of diffused leader nodes) to all the nodes in the network. The diffusion procedure takes place according to the following sequence of events. First, the master node sends a time-stamped message containing the standard deviation of the round-trip delay to its neighbors. Before transmission, the time stamp of the message is adjusted with half of the measured average round-trip delay (from the first phase) to account for the message delivery delay to its neighbors. Once the diffused leader nodes receive the time-stamped message, they set their clock according to the received time stamp and then broadcast their own time-stamped messages, containing their measured standard derivations of the round-trip times to their neighbors. Again, before transmission of the messages, the time stamps have to be adjusted with half of the measured average round-trip delay to their neighbors. For nodes that are not diffused leaders, if they only receive a message from one diffused leader node (e.g., node D in Figure 13.7), they just set their clock according to the time stamp they received. For the nodes that have received more than one time-stamped message originating from different diffused leader nodes (e.g., node E in Figure 13.7), they will use the standard deviations as weightings (the smaller the deviation, the larger the weighting) to combine the clock values and set their clocks according to the result.

The purpose of the third phase is to allow the sensor nodes to evaluate the stability of their local clock. First, the elected master nodes broadcast a number of time-stamped messages. The neighbor nodes receiving these messages calculate the two-sample Allan variance [32] of the local clock from the clock of the master nodes and send back these calculated Allan variances to the master nodes. Then the master nodes compute the average of all the Allan variances they received and send the result back to their neighbor nodes. By this procedure, all the neighbor nodes can evaluate the quality of their clocks with respect to those of their neighbors by comparing their calculated Allan variance with the average value. The above procedure is repeated, but with the elected diffused leader nodes broadcasting the time-stamped messages.

13.4.2.7 Synchronous and Asynchronous Diffusion Algorithms

In [33], two diffusion algorithms are proposed. The first one is called rate-based synchronous diffusion algorithm. The idea behind this algorithm is that in order for a network to achieve an equilibrium time, the clock at node i, denoted as c_i, should be adjusted according to the differences between its clock and its neighbors' clocks (assuming node i has exchanged clock readings with its neighbors). That is, the clock at node i should be set to

$$c_i - \sum_{j \neq i} r_{ij}(c_i - c_j),$$

where $r_{ij} > 0$ is the diffusion rate, $r_{ij} = 0$ if node i and node j cannot directly communicate, and the condition

$$\sum_{j \neq i} r_{ij} \leq 1$$

is enforced. The above algorithm can also be formulated using matrix notation. For a group of n sensor nodes, let $\mathbf{c}^t$ be the vector of length n containing the clock readings of all the sensor nodes at time t. The synchronous diffusion algorithm adjusts the clocks of different nodes using $\mathbf{c}^{t+1} = Rc^t$, where

$$\mathbf{R} = \begin{pmatrix} r_{11} & r_{12} & \cdots & r_{1n} \\ r_{21} & r_{22} & \cdots & r_{2n} \\ \vdots & \vdots & \ddots & \vdots \\ r_{n1} & r_{n2} & \cdots & r_{nn} \end{pmatrix} \tag{13.35}$$

and

$$r_{ii} = 1 - \sum_{j \neq i} r_{ij}.$$

It is shown in [33] that if the second largest eigenvalue of **R** is smaller than 1, the synchronous diffusion algorithm will converge, in the sense that all the elements in $\mathbf{c}^t$ will be equal.

The synchronous diffusion algorithm requires all the nodes to operate in an ordered manner. In order to remove this constraint, [33] proposed another algorithm, named asynchronous diffusion algorithm. In this algorithm, each node asks its neighbors about their clock readings and computes the average value. Then the average value is sent back to the neighbors so they can update their clocks. This algorithm gives a very simple averaging operation of a node over its neighbors, and the averaging operations by different nodes can be carried out at different times and in any order (thus the name asynchronous). It is shown in [33] that the clocks of sensor nodes at a sensor network converge to the average value by using this asynchronous algorithm.

13.4.2.8 Protocols Based on Pulse Transmissions

Recently, synchronization schemes that operate exclusively at the physical layer by transmitting pulses instead of message packets have been proposed in [34] and [35]. In [34], inspired by the synchronously flashing fireflies, the time synchronization problem in sensor network is modeled using pulse coupled oscillators (PCOs). In this scheme, each node (say node j) in the sensor network is associated with an increasing monotonic state function $x_j(t)$ taking values from 0 to 1. If a node is isolated, the state function $x_j(t)$ increases from 0 to 1 smoothly as a function of time, and the node emits a pulse when the state function achieves the unit value ($x_j(t) = 1$). After firing a pulse, the node immediately resets its state to zero. This results in periodic emission of pulses with period T. If a node is not isolated, it can receive pulses from other nodes. When a node receives a pulse, its state variable changes as follows:

$$x_j(\tau^+) = \begin{cases} x_j(\tau) + \varepsilon, & \text{if } x_j(\tau) + \varepsilon < 1 \\ 0, & \text{otherwise} \end{cases}, \qquad (13.36)$$

where τ is the time the node receives a pulse and ε is the advancement of the clock phase. This means that a node receiving a pulse either emits the pulse at the same time or shortens the waiting time for the next round of emissions. With the assumption that after a node fires a pulse, it enters a short refractory period, during which no signal can be received from other nodes (to avoid infinite feedback), it can be shown that only when the nodes emit the pulse simultaneously will they be insensitive to coupling, and therefore achieve synchronization.

In [35], a cooperative technique that constructs a sequence of pulses with equidistance zero-crossings is developed. The basic idea of this scheme is as follows. Assume there is a leader node and it emits a sequence of pulses with equidistance zero-crossings.

The surrounding nodes receive this pulse sequence, and based on the locations of the observed zero-crossings, the surrounding nodes predict when the next pulse will be transmitted. Then, these nodes emit pulses at their predicted times and an aggregate pulse sequence will be generated. It is shown in [35] that although the prediction at individual nodes may not be perfect, under certain conditions on the pulse and in asymptotically dense networks, the zero-crossings of the aggregate waveform sequence will be at the same positions as the zero-crossings of the original waveform sequence emitted by the leader node due to spatial averaging. This aggregate pulse sequence will be heard by the nodes lying further away from the leader node, and these nodes perform prediction as described before and emit their pulses at their predicted times. The procedure will be continued until all the nodes are synchronized.

Notice that the synchronization algorithms discussed in this section only provide a unified ticking rhythm across sensor nodes, but not the synchronization of clock time. A good analog is a group of people clapping together to get a rhythm. However, there exist applications in which a unified rhythm is enough, e.g., in distributed beamforming and reachback channel [36]. As another variation, a joint physical and network layer time synchronization scheme was proposed to overcome the effects of imperfect physical layer synchronization due to the nature of common wireless channels [37].

13.5 Adaptive Time Synchronization for WSNs

While all the above protocols in section 13.4 can achieve instantaneous synchronization among nodes, the timing of different nodes would drift apart as time passes; therefore, periodic resynchronization is needed to maintain long-term synchronization. Intuitively, less resynchronization needs less energy but leads to a larger synchronization error, while more frequent resynchronization leads to a smaller synchronization error but requires more energy. A natural question is what is the minimum resynchronization frequency (or equivalently, maximum resynchronization period) that can meet the desired synchronization precision. Therefore, adaptive algorithms are necessary to dynamically determine the resynchronization period, number of beacons to be used in each round of synchronization, synchronization accuracy, and so on. In this section, we will review three existing adaptive time synchronization algorithms proposed in the literature.

13.5.1 Rate-Adaptive Time Synchronization (RATS)

Consider the case where node A sends time-stamped messages to node B periodically with period τ, and node B records the receiving times of the messages. Based on a number of data points $(T_i^{(A)}, T_i^{(B)})$, where $T_i^{(A)}$ and $T_i^{(B)}$ are the time stamps made at node A and node B respectively, node B wants to determine the largest τ such that the synchronization error is smaller than a certain limit. The rate-adaptive time synchronization [38] is an algorithm that determines the optimal τ adaptively. Its idea can be summarized using the flowchart shown in Figure 13.8. First, node B calculates the optimal number of data samples for model parameters (e.g., clock offset and skew) estimation based on the current value of τ. Next, node B takes the required number of data points (stored

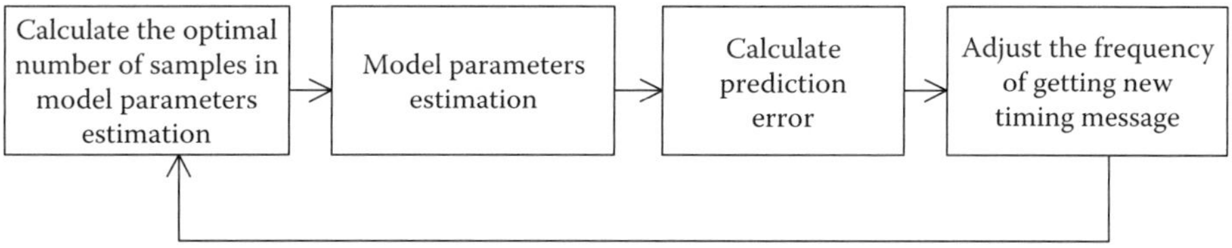

FIGURE 13.8 Flowchart of RATS run at the node receiving time-stamped messages from another node.

in memory) and estimates the model parameters. Then, node B computes the prediction error. Finally, using the calculated prediction error, node B adjusts the frequency of getting a new timing message from node A: If the prediction error is larger than the upper limit threshold E_u, then the timing message rate is not frequent enough from node A, and therefore τ should be decreased. On the other hand, if the prediction error is smaller than the lower limit threshold E_l, then there are fewer timing messages, and thus τ should be increased. Multiplicative increase and decrease strategies are used to enable fast convergence and quick response to the changing environment. After getting a new data point according to the new value of τ, the above process is repeated.

13.5.2 RBS-Based Adaptive Clock Synchronization

With the RBS setting, [39] extends the deterministic RBS protocol (discussed in section 13.4.1) to an adaptive probabilistic synchronization algorithm, allowing trade-offs between synchronization accuracy and resource requirement. It is based on the observation that if the relative clock skew error between two nodes ε, after applying RBS with one broadcast message, is a Gaussian RV with zero mean and variance σ^2, then the probability of error-free synchronization with N broadcast messages is given by

$$Pr\left(|\varepsilon|<\varepsilon_{max}\right)=2erf\left(\frac{\sqrt{N}\varepsilon_{max}}{\sigma}\right), \tag{13.37}$$

where ε_{max} stands for the maximum specified (allowable) clock offset for communications, and

$$erf(x)\triangleq 1/2\pi\cdot\int_0^x \exp\left(-t^2/2\right)dt.$$

From the above equation, it is clear that the performance criterion is a probabilistic measure since there is always a possibility that the clock offset is greater than some limit ε_{max}. However, one can reduce this probability to an arbitrarily small value by increasing N, the number of broadcasting messages in one round of RBS.

After application of RBS, we can bound the clock skew error with certain probability. However, since clocks from different nodes would drift apart as time passes, we need to reapply RBS periodically. Reference [39] proposes a formula to determine the maximum time between resynchronization $\tau_{\max}$ as

$$\tau_{\max} = \frac{\gamma_{max} - \varepsilon_{max}}{\rho} - d_{max}, \tag{13.38}$$

where γ_{max} denotes the maximum allowable clock skew at any time, ρ denotes the maximum drift of clock, and d_{max} is the maximum delay of time-stamp exchanges in RBS. With different synchronization precision requirements (specified by γ_{max}), one can determine the required resynchronization period $\tau_{\max}$.

13.5.3 Adaptive Multihop Time Synchronization

Adaptive multihop time synchronization (AMTS) [40] is based on a two-way message exchange mechanism similar to that used in TPSN. AMTS adaptively optimizes some crucial network parameters, such as the synchronization mode, the resynchronization period, and the number of beacons per pairwise synchronization, with respect to the current network status. AMTS consists of three functional phases:

- *Level discovery phase*: The same as that in TPSN, and used for generating a hierarchical structure in the network.
- *Synchronization phase*: Similar to the corresponding synchronization phase in TPSN. However, as opposed to TPSN, AMTS adjusts not only the current clock offset but also the clock skew to guarantee the long-term synchronization, while TPSN only estimates the clock offset. Hence, AMTS requires far less frequent resynchronization.
- *Network evaluation phase*: The reference node investigates the current status of network traffic in order to select the synchronization mode between *always on* (**AO**) (always maintain network-wide synchronization) and *sensor initiated* (**SI**) (synchronize only when it needs to). Besides, it optimizes the resynchronization period and the number of beacons per each pairwise synchronization.

The second and third phases (i.e., synchronization and the network evaluation phases) will be periodically repeated in order to minimize overall energy consumption with respect to the current network status. Since the level discovery procedure (section 13.4.2) and pairwise synchronization based on two-way message exchanges (section 13.3.1) have been detailed previously, we only focus on the network evaluation phase below.

13.5.3.1 Synchronization Mode Selection

The idea of selecting the synchronization mode between *always on* (**AO**) and *sensor initiated* (**SI**) is based on the observation that when network traffic occurs rarely and synchronization delay is not a critical problem, keeping all the sensor nodes synchronized all the time (**AO** mode) is not a good strategy since synchronization consumes a lot of energy. In addition, for some applications, the sensor clocks might be allowed to go out of synchronization unless sensing events happen. In this case, the **SI** mode, where only nodes participating in a particular multihop data transmission synchronize with each other, is a better choice.

In order to determine whether **AO** or **SI** mode is to be used, the following network parameters are introduced in AMTS:

- B: Number of branches (edges) in a spanning tree of the network. It can be obtained after the level discovery phase.
- τ: Resynchronization period, i.e., the time between resynchronizations.
- $\bar{h}$: Average of hops per unit time. In every sensing event, the destination node accumulates the number of hops that have occurred in that particular transmission to its storage. During the synchronization phase, the reference node collects the information about the total number of hops occurred in the last synchronization period and determines the average number of hops per unit time ($\bar{h}$) in the network. This information indicates how busy the network traffic is and can be included in timing messages with a small overhead.
- δ: Latency factor ($0 \le \delta \le 1$) reflecting the amount of allowed delay in data transmission. Higher latency factor means less concern for network delays. For example, δ is set to be 0 for sensor networks requiring network synchronization all the time. On the other extreme, for delay-independent networks, δ should be close to 1.
- N: Number of timing message exchanges per pairwise synchronization.

In the **AO** mode, the number of timing messages per unit time is given by $\bar{M} = 2BN/\tau$, while in the **SI** mode, $\bar{M} = 2\bar{h}N$. To minimize the number of timing messages per unit time $\bar{M}$, the synchronization mode should be selected as follows:

$$\frac{2BN\delta}{\tau} \underset{\text{SI}}{\overset{\text{AO}}{\lesseqgtr}} 2\bar{h}N, \tag{13.39}$$

or equivalently,

$$\tau \underset{\text{SI}}{\overset{\text{AO}}{\lesseqgtr}} \frac{B\delta}{\bar{h}}. \tag{13.40}$$

From (13.40), the synchronization mode changes from **AO** to **SI** when τ is smaller than $B\delta/\bar{h}$ and vice versa. In the **SI** mode, the reference node periodically inquires about the number of hops that occurred during the past time interval, and then it makes a decision whether or not to switch to the **AO** mode. Notice that AMTS is an iterative process, and the resynchronization period parameter in the mode selection will be from the last iteration of AMTS.

13.5.3.2 Determination of Resynchronization Period

As the resynchronization period τ increases, the network becomes more power efficient. Thus, τ should be chosen as large as possible. However, a too large value of τ induces a critical synchronization problem since the clock difference (offset) between nodes keeps generally increasing with time. Hence, there exists a maximum resynchronization period (τ_{max}) that is determined by the oscillator specifications and the accuracy of estimators.

Suppose that the clock timing mismatch ε between the two nodes is modeled as follows: $\varepsilon = \varepsilon_o + \varepsilon_s t$, where t denotes the reference time and ε_o and ε_s stand for the clock offset and skew errors, respectively. In general, it is difficult to determine any specific mathematical model for either clock offset or skew errors. Herein, we model both clock offset and skew errors by normal distributions based on the experimental results reported in [17] and [18]:

$$\varepsilon_{o,i} \sim \mathcal{N}(0,\sigma^2_{\varepsilon_{o,i}}) \quad 1 \le i \le N,$$

$$\varepsilon_{s,i} \sim \mathcal{N}(0,\sigma^2_{\varepsilon_{s,i}}) \quad 1 \le i \le N,$$

where $\varepsilon_{o,i}$ and $\varepsilon_{s,i}$ denote the clock skew and offset estimation errors after the ith message exchanges, respectively. Note that clock skew estimation is only available when there are multiple message exchanges. Hence, $\varepsilon_{s,1}$ stands for the clock skew error when no skew estimation occurred. Here, the maximum clock mismatch can be modeled as another normal distribution $\varepsilon \sim \mathcal{N}(0,\sigma_\varepsilon^2)$, where $\sigma_\varepsilon^2 = \sigma^2_{\varepsilon o,N} + \sigma^2_{\varepsilon s,N}\tau^2_{max}$, ($t = \tau_{max}$). Imposing the upper limit ε_{max} for the clock error via the probabilistic measure,

$$P_s = Pr\left(|\varepsilon| \ge \varepsilon_{max}\right) = erfc\left(\frac{\varepsilon_{max}}{\sqrt{2}\sigma_\varepsilon}\right),$$

where

$$erfc(x) \triangleq 2/\sqrt{\pi} \cdot \int_x^\infty \exp(-t^2)dt$$

and P_s denotes the synchronization error probability for pairwise synchronization. Thus, σ_ε can be determined when ε_{max} and the maximum allowable P_s are fixed. For instance, when P_s is limited to 0.1% and ε_{max} is 10 ms, then the standard deviation of clock mismatch (σ_ε) has to be smaller than 3.04 ms.

The maximum resynchronization period with N beacons can be written as

$$\tau_{max}^{(N)} = \sqrt{\frac{\sigma_\varepsilon^2 - \sigma^2_{\varepsilon_{o,N}}}{\sigma^2_{\varepsilon_{s,N}}}}. \tag{13.41}$$

Based on the lower bounds and asymptotic performance of the estimators, one can easily infer closed-form expressions of the variances $\varepsilon_{o,N}$ and $\varepsilon_{s,N}$ in terms of the variances $\varepsilon_{o,1}$ and $\varepsilon_{s,2}$, respectively. From the lower bounds derived in [40], $\sigma^2_{\varepsilon o,N}$ can be written with respect to N and $\sigma^2_{\varepsilon o,1}$ as

$$\sigma^2_{\varepsilon_{o,N}} = \frac{\sigma^2_{\varepsilon_{o,1}}}{N}.$$

Similarly, since the time differences between beacons are proportional to N and by far greater than the variance of delays, the following relationship can be obtained from the lower bound for the clock skew estimator derived in [25]:

$$\sigma^2_{\varepsilon_{s,N}} = \frac{\sigma^2_{\varepsilon_{s,2}}}{(N-1)^2}, \quad N \geq 2.$$

Therefore, for $N \geq 2$, $\tau^{(N)}_{\max}$ can be rewritten as

$$\tau^{(N)}_{\max} = (N-1)\sqrt{\frac{\sigma^2_{\varepsilon} - \frac{\sigma^2_{\varepsilon_{o,1}}}{N}}{\sigma^2_{\varepsilon_{s,2}}}}, \qquad N \geq 2. \tag{13.42}$$

Note that $\varepsilon_{s,1}$ can be obtained by the specifications of the crystal oscillator, and $\varepsilon_{o,1}$ and $\varepsilon_{s,2}$ can be determined by simple experimental tests. Therefore, the maximum resynchronization period is proportional to the number of beacons, and performing clock skew estimation will significantly increase $\tau^{(N)}_{\max}$ since $\sigma_{\varepsilon_{s,1}} \gg \sigma_{\varepsilon_{s,2}}$.

13.5.3.3 Number of Beacons Required for Each Pairwise Synchronization

The goal of AMTS is to minimize the average number of message exchanges ($\bar{M}$). Hence, from (13.41), finding the optimal number of beacons (N) resumes to solving the following optimization problem:

$$\hat{N} = \arg\min_{N} \bar{M}, \tag{13.43}$$

with

$$\bar{M} = \frac{2BN}{\tau^{(N)}_{sync} + \tau^{(N)}_{\max}} = \begin{cases} \dfrac{2B}{\tau^{(1)}_{sync} + \sqrt{\dfrac{\sigma^2_{\varepsilon} - \sigma^2_{\varepsilon_{o,1}}}{\sigma^2_{\varepsilon_{s,1}}}}} & N = 1 \\[3ex] \dfrac{2B}{\dfrac{\tau^{(N)}_{sync}}{N} + \dfrac{N-1}{N}\sqrt{\dfrac{\sigma^2_{\varepsilon} - \frac{\sigma^2_{\varepsilon_{o,1}}}{N}}{\sigma^2_{\varepsilon_{s,2}}}}} & N \geq 2 \end{cases},$$

where $\tau^{(N)}_{sync}$ denotes the synchronization time with N beacons and will be estimated at the reference node for different N values when the network is first established. Once N is estimated from (13.43), $\tau^{(N)}_{\max}$ can be obtained from (13.42).

Simulation results in [40] show that AMTS requires far less timing messages than TPSN when there exist multiple numbers of beacon transmissions. Moreover, the gap between the average number of required timing messages between AMTS and TPSN significantly increases as N increases, and thus AMTP is by far more energy efficient than

TPSN for large N values. Moreover, the adaptive features in AMTS make it applicable to various different types of sensor network applications.

13.6 Conclusions

In recent years, huge attention has been paid to WSNs due to their capability of serving a variety of purposes. Time synchronization is a significant part in WSNs, and a number of fundamental operations, like data fusion, power management, and transmission scheduling, require accurate time synchronization. Since the conventional time synchronization protocol for the Internet cannot be directly applied to WSNs, a number of synchronization protocols have been developed to meet the unique requirements of sensor network applications.

The importance of time synchronization also comes from the evolution of WSNs, which has been driven by technological advances in diverse areas. For instance, unlike the currently deployed WSNs, next-generation sensor networks may consist of dynamic mobile sensors or a mixture of static and dynamic sensors. In this scenario, far more sophisticated time synchronization protocols that efficiently deal with mobility of sensors will be required. Indeed, as the network becomes more complicated, the role of time synchronization becomes much more important.

In this chapter, basic features and theoretical backgrounds of the time synchronization problem in WSNs have been introduced, and three different basic approaches are analyzed and compared to reveal the general ideas and characteristics of time synchronization protocols in WSNs. In addition, a survey of existing time synchronization protocols in the literature has been provided, including the most recent results. The material of this chapter is presented from a signal processing viewpoint, which makes it distinguishable from other existing surveys. Furthermore, we shed light on adaptive time synchronization schemes for WSNs because of their huge benefits and flexibility to topology changes. We believe that the analysis and summary in this chapter will assist researchers in this area to select and develop more powerful synchronization protocols tailored specifically to the needs of their applications.

References

[1] I. F. Akyildiz, W. Su, Y. Sankarasubramaniam, and E. Cayirci. 2002. Wireless sensor networks: A survey. *Comput. Networks* 38:393–422.

[2] N. Bulusu and S. Jha. 2005. *Wireless sensor networks: A systems perspective*. Norwood, MA: Artech House.

[3] D. L. Mills. 1991. Internet time synchronization: The network time protocol. *IEEE Trans. Commun.* 39:1482–93.

[4] B. Sundararaman et al. 2005. Clock synchronization for wireless sensor networks: A survey. *Ad Hoc Networks* 3:281–323.

[5] J. Elson and K. Romer. 2003. Wireless sensor networks: A new regime for time synchronization. In *ACM SIGCOMM Comput. Commun. Rev.* (CCR) 33(1):149–154.

[6] F. Sivrikaya and B. Yener. 2004. Time synchronization in sensor networks: A survey. *IEEE Networks* 18:45–50.

[7] W. Su. 2005. Overview of time synchronization issues in sensor networks. In *Embedded systems*, ed. R. Zurawski. Boca Raton, FL: CRC Press, pp. 35–35-10.

[8] S. Ganeriwal, J. Elson, and M. B. Srivastava. 2005. Time Synchronization. In *Wireless sensor networks: A systems perspective*, ed. N Bulusu and S. Jha. Norwood, MA: Artech House, pp. 59–74.

[9] B. M. Sadler and A. Swami. 2006. Synchronization in sensor networks: An overview. In *Proceedings of the Military Communications Conference 2006*, Washington, DC, pp. 1–6.

[10] J. R. Vig. 1992. *Introduction to quartz frequency standards*. Technical Report SLCET-TR-92-1. Army Research Laboratory Electronics and Power Sources Directorate.

[11] D. A. Howe, D. W. Alan, and J. A. Barnes. 1981. Properties of signal sources and measurements methods. In *The 35th Annual Symposium on Frequency Control*, Philadelphia, PA, pp. 669–716.

[12] J. Van Greunen and J. Rabaey. 2003. Lightweight time synchronization for sensor networks. In *Proceedings of the 2nd ACM International Conference on Wireless Sensor Networks and Applications (WSNA)*, pp. 11–19.

[13] G. J. Pottie and W. J. Kaiser. 2000. Wireless integrated network sensors. *Commun. ACM* 43:51–58.

[14] H. Kopetz and W. Ochsenreiter. 1987. Clock synchronization in distributed real-time systems. *IEEE Trans. Comput.* 36:933–39.

[15] Mica2 and Mica2Dot. Accessed January 2007 from http://www.xbow.com/Products/Wireless_Sensor_Networks.htm.

[16] M. Maroti, B. Kusy, G. Simon, and A. Ledeczi. 2004. The flooding time synchronization protocol. In *Proceedings of the 2nd International Conference on Embedded Networked Sensor Systems 2004*, pp. 39–49.

[17] S. Ganeriwal, R. Kumar, and M. B. Srivastava. 2003. Timing-sync protocol for sensor networks. In *SenSys 03*, pp. 138–149.

[18] J. Elson, L. Girod, and D. Estrin. 2002. Fine-grained network time synchronization using reference broadcasts. In *Proceedings of the Fifth Symposium on Operating System Design and Implementation*, Boston, MA, pp. 147–163.

[19] K.-L. Noh and E. Serpedin. 2007. Pairwise broadcast synchronization for wireless sensor networks. In *Proceedings of IEEE International Workshop: From Theory to Practice in Wireless Sensor Networks (T2PWSN 2007)*, Helsinki, pp. 1–6.

[20] A. Papoulis. 1991. *Probability, random variables and stochastic processes*. 3rd ed. New York: McGraw-Hill.

[21] A. Leon-Garcia. 1993. *Probability and random processes for electrical engineering*. 2nd ed. Reading, MA: Addison-Wesley.

[22] H. S. Abdel-Ghaffar. 2002. Analysis of synchronization algorithm with time-out control over networks with exponentially symmetric delays. *IEEE Trans. Commun.* 50:1652–61.

[23] D. R. Jeske. 2005. On the maximum likelihood estimation of clock offset. *IEEE Trans. Commun.* 53:53–54.

[24] V. Paxson. 1998. On calibrating measurements of packet transit times. In *Proceedings of the 7th ACM Sigmetrics Conference*, vol. 26, pp. 11–21.

[25] K.-L. Noh, Q. Chaudhari, E. Serpedin, and B. Suter. 2007. Novel clock phase offset and skew estimation using two-way timing message exchanges for wireless sensor networks. *IEEE Trans. Commun.* 55:766–77.

[26] S. M. Kay. 1993. *Fundamentals of statistical signal processing: Estimation theory*. Vol. I. Englewood Cliffs, NJ: Prentice Hall.

[27] Q. Chaudhari and E. Serpedin. 2007. On maximum likelihood estimation of clock offset and skew in wireless sensor networks with exponential delays. *IEEE Trans. Signal Processing*, 56(4):1685–1697.

[28] M. L. Sichitiu and C. Veerarittiphan. 2003. Simple, accurate time synchronization for wireless sensor networks. In *Proceedings of the IEEE WCNC 2003*, pp. 1266–73.

[29] M. Lemmon, J. Ganguly, and L. Xia. 2000. Model-based clock synchronization in networks with drifting clocks. In *Proceedings of the 2000 Pacific Rim International Symposium on Dependable Computing*, Los Angeles, pp. 177–85.

[30] B. M. Sadler. 2006. Local and broadcast clock synchronization in a sensor node. *IEEE Signal Proc. Lett.* 13:9–12.

[31] I. Sari, E. Serpedin, and B. Suter. 2006. On the joint synchronization of clock offset and skew in RBS-protocol. In *Proceedings of the Military Communications Conference 2006*, Washington, DC, 1–4.

[32] W. Su and I. F. Akyildiz. 2005. Time-diffusion synchronization protocol for wireless sensor networks. *IEEE/ACM Trans. Networking* 13:384–97.

[33] Q. Li and D. Rus. 2006. Global clock synchronization in sensor networks. *IEEE Trans. Comput.* 55:214–26.

[34] Y.-W. Hong and A. Scaglione. 2005. A scalable synchronization protocol for large scale sensor networks and its applications. *J. Selected Areas Commun.* 23:1085–99.

[35] A.-S. Hu and S. D. Servetto. 2006. On the scalability of cooperative time synchronization in pulse-connected networks. *IEEE Trans. Information Theory* 52:2725–48.

[36] A.-S. Hu and S. D. Servetto. 2003. Asymptotically optimal time synchronization in dense sensor networks. In *Proceedings of the 2nd ACM International Conference on Wireless Sensor Networks and Applications (WSNA)*, pp. 1–10.

[37] Z. Tian, X. Luo, and G. B. Giannakis. 2004. Cross-layer sensor network synchronization. In *Proceedings of the 38th Asilomar Conference on Signals, Systems and Computers*, vol. 1, pp. 1276–80.

[38] S. Ganeriwal, D. Ganesan, H. Shim, V. Tsiatsis, and M. B. Srivastava. 2005. Estimating clock uncertainty for efficient duty-cycling in sensor networks. In *Proceedings of SenSys 05*, San Diego, pp. 131–141.

[39] S. PalChaudhuri, A. K. Saha, and D. B. Johnson. 2004. Adaptive clock synchronization in sensor networks. In *IPSN 04*, Berkeley, CA, pp. 340–348.

[40] K.-L. Noh and E. Serpedin. 2007. Adaptive multi-hop timing synchronization for wireless sensor networks. In *Proceedings of ISSPA2007*, Sharjah, UAE, pp. 1–6.

14

Adaptive Interference Nulling and Direction of Arrival Estimation in GPS Dual-Polarized Antenna Receiver

Moeness G. Amin
Villanova University

14.1 Antijam GPS Receiver Arrays

14.1.1 Interference Nulling

Four fundamental approaches based on spatial discrimination are commonly adopted in antijam Global Positioning Systems (GPS). The application of these approaches requires the GPS receiver to be equipped with an antenna array. The choice between which approach to undertake depends on various parameters, including the number of available satellites in the field of view (FOV), the nature of the jamming environment, the antenna polarization characteristics, and the ability to utilize the satellite direction information. The four approaches, placed in broad categories, are:

Null-steering arrays: These antenna arrays adjust their reception pattern to have low gain in the direction of the interferences, without considering the GPS satellite direction of arrival (DOA). This is the case for the power inversion technique, also known as minimum variance [1]. The power inversion (PI) approach for jammer nulling may unintentionally allow satellite signals to be suppressed along with the jammer.

Beam-steering arrays: These arrays make full use of the GPS satellite DOA information in the adaptive algorithm. They adjust the array reception pattern to have high gain in the direction of the GPS satellites and to form nulls in the direction of the interferers. This is the case of the minimum-variance distortionless response (MVDR) beamforming approach [2]. The MVDR-based interference nulling technique may inadvertently allow jammers to be passed on, along with the GPS signals, to the correlation loops of the receiver.

GPS signal-dependent arrays: These arrays utilize the GPS coarse/acquisition (C/A) temporal and periodic structure, and maximize the cross-correlation between the received signal and its one code-length delayed version. By this action, beams are formed toward the GPS satellites in the field of view and nulls are placed toward the jammers. This approach is known as a self-coherence antijamming GPS receiver [3]. It has a shortcoming of treating period jammers, which have the same fundamental period as the GPS code period, as desired signals.

Dual-polarized arrays: These arrays replace the circularly polarized GPS antennas by dual-polarized patch antennas. This approach requires a dual-feed dual-polarized antenna array, and can implement either beam-steering or nulling arrays. Including the polarization diversity in GPS adaptive antenna arrays increases the maximum number of possible jammer cancellations and gives more flexibility in the choice of the optimality criteria that can be applied to reduce jammer contaminations [4, 5].

14.1.2 Interference Direction Estimation

Interference nulling can be pursued jointly with the equally important task of estimating the interference directions of arrival. The latter can be achieved by implementing high-resolution subspace methods operating on the data covariance matrix, as is typically the case in commercial and military radar and communication receivers. Particular to navigation GPS receivers is the capability of providing simultaneous multiple beams, each oriented toward one satellite. For this reason, the sets of adaptive steering weights corresponding to the different satellites embed the interference spatial information and, as such, can be directly operated on by the eigndecomposition methods.

This chapter deals with interference nulling and DOA estimation for GPS receivers. Both tasks are performed prior to GPS signal dispreading, which is executed via the correlation loops of the receiver, utilizing the signal, jamming, and noise relative power underlying satellite geolocation application, and incorporating the navigation receiver information on polarization and spatial signatures of the satellites. Although focused on GPS, the approaches presented are applicable to Global Navigation Satellite Systems (GNSS), including Galileo, GLONASS, and Beidou [6]. All GNSS share the same operating principle. The receiver position is computed based on the distances between the receiving antenna and a set of satellites, and the receiver determines these distances by measuring the propagation time of the signals transmitted by the satellites. This propagation time can be obtained from the delay (referred to as pseudorange or code phase) of the complex envelope and from the carrier phase.

Dual-polarized beam-steering antenna array receivers are considered. We discuss the modeling and performance of optimum and adaptive antijam receivers implementing constrained minimization techniques.

14.2 Background

The Global Positioning System (GPS) is a satellite-based system used in localization and navigation for both military and civilian applications [7, 8]. It employs direct-sequence spread spectrum (DSSS) on two carriers, L1 at 1575.42 MHz and L2 at 1227.6 MHz. Each GPS satellite broadcasts a C/A code with chip rate at 1.023 Mchips/s or a precision (P) code at 10.23 Mchips/s. New navigation signal and code structures are considered, aiming at improving receiver positioning and combating multipath [6].

Depending on the operating environment, the signal-to-noise ratio (SNR) at the GPS receivers can be as low as –30 dB. For a strong interference, the jammer-to-noise ratio (JNR) may exceed 30 dB, which will severely degrade the GPS performance and the code synchronization process. Although the spread spectrum modulation provides some antijamming protection to weak signals, a high power interference (e.g., more than 5 dB above the noise) cannot be suppressed adequately by the spreading gain, causing inaccurate tracking and positioning errors. Many antijam mitigation techniques have been proposed, which are based on temporal processing [9–11], spectral-based processing [12–14], subspace projection [15], and spatial signal processing [1, 16, 17]. Combinations of the above techniques, such as time-frequency processing [18] and space-time processing [19, 20], provide superior jammer suppression compared to single-antenna or single-domain processing.

Adaptive techniques are commonly adopted for fast antijam implementations [21–30], and shown to be effective in spatial discriminations and interference nulling. The array response is continuously adjusted so that the interferers have low receiver gains. For spatial-only processing, each antenna is allocated a complex weight, which is adaptively chosen to minimize a desired cost function subject to satisfying a single or a set of desirable constraints. The two main types of adaptive antenna arrays are the nulling arrays [1, 4] and the beam-steering arrays [2]. In both arrays, interference cancellation can be achieved, irrespective of the signal temporal properties. It is noted that the MVDR method incorporates knowledge of the directions of arrival (DOAs) of the GPS signals to form beams toward the GPS satellites, while placing nulls toward the directions of the interferers. Although this method provides significant cancellation of the interference, it has high computational complexity, and its performance relies on accurate and prior knowledge of the DOAs of the desired signals. This knowledge is obtained from the navigation data and, therefore, could be difficult to acquire at "cold" start when the receiver has not yet acquired the signal and is already subjected to jamming. The power inversion method works in a different way. Because the GPS signals are typically –20 ~ –30 dB below the noise, the interferers are the dominant data components and, as such, are removed by suppressing the high-power-signal arrivals. In essence, the array weights are organized such that deep nulls are obtained at the directions of the high power signals, irrespective of their angular positions relative to the GPS DOAs. As the GPS signal directions are not taken into account in forming the array weights, it is possible that the GPS receiver array, in the power inversion method, nonintentionally offers a deep null or low gain toward one or more satellites in the field of view.

Equally important to jammer nulling is jammer direction estimation. High-resolution direction-finding techniques are widely used for the interference DOA estimation. These techniques are signal dependent and include the MVDR method as well as eigenstructure-based methods [31–35]. The latter include the multiple signal classification (MUSIC) algorithm [31], estimation of signal parameters via rotational invariance techniques (ESPRIT) [36–38], and Pisarenko harmonic decomposition (PHD) [39]. The eigenstructure methods utilize the property that the noise subspace eigenvectors of the data covariance matrix are orthogonal to the signal vectors. The DOA estimates are extracted from a low-dimension subspace of the estimated data covariance matrix via standard eigendecomposition or singular value decomposition (SVD).

Two important features separate the multiantenna GPS receiver from other receivers used in different applications. First, the GPS signal is of extremely low power compared to noise and potential interference, which makes the desired signal negligible prior to dispreading. The latter removes the spreading PN C/A code and compresses the signal to a narrow bandwidth. Second, the GPS in utilizing the direction of the satellites in the FOV can implement simultaneous beamformers, each with an adaptive set of weights. The above two features apply independent of the navigation satellite system and transparent to the navigation data signal structure [6].

Interference nulling and interference DOA estimation are typically two separate problems; each operates on the incoming data or the estimate of its covariance matrix. The two problems, however, can be jointly addressed by applying eigendecomposition to a matrix of the adaptive weights, rather than the covariance matrix. In this case, DOA estimation follows adaptive nulling and its accuracy, therefore, depends on the interference cancellation performance of the receiver array.

We show that the interference subspace information is embedded in the weight matrix, and thereby, interference DOA can be obtained directly from the nulling weights. The columns of the adaptive weight matrix pertain to the beamformers corresponding to different satellites. Each beamformer weight vector attempts to minimize the output power subject to a unit-gain constraint, i.e., a single constraint MVDR. In essence, instead of obtaining the data spatial covariance matrix for DOA estimation, the signal and noise orthogonal subspaces can be directly provided from the eigendecomposition of the weight matrix. The beamformer weights can be obtained using the least mean square (LMS) or other adaptive gradient algorithms [40, 41, 42]. The adaptive algorithms can be applied in baseband or instantaneous frequency (IF), and if required, the weights can even be adjusted using analog adaptive loops [43]. This approach was introduced in [43] and was referred to as concurrent adaptive nulling and localization (CANAL). The work in [43], however, is very sensitive to the noise power estimate, has not been applied to GPS, and did not consider a receiver with dual-polarized antennas.

Interference nulling and DOA estimation for GPS receivers can be pursued using dual-polarized antenna arrays. This improves the receiver capability for counteracting different types of the jammers, such as vertically polarized, horizontally polarized, and left-hand circularly polarized (LHCP) interferers. Further, the dual-polarized antenna array has around twice the number of degrees of freedom as its single-polarized antenna array counterpart.

14.3 GPS Receiver Model with Dual-Polarized Antenna Array

A GPS antenna array with N dual-polarized antennas is depicted in Figure 14.1. The dual-polarized antenna can be viewed as the linear combination of the vertical and horizontal elements. At each antenna, two received signals corresponding to the vertical and horizontal polarizations are collected. Each signal is assigned a separate weight. In baseband processing, an adaptive beamformer with complex weights, operating on the in-phase and quadrature signal components, is applied to linearly combine the dual-polarized antenna outputs.

Assume an N uniform linear array (ULA) with interelement spacing d. Consider D GPS signals incident on the array from the directions $\theta_1, \theta_2, \ldots, \theta_D$, and M interferers

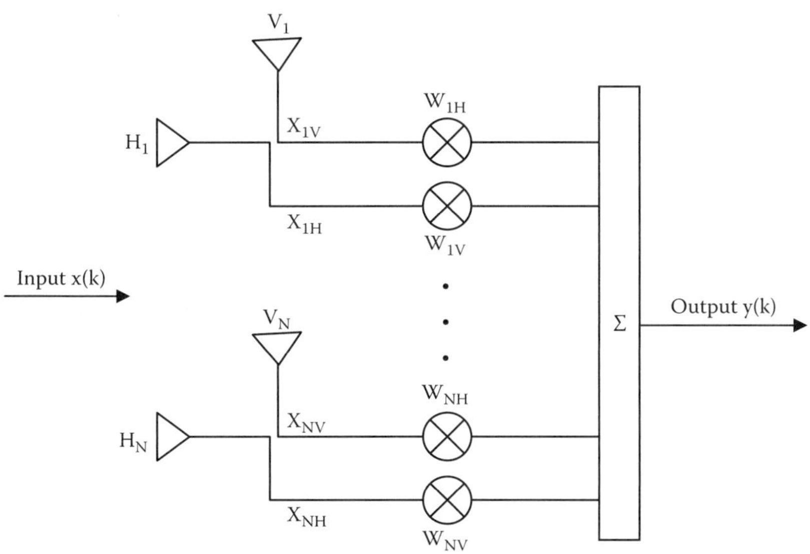

FIGURE 14.1 GPS receiver with dual-polarized antenna array.

arrive from the directions $\varphi_1, \varphi_2, \ldots, \varphi_M$. The channel is an additive white Gaussian noise (AWGN). We assume that the GPS signal is a direct line-of-sight signal with no reflection or diffraction components [44]. The kth data samples received at the horizontal element and the vertical element of the ith antenna are denoted as $x_{iH}(k)$ and $x_{iV}(k)$, respectively. Thus, the $2N$-by-1 data vector $\mathbf{x}(k)$ is given by

$$\mathbf{x}(k)=[x_{1H}(k) \quad x_{1V}(k) \quad \ldots \quad x_{NH}(k) \quad x_{NV}(k)]^T, \tag{14.1}$$

where $(.)^T$ denotes transpose. Let a D-by-1 vector $\mathbf{s}_D(k)$denote the D complex GPS signals at the kth sample,

$$\mathbf{s}_D(k)=[s_1(k),\ s_2(k),\ \ldots,\ s_D(k)]^T. \tag{14.2}$$

Similarly, the M-by-1 interference vector $\mathbf{i}_M(k)$ represents the M complex interferers at the kth sample,

$$\mathbf{i}_M(k)=[i_1(k),\ i_2(k),\ \ldots,\ i_M(k)]^T. \tag{14.3}$$

Let $\mathbf{A}_D(\theta)$ denote the GPS signal $2N$-by-D steering matrix,

$$\mathbf{A}_D(\theta)=\left[\mathbf{a}(\theta_1) \quad \mathbf{a}(\theta_2) \quad \ldots \quad \mathbf{a}(\theta_D)\right], \tag{14.4}$$

where $\mathbf{a}(\theta_i)$ is the $2N$-by-1 steering vector of the ith GPS signal incident on the antenna array from direction θ_i,

$$\mathbf{a}(\theta_i)=[a_{1H}(\theta_i) \quad a_{1V}(\theta_i) \quad \ldots \quad a_{NH}(\theta_i) \quad a_{NV}(\theta_i)]^H, \tag{14.5}$$

where $(.)^H$ denotes complex conjugate transpose. $\mathbf{A}_I(\varphi)$ represents the interference $2N$-by-M steering matrix,

$$\mathbf{A}_I(\varphi)=\left[\mathbf{a}(\varphi_1) \quad \mathbf{a}(\varphi_2) \quad \ldots \quad \mathbf{a}(\varphi_M)\right]. \tag{14.6}$$

In the above equation, $\mathbf{a}(\varphi_i)$ represents the $2N$-by-1 steering vector of the ith interference,

$$\mathbf{a}(\varphi_i)=[a_{1H}(\varphi_i) \quad a_{1V}(\varphi_i) \quad \ldots \quad a_{NH}(\varphi_i) \quad a_{NV}(\varphi_i)]^H. \tag{14.7}$$

In GPS, two binary phase shift keying (BPSK) codes are applied to the GPS information symbols. The coarse acquisition (C/A) code is a repeating 1 MHz pseudo random noise (PRN) code (1,023 bits) transmitted with the carrier frequency L1 = 1,575.42 MHz by the satellite vehicles (SVs). The precise (P) code is a very long 10 MHz PRN code (7 days) transmitted with both L1 = 1,575.42 MHz and L2 = 1,227.60 MHz. Over one navigation data symbol, the received GPS signal in the absence of noise and interference is given by

$$s_{\mathrm{d}}(t) = Ag(t-\gamma)\exp\left\{j[\omega_c t+\phi]\right\}, \tag{14.8}$$

where A is the signal amplitude, $g(t)$ is the ±1 valued C/A code, γ is the time delay of the original signal, ω_c is the carrier frequency, and ϕ is the carrier phase. After downconversion and phase tracking, the baseband signal is expressed as

$$s_{db}(t) = Ag(t-\gamma)\exp[j(\phi-\hat{\phi})], \tag{14.9}$$

where $\hat{\phi}$ is the estimated carrier phase. The code replica, generated by the receiver, is therefore

$$s_{\mathrm{D}}(t) = g(t-\hat{\gamma}), \tag{14.10}$$

where $\hat{\gamma}$ is the estimated time delay prior to processing through the GPS correlation loops. The received data vector $\mathbf{x}(k)$ is the superposition of the GPS signals, interferers, and AWGN noise, and can be expressed as

$$\mathbf{x}(k) = \mathbf{A}_D(\theta)\mathbf{s}_D(k) + \mathbf{A}_I(\varphi)\mathbf{i}_M(k) + \mathbf{n}(k), \tag{14.11}$$

where the $2N$-by-1 vector $\mathbf{n}(k)$ represents the AWGN noise at the $2N$ elements. The SNR of the GPS is in the vicinity of –20 dB, and the JNR is typically 20 dB. It is noted that with RHCP signals, $a_{iV} = -j\,a_{iH}$ $(i = 1, 2, \ldots, N)$. In this case, assuming all the antennas are omnidirectional, the steering vectors $\mathbf{a}(\theta_i)$ and $\mathbf{a}(\varphi_i)$can be expressed as

$$\mathbf{a}(\theta_i) = \begin{bmatrix} 1 & -j & \ldots & e^{j2\pi\frac{d}{\lambda}(N-1)\sin\theta_i} & -je^{j2\pi\frac{d}{\lambda}(N-1)\sin\theta} \end{bmatrix}^H, \tag{14.12}$$

$$\mathbf{a}(\varphi_i) = \begin{bmatrix} 1 & -j & \ldots & e^{j2\pi\frac{d}{\lambda}(N-1)\sin\varphi_i} & -je^{j2\pi\frac{d}{\lambda}(N-1)\sin\varphi_i} \end{bmatrix}^H, \tag{14.13}$$

where λ is the wavelength. For a horizontally polarized interference, $a_{iV} = 0$ $(i = 1, 2, \ldots, N)$, whereas for a vertically polarized interference, $a_{iH} = 0$ $(i = 1, 2, \ldots, N)$, and for LHCP interference, $a_{iV} = j\,a_{1H}$ $(i = 1, 2, \ldots, N)$. Let $\mathbf{R}$ represent the data spatial correlation matrix,

$$\mathbf{R} = E[\mathbf{x}(k)\mathbf{x}^H(k)], \tag{14.14}$$

where $E[.]$ denotes statistical expectation. In practice, $\mathbf{R}$ is replaced by its time-average estimates, $\hat{\mathbf{R}}$, and can be expressed as

$$\hat{\mathrm{R}} = \frac{1}{T}\sum_{k=1}^{T} \mathrm{x}(k)\mathrm{x}^{H}(k). \tag{14.15}$$

Denote the $2N$-by-1 complex beamformer weight vector for the N dual-polarized antennas as

$$\mathbf{w} = [w_{1H}\ w_{1V}\ w_{2H}\ w_{2V}\ \dots\ w_{NH}\ w_{NV}]^{T}. \tag{14.16}$$

The corresponding antenna array output $y(k)$ at the dual-polarized antenna array is given by

$$y(k) = \mathbf{w}^{H}\mathbf{x}(k). \tag{14.17}$$

Equations (14.11)–(14.17) and the analyses hereafter are also valid and applicable to the European, Russian, and Chinese satellite navigation systems [6].

14.4 Single versus Multiple Antijam Beamformers

The minimum variance distortionless response (MVDR) technique is an efficient tool to mitigate interference without compromising the desired signals. The array weights are computed to form unit gains toward the DOAs of the GPS signals, and to place deep nulls toward the directions of the interferers. This is achieved by minimizing the output power subject to unit-gain constraints,

$$\min_{w} \mathbf{w}^{H}\mathbf{R}\mathbf{w} \text{ subject to } \mathbf{C}^{H}\mathbf{w} = \mathbf{f}, \tag{14.18}$$

where the constraint matrix $\mathbf{C}$ represents the GPS steering matrix $\mathbf{A}_D(\theta)$. The optimal weights for the above constrained minimization problem can be obtained as [40]

$$\mathbf{w}_{opt} = \mathbf{R}^{-1}\mathbf{C}(\mathbf{C}^{H}\mathbf{R}^{-1}\mathbf{C})^{-1}\mathbf{f}, \tag{14.19}$$

where $\mathbf{f}$ is a D-by-1 vector of unit values, $\mathbf{f} = [1\ \ 1\ \ \dots\ \ 1]^{T}$. Applying the optimal weights to the received data vector $\mathbf{x}(k)$ yields the interference beamformer output

$$y(k) = \mathbf{w}_{opt}^{H}\mathbf{x}(k). \tag{14.20}$$

It is noted that the power inversion method is a special case of equation (14.19), where the constraint matrix is a vector of zero values except the first element, which has a unit value. Also, $\mathbf{f}$ is a scalar equal to 1.

In minimizing $E\{|y(k)|^2\}$, the dual-polarized antenna array will attempt to reduce the output power of the interference, which could be horizontally polarized, vertically polarized, right-hand circularly polarized (RHCP), or LHCP. Each antenna may act alone, by setting one of the two values of its corresponding complex weights to zero, or by forcing the two polarization weight components into a complex conjugation relationship. This would be clearly the case for cancelling any large number of horizontally polarized, vertically polarized, or LHCP interferers. The antenna may also act in conjunction with other receiver antennas, in which case the interference DOA plays a role in determining the weight values.

Since GPS signals are RHCP, the number of RHCP interferers that can be suppressed with N dual-polarized antennas is limited to $N - D$. The cancellation of a large or infinite number of horizontally polarized interferers is achieved by setting the weights of the horizontally polarized elements to zero. In this case, the total number of degrees of freedom is reduced to N, out of which $N - D$ degrees can be used to cancel, through spatial nulling, $N - D$ types of jammers, such as RHCP, LHCP, or vertically polarized interferers. Similarly, an infinite number of vertically polarized interferers can be cancelled along with a maximum of $N - D$ RHCP, LHCP, or horizontally polarized interferers. Further, an infinite number of LHCP interferers can be cancelled by the RHCP antenna array, which can additionally cancel $N - D$ RHCP, horizontally polarized, or vertically polarized interferers through spatial nulling.

It is noted, however, that depending on the number of dual-polarized antennas and the number of satellites in the field of view (FOV), D, one set of array weights may not be sufficient to cancel all interferers in dense jamming environments, even with the use of dual-polarized antenna array. This problem can be mitigated by using multiple MVDR beamformers, each corresponding to one satellite. In this case, several sets of weight vectors are applied; each weight vector is associated with one satellite and designed to satisfy a single unit-gain constraint. As a result, up to $N - 1$ RHCP interferers can be cancelled by the multiple beamformer approach, compared to only $N - D$ interferers in the case of the single beamformer approach. In addition to increasing the number of degrees of freedom, another important advantage of the multiple MVDR beamformers is in achieving regular array patterns. It is noted that in order to keep unit gain at all GPS satellite directions, a single MVDR beamformer will encounter difficulty suppressing the undesired signal, if it is close in angle to any one of the GPS satellites. With close angular separation between one satellite and one jammer position, the array response will be highly irregular, giving rise to several lobes in a random manner that makes the receiver vulnerable to newly borne interferers or on-off interferers with a long duty cycle. Insufficient interference cancellation compromises the receiver delay lock loops and provides undesirable tracking and positioning errors. With multiple MVDR beamformers, however, each of the $D - 1$ beamformer responses will null all interferers impinging on the GPS receiver, subject to the available degrees of freedom. Only the one response associated with the satellite that is aligned or close to an interference DOA will fail and provide an irregular pattern. In this respect, loss of acquisition, if it occurs, will be confined to only one satellite. With typically more than four satellites in the FOV, the loss of one satellite's information is not detrimental to the receiver pseudo-range estimate calculations. The multiple MVDR beamformers approach is shown in

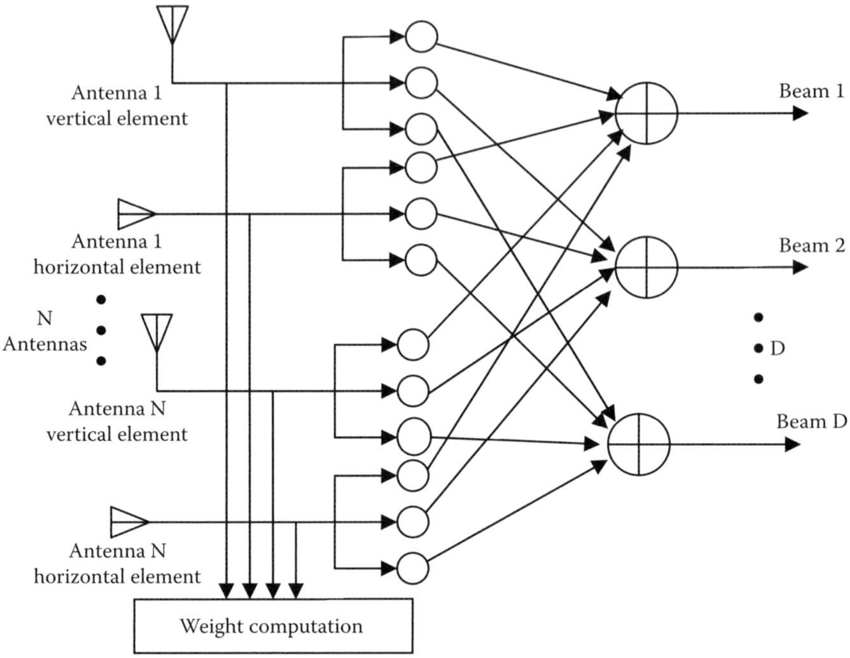

FIGURE 14.2 Block diagram of multiple MVDR beamforming.

the block diagram of Figure 14.2. The array response of each weight vector provides unit gain toward one direction of the GPS satellites and places nulls toward the directions of the interferers, irrespective of their temporal characteristics.

For the ith satellite, the $2N$-by-1 optimum weight vector in the multiple MVDR beam receiver is denoted as

$$\mathbf{w}_{i\_opt} = \mathbf{R}^{-1}\mathbf{c}_i / (\mathbf{c}_i^H \mathbf{R}^{-1}\mathbf{c}_i), \tag{14.21}$$

where c_i represents the steering vector toward the direction of the ith GPS signal. The corresponding array output is

$$y_i(k) = \mathbf{w}_{i\_opt}^H \mathbf{x}(k). \tag{14.22}$$

For illustration, we consider eight dual-polarized antennas. We compare the performance of single versus multiple MVDR beamformers in various jamming environments. Figure 14.3 shows the RHCP array response and horizontally polarized array response for twelve interferers, each with JNR of 20 dB. Five RHCP interferers are incident on the array with angles [–65, –50, –30, 40, 65] degrees, along with seven horizontally polarized interferers with angles [–80, –40, –15, 0, 30, 50, 80] degrees. Four GPS signals arrive from [–5, 5, 20, 27] degrees of elevation angles with an SNR of –20 dB. Figure 14.3(a) demonstrates that the single MVDR beamformer (solid curve) fails to cancel all five

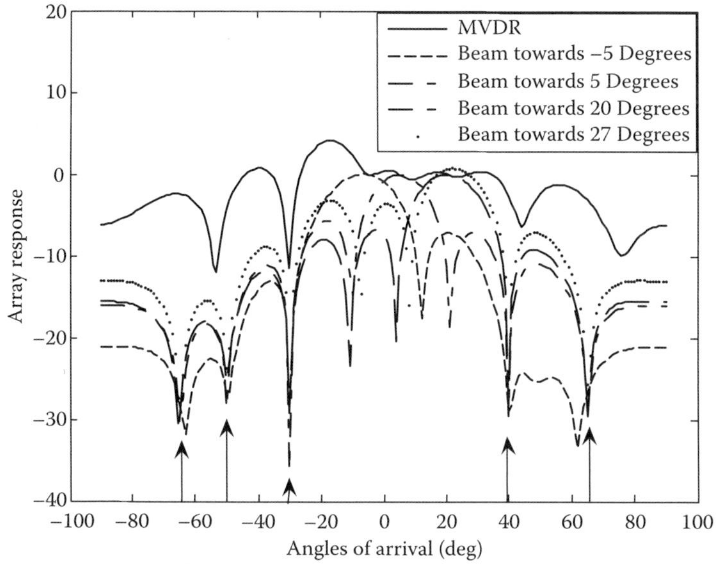

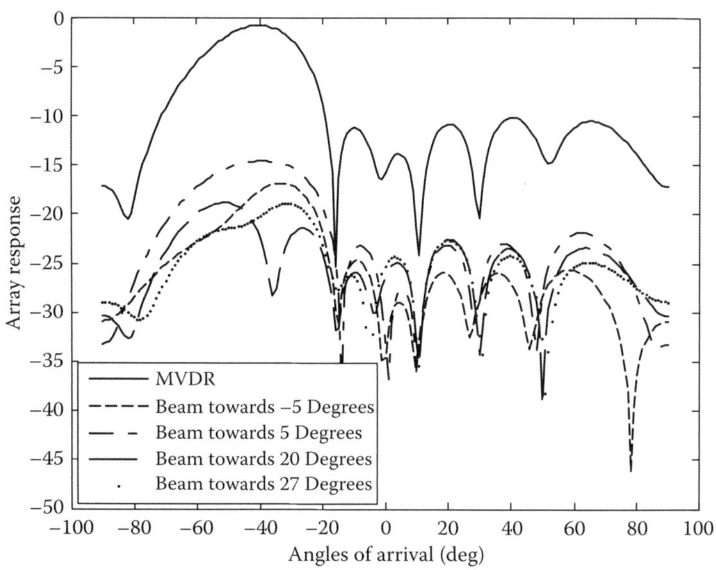

FIGURE 14.3 (Top) RHCP array response with single and multiple MVDR methods with five RHCP and seven horizontally polarized interferences. (Bottom) Horizontally polarized array response with single and multiple MVDR methods with five RHCP and seven horizontally polarized interferences.

RHCP interferers, whereas the multiple MVDR beamforming method, due to the existence of a sufficient number of degrees of freedom, places deep nulls at all interferer DOAs. Figure 14.3(b) demonstrates that both single and multiple MVDR methods null

TABLE 14.1 Weight Vector in Single and Multiple MVDR Methods

		Multiple Output Beamforming	
	MVDR	Beam 1	Beam 2
W_{1H}	0.0182 + 0.0395i	−0.0013 + 0.0010i	0.0011 + 0.0019i
W_{1V}	0.7563 − 0.2890i	0.0435 − 0.0011i	0.0831 − 0.0081i
W_{2H}	−0.0518 + 0.0866i	−0.0010 − 0.0009i	0.0008 − 0.0006i
W_{2V}	0.5980 − 0.0485i	0.1228 + 0.0152i	0.1466 − 0.0469i
W_{3H}	−0.0320 + 0.0181i	0.0013 − 0.0022i	−0.0016 + 0.0029i
W_{3V}	0.0921 + 0.2146i	0.1681 + 0.0541i	0.1335 − 0.0646i
W_{4H}	0.0566 − 0.0037i	0.0028 + 0.0015i	−0.0046 − 0.0041i
W_{4V}	−0.0948 + 0.1721i	0.1371 + 0.1086i	0.0906 − 0.0839i
W_{5H}	0.0964 − 0.0094i	−0.0028 + 0.0023i	0.0066 − 0.0027i
W_{5V}	−0.1696 + 0.0930i	0.0606 + 0.1661i	0.0521 − 0.1188i
W_{6H}	0.0268 + 0.0249i	−0.0015 − 0.0032i	−0.0000 + 0.0079i
W_{6V}	−0.1854 − 0.1926i	−0.0099 + 0.1735i	0.0256 − 0.1356i
W_{7H}	−0.0652 − 0.0562i	0.0030 − 0.0006i	−0.0065 − 0.0023i
W_{7V}	−0.0871 − 0.2303i	−0.0281 + 0.1067i	−0.0131 − 0.1407i
W_{8H}	−0.0167 − 0.1025i	−0.0004 + 0.0024i	0.0038 − 0.0030i
W_{8V}	0.1837 + 0.0734i	−0.0131 + 0.0330i	−0.0267 − 0.0809i

all the horizontally polarized interferers. In both cases, this is achieved by reducing the weights of the horizontally polarized antennas to approximately zero values. The weight vector values are given in Table 14.1. It is clear that while the horizontal weight at each antenna assumes a small value to suppress the horizontally polarized interferers, the vertical weights are organized as a beamformer to spatially cancel the vertical components of the RHCP interferers and to receive the GPS signals with unit gain.

In the next example, one of the interferers arrives from the same direction as one of the GPS satellites, e.g., 27 degrees of elevation angle. In this case, the optimum antenna array has to maintain a unit gain toward this direction, while trying to null a very close interference. Figure 14.4 depicts the array responses at the eight dual-polarized antenna arrays when two 20 dB JNR interferers arrive from 50 and 27 degrees. In this case, the single MVDR method will maintain a unit gain at 27 degrees, permitting the nonintentional GPS signal along with the interference to be received with equal sensitivity. Further, the null placed by the beamformer at 50 degrees appears shallow, with only −10 dB depth, and as such, the respective jammer may not be fully suppressed. On the other hand, with the multiple MVDR beamforming method, only the beamformer toward 27 degrees acts similar to the single MVDR beamformer and fails to cancel the interference. The other three beamformers show regular array responses and successfully suppress the two interferers with approximately −30 dB null.

In space-time processing, where spatial and temporal processing are applied jointly or sequentially, jammers that escape the spatial nulling can be dealt with in the temporal domain using excision filters [19, 40]. However, depending on the jammer signal characteristics, temporal and spatial filtering may not be able to provide proper jammer mitigation without compromising the desired signal.

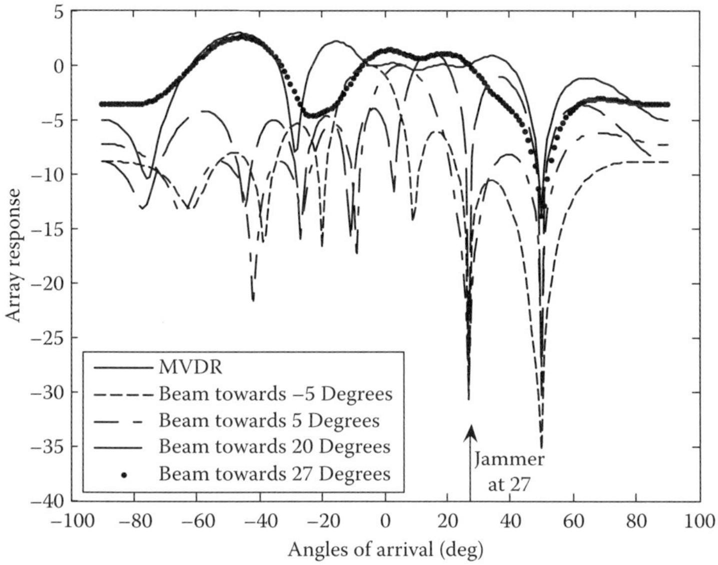

FIGURE 14.4 Array response performance at the eight RHCP antenna arrays with single and multiple MVDR methods with two RHCP interferers at 27 and 50 degrees.

14.5 Combined Nulling and DOA Estimation

Interference DOA can be obtained using subspace methods such as the MUSIC algorithm. Eigendecomposition is typically applied to the covariance matrix estimated from the data collected over an observation period,

$$\hat{\mathbf{R}} = \mathbf{U}_1 \mathbf{\Sigma}_1 \mathbf{U}_1^{\mathrm{H}}, \tag{14.23}$$

where $\mathbf{\Sigma}_1$ is a $2N$-by-$2N$ matrix representing the eigenvalue diagonal matrix, and $\mathbf{U}_1$ is a unitary matrix whose columns represent the corresponding eigenvectors. The signal subspace is spanned by the eigenvectors associated with the dominant eigenvalues of $\hat{\mathbf{R}}$. The MUSIC spectrum is given by [31]

$$\mathbf{P}_{MUSIC}(\theta) = \frac{1}{\mathbf{a}^H(\theta)\mathbf{U}_n\mathbf{U}_n^H\mathbf{a}(\theta)}, \tag{14.24}$$

where $\mathbf{U}_n$ is a matrix of the eigenvectors corresponding to the noise eigenvalues. The MUSIC algorithm as well as other eigenstructure methods require proper sampling of the data for covariance matrix estimation. These methods can be computationally demanding for GPS receivers, often call for large memory, and cannot be used effectively in simple analog adaptive loops. An alternative way is to build on the fact that multiple beamformers produce different sets of weights, and each carries interference information. One may therefore proceed to estimate the interferer DOAs by applying subspace methods, such as the MUSIC technique, to a matrix made of the weight vectors corresponding to different satellites. These weights are already generated for interference nulling purposes and can be obtained recursively through digital, or even analog, implementations. It is noted that this approach is different from beamspace MUSIC, where the covariance matrix of the beam data outputs is considered for eigendecomposition [45].

To elaborate on the combined cascade implementation of interference nulling and DOA estimation, we consider the optimum nulling weight matrix solution, given by

$$\mathbf{W}_a = \mathbf{W}_{opt}\boldsymbol{\alpha} \tag{14.25}$$

where

$$\mathbf{W}_a = \mathbf{R}^{-1}\mathbf{A}_D(\theta). \tag{14.26}$$

The $2N$-by-D matrix $\mathbf{W}_{opt} = [\mathbf{w}_{1\_opt}\ \mathbf{w}_{2\_opt}\ \cdots\ \mathbf{w}_{D\_opt}]$ consists of the weight vectors in the multiple MVDR beamforming approach, and

$$\boldsymbol{\alpha} = diag\left\{\frac{1}{\alpha_1}\ \frac{1}{\alpha_2} \cdots \frac{1}{\alpha_D}\right\}$$

is a diagonal D-by-D matrix, where α_i ($i = 1, 2, \ldots, D$) is the output power at each beamformer, $\alpha_i = y_i y_i^*$.

The spatial covariance matrix of the received data can be written as

$$\mathbf{R} = \mathbf{x}\mathbf{x}^H = \mathbf{A}_D\mathbf{P}_s\mathbf{A}_D^H + \mathbf{A}_I\mathbf{P}_I\mathbf{A}_I^H + \sigma^2\mathbf{I} = \mathbf{R}_s + \mathbf{R}_I + \mathbf{R}_n, \tag{14.27}$$

where $\mathbf{P}_s = diag(P_{si})$ ($i = 1, 2, \ldots, D$) and $\mathbf{P}_I = diag(P_{Ii})$ ($i = 1, 2, \ldots, M$) represent the diagonal power matrix of the GPS signals and the interferers, respectively. Due to the low power of the GPS signals, matrix $\mathbf{R}_s$ is negligible compared to the interference and noise covariance matrix. Therefore, equation (14.27) can be simplified to

$$\mathbf{R} = \mathbf{A}_I\mathbf{P}_I\mathbf{A}_I^H + \sigma^2\mathbf{I}. \tag{14.28}$$

The eigendecomposition of (14.28) is

$$\mathbf{R} = \mathbf{E}\boldsymbol{\Lambda}\mathbf{E}^H, \tag{14.29}$$

where $\boldsymbol{\Lambda}$ is the diagonal eigenvalue matrix, $\boldsymbol{\Lambda} = diag\{\sigma_1^2\ \ldots\ \sigma_M^2\ \sigma_n^2\ \ldots\ \sigma_n^2\}$, and $\mathbf{E}$ is the corresponding eigenvector matrix. In the above equation, σ_i^2 ($i = 1, 2, \ldots, M$) represents the ith significant eigenvalue, and σ_n^2 represents the noise eigenvalue. The first M columns of $\mathbf{E}$ span the interference subspace, i.e.,

$$span\begin{bmatrix}\mathbf{i}_1 & \mathbf{i}_2 & \dots & \mathbf{i}_M\end{bmatrix} = span\begin{bmatrix}\mathbf{e}_1 & \mathbf{e}_2 & \dots & \mathbf{e}_M\end{bmatrix}, \tag{14.30}$$

where $\mathbf{e}_i$ represents the ith column of **E**. If we adopt the approach in [43] for interference DOA estimation, then we define

$$\mathbf{W}_c = \frac{1}{\sigma^2}\mathbf{A}_D(\theta) - \mathbf{W}_a, \tag{14.31}$$

where σ^2 represents the power of the AWGN. Substituting equations (14.26) and (14.29) into (14.31), the cancellation weight vector for the kth satellite can be expressed as

$$\begin{aligned}\mathbf{w}_{ck} &= \frac{1}{\sigma^2}\mathbf{a}_k(\theta) - \mathbf{E}\Lambda^{-1}\mathbf{E}^H\mathbf{a}_k(\theta) = \frac{1}{\sigma^2}\mathbf{I}\mathbf{a}_k(\theta) - \mathbf{E}\Lambda^{-1}\mathbf{E}^H\mathbf{a}_k(\theta) \\ &= \left(\frac{1}{\sigma^2}\mathbf{I} - \sum_{i=1}^{M}\frac{1}{\sigma_i^2}\mathbf{e}_i\mathbf{e}_i^H - \sum_{j=M+1}^{2N}\frac{1}{\sigma_n^2}\mathbf{e}_j\mathbf{e}_j^H\right)\mathbf{a}_k(\theta),\end{aligned} \tag{14.32}$$

where **I** is the identity matrix. Typically,

$$\frac{1}{\sigma_i^2} \ll \frac{1}{\sigma^2}, \sigma^2 = \sigma_n^2,$$

and equation (14.32) can be simplified to

$$\mathbf{w}_{ck} = \left(\frac{1}{\sigma^2}\mathbf{I} - \sum_{j=M+1}^{2N}\frac{1}{\sigma^2}\mathbf{e}_j\mathbf{e}_j^H\right)\mathbf{a}_k(\theta) = \left(\sum_{i=1}^{M}\frac{1}{\sigma^2}\mathbf{e}_i\mathbf{e}_i^H\right)\mathbf{a}_k(\theta) = \sum_{i=1}^{M}\frac{1}{\sigma^2}\left(\mathbf{e}_i^H\mathbf{a}_k(\theta)\right)\mathbf{e}_i. \tag{14.33}$$

The above vector lies in the interference subspace. If $D > M$, then the columns of the cancellation weight matrix span the M interference subspace, i.e.,

$$span\begin{bmatrix}\mathbf{i}_1 & \mathbf{i}_2 & \dots & \mathbf{i}_M\end{bmatrix} = span\begin{bmatrix}\mathbf{w}_{c1} & \mathbf{w}_{c2} & \dots & \mathbf{w}_{cD}\end{bmatrix}. \tag{14.34}$$

It is noted that if the interferers are LHCP, the eigenvectors of the covariance matrix in the presence of weak GPS signals are presented as LHCP. Accordingly, the coefficients $\mathbf{e}_i^H\mathbf{a}_k(\theta)$ in (14.33) become zero, and the corresponding weight vector $\mathbf{w}_{ck}$ becomes zero as well. As a result, the DOAs of the interferers cannot be estimated from $\mathbf{W}_a$ or $\mathbf{W}_c$. This can be viewed as a drawback of using the adaptive weights for DOA estimation. If the interferers are RHCP, horizontally polarized, or vertically polarized, the above coefficients are nonzero, and equation (14.34) is valid.

If the exact order statistics are available, then one would generate the nulling weights, proceed with eigendecomposition of $\mathbf{W}_c$, and obtain a corresponding MUSIC spectrum, followed by peak finding and non-LHCP interference DOA estimation. This technique is known as concurrent adaptive nulling and localization (CANAL) [43] (although it performs nulling and localization in a sequential manner). In practice, however, only a short observation interval of the data is available, and as such, the noise perturbation and estimation inaccuracies will violate the interference subspace in equation (14.33), rendering equation (14.34) nonapplicable. For illustration, define

$$\lambda_i = \frac{1}{\sigma^2} - \frac{1}{\sigma_i^2}$$

for $i = 1, 2, \ldots, M$, and

$$\lambda_j = \frac{1}{\sigma^2} - \frac{1}{\sigma_{nj}^2}$$

for $j = M + 1, \ldots, 2N$, where σ_{nj}^2 represents the jth noise eigenvalue obtained from the estimated data covariance matrix $\hat{\mathbf{R}}$, and it is a perturbed version of σ^2. If $\sigma^2 \neq \sigma_{nj}^2$, then equation (14.32) can be rewritten as

$$\begin{aligned} \mathbf{w}_{ck} &= \left(\frac{1}{\sigma^2} EE^H - \sum_{i=1}^{M} \frac{1}{\sigma_i^2} \mathbf{e}_i \mathbf{e}_i^H - \sum_{j=M+1}^{2N} \frac{1}{\sigma_{nj}^2} \mathbf{e}_j \mathbf{e}_j^H \right) \mathbf{a}_k(\theta) \\ &= \sum_{i=1}^{M} \lambda_i \left(\mathbf{e}_i^H \mathbf{a}_k(\theta) \right) \mathbf{e}_i + \sum_{j=M+1}^{2N} \lambda_j \left(\mathbf{e}_j^H \mathbf{a}_k(\theta) \right) \mathbf{e}_j. \end{aligned} \tag{14.35}$$

It is clear that the cancellation weight vector is a linear combination of all interference and the noise subspace vectors $\mathbf{e}_i$ ($i = 1, 2, \ldots, 2N$) pertaining to the spatial covariance matrix. Proceeding with the singular value decomposition (SVD) for $\mathbf{W}_c$ yields

$$\mathbf{W}_c = \mathbf{U \Sigma V}, \tag{14.36}$$

where $\mathbf{U}$ is a $2N$-by-$2N$ unitary matrix, $\mathbf{V}$ is a D-by-D unitary matrix, and $\mathbf{\Sigma}$ is a $2N$-by-D diagonal matrix with singular values, γ_i, $i = 1, \ldots, D$,

$$\mathbf{\Sigma} = \begin{bmatrix} \gamma_1 & 0 & 0 \\ 0 & \cdots & 0 \\ 0 & 0 & \gamma_D \\ 0 & \cdots & 0 \\ 0 & \cdots & 0 \\ 0 & \cdots & 0 \end{bmatrix}. \tag{14.37}$$

If $\mathbf{W}_c$ is optimally calculated, and providing $D > M$, then its rank is M. However, with adaptive processing, the matrix rank is no longer M. In this way, the first D columns of $\mathbf{U}$ are the column subspace, containing the interference information ($M < D$), and the last $2N - D$ columns of $\mathbf{U}$ are the noise subspace. That is,

$$\mathrm{a}_k^H(\theta)\mathrm{u}_i = 0, \; (i = D + 1, \ldots, 2N) \tag{14.38}$$

for $k = 1, \ldots, M$. The DOAs of the interferers can be obtained by searching the peaks of the CANAL spectrum

$$P(\theta) = \frac{1}{\sum_{i=D+1}^{2N} \left\| \mathrm{a}^H(\theta)\mathrm{u}_i \right\|^2}. \tag{14.39}$$

It is noteworthy that the rank of the cancellation weight matrix $\mathbf{W}_c$ is $min(D, N)$. That is, only up to $D - 1$ RHCP interferers can be considered, compared to $N - 1$ by the MUSIC algorithm when operating on the data covariance matrix. By increasing D ($D < N$), the interference information will be further represented by a larger column subspace, producing enhanced interference DOA estimation performance. Accordingly, it is desirable to add $N - D$ beamformers to achieve a total number of N beamformers. For simplicity, we define the number of beamformers as L. If $D < N$, then the L beamformers consist of D beamformers toward the GPS directions and the $L - D$ beamformers toward arbitrary directions. In this case, equation (14.26) can be expressed as

$$\mathbf{W}_a = \mathbf{R}^{-1}\mathbf{T}(\theta), \tag{14.40}$$

and equation (14.31) becomes

$$\mathbf{W}_c = \frac{1}{\sigma^2}\mathbf{T}(\theta) - \mathbf{W}_a. \tag{14.41}$$

Therefore, equation (14.39) is rewritten as

$$P(\theta) = \frac{1}{\sum_{i=L+1}^{2N} \left\| \mathrm{a}^H(\theta)\mathrm{u}_i \right\|^2}. \tag{14.42}$$

The optimal performance will occur when $L = N$, and a maximum number of $N - 1$ interference DOAs can be estimated.

The multiple MVDR beamforming method can null a large number of different types of the interferers as presented in section 14.4. However, not all interferers can be considered with the CANAL method. For instance, if the interferers are LHCP, the inner product $\mathbf{e}_i^H \mathbf{a}_k(\theta)$ $(i = 1, 2, \ldots, M)$ corresponding to the ith signal subspace becomes zero. This is because $\mathbf{e}_i$ is LHCP and $\mathbf{a}_k(\theta)$ is RHCP. The eigenvectors $\mathbf{e}_j$ corresponding to the noise eigenvalues do not assume specific polarization. As a result, the cancellation weight vector $\mathbf{w}_{ck}$ no longer carries the interference DOA information, and is expressed as the linear combination of the noise subspace. Another important case occurs when the number of vertically polarized or horizontally polarized jammers exceeds the number of dual-polarized antennas N. In this case, the vertical elements or horizontal elements of the nulling weight matrix, respectively, will become zero, as this is the only option available to the array to suppress all jammers. The interferers' DOAs cannot be estimated based on the zero weight values. In other words, for the interference DOA to be estimated from the cancellation weight matrix, the interferers must have been already suppressed by spatial nulling.

Several simulation results of the interference localization in GPS are presented. We examine the interference localization performance as a function of the number of look directions L. Assume eight dual-polarized antennas, uniform linear antenna array with the interelement spacing of a half wavelength, are employed. Four RHCP GPS signals arrive at –20, 0, 15, and 45 degrees of elevation angles with SNRs of –20 dB. The number of snapshots is set to 1,000. Eight look directions are considered. Eight beamformers are formed, four of those toward the four GPS directions and the rest to –74, –55, 60, and 85 degrees of elevation angle. The latter were arbitrarily chosen. Seven $(M = 7)$ 20 dB RHCP interferers incident on the GPS receiver from –40, –30, –10, 10, 35, 40, and 70 degrees were generated. In Figure 14.5, the MUSIC spectrum and CANAL spectrum are presented and compared. The dashed line represents the spectrum with the CANAL approach applied to the cancellation weight matrix obtained in equation (14.41), and the solid line represents the spectrum with the MUSIC technique applied to the estimated data covariance matrix, presented by equation (14.24). Both techniques show seven clear peaks at the interferers' directions.

If the number of beamformers L reduces to six, toward –74, –20, 0, 15, 45, and 60 degrees, the rank of the cancellation matrix $\mathbf{W}_c$ reduces to six, and subsequently, the maximum number of RHCP interferers that can be cancelled is six. The performance is shown in Figure 14.6. It is clear that in this case, the covariance matrix-based MUSIC algorithm can estimate the DOAs of the interferers, while the CANAL method fails, since $M > L$.

In order to examine the sensitivity of the choice of the noise subspace dimension on the CANAL performance, we assume two interferers arrive from –40 and 35 degrees. We use the same eight look directions as in the previous example. We choose the dominant singular vectors $\mathbf{u}_1$ and $\mathbf{u}_2$ to represent the interference subspace, instead of $\mathbf{u}_1, \mathbf{u}_2, \ldots, \mathbf{u}_8$. The spectrum is shown in the solid curve in Figure 14.7. This figure demonstrates that the interference subspace is not properly captured in $\mathbf{u}_1$ and $\mathbf{u}_2$, but rather has a nonzero projection on all singular vectors $\mathbf{u}_1, \mathbf{u}_2, \ldots, \mathbf{u}_8$. The example below further underscores the above point.

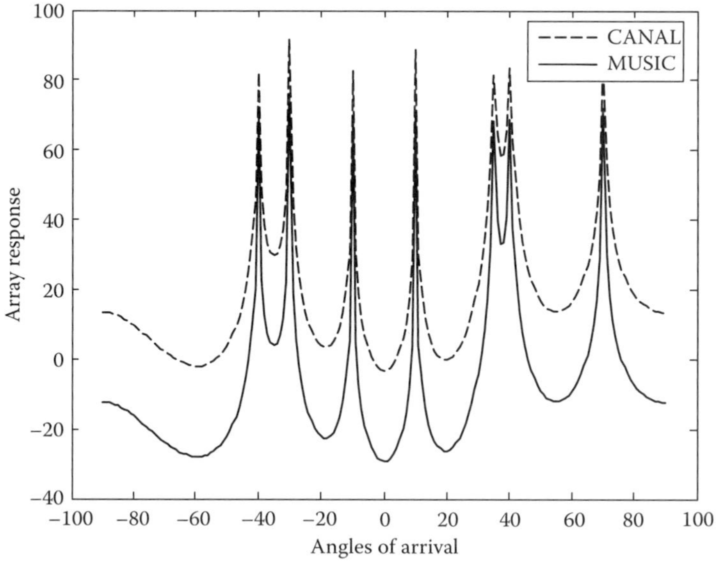

FIGURE 14.5 CANAL and MUSIC spectra with eight beamformers toward –74, –55, 60, 80, –20, 0, 15, and 45 degrees, where seven jammers arrive from –40, –30, –10, 10, 35, 40, and 70 degrees.

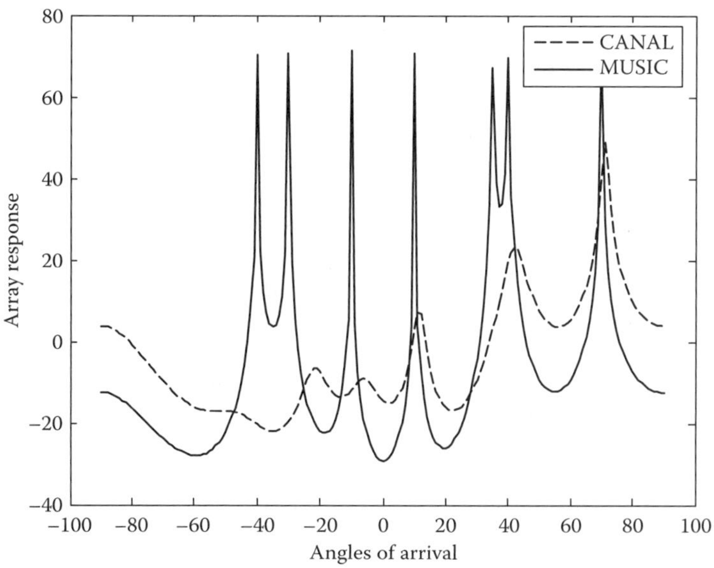

FIGURE 14.6 CANAL and MUSIC spectra with six beamformers toward –74, 60, –20, 0, 15, and 45 degrees, where seven jammers arrive from –40, –30, –10, 10, 35, 40, and 70 degrees.

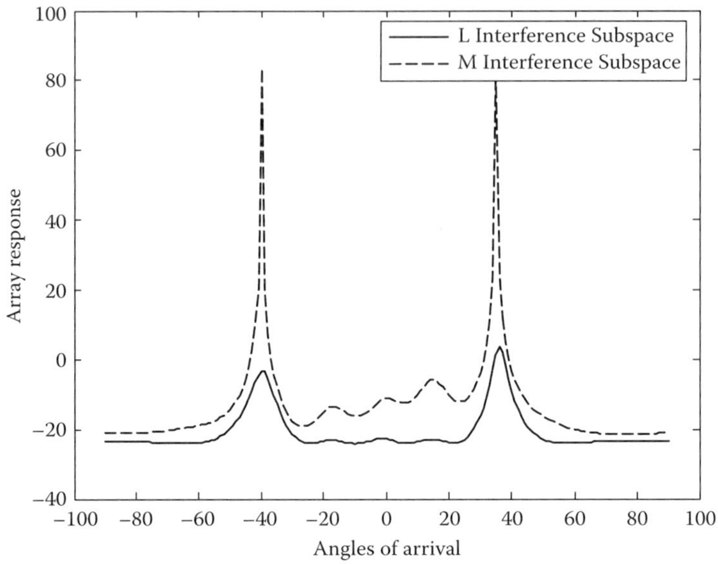

FIGURE 14.7 CANAL spectra where the noise subspace is of dimension $2N - M$ and $2N - L$ with eight dual-polarized antennas. Two interferers are located at −40 and 40 degrees.

We examine, in Figure 14.8, the normalized eigenvalues of the estimated data covariance matrix in the MUSIC technique as well as the normalized eigenvalues of the cancellation weight covariance matrix in the CANAL technique. The latter is obtained by the product $\mathbf{W}_c\mathbf{W}_c^H$. The solid line represents the eigenvalues in the MUSIC technique, where the first two eigenvalues are the interference eigenvalues, and the rest are small values representing the noise eigenvalues. The dotted line represents the eigenvalues in the CANAL technique. It is clear that the zero eigenvalues start from the eighth, and not from the third, entry, as expected in the ideal case.

In Figure 14.9, we compare the CANAL spectra under different numbers of MVDR beamformers, L. Two RHCP interferers arrive from −40 and 35 degrees of elevation angle with JNR of 20 dB. The number of beamformers is set to 5, 8, and 11. The number of dual-polarized antennas, N, is eight. The look directions for the three cases are the four GPS directions and other arbitrary directions, namely, [−20, 0, 15, 45, −55], [−20, 0, 15, 45, −55, 60, −74, 85], and [−20, 0, 15, 45, −55, 60, −74, 85, −10, 20, 70] degrees, respectively. It is clear that when $L < N$, the CANAL method estimates the interferer DOAs with poor resolution. On the other hand, when $L > N$, the rank of the cancellation matrix is N, and the CANAL spectrum performance will not be improved significantly.

In the next example, nine interferers arrive from [−40, 35, 70, −30, 40, −10, 10, 50, −60] degrees with JNR of 20 dB, and eight beamformers are formed toward the directions of 85, −74, 60, −55, 45, −20, 0, and 15 degrees. The first interferer from −40 degrees is RHCP, and all the others are horizontally polarized. With eight antennas, the multiple MVDR method will suppress the eight horizontally polarized interferers by placing zero weight values at the horizontal elements of the nulling weight matrix, and cancel the RHCP

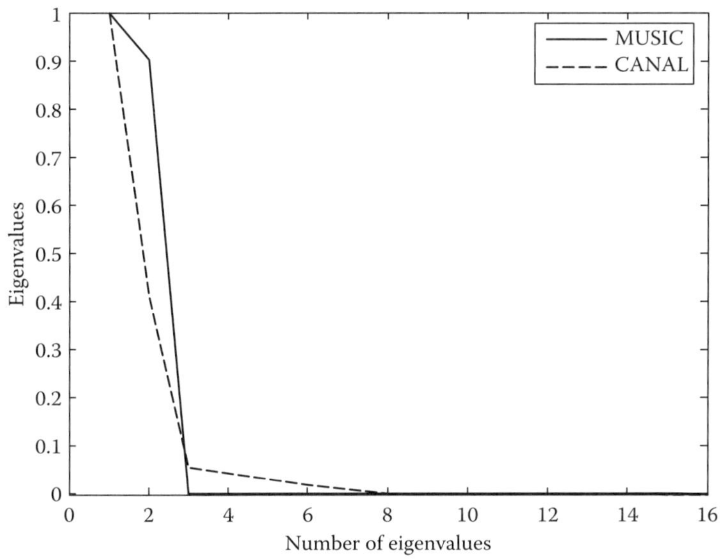

FIGURE 14.8 Eigenvalues of data covariance matrix and the cancellation weight covariance matrix in the MUSIC and CANAL techniques, respectively.

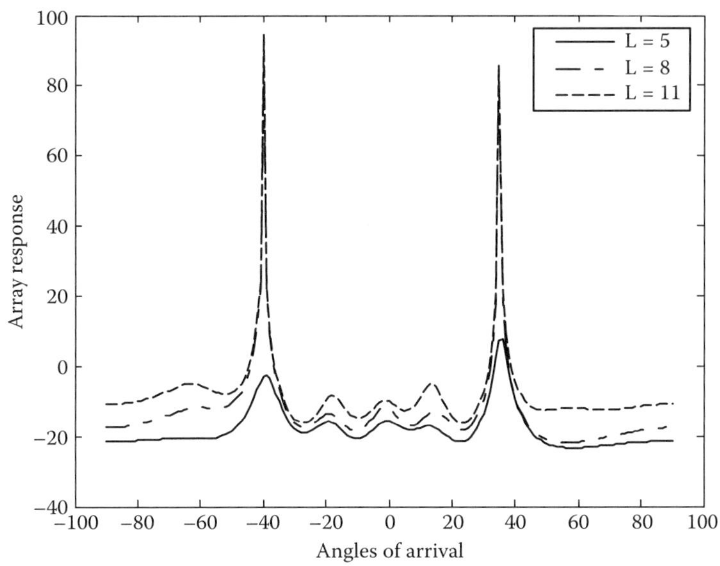

FIGURE 14.9 CANAL spectra with five, eight, and eleven beamformers, where two interferers are from −40 and 35 degrees.

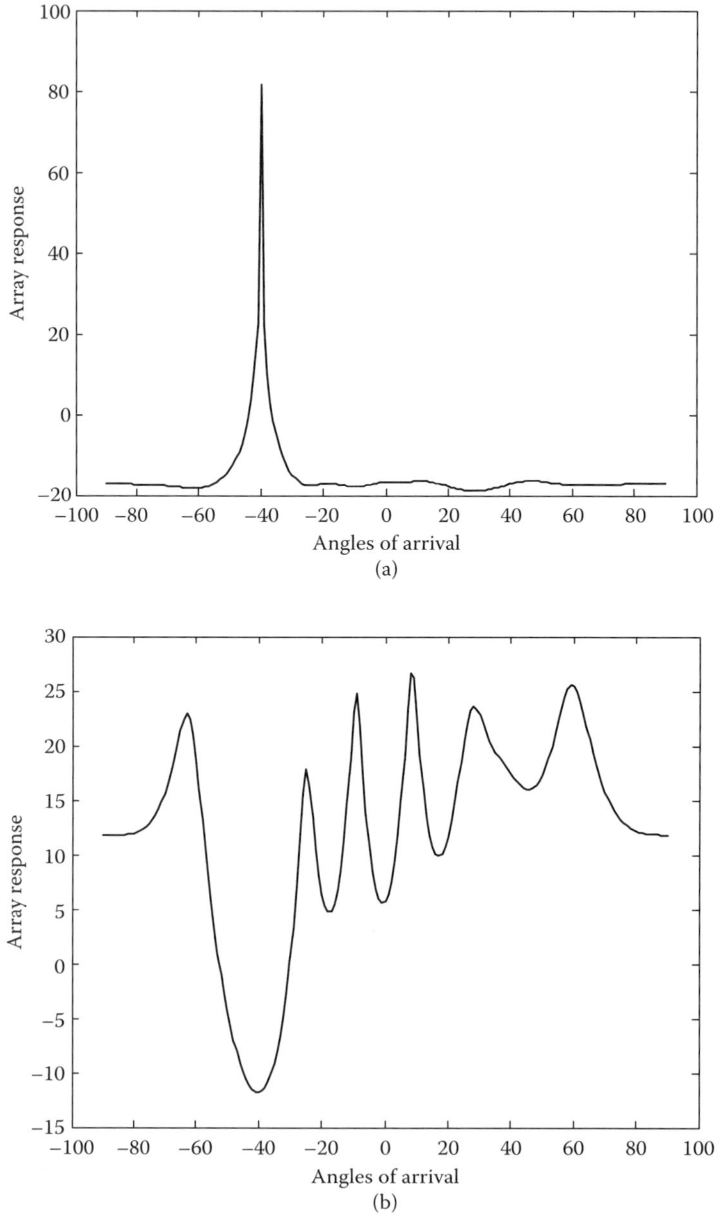

FIGURE 14.10 (a) CANAL spectrum with one RHCP jammer and eight horizontally polarized jammers. (b) CANAL spectrum with one RHCP jammer and eight horizontally polarized jammers.

interferer spatially using the MVDR method. The CANAL spectrum of the RHCP interferer and the horizontally polarized interferers are presented in Figure 14.10(a) and (b), respectively. The RHCP interferer is clearly located at −40 degrees. The CANAL spectrum for the horizontally polarized interferers shows random peaks. This demonstrates that it fails to perform direction finding of the interferers that are suppressed by the array orthogonal dual-polarization property.

14.6 Constrained Adaptive Algorithm

In this section, we recall a simple adaptive algorithm (constrained LMS algorithm) and use it to provide the cancellation weight matrix without the need to estimate the data covariance matrix. The constraint LMS algorithm is based on the gradient-descent algorithm. It iteratively adapts the weights of an antenna array to minimize the noise power at the output while retaining a unit response at the look directions. The multiple MVDR beamforming method solves for **w** according to

$$\min_{w} \mathbf{w}^{H}\mathbf{R}\mathbf{w} \text{ subject to } \mathbf{c}^{H}\mathbf{w}=1. \tag{14.43}$$

Denote $\mathbf{F} = \mathbf{c}(\mathbf{c}^{H}\mathbf{c})^{-1}$ and $\mathbf{P} = \mathbf{I} - \mathbf{c}(\mathbf{c}^{H}\mathbf{c})^{-1}\,\mathbf{c}^{H}$. The weight vector can be recursively updated as [40]

$$\begin{aligned} \mathbf{w}(k+1) &= \mathbf{P}\mathbf{w}(k)-\mu\mathbf{P}\mathbf{R}\mathbf{w}(k)+\mathbf{F}=\mathbf{P}(\mathbf{w}(k)-\mu\mathbf{R}\mathbf{w}(k))+\mathbf{F} \\ &= \mathbf{P}(\mathbf{w}(k)-\mu y(k)\mathbf{x}(k))+\mathbf{F}. \end{aligned} \tag{14.44}$$

The constraint LMS algorithm at the kth time sample can be expressed as

$$\begin{aligned} &\mathbf{w}(0)=\mathbf{F} \\ &\mathbf{w}(k+1)=\mathbf{P}(\mathbf{w}(k)-\mu y(k)\mathbf{x}(k))+\mathbf{F}, \end{aligned} \tag{14.45}$$

where μ is the adaptation step size. The output power of the antenna array is given by

$$\begin{aligned} P\_out &= yy^{H} = \mathbf{w}^{H}\mathbf{x}\mathbf{x}^{H}\mathbf{w} = \mathbf{w}^{H}\mathbf{R}\mathbf{w} \\ &= \left(\frac{\mathbf{R}^{-1}\mathbf{c}}{\mathbf{c}^{H}\mathbf{R}^{-1}\mathbf{c}}\right)^{H}\mathbf{R}\frac{\mathbf{R}^{-1}\mathbf{c}}{\mathbf{c}^{H}\mathbf{R}^{-1}\mathbf{c}} \\ &= \frac{\mathbf{c}^{H}\mathbf{R}^{-1}}{\mathbf{c}^{H}\mathbf{R}^{-1}\mathbf{c}}\mathbf{R}\frac{\mathbf{R}^{-1}\mathbf{c}}{\mathbf{c}^{H}\mathbf{R}^{-1}\mathbf{c}} = \frac{\mathbf{c}^{H}\mathbf{R}^{-1}\mathbf{c}}{(\mathbf{c}^{H}\mathbf{R}^{-1}\mathbf{c})(\mathbf{c}^{H}\mathbf{R}^{-1}\mathbf{c})} = \frac{1}{\mathbf{c}^{H}\mathbf{R}^{-1}\mathbf{c}}. \end{aligned} \tag{14.46}$$

Details of the derivations of equations (14.44)–(14.46) can be found in [40]. The optimum nulling weight vector in equation (14.25) then becomes

$$\mathbf{w}_{ai} = \mathbf{w}_{i} / P_{out\_i}. \tag{14.47}$$

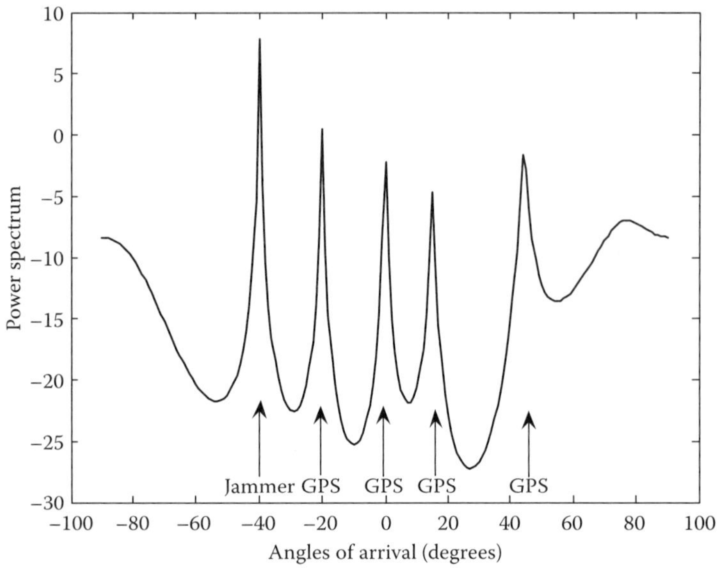

FIGURE 14.11 CANAL spectrum by LMS algorithm with eight dual-polarized antenna arrays, when one RHCP interferer arrives at -40 degrees.

In the following examples, the interference localization performance based on adaptive cancellation weight vector using the constraint LMS algorithm is examined. We consider eight dual-polarized ULAs with interelement spacing of a half wavelength. The four RHCP GPS signals arrive from –20, 0, 15, and 45 degrees with SNRs of –20 dB. Eight look directions include the four GPS directions and the additional directions of –74, –55, 60, and 85 degrees. The number of snapshots is set to 4,000, and the step size is 0.000005. In Figure 14.11, the CANAL spectrum is presented when one RHCP interferer arrives from −40 degrees with 20 dB JNR. It is evident that the directions of both the interferer and the GPS signals are presented by clear peaks, when applying the constraint LMS algorithm. This is the result of equation (14.35) and the mismatching of the true noise power value and the one offered implicitly by the converged weights. With this mismatch, the subtraction gives rise to the GPS signal subspace, which can no longer be ignored in the equation.

Figure 14.12 shows the output powers at the GPS receiver for each iteration with step sizes of 0.00005 and 0.000005. The output power is calculated by $\mathbf{w}_1^H(k)\mathbf{R}\mathbf{w}_1(k)$, where $\mathbf{w}_1(k)$ represents the nulling weights of the first beamformer at the kth iteration. The dashed line represents the output power with step size of 0.00005, whereas the solid line represents the output power with step size of 0.000005. It is obvious that the LMS converges faster when increasing the step size. In Figure 14.13, the first weight at the first beamformer is presented. The dashed line represents the optimum weight value obtained by the covariance matrix, whereas the solid line represents the adaptive weight. It is clear that the adaptive weight converges to the optimum value.

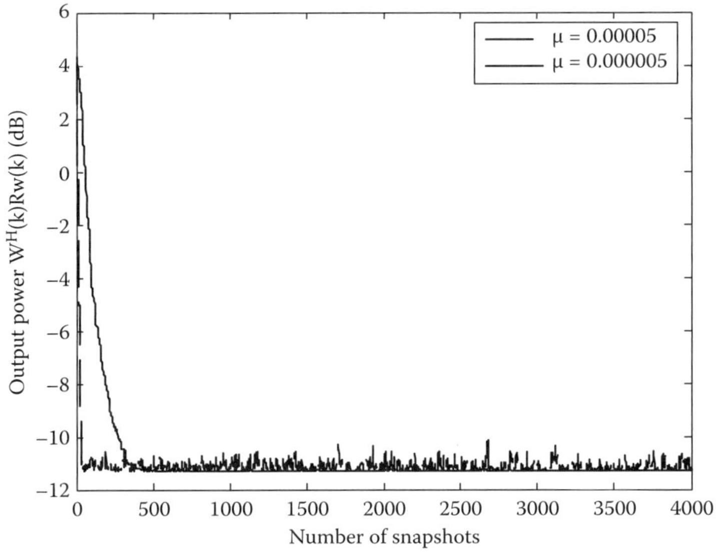

FIGURE 14.12 Output power at the GPS receiver for each snapshot with different step sizes.

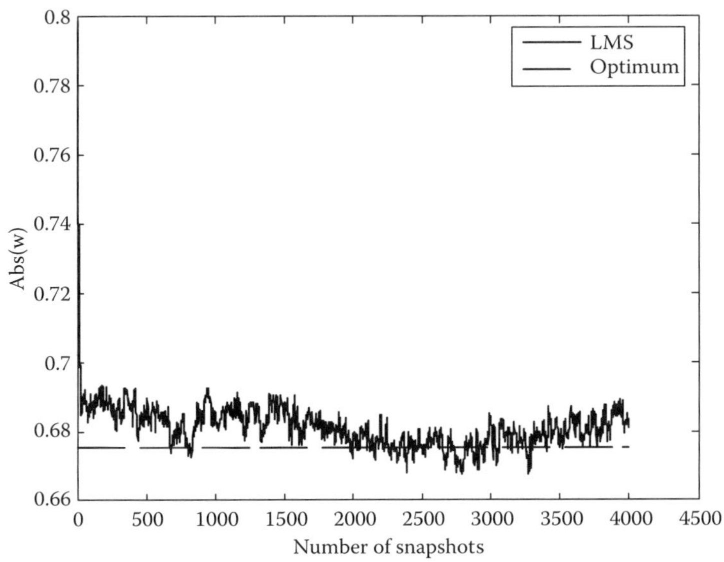

FIGURE 14.13 The adaptively obtained and optimum weight at the first beamformer for each iteration.

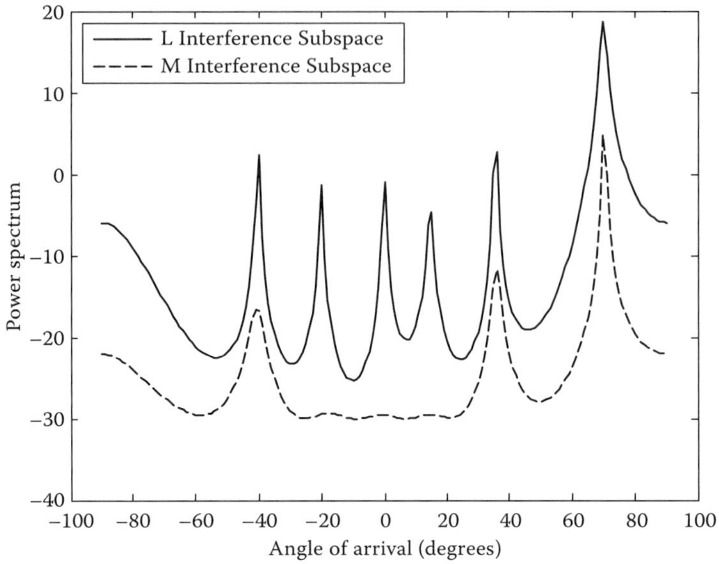

FIGURE 14.14 CANAL spectra by LMS algorithm where the noise subspace is of dimension $2N - M$ and $2N - L$ with eight dual-polarized antennas. Three RHCP interferers arrive at −40, 35, and 70 degrees.

In the next example, we consider three RHCP interferers arriving from −40, 35, and 70 degrees of elevation angle. We choose the dominant singular vectors $\mathbf{u}_1$, $\mathbf{u}_2$, and $\mathbf{u}_3$ to represent the interference subspace, instead of $\mathbf{u}_1, \mathbf{u}_2, \ldots, \mathbf{u}_8$. The CANAL spectrum is shown in Figure 14.14. When confining the subspace to dimension three, only the interference DOAs become visible.

In Figure 14.15, one horizontally polarized interferer arrives at −40 degrees with 20 dB JNR. We use the same GPS directions, look directions, and step size as in the previous example. It is evident that the horizontally polarized interferer has a clear peak at −40 degrees.

14.7 An Alternative Cascade Processing Approach

It is evident from the examples provided in this section that the CANAL method is sensitive to subspace dimension and the cancellation weight perturbation. Instead of applying the cancellation weight matrix, the nulling weight matrix itself can be used to obtain the DOAs of the interference. The nulling weight matrix is expressed in equation (14.40) as

$$\mathbf{W}_a = \mathbf{R}^{-1}\mathbf{T}(\theta). \tag{14.48}$$

The nulling weight vector for the kth look direction is

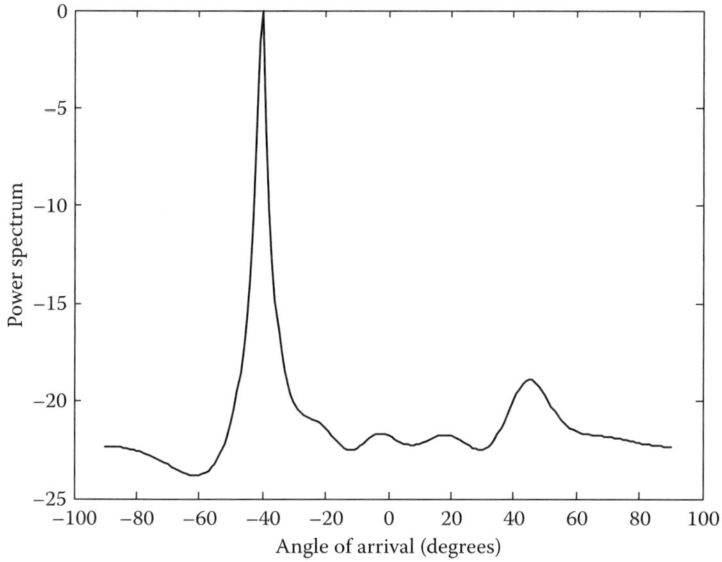

FIGURE 14.15 CANAL spectrum, by LMS algorithm for one horizontally polarized interferer from −40 degrees.

$$\mathbf{w}_{ak} = \mathbf{E}\mathbf{\Lambda}^{-1}\mathbf{E}^{H}\mathbf{a}_k(\theta) = \left(\sum_{i=1}^{M}\frac{1}{\sigma_i^2}\mathbf{e}_i\mathbf{e}_i^H + \sum_{j=M+1}^{2N}\frac{1}{\sigma_n^2}\mathbf{e}_j\mathbf{e}_j^H\right)\mathbf{a}_k(\theta). \tag{14.49}$$

For large JNR, $\sigma_i^2 \gg \sigma_n^2$ and equation (14.49) can be simplified as

$$\mathbf{w}_{ak} = \sum_{j=M+1}^{2N}\frac{1}{\sigma_n^2}\mathbf{e}_j\mathbf{e}_j^H\mathbf{a}_k(\theta) = \frac{1}{\sigma_n^2}\mathbf{E}_n\mathbf{E}_n^H\mathbf{a}_k(\theta) = \frac{1}{\sigma_n^2}(\mathbf{I}-\mathbf{E}_j\mathbf{E}_j^H)\mathbf{a}_k(\theta), \tag{14.50}$$

where $\mathbf{E}_n$ and $\mathbf{E}_j$ are long matrices representing the $2N$-by-$(2N - M)$ and $2N$-by-M eigenvector matrices corresponding to the noise and the jammer eigenvalues, respectively. It is noted that the nulling weight matrix is orthogonal to the jammer subspace, and the MUSIC spectrum can be directly obtained as

$$P(\theta) = \frac{1}{\sum_{i=1}^{L}\left\|\mathrm{a}^H(\theta)\mathrm{w}_{ai}\right\|^2}. \tag{14.51}$$

If $L = 1$, then the problem simplifies to looking for peaks by displaying the inverse spectrum. If $L > 1$, the common nulls are emphasized, whereas spurious system nulls are eliminated. In practice, however, the adaptive weights, at convergence, are perturbed, especially for low-power jammers and under fast convergence. The SVD of the nulling weight matrix can be expressed as

$$\mathbf{W}_a = \mathbf{U}_a \mathbf{\Sigma}_a \mathbf{V}_a, \tag{14.52}$$

where $\mathbf{U}_a$ is a $2N$-by-$2N$ unitary matrix, $\mathbf{V}_a$ is an L-by-L unitary matrix, and $\mathbf{\Sigma}_a$ is a $2N$-by-L singular value matrix. The matrix $\mathbf{U}_{an}$ of the first $L - M$ singular vectors corresponding to the nonzero singular values in $\mathbf{\Sigma}_a$ is orthogonal to the jammers' subspace, and the spectrum can be expressed as

$$P(\theta) = \frac{1}{\sum_{i=1}^{L-M} \left\| \mathrm{a}^H(\theta)\mathrm{u}_{an\_i} \right\|^2}. \tag{14.53}$$

In the following examples, four RHCP GPS signals arrive at –20, 0, 15, and 45 degrees with SNR of –20 dB. The eight look directions considered are the four GPS directions and additional directions of –74, –55, 60, and 85 degrees. The number of snapshots is set to 4,000. The nulling weight matrix is adaptively obtained by the constraint LMS algorithm with the step size of 0.000005. In Figure 14.16, the spectrum with one RHCP jammer is examined. The jammer signal is incident on the array with –40 degrees of

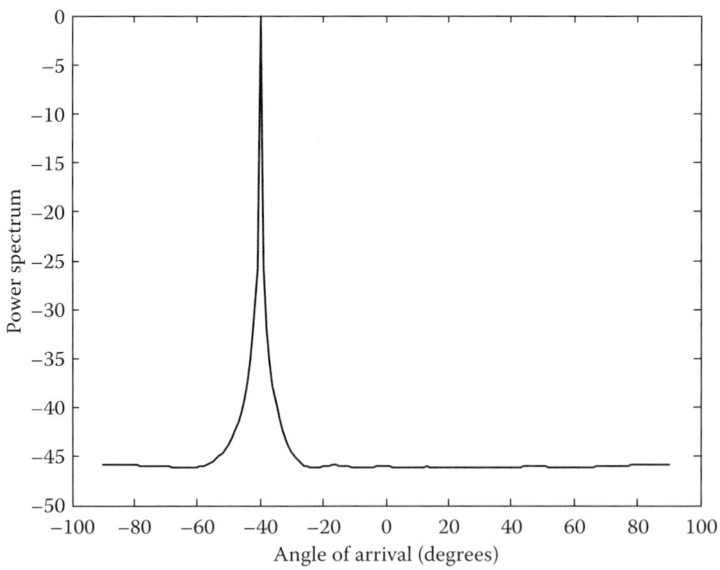

FIGURE 14.16 Spectrum based on the nulling weight matrix obtained by the LMS algorithm with eight dual-polarized antenna arrays, when one RHCP interferer arrives at –40 degrees.

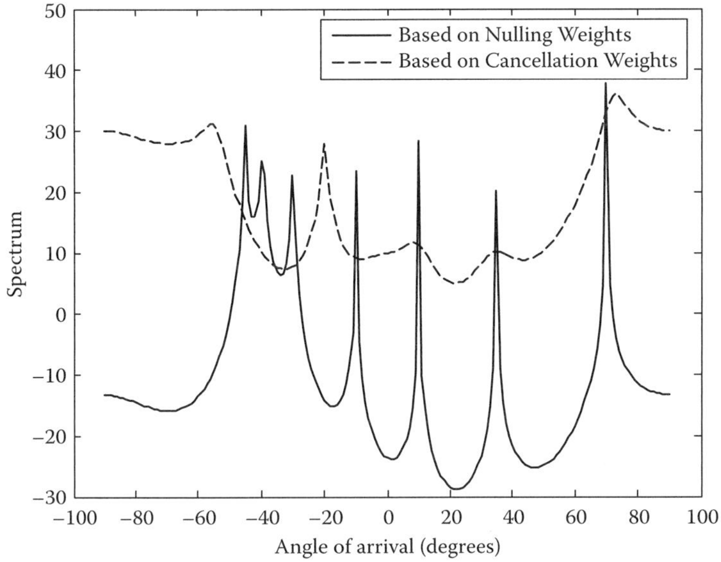

FIGURE 14.17 Spectrum based on the nulling and cancellation weight matrix obtained by the LMS algorithm with eight dual-polarized antenna arrays for seven RHCP interferers.

elevation angle and 20 dB JNR. The number of nonzero singular values is seven. The spectrum is obtained using equation (14.53).

Figure 14.17 compares the spectra based on the nulling weight matrix and the cancellation weight matrix for seven RHCP jammers, arriving from –40, 35, 70, –30, –45, –10, and 10 degrees. It is clear that the jammers can be properly localized if we apply the nulling weight matrix, whereas the cancellation weight matrix-based spectrum fails to exhibit the correct peaks at the jammer DOAs.

14.8 Summary

In this chapter, we addressed interference nulling and interference DOA estimation for GPS receivers with dual-polarized antenna arrays. It is shown that interference nulling can be achieved by invoking the orthogonal polarization properties of the array or by coherently combining the antenna outputs through optimum and adaptive weight values. For antijam GPS, the GPS receiver may be designed to employ one set of weights that forms one beamformer and places low gain toward the interference DOAs and unit gain toward satellites. It is shown that this approach has shortcomings, as it offers a limited number of degrees of freedom and provides poor nulling capabilities when the jammers and the satellites have close angular separations. Multiple weight vectors corresponding to multiple beamformers mitigate the aforementioned problems at the expense of increased receiver complexity. We have also presented a DOA estimation approach that operates on the adaptive weights rather than the estimated covariance matrix. Two different techniques under this approach are represented. The first deals with cancellation weight vectors that span the interference subspace, and the other is based on the adaptive nulling weight vectors that are orthogonal to the interference subspace. We provided simulation results comparing the two techniques under various jamming environments and using different interference signal polarizations.

Acknowledgment

The author thanks Ms. Jing Wang for providing the computer simulations.

References

[1] Zoltowski, M. D., and Gecan, A. S. 1995. Advanced adaptive null steering concepts for GPS. In *Proceedings of the 1995 Military Communications Conference (MILCOM 1995)*, vol. 3, p. 1214.

[2] Van Veen, B., and Buckley, K. 1988. Beamforming: A versatile approach to spatial filtering. *IEEE ASSP Mag.* 5:4.

[3] Sun, W., and Amin, M. 2005. A self-coherence based anti-jam GPS receiver. *IEEE Trans. Signal Processing* 53:3910.

[4] Trinkle, M., and Cheuk, W. C. 2003. Null-steering GPS dual-polarized antenna arrays. In *6th International Symposium on Satellite Navigation Technology.*

[5] Fante, F. L., and Vaccaro, J. J. 2001. *Jammer cancellation performance of a fully-polarimetric GPS array.* MITRE technical report.

[6] GNSS. 2006. GNSS album images and spectral signatures of the new GNSS signals. In *GNSS: Engineering solutions for the global navigation satellite system*, pp. 46–56.

[7] Kaplan, E. D. 1996. *Understanding GPS, principles and application*. Norwood, MA: Artech House Publishers.

[8] Grewal, M. S., Weill, L. R., and Andrews, A. P. 2001. *Global positioning systems, inertial navigation, and integration*. New York: Wiley.

[9] Ketchum, J. W., and Proakis, J. G. 1982. Adaptive algorithms for estimating and suppressing narrow-band interference in PN spread-sprectrum systems. *IEEE Trans. Commun.* 30:913.

[10] Milstein, L. B. 1988. Interference rejection techniques in spread spectrum communications. *Proc. IEEE* 76:657.

[11] Amin, M. G., and Zhang, Y. 2002. Interference suppression in spread spectrum communication systems. In *The Wiley encyclopedia of telecommunications*, ed. J. G. Proakis. New York: John Wiley.

[12] Badke, B., and Spanias, A. 2002. Partial band interference excisions for GPS using frequency-domain exponents. In *IEEE International Conference on Acoustics, Speech, and Signal Processing (ICASSP 2002)*, vol. 4, p. 3936.

[13] Capozza, P. T., Holland, B. J., Hopkinson, T. M., and Landrau, R. L. 2000. A single-chip narrow-band frequency domain excisor for a global positioning system (GPS) receiver. *IEEE J. Solid State Circuits* 35:401.

[14] DiPietro, R. C. 1989. An FFT based technique for suppressing narrow-band interference in PN spread spectrum communications systems. In *IEEE International Conference on Acoustics, Speech, and Signal Processing (ICASSP 1989)*, vol. 2, p. 1360.

[15] Amin, M. G., Zhao, L., and Lindsey, A. R. 2004. Subspace array processing for the suppression of FM jamming in GPS receivers. *IEEE Trans. Aerosp. Electronics Syst.* 40:80.

[16] Subbaram, H., and Abend, K. 1993. Interference suppression via orthogonal projections: A performance analysis. *IEEE Trans. Antennas Propagation* 41:1187.

[17] Fante, R. L., and Vaccaro, J. J. 2000. Wideband cancellation of interference in a GPS receive array. *IEEE Trans. Aerosp. Electronic Syst.* 36:549.

[18] Zhang, Y., Amin, M. G., and Lindsey, A. R. 2001. Anti-jamming GPS receivers based on bilinear signal distributions. In *Proceedings of the 2001 Military Communications Conference (MILCOM 2001)*, vol. 2, p. 1070.

[19] Xiong, P., Medley, M. J., and Batalama, S. N. 2003. Spatial and temporal processing for global navigation satellite systems: The GPS receiver paradigm. *IEEE Trans. Aerosp. Electronics Syst.* 39:1471.

[20] Myrick, W. L., Goldstein, J. S., and Zoltowski, M. D. 2001. Low complexity anti-jam space-time processing for GPS. In *International Conference on Acoustics, Speech, and Signal Processing*, vol. 4, p. 2233.

[21] Vaccaro, F. R. 2002. Evaluation of adaptive space-time-polarization cancellation of broadband interference. In *Proceedings of the IEEE Precision Location and Navigation Symposium*, p. 1.

[22] GNSS. 2006. GNSS solutions: Adaptive antenna arrays, multi-GNSS tropospheric monitoring, and high-dynamic receivers. In *GNSS: Engineering solutions for the global navigation satellite system*.

[23] Seco, G., Fernández-Rubio, J. A., and Fernández-Prades, C. 2005. ML estimator and hybrid beamformer for multipath and interference mitigation in GNSS receivers. *IEEE Trans. Signal Processing* 53:1194–1208.

[24] Moelker, D., van der Pol, E., and Bar-Ness, Y. 1996. Adaptive antenna arrays for interference cancellation in GPS and GLONASS receivers. In *Proceedings of the IEEE Position Location Navigation Symposium*, p. 191.

[25] Hatke, G. 1998. Adaptive array processing for wideband nulling in GPS systems. In *Proceedings of the 32nd Asilomar Conference on Signals, Systems, and Computers*, vol. 2, p. 1332.

[26] Gecan, A., and Flikkema, P. 1996. Jammer cancellation with adaptive arrays for GPS signals. In *Proceedings of the IEEE Southeast Conference: Bringing Together Education, Science, and Technology*, p. 320.

[27] Mingjie, D., Xinjian, P., Fang, Y., and Jianghong, L. 2001. Research on the technology of adaptive nulling antenna used in anti-jam GPS. In *Proceedings of the CIE International Conference on Radar*, Beijing, p. 1178.

[28] Zoltowski, M. D., and Gecan, A. S. 1995. Advanced adaptive null steering concepts for GPS. In *Proceedings of the Military Communication Conference*, vol. 3, p. 1214.

[29] Moelker, D., van der Pol, E., and Bar-Ness, Y. 1996. Adaptive antenna arrays for interference cancellation in GPS and GLONASS receivers. In *Proceedings of the IEEE Position Location Navigation Symposium*, pp. 191–98.

[30] Hatke, G. 1998. Adaptive array processing for wideband nulling in GPS systems. In *Proceedings of the 32nd Asilomar Conference on Signals, Systems, and Computers*, vol. 2, p. 1332.

[31] Schmidt, R. O. 1986. Multiple emitter location and signal parameter estimation. *IEEE Trans. Antennas Propagation* 34:276.

[32] Schweppe, F. C. 1968. Sensor array data processing for multiple signal sources. *IEEE Trans. Inform. Theory* 14:294.

[33] Krim, J., and Viberg, M. 1996. Two decades of array signal processing research: The parametric approach. *IEEE Signal Processing Mag.* 13:67.

[34] Kumaresan, R., and Tufts, D. W. 1983. Estimating the angles of arrival of multiple plane waves. *IEEE Trans. Aerosp. Electronics Syst.* 19:134.

[35] Grover, R., Pados, D. A., and Medley, M. J. 2007. Subspace direction finding with an auxiliary-vector basis. *IEEE Trans. Signal Processing* 55:758.

[36] Roy, R., and Kailath, T. 1989. ESPRIT—Estimation of signal parameters via rotational invariance techniques. *IEEE Trans. Acoust. Speech Signal Processing* 37:984.

[37] Swindlehurst, A., Ottersten, B., Roy, R., and Kailath, T. 1992. Multiple invariance ESPRIT. *IEEE Trans. Signal Processing* 40:867.

[38] Yu, S. J., and Lee, J. H. 1999. Efficient eigenspace-based array signal processing using multiple shift-invariant subarrays. *IEEE Trans. Antennas Propagation* 47:186.

[39] Fuchs, J. J. 1996. Rectangular Pisarenko method applied to source localization. *IEEE Trans. Signal Processing* 44:2377.

[40] Frost, O. L., III. 1972. An algorithm for linearly constrained adaptive array processing. *Proc. IEEE* 60:926.

[41] Griffiths, L. J. 1969. A simple adaptive algorithm for real-time processing in antenna arrays. *Proc. IEEE* 57:1696.

[42] Widrow, B., Mantey, P. E., Griffiths, L. J., and Goode, B. B. 1967. Adaptive antenna systems. *Proc. IEEE* 55:2143.

[43] Amin, M. G. 1992. Concurrent nulling and locations of multiple interferences in adaptive antenna array. *IEEE Trans. Signal Processing* 40:2658.

[44] Ercek, R., De Doncker, P., and Grenez, F. 2005. Study of pseudo-range error due to non-line-of-sight-multipath in urban canyons. In *Proceedings of ION GNSS 2005*, Long Beach, CA, pp. 1083–1094.

[45] Xu, X., and Buckley, K. 1993. Analysis of beam-space source localization. *IEEE Trans. Signal Processing* 41:501–509.

15

Reconfigurable Baseband Processing for Wireless Communications

André B. J. Kokkeler
University of Twente

Gerard K. Rauwerda
University of Twente

Pascal T. Wolkotte
University of Twente

Qiwei Zhang
University of Twente

Philip K. F. Hölzenspies
University of Twente

Gerard J. M. Smit
University of Twente

15.1 Introduction

Within wireless communications, advances in digital processing are pushing system performance to higher levels, providing increasingly larger bandwidths and introducing more complex coding and modulation schemes. Increasingly larger parts of processing of both the radio access network (RAN) and user equipment (UE) are mapped onto digital computing devices. Looking at this mapping process, one can distinguish different classes of processes (or tasks) and different types of computing devices.

A classification of processing is general purpose (GP processing), domain specific (DS processing), and application specific (AS processing). GP processing is generally characterized by a nonpredictable execution flow, whereas AS processing has a predictable execution flow. A characteristic of DS processing is that several similar tasks or algorithms are to be executed on a single computing device.

Computing devices or processors are generally classified as general-purpose processors (GPPs), digital signal processors (DSPs), reconfigurable processors (RPs), and

TABLE 15.1 Overview of Types of Processing and Computing Devices

	GPP	DSP	RP	ASIC
GP processing	*	(*)		
DS processing	*	*	*	
AS processing	*	*	*	*
Power efficiency	– –	–	+	++
Area efficiency	– –	–	+	++
Engineering efficiency	++	+	–	– –
Flexibility	++	+	–	– –

application-specific integrated circuits (ASICs). The ability to execute different types of processing for the different types of computing devices is indicated in Table 15.1.

GPPs are able to execute all types of processing. DSPs are processors that are tailored toward the signal processing domain. This is a relatively broad application domain, and DSPs therefore resemble GPPs to a large extent; both GPPs and DSPs are typically von Neumann machines [48]. Reconfigurable processors can be adapted in such a way that different types of processing within a specific application domain can be executed. ASICs generally are only suited for AS processing.

When mapping computing tasks onto computing devices, performance criteria of computing devices are evaluated. Performance criteria are presented at the lower part of Table 15.1. In general, the performance criteria are power, area, and engineering efficiency and flexibility. Because ASICs basically have no control overhead, most dissipated energy contributes to the execution of the tasks at hand. On the other hand, since GPPs are based on the von Neumann principle, substantial energy is dissipated for fetching and decoding instructions. From a power consumption perspective, this can be considered as overhead leading to low power efficiency. DSPs are also based on the von Neumann principle but are more power efficient. RPs show more ASIC-like behavior and are therefore more power efficient than DSPs. Area efficiency relates to the area required on an integrated circuit (IC) for realizing a specific task. ASICs are far more area efficient than GPPs. Engineering efficiency refers to the relative engineering effort required to have the computing device executing the demanded processing. For GPPs and DSPs, in general, compilers are available to efficiently compile high-level sequential code into machine instructions. Compilers for mapping applications onto RPs are not yet as mature as for GPPs and DSPs. For ASICs, functionality can be specified by means of schematic capture or a hardware specification language. However, the trajectory from this specification to an integrated circuit is time-consuming, risky, and expensive. Flexibility basically refers to the range of applications for which a computing device can be used. It is clear the GPP offers the most flexibility and ASIC the least.

Because of the diminishing feature size of integrated circuits, more and more functionality can be packed into a single chip. This has led to the creation of heterogeneous system-on-chips (SoCs), where multiple processing cores of different types are interconnected by means of a network-on-chip (NoC). An example can be found in [47].

The developments described above have their impact on wireless communications. If the RAN is considered, (chip) area and power consumption are not of utmost importance. Engineering efficiency is important because RAN equipment is sold in relatively low quantities. Flexibility is necessary in case new services have to be offered by the RAN. Therefore, its functionality is implemented in GPPs and DSPs as much as possible. For example, in UMTS base stations, chip-level processing is done by means of DSPs [54]. In UE, low power consumption has been the driving force, and large parts of functionality are traditionally implemented in ASICs. However, nonrecurrent engineering costs and flexibility become more and more important. This has led to a trend where, also in the UE, reconfigurable devices are becoming feasible.

In the remaining part of this chapter we will first discuss reconfigurable platforms since they are attractive for both RAN and UE. Second, the implementation of wireless communication applications on a reconfigurable platform will be described where we separately deal with licensed communication and a form of unlicensed communication called cognitive radio. Third, we zoom into the process of mapping applications onto a reconfigurable platform. Both the design time mapping and runtime mapping are discussed.

15.2 Reconfigurable Platforms

In [11], a definition of reconfigurable platforms is given: "Systems incorporating some form of hardware programmability—customizing how the hardware is used using a number of physical control points. These control points can then be changed periodically in order to execute different applications using the same hardware." Based on this definition, reconfigurable platforms can consist of fine-grained or coarse-grained functional units [45]. Fine-grained functional units implement a function on a single bit or a small number of bits, e.g., small lookup tables in a standard field programmable gate array (FPGA). Coarse-grained functional units are typically much larger and may consist of arithmetic and logic units (ALUs) and a significant amount of storage. Fine-grained and coarse-grained reconfigurable devices are described separately.

For the wireless communications domain, conventional computing architectures will be replaced by reconfigurable multiprocessor system-on-chips (MP-SoCs). On MP-SoCs, several instances of a single type of processing device can be implemented. These types of MP-SoCs are referred to as homogeneous tiled architectures. MP-SoCs with several instances of a few different types of processing devices are referred to as heterogeneous tiled architectures. We will give examples of both types of architectures. To interconnect different processors on an MP-SoC, we use a NoC. NoCs are described at the end of this section.

15.2.1 Fine-Grained Reconfigurable Devices

Recent advances in reconfigurable devices are based on the success of FPGAs. Within an FPGA, static random access memory (SRAM) bits are connected to configuration points. For example, writing to these SRAM bits, routing structures are configured and

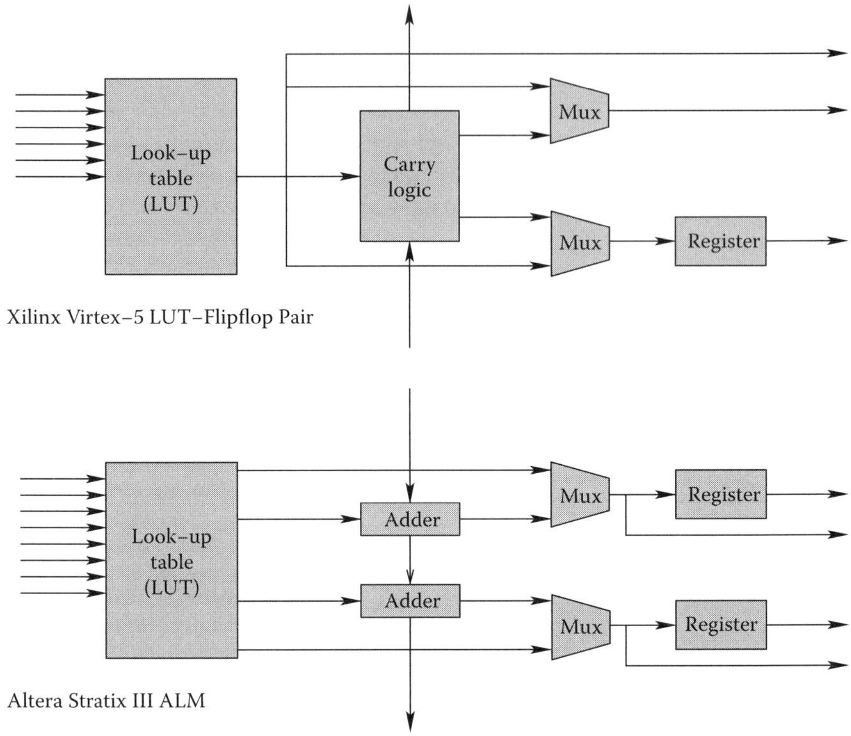

FIGURE 15.1 Basic computational elements of two state-of-the-art FPGAs.

(de)multiplexers can be set. Routing structures and (de)multiplexers are used to interconnect the basic computational elements of an FPGA. The basic computational elements of two state-of-the-art FPGAs are presented in Figure 15.1.

The computational elements are from recent FPGA families of the two dominant manufacturers: the adaptive logic module (ALM) of the Stratix III family of Altera and the combination of a lookup table (LUT) and flipflop of the Virtex-5 family of Xilinx. The ALM has eight inputs, connected to a LUT. At most two single-bit functions can be realized by the LUT. Furthermore, two adders and two registers are available. The unit of a Virtex-5 device that is comparable to an ALM is a LUT-flipflop pair. The LUT has six inputs to which carry logic and a single register are added.

In Table 15.2 an overview of the characteristics of two devices out of the Stratix III and Virtex-5 families is given. These devices are the ones with the maximum number of LUTs within their family.

The numbers that are given for features of the devices cannot be compared directly. A slice within the Virtex-5 series consists of four six-input LUT-flipflop pairs. Furthermore, the basic element within a Virtex-5 DSP slice is an 18-by-25-bit multiplier. Memory is organized in block RAMs. ALMs of the Stratix III series are comparable with a LUT-flipflop pair of the Virtex-5 series. An ALM has eight inputs. The multipliers within

TABLE 15.2 Characteristics of Two State-of-the-Art Fine-Grained Reconfigurable Devices

XilinxVirtex-5 XC5VLX330T		AlteraStratix III EP3SC340	
Slices ⇒ Number of LUTs	51,840 207,360	Adaptive logic modules ⇒ Number of LUTs	135,300 135,300
DSP slices (18 × 25 bit multipliers)	192	Multipliers (18 × 18 multipliers)	576
Clock frequency	550 MHz	Clock frequency	600 MHz
Memory In block RAM	11.664 Mbit	Memory In memory blocks	17.208 Mbit

the Stratix III device have two 18-bit input operands. Memory is organized in memory blocks. The frequencies given for both devices are the maximum internal clock speed.

Fine-grained reconfigurable devices are in principle only used in the RAN, for example, in base stations. Fine-grained devices offer a large flexibility at the cost of a reduced (chip) area efficiency and power efficiency. Area and power are of less importance for the RAN but are of utmost importance in the UE. An example of a HiperLAN/2 base station implemented on an FPGA is given in [37]. Furthermore, there are numerous examples of the implementation of parts of UMTS transmitters and receivers on an FPGA. Recent examples can be found in [5], [7], [34], and [43]. FPGA-based platforms are also used for the design and analysis of communication systems. An example is the BEE2 platform [8]. This is a modular platform, consisting of multiple FPGAs used to speed up the emulation and design of wireless communications systems.

15.2.2 Coarse-Grained Reconfigurable Devices

Within wireless communication systems, the computational data paths have path widths greater than a single bit, and for that reason, fine-grained architectures are much less efficient because of the routing area overhead. Coarse-grained architectures have word-level data paths and are therefore more efficient. While the basic elements within FPGAs are slices or LUT-flipflop pairs, the basic elements within a coarse-grained architecture are reconfigurable data path units (DPUs). Different coarse-grained reconfigurable devices distinguish themselves through the functionality provided by the DPUs and the structure of interconnections between different DPUs. A DPU can be as simple as a single arithmetic and logic unit (ALU) or relatively obese, containing multiple ALUs, multiple memories, and control/configuration logic. In case relatively small DPUs are used, FPGA-like structures interconnect the different DPUs. In case of multiple larger DPUs, usually a NoC interconnects different processors.

15.2.2.1 Coarse-Grained Reconfigurable Architectures

Recently, several word-level reconfigurable architectures have been proposed, very often in the context of a tiled architecture template. The Pleiades architecture [3] is a template to create an instance of a domain-specific processor. In the template, a control processor is surrounded by an array of heterogeneous autonomous special-purpose satellite processors. All processors are interconnected via a reconfigurable communication network. The adaptive computing machine by QuickSilver [25], is a heterogeneous system-on-chip. Nodes with different flexibility (adaptive nodes, domain-specific [ASIC-like] nodes, and programmable nodes) can be interconnected by means of a mesh interconnect. The XPP extreme processor platform [4] consists of clusters of computing elements and a packet-oriented communication network. A cluster consists of a set of parameterizable tiles. There are two types of tiles: memory tiles and ALU tiles. A memory tile has a capacity of 256 or 512 words. The word size can be either 16, 24, or 32 bits. The architecture uses a large number of processors, which makes the chip suitable for high-end applications. Silicon Hive provides reconfigurable accelerators designed according to a hierarchical approach. At the lowest level, the basic component is a VLIW-like processing and storage element (PSE). At the next level, multiple PSEs form a cell. A cell is a processor, capable of executing complete algorithms. One level higher, multiple cells can be combined. The accelerators are to be integrated onto a SoC. The ADRES architecture template [6] consists of a tightly coupled VLIW processor and a coarse-grained reconfigurable array. The reconfigurable array is intended to process computationally intensive kernels of applications. The VLIW host processor uses parts of the reconfigurable array to execute its instructions. The host processor and reconfigurable array are therefore tightly coupled and share resources. This architecture is dedicated to applications that require tight control of data flow operations. A similar approach is used in the Chimaera architecture [24]. In this architecture, the reconfigurable part has direct access to the host processor's register file. The Kilocore KC256 chip is a commercial version of the PipeRench chip [21]. The chips are characterized by a multicore computing kernel where cores can be cascaded to constitute multiple processing pipelines. Besides the configurable ALU, a processing core only contains a register file and no memory. Configuration and information data are stored in separate SRAMs. Because PipeRench is specially designed for pipelined applications, best performance is achieved when pipeline stages are identical or perfectly balanced. A specific coarse-grained reconfigurable architecture, developed at the University of Twente, is the Montium processor. The Montium will be described in more detail below and will be used throughout this chapter in the examples of coarse-grained reconfigurable processors.

15.2.2.2 The Montium

The Montium is described in detail in [27], and in this section the general structure is discussed. A single Montium processing tile is depicted in Figure 15.2.

The lower part of Figure 15.2 shows the communication and configuration unit (CCU), which deals with the off-tile communication and configuration of the upper part, the reconfigurable tile processor (TP). The TP is the computing part that can be dynamically reconfigured to implement a particular algorithm. At first glance the TP has a VLIW structure. However, the control structure of the Montium is very different. For (energy) efficiency it is imperative to minimize the control overhead. This is, for example, accomplished by scheduling instructions statically at compile time. A relatively simple sequencer controls the entire tile processor. The sequencer selects configurable tile instructions that are stored in the instruction decoding block (see Figure 15.2).

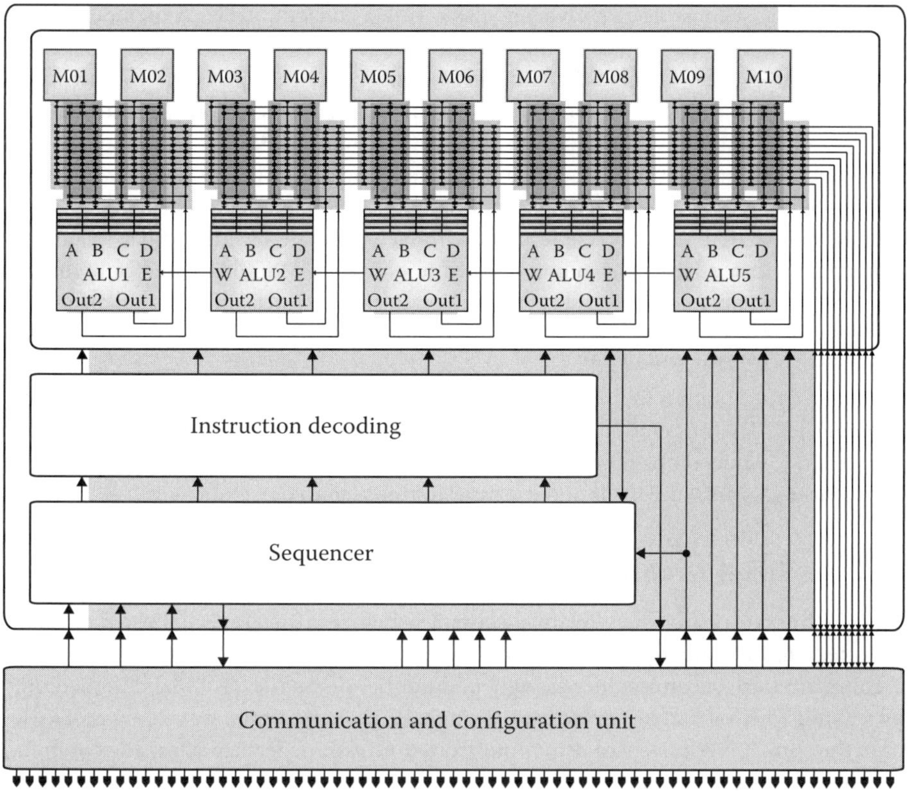

FIGURE 15.2 The Montium tile processor and network interface.

Furthermore, there are multiple ALUs (ALU1 … ALU5) and multiple memories (M01 … M10). A single ALU has four inputs (A, B, C, D). Each input has a private input register file that can store up to four operands. The input register file cannot be bypassed, i.e., an operand is always read from an input register. Input registers can be written by various sources via a flexible interconnect. An ALU has two outputs (OUT1, OUT2), which are connected to the interconnect. The ALU is entirely combinational, and consequently, there are no pipeline registers within the ALU. Neighboring ALUs can also communicate directly: the west output (W) of an ALU connects to the east input (E) of the ALU neighboring on the left.

The ALUs support both signed integer and signed fixed-point arithmetic. The five identical ALUs in a tile can exploit spatial concurrency to enhance performance. This parallelism demands a very high memory bandwidth, which is obtained by having ten local memories in parallel.

An address generation unit (AGU; not shown in Figure 15.2) accompanies each memory. The AGU can generate the typical memory access patterns found in common DSP algorithms, e.g., incremental, decremental, and bit-reversal addressing. It is also possible to use the memory as a lookup table for complicated functions that cannot be calculated using an ALU such as sine or division with a single constant value. A memory can be used for both integer and fixed-point lookups.

The reconfigurable elements within the Montium are the sequencer, the instruction decoding block, and the AGUs. Their functionality can be changed at runtime. The Montium is programmed in two steps. In the first step, the AGUs are configured and a limited set of instructions is defined by configuring the instruction decoding block. The sequencer is then instructed to sequentially select the required instructions. A predefined instruction set is available using an assembly type of mnemonics (Montium assembly). A compiler has been constructed to convert a Montium assembly program into configuration data for both the instruction decoding block and the sequencer.

15.2.2.2.1 Energy Consumption

Using power estimation tooling, the dynamic power consumption of a typical multiply-accumulate (MAC) operation in the Montium is estimated to be about 0.5 mW/MHz, realized in 130 nm complementary metal oxide semiconductor (CMOS) technology. The area of a single Montium TP is about 2 mm^2 in this technology [26].

15.2.3 Tiled Architecture

Tiled architectures are where relatively complex elements (tiles) are replicated on a single integrated circuit (IC). The tiles are interconnected via an on-chip network. Tiled architectures are becoming increasingly popular because a tile has to be designed only once, after which it can be copied onto a single IC multiple times. By adding more tiles onto the chip, it is relatively easy to profit from diminishing feature sizes. The computation model, programming model, interconnection structure, and memory organization can stay the same. Below, the following prominent tiled architectures are discussed: the RAW processor [49], the cell processor [29], the Polaris processor [47], and the Tile64 processor (see www.tilera.com). Furthermore, the Chameleon heterogeneous tiled architecture is introduced.

15.2.3.1 The RAW Processor

The RAW processor is one of the earliest tiled architectures. A RAW processor consists of a set of relatively simple tiles interconnected by a set of switches. Each tile contains instruction memory, data memories, an ALU, registers, configurable logic, and a programmable switch with an associated instruction memory. The general idea is that the internal hardware structure of both the tile and the switches is exposed to the compiler. This way there are two sets of control logic: operation control for the processor and sequencing routing instructions for the switches. A consequence is that the burden on the compiler is high, which leads to relatively long compile times. The configurable logic in each tile supports a few wide-word or many narrow-word operations and is coarser than FPGA-based processors.

15.2.3.2 The Cell Processor

The cell processor consists of eight replicas of a synergistic processor element (SPE) and a (single) power processor element (PPE) with a power core. An SPE consists of a division multiple access (DMA) unit, a local store memory (256 KB), execution units, and memory and bus interface controllers. The PPE is accompanied by first- and second-level caches. The nine cores are interconnected by a coherent on-chip bus. The SPEs and PPEs are SIMD processors where the element width of the SPE can range from 2 × 64 bits to 128 × 1 bits.

15.2.3.3 Polaris

Intel has recently released information on an 80-tile, 1.28 TFLOPS processor in 65 nm CMOS. The tiles are arranged as a 10-by-8 mesh network. Each tile consist of a processing engine connected to a five-port router that forwards packets between tiles. The PE contains two independent fully pipelined single-precision floating-point multiply-accumulate units, 3 KB single-cycle instruction memory, and 2 KB data memory. The NoC connects the PEs using packet-based communication, and the router in each PE can send and receive packets from any of the other tiles on the network.

15.2.3.4 Tile64

Tile64 is a processor based on the mesh architecture, originally developed for the RAW machine. The chip consists of a grid of processor tiles arranged in a network, where each tile consists of a GPP, cache, and a nonblocking router that the tile uses to communicate with the other tiles on the chip. Each processor has a register file and three functional units: two integer ALUs and a load-store unit. A processor also has a split L1 cache and an L2 cache. When there is a miss in the L2 cache of a specific processor, the L2 caches of the other processors are searched for the data before external memory is consulted. This way, a large L3 cache is emulated. The delay that is caused by the on-chip mesh network is exposed to the programmer, and during compilation, the compiler takes the delays into account when scheduling different tasks on different tiles.

15.2.3.5 The Chameleon Tiled Architecture Template

Trends in wireless communications show that many standards with the same purpose coexist. On one hand, new standards are being developed, but on the other hand, revisions of standards evolve quickly. Different contradicting requirements are identified for the platform that implements wireless communications systems:

- High performance: Due to the increasing complexity of signal processing in digital receivers, high-performance signal processors are needed.
- Flexibility: Radio receivers need to switch quickly between different standards, as they have to support a wide variety of standards. Within standards, fast switching between different configurations is required.
- Low power: Portable receivers need to be very energy efficient because they are battery powered.

To meet contradicting requirements for high performance, flexibility, and energy efficiency, a heterogeneous SoC template is designed. A heterogeneous SoC contains different types of processing parts, like general-purpose, fine-grained reconfigurable, coarse-grained reconfigurable, and dedicated hardware parts. Multiple processing parts are interconnected by means of the NoC. The idea behind a heterogeneous SoC is that

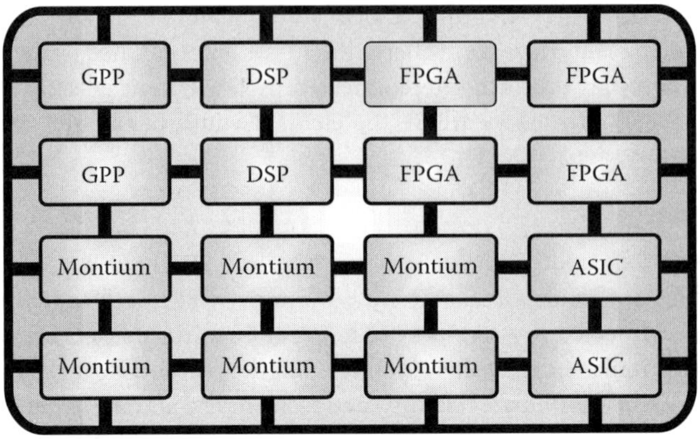

FIGURE 15.3 The Chameleon SoC template.

all parts of an algorithm can run on a processing part on which it can be implemented most efficiently. The ideas above translated into the definition of the Chameleon SoC template. This template is illustrated in Figure 15.3.

A Chameleon SoC that contains sixteen processing tiles is depicted. The black grid in the figure represents the NoC. The Chameleon SoC attempts to combine the best of all worlds: the performance and energy efficiency of ASICs, the bit-level flexibility of FPGAs, the domain-specific flexibility of the Montium and DSPs, and the general-purpose applicability of GPPs. Information exchange between the different tiles is realized by means of the NoC.

15.2.4 Network-on-Chip

A key element within the Chameleon SoC template is the NoC. In a NoC, a processing tile is connected to a router. Routers of different processing tiles are interconnected. Communication between two processing tiles involves at least the two routers of the corresponding processing tiles, but other routers might be involved as well. A NoC that routes data items in a SoC has a higher bandwidth than an on-chip bus, as it supports multiple concurrent communications. The well-controlled electrical parameters of an on-chip interconnection network enable the use of high-performance circuits that result in significantly lower power dissipation, shorter propagation time, and higher bandwidth than is possible with a bus (see also [52]). To describe the network traffic in a system, multiple types are identified [22]. According to the type of service required, the following types of traffic can be distinguished in the network:

- Guaranteed throughput (GT) is the part of the traffic for which the network has to give real-time guarantees (i.e., guaranteed bandwidth, bounded latency).
- Best effort (BE) is the part of the traffic for which the network guarantees only fairness but does not give any bandwidth and timing guarantees.

Within wireless communications, most traffic is in the GT category. Besides the mainstream of GT communication, a minor part (assumed to be less than 5%) of BE communication is foreseen, e.g., control, interrupts, and configuration data.

Several NoC solutions have been proposed that use different techniques to provide guarantees. The Æthereal NoC [23] combines a global time division multiplexing (TDM) schedule to provide contention free (i.e., guaranteed throughput) routes to network streams. The best-effort streams are handled in the slack time of the schedule via wormhole routing. The Nostrum NoC [38] uses a TDM-related technique called temporally disjoint networks. Containers are used to route the GT traffic. A third packet-switched solution uses virtual channels and wormhole routing to provide guarantees [33]. Using deterministic local arbitration mechanisms (e.g., round-robin) results in a guaranteed throughput per virtual channel. Assigning a single stream of traffic per virtual channel will give network-wide guarantees.

The performance of this virtual channel network is illustrated by a HiperLAN/2 case study. For a 6 × 6 network running at 333 MHz, the processes of multiple HiperLAN/2 receivers are placed on the processor tiles of the platform. Each process in the HiperLAN/2 process graph sends GT traffic at a rate of 256 bytes (i.e., one OFDM symbol) per 4 μs. Extra BE messages are offered to the network to control the receivers and emulate other applications running on the system concurrently. Figure 15.4 shows how the latency of the GT and BE messages depends on the offered BE load.

For the GT traffic, the mean and maximal latency of packets are given. When the offered BE load is low, the latency of the GT packets is lower than the guaranteed (or

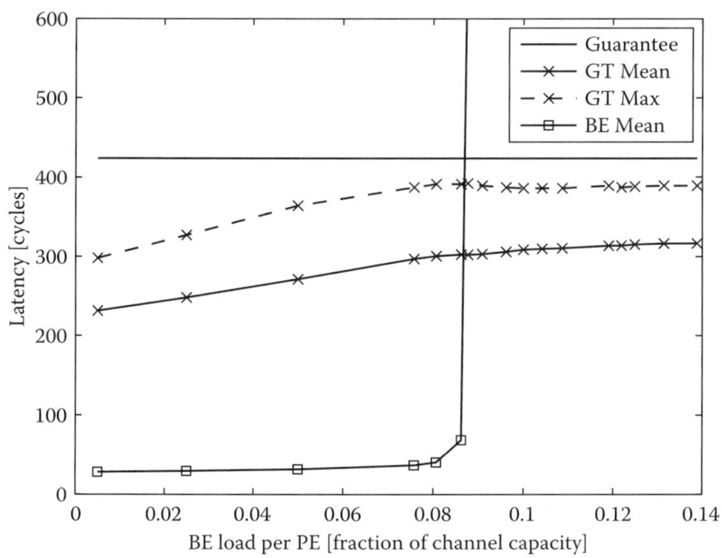

FIGURE 15.4 Message latency of the GT and BE traffic versus BE load for a 6-by-6 network (queue size 2 flits).

allowed) latency. The reason is that the GT traffic utilizes the bandwidth unused by the BE traffic. The latency of the GT packets is higher than the latency of the BE traffic because the GT packets are larger (256 bytes compared with 10 bytes for BE packets). With the increase of the BE load, the latency of the GT traffic increases too until the maximum delay reaches the limit. Further increase of the BE load increases the GT mean latency, but the GT maximum latency never exceeds the guaranteed latency.

Combining guaranteed traffic with best-effort traffic is hard [42]. When using dedicated techniques for both types of traffic, it is possible to reduce the total area and power consumption. The reasons for reconsidering circuit switching are that the flexibility of packet switching is not needed because a connection between two tiles will remain open for a long period (e.g., seconds or longer). Furthermore, large amounts of the traffic between tiles will need guaranteed throughput, which is easier to satisfy in a circuit-switched connection. Circuit switching also eases the implementation of asynchronous communication techniques, because data and control can be separated and circuit switching has a minimal amount of control in the data path (e.g., no arbitration). This increases the energy efficiency per transported bit and the maximum throughput.

Further, scheduling communication streams over non-time-multiplexed channels is easier because, by definition, a stream will not have collisions with other communication streams. In contrast with this, the Æthereal [23] routers are using time-multiplexed channels, and therefore have large interaction between data streams and have to guarantee contention free paths. Determining the static time slots table for these systems requires considerable effort. Because data streams are physically separated in a circuit-switched NoC, collisions do not occur. Therefore, no buffering and arbitration is required in the individual router. An example of a circuit-switched NoC is described in [51]. The number of parallel physical channels between routers is increased to increase the amount of simultaneously active circuitries. This solution utilizes the huge amount of wire resources provided by current and future silicon technologies.

15.3 Applications

In this section, we will discuss the mapping onto a heterogeneous SoC of different licensed communication standards and the mapping of the emerging cognitive radio application. Concerning licensed communication standards, first the realization of a UMTS receiver is given. Next, realizations of an OFDM receiver are presented. Finally, several digital broadcasting systems are discussed.

15.3.1 Licensed Communications

15.3.1.1 UMTS Receiver on Reconfigurable Hardware

The Universal Mobile Telecommunications System (UMTS) standard, defined by European Telecommunication Standards Institute (ETSI) [2], is an example of a third-generation (3G) mobile communication system. The communication system has an air interface that is based on direct-sequence code division multiple access (DS-CDMA)

[30]. The bit rate of UMTS at the physical level depends on the modulation type and the spreading factor (SF). Table 15.3 depicts the general characteristics of the UMTS communication system. Figure 15.5 depicts the basic operation principle of CDMA.

TABLE 15.3 Downlink UMTS Properties in the FDD Mode

Chip rate	3.84 Mcps
Chips per frame	38,400
Chips per slot	2,560
Slots per frame	15
Frame period	10 ms
Slot period	666.67 μs
Scrambling code length	38,400 chips
Scrambling code period	10 ms
SF	4–512
Symbol rate	7.5–960 ksps
Modulation	QPSK, QAM-16

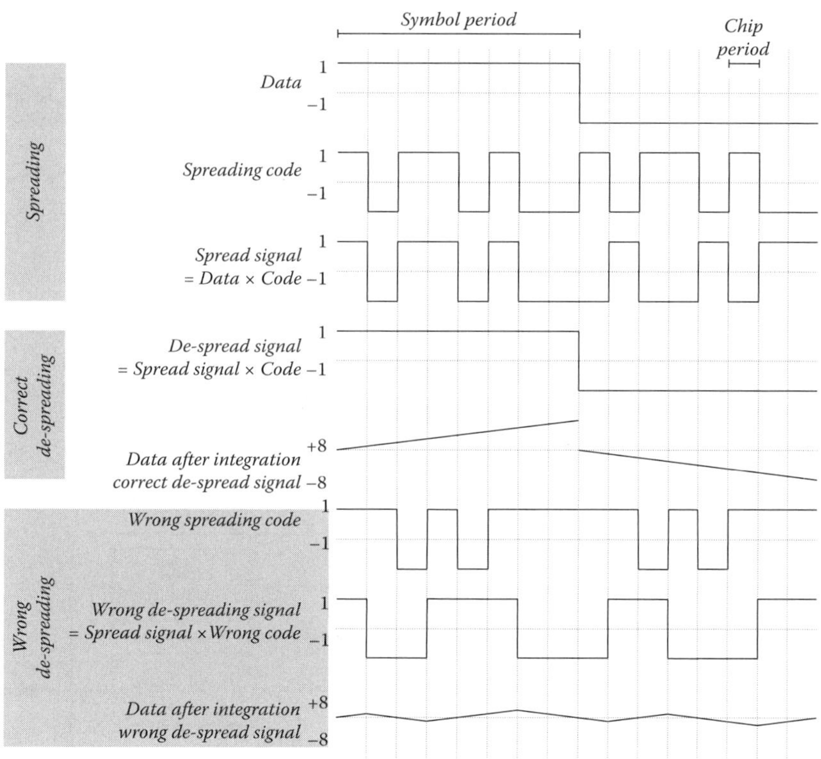

FIGURE 15.5 The CDMA operation principle.

We investigate the possibilities for implementing the digital signal processing functionality of a UMTS receiver on the Chameleon SoC template (see Figure 15.3). We only focus on the downlink of the UMTS receiver at the mobile terminal in the FDD mode.

15.3.1.2 Rake Receiver Implementation

Figure 15.6 shows the baseband processing, performed in the WCDMA receiver using a Rake receiver with four so-called Rake fingers. This WCDMA receiver has been implemented on an implemented platform, based on the Chameleon SoC template. This platform consists of four Montiums, two GPPs, and an FPGA part. Most of the computationally intensive baseband functionality has been implemented in coarse-grained reconfigurable hardware (Montiums), whereas fine-grained reconfigurable hardware (FPGA part) and GPPs are supposed to be used for additional control functionality.

Figure 15.7 shows the functional blocks of the WCDMA receiver that are intended to be implemented in the tiles of the heterogeneous reconfigurable SoC.

The first receive filter is commonly implemented as a pulse-shape filter. This pulse-shape filter is implemented on Montium tile 1. The output streams of the pulse-shape filter are

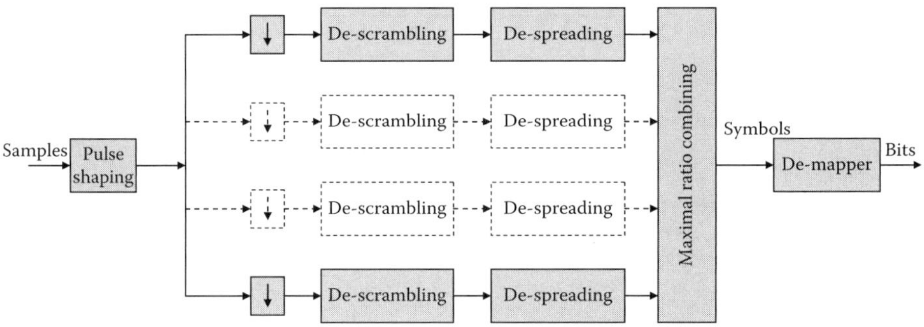

FIGURE 15.6 Baseband processing in the Rake-based WCDMA receiver.

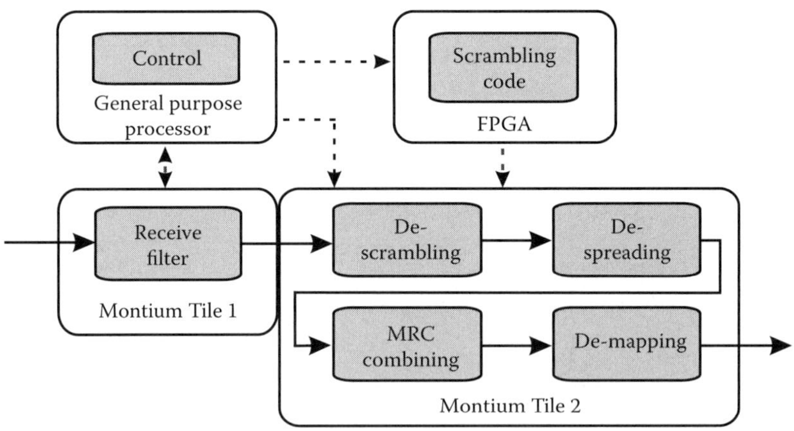

FIGURE 15.7 WCDMA receiver mapped to the heterogeneous reconfigurable SoC.

the input for the Rake receiver, which has been implemented on Montium tile 2. Generated scrambling codes can be mapped efficiently on an FPGA. The three blocks, Montium tile 1, Montium tile 2, and FPGA, are controlled by a general-purpose processor.

15.3.1.2.1 Block versus Streaming Communication

Two different mechanisms to exchange data with a Montium can be distinguished: block mode and streaming mode. Some applications require all the input data to be stored in local memories before execution can be started. This operation mode is called block mode. The most important feature of this communication mode is that, during the data transfers, the Montium is halted to make sure the execution is not started until it is sure that the data are valid. In streaming mode, the Montium processes data while simultaneously performing data transfers.

The timing properties of the UMTS communication system concern both the data processing and the control part of the receiver. It seems natural that data processing is performed according to the streaming communication principle, while control-oriented functions are partly done according to the block communication. The implemented WCDMA receiver has been realized according to the streaming communication principle because for block communication too many resources (memory for storing the scrambling code) are required during the baseband processing. According to Table 15.3 the scrambling code sequence consists of 38,400 samples. Hence, if the WCDMA receiver operates in block mode, 38,400 samples have to be stored in the local memory of the Montium. Even when data would be processed on a slot basis instead of a frame basis, still 2,560 data samples would have to be stored in local memory. Therefore, we may conclude that the block communication principle in the WCDMA receiver is not efficient, since blocks are too large.

15.3.1.2.2 Communication Requirements

The implemented WCDMA receiver thus operates according to the streaming communication principle. The implemented receiver can process four individual paths of the received signal (see also Figure 15.6). Consequently, the receiver requires four complex-number data streams for the four implemented fingers. All implemented fingers require the same scrambling code. The scrambling code can be generated using, e.g., an FPGA tile. The implemented receiver takes the complex-number scrambling code stream as an input. Before de-spreading starts, the appropriate spreading code is stored in the local memory of the Montium. The spreading code is stored in local memory because the code has a maximum length of 512 samples. Thus, a relatively small amount of data has to be stored in contrast to the scrambling code. Furthermore, the spreading code is assigned to a particular user in the UMTS communication system, and therefore, the spreading code will not change frequently. After de-spreading and before de-mapping, the received symbols of the individual signal paths are combined. During this combining phase, each symbol is scaled according to a complex-number coefficient before summation. These complex-number coefficients are provided by the channel estimator. The receiver outputs a bit stream with the received data. The characteristics of all the input and output streams are given in Table 15.4.

TABLE 15.4 Signal Stream Characteristics of the Implemented WCDMA Receiver

Signal Stream	Direction	Data Rate (Msps)
Data finger 1	Complex input	3.84
Data finger 2	Complex input	3.84
Data finger 3	Complex input	3.84
Data finger 4	Complex input	3.84
Scrambling code	Complex input	3.84
MRC coefficient finger 1	Complex input	3.84/SF
MRC coefficient finger 2	Complex input	3.84/SF
MRC coefficient finger 3	Complex input	3.84/SF
MRC coefficient finger 4	Complex input	3.84/SF
De-mapped bits	Real output	3.84/SF

15.3.1.2.3 Dynamic Reconfiguration

The complete Rake receiver is implemented in the Montium. This receiver includes the de-scrambling and de-spreading of four individual fingers, combining the results of the four fingers, and de-mapping (according to Figure 15.6). The configuration size of the complete Rake receiver in the Montium is only 858 bytes. Since two bytes per clock cycle can be loaded in the configuration memory, one Montium can be configured for Rake receiving in 429 clock cycles. With a configuration clock frequency of 100 MHz, a Rake receiver with four fingers can be configured in 4.29 μs.

In case the spreading code changes, and so the SF, only the new spreading code has to be loaded in the local memory of the Montium. Loading a particular spreading code into the local memory costs SF clock cycles; furthermore, one constant in the sequencer program (i.e., the loop counter) has to be changed.

The signal streams for the different fingers are buffered in local memories inside the Montium. When the delay of one of the paths changes, the buffering strategy of the local memories has to be changed (i.e., the AGU instructions are reconfigured). The buffering strategy of the memories is configured with 24 bytes. These 24 bytes can be reconfigured in twelve clock cycles. Consequently, the Rake receiver can update its complete path delay profile in 120 ns.

In total, for reconfiguring the number of fingers from, e.g., four to two, only 24 bytes have to be reconfigured in the configuration memory of the Montium. So, the Rake receiver can be reconfigured in 120 ns, which corresponds to twelve clock cycles. During this reconfiguration process from four to two fingers, the Montium sequencer program is changed as well as the ALU instructions.

15.3.1.2.4 Dynamic Power Consumption

Voltage and frequency scaling are important measures to control the dynamic power consumption of embedded systems. Because of the modular, regular structure of the Rake receiver, frequency scaling can easily be applied. The clock frequency of the Montium during Rake processing of four fingers is 20 MHz. Moreover, when the Rake

TABLE 15.5 Average Dynamic Power Consumption of the Montium Rake Receiver

	Clock Frequency Montium (MHz)	Dynamic Power Consumption Montium (mW)
2 fingers	10	5
4 fingers	20	10

receiver is reconfigured to process two fingers, the clock frequency of the Montium can be reduced to 10 MHz.

Table 15.5 summarizes the average power consumption figures for the implemented Rake receiver.

15.3.1.2.5 UMTS Performance Verification

The functional behavior of the implemented Rake receiver in the Montium has been verified by means of hardware/software co-simulations. Using a co-simulation environment, we can simulate the baseband functionality of the UMTS communication system in software (i.e., using Matlab and SaSUMTSSim [41]) and partly in hardware (i.e., using a VHDL simulator). SaSUMTSSim has been developed to evaluate the performance of the de-spreading and demodulation functions of the UMTS downlink [41].

The performance and functional behavior of the Montium-based Rake receiver have been evaluated under different propagation conditions. The propagation conditions for performance measurements in a multipath fading environment are defined in [1] for the UMTS communication system and are called cases. One of these cases, case 4, has been used to evaluate the Montium-based Rake receiver with four fingers.

Figure 15.8 shows the BER versus SNR performance for the Rake receiver with four fingers under case 4 conditions. The reception of one UMTS frame is simulated. The individual results of five simulation runs are given in Figure 15.8 for a reference Rake receiver (the SaSUMTSSim receiver) and for the Montium-based Rake receiver, which are labeled "Reference" and "textscMontium," respectively. For both receivers, the average of the five simulation runs is presented as well. The simulation results show that the performance of the Montium-based Rake receiver and the SaSUMTSSim Rake receiver are almost equal. Only for bad channel conditions (i.e., low E_c/N_0) does a small performance gap arise. This gap is caused by saturation in the ALUs. Scaling the input data of the Montium reduces the effect of saturation. The effect of proper input scaling is depicted in Figure 15.9 by the additional simulation results (labeled "Additional input scaling, Montium"). In Figure 15.9, only the averaged simulation results are shown, not the five individual results as in Figure 15.8. The additional simulation results are obtained by more rigorously scaling the input data of the Montium in bad channel conditions.

15.3.1.3 OFDM Receiver on Reconfigurable Hardware

The concept of orthogonal frequency division multiplexing (OFDM) has been known since 1966 [9]; however, due to implementational complexity, it has only been adopted since the 1990s. Currently OFDM is widely used as a modulation technique in wireless communication systems. OFDM is a special multicarrier modulation technique that

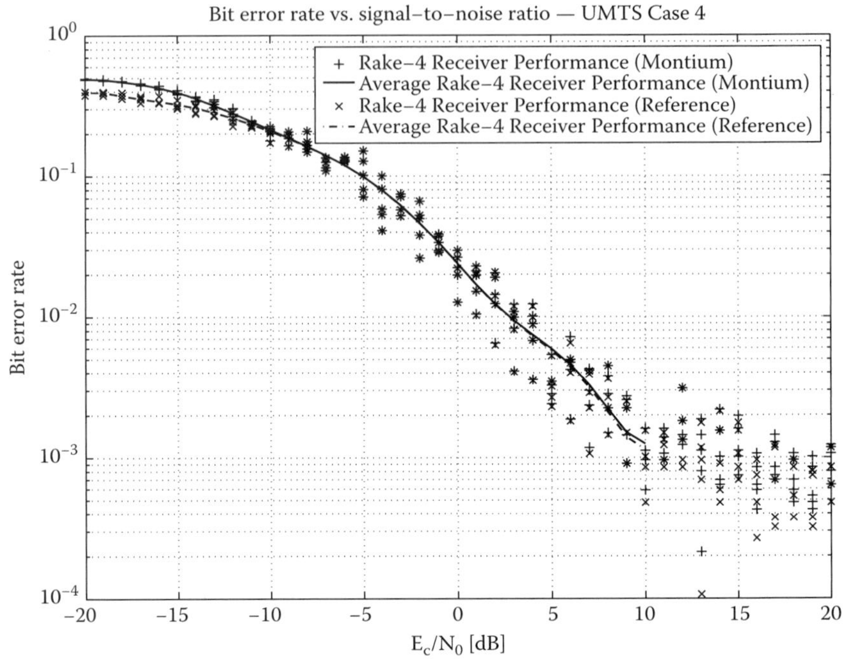

FIGURE 15.8 The BER before error correction of the Rake-4 receiver under case 4 propagation conditions with ideal channel estimation (+ and × indicate individual simulation points).

utilizes multiple subcarriers within a single channel. The modulation technique divides the high-data-rate information into several parallel bit streams, and each of these bit streams modulates a separate subcarrier.

OFDM-based communication systems are all designed according to a generic framework. Figure 15.10 shows the generic OFDM framework of an OFDM receiver. In this framework the characteristic properties are based on specific OFDM standards. This means that, for example, the number of subcarriers and the length of the guard interval may differ for different OFDM standards. The characteristics of an OFDM receiver for a single standard can even differ in the case where it has different modes defined. Characteristics of some OFDM-based standards, HiperLAN/2, DAB, and DRM are summarized in Table 15.6 [13–15].

15.3.1.4 OFDM Case Study: HiperLAN/2 Receiver Implementation

Parts of the baseband processing of a HiperLAN/2 receiver are implemented in the heterogeneous reconfigurable SoC of Figure 15.3. The coarse-grained Montium architecture [26, 28] is used as the target architecture for mapping the baseband DSP algorithms. The physical layer of the HiperLAN/2 receiver [13] is implemented on Montiums in combination with a GPP. Figure 15.11 shows the baseband processing blocks of the receiver that are implemented in the SoC. The solid arrows in Figure 15.11 indicate the data, which is processed in consecutive processing tiles. The input consists of data samples,

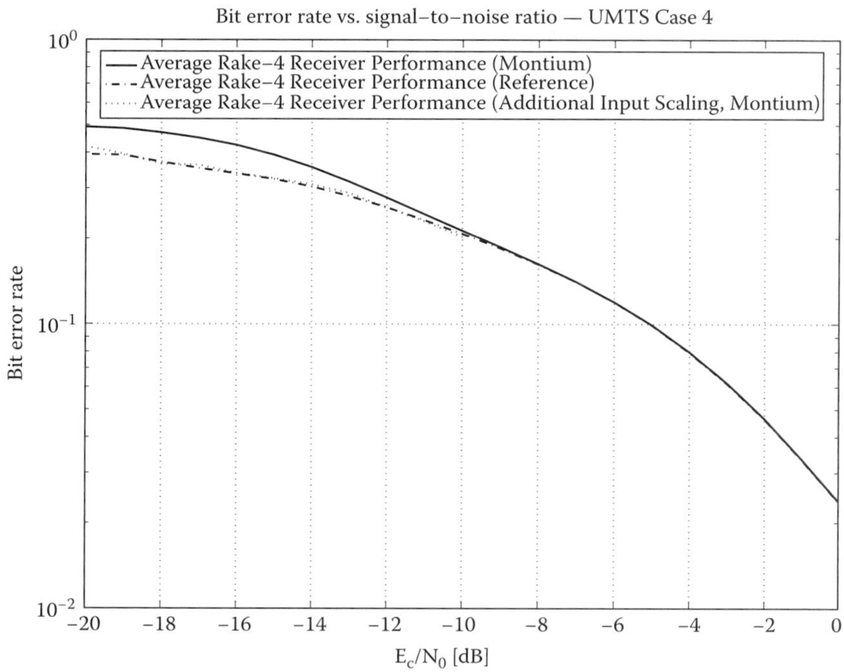

FIGURE 15.9 The influence of additional input scaling on the average BER before error correction under case 4 propagation conditions (additional results to Figure 15.8).

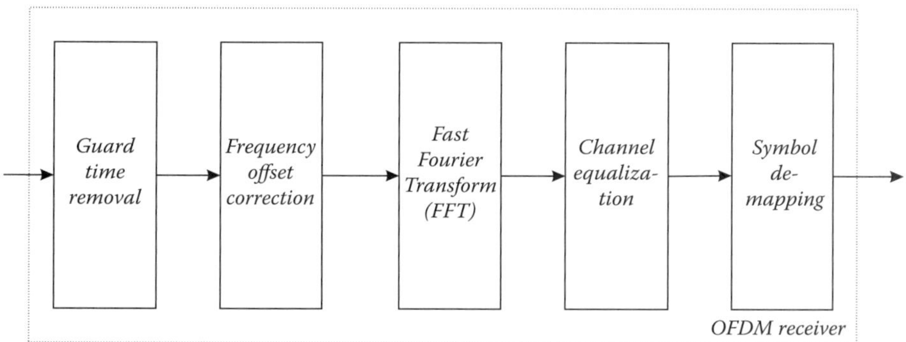

FIGURE 15.10 Generic OFDM receiver framework.

which form OFDM symbols. The output of the baseband processing part consists of encoded channel bits. The baseband processing functions are supported by a GPP, which is used for control purposes. The control streams between the processing tiles are shown as dashed arrows in Figure 15.11. The data streams between the processing tiles are mapped on channels of the NoC. The NoC provides the on-chip communication network in the SoC.

TABLE 15.6 Characteristics of Different OFDM Standards from [13–15]

		Hiper LAN/2	DAB				DRM			
			I	II	III	IV	A	B	C	D
Bandwidth	(MHz)	16.25	1.536	1.536	1.536	1.536	0.01	0.01	0.01	0.01
No. of carriers (FFT size), N		64	2,048	512	256	1 024	288	256	176	112
No. of modulated carriers, K		52	1,536	384	192	768	225	205	137	87
Symbol time, T_{OFDM}	[μs]	4	1,246	312	156	623	26,667	26.667	20,000	16,667
Guard time, T_g	[μs]	0.8	246	62	31	123	2,667	5,333	5.333	7.333
Useful time, T	[μs]	3.2	1,000	250	125	500	24,000	21,333	14 667	9,333
Subcarrier spacing, Δf	(kHz)	312.5	1.0	4.0	8.0	2.0	0.0417	0.0469	0.0682	0.1071

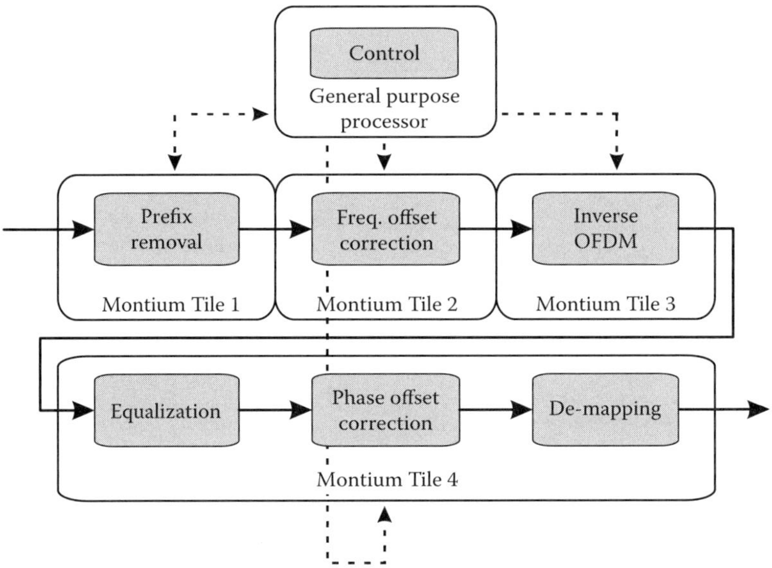

FIGURE 15.11 HiperLAN/2 receiver mapped on a tiled reconfigurable SoC.

Irregular tasks, which are outside the algorithm domain of the Montium, are performed in software on the GPP. The irregular processes in the HiperLAN/2 receiver are the channel estimation functions like frequency offset estimation and computation of equalization coefficients. Because of its relatively large coherence time, the channel estimates have to be updated only once per MAC frame, i.e., once per 2 ms. Table 15.7 shows the results of partioning the receiver's functionality over the Montium and the GPP.

The characteristics of the DSP kernels in the HiperLAN/2 receiver that are mapped on the tiled reconfigurable SoC (according to Figure 15.11) are summarized in Table 15.8.

TABLE 15.7 Reconfigurable Hardware/Software Partitioning of the HiperLAN/2 Functionality

	Implemented in	Multiplies per MAC Frame	Additions per MAC Frame
Determine frequency offset	Software	64	64
Determine equalizer coefficients	Software	0	0
Frequency offset correction	Montium	127,744	95,309
Inverse OFDM	Montium	383,232	574,848
Equalizer, phase offset, de-mapper	Montium	203,184	104,082

TABLE 15.8 Properties of the HiperLAN/2 Receiver Implementation

		Frequency Offset Correction	Inverse OFDM	Equalizer, Phase Offset, De-Mapper
Execution time	[Cycles]	67	204	110
Block mode				
Communication time				
(input + output)	[Cycles]	128	116	<100
Minimum Montium + NoC clock streaming communication	[MHz]	17	51	28
Minimum Montium + NoC clock block communication	[MHz]	49	80	53
Minimum Montium clock with block communication (NoC @ 100 MHz)	[MHz]	25	72	37
Configuration size	[Bytes]	274	946	576
Configuration time	[Cycles]	137	473	288

The table shows the impact of the communication overhead in the NoC, which can be performed by streaming or block mode communication. Furthermore, the configuration overhead of the DSP kernels implemented in the coarse-grained Montium is given.

15.3.1.4.1 Configuration

The configuration sizes of the Montium are small for the different functions (Table 15.8). Montium tile 3 (see Figure 15.11), on which the inverse OFDM is performed, requires the largest configuration. The configuration of tile 3 contains less than 1 kB of configuration data. The configuration data are written into the configuration memory of the Montium in 473 clock cycles, as every clock cycle 16 bits are written. So, tile 3 can be configured in 4.73 μs, which is dominating the time to switch from receive to transmit mode. Notice that the maximum radio turnaround time* of the HiperLAN/2 communication system

* The radio turnaround time indicates the time to switch from transmit to receive mode in a communication transceiver, and vice versa.

is 6 μs [12], so the Montium HiperLAN/2 implementation can be considered a real-time dynamically reconfigurable HiperLAN/2 system.

15.3.1.4.2 Frequency Scaling

Since the DSP kernels in the HiperLAN/2 receiver are mapped on different Montiums in the SoC architecture, frequency scaling can easily be applied. Frequency scaling in combination with voltage scaling is an important means to control the power consumption of embedded systems. The idea of dynamic voltage scaling is to keep the supply voltage as low as possible. The maximum operating frequency is tightly coupled to the supply voltage level. This means that by downscaling the clock frequency of hardware, the supply voltage can be lowered as well, resulting in a quadratic decrease of the power consumption.

All DSP operations in the physical layer of the HiperLAN/2 communication system are performed on OFDM symbols. Every OFDM symbol has a time period of 4 μs. So, it should be ensured that every 4 μs a new OFDM symbol can be processed by the receiver.

Typically, the clock frequency of the NoC is fixed and the clock frequency of the tiles can be varied. In case of block mode communication and under the assumption that the clock frequency of the NoC is fixed at 100 MHz, the clock frequency of the Montium tile for frequency offset correction has to be at least 25 MHz.

In case of block mode communication, the clock frequency of the NoC is equal to the clock frequency of the reconfigurable processor tile and the clock frequency can be adapted, and the clock frequency of the entire SoC (i.e., processor tiles and NoC) can be scaled to its minimum value. The minimum clock frequency of the system would in this case be reduced to 49 MHz for frequency offset correction. However, this situation is fairly unlikely to happen because managing the adaptable clock frequency of the common NoC is rather complex.

Introducing the streaming communication mode variant of the DSP kernels in the HiperLAN/2 receiver provides a situation where the data processing and input and output communication are performed in parallel. In this case the data words are processed immediately as they become available from the on-chip network, while previously processed data are sent to the next stage of processing. Consequently, the communication time is not a bottleneck. For example, the data processing for frequency offset correction is performed during sixty-seven clock cycles, and because input and output are done simultaneously, communication is reduced to sixty-four clock cycles. Processing needs to complete in 4 μs. Hence, the minimum clock frequency of the Montium tile is 17 MHz for frequency correction. In case of streaming communication, the clock frequency of the NoC should be at least 17 MHz. Typically, the clock speed of the NoC is fixed to the maximum operating frequency of the processing tiles.

15.3.1.5 Digital Broadcasting Systems

Many digital replacements of current analog broadcasting systems are available nowadays. It is expected that within a few years all analog broadcasting services will be switched off and replaced by digital standards like Digital Audio Broadcasting, Digital Video Broadcasting, and Digital Radio Mondiale [14–16]. For example, in the United States, television broadcasters have to switch from analog to digital before 2009. In the Netherlands, almost 98% coverage of DVB-T has been achieved and analog television broadcast has been completely turned off. Additionally, 88% of the United Kingdom is already covered with DAB.

One common characteristic of the digital radio systems is that they all employ orthogonal frequency division multiplexing (OFDM). Although digital radio standards are all OFDM based, many differences exist. Characteristic OFDM properties vary between the different standards, but the characteristics of the OFDM receiver can even change within one standard, because different modes are defined (see Table 15.6). Moreover, for all digital radio standards different source coding techniques are applied, i.e., MPEG-1 Layer 2 Audio Coding (MP2) or Advanced Audio Coding (AAC).

The identified DSP kernels in digital broadcasting receivers are based on the generic OFDM receiver framework (Figure 15.10). Mapping digital broadcasting receivers on the heterogeneous reconfigurable SoC can be done in a manner similar to the HiperLAN/2 implementation approach. The DSP kernels in, e.g., the DAB, DRM, and DVB receiver are similar to those in the HiperLAN/2 receiver and can be mapped to the SoC in the same manner as in Figure 15.11.

15.3.1.5.1 Digital Audio Broadcasting

The Digital Audio Broadcasting standard has been in existence since the early 1980s. The standard was initiated to replace the analog FM radio services and has already been adapted by many countries all over the world. DAB is capable of delivering data and audio services to end users. Data services are performed via packet transfers.

Figure 15.12 depicts the basic DAB receiver structure. After frequency and gain correction (AFC/AGC) of the DAB signal, the received signal is OFDM demodulated by means of a fast Fourier transform (FFT) (referred to as inverse OFDM) and differential quadrature phase shift keying (QPSK) demodulation. Then, the FEC decoding is applied by means of deinterleaving and Viterbi decoding. In the last stage of the receiver, the user data are demultiplexed and source decoded using the MP2 codec to audio.

15.3.1.5.2 Digital Radio Mondiale

Digital Radio Mondiale (DRM), approved as an ETSI standard in 2001, is proposed to replace the analog AM radio service for frequencies below 30 MHz. The DRM system delivers near-FM quality sound over short-wave, medium-wave, and long-wave radio channels.

Figure 15.13 shows the basic structure of the DRM receiver. In the first stage of the receiver, the FFT is applied to the received OFDM signal. All separate OFDM carriers are demodulated using the appropriate hierarchical constellation (BPSK, QPSK,

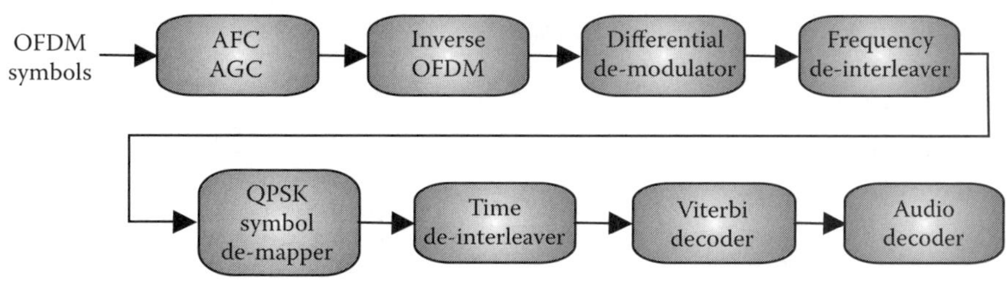

FIGURE 15.12 DAB receiver structure.

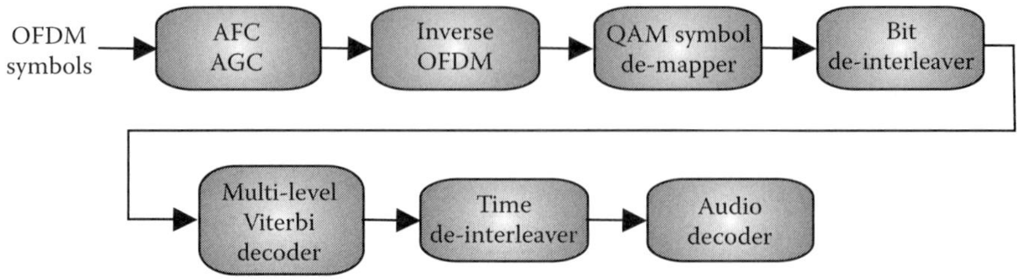

FIGURE 15.13 DRM receiver structure.

16-QAM [quadrature amplitude modulation], or 64-QAM). The multilevel coded data are decoded using an iterative Viterbi decoder approach. The last stage of the receiver comprises demultiplexing of the user data and decoding using the AAC codec to audio.

15.3.1.5.3 Digital Video Broadcasting

Digital Video Broadcasting–Handheld (DVB-H) is a new digital broadcast standard for the transmission of audio and video content to handheld terminals. DVB-H has been based on the standard for digital terrestrial television, Digital Video Broadcasting–Terrestrial (DVB-T). The DVB-H standard is similar to DVB-T, but it considers properties specific for mobile terminals that are portable, small, and battery powered.

In contrast to the digital audio counterparts, the DVB standard defines more robust forward error correction (FEC). DVB has been implemented with cascaded FEC; the convolutional inner coder is cascaded with an outer coder. The FEC outer coder is implemented using Reed-Solomon coding. Reed-Solomon proved to be a very powerful algorithm to solve (burst) errors. The inner decoder step is usually implemented by the Viterbi decoder. The outer decoding step, implemented by the Reed-Solomon decoder, should result in a quasi-error-free output, containing less than one uncorrected erroneous event per hour. The user data have to be decoded by the MPEG-2 codec to video (see Figure 15.14).

In future mobile platforms, one or more of these standards have to be supported. Therefore, reconfigurable multicore SoC architectures are needed to support all these and future OFDM standards.

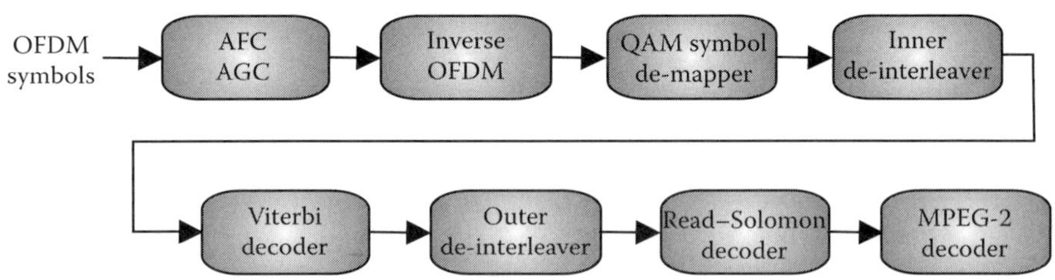

FIGURE 15.14 DVB receiver structure.

15.3.2 Cognitive Radio

In November 2002, the Federal Communications Commission (FCC) in the United States released a report [18] aimed at improving the management of spectrum resources in the United States. The report concluded that the current spectrum scarcity problem is largely due to the strict regulation on spectrum access. Spectrum measurements conducted by the FCC indicated that only small portions of the spectrum are heavily used, while other frequency bands are either partially used or unoccupied most of the time. So, spectrum utilization can be improved by making it possible for an unlicensed user (secondary user) to access the spectrum that is not occupied by the licensed user (primary user). The unlicensed user has the awareness of the spectrum and adapts its transmission accordingly on a noninterference basis. This spectrum access and awareness scheme is referred to as cognitive radio by the FCC.

The idea of cognitive radio was first presented by Joseph Mitola III in his paper [40], where he proposed that cognitive radio can enhance the personal wireless service by a radio knowledge representation language (RKRL). This language represents knowledge of radio at all aspects, from transmission to application scenarios, in such a way that automated reasoning about the needs of the user is supported. Cognitive radio thus is able to autonomously observe and learn the radio environment, generate plans, and even correct mistakes. A comprehensive conceptual architecture of cognitive radio was later presented in [31], where cognitive radio was thought to be a final point of the software-defined radio platform evolution: a fully reconfigurable radio that changes its communication functions depending on network and user demands. Recently, cognitive radio has become a very hot topic due to its impact on future spectrum policy, which could fundamentally change the current status of radio communication. At the Berkeley Wireless Research Center (BWRC), a dedicated Cognitive Radio Research (CRR) project is in progress. Their motivation is to improve the spectrum utilization by opportunistic use of the spectrum, which is the same as the FCC's initiative. Spectrum pooling is investigated in [50]. The basic idea is that a secondary user can dynamically access the licensed band by switching on and off OFDM subcarriers to avoid interference to the licensed user (primary user). The cognitive radio project at Virginia Tech does not specifically aim to improve spectrum utilization. This project is based on the observation that cognitive radio distinguishes itself by awareness and learning. In [10], a genetic algorithm-based cognitive engine is proposed to learn its environment and respond with an optimal adaptation. The European Union 6th Framework End-to-End Reconfigurability (E2R) project studies reconfigurability, software-defined radio, and cognitive radio. The key objective of the E2R project is to devise, develop, and test the architectural design of reconfigurable devices and supporting system functions for users, application and service providers, operators, and regulators in the context of heterogeneous systems. Although the project does not specifically address cognitive radio, dynamic spectrum allocation and evolution from software-defined radio to cognitive radio have been envisioned. In parallel with the ongoing research projects around the world, international standardization organizations also have proposals to improve the spectrum utilization. An example is IEEE 802.22, which is a new standard for a cognitive point-to-point (P2P) air interface for spectrum sharing with television bands.

The AAF project [53] focuses on spectrum awareness and access. The objective is to demonstrate a cognitive radio system to improve current emergency networks, which work in limited frequency bands. Each radio node works in an ad-hoc-based network. The node adopts the AAF protocol stack, which is consistent with the five-layer protocol reference model (physical layer, data link layer, network layer, transport layer, and application layer). In the physical layer, free spectrum is discovered. Whenever free spectrum is found, the AAF system will create an infrastructure using this spectrum. In cognitive radio, the first step is to focus on identification of free resources in the frequency domain by spectrum sensing. An OFDM-based system can approach the Shannon capacity in a segmented spectrum; the capacity to nullify individual carriers poses interesting opportunities for cognition, as was also observed in [50]. Cognitive radio has to operate in multiple bands and under different channel conditions and supports various multimedia services. A heterogeneous reconfigurable platform supports the reconfigurability of the physical layer of cognitive radio.

15.3.2.1 Spectrum Sensing

In order to identify the licensed user and locate unused spectrum, the system has to sense the spectrum. Spectrum sensing is not a new topic since a lot of research has been done in the area of signal detection and estimation. Three signal processing techniques are commonly used for signal detection: matched filtering, energy detection, and cyclostationary feature detection.

Matched filtering is an optimal way for signal detection in communication systems. It correlates the received signal with a known signal pattern that maximizes the received signal-to-noise ratio. However, cognitive radio may not have prior knowledge of the licensed user signal, and strict timing and frequency synchronizations are required for coherent detection. Therefore, matched filtering is not an option for spectrum sensing.

In situations where not much knowledge concerning the signal is available, energy detection [46] is often used to determine the presence of the signal. It measures the signal power within a certain time interval and frequency band. The detection decision is based on a noise threshold. However, limitations for the energy detection are: (1) the decision threshold is subject to changing signal-to-noise ratios; (2) it cannot distinguish interference from signals; and (3) it is not effective for spectrum spreading signals whose power has been spread. Therefore, energy detection is not always adequate.

Cyclostationary feature detection [19] is used to extract signal features in the background of noise. Since the modulated signal can be modeled as a cyclostationary process in which a signal varies in time with certain periodicities, it contains spectral redundancy information that can be exploited by analysis of the cyclic spectrum. The advantages of cyclostationary feature detection over energy detection have been recognized in [20]. Within the AAF project, the focus is on energy detection and cyclostationary feature detection. A system-level architecture for spectrum sensing is presented in Figure 15.15. The detailed theory can be found in [46] and [19]. At system level, energy detection can be implemented as an FFT algorithm, which has a computational complexity of $O(\frac{N}{2}\log_2 N)$, where N is the size of FFT. Cyclostationary feature detection is a combination of an FFT and spectral correlation, which has a computational complexity of $O(N^2+\frac{N}{2}\log_2 N)$. When a large N is used, the processing of cyclostationary feature

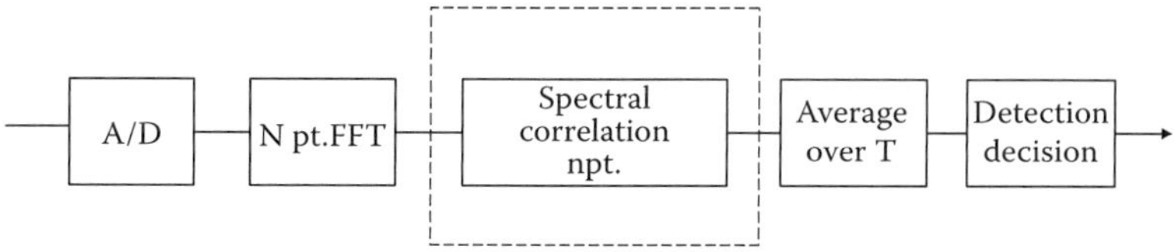

FIGURE 15.15 System-level architecture of spectrum sensing with the cyclostationary feature detection option.

detection can be prohibitive in terms of performance and computational power. It means that building a dedicated cyclostationary feature detector is simply too expensive. Therefore, cyclostationary feature detection is a complementary option when energy detection fails. So, energy detection can be switched to cyclostationary feature detection by turning on the spectral correlation functional module (see the dashed box in Figure 15.15). This option can be supported by a reconfigurable platform where the processing elements for spectral correlation can be switched on and off.

Based on the spectrum sensing result, an individual node needs to make a local decision on whether or not the band under consideration is empty. However, this local decision may not be reliable due to fading and shadowing. Therefore, collaborative sensing is proposed [39] to improve the quality of the licensed user detection.

15.3.2.2 OFDM-Based Baseband

The benefit of OFDM is that the high data rate of the whole system is transformed into relatively low-data-rate streams on each subcarrier that is more robust to intersymbol interference (ISI) caused by multipath delay spread. In the hardware design, an OFDM radio is also easy to integrate with spectrum sensing because both use FFT cores. The system robustness and hardware resource sharing make OFDM a good choice for cognitive radio baseband systems. More importantly, an OFDM system can optimally approach the Shannon capacity in the segmented spectrum by adaptive resource allocation on each subcarrier, which includes adaptive bit loading and adaptive power loading [53]. In an OFDM-based cognitive radio system, information bits are loaded as different modulation types onto each available subcarrier depending on the subcarrier's signal-to-noise ratio, while the subcarriers currently not available to cognitive radio are switched off. Two optimization methods for the adaptive resource allocation can be used for cognitive radio: using the power constraint or using the data-rate constraint.

We could maximize the data rate of the system under a certain power constraint. It is formulated as follows:

$$Max\ R = \sum_{k=1}^{K} \frac{F_k}{K} \log_2\left(1 + \frac{h_k^2 p_k}{N_0 \frac{B}{K}}\right) \tag{15.1}$$

$$\text{Subject to: } \sum_{k=1}^{K} p_k \le P_{total}$$

$$F_k \in \{0,1\} \text{ for all } k$$

$$p_k = 0 \text{ for all } k \text{ which satisfy } F_k = 0,$$

where R is the data rate, K is the number of the subcarriers, N_0 is the noise power density, B is the band of interest for cognitive radio, h_k is the subcarrier gain, and p_k is the power allocated to the corresponding subcarrier. F_k is the factor indicating the availability of subcarrier k to cognitive radio, where $F_k = 1$ means the kth carrier can be used by cognitive radio.

The system power minimization can also be applied under the constraint of a constant data rate. We formulate it as follows:

$$Min \sum_{k=1}^{K} p_k = P_{total}$$

$$\text{Subject to: } R = \sum_{k=1}^{K} \frac{F_k}{K} \log_2\left(1 + \frac{h_k^2 p_k}{N_0 \frac{B}{K}}\right) \tag{15.2}$$

$$F_k \in \{0,1\} \text{ for all } k$$

$$p_k = 0 \text{ for all } k \text{ which satisfy } F_k = 0.$$

A functional diagram of the system is presented in Figure 15.16. A bit allocation vector indicates how many bits are loaded on each subcarrier. The number of bits corresponds to the different modulation types used for each subcarrier. The bit allocation vector is determined by the spectrum occupancy information from spectrum sensing and the SNR of subchannels. The bit allocation vector is disseminated via a signaling channel so that both transmitter and receiver have the same information. The bit allocation vector does not change frequently, for instance, only after several frames. The basic idea is to load more bits on good subcarriers and zeros on carriers that cause interference to the licensed user or lead to poor transmissions.

15.3.2.3 Reconfigurable Physical Layer

As already foreseen by Mitola [31], a cognitive radio is the final point of software-defined radio platform evolution: a fully reconfigurable radio that changes its communication functions depending on network and user demands. His definition on reconfigurability is very broad, but we focus on the physical layer reconfigurability. We will discuss some possibilities for reconfiguration.

The proposed architecture for spectrum sensing is shown in Figure 15.15, where cyclostationary feature detection is used as a complementary option. This option has to be supported by a reconfigurable platform that can efficiently perform the spectrum correlation. In [35], a two-step methodology to analyze the mapping of cyclostationary feature detection (CFD) onto a Montium-based multicore processing platform is proposed. In the first step, the tasks to be executed by each core are determined in a

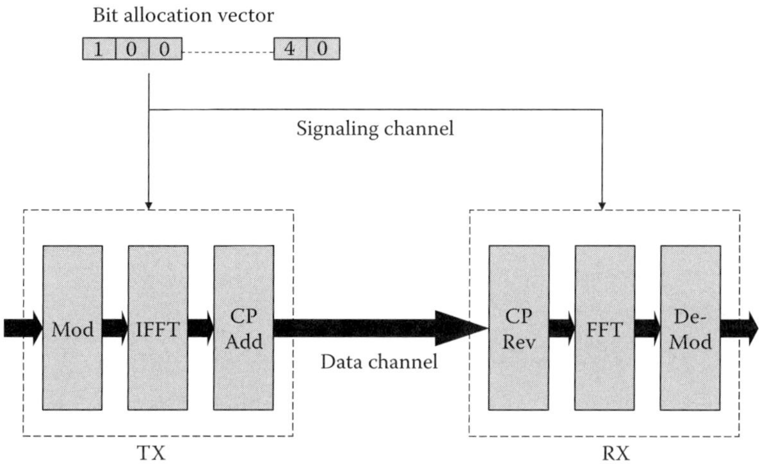

FIGURE 15.16 OFDM for cognitive radio.

structured way using techniques known from the design of array processors. In the second step, the implementation of the tasks onto the Montium is analyzed. It is shown that calculating a 127 × 127 discrete spectral correlation function requires approximately 140 μs on a system-on-chip (SoC) with 4 Montium cores. For energy detection, spectrum sensing can be done in different frequency resolutions with which the trade-offs between performance and computational power can be made. This can be achieved by a size-reconfigurable FFT on the Montium.

OFDM baseband processing for cognitive radio is a parameterizable OFDM processing chain that is configured by a configuration manager (see Figure 15.17). By applying different parameter settings in each task, the OFDM system is adaptive to various channel conditions and provides various data rates.

Relevant parameters are shown in Table 15.9, but are not limited by this table and can be extended to add more flexibility to the system. The number of OFDM symbols per frame (N_{sym}) is limited by the channel coherence time, the time during which the channel characteristics are constant. The number of guard samples (N_g) is chosen to deal with different channel delay spreads. Generally not all data are used to carry user information. A part of the data (e.g., pilots) is used to guarantee reliable transmissions. Different pilots are used for different purposes, such as channel estimation or phase offset estimation. They can be placed in the preamble section prior to each frame or embedded in the

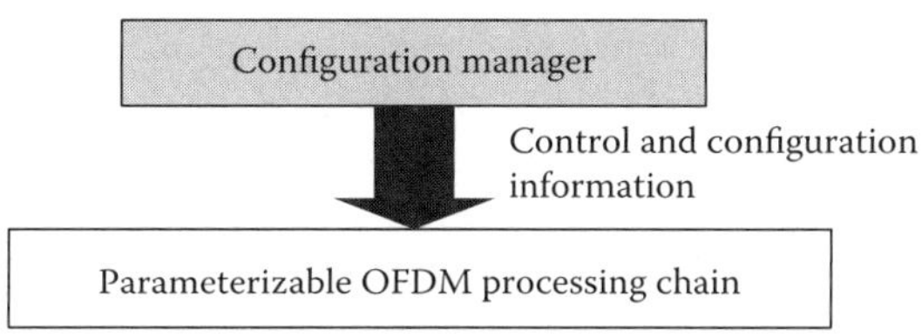

FIGURE 15.17 OFDM for cognitive radio.

TABLE 15.9 The Parameter Set for the Parameterizable OFDM

B	Bandwidth of OFDM system
N_{sym}	Number of OFDM symbols per frame
$N_{preamble}$	Preamble length per frame
N	Number of OFDM samples per symbol
N_g	Number of guard samples per symbol
N_{data}	Number of useful data per symbol
mod	Modulation modes
tab_{pilot}	Table for pilot information
tab_{format}	Table for format information

OFDM symbol. Modulation modes (mod) indicate the modulation type for the OFDM samples that carry useful information. The modulation mode in one OFDM symbol can differ on a subcarrier basis. In the cognitive radio case, the modulation mode can be set to zero to nullify carriers.

15.4 Mapping Applications to Multicore Architectures

In the most general terms, spatial mapping is the allocation of spatial resources to applications. As such, spatial mapping is to space what scheduling is to time. In the context of a multiprocessor system-on-chip (MP-SoC), the spatial resources to allocate are tiles and interconnections between tiles. These tiles may be processors or memory (controllers). Although the allocation of columns of fabric inside an FPGA could also be described as a spatial mapping problem, the system-on-chip context is considerably more coarse grained.

Baseband processing applications in wireless communications are streaming applications. Such streaming applications can be decomposed into multiple communicating computational kernels. Most streaming applications have intuitive decompositions into chains of such kernels, which is illustrated by the fact that most standards contain block diagrams of the components that make up the entire application [e.g., 17]. This decomposition can be as general as a Kahn process network (KPN) [32], or if more detailed information is available, some specialized form of a data flow graph (DFG).

A commonly used specialized DFG is the synchronous data flow graph (SDFG) [36]. The nodes of an SDFG are generally referred to as actors. The edges between actors are annotated with tokens. Tokens are an abstract notion of data that is communicated between two actors.

Actors are annotated for every incoming and outgoing edge. On incoming edges, the annotation is the number of tokens required on that specific input to fire the actor once, i.e., to execute the corresponding computational kernel. On outgoing edges, the annotation is the number of tokens the actor will produce after firing. Actors in an SDFG are self-timed, i.e., there is no global arbitration of what actor fires when, but rather a simple local firing rule: an actor may fire when all incoming edges have at least as many tokens on them as annotated.

Decomposing an application into actors and their (data) interdependence allows for more fine-grained resource management. Now, actors are the atomic units for which tiles need to be allocated. Whenever two different tiles are allocated to two interdependent actors, the interactor communication requires services from the interconnect between the tiles of the MP-SoC. In other words, communication needs to be routed through the network-on-chip (NoC).

If a tile supports multitasking, it can be allocated to multiple actors. In order to guarantee nonfunctional properties of the application (e.g., throughput and latency), all multitasking tiles must have deterministic schedulers. More formally, a tile's scheduler must be a latency-rate server [44]. This guarantees an upper bound on the latency between the first request for service and the first service response and a lower bound on the throughput (rate) from that point onward, provided it is kept busy. This restriction on multitasking tiles makes it possible to make allocations to actors independent of each other (composability), i.e., when a tile is already allocated to an actor, allocating it to another actor does not influence the nonfunctional properties of the first (if the sum of the actors' rate requirements do not exceed the rate of the scheduler).

To be able to perform the mapping of an application to tiles, a spatial mapping algorithm needs:

- A model of the application
- A model of the (MP-SoC) platform
- The constraints on the nonfunctional properties of the application
- Data on the resource requirements of process implementations (e.g., time, memory, and energy)

The constraints on the nonfunctional properties of the application can only be checked after it has been mapped. Thus, when a spatial mapping has been determined and latencies and throughputs of processes running on tiles are known, the constraints can be checked. A spatial mapping that lets the application meet its constraints on its nonfunctional properties is considered feasible.

15.4.1 Design Time Mapping

The spatial mapping of applications can be performed at design time, if all applications are known and the availability of the MP-SoC is guaranteed (what happens when this is not the case is described in section 15.4.2). This is a (quite often manual) optimization problem, in which the best allocation of tiles to actors should be found. A valid solution must work for all possible user scenarios. When one application is already running on an MP-SoC and a user starts another, then—assuming the combination of applications is allowed—the new application should be runnable with the remaining resources.

For sufficiently small applications and MP-SoCs, exhaustive search techniques can be used to find optimal allocations for all possible scenarios. The mapping of slightly larger applications is generally tackled by heuristic search methods that try to direct the search for a solution, possibly missing the optimum, but yielding an acceptable result.

Very large applications or failed heuristic searches are often tackled with manual help from a designer.

There is no general method for spatial mapping to all kinds of MP-SoC, because what constitutes a good mapping depends heavily on the available resources of the MP-SoC and the resource demands of the application. For example, when the number of processors per NoC node (multiple processors can be connected to a single router) is relatively high, communication is likely to become a problem in the mapping. In such a case, searches generally start off with determining the minimum bandwidth clustering of the process graph into clusters that fit on a tile.

When using SDFGs, a great advantage of performing a spatial mapping at design time is that a so-called static-order schedule can be calculated for actors mapped onto the same tile. This means that the order in which the actors fire is known and the actors are ordered such that they fire in sequence. If a static-order schedule is found, the schedule can be implemented by grouping the actors together and compiling them into a single piece of binary code for the tile, thereby eliminating the need for scheduling at runtime and the overhead it would induce.

It is worth noting that tiles of architectures that do not have support for multitasking can still be allocated to multiple actors, if such a static-order schedule can be found. This follows directly from the fact that actors in a static-order schedule can be compiled into a single binary executable.

15.4.2 Runtime Mapping

Performing the spatial mapping at runtime is necessary, whenever the application set is not known completely at design time. This happens for a wide variety of reasons; for example, when the platform allows the user to use software from any vendor, developed for that platform. When different software vendors produce software for the same platform independently, no one knows the complete application set. Also, when the resource availability of the MP-SoC is not known beforehand, the spatial mapping must be found at runtime. This can happen when tiles can be broken (by wear or by faults in production), or when applications are developed, not for a specific MP-SoC, but for a class of MP-SoCs that have the same types of tiles and the same type of interconnect, but in a different configuration. Runtime spatial mapping in this sense is inherently ad hoc, i.e., only when an application is started will the system know what resources are available.

When mapping to a heterogeneous MP-SoC, having multiple implementations of computational kernels for different types of tiles increases the options for the spatial map. Performance figures—(expected) worst-case execution time, energy consumption, etc.—of these implementations can be determined at design time. However, some figures of the applications simply cannot be determined at design time. Interprocess communication parameters (e.g., estimated latency), for example, need to be determined at runtime, as these are dependent on the specific mapping of the communicating processes. Likewise, it is only known at runtime on which tile a process will be executed and which processes are already running on this tile, so the actual execution time (often referred to as response time in the synchronous data flow world) of a process is only known at runtime.

15.5 Conclusion

Future wireless communication devices will contain integrated circuits where multiple processing cores are implemented on one chip. These multiprocessor systems-on-chip (MP-SoC) will be heterogeneous; they will consist of multiple types of processing cores. Bit-level functions will be executed by FPGA-type processors, word-level functions by coarse-grained reconfigurable processors, and control-oriented functions by GPPs. In wireless communication systems, processing is dominated by word-level signal processing that is efficiently implemented on coarse-grained reconfigurable processors. The Montium coarse-grained reconfigurable processor is an example of such a processor, enabling energy-efficient implementation of wireless communication receivers. This is illustrated for three classes of receiver baseband processing: UMTS Rake receivers, OFDM baseband processing (HiperLAN/2), and digital broadcasting receivers (DAB, DRM, and DVB). Furthermore, an MP-SoC, containing multiple Montiums, is a suitable platform for the implementation of cognitive radio (CR). The flexibility offered by coarse-grained reconfigurable processors supports the efficient implementation of new communication paradigms.

Because of the large number of transistors envisaged on future integrated circuits, a processing platform will not be fixed during its lifetime due to faults in production or wear. Furthermore, the applications are not fixed. Quality-of-service requirements and environmental conditions change, and cognitive radio has changing requirements by definition. Consequently, applications need to be mapped onto the processing platform at runtime.

Realizing that wireless communication applications are streaming applications explains the use of data flow graphs to model these applications. Data flow graphs naturally fit platforms with multiple processing cores where nodes of the graph are mapped onto the processing cores and edges are mapped onto the interconnecting network-on-chip. The use of data flow graphs supports the runtime mapping.

References

[1] 3rd Generation Partnership Project. 2004. Base station (BS) radio transmission and reception (FDD). 3GPP TS 25.104 v6.4.0.

[2] 3rd Generation Partnership Project. 2004. Physical layer—General description. 3GPP TS 25.201 v6.0.

[3] A. Abnous, H. Zhang, M. Wan, G. Varghese, V. Prabhu, and J. Rabaey. 2002. The Pleiades architecture. In *The application of programmable DSPs in mobile communications*, ed. A. Gatherer and A. Auslander, 327–60. New York: Wiley.

[4] V. Baumgarte, G. Ehlers, F. May, A. Nuckel, M. Vorbach, and M. Weinhardt. 2003. Pact xpp—A self-reconfigurable data processing architecture. *J. Supercomputing* 260:167–84.

[5] L. Bissi, P. Placidi, G. Baruffa, and A. Scorzoni. 2006. A multi-standard reconfigurable Viterbi decoder using embedded FPGA blocks. In *Proceedings of the 9th Euromicro Conference on Digital Systems Design: Architectures Methods and Tools* (DSD2006), pp. 146–54.

[6] B. Mei, S. Vernalde, D. Verkest, H. D. Man, and R. Lauwereins. 2003. Adres: An architecture with tightly coupled VLIW processor and coarse-grained reconfigurable matrix. In *Field-Programmable Logic and Applications*, pp. 61–70.

[7] J. Chandran, R. Kaluri, J. Singh, V. Owall, and R. Veljanovski. 2004. Xilinx Virtex ii pro implementation of a reconfigurable UMTS digital channel filter. In *Proceedings of the Second IEEE International Workshop on Electronic Design, Test and Applications*, pp. 77–82.

[8] C. Chang, J. Wawrzynek, and R. W. Brodersen. 2005. Bee2: A high-end reconfigurable computing system. *IEEE Des. Test. Comput.* 220:114–25.

[9] R. W. Chang. 1966. Synthesis of band-limited orthogonal signals for multi-channel data transmission. *Bell Syst. Tech. J.* 46:1775–96.

[10] C. J. Rieser. 2004. *Biologically inspired cognitive radio engine model utilizing distributed genetic algorithms for secure and robust wireless communications and networking*. PhD thesis, Virginia Tech.

[11] K. Compton and S. Hauck. 2002. Reconfigurable computing: A survey of systems and software. *ACM Comput. Surv.* 340:171–210.

[12] European Telecommunication Standards Institute (ETSI). 2000. Broadband radio access networks (BRAN); HiperLAN type 2; Data link control (DLC) layer. Part 1. Basic data transport functions. ETSI TS 101 761-1 v1.1.1.

[13] European Telecommunication Standards Institute (ETSI). 2001. Broadband radio access networks (BRAN); HiperLAN type 2; Physical (PHY) layer. ETSI TS 101 475 v1.2.2.

[14] European Telecommunication Standards Institute (ETSI). 2001. Digital Radio Mondiale (DRM); System specification. ETSI TS 101 980 v1.1.1.

[15] European Telecommunication Standards Institute (ETSI). 2001. Radio broadcasting systems; Digital Audio Broadcasting (DAB) to mobile, portable and fixed receivers. ETSI EN 300 401 v1.3.3.

[16] European Telecommunication Standards Institute (ETSI). 2004. Digital Video Broadcasting (DVB); Framing structure, channel coding and modulation for digital terrestrial television. ETSI EN 300 744 v1.5.1.

[17] European Telecommunication Standards Institute (ETSI). 2003. Digital Radio Mondial (DRM); System specification. ETSI ES 201 980 v1.2.2.

[18] Federal Communication Commision. 2002. Spectrum policy task force. Technical report 02-135.

[19] W. Gardner. 1988. Signal interception: A unifying theorectical framework for feature detection. *IEEE Trans. Commun.* 36:897–906.

[20] W. Gardner and C. M. Spooner. 1992. Signal interception: Performance advantages of cyclic-feature detectors. *IEEE Trans. Commun.* 40:149-159.

[21] S. C. Goldstein, H. Schmit, M. Budiu, S. Cadambi, M. Moe, and R. R. Taylor. 2000. Piperench: A reconfigurable architecture and compiler. *IEEE Comput.* 33:70–77.

[22] K. Goossens, J. van Meerbergen, A. Peeters, and P. Wielage. 2002. Networks on silicon: Combining best-effort and guaranteed services. In *Proceedings of the Design, Automation and Test in Europe Conference and Exhibition (DATE)*, pp. 423–25.

[23] K. Goossens, J. Dielissen, and A. Rădulescu. 2005. The Æthereal network on chip: Concepts, architectures, and implementations. *IEEE Des. Test Comput.* 220:414–21.

[24] S. Hauck, T. W. Fry, M. M. Hosler, and J. P. Kao. 2004. The Chimaera reconfigurable functional unit. *IEEE Trans. VLSI Syst.* 120:206–17.

[25] G. Heidari and K. Lane. 2001. Introducing a paradigm shift in the design and implementation of wireless devices. In *Proceedings of Wireless Personal Multimedia Communications*, Aalborg, Denmark, pp. 225–30.

[26] P. M. Heysters. 2004. *Coarse-grained reconfigurable processors—Flexibility meets efficiency.* PhD thesis, University of Twente, Enschede, The Netherlands.

[27] P. M. Heysters, G. K. Rauwerda, and G. J. M. Smit. 2004. Implementation of a HiperLAN/2 receiver on the reconfigurable Montium architecture. In *Proceedings of the 11th Reconfigurable Architectures Workshop* (RAW 2004), Santa Fé, NM, p. 1476.

[28] P. M. Heysters, G. J. M. Smit, and E. Molenkamp. 2003. A flexible and energy-efficient coarse-grained reconfigurable architecture for mobile systems. *J. Supercomputing* 260:283–308.

[29] H. Peter Hofstee. Power efficient processor architecture and the cell processor. In *Proceedings of the 11th International Symposium on High-Performance Computer Architecture*, pp. 258–262.

[30] H. Holma and A. Toskala. 2001. *WCDMA for UMTS: Radio access for third generation mobile communications.* New York: John Wiley & Sons.

[31] J. Mitola III. 2000. *Cognitive radio: An integrated agent architecture for software defined radio.* PhD thesis, Royal Institute of Technology, Sweden.

[32] G. Kahn. 1974. The semantics of simple language for parallel programming. In *IFIP Congress*, pp. 471–75.

[33] N. K. Kavaldjiev. 2007. *A run-time reconfigurable network-on-chip for streaming DSP applications.* PhD thesis, University of Twente, Enschede, The Netherlands.

[34] I. O. Kennedy and F. J. Mullany. 2004. Design of a reconfigurable UMTS channel processing engine. In *Proceedings of the 59th IEEE Vehicular Technology Conference (VTC 2004)*, vol. 3, pp. 1300–4.

[35] A. B. J. Kokkeler, G. J. M. Smit, and T. Krol. 2007. Cyclostationary feature detection on a tiled-SoC. In *DATE2007 Proceedings*, pp. 171–176.

[36] E. A. Lee and D. G. Messerschmitt. 1987. Synchronous data flow: Describing signal processing algorithm for parallel computation. In *COMPCON*, pp. 310–15.

[37] K. Masselos and N. S. Voros. 2007. Implementation of wireless communications systems on FPGA-based platforms. *EURASIP J. Embedded Syst.*, article ID 12192, doi:10.1155/2007/12192.

[38] M. Millberg, E. Nilsson, R. Thid, and A. Jantsch. 2004. Guaranteed bandwidth using looped containers in temporally disjoint networks within the Nostrum network on chip. In *Proceedings of Design Automation and Test Europe Conference*, p. 20890.

[39] S. M. Mishra, A. Sahai, and R. Brodersen. 2006. Cooperative sensing among cognitive radios. In *Proceedings of IEEE ICC*, pp. 1658–1663.

[40] J. Mitola. 1999. Cognitive radio: Making software radios more personal. *IEEE Pers. Commun.* 6:13–18.

[41] J. Potman, F. Hoeksema, and K. Slump. 2003. Tradeoffs between spreading factor, symbol constellation size and Rake fingers in UMTS. In *Proceedings of ProRISC 2003*, Veldhoven, The Netherlands, pp. 543–48.

[42] J. Rexford and K. G. Shin. 1994. Support for multiple classes of traffic in multicomputer routers. In *Proceedings of the First International Workshop on Parallel Computer Routing and Communication*, pp. 116–30.

[43] A. La Rosa, L. Lavagno, and C. Passerone. 2005. Implementation of a UMTS turbo decoder on a dynamically reconfigurable platform. *IEEE Trans. Computer-Aided Des.* 240:100–6.

[44] D. Stiliadis and A. Varma. 1998. Latency-rate servers: A general model for analysis of traffic scheduling algorithms. *IEEE/ACM Trans. Netw.* 60:611–24.

[45] T. J. Todman, G. A. Constantinides, S. J. E. Wilton, O. Mencer, W. Luk, and P. Y. K. Cheung. 2005. Reconfigurable computing: Architectures and design methods. *IEE Proc. Comput. Digital Tech.* 1520:193–207.

[46] H. Urkowitz. 1967. Energy detection of unknown deterministic signals. *Proc. IEEE*, 55(4):523–531.

[47] S. Vangal, J. Howard, G. Ruhl, S. Dighe, H. Wilson, J. Tschanz, D. Finan, P. Iyer, A. Sing, T. Jacob, S. Jain, S. Venkataraman, Y. Hoskote, and N. Borkar. 2007. An 80-tile 1.28 tflops network-on-chip in 65 nm cmos. In *Proceedings of the ISSCC*, pp. 98–99.

[48] J. von Neumann. 1993. First draft of a report on the edvac. *IEEE Ann. Hist. Comput.* 15:11–21.

[49] E. Waingold, M. Taylor, V. Sarkar, W. Lee, V. Lee, J. Kim, M. Frank, P. Finch, S. Devabhaktuni, R. Barua, J. Babb, S. Amarasinghe, and A. Agarwal. 1997. Baring it all to software: Raw machines. *Computer.* 30:86–93.

[50] T. A. Weiss and F. K. Jondral. 2004. Spectrum pooling: An innovative strategy for the enhancement of spectrum efficiency. *IEEE Commun. Mag.* 24:S8–14.

[51] P. T. Wolkotte, G. J. M. Smit, G. K. Rauwerda, and L. T. Smit. 2005. An energy-efficient reconfigurable circuit switched network-on-chip. In *Proceedings of the 19th IEEE International Parallel and Distributed Processing Symposium (IPDPS'05)—12th Reconfigurable Architecture Workshop (RAW 2005)*, Los Alamitos, CA, p. 155.

[52] H. Zhang, M. Wan, V. George, and J. Rabaey. 1999. Interconnect architecture exploration for low-energy reconfigurable single-chip dsps. In *WVLSI'99: Proceedings of the IEEE Computer Society Workshop on VLSI'99*, Washington, DC, pp. 2–8.

[53] Q. Zhang, A. B. J. Kokkeler, and G. J. M. Smit. 2006. A reconfigurable radio architecture for cognitive radio in emergency networks. In *Proceedings of European Conference on Wireless Technology*, pp. 35–38.

[54] L. Zhigang and G. Luoning. 2004. Design and implementation of umts-fdd/hsdpa baseband. In *Proceedings ICSP '04*, vol. 3, pp. 1857–60.

Index

B

D

F

G

H

I

N

Q

R

S

X

Z